THE HISTORY OF INFORMATION SECURITY

A Comprehensive Handbook

THE HISTORY OF INFORMATION SECURITY

A Comprehensive Handbook

Edited by

KARL DE LEEUW

and

JAN BERGSTRA

Informatics Institute, University of Amsterdam
Amsterdam, The Netherlands

ELSEVIER

Amsterdam – Boston – Heidelberg – London – New York – Oxford – Paris
San Diego – San Francisco – Singapore – Sydney – Tokyo

Elsevier
Radarweg 29, PO Box 211, 1000 AE Amsterdam, The Netherlands
The Boulevard, Langford Lane, Kidlington, Oxford OX5 1GB, UK

First edition 2007

Notice
No responsibility is assumed by the publisher for any injury and/or damage to persons or property as a matter of products liability, negligence or otherwise, or from any use or operation of any methods, products, instructions or ideas contained in the material herein. Because of rapid advances in the medical sciences, in particular, independent verification of diagnoses and drug dosages should be made

Library of Congress Cataloging-in-Publication Data
A catalog record for this book is available from the Library of Congress

British Library Cataloguing in Publication Data
A catalogue record for this book is available from the British Library

ISBN: 978-0-444-51608-4

For information on all Elsevier publications
visit our website at books.elsevier.com

Printed and bound in the United Kingdom

Transferred to Digital Print 2011

PREFACE

The two editors had different roles in the development of this Handbook. The idea to produce it came from Jan. Karl made the design and found most of the authors. Always keeping the design in mind Karl had extensive interactions with the authors about the content of their contribution. Besides reviewing all papers Jan took care of project management and funding issues. Karl wrote an extensive introduction that serves as a justification of the approach taken.

The production of Handbooks is at times difficult because it is probably impossible to achieve 100% of one's objectives. Admittedly some authors who we would have welcomed very much were either unable or disinclined to provide a contribution. Some potential authors could not get permission to provide vital information and some papers were left unfinished and could not be included for that reason. Nevertheless we are confident that the 28 contributions in this volume collect a wealth of information that may be of use and interest to many readers.

Of course we are extremely grateful to all authors who spent time and energy on the writing and rewriting of their contributions of which a number are very laborious indeed.

We would like to thank the members of the advisory board, Gerard Alberts, Hans van der Meer, Donn Parker and Martin Rudner, for their valuable comments, and for reviewing a number of papers.

We would also thank Arjen Sevenster from Elsevier deserves mention for many years of detailed interest, support and cooperation, as well as an unshakable faith in the positive outcome of the project, even when the editors saw deadlines being broken time and again.

One author has expressed his concern that the project took too long to complete and we can only apologise for this and acknowledge the fact that his contribution could still be included.

Karl de Leeuw
Jan Bergstra
Amsterdam, 21 May 2007

ADVISORY BOARD

CONTENTS

PART 5
PRIVACY- AND EXPORT REGULATIONS

PART 6
INFORMATION WARFARE

The History of Information Security: A Comprehensive Handbook
Karl de Leeuw and Jan Bergstra (Editors)
© Published by Elsevier B.V.

1

INTRODUCTION

Karl de Leeuw

Informatics Institute, University of Amsterdam
Amsterdam, The Netherlands

Contents

Abstract

This introduction gives an overview of the topics dealt with in this Handbook, and reaches the conclusion that society at large has to come depend increasingly on a civilian deployment of security tools. This unprecedented dependence entails risks of its own which are insufficiently weighted in current accounts of the impact of the digital revolution.

Keywords: communication security, identity management, intellectual ownership, cryptography, computer security, privacy, information warfare.

1.1 AN EXAMPLE FROM DUTCH HISTORY

The history of information security does not begin with the advance of the telegraph, the wireless, or the Internet. This claim is amply illustrated by an anecdote from Dutch history.

In January 1684 the Dutch Republic and France were on the brink of war. The French had attacked the Southern Netherlands which at that time was still a part of the Spanish Empire, and it seemed unlikely that the French would halt should the Spanish have been defeated. They had tried to conquer Holland twelve years before and there was little reason to trust their declarations of good will. Therefore, Stadholder William III, a semi-hereditary commander of the army and navy as well as chief executive of the foreign policy of the Republic was willing to fight the French in the Southern Netherlands.

The French ambassador in the Dutch Republic, Count D'Avaux, had no inclination to wait until William had succeeded. He entered direct negotiations with the city council of Amsterdam in order to prevent that city from raising money for troops as the stadholder had requested. Count D'Avaux's negotiations could easily be construed as a direct interference in the internal affairs of the Dutch State, but it was not uncommon for a French ambassador to do a thing like that. The stadholder, however, wanted to make a point of the counts activities and had the ambassador's messenger shadowed at a time when he was most likely carrying a letter on the negotiations to the King.

1

The courier was captured just after he crossed the border near Maastricht, by horsemen clearly belonging to the Maastricht garrison. The courier, robbed of all of his belongings except for his boots and his jacket, returned to his master with the story of what happened. The affair caused much distress to the members of the Amsterdam town council, but D'Avaux reassured them that the letter was fully coded and that no one could read it without a key. On 16 February 1684, however, the stadholder entered a meeting of the Amsterdam city council with a decoded copy of the letter, accusing its senior members, Hooft and Hop, of treason.

William claimed that he had received the copy from the governor of the Spanish Netherlands, De Grana, which made D'Avaux laugh. The only thing D'Avaux still could do to help his friends at the town council was to pretend that William's crypt-analyst had interpreted the letter wrongly and to support his claim be provided a 'genuine' copy for circulation. In turn then, William III released a copy, at first leaving out many of the unsolved code groups. Somewhat later, he released a full version in plain text, accompanied by the solutions from other letters which were also intercepted at the same time and which were even more compromising (Kurz [18]).

This incident made William III, who was to ascend to the British throne in 1689, well aware of the benefits of intercepting mail. In this case, he had used the information immediately and without effort to hide its source in order to provoke a crisis. As King of Britain, a few years later, he would commence with intercepting mail on a regular basis for purpose of strategy planning. Thus a trend began: during the 18th century, most countries employed code-breakers, linguists, and clerks to intercept the mail of foreign diplomats on a regular basis.

This example serves to illustrate the core concepts of information security. The ambassador's courier, travelling back and forth to the Court in Versailles, was a component of the French diplomatic information system which assured that the French emissaries abroad would keep the King informed in order to carry out his will. The loyalty of the courier, often a personal servant of the ambassador, was of vital importance for the system, as was his good health and general fitness of the horse. The encryption of the letters he carried, acted as a second line of defence. Thus, for this example, the protection of communication security depended upon three components: psychology, physical integrity, and encryption. A fourth element that is, diplomatic immunity provided the legal context for the French information system. It was significant that the robbing of the courier did not take place in Holland but rather just across the border. The Spanish were already at war in the Southern Netherlands; whereas the Dutch were not. Consequently exposing French interference in the internal affairs of Amsterdam as well as the treason attempt by members of the town's council was an excellent instrument employed in removing political opposition and turning public opinion against the French. It is information warfare at its best.

1.2 DEFINITIONS, TOPICS, AIM

This story from the 17th century remains a surprisingly relevant one for a proper understanding of the core concepts of today's information security, which is usually defined as: "the protection of information and information systems against unauthorised access or modification of information, whether in storage, processing, or transit, and against denial of service to authorised users. Information security includes those measures necessary to detect, document, and counter such threats" (Jones, Kovacich and Luzwick [16]).

Information system security, deals with "the entire infra-structure, organisation, personnel, and components that collect, process, store, transmit, display, disseminate, and act on information". This definition is independent of any stage of technological development, which means that it can be applied both to the courier services of the early modern world and to computer networks of today.

A similar approach is taken by Bruce Schneier in *Beyond Fear* [26, 6]. He argues that security concepts apply, or should apply, equally to computers and to the world at large. He defines a security system as a "set of things put in place, or done, to prevent adverse consequences. It is concerned with intentional actions which are unwarranted from the point of view of the defender, not necessarily illegal" [26, 12]. A security system presupposes a security policy requiring someone who defines it which generally is the asset owner [26, 34–36]. Those security policies reflect the political agenda of the owner(s) or a trade-off among various interested parties who move power in varying degrees from one set of players to another and who are anything but value-neutral. Taken in this manner, information security consists of the entire range of instruments available to limit the flow of information, including those theories, practises and insights that allow us to put them to good use. In a sense, we are dealing with an armoury for the exercise of social control [14].

Of course in information security, political and ethical considerations are particularly relevant. Information security consists of the entire range of constraints deliberately built into any information system in order to restrict its use. These constraints may encompass legal measures, institutional frame-works, social practises, or instruments, including devices or machines developed especially for that purpose.[1] These constraints are expressions of the wishes and perspectives of the system owners. Thus, the history of security technologies may well serve to illustrate MacKenzie's and Wajcman's thesis that the actual shape new technologies take on reflects the interests of the parties involved in the design process [22].

The aim of this book is to gather materials from different disciplines for a reconnaissance tour of the security domain of information systems. It is intended first as a field-survey and consists of twenty-nine contributions, dealing with episodes, organisations, and technical developments which are, in one way or another, exemplary or which have played an important role in the development of information security. The chapters are written by experts in such diverse fields as computer science, law, history, and political science and are

arranged into six sections. The sections on computer security, and on cryptology and communication security, constitute the core of the book and stay firmly within the traditional scope of information security. The section about intellectual ownership may surprise some readers. However, its inclusion is justified by the fact that copyright and patent registration systems are important instruments to define intellectual ownership while also providing organisational and legal means of enforcement.

Moreover, it provides a background for the debate about the propriety of software. The main argument in favour of Open Source software is that without it no software transparency, and therefore no computer security, can ever exist. The chapters in this section show how the Open Source movement borrowed key concepts from scientific publishing and also that alternatives do exist in the form of protecting software through patents or copyright. The section about identity-management includes chapters about biometrics, the collection of data about state subjects, and security printing. It aims at a better understanding of this topic through a multi-faceted approach. The fifth section deals with privacy and export regulations. These regulations are particularly relevant for the way information security is practised and have constituted hotly debated political issues in both Europe and the US. The sixth section contains only one contribution about information warfare. This part explores the strategic meaning of information security and addresses the question whether the information society of today encounters unprecedented threats.

Three questions will recur throughout the book. How did political or ethical motives influence the choice or shape of security instruments? Did scientific research, or new technological developments, influence legal practises or the execution of certain policies? Did technological or scientific considerations ever act as constraints for the formulation of political ambitions? In which ways can the availability or lack of security instruments be seen to act as a strategic consideration in the political debate?

1.3 HISTORIOGRAPHY

The origin of information security can be traced back to the rise of hierarchical command and

---

[1]There is some debate about the question whether the legal frame-work should be considered part of the security system. I like to follow Donn Parker [24, 36] who says it does.

control structures in administration and warfare from civilisations of the ancient world. Plenty is known about administrative procedures or command structures in western history, but there has been little effort to highlight the elements of confidentiality, integrity, and availability that figure so prominently in contemporary literature about information security.[2] Unfortunately, the extensive literature dealing with the rise of the Information Society does not pay any attention to security issues either (Toffler [28]; Beniger [7]; Agar [1]; Bijker, Hughes and Pinch [8]; Machlup [21]).

There are, however, very substantial contributions to the history of sub-areas. The history of cryptology has enjoyed considerable attention, for instance, particularly after Kahn published his epoch-making *Codebreakers* in 1967 [17]. The focus of the research since then is mainly on code-breaking and signals intelligence during World War II which was dominated by Anglo-American perspectives although Cold War research has also gained importance in recent years (Aid and Wiebes [2]). Moreover, military historians have paid considerable attention to command and control structures.[3] Contemporary cryptology is largely uncovered, with the exceptions of Simon Singh's *Codebook* [27] and Stephen Levy's *Crypto* [19].

The history of computer security has not been a topic for research thus far but, to a certain extent, hackers and free software have been researched.[4] There is a vast literature about the history of patents, and to a lesser degree, about the history of scientific publishing and copyright. The history of identity-management as such has never been investigated, but parts of it have been, such as the issuing of passports (Caplan and Thorpey [11]) and the use of biometrics (Breckenridge [10]), albeit in a forensic context mostly (Beavin [4]). Some case studies have also been written, and Edwin Black's

controversial and disquieting *IBM and the Holocaust* [8] may serve as an example. The privacy debate is closely related to identity-management issues, but has not resulted in historical research of any substance. The contributions of Bell [5;6] and Westin [29], do show a strong concern for the societal implications of the application of new information technologies, which may prove of heuristic value. Needless to say, this list is far from complete. I refer to the contributions in this book for further reading.

1.4 LIMITATIONS

This book may be a little disappointing for those wanting to know more about industrial espionage or more generally about information security in the context of corporate business, and this author is well aware that most practitioners today are keenly interested in protecting company assets. The focus of information on security in the corporate world, consequently, would be on assuring the continuity of business processes, industrial espionage, or the disruption of communication structures with criminal intent; these subjects certainly are part of a much larger portfolio. Unfortunately, this author has not been able to locate writers for this important subject. It should be noted, however, that safety, or the protection of assets from unintentional hazards, is not a proper part of this book since its intent is to cover security issues intentionally inflicted.[5]

A similar remark must be made for those wanting to know more about information security in the context of bureaucracy. The concern for confidentiality, integrity and availability of information has been an integral part of administrative practises for centuries. Unfortunately, in as much as the history of administrative practises has been written, there has been no effort to link it with the history of information system security.[6]

----

[2] For an introduction see for instance Gollmann [15] or Anderson [3].

[3] See for instance: John Ferris, *Airbandit C31 and strategic air defence during the first battle of Britain, 1915–1918* [13].

[4] See for instance: Steven Levy, *Hackers* [20]; Eric S. Raymond, *The Cathedral and the Bazaar* [25]; Sam Williams, *Free as in Freedom* [30].

[5] For the difference between safety and security see Schneier [26, 12].

[6] Angelika Menne-Haritz [23] has written extensively about the development of administrative procedures in Germany, but she did not provide a link with security.

This book then focuses on western history since the Renaissance. The basic ideas about intellectual ownership and the accessibility of knowledge were all formed during the 16th century in the western hemisphere as were core concepts in administrative organisation and diplomacy. This emphasis does not mean that the history of the empires of the ancient world, or the history of the great empires of the eastern hemisphere, will not provide interesting parallels, but these subjects necessarily fall outside the scope of this book.

1.5 INTELLECTUAL OWNERSHIP

In the first contribution examined, Jack Meadows has taken a close look at the way scientific publishing has evolved in early modern Europe. Before the 16th century, the idea of scientific publishing was virtually unknown. Scientific research took place in the context of alchemy or of secret societies, given to theosophical speculation, and scholars tended to adhere to secrecy if only to avoid being burned at the stake by the Papal Inquisition. The publication of Copernicus' *De Revolutiones Orbium Coelestium* by a protestant scholar, defying the authority of the Roman Catholic church in 1543; the publication of *De Humani Corporis Fabrica* by Vesalius in the same year; and two years later Cardano's *Ars Magna* in 1545 all marked a sudden change in scientific publishing. Part of the reason for the bold change was that the printing and distribution of complex texts and drawings had become easier. The fact that the career perspectives of scientists and scholars became increasingly dependent on personal fame also mattered. Thus Vesalius owed his appointment as personal physician of the Emperor Charles V directly to the publication of his book.

During the 17th century the cumulative structure of scientific knowledge became widely acknowledged and along with it a principle of scientific priority. The need for secrecy did not vanish, however; but it did become tied to the need of conducting verifiable experiments to support theories. Initially, priority claims were safeguarded by hiding a short reference to a discovery in an anagram and putting this at the disposal of a well-known colleague. Later, the depositing of a sealed and dated manuscript at a scientific academy came into vogue which had the advantage that much more detailed claims could be substantiated. During the 19th century, this depository practise vanished because by then it was commonly believed that the priority should go to whoever first published a discovery not to whomever the thought first occurred.

The 19th century, with its quest for scientific progress, marked the heyday of the scientific journal, and this pattern of scientific publishing is still with us today. The point of departure today is that the progress of science is best served by the free exchange of ideas even though this concept is not taken too literally. Scientific journals themselves act as a bottleneck since each is free to accept or reject the publication of a scientific paper. An initial rejection by one journal and a subsequent acceptance by another may result in a delay of years. The system of peer review used by most journals to rule out idiosyncrasies of the editing staff can turn counterproductive when a referee wants to avoid publication of anything threatening his authority.

Moreover, political pressures may also still be operating within this system of scientific publication. The arrest and trial of Galileo in 1632 is probably the best known example, but in the totalitarian regimes of the 20th century similar events have happened. For example, the purging of Jewish elements in physics by the Nazis which led to a discrediting of quantum physics and relativity theory is one example, and the ban on the publication of research based on Mendelian principles in Soviet Russia is another since traditional genetics did not fit communist beliefs. The western world is not exempt from restraints either. The classified research on behalf of the defence industry or contract research for commercial purposes may easily clash with the scientists' needs to make advancements known. The electronic publication of preprints has added to the armoury of the scientific author, but this solution does not remove any of the restrictions already mentioned. Scientific publishers may demand removal of the preprint from the Internet as soon as it is published in a magazine in order to make sure that revenues are not lost.

In the second contribution, Kees Gispen deals with the history of patent law with particular reference to the German patent system in the early 20th century. The first patents were granted the in late 15th century in Venice and in the early 17th century in England to 'men of exceptional genius' or 'first and true inventors' for accomplishments which would greatly benefit the 'common good'. These patents were privileges, however, not rights so there was no concept of intellectual property at stake either since the primary motive of the accomplishment was to benefit the common good of a state.

The concept of intellectual property itself waited until John Locke's 'natural right of proprietorship' was written. This concept lay at the root of the patent system which was introduced during the French Revolution in 1791, but it also affected practises in Britain where the concept of 'societal utility' acted as a counterbalance. By the end of the 18th century, both France and England were ready to dispose of state-granted monopolies used in the past. It was not self-evident, however, that patents and free trade could go together. Scottish philosophers, such as Adam Smith, believed they could, but exponents of the Manchester School later believed patents and free trade did not go well together.

The Manchester School sparked off an anti-patent movement with considerable influence in The Netherlands, Switzerland and Germany which lasted until the last quarter of the 19th century. The United States remained unaffected, however, since the concept of 'natural propriety rights' is firmly rooted in the US constitution. Following the French example, the United States adopted a patent regime based on the 'first-to-invent-principle', or to put it differently, based on the 'authorial principle'. In this way, an inventor's rights were well protected even though this system worked only when an innovation occurred within the context of small-scale enterprises.

By the end of the 19th century, the inventor-entrepreneur was no longer leading the way. By then, most innovation took place in the context of corporate businesses by employees more often than not working as a team. The basic ideas behind

the US patent system were not abandoned, however; but the US law did allow employers to claim the ownership of the inventions of their employees through an explicit clause in the employment contract. In reality this approach meant that inventors had little control over their inventions, but they did retain the non-material claim to the invention and were free to change employers or to start their own businesses.

Germany took a different course. After its unification in 1871, the Second German Empire was confronted with a bewildering diversity of patent regimes in the various states as well as an abolitionist movement inspired by the Manchester School. The leader of the pro-patent movement was Werner Siemens, a brilliant inventor-entrepreneur and head of the Siemens electrical company. He argued that German industry could not do without the legal protection of its inventions. The patent system should defend the interests of the industry not the interests of the inventors who were unable to act without the support of industry. With his influence, then, Germany adopted the first-to-file principle, meaning that whoever filed a patent first legally owned it. The German patent system allowed companies to claim that the inventions were made collectively, and therefore, the intellectual property of the company which, quite naturally, caused resentment among German engineers and which, in turn, resulted in political strife which was finally settled in 1957. Gispen argues, then, that the German system did not encourage inventors to start their own businesses as happened in the United States. Rather German patent laws encouraged conservative inventions within the context of existing large research establishments which, in the long term, was detrimental to innovation.

In the third contribution, Chris Schriks and Rob Verhoogt explore the emerging copyright protection in particular reference to The Netherlands. During the early modern period, The Netherlands had been the preeminent publishing house of Europe because of an absence of censorship and because of the presence of highly skilled immigrant communities who were on the run for religious and political oppression elsewhere. In countries like England and France, copyright protection had been part of the exercise of state and religious censorship, but in the Dutch Republic copyright protection was granted by way of privileges by requests of printers' guilds where the motive was purely economical, that is to say to protect the interests of an important industry.

The concept of intellectual ownership was introduced during the French occupation of The Netherlands under the influence of Enlightenment thinking and actions taken abroad by authors such as Pope, artists such as Hogarth, and composers such as Verdi. This period marked the beginning of the emancipation of the author from a publisher. The idea that the copy needed protection but the original work did not remained present not only in The Netherlands but also elsewhere. This preference of copy protection over original work protection can be seen when copyright protection was extended to images, meaning that engravings were protected yet the original images were not. This preference for copy protection resulted in heated debates which lasted for almost a century between authors, artists and composers on the one side and vested economic interests on the other. As late as 1877, the Dutch jurist H. Viotta argued, as a matter of principle, that copyright protection related exclusively to the multiplication of material objects and not to the spread of ideas.

The invention of new reproduction techniques, such as photography, sound recording, and in their wake film made a redefinition inevitable. The Berne Convention of 1886, and its revisions of 1908 and of 1928, gave due weight to the role of creative genius, and more or less enforced conformity of copyright regimes between western countries allowing for national differences to fade. The rise of the new distribution media which reached into people's homes through radio, television, tape recorders, and even copying machines could be used for limited scale multiplication made a revision of copyright rules necessary. The concepts of publishing and multiplying diverged, and the difference between private and public use was no longer always easy to make. The Internet, and the accompanying concept of Digital Rights Management, marks a new stage in the development of copyrights through the centuries which includes

the idea that technological changes might force changes in legal areas which does not necessarily mean that guiding principles of the past have become obsolete. The emergence of concepts such as Creative Commons or Free Source, which focuses on the idea of freely publishing technical information for every one to use and with no profit intended, may be taken a step further in the emancipation of the author from a publisher.

In the fourth and fifth contributions, Madeleine de Cock Buning writes about copyright protection for software and Robert Plotkin writes about the protection of software through patent law. Both subjects may traditionally not have been counted as security-related topics but in reality they very much are so related since they exemplify alternative ways of protecting the ownership of software without making source code unavailable for research purposes. An extension of patent and copyright law regimes to software may reconcile the interests of proprietary software developers and the need for code inspection in the name of public welfare. De Cock Buning explores the experiences with copyright law made after the 'unbundling' of soft- and hardware by IBM in 1970. The first efforts to extend a copyright regime to the protection of software were made shortly thereafter and proved less straightforward than one might think since both the algorithm and the actual expression in programming language need protection and copyright law traditionally only addressed the latter issue.

The tendency, then, both in Europe and in the United States has been to develop a new protection regime. In the US this resulted in a demand for a proof of arbitrariness of computer programs to ensure that the algorithm was not protected as well. In Germany, legislation required an 'exceptional originality' to justify the invocation of copyright protection. Needless to say neither solution has proven to be fully satisfactory.

Robert Plotkin explores the history of software patents by reference to a number of particular controversial cases in the US and to actual and proposed legal reforms in the US and Europe. He concludes that the ambiguous nature of software, described only in abstract or logical terms but always designed for a specific application, has made it difficult for patent law to formulate suitable criteria for patentability.

Jurisprudence has succeeded in fleshing out arguments in such a way that patent lawyers now know with reasonable certainty which requests are likely to be granted. Legislation has not, however, been able to resolve the root problems represented by the tensions between competing legal rules. Plotkin signals the rise of new problems due to the granting of patents for automated design processes; this new difficulty makes the line between 'idea' and 'application' even more difficult to delineate. Moreover, the granting of patents for fully automated business processes has torn the wall between industry and the liberal arts which will make the enactment of new legislation inevitable, even though policy makers are reluctant to take up the responsibility for this new legislation. This situation is slightly better in Europe than in the US as the aborted European Software Directive indicates.

1.6   IDENTITY MANAGEMENT

The second part of this book deals with identity management. The modern state cannot exist without knowing who its citizens are, and neither can banks exist without knowing who its customers are. The registration of ownership of land, houses, and movable properties is a prerequisite for both economic transactions and tax levying. Pieter Wisse develops a semiotics of identity management in three parts. In the first part (Identifying assumptions) he introduces how pervasive issues of identity really are. Philosophy, science and, not to be neglected, religion may indeed be viewed as attempts that come to terms with identity where ownership is concerned. In the second part (Identity in enneadic dynamics) Wisse presents an encompassing framework, inspired by C.S. Peirce, which is an enneadic model of semiosis which in turn is then applied to a pragmatic design, or a behavioural, orientation for identity management.

Communication and identity management are largely synonymous; identity management is essentially dialogical. Illustrations from natural history serve to emphasise the generally valid behavioural orientation. In part 3 (Social practises in identity management) Wisse illustrates how the semiotic framework helps to provide an overview of perspectives and discusses selected developments in identity management at the cultural level, that is as practised by human communities and societies. The partly electronic future of identity management is proposed as a program for open security in an open society.

Karel Schell treats the subject of document security, and more notably the security of banknotes, from their introduction in the 17th century until now. The history of banknotes provides us with a classical example of identity management. The banknote was introduced at various times and by various agents in order to facilitate money circulation, but it only proved successful after the emergence of a reliable system of central banks able to conduct a sound monetary policy. This system finally took shape around the year 1800 after two centuries of rigorous efforts, many of which went wrong.

The development of security printing was dependent on the existence of an authority willing to vouch for the authenticity of the document at hand which also meant that the document had to contain features that made reproduction difficult. Initially this purpose was served by the application substrate marks such as watermarks and security threads, or the use of unique type fonts, for instance the type fonts for the printing of music, invented by Johan Enschedé around 1780. The invention of photography around 1840 added new features to the armoury such as colour and eventually guilloche printing. Guilloche printing consists of complicated, engraved line patterns that are impossible to reproduce because of the subtle variety in distances between the lines.

The introduction of offset printing and colour scanners between 1920 and 1950 made the development of new deterrents necessary. The vulnerability of existing practises is exemplified by a number of cases of successful counterfeiting between and during the world wars which were mostly, but not always, politically motivated. The counterfeiting of British Pound notes by the SS is the most famous example. The deliberate introduction of moiré patterns, occurring when two periodic patterns, such as an array of straight lines or a series of concentric circles, are overlapped with imperfect alignment which would become visible when reproduced proved to be a helpful answer.

The invention of the laser colour copier and the digital revolution during the 1980s added optically variable printed images, holograms and covert images to the armoury. Interestingly, some substrate-based security features, such as watermarks and security threads, present for the last two centuries, proved far less vulnerable; whereas, the numbering of banknotes continued to be an indispensable means of identification. Moreover, the general requirements for banknotes have always remained roughly the same and were already listed in a report written by Sir William Congreve for the British Government in 1819. His point of departure was that security features should be "immediately recognised and understood by the most unlearned persons".

The use of micro text, incorporation of the serial number in the watermark, and guilloche structures were advocated by him as deterrents against counterfeit decades if not a century or so before their actual applications. The history of banknotes provides valuable clues about the history of identity management in general. Firstly, the need for 'virtualisation' is older than the digital age hence precedents can be found in the quest for security technologies. Secondly, the state of technology limits both possibilities and threats, which also means that technological progress does not allow more than temporary preponderance. Thirdly, general principles or guidelines can be established that may outlive a particular state of technology provided that the goals are sufficiently clear.

Edward Higgs depicts the development of identification techniques and indeed the various transformations in the concept of identity in England from the late middle ages onwards. England is a particularly interesting example since at various points in its history it played a leading role in the introduction of new identification techniques, such as a system for fingerprinting in 1901 and DNA-profiling in the 1980s. Higgs discerns three broad stages. During the early modern period, identification took place within the community in which a person lived or by lineage. During this period, the state was not the centralised, almost omnipotent entity it is today but rather a sphere of governance. The state formed a national community of those in every locality who were willing and able to be involved in the exercise of authority. For the lower strata of a society this framework meant that identification was a local matter. For the aristocracy and gentry, identification was bound up with proving descent through genealogies and involved the use of antiquarians to prove that one had the right to a name, a title, or land.

Characteristically, the Poor Laws of the 16th and 17th centuries intended to encourage and rationalise charity focused entirely on proving that the poor really belonged to a parish. Therefore, wandering beggars or vagabonds were flogged in public, put in jail, and finally physically branded. During the 19th century, urbanisation and industrialisation came with other more centralised forms of control. Local communal forms of governance proved unable to cope with the creation of vast, urban conglomerations while local elites ceased to be interested in the locality as their wealth increasingly came from positions or shares in national or international companies.

The individual increasingly had come to be captured in a nexus of centralised state registrations which identified him as a property-holding citizen. At the same time, those who did not take part in civil society had to be put under surveillance. The New Poor Law of 1834 ruled that the poor had to submit to imprisonment in all but name in so-called workhouses. Various criminal registries were set up, at first with limited success and from 1871 onwards improvements were made with the introduction of photography, body measurements, and fingerprinting. The increasing speed of transport added to pressures for more national forms of identification. Typically, the first sign was the introduction of the driving licence in 1903. Finally in 1915, the first modern UK passport was introduced with a description, a photograph, and a signature. In itself this identification use marked the beginning of the third stage in which bodily forms of identification were extended to the citizen instead of being reserved for the alien and anti-social elements in society.

The growth of the welfare state, beginning with the National Insurance Act in 1911, and the increasing role of banks and other financial institutions made some form permanent identification necessary. Efforts to introduce a British identity card, however, met with much resistance, except in war time, particularly from the conservative strata of society. The resistance continued well into the 1970s until it became clear that welfare benefits were often unjustly claimed. The government, then, tried to curb the abuse by verifying and cross-checking personal data in computerised databases. To that end the national insurance number was introduced. In 1989, 1994 and 1997, debates about a national registration and national identity cards surfaced without any tangible results. In 2004 the labour government tried to solve this issue once and for all by submitting its Identity Card Bill to Parliament. This bill called for the introduction of a national identity card with various biometric features including DNA-sequences so that this card was intended to be processed by machines.

Jim Wayman focuses on the rise of automatic recognition systems since the 1960s. He defines biometrics as "the automatic recognition of living individuals based on biological and behavioural traits". This idea is a clear departure from the traditional definition, which includes the fingerprinting, photographing, and measuring techniques done by hand as developed around 1900 by people like Galton and Bertillon. Wayman shows that during the 1960s, expectations were high. In 1963 Michael Trauring wrote that, biometrics in this sense of the word promised, "personal identity verification in seconds, requiring almost no trained or skilled operating personnel, with the advantage of not being susceptible to forgery or theft".

Technological developments for these processes fell far behind however. Early experiments with fingerprinting proved that they could easily be forged; whereas, facial recognition still lacked a sound technological basis to be of any use. Speaker recognition and automatic signature recognition performed slightly better, but computer technology was still insufficiently advanced to allow for any large scale applications. In 1970 biometrics moved more to the foreground, due to the start of a formal testing of fingerprint systems by the US National Bureau of Standards and included, in particular, holographic optical techniques for forensic purposes. In 1971, the US National Academy of Science organised a working group on speaker verification and the Acoustical Society of America held a meeting devoted to spoofing speaker recognition systems through mimicking.

In 1974, both the Stanford Research Institute and the National Physical Laboratory in the UK began working on signature recognition systems, and hand geometry verification was actually put to practise by the University of Georgia. In the same year, the US air force announced a program for a military-wide biometric identification system for access control that used a combination of biometric features to reduce error rates. The system was intended for production in 1980, but it was never deployed on the hoped for world-wide scale. In 1975 the Mitre Corporation followed with a formal test program for fingerprint, voice and signature recognition, and in 1977 the National Bureau of Standards (NBS) published a Federal Information Processing Standard for Automatic Personal Identification (API).

The twelve criteria for device evaluation, listed in this document are still applicable today. In 1983, formal testing programs were launched by the US Department of Energy and the US Department of Defense. The decision by California, to require fingerprints for driving licences met with resistance from congress so that this example was not followed by other states. Generally speaking, government agencies showed a greater interest in biometrics both in development and application than did businesses. This tendency continued during the 1990s. In 1994, a National Biometric Test Centre (NBTC) was launched by the Biometric Consortium, under joint control of NBS and National Security Agency (NSA). At approximately the same time, the European Commission founded the BIOTEST program at the UK National Physics Laboratory.

During the last decade of the 20th century sales of biometric devices increased a hundred fold whereas revenues increased ten fold. The actual application of biometrics systems remained limited, however, and was, or is, still largely experimental in nature. During the first years of the 21st century, pilot projects for frequent air travellers have been, and still are, undertaken in The Netherlands, the UK, the US, Germany, Malaysia, and Australia. The results seem to indicate notwithstanding the technological progress that has been made over the years that biometrics is still insufficiently reliable to be implemented on a large scale. In that sense, the call for application of biometric features on national identity cards by the US congress, with other countries voluntarily or involuntarily following suit, seems premature.

1.7 CRYPTOLOGY AND COMMUNICATION SECURITY

The third part of the book deals with cryptology and communication security. Gerhard Strasser depicts the development of cryptology in Europe during the Renaissance and thereafter. Cryptology was already being practised by the Greeks and the Romans during ancient times, but there is little proof of anything beyond mono-alphabetic substitution ciphers, let alone any form of cryptanalysis. During the middle ages most of what the ancients did or knew appears to have been forgotten, but Jewish mysticism encouraged an interest in the hidden meaning of biblical texts through assigning number values to letters and words. Jewish scholars engaged in such speculation had no interest in improving communication security, but their work became a source of inspiration for leading cryptologists during the Renaissance, such as Trithemius and Vigénère.

The first Treatise on cryptanalysis, however, was written during the 9th century by the Christian Arab philosopher Al Kindi. Unfortunately, his contribution remained largely unknown in the west, notwithstanding the inclusion of a vast amount of cryptological knowledge in an encyclopedia of Islamic knowledge by Ibn Ad-Durayhim in 1412. The real origin of cryptological thinking in the west lay with the rise of the Italian city states during the 14th century and the invention of standing diplomacy. As a result, diplomatic dispatches had to be ciphered on a daily basis which gave rise

to the emergence of a new type of official: the cipher clerk or cipher secretary. People like Argenti in Rome, Soro in Venice, and Simonetta in Milan wrote their treatises on cryptanalysis in secrecy for the dual purpose of improving the ciphers and codes of their masters and the training of new colleagues or successors.

During the 15th and 16th centuries, many cryptographic inventions were published by scholars such as Alberti, Bellaso, Porta and Cardano. Virtually all important cryptographic principles and devices were described in their books, ranging from the cipher disk and cipher square to the autokey and super enciphered codes. In 1624 this whole body of knowledge was comprehensively summarised by August Duke of Brunswick-Wolfenbuttel, writing under the pseudonym Gustavus Selenus. His book, *Cryptomeytices et Cryptographiae Libri IX*, became the standard reference work for the next two hundred years. The late 16th and early 17th centuries, an age of civil and religious warfare, saw a gradual proliferation of cryptological knowledge throughout the rest of Europe, with outstanding contributions in the field of cryptanalysis by mathematicians such as Viete in France, Wallis in England and Rossignol in France.

After 1650 the focus of many authors shifted to the search for universal language schemes, usually to facilitate the conversion of exotic peoples to Catholicism. Jesuit scholars such as Athanasius Kircher, Pedro Bermudo, or Kaspar Schott practised cryptography more or less as a side show of their ambition to spread Christianity through scholarly means which does not mean that their schemes and systems lacked practical value. Kircher, for example, invented a kind of cipher slide, the so-called *Arca Steganographica*, which proved a rather useful encryption tool for many courts in Europe. Generally, the popularity of universal language schemes in literature does not tell much about prevailing cryptological practises.

After the Peace of Westphalia in 1648, a system of standing diplomacy was adopted in all of Europe and was similar to the system that had existed in Italy before. This adaptation stimulated the diplomatic use of cryptography and lead to the emergence of a class of professional cryptologists throughout Europe, as had happened in

Italy earlier. In some countries, this effort resulted in the establishment of so-called black chambers which engaged in the intercepting and decrypting of diplomatic dispatches on a daily basis. In England and France, these black chambers were operated by members of the same family on a semi-hereditary basis and posing as post officials. Strict loyalty and utmost secrecy were of the essence and there are very few reports of cipher clerks revealing their secrets. Any cryptanalytic expertise was considered a trade secret which was vital for the continuity of the business and therefore jealously guarded.

The undersigned writes about the black chamber in the Dutch Republic. As early as the revolt against Spain at the end of the 16th century, the Dutch had been intercepting Spanish military communications. They were proven successful in decoding partly through the bribing of a Spanish code clerk and partly through cryptanalytical effort. The first half of the 17th century saw various code breakers at work, mainly focusing on military communications. The most famous of these was the playwright Constantijn Huygens the Elder, who followed Stadholder Frederick Henry in battle as his personal secretary. Too little is known about possible contacts between these cryptanalysts, but it seems likely that no effort was made to ensure transfer of knowledge which may be explained by the relatively open political structure of the Republic which made keeping secrets difficult.

Therefore, in secret negotiations with foreign diplomats stadholders and grand pensionaries had to act on their own, which entailed the risk of being accused of high treason when things did not work out as expected. The political machinery afforded little opportunity for the long-term allocation of secret funds for a black chamber which implied that code-breakers were hired on an ad hoc basis at the discretion of individual statesmen. This code-breaker use is further exemplified by the temporary emergence of a black chamber between 1707 and 1715, during the War of the Spanish Succession and its aftermath. This chamber was manned by the personal secretary of Grand Pensionary Heinsius, Abel Tassin d'Alonne. D'Alonne, who had become familiar with code breaking while serving

under his half brother Stadholder William III, focused on material intercepted at the post office in Brussels, which at the time was occupied by a combined Anglo–Dutch force.

Letters intercepted in the Republic proper were sent to Hanover to be decoded there. The reason for this transfer was to aid a political divergence with both the British and Hanoverians about the course taken in the Southern Netherlands. The important point to note is that only a few people were acquainted with d'Alonne's activity as a code-breaker, as the solutions never left the grand pensionary's office since both men worked in the same room. In the second half of the 18th century, another black chamber emerged in the Hague on the initiative of Pierre Lyonet, a cipher clerk in the service of the states general who excelled as a microscopist and an engraver of insects. Lyonet had been copying and storing French and Prussian diplomatic intercepts in order to have them decoded in London and he could not resist the temptation to try his luck on his own. He firmly believed that he was the first in the Dutch Republic to have ever been engaged in such activities and he was unaware that bishop Willes in London was assisted by a whole staff of translators and cryptanalysts.

Lyonet's work could easily go unnoticed since he already had been in the service of the states general since 1738. His cousin S.E. Croiset was called upon to assist him. Officially, Croiset was in the service of the post office. This formula lasted until the break-down of the Republic in 1795, with Croiset taking the place of Lyonet after his demise in 1789. Interestingly, Lyonet's activities as a code breaker also inspired him to improve the codes for the states general. Before that, Dutch cryptography was highly influenced by examples taken from literature, as can be seen in the application of Kircher's *Arca Steganographica* in the ciphers of the States General from 1658 onwards.

Jeremy Black writes about mail interception in 18th century Britain and the efforts to influence public opinion by the government. Mail interception focused primarily on foreign diplomats. It was carried out by a group of professional translators and code breakers working under the supervision of the secretary of the post office out of Secret

Service money after 1782 when the Foreign Office was created. The English black chamber originated in the 17th century when the mathematician John Wallis intercepted Stuart, French, and Spanish communications on behalf of parliament and later on behalf of William III. In 1715, the Hanoverian rulers took some of their own personnel with them from their black chamber at Celle. The collaboration with this black chamber remained in place until the personal union between Britain and Hanover ceased.

The appointment of the Reverend Edward Willes in 1716 marked the beginning of a veritable dynasty of code-breakers capable of reading the diplomatic traffic of all major and many lesser powers. Until 1746 the threat of a Jacobite overthrow was very real and it was supported by the threats of French and Swedish invasions. Therefore, domestic communications had to be monitored as well particularly in as much as members of the opposition were concerned and counter-insurgence strategies would have to include the influencing of public opinion, which was no straightforward matter. Censorship had been abolished and the government did not posses a newspaper of its own. Instead, the authorities had to rely on the releasing of favourable reports in both the foreign and the domestic press as well as the prevention of the publication of contentious information through lawsuits, particularly with reference to foreign policy.

The threat of legal action by the government had a pre-emptive impact. A case in point is the refusal of the Tory *Post Boy* in 1716 to print material offered by the Swedish envoy Gyllenborg about the British policy in the Baltic for fear of prosecution: a matter known to the government through the interception of Gyllenborg's reports. In the second half of the 18th century the threat of invasion and uprising waned, but the British Government remained dependent on public support for its foreign policy which was in contrast with the absolutist regimes in other states where foreign policy was considered the exclusive prerogative of the king and therefore never publicly debated. Of course, in this context, the need for reliable information, gained in importance.

Friedrich Bauer depicts the development of the rotor machine, a cipher machine based on a mechanisation and coupling of cipher disks. This machine was invented more or less simultaneously in The Netherlands, Germany, the US, and Sweden during or shortly after the First World War. In 1935 a modified version was adopted by the German military under the name ENIGMA and the machine was to play a key role in the German conquer of Europe five years later.

On introduction, the machine was considered practically unbreakable, due to its unprecedented key length. The Polish code-breakers nevertheless managed to penetrate German army traffic based on the exploitation of certain weaknesses in the way the machine was deployed and the use of key lists obtained through bribery. The Poles even devised a machine aimed at simulating the behaviour of the ENIGMA with various key settings, the BOMBA. After the Polish defeat in September 1939 this machine was transferred to England, along with all relevant documentation in order to assist the British cryptanalytical effort in Bletchley Park, which was to last throughout the war.

British mathematicians and engineers succeeded in making further improvements on both the machine and on cryptanalytical techniques which along with the capturing of key lists and a cipher machine on board of two German vessels allowed them to cope with the challenge posed by the introduction of a fourth rotor by the German navy in 1941. Bauer stresses that rotor machines, such as the SIGABA and the KL-7 or ADONIS did not share the design flaws of the ENIGMA and, consequently, were not penetrable in the same way. No rotor machine was ever theoretically unbreakable, however, because it was based on key repetition. A truly unbreakable cipher would have the characteristics of randomness and incompressibility, as may be the case with a one-time pad.

Jack Copeland fulfils a similar exercise for the Lorenz SZ 42, which was also broken at Bletchley Park, giving rise to the first computer. This machine, a teletypewriter used for communications between strategic command centres of the German military across Europe, was much more sophisticated than the ENIGMA and was only used for exchanges at the highest level of secrecy. With the

Siemens and Halske T52 and the Siemens T43, it belonged to a family of cipher machines devised for the protection of teleprinter traffic. They were not dependent on the rotor principle but rather on the use of a tape with additives: in the case of the T43 a one-time pad.

The Lorenz was eventually broken at Bletchley Park by building the Colossus: the first computer and usually seen as the result of Alan Turing's interest in electronics and his all consuming strife to build a 'decision machine'. Copeland shows that Turing, who interestingly was on a trip to the US when the actual work was done, played a more modest role than is generally believed. The decisive step from an engineering point of view that is, the large scale and controlled use of electronic valves as high-speed switches was first envisaged and implemented by Thomas H. Flowers.

Flowers, an engineer employed by British Telecom had been experimenting with electronics for over a decade. Moreover, the machine was programmed for the purpose of statistical calculations and devised by William Tutte. Turing had been involved in the breaking of 'Tunny', as the Lorenz was called, but at an earlier stage and his input had restricted itself to a pen and pencil method, involving a lot of guess work. Turing's ambition to build a 'logical engine' or 'decision machine' preceded the war by five years, but it was not instrumental in bringing about Colossus. The Colossus did serve for Turing and others as an example of how computers were to be built. Turing's professed interest in electronics, for instance, only occurred shortly after the war.

Silvan Frik writes the history of the only vendor of cipher machines that was commercially successful and remained so over the years: Crypto AG in Switzerland. This company was founded 1922 in Sweden as AB Cryptograph by Boris Hagelin with money from the Nobel organisation with the sole purpose of exploiting Arvid Damm's cipher machine. The success of this company was due to a large order placed by the Swedish army in 1926. Hagelin's key to success was his susceptibility to market demands. In 1930, the B-211 was launched, a portable electrical machine suitable for office-use. This machine was relatively expensive

and was mainly sold to diplomatic services and big business.

In 1934, a small almost pocket-sized machine was designed for military purposes, initially known as C-34. This machine was hand-driven and was designed for use under harsh conditions. The rearmament ensured large sales in Europe, but Hagelin's break-through as a primary supplier of cryptographic devices came with the licensing of the C-machine to the US military in 1941. The American twin of the C-machine, the M-209, was manufactured at the Corona typewriter factory. With the navy, the army and the air force as customers, sales quickly reached fifty thousand.

In 1942, the Office of Strategic Services (OSS), ordered a new Hagelin machine: the BC-543 with a keyboard and an electric drive intended for office use. This machine would continue to play an important role in peacetime. Nevertheless, the first years after the war saw a steep decline in demand, but the Cold War was able to revive the business. In 1952 Hagelin transferred his company to Switzerland because of its tradition in precision engineering, its favourable tax climate, and the absence of export control. Initially, the attention was focused on the development of a telecipher machine in collaboration with Gretener. Later, much effort was put in the upgrade of the C-machine, the CX-52, which did not perform well.

Another device, a one time pad generator called CBI-53, performed better but was too expensive to be a commercial success. In 1955, however, the CD-55 was patented, a genuine pocket-size machine originally designed for the French police but later sold in large quantities. This machine as well as the rising demand of telecipher machines for diplomatic use proved a sound basis for expansion of the factory. The development of a small radio device for telecipher machines, equipped with a small antenna and using little electricity, was another asset since it suited the needs of the French Foreign Office, for instance, which wanted its embassies to be able to operate independently of the power supply or the communication facilities of host countries.

It also paved the way for further diversification, which became increasingly important for the firm as the age of the cipher machine was drawing to a close. In 1970 Crypto AG entered the age of electronics under a new director, Sture Nyberg, who had worked with the company for many years. Shortly after the T-450 was launched, the first fully electronic online text encryption unit came onto the market. This machine was widely used for both civilian and military purposes.

Another innovation was the CRYPTOVOX CSE-280: a device for digital voice encryption. The continuing success story of Crypto AG is ample illustration of Kerckhoff's maxim that the secrecy of a communication line solely depends on the secrecy of its keys. The cipher system or device may well be known to enemies. This holds true for the age of cipher machines, as well as for the age of electronics and computers.

Matthew Aid writes about the deployment of Soviet signals intelligence (SIGINT) capabilities, mainly during the Cold War. He traces the origin of the Soviet code-breaking back to 1921, when the Cheka formed a cryptologic department, drawing its personnel mainly from the Tsarist black chamber that had rendered good service before the Revolution. The new department focusing on diplomatic traffic gained its first successes against German and Turkish diplomatic ciphers soon afterwards and was reading the diplomatic traffic of fifteen nations by 1925. Stalin's 'Great Terror' led to the execution of most of its staff during the latter part of the 1930s, including the successful founder of the department Gleb Bokiy who was considered politically unreliable.

Due to the outbreak of the war with Germany, efforts were made to regain strength mainly by recruiting academics. The Red army and navy had SIGINT organisations of their own, which did not collaborate well. Therefore, Soviet SIGINT capabilities were brought under direct control of the Peoples Commissariat of State Security 5th Directorate from November 3, 1942 onwards under General Shevelev. The result was a successful monitoring of German military activities in the air, sea, and land during the latter part of the war.

The military branch regained independence after the end of hostilities, resulting in a revival of much of the animosity between civilian and military intelligence that had existed before. Unfortunately, expenditure on civilian SIGINT was cut down and little efforts were made to invest in research and development, as was the case in the US which resulted in a technological backlog that was to last throughout the Cold War. Considerable results were gained, however, through conventional espionage, assisting SIGINT operations, such as the burgling of embassies in order to steal code books, or the bribing of American SIGINT personnel.

The most spectacular example of the last was the so-called Walker spy ring which provided the Soviets with cryptographic material for nearly twenty years in such quantities that it could easily have led to serious losses for the US Navy. Another important asset was the fielding of a new generation of domestically produced high-frequency intercept systems shortly after the end of the war based on a captured German design. This version allowed the Soviets to monitor virtual all NATO activity, mainly through traffic analysis and provided the Soviets with an early warning system in case of a pending nuclear war. Therefore, the Soviets were far from powerless, notwithstanding their inability to keep up with the pace of technological progress in the west.

Joseph Fitsanakis gives a brief account of the history of it National Security Agency, the American counterpart of the KGB's 5th Directorate in a manner of speaking. Unlike Tsarist Russia, the US had no long-standing tradition of code-breaking and, consequently, had been poor in the deployment of ciphers and of codes. By 1890 both the US army and navy had established small intelligence organisations, however, which were thoroughly under-funded and understaffed when World War I broke out in April 1917.

A newly-created intelligence section of the general staff was set up five days after America's entry into the war in order to fill in gaps of intelligence gathering. One of its branches was MI8, a cable and telegraph section, which was soon to become America's first full-grown civilian code-breaking organisation. Under the inspiring leadership of Herbert O. Yardley, the section decrypted eleven thousand foreign cable telegrams in nearly thirty months, mainly from Germany and from Central and South America.

There is no indication that the intelligence provided by MI8 greatly influenced American diplomacy during the negotiations at Versailles, and after the war Yardley's organisation was cut back from over three hundred in 1918 to less than twelve in 1924, which did not prevent Yardley from being successful against Japanese codes during the Washington Conference in 1921. However other major powers were untouched by these efforts in part because telegraph companies were no longer willing to supply the necessary intercepts. Not very surprisingly, Yardley's black chamber was closed in 1929 by the State Department, leaving the US without any code-breaking expertise.

The greatest damage was done to the army's Signal Corps, responsible for the provision of up to date cryptographic information. Therefore shortly after, a new branch was established, the Signals Intelligence Service (SIS), with cryptanalytic research being one of its functions. Under direction of William Friedman, the SIS elevated cryptographic practises in the army, and its methodical training program was at the root of the successes against enemy diplomatic codes at the outset of the war against Japan: better known as Purple. In 1949 the SIS assumed the co-ordination of the strategic activities of all US SIGINT organisations under the name of Armed Forces Security Agency (AFSA), and thus became a direct forerunner of the NSA.

In 1950, AFSA was still struggling to assert its authority and proved unable to predict the Korean War or to contribute much to it. The creation of the highly centralised NSA in 1952 was a direct response to this disappointing performance of the American SIGINT agencies which was meant to support well co-ordinated global policies in accordance with America's role as a superpower. Unlike its Russian counterparts, the NSA, as a matter of policy, heavily invested in computer research ultimately leading to the deployment of the Stretch supercomputer in 1962.

During the 1950s, the NSA also accomplished a spectacular augmentation of its intercept capabilities. These efforts were almost exclusively directed against the Soviet Union, leaving room for blind spots in other parts of the world, such as in the Middle East, where the Suez crisis would occur in 1956. The 1960s saw a further expansion of the agency, which by 1969 was to become America's largest and costliest intelligence agency with one hundred thousand employees and an estimated budget of two billion dollars.

The 1970s saw an increasing importance of satellite surveillance systems which were brought into orbit in collaboration with NASA and the National Reconnaissance Office (NRO). The satellites were a key component of a global, automated interception and relay system, codenamed ECHELON, fielded in collaboration with the UK, Canada, Australia, and New Zealand on the basis of an agreement stemming from the Second World War. Except from a comprehensive successful counterintelligence operation against the Soviets, known as Venona, little is known about the NSA's results and possible failures and even the scope of Echelon remains clouded in mystery.

The collapse of the Soviet Union in 1990 left the NSA without a mission, except for economic espionage against friendly countries which resulted in political tensions with the European Union and, in the US itself, to a revival of the debate about the NSA's possible role in the surveillance of American citizens. The organisation's budget was reduced substantially, and efforts were made to redirect priorities, nuclear proliferation being one of them. The future of the NSA is far from clear. The spread of fibre optic systems, combined with the increasing prevalence of strong cryptography, severely reduces its capability to monitor and the sheer increase in e-mail traffic in recent years seems to have a similar effect.

Bart Preneel investigates the development of contemporary cryptology, from the predecessors of the Digital Encryption Standard (DES) during the 1950s until today. Cryptology plays an essential role in the protection of computer systems and communication networks and civilian applications have increasingly gained in importance since the 1960s. For over thirty years, cryptologic research has slowly developed into a full-grown academic discipline. Progress depends on the publication of algorithms and proofs, leaving less and less room for the secrecy that has accompanied military research over the past.

Preneel dates the origin of this development to 1949 when C. Shannon published a mathematical proof that a perfectly random key sequence renders perfect security if the key is exactly as long as the message. This so-called Vernam scheme had been applied during and after the Second World War in various Soviet diplomatic cipher systems, which proved far from unbreakable for US cryptanalysts in the context of Venona. These results were due, however, to a reuse of tapes proving that errors in the use of a cryptosystem can be fatal even if the system is theoretically secure. This error pointed to a serious weakness in the Vernam scheme, however, because application depended on the exchange of huge volumes of cipher tapes which puts a strain on logistics.

The problems of key management and key exchange moved to the fore with the advent of the computer. Encryption was either left to stream ciphers, operating on the plaintext character by character or on block ciphers, operating on plaintext strings, which were sufficiently strong to defy attacks given a certain amount of computing power. This approach was firmly rooted in tradition, but had flaws that were particularly troublesome in computer networks. First of all, there was no possibility of exchanging keys over a network without taking the risk of interception. To make things worse establishing the identity of the other party was not straightforward because these so called 'symmetric' keys could easily be used to impersonate the original sender.

Public Key Cryptography, independently invented in 1976 by Diffie and Hellman and Merkle, was devised to solve both problems at once. In Public Key Cryptography, different keys were used to encrypt and decrypt messages. The public key was deposited somewhere on the web and could be used by anyone who wanted to send a message, but this message could only be decoded with the private key of the owner which never left the original location. The idea behind this method was based on an observation from number theory which says that it was much easier to multiply primes than to decompose them if these primes were of sufficient length.

Public Key Cryptography turned out to be the most important innovation in cryptology since the advent of the computer and it took only a decade to become an indispensable technology for the protection of computer networks. The irony of it all is that cryptology as a science has been of limited importance in achieving this breakthrough even as it successfully developed various models or approaches to think about the security of encryption algorithms. The information theoretic approach of Shannon has already been mentioned. The so-called 'complexity theoretic' approach departed from an abstract model of computation, for instance a Turing machine, and assumes that the opponent has limited computing power within this model. These approaches exist side by side with a more practical, or 'systems-based' approach, focusing on what is known about weaknesses and strengths of any algorithm in the context of a given configuration. The science of cryptology has never been able to prove that the factoring of primes is indeed as hard as it is supposed to be.

1.8 COMPUTER SECURITY

The fourth part of this book is on issues of computer security. Jeff Yost writes about the sense and nonsense of computer security standards, such as the *Orange Book*. He traces the origin of the quest for security standards back to the US Department of Defense's worries about security in time-sharing and multitasking environments during the 1960s. The DoD had played an important role, along with certain privileged companies such as IBM as well as high-level research institutions such as MIT, in the early development of computer technology during the 1940s and 1950s.

This development had taken place in private and relatively secure environments which were mainly protected by controlling physical access. The introduction of time sharing and multi-tasking made clear that there was no way of enforcing the strict classification of DoD-documents in degrees of confidentiality in a sophisticated computer environment. In October 1967, the Defense Science Board commissioned a task force, under the chairmanship of Willis Ware, to describe the problem. On February 11, 1970, this task force published its report, which stated among other things that providing satisfactory security control was in itself a system design problem.

In 1972, the US air force followed suit with a report written by James Anderson, developing the concepts of the security kernel and the reference monitor. In 1973, a first model for developing security policies was by Bell and LaPadula, focusing on the needs of the military and bureaucracy. In 1983, these and other initiatives acted as building-blocks for the *Orange Book*: the DoD's first rating system for establishing the level of security of any given computer system. In later years, there were an entire 'Rainbow Series' of security standard books that were being followed both in the US and in Europe and ending in 1996 with the *Common Criteria*.

These more recent rating systems were developed to accommodate the needs of network-computing and business life and privacy-sensitive civilian institutions, such as health care. The adoption of rating-systems like these outside of government use has been growing in recent years, notwithstanding the reluctance of organisations not bound by law. Interestingly, the US government has been far less successful in enforcing standards for cryptography. The introduction of DES in 1976 by the National Bureau of Standards happened with the specific purpose of promoting the use of weak cryptography for civilian use by limiting the key size. This effort backfired because of the rise of Public Key cryptography: a cryptographic algorithm of unprecedented superiority developed by academics unrelated to any government institution and marketed by private enterprise. This invention came to be the de facto world-wide standard for secure communication during the late 1980s and early 1990s against all efforts to block the export of sophisticated encryption technologies.

Dieter Gollmann writes about security models. A security model is a formal description of a security policy that a system should enforce. It is used as a yardstick in the design process of a computer system for determining whether it fulfils its security requirements. Security models are based on the state machine model. A state is an abstract representation of a computer system at any given moment in time. A security model represents a machine in an initial, secure state. The possible state transitions are specified by a partial state transition function defining the next state depending on current state and input. A "Basic Security Theorem" can be formulated when it is clear that all state transitions preserve the initial security of a particular system.

Security models are already recommended in the aforementioned report J. Anderson has written for the USAF in 1972. Various models were developed, but the Bell–LaPadula model from 1973 remained the most influential because it was recommended in the *Orange Book* and because it was applied in Multics. The BLP model reflects the multi-level security policy of the military, using labels such as 'unclassified', 'confidential', 'secret' and 'top secret'. The model introduces a reference monitor deciding whether to deny or to grant access on the basis of a user's clearance. In 1987 the BLP model was severely criticised by J. Maclean because it allowed a state transition which downgraded all subjects and all objects to the lowest security level and because it entered all access rights in all positions of the access control matrix and would still label a system as 'a secure state'.

Bell's counter argument was that if the requirements called for such a state transition, it should be possible to express it in security model. At the root of this disagreement is a state transition that changes access rights. Such changes would be possible within the general framework of BLP, but the model was intended for systems that are running with fixed security levels or to put it differently, it is intended for systems that are in a state of 'tranquility'. The BLP model had other limitations as well. It did not sufficiently capture multi-level security requirements for relational databases and neither did it capture the security requirements for commercial security policies. This flaw was pointed out by David Clark and David Wilson who argued that in business applications the unauthorised modification of data constituted a much bigger problem than a possible breach of confidentiality. They advocated the use of a specific set of programs for the manipulation of data items by privileged users working, if possible, under the condition of separation of duties.

Behind this debate, a technological shift can be noticed. BLP clearly refers to a world of general purpose, multi-user operating systems; Clark and Wilson refers to networks and application oriented IT systems. In parallel with the development of models for existing security policies, another line of work developed which explores the theoretical limitations of security models. Its purpose is the analysis of the intrinsic complexity of access control or the provision of generic frameworks for analysing information flow within computer systems. Early examples of these models are the Harrison–Ruzzo–Ullman model, the Take–Grant model from 1976, and a more recent example is the research on Execution Monitoring by Fred B. Scheider dating from 2000. Current work focuses on decentralised access control and the defining of flexible languages for expressing security policies without prescribing the types of policy that can be enforced.

Bart Jacobs, Hans Meijer, Jaap-Henk Hoepman and Erik Poll deal with the subject of programming transparency as a way of achieving of computer security. They trace the efforts to investigate computer security back to 1961, when Vyssotsky and McIlroy conceived a game for the IBM 7090 in which self-replicating computer programs tried to destroy each other. Starting out as a merely academic pastime, exercises like this gained importance because of the rise of multi-programming during the late 1960s. Efforts to improve programming transparency, as manifesting itself after the proclaiming of the software crisis in 1968, did not address security issues separately because security was perceived as a by-product of correctness. Ironically, computer scientists who did address such issues, such as the authors of the *Orange Book* occupied themselves mainly with security policies and access control.

The inadequacy of this approach became obvious only during the 1990s due to the spread of viruses and worms on the Internet and other exploiting bugs in programs. Jacobs and his colleagues believe a multi-faceted approach was necessary, and they consider the use of open software

to be an important pre-condition for improving computer security. Surprisingly, they recommend bringing the development of 'secure' software libraries into the hands of academic institutions under supervision of the state.

Laura de Nardis explains how the Internet protocol TCP/IP was initially designed for a closed community, such as the military and academia. Security hazards came to the fore after the adoption of the Internet by the general public. The sudden occurrence of the Morris-worm on 2 November 1988, infecting thousands of Unix-computers mainly in the United States, heralded a new era of public awareness. The attack was well planned by Robert T. Morris, son of a chief scientist at the National Computer Security Center to illustrate inherent Internet vulnerabilities and the inadequacy of prevailing network security measures.

Shortly afterwards, the Computer Emergency Response Team was founded to monitor irregularities on the Internet. Other, less benign attacks were to follow, leading to a rise of commercial security products, such as firewalls during the 1990s. The introduction of Virtual Private Networks to protect businesses and other organisations from intrusion was another approach which required strong encryption which, however, led to a fierce debate between policy makers who feared proliferation of military sensitive technologies to hostile countries and providers of encryption technology, lingering on until the turn of the century. This objection did not withhold viruses and worms from becoming legitimate plagues, causing serious disruptions of the Internet in 2001 (Code Red, Nimbda), 2002 (Klez), 2003 (Slammer, Blaster) and 2004 (My Doom, Bagle, Sasser).

Additionally, fear of cyber terrorism led to the development of a "National Strategy to Secure Cyberspace" by the US government, stressing the nation's dependency on information infrastructures. The recent growth of wireless Internet access has added a new dimension to the vulnerability of Internet use, and the numbers show that security incidents are rising annually. DeNardis does not rule out, however, that the introduction of a new Internet Protocol, IPv6, may result in an improvement in the near future.

Susan Brenner covers the issue of computer crime. She describes a development starting with the introduction of mainframe computers from conventional crimes, committed by insiders, often disaffected employees, to true computer crimes mainly committed by outsiders in the era of network computing and the Internet. The first wave of such activity emerged during the 1980s from the hacker community to make clear that legislation was lagging behind technological developments; the main problem shown was the 'virtual' character of gaining access which had never before been demonstrated.

Initially, these activities were relatively benign. During the 1990s, however, organised crime became increasingly involved on a world-wide scale, engaging in such diverse activities as online theft, online fraud, identity theft, and the use of malware for the purpose of extortion. These crimes were difficult to combat legally not only because of the unprecedented scale but also because of the international character of computer crime which demanded the harmonising of legislation and the collaboration of criminal investigation officers in the countries involved. The chapter ends on a pessimistic note. Brenner signals a lack of energy in the arena of international policy-making and law to combat cyber-crime. Unfortunately, the international crime scene is more energetic. There are now new types of organised crime, less hierarchically structured and less rooted in local folklore, or to put it differently more fluid and harder to combat than ever before.

Margaret van Biene-Hershey writes about IT-security and IT-auditing. IT-auditing came into existence around 1960 when accountants were initially confronted with important financial information that was being digitally stored. New, computer literate auditors were needed to assist the accountant in obtaining the evidence necessary to approve the financial statements made by an enterprise. They were called Electronic Data Processing auditors (EDP auditors) because during those early days of ubiquitous computer use the main issue was about the storage of data. No major changes were made in the structure of the work processes within any organisation; but the output increased

with fewer people. Input was made through punch cards, and magnetic tapes were used for back-up during the night.

The introduction of the IBM 360 in 1964 meant that computers became much faster, through the added features of multitasking and paging which made use of core memory which was suddenly much more flexible and efficient then ever before. The control on completeness and correctness could easily be done by the business unit responsible for the process. It was simply a matter of error detection. Audit-software programs were written to check the integrity of the master files through hash totals.

IT-auditing is primarily an audit of the computer centre's organisational procedures for carrying out production processes. In approximately 1970, this situation changed dramatically through the introduction of storage on disk, on-line data entry, and data base management systems which meant that the same data sets became accessible throughout a company and that the integrity of data could no longer be checked on a business unit level. Data base administrators assumed company-wide responsibilities and IT-auditors had to occupy themselves with procedures for data entry and data format which lay far beyond the scope of the computer centre.

Between 1972 and 1974, research initiated by IBM and executed by MIT, TRW systems, and IBM's Federal Systems Division itself resulted in a trail blazing series of technical reports about data security. The scope of IT-auditing was broadened with the audit of development of business systems and the audit of business systems controls. The IT-auditor reports directly to the management of the company and participates in the design of data communication processes as well.

The arrival of the PC and local area networks during the 1980s caused new data integrity problems, however, because programs and data could now, once again, be stored on the business unit level. The rise of the Internet added new security threats because it rendered the business networks vulnerable for attacks from outside. From 1990, the IT-auditor had to be concerned with new areas such as encryption and key management and also with the enforcement of privacy laws. The IT-auditor, then, increasingly assumes the role of an advisor to the management on security policy issues which does not mean that he is fully capable of fulfilling this new task. The poor quality of current Internet protocol, which makes it impossible to establish where a message comes from, prohibits the staging of fully effective security measures. The poor, or even absent, security-architecture in PC's adds to the problems. It is Van Biene's firm conviction that the computer industry should take responsibility for this issue.

1.9   PRIVACY AND EXPORT REGULATIONS

The fourth section consists of three contributions about privacy and export regulations. The first chapter in this part is written by Jan Holvast and deals with privacy or 'the right to be left alone' as it was first defined by Samuel Warren and Louis Brandeis in a famous article written in 1891. Privacy is a basic human need second only in importance to the need for safeguarding physical integrity and privacy issues can be traced back to biblical times.

The American constitution refers to privacy without using the term but by way of limiting the right of the state to enter citizens' homes. The concept acquired a broader meaning around the turn of the century with the rise of gossip journalism and the taking of pictures without consent. In 1928 a further dimension was added when federal investigators started wiretapping communications of citizens without a warrant under the pretext that wiretapping did not involve physical trespassing and was therefore fully constitutional. This reasoning did not hold and thereby, the right of the citizen to prevent the collection of data pertaining to him was firmly established.

The application of computer technology and the amassing of large volumes of data both by government agencies and by big business that goes with this area gradually changed the equilibrium that had existed before the war. Citizens have had no way of accessing data collected without new legislation. This issue sparked a political debate which resulted in a federal privacy act in 1974.

Unfortunately, efforts to curb the collection of data by private enterprise were hampered by the fact that the Constitution protected only the rights of the citizen against the state, but not against businesses, large or small. The development in Europe was somewhat different because the basis for future privacy acts was laid in 1950 with the European Convention for the Protection of Human Rights and fundamental freedoms. The information society resulted in a similar infringement of privacy as that found in the US and a similar effort to mend affairs through privacy acts in various member states culminated in a European Directive on Data Protection in 1990.

This Directive established a set of guiding principles for data collection, applicable on both private enterprise and government agencies. The core idea is that data collection should be purpose-bound, limited in the sense that these data should not be made available to third parties without consent, and that the individual should have the right of access without excessive costs. Regulatory agents, or data commissioners, are put in place to ensure that governments and business practises are in accordance with this law. In 1997, a second European directive was adopted, specifically related to the protection of privacy in the telecom sector which was a response to the emergence of spyware and the increase in unsolicited communications for marketing purposes, both occurring on the Internet.

In the US similar concerns had taken the form of self-regulating efforts by industry. The effect of these privacy laws seems limited, however, and insufficient to end further infringements. The driving forces behind the ever growing amassing of data are, in terms of money and power, too massive to be halted by legislation. A second, more abstract problem is the difference between information and data. Data can easily be protected, but information, that which is derived by putting data together, cannot be easily protected. In the end, it is information that constitutes the threat to privacy, not data.

Whitfield Diffie and Susan Landau explain why the tight US export control regulations for cryptography, valid for over forty years, were finally released at the end of the Clinton administration very much against the declared interest of the NSA.

Shortly after the Second World War the export of cryptography was put under the surveillance of the Munitions Control Board, a section of the state department which monitored the export of military goods with the objective of protecting national security. This Board acted on the instigation of third parties, in the case of cryptography, on the advice of the NSA or its predecessor.

During the early part of the Cold War, this method made sense because the demand for cryptography came almost exclusively from foreign governments in order to serve either diplomatic or military purposes. The situation changed with the rise of network computing during the last quarter of the 20th century. The importance of civilian use of cryptography increased, especially in the field of banking and commerce. The export control was retained nevertheless because it was considered an effective instrument to promote the deployment of weak cryptography abroad but also for civilian use in the US for reasons of crime investigation.

This use was detrimental to the interest of the US computer industry, increasingly challenged by foreign competition. After a public debate that continued for over a decade, control measures were relaxed in 1996 and finally abolished in 1999 as part of the election campaign of Vice-President Al Gore who sought support from Silicon Valley. Diffie and Landau do not believe that expediency tipped the coin altogether. The proven inability to prevent export of cryptographic algorithms in written form, such as Pretty Good Privacy (PGP), because of the Fifth Amendment, indicated that the existing export policy found little justification in the constitution and, therefore, had to change.

Finally, the rise of Open Source Software provided the rest of the impetus. Open Source Software is generated by a conglomerate of loosely associated programmers, who are scattered around the world. This allows anyone to add strong encryption modules to existing programs whenever the need requires the additions. Export control measures had always been targeted at companies, treating software as a finished and packaged product. The Open Source movement managed to make this approach largely ineffective.

Andrew Charlesworth writes about the consequences of the crypto controversy for the privacy debate predominantly in US and in Europe. Until 1970 there had been little public or academic interest in cryptology and organisations such as NSA have found slight difficulty in controlling cryptological research since previously this concern took place in the context of military institutions or government laboratories. All is changed now with the spread of personal computing which creates a need for cryptography as a privacy tool which in turn has stimulated a new wave of a cryptological research taking place in the context of computer science departments and practised as an academic discipline.

The watermark of academic research is the publication and subsequent scientific debate of results which potentially undermines any government monopoly in this field. Initially, the NSA focused on discouraging the publication of cryptological research in the name of national security and prohibiting export. The collapse of the eastern block in 1989 made this approach obsolete, however. At the same time, the rapid expansion of the Internet made the public availability of strong encryption tools commercially viable and a new political issue connected to the privacy debate. In 1994, the US Government responded by removing encryption software from the list of military goods that could only be exported with the consent of the state department. At the same time it also introduced the principle of key escrow. Key escrow entailed the obligation of commercial encryption suppliers to deposit keys with the law enforcement agencies or, worse still, the obligation to install so-called trapdoors in any encryption device put for sale thus allowing NSA to read encrypted communications.

This policy change was defended in terms of law enforcement and counter terrorism, but it also brought all communications under surveillance of the state which meant a huge encroachment on the privacy of American citizens. Unfortunately, key escrow made American encryption products unattractive for buyers outside the US who had something to hide from American authorities, such as governments with independent views of international affairs or businesses with American competitors. US efforts to find general acceptance of these measures through negotiation in international bodies, such as OECD failed because most countries believe that the NSA was using its large intercept capabilities mainly against its former allies for economic espionage.

Interestingly, the opposition to the American proposals led to a liberalisation of encryption market throughout most of Europe and even in countries that had been extremely restrictive in the past, such as France. After 2000 the effort to control the proliferation of encryption technologies dwindled in most countries, but this lessening does not mean that the concern for privacy is no longer warranted. With the introduction of its so-called Magic Lantern, the FBI is now able to monitor the on-line behaviour of any suspect in the name of law enforcement which certainly illustrates a general tendency of law enforcement agencies to evade encryption through the instalment of spyware. Another and perhaps more important threat is the use of strong encryption to protect intellectual ownership, as is the case in the Digital Copyright Millennium Act which makes it virtually impossible to access information without being known to the owner, which if it were possible to do would be an encroachment of privacy without precedent.

1.10 INFORMATION WARFARE

The sixth part of the book consists of one contribution that really does not fit into any of the other categories and that is Dan Kuehl's chapter about Information Warfare. Kuehl shows how military analysts take into account the growing strategic importance of information systems, both civilian and military, for warfare. He sees three vital areas in which the information revolution is shaping the future of national security. The first area is about protecting computer-controlled critical infrastructures from malevolent acts and attacks. The second is about the growing importance of information operations to military affairs, and the third is about the struggle for global influence in a political and ideological sense.

The threat to critical infrastructures or, more specifically, command and control centres is not new. The bombing campaign against Germany was

an effort to disrupt industrial plants, and the railway system served as an example of the former. The bombing of Turkish telegraph nodes by General Allenby during the campaign in Palestine in 1918 illustrates the latter. What is new, however, is the growing use of supervisory control and data acquisition technologies to monitor the status and to control the operation of a segment of this infrastructure, such as a rail network or an electric grid.

These can be attacked without much effort, from anywhere in the world by only a handful of people who do not even have to be in the same place. Information operations aim at exploiting the vulnerabilities of the electronic infrastructure. Psychological warfare, operations security, military deception, electronic warfare, and computer network operations are all part of information operations. The operations theatre itself merits a closer examination, however. The physical dimension of an information environment is made up of the infrastructures, networks, and systems that store, transfer, and use information. The information that is, the content is itself the second dimension.

The third dimension is the cognitive that is, how the information is perceived. Information operations aim at the manipulation of data, for instance for the purpose of military deception or psychological warfare which means that they focus on content but not on infrastructure. The third dimension comes closer to the traditionally known propagandistic aspects of warfare with the notable difference that the propagation of ideas, views or perspectives through the Internet is hard, if not impossible, to control. Information operations have acquired a prominent role in military theory because it may result in a rapid disruption of enemy command and control centres or supply lines. The Chinese in particular seem to envisage information operations in the context of asymmetric warfare against the West by allowing poorly equipped parties to gain military successes against an otherwise superior enemy. Kuehl argues that existing strategic concepts, such as those developed by Von Clausewitz are still viable to come to grasp with these new threats. He does emphasise, however, that the military cannot do its job alone; any strategic effort to protect a national infrastructure should include the civilian sector as well.

1.11 CONCLUDING REMARKS

The information society is based on an unprecedented civilian deployment of security tools and technologies which is insufficiently weighted in the current accounts of the impact of the new information and communication technologies, such as the Internet. On the contrary, the advent of the Internet has inspired many almost utopian visions of society. The Internet is supposed to bring an unrestricted flow of information. Eventually this influx will lead to the erosion of the role that traditionally has been hierarchically structured within organisations. Fairly typical is a statement by Douglas Cale, National director of Computer Assurance Service Practice of Deloitte & Touche who says: "The first characteristic of sound knowledge management is collaboration as opposed to organizational hierarchies. A second is openness as opposed to a separation of functions (...). Traditional structures such as entry controls, data security, and information classification are barriers to creating knowledge and managing it well".[7] This statement may all too well be true but it does not make such barriers superfluous.

Security tools have a long history but they have always been applied on a limited scale, in the context of hierarchically structured organisations, such as the military, within bureaucracy, or within corporate structures. Previously the deployment of security tools was in the hands of organisational apparatus that already obeyed rules of confidentiality, integrity, and availability. This may no longer be the case and it seems unlikely that technology can fully compensate for that lack. The spread of malware through the Internet proves that present computer security practises are insufficient. The quest for provable correctness of software has not been able to turn the tide thus far, perhaps because Open Source software is insufficiently backed by government policies.

Cryptographic protocols seem to be functioning well, but the scientists have not been able to prove why this is the case. Biometrics systems lag far behind the expectations of policy makers but improvement is likely due to the massive deployment

of inferior technologies that is now underway and the problems that this situation will cause in the nearby future.

The capacity of policy makers to steer technological developments is limited which can be learned, for instance, from the relaxation of export regulations for encryption software after many years of political strife. This policy change was as much a result of political expediency as of the advent of open software, which made it possible to export unfinished products and include encryption modules abroad. The liberalisation of the use of encryption software did not eliminate all threats to privacy, however. The Digital Copyright Millennium Act for instance, makes it virtually impossible to access any information without being known to the owner, which of course would be an encroachment of privacy without precedent.

More generally, policy makers all over the world have been reluctant to develop a legal regime that meets the necessities of the information age. This short-coming can be seen, for instance, in the fight against computer crime especially in as much as this requires international co-operation. Privacy laws have been proven unable to prevent the occurrence of massive data-warehousing for commercial purposes because the enforcement agencies that have been created everywhere are not strong enough to turn the tide.

Moreover, the continued existence of large signals intelligence establishments as well as the rumours of economic espionage seem to suggest that governments, even those that ascribe to the rules of fair competition, are not always playing legally and fair. Under these circumstances, a co-ordinated effort to protect the information infra-structure of democratic countries seems unlikely.

REFERENCES

[1] J. Agar, *The Government Machine*, MIT Press, Cambridge, MA (2003).
[2] M.M. Aid, C. Wiebes (eds), *Secrets of Signals Intelligence during the Cold War and Beyond*, Frank Cass, London (2001).
[3] R. Anderson, *Security Engineering. A Guide to Building Dependable Distributed Systems*, Wiley, New York (2001), pp. 200–203.

[7]Quoted in: Donn H. Parker [24, 7,8].

[4] C. Beavan, *Fingerprints. The Origins of Crime Detection and the Murder Case that Launched Forensic Science*, Hyperion, New York (2001).

[5] D. Bell, *The End of Ideology: On the Exhaustion of Political Ideas in the Fifties*, Free Press, New York (1965).

[6] D. Bell, *The Coming of Post-Industrial Society. A Venture in Social Forecasting*, Basic Books, New York (1973).

[7] J.R. Beniger, *The Control Revolution. Technological and Economic Origins of the Information Society*, Harvard University Press, Cambridge, MA (1986).

[8] W.E. Bijker, T.P. Hughes, T. Pinch (eds), *The Social Construction of Technological Systems: New Directions in the Sociology and History of Technology*, MIT Press, Cambridge, MA (1987).

[9] E. Black, *IBM and the Holocaust*, Crown Publishers, New York (2001).

[10] K. Breckenridge, *The biometric state: the promise and peril of digital government in the New South Africa*, Journal of Southern African Studies 31 (2005), 267–282.

[11] J. Caplan, J. Torpey (eds), *Documenting Individual Identity: The Development of State Practices in the Modern World*, Princeton University Press, Princeton, NJ (2001).

[12] A. Chandler, *The Visible Hand: The Managerial Revolution in American Business*, Harvard University Press, Cambridge, MA (1977).

[13] J. Ferris, *Airbandit C31 and strategic air defence during the first Battle of Britain, 1915–1918*, *Strategy and Intelligence. British Policy during the First World War*, M. Dockrill and D. French, eds, Hambledon, Condom (1996).

[14] M. Foucault, *Discipline and Punish: The Birth of the Prison*, Penguin, Harmondsworth (1980).

[15] D. Gollmann, *Computer Security*, Wiley, Chichester (1999), pp. 5–11.

[16] A. Jones, G.L. Kovacich, P.G. Luzwick, *Global Information Warfare. How Businesses, Governments, and Others Achieve Objectives and Attain Competitive Advantages*, Auerbach, Boca Raton, FL (2002), p. 20.

[17] D. Kahn, *The Codebreakers*, Macmillan, New York (1967).

[18] G.E. Kurtz, *Willem III en Amsterdam, 1683–1685*, Kemink, Utrecht (1928), pp. 95, 98, 107–109.

[19] S. Levy, *Crypto. How the Code Rebels Beat the Government-Saving Privacy in the Digital Age*, Penguin, Harmondsworth (2001).

[20] S. Levy, *Hackers: Heroes of the Computer Revolution*, Penguin, Harmondsworth (2001).

[21] F. Machlup, *The Production and Distribution of Knowledge in the United States*, Princeton University Press, Princeton, NJ (1962).

[22] D. MacKenzie, J. Wajcman (eds), *The Social Shaping of Technology*, Open University Press, Buckingham, Philadelphia (1999), pp. 10–11, 142–143.

[23] A. Menne-Haritz, *Business Processes: An Archival Science Approach to Collaborative Decision Making, Records, and Knowledge Management*, Kluwer Academic Publishers, Boston, Dordrecht, London (2004).

[24] D.H. Parker, *Fighting Computer Crime, a New Framework for Protecting Information*, Wiley, New York (1998).

[25] E.S. Raymond, *The Cathedral and the Bazaar: Musings on Linux and Open Source by an Accidental Revolutionary*, O'Reilly, Sebastopol, CA (1999).

[26] B. Schneier, *Beyond Fear*, Copernicus Books, New York (2003).

[27] S. Singh, *The Code Book. The Science of Secrecy from ancient Egypt to Quantum Cryptography*, Fourth Estate, London (1999).

[28] A. Toffler, *The Third Wave*, William Morrow, New York (1980).

[29] A. Westin, *Privacy and Freedom*, Atheneum, New York (1967).

[30] S. Williams, *Free as in Freedom: Richard Stallman's Crusade for Free Software*, O'Reilly, Sebastopol, CA (2002).

PART 1

INTELLECTUAL OWNERSHIP

2

LIMITATIONS ON THE PUBLISHING OF SCIENTIFIC RESEARCH

Jack Meadows

Information Science Department
Loughborough University
Loughborough, Leics LE11 3TU, UK

Contents

Abstract

Scientific research is usually thought of as a field where open publication is the norm. In the early days of science, secrecy was much commoner. Journals have played an important part in encouraging greater openness by allowing researchers to substantiate their claims to priority. However, journals also impose restrictions on publication, since editors and referees combine to decide what research should be published. Equally, there are restrictions on journal access and use due, for example, to financial limitations and copyright. Journals are not the only outlets for research, but other outlets – such as books – are subject to parallel restrictions. Report literature is often particularly difficult to access. Developments in electronic communication should, in principle, improve access to research information, but progress in this respect depends on the field of research.

Keywords: authors, classified research, copyright, editors, grey literature, journals, patents, peer review.

2.1 INTRODUCTION

The essence of publishing is to make information public. In other words, it is the opposite of secrecy. Scientists, in particular, have proclaimed the im-portance of publishing their work in the open literature. What then is there to say about hidden information in the sciences? The first point to make is that open publishing of scientific information is something that has developed over recent centuries. So one needs to ask why it has been accepted. Then there is the question of how well it works. In other words, how 'open' is open publishing in the sciences? Are there limitations on the freedom of publication of scientific research, which are acting beneath the apparently 'open' surface? Some scientific information – for example, commercial or military – is clearly not published openly. How does this fit in with the accepted process of scientific communication, and does it lead to significant tensions? These are the kind of queries that will be discussed here.

2.1.1 Background

In Renaissance Europe, new knowledge was often regarded as something not to be published. An obvious example is the work of the alchemists. At that time, 'alchemy' and 'chemistry' were used as interchangeable words, but alchemy was mainly concerned with the production of valuable metals from less expensive materials, or with the preparation of pharmaceutical remedies. Alchemy dates back to the Hellenistic culture of the first few centuries AD (though the word 'alchemy' itself comes from the Arabs, who followed up the ideas of the Greeks in this field). It was from the start secretive, and alchemists used allegorical or allusive language to describe what they were doing. For example, in an alchemical text of the early seventeenth century, we read that: "The grey wolf devours the King, after which it is buried on a pyre, consuming the wolf and restoring the King to life". This is actually a description of extracting gold from its alloys. The extraction was done by removing metal sulphides formed by a reaction with antimony sulphide, followed by a roasting of the resultant gold–antimony alloy until only the gold remained [3]. Some developed their own ciphers in order to hide their thoughts. For example, Leonardo da Vinci developed his own system of shorthand for writing down his ideas: he also employed 'mirror writing' – that is, writing from right to left. Practitioners often felt that keeping important innovations secret would enhance their prestige in the eyes of potential patrons. For example, the solution of certain types of equation in algebra was a matter of interest. In 1545, Cardano published *Ars Magna*, a book which

described how to solve these equations. In fact, the first steps towards a solution had been made around 1500 by del Ferro, but he never published his solution, only explaining it to one of his students. He was followed in 1535 by Tartaglia, who kept the solution secret in the hope of gaining more patronage. Cardano managed to extract the solution from Tartaglia, swearing an oath that he would not reveal the secret. Hardly surprisingly, when Cardano did publish, there was a major dispute.

2.1.2 The decline of secrecy

Ars Magna has been called the founding text of modern mathematics. Two years before it appeared, Copernicus published *De Revolutionibus Orbium Coelestium*, usually regarded as the founding document of modern astronomy and, in the same year – 1543 – Vesalius published *De Humani Corporis Fabrica*, the founding text of modern anatomy. There seems to have been upsurge around this time not only in the rate of innovation, but, more importantly, in publishing information about the progress that had been made. Part of the reason was certainly that the printing and distribution of complex text had become easier. This meant that, for example, the detailed anatomical drawings that Vesalius needed for his book could now be produced and printed. At the same time, such publications became an important way of bringing the author's work before the attention of possible patrons. Thus the success of Vesalius' book was a major factor in his appointment as personal physician to Charles V. More generally, as the sixteenth century gave way to the seventeenth, it came to be seen that science was a cumulative structure. For new developments to occur, they must be built on what was already known. Under these circumstances, keeping things secret would harm not only the scientific community as a whole, but also ultimately the work of each individual scientist. This belief was summed up by Newton (though he was neither the first nor the last to use this form of words) – "If I have seen farther than others, it is because I have stood on the shoulders of giants".

This is not to say that all secrecy departed from scientific innovation at that point. Methods were sought whereby claims for priority in making important advances could be established without actually making the innovation public. One device was the anagram. A sentence was composed describing the new discovery, and then the letters were jumbled up to disguise the original meaning, and the result published. Having established a claim to priority, the scientist could then go on to develop the work further until it was ready to be published. If, in the interim, someone else announced the discovery, the meaning of the anagram could be revealed and priority claimed. For example, Galileo wrote to Kepler in 1610 and his letter contained the following sequence of letters: smaismrmilmepoetalevmibunenugttaviras. Galileo subsequently told Kepler that the letters could be rearranged to give the sentence: *altissimum planetam tergeminum observari* [meaning – I have seen the uppermost planet triple]. In other words, Galileo had discovered the rings of Saturn; but he wished to have time to make more observations before publishing his account [21].

The use of anagrams had severe limitations. They had to be short, otherwise opponents could claim that the letters might be rearranged to give more than one meaning. More importantly, a short sentence could not include all the evidence for a claim, which might therefore be just be a lucky guess. For these reasons, anagrams were soon discarded. Another method of providing secrecy continued longer. This was the practise of depositing a sealed letter, carefully dated, with a learned society. When requested by the author, the officers of the society could open the letter and verify the contents, so confirming the author's priority. Since the manuscript could be of any length and so could contain all the evidence necessary, this method was not subject to the same objections as anagrams. It was used from time to time during the eighteenth century, but, by the nineteenth century, it was generally accepted that priority for a discovery should usually go to the person who made it public first, not to the one who first thought of it. This led to the situation that exists today where the emphasis is on priority of having a scientific paper accepted for publication. But it was still felt during the nineteenth century that discoverers should be

allowed some lee way to develop their ideas. This sense of 'property' in scientific work paralleled the nineteenth-century development of patents for technical innovation. An eminent scientist at the end of the nineteenth century commented:

> A habit has grown of recent years among certain scientific men, which many of those with whom I have discussed the subject join with me in regretting. It is this: after the announcement of an interesting discovery, a number of persons at once proceed to make further experiments, and to publish their results. To me it appears fair and courteous, before publication, to request the permission of the original discoverer, who probably has not merely thought of making identical experiments, but who has in many instances made them already, and has deferred publication until some grounds exist for definite conclusions.

> An analogy will perhaps help. If a patent has been secured for some invention capable of yielding profit, and some person repeats the process, making profit by his action, an injunction is applied for and is often granted. Here the profit of the business may be taken as the equivalent of the credit for the scientific work completed; no original idea, undeveloped, is of much value; before it produces fruit, much work must be expended; and it is precisely after publication of the original idea, that sufficient time should be allowed to elapse, so as to give the author time to develop his idea, and present it in a logical and convincing form [24].

This view was already becoming untypical when the author was writing. It has since more or less disappeared. Interestingly enough, the decline of the idea of research as a form of personal property has been accompanied by a decrease in the frequency with which published research leads to disputes over priority. This fall is partly a result of the success that journals have had in regularising procedures for claiming priority. But it also reflects an acceptance by scientists of the inevitability that some discoveries will be made simultaneously by two or more different people. The bitterness that early priority disputes could engender is illustrated by one of the most famous – the dispute between Newton and Leibniz over the invention of the calculus. The basic problem was that Newton developed the relevant mathematical concepts in the 1670s, but never published. Leibniz developed his own approach in the 1680s, and did publish. In the interim, Leibniz and Newton had corresponded about matters relating to the calculus. The result was that, in the early eighteenth century, Newton's supporters – with Newton's agreement – began to claim that Leibniz had borrowed his ideas from Newton. The resulting dispute involved not only the Royal Society, but foreign diplomats as well, and continued even after the death of Leibniz in 1716. This polarisation of views had an unfortunate effect on British mathematics. Continental mathematicians followed the approach developed by Leibniz, which proved to be more helpful than the Newtonian route followed by British mathematicians. As a result, British mathematics lagged behind developments on the Continent throughout the eighteenth century – a heavy debt to pay for not accepting the possibility of independent discovery.

In terms of making scientific research rapidly available, the rush to seize on published results, and to follow them up by further publication as soon as possible, has an obvious value. It does, however, also have a down side. An eminent scientist writing in the mid-twentieth century noted that the traditional scientific process of collecting data and developing theories was increasingly being circumvented. Instead, scientists were trying to make a best guess at what was happening and then collecting just sufficient data to see if their guess could be supported. This has certainly speeded up the growth of science, but has also made more likely the publication of research that would later need some revision. The wish of researchers to establish their priority in a discovery has been the main driving force behind this approach, but it can actually lead to problems in assigning priority. For example, some years ago a new method for testing Einstein's theory of general relativity became available. The possible application was taken up by two scientists – let us say M and N – who set to work independently. M published first – a letter explaining how he intended to carry out the experiment. A few weeks later, the front page of the New York Times carried a picture of M with an article about the experiment that he now had operating. Six weeks later still, M described his work at a scientific meeting, where he reported that he

had not yet obtained any definite results. However, N was present at the same meeting. He then reported that he had followed the same path as M, but had already obtained results agreeing with theory. He published his results not long afterwards, apparently pre-empting M. But two months later M reported not only his own confirmation of the theory, but also his discovery that reliable results required a much stricter control of temperature than N had used. So, after all this rapid release of information, who had priority? Contemporaries found it hard to decide, but the story underlines the point that the progress of research activity, which would have formerly preceded publication, may now be part of the publication process. It is to this latter that we now need to direct our attention.

2.2 AUTHORS AND PUBLISHING

The idea of unrestricted publishing of results grew up predominantly in the field of pure science and, since the nineteenth century, primarily in the world of academic research. The typical author of a paper in a scientific journal nowadays is most likely to be a member of staff at a science department in a university. The question, then, is what limitations may there be on such a person publishing a piece of research? The easiest way of examining this question is to look at the various stages of the publishing process and to ask what restrictions may operate at each stage. Publishing can be divided into three basic steps. It starts with the author, or authors, who produce the information. This information then passes through an editorial handling phase at a publisher, before going on to production and dissemination. The final stage is the receipt of the information by a reader (who may use it to help produce further information). So the first step in our examination of restrictions on the dissemination of information must begin with those faced by authors. They fall into two main categories – limitations due to material circumstances and limitations due to external pressures.

2.2.1 Restrictions on authors

Obviously, material limitations can prevent a scientist from producing research in the first place.

For example, experimental scientists need appropriately equipped laboratory facilities to do their work. But, even if results are produced, there can be difficulties in publishing them. The journals in which most scientists would like to publish are international in terms of circulation. Many of them expect papers to be submitted in English. Not all scientists know English well enough to write a scientific paper in the language. In a rich country – such as Japan – this can be overcome by employing expert translators. A poor country may be lacking in such facilities. Moreover, journals now increasingly ask for papers to be submitted electronically in a particular format, which can create problems for scientists in poorer countries. But if the research is published in a local journal and in the local language, there is a strong chance that it will be overlooked. There is the famous example of Gregor Mendel, who published the results of his genetics experiments in the *Proceedings* of his local Moravian natural history society during the 1860s. Although, in this case, the results were presented in the then dominant scientific language of German, their importance remained unrecognised until the early years of the twentieth century. The essential point here is that publication, as such, is not necessarily the same as making one's results readily available to the scientific community. As one scientist commented at the end of the nineteenth century: "It ought to be clear that mere printing in a half-known local journal is not proper publication at all; it is printing for private circulation" [16]. The problems of identifying and accessing published material will be looked at later under the section on readers.

Though scientists may differ in terms of the facilities and support they require, all need financial backing of some sort. Increasingly, the acquisition of grants to carry out research work has been seen not merely as a support for the research, but also as an assessment of the value of the research. Universities now proclaim how much external funding their science departments have obtained as a measure of how good the departments are. The resultant competition for grants means that many applications go unfunded. There is a knock-on effect here. It becomes more difficult under these circumstances to fund research that is speculative and may lead to little by the way of results, when there are applications that clearly will produce results. So the successful applications may often be the safer ones. Moreover, the policies of the grant-giving agencies often emphasise particular types of research. For example, in recent years, many countries have seen an emphasis on research that leads to useful applications. This move towards applicable research has implications for open publishing: both the grant-giving agency and the scientist's institution may impose limitations on this. The problem is that the views of these bodies may conflict with the views of the individual researchers on what it is reasonable to publish, and what is not. The effects of this potential clash will be looked at later. At this point, it is simply necessary to note that external pressures can limit an author's willingness or ability to publish research results.

Such pressures have existed throughout the development of modern science. For example, the concept that the Earth revolved round the Sun was contrary to the teaching of the Roman Catholic church in the sixteenth and seventeenth centuries. Copernicus, who first presented a detailed argument for the heliocentric theory, only had his work published when he was on his deathbed (and it was put through the press by a Protestant scholar). In the early seventeenth century, Galileo came to support the heliocentric picture, but did not publish on the topic after receiving a private warning from the Inquisition. The election of a new Pope led him to believe that a discussion of the topic would now be acceptable. However, his resulting publication of 1632, *Dialogue on the Two Chief World Systems*, led to his arrest and trial in Rome. He was forced to recant, his book was banned, and he was placed under house arrest for the remainder of his life. Two further examples come from the two great dictatorships of the twentieth century. Some physicists in fascist Germany believed that it was possible to create an Aryan physics, which would be purged of 'Jewish elements'. Since Jewish scientists had played a fundamental part in the development of relativity and quantum mechanics, this meant that both these topics were to be regarded with suspicion. Work on them continued, but some caution had to be exercised when it came to publication.

More serious restrictions occurred in Communist Russia – partly because the communists ruled for much longer than the fascists, and partly because decisions about what was acceptable in science came from the top. Stalin claimed that 'bourgeois science' – what was practised in the West – was of only limited potential compared with the scientific approach backed by communists. Part of the objection was again to 'Jewish science' in such areas as physics. But the main brunt of the attack was in the field of biology. Western understanding of genetics over the past century has been based on the approach pioneered by Mendel. Genetic material, according to Mendelian inheritance, changes only slowly from one generation to the next, and modifications are not directly due to the environment. Lysenko in the USSR argued, on the contrary, that characteristics produced by environmental conditions could be inherited. If this were true, then changes in genetic characteristics could be made to occur relatively quickly. This fitted in much better with communist beliefs and needs than the concepts of traditional genetics. By the 1940s, Lysenko had obtained the backing of Stalin. Publication of research based on Mendelian principles stopped, and its proponents were silenced. Indeed, the leading Soviet geneticist of the period, Vavilov, who strongly opposed Lysenko, died in prison in 1943. Support for Lysenko faded after Stalin's death, and publication of research based on Mendelian genetics once again became possible.

Though state-supported suppression of ideas is usually the most effective way of driving research underground, other pressure groups can have the same effect. For example, environmental pressure groups who oppose genetically modified crops have led to a more secretive approach to work in this area. Likewise, animal-rights activists who attack people working on laboratory animals ensure that there is greater secrecy about research in this area than would otherwise be the case. A number of similar cases can be found – for example, the influence of feminist extremists who attack research into male/female differences. Pressure of this sort has two effects. The first is to make communication relating to the research more secretive. The second is to deter researchers from entering the field

concerned. In fact, since scientists are, to some extent, geographically mobile, such pressure can lead to them moving to another country where the pressure is less. So the Jewish scientists who left Nazi Germany continued their research – mainly in the UK and USA. Even if researchers do not move physically, the different pressures in different countries mean that work on a particular topic will go on somewhere. Thus, research on laboratory animals is a more sensitive subject in the UK than in the USA, whereas the reverse is true of research into gender differences. This affects the caution of researchers in the two countries in terms of discussing their work. The overall result, in terms of global science, may be that the secrecy enforced by pressure groups has only a limited effect, since what is not investigated in one country may be investigated and published about in another.

2.2.2 Competition and access to results

External pressures obviously vary from subject to subject and from time to time. More generally, the scientists, themselves, though they wish to publish their research, worry about revealing too much about their work before it is ready for publication. The reason is that they are concerned that someone else may publish on the topic first, and they will lose their priority as the discoverers. A substantial minority of scientists – often those involved in research in a highly competitive field – are therefore unhappy about sharing the details of their research plans with others. A study of scientists in Croatia suggests that all scientists are aware of such secrecy, but senior scientists see this as a greater problem than their junior do [23]. However, the importance of priority varies to some extent from field to field. For example, direct competition in some areas of mathematics is relatively low, but, since the second proof of a proposition is not usually required, if a competitor publishes first, it can completely wipe out the value of a piece of work. In an area such as high-energy physics, confirmation of a result is often important. So, though priority may have been lost, publication of additional results is valuable. The worry about sharing ideas is, of course, that the work being done will be rapidly copied and published elsewhere. This probably occurs considerably less often than is feared, but it can happen. Here is a story about an English chemist related by an American:

> *X* has been known to wipe you out if he finds the direction of your research. He will quickly get into the area and publish the most important results. He once scooped Professor *Y* of this department. *Y* had talked freely about his research plans, and we are quite sure that someone he talked to visited England and got the news to *X*. *X* does not usually write up research notes, which are quickly published, but rather substantial articles, which take longer. On this occasion he did, however, lending substance to the idea that he knew *Y* was working on the same thing [8].

The point at issue here is speed of publication. Different journals publish at different rates. The result is that, if two different scientists make a discovery simultaneously, the one who publishes in the journal that appears first is likely to claim priority. This has been a problem for many years; consider the following complaint from a century ago:

> It would naturally be supposed that the *Trans. Zool.* is preeminently the place for such publications [on comparative anatomy]. But it will scarcely be credited that a wealthy Society like this, for some unknown reason, should allow in some cases as long as two-and-a-half years to elapse before publication of material received.... During such a long period I have found it necessary to keep pace with much more literature bearing upon the subject; but more than this, I have just suffered the chagrin of seeing a paper embodying a large slice of my results published by an Italian journal [34].

The opposite problem can arise with researchers who are perfectionists. They may sit on their results and refuse to publish when they should do. One of the notorious examples of this in the history of science is John Flamsteed, the first Astronomer Royal in England at the end of the seventeenth century. Flamsteed was appointed to make the measurements of stars and the Moon, as these were required to develop methods of navigation at sea. They were also needed for detailed testing of the ideas on motion and gravitation recently formulated by Newton. But Flamsteed refused to be

hurried. After many years of waiting, Newton's friend Halley finally gained access to some of the observations, and published them without Flamsteed's agreement. Flamsteed responded by buying as many copies of Halley's report as he could and burning them. Only a few years later Flamsteed, himself, died with the results still unpublished. They were finally published by Flamsteed's assistant, nearly half a century after Flamsteed began work.

Slow publication, whatever the cause, is not only capable of upsetting fellow-scientists: it can also affect the free flow of scientific information. Since it is one of the problems arising from the publishing process, we turn to this next.

2.3 EDITORIAL ACTIVITIES

The important step in making their work public for most scientists working on pure research is the submission of their results to a journal. Their selection of which journal to use is based mainly on three things – the prestige of the journal, the expected readership of the journal, and the speed with which it publishes. Thus journals such as *Science* or *Nature* have high prestige because of the quality of the research they publish, because they have a large audience world-wide, and, because, since they appear on a weekly basis, they can publish new research very rapidly. Concern about speed of publication tends to be dependent on discipline. In disciplines where speed is regarded as important, it has been customary for some decades to have 'letters' journals. Their purpose is to publish short accounts of research appreciably more quickly than the standard journals in the subject can manage. It was initially supposed that these short announcements would be followed up at leisure by fuller accounts in the standard journals. Instead, 'letters' have frequently become ends in themselves and letters journals have been used for the speedy publication of results in areas of research that are developing rapidly.

2.3.1 Editors and referees

A journal can be seen as a kind of bottleneck, which determines whether research is published, and, if so, how quickly. The way in which journals handle submitted material is therefore highly significant in terms of providing public access to scientific information. So it is important to examine how material submitted to a journal is actually treated. The process begins when a manuscript from a scientist arrives at a journal. It is first looked at by the editor, who makes the initial decision on how to deal with it. One option is immediate rejection. Journals which are regarded by researchers as having a high prestige typically receive many more articles than they can publish. In consequence, their editors can, and do, reject articles on their own individual initiative. A study of one major medical journal, for example, found that over half the incoming material never progressed beyond the editorial office. There is some indication that, besides deciding which material can be immediately rejected, editors are also more likely to assess – more favourably – articles from well-known researchers, since they are expected to submit material that will prove acceptable for publication. Editors of the most important journals can directly influence what is accepted, and what rejected, simply by favouring particular topics or methodologies at the expense of others. Even editors of journals where the pressure for publication is less can influence the contents of a journal in this way.

In times past, editors often relied on their own opinions considerably more than they do today. Growing specialisation has made it more difficult for an individual to encompass the whole of a field. Sometimes, reliance is placed on the experience of the authors. It is argued that leading scientists would not wish to sully their reputation by publishing substandard work. Consequently journals aimed at such authors can be edited more lightly. For example, members of the National Academy of Sciences in the USA can publish their research in the Academy's *Proceedings* after only a fairly informal review. But, for the most part, editors call on specialist helpers – 'referees' – to help them assess the value of submitted material that is not obviously unacceptable. The process of using referees has become so common that publication in a 'properly refereed' journal is now regarded as being virtually synonymous with publishing research

that is acceptable to the scientific community. This does not remove the editor's final responsibility for publishing a paper. A notorious example in recent years was the publication of a paper on the MMR (measles, mumps and rubella) vaccine that is routinely given to young children in the UK. An article appeared in *The Lancet* claiming that administration of the triple vaccine might be linked to autism. The result was that large numbers of parents refused to allow their children to be vaccinated. Further work in recent years, indicates that there is no such link. A number of people working in the field think that the execution and interpretation of the research was sufficiently open to question when the article was submitted that it should have been rejected. Whether fairly, or not, the editor has received some of the author's blame for publishing the material.

The obvious hope in appointing referees is that they will be knowledgeable and experienced researchers in the field of the articles they are sent. The problem is that such researchers are usually busy people. Refereeing is essentially an unpaid activity which has to be fitted as best as possible into the referee's schedule. In consequence, referees can often only afford a few hours in which to read and assess an article. This allows some margin for error, especially in fields, such as mathematics, where careful following of the argument can be time-consuming. At the simplest level, a referee's report to an editor can contain significant factual errors. It has been estimated in one biomedical field that up to a fifth of referees' reports may have this problem. Since editors do not necessarily pass on all the referee's comments to the author, this may lead to errors in terms of acceptance or rejection of the paper. The other side of the coin is that referees may miss errors that actually do occur in submitted articles. To the extent that this does not prevent the article from being published, it is less negative. However, if the error is well hidden, it may lead subsequent researchers astray. In fact, articles are typically looked at by the editor plus two referees, so obvious errors should normally be spotted. Nevertheless, some articles containing errors do get through, and have to be corrected subsequently. The usual response is to publish a notice of the correction in a later issue of the journal.

Unfortunately, this does not help people who read the original article at a later date. Most such corrections are fairly trivial, but on rare occasions it may be necessary to retract the whole article. Since this is likely to involve differences of opinion, either between the authors (if there is more than one), or between the authors and their institutions, there may well be legal implications. Editors therefore tend to be cautious, which means that major errors may be less well publicised than minor errors.

Apart from actual errors, referees can be biased by the varying standards they apply to refereeing. The invitation to referee tends also to be seen as an invitation to criticise, but some referees are more critical than others. For example, referees of articles submitted to a radiology journal were asked to grade the manuscripts they looked at on a scale running from 1 (= outstanding) to 9 (= unacceptable). The mean score averaged over all the referees was 4.8. Some referees deviated considerably from this average, and it appeared that 5–10 percent were consistently high in their gradings, with a similar percentage consistently low [29]. In other words, most referees seemed to have a roughly similar approach to standards, but a few were consistently stricter and a few consistently more lenient. It might be supposed that editors would discard referees whose opinions differ significantly from the norm. Though this does happen, it is not always the case. Editors learn the habits of regular reviewers, and can allow for them. Indeed, editors can use such referees to help support their own attitude to articles. An article that an editor particularly wants to publish can be sent to a lenient referee, while one that the editor regards as suspect can be sent to a strict referee. (The situation is usually more complex than this when more than one referee is employed.)

2.3.2 Refereeing problems

Because refereeing an article is often seen as a low priority, an article can sit around for an appreciable time before the referee gets down to looking at it. Editors have to make allowance for this; so, for example, a journal that tries to take six months from the receipt of an article to its publication may allow up to half that time for the refereeing. This is if the article presents no problems: manuscripts that are in any way controversial can be delayed much longer. It has to be said that as much, or more, delay can be introduced by the authors themselves. By the time their paper is returned with the referees comments, they have often become involved in other work, and may take an appreciable time before they respond.

Referees can make three types of recommendation – accept, revise, or reject. In terms of their effect on information flow, these correspond to making available, delaying availability, and not making available. Consequently, it is the two latter types of decision that need looking at here. If the author disputes the comments of a referee, the delay can be lengthy, as the following comment shows:

> The acceptance of a recent paper of mine has been delayed over nine months because of a disagreement with a referee. I agree that strict refereeing is necessary to maintain the standards of a journal. Unfortunately, the main disagreement was at a quite fundamental level of how one can approach the relevant physical problems. It is by the public exchange of conflicting ideas that progress is often made. Surely it should not be the function of the referee to try to suppress this discussion. In any case, there must be a point when a referee should say, 'I disagree completely; publish the paper' [9].

Disagreements over how to approach a question can lead to long delays in publication or, indeed, to outright rejection. An example of this can be found in the earlier part of the twentieth century. The two leading statisticians in England at the time, Pearson and Fisher, were involved in a long dispute over the fundamental criteria to be applied in their subject. Pearson was editor of the only specialist statistics journal in the country, and consistently refused to accept Fisher's papers for publication. This sort of problem merges in with delays due to overlap with the referee's work. There is nothing new in this, as witness remarks by T.H. Huxley in the nineteenth century:

> I know that the paper I have just sent in is very original and of some importance, and I am equally sure that if it is referred to the judgement of my 'particular friend', that it will not be published. He won't be able to say a word against it, but he will pooh-pooh it to a dead certainty. You

will ask with some wonderment, Why? Because for the last twenty years, he has been regarded as the great authority on these matters, and has had no one to tread on his heels, until at last, I think, he has come to look on the Natural World as his special preserve, and 'no poachers allowed'. So I must manoeuvre a little to get my poor memoir kept out of his hands [12].

Editors sometimes try to get round this problem by allowing authors to suggest possible names for at least one of the two referees. It is less easy to guard against referees who use their role to aid their own research. This is certainly less common than aggrieved authors are likely to think. However, it happens occasionally, and can lead not only to delay in publication, but also to preemption, as the following story claims:

One referee was concerned with a paper finally published in the *Journal of Clinical Investigation*. He caused a delay of over a year in the publication of this paper, requiring first prescribed revision and then a revision of the revisions prescribed by him. In the meantime he had published a paper which took some of the sting out of our paper [37].

Occasionally, such claims have led to formal investigations. For example, one biomedical dispute resulted in the creation of an adjudicating panel. The panel decided that the referee concerned had been guilty of, "patterning his experiments after those detected in a manuscript sent to him for review, using the information to publish his own work and falsifying his records to claim credit" [13].

Authors are often unhappy at having to revise their work, but they are a good deal unhappier when it is rejected altogether. Clearly rejection raises the most significant questions in terms of access to information. The first point to note is that rejection rates depend on the discipline concerned. Studies of leading journals in different fields indicate that physics journals usually have the lowest number of rejections (perhaps a quarter of the submitted articles). They are followed by journals in chemistry and biology. Subjects that lie on the borderline of science and social science, such as psychology and economics, often have rejection rates of a half or more. The main factor at work in these

differences appears to be the level of agreement between practitioners in the subject concerning the conceptual framework. The greater the uncertainty over how research should be formulated and presented, the more difficult it becomes to agree on what represents an acceptable article. An interesting experiment was tried in psychology to try and assess this effect. Referees used by an American psychology journal were sent articles that had a basically similar content, but with slightly different forms of presentation. The data sections were adjusted to give results that were thought likely either to agree with, or to contradict, the theoretical standpoint each referee was believed to have. It was found that articles that agreed with the presumed conceptual framework of a referee received higher gradings than those that contradicted it [17].

If there are disagreements between authors and referees over the approach to be taken to a particular piece of research, then there are also likely to be disagreements between referees. In other words, disagreements between referees should be statistically similar to those between authors and referees. The simplest way of testing this is by noting the frequency with which one referee recommends acceptance while the other recommends rejection. Even this simple test has problems. Editors may choose referees because they have complementary skills. For example, one may be an expert on the apparatus used, while the other is an expert in the statistical handling of data. In this case, there need be no contradiction in one approving the article while the other rejects it. Nevertheless, referees, whatever their expertise, are supposed to judge the article as a whole. So a comparison of judgements is a reasonable test. One study of the refereeing of psychology journals measured the level of agreement between referees on a scale running from -1 (totally opposed recommendations) to $+1$ (complete agreement). The average score for the journals examined came out at $+0.27$. In other words, there was moderate, but far from perfect agreement. As expected, agreement in other areas of science is better. An examination of the decisions made to accept or reject articles submitted to a leading physics journal found there was a 97 percent level of agreement between the referees, as compared with a 75 percent level for the referees of two biomedical journals [18].

The level of disagreement may be inherent in a subject, but are there also systematic biases in judgement that affect referees? For example, younger researchers tend to think that their work is more likely to be rejected than the work of eminent researchers in their field. Is it true? One detailed study of physics actually found that younger researchers were somewhat likelier to have their articles accepted than older researchers [38]. Though physics may not be the best example – it is a subject where young researchers have often made outstanding contributions – bias relating to the age or prestige of the author does not seem to be a major problem in science refereeing. The same seems to apply to articles sent in by female authors, though female authors sometimes feels that such a bias exists. Equally, there does not seem to be a bias against papers from developing countries. It is certainly true that papers submitted from such countries typically have a higher ejection rate than papers from developed countries. But, on investigation, the difference is usually attributable to factors such as out-of-date research, or incomprehensible writing in a language foreign to the authors. The only instance where a significant indication of bias has been suggested relates to the author's home institution. There is anecdotal evidence that researchers who move from a little-known institution to a well-known one find it easier to get their research published, but surveys find any such effect difficult to pin down. One exception is a survey of publishing in the physical sciences in the UK. This divided both referees and submitted articles into two groups, depending whether they were from 'major' or 'minor' universities. Referees from the minor universities graded manuscripts at much the same level regardless of source. Referees from major universities graded manuscripts from other major universities significantly more highly than those from minor universities [7].

One method that is often used to try and avoid individual or institutional bias is to delete the author's name and affiliation from the copies sent out to referees. (This does not always work. Most authors refer to their own work more often than other researchers do; so a scan of the references attached to an article can give some hint as to the identity of the author.) Authors, on the contrary, sometimes suggest that they should know who is refereeing the work they have submitted. This, they feel, would allow them to detect any conflict of interest. Referees tend to be less enthusiastic, as they foresee arguments with irate authors. Since the referees are donating their time free of charge, editors for the most part feel that they must go along with what referees want. One compromise that is used is to publish a list of the people who referee for a journal without specifying which articles they have refereed.

Refereeing becomes particularly important for articles that contain major advances. Since such articles may challenge the currently accepted conceptual framework, they face more problems in being accepted than is the case for more run-of-the-mill contributions. Back in the nineteenth century, Lord Rayleigh was looking through the archives of the Royal Society when he came across a paper that had been submitted nearly fifty years before. It had been turned down for publication, but actually contained ideas which had since become of major importance in physics. In the first half of the twentieth century the leading British journal *Nature* rejected, on the advice of referees, papers reporting three of the major scientific advances of the inter-war years. (It is said that these blunders made the editors of *Nature* suspicious of referees' opinions for a number of years afterwards.) Reliable surveys of the topic are not easy to carry out. However, one study has looked at the publication process that preceded the appearance of some two hundred of the papers that subsequently became very highly cited. (The extent to which a paper is cited by one's peers is usually taken as the best measure of its importance.) It was found that there had been problems with the acceptance of around ten per cent of these papers. Another survey looked at the evaluation of chemical papers by referees, and compared their gradings with the assessments of readers (as measured using citation counts and informed opinion). It concluded that: "Our results are not favorable to the referees. A significant relationship is found, but one which would indicate that papers that become highly cited received generally lower referee evaluations than papers which were cited less frequently" [30].

Authors often try to compensate for rejection by sending their contribution to another journal. Whether they follow this path, or not, tends to depend on the nature of their work. Applied scientists and engineers may be judged more by the progress of their projects than by publications in journals. So such people may have little motivation to resubmit a rejected paper. But most others try again. For example, a study of articles rejected by the *British Medical Journal* found that three-quarters of those whose subsequent history could be tracked were eventually published in other journals. It might be supposed that their chance of acceptance by another journal had been enhanced by the feedback they had received from the referees of the first journal. In fact, only about a fifth of the articles had been revised before they were resubmitted [15]. This low revision rate seems to be quite common for articles resubmitted to another journal. A survey across science and technology found that some 60 percent of authors whose manuscripts were not accepted by their first-choice journal subsequently submitted them to another journal. Only about half had carried out any revision, yet some 90 percent of the resubmissions were accepted by the second-choice journals [6]. An interesting example that has been examined in some detail relates to a biomedical article dealing with the treatment of patients who have eaten a poisonous fungus. It was rejected successively by four journals, only the first of which gave detailed reasons. It was finally accepted by the fifth journal to which it was sent: it was still in the identical form which had been used for submission to the first journal [28]. Most resubmissions are to journals which have less prestige than the original choice, so that competition for acceptance and, correspondingly, the rejection rate is less. However, a small number of papers are resubmitted to, and accepted by, journals whose standing in the eyes of the community is at least equal to that of the original journal. These are presumably cases where the authors are backing their own opinion of their work against the opinions of the original referees.

The overall result of such resubmissions is that less material is lost to the public domain than might be expected from the raw rejection rates of journals. What does happen is a slowdown in the transmission of the information contained in the papers. For example, the biomedical paper mentioned above, would have appeared nearly two years earlier if it had gone straight through the refereeing process of the first journal. It is also true that the journals used for resubmission often have a smaller readership than the journal originally selected.

2.4 OTHER FORMS OF PUBLICATION

2.4.1 Books and other outlets

Most new scientific research nowadays is published in journals. But some material, especially in the biological sciences, still appears in book form. The likelihood of having a scholarly book accepted for publication is considerably less than having a research paper accepted. There are various reasons for this, but the major one is financial. Because journals contain a mix of articles, the cost of publishing any individual article is relatively low. Moreover, journals are subscribed to beforehand, so that the publisher may have the money in hand to publish the article even before it is submitted. Books, on the contrary, are relatively expensive to produce, and their cost may only be recouped over an extended period of time. Books, unlike most journal articles, are also often the result of an initiative on the part of the editor. Indeed, editors and editorial bias is more important in the publication of books than in the publication of journal articles, though external reviewers are almost always used for scholarly texts. For example, we have seen that there is rather little bias in the journal refereeing system when it comes to the author's institutional affiliation. The same is not true of books. Scholarly publishers typically target the elite universities or research centres when then are looking for authors. They also tend to prefer experienced authors, which usually means authors who have already published a successful book. Finally, books take longer to write and to publish than a research article. The overall result of these limitations is that publishing research in book form tends to be more restrictive than publishing in journals.

Scientists do not restrict the dissemination of information about their work to formal publication in journals or books. Apart from informal presentations within their institutions, researchers often present accounts of their research at conferences. The most important of these conferences are highly selective, rejecting a number of the proposals submitted. The proceedings of such conferences are usually subsequently published, though sometimes only after a long delay. Researchers in fields such as computing or engineering may regard the publication of their presentations at such conferences as being sufficient. So they may not go on to publish their work in a standard journal. Scientists in other fields are more likely to see conference proceedings as aimed at a restricted audience, and therefore go on to publish a more polished version in a journal. Access to conference proceedings can sometimes be difficult. The same is true of access to research theses, though copies are always housed in the institutions where the students have obtained their degrees. Fortunately, the main results contained in research theses are usually published in regular journals, though there may be some delay, and in some fields, such as engineering, such further publication is not always the norm.

Media reports of scientific advances face a different problem. They may receive very wide publicity before the scientific community has had time to assess their validity. For example, the discoverers of cold fusion (a possible way of producing nuclear fusion at low temperatures) decided to obtain maximum publicity for their work. They therefore announced their results at a press conference, and their scientific peers first heard about it via the media. However, the paper about it that they had submitted to *Nature* was rejected by the referees, with the result that other scientists were left in the dark for several months over how the experiments were actually carried out. As this incident might suggest, the scientific community as a whole is unhappy about premature revelation of results via the media. Some journals, indeed, refuse to publish papers that have had prior exposure in the media.

Publicity for an article does not necessarily stop when it has been published. Further publicity can come from mentions in articles reviewing the branch of science concerned, and by inclusion in specialist databases. One method of bringing a paper to the attention of a wider audience is to distribute offprints (copies of the paper bought by the author from the publisher, and usually sent to the author when the journal containing the article is published). Offprints are particularly popular in the biomedical sciences where they can generate considerable income for a journal. A drug company may sometimes spend over a million dollars on buying offprints of a single study. These large sums of money can, however, lead to pressure being exerted on the refereeing process. An editor of a leading medical journal cited the following example of a paper that was being considered for publication.

> When the journal raised questions with the authors over the paper, the drug company sponsoring the research telephoned [the editor], asking him to 'stop being so critical'. The company told him, 'If you carry on like this we are going to pull the paper, and that means no income for the journal' [11].

It has to be remembered though that access to a publication does not necessarily mean that a reader has all the information that is needed. It was originally expected that published articles dealing with experiments would contain enough information for a competent reader to repeat the experiment. The growing complexity of experimental techniques and equipment, along with the need to keep papers short so that a journal could afford to publish them, means that someone wishing to repeat an experiment may not find it easy. Even theoretical papers often telescope a few steps in the mathematics in order to shorten the material. Readers who encounter a phase such as, 'It can be seen that ...', are likely to heave a sigh and get down to filling in the gap.

2.4.2 The impact of new developments

The editorial and refereeing system is intended to ensure that science published in journals has been subject to quality control. In general, this requirement is fulfilled, but the system also acts to slow down the passage of scientific information, and sometimes prevents its appearance altogether. One strongly expressed opinion of the peer review process is that it is: "slow, expensive, profligate of academic time, highly subjective, prone to bias, easily abused, poor at detecting gross defects, and almost useless in detecting fraud" [31]. The discussion has so far concentrated on the system as it operates when the end-product is printed on paper. It is interesting to compare the restrictions imposed by this format with what happens when the end-product is information in electronic form. Two of the things that interest authors are the readership for their work and the speed with which their work can be published. Publication online always holds the promise of increasing readership. Anyone, anywhere in the world, who has a networked computer can, in principle, access the paper. Equally, it takes much less time to put up a paper on the network than to produce and distribute a printed journal. So publishing via a network has more potential for rapid world-wide circulation of information than print-on-paper publishing. Networking also has a number of ancillary attractions for increasing information flow. For example, there are effectively no space limitations, so as many details of the work as desired can be included. Moreover, papers can be made multi-media, including sound and moving pictures, in a way that is impossible for printed material. The references at the end of a paper can be linked directly to the original publications mentioned, and any corrections to a paper can be incorporated in it directly even post-publication.

As a result, electronic publishing is seen as having a lot to offer in terms of wider dissemination. It might be expected therefore that authors would seize the chance of publishing in that way. In fact, the transfer from print publication to electronic publication is occurring, but at a moderate pace. The main reason relates to the other significant factor, mentioned above, in determining authors' choice of journals – the prestige of the journal in the scientific community. New journals take time to establish themselves, and a new journal in a new medium has a particular problem in becoming widely accepted. However well-designed a journal is, it will get nowhere if authors fail to send it their papers for publication. Consequently, we are currently at an intermediate stage. Established journals are appearing in both printed and electronic forms. This satisfies the authors so far as the acceptability of the journal is concerned, but allows some – though not all – of the freedom of electronic handling. However, it already brings advantages. The great majority of science papers are now submitted in electronic form, and are increasingly handled this way by both editors and referees. This not only saves on costs, it is already speeding up the refereeing process.

An interesting example of the direction in which electronic publishing of research may go is provided by electronic preprint archives. Preprints have existed for many years as print-on-paper. They consisted of copies of papers which had been sent to journals, or had been accepted but were awaiting publication. They had the drawback that they were only distributed to people whom the authors knew to be interested in their work. There were always others with a potential interest whom the authors did not know. In recent years, a number of electronic preprint archives have been established. They have the great advantage that they are well known within their particular communities, and the preprints are therefore available to everyone in the discipline who has a networked computer. The person in charge of the website typically scans incoming material, and throws out anything that is obviously unacceptable. Otherwise, there is no refereeing. Since authors have the prime responsibility for mounting the paper on the Web, delays in a submitted paper appearing are minimal. The original intention was that the electronic files would simply be of use until the properly refereed paper appeared in a journal. However, it has become evident that an appreciable fraction of the material submitted is not being published elsewhere. This raises the question of how well a non-refereed form of publication works.

The first electronic preprint service was established in the early 1990s, and was devoted to high-energy physics. It clearly filled a definite need in the community: by the end of the decade, the site was recording up to 150,000 visits a day by potential readers. Factors that make the site attractive are easy enough to list. Rapid publication was always regarded as important in this community, and, in consequence, it had a well-organised system of preprint circulation before the advent of online communication. All members of the community are computer literate, and are tightly bound, both in the sense that most readers are also authors and because they are often members of research groups. As a result, there is a fair level of agreement on what constitutes an worthwhile research paper. In other words, the community tends to be self-monitoring as regards research standards, and this makes the electronic preprints (often called *e-prints*) acceptable. Indeed, recent studies confirm that papers that appear first in preprint form are subsequently more often referred to than papers that appear in print only. For example, a recent study of astrophysics papers found that, on average, papers that had been mounted as electronic preprints were cited more than twice as often as those that had not been posted [26].

Other research communities do not necessarily possess the same characteristics as the physics community, but electronic preprints sites are increasingly appearing across a variety of disciplines. One reason may be a slow drift in the method of assigning priority in a discovery: from the person who publishes first in a properly refereed journal, to the person who publishes first via any channel acceptable to the community. Since an electronic preprint is about as rapid as you can get, this obviously encourages their use. But it can create problems in terms of publication in ordinary journals. Some editors accept that electronic preprints have their old function of simply alerting researchers to a piece of work before it is published through traditional channels. Others, however, see an electronic preprint server as an alternative means of publishing, and therefore refuse to publish papers that have already appeared as electronic preprints. From the way scientists cite e-prints in their papers, it seems that they are increasingly seeing an e-print server as a legitimate publication channel. But it does depend on subject. In a subject such as physics, it usually becomes evident fairly soon if the contents of

a paper are wrong. Though incorrect contents can waste researchers' time, they do not usually cause other major problems. Contrast this with medicine, where wrong information can lead to incorrect treatment of patients. Hardly surprisingly, medical researchers tend to more unhappy than physicists at the publication of un-refereed research.

One of the virtues of publication of an e-print is that readers can send in comments on a paper, which can be linked to the original for other authors to see. This obviously raises the possibility of a new approach to refereeing, turning it into an open process that operates after the material has been made publicly available. The main problem is that the people who send in comments are self-selected, and need not be those who are best fitted to referee the paper. Various experiments have been tried to combine the old with the new. For example, one journal has allowed open comment on the electronic version followed by the appointment of referees in the ordinary way. These referees make use of readers' comments in drawing up their recommendations. Presuming the paper is accepted, its final version appears in both print and electronic form with the latter still linked to the past discussion [19]. This sort of approach has the problem that papers that are not accepted have to be deleted from the server, so material appears for a time and then disappears. However, alternative approaches are possible. One suggestion is that both readers' comments and online usage of preprints should be monitored. Those reprints that arouse enough interest over a certain period of time should be subjected to a normal refereeing process. If they pass it, they should be noted to readers as having a higher status than other papers in the database, but the latter are still available.

2.4.3   Copyright

The ultimate example of open publishing occurs when all authors have their own websites on which they can put their papers. The idea of 'open access' has been widely debated in recent years. It is easy enough for authors to mount their work online (usually on the website of their host institution). The problem is allowing others who might be interested to know that the paper is there. The most important development here has been the growth of the Open Archive Initiative. This lays down that papers appearing on an institutional website should be tagged in a standard way. Retrieval services can then search across all the websites as though they formed a single server. It is envisaged that most authors will put up their work initially as a preprint. Then, presuming that the preprint is accepted by a journal, it will be replaced by the final version of the paper. It is at this point that a legal query arises. Who owns the rights in an article that has been published in a journal?

For much of the history of journal publication it was assumed that authors retained the copyright in articles that they published. In the last half-century, publishers (and especially commercial publishers) have come to claim that the copyright should remain with them. Indeed, they commonly ask authors to sign a form handing over the copyright. In the past, this has not usually created problems for the individual author, but it could do for the host institution. Such an institution may have paid for the research and for the time required by their employees to write and publish their work, only to find at the end that they had to pay a fee in order to make multiple copies of their employees' publications. With the advent of online publication, authors, themselves, are encountering problems in terms of their freedom to publish. There is nothing to stop them putting up a preprint on their website, other than the chance that it might not be liked by the editor who receives the paper for consideration. But once the paper has been published, the publisher can claim copyright, and require authors to remove the final versions of their papers from their websites. The initial response of an US society publisher to the new situation was to lay down the following policy:

- If a paper is unpublished, the author may distribute it on the Internet or post it on a website but should label the paper with the date and with a statement that the paper has not been peer reviewed.
- Upon submitting the paper for publication, the author is obligated to inform the editor if the paper has been or is posted on a website. Some editors may consider such a Web posting to be prior publication and may not review the paper.

- If a paper is accepted for publication and the copyright has been transferred to the [publisher], the author must remove the full text of the article from the website. The author may leave an abstract up and, on request, the author may send a copy of the full article (electronically or by other means) to the requestor [33].

Some publishers have insisted on more stringent limitations than this. Recent proposals in both the USA and the European Union relating to copyright in electronic material have also tended to strengthen the hands of publishers. It has been suggested that enforcement of publishers' copyright is actually holding back scientific research in some fields. For example, research in biodiversity depends on the collection and subsequent collation of vast quantities of information concerning species. Refusal of permission to include some of the existing data into a central database means there are problems in putting together a complete picture. In view of the long period for which copyright continues, holes in the data may create a long-term problem. Hardly surprisingly, authors and their institutions are opposed to such restrictions. One response has been to refuse to sign copyright transfer forms: after all, government departments will not allow the transfer of copyright, but papers from government establishments are still published. The number of papers archived on institutional websites is certainly increasing. One problem is that a handful of publishers dominate the world of science journal publishing. This puts them in a strong position in terms of negotiating with authors. However, it does seem that publishers are becoming more flexible in their responses to the activity of self-archiving.

2.4.4 Access by readers

There is one further group that needs consideration – the readers. An item of information may be published, but is it available? The rapidly rising prices of journals in recent decades have meant that the number of titles available locally has fallen almost everywhere. This has been offset to some extent by national provision – for a fee – of photocopies on request, but clearly the great hope lies in online provision. A recent survey of 174 medical journals has found that a quarter of them have all, or some, of their recent content freely available on the journal's website [20]. If this sort of trend continues, readers should have an increasingly greater access to the literature. This is particularly important for researchers in developing countries. Scientists in such countries have had major problems in finding out about research under a print-on-paper regime. The rising prices of journals has made it increasingly difficult for their institutions to subscribe, especially since the most significant journals in any field are often also the most expensive. In addition, problems with the mail meant that copies usually took appreciably longer to reach recipients in developing countries than their colleagues elsewhere. Online availability of journals and articles is already improving access to scientific information in developing countries. The proviso, of course, is that the scientist has a computer attached to the Web. Those who do not are actually more disadvantaged relative to their peers than under a print-on-paper regime. Countries round the world are therefore giving priority to Web networking.

The increasing use of the Web for publishing scientific work means that readers, whether in developing or developed countries, must become adept at tracking down what is available. Unfortunately, scientists are often not very good at online searching. In industry, this does not matter, because much of the searching is delegated to trained specialists. In universities, it often depends on whether the scientific field concerned has a specialist database, or not. If it does, then scientists in that field will soon learn to rely primarily on it. For example, NASA maintains a database for space science, astrophysics and geophysics that covers nearly all the published sources in these fields. Where there is no single database, the situation is likely to be less satisfactory. Scientists then often go to a general search engine, such as Google, which will only return a part of the material they seek. (Attempts are being made to improve the situation. For example, the new search engine, Google Scholar, covers material aimed at researchers.) The basic problem is that it is difficult to know just what information

a particular search engine is covering. For example, search engines naturally concentrate on keeping their most-used material up to date. This may mean that they neglect sources of less-used information. Some search engines order their responses partly by commercial criteria, which leads to some sites appearing in the search list ahead of others. Since many users only look at the first few sites in the list, this can lead to the loss of potentially useful information.

2.4.5 Patents

Patent specifications, like journal articles, are there to make things public, not to keep them private. Indeed, the word 'patent' derives from the Latin *litterae patentes*, which means 'open letters'. Of course, the main function of a patent is to impose limitations, but on actions rather than information. They are intended to prevent others from exploiting an invention or innovation without permission. From this viewpoint, they act rather like copyright does for text or pictures, but they have a much shorter lifetime than the copyright laws demand. The original English lifetime of a patent, established in the early seventeenth century, was for fourteen years. This was derived from the system of apprenticeship then in use for technical training. An apprentice was indentured for seven years, so fourteen years allowed for two successive groups of apprentices to be trained with the new equipment or technique. Nowadays, a somewhat longer time is allowed, but, depending on the country, it is usually no more than 15–20 years. Drug companies typically complain that, because it can take eight years or more after receiving a patent to carry out the necessary trials with patients, they effectively have a much shorter time during which they can enforce their rights.

Obtaining a patent requires a fair degree of patience along with a certain amount of money. For example, obtaining a European patent can take several years along with several thousand euros. This means that large firms are often in a better position to cope with patent requirements than small firms or individuals. In terms of secrecy, a patent specification keeps less back than a journal article. It is expected to contain sufficient information for someone else to copy the innovation without requiring further information. For this reason, industry is often cautious in terms of when applications for patents are made. Some firms prefer to maintain trade secrets – as in the formulae for certain drinks – rather than apply for a patent. Some patent applications can actually be kept secret legally, as when they bear on matters of national security. US law differs from most in allowing a grace period of one year between the time when information about an innovation is first published and the date of filing an application for a patent. The idea is to encourage earlier publication of information about innovations and to do away with some of the secrecy which, in other countries, precedes the application. The downside is that it encourages litigation over who was first to come up with an innovative idea.

Universities are under great and increasing pressure to create innovative applications in order to provide themselves with additional sources of funding. This means that they are now much concerned with generating their own patents. There is an immediate tension here between the tradition of open publication by university staff and the commercial need for secrecy prior to seeking a patent. There have been examples in the past of individuals foregoing patent rights in their work for humanitarian reasons. For example, Jonas Salk was at the University of Pittsburgh when he produced his polio vaccine in 1954. He refused to take out a patent. Instead, he published his work in the open literature, so that the vaccine could be taken up widely and rapidly. At the same time, it has been argued that taking out a patent may be the best way of making a technique available. For example, an eminent biochemist at Stanford University was reluctant to apply for a patent for his new gene-splicing technique. He was persuaded to do so by his university on the grounds that making it available at a reasonable price would make it unnecessary for users to develop their own approaches. Were they to do this, it could lead to a large number of different techniques being used, each kept as a trade secret. So far as the United States are concerned, the matter was effectively settled by the passage of the Bayh–Dole Act through Congress

in 1980. This not only gave universities the title to inventions made with Federal funding, but also obliged them to commercialise such innovations whenever possible. During the 1990s and partly as a response to this, the number of patents owned by US universities (mostly in the biomedical field) more than doubled [1]. The mix of university and commercial patents in a particular field can lead to problems in terms of communicating research. For example, one major American corporation has offered licences to academic researchers to use its patented products, but the terms seem to require licensees to consult the corporation before any results are publicised or used for commercial purposes [10].

2.5   GREY LITERATURE AND RESTRICTIONS ON AVAILABILITY

Journals, books and patents are well-regulated publications, which can usually be searched for, and tracked down relatively easily. They are supplemented by a whole host of documents and reports produced as spin-off from an organisation's activities. Publications of this latter type are usually referred to as 'grey literature', because their exact status as publications is often indeterminate. Some of this material may be publicly available, but it can be hard to track down and is not always easy to access. Electronic grey literature is more readily detected than printed material. So the move to online access has helped open up previously unrecognised sources. However, grey literature continues to offer problems to scientists seeking information, in some important instances because the contents are only made available to a particular group of people.

2.5.1   Classified research

Reports important for military or defence purposes provide an obvious example of limited-circulation material. In many cases, the existence of the information may be generally known, but access is limited. An example is provided by early high-resolution pictures of the Earth's surface. The quality of samples on general release was degraded in order not to show the level of detail that could be attained. During the Cold War years, a significant fraction – perhaps a quarter – of the research produced by U.S. National Laboratories was classified [35]. However, the classified material was circulated between the laboratories and its existence, though not its contents, guessed more widely. On occasion, scientists have themselves imposed limits on what should be revealed of research with potential military applications. As far back as the end of the eighteenth century, Lavoisier and other French chemists voluntarily kept secret work they were doing on new explosives during the French Revolution. More recently, a leading physicist proposed that research into nuclear reactions should be kept secret in the period just before the Second World War. This attempt at self-censorship failed, mainly because the leading French researcher in the field insisted on publishing his work.

The question of what level of secrecy is required for research information relating to defence is hotly debated, especially in the United States. One example is the exposure of the American armed forces to substances that might endanger health. A Senate investigation in 1994, claimed that the Department of Defense had deliberately exposed military personnel to potentially dangerous substances, and had often kept both the action and the results secret [25]. Accusations of secrecy about this and other defence activities continue to be made; yet the need for classifying much information was queried by the Department's own Defense Science Board in 1970. They made four main points:

1. It is unlikely that classified information will remain secure for periods as long as five years, and it is more reasonable to assume that it will become known to others in periods as short as one year.
2. The negative aspect of classified information in dollar costs, barriers between US and other nations and information flow within the US is not adequately considered in making security classification determinations. We may gain far more by a reasonable policy of openness because we are an open society.
3. Security classification is most profitably applied in areas close to design and production, having

to do with detailed drawings and special techniques of manufacture rather than research and most exploratory development.
4. The amount of scientific and technical information which is classified could profitably be decreased perhaps as much as 90 percent by limiting the amount of information classified and the duration of its classification [27].

Hardly surprisingly, US academic scientists tend to object to excessive secrecy which affects the publicising of their work. The consequent potential for clashes has been the subject of comment for some time past. At the end of the 1960s, Senator Fulbright in a speech to the US Senate, suggested that government funding of projects was:

> ... purchased at a high price. That price is the surrender of independence ... and the distortion of scholarship. The corrupting process is a subtle one: no one needs to censor, threaten or give orders to contract scholars; without a word of warning or advice being uttered, it is simply understood that lucrative contracts are awarded not to those who question their government's policies, but to those who provide the government with the tools and techniques it desires [14].

At the time Fulbright was speaking, the vast majority of external research funding in US universities was coming from government agencies – almost half from the Department of Defense. The end of the 1960s was, of course, the time of the Vietnam war, and cases did occur of scientists who published anti-war comments being warned of possible loss of grants. However, the pressure was not specifically aimed at restricting the publication of research results. There is a continuing worry, however, that groups of voters may bring pressure to bear against the Federal support of particular types of research (e.g., studies of homosexuality).

2.5.2 Industry and secrecy

The amount of industrial funding of research in universities, both in the USA and elsewhere, has been growing rapidly in recent decades. So, correspondingly, have questions about the publication of sponsored research. As an example of this growth, over a thousand university-industry R&D centres existed in the USA in the early 1990s, more than half of which had been set up in the previous decade. The income received by these centres was over twice as much as the total support for all academic research provided by the National Science Foundation. The rapid growth in joint university-industry research is also reflected in the awards of patents. In 1974, the leading hundred US research universities were awarded less than 200 patents. Ten years later, the annual number had risen to over 400; by 1994, it had reached nearly 1,500 [36].

Industry-funded research can raise – sometimes in an acute form – problems of secrecy. Tension is particularly likely when university research is actually carried out in close collaboration with the industrial firm. Researchers in industry are usually not greatly concerned with the publication of their results. They work in industry because they like the research environment and the pay that industry offers. Academic researchers, on the contrary, typically rely on publication to establish and further their careers. The difference can affect research collaboration between the academic and industrial worlds in a number of ways. Thus a study carried out by the Parliamentary Office of Science and Technology in the UK suggested that the commercialisation of research in British universities could lead to a conflict of opinions in at least three ways:

- *The priority for disclosure* where academics wish to publish their research results, while industrial partners wish disclosure to be delayed until patents can be taken out.
- *The degree of formality in academic/industrial collaboration* where industry wishes to protect their resources invested in the collaboration on a formal basis which can constrain academics' wish to publish openly.
- *The appropriate phase of research at which to patent* where patenting 'up-stream' research can restrict the flow of even early research results, which could jeopardise the historical openness with which experimental use exemptions have been granted [22].

The pressures are particularly severe in biomedical research – currently one of the major areas of growth in science – especially when it affects the pharmaceutical industry. In some cases, the tension has led to highly publicised clashes, as in the following two incidents:

> Nancy Olivieri, at Toronto's Hospital for Sick Children, became convinced that use of the drug deferiprone on children with thalassemia had dangerous side effects. The manufacturer of the drug that funded Olivieri's study cancelled her research and threatened her with a breach of contract suit should she inform her patients or publish her negative findings. When Betty J. Dong, a researcher at University of California, San Francisco, and her colleagues found the effectiveness of a new synthetic thyroid drug to be equivalent to three other thyroid drugs, the drug's manufacturer refused to allow publication based on a contractual agreement that Dong had signed prohibiting disclosure of proprietary information. Instead, the company reanalyzed the data and published an opposite conclusion, without acknowledging the experiments carried out by Dong. Having signed nondisclosure agreements with the companies, both Olivieri and Dong found that their universities, daunted by monetary and legal stakes, would not support them. However, they did find support from colleagues who argued that the professional responsibility to inform the public of their research findings outweighed the contractual obligations [5].

Commercial pressures can also affect the author-reader relationship for published material. A common problem here is the existence of a conflict of interest. It typically occurs when authors are receiving funding from companies whose products they are examining. The leading medical journals – such as the *British Journal of Medicine* – now have strict policies regarding declarations of potential conflicts of interest. However, not all journals are equally stringent, and, in any case, it depends on the authors deciding whether there is a conflict of interest. So one study of scientific papers published in the early 1990s found that, at that time, about one third of the papers examined had authors with financial interests in the research. Many of these were not mentioned in the papers. Here is an example of the sort of problems that can arise:

> More than a year before the appetite suppressants known as 'fenphen' were pulled from the market, Stuart A. Rich and others published a study in *The New England Journal of Medicine* showing that such drugs could lead to a potentially

fatal lung condition. In that same edition, the journal also ran a commentary by two other academic physicians that minimized the study's conclusions about the health dangers of the drugs. What readers were not told was that each of the authors of the commentary had been paid consultants to companies that make or distribute similar drugs [2].

There is some evidence that government agencies tend to be over-cautious in their assessment of what information on drugs can be released. They are, of course, under pressure from manufacturers, who prefer information about the merits, or otherwise, of their products not to be reported publicly by an independent body. In fact, excessive secrecy can sometimes be counterproductive. An international body looking into this question observed:

> Drug agencies and inspectorates often maintain secrecy to a much greater extent than law or logic actually demand. Some laws, for example, only strictly require secrecy as regards personal data and the method of preparation of a drug, yet one often sees that no part of the regulatory file is accessible, and that reports about adverse reactions or poor manufacturing standards are sealed [32].

Pressures for the commercialisation of research are clearly affecting open communication in academic research. A study of academic researchers in the life sciences in the USA found that about a fifth had experienced delays of six months or more in the publication of their results, mostly in order to protect their monetary value. A third of those surveyed said that they had been denied access to research information relating to products by academic colleagues. Those with industrial support were appreciably more likely to refuse access than those who did not have such support [5].

It is hardly surprising that a survey published by the U.S. National Academy of Sciences notes:

> To researchers and educators in the natural sciences, this pressure toward privatization and commercialization of scientific data is of great concern. Many fear that scientific data, the lifeblood of science, will be priced beyond their means, especially in less developed countries. It is argued, correctly, that the conduct of scientific research, including the maintenance and distribution of scientific data, is a public good, provided for by government funding. This traditional model has worked well in the past, and

many scientists are of the view that privatizing the distribution of scientific data will impede scientific research [4].

2.6 CONCLUSION

In the early days of modern science, researchers often maintained some degree of secrecy regarding their research activities. But, by the end of the seventeenth century, the value of openly publishing the results of research was already coming to accepted. Scientists did not, however, feel under great pressure to do so. At the end of the seventeenth century, Newton had to be persuaded by his friend Halley to publish his work on gravitation and motion. At the end of the next century, Cavendish left unpublished a number of the major experiments he had carried out. Cavendish was a wealthy aristocrat, and was consequently under no pressure to publish. When his manuscript notes were examined in the latter part of the nineteenth century, it was found that he had anticipated a number of the major advances that had subsequently been made concerning electricity. In the early nineteenth century, Gauss notoriously failed to publish many of the significant advances in mathematics that he had made. Gauss, unlike Cavendish, was not a wealthy man, but he was so prolific in research that he could select what he published. When his notes were examined after his death, they were found to contain discoveries that later brought fame to a range of other mathematicians. For example, he had already begun to develop ideas on non-Euclidean geometry while still a boy.

The situation changed during the nineteenth century. As science became increasingly professionalised, it became increasingly necessary to publish work in order gain recognition and so to obtain worthwhile positions (especially in the academic world). During the twentieth century, open publishing of research, especially in journals, became accepted as the academic norm.

Scientists accepted that some research, not least that involving the defence of their country, might need to be kept secret for a time. They also accepted that a degree of secrecy could be necessary for some industrial research results. However, much of this research was carried out by government or industrial scientists, and did not usually affect the flow of academic science. The Second World War marked a watershed in this respect. Academic scientists were caught up in military developments, to which their contribution proved vital. As a result, governments after the war, recognising the national importance of scientific research, increasingly provided funds for civilian research. Industrial firms also recognised the value of research carried out in academic environments, and, especially with the recent growth of biotechnology, pumped funding into academic institutions. The down side of all this has been the growth of demands for secrecy concerning some of the research that has been supported. Academic scientists in some research areas now find themselves caught between their own wish to publish and the requirements of the bodies supporting their research.

The growing number of scientists doing research has been balanced by a growth in the number of the journals where they can publish their results. The assessment system that has evolved in an attempt to control the quality of published research in these journals is widely accepted, but has some flaws. In particular, it slows down the flow of information. The development of electronic publishing has opened up the possibility both of more rapid publication and of wider circulation of research results. In order to take advantage of these possibilities, publication will need to become more flexible, not least in terms of quality assessment. Experiments along these lines face the need to resolve questions of copyright. However, in general terms, online publishing seems to be improving the dissemination of research information, while external funding seems to be hindering it.

REFERENCES

[1] L. Belsky, E.J. Emanuel, *Conflicts of interest and preserving the objectivity of scientific research*, Health Affairs 23 (2004), 268–270.

[2] G. Blumenstyk, **Conflict-of-Interest Fears Rise as Universities Chase Industry Support**, Chronicle of Higher Education (1998), available at http://chronicle.com/data/articles.dir/art-44.dir/issue-37.dir/37a00101.htm.

[3] W.H. Brock, **The Fontana History of Chemistry**, Fontana Press, London (1992).

[4] Committee on Issues in the Transborder Flow of Scientific Data, **Bits of Power: Issues in Global Access to Scientific Data**, National Academy Press, Washington, DC (1997).

[5] A. Crumpton, *Secrecy in science*, Professional Ethics Report 12 (1999), available at http://www.aaas.org/spp/sfrl/per/per16.htm.

[6] W.D. Garvey, **Communication: The Essence of Science**, Pergamon Press, Oxford (1979).

[7] M.D. Gordon, *The role of referees in scientific communication*, **Technology and Writing: Readings in the Psychology of Written Communication**, J. Hartley, ed., Jessica Kingsley, London (1992), pp. 263–275.

[8] W.O. Hagstrom, **The Scientific Community**, Basic Books, New York (1965).

[9] A.G. Hearn, *What are referees for?*, Observatory 90 (1970), 263.

[10] M.A. Heller, R.S. Eisenberg, *Can patents deter innovation? The anticommons in biomedical research*, Science 280 (1998), 698–701.

[11] R. Horton, *Editor claims drug companies have a "parasitic" relationship with journals*, British Medical Journal 330 (2005), 9.

[12] L. Huxley, **Life and Letters of Thomas Henry Huxley**, Vol. 1, Macmillan, London (1900).

[13] M.C. La Follette, **Stealing Into Print**, University of California Press, Berkeley, CA (1992).

[14] S. Lang, *The DOD, government and universities*, **The Social Responsibility of the Scientist**, M. Brown, ed., The Free Press, New York (1971), pp. 51–79.

[15] S. Lock, **A Difficult Balance: Editorial Peer Review in Medicine**, Nuffield Provincial Hospitals Trust, London (1985).

[16] O. Lodge, *Letter*, Nature 48 (1893), 292.

[17] M.J. Mahoney, **The Scientist as Subject: The Psychological Imperative**, Ballinger, Cambridge, MA (1976).

[18] H.W. Marsh, S. Ball, *The peer review process used to evaluate manuscripts submitted to academic journals: Interjudgmental reliability*, Journal of Experimental Education 57 (1989), 151–169.

[19] G. McKiernan, *Peer review in the internet age: five (5) easy pieces*, Against the Grain 16 (2004), 52–55.

[20] M.E. McVeigh, J.K. Pringle, *Open access to the medical literature: how much content is available in published journals?*, Serials 18 (2005), 45–50.

[21] A.J. Meadows, **Communication in Science**, Butterworths, London (1974).

[22] Parliamentary Office of Science and Technology, **Patents, Research and Technology – Compatibilities and Conflicts**, House of Commons, London (1996).

[23] K. Prpic, *Generational similarities and differences in researchers' professional ethics: an empirical comparison*, Scientometrics 62 (2005), 27–51.

[24] W. Ramsay, *Letter*, Nature 53 (1896), 366.

[25] J.D. Rockefeller IV, Is military research hazardous to veterans' health? Lessons spanning half a century, United States Senate Report, Washington, DC (1994).

[26] G.J. Schwarz, R.C. Kennicutt Jr., *Demographic and citation trends in Astrophysical Journal papers and preprints*, Bulletin of the American Astronomical Society 36 (2004), 1654–1663.

[27] F. Seitz, Report of the defense science board task force on secrecy, Office of the Director of Defense Research and Engineering, Washington, DC (1970).

[28] D.A.E. Shephard, *Some effects of delay in publication of information in medical journals, and implications for the future*, IEEE Transactions on Professional Communication 16 (1973), 143–147.

[29] S.S. Siegelman, *Assassins and zealots: variations in peer review*, Radiology 178 (1991), 637–642.

[30] H.G. Small, **Characteristics of Frequently Cited Papers in Chemistry**, National Science Foundation, Washington, DC (1973).

[31] R. Smith, *Opening up BMJ peer review*, British Medical Journal 318 (1999), 4–5.

[32] Statement of the International Working Group on Transparency and Accountability in Drug Regulation (1996), available at http://www.haiweb.org/pubs/sec-sta.html.

[33] N.G. Tomaiuolo, J.G. Packer, *Preprint servers: pushing the envelope of electronic scholarly publishing*, Searcher 8 (2000), available at http://www.infotoday.com/searcher/oct00/tomaiuolo&packer.htm.

[34] S. Vincent, *Letter*, Nature 55 (1896), 79.

[35] P.J. Westwick, *Secret science: a classified community in the national laboratories*, Minerva 38 (2000), 363–391.

[36] L.J. White, University–industry research and development relationships: the university perspective, New York Academy of Sciences Workshop Series on University–Industry Relationships (1998), available at http://www.stern.nyu.edu/eco/wkpapers/workingpapers99/99-12White.pdf.

[37] R.D. Wright, quoted in: A.J. Meadows, **Communication in Science**, Butterworths, London (1996), p. 42.

[38] H. Zuckerman and R.K. Merton, *Patterns of evaluation in science: institutionalisation, structure and functions of the referee system*, Minerva 9 (1971), 66–100.

3

INDUSTRIALISTS, INVENTORS AND THE GERMAN PATENT SYSTEM, 1877–1957: A COMPARATIVE PERSPECTIVE[1]

Kees Gispen

Croft Institute for International Studies and
Department of History, University of Mississippi
Mississippi, MS, USA

Contents

Abstract

The modern German patent system came into existence in the last quarter of the nineteenth century in reaction to a perception of excessive authorial or inventor rights in French, British and American patent law. The German system also anticipated systematic R&D in the modern research laboratory. It was based on the so-called first-to-file system and developed the concept of the 'company invention' to deprive inventors, especially employee inventors, of any residual rights in their inventions. Inventor discontent resulted in numerous failed patent reform attempts, which finally succeeded in the 1930s and early 1940s. The National Socialist regime made the company invention illegal, changed the theoretical underpinnings of the patent code to introduce the inventor's right principle, and adopted statutory inventor compensation. These reforms survived the Third Reich and became an integral part of the Federal Republic's consensus-based political economy.

Keywords: Academy of German Law, Act on employee inventions of 1957, anti-patent movement, Ardenne, Manfred von, assignment, authorial principle, Bosch, Carl, bureaucratization of industry, company invention, captive research, compulsory licensing, compulsory working, conservative inventions, corporate management, creative destruction, disclosure, Duisberg, Carl, employed inventor, employer, fair compensation, first-and-true inventor, first-to-file system, first-to-invent principle, Fischer, Ludwig, Frank, Hans, free inventions, free-trade ideology, Gareis, Karl, German Association for the Protection of Industrial Property, Hitler, Adolf, IG Farben, incentive, independent inventors, innovation, intellectual property, invention-award system, inventor reward, inventor's right, labor law, monopoly, National Socialism, Nazi inventor policy, organized inventing, patent agreements, patent examination, patent fees, patent office, radical inventions, reactionary modernism, registrant's right, research and development, routine invention, royalties, service invention, shop right, Siemens, Werner von, statutory inventor award, technological culture, Weimar Republic, work-for-hire doctrine.

3.1 THE CASE FOR PATENTS: ORIGINS AND THEORIES

"To promote the progress of science and useful arts", nations have long used incentives known as patents [117]. The World Intellectual Property Organization (WIPO) defines a patent as "an exclusive right granted for an invention, which is a product or a process that provides a new way of doing something, or offers a new technical solution to a problem. A patent provides protection for the invention to the owner of the patent for a limited period ..." [25: 9; 128]. The earliest known patents were special privileges that governments granted to individuals in exchange for the introduction of new processes or techniques that were expected to stimulate economic activity. New processes and techniques were understood at the time to include both inventions in the modern sense of the word and existing technologies imported from abroad because they were unknown in the home country (technology transfer). These early patents were simply one instance of a more common practice in medieval and early modern Europe, whereby rulers issued public 'letters patent', 'liberties' and other privileges to corporate bodies or individuals in return for some tangible benefit to the state or to themselves.

The first actual patent law was the Venetian statute of 1474, which promised exclusive rights to "great men of genius" who would introduce hitherto unknown machinery that was useful and beneficial to the 'commonwealth' [9: 139–142; 71: 166–224]. The next known patent law was the English Statute of Monopolies of 1624, which originated in the context of the historic power struggle between King and Parliament over constitutional government. The chief purpose of the English statute was to abolish the monarch's right to grant special monopolies, which he was accused of abusing, thereby causing "great grievance and inconvenience" to his subjects. Accordingly, the main section of the law stipulated that "all monopolies and all commissions, grants, licenses, charters

and letters patents to any person ... bodies politic or corporate whatsoever ... are altogether contrary to the laws of this realm, and so are and shall be utterly void and of none effect ...". But the statute specified one crucial exception. Prohibition of temporary monopolies would "not extend to any letters patents and grants of privilege for ... the sole working or making of any manner of new manufacture within this realm to the first and true inventor or inventors of such manufactures". So long as it was granted for something truly new, the reasoning went, the temporary monopoly (a maximum of fourteen years) would not raise prices of existing commodities or processes, which meant that it would not harm but rather benefit the public good [70: 10–19; 113: 434–435].

In light of subsequent developments and later theories justifying patents, the rationales of the Venetian and English patent statutes merit careful scrutiny. These laws granted special privileges to "great men of genius", or to the "first and true inventor", for the purpose of encouraging them to introduce new technologies, which would benefit the common good. Inventors might be especially talented individuals, but that did not mean they had a *right* to a patent. The notion of authorial property rights was as yet weakly developed at this time, which meant that there was also little or no concept of the invention as the inventor's 'intellectual property'. All the emphasis was on utility for the state and encouragement of industrial progress. The inventor is mentioned, not for his own sake, but only because it was impossible to conceive of invention and innovation apart from the special individuals who introduced them.

In the last decades of the seventeenth century and especially during the eighteenth century the rationale for patents began to change. The spread of Enlightenment ideas brought about subtle but important shifts in the assessment of the inventor's position. Based in large measure on the theories of John Locke, the principle of individual rights, especially individual property rights, came to occupy center stage, gradually giving rise to a concept of intellectual property, or the notion that the author or inventor possessed a fundamental, 'natural right' of proprietorship in his or her intellectual creation. The rise of Romanticism with its discourse of personality, originality and genius by the

---

[1]This contribution is an expanded and revised version of Chapter 1 in the author's *Poems in Steel* [37], which includes textual comments in the endnotes.

end of the eighteenth century strongly reinforced this concept of the author's and the inventor's property right [70: 202–222; 98; 127: 425–448].

The concern with authorial property rights significantly strengthened the inventor's hand in the British patent system, though it never went so far as to cause the original principle – that patents were temporary monopolies granted by the state to the inventor for reasons of public utility – to be overthrown. Still, the focus on authorial rights gave rise to a new argument for patents that recognized inventors' rights and tried to balance these against the needs of society. Patents now appeared as justified by a blend of the economic principle (incentives for making inventions that promote technological progress) and a principle of justice (the inventor, as original author and benefactor of humanity, is entitled to fair compensation and a just reward for handing over to society the fruits of his labor). When the United States passed its first patent law in 1790, it essentially adopted this approach as well, emphasizing the advantage to society of developing new inventions but also "securing for limited times to authors and inventors the exclusive right to their respective writings and discoveries" [69: 1–29, esp. 10ff.; 117].

The authorial principle became especially strong in France, which based its 1791 patent law entirely on the principle of the inventor's right. As in England a century and a half earlier, French lawmakers at the time of the Revolution chose to correct the monarchy's abuses by rigorously eliminating all monopolies. Recognizing the unique importance of inventions for French power and the economy, however, the National Assembly voted to retain patents of invention. It justified this exception to an otherwise doctrinaire political and economic liberalism on the grounds of the "universal rights of man", one of which was said to be the author's property right in his invention. Section I of the French code proclaimed, "Every discovery or new invention, in all branches of industry, is the property of its author; the law therefore guarantees him the full and entire enjoyment thereof . . ." [28: 7; 48; 69: 8–19; 92: Preface]. The emphasis here was totally on the rights of the inventor, the good of society playing a minor, nearly invisible role.

France retreated from this extreme position in 1844, when it adopted a new patent law founded on a more reasonable balance between societal utility and individual rights. But it retained a strong dose of the latter in its revised case for patents: that they provided the best motive for inventors to disclose their inventions to the public instead of keeping them secret. By keeping valuable technological knowledge out of the public domain, secrecy was said to harm society as well as the inventor – society by preventing valuable technological knowledge from becoming more widely known, the inventor by rendering his ownership rights precarious and insecure. "The 'incentive-to-disclose' theory of patent protection", Fritz Machlup and Edith Penrose pointed out long ago, "was often formulated as a social-contract theory" based on the social theories of Jean-Jacques Rousseau. The patent was represented as a contract between the inventor and society, which did justice to both and benefited them equally [52; 69: 25–26; 92: Preface].

Incentive, reward, disclosure, inventor's right – to this day the case for patents revolves around these four principles, either in a form that combines one or more of theories, or as a conscious and deliberate choice among them. In both cases, however, the theoretical underpinnings of nations' patent laws have important consequences for the ways in which they administer their patent systems in practice. To illustrate this point, we will next examine the treatment of inventors in two contemporary but conceptually very different patent regimes.

3.2   INVENTOR RIGHTS IN THE UNITED STATES AND GERMANY

As far as basic organization is concerned, today's patent systems generally belong to one of two types. The one most widely used is the so-called first-to-file system, in which a patent is issued, not to the 'first and true inventor', but to the person who is the first to apply for the patent. The invention, rather than the inventor, is the primary focus of attention in this system. It encourages rapid disclosure, reduces the likelihood of protracted and expensive litigation over authorship, and is therefore supposed to result in the speediest possible innovation. Most nations, including Germany, administer patents under some version of the first-to-file system [8].

The other, less common type of patent system is based on the first-to-invent principle, which entitles the first and true inventor to a patent. The rights of the inventor weigh much more heavily in this system than in the first-to-file system. No one can file for a patent who is not the original inventor, and the claim of a second inventor who has made the same invention later but applies for a patent earlier than the first inventor (known as 'interference') is invalid. The United States is the most important of the few countries whose patent systems are still based on the first-to-invent principle.

The first-to-invent system is older and used to be more popular than the first-to-file system. In addition to governing the United States code, it served as the foundation of French patent legislation from the eighteenth centuries until after World War II. The same was true for the British patent system.[2] Like the American code, the original British and French patent systems dated from the time when inventing was still mostly a matter of individual entrepreneurship, private effort and, by today's standards, small proportions. Geared toward those economic–technological circumstances, the first-to-invent system also reflects the political and cultural context of eighteenth-century and early nineteenth-century liberalism, including the authorial principle discussed above. The first-to-invent principle treats inventions as the product of individual authorship and grants the inventor broad li-

cense to maximize profits and litigate his claims in court.

Despite the extensive protection that the first-to-invent system offers in theory, for the vast majority of inventors in the United States it does not do so in practice. Nor, as far as most inventors are concerned, does the US patent system act any longer as the incentive that it was meant to be. The bureaucratization of industry and the emergence of organized inventing in corporate and government laboratories since the last quarter of the nineteenth century have turned the majority of inventors into salaried employees. The employer now comes between the inventor and the invention and lays claim to the rights of the employee's intellectual product.

The first-to-invent system does not readily entitle the employer to the results of the employee's inventive genius as part of the general employment contract. American law "considers an invention as the property of the one who conceived, developed, and perfected it, and establishes, protects, and enforces the inventor's rights until he has contracted away those rights" [3: 984]. Without such a contract, only the rights to those inventions the employee was hired to invent (work-for-hire doctrine) pass directly to the employer. Other work-related inventions the employee makes on his own initiative, even if using the employer's time and the employer's materials, equipment or labor, remain his property subject to a 'shop right', which gives the employer a non-exclusive license to use the invention. Inventions made at home and unrelated to the job also belong to the employee, even if they fall within the scope of the employer's business operations [86: 60–64]. To circumvent the obstacle represented by the patent code, employers routinely require scientists and engineers to sign special patent agreements. In these separate, pre-employment contracts it is usually expressly agreed that as a condition of employment any inventions or improvements made by the employee during the employment shall belong to the employer [3: 995; 80; 82: 84–109, esp. 90, 97–101; 83: 648–662, 719–745; 109; 112: 7].

The permissible scope of pre-employment patent agreements in the United States has always been extremely broad. An employee can bargain away the right to ideas, concepts, inventions,

---

[2]France retained the first-to-invent principle until 1978, when in the context of the European Union's patent harmonization efforts it adopted a new patent code based on the first-to-file system. Likewise, the United Kingdom only abandoned the first-to-invent principle going back to the 1624 Statute of Monopolies in 1977, with the Patent Act of that year. Canada retained the first-to-invent system until passage of the 1987 patent reforms. The United States patent law of 1790, first revised in 1793, came of age with the reforms of 1836. Despite recent pressures for global harmonization of patent systems, none of the subsequent revisions of the American patent code, through the Intellectual Property and Communications Omnibus Reform Act of 1999, has tinkered with the first-to-invent system.

know-how, proprietary information, or patents in exchange for employment with the employer. It is not uncommon for such contracts to require assignment of all inventions the employee makes, regardless of time, location, or subject matter, and extending for six months or more beyond termination of employment. Only a few states – Washington, California, Illinois and Minnesota – have statutes restricting the employer's right to an employee's invention that is unrelated to the employer's business. Some companies have voluntary invention-incentive plans, offering employees a nominal sum or special recognition when a patent application is filed for one of their inventions or when it is patented. Sometimes they have special invention-award systems or profit-sharing plans. Such programs, however, are by no means the norm and rarely very generous. While most universities (and under certain conditions the federal government) typically offer the employee attractive royalty arrangements, most companies in the private sector pay their employees little, if anything, for their inventions [80: 425–496; 34: 33]. In sum, as far as the salaried inventor in industry is concerned, the intention of the first-to-invent principle at the heart of the United States patent code has for all practical purposes been turned on its head.

In contrast to the United States, most European nations and Japan have legislation to protect the salaried inventor. Sometimes these laws merely restrict the scope of patent agreements, preserving the rights of employees by making it illegal for employers to require the wholesale assignment of all inventions. In other cases, the restrictions go a step further by mandating compensation for certain employee inventions assigned to the employer. Some nations, including Germany and Sweden, prohibit patent agreements altogether and have comprehensive statutes determining all the rights in all possible types of employee inventions [79; 86: 60–64; 89: 74]. There have been sporadic attempts – all of them unsuccessful – to introduce similar legislation in the United States since the early 1960s. In 1962, Congressman George Brown of California introduced bills that would have outlawed pre-employment patent agreements. At about the

same time, Senator John McClellan of Arkansas held hearings to determine whether legislation was needed to rekindle the salaried employee's inventive energies and accelerate technological progress. In the mid-1970s, Senator Philip Hart of Michigan proposed legislation that would have banned patent agreements unless the employer agreed to pay the employee a minimum of two per cent of the profits or savings generated by the latter's invention. In 1981, Congressman Robert Kastenmeier of Wisconsin sponsored a bill restricting the permissible scope of pre-invention patent agreements. During the 1970s and into the early 1980s, Representative John Moss of California sponsored various resolutions – all modeled on the West-German employee-invention statute – that distinguished between free inventions, which would belong to the employee outright, and service inventions, which would be assigned to the employer subject to compensation [2; 85: 28–32; 83; 86: 60–64].

None of these proposals was ever enacted into law, for reasons that concern broad cultural differences between the United States and Europe. One factor is the pervasive influence of capitalist ideology in the United States and disproportionate power of corporate management relative to that of labor. This imbalance is especially pronounced with regard to the relations between management and salaried professionals such as engineers, who have been more heavily influenced by the individualist ethos of American-style professionalism than their European or Japanese counterparts. The degree of militancy, solidarity and unionization among American engineers, who typically see themselves as future managers, has always been low. Salaries and living standards have also been sufficiently high to undercut the temptation of collective action. Despite widespread discontent with patent agreements, American engineering associations have never pursued the issue with any real passion [66; 72: 403–421; 87; 88; 96; 122; 130]. Meanwhile, employers have consistently and effectively lobbied against reform of the status quo [84: 34].

A second consideration is that the technological culture of the United States remains otherwise quite favorable to the inventor and innovation, in spite of the invention-stifling potential of corporate patent practices. Unlike the procedure in many other countries, the patent always goes first to the actual inventor. The employee may have to assign his rights to the employer but still receives the honor and name recognition that are important for career mobility and self-esteem. Large bureaucratic companies may waste much of their engineering talent, but it is not uncommon for engineers and industrial scientists in their most creative years to switch jobs or start their own business. There has also long existed an elaborate network of private, semi-private and government research institutes, as well as contract and independent research by universities and special agencies. Academics typically receive a very generous share of the royalties generated by their inventions. United States patent fees have always been low compared with those of other countries, and venture capital for promising technologies has been readily available. There exists a thriving culture of small, entrepreneurial firms nurturing innovation in the United States, supported by capital markets much more flexible and aggressive than in other countries, including Germany. Stock options are a common form of incentive and compensation for critical employees not readily available in Germany – certainly not in the decades before World War II [17]. While the percentage of patents issued to independent inventors in the United States has long been declining as a proportion of the total, historically the rate has been significantly higher than in Germany. In the early 1950s, for instance, the number of patents issued to individuals in contrast to corporations or government was above forty percent of the total in the United States and less than twenty-five percent in Germany. By the early 1970s those figures had dropped to 23.5 percent and to somewhere between ten and twenty percent respectively. While German percentages have continued to shrink, to some 13 percent of the total in 2001, the American figures appear to have stabilized or slightly increased, comprising some 36 percent of all filings in 1987 and almost 27 percent in 1999 [8: 4; 23; 39: 41–43; 42; 51: 124–125; 82: Chapter 7; 85: 28; 83; 93: 80; 111: 4–5].

Finally, the United States has a record of technological achievements that until quite recently seemed to obviate any need for special measures to stimulate inventiveness from within the large corporation. America's many technological triumphs in the two or three decades since World War II made it easy to believe that the systematization and routinization of invention by the large organization had fully succeeded in supplanting the older mechanism of generating innovation through individual incentive and the independent inventor. As one particularly bold exponent of this view put it in 1947, "invention is a term that is not essential" for modern technological progress. The "orderly process of basic research, technical research, design, and development, or approximately this process as conducted in organized laboratories has become so powerful that the individual inventor is practically a thing of the past". According to the same author, "the vast majority of new things today come from organized scientific effort, not from a single inventor. Consequently, invention is not considered to have a place in a critical examination of... research and development" [110: 43–44]. It held that division of labor and bureaucratization make invention predictable and that they rationalize innovation. In this system, most engineers and industrial scientists are reduced to obedient automatons, 'captives' who are "told by business executives what problems to work on" [40: 252; 58: 34–39; 82: Chapter 6]. These specialists, in the words of one concerned insider, "are primarily hired for their competence in certain limited fields, outside of which they are not encouraged to go, or even to satisfy their curiosity. This is partly to prevent them from wasting their time" [126: 81–83, 89–91, 96, 145]. It was the natural corollary of such a system that special incentives to encourage creativity were unnecessary, even counterproductive.

When critics of corporate inventor policy in the United States look for models of a more rational system of inventing and encouraging innovation, they invariably point to Germany. Of all the countries that have intervened in the relations between management and employed inventors, Germany has wrestled with the problem longest and come up with the most comprehensive solution. By many accounts it is also the best solution. The Federal Republic's Act on Employee Inventions of

1957 reconciles the conflicting principles of labor law and patent law by treating the employee as the creator and therefore original owner of the invention but granting the employer the right to acquire the invention, subject to compensation and name recognition of the employee in the patent documents. The Act covers all patentable inventions by all employees (except high-level management) in the private and public sectors, including all military personnel and civil servants. Academics are included as well, though they are subject to separate, more advantageous regulations. A permanent arbitration board settles unresolved disputes between employees and employers.

The German 1957 statute revolves around two key issues: which inventions the employer can claim and how the employee is remunerated for inventions assigned to the employer. As to the first point, the employee must report all inventions made during the time of employment. The law divides them into 'service inventions', which the employer may claim subject to monetary compensation, and 'free inventions', which belong to the employee. The scope of service inventions is substantially wider than what an employer would be able to claim without a pre-employment patent agreement under US common law [86: 61]. Service inventions are defined as inventions that are made during the period of employment and that in addition are either (a) related to the employee's responsibilities *or* (b) based in some measure on the company's in-house experience or activities. All other inventions are free inventions, the only restriction being that the employer must be given the opportunity to purchase a shop right in them [90: 31–55, esp. 37–44; 100; 102: 57–79].

The second focus of the 1957 Act is the remuneration mechanism. By claiming a service invention, the employer assumes an obligation to pay the inventor special compensation and must file a patent application. The amount of compensation is calculated by taking into account two main criteria: the invention's economic value and the employee's share in making the invention. The economic value of the invention is usually determined according to the so-called license analogy: how much it would cost to obtain a license had the invention been classified as free or belonged to an outside inventor. The employee's share in the invention is a function of three factors: degree of initiative in formulating the *problem*, level of achievement in finding the *solution*, and creative responsibilities and *position* in the company. The more articulation of the problem falls within the scope of professional responsibility, the smaller the factor. The more conceptualization of the solution is a function of standard professional skills or the company's state of the art, the smaller the factor. The higher the rank and the greater an employee's responsibilities, the smaller, once again, is the factor. The sum of the three factors represents the employee's share, which can range from zero to one hundred percent. The amount of financial compensation is the product of the employee's share and the invention's economic value, and ranges in practice from two to seven percent of the employer's profit from the invention [102: 57–59; 85: 30–31].

Despite occasional grumbling about the financial burden and large amount of paperwork, the overall attitude of German employers toward the Act is said to be positive. There is broad agreement that the law's compensation mechanism fosters invention and creativity, and that direct financial remuneration is more effective than any other incentive to spur invention. On balance, therefore, the benefits of a broad-based impulse to innovate outweigh the disadvantages of somewhat higher costs. As far as employees are concerned, the most common complaint has to do with supervisors being unjustly named as co-inventors. Another criticism is lack of information about the law and insufficient familiarity with it. The vast majority of employees, however, looks upon the statute with favor and, with some luck, can earn handsomely by it [39: 41–42, 49, 63; 85; 84: 33].

3.3 WERNER SIEMENS AND THE PECULIARITIES OF THE GERMAN PATENT CODE

It is ironic that the favorable situation of German salaried inventors today, like the unsatisfactory one of their American counterparts, is the outcome of an historical development that originated in exactly the opposite condition. In the United States, the pro-inventor thrust of the patent statute's first-to-invent system has effectively been reversed as far as employees are concerned. In Germany, enlightened inventor policies are the legacy of a patent system that was expressly designed, in the last quarter of the nineteenth century, to eliminate inventor rights based on the authorial principle.

Germany's patent code dates from 1877, a time when the industrialization of invention, bureaucratization of industry, and transformation of entrepreneurial into organized capitalism were taking shape. Classical liberalism with its concern for individual rights had been dethroned as the reigning ideology, and collectivist social theories, conservative as well as socialist, were moving center stage. Tariffs and other forms of protection for industry and agriculture were replacing the free-trade principles that had guided economic policy until the mid-1870s [31; 65; 99; 124]. The 1877 code reflected these intellectual and socioeconomic changes when it replaced the poorly conceived and ineffective patent systems of the individual German states that preceded it.

The greatest flaws of the pre-1877 patent situation in Germany were fragmentation, lack of uniformity, and inconsistency. The nature of patent protection and theory differed from state to state, ending at the border of jurisdictions that no longer bore any relationship to the patterns of commerce in the unified market of the German customs union. Patent laws were an odd mixture of French ideas of the natural rights of the inventor, anti-patent sentiment derived from the free-trade principles of the Manchester School of Economics, and the mercantilist tradition of granting special privileges to spur economic activity [38]. Some states followed the French practice of registering and approving all patent claims without examination, which resulted in numerous overlapping and conflicting legal claims. Other states had extraordinarily strict or impenetrable examination procedures and rejected the overwhelming majority of claims. There was no uniform period of protection from state to state or even within states. In Prussia, for instance, the maximum duration of a patent was fifteen years, but the typical length was three years, which in most cases was far too short to move an

invention from conceptualization to market. Following French practice, the majority of early German patent laws did not call for public disclosure, exposing inventors and businessmen to the danger of involuntary patent infringement and perpetuating industrial secrecy. Patents also constituted an obstacle to inter-state commerce, as different monopolies of manufacture and exclusion in the different German states prevented the free flow of goods [6: 126–129; 28: 9–19; 42: 19–68].

Partly as a consequence of these particular problems and partly because of the general, European-wide anti-patent movement, which had grown in tandem with the rise of free-trade ideology in the first seven decades of the nineteenth century, a powerful anti-patent lobby had emerged in Germany about 1850. There were similar anti-patent movements in other countries, but with the exception of the Netherlands, which in 1869 discarded patent protection altogether, and Switzerland, which held out against adopting proper patent legislation until 1907, none was as strong as that in Germany [97: 463–496; 103]. The leading adherents of the German anti-patent movement were political economists and high-level administrators in the civil service. Its ideological underpinnings were the Manchesterian free-trade doctrines that controlled German economic policy from mid-century to the early 1870s. Adam Smith, Jeremy Bentham and John Stuart Mill had all condoned patenting as sound economic policy and a natural complement to other aspects of liberal economics. Their German disciples, however, increasingly came to see patents as harmful monopolies, unwarranted privileges for inventors, and holdovers from an old regime that restrained trade and impeded economic growth. German patent opponents, like those in Switzerland, also argued that their country was an economic follower nation, which acquired much of its technology by unlicensed borrowing and copying from abroad and was therefore better off without patent protection.

Along with this denial of the economic justification for patents, the natural-law theory of the inventor's inherent right in his invention came under attack. Logical difficulties – such as the contradiction between the theory of an original property right in inventions and the requirement of imposing time limits on patents, or granting a patent to the first inventor but depriving all subsequent, including genuine, inventors – were advanced to discredit the natural-law theory of intellectual property. It was argued that the purpose of government policy was not to enrich individual inventors but to benefit society as a whole; that patents harmed technological advancement and increased prices; that inventing was not analogous to artistic or literary creation but rested on scientific discoveries, which were objective and accessible to all and explained why the same invention was often made by different inventors at approximately the same time in different places [28; 42: 69–85; 103: 3–15].

The anti-patent movement was especially strong in Prussia, where it succeeded in reducing to a trickle the number of patents issued. If the depression of 1873–1879 had not intervened, patents would likely have been eliminated altogether. But the deep economic crisis of the 1870s turned the tide and handed victory instead to a pro-patent, industrialist lobby that had emerged in the 1860s. The leading figure of the pro-patent movement was Werner Siemens, the brilliant inventor-entrepreneur at the head of the Siemens electrical company and the individual who did more than anyone else to shape the 1877 patent code.

In contrast to the free traders, Siemens was convinced that German industry could not survive without patent protection. However, he shared their views about the shortcomings of the existing patent system, especially concerning the natural-law theory of the inventor's right. He agreed that by not requiring disclosure and implementation and by interpreting inventions as natural property rights entitled to secrecy, the old statutes gave inventors excessive privileges, which harmed economic growth. This and similar arguments did much to narrow the gap between the two sides and helped prepare the ground for passage of the 1877 statute [28; 42].

Although he was Germany's greatest living inventor, in the campaign for a new patent code Siemens acted first and foremost as an industrialist. Siemens was determined to secure a patent system that would protect industry – not inventors – and

that would facilitate the acquisition of the rights in inventions by industrialists and investors. As he put it in a speech to the Association of German Engineers in 1883: "The interests of inventors and the interests of industry are not always the same. The interest of the inventor may only be furthered to the extent that it promotes the interest of industry, and when both interests come into conflict, the law must always put the latter first" [41: 169]. Siemens justified this position with a patent philosophy that was at once old and new: old in that it revived the economic principle, and new in that this principle entirely replaced the reigning authorial principle. The purpose of patents, according to Siemens, was to make it possible for industry to develop, "and not for inventors to make a lot of money" [41: 169]. Neither the Anglo-American blend of economic motives and copyright analogies protecting the rights of the individual inventor, nor French theories of natural law and intellectual property should therefore form the foundation of Germany's patent system. The only legitimate focus of patent legislation was the interest of the national economy – specifically, the speediest possible disclosure and dissemination of new ideas and the willingness of industry to invest in the expensive process of developing new technology.

A patent system based on this purely economic principle would have at least four features, according to Siemens. First, it should have a patent examiner – lacking in French and much current German law – to make a technical and historical investigation of the patent claim and determine its merits. This would prevent the accumulation of worthless, identical or contradictory patents and spur industrial activity by reducing the fear of litigation. Second, it should have high and steeply rising annual fees, to discourage inventors from holding on to monopolies that were not financially rewarding. Inventors should either realize and market their invention or get out of the way and let someone else try. In either case, high and progressive fees meant that there would be no accumulation of useless patents. Third, patent rights should be limited not just in time but also by the principles of compulsory licensing and compulsory working. Thus, if the patent holder did not attempt to bring the invention to market in a meaningful way, someone else should have the right to force the granting of a license to effect commercialization.

Finally, the patent system should be based on the first-to-file principle (which became known in Germany as the principle of the 'registrant's right'), rather than the first-to-invent, or the 'inventor's right' principle. As Siemens envisaged it, moreover, the first-to-file principle did not just mean that the patent would go to the first *inventor* to file, but rather to the first *person* to file, regardless of whether that person was the inventor or not. Siemens and his supporters argued that this rule would accomplish a number of important objectives. It would get rid of the destructive practice – associated with the theory of the inventor's intellectual property – of keeping technological advances secret. It would eliminate the problem of inventor procrastination and force the invention's fastest possible public disclosure and dissemination by rewarding the first person to report the invention. It would free the patent office from the difficult and time-consuming task of determining priority and reduce litigation on that point. Finally, Siemens believed – incorrectly, it turned out – that the first-to-file system would obviate the need for assignment of an invention by an employed inventor to the employer and thus eliminate a potential source of conflict over this question.

When the new patent statute went into effect in 1877, it incorporated all the above features. Sometimes described as the "Charta Siemens", the German code pioneered the economic or incentive principle that is the basis of most current patent systems and theories [5: 26; 28: 5; 44: 104; 68: 19–22, 76–77; 103: 3–4]. The new system did, in fact, cause German patents to become highly regarded; it forced rapid disclosure and dissemination, it cut down on secrecy and litigation, and it spurred investment in technological innovation. In this sense, adoption of the patent statute was a forward-looking and very modern development. It created a stable legal platform that promoted predictability and calculability. The new code was therefore well suited to the dawning age of corporate capitalism; to the large technological systems based on electricity, chemistry, physics and engineering that are the mainstay of modern life; and

to the large investments of capital needed to realize them. The code's role in making Germany an industrial powerhouse has been stressed time and again [7: 149–155; 21: 7; 39; 42: 136–142; 68: 105]. At the same time, it is clear that the new German patent system purchased its forward-looking qualities at least in part at the expense of inventors' rights. In this sense – by eliminating and denying the legitimacy of the authorial principle – the statute was an illiberal, even a reactionary piece of legislation, in which the needs of industry were a euphemism for the interests of corporations and industrialists, and social power was redistributed to the detriment of the individual [42: 137–142; 56: 268–284; 82: Chapter 6; 114].

The code's steeply progressive fee schedule and the principle of compulsory working and licensing were major obstacles to independent inventors and small business [42: 147–149; 60: 11–12]. Both groups usually lacked the resources to maintain a patent for the full fifteen years, or to hold out against large corporations and capitalists with deep pockets. This had consequences for the nation's political and technological cultures. It helps explain, for example, why Wilhelmine and Weimar Germany could generate so much discontent among engineers, inventors, small entrepreneurs, and other affected groups. Likewise, it sheds light on the causes of industrial concentration and clarifies why Germany was less hospitable to the types of technological innovation associated with independent inventors than the contemporary United States [35: 223–336; 57: 27–114; 63: 271; 67: 42–54; 119].

The first-to-file system affected employed inventors much more than independents. As far as the latter were concerned, inventor and filer ordinarily were the same person, which mooted the issue of entitlement to the patent. But this was not true for employed inventors. Employees lost the right to file for a patent – and with it all other claims and rights in the invention, such as the honor of authorship, professional recognition and any claim to special rewards. Much like today's technical writers and the authors of computer manuals, salaried inventors received no public credit or acknowledgment and remained anonymous operatives behind the company label. The only name on the patent document was that of the company or the employer.

The exact way in which the patent code's first-to-file system ended up neutralizing employee-inventors is not immediately obvious. Understanding the circuitous route by which it did so, however, is important for a proper appreciation of the struggles and developments – discussed more fully below – that followed in the wake of the code's adoption. Since the patent code ignored authorship, the threshold question was not: who is the first inventor, but who has the right to file – the employer or the employee? The statute was silent on this question, except for voiding the claims of filers who had acquired the invention illegally [42: Appendix]. Since the concept of illegal acquisition was open to interpretation, it invited a great deal of legal wrangling and controversy. In spite of Siemens' efforts to preempt the issue, the right to file therefore became the subject of frequent litigation between employers and employees and a matter of dispute among patent experts [26].

Conservative legal scholars and corporate lawyers argued that employees owed the whole of their physical and mental powers to the company for which they worked. In all cases the employer was therefore entitled to file for the patent. This conclusion was based on labor law, according to which the product of an employee's work automatically belonged to the employer. It was further justified by extending to salaried employees in the private sector the theory of the relationship between the state and the civil servant, according to which the civil servant merged his entire being into the state in return for a guarantee of life-time employment and old-age pension [45; 61; 62; 115; 123: 196–244].

The opposing, liberal school of thought continued to recognize the inventor's authorial rights, even though the patent code did not mention them. Scholars in this tradition pointed out that labor law had only limited applicability to employees' intellectual products, the original rights to which always resided in the author. Analogously, and since only individuals – and not companies or other abstract, legal entities – could invent, the act of inventing created the inventing employee's right to

the invention and with it the right to file. Karl Gareis, future president of the University of Munich and the author of an early commentary on the patent code, argued precisely this point in an 1879 treatise on employee inventions: "a legal person or a corporation never discovers something itself; it is always the latter's employees, etc. who do so, thereby creating a personal right to which they themselves are entitled" [33: 6].

This conflict was never fully resolved, but the inventor's right remained firmly anchored in legal precedent and despite the new patent code survived many legal challenges [13: 6; 16; 29]. In theory, therefore, the employee's right to file and acquire the patent remained quite broad. It was limited by only two conditions. First, contractual obligations could require the employee to assign his invention to the employer. If the employee's duties included making inventions and technical improvements, the employer was automatically entitled to file for the relevant patent applications. Just as in the United States, however, an employer's right did not automatically extend to inventions unrelated to the employee's responsibilities or those made at home [91: 185–186; 125: 2–9]. Companies therefore routinely required employees to sign separate patent agreements, stipulating that any and all inventions they made would belong to the employer. Science-based industry in Germany used patent agreements just as frequently as its American counterpart. The Siemens company, for example, introduced its *Patentrevers* (literally, patent declaration) almost immediately upon enactment of the patent law in 1877. The document in question stated that "passage of the German patent law, according to which the first person to report an invention is entitled to the patent right so long as illegal acquisition cannot be shown, makes it necessary to establish certain principles for protecting the company's interests, … the acceptance of which every technical employee must declare by his signature". Article I, which remained in effect until the late 1920s, stipulated that,

> the company can claim as its exclusive property inventions or improvements of any kind made by the employees. Special evidence that the employee has made the invention or the improvement in the course of his professional obligations

or while using the means and experience of the company is not required.

In the United States, a patent agreement like this would have been perfectly legal. In Germany, however, it was not: sometimes the courts sided with the employee, on the grounds that a patent agreement was so broad as to violate the principle of 'common decency' established in article 138 of the Civil Code [26: 46; 33; 73: 3–8; 106].

The second limitation of the employee's rights in Germany was constructed on the basis of the patent code's clauses pertaining to illegal acquisition of the invention. Inventing in employment typically involved collaboration among different employees or use of the company's equipment, in-house knowledge or experience. This fact gave rise to the concept of the 'establishment invention' (*Etablissementserfindung*), known after World War I as 'company invention' (*Betriebserfindung*). The precise meaning and scope of the company invention were never entirely clear. Liberals such as Gareis defined it narrowly, as an invention that resulted from so many different contributions by different collaborators and prior experiences that the individual(s) who had made the decisive conceptualization was no longer identifiable [32; 33: 7; 102: 80; 125: 10]. Scholars in the industrial camp defined the company invention broadly, applying it to most inventions that resulted from the organization's division of labor and employer inputs. In either case, the company was said to be the original inventor and therefore the entity entitled to file. If an employee filed instead, the employer could block granting of the patent or get the employee's patent voided, on the grounds that the invention had originated in collaborative work and was based on company know-how and materials. The invention was then said to be original company property, which the employee had acquired illegally. By making the company the original inventor, the concept of the company invention eliminated the employee's special, creative input and thus his capacity as inventor altogether [27: 3–32, 43–58; 33; 35: 264–287; 125: 9–10, 14, 21–27].

The concept of the company invention was a direct consequence of the first-to-file system applied to inventing in employment. It became German employers' most effective weapon in their arsenal of legal tools to acquire employees' inventions without financial compensation and moral recognition of authorship. The company invention proved more reliable than patent or employment contracts, although much still depended on its exact, legal meaning. Industrialists always strove to make the company invention as broad as possible. The most aggressive employers defined essentially all inventions that way, while others somewhat less extreme extended it merely to include all ordinary professional achievements. Conversely, employees attempted to get rid of the concept altogether, demanding fundamental reform and adoption of a patent code based on the first-to-invent system. Failing this, employed inventors constantly struggled to keep the company invention as narrow as possible.

In time, this conflict gave rise to the emergence of informal conceptual refinements. On the eve of World War I, a tentative distinction had evolved between 'company inventions', 'service inventions' and 'free inventions'. Free inventions were mostly uncontroversial. They were defined as inventions unrelated to the employer's business and therefore the unrestricted property of the employee. Service inventions (*Diensterfindungen* or *dienstliche Einzelerfindungen*) originated with one or more individual employees, but the ownership rights went to the employer, in exchange for which the employee(s) received name recognition in the patent literature and perhaps a financial reward as well. Essentially, service inventions were all those employee inventions that were neither free inventions nor company inventions, on whose definitions they therefore depended.

Company inventions originated in the collaborative research and development effort organized by the company. As collective inventions, they belonged to the employer outright and did not entitle the employee(s) to name recognition or to financial compensation. The exact definition of the company invention, however, remained a matter of endless dispute and controversy, both before and after World War I. For instance, the 1914 Augsburg meeting of the German Association for the Protection of Industrial Property (Deutscher Verein für den Schutz des gewerblichen Eigentums, or DVSGE), which was dominated by big business, defined company inventions as all those inventions "whose genesis is conditioned in an essential way by the experiences, means of support, suggestions, or preliminary activities...of the company". This was so broad that for all practical purposes it would have eliminated service inventions. In contrast, the imperial goverment's 1913 patent reform bill, which favored employees, used a narrow definition: it limited company inventions to those that "cannot be traced back to specific individuals as inventors". This was so restrictive as to make almost everything a service invention [8: 18–23; 20: 1–6; 21: 10–17; 22: 1–18; 26: 10–15, 24–27; 35: 285–287; 78: 13–16; 120: 29–35; 125: 10–29].

The outbreak of war in August of 1914 prevented passage of the 1913 bill. As a consequence, the definitional question remained unsettled. The conceptual distinction between free, service and company inventions remained fuzzy, and the anti-employee thrust of the German patent system remained in effect. The result was that the company invention emerged as the single most contentious issue in management–labor disputes about inventing and patent rights in Germany – until the National Socialist dictatorship outlawed it in 1936. For almost six decades, therefore, Germany's patent code was a Janus-faced institution, in which economic modernity and sociopolitical reaction constituted two sides of the same coin. In this respect the code was very much like other institutions of Imperial Germany, such as the country's social insurance system, its feudalized industrial elite and, in general, the various other amalgams of reaction and modernity that have long fascinated students of German affairs as a source of political dynamism [19; 42: 121–24, 136–142; 43: 121–124, 136–142; 77; 75; 81; 101; 116: 210–222; 118].

3.4 TECHNOLOGICAL CULTURE AND THE INVENTOR'S REWARD

If one accepts the basic premise of patents, that providing incentives for invention and innovation is good policy because it spurs technological progress and economic growth, it follows that the incentives should be designed so as best to achieve the desired outcome. Prior to the age of industrial bureaucracy and the large corporation, this appeared to be easy. Inventor and innovator were usually the same person, and it made sense to award the patent to the inventor. With the growing division of labor between entrepreneurial-managerial and technical-inventive functions, however, this approach became increasingly problematical. If innovation – bringing new ideas to market – was judged to be more challenging than invention – coming up with the new ideas in the first place – for instance, it made sense to award the patent to the company or employer rather than the inventor. This was the path chosen by Germany, while the United States retained the old patent system but found other ways to accomplish the same goal.

How did the change affect actual inventors – the people in the laboratories, in the design offices, and on the shop floor? More specifically, what happened to their incentive to make new inventions? There is no single or easy answer. Over the years and decades, different experts have given different answers. Generally speaking, however, corporate management worried little about the problem, confident in its ability to keep the flow of new ideas going. Many others – engineers, scientists, scholars, lawyers, publicists, legislators, policymakers – however, became concerned that depriving employees of their patent rights or other meaningful rewards for their inventions would kill the necessary incentive to fire their genius, which in turn would have negative consequences for technological development. Such reasoning played a major part in the pro-inventor measures first adopted by the Nazi regime and subsequently continued by the Federal Republic's Employee Inventions Act of 1957.

How real were such concerns? In theory, the privileging of the employer over the inventor in Germany's 1877 patent code may well have retarded the nation's inventive energies and thus harmed the prospects for technological advancement. But it is by no means obvious that it did so in practice. The late nineteenth century and early twentieth century were among the technologically most dynamic and creative periods in modern German history. However much the patent code may have reduced employed inventors' incentives, the negative effects appear to have been counterbalanced by other factors.

Technological progress is multifaceted and depends on many different factors, some of which cancel each other out. Any reduction in inventive creativity caused by the anti-inventor thrust of the 1877 patent code, for instance, may have been offset by improvements the statute made in the conditions for industry's commercial exploitation of new ideas. The law also spared important categories of inventors such as university-based researchers and scientists in research institutes, who constituted a crucial source of inventive talent in nineteenth- and twentieth-century Germany. Nor do patent statistics – whatever their worth as a measure of technological progress – support a view of flagging inventiveness, showing instead a pattern of explosive growth from 1877 to 1930, when patent filings reached a peak that has not been equaled since. Despite such difficulties and the appearance of unbroken inventive dynamism, it is worthwhile to inquire further into the problem of the 1877 patent code's impact on Germany's technological culture.

A first step is to consider more closely the concept technological culture itself. Technological culture may be defined as the configuration of attitudes, practices, institutions, technical artifacts and systems that set apart one society's or one period's approach to technology from that of another [54; 64; 95]. The leading cause of variation in this regard is the fact that technology, as Wiebe Bijker, Trevor Pinch and Thomas Hughes have shown, is 'socially constructed'. That is, technology is shaped by the different natural environments and different historical contexts – social, economic, political, cultural, legal – of different societies. Because those exogenous factors change over time, technological culture is not a static concept but a dynamic one and itself subject to change. As a consequence, one can delineate periods or epochs in the development of a society's technological culture, even though patterns, once established, tend to survive long after the conditions that gave rise to them have disappeared.

The concept technological culture is not merely descriptive of a society's various technological peculiarities but also serves analytical purposes. It seeks to explain the 'why' of those peculiarities in terms of the reciprocal influences and the functional relationships between technology and other societal factors. Perhaps the most striking feature of Germany's technological culture in comparative historical perspective is its precocious and passionate professionalism and its high degree of formalism. Since the first half of the nineteenth century, German engineers have shown a disproportionate interest in mastery of a discipline's theoretical foundations and abstract perfectionism, accompanied by a certain disdain for empiricism and a hunger for credentialing. Another distinctive quality, according to Joachim Radkau, is the caution and conservatism that, relative to the United States, Germany has exhibited in its adoption of new technologies during the past two hundred years. Both characteristics can be linked to Germany's bureaucratic tradition and to the early emphasis on school-based professionalization and specialized scientific training of the majority of its technologists. All of this generated a certain degree of intellectual compartmentalization and inventive conservatism. That is, organizational and social constraints tended to produce inventors firmly anchored in one single 'technological frame' and therefore likely to channel their creativity into creating improvements or variants of established technologies [10; 35].

To characterize the pattern of inventing that strengthens and develops existing technologies, Hughes has coined the phrase 'conservative inventions'. Hughes associates conservative invention primarily with the process of inventing in industrial research laboratories by industrial scientists and with the economic interests of the corporations heavily invested in those technologies. The counterpart to conservative invention is 'radical inventions', by which Hughes means 'system-originating' advance that potentially threatens existing technology. Radical inventions, according to Hughes, are disproportionately associated with independent as opposed to salaried inventors. The great independents were (and are) not inventors in

...the narrow sense of the word nor specialized researchers, but more like visionaries, individuals such as Henry Ford, Walt Disney, Thomas Edison, Werner Siemens, Rudolf Diesel, Carl Bosch, Emil Rathenau and, one might add, Steven Jobs or Bill Gates, pioneering whole new industries and technological systems and ways of life [53: 53–95, 180–193; 54: 57–66]. Conceptually, Hughes' independent inventors are closely related to Joseph Schumpeter's capitalist entrepreneurs ('creative destruction') and Max Weber's charismatic religious leaders, innovators such as Jesus or Buddha, whose power is so great that they successfully challenge everyday reality and inspire new patterns of social organization. Sociologically, the inventive creativity and innovative power of independents can be linked to their being more marginal, less professionalized and less committed to the traditions of a discipline than industrial scientists. Independents are more likely to be acquainted with what Bijker calls 'multiple technological frames' and to possess greater 'interpretive flexibility' than the average research scientist in a bureaucratic setting. Historically, independents have played a major role in the technological culture of the United States, whose most notable trait is probably the abandon with which it has been willing to innovate and discard older technological systems in favor of newer ones [10; 36; 76: 151–155].

The last quarter of the nineteenth century and the first two decades of the twentieth century were the golden era of the independents, according to Hughes. After World War I, salaried professionals and specialized researchers began to move center stage, owing to the continuing growth of large corporations and aided by patent practices inimical to the independent inventor. The corporations bureaucratized inventing in the industrial research laboratory and created the modern system of scientific research and development in their quest for stable and controllable technological systems. The decline of independent inventors meant that radical inventions became less common, though they did not disappear. Technological progress became characterized by the predominance of conservative inventions, rationalization, consolidation and the systematic elimination of what Hughes calls 're-verse salients', problems that emerge as a consequence of advances in other parts of the system. In the end, a world emerged that is dominated by huge technological systems possessing, if not their own unstoppable dynamic and autonomy, a great deal of 'momentum'.

Hughes' concept of momentum may be paraphrased as the tendency of a technological system to continue to grow or persist far beyond the point at which the conditions that gave rise to it continue to operate. Technological systems acquire momentum because of the visions, the material interests, the discourse, the know-how – entire ways of life – of the people and institutions involved in them [54: 71–80]. Momentum also is a leading cause of distrust and critique of large technological systems. In part, this is because momentum tends to suppress new and competing technology along with the advocates of change – a tendency that, in turn, may result in the loss of the system's own technological dynamism and in technological freezing, turning large systems with momentum into 'dinosaurs'. Another source of apprehension is that adherence to the instrumental rationality of a large technological system easily becomes irrational or destructive from a human perspective, just as modern bureaucracy, in Max Weber's view, is both humankind's greatest organizational achievement and the attainment of complete dehumanization, the "iron cage of future serfdom". Powerful versions of both kinds of technology critique emerged in the United States in the early 1970s, closing a chapter of technological enthusiasm that had started in the 1870s [53: 1–12, 443–472, esp. 462].

In the 1970s and 1980s, several highly publicized technological disasters and a steep decline in American competitiveness in industries such as automobiles, consumer electronics and machine tools brought the problems of large technological systems to the attention of a wider public. Long before then, however, scholars and intellectuals had begun to reexamine the technological culture created by the momentum of large technological systems. The questioning included a critical reexamination of the reigning ideology of inventing, which held that the large organization's scientific research laboratory and systematic R&D were vastly superior to the amateurish and unsystematic fumbling of the independent inventor. This ideology counted many eloquent adherents, among them John Kenneth Galbraith, who in 1952 had written about a "benign Providence ..., [which] has made the modern industry of a few large firms an almost perfect instrument for inducing technical change ...". Galbraith continued:

> There is no more pleasant fiction than that technical change is the product of the matchless ingenuity of the small man forced by competition to employ his wits to better his neighbor. Unhappily, it is a fiction. Technological development has long since become the preserve of the scientist and the engineer. Most of the cheap and simple inventions have, to put it bluntly, been made [32: 92; 58: 35–36].

In a study first published in the late 1950s and reissued in an updated edition in the late 1960s, John Jewkes, David Sawers and Richard Stillerman confronted the view that "originality can be organized". They investigated the origins and development of a large number of recent inventions and innovations, showing that independent inventors, individual genius, the 'uncommitted mind', and the quality of being an outsider remain among the most important sources of invention. They also concluded that large corporate research organizations, rather than being agents of change, frequently become centers of resistance to change. Captive research within the large company is less creative than that by independents, universities and government research institutes. Within the organization it is usually the quasi-autonomous individual who is most creative. In light of their findings, the authors posed this fundamental question:

> Whether Western societies are fully conscious of what they have been doing to themselves recently; whether, even although there now may be more people with the knowledge and training that may be a prerequisite for invention, they are used in ways which make it less likely that they will invent; whether there is truth in the idea that, although the institution is a powerful force in accumulating, preserving, discriminating, rejecting, it will normally be weak in its power to originate and will, therefore, carry within itself the seeds of stagnation unless the powers and opportunities of individuals to compete with, resist, challenge, defy and, if necessary, overtopple the institutions can be preserved [58: 182; 126].

To counter stagnation and "offset the dangers of rigidity", Jewkes, Sawers and Stillerman suggested the possibility of introducing measures to help independent inventors, such as changes in the patent system, public rewards or tax credits [58: 186–193]. They did not mention special programs or incentives for salaried inventors, evidently because they had so little confidence in the ability of corporations to be creative in the first place. The authors did concede that large organizations and corporate R&D play an important role in the development, engineering, production, and marketing phases of the innovation process, but that was not their central concern.

One can make a good case, however, that assistance for the independent inventor, though by no means misplaced, is at best a partial solution. The greatest problem of the technological culture of the United States during its most recent technological slowdown, for instance, was not a lack of original ideas and inventions, nor an inability to develop and introduce new products, but failure to generate the steady stream of improvements that sustain technological systems, conquer foreign markets, and keep products competitive in the long run. Why did this happen? And why do large technological systems and corporations develop what Hughes calls 'conservative momentum', i.e., become rigid, inflexible, and stagnant? The answer has much to do with the nature of bureaucracy and the vested interests of all those who develop a stake in the technological status quo. Perhaps those impersonal forces are powerful enough by themselves to overwhelm the kind of technological dynamism that springs from the principle of conservative inventing and gives systems their evolutionary potential. But it seems obvious that any tendency toward stagnation and momentum is powerfully reinforced by corporate patent practices that deprive industrial scientists of the incentive to bring forth their best efforts and other labor policies that disregard the psychology of human motivation [2; 39: 41–43; 85; 86; 89; 111: 12–37; 112: 89–126]. Reversing those policies by introducing award schemes for salaried inventors might not necessarily break momentum. However, it can reinvigorate conservative inventing and therefore regenerate otherwise moribund technological systems. To that extent the possibility of a social (re-)construction of technology, which Hughes posed as a question in the final pages of *American Genesis*, is very real [53: 459–472; 55].

The validity of this conclusion is supported by the experience with employee-inventor schemes of nations such as Japan and Germany. These countries have managed to best the United States, not in the newest and most glamorous technologies, nor in inventive genius, but more typically in those industries – many of them older – whose vitality depends on the gradual perfection and ongoing accretion of many small changes, i.e., on the dynamism of conservative inventing in the large corporation. It was largely this consideration that in the 1970s and 1980s prompted the reform initiatives by members of the U.S. House of Representatives and others to remedy the incentive-killing effect of the employer's interposition between private inventor and public inducement.

Corporate management has framed its opposition to those efforts in terms of general philosophical considerations about restrictions on free enterprise, but also marshaled specific counter arguments about the nature of modern inventing and inventing in employment. In a recent comparative study on employees' inventions, British legal scholar Jeremy Phillips has grouped those arguments into five different categories. The first one centers on the idea that statutory awards for salaried inventors are unnecessary. This is so because the market is said to be a fair mechanism of determining the rewards for work performed. Phillips quotes inventor and employer Jack Rabinow, who argued in the 1960s that, "In this society, anyone who wants to change employers can, as I left the Government ... I think inventors should get all they can in a competitive society, such as ours. And if all they can get is a good salary, then that is all they deserve" [89: 29–30]. Other reasoning of this type includes the argument that inventors are a special breed of people who invent anyway, so that "there is no point in stimulating them to do what they already do"; and that "so few really

important creative inventions are made that there is very little point in establishing the machinery of a statutory scheme to encourage them" [89: 31].

A second type of argument revolves around the position that rewarding salaried inventors is impracticable. "It is difficult, if not impossible", one strand in this line of reasoning holds, "in these days of teamwork and group invention to isolate the person who made the invention from those around him who assisted in it" [89: 32]. This thesis, which German employers also used frequently in the first part of this century, still has considerable scholarly support in the United States today. Thus, Hugh Aitken in his 1985 study of the development of American radio maintains that invention is a "process ... essentially social and cooperative[,] ... with considerable duration in time, one to which many individuals contribute in a substantial way, and in which the conception of the thing invented or discovered changes". The 'eureka moment', according to Aitken, is largely an arbitrary decision and a romantic fiction resting on the "bias built into our patent laws" [1: 548–549].

Arguments in the third group maintain that rewarding salaried inventors actually harms invention, because fear of losing credit discourages employees from sharing their ideas with colleagues, encourages them to consider only their own narrow self-interest rather than the well-being of the company, and disrupts teamwork [82: 100–101; 89: 32–34]. The two final types or argument rest on the notion that special awards are unfair to both employees and employers. Awards are said to be unfair to employees because they single out inventors when other employees in the organization are just as important to successful innovation or efficiency but get no special rewards. Awards are unfair to employers because they should be entitled to the employees' labor as a matter of course. Instead they will be flooded by worthless ideas and harassed by frustrated employees demanding the marketing of their inventions [89: 35–36].

Industrial scientists, engineers and the advocates of inventor rewards have good answers to each of these arguments. Like management, the inventor side typically presents its case as an analysis of the inventive process and a discussion of the inner workings of corporate organization. As a consequence, the issues of modern technological progress, the nature of inventing, the innovative potential of large industrial corporations, and the achievements of salaried professionals tend to assume the guise of arcane technical discussions and seemingly trivial disagreements among patent experts. This appearance is very deceptive. At bottom, the debate about inventor rewards is about the relationship between the individual, society and technology, about the reciprocities among technological progress, industrial organization and politics. It is a debate in which fundamental questions of the division of labor in modern capitalist society, economic competitiveness, creativity, bureaucracy, power, and social policy are inextricably intertwined.

3.5 THE INVENTOR DEBATE IN GERMANY AND NATIONAL SOCIALIST INVENTOR POLICY

Conservative inventing by salaried employees and the kind of innovation with which it is associated came to define Germany's technological culture earlier than in the United States. In the decade before World War I, the large corporations in electrical and mechanical engineering and chemistry – companies such as Siemens, AEG, MAN, Hanomag, Deutz, Bayer, BASF, Agfa and Hoechst – already accounted for the bulk of German innovation. The reasons centered on the synergies of Germany's bureaucratic tradition, its highly developed and formalized system of technical education, the emergence of organized capitalism in the late nineteenth century, and the 1877 patent code. General economic, geopolitical and natural-resource conditions, which made life harder for independents in Germany than in the United States, played a role as well. Combined with the intense expectations of professional status and privilege among German engineers, those factors made the patent code, more specifically the rights of salaried inventors and high patent fees, into a political issue as early as the first decade of the twentieth century [35: 264–287; 74].

By 1905 unionized engineers, progressive lawyers and reformist politicians had launched a campaign to revamp the patent code. The reformers, who attracted considerable public attention and support, demanded that the first-to-file system be scrapped in favor of the Anglo-American first-to-invent system. They insisted that employers be prohibited from acquiring the patent rights of employees' inventions, except in exchange for steep royalty payments. Intimidating or browbeating employees into accepting token royalties or rewards should be forbidden. Reformers also pressed for abolition of the progressive patent fee schedule in favor of a low, one-time payment such as existed in the United States. German big business in the science-based industries put up a forceful defense of the status quo but was unable to hold the line. By 1913, the Imperial government had drafted a reform bill that steered a middle course between the two sides. It adopted the first-to-invent principle, albeit without burdening the patent office with investigating claims of priority and interference; it conceded name recognition and the right to a claim of special compensation for service inventions by employees, but not statutory rewards; and it envisaged mild fee reductions, though not surrender of the 'self-cleansing' principle of progressive fees. Passage of this bill was postponed indefinitely by the outbreak of World War I, but in July of 1914 not even the most determined opponents of reform doubted that it would become law [35: 241–242, 264–287; 107].

The details of the inventor conflict before World War I need not concern us here, but it is important to emphasize that before 1914 the issue was framed in terms of social policy, fairness and equity for inventors – and not as a question of technological momentum or dynamism [35: 264–287]. As the official government commentary on the 1913 bill put it:

> Complaints about the lack of social sensitivity and fair consideration for the economically weaker party, and about the high fees that burden patent protection, have played a large role in the public discussion. There are vigorous demands for recognizing the rights of the intellectually creative inventor as such and establishing a

legal basis for the moral and economic empowerment of the economically dependent inventor [21: 7].

The imperial government justified the bill on the same grounds of political morality and expressly discredited the first-to-file theory that had guided the framers of the 1877 code as illiberal and immoral. "The circumstance that the current law does not even mention the inventor contradicts to a certain extent the fact that in the final analysis all cultural progress stems from the knowledge, the will and the deed of individuals". Official recognition of this fact was the main reason for changing patent systems and adopting the first-to-invent principle. In the language of the official commentary, "being the conceptual author of the invention should suffice by itself to establish the right to the patent" [21: 10].

In the Weimar Republic, severe economic problems and manifestations of technological momentum brought a new dimension to the struggle. The conflict expanded into a larger problem, which compounded the original social issue with fears about technological stagnation. Critics of the status quo began arguing that the patent code and corporate inventor practices were not merely unfair, but also harmed the vitality of Germany's technological culture and imperiled its destiny as a nation. Regardless of motives, however, debate over the merits of changing the patent system and adopting inventor rewards were always articulated as disagreements about the nature of inventing. The issue was framed in terms of conflicting interpretations of the inventive process in modern society. This was the case both before World War I and after.

The arguments of German industrialists often were variations of the positions that Phillips encountered among American management in the 1980s. There was, however, one crucial difference. In most countries, Phillips points out, the question of the "initial ownership and control of inventions" was not contested. Typically, industrialists "swiftly conceded that, by virtue of the mere fact of making an invention, the employee was entitled to be recognized as the inventor, notwithstanding any part played by the employer in financing or assisting in the making of that invention". Employers

merely contested the employee's right to compensation for the use of his invention [89: 25]. That was not true in pre-Hitler Germany, where employers went one step further and relied on the company invention to deprive employees of their status as inventors as well. This is how Carl Duisberg, the head of Bayer and the intellectual father of IG Farben, in 1909 described inventing in the synthetic dye industry:

> A given scientific theory is simply put to the test, either at the instigation of the laboratory's supervisor or at the initiative of the respective laboratory chemist. The theory tells us that the product must possess dyeing properties, but that matters less than finding out whether the new dye can do something new.... The chemist therefore simply sends every new product he has synthesized to the dye shop and awaits the verdict of the dyeing supervisor.... Not a trace of inventive genius: the inventor has done nothing more than routinely follow a path prescribed by the factory's method [24].

Some scholars believe that Duisberg's blunt statement accurately reflected the realities of industrial exploitation of Azo-dyes chemistry before World War I, and that inventing in this field had indeed changed to simple, industrial routine [7: 149–155]. It is true that the underlying 'coupling process' had been known since the 1880s and the role of individual inventive creativity reduced with respect to the basic chemical principles. But it is also true that ongoing profitability in the commercial production of Azo-dyes called for continual inventiveness and rationalization in the manufacturing process [121]. Moreover, Duisberg made his remarks in the midst of the prewar struggle over inventor rights as a polemic against the pro-inventor lobby. His words did not reflect management's private thinking or inventor policies, which held that routine inventing – even in dye making – was the exception [48]. Nor did the alleged absence of inventiveness prevent the Bayer executive from making a complete about-face when it came to defending the grounds on which such 'non-inventions' should be patented. In 1889 Duisberg had succeeded in preserving the patentability of 'routine inventions' by convincing the *Reichsgericht* to widen the criterion of patentability in

chemical invention from novelty of process to encompass novelty of usage as well. Subsequently
this became known as the doctrine of the 'new
technical effect'. Inventiveness and patentability,
that is, were said to reside in having produced new
and unexpected technical properties, a concept that
figured prominently in the 1891 revision of the
patent code and made possible continued patenting of inventions whose underlying process was
no longer judged to be original [35: 230, 264–287;
48]. In other words, the experts agreed that mere
knowledge of the general paradigm did not eliminate inventiveness in producing specific results.
Of course, this smuggled the human factor right
back in. Finally, there can be no doubt that despite a tendency toward routinization, some Azo-
dye chemists were a great deal more creative than
others, and that the successful ones tended to climb
the corporate ladder by accompanying their inventions through all the stages from the laboratory to
market. On balance, it seems prudent to treat Duisberg's rhetoric about automatic and routine inventions with considerable circumspection.

In 1921, Ludwig Fischer, director of the patent
division of the Siemens company, published *Company Inventions*, a slim but highly controversial
volume, in which he took the employers' position
to its logical conclusion. Fischer began by defining
inventing as "subjugating the forces of nature to a
purpose" (*Bindung von Naturkräften nach einem
Zweckgedanken*) [27: 4]. Seemingly unobjection-
able, this definition was in fact quite misleading.
It ignored the seminal, creative, groping stage of
invention, when answers and questions are not always clearly distinguishable and the world as taken
for granted is first challenged [126]. Instead, Fischer posited a later phase and tamer dimension of
the inventive process as its totality, by defining it as
an activity in which the desired technological outcome or function was known at the outset. His definition therefore reduced inventing to a question
of traveling an unfamiliar road strewn with 'obstacles'. The most important means to overcome
those obstacles, or the actual process of inventing,
according to Fischer, were the kind of scientific-
technical training, experience, systematic activities, and rational planning that are within reach of

"professional technicians with wholly average tal-
ents" and that accounted for "ninety-nine per cent
of the countless inventions made in the past twenty
or thirty years". The thing that

> fertilizes inventing more than anything else,
> however, and that makes it into rational activ-
> ity, is the teamwork of many, who complement
> and stimulate each other; the sensitive use of the
> experience of others; time and freedom for spe-
> cialization in one area, ...; and finally and most
> of all: possession of the necessary tools to do the
> work [27: 10].

Few if any of those conditions were within reach
of the independent inventor. It was the company
that "create[d] the most important conditions for
this rationalization" of inventing. It was the company that identified the thing to be invented; the
company that provided the employee with the necessary tools, the laboratory, the time and money
for experiments, the library, the division of labor, the experience, the state-of-the-art knowledge,
the patent searches, and the overall climate of
intellectual stimulation. The employee, therefore,
was "exposed to a thousand influences, which are
of the greatest significance for the gradual, step-
by-step elimination of the obstacles", and which
collectively may be designated as the 'company
spirit'. According to Fischer, it was the 'company
spirit' that was the real inventor, the agency that
"steers the activity of the employee in certain, rational channels and eliminates the obstacles, so
that inventions must come about". Conversely, an
"employee influenced by the company spirit who
would never be able to find anything that might
be patented as an invention... would be a bun-
gler" [27: 11–13]. Fischer concluded that the "employee invention as a rule is a company invention. The achievement of a company inventor as a
rule does not exceed the measure of what can be
expected from a competent full-time professional
(*Voll-Techniker des Faches*)" [5: 14–35; 27: 15].

Ironically, it was not only the right and big business that assailed the individualistic dimension of
inventing, but also the blue-collar working class.
The left did so, however, for different reasons. In
1923 the socialist *Metallarbeiter-Zeitung* attacked
the notion that inventors ought to receive special
compensation for their creativity. To support its

view, the journal quoted the famous inventor and physicist, Ernst Abbe, who reportedly had turned down a financial interest in an invention he had made for the Zeiss firm on the grounds that accepting it would be an injustice. "It is true that the invention originates with me", Abbe is quoted as saying, "but its reduction to practice and its development is the work of my numerous collaborators. I would not have been capable to market the invention profitably by myself. I owe this success to the collaboration of the firm's employees", who therefore shared the right to its financial profits in common. "So it is with every collective endeavor", commented the metal workers' publication approvingly. "All those who are involved in it have a claim to it, and there is no possible way of determining how much the individual is entitled to by law" [108].

Similarly, in an article entitled "Invention or Development", the Social Democratic organ *Vorwärts* argued in 1921 that "in the currently prevailing conception of history the significance of the individual human being for historical development is vastly exaggerated. In the bourgeois conception of history, able field marshals figure as the motor of world history". It was the same, said the paper, with bourgeois interpretations of technological development, which looked only to the great inventors. This perspective was wrong because it ignored the more fundamental, economic forces that drove historical development. "Progress", concluded the paper, "does not depend on the genius and creative powers of individual human beings, but on the industrious collaboration, the enthusiasm of countless people" [108].

This convergence of the interpretations of inventing by big business and labor is a good example of why the German middle class felt threatened by the political constellation of the Weimar Republic. The attack on individualism and embrace of collectivist arguments by society's big battalions came across as a repudiation of bourgeois culture and values. This helps explain why in defense of their world the middle classes could embrace, first, extreme ideas and, later, the politics of extremism. Their reaction is demonstrated with great clarity in the response of industrial scientists to the collectivist view of inventing, especially its employer

version. Outraged by Fischer's orderly inventing machine, professional employees denounced *Company Inventions* as the work of a hired hand and attacked its author in other personal ways. More importantly, they countered with references to bourgeois cultural icons such as Goethe and Max Eyth, the well-known nineteenth-century engineer–poet. Both Goethe and Eyth had portrayed the inventor as the genius whose flashes of insight propel human history, as the archetypal Promethean, made in the "image of the Creator, a being in which God has placed a spark of His own creative power" [13; 104: 178].

Many other Germans agreed that technological originality was not simply a matter of corporate organization. For instance, Carl Bosch, IG Farben's chief executive and winner of the 1930 Nobel prize for inventing high-pressure catalytic hydrogenation, commented in 1921 on this engineering achievement by equating it with the creativity of the artist. Neither the artist nor the engineer, Bosch maintained, "is in the final analysis master of his thoughts and ideas". It was "false to assume that everything has been calculated, everything figured out". The crucial idea comes to the inventor "at the right moment, just as it does for the artist in his creative urge" [49: 95]. Adolf Hitler, too, made his contribution to the debate. In *Mein Kampf* (1924) Hitler wrote, "the inventions we see around us are all the result of the creative forces and abilities of the individual person". It was "not the collectivity that invents and not the majority that organizes or thinks, but always, in all circumstances, only the individual human being, the person". Only that society was "well organized", according to the future dictator, that does "all it can to facilitate the work of its creative forces and to employ them to the advantage of the community. The most valuable part of the invention ... is first of all the inventor as an individual" [47: 496–498; 59: 1–3].

To be sure, Hitler's central concerns were not lower patent fees or royalties for salaried inventors. But, as Rainer Zitelmann has argued, Hitler did think of himself as a revolutionary, a fighter for social equality and equal opportunity within the racial community, and the historical agent who would restore charisma and individual dynamism to their rightful place in a nation he maintained was collapsing under the weight of Jewish conspiracy, socialist collectivism and bureaucratic capitalism. Hitler was also fascinated by technology and infatuated with the idea of the modern 'Aryan' hero, the individual inventor as the source of vitality and dynamism in industrial society [129: 49, 83–84, 92–125, 259, 264, 321–325, 347, 401]. Given the general climate of despair, the manifestations of technological momentum and stagnation, and the particular frustrations generated by the patent code in Weimar, such views established an elective affinity with certain right-wing intellectuals and with many engineers and inventors.

Jeffrey Herf has studied those affinities in detail and coined the term 'reactionary modernism' to describe the peculiar blend of pro-technology and anti-Enlightenment values in the ideology of engineers and right-wing intellectuals in the Weimar Republic and the National Socialist regime [43]. It is true, of course, that the hyperbole inventors and engineers frequently employed when they described their work was a kind of applied, *völkisch* romanticism: a transfer of the categories German idealism to the world of twentieth-century business and technology. But it is true as well that such rhetoric, even if exaggerated, sprouted from more than a grain of truth. Above all, it was a reaction against the systematic denigration of intellectual creativity by German employers using the modern but perverse concept of the company invention. The label of reactionary modernism, therefore, should not be restricted to the ideology of Nazis and the attitudes of discontented engineers and right-wing intellectuals after World War I. The term may be applied with equal if not greater justification to institutions such as the 1877 patent code, which the Weimar Republic inherited from Werner Siemens and the many other bourgeois authoritarians who helped make Imperial Germany.

To describe the company invention and the German patent code as examples of reactionary modernism is not to argue that employers' insistence on the reality of collective inventing in the corporate setting was nothing but a clever scheme to enhance their own power vis-à-vis employees. It is obvious that both sides were partially correct. The reality of modern inventing comprises both the organizational dimension and the individual one. Inventing in the industrial research laboratory is a combination of what David Hounshell calls 'individual act' and 'collective process' [50: 285]. Moreover, the collective or teamwork dimensions, as Erich Staudt and his colleagues at Bochum University's Institute for Applied Innovation Research point out in a study of recent developments, are "gaining ever greater significance in industrial research and development. Growing technical complexity as well as the general transition from 'Little Science' to Big Science' increasingly demand knowledge that integrates various specialties and is interdisciplinary". More than half of all German inventions in 1990 were so-called team inventions (two or more co-inventors), and in the chemical industry the figure was about eighty percent [111: 38–39].

Teamwork, however, does not mean the absence of individual intellectual creativity, nor the disappearance of the independent inventor. Rather, research teams and the independent operate simultaneously and along a spectrum, working on different problems in different fields, in pulses of radical breakthrough and conservative consolidation of different technologies. Manfred von Ardenne, the scientific entrepreneur and gifted independent who in 1937 at age thirty invented an electron-scanning microscope, refers to this fact in his autobiography. Reflecting in the early 1970s on the story of his own creative spark and feverish capture of the essential conceptualization that led to the new microscope, Ardenne writes that, "even under today's circumstances, in the pioneer stage of a young science the decisive impulse can come from a single researcher". According to Ardenne, who opted for the Soviet Union and the German Democratic Republic after World War II, this is true even though the independent stands on the shoulders of his predecessors. In the later stages, however, "especially when important 'developments' take place, it is rarely the case anymore that the decisive impulses stem from a sole researcher by himself. Usually, those impulses are vectors of components in which several partners, for example a collective, participate" [4: 139–140; 12: 317–324].

In 1934, the chemist Professor H.G. Grimm, director of IG Farben's Ammonia Laboratory in Oppau, made a similar observation. In a lecture on inventive creativity to the German Research Institute for Aeronautics that year, Grimm sketched the organization and management of inventing in large chemical research laboratories. The chemical 'inventor factories', he pointed out, were typically organized in different research teams, some of which were as small as one or two members. The latter were often "specialists – for instance, mineralogists, bacteriologists, etc. – who serve the entire lab, or also solo inventors, 'one-horse carriages', who have difficulty fitting in with team work, but who are frequently particularly valuable because of their original ideas". To ensure that large research laboratories did the things for which they were designed, namely "create new products and processes, i.e., economically usable inventions", according to Grimm, two conditions must be met. First, there had to be a "good atmosphere inside the laboratory, and second, its internal organization had to function with so little friction that the individual inventor hardly notices it". Creating this good atmosphere, said Grimm, consisted in establishing the "optimal mental and material conditions of work for the individual inventor, whose importance for the entire operation may be compared to that of the soldier at the front". The inventor must get the feeling that, "although he must fit into the larger whole and his research mission has been set by the needs of the factory, he can be free in his creativity and, within the given framework, his creative talent can roam freely and without hindrance". Citing Carl Bosch, Grimm cautioned his audience not to forget "that for the technological inventor too, completely analogous to the artist, everything depends on the creative idea and the right mood. In neither case can one summon these by command or produce them artificially. All one can do is to establish the appropriate conditions for their appearance" [46].

In the Weimar Republic, too, cooler heads recognized that the problem was not one of mutually exclusive alternatives but one of complementarity and finding the common ground on which productive research in the corporation could be erected. The chemical industry was one sector where such a compromise came into being as early as 1920. Employers and employees worked out a formula to differentiate among the various types of inventions and reward a majority of them with royalties and author recognition. But the chemical industry was an exception, and after 1924 even that industry's leaders tried to reduce costs and rationalize the innovation process by rolling back the financial gains industrial scientists had made earlier. Meanwhile, efforts to reform the patent code and pass labor legislation with compulsory rewards for employee inventors failed. On the whole, therefore, and despite recognition of its importance, the problem continued to defy solution in the Weimar Republic. This deadlock helped set the stage for the Nazi takeover of 1933 and for the decisive break the dictatorship's pro-inventor reforms made with the past.

In the Third Reich, the inventing debate continued, though the balance of power now tipped in favor of the pro-inventor party. Overriding the resistance of big business, the Nazi government eliminated the company invention and introduced the principle of the inventor's original right in the invention. (For pragmatic reasons, however, it retained a procedure before the Patent Office that continued the first-to-file system.) In addition to a reduction in patent fees and expanded government powers to compel the working of patents, these and other changes were all contained in the new 1936 patent code.

Despite the usual Nazi rhetoric about subordinating the interests of the individual to those of the community, the new patent law in fact represented a kind of emancipation act for employed inventors. First, it accorded them the important symbolic recognition as inventors that they had sought for decades. Second, it created the legal foundation for statutory compensation for employee inventions, which was a crucial modification of the original patent code. Adoption of the first-to-invent principle and elimination of the company invention meant that employers could no longer deprive employees of authorship and block inventor rewards by arguing that the company was the real inventor. This was tantamount to conceding that employee inventions in principle were special accomplishments deserving of special compensation. Of course, the law left open the all-important practical aspects of the question, which took a great deal of additional wrangling and the pressures of total war to be solved. But by 1936 Germany had finally reached the point where, in most other countries, the debate about employee inventions began [89: 25].

Between 1936 and 1942, when the National Socialist regime imposed statutory compensation as part of a comprehensive system of inventor assistance, the inventor debate continued in a different form. It assumed the guise of discussions and negotiations about the technical details of whatever compensation mechanism would be adopted. Employers now argued their case in terms of company contributions, employee responsibilities and the methods of calculating an invention's value to establish the principle of the zero option: that there should be cases in which the invention was so minor, or so obviously within the scope of the employee's duties or a function of company state of the art, that the compensation formula would yield no reward at all. It was in the context of these discussions that the regime's inventor-policy experts, abandoning their sometimes naive faith in the value of amateur and little inventors, developed the highly articulated and nuanced system for employee compensation that became the basis for the 1957 Act on Employee Inventions.

What is the larger historical significance of the inventor debate, and how does it relate to the evolution of Germany's technological culture? With regard to the Weimar Republic, there is only sporadic evidence that problems of technological momentum and stagnation were the immediate consequence of employers' inventor policies in those years. In as much as the patent code was one factor in causing Germany's technological culture to deemphasize radical inventing before World War I, it is plausible that the employers' hard line toward inventors during Weimar increased that tendency, but it is impossible to say by how much. There can be no doubt, however, that problems associated with momentum, bigness, rationalization, the treatment of inventors, and reluctance to introduce new technologies (e.g., the automobile, telephone or radio) unnecessarily delayed the transition to a mass

consumer society. In turn, this created a political climate in which the politics of inventing helped wreck the Weimar experiment as a whole. The conflict over inventor rights and patent code reform thus contributed in its own small way to the polarization and political stalemate that led to Hitler [11]. Considering that employers on the eve of World War I had already resigned themselves to reform of the patent system, the failure to recreate the conditions for compromise in the Weimar Republic is particularly tragic. At the same time, however, Weimar was the crucible from which Nazi inventor policy emerged. The dictatorship's introduction of measures such as free assessment of an invention's economic viability, inventor trustees in the factories, statutory compensation, patent-fee reductions, the first-to-invent principle, compulsory licensing, and, in general, its emphasis on the inventive genius and creative energy of the individual, were a backlash against the inventive conservatism and corporate intransigence of the Weimar period.

In the Third Reich, inventor policy became embedded in the comprehensive racism that was the regime's dominant organizing principle. Inventor assistance and financial rewards, for instance, did not extend to individuals categorized as *Ostarbeiter*, Poles, Jews and Gypsies. Likewise, the sizeable contingent of Jewish patent attorneys, consultants and participants in inventor discussions during Weimar disappears from the sources virtually overnight from 1933. We do not have to ask what happened to them. In 1944, Erich Priemer, head of the NSDAP's Office for Technical Sciences, spoke with pride of the successful cleansing operation initiated in 1933, when "the profession of patent attorneys, governed at the time by Jews, itself consisted largely of Jews. The ranks of independently practicing invention promoters had been infiltrated by elements corrupted both professionally and morally, which discredited the entire profession" [94: 3–5]. Nazi inventor policy was, therefore, an integral part of the Third Reich's descent into barbarism.

With regard to the community of racial 'citizens', however, there can be no doubt that the regime's inventor measures constituted a socially modernizing and progressive development [14: 1–4, 8–22, 304–307; 15: Preface; 18]. Nor is there any doubt that this was the intention. As Hans Frank, head of the National Socialist Jurists Association and President of the Academy of German Law, boasted following enactment of the new patent code in 1936, "the Reich government openly and freely presents the new Patent Code for inspection, in the consciousness of having established a modern and progressive system, which will not remain without influence on the patent codes of other civilized nations (*Kulturnationen*)" [30: 7]. Ironically, there was more than a grain of truth in this. The regime's inventor measures, moreover, appear to have had a positive effect on Germany's technological culture, even if the motives of policy makers rooted in an exaggerated technological romanticism and their expectations of a wave of radical inventions remained mostly unfulfilled. As a dialectical counterpoint to the reactionary modernism of corporate inventor policies in Weimar, however, the Nazi measures produced a synthesis that energized conservative inventing in the large firm, both during the Third Reich and after.

The Allied victory of 1945 destroyed the racist shell in which the regime's inventor policies – and all Nazi policies – had been encased. Once the ideological and race dimensions were excised, however, the substantive components of those measures gained a new lease on life. Thus, while all references to National Socialism were deleted and Nazi Party agencies such as those for compulsory inventor counseling and assistance dismantled after the war, the core of the new policies survived in the Salaried Inventor Act of 1957 and current law. The principle of inventor assistance was considered important enough to result in the establishment of several non-profit R&D institutes and voluntary inventor support organizations (e.g., the Fraunhofer Gesellschaft) in the Federal Republic. Moreover, the regime's measures indirectly raised the status and the attractiveness of the engineering profession, strengthening the opportunities for financial gain and rights of practitioners vis-à-vis management. Finally, by forcing resolution of a long-standing conflict in management-labor relations, the National-Socialist inventor policies laid part of the groundwork for a functional social order after the war. This outcome, it seems to me, invites us to reflect on history's profound ironies and dialectical turns.

REFERENCES

[1] H.G.J. Aitken, *The Continuous Wave: Technology and American Radio, 1900–1932*, Princeton University Press, Princeton, NJ (1985).

[2] K.R. Allen, *Invention pacts: Between the lines*, IEEE Spectrum, March (1978), 54–59.

[3] American Law Reports, Annotated, Vol. 153 (1944).

[4] M. von Ardenne, *Ein glückliches Leben für Technik und Forschung: Autobiographie*, Kindler-Verlag, Zurich and Munich (1972).

[5] L. Beckmann, *Erfinderbeteiligung: Versuch einer Systematik der Methoden der Erfinderbezahlung unter besonderer Berücksichtigung der chemischen Industrie*, Verlag Chemie, GmbH, Berlin (1927).

[6] K.F. Beier, *Wettbewerbsfreiheit und Patentschutz: zur geschichtlichen Entwicklung des deutschen Patentrechts*, Gewerblicher Rechtsschutz und Urheberrecht 80 (3) (March 1978), 123–132.

[7] H. van den Belt, A. Rip, *The Nelson–Winter–Dos model and synthetic dye chemistry*, **The Social Construction of Technological Systems: New Directions in the Sociology and History of Technology**, W.E. Bijker, T.P. Hughes, T.J. Pinch, eds, MIT Press, Cambridge and London (1989).

[8] W. Belz, Die Arbeitnehmererfindung im Wandel der patentrechtlichen Auffassungen, diss., Hochschule für Wirtschafts- u. Sozialwissenschaften Nürnberg (1958).

[9] E. Berkenfeld, *Das älteste Patentgesetz der Welt*, Gewerblicher Rechtsschutz und Urheberrecht 51 (5) (1949), 139–142.

[10] W. Bijker, *The social construction of bakelite: Toward a theory of invention*, **The Social Construction of Technological Systems: New Directions in the Sociology and History of Technology**, W.E. Bijker, T.P. Hughes, T.J.Pinch, eds, MIT Press, Cambridge and London (1989).

[11] M. Broszat, *Hitler and the Collapse of Weimar Germany*, transl. and foreword by V.R. Berghahn, Berg Publishers, Ltd, Hamburg, Leamington Spa, Dover (1987).

[12] W. Bruch, *Ein Erfinder über das Erfinden*, **Hundert Jahre Patentamt**, Deutsches Patentamt, Munich (1977), pp. 317–324.

[13] Budaci (Bund angestellter Chemiker und Ingenieure), *Denkschrift zum Erfinderschutz*, Sozialpolitische Schriften des Bundes Angestellter Chemiker u. Ingenieure E.V., 1st series, no. 6, September (1922), 45–55.

[14] M. Burleigh, W. Wippermann, *The Racial State: Germany 1933–1945*, Cambridge University Press, Cambridge and New York (1991).

[15] T. Childers, J. Caplan (eds), *Reevaluating the Third Reich*, Holmes & Meier, New York (1993).

[16] Correspondence between MAN patent experts Martin Offenbacher and Emil Guggenheimer (January 22, 25,

and February 2, 6, and 26, 1924) in MAN, Werksarchiv Augsburg, Nachlass Guggenheimer, K65 (1924).

[17] Correspondence of October 12 and 17, 1922, between Vereinigung angestellter Chemiker und Ingenieure der Farbwerke vorm. Meister Lucius & Brüning, E.V., Höchst a/M and management board of same, Hoechst Historical Archive, 12/35/4 (1922).

[18] D. Crew (ed.), *Nazism and German Society 1933–1945*, Routledge, London and New York (1994).

[19] R. Dahrendorf, *Society and Democracy in Germany*, Doubleday and Company, Inc., Garden City, NJ (1969).

[20] Deutscher Verein für den Schutz des gewerblichen Eigentums, *Beschlüsse des Augsburger Kongress*, Julius Sittenfeld, Berlin (1914).

[21] Deutscher Verein für den Schutz des gewerblichen Eigentums, *Vorläufige Entwürfe eines Patentgesetzes, eines Gebrauchsmustergesetzes und eines Warenzeichengesetzes nebst Erläuterungen*, Carl Heymanns Verlag, Berlin (1913).

[22] Deutscher Verein für den Schutz des gewerblichen Eigentums, *Vorschläge zu der Reform des Patentrechts: Denkschrift der Patentkommission und der Warenzeichenkommission*, Carl Heymanns Verlag, Berlin (1914).

[23] Deutsches Patent- und Markenamt, press release of Mar. 21 (2001), available at http://www.dpma.de/infos/pressedienst/pm010320.html.

[24] C. Duisberg, Comments at Stettin Kongess für gewerblichen Rechtsschutz, 1909, quoted in Zeitschrift für angewandte Chemie 22 (1909), 1667.

[25] H.I. Dutton, *The Patent System and Inventive Activity During the Industrial Revolution 1750–1852*, Manchester University Press, Manchester (1984).

[26] K. Engländer, *Die Angestelltenerfindung nach geltendem Recht: Vortrag vom 24. Februar 1925*, A. Deichertsche Verlagsbuchhandlung Dr. Werner Scholl, Leipzig and Erlangen (1925).

[27] L. Fischer, *Betriebserfindungen*, Carl Heymanns Verlag, Berlin (1921).

[28] L. Fischer, *Werner Siemens und der Schutz der Erfindungen*, Verlag von Julius Springer, Berlin (1922).

[29] L. Fischer, *Patentamt und Reichsgericht*, Verlag Chemie, GmbH, Berlin (1934).

[30] H. Frank, preface to *Das Recht des schöpferischen Menschen: Festschrift der Akademie für Deutsches Recht anlässlich des Kongresses der Internationalen Vereinigung für gewerblichen Rechtsschutz in Berlin vom 1. bis 6. Juni 1936*, Reichsdruckerei, Berlin (1936).

[31] M. Fulbrook, *A Concise History of Germany*, Cambridge University Press, Cambridge and New York (1990).

[32] J.K. Galbraith, *American Capitalism: The Concept of Countervailing Power*, Houghton Mifflin, Boston (1952).

[33] J. Gareis, *Ueber das Erfinderrecht von Beamten, Angestellten und Arbeitern: Eine patentrechtliche Abhandlung*, Carl Heymanns Verlag, Berlin (1879).

[34] G.S. Geren, *New legislation affecting employee patent rights*, Research & Development, January (1984), 33.

[35] K. Gispen, *New Profession, Old Order: Engineers in German Society, 1815–1914*, Cambridge University Press, Cambridge and New York (1989).

[36] K. Gispen, *Bureaucracy*, Encyclopedia of Social History, P.N. Stearns, ed., Garland Press, New York and London (1994), pp. 86–90.

[37] K. Gispen, *Poems in Steel: National Socialism and the Politics of Inventing from Weimar to Bonn*, Berghahn Books, New York and Oxford (2002).

[38] W.D. Grampp, *The Manchester School of Economics*, Stanford University Press, Stanford, CA (1960).

[39] K. Grefermann, *Patentwesen und technischer Fortschritt*, Hundert Jahre Patentamt, Deutsches Patentamt, Munich (1977), pp. 37–64.

[40] W. Hamilton, I. Till, *Law and Contemporary Problems*, vol. 13 (1945).

[41] K. Hauser, *Das deutsche Sonderrecht für Erfinder in privaten und öffentlichen Diensten*, Die Betriebsverfassung 5 (9) September (1958), 168–175, 169.

[42] A. Heggen, *Erfindungsschutz und Industrialisierung in Preussen 1793–1877*, Vandenhoeck & Ruprecht, Göttingen (1975).

[43] J. Herf, *Reactionary Modernism: Technology, Culture and Politics in Weimar and the Third Reich*, Cambridge University Press, Cambridge and New York (1984).

[44] E. Heymann, *Der Erfinder im neuen deutschen Patentrecht*, Das Recht des schöpferischen Menschen: Festschrift der Akademie für deutsches Recht anlässlich des Kongresses der Internationalen Vereinigung für gewerblichen Rechtsschutz in Berlin vom 1. bis 6. Juni 1936, Akademie für Deutsches Recht, Reichsdruckerei, Berlin (1936).

[45] O. Hintze, *Der Beamtenstand*, Soziologie und Geschichte: Gesammelte Abhandlungen zur Soziologie, Politik und Theorie der Geschichte, G. Oestreich, ed., Vandenhoeck & Ruprecht, Göttingen (1964), pp. 66–125.

[46] Historisches Archiv, BASF, Frankfurt a. M. B4/1978, Grimm to Brendel, 25.10.1934.

[47] A. Hitler, *Mein Kampf*, Franz Eher Nachfolger, Munich (1925–1927).

[48] Hoechst Historical Archive, 12/255/1 (Arbeits u. Sozialverhältnisse 1909–1934. Erfinderrecht).

[49] K. Holdermann, *Im Banne der Chemie: Carl Bosch Leben und Werk*, Econ-Verlag, Düsseldorf (1953).

[50] D.A. Hounshell, *Invention in the industrial research laboratory: Individual act or collective process*, Inventive Minds: Creativity in Technology, R.J. Weber, D.N. Perkins, eds, Oxford University Press, New York and Oxford (1992).

[51] http://www.invention-ifia.ch/independent_inventors_ statistics.htm.

[52] http://www.progexpi.com/; esp. http://www.progexpi. com/htm7.php3.

[53] T.P. Hughes, *American Genesis: A Century of Invention and Technological Enthusiasm 1870–1970*, Penguin, New York (1989).

[54] T.P. Hughes, *The evolution of large technological systems*, **The Social Construction of Technological Systems: New Directions in the Sociology and History of Technology**, W.E. Bijker, T.P. Hughes, T.J. Pinch, eds, MIT Press, Cambridge and London (1989), pp. 51–83.

[55] T.P. Hughes, *Rescuing Prometheus: Four Monumental Projects that Changed the Modern World*, Vintage Books, New York (1998).

[56] K.H. Jarausch, *Illiberalism and beyond: German history in search of a paradigm*, Journal of Modern History 55 (2) (1983), 268–284.

[57] K.H. Jarausch, *The Unfree Professions: German Lawyers, Teachers, and Engineers, 1900–1950*, Oxford University Press, Oxford and New York (1990).

[58] J. Jewkes, D. Sawers, R. Stillerman, *The Sources of Invention*, 2nd edition, W.W. Norton & Company, Inc., New York (1969).

[59] R. Kahlert, *Erfinder-Taschenbuch*, Verlag der Deutschen Arbeitsfront, Berlin (1939).

[60] R. Koch, Forderungen an ein nationalsozialistisches Patentrecht, unpublished memorandum, October 12 (1934), Bundesarchiv Potsdam 2343 (Reichsjustizministerium), Bl. 20, pp. 2, 11–12.

[61] J. Kocka, *Angestellter*, **Geschichtliche Grundbegriffe: Historisches Lexikon zur politisch-sozialen Sprache in Deutschland**, Vol. 1, O. Brunner, W. Conze, R. Koselleck, eds, Ernst Klett Verlag, Stuttgart (1972), pp. 110–128.

[62] J. Kocka, *White Collar Workers in America 1890–1940: A Social-Political History in International Perspective*, Sage Publications, London (1980).

[63] W. König, W. Weber, *Netzwerke Stahl und Strom 1840 bis 1914*, Propyläen Technikgeschichte, Vol. 4, Propyläen Verlag, Berlin (1990).

[64] W. König, *Künstler und Strichezieher: Konstruktions- und Technikkulturen im deutschen britischen, amerikanischen und französischen Maschinenbau zwischen 1850 und 1930*, Suhrkamp Taschenbuch, Frankfurt a. M. (1999).

[65] I.N. Lambi, *Free Trade and Protection in Germany, 1868–1879*, F. Steiner, Wiesbaden (1963).

[66] E. Layton Jr., *The Revolt of the Engineers: Social Responsibility and the American Engineering Profession*, The Press of Case Western Reserve University, Cleveland, OH (1971).

[67] R. Linde et al., Discussion of patent fees, Technik und gewerblicher Rechtsschutz: Zeitschrift zur Förderung des Erfindungs- und Markenschutzes, Mitteilungen des Verbandes Beratender Patentingenieure, No. 4–6 (October–December 1928), pp. 42–54.

[68] F. Machlup, *An Economic Review of the Patent System*, Study of the Subcommittee on Patents, Trademarks, and Copyrights on the Judiciary, US Senate, Study no. 15, Washington, DC (1958).

[69] F. Machlup, E. Penrose, *The patent controversy in the nineteenth century*, Journal of Economic History 10 (1) (1950), 1–29.

[70] C. MacLeod, *Inventing the Industrial Revolution: The English Patent System, 1660–1800*, Cambridge University Press, Cambridge (1988).

[71] G. Mandich, *Venetian patents (1450–1550)*, Journal of the Patent Office Society 1 (3) (1948), 166–224.

[72] P. Meiksins, *Professionalism and conflict: the case of the American association of Engineers*, Journal of Social History 19 (1986), 403–421.

[73] W. Meissner, Unser Patentrevers, unpublished ms. dated June 27 (1899), Siemens Historical Archive, SAA 4/Lk 78, Nachlass Wilhelm v. Siemens.

[74] G. Meyer-Thurow, *The industrialization of invention: a case study from the German chemical industry*, Isis 73 (256) (1982), 363–381.

[75] W.J. Mommsen (ed.), *The Emergence of the Welfare State in Britain and Germany, 1850–1950*, Croom Helm, London (1981).

[76] W.J. Mommsen, *The Political and Social Theory of Max Weber: Collected Essays*, University of Chicago Press, Chicago (1989).

[77] W.J. Mommsen, *Der autoritäre Nationalstaat: Verfassung, Gesellschaft und Kultur des deutschen Kaiserreiches*, Fischer Taschenbuch, Frankfurt a. M. (1990).

[78] H. Müller-Pohle, *Erfindungen von Gefolgschaftsmitgliedern*, Deutscher Rechtsverlag GmbH, Berlin, Leipzig, Vienna (1943).

[79] F. Neumeyer, *The Law of Employed Inventors in Europe*, Study No. 30 of the Subcommittee on Patents, Trademarks, and Copyrights of the Committee on the Judiciary, 87th Congr., 2nd sess., U.S. Government Printing Office, Washington, DC (1963).

[80] F. Neumeyer, *The Employed Inventor in the United States: R&D Policies, Law, and Practice*, MIT Press, Cambridge, MA, and London (1971).

[81] R. Nirk, *100 Jahre Patentschutz in Deutschland*, **Hundert Jahre Patentamt**, Deutsches Patentamt, Munich (1977), pp. 345–402.

[82] D. Noble, *America by Design: Science, Technology, and the Rise of Corporate Capitalism*, Oxford University Press, New York (1979).

[83] N. Orkin, *The legal rights of the employed inventor: new approaches to old problems*, Journal of the Patent Office Society 56 (10) (1974), 648–662; and 56 (11), (1974), 719–745.

[84] N. Orkin, *Innovation; motivation; and orkinomics*, Patent World, May (1987), 34.

[85] N. Orkin, M. Strohfeldt, *Arbn Erf G – the answer or the anathema?*, Managing Intellectual Property, October (1992), 28–32.

[86] G.P. Parsons, *U.S. lags in patent law reform*, IEEE Spectrum, March (1978), 60–64.

[87] G.P. Parsons, letter to Robert W. Bradford of September 4, 1979, in SPEEA (Seattle Professional Engineering Employees Association) File on Patent Agreement Negotiations (1979).

[88] R. Perrucci, J.E. Gerstl, *Profession without Community: Engineers in American Society*, Random House, New York (1969).

[89] J. Phillips (ed.), *Employees' Inventions: A Comparative Study*, Fernsway Publications, Sunderland (1981).

[90] R. Plaisant, *Employees' inventions in comparative law*, Industrial Property Quarterly 5 (1) (1960), 31–55.

[91] H. Potthoff, G. Jadesohn, S. Meissinger (eds), *Rechtsprechung des Arbeitsrechtes 1914–1927: 9000 Entscheidungen in 5000 Nummern in einem Band systematisch geordet*, Verlag von J. Hess, Stuttgart (1927).

[92] E. Pouillet, *Traité théorique et pratique des brevets d'invention et de la contrefaçon*, Marchal et Billard, Paris (1899).

[93] K. Prahl, *Patentschutz und Wettbewerb*, Vandenhoeck & Ruprecht, Göttingen (1969).

[94] E. Priemer, Begrüßungsansprache und Rückblick auf 10 Jahre Erfinderbetreuung durch die NSDAP, Tagung der Reichsarbeitsgemeinschaft Erfindungswesen im Hauptamt für Technik der Reichsleitung der NSDAP, Amt für technische Wissenschaften am 11. January 1944, Munich (1944).

[95] J. Radkau, *Technik in Deutschland: Vom 18. Jahrhundert bis zur Gegenwart*, Frankfurt a. M., Suhrkamp, Frankfurt (1989).

[96] T.S. Reynolds, *The engineer in 20th–century America*, **The Engineer in America: A Historical Anthology from Technology and Culture**, T.S. Reynolds, ed., University of Chicago Press, Chicago and London (1991).

[97] D.S. Ritter, *Switzerland's patent law history*, Fordham Intellectual Property, Media & Entertainment Law Journal 14 (2) (2004), 463–496.

[98] M. Rose, *Authors and Owners: The Invention of Copyright*, Harvard University Press, Cambridge, MA, and London (1993).

[99] H. Rosenberg, *Grosse Depression und Bismarckzeit: Wirtschaftsablauf, Gesellschaft und Politik in Mitteleuropa*, Walter de Gruyter & Co., Berlin (1967).

[100] M. Ruete, *The German employee-invention law: An outline*, *Employees' Inventions: A Comparative Study*, J. Phillips, ed., Fernsway Publications, Sunderland (1981), pp. 180–212.

[101] W. Sauer, *Das Problem des deutschen Nationalstaates*, **Moderne deutsche Sozialgeschichte**, H.-U. Wehler, ed., Kiepenheuer & Witsch, Cologne and Berlin (1968), pp. 407–436.

[102] H. Schade, H. Schippel, *Das Recht der Arbeitnehmererfindung: Kommentar zu dem Gesetz über Arbeitnehmererfindungen vom 25. Juli 1957 und deren Vergütungsrichtlinien*, 5th edition, Erich Schmidt Verlag, Berlin (1975).

[103] E. Schiff, *Industrialization without National Patents: The Netherlands, 1869–193; Switzerland, 1850–1907*, Princeton University Press, Princeton, NJ (1971).

[104] H. Schmelzer, *Erfinder oder Naturkraftbinder?*, Der leitende Angestellte 3 (23) December (1921).

[105] J. Schmookler, *Invention and Economic Growth*, Harvard University Press, Cambridge, MA (1966).

[106] Siemens Historical Archive, SAA, 4/Lk78.

[107] Siemens Historical Archive, SAA, 4/Lk153 (Nachlass Wilhelm von Siemens).

[108] Siemens Historical Archive, SAA, 11/Lf36, 11/Lf 352, and 11/Lf364 (Koettgen Nachlass).

[109] J. Silva Costa, *The Law of Inventing in Employment*, Central Book Company, Inc., New York (1953).

[110] L.E. Simon, *German Research in World War II: An Analysis of the Conduct of Research*, Wiley, New York (1947).

[111] E. Staudt et al., Der Arbeitnehmererfinder im betrieblichen Innovationsprozess, Institut für angewandte Innovationsforschung der Ruhr Universität Bochum, Study No. 78, Bochum (1990).

[112] J.R. Steele Jr., *Is This My Reward? An Employee's Struggle for Fairness in the Corporate Exploitation of His Inventions*, Pencraft Press, West Palm Beach (1968).

[113] C. Stephenson, F.G. Marcham (eds), *Sources of English Constitutional History*, Harper & Row, New York (1937).

[114] F. Stern, *The Failure of Illiberalism*, University of Chicago Press, Chicago (1975).

[115] B. Tolksdorf, *Das Recht der Angestellten an ihren Erfindungen*, Zeitschrift für Industrierecht (1908), 193–200.

[116] E. Troeltsch, *The ideas of natural law and humanity in world politics*, **Natural Law and the Theory of Society 1500 to 1800**, O. Gierke, ed., transl. E. Barker, Cambridge University Press, Cambridge (1950).

[117] United States Constitution, Art. I, sec. 8, cl. 8. available at http://www.archives.gov/national-archives-experience/charters/constitution_transcript.html.

[118] T. Veblen, *Imperial Germany and the Industrial Revolution*, The University of Michigan Press, Ann Arbor, MI (1968).

[119] B. Volmer, Entwurf eines Gutachtens über die Möglichkeiten einer wirksamen Förderung der Erfinder im Bundesgebiet, December 31, 1951, pp. 10–12, in Bundesarchiv-Zwischenarchiv Sankt Augustin-Hangelar, B141/2761.

[120] B. Volmer, D. Gaul, *Arbeitnehmererfindungsgestz: Kommentar*, 2nd edition, C.H. Beck'sche Verlagsbuchhandlung, Munich (1983).

[121] J. Walter, *Erfahrungen eines Betriebsleiters (Aus der Praxis der Anilinfarbenfabrikation)*, Dr. Max Jänicke Verlagsbuchhandlung, Leipzig (1925).

[122] R.E. Walton, *The Impact of the Professional Engineering Union*, Harvard University Press, Cambridge, MA (1961).

[123] M. Weber, *Bureaucracy, From Max Weber: Essays in Sociology*, H.H. Gerth, C. Wright Mills, transl. and eds, Oxford University Press, New York (1946), pp. 196–244.

[124] H.-U. Wehler, *The German Empire 1871–1918*, K. Traynor, transl., Berg Publishers, Leamington Spa/Dover, UK (1985).

[125] P. Wiegand, Die Erfindung von Gefolgsmännern unter besonderer Berücksichtigung ihrer wirtschaftlichen Auswirkungen auf Unternehmer und Gefolgsmänner, Dr.-Ing. Diss., Technische Hochschule Hannover, 1941.

[126] N. Wiener, *Invention: The Care and Feeding of Ideas*, S.J. Heims (intr.), MIT Press, Cambridge, MA, and London (1993).

[127] M. Woodmansee, *The genius and the copyright: Economic and legal conditions of the emergence of the 'author'*, Eighteenth-Century Studies 17 (4) (1984), pp. 425–448.

[128] www.wipo.org/about-ip/en/studies/publications/ip_definitions.htm.

[129] R. Zitelmann, *Hitler: Selbstverständnis eines Revolutionärs*, Berg Publishers, Hamburg, Leamington Spa, Dover (1987).

[130] R. Zussman, *Mechanics of the Middle Class: Work and Politics Among American Engineers*, University of California Press, Berkeley, CA (1985).

The History of Information Security: A Comprehensive Handbook
Karl de Leeuw and Jan Bergstra (Editors)

4

REFLECTING MEDIA: A CULTURAL HISTORY OF COPYRIGHT AND THE MEDIA

Robert Verhoogt and Chris Schriks

Ministry of Education, Culture and Science
The Hague, The Netherlands

Contents

Abstract

Intellectual property is a fascinating historical amalgam of concepts of individuality, property and technological innovation. Media revolutions in the past, like the introduction of the printing press, phonographs, television and computer, offered new means of production, distribution and consumption of information. The introduction of new media constantly resulted in new legal questions of intellectual property. Or should we say the same legal questions over and over again? We investigated the cultural history of intellectual property law in relation to the introduction of new media and its legal consequences in The Netherlands. Many aspects of modern intellectual property law can be better understood through the lens of the past. Therefore, we presented some headlines of the cultural history of the media and intellectual property in the successive fields of text, visual, audio, audio-visual to multimedia and their mutual reflections.

The issues of reproduction, publication and fair private use were repeatedly evaluated in relation to another medium: text, images, sound and the combination of them. Nevertheless, the approach of this issue changed over the ages. Initially, the system of legal protection was orientated on the reproduction that was protected by privileges. The book was protected before its original manuscript, and the reproductive engraving before the original painting. During the eighteenth century the legal system of protection slowly changed to focus on the original work, rather than the reproduction of it, thanks to effort of writers, painters and composers. From the beginning of the twentieth century intellectual property was definitely focused on protecting the originality of creative products, whether in the form of text, images, sound, film, television or computer games.

In general, the protection of texts paved the way for protecting visual media like graphic and photographic images, which in their turn influenced the protection of audio-visual media, like film and television. They all created their own niches in copyright protection during the course of time. The protection of these analogue media founded the way in which we approach and protect intellectual property in the digital era today. The production, distribution and consumption of copyright protected material are radically changed in our age. One thing seems not to have changed at all: the unique position of the creative author and his special relation with his work, whether it's designed in text, images, sounds or a sophisticated mix of it. The primacy of the author still seems to be an important normative point of departure for the recognition and protection of the author's intellectual property. The traditional system of copyright protection showed to be flexible enough to react and incorporate new media revolutions in nineteenth and twentieth century and will probably survive the twenty first century, supported by new approaches of Copyleft, Creative Commons and Open Source.

Keywords: intellectual property, privilege, printing press, visual arts, photography, radio, gramophone, cassette-tape recorder, music, film, television, video recorder, computer, multimedia.

4.1 INTRODUCTION

The introduction of the printing press, the phonograph, television and the computer have offered new methods of production, distribution and consumption of information. They have captured literature for different cultures, preserved music long after the last note was played, or broadcasted sounds and images around the world. The introduction of new media constantly resulted in new legal questions about intellectual property. However, are these really *new* legal questions, or repeatedly the *same* questions of intellectual property?

Intellectual property is a fascinating mixture of concepts of individuality, property rights and technological innovation. In the essay *What Is an Author?*, Michel Foucault confirms that 'the author' is not a constant universal concept but rather must be considered as a social-cultural notion, constructed both by cultural and legal influences.[1] The same point can be made about the legal system protecting 'the author'. Intellectual property

provides the author the exclusive right to reproduce and to publicize his work, as stated for example in article 1 of the Dutch Copyright Act 1912. In the context of intellectual property, this work can be defined as a personally designed human thought (P.B. Hugenholtz, *Auteursrecht op Informatie. Auteursrechtelijke Bescherming van Feitelijke Gegevens en Gegevensverzamelingen in Nederland, de Verenigde Staten en West-Duitsland. Een Rechtsvergelijkend Onderzoek*, Kluwer, Deventer (1989).) This personal design expresses the special bond between the author and her or his ideas, which constitutes the special legal protection of intellectual property. Or, as Rose, rightly states in *Authors and Owners, The invention of Copyright* (1993): "Copyright is founded on the concept of the unique individual who creates something original and is entitled to reap a profit from those labors" (M. Rose, *Authors and Owners. The invention of Copyright*, Cambridge (Mass), London (1993), p. 2). The Copyright Act 1912, although still valid and useful today, is in itself a product of cultural and legal processes and still bears its interesting historical marks. Examining the law we encounter references that remind us of media revolutions in the past, such as the introduction of the printing press and photography, of radio and records, of television and computers. We will investigate those traces from a cultural historical point of view. This historical approach can provide interesting changes and continuities in the development of intellectual property law and puts the actual legal system into another perspective (D. Saunders, *Authorship and Copyright*,

---

[1] "The 'author function' is tied to the legal and institutional systems that circumscribe, determine, and articulate the realm of discourses; it does not operate in a uniform manner in all discourses, at all times, and in any given culture; it is not defined by spontaneous attribution of a text to its creator, but through a series of precise and complex procedures; it does not refer, purely and simply, to an actual individual insofar as it simultaneously gives rise to a variety of egos and to a series of subjective positions that individuals of any class may come to occupy, Michel Foucault, *What Is an Author?* (D. Bouchard, *Language, Counter-Memory, Practice: Selected Essays and Interviews*, Cornell University Press, Ithaca (1977), pp. 130, 131.)

Routledge, London New York (1992); B. Sherman, A. Strowel (eds), *Of Authors and Origins. Essays on Copyright Law*, Clarendon Press, Oxford (1994).)

In this review of the cultural history of intellectual property law in relation to new media and its legal consequences, we will focus on The Netherlands. The invention of the printing press, photography, the phonograph, film, radio and television created important media revolutions in production, distribution and consumption of information. Nevertheless, the word 'revolution' can be misleading. Firstly, it assumes radical and instantaneous cultural changes within society. In reality, it has taken quite some time and effort before the media realized their character as a mass media. More than a radical presentation, the media were the result of complex technical, economical and cultural processes. Their changing status from an expensive apparatus for exclusive professional use to a mass medium affected their relation to copyright law, as will be explained.

Secondly, the word 'revolution' seems to indicate a unique chain of events introduced by each new medium. However, more than once the new media were closely related to existing media. For example, visual image and music became legally protected because of their similarities with the written word. Film was protected because of the parallel with standstill photographic images and digital media because of their correspondences with analogue media. So regardless of intrinsic unique qualities, these new media reflected qualities of pre-existing media, sometimes of a technical nature and sometimes of a conceptual nature. In legal discussions, therefore, these changes or continuities reflected in the new medium, appeared mostly as rhetorical arguments for legitimizing why, or why not, a new medium should gain legal protection.

In this digital era, it is interesting to analyze the historical development of analogue media culminating in the modern topic of multimedia. Many aspects of modern intellectual property law can be better understood through the lens of the past (B. Sherman, L. Bently, *The Making of Modern Intellectual Property Law. The British Experience,* *1760–1911*, Cambridge University Press, Cambridge (1999).) Therefore, we will present some headlines of the cultural history of the media and intellectual property in the successive fields of text, visual, audio, audio-visual to multimedia and their mutual reflections.

4.2 TEXT

4.2.1 The printing press

In 1999 *The New York Times* selected Gutenberg as the 'man of the millennium', thanks to his invention of printing press in approximately 1450.[2] Gutenberg's invention made it possible to produce and reproduce texts much faster than writing them. The principle of printing, although earlier used in China and Korea, appeared to be the right invention for the right time in Europe to answer the growing need of the reading public. Within a few decades the invention was widespread in Europe and by 1500, there already were 250 cities with active printers and publishers (Fig. 4.1).

Specialized printers, engravers, booksellers and publishers worked efficiently together for the production and distribution of printed matters. Publishers in Paris, Venice, Antwerp, Amsterdam and London paved the way for the International Republic of Letters of the Renaissance and the Enlightenment. Hieronymus Cock, Plantin, Elsevier, or the Remondini family produced enormous quantities of religious texts, almanacs, scientific essays, maps, prose and poetry, bound or not [15; 27]. These items could be found across at the book fairs of Frankfurt, Dresden and Leipzig, the bookstalls along the Seine in Paris, or in the bookshops on the Kalverstraat in Amsterdam.

Historical documents show that in 1522 the preacher Martin Luther was upset by the reprinting of his work. The German translator of the Bible was, more than anyone else, aware of the significance of his texts, so he was especially disturbed to see his very own texts printed by others when his words were twisted, misrepresenting his precisely worked out ideas [79: 37]. According to Luther the

---

[2]The section *Text* is an adaptation of Schriks, *Het Kopijrecht* [79].

Figure 4.1. Early example of a printer's mark of the printer Jacob van Breda of 1515.

irresponsible reprinting of his work not only infringed upon him as the author but also his readers who consequently were falsely informed. Even the printer was financially harmed.

One of the consequences of printing was the *reprinting* of books. Reprinting certainly was less expensive than original printing and could be a serious source of offending a fellow printer. The original printer would have taken considerable investment risk; a less expensive reprinting process, then, could be seen as a form of piracy of theft. However, the re-printers pointed out that no demonstrable legal justification could be found for the protection of written knowledge. The owner of a purchased book was free to use it as he chose, including reprinting. Cheap reprinting was in the interest of learning and literature, and it was undesirable for one person to hold a monopoly on a work. Despite

this argument, printers and booksellers of original editions appealed to natural, civil, and/or common law. They argued that their acquired ownership rights had been infringed upon when their more expensive editions had been reprinted without permission. There was considerable confusion in terminology and interpretation about what was considered as piracy or theft in these cases. Attempts were made to distinguish between permissible and *not* permissible reprinting. Not-permissible reprinting was taken to encompass different types of unfair competition and unfair acts towards those in the same profession. Due to a lack of a clear concept of the term piracy, reprinting remained a problem for ages.

Luther's contemporary Desiderius Erasmus proposed a solution to Luther's problem. What if the emperor declared a ban on reprinting a text in a particular region and during a particular period of time, for example for two years? By way of example, Erasmus pointed to the trade of cotton: if selling English cotton for Venetian Cotton was seen as fraud, then the reprinting of another printer's text should also be seen as theft. Governments were necessarily involved in these decisions. It led to serious discussions about the protection of an acquired ownership for original editions and complaints about the unfair competition, haste, and accompanying inaccuracies found in reprints. Governments hesitated to make judgments out of a desire not to slow the book trade and, instead, considered each complaint on a case-by-case basis; governmental legislation between what was permissible and what was not was most difficult.

Ecclesiastical and secular governments recognized that unrestricted printing and publication of writings could endanger civil society, as well as the unity between the church and the state. Consequently unwanted activities within the printing industry had to be prevented. So the emperor, the pope, feudal lords and magistrates utilized the possibilities afforded to them by the ius regale, from which they derived their prerogatives. They had the power and the authorization to ban certain publications, but also to extend printing and book privileges prohibiting piracy of a publication during a certain period of time under penalty of sanctions. This practice of book privileges started in Venice as early as 1469. Book privileges fit well within the older legal tradition of patent law of industrial property which had begun more than a century earlier.[3]

Due to the opportunistic application of this system, book privileges began to take on the nature of favors or arbitrarily extended privileges. However, as more privileges were granted, customs developed around them as well. They could be transferred, sold, split up, lent, or distributed. In principle, it was possible for anyone in the Dutch

Republic to apply for a privilege. Obtaining one was not considered a special favor. The authorities in France, Italy and the Dutch Republic who granted privileges copied from each other as to how and in what form privileges were granted. Whether a printer or bookseller could successfully take action against a violation of his privilege depended on the current legislation, on precedent, case law, or whether the juridical authorities possessed the necessary expertise and were prepared to take action.

The system of privileges was commonly linked to censorship. Especially in France, discussions about piracy and literature had been part of the cultural politics of the government from very early on. From 1534 printing anything without permission from the authorities was prohibited. Every work was to be examined before going to press, preventing publications containing false information about the Christian faith. In the Dutch Republic, the situation was different, because the exercise of censorship was limited to the regulation of trade and to counteract unfair competition. There were various types of privileges, such as those of commissions, of general privileges attached to the protection of a written work, for a specific work, or for the collected works of one author. The duration of the protection, which initially differed considerably, was slowly extended and from around 1650 amounted to a standard period of fifteen years. In the Dutch Republic, between twenty-five hundred to three thousand privileges were granted during the 17th and 18th century: no more than approximately one percent of all publications published in this period. These publications mainly consisted of small scale brochures, schoolbooks and small almanacs. From the viewpoint of production cost however, the privileged (more expensive) publications were a substantial part of the total production.

The legal basis for the privilege of protection from piracy was found in the Dutch 'kopijrecht': the right to copy. This right existed as a custom giving the rightful possessor of a manuscript the right to have it appear in print. The owner of the right of copy had reasonable grounds on which to fight a pirating printer from reprinting a work. Once acquired, the right of copy was no longer connected

---

[3] One of the first patents even was granted in 1331 to a Flemish weaver who wanted to work in England, see: D.I. Bainbridge, *Intellectual Property* [10: 244]. See also about patents in general: MacLeod, *Inventing the Industrial Revolution, The English Patent System 1660–1800* [63].

with the author but rather with the activities of a printer or bookseller. The term '*kopijrecht*' was considered synonymous with a complete right of ownership without nuances. It can be related to foreign terms, such as '*droit de proprieté*' and '*Verlagseigentum*' even though each of these terms had a different meaning linked to the ideas of natural law, common law, and custom. Anything which could be printed could be made subject to right of copy, be it a single book, bulk printings, complete funds, stocks, blank sheets, works in the making, and also privileges or parts of privileges and translations. If it could be copied, it could be protected against unauthorized reprinting.

The system of privileges was vertically directed between the public state and its citizens, but also horizontally among citizens as private parties themselves. As the number of printers and booksellers increased, specific problems required adequate solutions with respect to piracy, privileges, auctions, price-cutting, and other such matters. As a reaction to this need, the most important printing cities in the Dutch Republic established their own bookseller guilds – Middelburg (1590), Utrecht (1599), Haarlem (1616), Leiden (1651), Amsterdam (1662), Rotterdam (1699) and The Hague (1702). The bookseller's guilds primary tasks consisted of providing advice regarding applications for privileges. The guilds wanted nothing to do with general and monopolistic privileges but rather preferred privileges extended only to the protection of a specific work. The guilds promoted internal agreements against piracy and attempted to persuade their members of the fact that customs had to be respected. However, their requests to the city magistrates of state's governments to provide for a statute countering piracy of non-privileged books were repeatedly denied. The customary preferential right to pirated versions and translations of foreign works, on the other hand, would continue for a long time.

Anyone could ask for a privilege, including the author himself. Compared to the printer, publisher, or bookseller, the author held no special position in relation to the publication. The legal and economic emancipation of the author was to start in the 18th century, influenced by the Enlightenment and the impetus of John Milton, John Locke, Jean Jacques Rousseau and Adam Smith [30; 96]. Nevertheless, already since the 16th and 17th centuries, authoritative and scientifically educated authors of the Renaissance began requesting privileges for publication of their own works, as did inventors. Guilds, then, were afraid that the customs used in the book trade would be affected. In their opinion, only professional practitioners of the book trade, who were qualified by guild statutes, were entitled to right of copy and privileges. The motto of guilds remained as before: no membership, no privilege.

As soon an author's privilege showed general traits, it acquired the nature of an author's copyright to some degree, which was the case with the preacher D'Outreyn from Dordrecht who in 1707 asked for a privilege to protect his entire *oeuvre* from piracy. His efforts echoed the experiences of his colleague Luther nearly two centuries before. With this effort, he wanted to prevent a privilege which had already been granted to his bookseller from being prolonged after his death. The magistrate in Dordrecht was in favor of D'Outreyn, but the guilds were against. The States of Holland refused to grant privileges for books for which a privilege had already been issued. Another example was the famous professor Boerhaave at Leiden University. Displeased with the many poor pirated copies, he submitted a request to the States of Holland to take measures. It resulted in a long discussion with the guilds. The primary concern of Boerhaave and the other professors was that the reprinting and pirating of their work without their permission damaged their reputation. The essence of the proclamation of 1728 was that nobody in Holland was allowed to print a new work by the professors and other members of the University if they – or their heirs – had not granted them written permission to do so. It was the first time that a right was granted to a group of authors. However, this occurred not because they were authors but because they had asked for it.

The aversion to the monopolistic nature of the guild began to be expressed more and more in the eighteenth century. As a consequence of guild monopoly, English legislation – earlier than the French or the Dutch Republic – looked to another form of literary ownership which resulted in *The Statute of Queen Anne* of 1710. This statute was not intended to give the author an authorial property, but rather to protect the copyright holders against piracy and to restrict perpetual monopolies. It regulated ownership of copies, penalties for violation registration, consulting the register, measures against prices being too high or unreasonable, and the import of foreign copies. Of greater importance was that the subsequent act stipulated that an exclusive right of publication fell to the author of a work which was already in print, if he had not transferred his copyright, or to the bookseller who had acquired the copyright from the author, either for a period of twenty-one years. The author, or his heirs, of a not yet printed or published work was given the same right for a period for fourteen years. In practice, not much became of the author's rights because they continued to be dependent upon guilds for the printing and distribution of their work. It was not until later that booksellers realized that with this act, the system of book privileges was formally abandoned and that extensions of privileges were no longer possible.

The English poet Alexander Pope was one of the first to attempt control over the production and distribution of his work using the Statute of Queen Anne. He carefully arranged his contracts with printers and booksellers and frequently restricted a contract only to the first printing of his book, for which he personally designed the layout, letter types, and format as a printer or publisher would. In 1723 he even went so far to sign a contract with a printer for the publication of *Works*, which reserved his right to print parts of the book with other printers. He made personal detailed agreements and never hesitated to go to court to guarantee compliance. In his mind, as far as he did not succeed as a writer, he considered himself to be a printer or even a publisher if necessary. Pope's approach to authorship was supported by several court cases in between 1760 and 1770. In the case Millar v Taylor (1769) the term copyright was defined as the 'sole right of printing, publishing, and selling', which in the opinion of the judges according to common law belonged to the author, who subsequently lost that right should he transfer his copyright. In 1774 the House of Lords ruled that an author did not lose his rights after the publication of his book nor were third parties allowed to reprint and sell an author's book against his will. The frustration on the part of booksellers was great. After futile attempts on the part of the London booksellers to get a new act passed through parliament, more realistic ideas began to dominate although the author remained economically and organizationally dependent upon his publishers [75: 59–66].

Pope was not the only writer devoted to his rights as an author. In Germany Goethe was struggling with the same issues regarding the reprints of his works. As a successful writer he was well aware of his market value. He even tried to start his own 'Selbstverlag', although not very successfully. He published *Götz von Berlichingen* on his own but it was a financial loss soon afterward with the publication of five pirated versions. After Goethe many writers tried to publish their own works in order to stay in control of the production and the distribution of their works, and there was a growing annoyance between writers and re-printers or what the authors considered to be pirating their '*Geistige Eigentum*'.

In France every work, except for the classics, was subject to censorship and to protection from piracy, officially at least (Fig. 4.2). The reality proved very different. The increasing number of printing houses brought about an increase in the frequency and size of enactments. The *Code de la librairie et Imprimerie de Paris of 1723* aimed to reserve the entire book trade only for admitted, registered, and patented printers and booksellers. Under these regulations, authors received the freedom to publish and sell their own works, which was perceived as the first form of an author's copyright. The concept that intellectual creations were inalienable was first recognized in the Enactment of 27 March 1749. With the enactment of 1777 the French government demonstrated that it championed a more fundamental approach to the literary right of ownership than was the case in other countries, with the exception of Great Britain. The book privilege was still a favor, but it was granted from the point of view of justice. It entailed the recognition of a legal standard which constituted a right

Figure 4.2. The Galerie du Palais was one of the places where the French reading public bought their books in the seventeenth century.

for the author who had the right to publish and sell his work without losing his ownership. The French Revolution of 1789 ended the 'ancient regime' of the book privilege. In March 1790 the last privilege was granted.

The Copyright Act of 1793 specified the rights of authors: writers, composers, painters, illustrators, designers and engravers were granted exclusive rights for the duration of their lives (plus seven years for heirs and successors) to sell their work, have them sold, and distributed in the French Republic. The law no longer spoke of a *droit de copie*, which was now given a more limited meaning; henceforth, the right of copy concerned the work itself. Although objections were raised against this

enactment, it does not detract from the fact that the act was a giant step in the direction of a modern author's copyright. It still proved to be a difficult course of development, however. During the Napoleonic period censorship was by worse than even. The enactment of 1810 regulated the printing and publishing business exhaustively and in detail and was reminiscent of what the English had tried to do almost two centuries before as well as of the French decrees of 1686 and 1723. The Code Penal of 1810 replaced the penal measures in previous acts against pirates and importers of pirated copies. After the Napoleonic period, the decrees were supplemented with a number of acts and new decrees, including the Act of July 1866 which increased the limit of protection to fifty years after the death of an author.

In the Dutch Republic the tradition of the privilege-system existed throughout the 18th century. During the last decade of that century, it became clear that the vested order in The Netherlands was no longer self-evident. Following the French example, a centrally governed state was pursued in which there was to be a place for human rights. The Revolt of 1795 took place within a few weeks. The abolition of the privileges was the most important event for the printers and booksellers although they immediately looked for a substitute to secure their rights. Printers and booksellers maintained that the right of copy was a right of production and trade, which had to be protected by law. A book in which investments had been made should be treated as any other capital put out to generate interest. Translations too, should be protected and it should be prohibited to publish alternative translations or retranslations. The decree of 1795 stated that no more privileges would be granted since everyone had the right to protect ownership of a work. Consequently, a right was constituted for the bookseller who had obtained the copy of an original work and his heirs and successors to publish that work. This right was a right of ownership which nobody was allowed to infringe, either by producing pirated copies or by importing and selling them. The essence of this decree, and the following one of 1796, was more of a household regulation for the book trade than of an act that protected an original work.

In 1808, during the reign of Louis Napoleon, the guilds were abolished. Henceforth, everybody had the right to operate a business. Since the time of Gutenberg, publishers and printers reproduced texts in books, newspapers and periodicals. The production of text originally was a handicraft, as described by the famous publisher Christoffel Plantin in 1567, but descriptions from the 17th and the 18th century provided a similar impression of the printer's workshop (Figs 4.3 and 4.4).

Remarkably, then, little seemed to have changed in the daily production of texts. Therefore, the expanded production of printed material seemed to be more the result of the increasing number of printing presses rather than the improvement of the presses themselves. Generally speaking, the actual production of printed matters by the end of the Enlightenment had barely changed compared with Gutenberg's production at the end of the Middle Ages [37].

The 'ancient regime' of the privilege system disappeared by the end of the eighteenth century. In the same period, the traditional techniques of the printing press were altered when the Industrial Revolution brought important changes to the production of printed materials. The traditional manual wooden printing press was exchanged for a new machine made of iron, the first built by Charles Earl of Stanhope in 1804. The new presses of Stanhope and the rotating presses of Friedrich Koenig fueled tremendous increase of the production of printed matter. By the end of the nineteenth century, presses produced about 1500–5000 sheets per hour [44], which production was made possible because of the innovation of paper and ink. In 1799 the Frenchman Louis Robert invented his paper machine to produce endless sheets of paper ('papier sans fin'). In addition, ink was chemically improved so that it could more easily be used for printing a greater quantity of letters than ever before [37].

After Louis Napoleon relinquished his throne, the Kingdom of Holland became a French province where legislation and administration were quickly made to fit the French mold. French legislation was effective from 9 April 1811 to 20 November 1813, when The Netherlands regained its independence. The Act of 19 July 1793 also came into effect in 1811. It was the only act in The Netherlands, together with two articles from the Regulation of 1810, in the period up to 1881 that favored authors. The idea was to give an author priority over a printer or bookseller, although it remained the question as to whether an author could actually publish his own work at his own risk considering the many regulations involved.

When the Kingdom of Holland was reunited with the southern part of The Netherlands in 1814, it turned out that the regulations in Belgium were completely different from those in the northern Netherlands. The northern Netherlands protected

IMPRESSIO LIBRORVM.
Poteſt vt vna vox capi aure plurima: Linunt ita vna ſcripta mille paginas.

Figure 4.3. This copper engraving by the engraver–publisher Philip Galle shows a visual impression of daily business of the book production in the sixteenth century workshop.

the printer and bookseller from piracy, while the southern Netherlands protected the author, his widow, and the heirs of the first generation, all of whom enjoyed a life-long right of reproduction and were in a more independent position than their Dutch colleagues. The Dutch decree contained a preferential right to translate foreign works which could not be found in the Belgian decree. It was a curious phenomenon that in one state and under one sovereign, two conflicting acts existed simultaneously. Consequently, the north and the south did not acknowledge each other's book act causing conflicts to arise which generated an acute need for a unified act and the need for the government to choose between an author's copyright or a publisher's right.

The Act of 1817 regulated the printing and publishing rights of literary works for all of The Netherlands. Its purpose was the right to multiply by means of print, to sell or have sold the literary works or works of art, entirely or partially, abridged or reduced in size, without distinction in form or presentation, in one or more languages. The definitions of the act show that there was no '*communis opinio*' regarding the intellectual rights of the creator nor was this act free of subtleties either.

The author's copyright remained minimal. In 1864 the leading Catholic conservative spokesman Alberdingk Thijm still was not convinced of its correctness: 'The most annoying inconsequences of today's Liberalism, is for sure the increasingly strong protection of the so called 'literary property'

Figure 4.4. Printing business ca 1740. Boetius after J.A. Richter copperengraving, Gutenberg Museum, Mainz.

[1: 75]. He considered copyright to be a revolutionary product of the Enlightenment and mostly as a 'crusade against reproduction'. In his view boundless reprinting of printed matters was in the interest of the general public. On the other hand, the critical writer Multituli alias Eduard Douwes Dekker was a strong fighter for the individual rights of the author. He gave his manuscript of his book *Max Havelaar* to Jacob van Lennep to be printed. Nevertheless, he wanted to have the final say in the printing of his book as he stated on 12th of October 1860: "the treatment of the Max Havelaar *I* must... judge. That right has not been sold, nor paid for. That right is not for sale, and can't be paid" [18: 93]. He wrote in a similar way to his publisher Funke on 16 august 1871: "Passing by the author in respect to the reprinting of his work – inconceivable in foreign countries – is a mistake" [18: 115; 95]. Multatuli lost his case against Van Lennep and had scant support from the Dutch Copyright Act of 1817. A cold consolation was the fact that painters enjoyed even less protection.

Intellectual property remained fragile in The Netherlands during the nineteenth century [32; 54; 74]. For some members of the Dutch parliament their resistance against the Copyright Act of 1881 was still a matter of principle. Some still judged copyright as a 'tax on reading' which raised the prices of printed materials and excluded the public from knowledge [58]. Although copyright in general was accepted in the surrounding countries, the Dutch still hesitated to accept this principle. The strong influence of the tradition of publishers was probably the main reason for this resistance (Fig. 4.5). The illustrious history of the publishers of the Dutch Republic was still known and honored [66].

Transition to a new legal standard for intellectual ownership was also influenced by international developments, such as the Trade and Ship-

Figure 4.5. The relation between the poor author and the powerful publisher is strikingly characterized by Thomas Rowlandson in this aquatint of 1797.

ping Treaty of 1841 between The Netherlands and France. It stipulated that a further treaty would mutually guarantee literary ownership, the treaty between The Netherlands and France of 1855, together with the additional agreement of 1860. In the meantime opinions were ripening ready for the Act of 1881, followed by the Berne Convention of 1886. The different national traditions of intellectual property law in the Anglo-American countries, Germany and France, developed during the eighteenth century, slowly fused during the nineteenth century, as the result of international law and the international Berne Convention 1886 [40].[4] To the annoyance of the Dutch lawyer Robbers his country was not part of the convention of 1886 [74]. But The Netherlands could not resist the influence of the Berne Convention and around 1900 the admission to the Convention was a hot legal issue.

[4]F.W. Grosheide, *Auteursrecht op maat* [40: 274–278]. See also for the interesting comparison with the development of intellectual property tradition in developing countries: F.W. Rushing, C.G. Brown, *Intellectual Property Rights in Science, Technology, and Economic Performance* [76] and A. Firth, *The Prehistory and Development of Intellectual Property Systems* [31].

Entering the Convention presupposed a Copyright Act that protected the author without restrain. It resulted in the Copyright Act of 1912 (Auteurswet 1912: Aw). It was not until than that the right of the author or his successors-in-title to publish and multiply each creation in the areas of literature, science or art, in whatever manner of form it was reproduced was provided for. The Berne Convention strongly reduced piracy in those countries which were party to the treaties, including The Netherlands.

4.2.2 Stencil machine and photocopier

Henceforth, the modern author possessed the exclusive right to reproduce his text and was increasingly able to do so. In 1874 the Remington Company introduced the type machine which could rapidly type the individual letters on paper. The text could even be easily reproduced with a simple sheet of carbon paper. The invention of the linotype machine a few years later, 1885, provided a fast means for the production and reproduction of huge amounts of text, very useful to the production of newspapers. In this same period David Gestetner presented his stencil machine to reproduce texts even easier. During the first half of the twentieth century this machine was widely used in public offices, community centers and universities. Meanwhile, others were already experimenting with photocopy, using photography for the reproduction of text. The photocopier of Rank Xerox was probably the most successful. During the 1960s and 1970s the photocopier replaced the old-fashioned stencil machine. The type machine, stencil machine, and photocopier made it possible to produce and to reproduce text easily, even without the help of the printer or publisher.

Copyright protected material from books and newspapers was widely reproduced, to the great annoyance of authors and publishers. These newest ways of reproduction also required an adaptation of copyright protection. Since 1912 the Dutch Copyright Act offered the opportunity of reproduction for exercise, study, and private use of texts without the explicit approval of the author. The introduction of the stencil machine and photocopier raised the issue of the limits of private use of text [38;

40; 82]. In 1972 the act resulted in the revision of the law, particularly of article 16B Aw and article 17 Aw. Article 16B Aw offered the opportunity to make a few copies of texts for exercise, study, or private use without the author's consent. The exact size of this material is difficult to determine. Article 17 Aw, however, described the possibility for reproducing small parts of texts for companies, organizations, and institutions, which would be compensated by a reasonable fee. This concept was further developed by the special regulation of 20 June 1974 (Stb.351) for the reproduction of texts [82: 240–241]. With these developments in mind, authors and publishers in The Netherlands established Reprorecht in 1974. Since 1985 this is the official organization to receive and divide the collected fees to authors for the reproduction of their texts.

To summarize: the invention of the printing press resulted in the strong increase of the production and reproduction of texts. A special system of book privileges was needed to protect the individual commercial interests of publishers. At first an author of a text did not have a special position in the protection of published texts. During the eighteenth century the Enlightenment concept of 'the author' emerged in the sense that the writer of a text should be the first to honor and profit from publication. The idea evolved in different countries with different meanings and at different paces. Separate traditions of intellectual property traditions appeared and then slowly disappeared with the creation of the Berne Convention in 1886. In the meantime new means for (re-)production of text were invented which in turn meant that reproduction of text for private use became an issue to be resolved as well. The intellectual property law was revised to guarantee the balance between the author's exclusive rights to reproduce his work and the free flow of information and its private use. The protection of intellectual property was designed in the first place for the author of written words, but soon it inspired authors of other media, for example images, to protect their works as well.

4.3 VISUAL

4.3.1 Graphic

The invention of graphic art coincided with the invention of the printing press. Woodcuts and engraving offered means to reproduce images like text as 'exactly repeatable statements' [51]. The artists Albrecht Dürer, Andrea Mantegna and Lucas van Leyden experimented with graphic techniques. Within a few decades graphic specialists, such as Marcantonio Raimondi who engraved the work of Raphael and Michelangelo, started reproducing their works. A fast growing production of woodcuts, copper engravings, and etchings appeared during the sixteenth and seventeenth century. Copper engraving turned out to be an especially successful technique for the reproduction of images of nature, mankind and art [45].

Once Dürer found that prints of his works were being produced in Italy without his permission, he left for the south of Europe. The offender was the well known engraver Marcantonio Raimondi himself; he had engraved Dürer's compositions and even signed them with the monogram of the German master. Dürer remained powerless. The only thing he could get Raimondi to do was to cease using his [Dürer's] monogram [97]. Luther's contemporary Dürer is an early example of an artist devoted to the legal protection of his work, although without success. He could not refer to his copyrights as an author, and apparently he did not possess a privilege to print his work. During the sixteenth and seventeenth century it became common use by artists to gather privileges to print their work. In 1607 the artist M.J. van Mierevelt received a privilege for prints for one of his paintings; while, a few years later the engraver Jacob Matham received a privilege for a print after a composition of Mierevelt. The famous Peter Paul Rubens too, was well aware of the importance of privileges when he negotiated with the representatives of the Dutch Republic about the reproduction of his work [12]. The reproduction of images was strongly related to the system of privileges and was comparable with the situation of books [34; 35] since it offered publishers, engravers, and artists the exclusive right to produce and distribute prints in a certain region during a certain period of time. It should

be noted that painters and engravers could be protected, however *not as an author* but rather as publishers.

The English copyright law was founded on the previously mentioned Queen Anne's Act of 1710 which recognized the rights of the literary author for the first time.[5] The influential painter William Hogarth extended this vision to the visual arts. Like his contemporary Alexander Pope, Hogarth fought the power of publishers with his conviction that the author deserved the exclusive right to exploit his work. His struggle succeeded in the Engraving Copyright Act of 24 June 1735, also known as the Hogarth Act:

> every person who shall invent and design, engrave, etch or worked, in *mezzotinto* or *chiaro oscuro*, or from his own works and invention shall caused to be designed and engraved, etched or worked, in *mezzotinto* or *chiaro oscuro*, any historical or other print or prints, shall have the sole right and liberty of printing and reprinting the same for the term of fourteen years, to commence from the day of the first publishing thereof, which shall be engraved with the name of the proprietor on each plate, and printed on every such print or prints.[6]

This law offered protection to the engraver who designed and engraved his own prints for the term-limit of fourteen years after the first publication [69]. It was possible to take action against publishers or others who without permission copied, printed, traded, published, exposed prints, or others incited to do so. After his death Hogarth left all his possessions – including his copyrights – to his widow Jane Hogarth. Therefore the Hogarth Act was changed in 1767 which stated:

> [...] that Jane Hogarth, widow and executrix of the said William Hogarth, shall have the sole right and liberty of printing and reprinting all the said prints, etchings, and engravings of the design and invention of the said William Hogarth, for and during the term of twenty years [...].[7]

Thenceforth, the author's rights were extended after his death and transferred to his legal successors, such as widow(er)s and children. The Hog-

---

[5] Act 8 Anne, c. 19.
[6] Act 8 Geo. 2, c. 13.
[7] 7 Geo. III. c. 38.

arth Act provided a remarkable copyright protection for images, especially when compared with the situation in Europe and the United States. In 1800 the writer Joseph Farrington wrote in his diary about the Hogarth Act: "This was an important step as it encouraged speculations. Before such security was obtained plates intended to be published were through the intrigues of Artists copied & the Copies preceded the Originals".[8]

In France also there existed the system of privileges since the sixteenth century, not only for text but also for printed images. The system remained largely unaffected till the French Revolution. The recognition of the individual citizen was soon translated to the individual visual artist, stated by the Copyright Act of 1793, article 1: "The authors of manuscripts in all kind, the music composers, the painters and designers that will do engrave pictures or drawings, will enjoy, during their entire life, exclusive right to sell, to do sell, and distribute their works in the territory of the Republic [...]".[9] The law protected the author's property or his work. Prints produced without permission could be confiscated and were considered property of the author.[10] Since the Law of 1793 artists were recognized as author with accompanying rights. Engravers, however, were probably not yet regarded as authors by this law, but they were given equal rights with the law of 1810: "The authors, be national or foreigners, of all work printed or engraved, can yield their right to a publisher or bookseller, or to all other person".[11] This new law also changed the duration of the copyright protection to twenty years after the death of the author.[12] Nevertheless, the author had to remain alert. If he did not explicitly claim his rights when selling his work, he still might have lost them immediately [2], to the great annoyance of the painter Horace

Vernet, for example, as he stated in his essay *Du droit des peintres et des sculpteurs Sur leurs ouvrages* (1841) [91].

The Dutch protection of intellectual property can be situated between the English and French jurisdictions. During the Dutch Republic the Dutch publishers acted as the bookshop of the world, mainly through the lucrative business of reprinting. The privilege system was an essential element in this system, not only in respect to text but also to images. This 'ancient regime' of the history of copyright remained largely intact till the end of the 18th century. In 1808 the Dutch director-general of education strongly insisted that engravers needed a stronger legal protection for their prints and maps. Therefore the king declared by decree of 3 December 1808 that any engraver would have the same rights as writers, which forbid the re-engraving, publishing and selling of prints without the permission of the original engraver [79].

It must be emphasized that the Dutch decree of 1808 only intended to protect *graphic* images, not original paintings or drawings. If the painter engraved his own painting, than the engraving was protected but not his original painting. In that sense the painter could do nothing against prints made by a professional engraver. The remarkable result was that reproductions were protected, but the original was not. In a way the position of the artist was subordinate to the public interest. The author had no more rights than the property of his painted canvas [3; 6]. Contrary to the modern approach of intellectual property, it was not the *originality* of the painted or drawn image that gave cause to protect the art, but on the contrary it was the *reproducibility* that needed protection, as was the case for graphic prints. This effect is a remnant of the traditional approach of copyright of texts in which the printed book was protected before its original manuscript. In the same sense the image was protected too as far it could be reproduced, so in that sense the original painting or drawing didn't need legal protection.

4.3.2 Photography

The Industrial Revolution brought forth many new products and techniques, also new ways of reproducing images. Besides the traditional copper

---

[8]Farrington quoted in Friedman 1976 [33: 35].

[9]Decret des 19–24 juillet 1793 (an II de la République), art. 1.

[10]Decret des 19–24 juillet 1793 (an II de la République), art. 3.

[11]Decret du 5 février 1810, art. 40.

[12]Decret du 5 février 1810, art. 39. The issue of the droit de reproduction was soon settled in the French Art world as can be read in the French Art Journal, *L'Artiste* of 1839, see: F. Lecler [59: 142].

engravings and etchings, the mezzotint was developed. During the second half of the eighteenth century, the following new graphic techniques came into use: crayon-manner, stipple engraving and aquatint. Ernst Rebel showed in *Faksimile und Mimesis. Studien zur Deutschen Reproduktionsgrafik des 18. Jahrhunderts* (1981) the relation between these new methods and the new concept of the facsimile, described as "the technical reproduction of an image or text that claims to be the best faithful copy" [translated R.V.] [72: 131]. The goal of achieving an identical image to an original set new standards for realism as William Ivins stated: "Up to this time engravings had looked like engravings and nothing else, but now, thanks to the discovery of new techniques, the test of their success began to be the extent to which they looked like something else" [51: 170]. This flow of new graphic reproduction techniques continued into the 19th century [11].[13] After Aloys Senefelder presented his lithographic technique, the engraver was released from the difficult and tiring task of engraving. The image could easily be drawn onto a stone, but more importantly a large number of strong prints could be made. After these early experiments, lithographers could easily print some tens of thousands of prints. Lithography resulted in the first large growth in 19th century production of images. Since the Hogarth Act printmakers of engravings, etchings, and mezzotints were protected and so were lithographs.[14]

The request for exact reproducible images already existed in the 18th century, but it was to be fulfilled in the 19th century by the invention of photography. The first photographic experiments of Nicephore Nièpce date from the 1820s, but the medium was further shaped in the 1830s. The method of Daguerre was an immediate success, and thanks to the close contacts between the French and British pioneers of photography the medium was soon internationally widespread [42]. At the beginning the reproduction of photographic images remained a difficult problem for Daguerre. In the meantime, however, William Henry Fox Talbot (1800–1877) and John Herschel (1792–1871) were already practicing with the concept of the negative [64]. This principle of the negative showed to be critical for the reproduction of images. In 1847 the French photographer Louis-Désiré Blanquart-Evrard (1802–1872) received his patent for his version of negatives. Talbot proved the success of the photographic negative although it remained fragile. By the 1850s this fragility changed with introduction of the glass negatives and the collodium process. The glass negatives resisted wear and tear of the reproduction process much better then its predecessor. In addition, the glass negative reduced the exposure time and produced higher quality images [42; 50]. For light exposure, photographers were still completely dependent upon sunlight, so the quality of photography was as changeable as the weather.

The development of photography resulted in the question of whether a photographer deserved the same legal protection as a printmaker. In several court cases in France during the 1850s it was decided that photographical reproduction of works of art without the permission of the artist was no longer acceptable.[15] But was the photographer himself protected? The problem was still a complex one because a precise definition of photography was not clear. *The* photography did not exist; rather, in reality there existed a motley collection of methods and techniques more or less based on a commonly used and basic photographic principle. In one way of looking, photographic technique was considered to be an art form, but in another way of looking merely a technique. As a consequence photography was sometimes protected as

---

[13]See about the topic of art reproduction in the nineteenth century: R.M. Verhoogt, *Kunst in reproductie. De reproductie van kunst in de negentiende eeuw en in het bijzonder van Ary Scheffer (1795–1858), Jozef Israëls (1824–1911) en Lawrence Alma Tadema (1836–1912)* [89]. This book will be published in 2007 in English.

[14]"It is hereby declared that the provisions of the said Acts are intended to include prints taken by lithography, or any other mechanical process by which or impressions of drawings or designs are capable of being multiplied indefinitely, and the said Acts shall be construed accordingly.". See: 15 Vict. c. 12 art. XIV [69].

[15]Besides permission of the artist, it was usually necessary to have permission of the actual owner of the work, museum of exhibition committee. See about photographing works of art in the Louvre: McCauley [65].

intellectual property, sometimes as industrial property by patent laws. It resulted in a complex discussion about which photographer should be protected according to different legal systems. In France the legal problem of photography was solved by victories of the photographic company Mayer and Pierson in 1862 and 1863. From than on, photographers and their work were protected on the same basis of the Copyright Act of 1793, as painters and engravers were [65: 30–34].

In 1863 the famous art dealer and print publisher Ernest Gambart stated in his pamphlet: *On Piracy of Artistic Copyright* (1863): "It is now a question for the legislature and the public to decide whether or not the school of English line-engraving, once occupying so high a position, shall perish or be maintained" [62: 21]. In reaction to the flow of pirated art reproductions the dealer pleaded for a better copyright protection of works of art. Popular paintings were immediately photographed which could seriously damage the business of traditional engravings [88]. Gambart insisted: "It is not, [...] against competition that protection for copyright in art-works is demanded, [...] but against robbers" [36: 4].[16] He continuously went to court to fight against illegal photographic reproductions during the 1850s and 1860s.[17] Holman Hunt's famous painting *The Light of the World* alone resulted in more than twenty court cases [42; 62]. Pirate photographers defended their actions by referring to the Hogarth Act which did not mentioned photography at all. However, in 1863 the judge confirmed that photographers could not refuse any longer the generally accepted meaning of 'intellectual property' [69]. From then on photographers, like engravers, needed permission to reproduce works of art. This was an important statement that underlined the importance of intellectual property of images, regardless the used reproduction technique.

Meanwhile, there still was no copyright protection for original works of art in Great Britain. In protest against this injustice the Society of Arts and the Royal Institute of British Architect presented a petition to the House of Lords in 1858. It resulted in the Copyright Act of 29 July 1862 which stated:

The author, being a British subject or resident within the dominions of the Crown, of every original painting, drawing and photograph which shall be or shall have been made either in the British dominions or elsewhere, and which shall not have been sold or disposed of before the commencement of the Act, and his assigns shall have the sole and exclusive right of copying, engraving, reproducing, and multiplying such painting or drawing, and the design thereof, or such photograph, and the negative thereof, by any means and of any size, for the term of the natural life of such author, and seven years after his death.[18]

It protected the author during his life against the "[...] repeat, copy, colorable imitate, or otherwise multiply for sale, hire, exhibition, or distribution, any such work or the design thereof [...]".[19] Taking action against illegal reproductions nevertheless supposed the registration of the copyright and the possession of the copyright. After selling his work including its copyright, the artist was not able to act against pirated reproductions. Therefore, it was important to have *and to keep* the copyright as well.[20] The copyright Act of 1862 was an important step in the development of the English system of intellectual property. More than a century before, Hogarth set the standard that was still valid in the 19th century. Besides the law of 1862 there were several laws of copyright effective in Great Britain. Engravers had to refer to the Hogarth Act and the Amendment of 1767. Prints were protected for a term of twenty-eight years; paintings, drawings and photographs during the life of the author plus seven years; and sculptures for fourteen years. In 1891 the jurist G. Haven Putnam remarked ironically: "We do not think it desirable that these distinctions should continue" [71]. In Great Britain, and like elsewhere, the copyright laws were regulated by the development of international copyrights, especially by the Convention of Bern.

In The Netherlands graphic images were protected, but the painted and drawn picture was not. In reaction to this situation a group of well known artists under leadership of Jozef Israëls presented a petition to members of the Dutch Parliament, as was earlier done in England. The petition was signed by twenty contemporary artists:

---

[16]See about Gambart's struggle for a better protection of intellectual copyright [6; 7].

[17]See about copyrights in the Victorian Art world [4–9; 42].

[18]Act 25 & 26 Vict c. 68, art. 1.

[19]Act 25 & 26 Vict. c. 68., art. VI.

[20]Act 25 & 26 vict. c. 68, art. I.

D.A.C. Artz, H.W. Mesdag, P. van der Velden, O. Eerelman, J. Vrolijk, B. Höppe, J.J. van de Sande Bakhuyzen, J. Maris, J.A. Neuhuijs, F.P. ter Meulen, J. van Gorkum jr., J.M. Schmidt Crans, T. Offermans, J.H. Neuhuijs, E. Verveer, F.J. van Rossum du Chattel, H.J. van der Weele, C.L.Ph. Zilcken and J. Bosboom [60; 87]. The artists claimed attention for the fact that their work was still not protected. They were supported by the artist, critic and jurist Carel Vosmaer who stated in the influential periodical *De Nederlandsche Spectator* in 1879 that the copyright belonged to the artist: "he must know and decide whether his work will be reproduced or not" [94]. The Copyright Act of 1817 did not protect the artist. Vosmaer underlined the similarities between artists and writers, between word and image, although others mostly pointed to the differences between them. The jurist Fresemann Viëtor stated in his advice for the Copyright Act of 1881 that a painting could not be mechanically reproduced; therefore, there was no reason for protecting painters. Graphic works like engravings, lithographs, or photographs on the contrary could be easily reproduced and did deserve protection. The claim of the artists was not awarded. The Copyright Act of 1881 focused on graphic printed works [73]. Authors of prints were protected against pirated reproduction for the term of fifty years after registration. Creators of paintings, drawings, and sculptures remained unprotected.

Nevertheless, the claims of the artists were also taken into account. A new copyright act was promised, especially for artists and their visual works [73]. Article 1 recognized the author's exclusive right to copy his work by another artistic of mechanical means. Supplement to this, article 4 stated that the author of a photographic reproduction would be recognized as the original author of the original work. The special copyright act for visual arts was inspired by a similar German law of 1876.[21] Due to the critical response to copyright law for the visual arts in The Netherlands, the special regulation never succeeded.[22] Unlike France

and Great Britain, paintings, drawings, and sculptures were still not protected in The Netherlands by the turn of the century. This neglect lasted until the Copyright Act of 1912 which finally recognized artists on the same legal basis as writers. From that moment onward, the visual image received the same protection as the written word [20; 53; 61].

To summarize: inspired by copyrights on texts, intellectual property protection was extended to the protection of images. Because of its graphic similarities, this type of protection was firstly translated to graphic images, such as engravings, etchings, lithographs and photographic images. Only after that initial protection did original paintings, drawings, and sculptures receive appropriate protection as well in The Netherlands and in Great Britain. Interestingly though, it was the reproduction that was the starting point for copyright protection rather than the originality of the image. At the same time that photographers tried to catch a visual impression of reality, other artists were already experimenting with techniques to record the sounds of reality. Soon the discussions about intellectual property moved into the field of audio creations.

4.4 AUDIO

On 3 January 1873 the composer, Giuseppe Verdi wrote to his publisher Tito Ricordi:

> I will not go any further into the matter of the *droit d'auteur*. I can see clearly now that you won't like it, when I will claim it as my right, and you will only receive a small part of the agreement and the publication, that as you said, will produce that little that your profit will be minimal [86: 141].

Like writers and other artists, composers were well aware of their rights and of the injustice of pirated publications printed often played without their permission. Thanks to the system of copyright, composers were also protected. In 1877 Viotta wrote about the copyright of music: "The reprinting of music in general, is regarded along the same rules as the reprinting of literary works" [92: 42]. Actually, it was not the music that was protected but rather the printed scores and the performances of it. At first it was not possible to

---

[21] Handelingen Tweede Kamer Staten-Generaal 1876–1877, bijlagen nr. 202, p. 4.

[22] Handelingen Tweede Kamer Staten-Generaal 1884–1885, bijlage 72 no. 3.

record and reproduce music itself. As long as it could not be reproduced, it did not need to be protected.

4.4.1 Barrel organ and pianola

By the end of the 18th century, there was a rage to reproduce music mechanically for the music box. It was the first time that it was possible to play music without musicians. The music patterns of folksongs were soon translated into ingenious patterns on a wooden spool. Turning the cylinder moved a complex mechanism that produced the sound of music. In the 19th century the spool was rotated by hand which gave rise to the barrel organ. The wooden spools were changed into cardboard books of music. The barrel organ became extremely popular especially in France and the German states in the second half of the 19th century. At the same time others inventors were already experimenting with a related mechanism: the mechanical piano, which in 1895 was known as the pianola. Music pieces translated into paper rolls could be played like the barrel organ, but much more subtly.

The arrival of the pianola resulted in an interesting new legal question about artistic copyright on music. Composers insisted on the extension of their copyright from printed scores to the mechanical music carriers. The publishers of the pianola rolls did not agree and were seriously offended by this increase of a composer's rights. At first, they were supported by legal statements of American lawyers, for example in the case White Smith Music Publication Co v Apollo Co (1908). The American producer of pianolas, Apollo, had translated two popular songs for the pianola. The publisher of the music was offended by this situation and described it as the infringement of his copyright. The Supreme Court judged that the translation for the pianola was not against the law and agreed with the pianola producer stating:

> The definition of a copy, which most commends itself to our judgment is ... "a *written or printed record* [curs R.V.] of a musical composition in intelligible notation". The musical tones which are heard when a perforated roll is played are not a copy which appeals to the eye. Even those skilled in making rolls are unable to read them as musical compositions. There is some testimony that with great skill and effort an operator might read rolls as he could read music written in a staff notation, but the weight of testimony is the other way, and a roll is not intended to be read like sheet music [21: 110].

The judge approached music essentially in a textual way. According to the judge the pianola roll was illegible for (almost) everyone; therefore, it did not need protection. In a strict sense, music was protected as far it was printed and could be read, but music itself was not protected. The development of the barrel organ and pianola ceased, however, because of the invention of a new carrier of music: the gramophone.

4.4.2 Gramophone

In 1877 the inventor Thomas A. Edison presented his phonograph which could record and reproduce sound. At first, Edison was particularly interested in recording the sound of the human voice, more than of musical instruments. The phonograph reproduced sound although the phonogram itself was difficult to multiply. In that sense Edison's inventions reminds to the invention of Daguerre which was not suitable for reproduction either. This lack changed after the replacement of the wax cylinder by a rotating record which was invented by Emile Berliner (1851–1921) in 1888. Initially, the gramophone was made for the recording of the human voice. After a time the record appeared to be useful for the reproduction of music, although many musicians hesitated to record their music. The famous tenor Enrico Caruso agreed to record his music only for a lucrative financial compensation. In 1901 he made his first record and in 1904 he had already sold over a million copies. Thereafter he made numerous recordings which made him world famous [19; 21]. From 1915 the fragile records turned seventy-eight rotations per minute; by 1948 these records were replaced by the vinyl records that went at thirty-three rotations per minute and were made by for example Columbia Broadcasting System (CBS). In addition to this change, the singles became an immediate success because of the production of jukeboxes. At first the music itself was played in mono, but since 1958 in stereo.

In Germany composer's rights were soon extended to mechanical music carriers. Instead of taking the visual readability as a point of departure, like the American judge did, the German judge considered its economic value of overriding importance. If the pianola roll was traded as an independent product, than the author was the first that deserved to profit from it, which was a progressive but short-lived interpretation of the copyright law. The growing interests of the record business were not always for the benefit of the composer. To the contrary, thanks to the lobby of the record companies the composer was not protected from 1901 against mechanical reproduction. He could act against illegal pianola rolls but again remained powerless against gramophones. The Dutch Copyright Act did not intend to protect music. The law was not explicit, but according the copyright expert De Beaufort the statement 'by print made public' could not have been written for music, because that was still economically impossible to do [21: 120–121]. The copyright of the composer supposed a market to benefit from this right, so the composer had no right without a profit.

The record business was booming during the first decades of the 20th century and then economic changes forced a reconsideration of the rights of composers. In 1909 the U.S. Copyright Law was changed to reflect an interest for composers. According to article 1 the author had the exclusive right: "to make any arrangement or setting of it, or of the melody of it, in any system of notation, or any form of record in which the thought of an author may be recorded and from which it may be read or reproduced" [21: 111]. The author of music was now protected against reproduction without his permission. However, the record industry took care that this was only the case for music made after 1909. Music dated before remained free for use. In addition composer protection was for a single time per composition. After permission was given, the music was free to all to reproduce. The composer also had to register his rights like he had to do in relation to texts and images. Interestingly, the composer even had to register the music score in order to protect the music. In that respect the protection of music in the United States was still related to the text of music.

This change can also be observed in Europe. The revision of the Berne Convention in 1908 recognized the author's rights of a composer in relation to mechanical reproduction. The national laws were changed in this direction too. Remarkably, the record companies too claimed their copyright for the individual recording itself, besides the composer's rights to his music. It was not a strange idea. Photographs of paintings of prints were protected too as individual works, regardless of the rights of the painter or engraver. The music companies did not succeed in their claims. Nevertheless, the film companies much later did receive their individual rights as producers, as will be shown below. The Dutch Copyright Act of 1912 also offered the exclusive copyright to composers for mechanical reproduction of their music [52]. Some members of Parliament suggested to present also copyrights to producers of pianola roll and records, but could not convince the Dutch government. The proposal to exclude the very popular barrel organ from the system of copyright for music also failed to succeed.

4.4.3 Radio

The Italian tenor Caruso became famous by making and distributing his recorded voice and he was one of the first to be heard on the radio in 1910. This new medium for sound was introduced in 1896 by Guglielmo Marconi (1871–1937). The basis of the radio was the combined inventions of Morse's telegraph and Bell's telephone of 1876 which made it possible to sent sounds over long distances [19]. Like photography, the telephone was first of all a source of amusement. After some time, it was realized that telephone could connect people for communication. At first, the laying of bundles of cables by AT&T made this even more available in the USA than in Europe. Around 1900 in the USA, there was one telephone to every 60 people, in France one to every 1216, and in Russia one telephone to every 6958. Telegraph and telephone were combined to send information by cable. Marconi's innovation made it possible to send sounds over long distances without cables via the ether. Following Caruso's performance on the radio in 1910, radio broadcasting was programmed in the 1920s. The British Broadcasting Corporation (BBC) started in 1922, the Dutch radio telephonic company or "*radiotelefonische omroep der Nederlandsche Vereeniging voor radiotelegrafie*" one year later. Within a few years new radio organizations which were established in The Netherlands were mostly built on cultural-religious foundations: NCRV, KRO, VRPO, AVRO and the VARA. Radio could reach a mass audience of millions all over the world. On 23 January 1923 the first signals of the Dutch Radio Kootwijk were received in Bandoeng in Indonesia. Radio could now reach everyone, and soon everyone could reach for a radio. A fast growing number of households could afford a radio. The possession of radios increased greatly despite the suppression of the German occupation during the Second World War. After the war, radios became better, smaller and cheaper. In the 1950s Philips Corporation introduced the portable transistor radio so that people could listen to the radio at the park, on the beach, or in your car while driving.

Consequently radio could reach a huge audience, which then raised the question: how huge? What was the meaning of radio in the publication of information? This question was put forward in an important court case in the Dutch history of copyright: the Caferadio case of 1938. The cause of conflict was the radio broadcasting of a concert of Franz Lehar's *Der Zarewitch*. The café owner turned on the radio so that everyone in the café could enjoy the concert. The composer than addressed the café owner claiming that this was an infringement of his copyright. The café owner defended his conduct that the music piece was performed in the studio and that anyone could listen to it on the radio. The High Court judged otherwise on 6 May 1938.[23] Turning on the radio in public was indeed a new way of making a performance public. The composer's permission for the performance in the studio did not imply that the permission for making this public by other ways was simultaneously granted. In that sense the High Court recognized the possibility of separate ways of a performance being made available to the public, which supposed separate permissions to do so.

This judgment was an important break-through in the Dutch history of copyright on sound. It offered the author the opportunity to exploit his work at different markets [82: 170–171, 175].[24] In 1938, Oranje already suggested the idea of a special tax in order to compensate copyright owners for the private use of radio [67]. Although this idea was never realized for radio, it was used in relation to the tape and cassette recorders as will be shown below.

The new possibilities of radio offered many kinds of uses and misuses, for example by radio pirates. The first illegal radio stations appeared in the 1960s, such as the illustrious *Radio Caroline* (1964) [19]. These radio stations broadcasted from oilrigs and ships at sea, like the Dutch *Radio Veronica* and attracted a huge loyal audience of listeners. Besides these commercial and professional stations, there existed a large number of amateur radio broadcasters with 27MC capacity that reached regional and local listeners. The pirate radio stations played mainly pop music records and tried to stay out of reach of the regular legal copyright systems. In resistance to these illegal stations new similar legal ones were promoted like BBC's Radio 1 in 1967 and the Dutch broadcast organization TROS. Radio pirates are notorious for illegal broadcasting of records and are typical to the tensions that emerge when new technical possibilities of broadcasting music arise and the increasing need of artists to regulate their creative products. Radio pirates tried to slip through the legal net of copyright protection. This was an important test of the legal system set up for broadcast regulation, including copyrights; and, in that sense of testing legality the radio pirates had a similar influence on the system of copyright as did the commercial television stations and Internet providers later.

4.4.4 Tape and cassette recorders

Besides music records, inventors were experimenting with another important recording system for music: the tape recorder. The Danish engineer Veldemar Poulsen presented in 1900 the first system to record sound on a magnetic tape. It re-

---

[23]HR 6 mei 1938, NJ 1938, 635.

[24]This opinion was confirmed by the important Draadomroepcase, HR 27- juni 1958, NJ 1958, p. 405.

sulted in the popular Dictaphone in the beginning of the 20th century. New possibilities emerged in the 1920s and 1930s due to German engineers and the development of plastic for the use of recording and the playing of sound. More than the gramophone, the tape recorder provided effective means of recording sounds. During the first half of the 20th century, the tape recorder was available only for professional music studios, but after World War II amateurs could also get their own tape recorder. Special self-made tape recorders were made in the 1950s to be built at home. From then on, American and German producers offered inexpensive and successful machines like the Record-o-Matic recorder. In the 1960s smaller and even less expensive recorders were produced by Japanese firms, such as AKAI and AWAI. The cassette player invented by the Philips Company in 1963 also fits in this tradition.

The coming of the tape cassette made it possible for more people to record sound easily, especially music which increased the habit of copying music at home for private use. The record industry was seriously affected by this private capturing and use of music. Especially in the 1970s, sale of recordable cassettes, to be used at your own tape recorder at home, were widely available. This private use strongly increased in the 1980s after a Japanese student started to lend out his record collection. Within one decade there were six thousand shops in Japan lending records; soon this concept was employed worldwide [68]. A conservative estimate shows that in 1982, sixty percent of the households in the European Union possessed tape recorders. By 1989, each household in the EU possessed three tape recorders and over 444 million hours of recorder cassettes was sold and 562 million hours of recordable cassettes. Given the fact that empty cassettes can be reused for recording two to three times, these numbers add up to approximately 1.405 hours of music [23; 41]. More than fifty percent was pop music compared to ten percent for classical recordings [23].

The system for regulating intellectual property for sound recordings held for half of the 20th century. Nevertheless the new technical means for private use, such as the tape recorder, has forced the industry to reconsider copyright legislation [68]. As Cock Buning has shown, copyright protection of music in the United States, Germany and The Netherlands has changed. Instead of a limited selection of reproduction techniques, legislators have decided on a more general approach: every fixation of music is considered to be a mechanical reproduction which supposes the composer's permission. In 1967, the Berne Convention was revised to reflect this idea. In 1978, US legislation described a recording as: "material object in which sounds, other than those accompanying a motion picture or other audiovisual work, are fixed by any method now known or later developed, and from which the sounds can be perceived, reproduced, or otherwise communicated, either directly or with the aid of a machine or device".[25]

The protection against private use of copyright protected material changed the Dutch copyright law as well. The Copyright Act of 1912 article 17 Aw offered the opportunity of reproduction for private use. At first this allowance covered the copying of texts and images for private use and for pleasure. Until the 1970s, this legal provision was of little importance; however, with the advent of new technical means, such as the copier, the stencil machine, tape recorder, cassette recorder, and the video recorder all leeway has now changed. Article 17 Aw was revised in 1974 to article 16B Aw which states that it is not an infringement on the copyright of a work of literature, science or the arts when the reproduction is strictly made for private practice, study, or use. Article 16B part 4 Aw adds, however, a remarkable addition for music. It prescribes that this reproduction for private use must have been personally made. This rule was written at the time when tape recorders were still expensive complex machines and only available to professional users. The legislature was alert for the possibility that professional companies would produce reproductions on a large scale for the private use of others. In order to prevent this misuse, it became essential that reproductions for private use would need to be personally made. In reality, the technological innovation of the tape and cassette recorder

---

[25]Quoted by Cock Buning [21: 114].

made it easy for anyone to reproduce music. Thus, soon this rule had lost its meaning.

Since the 1970s there were suggestions on compensating composers for large scale reproductions of their work for private use. It seemed a reasonable idea, but a complex one to realize. In addition, how to set and to maintain general rules for the public in order to regulate these actions in private? So instead of regulating individual actions, legislatures have focused on the recording equipment. The first recording tax regulation on recording machines went into effect in Germany in 1965 [23]. Germany was leading in the production of tape and cassette recorders and was the first to compensate composers with a special financial regulation. The idea of a special tax in order to compensate private use was already suggested in 1938, by Oranje, as mentioned before [67]. Afterwards similar regulations also became valid elsewhere in the 1980s and 1990s [23].[26]

In The Netherlands article 16C Aw of the Copyright Act was introduced in 1990 [82]. It offered the possibility to reproduce images and sounds for private use in return for financial compensation for the composer. It was not the buyer or user of the equipment who was obliged to pay the creator, but rather the producer or importer of the tapes or cassettes. However, the regulation supposed the passing on of these expenses to the consumer who reproduces the material for his private use. In that sense it is a practical and effective regulation of private use, although it breaks up the direct relationship between the reproduction and the related compensation for it.

In the shadows of the official commercial music business there emerges an interesting phenomenon of the bootleg recording from the 1960s. Initially, the bootleg was about non-published studio recordings, retrieved from the garbage cans of music studios. In that manner, rare recordings of Bob Dylan, The Beatles, or of Jimmy Hendrix could be distributed and was extremely popular since the unpolished bootleg recordings were as attractive as a personal sketch of a visual artist. It was a most original impression of the artist himself. Since the tech-

nology provided inexpensive yet effective recording equipment, anyone could make his or her own bootleg recording of a favorite artist performing live in concert. This type of bootleg copy resulted in an even larger production of bootleg recordings and sought after by a huge and curious fan base. Thus live concerts were recorded on the spot and then reproduced by fans for sale to other fans. This method greatly annoyed the musicians themselves. For example when Phil Collins performed his *Live and Alive Tour* in 1983 the concert was recorded and later published in Germany without his permission. It took him years of legal action before he succeeded in his claims of copyright at the European Court [41].

To summarize: at first, the copyright of sound was protected because of its similarities with text. However, within a relatively short period, the composer held the exclusive rights for mechanical reproduction of the original music pressed onto records. Radio broadcasting offered a new a way of presenting music that reached a huge audience. The tape recorder and the cassette recorder provided even newer possibilities on the use and abuse of reproducing music for private use. It forced reconsideration of legal rights in order to protect the intellectual property of the composer or performer. Meanwhile, others were experimenting with the combination of sounds and images that is, of audio-visual media.

4.5 AUDIO–VISUAL

4.5.1 Motion pictures

Fascinated by motions of man and animals, the photographer Eadweard Muybridge photographed his famous series entitled *Animal Locomotion*. New techniques by the end of the 1870s offered new means of reducing shutter time and consequently being able to catch movement as separate photographic pictures. A jumping horse, a man running, or a child crawling were photographed by Muybridge. For the first time in history it became possible to visualize and analyze motion in individual images. The photographic series seem to be similar to series of film frames that when

---

[26]See about an extensive international overview in this matter Davies and Hung [23: 116–214].

run produce the illusion of movement. But the appearances can be deceptive. Apparently Muybridge did not intend the illusion of movement, but rather the illusion of a falling cat or a jumping horse to be caught in a frozen stand. His pictures are the ultimate attempt of a photographer to freeze a moment of reality forever. From that perspective, Muybridge's photographs are typical for *standstill* photography, more than they represent the beginning of the *motion* picture industry. Despite this difference of intent, there is an important relation between photography and film as related to copyright, as will be explained below.

The beginning of the film industry is founded on the idea of the Belgian Scientist J.A.F. Plateau who had a collection of separate yet related standstill images which suggests movement when they are rapidly and successively shown. The remarkable machine, the Phenakistiscope, first presented the illusion of motion in the 1830s. From that exhibition, several machines designed for this effect have been produced, such as the praxinoscope (1877) and the kinetoscope (1891), brainchild of the influential inventor Edison. In 1895, the Lumière Brothers presented their invention: the cinematograph and it was an immediate success. Many curious people were amazed by their demonstration. By 1897, there was already a permanent installation of the cinematograph, a cinema, where various types of films were shown, mostly short documentaries with simple stories but accompanied by moving pictures. Soon thereafter, real life situations were captured in simple plays and shown on the cinematograph. Initially, however, there was more to see than to listen to since these were silent films.

Soon after the brothers Lumière, the brothers Pathé founded their film company, in September 1896. The firm was established in Paris but quickly developed to an international company with branches in Amsterdam, Moscow, St Petersburg, New York, Brussels, Berlin, Kiev, Budapest, London, Milan, Barcelona, Calcutta and Singapore by 1907. The firm produced a variety of movies: comedies, historical films, drama, and sport. The distribution of movies expended after the company started renting (1907) as well as selling films [14]. Pathé provided a constant supply of new movies which made it possible for cinemas to update their programming of movies. Consequently the Pathé firm dominated the early production and distribution of movies. But soon there were new film producers. In Great Britain, Germany, and especially in the United States, several film companies were established in a remarkably short period of time. At first on the east coast of the U.S.A., film activities soon moved to the west coast. In 1912, Universal Pictures was established in Hollywood, California. William Fox started too his Fox Film Foundation which was the parent company of other successful Hollywood film companies, including Metro–Goldwyn–Mayer and Paramount. In this same year (1912), the British actor Charlie Chaplin played in his first movie [19]. A few years later (1914) he acted his famous character of the little tramp in the movie *Kid Auto Races in Venice*. He became world renown for this character, which reflects both his talent as a comedian and the expanding influence of the medium he used. One of his first classics was the movie *The Gold Rush* (1925), the same year that other classics as Eisensteins *Battleship Potemkin* and MGM's *Ben Hur* were released.

In 1904 a French judge decided that film was to be considered as an 'oeuvre purement mecanique' which did not deserve copyright protection [90]. A few years later the French Court repeated that the illusion of motion was the machine's merit, not the author's. Therefore there was no reason for protecting the author. However, the film industry was booming and in a few years discussions of copyright protection for films began. Or rather the protection copyright *against* films, since movies were soon considered as reproductions of copyright protected works, such as plays, books, pantomimes, or photographs. This borrowing seemed to endanger the already existing copyright protected works that movies used. In that sense authors were soon offered copyright protection of their works in the form of films. At first, this did not imply the recognition of the individual artistic quality of motion pictures themselves, let alone the legal protection of it. This approach can also be observed in the revision of the Berne Convention of 1908. According to the convention, film was to be considered as a technical method for reproduction which could not be compared with the 'oeuvres litteraires et artistique' as stated in article 2. Nevertheless, article 14 part 2 offered the opportunity for the protection of certain types of movies: "are protected as literary or artistic works when, by the devices of the presented or the combinations of the represented scenes, the author will have given to the work a personal and original character".

The question soon arose as to whether all movies were to be protected, or only certain ones. There was no confusion about the directed feature films, but the situation of popular documentaries was not clear at all. Some proponents disputed the original quality of documentaries since they were more or less direct recordings of reality. According to the jurist Vermeijden in 1953, documentary films were also protected by the Berne Convention, but not under article 14 part 2 but rather under article 3 of the convention which stated: "The present convention applies to the photographic works and to the works obtained by an analogous procedure". This rule was originally intended for photography copyright protection, but in certain ways photography could include documentary films, as Vermeijden and others underlined. Without putting too fine a point on this analysis, this discussion did reflect the idea that films might be protected because of its similarities to literary works, plays and photography rather than because of its own inherent qualities [13; 90].

Inspired by the revision of the Berne Convention of 1908, copyright protection of movies was incorporated into the Dutch Copyright Act of 1912. According to article 10 Aw works of literature, science and the arts explicitly included photographic or cinematographic works. With respect to the visual arts, the Copyright Act of 1912 was rather late off of the starting block; however, in relation to motion pictures it was timely. Nevertheless, the definition of cinematographic works was not always clear. The Dutch legislature considered film in general in relation to photography with motion as the essential character. The rapid successively shown images produced the illusion of motion, which was the main distinction from standstill photography [90]. Vermeijden underlined the absence of sound of the early film, although the soundless film was

usually accompanied by piano, organ, or gramophone music. But the music varied with every performance and was not directly linked to the moving images. Neither was there any optical–acoustic unity to these early films; the author of the movie could not influence any of the incidental music that might accompany his film. In fact, the publication of images and sounds was parallel but unrelated [90]. It is clear that the Dutch Copyright Act of 1912 had only silent films in mind [90].[27]

4.5.2 Movies with sounds

Initially, film was a purely visual medium. From the beginning of the roaring 1920s there were experiments that technically combined film with sound: spoken words and music. These techniques were innovative and within a decade films with sound accompaniment were successful and popular. The first of these films was Warner Brothers' *The Jazz Singer* (1927) [19: 139]. Within a year, new popular film genres appeared with sound as an essential element. Walt Disney's *Steamboat Willy* (1928), featuring Mickey Mouse was the first cartoon picture with sound. Alfred Hitchcock's *Blackmail* (1929) was his first thriller motion picture. And the musical film also became a rage by the end of the 1920s, with for example *The Singing Fool* (1928), including the first successful soundtrack ever: Al Jolson's *Sonny Boy*. Henceforth movies were a true audio–visual medium. The silent movie had already passed its peak by 1930 and had been replaced by the modern audio–visual film. In the same period the traditional black-and-white movies changed into the modern colored pictures. It brought the golden age of the Hollywood film industry, only temporarily tempered at the beginning of World War II. The productions continuously increased as did the number of cinemas showing the films; even drive-in cinemas in the 1950s became popular.

---

[27]Despite the numerous court cases about films at the beginning of the twentieth century in foreign countries, there were no cases in The Netherlands before the thirties. The early film entrepreneur Desmet paved the way for the Dutch cinema from 1910, but was apparently never forced to go to court to protect his business. See extensively about Desmet and the early Dutch cinema: Ivo Blom, *Jean Desmet and the early Dutch Film Trade* [14].

Sound and color changed the status of films during the 1920s from a legal point of view. The revision of the Berne Convention in 1928 changed article 14 part 2. Films were protected as literary or artistic works the film productions when, the author will have given to the work an original character. If this character was lacking, the film production enjoyed the protection as a photographic works. Interestingly enough is the disappearance of the reference to the presentation of (combined) scenes. An important consecutive element for legal protection was no longer used; consequently the special relation between film and the theatre play ended. Paradoxically, the disappearance of this element from films opened the way to a more independent legal protection based on its intrinsic qualities rather than because of a films similarity to other media [90]. By now, the modern audiovisual movie was thoroughly developed. Remarkably enough the revision of the Berne Convention of 1928 did not mention the introduction of sound to the media of film. The Dutch Copyright Act was revised in 1931 with respect to the protection of photographic and cinematographic works. But here, too, the introduction of sound was not an issue. The traditional definition of cinematographic works remained unchanged. Nevertheless, according to Vermeijden, the well-used open-ended definition was not an obstruction for the copyright protection of a film including its sound [90]. Film with sound easily fitted into the legal system once developed for the silent movie.

The arrival of the audio–visual film raised interesting legal questions about the relation between sound and image. A Dutch judge regarded film as a 'collective work'. A judge of the court in The Hague considered on 24 September 1934 that film is to be considered as a collective work, the musical part included, and which intellectual property is to be protected, notwithstanding the copyright of the film in general. Contrary to this opinion, the court of Amsterdam considered a well-known Dutch case about the film *Fire over England* on 25 January 1940 and concluded that the film's music is not an individual element of the movie but rather that all sound, spoken and played,

is an integral part of the movie.[28] The High Court refused this argumentation and continued to consider film as a collective work of which the separate elements, like music, could be separately protected in addition to the film as a whole. This judgment remained the leading approach in The Netherlands, but in Great Britain, the United States and France the dominating legal opinion was of the optical–acoustic unity of film. Whether a film was considered as a collective work or as a unique single work, the fact is that film was developed into a separate and independent art medium between World War I and II, which was confirmed by the revision of the Berne Convention of 1948. It placed films on the same legal basis as other works of literature, science and the arts [90]. The legal differences between the individual film genres, such as documentaries, commercials, or feature films also faded away. Still there was an interesting and on-going discussion about the status of cartoons. The copyright expert De Beaufort regarded cartoons neither as films nor photography, but as types of drawings. Vermeijden however underlined in 1953 the similarities between cartoon and other kinds of movies: moving images in combination with sound. In that sense he refused to exclude cartoons from the copyright protection afforded to films [90].

The medium of film has been regarded on the same legal basis as works of literature, science and the arts for some time, but evidently that standing has not been enough. In the 1960s ideas were put forth about an independent legal system for film. The production of film supposed a complex organization of many different specialists, contrary to single authors of books, paintings or recorded music. This complexity of effort is underlined by the credit titles at the end of a film. Therefore, article 14 bis was formulated for additional protection of film at the revision of the Berne Convention of 1967 in Stockholm. The main goal was special protection for the producer of the film who managed the actors, composers, designers of special effect

---

[28]The Court regarded music specially composed for the movie "as a part of the independent artwork, that the film with sound is", Hof Amsterdam 25 januari 1940, N.J. 1940, 740. Vermeijden [90: 59].

while each made their contributions to the film production. Strictly speaking, the producer is not an author with an artistic contribution, but at the same time a director is essential for the original composers involved in film production to deliver their artistic contribution. The producer plays an essential part in providing the needed financial sources, running risks, and engaging necessary authors for each film produced. Special protection of the producer, then, seemed also to be fair. This complexity of protection can be reminiscent of similar claims of publishers, guilds, and record companies from the past. Therefore article 14 bis of the convention offered the producer a special protected right in film production even if special contracts with the authors involved would be missing. This special producer's right was introduced in The Netherlands by the revision of the Dutch Copyright Act 1912 in 1985 [82]. The producer's copyright is protected for various types of movies, from long feature films, documentaries, videos, television films, and programs for live television.

4.5.3 Television

During the 1950s rivalry emerged between film and another audio-visual medium: television. The development of television from the first experiments to a successful mass medium took quite some time, just like photography and radio. The German engineer Paul Nipkov experimented with television around 1900 but the next important steps forward were taken after World War I. Engineers in Germany, Great Britain, the USA and Japan seriously improved the technique for television in the twenties and thirties, interrupted for a few years during World War II [19]. In The Netherlands the Philips Company started its experimental broadcasting in 1948. The experimental exclusive medium now quickly transformed to the modern mass medium capable of reaching millions of viewers all over the world.

Television was an immense success that soon affected the success of cinema and even society in general [17]. In 1953 President Eisenhower wrote: "if a citizen has to be bored to death, it is cheaper and more comfortable to sit at home and look at

television than it is to go outside and pay a dollar for a ticket".[29] Eisenhower was not the only one with such an opinion. In the United States, cinema's audiences reduced from ninety million in 1948 to forty-seven million in 1956, almost fifty percent within a decade. At the same time millions of viewers were watching the hearings of the McCarthy and the shows of Ed Sullivan and Lucy Ball. Film companies increasingly changed their focus to the production of programs for television in order to compensate for losses sustained by their film companies. The advantage of colored movies instead of black-and-white television faded too. The television pioneer David Sarnoff devoted himself to the development of color television. Technically color television was already possible in the 1940s, but it was not until the 1950s that color television was economically feasible. Cartoons, especially *The Flintstones*, were very popular and important for the introduction of color television to general audiences. Viewers increased due to the addition of TV satellites for transmitting TV signals after 1962. When Neil Armstrong set foot on the Moon in 1969, more than 850 million watched him on TV [19].

Film and television competed for the same audiences yet influenced each other too. It took years before films were recognized as an independent work and an intellectual property. When this recognition was finally realized, television was the first to profit from this awareness and provided new opportunities for broadcasting many types of previously existing works. In particular, the advent of cable television networks raised even more legal questions about regulation of the publication of information by broadcasting. Centralized cable networks were developed during the 1950s for improving access and quality of television for consumers. In 1952 in the United States, there were seventy central cable networks that reached fourteen thousand subscribers; whereas by 1965 there were already over fifteen hundred systems that were reaching three and one-half million people. In The Netherlands there were experiments with cable networks. Queen Juliana, Queen of the Netherlands, announced in her 1964 Queen's speech the

---

[29]Quoted by Briggs & Burke [19: 189].

ambitious plan to provide one national cable network for television [55]. This was not actually realized immediately; but once it did happen, the national cable infrastructure covered nearly the entire country. The first Dutch program by way of cable television was broadcasted in the small village Melick Herkenbosch by 18 October 1971. Afterwards many communities followed this example [55], and today, The Netherlands is one of the most densely cabled countries in Europe with almost ninety percent of its households connected to a cable television network. Cable television networks have made it possible to regulate the supply of television on a national, regional, and a local scale. Cable television networks, therefore, were an important extension of broadcasting television by air supported by satellites.

As with radio broadcasting, television broadcasting provided a new way to publish copyrighted materials. In 1938, Oranje pointed to the similarities between radio broadcasting and a new innovation called television in relation to copyright: "In this book we shall not refer separately to television, since we cannot at the present moment (1938) fully realize what new questions will arise in connection with this subject. This does not alter the fact that probably most of the conclusions which we reach in this book will also be applicable to the protection of the broadcaster of television" [67: 2]. Initially, the government argued that cable television networks did not reach a *new* public of viewers, only supported the already existing public with higher quality television. Therefore, it was not a new way of making information public. Nevertheless, the high court did not agree with this argument and refused the criterion of a new public in 1981 and 1984 cases. Passing of television signals through cable television networks was to be regarded as a new way of making public that which needed permission and compensation of the authors. In that sense the Dutch high court followed the argumentation of the revised Berne Convention in this respect. Indeed, even broadcasting by 'cable pirates' who illegally used the cable networks was regarded as publications by the legitimate cable companies [82]. Cable networks were installed for television, but are today heading to a new future through the use of the Internet.

4.5.4 Videorecorder

Live television offered the viewer the excitement of being present to a performance. Not long afterwards new technological means were available to record this experience on tape and keep the memory alive. In line with the tape recorder engineers presented in 1969 the first video recorder to record sound *and* image, and in 1971 the first examples were supplied to the European market. The video recorder became popular, especially in Japan and, in 1989 one of three television owners also owned a video recorder [23; 41]. Like the earlier tape recorder, the video recorder offered new opportunities for reproduction for private use. A videorecorder turned one's individual television into a home cinema or 'armchair theatre'. The introduction of the video recorder from the start was supported by the renting of video tapes. However, more than playing prerecorded films on television, the video recorder was mostly used for recording television programs to be viewed at later on a more suitable moment, called 'time shifting'. The introduction of the double-deck video recorders in 1990 increased possibilities even more. In 1988–1989, research into the use of video recorders showed that the equipment was mostly used for recording television p rograms for time shifting (96.3%) and to play rented videocassettes (66.6 %) [23].

The effect of the video recorder on copyright protected material was similar to that with tapes and cassette recorders which raised questions about the limits of private use. In February 1977 in order to deal with these issues, a special committee was formed and supervised by United Nations Educational, Scientific and Cultural Organization (UNESCO) and the World Intellectual Property Organization (WIPO): "The Working Group considered that, in view of the ease of reproduction video grams in the form of videocassettes, it was probable that such a mode of reproduction would not satisfy the restrictive conditions laid down by the above mentioned provision [article 9 (2) Berne Convention, RV] and that, consequently, such reproductions were subject to the exclusive right of reproduction, under the Berne Convention" [23: 70]. On this basis the committee considered that "as long as the state of technical progress did not allow copyright owners effectively to exercise the prerogatives of the exclusive right in the event of the private reproduction of video grams, the only solution seemed to be the establishment of a global compensation for authors or their successors in title. It was pointed out that such payment would be in the nature not of a tax or other monetary imposition, but rather of an indemnification for being deprived of the opportunity to exercise exclusive rights" [23: 76]. Afterwards several new committees were established to look at these ideas for further recommendations.

A special UNESCO/WIPO Committee underlined the principle again in 1988 quite clearly: using video recorders for private use directly affected the private interests of copyright owners. They deserved reasonable financial compensation by way of special taxes on blank video cassettes and other video recording equipment [23]. This idea of a special tax as financial compensation for private use was mentioned before in relation to the recording of sound. A special regulation by article 16C Aw was introduced in the Dutch Copyright Act 1912 in 1990 for audio *and* for video cassettes with a special tax for compensation for private use [82]. With that move, the regulation of private use was easily extended from audio to audio-visual material.

To summarize: at first, the pictures started to move, than they were accompanied with sound forming new audio-visual media. The movies were to be seen in local cinemas, but soon people did not have to leave their homes and could watch films on televisions. Radio brought sound to a large at-home audience, and a few decades later television added visual motion to this experience, first in black and white and then in color, and finally supported by cable television and satellites. Audio-visual media like films and television raised interesting legal questions about the relation between sounds and images. In line with the private use of sounds on tape and cassette recorders, the video recorder seriously changed the private use of motion pictures. Looking back, however, this alteration seemed only to be the quiet before the storm of the most revolutionary digital multimedia.

4.6 MULTIMEDIA

During World War II the British engineers tried with help of the Colossus machine to crack the secret codes of the Germans. This technological monster with all its tubes and wires looked impressive. Actually, its impressive size was the result of the lack of the smallest part of the machine: the chip. The production of chips was started in 1954 by the Texas Instruments company and extended by the Intel company in the 1960s [19]. Meanwhile the International Business Company (IBM), another pioneer in the digital field, presented its first computers. Initially these computers were not *personal* computers. Instead they were expensive super calculators available only to a small number of professional users [28].

Like photography, radio and film, computers were introduced to the general public as a new form of entertainment. At the end of the 1970s, the Atari Company presented the first computer games, which were to be played on television sets. The compact disc (CD) was another important digital medium introduced in Japan in October 1982. A few months later this digital invention was presented to the European market (March 1983) and in the United States (June 1983). Since 1985 the CD has been a huge commercial success that seriously affected the selling of traditional records. Like the addition of the single to traditional record, the CD was supplemented by the CD single in 1987. At the end of the 1980s the selling of CDs exceeded the selling of records and music cassettes [23]. After the basic music CD, other variations were produced, such as the CD ROM (compact disc, read only memory) for images, CD-I (CD Interactive) for games, and the DVD (Digital Video Disc) for audio-visual works like films.

New digital products, like video games, raised interesting but complex legal questions of copyright, as stated by Salokannel: "Literary works, audiovisual works, art works and musical works have been merged together in protected multimedia products, thus producing, at least at first glance, a supposedly 'new' category of works" [77: 20].[30] In considering the legal protection for these types

---

[30]Salokannel, p. 20. See also [25; 70; 98].

of works, a Dutch judge stated in 1983 that the computer game *Pacman*, produced by the Atari Company, deserved to be regarded as a work in order to protect its copyright because of its creative structure and design.[31] A few years later, in 1986, there was again an important court case about Atari, now in France in which the illusive, interactive character of the computer game was discussed. As Majut Salokannel stated: "The Court stated that the legal protection covers all works of original intellectual creation regardless of any aesthetic consideration. Neither the interactive nature of the video game nor the fact that the game functions with the aid of a computer program has any relevance when determining the protectability of the work" [77: 76–77]. It is difficult to point to the single author of a computer game since it supposes a complex cooperation of several specialists working on one game. In this respect, games were soon compared with traditional media, such as film and TV; for example, in the Belgium Nintendo case: "a video game is qualified as an audiovisual work defined as an animated mixture of sounds and images which is destined to be shown in front of a public [. . .] a video game is a film in terms of Belgium law and Nintendo and co is to be deemed as the producer of a first fixation of the film" [77: 78]. Like film, video games are the creation of many creative authors. In that sense extensive credit titles at the end of your game seemed to be reasonable.

Apple, IBM and Microsoft brought computers into reach of many consumers by the mid-1980s. During this period, computers became *personal* computers affordable for a lot of people. The successful software of Microsoft Windows offered a first impression of a virtual world, and that virtual digital world was realized within one decade when computers were connected to the worldwide web of Internet. Search engines like Netscape, Navigator and Internet Explorer have prevented consumers from getting lost in virtual space. In the beginning, there were mostly standstill images, but soon enough the images started to move in fascinating audio-visual animations and in beautiful colors.

In the digital era boundaries between text, visual, sound, and audio-visual works are fading away. Digital text, image, sound, and audiovisual information are based on the uniform binary codes of 0s and 1s which is solely responsible for adapting and exchanging information so easily. Due to simple functions, such as "copy and paste" of text, sound, and images every amateur can now produce works on his personal computer that many professionals only dreamed of ten years ago. Because of the World Wide Web, there is an almost infinite supply of information available to use and to adapt. Digital media presents a new dimension for clarification with regards to the private use of information. Because of the Internet, the modern user can reach far more copyright protected material than ever before. Texts, sounds, and images can easily be downloaded and sampled in any form a user wishes. The new digital creation can be copied to a hard disk, CD or USB (universal serial bus) stick. More importantly, the publication of a new work on the Internet is global available to everyone. The amateur radio was used to reach a local or regional audience. The modern user of the Internet can easily reach a world wide audience.

The modern digital era provides a fascinating combination of content and means to adapt and publicize creative works to a large public audience; consequently, digital media has challenged the traditional system of copyright protection more than ever before. For example in the field of music, new initiatives like Napster and Kazaa seriously affected the production and distribution of music. Like the introduction of the tape and the cassette recorder, it was pop music in particular that was mostly downloaded and pirated. The huge reproduction for private use was soon seen as private abuse of copyright protected music and resulted in a flow of legal cases about the repeating issue of the limits of private use. Without too much analyzing, this private use of digital media still seems acceptable by the industry, although the limits are in on-going dispute. The reactions to the possibility of large scale private use can be categorized as follows.

In the first place the Dutch legislature acted on changing the copyright law, influenced by the international law of WIPO and the European Union (Directive 2001/29/EU). Regulations for private use

---

[31] Hof Amsterdam 31 maart 1983, see about this case: P.B. Hugenholtz, J.H. Spoor, *Auteursrecht op Software* [46: 38–39].

were also applied to new digital works in order to guarantee reproduction of digital works for private practice, education, or pleasure. In that respect there no longer is a difference between 'offline works' and 'online works'. In the tradition of earlier discussions about special taxes for private use of the photocopier, tape and videorecorder, this discussion had been started in relation to MP3 music player and the use of USB sticks [39]. Publishing information on your website is regarded as a new way of publication which supposes the permission of the author. In addition, article 45 N of the Dutch Copyright Act 1912 states that the opportunity of private use through articles 16 B and 16 C is not valid in respect to computer programs. Thereby, the legislature blocked the easy reproduction of computer programs and restricted private use. These revisions of the Dutch Copyright Act 1912 were the result of international developments and discussions. It remains to be seen whether this *national* legal protection will be efficient and sufficient in the global *international* context of the Internet.

Secondly, the producers of digital information reacted to the possibility of extensive private use with the technological development of Digital Rights Management (DRM) [48]. Technological cryptography must screen off information and protect it for a selected group of users [57]; nevertheless, the better the protection, the greater the challenge to break it. The consequence is the permanent arms race between producers and consumers about the technological protection of information. In this rat race the producers are supported by the copyright legislation article 29 A Aw of the Dutch Copyright Act forbids passing by technological measures. The limits of the producers DRM and the consumers private use are still disputed, for example, in the recent French court case *Christophe R.v Warner Music* (10 January 2006) in which the French court underlines to make sure that private use is still possible, despite technical measures of DRM [43]. Of course, the topic of DRM is closely related to the history of cryptography in which other essays in this book are addressing.

Thirdly, copyright owners more and more release their works free for use by everyone based on concepts of 'Creative Commons' or 'Open Source'. This concept to publish information without restraint is, interestingly, as old as the Internet itself. Rooted in the hippy culture of the 1960s, this idea was especially promoted by institutions like the University of Berkeley in California [56]. Closely related ideas of 'Open Source', 'Creative Commons' or 'Copyleft', are all founded on the idea of unrestricted publishing of information. These concepts are often presented as opposed to the traditional protection of intellectual property; 'Copyleft' as the opposite of Copyright. However, this apparent opposition does not hold because the freedom of an author to publish his work without restriction can be regarded as the essence of the traditional copyright system. An author's choice to publish his work with no restrictions, looks like giving up copyright, but can also be considered as the ultimate expression of the author's copyright. The general call of authors to their colleagues to publish work freely is not against the idea of intellectual property. In that sense the development of 'Creative Commons' and 'Open Source' seems no alternative for the traditional system of intellectual property, but rather a new expression of it. Therefore, the 'Creative Commons' movement even fits in the traditional copyright system which can also be shown with a recent Dutch court case about the limits of the Creative Commons license of 9 March 2006.[32] The case was about photographs published on the Internet with reference to the license used by the organization *Creative Commons*. The photographs were, in time, published by a glossy magazine. The claimant stated this was illegal because this was against the conditions of the referred license. One of the conditions forbids the commercial use of the published photographs like the magazine did. The glossy magazine defended the use of the photograph referring to the idea of 'Creative Commons'. The Amsterdam judge agreed with the argumentation of claimant. Information published on the site of 'Creative Commons' is not completely free to use, though restricted by the conditions of the used license of Creative Commons and the copyright law in general.

---

[32]Rb Amsterdam 9-3-06, 334492/ KG 06-176 SR.

The era of electronic multimedia presents a revolution in the production, reproduction, and publication of texts, images, sounds, or the audiovisual combination of these aspects. Despite important changes, the traditional system for the protection of intellectual property still seems useful and valuable for the protection of digital works. The essential articles of the Copyright Act 1912 are generally formulated, abstracted from individual techniques and media. Compared to the copyright laws abroad, the law of 1912 was quite late, but it showed itself to be flexible in adapting to the media revolutions of the 20th century. Some specialists predict that the traditional copyright protection has an uncertain future; others show more confidence in the meaning and organization of intellectual property protection in the digital era [25; 26; 49]. The main reason for protecting the intellectual property is the author and his bond held with his creative work. From a technological perspective this special relationship can easily be broken with the original author easily being replaced. From a normative legal perspective this separation might affect the original author more now than ever before. The modern Internet artist still holds a clear relationship with his digital creation similar to the relationship that the writer Alexander Pope, the artist Horace Vernet, or the composer Giuseppe Verdi had with their creative works.

Salokannel rightly states: "A multimedia application, however complex in structure, is always, at least at the present stage of technological development, based on the creative thought of the human intellect; we still have authors who create the works, who design and compose the structure of the work, all activities which set limits to user interactivity, regardless of the multitude of choices, or even learning abilities of the multimedia application. It is always the authors, the physical persons who have created and pre-determined the choices available to users to interact with the work. Even when we speak of the so-called computer-generated works, the computer program generating the work is a result of the creative input of human minds" [77: 89]. Strictly speaking, it does not seem to be necessary to provide a separate protection for a work of digital multimedia. The definition of work still seems valuable for the protection

of texts, images, sounds, and their combination in this digital era. The interactive dimension, the new possibilities of private use, or the combined creation of several authors makes the protection of the digital work complex but not principally different than works before the digital revolution [77]. The digital multimedia work still reflects its author's personal mark signifying the special relation between the author and his own creation.

4.7 CONCLUSIONS

Successively introduced media over the past centuries has resulted in questions being raised about protection of intellectual property. The questions raised seem to have a similarity with regards to the reproduction of text, images, sound, and the combination of these factors. Nevertheless, the approach to intellectual property protection has changed over the centuries. Initially, the system of legal protection was orientated to the reproduction that was protected by privileges. The book was protected before its original manuscript, and the reproductive engraving before the original painting.

During the eighteenth century the legal system of protection was slowly changed to focus on the original work, rather than the reproduction of it because of the effort of writers, painters, and composers. However, this approach was a gradual cultural and legal transformation. In 1877, the Dutch jurist Viotta still took the reproduction of a work as the point of departure in the 19th century copyright jurisdiction in The Netherlands: 'The object of intellectual property is not based on the idea, nor on it's originality of the product, but on the *reproduction* of it. Instead of in the intellectual sphere, the intellectual property realizes itself in the material world' [92: 12].

From the beginning of the 20th century, intellectual property was definitely focused on protecting the originality of creative products, whether in the form of text, images, audio, films, television or digital products. Traditions of similar legal questions can be observed in the history of intellectual property, such as the issue of fair private use. The stencil machine and photocopier, tape and cassette recorder, the video recorder and the personal computer all resorted to the same issue of private use.[33] In general, the protection of texts paved the way for protecting visual media like graphic and photographic images, which in their turn influenced the protection of audio–visual media, such as film and television. They all created their own niches in copyright protection over the course of time. The protection of these analogue media founded the way in which we approach and protect intellectual property in the digital era today.

We successively discussed text, image, sound, and audio-visual media in relation to copyright laws. These media often combine together. Although the term 'multimedia' is commonly used for the digital era, it would be interesting not to restrict the meaning of it to the current affairs. It suggests that media in history used to be completely singular and separated of each other. Which was not always the case. For example texts and images were fused by 19th century wood engraving used in popular, illustrated magazines; silent movies also mixed moving images with text; and television gave way to Teletext (textual information broadcasted by TV signals), started by the BBC in 1974.[34] In that sense media have combined for centuries. Contrary to this, the audio-visual works of film and television were often separated into new products. Films provided popular soundtracks, pulled loose from the original picture [55; 68]. The soundtrack became a new work in itself, just as the photographic stills of the movie

scenes became single units of creative works as well. Even before the era of the multimedia, media were creatively combined and separated again. From a theoretical perspective too, media in history has showed impressions of interesting combinations of media. The understanding and legal approach of a new medium was often based on similarities with already existing media. Engravers and photographers referred to writers in their request for copyright protection. Sound was protected as far it could be read and film because it's relation to photography and plays. Therefore, even singular media reflected dimensions of multimedia.

The relation between multimedia and copyright issues has been discussed for decades. In 1984, UNESCO predicted the need to adapt traditional copyright protection in the light of new digital multimedia. Their Second Medium-Term Plan (1984–1989) stated: "Moreover the traditional form of copyright, developed essentially in order to protect printed words, needs to be adapted to the present day, now that the emergence of revolutionary techniques – reprography, disks and other forms of magnetic recording, cable transmission, communications, satellites and computers – has completely transformed the conditions in which texts, images and sounds are reproduced and disseminated, suggesting that in the future works may be disseminated instantaneously and universally, leaving no real possibility of ever learning or even estimating the number of users or the volume of material involved. One of the tasks to be faced in order to cope with these technical changes is to find a way to protect the works which are carried by these new techniques or new copyright media and to protect the techniques or media themselves, either through copyright or through rights related to copyright" [23: 81].

The question is whether this prediction will turn out to be a reality. Of course the production, distribution, and consumption of copyright protected material have radically changed in our age. Nevertheless as Salokannel already stated: "Digital technology does not make an artist of every citizen, just as the invention of the typewriter did not make all of us writers, nor did the advent of the video camera, cause a drastic increase in the number of cinematographers" [77: 28]. One thing seems not to

---

[33]Remarkably, this was not the case for images. Of course, photo cameras are very popular with a large public using them for taking holiday pictures for more than a century. However, instead of the recording equipment for sound and film, the camera is less used for reproduction of copyright protected material. People used the camera much more for the production of images of reality itself, rather than reproduction of existing works. Therefore, photography never caused the question of fair private use, let alone the revision of the copyright protection. This will all change by the easy access to copyright protected images on Internet and software like photoshop to manipulate existing images.

[34]Another crossover in the context of the Dutch radio and television was the complex discussion about the copyright protection of the text of radio and television guides, see: L. van Vollenhoven, *De Zaak TeleVizier. In het bijzonder in auteursrechtelijk verband* [93: 109–110].

have changed at all: the unique position of the creative author and his special relation with his work, whether it's designed in text, images, sounds, or as a sophisticated mix of these elements. The primacy of the author still seems to be an important normative point of departure for the recognition and protection of the author's intellectual property. The traditional system of copyright protection has shown itself to be flexible enough to react and incorporate new media revolutions from both the 19th and the 20th centuries. It will probably survive the 21st century as well, supported by new approaches of 'Copyleft', 'Creative Commons' and 'Open Source'.

The digital era provides new fascinating opportunities of which the (legal) consequences are still unclear. Digital information can easily be reproduced, but its authenticity and reliability can not. Searching for a reliable provenance of information on the Internet seems to be like looking for the starting point of a circle. More than traditional knowledge institutions like libraries, archives, and universities, the search engines like Google seem to be the new gateways for the digital republic of letters of tomorrow [29]. Finding information is surprisingly simple and most often based upon association and intuition rather than on rational systematic analyses. The widely used Google system has already been turned into a verb, that is 'to google'; and it will probably soon be used to identify a whole young generation: the 'Google generation'. New products can easily be (re)produced and read by everyone around the world. According to copyright legislation, an author's work is protected during a lifetime and up to seventy years after death. However, the digital techniques can not give any guarantee for such a period. It is quite questionable whether any author can still read his own multimedia work in even ten or twenty years after its first publication. In that respect it is not a question of whether copyright can resist the digital challenges but rather if the digital media can fulfill an author's expectations in the future.

One sign of the digital era is the mobile phone. As a real personal digital assistant, cell phones provide multiple features; calling is just one of those. It can be used as photo camera, to play music

through its ring tones, to send short textual messages (SMS), or to play movies, games, or even to watch television or surf the Internet. And all wireless. It releases the Internet of its physical cable network and uses the digital transmission by air, like radio and television. The mobile phone offers endless new opportunities for all kinds of private use. No concert, film, or performance is safe from bootlegged recordings as long as mobile phones are present. The mobile phone can be considered as a fascinating culmination of the media of text, visual, audio and audio-visual and as the ultimate expression of the multimedia era. Interesting discussions have already started about this new medium in relation to intellectual property, creating feelings of *déjà vu* [83].

REFERENCES

[1] J.A. Alberdingk Thijm, *Kopijrecht*, De Dietsche Warande (1864), 75.
[2] Anonymous, *regtzaken*, De Kunstkronijk (1842–1843), 42.
[3] Anonymous, *Property in art*, The Art Journal (1849), 133.
[4] Anonymous, *Copyright in art*, The Art Journal (1862), 241.
[5] Anonymous, *Photo-Sculpture*, The Art Journal (1863), 59.
[6] Anonymous, *Copyright in Sculpture*, The Art Journal (1863), 59.
[7] Anonymous, *Reviews*, The Art Journal (1863), 128.
[8] Anonymous, *Infringement of Copyright*, The Art Journal (1863), 103.
[9] Anonymous, *Infringement of Copyright*, The Art Journal (1863), 210–211.
[10] D.I. Bainbridge, *Intellectual Property*, Blackstone, London (1992).
[11] S. Bann, *Parallel Lines. Printmakers, Painters and Photographers in Nineteenth-Century France*, New Haven, London (2001).
[12] H.L. De Beaufort, *Het Auteursrecht in het Nederlandsche en internationale recht*, Den Boer, Utrecht (1909).
[13] H.L. De Beaufort, *Auteursrecht*, Tjeenl Willink, Zwolle (1932).
[14] I. Blom, *Jean Desmet and the Early Dutch Film Trade*, Amsterdam University Press, Amsterdam (2003).
[15] A.W.A. Boschloo, *The Prints of the Remondinis. An attempt to Reconstruct an Eighteenth-Century World of Pictures*, Amsterdam University Press, Amsterdam (1998).

[16] D. Bouchard, *Language, Counter-Memory, Practice: Selected Essays and Interviews*, Cornell University Press, Ithaca (1977).

[17] P. Bourdieu, *Sur la television*, Liber-Raisons d'agir, Paris (1996).

[18] H. Brandt Corstius, *Brieven van Multatuli gekozen door H. Brandt Corstius*, Querido, Amsterdam (1990).

[19] A. Briggs, P. Burke, *A Social History of the Media. From Gutenberg to the Internet*, Polity Press, Cambridge (2005).

[20] D. Cassell, *The Photographer & the Law*, BFP, London (1984).

[21] M. de Cock Buning, *Auteursrecht en informatietechnologie. Over de beperkte houdbaarheid van technologiespecifieke regelgeving*, Otto Cramwinckel, Amsterdam (1998).

[22] K.D. Crews, *Copyright, Fair Use, and the Challenge for Universities. Promoting the Progress of Higher Education*, University of Chicago Press, Chicago (1993).

[23] G. Davies, M.E. Hung, *Music and Video Private Copying. An International Survey of the Problem and the Law*, Sweet and Maxwell, London (1993).

[24] E.J. Dommering, *Het auteursrecht spoelt weg door het electronisch vergiet. Enige gedachten over de naderende crisis van het auteursrecht*, Computerrecht 3 (1994), 109–113.

[25] E.J. Dommering et al., *Informatierecht. Fundamentele Rechten voor de Informatiesamenleving*, Otto Cramwinckel, Amsterdam (2000).

[26] E.J. Dommering, *Nieuw auteursrecht voor de eenentwintigste eeuw?*, Computerrecht 3 (2001).

[27] B.P.M. Dongelmans, P.G. Hoftijzer, O.S. Lankhorst (eds), *Boekverkopers van Europa. Het 17de-eeuwse Nederlandse uitgevershuis Elsevier*, Walburg, Zutphen (2000).

[28] D. Draaisma, *De Metaforenmachine. Een geschiedenis van het Geheugen*, Historische Uitgeverij, Groningen (1998).

[29] N. van Eijk, *Zoekmachines: Zoekt en Gij Zult Vinden? Over de Plaats van Zoekmachines in Het Recht*, Otto Cramwinckel, Amsterdam (2005).

[30] J. Feather, *Publishing, Piracy and Politics. An Historical study of copyright in Britain*, Mansell, London, New York (1994).

[31] A. Firth, *The Prehistory and Development of Intellectual Property Systems*, Sweet and Maxwell, London (1997).

[32] J.F. Viëtor, *Het auteursrecht. Kantteekeningen op Het Ontwerp Wet Tot Regeling van Het Auteursrecht*, J.L. Beyers, Utrecht (1877).

[33] W.H. Friedman, *Boydell's Shakespeare Gallery*, Garland Publishers, New York, London (1976).

[34] P. Fuhring, *The Print Privilege in Eighteenth-Century France-I*, Print Quarely 2 (1985), 175–193.

[35] P. Fuhring, *The Print Privilege in Eighteenth-Century France-II*, Print Quarely 3 (1986), 19–33.

[36] E. Gambart, *On Piracy of Artistic Copyright*, William Tegg, London (1863).

[37] P. Gaskell, *A New Introduction to Bibliography*, Oak Knoll Press/St. Pauls Bibliographic, Oxford (1995).

[38] S. Gerbrandy, *Kort Commentaar op de Auteurswet 1912*, Quint, Arnhem (1988).

[39] S. van Gompel, *De vaststelling van de thuiskopievergoeding*, AMI. Tijdschrift voor Auteurs-, Media & Informatierecht 2 (2006), 52–62.

[40] F.W. Grosheide, *Auteursrecht op maat. Beschouwingen over de grondslagen van het auteursrecht in een rechtspolitieke context*, Kluwer, Deventer (1986).

[41] J. Gurnsey, *Copyright Theft*, Ashgate Publishing, Hampshire, Vermont (1995).

[42] A.J. Hamber, *"A Higher Branch of the Arts" Photographing the Fine Arts in England, 1839–1880*, Gordon and Breach, Amsterdam (1996).

[43] N. Helberger, *Christophe R. v Warner Music: French Court Bans Private-Copying Hostile DRM*, INDICARE Monitor, 24 February (2006).

[44] J. Hemels, H. Demoet (eds), *Loodvrij en Digitaal. Visies op Innovatie in Grafische Communicatie*, Alfa Base, Alphen aan de Rijn (2001).

[45] A.M. Hind, *A History of Engraving & Etching. From the 15 Century to the Year 1914*, Dover, New York (1967).

[46] P.B. Hugenholtz, J.H. Spoor, *Auteursrecht op Software*, Otto Cramwinckel, Amsterdam (1987).

[47] P.B. Hugenholtz, *Auteursrecht op Informatie. Auteursrechtelijke Bescherming van Feitelijke Gegevens en Gegevensverzamelingen in Nederland, de Verenigde Staten en West-Duitsland. Een Rechtsvergelijkend Onderzoek*, Kluwer, Deventer (1989).

[48] P.B. Hugenholtz, *Het Internet: Het Auteursrecht Voorbij?*, Preadvies vo de Nederlandse Juristenvereniging (1998).

[49] P.B. Hugenholtz, *Haalt de Auteurswet 2012?*, Jaarverslag Nederlandse Uitgeversbond (2000), 56–61.

[50] R. Hunt, *Photogalvanography; or engravings by light and electricity*, The Art Journal (1856), 215–216.

[51] W.M. Ivins, *Prints and visual communication*, MIT Press, Cambridge (USA), London (1996).

[52] R.A. de Jonge, *Muziekrecht. Theorie en Praktijk van de Nederlandse Muziekindustrie*, Strengholt, Naarden (1991).

[53] J.J.C. Kabel, *Beeldende kunst en auteursrecht, Kunst en Beleid in Nederland 5*, Amsterdam (1991), pp. 67–124.

[54] J. Van de Kasteele, *Het Auteursrecht in Nederland*, P. Somerwil, Leiden (1885).

[55] F. Klaver, *Massamedia en Modern Auteursrecht*, Wetenschappelijke Uitgeverij, Amsterdam (1973).

[56] K.J. Koelman, *Terug naar de bron: open source en copyleft*, AMI. Tijdschrift voor Auteurs-, Media & Informatierecht 8 (2000), 149–155.

[57] K.J. Koelman, *Auteursrecht en technische voorzieningen. Juridische en rechtseconomische aspecten van de*

bescherming van technische voorzieningen, SDU Den Haag, Amsterdam (2003).

[58] L. Kuitert, *Het ene Boek in Vele Delen. De Uitgave van Literaire Series in Nederland 1850–1900*, Amsterdam University Press, Amsterdam (1993).

[59] F. Lecler et Léon Noël, *Revue des editions illustrées, des gravures et des lithographies*, L'Artiste I (1839), 142.

[60] A. Ligtenstein, *Het ontwerp van wet tot regeling van het auteursrecht op werken van beeldende kunst. Meer dan kunstenaars alleen...*, Pro Memorie (2005), 297–308.

[61] T. Limperg, *Auteursrecht in de Hortus der Beeldende Kunsten*, Phaedon, Culemborg (1992).

[62] J. Maas, *Gambart. Prince of the Victorian Art World*, Barrie and Jenkins, London (1975).

[63] C. MacLeod, *Inventing the Industrial Revolution. The English Patent System 1660–1800*, Cambridge University Press, Cambridge (1988).

[64] G.H. Martin, D. Francis, *The Camera's Eye*, *The Victorian City: images and realities*, H.J. Dyos, M. Wolff, eds, London (1973), pp. 227–245.

[65] E.A. McCauley, *Industrial Madness. Commercial Photography in Paris, 1848–1871*, Yale University Press, New Haven, London (1994).

[66] R. van der Meulen, *Boekhandel en Bibliographie. Theorie en practijk*, A.W. Sythoff, Leiden (1883).

[67] J. Oranje, *Rights Affecting the Use of Broadcasts*, A.W. Sythoff, Leiden (1938).

[68] D. Peeperkorn, C. van Rij (eds), *Music and the New Technologies. Soundtracks, Cable and Satelite, Rentals of CDs, Computers and Music, Sounds Sampling, Cover Versions*, Maklu, Antwerpen (1988).

[69] C.P. Phillips, *The Law of Copyright in Works of Literature and Art and in the Application of Designs*, London (1863).

[70] E.W. Ploman, L.C. Hamilton, *Copyright. Intellectual Property in the Information Age*, Routledge/Kegan Paul, London, Boston (1980).

[71] G.H. Putnam, *The Question of Copyright. A Summery of the Copyright Laws at the Present in Force in the Chief Countries in the World*, G.P. Putnam's Sons, New York, London (1891).

[72] E. Rebel, *Faksimile und Mimesis. Studien zur Deutschen Reproduktionsgrafik des 18. Jahrhunderts*, Mittenwald, München (1981).

[73] M. Reinsma et al., *Auteurswet 1881. Parlementaire geschiedenis wet 1881 – ontwerp 1884*, Walburg, Zutphen (2006).

[74] J.G. Robbers, *Het Auteursrecht. Opmerkingen en Beschouwingen*, J.G. Roblers, Amsterdam (1896).

[75] M. Rose, *Authors and Owners. The invention of Copyright*, Cambridge, MA, London (1993).

[76] F.W. Rushing, C.G. Brown, *Intellectual Property Rights in Science, Technology, and Economic Performance. International Comparisons*, Westview Press, San Francisco, London (1990).

[77] M. Salokannel, *Ownership of Rights in Audiovisual Production. A Comparative Study*, The Hague, London, Boston (1997).

[78] D. Saunders, *Authorship and Copyright*, Routledge, London, New York (1992).

[79] C. Schriks, *Het Kopijrecht. 16de tot 19de eeuw. Aanleidingen tot en Gevolgen van Boekprivileges en Boek-Handelsusanties, Kopijrecht, Verordeningen, Boekenwetten en Rechtspraak in Het Privaat-, Publiek- en Staatsdomein in de Nederlanden, met Globale Analoge Ontwikkelingen in Frankrijk, Groot-Brittannie en het Heilig Roomse Rijk. Completed with an Ample Summary in English*, Walburg, Zutphen (2004).

[80] B. Sherman, A. Strowel (eds), *Of Authors and Origins. Essays on Copyright Law*, Clarendon Press, Oxford (1994).

[81] B.Sherman, L. Bently, *The Making of Modern Intellectual Property Law. The British Experience, 1760–1911*, Cambridge University Press, Cambridge (1999).

[82] J.H. Spoor, D.W. Verkade, *Auteursrecht*, Kluwer, Deventer (1993).

[83] S.K. Sukhram, P.B. Hugenholtz, *Het mobiele Internet*, AMI. Tijdschrift voor Auteurs-, Media & Informatierecht (2003), 161–168.

[84] A.G.N. Swart, *Opmerkingen Betreffende Auteursrecht op Werken van Beeldende Kunst*, S.C. Van Doesburg, Leiden (1891).

[85] J.D. Veegens, *Het Auteursrecht Volgende Nederlandse Wetgeving*, Gebr Belinfante, 's-Gravenhage (1895).

[86] G. Verdi, *Autobiografie in Brieven. Gekozen, Bezorgd en Vertaald Door Yolanda Bloemen*, Arbeiderspress, Amsterdam (1991).

[87] R.M. Verhoogt, *En nu nog een paar woorden business. Reproducties naar het werk van Alma Tadema*, Jong Holland 12(1) (1996), 22–33.

[88] R.M. Verhoogt, *Kunsthandel in prenten. Over de negentiende-eeuwse kunsthandels van Goupil en Gambart*, Kunstlicht 20(1) (1999), 22–29.

[89] R.M. Verhoogt, *Kunst in Reproductie. De Reproductie van Kunst in de Negentiende Eeuw en in het Bijzonder van Ary Scheffer (1795–1858), Jozef Israëls (1824–1911) en Lawrence Alma Tadema (1836–1912)*, unpubl. Diss. Universiteit van Amsterdam (2004). Will be published in English in spring 2007.

[90] J. Vermeijden, *Auteursrecht en het Kinematographisch Werk*, Tjeenk Willink, Zwolle (1953).

[91] H. Vernet, *Du Droit des Peintres et des Sculpteurs sur les Ouvrages*, Imp. d'E Proux cf ce, Paris (1841).

[92] H. Viotta, *Het Auteursrecht van den Componist*, L. Roothaan, Amsterdam (1877).

[93] L. van Vollenhoven, *De Zaak TeleVizier. In het Bijzonder in Auteursrechtelijk Verband. Een Documentaire Studie*, Kluwer, Deventer (1970).

[94] C. Vosmaer, *Het adres der kunstenaars over het eigendomsrecht van hunne werken*, De Nederlandsche Spectator (1879), p. 113.

[95] J.A. de Vries, *Enige beschouwingen over het 19e eeuwse auteursrecht tegen de achtergrond van het incident Douwes Dekker/Van Lennep*, AMI. Tijdschrift voor Auteurs-, Media & Informatierecht 9 (1996), 180–184.

[96] M. Woodmansee, **The Author, Art and the Market. Rereading the History of Aesthetics**, Columbia University Press, New York (1994).

[97] C. Zigrosser, **The Book of Fine Prints. An Anthology of Printed Pictures and Introduction to the Study of Graphic Art in the West and East**, Crown Publishers, New York (1956).

[98] M. de Zwaan, **Geen Beelden Geen Nieuws. Beeldbeperkingen in Oud en Nieuw Auteursrecht**, Cramwinckel, Amsterdam (2003).

5

THE HISTORY OF COPYRIGHT PROTECTION OF COMPUTER SOFTWARE

THE EMANCIPATION OF A WORK OF TECHNOLOGY TOWARD A WORK OF AUTHORSHIP

Madeleine de Cock Buning

University of Utrecht, The Netherlands

Contents

Abstract

Driven by the technological and economic reality of the expansion of the computer programs market in the last half of the 20th century, software has been emancipated from work of technology towards work of authorship. This contribution will elucidate on this process of emancipation which has given rise to fundamental discussions at the core of Intellectual property law.

Keywords: computer software, intellectual property, software industry, Etienne Pascal, IBM, Mark 1, Remington Rand, software pool, unbundle, World Intellectual Property Organization, computer program, Patent and Trademark Office (PTO), Copyright Act, pianola roll, ROM-incorporated, National Committee on New Technological Uses of Copyrighted Works (CONTU), Office of Technology Assessment (OTA), look and feel, reverse engineering, Patent Gesetz, Inkassoprogramm, Oberdurchschnittlichkeits, Software Directive, programming idea, sui generis regime, Berne Convention, European Patent Treaty, European Software Directive, European Commission, back-up copies, decompilation, General Agreement on Tariffs and Trade (GATT), literary works, Trade-Related Aspects of Intellectual Property Rights (TRIPS), World Trade Organization.

5.1 INTRODUCTION

Computer software was one of the first electronic information technology products which penetrated the domain of intellectual property. The question of adequate protection arose concomitant with the large-scale expansion of the software industry. Producers have emphasised to legislators that software development is costly and that copying without the approval of the rightful owner is quite easy.

Due to the technical character of software, more so than with other information technologies, the question arose as to which area of the law should provide for computer software protection and specifically whether the protection should come from within patent law, which has a high threshold and a broad scope of protection or from copyright law, which has a low threshold and a narrow scope of protection. The very nature of this question indicates that computer software is a hybrid which deserves serious legal consideration.

This article will outline how the protection of computer software has evolved, and it will not only outline the historical development as found in the United States of America, the historical mecca of information technology, but in particular it will address The Netherlands and Germany as exponents of European developments. Within these two countries noteworthy and fundamental discussions have arisen in regards to software as part of an intellectual property arena. Attention will also be devoted to European and International legislation that has surrounded the processes of emancipation. To begin with a brief description of the history of the development of software, however, is essential to presenting the framework of the legal developments.

5.2 FRAMEWORK: TECHNOLOGICAL EVOLUTION

Etienne Pascal's adding machine from the 17th century can well be considered the earliest forerunner of the computer. In 1890 Herman Hollerith designed a machine to mechanically take the entire census of the American population. It is from this project that the subsequent IBM computers were derived [3]. The breakthrough for computer technology came with World War II. The Mark 1 from 1944 was the first large mechanical computer. This computer came about thanks to the collaboration of the American Marines, Harvard University and IBM. The American Army commissioned the construction of the first electronic computer [81], which could be programmed by changing cords on the 'control board' much in the same way as on a telephone switchboard of earlier times [55]. Von Neumann developed a method of programming to replace the time-consuming repositioning of cords. The data and the software could thus be entered externally, which shortened programming time considerably [31]. The very first computer for commercial use was introduced by Remington Rand in 1951. Three years later IBM brought another machine on the market which had more possibilities [3]. From that moment on the computer industry has been growing at a very rapid pace. The computer has permeated all facets of organisations and companies. The first integrated connector (chip) appeared on the market in the 1960s, and with the invention of the chip, the computer became an affordable and more manageable product for everyone.

In the beginning of the computer era, software developments and distributions were in the hands of the producers. Large hardware producers generally provided computer programs free of charge as part of the hardware sold ('bundle'). Users hardly ever needed to copy the accompanying software, and there was not a market for independently developed software. Hardware producers encouraged users to form collective software banks, thus thinking to create greater sales of hardware. Specifically developed programs were deposited in these 'software pools' which everyone could draw on free of charge. After a number of years it became clear that the 'software pools' were functioning below expectations. Businesses began realising that the value of the software used in conducting their business was increasing and they became less prepared to place this self-developed software free of charge at their competitor's disposal via the pool [31].

5.2.1   Unbundling

IBM decided to 'unbundle' software and hardware in 1969, due to competitive considerations and this decision was the first impetus toward an independent software industry [39]. This process was hastened when chip-driven mini- and microcomputers appeared on the market. IBM's mainframe computers were now being replaced on large scale by personal computers, and 'tailor-made' computer programs for these mainframes were no longer required. The hardware-independent production of software, the increase in scale, and the simple manner in which software – for businesses or for personal use – could be copied all joined to create a strong need for legal protection of intellectual property rights among software producers.

Many definitions of a computer program have been formulated in the technical and legal literature [70]. The World Intellectual Property Organization (WIPO), that is part of the United Nations which is responsible for intellectual property, gave the following definition in 1978: "a set of instructions capable, when incorporated in a machine-readable medium, of causing a machine having information processing capabilities to indicate, perform or achieve a particular function, task or result".[1] In Article 101 of the US Law on Copyright can be found: "A computer program is a set of statements or instructions to be used directly in a computer in order to bring about a certain result". Framed in the Green paper from 1988, the European Commission described a computer program as: "a set of instructions which has as an aim to allow an information-processing machine – the computer – to carry out its functions".[2]

The computer can only process tasks which are given in binary form: the object code. Various symbol languages have been developed to simplify the programming of a computer. Most computer programs have been written in higher program languages, such as Pascal, C, Prolog, Lisp and APL. The code in a higher program language is named the source code. The source code is converted (compiled) to a particular computer-readable object code with the assistance of a separate computer program (the compiler). During decompilation attempts are made to translate the compiled code back to a semi-source code. Decompilation is part of a broader process of analysis of a computer program: reverse engineering. The analysis of a computer program can serve various purposes. It can provide information about the parts of a program, which take care of the interaction between the components of a system (the interface); and it can be used for the production of a computer program that can work together with a decompiled program (an interoperable program). So too can decompilation assist the user of a program in making a code appropriate for a specific application or in removing errors. The least valued aim of decompilation by the software producer is the production of a competing program.

This brief account of the development of the independent software market serves as a framework for that which will be described below. How has legal protection evolved with the passage of time?

5.3   THE UNITED STATES OF AMERICA

Although computer programs were being developed in the United States at universities and within governmental institutions for fifty years, the question of legal protection only became apparent when commercial software was being exploited in the 1970s [70]. After an initial demonstration that patent law was the most appropriate means of protection for computer programs due to the exceptionally technical nature of the material, an interest in copyright law protection also grew [80]. Nonetheless, patenting software-related inventions was frequently not allowed until the 1980s [70]. The basis for this was the 1968 guidelines of the US Patent and Trademark Office (PTO) in which it was stated that: "The basic principle to be applied is that computer programming *per se* (. . .) shall not

---

[1]WIPO/IP/ND/87/3.
[2]COM (88) 172 def.

be patentable".[3] Thus it remained a relatively uncertain matter, even when the Supreme Court in 1981 began paving the way for the patenting of software [6].[4] In addition, a number of practical difficulties occurred at the beginning of the software patenting process. When it was time to gain an extension of a patent, there was an additional requirement that the invention be made public in exchange for strong protection rights [79]. Furthermore, it was a costly affair to acquire a patent which had become the special reserve of large software industries. As a result of these complications, copyright, then, became an attractive alternative.

5.3.1 Shortly after the introduction of computer programs

The question of whether software is protected under copyright law first appeared under the effect of the Copyright Act of 1909. It was disputed whether a computer program could be registered with the Copyright Office [70]. After a period of uncertainty the Copyright Office officially announced in 1964 that the registration of computer programs was possible under the 'rule of doubt' but expressed reservations at that time about the practical value of a registration process [16]. People at the Copyright Office had questions about a few issues, such as could a computer program be considered 'writing' as defined in the US Constitution. Was the reproduction of this computer programme in the memory of a computer to be seen as a 'copy' that could be registered?

In a study carried out while the US Government was in the process of revising the Copyright Act of 1909, an answer to the first question occurred [70]; under the law 'writings' were considered the fruit of intellectual labour and 'authors' of these writings were to be seen as having provided the creative contribution. Computer programs could, therefore, be considered to fall under copyright law [2]. The second question which was formulated by the Copyright Office in 1964 was the establishment of a computer programs in the memory of a computer to be seen as 'copy', which came from a 1908

decision on *White-Smith Music Pub. Co v. Apollo Co*.[5] The Supreme Court ruled in this case that the producers of a pianola roll had not infringed on the copyright of the maker of the melody which had been imprinted there because the music on the roll was not aurally detectable without the aid of a mechanical device. In 1909 the Copyright Act was amended in order to consider pianola rolls as reproduction in the sense of the Copyright Act. This Act was not redefined with a technology neutral concept of a 'copy' but instead included a specific article concerning sound recordings.[6] The most important obstacle for the protection of computer programs thus remained under the teachings of *White-Smith*. Computer programs were established in a manner which people did not immediately perceive without the aid of an instrument. Registration was therefore not possible with the Copyright Office. It is for this reason that relatively few programs were offered for registration with the Copyright Office [16].

A tremendous resistance existed in the 1960s against the copyright protection of software [79]. Furthermore the necessity of this protection was questioned. According to legal scholar Breyer, no need existed for copyright protection since most software was provided with its required hardware. In 1970, Breyer was already predicting a worldwide computer network which would provide for sufficient technical protection:

> Copyright protection is not likely to be needed if it is relatively easy to identify, organize, and bargain with groups of potential program buyers [9].

He himself felt copyright protection was undesirable because expenses for recovery of rights were disproportionately high. The fear of a strong expansion of the copyright domain was also a reason used to argue against the copyright protection of software. Legislators, in the Copyright Act of 1976, still declined to decide on new material. Software was not included in the list of the copyright-protected works of article 102(a).

---

[3]Examination of patent applications of computer programme, 33 Fed. Reg. 15609., 855 Pat. Off. Soc. 581 (1969).
[4]Diamond vs. Diehr 450 U.S. 175.

[5]209 U.S. I, 52 L. Ed. 655, 28 S.Ct. 319 (US Supreme Court 1908).
[6]U.S. CA 1909, art 26.

In one of the earliest important cases after 1976 in which copyright protection for software was invoked, *Synercom Technology v. University Computing*,[7] dimensions of protection, albeit restricted, were allowed and assumed. The district court did not allow any protection on the 'ordering and sequence' of important components of the computer program. The structure of these components was, according to the decision of Judge Higginbotham, nothing more than the reproduction of an idea. He thus differentiated between literary works, in which the 'ordering and sequence' was traditionally well protected, and computer software.

Because the legislators in the Copyright Act of 1976 explicitly distanced themselves from the *White-Smith* doctrine, the path was open for copyright protection of software. Works which could only be discerned with the assistance of an instrument were offered copyright protection [56]. Copyright protection of ROM-incorporated computer software remained, as a result, but was not yet guaranteed. Thus the District Court of Illinois in *Data Cash Systems v. JS. & A*[8] decided that software solely in the form of a flow diagram and source code was protected. The chess game that was incorporated into a ROM chip and that literally was copied, according to Judge Flaum, was not protected against copying since it was not visually discernible. A few years later the District Court of Northern California, in *Tandy v. Personal Micro Computers*, distanced itself from the *Data Cash* decision. The teachings of *White-Smith* were no longer applicable. Based on article 102 of the new law, a chip could be considered as a tangible medium of expression: software in a ROM could now be protected.

The US Congress instituted the National Committee on New Technological Uses of Copyrighted Works (CONTU) in 1974 [16]. This committee was assigned the task, among others, of immersing itself in the copyright protection of software. There were proponents of copyright protection on the committee, as well as harsh opponents. Committee member Nimmer proposed that copyright protection should be restricted to works which were

automatically generated with the aid of software generating computer programs; howsoever, computer programs in and of themselves should be excluded from copyright protection. The chairman of the US Copyright Association, Hersey, proposed that copyright law was principally unsuited for the protection of what he called "a mechanical device, having no purpose, beyond being engaged in a computer to perform a mechanical work".

The majority of the CONTU members did not have any objections, in principal, against introducing these creations into the domain of copyright law. As a result copyright protection would encourage the development of computer software in the United States, and the CONTU report provided a number of recommendations on how to secure this protection [16].

5.3.2 Further development of the technology and expansion of the software market

By 1980 CONTU's recommendations eventually led to the incorporation of software into US copyright legislation[9] making the United States the first country to included software under the protection of the law. The Software Copyright Act of 1980 changed the Copyright Act of 1976 and categorised computer programs as 'literary works' under 102(a)(l). Article 117 made a specific exception to the existing exclusive rights – to be excluded were those actions which were required in order to use a computer program. The other actions which were considered in article 117 as exceptions to the exclusive rights were the conversion into a different programming language, the addition of program components, and the placement of small changes in the program. However, making back-up copies was still subject to the approval of the rightful owner.

In 1980 the Office of Technology Assessment (OTA) published a critical assessment of the Software Protection Act of 1980.[10] Software was a hybrid creation with features shared with patented

---

[7]462 F. Supp. 1000 (U.S. Distr. Crt. N.D. Tex., 1978).

[8]480 F. Supp. 1063 (U.S. Ditsr. Crt. N.D. III, 1979).

[9]Software Copyright Act, Pub. L. no. 96-517, 94 Stat. 3028 (1980).

[10]U.S. Congress, Office of Technology Assessment, Intellectual Property Rights in an Age of Electronics and Information, OTA-CIT-302, Washington, DC, April 1986.

and copyrighted objects, and according to the OTA there was a danger of limiting a free interchange of due to the low threshold and long-term protection of copyright law. The OTA proposed the introduction of a new protection regime. The object of the regime would be all 'functional works' such as computer chips, expert systems, and recombinant DNA. The OTA further proposed the creation of special software institutes to collect royalties and to settle disputes.

The judiciary had relatively little difficulty with the application of the Software Protection Act of 1980 [26]. The originality required in interpreting the law did not pose any appreciable issues. The presence of a number of choices in the creation of a computer program did make it susceptible to copyright protection from the beginning.

5.3.2.1 *The extent of protection.* Determining the extent of protection as well as the application of the traditional idea or the expression dichotomy of computer programs presented the judiciary with some larger difficulties. Which elements of a program should be considered as protected expressions and which existing parts should be deprived of copyright protection? The question was applicable in a series of court decisions.

In the 1978 decision on *Synercom Technology v. University Computing*,[11] the extent of the protection of computer programs allowed was scant. In that decision the structure of a computer program was deprived of protection while the works of literature were allowed protection. On the other hand, in 1986, an unusually large extent of protection was allowed in *Whelan v. Jaslow*.[12] This protection in itself was noticeably larger than usual for traditional works of literature. It was proposed that all software elements – except the general aim of the purpose of the program – should be considered as protectable. In the law journals, the *Whelan–Smith* decisions projected an extent of protection that was generally criticised [10].

The Court of Appeals in 1992, in *Computer Associates v. Altai*[13] took a position contrary to the

extent of protection granted in *Whelan v. Jaslow* [82].[14] The Second Circuit Court dismissed the position that all elements should be considered as protected design, except for the function or purpose of the program. In the *Altai* case, the Court of Appeals developed a three-pronged 'abstraction, filtration and comparison' test. Using this, it could be determined whether there was 'substantial similarity of non-literal elements of computer programs'. The approach adopted in the *Altai* case was increasingly employed in the 1990s.

The 'look and feel' court decisions of the 1990s also distanced themselves from the broad extent of protection that resulted from *Whelan* [41]. After a broad extent of protection had been allowed in the court in a series of decisions on the protection of user-interfaces, the protection requested in *Lotus Development Corp. v. Borland International Inc.* was dismissed.[15] The First Circuit Court considered Lotus' menu structure, as according to article 102(b)USCA as an unprotected 'method of operation' [5]. Thus the possibility to use (parts of) the menu structure of an existing computer program for a (competing) program was not ruled out.

5.3.2.2 *Reverse engineering.* Except for the extent of protection, one exception to the exclusive right formulated in 117 was further interpreted [10]. Should the research on the program code, for which duplication of the Code was needed, be allowed? The law was silent on reverse engineering. After the court at first considered software research as an infringement on the exclusive *rights of the rightful* owner,[16] the lower courts began cautiously to allow reverse engineering.

Reverse engineering was deemed admissible for the first time in *Johnson v. Uniden*.[17] In *Vault v. Quaid*[18] and *NEC v. Intel*[19] the reproduction and functioning of a program, necessary for reverse engineering, were indeed not considered as infringement. In two decisions from the beginning of the

---

[11]462 F. Supp. 1000 (U.S. Distr. Crt. N.D. Tex., 1978).

[12]797 F. 2d 1222 (3rd Cir. 1986).

[13]23 USPQ 2d 1241 (2nd Cir. 1992).

[14]797 F. 2d 1222 (3rd Cir. 1986).

[15]49 F. 3d 807 1995; 288 SC 94-2003 (Supreme Court Jan 16, 1996).

[16]For example in Apple Computer v. Franklin Computer Inc. 714 F. 2d 1240 (3rd Cir. 1983).

[17]623 F. Supp. 1485 (D. Minn. 1985).

[18]847 F. Supp. 255 (5th Cir.)

[19]10 USPQ 2d (BNA) 1177 (N.D. Cal. 1989).

1990s, the criteria for admissibility of reverse engineering were precisely stated. In *Sega v. Accolade*[20] and *Atari Games v. Nintendo* the reproduction of a computer program for the purpose of researching whether unprotected elements could be used in a compatible computer program was considered as 'fair use'. Thus was the path opened up for later programmers to build a computer program without damaging the rightful interests of the rightful owner [41]. Additionally, in *Atari v. Nintendo*, a halt was called to software producers who were aspiring to patent-like protection via amorphous copyright protection:

> An author cannot acquire patent-like protection by putting an idea, process, or method of operation in an unintelligible format and asserting copyright infringement against those who try to understand that idea, process, or method of operation.[21]

5.3.3 Long term: normalisation

Software was admitted to the domain of copyright during the 1980s. A call was sounded now and then during the 1990s for a separately defined protection for computer software. Four prominent writers issued a manifesto in 1994 in which they proposed a specific protection regime [61]. The introduction of a short-term know-how protection was advocated. This protection would need to be complemented with a registration system. Industries wishing to make use of this knowledge would know from the registered information whether the know-how protection had expired. It was truly improbable that US legislature would take any type of legislative action [27]. What was probable was that, in the long term, an amendment to a law would occur for the codification of judiciary-created norms and for the advancement of legal certainty. With this idea, the creation of a protection legislation separately defined, as opposed to the long-term practise of protection, would come into effect. In the meantime, the necessity of technologically specific regulations for software protection appeared to have decreased. Due to the fact that US copyright legislation, at least in regard to software, recognised

fewer and less technology-bound legal restrictions than continental European legislation, the system has remained reasonably flexible. The *judiciary* has taken the room to clarify, in a step-wise manner, the still existing grey areas [41]. In *Sega v. Accolade*[22] and *Atari v. Nintendo*[23] a solution to the question of admissibility of reverse engineering was offered. In *Lotus v. Borland*[24] the issue of user-interface protection was clarified, and the extent of protection under copyright law was returned to a more reasonable level in *Computer Associates v. Altai* [8].[25]

5.4 GERMANY

In Germany, it was as in the United States in that at first special attention was given to the patent-like protection of software. Articles which appeared in law journals were however initially critical of patenting software. The German Supreme Court disallowed the patenting of various software-related inventions because of its lack of a technical character.[26] The Patent Gesetz, in 1976, adopted the ruling of article 52 of the European Patent Treaty. With this move, 'computer programs as such' were excluded from patent-like protection. In 1986 the German Patent Commission issued guidelines in which the patenting of software-related inventions was allowed under limited circumstances. As of 1991 software-related inventions were also definitively deemed patentable by the German Supreme Court.[27] A lack of technical character was no longer considered an obstacle to patenting. In the 1990s truly a large number of requests for the German newness and inventiveness requirements failed.

It took time in Germany not only for software to be protected by patents but also for the copyright

---

[20]977 F. 2d 1500 (9th Cir.).

[21]975 F. 2d 832 (Fed. Cir. 1992).

[22]977 F. 2d 1500 (9th Cir.).

[23]975 F. 2d 832 (Fed. Cir. 1992).

[24]49 F. 3d 807 1995; 288 SC 94-2003 (Supreme Court Jan 16, 1996).

[25]23 USPQ 2d 1241 (2nd Cir. 1992).

[26]Dispotitionsprogramm, GRUR 1977, 69–99; Staken, GRUR 1977, 657–659; Prüfverhfahren, GRUR 1987, 101–103.

[27]Settenpuffer, Computer und Recht 1991, 568; Tauchcomputer, Computer und Recht, 600.

nature of software to be recognised. Early court decisions did not allow for copyright protection of software since software could not meet the original criterion of protection required, among other reasons.

5.4.1 Shortly after the introduction of computer programs

As early as 1965 some authors emphasised how important the protection of computer programs was for innovation in the software industry. He stated, to his own regret, that the existing copyright system was unsuitable. Through copyright protection, only the copying of the expression of a program can be prevented, but not the underlying idea [57]. He therefore proposed an alternative form of protection. Each and every computer program would be tried after it had been filed with one of the as yet to be established bodies. Those interested parties would be able to buy or borrow these programs. It would have been, according to Ohlschegel, no more than rewarding an idea without an approval to acquire it. Ohlschegel came up with his solution at the time when the independent software market did not truly exist. Until 1969, software was written principally for a specific hardware environment and software was provided free of charge with the hardware.

In the first judgement in which copyright protection for software became a focus, protection was not allowed by LG Mannheim. The program was not perceptible to the senses and, in addition, lacked the 'aesthetic content' required for copyright protection.[28]

This decision incited a storm of criticism in law journals. Some legal scholars were of strong opinion that almost every program – with the exception of simple, straightforward programs – has its own personal character [48]. This position was used in decisions of the lower courts in Germany. In two subsequent decisions copyright protection of computer programs was granted.[29] The decision of Landesgericht (LG) Mannheim[30] to dismiss

protection on the grounds of a lack of 'aesthetic content' was overturned on appeal to Oberlandesgericht (OLG) Karlsruhe.[31]

In the 1980s, after the lower courts had recognised copyright protection of software in a large number of instances [18], the decision of the Germany Supreme Court in May 1985 recognised this protection with the *Inkassoprogramm* case.[32] The German Supreme Court explicitly distanced itself from that of the 1981 LG Mannheim-formulated discernibility requisites but strengthened the originality test considerably.

According to the German Supreme Court, computer programs, in principle, were protectable as 'Schriftwerke' or 'Darstellungen wissenschaftlicher oder technischer Art' under article I UrhG or that is if they could not be directly discerned. In *Inkassoprogramm*, the German Supreme Court developed a two-step test to be able to determine the originality of a computer program. Firstly the program had to be compared with other computer programs which performed comparable functions. Secondly the comparison had to determine whether the program was of above-average quality [19], that is programs were clearly required to surpass the average generally produced by programmers vis-à-vis choice, collection, ordering, and division of particulars and instructions. The German Supreme Court thereby honed the German originality criterion significantly by requiring above-average creativity [63].

With this decision the German Supreme Court deviated from the generally accepted position in copyright doctrine. Legal scholar Haberstumpf blamed the German Supreme Court for placing the *Oberdurchschnittlichkeits* criterion somewhere between the level of invention of patent law and the originality of copyright law. According to Haberstumpf, this ambiguous position could only lead to making an already complex question of software protection even more complicated [29]. Legal scholar Bauer described the decision as "die grundsätzliche Bejahung und praktische Verneinung des Urheberrechtsschutzes für Computerprogrammen"

---

[28]LG Mannheim 12 June 1981, Betriebsberater 1981, 1543.

[29]LG Mosbach 13 July 1982 GRUR 1983, 70; LG Munchen 21 December 1982, GRUR 1983, 175.

[30]LG Mannheim 12 June 1981, Betriebsberater 1981, 1543.

[31]LG Karlsruhe 9 February, GRUR 1983, 300.

[32]BGH 9 May 1985, Computer und Recht 1985, 22.

[4]. ("The theoretical acceptance and the practical denial of copyright protection for computer programmes".) Bauer concluded that most programs would be denied protection with this decision. They thought more effective protection of computer programs could only be guaranteed under newly developed specific legislation [63].

5.4.2 Further development of the technology and expansion of the software market

In April of 1985, a month before the German Supreme Court handed down its decision on *Inkassoprogramm*, the German legislature explicitly established the possibility of copyright protection of software in the German Copyright Act (UrhG). Starting in 1986 computer programs were, in so many words, copyright protected by the introduction of the 'Programme für die Datenverarbeitung' in the category of literary works listed in the catalogue of works of article 2.1 UhrG. In principle the standard copyright rules thus became applicable to computer programs. Each program which could be considered as a 'persönliche geistliche Schöpfung' would therefore need to be considered for protection [21]. In the above-mentioned *Inkassoprogramm* judgement of 9 May 1985, the German Supreme Court had established a truly higher threshold for protection than for the other products. In the second software decision, *Betriebssystem*, handed down in 1991, the German Supreme Court also employed an even more rigorous test to assess the originality of a computer program.[33]

With the amendment of the law in 1995, a technology-specific condition was included in article 53 UrhG for situations concerning individual practise, study, or use of software. Each duplication, including a reserve copy, required the prior approval of the rightful owner. An exception was made for actions which were necessary for the normal use of the computer program. What should be understood by the term 'normal' was not further elaborated. What was also ambiguous was whether decompilation research was permitted. Because of this ambiguity, an extensive polemic arose in the law literature [30; 46; 49; 51].

The proposal for the accompanying neighbouring rights[34] protection of software of 1989 is also worthy of note. The German government proposed creating a new neighbouring right in order to position the Software Directive then being created within national legislation. This proposal was created in order to prevent the German practical understanding of this protection from becoming over-extended. The legislative bill was strongly criticised because it was not in keeping with either international developments or with the European Directive which did not have neighbouring protection in mind but rather copyright protection [45].

With the implementation of the Software Directive, it was not neighbouring rights protection that was chosen but rather a special section (article 69a–69b UrhG) within copyright legislation which was set up for computer programs [67]. The concern, then, continued to exist, in the event that the provisions were placed within the existing law [22], that the traditional German copyright laws would be influenced by the software-focused detailed provisions of the Directive.

In contrast to the German legislation, The Netherlands, England and France, for example, adopted a definition of originality, as found in article 69a lid 3 UrhG. This article contained not only a definition of originality, such as described in article 1 lid 3 of the Directive, but also an elaboration on the explanatory notes, which were non-binding. This legislation stated that "no criteria may be derived in respect of the qualitative or aesthetic value of a program". This addition was intended to render harmless the onerous originality test of the German Supreme Court's judgement on *Inkassoprogramm* [34].

5.4.3 Long term: normalisation

In Germany at the beginning of the 1990s, the initial level for the software-raised copyright protection threshold was lowered. In 1993, the German Supreme Court in *Buchhaltungsprogramm* considerably relaxed the originality test in conformance

---

[33]BGH 112/264, GRUR 1991, 451.

[34]A neighbouring right is a right that is granted to e.g. a performing artist for its performance. It is partly similar to copyright.

with the European Software Directive.[35] In case law and in the law journals there was still more discussion to come about the content and extent of protection [22].

Specifically the implementation of the detailed and technology-bound provisions of the Software Directive in article 69a–69g UrhG required further interpretation [1]. An inroad in this direction was made in the *Holzhandelprogramm* decision of the German Supreme Court.[36] The court emphasised that copyright law could not stand in the way of the use of a work [35]. The use of a computer program was, according to the German Supreme Court, not a relevant copyright infringing event [52].

The Court did not decide on whether the temporary loading of a program into the working memory ought to be considered as a copyrightable-relevant duplication. If this action is to be considered as belonging to an exclusive ownership right, the multiplication is then essential for the 'contemplated use', in any event, to be allowed in conformance with article 69d. 1 UrhG [59]. What is to be understood exactly by 'contemplated use' is, by and large, unclear.

In German case law it was determined that the removal of technical protection in order to use the program simultaneously on other computers without the approval of the rightful owner could not be considered as contemplated use. Under what circumstances decompilation could take place would also have to wait for future court decisions. The technology-specific provisions of the Software Directive did not establish the contemplated legal certainty.

5.5 THE NETHERLANDS

Legal scholar Verkade, in the first publication appearing in the Netherlands concerning the protection of software, questioned the practical value of copyright protection [71]. In his opinion a patent would be the ideal form of protection. A patent owner could then not only prohibit third parties from producing and selling the program but also

from using it. Verkade certainly was aware that there was no certainty that the Board of Patents would actively proceed to grant patents for software inventions. In 1970 the Appeals Department of the Board of Patents indeed denied a patent in its *Telefooncentrale* ruling.[37] It was the Department's opinion that the computer program in question was not a mode of operation as meant by the 1910 Patent Act (*Rijksoctrooiwet 1910*) because the accompanying process was not a material substance and was therefore judged not to be industrial.

After the Board of Patents' denial, Verkade came out in for the expansion of the system of patent rights in order to be able to include software [72]. Nothing occurred in the 1970s to expand this range of possibilities under the law. Undoubtedly the Board of Patents' amending of its policy of granting patents in the 1980s was influenced by industrial lobbyists, such as those found in the US and the European Patent (Octrooi) Bureau (EOB) which had achieved some success. In 1983, this change manifested itself clearly when the Appeals Department of the Board of Patents made an important first move in amending its policy for patenting software as a mode of operation in the *Tomograaf* decision [40].[38] In a decision of 1985 the Appeals Department deemed software a patentable product provided that the storage in the working memory of a computer which was directly addressable was equally patentable as a product.[39] With the *Streepjes-code* decision of 1987 the possibility of patenting software was further expanded.[40] The Appeals Department awarded patent rights to the deciphering of a bar code pattern considering its patentability as a mode of operation. Materialised information, regardless of the form assumed, was, according to the Department of Appeals, of a material nature and could then be an object for a mode of operation patent.

5.5.1 Shortly after the introduction of computer programs

The first public discussion on the protection of software took place in 1971 [71]. The general feel-

---

[35]BGH 14 Juli 1993, GRUR 1994, 39.

[36]BGH 20 January 1994, Computer und Recht 1994, 275–277.

[37]16 December 1970, BIE 1971, 54.

[38]19 January 1983, BIE 336.

[39]2 September 1985, BIE 1985, 435.

[40]11 May 1987, BIE 1987, 174.

ing was that "(...) not much is to be expected for the protection of computer software under copyright law". Copyright law could be invoked only in the instance of indiscriminate duplication. Legal scholar Cohen Jehoram felt that this would seldom or never arise in the area of computers [71]. If a serious lack in copyright law were to be shown, it would not be against protected use. The program could come into third-party hands through 'various leaks' (w.o. article 17 Aw) which could be viewed as not impeded by copyright in order to use the program. The trust in (professional) ethics of software producers and users put this problem into perspective. Legal scholar Helmer stated that: "it is gratifying to acknowledge that the larger part of computer users are normally honest, dependable people" [71]. Naturally, specific regulations could be formulated to give the understanding of norms additional strength. Helmer felt this should be indicated on the program for the intended user. Should copies be made or used by third parties, "then the people who saw this program should be surprised, with all that follows from this". Helmer was additionally a proponent of a formalised professional code of ethics and a disciplinary tribunal, "which as a tribunal does nothing more than letting people who feel cheated have a good cry, and will thus exist to make potential cheaters flinch" [71]. Verkade stated in 1972 that certainly what most important in a computer program, the 'programming idea' was not protected [72]. It was, as he said, incredibly easy, especially in developing software, to give the underlying idea another form [72]. In 1973 Tenbergen was even more pessimistic than Verkade had been: "copyright protection definitely rests – and in The Netherlands without any restrictions – on *all* computer programs but has (...) barely any appreciable practical value" [68].

5.5.2 Further development of the technology and expansion of the software market

The concern about the appearance of software piracy increased during the 1980s [36]. It was proposed with increasing frequency that the existing copyright law was considered to be insufficient and should be supplemented with specific protection.

There was some anxiety that if the software producer were not sufficiently protected then his motivation to develop software would flag. When legal scholar Borking, at the beginning of the 1980s, discovered, to his amazement, that there was a 'gap in the copyright law' he proposed an alternative form of protection [37].[41] The fact that the object code in ROM or the magnetic strip or drive was not readable meant, according to him, that this absence had remained unprotected from the outset [64]. Borking considered it more feasible to protect software on ROM as falling under the category of article 10 section 1 sub 9 of the Dutch Copyright Act (Auteurswet, 'Aw') as a photographic work. With part of the production process, it was discussed as if it were a photolithographic process. He thought the chip's mask could be considered as a photographic negative and the result of a photolithographic process, the chip, as a special photographic print.

In the law literature a polemic was carried on between the proponents and adversaries for copyright law protection. The objection to copyright law protection of software was that copyrighting itself did not allow for the area of technology since it had always been an instrument to protect art and literature. In this context, a number of authors expressed the concern that if copyright be made applicable to objects in the area of industry then the legislature might be moved to legally limit copyright [74].

Legal scholar. Quaedvlieg felt that copyright protection of a technical product was, in addition, fundamentally incorrect. The utilisation of the teachings of a subjective work had to lead to the exclusion of protection [58]. Because the programmer, when creating software, is concerned with technical functioning and not with design, the programmer would not feel that any work could thus stand as having an individual or personal character.

A counter-argument to copyright protection could also be heard regarding the intention of a work. A computer program had as its purpose the functioning of a machine whereas traditionally works protected by copyright served to communicate with humans, a work served if it were

---

[41]Borking, Informatie 1983, 28–31.

to be discernible by humans rather than by a machine. Only those parts of a program which were intended as communication with people, such as the design of a video game or program documentation, should be considered, according to Vandenberghe, as falling under copyright protection [70]. The aspect of discernibility had become relevant for the first time when new information carriers, such as gramophone records and pianola rolls were introduced. In court decisions and in the copyright tenets used in the United States, protection was disallowed because these information carriers did not permit immediate discernibility. Grosheide, who actually introduced the criterion of discernibility in The Netherlands, felt the object code had lost its characteristic of sensory discernibility through the irreversible conversion of the source code to object code. He further posited that an essential difference existed between gramophone records and an object code. The playing of a gramophone record gave the particular work inherent discernibility, while the object code consisted of instructions [28]. Verkade was convinced, on the other hand, that the software in the memory, though neither primarily intended nor suitable for people to read or discern, was also protected by copyright [74]. Verkade found wide support in various court decisions.

Besides the dogmatic objections to copyright protection of software, pragmatic objections were raised as well [62]. Vandenberghe was also of the opinion, based on considerations of competition law, that software ought not to be protected under traditional copyright law. He felt that because a computer program did not aim to be directly communicated to people such as books or paintings do, that a program was part of the technology [70]. For this reason software did not merit copyright protection. In the area of technology at least a fundamental right of freedom of imitation prevailed; whereas, intellectual property ought to remain an exception. Inspired by his dislike of copyright protection of software Vanderberghe fashioned his own regime for computer programs. The particular protection which he proposed, however, was still predominantly grounded on copyright law [70]. Only the duration of protection differed radically in his proposal. He argued for a duration of

protection of fifteen years after the creation of the program or ten years after its first use. Berkvens was another advocate for *sui generis* protection [7].

Van Hattum proposed allowing the understanding of 'technical effect' (from patent law) to play a role in the protection of computer programs [33]. By making an exception for the technical element for the protection of a copyrightable work, technical developments would not become unnecessarily inhibited.

Until the introduction of a *sui generis* regime for software nothing exceptional happened in The Netherlands in the 1980s. Nor were computer programs, in source or object code, added to the catalogues of works in article 10 Aw. The courts in The Netherlands in the meantime had surprisingly little difficulty with the copyright protection of software. Computer programs came to be considered as falling under copyright protection provided that they were original. In some event protection was denied because simple computer programs were not deemed to be sufficiently original.[42]

5.5.2.1 *Reverse engineering.*

Great ambiguity existed about the content of the copyright exploitations rights of computer programs [73; 77]. Should the loading of a program into a computer's memory be considered as duplication under copyright protection? Copyright law did not traditionally oppose the use of a work [58]. The admissibility of reverse engineering was – specifically as a result of the European Directive being established – a topic for discussion. The tenet argued that any copyright infringement could only be established if multiplication was carried out with the intention of making a profit, which meant that reverse engineering did not constitute infringement. Others felt, however, that intentions had no bearing on the factual act [75]. Adherents of this non-profit thinking concluded that reverse engineering which was carried out without the intention of making profit did not constitute infringement, whereas intention toward a commercial pursuit did [28]. According to opponents to this way of thinking, commercial intention had no bearing on the factual act and was thus a

---

[42]Rb Arnhem 21 February 1985, Computerrecht 1985/5.

principle always subject to the approval of the author [24; 65].

With the implementation of the Software Directive it was more explicitly defined under which circumstances reverse engineering was admissible and which duplication rights for individual use was allowable.[43] A new Chapter VI (article 45h–45n) of the Dutch Copyright Act introduced a 'fourth regime' for computer programs [76]. In this chapter the technology-specific provisions of the Directive were adopted in reference to exclusive rights, the limiting of these exclusive rights, and for decompilation [20]. Furthermore, article 10.1 Aw was expanded by adding a twelfth category: 'computer programs' and protection for pieces of writing was excluded [11].

5.5.3 Long-term effects

For some time in The Netherlands, computer programs have belonged to the domain of copyright law [38]. In Dutch case law the copyright protection of computer programs has been generally accepted without difficulty, so the inclusion of software in the catalogue of works of article 10 Aw in 1994 was thus merely a confirmation of a long-term practise of protection. With the implementation of the Software Directive the development of Dutch copyright fit neatly into place.

In addition the Dutch system for the implementation of the Software Directive has expanded its scope by adding a large number of technology-dependent provisions. These provisions were intended to clarify the area of protection but it also raised many new questions. The exceptions to exclusive rights remains to be further elaborated on in case law. For instance, one of the ambiguous areas is article 45j Aw, which states that the reproduction that is required for use of a computer program is allowed. Is the correction of errors use of a program allowed? Owing to the fact that technology requires physical duplication, case law has not been able to clearly express what is most reasonable: when correcting errors the owner of a program should be independent of the rightful owner.

5.6 DEVELOPMENTS IN THE INTERNATIONAL ARENA

A large number of international legislative initiatives employed in an attempt to affect harmonisation between national systems of protection of software. The first to be discussed here is the Berne Convention. Discussion of the European Software Directive then follows, with the Trips Treaty last.

5.6.1 The Berne Convention

The World Intellectual Property Organization (WIPO) was actively involved in 1971 in the subject of the protection of computer software. During a meeting of the countries united under the Berne Copyright Union (*Union* countries) WIPO posited that effective, legislative protection of software was lacking. A specially adapted form of protection was, according to WIPO, the optimal solution to be hoped for.[44] At the instigation of the intellectual property organisation, AIPPI research was conducted to find out whether computer programs might be protected via a patent-like approach of depository and registration. This idea was very quickly abandoned because software in the 1970s in many European Union countries was not deemed to be patentable. Specifically the rejection of patenting of 'computer programs as such' in the European Patent Treaty in 1973 cut off the path for this approach. In 1978, the International Agency of WIPO developed "Model Provisions for the Protection of Computer Programs".[45] This legislative model provided for protection against unauthorised publication, copying, sales, and use. In addition the legislative model included a condition on reverse engineering, which was allowed in as far as it served in the independent production of a program. The term of protection extended for 20 years after the first use or the sale of the licensing of the software and a maximum of 25 years after its creation. WIPO proposed that the EU countries set up an (optional) depository system for software. After a few rounds of discussions, the proposal failed. In the years following

---

[43]Law of 7 July 1994, 521.

[44]Copyright 1971, 34.
[45]WIPO/IP/78/3.

the legislative model proposal, software seemed to be protected completely under copyright law in most of EU countries, and WIPO felt compelled to withdraw its proposal [60]. It remained committed to finding appropriate instruments for protecting computer software and developed a "Draft Treaty for the Protection of Software". This draft treaty quickly faded from interest just as the legislative model had in 1978. EU countries did not wish for any sui generis regulation for the protection of computer programs; nonetheless, the need existed for certainty concerning the question of software falling within the context of the existing Berne Convention. Ricketson questioned the applicability of the Berne Convention on all forms of software. He was of the opinion that the source code of a program could well be considered as a copyrightable work, but felt that the object code fell outside the context of the article since it could not be readily discernible [60].

Other authors were certainly convinced of the applicability of the Berne Convention to computer programs [42; 43]. They further referred to the wording of article 2.1 BC. The description of the object of copyright protection was unusually broad and included "every production in the literary, scientific and artistic domain, whatever may be the mode or form of its expressions" since 1886. Because the catalogue of works was not exhaustive, software could be protected under the Berne Convention according to these authors [42].

Lack of clarity about the applicability of the Berne Convention motivated the International Agency of WIPO, in 1991, to develop a 'Possible Protocol to the Berne Convention'. With this instrument WIPO intended, among other things, to clarify and supplement the Berne Convention, in respect to a number of new objects under which software as well as those works generated by computer [12].[46] The protocol explicitly stated that computer programs in source, and object code, were protected under the Berne Convention. In response to the discussions surrounding the supplementing of the restrictions to the reproduction right (Article 9.2 BC), WIPO included a condition in the

protocol, which allowed the rightful owner to duplicate the program in so far as was necessary for normal use or for making back-up copies. The protocol also included an unusually detailed regulation for decompilation research which was strongly reminiscent of the European Software Directive.

In expectation of the TRIPS Treaty, in which the WTO would include an international copyright regulation, agreement on the protocol was frozen. When in 1994 the WIPO Meeting of Experts convened, both the United States and the European Commission did propose merely to confirm that the EU countries were required to protect applications as system software in source and object code as literary works as meant in the Convention. Article 9.2 BC would thus only need to be confirmed without the necessity of formulating technology-specific provisions in reference to back-up copies and compilation. It was formulated as follows:

> Computer programs are protected as literary works within the meaning of Article 2 of the Berne Convention. Such protection applies to computer programs, whatever may be the mode or form of their expression.[47]

5.6.2 The European Software Directive

The appearance of the protection of computer programs in the European member states varied greatly in the 1980s. Within the twelve member states of the European Community there was hardly any explicitly included computer software in its copyright legislation. The legislation of Germany (1985), England (1985), France (1985) and Spain (1987) provided for the copyright protection of software. The other member states did not. In law journals of these countries copyright protection of software was certainly recognised. In some countries legislation was in preparation. The European Commission, motivated by potential disturbance in the internal market and wanting to improve European competitiveness in the world market, particularly vis-à-vis the United States and Japan, brought out its Green paper *Copyright and the Challenge of Technology* in 1988. The EC once again urged the harmonisation of legislation

---

[46]BCP/CE/1/2, 38.

[47]CRNR/DC/94.

regarding software protection.[48] The majority of member states and special interest groups had, in the meantime, spoken out in favour of copyright protection. The Commission motivated the choice for form of protection as follows:

> ... copyright has already in the past proved its capacity to adapt to new technologies, such as film and broadcasts. Copyright protection does not grant monopolies hindering independent development. Copyright protects only the expression but not the underlying idea of a work. It does not therefore block technical progress or deprive persons who independently developed a computer program from enjoying the benefits of their labour and investment.[49]

The text of the Directive proposal was strongly criticised in the law journals and in the industry [13; 14; 50; 66]. The contents of the protection also provoked criticism. In respect to the exceptions to exclusive rights two situations were discussed in the proposal: those in which a program user acquires rights based on a written agreement and those in which this was not the case. If there was a written agreement then the user would have no other rights than those granted in the agreement. If no written agreement existed the rightful recipient may make reproductions only in so far as the use of the computer program requires. Furthermore, the making of back-up copies without the approval of the rightful owner would constitute infringement. Decompilation was allowed to the extent that the associated software was then made operative which worked in connection with the decompiled program.

The Directive proposal further introduced a new item: a regulation concerning the 'secondary infringement' of copyright. The person who in one way or another 'introduces, possesses, or sells' an unlawfully made copy of a program, knowingly or having a reasonable knowledge of its unlawfulness, would be infringing on the copyright of the rightful owner. The manufacturing, possession, or selling of articles intended to evade or remove the technical protection of a program would also constitute infringement.

The first reading of the Directive proposal by the European Parliament resulted in a large number of amendments [23]. After most of the amendments had been accepted by the Commission the Board of Ministers established the Directives on 14 May 1991 [17].[50]

A computer program, according to the Directive, would be considered protected if it is original in the sense that the result is the intellectual creation of the author. There is a departure here from the Anglo-Saxon understanding of originality such as was initially covered in the Directive proposal. With the recognition in the preamble to the Directive that no qualitative or aesthetic criteria may be applied, the German demand for *Überdurchschnittlichkeit* was defeated.

5.6.2.1 *Reverse engineering.*

In the Directive, the permanent or the temporary reproduction of a computer program were rights to be retained by the rightful owner, whether in translation or in processing or any other changes to the program. These exclusive rights were viewed by some as complex restrictions. The provision on the debated subject of reverse engineering was unusually complex because of the difficult compromise between various interests. A debate quickly arose in scientific circles specifically on this subject [25; 32; 54; 78], and bitter opposition flared up among industrial lobby groups. Lobby groups exerted strong pressure on the European Commission and the European Parliament. The Software Action Group for Europe (SAGE) and the European Committee for Interoperable Systems (ECIS) were diametrically opposed to each other. SAGE represented the largest software and hardware producers who wanted to reduce, as much as possible, any restrictions on their rights. ECIS was in favour of the smaller software industry which was dependent on interface information from the products of market leaders. ECIS contended that strong protection would be at the expense of a competitive market. The user organisation (CEU) banded together over this standpoint [15].

To strengthen the protection of software, member states of the Directive took further dedicated

---

[48]7 June 1988 (COM) 88 172 final.
[49]OJ C 91/7.

[50]EG 1991, L 122.

measures against 'secondary infringement'. The sale or possession of an illegal copy was not considered as infringement on the exclusive right of the author, as in the initial proposal, but rather as leading to sanctions within the framework of the preservation of copyright.

Member states of the European Union were required to implement the Directive before 1 January 1993. Most of the countries met their obligations in 1994. The most detailed provision of the Directive (article 6) did not create the much hoped for clarity. Decompilation research was allowed under article 6 in the situations in which research had (until then) only relative value, namely for achieving interoperability [11]. There was still ambiguity concerning the admissibility of correcting errors.

5.6.3 WTO: TRIPs

The United States and the European Community proposed creating a treaty, within the framework of the General Agreement on Tariffs and Trade (GATT) to strengthen bonds concerning the international pirating of intellectual property. The increasing piracy of computer software was also a cause of concern. Protection under the Berne Convention – to which the United States had not yet ascribed – was deemed as inadequate [47]. Not only was there ambiguity about whether computer software fell within the venue of the Berne Convention, but also whether the assertion of sufficient rights was guaranteed [69]. WIPO was rendered powerless by the disagreement of protection between member states [44].

Negotiations on the treaty began in 1986 [53]. In order effectively to proceed against piracy, there was the option of offering attractive trade tariffs to developing countries for their manufactured and agricultural products in exchange for the introduction and enforcement of the protection of intellectual property. This 'carrot and the stick' method was viewed, especially by Western industry, as an efficient solution to the international problem of piracy. Developing countries were originally opposed to including software in the treaty, but were finally won over. Countries which signed the treaty were obliged to protect computer programs

in source- and objectcode as 'literary works' under the Berne Convention. No technology-specific rules were included but possible restrictions to the right of exploitation by the rightful owner were set aside in general terms.[51]

The whole collection of intellectual property measures was named Trade-Related Aspects of Intellectual Property Rights (TRIPS) and became a part of the Agreement of Marrakech as established by the World Trade Organization in 1993.

5.7 SUMMARY AND CONCLUSIONS

In this contribution we saw that shortly after the first computer program appeared on the market the necessity of protection was frequently questioned. In the 1960s and 1970s software was mostly highly specific to and accompanied hardware. Contracts were able to provide producers with the provisions for the desired guarantees for protection. Intellectual property protection against infringement by third parties was not considered to be necessary. In addition, there was great trust in professional ethics in the beginning. Furthermore, many considered copyright protection of computer programs as undesirable since the traditional rightful owners, such as authors, artists, and inventors could be injured by this approach. The introduction of a new copyrightable object could have run counter to the rights of the existing rightful owner thus creating smaller portions of the pie for everyone. The requirements of originality as well as discernibility would additionally have stood in the way, in principle, of copyright protection. The lack of a technical or material character was raised in opposition to the protection of software under patent law.

With the large-scale expansion of the software industry in the 1980s, the concomitant request was raised by producers for unambiguous protection. The greatest market share was no longer held by made-to-order software but rather by standard software. After the need for protection had been generally recognised, the discussion shifted towards the

---

[51] Art. 13: "Members shall confine limitations or exceptions to exclusive rights to certain special cases which do not conflict with a normal exploitation of the work and do not unreasonably prejudice the legitimate interests of the rightholder".

appropriateness of various regimes. Two possibilities were at hand: the regulation of communication of cultural goods (copyright) on the one hand, and the protection of technological-industrial products (patent law) on the other. The two part nature of software lay at the very core of the debate. A computer program consists of an algorithm and of a set of instructions, employed by the computer to carry out its task [28].

At first the discussion was primarily focused on patent right protection of software. At first glance software seemed closer to the domain of patent right (technological) than to copyright. The essential value of a computer program rests less in its expressive manner than its problem-solving structure, that is the algorithm. The content of patent right protection is attractive because trading and production can be prohibited and because commercial use remains exclusively the property of the rightful owner. In reality, the patenting of software has remained a highly hazardous undertaking and it has resulted in great ambiguity of policy all over the world.

Against this complex background, positions shifted from a patent right to a copyright approach. WIPO formulated, in 1978, a copyright-inspired *sui generis* regime for the protection of software. An individual protection regime was chosen because protection under traditional copyright was considered inadequate. A user-right and a shorter time period of protection was proposed. The minimum term of fifty years after the death of the author that was stated in the Berne Convention was considered to be too long. In addition WIPO proposed a depository system but withdrew this proposal as it apparent from case law that software was protected under copyright law. A decision of the German Supreme Court formed the most conspicuous exception. In the *Inkassoprogramm* case, the German Supreme Court, in 1985, posed an above average originality requirement (*Überdurchschnittlichkeit*). In The Netherlands, computer software was frequently granted protection also without it being explicitly stated in the law. In the United States software was protected under copyright law as long as sufficient freedom of choice in the programming could be determined, as opposed to software that was purely dictated by technological requirements.

In 1980 copyright protection of computer software in the United States was codified in the Copyright Act. With this act, every doubt as to the applicability of the White-Smith doctrine was removed. In Europe, Germany was the first to change to copyright legislation. In the act of 24 June 1985, programs for data processing were lined up under *Sprachwerke, wie Schriftwerke und Reden*. France (3 July 1985) and England (16 July 1985) were soon to follow in the same year as were other countries somewhat later, among them Sweden, Denmark and Spain. Japan also opted for copyright protection.

In time software was indeed allowed under the domain of copyright but yet a growing dispute over the content of this protection arose. Technology-specific provisions were adopted to supplement the existing exploitation rights and restrictions. The most well-known example is the European Software Directive launched in 1991. One part of the recommendations for technology-specific legislation was finally not enacted in law. The Office of Technology Assessment in the United States proposed introducing copyright and patent protection regimes in parallel and a third protection regime for 'functional works'. The above-mentioned proposals of the international agency WIPO also amounted to technology-specific protection of computer software. The German government suggested a new neighbouring rights law for software. Japan considered a *sui generis* regime for computer programs [41]. Additionally various suggestions from the (international) law journals were made for technology-specific protection.

An interweaving of copyright and patent protection can certainly be detected in the protection regime of the Software Directive. Thus, on the one hand, was the underlying idea of a computer program explicitly excluded from copyright protection (article 5) but on the other admitted to protection by the formulation of exclusive rights, specifically by the prohibition of decompilation (article 6). The mere use of software in the Directive was, in addition, for the first time in the copyright era considered as an exclusive right (article 4).

In the beginning of US case law, there was a discernible tendency that patent and copyright infringements would be interwoven in this complex issue, which was why with the introduction of technology in the intervening years an unusually broad protection was granted. In the Whelan decision of 1986 the elements of a program not able to be protected were limited to the purpose and the function of the program. The other elements, including the algorithms, were protected under copyright. This was how copyright protection took on patent-like characteristics without the requirement of disclosure being posed and with a longer period of protection. In a 1986 decision, the French Court of Cassation, seemed to wish to replace the criterion of originality of software by a criterion of newness taken from patent right.[52] The protection of invested knowledge and science was at the forefront of this decision [41].

Inversely some advance of patent protection into copyright protection was also perceptible. During the 1980s the requirements of the technical character and the material form were relaxed to the benefit of the protection of software-related inventions. The U.S. Supreme Court, in *Diamond v. Diehr*, the German Supreme Court, in *Antiblockiersystem* and the German Patent Agency in their Guideline of 1986, The Netherlands Board of Patents Appeals Department in *Tomograafbeschikking*, and the EOP in *Vicom Systems* all appeared prepared to provide for the rising need of protection.

During the 1990s in the United States, the readiness to patent software was itself so great that agencies granting patents tested the requirements of newness and inventiveness inadequately. Previously recognised or trivial software were patented. The PTO proposed guidelines in 1996 in which this over-reactive tendency was reined in. In addition, in these guidelines the PTO adopted a more medium- and technology-independent approach towards a software patent than previously. In general the need for technology-focused legislation appears to be diminishing both in patent and in copyright law. Since computer programs have become part of our day to day economic and technological reality, the need for technology-focused legislation has diminished and computer programs have truly been emancipated to full blown works of authorship.

---

[52]Court de Cassation 7 March 1986, Expertises 1986.

REFERENCES

[1] R.B. Abel, *Zweites Urheberrechtsänderungsgesetz: neuer Rechtsschutz für Computerprogramme*, Recht der Datenverarbeitung (1993), 105–109.

[2] J.F. Banzaff III, *Copyright Protection for Computer Programs*, Copyright Law Symposium ASCAP' (1966), 119–179.

[3] C.J. Bashe et al. (eds), *IBM's Early Computers*, Cambridge, MA, London (1986).

[4] K.A. Bauer, *Reverse engineering und Urheberrecht*, Computer und Recht (1990), 89–94.

[5] D. Bender, *Lotus v. Borland appeal onscreen program menus not copyright-protected*, Computer Law & Practice (1995), 71–73.

[6] D. Bender, A.R. Barkume, *Patents for software-related inventions*, Software Law Journal (1992), 279–298.

[7] J.M.A. Berkvens, Congestie op de datahighways (oratie), Deventer (1991).

[8] W.B. Bierce, M.C. Harold, *New US patent guidelines offer hope to software developers in era of diminishing copyright protection*, Computer Law & Practice (1995), 94–96.

[9] S. Breyer, *The uneasy case for copyright: a study on copyright in books, photocopies and computer programs*, Harv. L. Rev. 84 (1970), 281–351.

[10] D.S. Chisum et al., *Last frontier conference report on copyright protection of computer software*, Jurimetrics J. (1989), 15–33.

[11] M. de Cock Buning, *Auteursrecht en "reverse engineering", techniek en theorie*, IER 5 (1993), 129–137.

[12] M. de Cock Buning, *Recente auteursrecht verdragen met onderhandelingsruimte*, Informatie Professional 2 (1997), 20–22.

[13] M. Colombe, C. Meyer, *Seeking interoperability: An industry response*, EIPR (1990), 79.

[14] M. Colombe, C. Meyer, *Interoperability still threatened by EC Software Directive: A status report*, EIPR (1990), 325–329.

[15] D.J. Conners, A. Westphal, *The European Community directive on the Legal Protection of computer programs: a comparison between European and U.S. Copyright Law*, Communication and the Law (1992), 25–55.

[16] CONTU, *Final Report of the National Commission on New Technological Uses of Copyrighted Materials*, Washington, DC (1979).

[17] B. Czarnota, R. Hart, *Legal Protection of Computer Programs in Europe, a Guide to the EC Directive*, London/Dublin/Edinburgh/München (1991).

[18] D. Dietz, *Das Problem des Rechtschutzes von Computerprogrammen in Deutschland und Frankreich. Die kategoriale Herausforderung des Uhreberrechts*, BIE (1983), 305–311.

[19] E.J. Dommering, *De software richtlijn uit Brussel and de Nederlandse Auteurswet*, Informatierecht/AMI (1992), 82–89.

[20] E.J. Dommering, *Noot bij Pres. Rb Breda 13 februari 1991*, Informatierecht/AMI (1992), 174–176.

[21] Th. Dreier, Rapport National Republique Fédérale d'Allemagne, General Report Software Protection ALAI 1989, Montreal (1990), 238–243.

[22] Th. Dreier, *Verletzung urheberrechtlich geschützter Software nach der Umsetzung der EG-Richtlinie*, GRUR (1993), 781–793.

[23] E.A. Dumbill, *EC Directive on computer software protection*, Computer Law & Practice (1991), 210–216.

[24] S. Gerbrandy, **Kort commentaar op de Auteurswet 1912**, Gouda Quint, Arnhem (1988).

[25] K. Gilbert-Macmillan, *Intellectual Property Law for reverse engineering computer programs in the European Community*, Computer & High Technology L.J. (1993), 247–264.

[26] J.C. Ginsburg, *No "Sweat"? Copyright and other protection of Works of information after Feist v. Rural telephone*, Columbia Law Review (1992), 338–388.

[27] P. Goldstein, Comments on *A manifesto concerning the legal protection of computer programs*, Columbia Law Review (1994), 2573–2578.

[28] F.W. Grosheide, Auteursrecht op maat (diss.), Kluwer, Deventer (1986).

[29] H. Haberstumpf, *Grundsätzliches zum Urheberrechtsschutz von Computerprogrammen nach der Urteil des Bundesgerichtshofes vom 9-5-85*, GRUR (1986), 222.

[30] H. Haberstumpf, *Die Zulässigkeit des Reverse Engineering*, Computer und Recht (1991), 127–141.

[31] H.W.A.M. Hanneman, The patentability of comptersoftware (diss.), Kluwer, Deventer/London (1985).

[32] R.J. Hart, *Interface, interoperability and maintenance*, EIPR (1991), 111–116.

[33] W.J. van Hattum, *Techniek is meer dan stof alleen*, BIE (1988), 143–148.

[34] Th. Heymann, *Softwareschutz nach dem EG-Richtlinienentwurf*, Computer und Recht (1990), 10.

[35] Th. Hoeren, *Look and Feel im deutschen Recht*, Computer und Recht (1990), 22.

[36] P.B. Hugenholtz, *Softwarebescherming, een tussenstand voor de thuisblijvers*, Auteursrecht/AMR (1984), 90.

[37] P.B. Hugenholtz, *Compterprogramma's een nieuwe categorie werk?*, Auteursrecht/AMR (1985), 32.

[38] P.B. Hugenholtz, Auteursrecht op informatie (diss.), Kluwer, Deventer (1989).

[39] P.B. Hugenholtz, J.H. Spoor, **Auteursrecht op Software**, Kluwer, Amsterdam (1987).

[40] J.L.R.A. Huydecoper, *Originaliteit of inventiviteit? Het technisch effect in het auteursrecht*, BIE (1987), 106–112.

[41] D.S. Karjala, *Recent United States and international developments in software protection*, EIPR (part I) (1994), 13–20; EIPR (part II) (1994), 58–66.

[42] M. Keplinger, *Authorship in the information age. Protection for computer programs under the Berne and Universal Copyright Conventions*, Copyright (1985), 119–128.

[43] M. Kindermann, *Computer software and copyright conventions*, EIPR (1981), 6–12.

[44] M. Kindermann, *Vertrieb und Nutzung von Computersoftware aus urheberrechtlicher Sicht*, GRUR (1983), 150.

[45] M. Kindermann, *Urheberrechtsschutz von Computerprogrammen*, Computer und Recht (1989), 880.

[46] M. Kindermann, *Reverse engineering von Computerprogrammen*, Computer und Recht (1990), 638.

[47] G. Kolle, *Computer software protection present situation and future prospects*, Copyright (1977), 70–79.

[48] G. Kolle, E. Ulmer, *Der Urheberrechtsschutz von computerprogrammen*, GRUR Int. (1982), 489.

[49] R. Kortsik, Der Urheberrechtliche Schutz von Computersoftware, Zugleich ein Beitrag zum urheberrechtlichen Inhaltsschutz wissenschaftlicher Werke (diss.), Mainz (1993).

[50] M. Lehmann, Th. Dreier, *The legal protection of computer programs: certain aspects of the proposal for an (EC) Council Directive*, Computer Law & Practice (1990), 92–97.

[51] M. Lehmann, *Das neue deutsche Softwarerecht, Novellierungsvorslag zum Urheberrecht*, Computer und Recht (1992), 324–328.

[52] M. Lehmann, *Vermieten und Verleihen von Computerprogrammen, Internationales, europäisches und deutschen Urheberrecht*, Computer und Recht (1994), 271–274.

[53] M. Lehmann, *TRIPS, the Berne Convention, and legal hybrids*, Columbia Law Review (1994), 2621–2629.

[54] C.R. McManis, *Intellectual property protection and reverse engineering of computer programs in the United States and the European Community*, High Technology L.J. (1993), 25–99.

[55] N. Metropolis, J. Howett, G.C. Rota (eds), **A History of Computing in the Twentieth Century**, New York/London/Toronto/Sydney/San Francisco (1980).

[56] C.J. Millard, **Legal Protection of Computerprograms and Data**, London/Toronto (1985).

[57] H. Öhlschelgel, *Sollen und können Rechenprogramme geschützt werden*, GRUR (1965), 465–468.

[58] A.A. Quaedvlieg, Auteursrecht op techniek. De auteursrechtelijke bescherming van het technische aspect van industriële vormgeving and computerprogrammatuur (diss.), Zwolle (1987).

[59] A. Raubenheimer, *Implementation of the EC Software Directive in Germany – Special provisions for protection of computer programs*, IIC/5 (1996), 609–648.

[60] S. Ricketson, **The Berne Convention for the Protection of Literary and Artistic Works: 1886–1986**, London (1987).

[61] P. Samuelson, R. Davis, M.D. Kapor, J.R. Reichman, *A Manifesto concerning the legal protection of computer programs*, Columbia Law Review (1994), 2308–2431.

[62] P.C. van Schelven, **Bescherming van software and chips in juridisch perspectief**, Boom, Lelystad (1986).

[63] M. Schultze, *Urheberrechtschutz von Computerprogrammen – geklärte Rechtsfrage oder blosse Illusion*, GRUR (1985), 997.

[64] J.H. Spoor, *Piraterij van computerprogramma's*, Auteursrecht/AMR (1985), 31.

[65] J.H. Spoor, D.W.F. Verkade, Auteursrecht, Deventer (1993).

[66] I.A. Staines, *The European Commission's proposal for a Council Directive on the legal protection of computer programs*, EIPR (1989), 183–184.

[67] B. Steckler, *Legal protection of computer programs under German Law*, EIPR (1994), 293–300.

[68] R.A. Tenbergen, *Juridische bescherming van computerprogramma's*, Informatie 5 (1973), 220–223.

[69] E. Ulmer, *Copyright problems arising from the computer storage and retrieval of protected works*, Copyright (1972), 37–59; RIDA (1972), 34–150.

[70] G.P.V. Vandenberghe, Bescherming van computersoftware (diss.), Antwerpen (1984).

[71] D.W.F. Verkade, *De software-beslissing van de oktrooiraad*, Informatie (1971), 170.

[72] D.W.F. Verkade, **Softwarebescherming**, Alphen aan den Rijn, Brussel (1972).

[73] D.W.F. Verkade, **Juridische bescherming van programmatuur**, Alphen aan den Rijn, Brussel (1986).

[74] D.W.F. Verkade, **Ongeoorloofde mededinging**, Zwolle (1986).

[75] D.W.F. Verkade, *Computerprogramma's en geschriften – een onverstandige Nota van Wijziging*, BIE (1988), 46–52.

[76] D.W.F. Verkade, *Computerprogramma's in de Auteurswet 1912: het vierde regime...*, Computerrecht (1992), 86–97.

[77] D.W.F. Verkade, *Intellectuele eigendom*, **Recht and Computer**, H. Franken, H.W.K. Kaspersen and A.H. de Wild, eds, Kluwer, Deventer (1997), pp. 178–229.

[78] Th.C. Vinje, *Die EG-Richtlinie zum Schutz von Computerprogrammen und die Frage der Interoperabilität*, GRUR Int. (1992), 250–260.

[79] M.R. Wessel, *Legal protection of computer programs*, Harvard Business Review 43 (1965), 97–106.

[80] R.W. Wild, *Comment: Computer program protection: the need to legislate a solution*, Cornell L. Rev. 54 (1969), 586–609.

[81] M.R. Williams, **A History of Computing Technology**, Englewood Cliffs (1985).

[82] L.J. Zadra-Symes, *Computer Associates v. Altai, The retreat from Jaslow*, EIPR (1992), 327–330.

6

A HISTORY OF SOFTWARE PATENTS

Robert Plotkin

Boston University School of Law
91 Main Street, Suite 204, Concord, MA 01742-2527, USA
E-mail: rplotkin@rplotkin.com

Contents

Abstract

A history of software patents is presented by reference to particular controversial software patents and to actual and proposed legal reforms that have attempted to address perceived problems with software patents. This history is divided into four stages: (1) pre-software; (2) early software patents; (3) recent software patents; and (4) the present. These stages are defined by legal events, such as the resolution of lawsuits over particular software patents, and most directly expose the development of legal standards that apply to software patents. Such legal standards have not, of course, developed in a vacuum. The legal history of software patents has influenced, and been influenced by, the history of computer technology and the history of economic forces driving the development of such technology and competition over legal rights in such technology. The interaction of these three histories – legal, technological and economic – is explored in an attempt to explain the difficulties faced at present and the challenges that lie ahead.

Keywords: patent, intellectual property, software, computer program.

6.1 INTRODUCTION

Software patents have been controversial for nearly as long as software has existed. Although some of the controversy has focused on whether particular software patents, such as Amazon.com's 'one-click shopping' patent [49], should have been allowed, the heart of the debate is over the more fundamental question whether patent protection should extend to software at all. Proponents of software patents contend that software is merely another technological development, like the cotton gin and internal combustion engine before it, to which patent law's clear mandate is intended to extend. Software patent critics argue that software, unlike machinery in the Industrial Age, is too intangible to be worthy of patent protection and that software patents will bring innovation in the software industry grinding to a halt.

The longevity of the debate over software patents is itself curious and possibly unique. Patent law by its very nature is intended to apply not only to individual new inventions but also to new fields of technology as they arise. Although the previous introduction of new technological fields, such as electrical circuitry, posed some theoretical and administrative difficulties for patent law, the law overcame them relatively easily and without a public outcry over whether the law should apply to inventions in such fields per se.

In contrast, the essential question whether software should be 'patentable subject matter',[1] often carried on within the confines of the legal profession, but just as often including members of the general public typically unconcerned with intellectual property policy, shows no sign of abating either in theory or in practice. The recent multi-year row over the so-called *European Software Patent Directive*[2] is perhaps the best example of the contention that there is no more consensus today about whether software should be patentable subject matter than when the commission convened by US President Johnson [23] in 1965 concluded that software should not be patentable for the eminently pragmatic, and hence theoretically unsatisfying, reason that the US Patent and Trademark Office (USPTO) lacked the resources to search effectively for existing patents in the software field.

One may conclude from the current state of software patents that we have made no progress over the last half-century. Software patent opponents would make an even stronger indictment: that software patents will continue to be granted worldwide despite the failure of the *European Software Patent Directive* and despite the lack of a compelling or even plausible justification for such patents, simply because multinational corporations and others with the resources to lobby for preferential legislation and court decisions have the power to influence the law in their favor.

The history of software patents reveals, however, that although the path of the law has been influenced by private entities seeking to expand the scope of legal rights that are available for them to use to their private advantage, naked economic power has not been the only or perhaps even the

most significant force steering patent law as it applies to software. Rather, unique and evolving features of computer technology have created tensions among competing social policies and legal principles, and at least some of the confusion and disagreement within legislatures and courts over software patents may reasonably be attributed to good-faith, if often misinformed, attempts to resolve such tensions.

These themes will be elaborated upon by presenting a history of software patents by reference to particular controversial software patents and to actual and proposed legal reforms that have attempted to address perceived problems with software patents. This history is divided into four stages: (1) pre-software; (2) early software patents; (3) recent software patents; and (4) the present. These stages are defined by legal events, such as the resolution of lawsuits over particular software patents, and most directly expose the development of legal standards that apply to software patents. Such legal standards have not, of course, developed in a vacuum. The legal history of software patents has influenced, and been influenced by, the history of computer technology and the history of economic forces driving the development of such technology and competition over legal rights in such technology. The interaction of these three histories – legal, technological and economic – is explored in an attempt to explain the difficulties faced at present and the challenges that lie ahead.

The history presented herein is intended not to be comprehensive, but rather to provide examples that illustrate the most significant historical trends in patent law as it has applied to software. Although the discussion focuses primarily on the US and secondarily on Europe, this represents not a value judgment but an attempt to confine the scope of this paper in the interest of concision.

6.2 PATENT LAW PRIMER

Although most of the discussion herein applies to patent law generally, independent of any particular legal system, it is worthwhile to provide a general overview of the patent systems in the US and

---

[1]Although the term 'statutory subject matter' typically is used in the US, I use the term 'patentable subject matter' throughout to eliminate any dependence on the US patent system in particular.

[2]The actual title of the proposed directive was the *Directive on the Patentability of Computer-Implemented Inventions.*

Europe before describing particular legal cases. Article I, Section 8, Clause 8 of the US Constitution, referred to herein as the '*Patent Clause*', provides the Congress with authority to implement patent policy through appropriate enabling legislation. More specifically, the Patent Clause states that "The Congress shall have power . . . to promote the progress of . . . useful arts, by securing for limited times to . . . inventors the exclusive right to their . . . discoveries".

For purposes of the present discussion, the essential feature of the Patent Clause is that it grants the Congress power to provide inventors with exclusive rights to their inventions ('discoveries') for a limited time, in order to promote the progress of the useful arts. One justificatory theory for the *Patent Clause* is that providing such limited rights will provide inventors with an incentive to invent and to disclose their inventions to the public, while requiring such exclusive rights to expire will enable the public to make free use of such inventions after such expiration, thereby striking a balance between private and public benefit that will maximize technological innovation – the "progress of the useful arts".

Congress first exercised its authority under the Patent Clause by enacting the first Patent Act in 1790. Although the Patent Act has undergone several major and many minor revisions since 1790, none of them is sufficiently relevant to the topic of this chapter to bear mentioning. Therefore, although the present discussion will use the current provisions of the Patent Act as a point of reference except where otherwise mentioned, such anachronistic usage will not detract from the accuracy of the conclusions drawn.

The European Patent Convention (EPC) takes a somewhat different approach, defining the subject matter of patent law as "inventions which are susceptible of industrial application" [15], rather than those that are merely "useful". As we shall see, although in many cases the US and European approaches result in the same outcome, in some of the most controversial areas relevant to software they do not.

To understand the US Patent Act it is useful to understand what a patent is and how it is ob-

tained and enforced. A patent is a document[3] that describes an invention and that confers upon its owner the legal right to exclude others from making, using and selling that invention during the term of the patent. If, for example, a patent describes a particular kind of hammer, the patent owner has the legal right to prohibit anyone[4] from making, using or selling the same hammer. In general, the patent owner may grant or withhold this right arbitrarily, just as he may grant or withhold the right to enter his house arbitrarily. This power to exclude others – the 'exclusive right' granted by the patent – enables the patent owner to attempt to profit from the patent in a variety of ways. The patent owner may, for example, choose to manufacture the hammer himself and exclude competitors from doing so, thereby extracting a premium price. Or he may, for example, allow others to manufacture the hammer in exchange for a payment referred to as a "royalty".

Someone who makes, uses, or sells a patented invention without permission of the patent owner is said to "infringe" the patent.[5] The patent owner, upon discovering an infringement, may sue and obtain monetary damages [76] or an injunction [75] (a court order prohibiting the infringer from further making, using, or selling the patented invention).

Whether a particular hammer is "the same" as the hammer described in the patent involves interpreting the patent to identify the precise contours of the legal rights it grants. Modern patents are divided into two distinct but interrelated sections that are intended in part to facilitate interpreting the patent. The first part, referred to as the "specification", "disclosure" or "description", includes a description of how to make and use "embodiments" (particular instantiations) of the invention. The specification typically includes both text and accompanying drawings. One important purpose of

---

[3]One might quibble that a "patent" is a set of legal rights created by a document referred to as "letters patent". This distinction, whether or not valid, is irrelevant to the purposes of this paper.

[4]This discussion will ignore for purposes of simplicity the limited geographic effect of patents.

[5]The US Patent Act [72] defines additional activities that constitute infringement, but these are not important for the purposes of this paper.

the specification is to enable the public to make and use the invention when the patent expires.

The second part of a patent is referred to as the "claims". These point out the particular features of the invention that have satisfied the requirements for patentability and that the patent owner has the legal right to exclude others from implementing. Returning again to the example of a hammer, assume that the hammer described in the specification of a particular patent differs from all previous hammers due to a particular easily-graspable handle. The claim(s) of such a patent would point out the novel features (such as the particular shape) of the hammer's handle, but need not reiterate the detailed technical description that was provided in the patent's specification. The claims, therefore, are intended to serve as a concise (albeit broad and abstract) description of the "inventive" features of the invention.

To obtain a patent, an inventor must write a patent application that contains a specification and claims. To draft such a document, the inventor must therefore attempt to identify what distinguishes her invention from previous inventions. The set of all relevant previous inventions is referred to as the "prior art". A sophisticated inventor will employ a patent attorney who is well-versed in drafting patent applications using the appropriate legal language and in a way that points out how the invention differs from the prior art.

Another common strategy in drafting patent applications, and one that is particularly relevant to the history of software patents, is to draft the specification and claims to characterize the invention as broadly as possible. Consider again the example of the hammer having a handle with a particular new shape. The inventor likely desires that the patent provide him with legal protection for *any* hammer having a handle with that shape, whatever the other features of the hammer may be. For example, the inventor would likely desire the ability to sue anyone who sells a hammer having the invented handle shape, regardless of the shape or size of the hammer's head, the material out of which the handle is made, or whether the handle has a hole to facilitate hanging it on a hook. A patent that is well-drafted from the point of view of the inventor, therefore,

distills the invention down to its essence and characterizes it as broadly as possible so as to provide the inventor with maximum ability to use the patent to exclude others from employing the same inventive concept.

The public shares, to a certain extent, the inventor's interest in enabling the inventor to obtain legal protection that covers embodiments of the invention other than the one(s) he has actually constructed, and even embodiments other than those specifically described in the patent specification. If the patent is viewed as a contract between the inventor and the public, and the inventor has truly invented a new and useful hammer handle that can be constructed of any material and used in conjunction with any hammer head, then to hold up its end of the bargain the public should grant the inventor a relatively broad right in the hammer handle. Failure to do so may discourage the inventor, and other inventors, from investing the resources necessary to produce further inventions and disclose them to the public in the form of patent applications.

The public also has an interest, however, in imposing limits on how broadly an inventor may obtain protection for a particular invention. In the case of the hammer, for example, it is important that the patent not be interpreted so broadly as to encompass prior art (previously-existing) hammers and thereby enable the inventor to prohibit people from using hammer handles that were already in the public's possession. The prior art therefore provides a *backward-looking* check on the scope of patents, by prohibiting inventors from effectively removing inventions that already are in the public domain.

Patent law also incorporates important *forward-looking* checks on the scope of patents. Considering again the hammer example, it is important that the patent not be interpreted so broadly as to cover any future hammer handle that one might invent. A patent interpreted so broadly could stifle innovation by imposing potentially prohibitive costs on other inventors who seek to build on, or design around, the particular hammer handle described in the patent. As we will see in cases such as *O'Reilly v. Morse* [53] and *Gottschalk v. Benson* [27], courts have often tried to impose forward-looking checks on the scope of patents for precisely this reason, but have had understandable difficulty doing so due to the vagaries of written language and the impossibility of predicting the future.

The final check, closely interrelated to the other two, inquires into what the inventor has invented presently. Although I have described the specification and claims as two distinct parts of a patent, they are interrelated, and when the claims are unclear they may be interpreted by reference to the specification, in which the inventor describes in detail how to make and use his invention. In this way, the scope of the claims may be limited more closely to what the inventor actually has invented, thereby better striking the previously-mentioned balance between private and public benefit.

Having discussed some of the policies and principles underlying patent law, I will now describe in general terms the legal requirements for patentability (the requirements that an invention[6] must satisfy to qualify for patent protection). In general, an invention must be new [17; 68], non-obvious [18; 69],[7] and either useful (in the US [67]) or susceptible of industrial application (in Europe [19]) to merit patent protection. I will refer to these as the 'substantive' requirements for patentability because they relate to the invention itself rather than to any particular way in which the invention is described in a patent application.

The requirement that an invention be new is referred to as the "novelty" requirement and implements the requirement that an invention distinguish in some way from the prior art. The novelty requirement establishes a relatively low hurdle for patentability in the sense that any difference between the claimed invention and the prior art will satisfy the novelty requirement. The novelty requirement is important, however, because it prevents the removal of inventions from the public domain.

Certain modifications to inventions in the prior art may satisfy the novelty requirement and yet still not be entitled to patent protection. Intuitively, for example, one might agree that a patent should not be granted on a hammer handle whose only difference from previous hammer handles is that it is longer by ten percent, even though such a handle did not exist in the prior art and therefore is novel. The non-obviousness requirement is an attempt to weed out modifications such as this, which do not constitute an advance over the state of the art. The non-obviousness requirement performs this function by requiring that an invention not have been obvious to "a person having ordinary skill in the art" to which the invention pertains. As one might expect, the reliance on a hypothetical "person having ordinary skill in the art" makes this requirement, although critical, particularly difficult to apply in practice.

The final requirement for patentability is difficult to define in a way that is consistent across the US and Europe because of differences in the approaches taken by the two jurisdictions. In the US, §101 of the Patent Act does not impose an express requirement that the invention fall within the 'useful arts', but instead imposes what traditionally have been interpreted as two distinct requirements, the "statutory subject matter requirement" and the "utility" requirement, which, in combination, serve the purpose of limiting the scope of patent law to the useful arts.

More specifically, §101 states in relevant part that "Whoever invents or discovers any new and useful process, machine, manufacture, or composition of matter, or any new and useful improvement thereof, may obtain a patent therefor". This "statutory subject matter" requirement requires that the invention be either a process, machine, manufacture, or composition of matter. This is an attempt to capture the categories of subject matter that should be susceptible to patent protection. Notably absent from the list are abstract ideas and works traditionally falling in the liberal arts, such as novels, poems, and songs.

The 'utility' requirement traditionally has been interpreted relatively narrowly, to require only that

---

[6]Although a strict usage of the term "invention" would limit it to refer only to those products and processes that have satisfied all of the requirements for patentability, I will use the term "invention" here more loosely to refer to any product or process for which patent protection is sought or has been granted.

[7]The EPC uses the term "involving an inventive step" to refer to the non-obviousness standard.

the claimed invention "be capable of some beneficial use in society" [9]. An invention would satisfy the utility requirement if it were more than "a mere curiosity, a scientific process exciting wonder yet not producing physical results, or [a] frivolous or trifling article or operation not aiding in the progress nor increasing the possession of the human race" [58]. In recent years, however, the twin requirements of statutory subject matter and utility have undergone twin metamorphoses, making their status as individual requirements and their interrelationship unclear. Although some details of these changes will be described below, rather than attempt to divine the precise contours of these shifting requirements, it is sufficient to note that they are an attempt to define a standard for determining whether a particular invention falls within the "useful arts".

The EPC, in contrast, takes a more explicitly unified approach, imposing the single requirement that an invention be "susceptible of industrial application" [15] (which means that it can be made or used in any kind of industry), [19] to qualify as patentable subject matter. The waters are quickly muddied, however, by the EPC's express statement that "programs for computers" "shall not be regarded as inventions" [16]. Further complicating matters is that there is an exception to this exception: only computer programs "as such" are excluded from subject matter patentability. Attempting to provide some clarity in the context of software, the European Patent Office (EPO) issued guidelines distinguishing between patentable and non-patentable subject matter in the context of software based on whether the claimed subject matter has a "technical character" or produces a "technical effect" [22].

In the end, determining whether any particular computer program is patentable subject matter under the EPC requires determining whether an attempt is being made to patent the computer program "as such". This roundabout definition of patentable subject matter effectively creates as much difficulty determining whether software is patentable subject matter in Europe as it does in the US, as we shall see from the cases described below.

The substantive requirements of patentable subject matter, novelty and non-obviousness are complemented by a set of formal requirements specifying the form that the patent application must take and the relationship of this form to the interpretation of the patent.

Patent law includes, for example, "disclosure" requirements which dictate the manner in which the inventor is to write the specification and claims. The fundamental purposes of these requirements are to ensure that the inventor has provided sufficient evidence of the existence of a patentable invention [20; 70], to ensure that the inventor has described that invention in sufficient detail to enable the public to make and use the invention, and to put the public on notice of what exclusive legal rights the patent grants to the inventor [21; 71].

To satisfy the "enablement" requirement, for example, the inventor must describe "the invention, and ... the manner and process of making and using it, in such full, clear, concise, and exact terms as to enable any person skilled in the art to which it pertains ... to make and use the same" [70]. The intent behind this requirement is that a scientist, engineer, or other practitioner in the relevant field be able to use the patent specification as a guidebook for constructing a working embodiment of the invention.

The inventor's claims must "particularly point [...] out and distinctly claim [...] the subject matter which the applicant regards as his invention" [71]. The intent behind this requirement is to ensure that the inventor's claims are clear and therefore put the public on notice of the inventor's legal rights.

The final formal standard is not so much a requirement as a rule for interpreting patent claims that the inventor chooses to write in a particular way. Although it is difficult to make the import of this rule clear without providing specific examples of its application, its importance for software patents cannot be overlooked. As codified in the US Patent Act, the rule states that "[a]n element in a claim for a combination may be expressed as a means or step for performing a specified function without the recital of structure, material, or acts in support thereof, and such claim shall be construed to cover the corresponding structure, material, or acts described in the specification and equivalents thereof"[8] [72].

This cryptic passage is likely incomprehensible to anyone other than a patent attorney. But it is important to obtain at least a basic understanding of its meaning before considering specific software patent cases. The motivation behind the rule embodied in the quoted text is that any particular invention – such as a hammer handle – may be described either in terms of its structure or its function. A "structural description" of the handle might describe the particular shape of the handle and the physical materials out of which it is constructed. A "functional description" of the same handle, in contrast, would describe the handle in terms of the function(s) it performs, such as "a hammer handle for facilitating gripping", without pointing out any particular physical structure for performing such function(s). The distinction between a structural description and a functional description is sometimes characterized as the difference between describing what a thing *is* and what the thing *does*. It is critical to recognize that both kinds of description, however different they may appear, may describe the same object.

From even this simplified example it may be understood why inventors might prefer to write claims using functional language rather than structural language. First, functional language may simply be easier to write. It is easier to describe an automobile as "a machine for transporting people, including wheels, an engine for driving the wheels, a steering wheel for controlling the direction of the wheels, an accelerator pedal for controlling delivery of gasoline to the engine", and so on, than it is to describe in detail the physical structure of each component. Second, and more importantly from a policy perspective, functional language is, in an important sense, more abstract and therefore broader than structural language that describes the same object. Inventors prefer to describe their inventions using broader language for the reason described above, namely that such language has the

potential to provide them with stronger legal protection.

The retort should by now be familiar: that the public has a corresponding interest in limiting the ability of inventors to obtain broader patent protection merely by describing their inventions using broader, functional, language. The intent of the statutory provision quoted above is precisely that – to 'cut back' on the scope of a patent claim that the inventor has written using functional language [40]. Returning to the hammer example, if an inventor describes a hammer having a particular shape in the patent specification and then writes a claim using broad functional language, such as "a hammer handle for facilitating gripping", such a claim is required to be interpreted to be limited to the specific hammer handle described in the specification. The inventor cannot, in other words, capture broader legal protection merely by the form of language in which he chooses to draft his patent claims.

Although patent law includes a variety of additional rules (such as those for identifying whether a product infringes a patent claim), the policies, principles, and rules just described provide the background necessary to understand the history of software patents that will begin to be provided in the following section.

6.3 A PRE-HISTORY OF SOFTWARE PATENTS

Software patents do not exist in a vacuum. The basic requirements for "patentability" – that an invention be novel, non-obviousness (involve an inventive step), and fall within the "useful arts" or have an "industrial application" – are intended to apply generally to all classes of inventions. The high-level standards that determine whether a particular invention is 'novel' are intended to apply, at least in theory, to all fields of technology. (For an argument that patent law both is and should be technology-specific in certain ways, see Burk and Lemley [7].)

As a result, when a new technological field emerges, legal actors (such as lawyers, judges and patent examiners) attempt to interpret the existing law in light of the new field. New legislation is

---

[8]Note that there is no similar provision in the Articles or Implementation Regulations of the European Patent Convention.

rarely enacted to modify the law in light of the new technology.[9]

To understand the application of patent law to software, therefore, it is necessary to understand the state of patent law as it existed when software came into being, namely at the advent of the modern computer at roughly the midpoint of the 20th century. The following discussion describes prominent cases involving legal questions that are most relevant to the patentability of software, namely: (1) whether, and under what circumstances, a process (in contrast to a product) should be patentable; (2) how broadly a patent for a process (or a machine defined in terms of its function) should be interpreted; (3) what kind of description an inventor should be required to provide of his invention in exchange for a patent; and (4) what the relationship is between the end use(s) of an invention and its patentability. The relationship between these questions and software patents will be made more clear in the description of actual software patents below.

In any field of law, there is often a single case that, although not a Rosetta Stone, provides insight into all that came after it. The English case of *Neilson v. Harford* [52] plays this role in the field of software patents. Although not technically binding precedent elsewhere in Europe or in the US, the court's opinion in *Neilson* has often been cited for the insights it provides. The case involved a patent for a blast furnace, in which "a blast or current of air must be produced by bellows or other blowing apparatus, and is to be passed from the bellows into an air-vessel or receptacle, by means of a tube, or aperture, into the fire". Although the patent was written to cover a particular blast furnace, the defendant in the case argued that the patent should be invalidated because it attempted more generally to cover an unpatentable scientific principle, namely that hot air feeds a fire better than cold air.

The court upheld the patent on the grounds that it claimed not a principle, but a specific machine (the blast furnace) embodying that principle, and

that the inventor had described in sufficient detail how to make and use that machine. Although the import of this case for software patents may not be clear at first glance, its relevance is that in the context of software it can be difficult to draw the line between an unpatentable principle and a patentable application of that principle. More generally, the dividing line between an (unpatentable) abstract idea and a (patentable) practical application of that idea is a recurring theme in most of the cases discussed in this and the following section.

The US Supreme Court addressed a patent on an improved machine for making pipe in *Le Roy v. Tatham* [43]. The inventors of the machine had found that lead was more suitable for making pipe than other metals because, once formed into a desired shape while in its molten form, it was easily cooled and held its shape without cracking or exhibiting other undesirable characteristics that had plagued pipes constructed of previous materials.

Le Roy, the patent owner, sued Tatham for making pipe using a device that allegedly was covered by Le Roy's patent. Tatham countered by arguing that the patent should never have been granted because it attempted to secure protection for an unpatentable principle, namely the principle that lead holds its shape when cooled. The argument, in essence, was that even if Le Roy had made some modifications to a pipe-making machine to make it more suitable for forming pipes out of lead, Le Roy's 'true' invention was a scientific principle, not a machine.

The Supreme Court considered both arguments and struck down Le Roy's patent, holding that the invention was not patentable subject matter because it merely covered an unpatentable principle. The oft-quoted phrase from this case is that "[a] principle, in the abstract, is a fundamental truth; an original cause; a motive; these cannot be patented, as no one can claim in either of them an exclusive right". The Court distinguished between unpatentable principles and patentable applications of such principles by explaining that "the processes used to extract, modify, and concentrate natural agencies [such as steam or electricity], constitute the invention. The elements of the power exist; the invention is not in discovering them, but in applying them to useful objects".

---

[9]One example of special legislation enacted specifically to protect semiconductor chip mask works, The Semiconductor Chip Protection Act, is generally recognized as a failure, as described in Kukkonen [41].

Both *Neilson v. Harford* and *Le Roy v. Tatham* adhere on their faces to the rule that principles are not patentable subject matter and that practical applications of such principles are patentable subject matter. The different outcomes in these two cases, both of which involved patents claiming improved machinery, demonstrates just how difficult it can be to apply this distinction between patentable and non-patentable subject matter in practice.

O'Reilly v. Morse [53] involved a dispute over a claim in Samuel Morse's patent on the telegraph. Although there is still debate about whether Morse is entitled to the label "first inventor of the telegraph", that historical question is outside the scope of this paper. The relevant portions of *Morse* are best understood not as a dispute about *whether* Morse should have been entitled to a patent on the telegraph at all, but rather as a dispute about *how broadly* the scope of his patent should extend, in light of what he had actually invented and how he had described the invention in his patent.

Morse's patent included many claims. The relevant claim, claim 8, read as follows:

> I do not propose to limit myself to the specific machinery or parts of machinery described in the foregoing specification and claims; the essence of my invention being the use of the motive power of the electric or galvanic current, which I call electromagnetism, however developed, for marking or printing intelligible characters, signs or letters, at any distances, being a new application of that power of which I claim to be the first inventor or discoverer.

To paraphrase, Morse sought patent protection not only for the particular telegraph that he had described in the patent specification, but more generally for *any* device (even if not yet invented) that used electricity to transmit messages over any distance and to mark them in a tangible form at the receiving end.

The US Supreme Court considered this claim to be too broad and held it to be invalid. Although the Court gave several reasons for its holding, the one most relevant to the present discussion is that:

> If this claim can be maintained, it matters not by what process or machinery the result is accomplished. For ought that we now know some future inventor, in the onward march of science,

may discover by means of the electric or galvanic current, without using any part of the process or combination set forth in the plaintiff's specification. His invention may be less complicated – less liable to get out of order – less expensive in construction, and in its operation. But yet if it is covered by this patent the inventor could not use it, nor the public have the benefit of it without the permission of this patentee.

The Court, in other words, expressed the concern that if Morse's claim 8 were allowed and given a broad interpretation, it would stifle innovation by future inventors. Because Morse's claim extended far beyond the telegraph that he had explained how to construct and use in the patent, the Court struck the claim down as too broad.

Although the Patent Act existed in a different form at the time of the *Morse* case, it is worth noting that if the same case were decided under the current incarnation of the Patent Act, claim 8 would likely be interpreted as a claim written using functional language (in so-called "means-plus-function form"), because the claim describes what the telegraph *does* rather than what it *is*. The rule codified in 35 USC §112 (6) would dictate, therefore, that the claim be limited in scope to the particular telegraph described by Morse in the patent specification. This outcome would allow the claim to remain valid, while limiting the scope of the claim so significantly as to address the Supreme Court's concern that it would unduly inhibit the work of future inventors.

All of the cases described so far involve patents for *products*. Patents on *processes*, however, have also long been recognized. Patent law has always had conceptual difficulty with patents on processes, however, due to their inherent intangibility in contrast with products. Two landmark cases involving processes are *Cochrane v. Deener* [10] and *Tilghman v. Proctor* [62].[10] *Cochrane* involved a patent on an improved process, spelled out in the patent in detail, for bolting (making) flour that Cochrane claimed could increase the production of high-quality flour. Although the

---

[10]The patentability of mechanical processes, in contrast to chemical processes, was expressly recognized in *Expanded Metal Co. v. Bradford* [24].

patent described particular machinery for performing the process, the dispute in the case was over the patentability of the process itself apart from the particular machinery for performing it. The Supreme Court held that "a process may be patentable, irrespective of the particular form of the instrumentalities used". The importance of this case for software patents is the recognition that processes may be patentable independent of the particular machinery for performing them.

The disputed patent in *Tilghman* "relate[d] to the treatment of fats and oils, and [was] for a process of separating their component parts so as to render them better adapted to the uses of the arts". In determining whether the claimed process was patentable subject matter, the Supreme Court determined that several other processes existed for achieving the same result, and therefore concluded that Tilghman was not attempting to obtain protection for "every mode of accomplishing this result", but rather "only claims to have invented a particular mode of bringing about the desired result". Comparing this to the outcome in *Morse*, one may conclude that whether a patent claim is too broad or 'abstract' to be patentable depends on whether the claim attempts to cover all possible ways of performing a particular function (as in the case of Morse's claim 8) or merely one possible way of performing that function (as in the case of Tilghman's claim).

The preceding cases represent attempts to draw a line between: (1) inventions that are in some sense too abstract to qualify as patentable subject matter; and (2) inventions that are practical applications of an abstraction and therefore qualify as patentable subject matter. Such 'abstractions' may come in many forms, such as scientific principles, laws of nature, and natural phenomena. The abstraction in dispute in *MacKay Radio & Telegraph Co. v. Radio Corp. of America* [44] was of yet another kind: a mathematical formula.

The patent at issue in *MacKay* was directed to a V-shaped antenna, wherein a particular mathematical formula dictated the angle of the V and thereby improved the antenna's performance. Interestingly, the inventor of the antenna did not claim to have invented the formula, which was well known at the time he invented the antenna. The question instead was whether the inventor's *application* of the formula to the design of an antenna qualified as patentable subject matter. The Supreme Court held that it did, because "[w]hile a scientific truth, or the mathematical expression of it, is not patentable invention, a novel and useful structure created with the aid of knowledge of scientific truth may be".

This holding bears repeating: a novel and useful structure created with the aid of knowledge of scientific truth may be patentable subject matter, even if a mathematical expression of the scientific truth is not patentable subject matter. When combined with the fact that "any new and useful improvement" of a new and useful structure is itself patentable subject matter [67], the stage was set for controversy over patents on software.

6.4 EARLY SOFTWARE PATENTS

Slee and Harris's Application [60], decided in 1965 in the UK, is the earliest software patent case of note. The patent application included both method claims and apparatus claims covering techniques for using a new algorithm to solve a problem in a field of computer science known as linear programming. The patent court upheld the apparatus claims, which covered a programmed computer and means for programming the computer, for the reason that they were directed to a machine modified in a particular way. The court rejected, however, the method claims for failing to define a "vendible product", the only identifiable end product being "intellectual information". This case is the first example we shall see of courts searching for a distinction between forms of software that are abstract (and hence non-patentable subject matter) and forms that are in some sense concrete (and hence patentable subject matter).

During the same year that *Slee and Harris's Application* was decided, US President Johnson issued an executive order [64] creating a Commission to perform a comprehensive study of the US patent system. The Commission, in its final report issued in 1966, recommended that patents not be granted on computer programs. Rather than provide a theoretical justification for this rejection of software patents, the Commission provided the eminently pragmatic reason that:

> [t]he Patent Office now cannot examine applications for programs because of the lack of a classification technique and the requisite search files. Even if these were available, reliable searches would not be feasible or economic because of the tremendous volume of prior art being generated. Without this search, the patenting of programs would be tantamount to mere registration and the presumption of validity would be all but nonexistent.

The report did, however, provide a policy-based reason for precluding patent protection for software: "the creation of programs has undergone substantial and satisfactory growth in the absence of patent protection and ... copyright protection for programs is presently available". In part as a result of the Commission's report, companies in the US pursued legal protection for software primarily by pursuing copyright and trade secret protection.

The first true software patent case heard by the US Supreme Court was *Gottschalk v. Benson* [27], decided in 1972. *Benson*, decided six years after the Commission's report, involved a patent on a method for converting numbers from one format (binary-coded decimal) to another (pure binary). The details of the method are not important for present purposes. Rather, what is important is the following. One of the disputed claims described a version of the method as carried out by the hardware of a computer. Another one of the disputed claims described another version of the method independent of any particular machinery for performing it. Both of these claims described a procedure for performing a function – converting numbers – that in some sense is "mathematical", and neither of them stated expressly *why* someone might want to perform this conversion. In other words, neither claim pointed out any particular *end use* of the numerical conversion method.

The Supreme Court held the claim invalid, based in large part on the factors just described. The Court considered the core issue in the case to be whether any of the claimed methods was a "process" within the meaning of the US Patent Act. Recall that the Act imposes a threshold requirement that a claimed invention be either a process,

machine, manufacture, or composition of matter. Reviewing cases such as *MacKay Radio*, *Le Roy*, and *Expanded Metal Co.*, the court concluded that:

> It is conceded that one may not patent an idea. But in practical effect that would be the result if the formula for converting BCD numerals to pure binary numerals were patented in this case. The mathematical formula involved here has no substantial practical application except in connection with a digital computer, which means that if the judgment below is affirmed, the patent would wholly pre-empt the mathematical formula and in practical effect would be a patent on the algorithm itself.

This short passage, which formed the basis for the Court's conclusion that neither of Benson's claims was directed to a patentable "process", is pregnant with meaning. First, it reflects the division between unpatentable ideas and patentable practical applications of ideas, as established in the line of cases described above. Second, it reflects the concern expressed in *Morse* that a patent which is not limited to any particular practical application will provide legal protection to the inventor that is in some sense too broad.

Third, it reflects a judgment, not expressed directly in previous cases, that a practical application of an idea may nonetheless *not* be patentable if that practical application is embodied in a digital computer and is not susceptible of being embodied in any other way. This conclusion was interpreted by many as a pronouncement that computer programs were excluded from the protection of patent law, despite the court's caveat that "[w]e do not so hold" that "[this] decision precludes a patent for any program servicing a computer". After *Benson*, however, it was difficult for many to imagine circumstances under which a computer program could qualify as patentable subject matter.

One effect of the decision in *Benson* was to further encourage software innovators to seek legal protection in the legal regimes of copyright and trade secret law. (For a description of the status of legal protection for software in the US in the 1970s, see Samuelson [59].) These mechanisms, however, proved inadequate. Trade secret law does not prevent a competitor from "reverse engineering" a product (learning how it works by experimenting with it) and then making and selling copies of it. Many kinds of software can be reverse engineered, often with the assistance of software that partially automates the process. As a result, companies who sold their software publicly risked losing the ability to exclude others from using that software.

Copyright law had its own limitations. The scope of a copyright in a computer program, for example, is limited relatively narrowly to the particular computer code in which the program is written. This created a risk that competitors could effectively copy a program without penalty by making relatively minor changes to it, akin to copying a novel by changing the names of the characters and some details in each scene. This left software developers unable to prohibit others from copying the combination of functions performed by their software, independent of the particular code in which such functions were implemented.

Driven by the potential economic benefits that patent protection for software could provide, and informed by the reasoning in *Benson*, high-technology companies continued to press for software patents, primarily by drafting patent applications in a way that stressed the practical real-world uses of their software and the tangible machinery in which such software was implemented.

Parker v. Flook [54], for example, involved a claim to a method for updating an alarm limit in a catalytic conversion process. A catalytic converter is a device now commonly used to remove pollutants from the exhaust of an internal combustion engine. It is important to measure the value of certain "process variables", such as temperature and pressure, during the catalytic conversion process to ensure that they do not rise too high or fall too low. Specific maximum and minimum "alarm limits" are assigned to each process variable, and an alarm is triggered if any process variable falls outside of the range of its alarm limits.

Flook's claim was directed to a method for updating such an alarm limit by performing four steps: (1) measuring the present value of a process variable; (2) identifying a new alarm base using a specified mathematical formula; (3) identifying an updated alarm limit for the process variable based on the formula used in step (2); and (4) setting the

alarm limit for the process variable to the updated alarm limit.

As in one of the claims in *Benson*, the claim in *Flook* did not limit itself to performance using any particular machinery (such as a computer). Rather, it merely characterized itself as a method and recited the steps comprising the method. In further similarity to *Benson*, the specification of the patent in *Flook* stated that the method could be carried out by software executing on a general-purpose computer.

In contrast to *Benson*, however, Flook's claim was not devoid of a description of its "end use". Rather, the claim stated expressly that it was a claim for "[a] method for updating the value of at least one alarm limit on at least one process variable involved in a process comprising the catalytic chemical conversion of hydrocarbons". In other words, the claim stated that it was limited to use in catalytic chemical conversion of hydrocarbons. It did not attempt to capture any use of the mathematical formula for any purpose.

Further distinguishing the case from *Benson*, however, was the fact that the sole novel aspect of Flook's claim was the mathematical formula provided in step (2). The court, pointing to this feature, viewed the claim as a "method of calculation" and held that "a claim for an improved method of calculation, even when tied to a specific end use, is unpatentable subject matter under §101". The caveat – "even when tied to a specific end use" – must have surprised the patent attorney who no doubt drafted Flook's claim in light of the *Benson* court's hint that Benson's claim may have survived had it specified a particular end use.

The Court, recognizing that its decision could be interpreted as a per se prohibition on software patents, issued the pre-emptive admonition that "[n]either the dearth of precedent, nor this decision, should ... be interpreted as reflecting a judgment that patent protection of certain novel and useful computer programs will not promote the progress of science and the useful arts, or that such protection is undesirable as a matter of policy". It was difficult for many, however, to interpret the decision in any other way. How, for example, could the rejection of Flook's claim to a method involving a *novel* mathematical formula be reconciled with the Court's previous acceptance of the V-shaped antenna claim in *MacKay Radio*, which included a mathematical formula that had been published thirty years earlier? A reasonable observer might conclude that the factor distinguishing the two cases was simply the presence of software in one and the absence of software in the other.

Not all important software patent cases in the US were decided by the Supreme Court. Until 1982, the Court of Customs and Patent Appeals (CCPA) heard appeals from decisions of the US Patent and Trademark Office. After *Parker v. Flook*, determined patent applicants continued to revise their strategies and began to obtain some favorable rulings in the CCPA. *In re Sherwood* [37] is an exemplary case. Sherwood's patent was directed to a purportedly novel method for "seismic prospecting" – identifying subsurface geological formations by transmitting sonic waves into the earth and measuring the reflections. Such prospecting enables detailed three-dimensional maps of the Earth's subsurface to be constructed, and thereby to identify potential mineral deposits or fossil fuels, without incurring the significant expense of drilling.

The process of producing such three-dimensional maps based on the reflected waves could be described in terms of mathematical operations. But Sherwood, apparently recognizing that the claims in *Benson* and *Flook* failed at least in part due to their mathematical character and lack of connection to a physical apparatus and physical results, chose instead to write at least some claims for a *machine* (rather than a method) that produces subsurface maps. The specification of Sherwood's patent described both how the claimed method could be carried out by an analog computer and by a digital computer running appropriate software.

The CCPA held these claims to qualify as patentable subject matter, rejecting the Solicitor General's argument that Sherwood was "really mathematically converting one set of numbers into a new set of numbers", despite having dressed his claims in the clothing of an electrical device. The court, while admitting that the claims implicitly involved mathematical operations, focused instead

on the fact that the claimed machine transformed electrical signals from one form into another in the course of carrying out those mathematical operations, thereby satisfying the requirements for patentable subject matter.

The difference in outcome between *Sherwood* on the one hand and *Flook* on the other draws out at least two different approaches that courts have taken in analyzing software patent claims: (1) analyzing the claim as a whole to determine whether it qualifies as patentable subject matter (*Sherwood*); and (2) dissecting the claim and analyzing part of it to determine whether it qualifies as patentable subject matter (*Flook*). The court in *Sherwood*, for example, did not find it determinative that the claim implicitly *included* mathematical operations because the claim *as a whole* was directed to more than just those mathematical operations; it was directed to a machine that performed a physical transformation on electrical signals from one form into another. This approach has its roots in *Neilson v. Harford*'s focus on the blast furnace as a machine rather than solely on the principle on which it was based, and was reflected in the affirmance of the V-shaped antenna claim in *MacKay Radio* despite the claim's inclusion of a mathematical formula. *Flook*'s reliance on the inclusion of a mathematical formula in the claim as the basis for holding the claim to constitute non-patentable subject matter traces its roots back through *Benson* and at least as far back as *Tilghman v. Proctor*, decisions which attempted to look beyond the inventor's chosen characterization of his invention to the "true" invention itself.

It is beyond the scope of this paper to perform a comparative analysis of the relative arguments for and against these two different approaches to claim interpretation, each of which has surfaced in various incarnations in different cases. The important point is that each approach leads to a very different conclusion about whether software is patentable subject matter. In general, courts that have adopted the "claim as a whole" approach have found software to constitute patentable subject matter, while courts that have adopted the "claim dissection" approach have found software to be lacking in patentability.

Although one may view this distinction in a purely formal sense as a difference in methods for interpreting the language that is used to describe software, an understanding of computer technology helps to provide a substantive explanation for this difference. A modern computer, as an instantiation of a Turing machine[11] having a von Neumann architecture,[12] has two essential components: hardware and software. According to one set of definitions, "hardware" is the physical set of components comprising the computer, and "software" is the information embodied in those physical components.[13] More specifically, software is a set of instructions that "instructs" the hardware to perform certain actions, much like the instructions in a cake recipe instruct a chef how to bake a cake.

Under this "hardware as tangible components" and "software as intangible instructions" view, it is clear why the two different approaches to claim interpretation described above lead to different answers to the question whether software is patentable subject matter. If a claim to a programmed computer is interpreted by dissecting out the software from the hardware, the software will fail to qualify as patentable subject matter because by definition it is intangible and therefore a kind of abstract idea lacking in practical application. If the same claim is interpreted by analyzing the combination of hardware and software as a whole – the hardware in which the software is tangibly embodied – then the hardware–software system will qualify as patentable subject matter because by definition the programmed hardware is a tangible machine and constitutes a practical application of the software.

Although the preceding description glosses over some potential difficulties (such as whether any physical instantiation of software necessarily constitutes a "practical application" of the software), it captures as a first approximation the reason why courts have had such difficulty deciding which of

two distinct interpretive approaches to apply to software patent claims. Although courts could have had this discourse over claims in previous fields of technology, the sharp distinction between computer hardware and software, and the interrelationship between them, has drawn out the tension between the two approaches in a way that is no longer avoidable. Although, as we shall see, the "claim as a whole" approach now holds sway (at least in the US), it is not clear that this is because it has proven itself to be superior as a matter of logic.

The US Supreme Court next addressed software patents directly in *Diamond v. Diehr* [13], which the court characterized as being about "whether a process for curing synthetic rubber which includes in several of its steps the use of a mathematical formula and a programmed digital computer is patentable subject matter". One would correctly conclude that the court's decision to characterize the case as one about "a process for curing synthetic rubber" hints at the Court's conclusion that the process is patentable subject matter. A court inclined to consider the process non-patentable might, for example, characterize it alternatively as a "process for calculating a mathematical formula which, as a byproduct, involves the curing of rubber".

In any case, the disputed claim stated that it was a "method of operating a rubber-molding press for precision molded compounds with the aid of a digital computer", and recited steps including calculating the well-known Arrhenius equation based on the measured temperature within the mold press, and opening the mold press at a time determined by the result of the Arrhenius equation. The Court, while recognizing that laws of nature, natural phenomena, and abstract ideas are not patentable subject matter, held the claim to be directed to patentable subject matter because the inventors sought "only to foreclose from others the use of [the Arrhenius] equation in conjunction with all of the other steps in the claimed process". The claim, in other words, left the public free to use the Arrhenius equation in isolation or in conjunction with steps other than those in the claimed process.

Although it is beyond the scope of this paper to analyze *Diamond v. Diehr* in light of all of the

---

[11] For background on Alan Turing and the origins of the *Turing machine*, see Hodges [28].

[12] For a description of the "von Neumann architecture", named after John von Neumann, see Davis [11].

[13] For an alternate definition, according to which hardware is the fixed part of a computer and software is the variable part of a computer, see Plotkin [56].

cases that came before it, particularly *Benson* and *Flook*, many observers have concluded that the reasoning in these cases is irreconcilable and reflects a change in judicial philosophy rather than justifiable factual distinctions among the cases. For better or worse, *Diamond v. Diehr* opened the door to widespread patenting of software by announcing that a method is not excluded from patentability merely because the method is implementable in software or involves the performance of mathematical operations. Furthermore, the fact that the claims in *Diehr* included "physical acts", such as the act of opening the mold press, encouraged future patent applicants to include descriptions of such acts in their claims to bolster their arguments for patentability.

Another trend that began around the time of cases such as *Sherwood* and *Diehr* is a shift away from attempting to characterize particular methods as "mathematical" or "non-mathematical" and basing the patentable subject matter determination on the outcome of that characterization. In the case of *In re Taner* [38], for example, which involved seismic prospecting technology similar to that at issue in *In re Sherwood*, the court agreed that the electrical signals used in the claimed method were "physical apparitions", and that the fact that those apparitions *may be expressed in mathematical terms* is in our view irrelevant" (emphasis added). The court here implicitly acknowledges that, in some if not all cases, one may *choose* to describe a particular product or process in either mathematical or non-mathematical language. If it is correct that whether a particular invention is "mathematical" depends not on any inherent feature of the invention, but rather on the kind of language in which the inventor chooses to describe the invention, then hinging the patentable subject matter determination on whether the invention involves mathematics is effectively to turn over the determination to the drafter of the patent application. Since this cannot be the correct outcome, courts began to decrease their reliance on the presence or absence of mathematical language in patent claims and to search for some deeper, more objective, test for determining whether claims satisfy the patentable subject matter requirement.

The European Patent Office (EPO) Board of Appeal decided *Vicom/Computer-Related Invention* [77] five years after *In re Taner*. The Board upheld claims directed to a process and product for digital image processing as satisfying the patentable subject matter requirement, reasoning that "if a mathematical method is used in a technical process, that process is carried out on a physical entity (which may be a material object but equally an image stored as an electrical signal) by some technical means implementing the method and provides as its result a certain change in that entity". Three years later, in *IBM/Data Processor Network* [30], the Board upheld as patentable subject matter claims related to the coordination and control of the internal communications between programs and data files stored in different computers connected as nodes in a telecommunications network because the invention was concerned with the internal workings of computers. The Board's focus in both of these cases on the physical nature of the machinery and electrical signals in which the claimed method was implemented matches closely the reasoning in the US cases of *Sherwood* and *Taner*.

The patent in dispute in *Arrhythmia Research Technology, Inc. v. Corazonix Corp.* [1] was directed to a method and apparatus for analyzing electrocardiographic signals in order to determine certain characteristics of heart function. The specification of the patent stated that the claimed method and apparatus could be implemented in software executing on a general-purpose digital computer. The court, in beginning its analysis of whether the claims satisfied the patentable subject matter requirement, quickly dismissed the possibility that the 'mathematical' nature of all computer operations excluded the claims from patentability, quoting from the earlier case of *In re Bradley* [34], which stated that although computers manipulate data by performing mathematical operations, "this is only *how* the computer does what it does. Of importance is the significance of the data and their manipulation in the real world, i.e. *what* the computer is doing".

Note the shift here from the manner in which a claimed method or apparatus is *implemented*, i.e. whether it operates by performing mathematical functions, to the *meaning* ("significance") of that implementation in the "real world". Had the Supreme Court adopted the same focus in *Flook*, the claim would likely have been upheld on the ground that it had significance for catalytic conversion – unquestionably a "real world" use – despite the fact that it involved the performance of mathematical operations. Similarly, the output of the method, while characterizable as a number, "is not an *abstract* number, but is a signal related to the patient's heart activity" (emphasis added). Applying this analysis to the disputed claims, the court found that the invention transformed electrocardiograph signals from one form to another, and that these signals "are not abstractions; they are related to the patient's heart function".

The court in *Arrhythmia*, like the courts in *Sherwood* and *Taner*, considered it relevant that the transformed signals were themselves physical. But the *Arrhythmia* court went one step further, relying additionally on the fact that the signals *represented information* that was useful in a "real world" context: identifying problems with heart function.

The EPO Board of Appeal also required something beyond the mere physical transformation of electrical signals within a computer in *IBM/ Computer Programs* [29]. The claims were directed to methods and apparatus for performing resource recovery in a computer system. Interpreting Article 52 of the European Patent Convention not to exclude from patentability *all* computer programs, the Board concluded that a claim to a computer program is not excluded from patentability if the program is capable of bringing about a technical effect beyond the normal physical interactions between hardware and software that are caused by *any* computer software executing on hardware. This "further technical effect" standard has turned out to be extremely difficult to apply in practice,[14]

and this author [57] has argued elsewhere that it is impossible to apply even in theory.

Compare this to *IBM/Document Abstracting and Retrieving* [31] decided by the EPO Board of Appeal in 1990. The case involved a claim directed to abstracting a document, storing the abstract, and retrieving it in response to a query. The board rejected the claim for lack of patentable subject matter for the reason that the electrical signals generated when implementing the claim did not represent a physical thing, but rather represented part of "the information content of a document, which could be of any nature". This approach is similar to that taken by the Federal Circuit in *Arrhythmia*, in the sense that heart function can reasonably be considered to be "physical" in a sense that the information content of a document is not. The UK Court of Appeal came to a similar conclusion based on similar reasoning in *Merrill Lynch's Application* [47], holding that a computer-implemented business method was not patentable subject matter because the end result was "simply 'a method of doing business', and is excluded" from patentable subject matter for that reason.

The cases described thus far lay the groundwork for the problems that more recent cases have grappled with, and that the law continues to grapple with today. In particular, *Arrhythmia* draws on the line of cases preceding it to recognize that identifying the presence or absence of mathematical language in a patent claim is likely not useful, even as a first step, in determining whether the claim is directed to patentable subject matter. Furthermore, *Arrhythmia* emphasizes that although an invention must be "physical" in some sense to qualify as patentable subject matter, "physicality" alone cannot be sufficient because such a standard would enable *all* computer-implemented methods and data to be patentable subject matter, even if such methods or data fell outside of the useful arts or were not susceptible of an industrial application. In *Arrhythmia* we see the beginning of a search for a standard

---

[14]Approximately 300 people attended workshops held by the UK Patent Office to attempt to define what constitutes a "technical contribution". The report [63] summarizing the workshops makes clear that there was little agreement about how to define "technical contribution". More specifically, the report stated that "[i]t is clear from the workshops that none of the definitions as they stood had wholehearted or even widespread support; tallied closely with the status quo when applied by

the attendees at the workshops (i.e. they would all make a significant change to the boundary between what is and is not patentable); was unambiguous (i.e. gave a near-unanimous answer in most cases)".

that incorporates a requirement of "real world" applicability into the patentable subject matter standard. The next section examines the cases that continued to struggle with the use of physicality and real-world usefulness as requirements for subject matter patentability.

6.5 RECENT SOFTWARE PATENTS

In the US, 1994 was "the year of the algorithm". Despite the holdings in the cases described above, many computer scientists and patent lawyers continued to believe that software was not patentable subject matter, or at least that obtaining a patent on software was sufficiently difficult to make it not be worthwhile. Several cases decided in 1994, however, reaffirmed the subject matter patentability of software in a variety of contexts, leading to a boom in software patent applications that has continued to date.[15]

The technology at issue in *In re Allapat* [33] was a machine for performing a kind of 'anti-aliasing' that could smooth the appearance of waveforms on a digital oscilloscope. Although the only disputed claim was directed to a machine, the claim was written in means-plus-function form and was defined in terms of a set of functions performed by the machine. Recall from the discussion of *Morse* above that such claims, although permissible in the US, raise concern about overbreadth and therefore are interpreted under 35 USC §112 (6) to be limited in scope to the particular structure described in the specification for performing the recited functions.

The court held, perhaps not surprisingly after cases such as *Sherwood* and *Arrhythmia*, that the claim satisfied the patentable subject matter requirement. What is noteworthy is not necessarily the novelty of the outcome, but the depth and breadth of the court's defense of claims written in means-plus-function form for methods that are carried out by software on a general-purpose computer – so-called "computer-implemented methods". Judge Newman, for example, in a concurring opinion, elaborated on the inappropriateness of rejecting claims simply for incorporating mathematical language:

> Devices that work by way of digital electronics are not excluded from the patent system simply because their mechanism of operation can be represented by mathematical formulae. The output of an electronic device or circuit may be approximated to any required degree as a mathematical function of its current state and its inputs; some devices, such as the transistor, embody remarkably elementary mathematical functions. Principles of mathematics, like principles of chemistry, are "basic tools of scientific and technological work". [Citation omitted.] Such principles are indeed the subject matter of pure science. But they are also the subject matter of applied technology.
> ... Mathematics is not a monster to be struck down or out of the patent system, but simply another resource whereby technological advance is achieved. Alappat's claim to a rasterizer that is characterized by specified electronic functions and the means of performing them no more preempts the mathematical formulae that are used to direct these functions than did Chakrabarty's bacterium preempt genetic theory.[16]

Here, Judge Newman makes the case that to maintain an exception for "mathematical" subject matter would be to create an exception that would swallow the rule, since *all* physical phenomena are susceptible of description in mathematical terms.

If this were not enough to secure the last nail in the coffin of a per se subject matter exclusion for software, *In re Warmerdam* [39] and *In re Lowry* [36] went one step further. Although the preceding cases involved computer implemented *methods* (whether claimed directly as methods or indirectly as computers embodying those methods), *Warmerdam* and *Lowry* involved claims for *data structures*. There is some ambiguity in the term "data structure": in some cases it refers to a particular set of data stored in a computer, such as data representing the address of phone number of an individual, while in other cases it refers to the *structure* or *format* of such data, such as the organization of the data into a distinct fields for storing the address and

---

[15]For historical statistics on US software patent filings, see Besson and Hunt [6].

[16]This refers to *Diamond v. Chakrabarty* [12].

phone number. Under either definition, it is reasonable to query whether a data structure might not be patentable subject matter even though a method is.

The court came to different conclusions in these two cases, holding that the data structure in *Warmerdam* was not patentable subject matter but that the data structure in *Lowry* satisfied the patentable subject matter requirement. Although this may at first glance appear to be a case of flat inconsistency, a simple explanation for the difference in outcome is attributable to a difference in the way in which the claims in the two cases were drafted. The patent in *Warmerdam* included several disputed claims, the relevant three of which were directed to: (1) a method for generating a data structure; (2) a machine having a memory which contains the data structure; and (3) the data structure itself. The court rejected claims (1) and (3) for lacking in patentable subject matter, while upholding claim (2). In contrast, the court upheld all of the disputed claims in *Lowry*, all of which involved a memory containing a data structure.

The lesson from *Warmerdam* and *Lowry* is that claims which do not expressly state that they require a computer or other physical apparatus – such as Warmerdam's claims for a computer-independent method or for a data structure in the absence of a memory for storing it – are likely to fail the patentable subject matter test, while claims involving data structures actually stored in a computer memory will at least not be rejected for constituting "abstract ideas".[17] Whether or not this is a rational distinction, patent applicants paid attention and began filing patent applications for data structures.[18]

---

[17]In another case, *In re Beauregard*, IBM argued for claims covering a computer-readable medium, such as a floppy diskette, containing a computer program. Although the Federal Circuit Court of Appeals never decided the case, the Patent Office backed down from its position that such claims did not satisfy the patentable subject matter requirement. As a result, so-called "Beauregard claims" covering stored computer programs have become commonplace.

[18]A search of the USPTO patent database for patents with "data structure" in the title reveals 330 such patents, and a review of the titles indicates that many of them are directed to data structures themselves rather than merely methods for manipulating data structures. This number likely underestimates the total number of US patents including claims to data struc-

In 1996 the USPTO published guidelines [65] for examining software patent applications that are still used today. Although these guidelines do not have the force of law, they are an attempt to interpret existing law and to provide patent examiners with procedures to follow in examining software patent applications. And they reflect a recognition by the USPTO that software can qualify as patentable subject matter.

Both the Federal Circuit Court of Appeals and the US Supreme Court were relatively quiet on the question of software patentability until the Federal Circuit's decision in *State Street Bank & Trust Co. v. Signature Financial Group, Inc.* [61] in 1998. The case is important both as a case about patentability of software and as a case about patentability of business methods. Although the law prior to *State Street* was unclear, many patent attorneys and businesses had concluded that the law prohibited patents on business methods because, for example, they constituted "abstract ideas" or failed to fall within the "useful" arts.

The technology at issue in *State Street* was a computer-implemented business method. More specifically, the single claim in dispute was directed to "a data processing system for managing a financial services configuration of a portfolio established as a partnership, each partner being one of a plurality of funds" – a kind of meta-mutual fund. The claim, written in the now-familiar means-plus-function format, defined the data processing system in terms of a set of functions to be performed for updating share prices in the meta-mutual fund. The court, following the dictate of 35 USC §112 (6), found that each of the recited "means" corresponded to a structure described in the specification and concluded that the claim therefore constituted a "machine", which is proper patentable subject matter under §101 of the Patent Act.

This did not end the inquiry, because the lower court had concluded that the claim fell into one of two judicially-created exceptions to patentable subject matter: the "mathematical algorithm" exception and the "business method" exception. The

---

tures, since such claims may be included in software patents not having the term "data structure" in the title.

court dealt swiftly with the "mathematical algorithm" exception, finding that mathematical algorithms are only non-patentable subject matter *to the extent that* they are merely "abstract ideas". Certain kinds of "mathematical subject matter", the court concluded, could constitute patentable subject matter if they produce "a useful, concrete and tangible result".

With respect to the facts of the case at hand, the court held that "the transformation of data, representing discrete dollar amounts, by a machine through a series of mathematical calculations into a final share price, constitutes a practical application of a mathematical algorithm, formula, or calculation, because it produces "a useful, concrete and tangible result" – a final share price momentarily fixed for recording and reporting purposes and even accepted and relied upon by regulatory authorities and in subsequent trades".

The court then eliminated the business method exception by announcing that it had never really existed, at least in practice. "We take this opportunity to lay this ill-conceived exception to rest", announced the court, holding that business methods should be "subject to the same legal requirements for patentability as applied to any other process or method". Contrast this approach with that taken by the EPO Board of Appeal in *IBM/Semantically Related Expressions* [32] nine years earlier. That case involved a claim to a text processing system for automatically generating semantically related expressions. The Board rejected the claim for the reason that the system related to the field of linguistics. Had the Federal Circuit applied this reasoning in *State Street*, it would at least have had to confront the question whether business is a field protectable by patent law. This is one example in which the distinction between the broad application of US patent law to the 'useful' arts and Europe's narrower focus on inventions susceptible of an "industrial application" makes a difference.

The ruling in *State Street* that business methods were not per se excluded from patentability caused an uproar that extended far beyond the legal profession.[19] Patent applicants, apparently unde-

terred by the public outcry, flooded the US Patent and Trademark Office with patent applications for business methods in light of the *State Street* decision.[20] Individual high-profile business method patents that were granted after *State Street*, such as Amazon.com's "one-click shopping" patent [49], and Priceline's "reverse auction" patent [48], were the objects of particular scorn.

Both business method patents and software patents generally were criticized not only for failing to claim patentable subject matter, but also for claiming techniques that either were not new or were trivial modifications to the prior art. A common complaint, for example, was that patents were being granted for software that simply automated techniques (such as reverse auctions) that had been well known for years or even centuries.[21]

In response to both the public criticism and the sheer volume of applications, Congress enacted a "prior user" defense to infringement of business method patents [74], and the USPTO launched a Business Methods Patent Initiative to undertake internal reforms in an attempt to ensure that business method patents were examined rigorously [8]. Business method patents continue to be controversial but have become entrenched in the US. Such patents have been granted in Europe and Japan despite the relatively higher hurdles that such patents must overcome in those jurisdictions.

As far reaching as the *State Street* decision may be, it was not the Federal Circuit's last word on software patentability. *AT&T v. Excel* [2] involved a patent on a method for assisting long-distance telephone carriers in billing telephone calls at different rates "depending upon whether a subscriber calls someone with the same or a different long-distance carrier". More specifically, when a long-distance call is placed, a message record is transmitted which contains information such as the caller and callee's telephone numbers. The claimed method involved adding to the message record an additional data field to indicate whether the call involves a particular primary interexchange carrier

---

[19]For academic criticisms of *State Street*, see, e.g., Melarti [46] and Dreyfuss [14]. For non-academic criticisms, see, e.g., McCullagh [45] and Gleick [26].

[20]For a discussion of patent floods in the business methods context, see Meurer [51].

[21]For a discussion of problems raised by "trivial" patents and recommendations for solving such problems, see Bergstra and Klint [4].

(PIC). The addition of this data field can be used to provide the differential billing mentioned above.

The court quickly concluded that the disputed claims qualified as "processes" under the Patent Act and that the lower court's rejection of the claims as covering "mathematical algorithms" could not stand under *Alappat* and *State Street Bank*.

The court then turned to the defendant's argument that a method claim only satisfies the patentable subject matter requirement if the claimed method works a "physical transformation" or conversion of subject matter from one state into another. While recognizing the grounding of the "physical transformation" requirement in previous cases, the court announced that "[t]he notion of physical transformation ... is not an invariable requirement, but merely one example of how a mathematical algorithm may bring about a useful application. The court justified this rejection of physicality as an absolute requirement by reference to *Arrhythmia*, *State Street* and *Warmerdam*.

The court's reasoning in *AT&T* is both incorrect and unfortunate, and reflects a fundamental misunderstanding of computer technology and the language used to describe it. No previous case, not even those specifically cited by the *AT&T* court, upheld a method claim lacking in a physical transformation. *Arrhythmia* involved transformation of physical signals representing heart function into other physical signals; *State Street* involved a physical computer transforming physical signals representing dollar amounts into other physical signals; and the claims in *Warmerdam* that were upheld involved physical signals stored in a physical memory.

More importantly, the claims at issue in *AT&T* itself involved physical transformations, albeit implicitly. Claim 1, for example, included a step of inserting a PIC indicator into a message record. Message records in telecommunications systems are embodied in physical signals, as is the PIC indicator recited in the claim. Perhaps this could have been made clearer to the court if the claim had been rewritten to recite "a message record *signal*" and "a PIC indicator *signal*", thereby making the analogy to the patentable signals in *Sherwood* and *Arrhythmia* more apparent. It is unfortunate that such

a formal difference in claim drafting had the substantive consequence that it had in *AT&T*.

6.6 THE PRESENT

There have been no significant court decisions regarding software patentability in the US since *AT&T v. Excel*. At present the law has reached an equilibrium, although one that is likely to prove unstable. Patent attorneys have learned how to write and prosecute software and business method patents patents, and the USPTO has revised its procedures and trained the examining corps sufficiently to reduce the high degree of unpredictability that once existed in the examination of software patents to a level that is tolerable to most applicants. Although representatives of high-tech industry and the general public continue to criticize patents on software and business methods, there is no current impetus for the USPTO or the patent bar to take action in response.

And although a major effort at legislative patent reform is underway in the US,[22] none of the proposed modifications to the law address the fundamental problems that are raised by software patents as described herein. Instead, the reform effort is motivated primarily by the desire to harmonize law in the US with that in other countries. Some previous recommendations that might have addressed some of the criticisms of software patents, such as those for changing the standard of non-obviousness, are no longer being considered.[23] Even those proposed reforms that are motivated in part by particular concerns about abuses of software patents, such as those directed at so-called "patent trolls", address symptoms rather than causes of the problems. In short, the current patent reform efforts in the US are not likely to change the status quo with respect to the problem of software patentability and the scope of software patent claims.

Attempts at legislative reform in Europe have been much more vigorous, as represented most

---

[22]Currently represented by [55].

[23]A summary of the currently-recommended provisions may be found in Berman [5].

vividly by the multi-year debate over the proposed European Software Patent Directive.[24] The Directive was originally proposed by the European Commission in a stated attempt to harmonize the legal standards that applied to software patents throughout Europe, based on perceived inconsistencies among the laws of member nations that had created confusion about the patentability of software under the European Patent Convention. The Directive was opposed by a coalition consisting primarily of small to medium-sized businesses and proponents of open source software. One particularly vocal voice against the Directive was the Foundation for a Free Information Infrastructure, which maintained a web site [25] used to document and advocate against the Directive. Critics of the Directive generally argued that software was too intangible to be patented and that software patents would stifle innovation in the software industry, particularly by providing an advantage to large companies over small and medium-sized companies.

Whatever the advantages and disadvantages of the Directive may have been are now moot, since the EU Parliament voted overwhelmingly on July 6, 2005 to reject it. Despite a common misperception to the contrary, this does not mean that software is now not patentable in Europe. Rather, it means that the status quo is maintained, according to which the patentability of software is governed by the standards adopted by the European Patent Office (EPO) and the interpretation by each nation of its own national law. This will likely lead to more uncertainty and controversy as differences in interpretation arise within and among the EPO and nations that are party to the European Patent Convention.

Although the lack of presently-ongoing projects for meaningful legislative reform in either the US or Europe may be dispiriting, new opportunities for progress continue to arise from what might appear to be unexpected sources. For example, the US Court of Appeals for the Federal Circuit recently decided the case of *In re Fisher* [35], involving a patent claim for five purified nucleic acid sequences that encode proteins and protein fragments

in maize plants. The court rejected the claim for failure of the corresponding specification to disclose at least one specific, substantial, and credible utility in accordance with prior precedent and as articulated in the USPTO's Utility Examination Guidelines [66]. Although *Fisher* is a biotechnology case, the Federal Circuit's requirement that a claimed invention have a "real world utility" that provides "some immediate benefit to the public" may represent an increased willingness of the court to put the burden of proving the utility of an invention on patent applicants in any field of technology. Those seeking reform of the system may need to be on watch for opportunities such as that provided by *Fisher* until there is a resurgence of the political will to engage in legislative reform.

6.7 CONCLUSIONS

The history of software patents presented herein, although incomplete, provides a basis for understanding the fundamental reasons for the law's difficulty in adapting to software. At bottom, the inability of courts to resolve the question of software patentability stems from several tensions that technological features of software draw out between different legal rules respecting patentability that cannot be resolved satisfactorily without resort to policy judgments that courts are not competent to make.

For example, one recurring tension that emerges is that resulting from the particular susceptibility of software to being understood and described in terms that are in some sense "abstract", and the contrasting ability of software to be implemented in a physical form and achieve concrete results.[25] The former feature lends support for the case that software is an "abstract idea" and therefore excluded from patentability, while the latter feature bolsters the case for software's practicality and therefore its patentability. Although, as Judge Newman recognized in her above-quoted concurrence in *Alappat*, all of the physical world is *susceptible* to being described in mathematical terms,

---

[24]For a history and the current status of the directive, see Basinski [3].

[25]I have argued elsewhere that this susceptibility is related to the fact that software is designed using a semi-automated process, but the validity of that argument is not necessary for the conclusions drawn herein. See Plotkin [55].

software has raised particular difficulties because it invariably is designed, described, and understood *solely* in mathematical (logical) terms. The law has not previously had to address an entire field of technology whose subject matter is implemented in a physical form yet customarily, and perhaps necessarily, characterized in "abstract" terms. Seen in this light, the courts' confusion about whether to focus on the physical form of the invention or the terms in which it is described is understandable, even if not justified.

Another recurring tension is that resulting from the ability of computers to automate processes in both the technological/industrial arts and the liberal arts. The implementation of an automated business method in a computer, for example, would appear to satisfy the patentable subject matter requirement because the result is an improvement to a "machine". On the other hand, business methods traditionally have not been considered to fall within the technological or industrial arts. Although machines have long been used to automate *parts* of business processes (such as performing calculations and sorting cards), computers make possible a new kind of end-to-end automation that almost entirely removes the human from the loop.

As a thorough reading of the cases cited herein would verify, the legal system has done a good job of fleshing out the arguments for and against software patentability. Case-by-case adjudication, however, has failed to generate a comprehensive approach for applying these arguments to the root problems represented by the tensions between competing legal rules just described. Although the existing system has produced a set of rules that patent attorneys, examiners, and judges are able to apply with reasonable certainty in the majority of cases, the rules at the margins represent plugs in the dike that will only last so long until the mounting pressure drives them out.

Increasing computer automation of design processes, for example, will result in an increasing number of inventions that were designed by providing a computer with an abstract description of a problem to be solved. Technologies such as genetic algorithms are increasingly being used to design not only software but electrical and mechanical devices. When the inventors of such products file patent applications claiming protection for them in terms of the abstract specifications that produced them, the legal system will once again have to address the patentability of inventions for which there is a gap between abstract description and concrete physical implementation. The problem will inevitably grow to a size that demands a comprehensive rather than piecemeal solution. The development and adoption of such automated design processes are being driven not only by the reductions in design time and cost they can produce, but also by increased funding for nanotechnology and other technologies having a degree of complexity that is beyond the ability of human designers to manage alone.

Similarly, patents on automated methods for addressing problems in the liberal arts will reach a point where a fundamental re-evaluation is required. In addition to the flood of patent applications for computer-implemented business methods, the USPTO has seen applications for non-machine-implemented methods for:

- obtaining a percentage of total profits from exclusive marketing arrangements with independent producers of different products;
- teaching a vocabulary word by providing a recording of a song containing the word to a student;
- facilitating conversation between at least two dining partners by providing them with printed open-ended questions to ask each other; and
- creating an expression of a positive feeling such as love toward another person by placing a note expressing the positive feelings in a package and providing the package to the person with specified instructions. (See Kunin [42].)

Lest one conclude that these patent applications are not worthy of concern because they have not yet been granted as patents, consider that a US patent [50] has been granted on a method of sharing erotic experiences by providing a building with a purportedly novel kind of peep show. Once the dividing line between the technological/industrial and liberal arts comes down, a new wall will need to be erected if the original purpose of the patent system is to be maintained.

Tensions such as those described are not susceptible of being resolved by courts as a matter of law because they require judgments about public policy, such as whether the economic justifications for patent law apply to computer-automated inventing and whether developments in the liberal arts, even if computer-automated, are best promoted by granting patent rights in them. In a representative democracy, such policy judgments are made by the legislature. Despite the recent failure of the EU Software Patent Directive for political reasons, additional efforts of the same kind are precisely what is needed for all patent systems. Even if the tensions described above are not logically reconcilable, a democratic process may at least recognize such tensions and make a choice about which imperfect set of outcomes is preferred. Such a pragmatic resolution, while perhaps lacking in the perceived intellectual purity of a judicial opinion announcing the "correct" rules for determining software patentability, would nonetheless represent democracy at its finest.

ACKNOWLEDGEMENT

I would like to acknowledge research assistance provided by Brandy Karl in the preparation of this paper.

REFERENCES

[1] Arrhythmia Research Technology, Inc. v. Corazonix Corp., 958 F. 2d 1053, Fed. Cir. (1992).

[2] AT&T v. Excel, 172 F. 3d 1352, Fed. Cir. (1999).

[3] E.J. Basinski, *Status of computer-implemented invention patents in Europe*, July 11 (2005), available at http://www.aippi-us.org/images/BasinskyCIIUpdate.doc.

[4] J.A. Bergstra, P. Klint, *About trivial software patents: The IsNot case*, Science of Computer Programming 64 (2007), 264–285.

[5] Hon. H.L. Berman, Congressional Record of June 8 (2005), at p. E1160.

[6] J. Besson, R.M. Hunt, *An empirical look at software patents*, available at http://www.researchoninnovation.org/swpat.pdf.

[7] D.L. Burk, M.A. Lemley, *Is Patent law technology-specific?*, Berkeley Technology Law Journal 17 (2002), 1155.

[8] Business Method Patents: An Action Plan, United States Patent and Trademark Office, available at http://www.uspto.gov/web/offices/com/sol/actionplan.html.

[9] D.S. Chisum, *Patents, A Treatise On The Law Of Patentability, §4.02*, M. Bender, Newark, NJ (2006).

[10] Cochrane v. Deener, 94 US 780 (1877).

[11] M. Davis, *Engines of Logic: Mathematicians and the Origin of the Computer*, W.W. Norton, New York (2000), 182.

[12] Diamond v. Chakrabarty, 447 US 303 (1980).

[13] Diamond v. Diehr, 450 US 175, 177 (1981).

[14] R.C. Dreyfuss, *Are business method Patents bad for business?*, Santa Clara Computer and High Technology Law Journal 16 (2000), 263.

[15] European Patent Convention, art. 52(1), Oct. 5 (1973).

[16] European Patent Convention, art. 52(2), Oct. 5 (1973).

[17] European Patent Convention, art. 54(1), Oct. 5 (1973).

[18] European Patent Convention, art. 56, Oct. 5 (1973).

[19] European Patent Convention, art. 57, Oct. 5 (1973).

[20] European Patent Convention, art. 83, Oct. 5 (1973).

[21] European Patent Convention, art. 84, Oct. 5 (1973).

[22] European Patent Office, *Guidelines for examination in the European Patent Office*, Pt. C, Ch. IV (2.1) June (2005).

[23] Exec. Order No. 11,215, 30 Fed. Reg. 4661 (1965).

[24] Expanded Metal Co. v. Bradford, 214 US 366 (1909).

[25] Foundation for a Free Information Infrastructure web site, available at http://www.swpat.ffii.org.

[26] J. Gleick, *Patently absurd*, N.Y. Times (Magazine), Mar. 12 (2000), 44.

[27] Gottschalk v. Benson, 409 US 63 (1972).

[28] A. Hodges, *Alan Turing: The Enigma*, Walker & Company, New York (2000).

[29] IBM/Computer Programs, T1173/97, Eur. Pat. Office (1998).

[30] IBM/Data Processor Network, T6/83, Eur. Pat. Office J. 1–2 (1990).

[31] IBM/Document Abstracting and Retrieving, T115/85, Eur. Pat. Office J. 1–2 (1990).

[32] IBM/Semantically Related Expressions, T52/85 Eur. Pat. Office J. R-8, 454 (1989).

[33] In re Allapat, 33 F. 3d 1526, Fed. Cir. (1994).

[34] In re Bradley, 600 F. 2d 807, 812, C.C.P.A. (1979).

[35] In re Fisher, Docket No. 04-1465, decided on September 7 (2005).

[36] In re Lowry, 32 F. 3d 1579, Fed. Cir. (1994).

[37] In re Sherwood, 613 F. 2d 809, C.C.P.A. (1980).

[38] In re Taner, 681 F. 2d 787, 790, C.C.P.A. (1982).

[39] In re Warmerdam, 33 F. 3d 1354, Fed. Cir. (1994).

[40] Johnston v. IVAC Corp., 885 F. 2d 1574, 1580, Fed. Cir. (1989).

[41] C.A. Kukkonen III, *The need to abolish registration for integrated circuit topographies under TRIPS*, IDEA: The Journal of Law and Technology 38 (1997), 105, 109.

[42] S.G. Kunin, *Patent Eligibility 35 USC §101 for non-machine implemented processes*, presentation given at PTO Day, December 6 (2004).

[43] Le Roy v. Tatham, 55 US 156 (1852).

[44] MacKay Radio & Telegraph Co. v. Radio Corp. of America, 306 US 86 (1939).

[45] D. McCullagh, *Are patent methods patently absurd?*, CNet News.com, October 15 (2002), available at http://news.com.com/Are+patent+methods+patently+absurd/2100-1023_3-962182.html.

[46] C.D. Melarti, *State Street Bank & Trust Co. v. Signature Financial Group, Inc.: Ought the mathematical algorithm and business method exceptions return to business as usual?*, Journal of Intellectual Property Law 6 (1999), 359.

[47] Merrill Lynch's Application, R.P.C. 561 (1989).

[48] Method and apparatus for a cryptographically assisted commercial network system designed to facilitate buyer-driven conditional purchase offers, US Pat. No. 5,794,207, issued August 11 (1998).

[49] Method and system for placing a purchase order via a communications network, US Pat. No. 5,960,411, issued September 28 (1999).

[50] Method of shared erotic experience and facilities for same, US Pat. No. 6,805,663, issued October 19 (2004).

[51] M.J. Meurer, *Business method patents and patent floods*, Washington University Journal of Law and Policy 8 (2002), 309.

[52] Neilson v. Harford, 151 Eng. Rep. 1266 (1841).

[53] O'Reilly v. Morse, 56 US 62 (1854).

[54] Parker v. Flook, 437 US 584 (1978).

[55] Patent Act of 2005, H.R. 2795, 109th Cong. (1st Sess. 2005).

[56] R. Plotkin, *Computer programming and the automation of invention: A case for software patent reform*, University of California and Los Angeles Journal of Law and Technology 7 (2003), Section II.C.

[57] R. Plotkin, *Software patentability and practical utility: What's the use?*, International Review of Law, Computers, and Technology 1 (2005), 23.

[58] W. Robinson, **Treatise on the Law of Patents for Useful Inventions**, Little, Brown, Boston, MA (1890), 463.

[59] P. Samuelson, *Contu revisited: The case against copyright protection for computer programs in machine–readable form*, Duke Law Journal 1984 (1984), 663.

[60] Slee & Harris's Application, R.P.C. 194 (1966).

[61] State St. Bank & Trust Co. v. Signature Fin. Group, Inc., 149 F. 3d 1368 (Fed. Cir. 1998).

[62] Tilghman v. Proctor, 102 US 707 (1881).

[63] United Kingdom Patent Office, Report on the Technical Contribution Workshops, available at http://www.patent.gov.uk/about/ippd/issues/eurocomp/.

[64] United States Executive Order No. 11215, 30 Fed. Reg. 4661, April 10 (1965).

[65] United States Patent and Trademark Office, Manual of Patent Examining Procedure §2106, 61 Fed. Reg. 7478-7492, February 28 (1996).

[66] United States Patent and Trademark Office, Utility Examination Guidelines, 66 Fed. Reg. 1092–1099, January 5 (2001).

[67] United States Patent Act, 35 USC §101 (2002).

[68] United States Patent Act, 35 USC §102 (2002).

[69] United States Patent Act, 35 USC §103 (2002).

[70] United States Patent Act, 35 USC §112 (1) (2002).

[71] United States Patent Act, 35 USC §112 (2) (2002).

[72] United States Patent Act, 35 USC §112 (6) (2002).

[73] United States Patent Act, 35 USC §271 (2002).

[74] United States Patent Act, 35 USC §273 (2002).

[75] United States Patent Act, 35 USC §283 (2002).

[76] United States Patent Act, 35 USC §284 (2002).

[77] Vicom/Computer-Related Invention, 2 Eur. Pat. Office Rep. 74 (1987).

PART 2

IDENTITY-MANAGEMENT

The History of Information Security: A Comprehensive Handbook
Karl de Leeuw and Jan Bergstra (Editors)

7

SEMIOTICS OF IDENTITY MANAGEMENT

Pieter Wisse

PrimaVera Research Fellow
Amsterdam University, The Netherlands

Contents

Abstract

This chapter develops a semiotics of identity management in three parts. Section 1 (Identifying assumptions) introduces how pervasive issues of identity really are. Philosophy, science and, not to forget, religion may indeed be viewed as attempts for coming to terms with identity. Section 2 (Identity in enneadic dynamics) presents the encompassing framework. An enneadic model of semiosis is sketched, which is then applied to design a pragmatic, or behavioral, orientation at identity management. Communication and identity management are largely synonymous; identity management is essentially dialogical. Illustrations from natural history serve to emphasize the generally valid behavioral orientation. Section 3 (Social practices in identity management) illustrates how the semiotic framework helps to gain overview. It discusses selected developments in identity management at the cultural level, i.e. as practiced by human communities and societies. The partly electronic future of identity management is proposed as a program for open security in an open society.

Keywords: identity and difference, identity management, information precision, open security, open society, security management, semiotics, semiotic ennead.

"Be yourself" is the worst advice you can give to some people [56: 265].

7.1 IDENTIFYING ASSUMPTIONS

7.1.1 Limits to access

Organizations in both the private and the public sector apply what is currently known as identity management, and their purpose is to control interests. From such a concern with security, however, any particular organization certainly does *not* put its own identity to the test when 'managing identities'. By way of analogy and more generally speaking, in the area of actors and their individual persons, it is only other actors who are considered problematic. Are they potentially harmful? Could they threaten interests?

Knowing who the other actor is, that is being able to verify another actor's identity upon contact, supports a policy of selective access [61: 14]. Where identity management is pursued within a security matrix of controlled process and property, it is essentially an area that is *identical* to access control.

Access, however, is not a clear-cut concept any longer, if it ever has been. An especially relevant trend is for access control systems increasingly to involve application of *digital* information and communication technology (IT). In fact, IT does not just replace traditional instruments for granting or denying physical access. It also establishes an additional access category that is, a virtual access. IT has the ability now to make information resources instantly available across distance in place and/or time.

When IT is deployed for controlling physical access and leaves the access at a certain stage, it is fitting to conclude, in general, that the new identity management is supporting an otherwise old security. A different lock may have been installed, but it is one that works with a correspondingly digitalized key.

Security, then, only becomes more specific in an informational sense with virtual access. It is with virtual access that identity management, in one of its popular meanings, actually meets *information*

security since virtual access is by definition connected to information resources. Regarding information security, identity management requires as seamless a modulation as possible into authorization for using resources.

> [F]irst, a reliable structure is needed under which authorizations to use resources are conferred (or revoked), and second, a reliable mechanism must exist to verify the authorization each time an access is attempted [67: 404].

7.1.2 An encompassing framework

Where identity management lacks requisite variety, risk rather than security results. Information security should benefit from identity management properly designed and executed. But how does semiotics, the theory of the sign, contribute? With a heritage dating back at least to fourth century thinkers and reinforced from the seventeenth century onward [23], semiotics abstracts from IT. It provides a perspective for recognizing that there is much more to take into account for identity management than merely virtual access control to digitally-coded information resources.

In this paper, the semiotics of identity management is developed from three parts. Section 1 (Identifying assumptions) is an introduction which aims for an appreciation of how pervasive identity issues truly are. Philosophy, science and, religion may indeed be viewed as attempts for coming to terms with this identity issue; and together, the paragraphs in Section 1 especially argue for the relative nature of assumptions. Whatever assumptions have been useful at one time may now require adjustments, and so on; and assumptions for identity and, subsequently, for identity management are no exception.

Section 2 (Identity in enneadic dynamics) presents the encompassing framework where an enneadic model of semiosis is sketched and then applied to design a pragmatic, or behavioral, orientation at identity management. Communication and identity management are largely synonymous; identity *management* is essentially dialogical. Illustrations from natural history serve to emphasize the generally valid behavioral orientation.

Section 3 (Social practices in identity management) applies the semiotic framework. It discusses selected developments in identity management at the cultural level, that is, as practiced by human communities and societies. The partly electronic future of identity management is proposed as a program for open security in an open society.

There are no direct conclusions or recommendations at the end. It is, rather, as a framework for further work that semiotics of identity management is outlined.

7.1.3 Security in sign accuracy

A general meaning that dictionaries supply for *identity* is "the condition, quality or state of being some specific person or thing that is, individuality and personality" [86: 674; 97: 578]. Identity is not mentioned as a condition for entry or for access. Then again, such application might be inferred from *identity* referring to "identification, or the result of it" [86: 674]. If so, what qualifies as the result of identifying-as-a-process on the basis of an identification-as-a-sign? After all, it is the unequivocal knowing of "some specific person or thing", that is, in its essential uniqueness. When *identifying*, security therefore primarily hinges on accuracy. What counts especially is the precision of the sign or the information provided. Identity management is thus a precondition of information security and, as will also be shown, security in general. Pragmatic philosopher John Dewey (1859–1952) places his behavioral point of experimental logic as follows:

> The judgment of what is to be done implies [...] statements of what the given facts of a situation are, taken as indications to pursue and of the means to be employed in its pursuit. Such a statement demands accuracy. [...A]ccuracy depends fundamentally upon relevancy to the determination of what is to be done [24: 345].

Who is the one individual, or actor, who seeks to know-as-a-unique-designation the other individual, be it another person and/or object? Why and when are identifications believed to be relevant? And, indeed, when are they not relevant? What underlies variety in the rigor of identification? This inquiry (re)establishes how a behavioral theory of identity management relates individuality with identification and (inter)action. Social complexity

demands that a comprehensive, productive theory should guide the practice of identity management.

> [T]o resolve [complexity] into a number of independent variables each as irreducible as it is possible to make it, is the only way of getting secure pointers as to what is indicated by the occurrence of the situation in question [24: 37].

Accuracy requires a formally rich approach to information modeling. This paper presents such a generalized, interdisciplinary theory. A *semiotics* of identity management suggests integrative directions in both historical investigations and future-oriented theory and practice of information security.

7.1.4 Taking time off

Dealing responsibly with identity, there simply is no escape from taking a stand on assumptions. Some people may not call collected assumptions a philosophical doctrine, but that doctrine is invariably what assumptions constitute. As it is,

> [t]he necessity of taking into account the best available non-philosophical work on space and time when formulating one's philosophical doctrines is [...] evident far back into the history of the subject [82: 850].

Among the most advanced theories in physics today are those on relativity which is actually where this paper's inquiry threatens to come to a halt before it has even begun. Radically applying the relativistic concepts of space–time may yield identity issues so problematic as to be beyond the control of information security. Quantum mechanics certainly offers no solution. Heisenberg (1901–1976), known for his uncertainty principle, mentions an "interference of probabilities" [40: 149]. A constant and ultimate ground for distinctions and therefore for identities is lacking. A constraint is definitely in order [71; 90], so for this work, space and time are considered as related but separate.

Epistemologically separating time from spatial dimension(s), however, does not preclude time from being seen as constituting other irreducible configurations, such as systems. Founding his intuitionism, Luitzen Brouwer (1881–1966) argues from a point that distinguishes between man and nature.

Man has the capacity for mathematical experience of his life, i.e. recognizing repeating sequences in the world, and causal systems through time. The connected primal phenomenon is the time intuition as such. It allows for repetition of *a thing in time and a thing yet again*, thus leading life's moments to become split into sequences of qualitatively different things [13: 81].

In other words, time intuition in Brouwer's sense is conditional for the essential mathematical practice of, say, particularizing. Time intuition is man's cognitive imperative for recognizing systemic individuals.

> The primal intuition of mathematics (and of all intellectual action) is the substrate when all observations of change have been abstracted from quality, i.e., that intuition operating from a unity of the continuous and the discrete, from the potential of joining several units in a single thought with such units connected by an *in-between* which is never exhausted because ever new units are joined up. As irreducible complements, the continuous and the discrete constitute that primal intuition; they have equal rights and are equally manifest. It is therefore ruled out to abstain from either one as an original entity. One cannot be constructed from the other, taking the other first as a self-contained entity; making such an independent assumption is already impossible [13: 49].

Assumptions, or axioms and postulates, are instrumental designs for optimizing mathematical cognition in Brouwer's wide intuitionist scheme. It should be noted that time intuition is not merely a *human* understanding (see Sections 7.2.6 and 7.2.7, on natural history). However, the point remains that axioms are artifacts, as has been increasingly acknowledged by members of the (Neo)Kantian movement [68; 88]. Assumptions arise in inquiry and serve to compress speculation.

> We may take a postulate to be a statement which is accepted without evidence, because it belongs to a set of such statements from which it is possible to derive other statements which it happens to be convenient to believe. The chief characteristic which has always marked such statements has been an almost total lack of any spontaneous appearance of truth [83: xix–xx].

The number zero is a primary example of this no-absolute-boundary idea; as an efficient borderline concept, it allows controlled crossing from the irrational to the rational and vice versa. As another example, so does Schopenhauer's (1788–1860) concept of the will [81].

Because of Immanuel Kant's (1724–1804) idea that boundaries preclude absolute rationality as the Enlightenment had proposed, by the beginning of the 20th century mysticism was the accepted label for practicing a strong concern for assumptions [96]. Most scientists would again call this area metaphysics, and even Bertrand Russell (1872–1970) admits to some limits.

> Human beings cannot, of course, wholly transcend human nature; something subjective, if only the interest that determines the direction of our attention, must remain in all our thoughts [77: 30].

However, Russell still tries to keep his logic as absolute as possible that is, opposing what he must have considered contamination from behavior. It leads him to disagree, for example, with Dewey on the latter's experimental logic. Russell argues it is "only personal interest" that might interfere somewhat with absolute knowledge. With Arthur Schopenhauer it is precisely the irreducibility of interest-driven action (read: will) why absolute knowledge is unattainable in principle.

Ideas that must be assumed that is, those ideas which cannot be proven are also the ones that ground mathematics, or as Schopenhauer, at an early period called these ideas: interpretations (German: Vorstellung). Man's particular state therefore also determines his mathematics through his particular choice of mysticism, say his first principle(s) [11; 52; 47]. Indeed almost a century has passed since Brouwer axiomatized time intuition. *The Mystery of Time* [48] remains ontologically as unresolved as ever. Through to today, time is still induced from an experience of order and, working from the opposite direction, measurements that imply time involve man in *man-aging* order, or at least in making an effort to do so.

Measuring time suggests time intuition is applied to itself that is, reflexively. With the constraint of a separate time dimension and for practical purposes, time is believed to develop linearly. Along a continuous time line, individual time points or, rather temporal intervals, may be marked. The precision varies with the instruments used for measuring. As Brouwer remarks,

> Similarity of instruments leads to the expectation of similar laws for different areas in physics. For our counting and measuring instinct is similarly affected by [different phenomena], when we submit them all to certain specific instruments; we can subsequently apply a single mathematical system, but it is only a lack of appropriate instruments which has prevented us so far from discovering other mathematical systems relevant for one particular phenomenon and not for another [12: 15–16].
>
> For example, the laws of astronomy are nothing but the laws of the instruments we use for measuring the course of celestial bodies [14: 30].

The word 'year' is used to name a certain interval of duration, and years are counted in order to sequence major events. What inhabitants of the western world take for the year zero actually relies on a particular system of mysticism, called the Christian religion. While it is far from 'mathematically' certain that there is an historical Jesus, let alone that his life and most notably his death are faithfully rendered by the New Testament gospels [22], the alleged year of his alleged birth was afterwards set as the beginning of an era.

More recently, in scientific publications, the politically more correct references BCE (before the common era) and CE (common era) which replaces BC (before Christ) and AD (anno domini). The calendar itself for establishing temporal order and supporting measurements has undergone several reforms in attempts to gain precision and periodicity [48].

7.1.5 Tangled identities

Along a temporal dimension that is, 'in' time, identity might be determined as what exists between birth and death. Then, is *nothing* what lies before or what comes after that period 'in' existence? As an assumption, it indeed seems too simple; the procedure actually allocates identity to the before-birth and the after-death, too.

So, what is real? An absolute, or perfect, identity in the idealist sense has often been proposed; for example, Plato's form, the Judeo-Christian god, and Hegel's spirit are some of the ideas on a metaphysical absolute which help to avoid an infinite regress in argument and explanation. Instead however, the concept of empirical, behaviorally responsible identity becomes increasingly muddled:

> [T]he speech (*logos*) of many philosophers (in reality, ideologues) is not human speech, but either silence, indicative of a world devoid of meaning, or ideology, the false speech of a superindividual: History [59: xiii].

From the western world's calendar's point of reference that is, the birth year of Jesus or the beginning of the Common Era, an illuminating example of tangled identities are derived. In general, of course, a long tradition exists of biblical criticism which aims at clarifying this lacking consistency, among others Erasmus and Spinoza. The story of Jesus' birth as it has been handed down [74] certainly displays an intriguing range of perspectives on both identity and its management; and, therefore, this cultural event of one man's birth exemplifies much of what is invariably problematic in identification with regard to hermeneutics that is, establishing a particular identity with adequate certainty.

Even when only *The Bible* is consulted, it is already the story of the birth of Jesus which does not have an unproblematic identity. For the New Testament offers not one but rather two birth narratives [4] – one attributed to Matthew and another to Luke. Immediately, the preliminary question of identity arises. Did Matthew really exist that is, is he an historical person? And what is known about Luke to ascertain his existence?

Questions abound, too, for example on language. Rendered in modern English, what counts now as the text(s) for the reader? Is s/he aware of historical change in the concept of authorship, for example? What are the limits of such awareness?

Luke's gospel opens with a claim at credibility: "I myself have investigated everything from the beginning". From this claim of an orderly account, Luke proselytizes that the reader "may know the certainty of the things that is, the reader has been taught". This emphasis on truthfulness raises yet another question of identity. The birth accounts by Matthew and Luke, respectively, demonstrate striking differences which could merely reflect different perspectives on the same events as well as several conflicting points. In doing so, does Luke more or less proclaim Matthew a liar?

So far, only circumstantial issues of identity have been referred to. What events surrounding the birth of Jesus do the evangelists write about? With Luke, it is to Mary that the angel Gabriel speaks on behalf of God. Mary inquires how she may give birth to a son while being a virgin. Gabriel announces, "[t]he Holy Spirit will come upon you, and the power of the Most High will overshadow you. The holy one to be born will be called the Son of God".

As if Mary needs convincing, Gabriel mentions that "[e]ven Elizabeth your relative is going to have a child in her old age, and she who was said to be barren is in her sixth month". When Mary visits her, Elizabeth (who is pregnant with John the Baptist) is filled with the Holy Spirit, too. Elizabeth now confirms Mary's conception: "But why am I so favored, that the mother of my Lord should come to me?" With the word "Lord" for Mary's baby, is Elizabeth emphasizing the unity of God-the-father and Jesus-the-coming-son?

Luke does not waste many words on Joseph. Regarding fatherhood, it might be a telling idea that Joseph is only introduced when it is time for the Roman census. He belongs to the house of David and is therefore directed to Bethlehem. From a Roman perspective, it makes perfect sense to let everyone register at their tribal headquarters, in a manner of speaking. First of all, such an arrangement prevents most people in the Jewish peasant, tribal society to have to travel far, if at all. Secondly, the probability of dependable identification is higher among an expected concentration of relatives, friends, and other acquaintances. Joseph is accompanied by "Mary, who was pledged to be married to him and was expecting a child". Please note, it says "a child", not Joseph's child.

Luke is subsequently silent on Mary and Joseph actually taking part in the census. "[T]he time came for the baby to be born". Bethlehem as the place of birth substantiates the claim to King David's lineage. Jesus is placed in a manger. As "there was no room for them in the inn", they seem to have forfeited registration at least by the Romans. They would surely have written Joseph down as Jesus'

father. Rather than having to explain multiple fatherhood to the secular authorities, fatherhood is only attributed to God. The initial birth announcement is made by an angel to shepherds. After verifying that "all the things they had heard and seen [...] were just as they had been told", the shepherds "spread the word". The early youth of Jesus is uneventful enough. Luke reports that "the child grew and became strong; he was filled with wisdom, and the grace of God was upon him".

Now consider Matthew's account. Mary, he writes, "was found to be with child through the Holy Spirit". Apparently, at one stage Joseph notices her pregnancy and quickly rules himself out as the biological father, so "he had in mind to divorce her quietly". Matthew explains that it is to Joseph that an angel divulges the nature of her conception: "do not be afraid to take Mary home as your wife". The angel leaves Joseph instructions for naming the child.

Bethlehem is also stated by Matthew as the place of birth and sets the scene for Herod's involvement. Herod rules as king over Judea, which includes Bethlehem. He meets some magi, or wise men, who have come from the east inquiring after "the one who has been born king of the Jews". Herod pretends an equal interest in worship: "[a]s soon as you find him, report to me". Under guidance from a star, the magi locate "the house [where] they saw the child with his mother Mary. [T]hey bowed down and worshipped him. Then they opened their treasures and presented him with gifts [...] having been warned in a dream not to go back to Herod, they returned to their country by another route".

Matthew again has an angel appear to Joseph, giving him a warning that "Herod is going to search for the child to kill him". The family escapes into Egypt. After receiving word of Herod's death from "an angel of the Lord", Josef and his family do not return to Judea. Instead, Matthew informs, "he withdrew to the district of Galilee, and he went and lived in a town called Nazareth". It is of particular interest that Matthew puts this forward as fulfilling the prophecy that Jesus "will be called a Nazarene".

So Luke starts from Nazareth and subsequently offers an explanation for the Judean affiliation of

Jesus. With Matthew it is exactly the other way around; starting from Bethlehem, he builds the case for Jesus as a Galilean.

Juxtaposing gospels, another concept of *identity* arises to provide a theme for discussion. It is "the state of being the same in substance, nature, qualities that is, absolute sameness" [86: 674]. As indicated, Matthew and Luke, or whoever authored the narratives carrying these names, have produced differing texts on the same subject which does not have to pose any problems. The requirement of identity-as-sameness is clearly nonsense when looking at separate items in the first place. The more exact question is whether such essentially different texts, or signs, identify one-and-the-same object, as in this case of the historical Jesus. A reliance on identity-as-uniqueness of space and time helps to draw out contradictions. Arguing from the assumption of a single place of residence at a particular point in time, Luke has Jesus spend his early youth in Galilee while Matthew places him at that time in Egypt. Therefore, honoring the constraint, both accounts cannot be accurate.

The birth story, or myth, of Jesus has been commented upon at such length here because it appears at odds with the identity management of this person. But is it, really? It should be possible to make evolutionary sense of the success from tangling identities. Is it not a reasonable hypothesis that the myth intertwines identities to such an extent that only the assumption of a single super-identity restores order? In such a scheme, Jesus figures the necessary reconciliation of every mortal being with an added super-identity. If so, as a movement with a persistent record of supporters for now two millennia, its ideology qualifies as identity management, not despite of but precisely because of the use it makes of tying up identities into an irresolvable knot by setting up the miraculous escape from it.

The general point this above seemingly out-of-place analysis tries to make is that there is more to identity management than one-to-one positive identification and verification. A semiotics of identity management provides the enlarged framework required for the requisite variety. Before that framework proper is addressed in Section 2, a last introductory paragraph continues to widen concerns for identity management.

7.1.6 A false security in management

Whenever an encounter with the popular use of the term *management* occurs, a healthy degree of suspicion might be in order. Another example, currently in vogue is, of course, *architecture,* with *system* now rapidly losing persuasive power. Such words are suggestive of a positively-valued qualification, that is, an advertisement. But what are actual, or real, benefits?

Also in the case of the word management, a false sense of security might easily be conferred onto the what-of-management, in this case *identity*. In other words, a reference to management attempts to argue, unconsciously more often than not, for an unproblematic status of the core concept presented which implicitly suggests the underlying promise of management's application is the assumption of identity's secure existence. Ah, yes. Quite so. Identity again is established and settled: security underlies the word *management*. The label might as well read *identity security*.

It is not that simple with *identity*. Indeed, identity especially escapes singular, absolute definition as required to ground the concept of identity in ant reliable way [9: 682–684; 18: 379–381; 35: 385–388]. Stewart Clegg summarizes Michel Foucault's position:

> Identity is never fixed in its expression nor given by nature. No rational, unified human being, or class or gendered subject is the focus or source of the expression of identity. [. . . It] is contingent, provisional, achieved not given. Identity is always in process, always subject to reproduction or transformation through discursive practices that secure or refuse particular posited identities [19: 29].

What about *individual*? Does it offer solid conceptual ground [31; 60; 85]? Does a particular meaning of identity include that which distinguishes one individual from another? But, then, is not *identity* actually a strange term, a misnomer even? *Distinction* or *difference* may be more suitable. A host of philosophical writers, such as Jacques Derrida apply this perspective in various ways. Debunking "neo-dogmatic slumber", Molnar argues:

Disengagement from the Hegelian *fascinatio* must [...] begin with the proposal of a different epistemology, one whose ideal is not fusion but *distinction*. It is through distinction that the mind operates, concepts are articulated, and being is understood. It is through distinction that we recognize the objective reality of the extramental world (as distinguished from the self), of the moral elements in human situations, and of the inherent limits of human beings in thought and action [59: xv].

Apart from *individual* being a problematic concept, the question of *difference management* also arises when more than one individual exists and one of those particular individuals needs to establish his/her identity. It seems logical to assume that a different meaning of *identity* is invoked in this matter and that synchronic identity of different individuals would be a contradiction. Identity, then, pertains to diachronic identification of a single individual.

With a diachronic identification of an individual what is meant is the *continuity* of an individual's existence or the recurrence of the appearance of an individual. Stated another way, diachronic identification is the *difference* not between separate individuals in a possibly single event, but rather between separate events in which a single individual is involved. This idea recalls Ludwig Wittgenstein's image of relatedness through family resemblance to mind [95: 32]. Following the life trail of an individual, diachronic identification leads back to *identity management*, as far as the label goes even though it is preferred to conceive of it as *i/d management*, with *i/d* standing for *identity & difference*. The concepts are essentially interdependent.

Identities are not absolute but are always relational: one can only be something in relation to some other thing. Identity implies difference, rather than something intrinsic to a particular person or category of experience [19: 29].

Such perplexities may find compact expression through contra-grammar [38]. A contra-gram playfully hypothesizes an equivalence of inverse relationships. It thereby immediately raises questions of identity and difference and may even point at reconciliation through an oscillation of concepts.

For example, with two terms x and y involved, a contra-grammar's formula is: the x of y is the y of x. So, for the (meta) terms identity and difference, the subsequent contra-gram would read: The identity of difference is the difference of identity.

Is it possible that difficulties of identification would not ever have arisen if the idea of identification had been phrased differently, for example as *identification management*? Certainly not, for sooner or later the *object* of identification, that is, the individual should have to be confronted and that is a moment of difference *and* identity of the objects or individuals in an identification process.

When refraining from emphasizing that individuals subjected to identity management are not limited to persons and may include objects, it is only because this issue has yet to be covered. Indeed, animals and even plants *actively* perform identity management (see below). And a human community settles on more than one plot of land; its members live in more than one house, drive more than one car, milk more than one cow, and on the variety multiples. The idea of community management (sic!), therefore, implies requirements for identity management, and it follows next that identity management includes relationships. For example, an individual house is built on an individual plot of land, and therein lies an identification. The 'plot' thickens when identification is recognized to rest largely on correlation. A person may seek to establish an identity by referring to a date of birth, for which this date becomes the individual's identification. Status symbol entails an even more elaborate attempt at manipulating identity. Identity, and therefore identification, turns out to consist of a dynamic structure of differences.

7.2 IDENTITY IN ENNEADIC DYNAMICS

7.2.1 Introducing Peircean semiosis

Extended identity management is ultimately grounded on semiotics with its core concept of semiosis. Semiotics is about signs, and semiosis will be explained in more detail shortly. A first issue is to determine who produces and interprets signs; and, whereas, it is obvious that a particular individual does the interpreting, a problem remains since a 'particular individual' is a concept that encompasses a variety of identity management possibilities.

> The term 'person' has a history of special use in legal and theological contexts. Apart from these, the term is often synonymous with 'human being'. The history of thought about persons is thus linked to changing legal, social, and theological trends as well as to more general reflection on the nature of the human subject, the 'I' that thinks, feels, reflects on itself and its doings, and carries responsibility for his previous actions. [...] The problem of personal identity resolves into two different questions, one concerning the unity of the self at a time, and the other with dealing with unity through time [9: 692].

Charles Peirce (1839–1914) was an eccentric scholar of a Renaissance character who produced an enormous volume of work on a great variety of subjects [10; 43]. His body of work cannot be divided into disjunctive philosophical classes, and there are at least two conflicting frames of reference, that is, transcendentalism and naturalism [34: 5–7]. While Peirce certainly shifted emphasis from one object of study to another, he also pursued a unifying approach, admittedly with various degrees of success. What concerns this paper is Peirce's overriding unification, for which his concept of semiosis is exemplary. Semiosis is

> an action, or influence, which is, or involves, a coöperation of *three* subjects, such as a sign, its object, and its interpretant, this tri-relative influence not being in any way resolvable into actions between pairs [64: 282].

More specifically,

> [t]he sign stands for something, its *object*. [... And it] addresses somebody, that is, creates in the mind of that person an equivalent sign, or perhaps a more developed sign. That sign which it creates I call the *interpretant* of the first sign [63: 99].

An abstracted schematic representation is the semiotic triad, as in Figure 7.1. Peirce's key idea is that interpretant is not directly related to an object. Experience and behavior are in a relationship that makes itself explicit as a third term: sign mediates. This additional, third element balances traditional

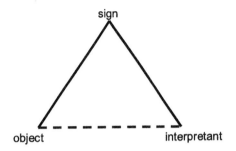

Figure 7.1. Peirce's semiotic triad.

idealism with realism. Peirce's innovative triad is a succinct expression of transcendentalism where he emphasizes its irreducible nature that is, any reduction of the transcendental compromises the integrity of a semiosis. Peirce's *irreducible* triad, of course, strongly establishes his semiotics also as a phenomenology. Interpretation as Peirce suggests, therefore, is closely related to the concept of a perceptual or a phenomenal field [20: 25].

A preliminary result may be derived from the triad. There is a much needed, formal distinction between identity and identification. The former, identity, entails the object itself. Identification, on the other hand, is the sign that *acts* as a reference. Therefore identification is *not* the identity; it stands for it. This observation lies at the heart of criticisms directed at the pre-semiotic logic of Aristotle [24; 25; 45], and as Peirce remarks:

> The Sign can only represent the Object and tell about it. It cannot furnish acquaintance with or recognition of that Object [63: 100].

A corollary entails different meanings for security. Actually, there are three main meanings as they correspond to the three semiotic triad elements. Interdependency rules between an object's security, a sign's accuracy (read: precision or rigor), and an interpretant's relevance. Please note that for practical purposes, "a sign's security" is not written.

7.2.2 Variety: enneadic semiosis

The triad lacks requisite variety [2] for reconciling identity with difference. Still, Peirce offers an important clue for development where he argued that a sign is something which stands to somebody for something in some respect or capacity. [...It] stands for that object, not in all respects, but in reference to a sort of *idea*, which I have sometimes called the *ground* of the [sign] [63: 99].

Elsewhere, each element is first 'grounded', thereby restructuring the triad into a hexad. Next and in the opposite direction, elements for precision are added; the final result is an ennead [90: 146]. Figure 7.2 reproduces the semiotic ennead. The 'identifications' (in enneadic terms, read: signatures) for the elements have undergone slight modification as compared to the original version [93].

The ennead retains Peirce's essential assumption on transcendentalism, so, it also *irreducibly* links all its elements in semiosis. The original three elements reappear as dimensions, each dimension now constituted by three more detailed elements. For example, the dimension of interpretation involves motive, focus and concept.

It is beyond the scope of this overview of the semiotics of identity management to elaborate at length on enneadic dynamics. The next paragraph explains movements along the sign dimension, only. Please note that along all dimensions, shifts are assumed to occur hierarchically. In one direction, for example the focus can shift into what was motivational before. Thus the concept broadens. In the other direction, the focus may shift down to a narrower concept than was considered previously. This is accompanied by a broader motive.

[A] contraction of reference accompanies an expansion of awareness, and an expansion of reference accompanies a contraction of awareness [83: 10].

In the semiotic ennead, formal correspondences rule between dimensions. A signature therefore mediates a focus on an identity. In similar fashion, other relationships within the ennead hold.

The semiotic ennead safeguards against conceptual simplification. It can be called upon whenever relevant rigor seems beyond currently applied differences [91; 94]. However, it is cumbersome to suggest semiosis in every separate sign that is, explicitly touching all dimensions and their elements. A practical emphasis usually operates. In modern

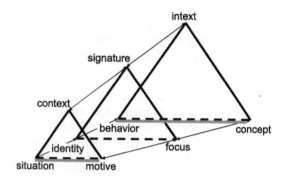

Figure 7.2. Semiotic ennead.

science, signs are supposed to be especially selected for abstraction from interpretational (read: psychological or subjective) variety. Active denial of subjectivity becomes a problem when real differences have a subjective (read: individual) basis in interpretation [72].

Until recently, science has largely ignored the sign dimension. This neglect certainly has changed with the linguistic turn in philosophy that is, adopting linguistic methods [75]. Regretfully though, Peirce's explicit prescription for irreducibility of elements/dimensions still seems largely overlooked and violated. Modern linguistic disciplines deny interpretational variety where they should include it. They have often proceeded to treat signs as objects, thereby losing sight of the relationship between sign and object as well. Symbolic logic is likewise often practiced at irrecoverable distance from empirical identification.

Drawing special attention to the way that the semiotic ennead encompasses motive and identity in a single 'system', is an interesting perspective. When the relativism included in motive is factored into an investigation, some puzzles of strictly objectivist logic dissolve. An example is the so-called sorites paradox [42]. Once again, what is essentially questioned as paradoxical is identity and, by consequence, difference, too. Soros (Greek: σωρός) means heap, as in sand, for example. How much sand is needed for an accumulation rightly to be called a heap? This question only leads to a paradox when considered from an absolutist perspective. Changing perspective, however, the question simply dissolves as a pseudo-problem when it depends upon both the particular subject and its particular situation of what constitutes a heap that is, refining the relevant concepts. There is nothing vague about the designation, in this case a heap as soon as its subjective situation is predicated.

7.2.3 Toward a structural concept of identity

One of the ennead's elements is labeled *identity*, and it occupies the pivotal position along the object dimension. Such a structural(ist) position dissolves any frustration with unattainable definition in an absolute sense. As it were, identity is unburdened. Conceptual load is shared between all elements of the semiotic ennead. Identity itself is even radically changed into a pivot, *only*. It should function unhampered as a hinge between situation and behavior. The current paragraph reproduces a demonstration [39] limited to the object dimension, that is, without taking the two other semiotic dimensions explicitly into account. (As argued before, the bias of expressing-by-sign is impossible to avoid; sign expression is a pleonasm.)

Social psychology instructs about the situational nature of behavior [17; 24; 25]. An actor or an agent (or object, or individual) is assumed to reside in various situations. Hence variety exists in an actor's behavior. In fact, a particular behavior corresponds with the actor *as far as a particular situation goes*. Adding situation and behavior therefore turns inside-out the treatment of an actor as entity/object. Only an actor y's barest identity remains necessary and sufficient for relating a situation x and a behavior z (Figure 7.3(a)). The whole of actor y is now reflected by his particular behaviors *across relevant situations* (Figure 7.3(b)).

Please note that juxtaposition of behaviors rests on repeating the reference to the actor's identity. An instance of identity exists for every relevant situation that is, where the actor has a particular behavior. Through repeated identities, differences (particular behaviors) are reconciled with unity (one actor).

Additional compactness and flexibility comes from the assumption that situation, (actor's) identity, and behavior are relative concepts. A rigorous set of modeling constructs applies throughout. Following the spatial orientation of Figure 7.3, decomposition can proceed both up and downward.

Upward, for example situation x_1 can itself be considered as constituted by several actors' identities, presumably all different from I_y, appearing in a correspondingly less determined situation. Introducing levels, the original situation x_1 may be designated as situation $x_{m,1}$. The result of one step of upward decomposition is shown in Figure 7.4. The limit to upward decomposition lies in ambiguity. It can proceed as long as identities at the next lower level, such as I_y for the step illustrated here, can connect only with a single identity in what previously was held as an undifferentiated situation.

Upward decomposition has a paradoxical ring to it. Its purpose is not to break *down* a whole into parts, but to discover the *up* structure in a situation. It can be extremely productive to attempt it from any instance of an identity as a starting point.

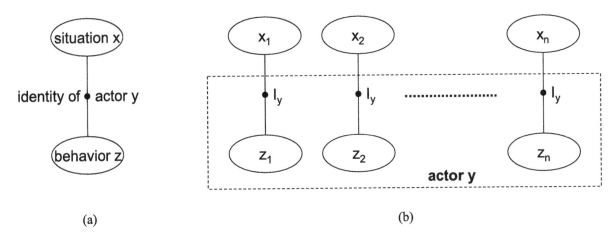

(a) (b)

Figure 7.3. The object dimension of situation, (actor's) identity and behavior.

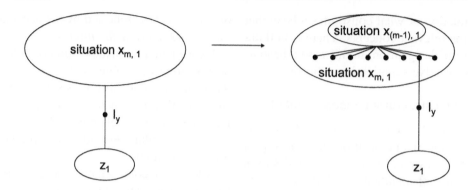

Figure 7.4. Upward decomposition.

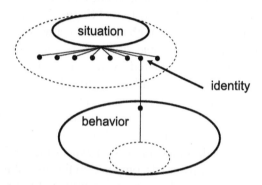

Figure 7.5. A shifted configuration of situation, identity and behavior.

Which situation is more applicable: a person's particular behavior correctly located in a city or a specific reference to the person's home within a city? Does it matter in which city the individual's home is located or was the original situational designation even irrelevant? In Figure 7.5, the right-hand side of Figure 7.4 has been adapted to indicate what is now significant with regards to situation, identity, and behavior after one step of upward decomposition.

Reading Figures 7.4 and 7.5 in reverse order gives an impression of downward decomposition; the original identity is encapsulated within a more determined situation while the original behavior is decomposed into identities each with situational behavior at a lower level of the conceptual model.

There are no limits to upward and downward decomposition, at least not theoretically. An upper boundary condition must be formally set, however, and reflect the model's horizon, corresponding to the most generally accounted for that is, the least

determined situation; in metapattern's visual language it is a thick horizontal line.

For a hint as to where downward decomposition may ultimately lead, it seems apt to mention connectionism which reflects the perspective, often associated with artificial intelligence, on meaning as a neural process [58: 166–167]. Then, brain activity is engaged in identity management as outlined here. The ennead provides a sophisticated metamodel for connectionism, too. A conceptual *model* itself lies along the sign dimension and takes on the shape of a lattice of nodes. Some nodes are connected to the 'horizon'. Different instances of equal identity may be connected laterally to indicate that an actor's behavior in one situation is invoked from his behavior in another situation.

How dynamics are accommodated also goes beyond the fixed, singular instance of traditional identity. On the assumption that each instance of an identity is supplied with a unique identifier, such instances can be moved about in a controlled manner. For example, the home address of person A was x_1 up to time t_1, and from then on it has been x_2.

The more radical the decomposition in both directions is, including finely grained management of temporal variety, the more an actual model moves toward a lattice consisting mainly of identities that is, they serve to *connect*. Only relatively little behavior remains to be specified at what has been designed as the extreme result of downward decomposition.

7.2.4 The systemic nature of identity management

Radical decomposition suggests an object's identity emerges from differences that is, consists of relationships with different objects. Likewise, every such different object also exists as an emerging identity. Identities mutually determine each other and for living beings dynamically do so [46].

The general implication is that identity cannot be managed in isolation, so other identities are necessarily involved. Absolute autonomy is an illusion. Identity management is systemic. Identity management *is* relationship management.

Peircean semiotics instructs that an object cannot be directly known. Identification mediates identity. Foucault (1926–1984) emphasizes identity management as power play by institutions that is, in (social) situations. His studies inquire into

> the power of normalization and the formation of knowledge in modern society [28: 308].

Etzioni also explores "the nature of compliance in the organization" in his case for arriving at "middle-range organizational theory" [26: xii].

Interdependency of identities in practice has not been applied to radically integrated management. There are many constraints. First and foremost, what is hardly recognized yet is the need to strike a social *balance* between identity and difference. For example, with dogmatic reference to privacy, identity is often argued for as an absolute value. However, identity taken to this extreme ends up at its opposite; absolute difference equals nonsense. In general, bias is a common source of divided identity management. Then, a particular interest leads only to considering a corresponding kind of object as carrying identity. Apartheid is the paradigm praxis [8: 195–225]. Properties are seen as internal to an object that is, encapsulated, rather than constituted by relationships with different objects in their own 'right'. However, one individual's *main object* may be another individual's *object property*, and several interests may include similar properties. It all results, then, in duplication.

Instead, a rational design for identity management should start from its systemic nature. Privacy, too, entails social relationships. Actual choices

must of course be politically made. Human interest is ultimately moral. In a similar vein, appreciation of identity as a system of interdependencies may guide historical research into management theories and practices. However, this paper will first radicalize the framework and present a concept of language that is consistent with the semiotic ennead. Language, too, must be properly situated with the cue coming from Schopenhauer.

7.2.5 Language use *is* identity management

Schopenhauer designed a transcendental concept of great practical value, the will, for dissolving the *Weltknoten*. The knot-of-the-world is his metaphor for the fundamental question of mind–body dualism. Associated with both the concept of an individual as a unique objectification of the will and an individual's capacity for empathy, he rigorously accounts for a large variety [80; 81] of difficulties. The semiotic ennead includes and thus enhances Schopenhauer's conceptual scheme, too, adding formal precision since motive now correlates with context and situation [90].

Every instance of semiosis is unique, and this assumption alone argues for the irreconcilable difference between an interpretation resulting in a sign-as-produced and an interpretation resulting from a sign-as-consumed. Significs already establishes the formal distinction between speaker's and hearer's meaning. Gerrit Mannoury (1867–1956) already stipulates the difference in his early work on the foundations of mathematics [51] that is, before he was introduced to significs, and in his later development of significs, it is a recurrent theme [52; 53]. The irreducible part motive plays in enneadic semiosis explains the nature of sign use that is, language: Every sign is a request for compliance. This slogan is my own [90], but I have subsequently learned that Brouwer already holds an equally radical motto for language use [79: 204–205].

The idea that every sign aims at compliance impinges on the concept of identity management which can and should now be approached from such a generalized language perspective. There are already extensive sign decompositions along the enneadic lines elsewhere presented (see [90: 235–291]), this paper will only provide a summary.

sign exchange

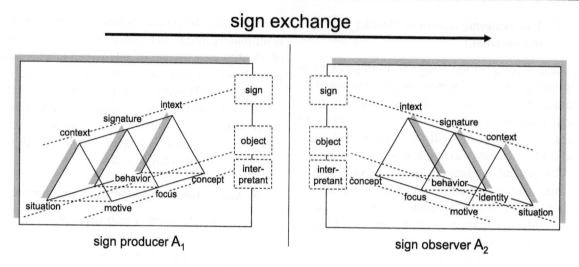

Figure 7.6. Dia-enneadic sign exchange.

Please note, such sign structures have a generative quality which is essentially pragmatic that is, they have no relationship with transformational grammar, for example. (Such attempts at structure do not include the interpretation dimension and even deliberately reduce, in case of syntax, sign structure from a relationship with the object dimension.)

For the sake of overview, a single sign exchange is limited to two participants: the sign producer A_1 and the sign observer A_2. Figure 7.6 reproduces the full dia-enneadic model for sign exchange [93].

For a particular exchange, A_1 acts as the sign producer, with A_2 as the sign observer or sign consumer. By definition, the initiative lies with A_1. A particular motive induces the sign producer to attempt to elicit behavior from A_2. A focus connects the producer's motive to a concept of A_2's projected behavior. On the surface, there seem to be three 'units' to identify with the sign:

1. A_2 is called upon to exhibit behavior.
2. The behavior for compliance needs to be specified.
3. A_1 is known as the requester, if not the beneficiary.

However, the process is not that simple when the enormous variety is factored into sign production. It is roughly dependent on

a. the relationship between A_1 and A_2 *as seen by* A_1, and

b. the behavioral request *as put 'up' by* A_1 for compliance by A_2.

Their relationship is far from constant, though. This particular sign, too, will influence it, and so on. It reflects the emerging nature of identity. A_1 will always integrate a strategic assessment. What are the opportunities and/or risks for his on-going relationship with A_2? As for identifying A_2 in his sign, he can choose from no identification at all to the most precise and explicit identification he can muster. Similar theoretical ranges exist for referring to the behavior required and for identifying himself as the sign producer. The practical options of course depend on A_1's actual *repertoire* for identity management. When language use in its widest sense is treated as identity management, previously unsuspected phenomena can be analyzed coherently. The practice of communication *is* identity management [32].

The dia-enneadic framework displays, by definition, eighteen interacting elements. Deployed fully, it provides a correspondingly rich concept of information and, from the assumptions for the framework, communication. Information concepts with reduced variety can simply be 'produced' as elements are removed. This procedure explains why several disciplines refer to information while completely bypassing issues of identity management. For example, an information concept that only 'fits' physical signal transmission is far from behaviorally relevant in any motivated sense. Although lacking a systemic approach that is, not from an overall formal framework such as paired enneads, Rafael Capurro already points at conceptual differentiation:

> I object to analogies being made, not to mention an equivalence assumed, from incommensurable horizons. On the contrary, a differential characterization of the concept of information is required for particular areas in order to establish fundamental boundaries as exist between, for example, genetic information and what is considered "human transmission of meaning" [16].

7.2.6 A natural history of identity management

Now what can be claimed is an origin and relevance for identity management beyond historical sources and strictly cultural explanation. The semiotic framework for identity management applies outside of human culture that is, as it is relevant for so-called natural behaviors. Whatever inter- or multi-disciplinary design may lose, the question can now be raised if anything is truly lost based on a 'professional' level and can be done so by making a rigorous reference to singular disciplines through innovative audacity and, hopefully, gaining general relevance. Schopenhauer and Peirce, for example, remained academic outsiders during their own lives. Synthesis is initially often dismissed as amateuristic. Therefore, some adventurous conjectures on ethology [54] or, more specifically, zoosemiotics [62] could prove academically risky, but still the case can be made that these processes help. A genuinely generalized theory of identity management should easily negotiate the traditional distinction between natural and cultural behavior, and also hold regardless of the species of the organisms [36]. This requirement for theoretical relevance fits Mannoury's principle of graduality (read: relativity).

> Nature is continuous. It doesn't entail acute differences, there are no sharp boundaries. Nature holds no absolute similarities, nor is anything absolutely stable. And when we try to apply words for referring to an object or an appearance that belongs to this nature, we find it impossible to overcome this indiscriminate character, that is, a relatedness of one to the other [51: 5].

An orientation at natural behavior draws attention to the *asymmetries* of sign exchange. There is no cultural superstructure where 'rules of order' may be mistaken for the 'order' itself. Yet, it is 'normal' that A_1 wants A_2 to comply with the former's rule as the latter's reality. What is 'cultural' in addition to 'natural' might not be the attempt at imposing a rule, which is already natural, but rather the motivated adaptability that is, the rules that may change by design. With propaganda invested in beliefs of human societies 'ruled' by equality, solidarity, and other motivations as well as real achievements in emancipation notwithstanding, stepping outside the human sign exchange for even a moment might secure theoretical validity. Of course, it is clearly impossible for a human being to step outside of its human sign exchange, which is an inevitable bias already established. On account of irreducibility in semiosis, neither the subject, nor the object of inquiry is independent from the propaganda system.

As to ethology, this paper argues from abstraction that is, sticking to A_1 and A_2 and suggesting intermediary examples for inspiration and/or popular illustration. Imagine an interaction where A_1 is the predator and A_2 is the prey. How A_1 manages its own identity depends. Suppose A_1 is hungry and simultaneously has actually spotted A_2. Or it could be that A_1 spots A_2 *because* A_1 is hungry. As long as the situation is the latter, A_1 must maneuvering into a promising attacking position while A_2 still has serious defense or escape options; and, consequently, A_1 would profit from hiding its *own* identity and remain unnoticed or even fake a different identity so that A_2 does not sense a threat.

Mimicry is usually reserved to indicate faking behavior to threaten off an opponent of superior strength, for example if A_1 has come within striking distance of A_2. When A_1 can overpower A_2 by surprise, up to the very moment of striking, A_1 would continue attempts to manage no identity or a different one. There really is no paradox when *no identity* is considered from the perspective A_1 applies to manipulating A_2's perspective. Or, on the assumption that experience of A_1's identity may undermine A_2's defense, A_1 can forcefully show it or even fake a different identity for *that* very purpose. Does a dog, just to mention two options, bark aggressively to scare another dog into an easier victim or does the dog bark to scare off an opponent dog which is assessed as actually being more powerful?

There is a solid case for a rich natural variety in identity management. It is obvious that when A_1 is the prey and A_2 the predator, other ranges of identity behaviors may hold for A_1. The situations are infinite and include areas such as mating and childrearing. The semiotic ennead, where identity relates situation to behavior, points to an equally infinite variety of behaviors where identity is involved. Please note that what is being referred to is the actor's *own* identity. The hypothesis must include that one actor, say the predator, *manages* its own identity when potentially offering differential behavior for influencing how other actors shall behave and that the issue of consciousness pragmatically is irrelevant. For example, an octopus does not have consciousness as humans do and this exploration is only concerned with the variety of appearance in which the octopus engages. By mastering, then, the ability to change skin coloring, an octopus certainly is an identity manager, and its influence is achieved not through direct physical impact or impulse but rather through sign [80: 62]. A sign draws effect, whichever way it is used, when the other actor is capable of sign consumption, i.e. semiosis, which is why a dia-enneadic model is required for adequately explaining sign exchange.

7.2.7 Security balance in the natural state

No matter how anecdotally this paper sketches the variety of ways of identity management, it must be acknowledged that identification is not a linear function with general validity. A reliable sign of identity is not *necessarily* good, let alone that a more reliable sign is *always* better. Again, it depends. Take the rabbit baby A_1 and his rabbit mother A_2. The mother wants optimal conditions for supervising her baby. For that 'situation' alone, the baby should carry markings marking him as clearly distinguishable. The baby's survival is promoted when the mother can easily establish control. The same markings, however, may threaten the baby's survival in a different 'situation' that is,

when a predator thereby can more easily recognize the baby rabbit as prey. So, it is a matter of ecological balance, a balanced capacity for self-identity management and constituents of evolutionary fitness. When new situations arise, the old capacities may not support survival. Yet, an evolutionary orientation introduces broader feedback constraints; successful self-security of a species at one stage may create the conditions for population decline and even extinction at another stage.

Orientation at the natural state shows that self-management of self-identity *is* also self-management of self-security. This may seem to be a contrived expression, but it helps to set the stage for dealing with *security* as a cultural, non-linear phenomenon. Instead, it is a matter of optimization. Security and risk should be a matter of balance within situations. Suggestive of such equilibrium, this balance would also be useful when it could be shown that social animal life exceeds self-management of self-identity.

However the question also arises as to when identity management become a group phenomena. Structurally what is meant is must be seen minimally as a tri-partite relationship. A third party enters the stage and such a third party relationships may also be brought onto the stage at the initiative of one or both of the original actors. Or, it may actually be a 'first party' who engages other actors for compliance, taking on the third party-role for that purpose. In such situations, the original actors A_1 and A_2 both, but of course each for his own motives, acknowledge the 'third party' as an intermediary for identity issues which depends upon the behavioral precision allowed. For example, with the matriarch of an elephant herd, which party determines mating partners? Such a question might be left open at this point, but perhaps the answer has already been suggested elsewhere through ethological research. Another example may raise the question as to which identity issues are at the discretion of the church patriarch within a religious community.

7.2.8 Individual relational experience

Before engaging a third party in the generalized theory of identity management, however, a more basic social concept should first be developed: relationship. It is seen as a corollary of methodological individualism [5; 30] that social relationships are essentially one-to-one that is, between A_m and A_n. Please note that the generalization is being taken another step by introducing symbols m and n more closely associated with variables. At that point then, A_m is an individual 'object' capable of semiosis. This requirement includes memory. Remaining with A_m's perspective, A_n reflects a pragmatic belief held as well and can therefore be anything imagined. What counts here is that A_m believes A_n exists and that there is a relationship that is, there is an engagement of sign exchanges.

Let A_m and A_m *both* be actors which emphasizes that the relationship is an individual experience for *each one* of the party. In other words, A_m's experience of relating to A_n is *different* from A_n's experience of relating to A_m. This short-hand notation, for the sake of convenience, is $a_n@A_m$, respectively $a_m@A_n$. A special case is $a_m@A_m$: self-experience.

A particular behavior by A_n can of course impact $a_n@A_m$, that is, act as a constituent of the relational experience A_m holds over A_n, *only* when A_m actually attributes it to A_n. It follows that all behaviors in one actor's particular relational experience have at least the other, related actor as their common attribute. For A_m, A_n appears both diachronically identical (the same individual 'other' is involved throughout) and different (at the level of the behaviors). It is therefore an identity attributed to A_n with which A_m connects separate behavioral experiences with A_n into an overall relational experience, underlying the experience of their relationship's continuity. The experience of other-identity provides the focus for relational experience. Identity management is conditional for relational experience. As has already been emphasized, modes of identity signification can vary widely. The potential prey A_n who applies camouflage, for example, tries to remain outside predator A_m's $a_n@A_m$, at least while the dangerous 'situation' lasts.

In the natural state, $a_n@A_m$ strictly resides inside A_m. In other words, the instrument for the achievement of relational experience is internal memory only [76]. Relational experience is instrumental for re-cognition. Ample opportunities for error exist, though. For example, A_m may identify another individual, but the wrong one. Then, $a_n@A_m$ is not updated, but $a_p@A_m$ inadvertently is. A direct survival issue is how such 'mistakes' impact on A_m's own security, and on A_n's, A_p's, and so on for that matter. Once again, it seems that the requirement for precision in identity management is contingent.

7.2.9 Subjective-situational identification requirements

To repeat, one actor's relational experience interprets behavioral exchanges with another actor. What one actor constructs and maintains as another identity serves as a continuous focus: by definition, A_n's identity acts (read: signs) as the pervasive attribute of behaviors occasioning A_m to form relational experience $a_n@A_m$.

This paper's concept of methodological individualism implies that when two individuals interact, they may have *different* needs for attitudes toward identification. It is important to emphasize this point since it lies at the basis of realistic identity management. To illustrate, a sketch of the space of variable identification is given in more detail.

Let A_m initiate an interaction. What stand might possibly be taken by the other actor's identity? An assistant working behind the counter in a bakery shop may not be particularly interested in positively identifying a customer who has walked in and orders a single loaf of bread. However, the assistant's interest may change when the stranger orders fifty apple pies and wants to take them immediately yet reports that s/he will be back later to pay for them. Or consider someone who requests to drive off in a brand-new car without leaving sufficient payment as 'security', which certainly would instigate additional scrutiny by the sales person.

Against the background of a dynamic relationship, identification requirements vary by situations that is, from interaction to interaction. At one end of the range, A_m becomes positively and completely certain about a co-actor's identity. So, A_m seeks full guarantees to 'know' the other actor as A_n and then updates $a_n@A_m$ with corresponding

confidence. The other end of the range for identification requirements consists of A_m refusing to
learn about the identity of the other actor who participates in that interaction. Of course, when the
other actor is perfectly anonymous, for all n, A_m's
relational experience $a_n @ A_m$ remains unchanged.
Even though the initiative for the interaction lies
with A_m, that position may be just what is wanted.

Still taking the cue from A_m, the attitude with
respect to A_n identifying him with a particular behavior will vary, too. A similar range applies. At
the one end, A_m would like to be as certain as possible that A_n correctly identifies A_m. At the other
end of the attitudinal range, when A_m does not
want any doubt left that involvement is to be secret, action is anonymous. With two such ranges,
a two-dimensional space covers the interactional
perspective on identity for one actor committing to
an interaction with another actor. See Figure 7.7.

Of course, every *other* actor just as well enter-
tains an attitude toward identifying all actors in any
interaction. (For overview's sake, the number of
actors is limited to two.) Figure 7.7 therefore holds
for both A_m and A_n that is, for A_n with n as an
even more general variable. Closer inspection reveals that at this level of generality it does not matter if one actor takes the initiative for an interaction. For example, when A_m wants to be explicitly
recognized, A_n can nevertheless stubbornly try to
avoid learning about A_m's identity.

The attitude of A_m, respectively A_n may be
thought of as occupying a point in two-dimensional
space, and these related representations express the
extent of symmetry or asymmetry in their attitude
regarding actors' identification. In Figure 7.8, the
axis for symmetry is added as a broken line. For
example, when a message arrives with the request
to pay a certain amount in taxes, does an actor accept at face value that it is a message sent by the
appropriate tax authority? Proportional 'amounts'
of trust must first be established.

Interactions for which actors apply symmetrical or matching perspectives are *by definition* conducted smoothly on identification which, again,
may include a demand for anonymity. Asymmetri-
cal attitudes make interaction problematic but can
be 'solved' in a variety of ways. One actor, or both,

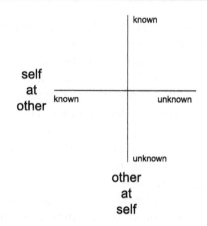

Figure 7.7. Single actor's interactional perspective on multiple actors' identities.

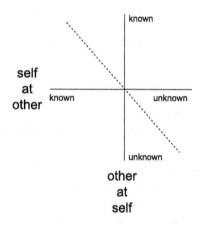

Figure 7.8. Symmetry or matching attitudes on identification.

may simply forgo the interaction. The withdrawal
of any one actor is the end of the exchange. Of
course, the actor should be free to 'act' in such
a way. One actor may, for example, consume the
other actor and thus mark the consummation of the
exchange. When freedom is illusory, as it often is,
the conflict is solved so as to facilitate the subsequent exchange. Actors may negotiate with symmetry and arrive at a consensus. Or, dependent on
their relationship, the powerful actor may simply
refuse to budge, forcing the weaker actor to change
attitude in order to comply. Power dictates symmetry; but, then, the weaker actor does not really comply so much as pretend symmetry, trying to hold
out from a position of less insecurity where he believes his interests are less liable to suffer and wait

for a more even working opportunity in order to shift the balance.

7.3 SOCIAL PRACTICES IN IDENTITY MANAGEMENT

7.3.1 In(s) and out(s) of institutions

Identity management occurs in relationships, and actors engage in interactions from their respective motives. In other words, an exchange is always a meeting of an individual's interests, or will. A general concept of security, then, is whatever promotes interests. As interests vary with a situation and a subject, so do the factors that contribute to, and may be experienced as, an individual's security. It follows, then, that a history of identity management is essentially a *political* history, and in terms of power and knowledge, Foucault draws a similar conclusion [28].

This historical sketch starts by examining several social institutions. What is an institution? It should be no surprise to consider an institution as a type of situation. More precisely, an institution is an(other) attempt at control which means it is actually an institution-*use* that is the helpful concept for social psychology. Wittgenstein, in a similar way, classified language when he emphasized language *use* [95].

Through establishing and maintaining a particular situation, participants are subjected to behavioral control. One person seeks compliance from another person on the alleged authority of an alleged institution (read: collective identity, which of course is a contradiction in terms). Whenever an exchange can be 'ordered' as taking place within a particular institution's sphere of influence, the actors involved are counted upon to exhibit corresponding behaviors.

Kinship is often regarded as the prototype of institutions.

> [T]he study of varieties of kinship arrangements is one of the oldest interests of anthropology. Like language, kinship studies demonstrate the power of culture to form systems of thought and behavior on the level of groups rather than on the level of individuals [37: 260].

> [T]wo principles are expressed in the organization of domestic life everywhere. The first is relatedness through *descent* or parentage. The second is relatedness through marriage or *affinity*. People who are related to each other through descent or a combination of affinity and descent are relatives, or kin. And the domain of ideas constituted by the beliefs and expectations kin share about one another is called *kinship*. [. . . M]arriage may [. . .] establish "parentage" with respect to children who are biologically unrelated to their culturally defined "father". [. . .] A similar distinction is necessary in the case of "mother" [37: 261].

The cultural character of kinship emphasizes its acquired nature. A person must first of all learn what kind of relative s/he is to another person. What is actually learned is a system, not so much of labels but of behavioral patterns for interaction. Kinship effectively organizes role learning by aspiring actors. Wittgenstein, arguing how much of education is basically rote training (German: *Abrichtung*) [95: 4], would probably consider kinship a language game as well. When a subsequent exchange classifies as a particular relative-ness, corresponding acquired behavior(s) should also be triggered.

The concept of relative draws attention to a characteristic that makes institution a key perspective on identity management. An institution resides in a larger culture-scape (such a 'scape' also being an encompassing institution, of course). Thus it becomes clear how an institution is occupied with entry control. Every institution divides a population. Insiders belong to an institution and outsiders do not. Note that institutional criteria are applied for classifying insiders and outsiders, respectively. An individual person may have different preferences. For example, no, I do not want to interact as the cousin to this aunt, but yes, I would like to join my favorite soccer club. As the semiotic ennead indicates, identity 'connects' such various behaviors across situations; metapattern supports formal modeling.

Membership control relies on identification. At the scale of a small band of hunter-gatherers, near continuous bodily proximity seems to virtually (pun intended) guarantee one member's confidence in another member's identity-as-member. Anyone else qualifies as a non-member. So scale, surely, is an important determinant for institutionalized identity management, that is, for mediated attempts at a secure division between members and non-members. Immediate, sufficiently 'familiar' acquaintance among members cannot practically extend beyond a small community.

The kinship institution may well have arisen to accommodate an increasing scale of human habitat. Two different persons A_m and A_n may not be kin, for example, but there just might be a person A_p, who is kin to both. Where kin still equals immediate familiarity, A_p can act as identification broker between A_m and A_n. Having networks of kin overlap provides an orally mediated device for identity management at the scale of the larger community/society. Of course, this conjecture stretches the relevance of kinship well beyond the domestic sphere.

Overlapping networks may no longer be efficient for identity management as, for example, the distance in kin between A_m and A_n exceeds a certain number of intermediaries. Rather than a single A_p remarking to either A_m or A_n, or to both, that A_p could vouch for their respective identities, it is now up to A_m and/or A_n to chart a path along several linking points. Even if that could be managed, the incentive for intermediaries to participate in such a chain of identification is heavily compromised. The question then arises as to what each member stands to gain from such small contributions compared to an effort which is not elastic.

Actors who aim to engage in exchange, but who find themselves at a significant distance in original kin, reflect a dimension of social change. In such a 'developed' society, often the concept of kin does not adequately support the identity aspect of relationships that actors require for interactions in order to secure interests. Actors can gain when they also become organized around their special interest. They actually engage in kin, too, albeit not of the 'relative' kind. Actually, the so-called special interest group nowadays is such a common form of an institution made up of affiliations that, at least in western culture, any orientation of an original kin, beyond the nuclear family, is exceptional.

Extending the concept of kin to special interest helps to explain why the original kinship system of

identity management continues to function. What was added, and still is, are ever new and changing personal networks: a person has traditional relatives, but for example also colleagues in civil engineering, neighborhood residents, and so on. The social variety operating on identity has correspondingly increased [3], and the ubiquity of such networks has kept identification brokerage as an efficient approach to identity management in daily life. A popular saying suggests that anyone can be reached through at three phone calls (which, of course, assumes that everybody involved can be contacted through the same medium; today that medium might now be email). However, the nature of identification may vary from one institution/network type to another.

Religion is also relevant for identity management which possibly can best be appreciated from a hypothesis about its origin. Humans become conscious of life's uncertainties and motives then drive each person to attempt to control events in order to secure results that benefit their own interests. While it is usually clear enough how to direct actions, a person can be left with uncertainties. Mysticism or religion, then, are identified targets for attention, and once a particular range of events can be explained through this lens, or at least be addressed, humans can start to behave in concert with that institution. A person, then, may try to promote events or prevent them from happening. Either way, the actor who a person now assumes to be 'responsible' for events happening in a particular way may also be influenced by offerings. A sign is an offering, too: humans request compliance from whomever or whatever they consider an addressable actor. Religion, therefore, is essentially an interest-based worldview.

It is impossible to hold unbiased views because motives pervade action which is precisely why Brouwer sets mysticism as a necessary limit to rational explanation and development. However, Brouwer's attitude should not be confused with irrationality; in fact, he meant quite the opposite that is, he recognized that rationality's natural limit is optimal [13].

A church is of a different order than religion. Originally, it may have been that a church helped to organize the requests a society's members direct at what they identify as axiomatized actors or deities. When all members participate, a church actually contributes to organizing the whole of that society. It can also be that an encompassing institution, such as a state orders that its inhabitants should participate without exception. A church is then directly applied for state interests (with the state, according to methodological individualism, like the church, being an institution serving interests of particular persons). As it happened, Christianity was proclaimed Rome's state religion in year 324 CE. From the perspective of identity management it is especially interesting to see how its corresponding institution that is, i.e. the Roman church borrowed from original kinship for prescribing behavior: father, son, and even mother.

For the western world, the middle ages coincided with church-mediated identity management. Societies were feudal; secular lordships (or ladyships) might change abruptly; and the church was a stabilizing force. The church also organized at the scale of immediate personal contacts. Among a largely sedentary and illiterate populace, not to mention how (il)literate the worldly rulers were, the church does not even have to develop institutionalized identity management in order to aid control.

7.3.2 The rise of pluriformity in power

The middle ages ended when the hierarchy, as constituted by a limited number of power institutions, disintegrated [6]. The mutual reinforcement of wide-ranging developments, such as book printing and overseas trading resulted both in the growth and the redistribution of wealth and consequently power. A significant number of persons became citizens. As recognized stakeholders in a particular society, these citizens demanded, and were increasingly awarded, legal rights. Nonetheless, specific rights still cannot be properly managed without a clear identity management.

An important distinction to make is that wealth in the age of emerging citizenship is fairly fluid. Money lies at the heart of identity management, and it provides a single dimension for expressing the value of exchanges. When money is identified

in its own right, so to speak, economic activity is promoted. As distance in trust between participants increases (and in tandem, the distance in power decreases through a lack of a common legal framework), a step toward securing rights was to record exchanges or transactions. It follows, then, that the legal framework more or less prescribes what aspects in a transaction need to be identified, and how rights are to be supported.

When a transaction occurred, ownership changed. Citizens acquired the basic right to be owners, also, and for particular classes of objects, ownership entailed subsequent rights as well as duties. For reasons of administration (read: protection and enforcement), ownership was registered according to the ruling legal framework.

The formation of nation states and the development of citizenship are mutually reinforcing aspects of overall change [50]. A sovereign state frees citizens from the instability of inherited rule and/or arbitrary change of rule that resulted from marriages within the noble class.

As it is, a state of citizens requires an explicit legal framework for social control.

Interestingly, though, early citizen registration and citizen-related ownership usually did not happen at the scale of the nation state. In many countries where citizen rights were formally established, churches kept the records of the main events in a person's life (birth, marriage, parenthood, death) while records of real estate property were kept by the local authority. Many other special-purpose registrations also originated in the centuries immediately following the middle ages. Often, as to be expected in states where citizens are increasingly participating in government, money was a key motivator. For example, the legal framework allowed for entities to participate with a joint stock in economic transactions (the concept of their limited liability was only added in the nineteenth century). This trend of recognizing actors with their possessions and contributions more individually soon permeated the wider culture. The new sciences of empiricism, for example, were also set on a course of identification that is, general laws of nature and detailed classification schemes.

7.3.3   We, the standard

By the end of the eighteenth century, an elite citizen force had already for some time ruled countries such as England and The Netherlands. Relatively speaking, some significant aspects of a democracy had developed.

An absolute monarch still ruled France, so when that change came about, it was radical. Yet in important ways, the French revolution attempted to redesign the single power hierarchy (a pattern which was repeated by the Russian revolution). The state was then placed at the pinnacle, but the practice of policing continued or even intensified [29: 153]. Institutions that had collaborated with and effectively shaped the ancient regimes were banned or at the least quietly subsided. The church, for example, lost its immediate political position, but there was a revolutionary logic of state control, which is why France set a new direction for itself in identity management.

In the wake of the French revolution, deliberate efforts at standardization were undertaken. Money was recognized as only one particular aspect of a rationalized (e)valuation. Other aspects were also explicitly brought to bear upon identification that had a state authority to reinforce it. Standardized measures for length and weight, for example, were scientifically designed and developed and politically enforced with diffusion of their application rapid throughout the Napoleonic empire.

State involvement both accompanied and supported by standardization entailed identity management as a genuine social infrastructure. How different states were involved varied widely, though. At one extreme, a state might limit itself to conditions for social dynamics according to a general democratic policy; but at the other extreme the infrastructure for identity management would be state-controlled for the purpose of day-to-day force over all inhabitants' behaviors. Major infrastructural components were largely invariant, however; which explained why divergent political systems could co-exist. Supranational standardization in measuring (read: identification) certainly promoted exchange. However, it also promoted state control. And constraints invariably ruled, for only a subset of variety was officially recognized. A bureaucracy

cannot differentiate, for that matter no one can, beyond the standards applied for identification [8].

Culturally, too, security is a matter of equilibrium. From the perspective of an individual citizen, infrastructure will be a positive value when it serves a citizen's interests and when there is no reasonable alternative available. On the other hand, infrastructure will hold a negative value when an individual citizen views it as an obstacle or as a threat.

This civil 'equilibrium of state' is dynamic, and many states pursue an orientation toward service. The citizen is primarily considered a tax-paying customer; infrastructure makes for efficient delivery of services. A comprehensive array of identifying registrations (person, address, house, car, etc.) is available and such information can also be used for different purposes. It is a legitimate question to ask whether or not citizens would consent to actual usage of this collected information. When a citizen is suspected of fraud, asking for permission to use personal information collected is hardly relevant, but it is relevant to ask where the state should set limits on surveillance of its citizens. The constituent aspects of civil equilibrium are so interwoven as to make an evaluation burdensome for the citizen whose purpose is to get on with life's most immediate concerns. On the other hand it very well might be precisely the impression of interests ill-served that makes citizens suspicious of their own 'state' whether or not such impressions are factually supported. The tension for citizens occurs from the very complexity of the situation that is, the lack of transparency of the state's intentions. In defiance then, some citizens will view the solution to be that of managing their personal identification(s) and the identifications of what is 'owned' by each citizen individually. However, that option has never been available. Identity management is always a relationship. What should be questioned, though, is the one-sidedness as civilly experienced that is, a state's formal domination of that very relationship with correspondingly biased rules for identification.

Of course it is true that digital information and communication technology has allowed for an increasingly networked human life, and perhaps a too quick response based upon previous conditions would not satisfy. Active participation of citizens in government now requires another fundamental effort at social (re)engineering. Security from variety should be balanced with security from standardization/uniformity. The civil demand for variety increases with the distribution of knowledge [84].

7.3.4 Attention management

A post-modern citizen is more or less safe from physical harm during protracted periods. Cannibalism is rare nowadays so not many people are going to be eaten, for example. Excluding for the moment types of physical abuse, the post-modern citizen nevertheless is still heavily targeted [7; 87]. A multitude of other actors are always attempting to engage others in various interactions, and these attempts are increasingly of a commercial nature [73]. In a crowded society, where spending power is limited and competition abounds, an actor therefore might shift energy to getting the attention of another actor in the first place. Specialized attention-getting interactions aim to set up and maintain a relational experience. Brands serve efficiently in attention management [1]. Identities for life styles are constructed, and products become transient properties of a continuing brand. The targeted actor should swiftly gain and maintain the impression that personal interests are optimally served by continuing the relationship, and for precisely that purpose, the individual is offered a chance to participate in subsequent interactions, in exchange for some financial resources, of course. The illusion of difference promotes uniformity [41; 44].

Attention management is natural, as ethological studies demonstrate. Political and commercial institutions, too, have always practiced attention control through identity management (if only by blurring their own identity or of ignoring the identity of the 'other'). So have individual persons in cultural settings (and to a large extent through their institutions). Adolescence, for example, is a young person's revolt against the institution of parental rule. Parents who completely ignore their child's claims of growing independence frustrate the teenager's

individual development (and in the process make their own lives miserable since a child will not stop claiming grounds for independence). Luckily, the damage is often limited. Usually parents learn to play their new roles in most cases behind cue but still close enough to accommodate some measure of dynamics of equilibrium. Consequently the adolescent grows up. No longer a child, one time or another s/he 'simply' leaves and ventures out autonomously into the wider world.

When it happens that an adolescent does not want to leave home and instead stays, there is still a compromise to be reached: the young adult will need sufficient space to pursue personal interests yet may still be under control of the home space. Such conflicts of interests between the parent and the young adult still living at home has always occurred, and both actors in this drama attempt to exploit asymmetries. One actor may ignore the other's request for compliance who in turn may feel frustrated with the young adult's stymied development. But then, that particular actor has apparently developed enough of an identity (read: motivated individuality) to state a request. In fact, it is the very ability to state a request that precisely constitutes a living being's individuality. This dynamic, then, also explains why a dictator meets insurgence with radical repression that is, murdering opponents and their relatives and friends. Ignoring resistance effectively fuels it and so does repressing resistance, for that matter. A dictator merely gains time as other people to raise the courage for a similar request directed against him (and the regime, of course). As murdering and other untoward events transpire, the dictator must also request compliance. Public executions or concentration camps are signs that a dictator deters other persons from offering resistance, forcing them into submission. Once again: $a_n @ A_m$.

A state may conveniently label activities as terrorism, and an analysis under the heading of identity management may help to clarify confusion over this social issue with Popper's (1902–1994) concept of the open society provides a perspective [69]. His concept details that a citizenship can change the government. 'The people' are given the opportunity to do so at set intervals when free elections are held. No physical force is required; the old government leaves; the new government enters. The difference resulting from a democratic change in government is usually small, though, as much of the base administration remains.

Much in the same vein, Etzioni conceives of the active society where actors opt

> for the uninhibited, authentic, educated expression of an unbounded membership [27: 13].

When citizens direct force against their *own* state's institutions, or social processes in general, something is wanted. The first point to establish is whether the target that citizens aim for under these circumstances is part of an open society. Suppose it is and that the citizens truly want to destabilize the current government. It is legitimate to wonder then if their interest lies with continuing an open society. Do they want people to vote for them, next time around? If so, are they drawing attention to their political program? Are they actually a political party? Do they properly manage its identity by committing violence? Are they really making a convincing statement for the open society? It seems reasonable to argue, from the perspective of the open society that is, for a case of severe maladjustment. They are criminals, perhaps medical patients and the inmates want to run the asylum.

It happens, though, that a citizen can be unfairly treated. Institutions in an open society cannot repress, but they certainly can ignore citizens and their legitimate needs. Open societies can also make mistakes so that in mounting frustration, citizens may well escalate their measures for getting the governments attention. By definition, the legal framework sets the limit for what is permissible as civil action, and the citizen who crosses it behaves criminally. However, his transgression should also be taken as a request that the identified institution may not be functioning as it should.

An open society represents political relativity and the potential for change can be a valid measure of the success of that open society. Some ideologies are closed in the sense of absolute rule, and violence committed in its pursuit of elimination of political change is dictatorial and therefore criminal.

At present, large scale violence mainly occurs at what seems an intersection of political systems.

One institution applies the label of terrorism in denial of institutionalism for another actor who resorts to violence precisely for the purpose of getting accepted as an institution. Mere acknowledgment of a political identity, however, does not result in halting violence, let alone guaranteeing a lasting peace. Struggle continues when actors find their interests are not yet given adequate attention by other actors. At the political intersection, a vacuum of legal framework operates and at that point identity management lacks a common ground. In its absence, actors continue with interactions to prepare such ground in their own, particular image. The semiotic ennead helps to explain how one and the same actor may display a wide variety of behaviors. The individual acts situationally.

7.3.5 Contingency management

Crime, terrorism, and war provide extreme threats to 'normal' citizens' security. Identity management in open societies requires a broader foundation. This paper argues that a defining characteristic of a citizen is his dependence on products and services provided to him which include ways of earning a living. A citizen is not at all self-sufficient but rather highly dependent, but of course, humans never were solitary animals. In citizenship, an individual partakes in a variety of chains of production and consumption. Economic parameters cause such chains to stretch; more and more often, citizens engage in specialized contributions. On the side of consumption, citizens reap benefits, such as choice, quality and price. There are also larger risks, though. As chains become more complex, control of both processes and results is getting difficult and sometimes impossible to achieve.

The vegetable specimen on a dinner plate, for example, may be from a genetically manipulated species, have its growth protected by pesticides, stimulated by fertilizers, and its shelf-life extended by yet other chemical treatments. Suppose a person gets ill from eating this vegetable, very ill. When more people suffer serious complaints and a medical pattern is established, this particular food chain becomes an official public health issue. Sooner or later, the 'product' is banned, the process chain redesigned, or whatever else it takes to make right the imbalance. Such reactive measures, if they are to be successful, depend of course on both extended and precise identification. Extension means that it must be possible to retrace the chain, including what actually happened and under whose responsibility for each step along the chain. Precision pinpoints the vegetables 'behavior' in the food-consumption situation, and it should of course also be possible to broaden the investigation to other situations in which the vegetable exhibits identified behavior(s).

Assume a system of registration is properly but minimally functioning for the food-consumption chain of the type of vegetables introduced in the above paragraph. To a point, all seems well that is, no harmful effects have occurred from consumption of this vegetable. The very existence of a formal registration may then cause insecurity, or at least have unintended negative consequences that bring on doubts. Variety control increases variety, and so on. Identity fraud entails, in this case, passing a different vegetable for the one that is certified. Such fraud may be committed for a variety of reasons. Bypassing certification should yield greater profit, and there does not actually have to be anything wrong with the substituted vegetable. In fact, it might qualify even better for certification than the previous vegetable. Then again, it might not and indeed constitute a risk to public health.

The general point for identity management is that a system of registration is always a sign in its own right, too. It was apparently judged necessary to establish trust through certification because some 'items' offered on the market fell short in consumer quality. The identification scheme may very well succeed in eliminating the offerings which originally caused concern. The dilemma consists in irreducibly constructing a basis for misuse of trust, too. However, trust is also what keeps fraud within certain bounds. When fraud undermines trust, or confidence, it effectively destroys the vehicle it requires to function at all. It is no coincidence that fraud is popularly called the confidence game or trick [97: 223].

7.3.6 Electronically mediated identity management

Like a coin, official identification has two sides. The aim may be to enhance security, but the con-

sequence may be that insecurity results. If a person can trust keeping belongings in a locked compartment, then it is the key that should be of concern. Not just the control side, but especially the risk side of identity management enters a new stage with the use of the electronic media that is, digital information and communication technology, and personal access is a good example.

The first idea to examine is physical access. A citizen may want to enter (or exit) an area which a particular institution controls. Traditionally, a guard often euphemistically called a receptionist stops the entrant. In most cases, this guard does not establish the visitor's identity at all. The visitor may be requested to register, but the information remains unchecked. Leaving citizen privacy aside, suppose the guard does check and for that purpose assume that the guard has information access to the state registration system containing standardized identifications of citizens. It is a legitimate point to wonder what level of certainty a check must provide, and that answer depends, of course. Suppose that, upon the guard's request, the citizen supplies a person's last name, the name of a town, of a street, and a house number. The state registration system responds to the search with zero, one or more facial pictures of persons. The guard can proceed with identification. If the guard believes that s/he can confirm a particular picture as an adequate representation of the person who wants to enter, that entrant has been positively 'identified' as the citizen.

This procedure can be repeated for areas contained within a larger area and the level of certainty required may be raised with each checkpoint, adding additional criteria at each check station, even to the point of asking what the purpose of the entrant's visit is, which in turn may be checked with the person the entrant is intent upon visiting.

In many organizations, employees physically enter a restricted area without interference by a human guard. They can pass after an automated check. For example, an employee can open a door by entering a code. Or a card he presents is scanned. And/or a biometric sample (fingerprint etc.) is taken, processed, and compared with stored templates.

When substituting a nation state for an organization and applying the identification process at a airport, it is easy to see that automated checks are always performed upon entry and exit by citizens. Indeed, the virtual guard controlling a particular state's border even has access to systems of identification registration of several other states. Rather than limiting an area for access, such supranational cooperation extends it. Prospects for identity fraud are correspondingly attractive.

Information access is increasingly disengaged from the restrictions for physical access. An employee does not necessarily have to travel to his office desk in the employer's building in order to view, change, and/or extinguish information. And with such e-governments, a citizen can handle transactions without physically having to visit the counterpart government institution and vice versa. The question of adequate levels of trust implemented is covered in the final paragraph outlined for a view of a possible future.

7.3.7 Open security

Trust does not necessarily correlate with security since the overriding concept is one of interest [33: 392]. An actor trusts what promotes his interest that is what consummates his motives. Therefore, usability can equally be correlated with trust, and a design issue then becomes balancing usability with security.

I believe the concept of an open society helps to focus trust. An open society secures a climate for non-violent creativity. A citizen must not only be free to design a policy in line with the basic idea of an open society, but also to present it in order to be elected for government in order to implement it.

How can an open society secure itself as it develops? An open society's essential focus must be the rights of its individual citizens and this focus comes with a serious obligation. An actor has the responsibility of identity as to how s/he stands for participation. It is a matter of a democratic government to determine which interactions are potentially of general interest to warrant a third-party memory of participation: $a_m @ A_p$. The third party is a particular government institution of the citizen's state.

The still growing density of social differentiation makes the modern distinction between a private and a public domain somewhat already outdated. It is now more apt to argue for a private and a public aspect of every single interaction even as the proportion varies, of course. At one extreme, it is to be determined what is completely private, and at the other extreme what is overtly public.

A sufficiently large public aspect (read: democratically-up-to-legally determined interest) requires proportional identification of both participating actors and objects otherwise referred to in their interaction. The demand for proportionality means that anonymity must also be safeguarded when deemed appropriate for extending the open society. There is, by the way, nothing against a person applying different identifications. Between interactions of sufficient public interest, though, it should be possible to relate the use of different identifications of a single person. A person may have an artist's name but be liable for income tax for all other identifications taken together. Rather than opposing multiple identifications, a state should invest in their management for its democratically allocated tasks. In the meantime, a citizen can more effectively guard privacy through different identifications where their connections are not allowed.

As private and public sectors integrate, a citizen can use state-guaranteed identification throughout, and the same idea applies to state-guaranteed object identifications, for example of cadastral plots. The increase in electronically mediated identification must, however, be compensated for and complemented with additional real-life verification. Admission to a hospital, state-funded or otherwise, could qualify. It would mean that the hospital is explicitly co-responsible for maintaining a quality register of personal identifications. For example banks and insurance companies could be given a similar status with respect to verification. An organization/institution would have to 'pay' for failure to comply. Then again, security may not compromise usability. The choice of 'natural' verification confrontations should benefit acceptance.

The freedom allowed a citizen to entertain different identifications will, in most cases, promote the general use of the state-guaranteed identification since a single 'key' is convenient. It works when a citizen can trust that records of relational experiences in which s/he figures are only shared between institutions when such an overview can reasonably be expected to increase security.

More and more interactions will be electronically mediated. At least technically, a single infrastructure or network of infrastructures will be available to carry the information traffic. Metapattern allows for conceptual integration through interdependent identifications [49; 89; 92] and such potential for open interconnection prioritizes authorization. An actor may have general access to and participation in certain interactions and information but may require permission for other interactions including specific information. Authorization involves yet another domain of explicit identification and combines the complexities of identifying actors, (other) objects, and process chains with separate activities. The unambiguous solution rests with individual situational behavior as the unit of authorization. For it has been argued that behavior is specific for a particular actor in a particular situation.

An open society, or civil society, as it is also called, democratically determines its risks. Every new government may set a different agenda where, for example, income redistribution fraud is considered a serious security risk or a part of taxation is simply not spent as it should be. Policies suffer while citizens are overtaxed and trust is undermined. Yes, democracy is especially at a dilemma against absolutist threats, supported by violence. However, the democracy of risk is the risk of democracy. An open society can only remain 'open' within a characteristic bandwidth of variety and its corresponding vulnerability. To try and operate outside such an open bandwidth would expose society to greater risks, such as that of rapid closure. Political closure is easy to step into but extremely difficult to recover from.

A pervasive, dynamic equilibrium in identity management is an achievement by participants in a particular culture. Being socially open and active demands a self-biased opposition against its 'other'. More regretfully, this demand can work

from the opposite direction as well as Foucault highlights [28]. A society which sees itself as open will also be in self-denial about its oppression; while, a dictatorship must be visible to be effective in its oppressions, it also holds the potential for being an open society or, as a contragram, the openness of oppression is the oppression of openness).

7.3.8 Identifying relevance

Within any scope, it is impossible to give an exhaustive treatment of identity management and security issues surrounding it. With Bertrand Russell's remark, albeit in a different context, it is best summarized: "[t]he net result is to substitute articulate hesitation for inarticulate certainty" [78: 9]. With this idea, what a semiotics of identity management loses in technical constraint might very well be gained in social–political relevance. A wide range of behaviorally relevant phenomena may be included in both historical analysis of and design (read: synthesis) for future identity management.

REFERENCES

[1] D.A. Aaker, *Managing Brand Equity*, The Free Press, New York (1991).

[2] W.R. Ashby, *An Introduction to Cybernetics*, Methuen, London, UK (1964), originally published in 1956.

[3] R.F. Baumeister, *Identity: Cultural Change and the Struggle for Self*, Oxford University Press, New York (1986).

[4] *The Bible*, New International Version, Hodder & Stoughton, London, UK (1989).

[5] P. Birnbaum, J. Leca (eds), *Individualism, Theories and Methods*, Oxford University Press, Oxford, UK (1990), originally published in 1986 in French.

[6] W.P. Blockmans, *Geschichte der Macht in Europa: Völker, Staaten, Märkte*, Frankfurt/M, Germany (1998), originally published in English in 1997.

[7] D.J. Boorstin, *The Image*, Penguin, Harmondsworth, UK (1963), originally published in 1962.

[8] G.C. Bowker, S.L. Star, *Sorting Things Out: Classification and Its Consequences*, MIT Press, Cambridge, MA (1999).

[9] A.A. Brennan, *Persons, Handbook Metaphysics and Ontology*, Analytica Series, H. Burkhardt and B. Smith, eds, Philosophia, Munich (1991), pp. 682–684.

[10] J. Brent, *Charles Sanders Peirce*, Indiana University Press, Bloomington (1993).

[11] L.E.J. Brouwer, *Leven, Kunst en Mystiek*, Waltman, Delft, Netherlands (1905). Translated into English as *Life, Art, and Mysticism*, Notre Dame Journal of Formal Logic 37 (3) (1996), 389–429. With an introduction by the translator, W.P. van Stigt.

[12] L.E.J. Brouwer, *Letter to D.J. Korteweg, November 13th, 1906*, *Over de grondslagen der wiskunde*, Mathematisch Centrum, Amsterdam, Netherlands (1981), pp. 14–18. The letter is only included in the 1981 edition.

[13] L.E.J. Brouwer, *Over de grondslagen der wiskunde*, Mathematisch Centrum, Amsterdam, Netherlands (1981), with materials added by the editor D. Van Dalen, originally published in 1907.

[14] L.E.J. Brouwer, *Unpublished fragments*, *Over de grondslagen der wiskunde*, Mathematisch Centrum, Amsterdam, Netherlands (1981), pp. 25–35. Fragments are only included in the 1981 edition.

[15] H. Burkhardt, B. Smith (eds), *Handbook of Metaphysics and Ontology*, Analytica Series, Philosophia, Munich, Germany (1991), two volumes.

[16] R. Capurro, *Heidegger über Sprache und Information*, Philosophisches Jahrbuch 88 (1981), 333–343.

[17] L.J. Carr, *Analytical Sociology: Social Situations and Social Problems*, Harper, New York (1955).

[18] K.C. Clatterbaugh, *Identity*, *Handbook Metaphysics and Ontology*, Analytica Series, H. Burkhardt and B. Smith, eds, Philosophia, Munich (1991), pp. 379–381.

[19] S. Clegg, *Foucault, power and organizations*, *Foucault, Management and Organization Theory*, A. McKinlay and K. Starkey, eds, Sage, London, UK (1998), pp. 29–48.

[20] A.W. Combs, D. Snygg, *Individual Behavior: A Perceptual Approach to Behavior*, Harper, New York (1959), revised edition, originally published in 1949.

[21] *Concise Routledge Encyclopedia of Philosophy*, Routledge, London, UK (2000).

[22] J.D. Crossan, *The Historical Jesus: The Life of a Mediterranean Jewish Peasant*, T&T Clark, Edinburgh, UK (1991).

[23] J. Deely, *Four Ages of Understanding: The First Postmodern Survey of Philosophy from Ancient Times to the Turn of the Twenty-first Century*, University of Toronto Press, Toronto, Canada (2001).

[24] J. Dewey, *Essays in Experimental Logic*, Dover, New York (1963), originally published in 1916.

[25] J. Dewey, *Logic, The Theory of Inquiry*, Rinehart and Winston, Holt, NY (1960), originally published in 1938.

[26] A. Etzioni, *A Comparative Analysis of Complex Organizations*, 2nd edition, The Free Press, New York, (1975), originally published in 1961.

[27] A. Etzioni, *The Active Society: A Theory of Societal and Political Process*, The Free Press, New York (1968).

[28] M. Foucault, *Discipline and Punish: The Birth of the Prison*, Penguin, Harmondsworth, UK (1979), originally published in French in 1975.

[29] M. Foucault, *The political technology of individuals*, *Technologies of the Self*, L.H. Martin, H. Gutman and P.H. Hutton, eds, Tavistock Press, London, UK (1988), pp. 145–163.

[30] M.P.M. Franssen, *Some Contributions to Methodological Individualism in the Social Sciences*, Amsterdam, Netherlands (1997).

[31] V. Gerhardt, *Individualität: Das Element der Welt*, Beck, Munich, Germany (2000).

[32] E. Goffman, *The Presentation of Self in Everyday Life*, Penguin, Harmondsworth, UK (1980), originally published in 1959.

[33] S. Gosepath, *Aufgeklärtes Eigeninteresse: Eine Theorie theoretischer und praktischer Rationalität*, Suhrkamp, Frankfurt/M, Germany (1992).

[34] T.A. Goudge, *The Thought of C.S. Peirce*, Dover, New York (1959), originally published in 1950.

[35] J.J. Gracia, *Individuality, Individuation*, *Handbook Metaphysics and Ontology*, Analytica Series, H. Burkhardt and B. Smith, eds, Philosophia, Munich (1991), pp. 385–388.

[36] P.P. Grassé, *Das Ich und die Logik der Natur*, List, Munich, Germany (1973), originally published in French in 1971.

[37] M. Harris, *Culture, People, Nature: An Introduction to General Anthropology*, 7th edition, Longman, New York (1997).

[38] J.D. Haynes, *Meaning as Perspective: The Contragram*, Thisone, Palmerston North, New Zealand (1999).

[39] J.D. Haynes, P.E. Wisse, *The relationship between meta-pattern in knowledge management as a conceptual model and contragrammar as conceptual meaning*, *Proceedings of the First Workshop on Philosophy and Informatics*, Deutsches Forschungszentrum für Künstliche Intelligenz, Kaiserslautern/Saarbrücken, Germany, Research Report 04-02 (2004).

[40] W.K. Heisenberg, *Physik und Philosophie*, Ullstein, Frankfurt/M, Germany (1959).

[41] M. Joseph, *Against the Romance of Community*, University of Minnesota Press, Minneapolis, MN (2002).

[42] R. Keefe, P. Smith (eds), *Vagueness: A Reader*, MIT Press, Cambridge, MA (1997).

[43] K.L. Ketner, *His Glassy Essence: An Autobiography of Charles Sanders Peirce*, Vanderbilt University Press, Nashville, TN (1999).

[44] N. Klein, *No Logo: Taking Aim at the Brand Bullies*, Picador, New York (1999).

[45] A. Korzybski, *Science and Sanity: An Introduction to Non-Aristotelian Systems and General Semantics*, International Non-Aristotelian Library Publishing Company, Lakeville, CT (1958), originally published in 1933.

[46] L. Krappmann, *Soziologische Dimensionen der Identität: Strukturelle Bedingungen für die Teilnahme an*

Interaktionsprozessen, Klett-Cotta, Stuttgart, Germany (2000), originally published in 1969.

[47] G. Lakoff, R.E. Núñez, *Where Mathematics Comes from: How the Embodied Mind Brings Mathematics into Being*, Basic, New York (2000).

[48] J. Langone, *The Mystery of Time: Humanity's Quest for Order and Measure*, National Geographic, Washington, DC (2000).

[49] S.B. Luitjens, P.E. Wisse, *De klacht van de Keten: een Erasmiaans perspectief op Stroomlijning Basisgegevens*, Ictu, The Hague, Netherlands (2003).

[50] C.B. Macpherson, *The Political Theory of Possessive Individualism*, Oxford University Press, Oxford, UK (1962).

[51] G. Mannoury, *Methodologisches und Philosphisches zur Elementar-Mathematik*, Van Gorcum, Assen, Netherlands (1909).

[52] G. Mannoury, *Mathesis en Mystiek: een signifiese studie van kommunisties standpunt*, Wereldbibliotheek, Amsterdam, Netherlands (1925).

[53] G. Mannoury, *Handboek der Analytische Signifika*, Kroonder, Bussum, Netherlands, vol. 1 (1947), vol. 2 (1948).

[54] P.R. Marler, editorial consultant, *The Marvels of Animal Behavior*, National Geographic Society, Washington, DC (1972).

[55] L.H. Martin, H. Gutman, P.H. Hutton (eds), *Technologies of the Self*, Tavistock Press, London, UK (1988).

[56] T. Masson, *Aphorism*, *Peter's Quotations: Ideas for Our Time*, L.J. Peter, Morrow, New York (1997), p. 265.

[57] A. McKinlay, K. Starkey (eds), *Foucault, Management and Organization Theory*, Sage, London, UK (1998).

[58] B.P. McLaughlin, *Connectionism*, *Concise Routledge Encyclopedia of Philosophy*, Routlege, London, UK (2000), pp. 166–167.

[59] Th. Molnar, *God and the Knowledge of Reality*, Basic, New York (1973).

[60] R. Müller-Freienfels, *Philosophie der Individualität*, Felix Meiner, Leipzig, Germany (1921).

[61] S. Navati, M. Thieme, R. Nanavati, *Biometrics: Identity Verification in a Networked World*, Wiley Computer Publishing, New York (2002).

[62] W. Nöth, *Handbook of Semiotics*, Indiana University Press, Bloomington, IN (1995), original publication of English-language edition in 1990, originally published in 1985 in German.

[63] C.S. Peirce, *Logic as semiotic*, compilation by J. Buchler from three selected manuscripts dated 1897, 1902 and 1910, respectively, *Philosophical Writings of Peirce*, J. Buchler, ed., Dover, New York (1955), pp. 98–119.

[64] C.S. Peirce, *Pragmatism in retrospect: a last formulation*, manuscript (1906), *Philosophical Writings of Peirce*, J. Buchler, ed., Dover, New York (1955), pp. 269–289.

[65] C.S. Peirce, *Philosophical writings of Peirce*, J. Buchler, ed., Dover, New York (1955).

[66] L.J. Peter, *Peter's Quotations: Ideas for our Time*, Morrow, New York (1997).

[67] C.P. Pfleeger, *Data Security*, *Encyclopedia of Computer Science*, A. Ralston and E.D. Reilly, eds, Petrocelli-Charter, New York (1976), pp. 403–406.

[68] K.R. Popper, *The Logic of Scientific Discovery*, Hutchinson, London, UK (1968), originally published in German in 1934.

[69] K.R. Popper, *The Open Society and its Enemies*, 4th edition, Harper & Row, New York (1962), originally published in 1945.

[70] A. Ralston, E.D. Reilly (eds), *Encyclopedia of Computer Science*, 3th edition, reprint, International Thomson Computer Press, London, UK (1976), originally published.

[71] N. Rescher, *Philosophical Standardism: An Empiricist Approach to Philosophical Methodology*, University of Pittsburgh Press, Pittsburgh, PA (1994).

[72] V. Riegas, Ch. Vetter, *Zur Biologie der Kognition*, Suhrkamp Frankfurt/M, Germany (1990).

[73] A. Ries, J. Trout, *Positioning: The Battle for Your Mind*, McGraw-Hill, New York (1981).

[74] J. Rogerson (ed.), *The Oxford Illustrated History of the Bible*, Oxford University Press, Oxford, UK (2001).

[75] R.M. Rorty (ed.), *The Linguistic Turn: Recent Essays in Philosophical Method*, University of Chicago Press, Chicago, IL (1967).

[76] B.M. Ross, *Remembering the Personal Past: Descriptions of Autobiographical Memory*, Oxford University Press, New York (1991).

[77] B. Russell, *Mysticism and Logic*, Unwin, London, UK (1963), originally published in 1917.

[78] B. Russell, *An Inquiry into Meaning and Truth*, Penguin, Harmondsworth, UK (1973), originally published in 1940.

[79] H.W. Schmitz, *De Hollandse Signifika: Een reconstructie van de geschiedenis van 1892 tot 1926*, Van Gorcum, Assen, Netherlands (1990), originally published in 1985 in German.

[80] A. Schopenhauer, *Über die vierfache Wurzel des Satzes vom zureichenden Grunde*, Felix Meiner, Hamburg, Germany (1957), reprint of 1847-edition, first edition originally published in 1813. Translated into English as *On the Fourfold Root of the Principle of Sufficient Reason*, Open Court, La Salle, IL (1997), translation originally published in 1974.

[81] A. Schopenhauer, *Die Welt als Wille und Vorstellung*, Diogenes, Zürich, Switzerland (1977), reprint of 1859-edition, 4 vols, originally published in 1818. Translated into English as *The World as Will and Representation*, Dover, New York, vol. 1, reprint 1969; vol. 2, reprint 1966, translation originally published in 1958.

[82] L. Sklar, *Space–time*, *Handbook of Metaphysics and Ontology*, H. Burkhardt and B. Smith, eds, Philosophia, Munich (1991), pp. 850–852.

[83] G. Spencer-Brown, *Laws of Form*, Dutton, New York (1979), originally published in 1969.

[84] N. Stehr, *The Fragility of Modern Societies: Knowledge and Risk in the Information Age*, Sage, London, UK (2001).

[85] P.F. Strawson, *Individuals: An Essay in Descriptive Metaphysics*, Methuen, London, UK (1964), originally published in 1959.

[86] D. Thompson (ed.), *The Concise Oxford Dictionary of Current English*, 9th edition reprinted with corrections, Clarendon Press, Oxford, UK (1998), originally published in 1911.

[87] C. Türcke, *Erregte Gesellschaft: Philosophie der Sensation*, Beck, Munich, Germany (2002).

[88] H. Vaihinger, *Die Philosophie des Als-Ob: System der theoretische, praktischen und religiösen Fiktionen der Menschheit auf Grund eines idealistischen Positivismus*, Felix Meiner, Leipzig, Germany (1918), originally published in 1911.

[89] P.E. Wisse, *Metapattern: Context and Time in Information Models*, Addison-Wesley, Boston, MA (2001).

[90] P.E. Wisse, *Semiosis & Sign Exchange: Design for a Subjective Situationism, Including Conceptual Grounds of Business Modeling*, Information Dynamics, Voorburg, Netherlands (2002).

[91] P.E. Wisse, *Multiple axiomatization in information management*, *PrimaVera*, Working Paper Series in Information Management, Nr. 2002-06, Amsterdam University, Amsterdam, Netherlands (2002).

[92] P.E. Wisse, *Stroomlijning tot informatiestelsel*, *PrimaVera*, Working Paper Series in Information Management, Nr. 2003-04, Amsterdam University, Amsterdam, Netherlands (2003).

[93] P.E. Wisse, *Dia-enneadic framework for information concepts*, Voorburg, Netherlands, available at *www.wisse.cc*, see publications/articles & papers (2003).

[94] P.E. Wisse, *Information metatheory*, *PrimaVera*, Working Paper Series in Information Management, Nr. 2003-12, Amsterdam University, Amsterdam, Netherlands (2003).

[95] L. Wittgenstein, *Philosophical Investigations*, English/German edition, MacMillan, New York (1968), originally published in 1953.

[96] R. Wright, *Three Scientists and Their Gods: Looking for Meaning in an Age of Information*, Times/Random House, New York (1988).

[97] H.C. Wyld (ed.), *The Universal Dictionary of the English Language*, 2nd impression, Routledge & Kegan Paul, London, UK (1952), originally published in 1932.

8

HISTORY OF DOCUMENT SECURITY

Karel Johan Schell

Schell Consulting BV
Noordwijk, The Netherlands

Contents

Abstract

Staying ahead of the currency counterfeiters is the recurring theme in the history of document security. Starting with the first banknote of the Mogul empire and ending with an example of a modern note overloaded with security features, the successive technological developments in reproduction and the reaction by the issuing authorities are elucidated in between. The historical background of the security features, the physics behind the features and the why and who at the turning points are explained, a few times larded with a personal view and recollection of the author.

The genes of paper currency are described and after the invention of typography, the birth of security printing. In the successive chapters, growth and maturity of security printing and its associated security features are elucidated. The digital revolution at the end of the 20th century made certain features obsolete and required a new approach. But the statement by Sir William Congreve in 1819 – "features to be immediately recognised and understood by the most unlearned persons" remains valid also in the 21st century.

Keywords: banknotes, counterfeiting, falsifying, Bank of England, political counterfeiting, medallion, guilloche, micro text, Congreve, Bradbury, CEF, Special Study Groups, Gutenberg, Coster, letterpress, typography, types, Johan Palmstruch, dry seal cachet, assignats, Johan Enschedé, Fleischmann, music types, Benjamin Franklin, intaglio, embossing, watermark, dry seal relief printing, four colour reproduction, photography, Le Blon, CMYK, screens, subtractive colour mixing, Guillot, MTF, electrotype, wiping, multi-colour, Hoe, Ganz, Gualtiero Giori, casual counterfeiter, professional counterfeiter, economic counterfeiting, super dollar, Reis, Waterlow & Sons, Operation Bernhard, economic counterfeiting, offset printing, colour scanners, Crossfield, Hell, laser, gravure, screen printing, analogue image, digital image, continuous tone, half tone, screen frequency, Fox Talbot, Moiré, scanners, screen trap, copier technology, frequency-modulated screens, Gualtiero Giori, Orlof intaglio, Koenig & Bauer, KBA, Simultan, RZ offset, Peoples Bank of China, Germann, see-through feature, inkjet, desktop, Microsoft, Windows, HP, PC, xerography, Carlson, Xerox, toner, electrostatic drum, RGB, CIE, colour triangle, Canon laser colour printers, FM screening, monet screen, amplitude modulation (AM), laser scanner, information density [bit/cm^2], dot size, first-line of defence, second-line of defence, third-line of defence, forensic, gamut, 3D devices, covert, taggants, tracers, DOVID, holography, latent image, transitory image, Merry, Hutton, van Renesse, metamerism, BEP, Four Nations Group, Bretler, Sicpa, San José conference, OVI, 60 Baht note, optically variable pigments, human perception, digital sampling, dpi, screen angle modulation, SAM, EURion, CBCDG, CDS, digital watermarking, substrate, RBA, Fabriano, paper endurance, fibre length, mold, dandy roll, electrotype, Dickinson, De Portal, Portals, Congreve, DNB, Napoleon, security threads, fibres, planchettes, Microperf, Orell Füssli, Swiss currency, perforations, rouble, Crane, sorting machines, polymer thread, polymer substrate, Stardust, Optiks, Varifeye, Tumba, Smidt van Gelder, Luminus, Enschedé, American Banknote Company, UV fluorescence, MM feature, Stokes, AQUS, infrared, barcode, transmission, dog-ears, circulation, not coin boundary, Cyanamid, Melamine, Dupont, Tyvek, non-woven, ink adherence, Caribbean, Isle of Man, Guardian, polymer substrate, bicentenary, Larkin, Coventry, transparent window, self-authentication, metameric colours, metameric image, New Zeeland, Y2K, μSAM, holography, Gabor, Leith, Upathnieks, three-dimensional image, coherence, laser, holograms, Denisyuk, Lippmann–Bragg, Benton, Polaroid, white light transmission holography, MasterCard, embossed holograms, plastic deformation, oblique illumination, ISARD, Landis & Gyr, Banknote Printers Conference, Machinery Committee, diffractive pattern, die, Numerota, coating, foil, kinegram, Kurz, Austria, Mozart, hot melt, SNB, Overhill, Reinhart, Kuhl, Moser, holographic patch, holographic grating, metallized, polyester carrier, ironing, foil application machine, Schaede, printable holograms, ink, pigment, vehicle, light fastness, drying mechanism, silkscreen, magnetic, de la Rue, up converting, lanthanide, Belgian national bank, OVI, NLG 250, phosphorescence, infrared transparent, KBA-Giori, Jules Verne, SPIE.

8.1 INTRODUCTION

Document security addresses the question of how to secure printed securities, such as banknotes or passports against counterfeiting and falsifying. Counterfeiting implies an illegal reproduction of a document where the falsification includes an illegal change of information on the document. With that definition in mind, an illegally reproduced bank-note is considered to be a counterfeit; a genuine passport where the photograph of the bearer has been replaced is a falsification.

The history of document security will cover all types of security documents, but from the past until the end of the 20th century, threats against document security and the technologies employed against these threats have, more or less, been identical for all types of security documents. The suggestion then is that the history of security features in paper currency will offer a reliable framework for the history of all document securities.

Governments of today are increasingly inclined to link an identification (ID) card number, tax number, or social security number with a unique biometric feature. Biometrics and biometric passports are gaining momentum and in the last decade of the 20th century a bifurcation of security technologies against counterfeiting and falsifying has become noticeable. By the mid-21st century, the history of document security will most probably not be based on the history of paper currency alone. For the time being, however, it will serve the purpose of this article to use paper currency as the focus of discussion.

Counterfeiting money is one of the oldest crimes in history and has been considered treasonous and punishable by death in its earliest history. In England for instance, counterfeiting was a hanging offence from 1697 until 1832. During that period, the Bank of England was considered reprehensible in issuing notes without sufficient security, which then was the principal reason for outrage over executions [6: 52].

The general public still traces the appearance of counterfeits to insufficient security and still considers central banks at fault. As in the past, there are people who might silently admire a lonely Robin Hood-type counterfeiter with his engraving tools rather than admire mighty centralised banks. On the other hand, there is no love lost when the mafia or drug dealers appear on the counterfeiting stage; there is little or no compassion for the serious crime of professionally organised counterfeiting.

However in retrospect, society's ordeal over political or banking counterfeiting is neither steady nor unanimous if someone also condemns the counterfeiting of English pound notes during the Second World War because burdens imposed as a result of the Nazi regime or because of the exploitation of concentration camp prisoners. Furthermore, production of today's counterfeited super dollar might also be condemned because of the suspect dictatorial regime of North Korea.

Further examples of political counterfeiting follow. In 1793 in an attempt to destabilise the young French republic, the English government of William Pitt supported French refugees in so-called 'assignats' counterfeiting. These assignats were initiated in the rebellious Vendee department. In 1806, Napoleon ordered the counterfeiting of Russian banknotes and financed a part of his wartime expenses by producing notes from printing plates confiscated in Vienna [24: 62]. Rumor had also suggested that England supported the production of counterfeited dollar notes during the American Revolution (1775–1883) [24: 62]. In 1918 after the First World War, England came seriously [24: 63] under suspicion for having plans to smuggle counterfeit Austrian banknotes into Austria.

However to return to England at the end of the 18th century, in 1797 the Bank of England invited the general public to submit suggestions for the improvement of its notes thus implicitly acknowledging the counterfeiting problem.[1] The results of this invitation which was presented in 1803 by a committee were meagre and consisted mainly of using new paper and keeping the design simple in order to make the money easily recognised by the general public [10: 57–59]. More then a decade later in 1817, parliamentary interest enforced a new investigation by a royal commission of inquiry, and in 1818 the Society for the Encouragements of Arts, Manufacture, and Commerce issued an inquiry into the state of the currency of the country, its defects, and possible remedies. By 1819, the inquiry resulted in *The report on the mode of preventing the forgery of banknotes.*[2]

For many reasons, this report deserves its place in the history of document security. It is possibly

---

[1] In 1811 the value of counterfeit notes rose to £16,000.

[2] The report can be found in the libraries of the Museum Enschedé and the Royal Society of Arts in London, the latter under the code SC/EL/5/314.

the first concentrated effort against counterfeiting and many of its suggestions are valid even after nearly two centuries. A few of the suggestions are as follows:

- Construct the large numerals of the denomination out of extremely small types, which is more or less the forerunner of today's micro text.
- Create a background of engraved Spiro graph patterns, which is more or less the forerunner of the medallion and guilloche structures.
- Incorporate the note serial number in its watermark.[3]
- Sir William Congreve suggested securing the paper substrate and made a still valid statement about security features: make them "as they may be immediately recognised and understood by the most unlearned persons".

In 1856 Henry Bradbury, the founder of the security printing house of Bradbury & Wilkinson stated: "Had however, the report of the committee been acted upon, it appears to me that this state of things[4] could never have arisen, for they arrived at those sound conclusions which are perfectly applicable at the present day".

In Vienna in 1923, a second concentrated effort took place with the inception of Interpol. Eliminating international counterfeiting is one of Interpol's assignments, and in 1924 under the supervision of Interpol, the first edition of *Contrefaçons et falsifications (CEF)* was issued.[5]

In the 1970s, the first threat of colour copiers prompted four English-speaking nations[6] to establish the Four Nations Study Group. Later the research efforts were pooled with others in an international Special Study Groups, with the imaginative names as SSG1 and SSG2. Since that time, it has been realised that at no since the introduction of the first banknotes in the 17th century have threats of counterfeiting been greater. Technology has managed to change the tools of individual counterfeiters from that of engraving to that

of digifeiting[7] so the pace for counterfeiting banknotes has not slowed any.

A reasonable question might be to ask if the banknote will last, and the answer is yes. Just as it was in the past, banknotes offer ease of use and transaction anonymity. In addition banknote use propels efforts that search for adequate answers to emerging newer technological threats. These past and emerging threats provide the very structure for this history on document security.

First, the period up until 1800 covers the early days and origins of paper currency. Secondly, the period 1800–1940 will detail the establishment of central banks, the regular issue and circulation of banknotes, and the upgrading and development of security features to deter counterfeiting. Thirdly, the period after the Second World War until 1990 will discuss how banknotes are secured on a more structural base with mature and proven counterfeit deterrents. Finally, the period from 1990 until today will examine how by the end of the 20th century a new threat has appeared because of the proliferation of digital technology which indicates that the armoury of existing security features is now insufficient and that the rise of new counterfeit deterrents must dominate.

8.2 THE FIRST PERIOD UNTIL 1800, BLOCK PRINTING, THE INVENTION OF TYPOGRAPHY AND THE GENES OF PAPER CURRENCY

It is said that in 1440, Johan Gutenberg in Mainz invented the letterpress technology and by 1456 produced the world's first printed book using movable types. The Dutch state that in 1455 Laurens Janszoon Coster assembled printing forms by using and reusing single types. In either case, the invention of typography triggered the start of a blooming craft, and even more importantly, it made printed matter easily available and seriously helped to reduce illiteracy. Prior to the 15th century, books

---

[3]This suggestion was considered highly impractical in those days, but in 2002 the US patent 6,395,191 B1, which underlies the same principle, was granted to this writer.

[4]Henry Bradbury refers to the rising number of counterfeits.

[5]CEF registers and identifies all the known counterfeits.

[6]Australia, Canada, UK and USA.

---

[7]Sarah Church of the Bureau of Engraving and Printing in Washington coined the word digifeiting for the counterfeiting of a banknote with a cheap scanner in combination with a PC and an inkjet colour printer.

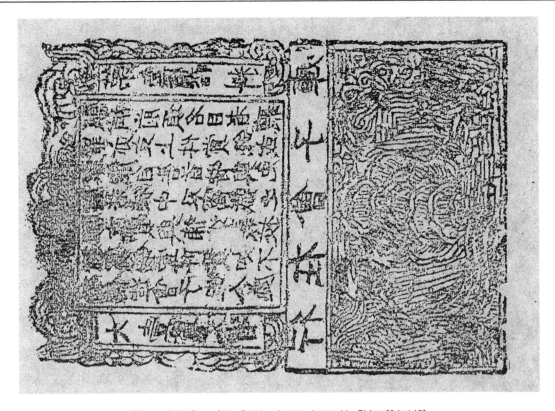

Figure 8.1. One of the first banknotes issued in China [24: 145].

were mainly handwritten with the exception of block-printed books as were produced in China.

The medieval traveller Marco Polo mentions a particular application of block printing circa 1300 [12: 182; 45: 20]:

> In this city of Kanbalu is the mint of the grand khan, who may truly be said to possess the secret of the alchemists, as he has the art of producing paper money.
> When ready for use, he has it cut into pieces of money of different sizes. The coinage of this paper money is authenticated with as much form and ceremony as if it was actually of pure gold and silver and the act of counterfeiting is punished as a capital offence.

In fact the Mogul empire adopted the use of paper money from China.[8] That the Chinese were the first to apply the innovation of paper money is plausible since they also invented both paper

and (block) printing. In his dissertation [47], Dr. Willem Vissering more precisely states that bartering may be the root for the introduction of coins and later the introduction of paper money as replacement for heavy coins. The appearance of the first Chinese banknote can be traced back to the year 940.

8.2.1   The origin of paper currency

The genes of paper currency are extensively described in many textbooks [7; 12; 45; 48]. Briefly paper currency's inception in the western hemisphere is in the 16th century with traders and goldsmiths issuing depositary notes in exchange for the valuables they had taken on deposit on behalf of the *noblesse* and of citizens. In general the goldsmith is favoured because of his secured storage options. In the beginning of this period, then, the valuable deposit is sealed by the owner, kept apart, and also registered. From time to time, the depositary notes were exchanged and so a transfer of the ownership

---

[8]China is by far today's largest producer and consumer of banknotes with an annual production of about twenty-five billion notes.

Figure 8.2. One of the first European banknotes issued by the Bank of Sweden in 1666. The dry seal cachet in the centre is a security feature against counterfeiting or as it was called in those days – a token of authenticity [7: 3].

of the valuables, mostly heavy coins, was effectuated.

By the beginning of the 17th century, traders in Amsterdam were receiving international payments in gold and silver coins of different weight and/or purity. Mutual trade was then eased by an exchange of vouchers and the exchange office was called the Bank of Amsterdam. The vouchers had a fixed value linked to an amount of precious metal. The system did not try also to scale down the privileges and profit margins of the guild of money-changers, but neither the depositary notes nor the vouchers were banknotes in today's sense of the term. These notes simply replaced the circulation of heavy coins.

When however the goldsmith or the custodian noticed that the deposits remained more or less stable, the idea arose to issue more notes with a promise-to-pay, and with this event there arose the beginnings of banknotes. Those promises-to-pay were considered a type of loan and were delivered to investors, merchants, craftsmen and other appropriate parties. The promise-to-pay could move from person to person as long as the general public trusted the issuer. In fact the banknote's origin is directly linked to the amount of promises-to-pay that would exceed an underlying amount of precious metals.

In 1652 the Dutchman Johan Palmstruch, born in Riga in 1611, presented a plan for the building of a bank in Stockholm [7: 3]. This bank issued depositary as well as credit notes as shown in Figure 8.2. The depositary note had a direct link with coins that were in storage. The credit notes were issued when it was considered unlikely that a party had to pay off all credit at one time.

Granted in 1656, this liberal charter was a milestone in monetary exchange and it authorised Stockholm's bank to accept deposits, grant loans and mortgages, and issue bills of exchange. By 1661 then this bank became the first European chartered bank issuing notes.

A few years later in the autumn of 1663, thoughtless increases of credits brought the bank in Stockholm into sufficient stormy situations that the Swedish government had to intervene [7: 4].

In approximately the same period, goldsmiths in England started the circulation of depositary notes that were not fully covered by stored valuables or coinage. But the use of these banknotes became institutionalised by 1694 with the foundation of the Bank of England [10: 11]. This bank also acted as a source of finance for the war against Louis XIV.

The same Louis XIV brought France to a financial abyss and after his death in 1715 recovery began by consolidating and clearing debt. In 1716 the

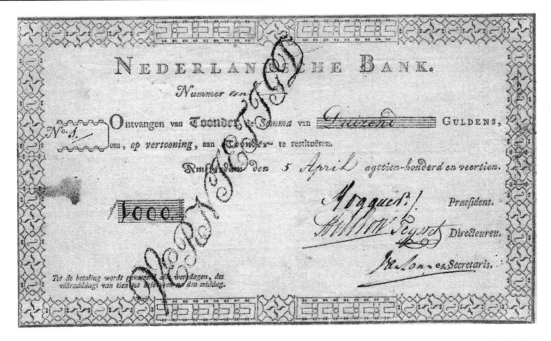

Figure 8.3. Decorative edges of music types as counterfeit deterrent in an early banknote. The date of issue is the 5th of April 1814.

Scotsman John Law was granted the privilege to found a national bank. He speculated with the procurement of a trading company and had to create extra banknotes in such an amount that the general public lost confidence in the enterprise. Panic broke out in 1720 and very few owners of paper currency did not suffer bankruptcy through this scheme.

Seventy years later the French began issuing assignats. The currency press ran freely in order to finance the Napoleonic wars, thus discrediting again paper currency in France and in its occupied countries.

8.2.2 The birth of security printing

By the end of the 18th century, the lowlands of Western Europe were under Napoleonic occupation and the worthless assignats were introduced there as well. Apparently some security was requested to prevent counterfeiting, and someone[9] at the printing house of Johan Enschedé in Haarlem, The Netherlands suggested using the in-house design and engravings for music types. Midway into

the 18th century, Enschedé [15] employed the German punch cutter Joan Michael Fleischman. His masterpiece was a design of music typefaces, rendering the engraving of music superfluous. Everything composable could also be typeset and the Dutch edition of the 'Violin school' of Leopold Mozart brought this marvel of craftsmanship to public attention. However the investment [44] was only later recouped when the unsold pieces of music types became decorative edges.

These music types were unique at that time so their application in letterpress[10] printing presented itself as a security feature.

The French occupation of the lowlands ended in 1813. The deplorable situation created by the many Napoleonic wars called for drastic measures in order to facilitate the ailing industry [46], and in The Netherlands a national bank was founded. The main task was supplying credit and buffering the procurement of liquid assets. The first banknotes as issued in 1814 by the Dutch Central Bank are displayed in Figure 8.3. Exclusive types used in

---

[9]Who is alas unknown!

[10]The letterpress printing technology makes multiple copies of text or images by direct impression of an inked and raised surface against paper [2: 17–28].

Figure 8.4. Colonial currency from the printing press of Benjamin Franklin [45: 110].

the worthless assignats proved their worth after all since they could be used as an effective deterrent against counterfeiting and are an early example of security printing.

The start of the printing of securities cannot be precisely traced. The first pound notes in England date from the beginning of the 18th century. Before then, colonists in North America issued paper money because of the shear lack of precious metals. Benjamin Franklin, printer by education and owner of a printing press [48: 133], printed some of the first paper money used in America. He surely recognised the counterfeit threat since he stated that counterfeiting deserved death, but his block printed currency, as seen in Figure 8.4, neither surpasses the quality of early Chinese notes nor provides sufficient protection against counterfeiting.

The samples shown in Figure 8.5 deserve a place in the history of printed currency. From the perspective of the history of document security, the counterfeit deterrents in the early pound notes are the engraved vignette with Britannia and the engraved value of the denomination. The printing from an indent or engraved image is called intaglio

printing. The image is engraved in a hard material, such as a steel plate. The engraved impressions are filled with ink and the excess ink is wiped from the surface. Subsequently, the plate is brought into direct contact with the paper substrate under such force that the paper is pressed into the engraving. The paper is then plastically deformed and the deformation is called embossing. In this way, a tactile relief is formed that is unique to the intaglio printing process and that is caused by a combination of ink thickness and paper embossing.

An even more important security feature is the visibility of the line watermark. After 1809, numbers printed in a letterpress replaced handwritten note numbers.

By the end of the 18th century, then, document security was based on a unique and/or engraved form printed in one colour, either from a letterpress or an intaglio method, plus a watermark. The dry seal relief printing as used in the 1666 issue of the Bank of Sweden is *de facto* an intaglio print without ink.

Counterfeiters in those days were acting more or less on a casual level; they were draftsmen and engravers and their tools [10: 43] are displayed in Figure 8.6.

8.3 FROM 1800 UNTIL 1940, SECURITY PRINTING MATURES

Currency printing reached its adolescence in the 19th century and matured in the 20th century. The economic growth, its related increased currency circulation, and the development of printing technology together were responsible for its growth and maturity. But with the development of printing technologies, the reproduction quality also increased. Before World War II, this quality increase brought four-colour reproduction into the hands of counterfeiters and by the end of the 20th century, office and home colour printing devices were in the hands of laymen. So the advantage of the development of reproduction technology on the other hand also improvised easier tools for counterfeiters. The growth of security printing into maturity passes through certain steps in technologies, which are also linked to counterfeit threats.

Figure 8.5. [7: 4] A promissory note of £50. Issued in 1771. The nominal value is printed and the line watermark mold [10: 38], as shown below, is clearly visible.

8.3.1 Photography

The core of the invention of typography in 1455/1456 either by Gutenberg in Germany or Coster in The Netherlands is the reusable single type [32: 131]. With single types, a printer can compose a letterpress printing form instead of having to print from a fixed block. Single types can be reused after each printing. The types are cast and typecasting took place in type foundries. The possession of exclusive casting forms made the type foundry[11] and its related printing house rather exclusive. Security and exclusivity are closely linked and the exclusive movable types were the building blocks for security in a letterpress. In the be-

---

[11]Charles Enschedé wrote at the beginning of the 20th century his *opus magnum* entitled *Type foundries of the lowlands*.

Figure 8.6. Forgers tools at the end of the 18th century.

ginning these exclusive type fonts, like the earlier mentioned music types, were printed in black and white. The security of this image was based on the exclusiveness of the movable types and disappeared with the introduction of photography.

The first practical application of photography is attributed to Louis Jacques Mandé Daguerre and the first images were called daguerreotypes. The actual invention was published on 7 January 1839. Thereafter Coe [11], Lilien [26: 21], William Henry Fox Talbot, Nicéphere Niepce, Sir John Herschel, Scott Archer, as well as many other people established their names in the ranks of pioneers in photography. The rise of photography endangered not only the jobs of portrait painters by putting portrait technology into the hands of the layperson it also reduced the security of printed matter. The printed form could be copied on a large scale and exclusive types could be reproduced. Document security responded with this new technology by introducing colour into the security process.

8.3.2 Colour

With the introduction of colour, photographic reproduction in black and white receded for the time being. But the development of the technology for reproduction in colour gained momentum in the 19th century and faced document security with the next challenge.

Letterpress printing in more than one colour and using separate printing forms was used almost as early as the printing press. There are examples from the 16th century, where closely related shades of colour – each of them printed with a separate printing form – are designed to imitate coloured drawings. The foundation of the four-colour reproduction technology can be traced back to the beginning of the 18th century, when the German painter and graphic artist Jacob Christoph Le Blon (1667–1741) made the first three-colour prints by the overlapping printing of three gravures [26: 12], with transparent yellow, red and blue inks.

There are many good textbooks [3; 9; 20; 28] on the theory of colours, its perception and reproduction. The most succinct however is the description by Rudolf L. van Renesse in the revised issue of Optical Document Security [33], given here in full with kind permission of the author.

Colour separation constitutes the basis of colour reproduction. This process decomposes the colours of an original object into the subtractive colour components cyan, magenta, yellow (denoted as CMY), and black, which for simplicity is further disregarded here. Before today's digital age, separate photographic negatives were made of the original colour object on black-and-white film through red, green and blue-violet colour filters. Continu-

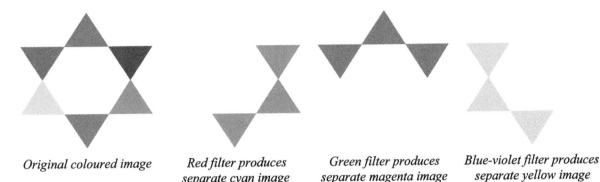

Original coloured image   *Red filter produces*   *Green filter produces*   *Blue-violet filter produces*
separate cyan image   *separate magenta image*   *separate yellow image*

Figure 8.7. Schematic explanation of colour separation and reproduction.

ous tone of the negatives[12] was achieved by placing suitable halftone screens in contact with the photographic film during exposure. These halftone screens break the image into dot patterns, the resulting dot size on the negative being a function of the brightness of the original colour. Separate printing plates are subsequently prepared from these negatives for the subtractive colours cyan, magenta, and yellow. Each separate colour layer is then printed in transparent inks, one on top of the other to reproduce the original colours by subtractive colour mixing. Because of the use of screens, this type of reproduction is referred to as screen reproduction.

The principle of colour reproduction by colour separation is further explained with reference to Figure 8.7, which shows an original colour image in the six basic colours of the colour wheel.

- A first photographic negative of the original image is made through a red filter so that the photographic film will receive only an exposure of the magenta, red and yellow coloured areas, which all reflect a substantial amount of red.
- Likewise, a second negative is made through a green filter so that the photographic film will only receive an exposure by the yellow, green and cyan coloured areas, which all reflect a substantial amount of green.
- Finally a third negative is made through a blue-violet filter. Only magenta, blue-violet and cyan

areas will expose the negative while it will remain unexposed in the red, yellow and green areas, which substantially absorb blue-violet.

Subsequently three printing plates with photosensitive coatings are exposed in contact with the three separation negatives. In the transparent areas of each negative the photosensitive layer is exposed and hardened thus becoming insoluble. Chemical processing then develops the plates and three printing plates are prepared respectively for the cyan, the yellow and the magenta image. Figure 8.7 separately shows the cyan, magenta and yellow images that are printed on top of each other on white paper. Obviously, due to the laws of subtractive colour mixing, the overlapping colours will reproduce the original colour image.

But not all the colours can be reproduced faithfully and it then goes without saying that this lack of faithfulness is used as a tool against (illegal) reproduction. Colours such as dark green, stone red, sienna brown and purple-violet are difficult to reproduce which explains the rationale behind the colour impression of many of the notes issued into the middle of the 20th century.

8.3.3   Guilloche

With respect to this history's time line, by the middle of the 19th century printed security features consisted of:

- Exclusive types as features in black and white printing.
- Colours, which are difficult to reproduce faithfully.

---

[12]An elucidation of the screening process is also presented in Section 8.4.3.

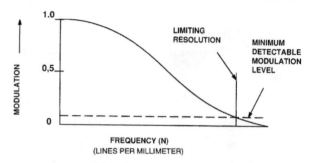

Figure 8.9. MTF. Parallel lines [32: 142] with a frequency above the limiting resolution will not be reproduced.

Figure 8.8. Guilloche and medallion pattern circa 1860.

Around that time another printed security device was introduced. A relief pattern as suggested by many parallel lines and then later still intricately interwoven patterns or guilloches.

The relief or medallion patterns are created on line engraving machines. A fine needle follows a relief image and the up and down movement is mechanically transformed onto a plane movement. The engraving machine produces complex patterns through a mechanical arrangement of gears and drives. The patterns are called guilloche, after its inventor the Frenchman Guillot. From 1850 onward, there were a group of famous suppliers for these sophisticated mechanical ruling-, medallion- and/or guilloche-machines, which included companies such as F.G. Wagner Jr. and Walter Kaempf at Berlin. Today the complex mechanical movements are replaced by mathematical formulas in computer programs.

The difficulty of re-origination of medallions and guilloches made them effective deterrents from the beginning. To explain why these fine line patterns still act as a deterrent against reproduction, the quality parameter for optical systems called MTF must be introduced. MTF stands for modulation transfer frequency and tells which spatial frequency of parallel running lines can be reproduced.

Figure 8.9 shows that reproduction of parallel lines with a frequency above the limiting resolution results in a diffuse image. But in addition, micro texts become unreadable in reproduction when a texture frequency is above a certain threshold and closely spaced line crossings of guilloche patterns will become diffuse and unreadable in reproduction. In short, a high quality reproduction system is needed for the reproduction of medallion structures, guilloches, or micro text, which is why these items act as counterfeit deterrents.

By the end of the 19th century and the beginning of the 20th century, the armoury of printed security devices against counterfeiting consisted of exclusive types and of difficult to reproduce colours and fine line structures. Letterpress was the dominant printing technology for general printing and the technology to reproduce colour in the letterpress process depended heavily on the level of craftsmanship and proficiency.

8.3.4   Printing technologies

The security features of the bank in Stockholm note of 1661 and the Bank of England note of 1794 were produced respectively by letterpress and intaglio. The unique music types as applied for the assignats and the first Dutch notes were based on letterpress technology and until the middle of the 19th century either intaglio or letterpress were considered to be the leading security printing technologies. This advance, however, changed with the arrival of mechanical engraving machines for the medallion and guilloche images, circa 1850. Engraving machines originate engraved line images directly onto steel, and these images are subsequently transferred into intaglio printing forms either by electrotyping or by transfer pressing (mysteriously called Siderography). Thereafter, intaglio has dominated the engraving process while the letterpress, although tolerated because of the invention of photography, ultimately had to be demoted

Figure 8.10. Ganz intaglio press.[14]

with the introduction of colour. In those earlier days, to produce multi-colour images by intaglio was difficult if not impossible but that limitation no longer applied.

In addition to the multi-colour difficulty of the intaglio printing process, there is another handicap worth mentioning. After inking a recessed engraving, the excess ink has to be removed by wiping. In the past, the ball of the thumb of a craftsperson or an intaglio printer performed this duty. The ball of the thumb was replaced by an industrial wiping process and the solutions for this wiping problem were all based, in those early days, on endless cotton tapes or cloths. Even after World War II this procedure was the case. The intaglio process, then, displayed unique features in the printing of securities but also displayed two major problems, i.e. the wiping process and the impossibility of printing multi-colour designs in one run or with one wiping cloth.

In the first half of the 20th century, the major machine supplier for the intaglio printing of securities were the Hoe Company from the UK and Ganz machines from Budapest,[13] which were designed by Dr. Fritz Heinrich.

In the archives of Ganz, or rather what has been left of it, a researcher can retrieve the list of customers that predate World War II and can find the printing houses of various national banks, such as Enschedé, Giesecke & Devrient, and the Impremera di Carte Valori in Milan. The latter, in those days, employed Gualtiero (Rino) Giori (1913–1992) who experienced, before World War II, the joys and the problems of being an intaglio printer. This history will return to the story of this remarkable man in Section 8.4.5.

8.4 THE EVOLUTIONARY YEARS FROM 1940–1980. CLASSIFYING COUNTERFEITERS AND THE ARRIVAL OF MATURED SECURITY FEATURES

8.4.1 Counterfeits

One of the achievements of the Interpol conference in 1923 at Vienna was the classification of counterfeiters into the categories of casual, professional and political.

A *casual* counterfeiter is an individual who periodically issues a reproduced security document

[13]The Ganz machines in Budapest have produced the Hungarian banknotes until 1996.

[14]Dr. Bela Egyed, the former manager of the Hungarian Banknote Printing Works kindly supplied me with this Ganz information.

such as a banknote. A popular example is the holi-daymaker illegally solving the budget deficit of his holiday. The frequency of the issue is low, as is, relatively speaking, the financial damage. Since the amount is small, so are the chances of getting caught. However the nuisance value is consider-able and with the proliferation of colour copiers, computers, scanners and four-colour printers, ca-sual counterfeiting has increased significantly over the last decades of the 20th century. The early counterfeiters were acting more or less casual.

The *professional* counterfeiter is linked to or-ganised crime. The number of notes brought into circulation is relatively large because this profes-sional counterfeiter employs a printing process. The general public will always admire the single-handed professional counterfeiter. He has an aura of heroism combined with craftsmanship and only a few of the numerous reports and stories about these heroic, tragic, and/or romantic lives are men-tioned in this history [6; 7: 81, 82; 24: 54; 37; 61].

Arthur Alves Reis is famous as the man who almost stole Portugal. In 1924, he contracted the security printer Waterlow & Sons in England to print[15] two hundred thousand notes of five hun-dred escudos for Portugal, supposedly on behalf of the bank of Portugal. He managed to get the notes delivered personally and proceeded to issue them. The discovery of this perfect forgery, as in many other cases, had occurred because of 'futile' de-tails, such as note numbers and note numbering se-quences.

Peter Ritter van Bohr (1772–1847) is known as one of the master counterfeiters of the world. Circa 1840, he expertly forged, and consequently lost the eyesight of one eye, ten and one hundred Austrian gulden notes. The police finally traced him through the procurement, with these counterfeited notes, of an unusual watch by his wife Mathilde. He was sentenced to death in 1846, granted grace by Em-peror Ferdinand I, and died in prison in 1847.

---

[15]Reis also approached security printer Enschedé in The Netherlands where there is a story that they refused the order because they became suspicious. However a detailed investi-gation of the minutes of the management meetings of those days will reveal that the order had to be refused by sheer lack of capacity.

Quiet another illustrious figure is Mr. Bojarski from France. He counterfeited French banknotes with such perfection that the *Banque de France* ac-tually granted compensation for the exchange, but only for this one occurrence. In 1964, Bojarski's cellar was discovered as his counterfeit produc-tion centre, and the police and other experts were amazed by the quality and the display of the simple instruments that he used!

A comparable affair was the issue of a few beau-tiful counterfeited Swiss one thousand Franc notes, known as the *affaire Fourmis*. The superb counter-feited Swiss thousand Franc notes were discovered in the 1980s and the trail finally led to a modest but able draftsman, printer, and engraver. He produced the front and reverse side of the notes separately and composed them together thus cleverly simu-lating the watermark. The front and reverse sides were registered by using a see through security fea-ture. Personally this writer has always found this innovation to be a genial and unforeseen applica-tion of a security feature.

The gamut of counterfeiters displays many other unique individuals, all being more tragic figures than villains, such as the Ukrainian Salomon Smo-lianoff, who was bankrupted in the October revolu-tion of 1917 and turned to counterfeiting. In Berlin from 1930–1935, he produced counterfeited bank-notes from a small print shop. Smolianoff, in the end, was also involved in the operation 'Bernhard' as described below. There is also the tragic figure of the Austro–Hungarian Prince Ludwig Windis-chgraetz who tried to disrupt the French economy with massive counterfeited banknotes as a revenge tactic after World War I. The quality however was so poor that his partners in crime were discovered and seized after issuing the first note.

Political counterfeiting involves a deliberate at-tempt to deregulate and destabilise the financial structure and economy of a state and is closely linked to war and terrorism. Under government supervision, dedicated and specialised currency printing equipment reproduces the currency of the enemy state.

The operation 'Bernhard' of the Second World War is best known and well documented. During the Nazi regime in Germany, Himmler ordered the production of perfect imitations of British banknote paper complete with watermark in a top-secret operation. The SS installed a fully equipped printing works in the concentration camp Sachsenhausen near Berlin and prisoners worked with highly qualified printers[16] to produce high quality British pound notes. The name of the operation had to do with the SS officer in charge Bernhard Krüger. It is said [17] that less than ten percent of the total production had been issued. By the end of the war, the counterfeit notes were dumped into Lake Toplitz. The discovery of the perfectly counterfeited pound notes actually helped to accelerate the implementation of the security thread feature in the English currency. Five and ten pound notes of the Operation Bernhard are still[17] available.

A more recent example is the discovery of the so-called 'super dollar' which is a US one hundred dollar counterfeit printed in intaglio and of such a quality that even experts have difficulty in discriminating counterfeits from genuine notes. Rumor has

it that the super dollar is produced in North Korea, but this idea has yet to be confirmed. At any rate, the 'super dollar' triggered and accelerated the decision for the upgrading of US dollar notes.

The Reserve Bank of India[18] suffers today from perfect counterfeits with intaglio printing on watermarked paper. Although in India usually from times past, unfortunate events have come from Pakistan, in this particular case a former USSR state close to the Indian border is suspected.

A peculiar new phenomenon of the last decade is the appearance of economic counterfeiting. Rumour has it that a paper[19] mill in Indonesia organised counterfeit notes of one hundred thousand Rupiah on polymer in order to prevent a further penetration of polymer as a currency substrate in Indonesia.

8.4.2   The need to refurbish and expand security features

Just before WW II the armoury of security against counterfeiting existed mainly of exclusive types, colours and fine line structures that were difficult to reproduce. Letterpress was the dominant printing technology for general printing and the technology to reproduce colour in letterpress depended heavily on the level of craftsmanship and proficiency.

By 1920, offset printing[20] technology had arrived and from 1950 onward it became so suc-

Figure 8.11. Forged £10 notes in WW II. Recovered in 1959 [10: 156], from Lake Toplitz [24: 64].

---

[16]I encountered one of these printers in my early Enschedé years.

[17]They are offered at the auction house *www.first-dutch.nl* for €45 each.

---

[18]Private communication.

[19]Private communication of Securency Pty. Ltd dated August 2004.

[20]The inked image in *offset printing* is offset to the substrate via an intermediate rubber cylinder. Printing in *wet offset* is based on the phenomenon that oil and water do not mix. Oily inks, therefore, can be deposited on hydrophobic printing areas of the plate, while such inks are rejected by hydrophilic, water holding non-printing areas. The offset plate is coated with a photosensitive layer that reacts on exposure. The unexposed real estate (non-printing areas) can be washed away and during the printing process, the printing plate first is moistened so that water is deposited on these non-printing areas. Then the plate is inked and the hydrophobic printing image areas accept the ink. The image is then conveyed to the rubber cylinder and is subsequently offset to the paper substrate. *Dry offset* is a combination of letterpress and offset and utilises an intermediate rubber cylinder for ink transfer from printing plate to substrate. However the inked surface is raised like in the case of letterpress [2: 222–229].

cessful that it ousted letterpress technology in general printing. Secondly the development of colour scanners moved colour reproduction from craft to technology. Offset printing and colour scanners became the standard tools for general printing and counterfeiters picked up the technique quite easily. In particular the continuous development of automatic colour scanners revolutionised the colour separation process and normalised the faithfulness and reproducibility of colour reproduction. Crossfield in the UK and Hell in Germany were leading the development of colour scanners. Both companies were fierce competitors, as were their founders in WW II. Sir John Crossfield and Dr. Rudolf Hell were both involved in the crucial development of radar.

And last but not least the laser appeared around the 1970s. In 1960, more than forty years after the theoretical prediction in 1916 of stimulated emission by Albert Einstein, Theodore Maiman presented experimental proof of the laser's existence. In its earliest years, even the Nobel Prize winner and co-inventor of the laser A.L. Schawlow referred to it as "a solution looking for a problem". Comic strip artists who drilled tunnels and shot science fiction monsters with laser beams, among other things, introduced the first application of lasers in the graphic arts. For these Sci-Fi applications, the capabilities of commercial lasers are pathetically inadequate. But laser light is monochromatic and coherent and these properties have more than adequately caused dramatic shifts in graphic technology. Hell in Germany launched the first laser colour scanner in the early 1970s. Thereafter, the search for new security devices had to be increased and central banks along with security printers strengthened their research by contracting with physicists.[21] At this point, this history must raise the issue of what makes offset printing combined with colour scanners, and in particular the laser colour scanner, so dangerous.

8.4.3 Screens

The printing techniques used by security printers are intaglio, gravure, offset, letterpress and screen-printing which all serve the purpose of placing ink

onto paper and looking at a small area to see if there is or if there is not ink on the paper. In today's terminology, printing is a digital process.

There are two kinds of image elements that have to be re-produced: continuous tone or half tone photographs and line elements or line art. It is taken for granted that a line or a typeface is bi-level and is printed on a white substrate.

The multilevel or analogue, continuous tone or half tone image is converted to a bi-level or binary or digital image by using a technique known as screening. Screening converts the halftones of an image into single dots. The mutual distance of the dots is constant and is called the screen frequency. The size of the dots is related to the density of the halftone, which means that large dots make for dark tones while small dots make for light tones.

William Henry Fox Talbot was the first [26: 19–21] to transfer an analogue or continues image into small dots. In 1852 he reported that it is pos-

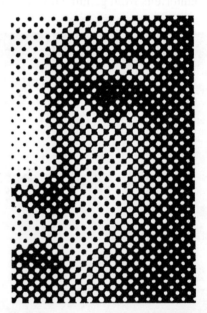

Figure 8.12. The dot size is modulated and the dot distance is determined by the screen frequency. A grid of cross hatched lines looks grey when viewed from a distance which is the principle of the halftone illustration, provided that the original image can be coded into black spots of varying sizes. The early method places an opaque screen, bearing a fine mosaic of tiny transparent squares, close to a high contrast photographic plate. The emulsion is of high contrast and behaves like a binary switch, which is the old-fashioned way of getting an analogue into a digital conversion [2: 194].

---

[21] I joined the security printer Joh Enschedé in 1977 after being employed for ten years by Philips in The Netherlands.

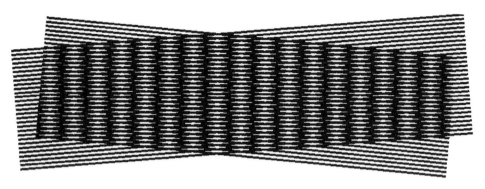

Figure 8.13. Moiré patterns[24] are optical illusions of apparent patterns that are not truly there.

sible to transform a continuous tone picture into a halftone by interposing a woven textile and thus make it suitable for the etching of a printing plate.

If colour printing is needed and as explained in Section 8.3.2, there are three basic printing forms: cyan, magenta and yellow, which have to be printed in close register. In addition the screens have to be set at different suitable angles; otherwise, the result is blotchy moiré[22] patterns.

Apparently, then, faithfulness and reproducibility of a coloured image depend upon the handling of screens or the craftsmanship of the dark room. In early days of colour printing, the quality of the output of the preliminary or photographic department of a printing house depended to a large extent on craftsmanship in the dark room.

Similarly the previous production of a reasonable counterfeit depended on a highly skilled dark room operator, and before the arrival of electronic and laser scanners, counterfeiters were sometimes admired for their ability to make good colour separations.

The electronic and laser scanners turned the mystique of the craft of the dark room into the skill of operating an expensive piece of equipment. The management of a printing house was able to gain better control over its preliminary department and accepted the downside of losing certain craftsmanship of the graphic arts. An operator could produce colour separations after a six-week training course. Simultaneously, however, the same equipment became a tool for the counterfeiter, and once again security printers were confronted with a need to look

for new security improvements in their reproduction technology.

8.4.4 Moiré[23]

When two periodic patterns, such as an array of straight lines, a mesh, or a series of concentric circles are overlapped with imperfect alignment, a large-scale moiré pattern appears.

Anyone can see this moiré pattern quite easily. A fine mesh can be found in a sheer curtain, a handkerchief, or any fine woven fabric by looking at a double layer of such fabric in transmission and noticing the large-scale wavy pattern where the lines of the two layers overlap.

Because large-scale patterns result from fine periodic arrays, moiré patterns provide a magnification effect which can be used, for example, to determine the spacing of the fine array or in the case of security printing as an indicator of a reproduction. If a regular pattern, called a screen trap, is introduced into the design of a document, it will generate a moiré pattern when the document is reproduced using scanner technology as described in the previous section. Although the screen trap and the scanning frequency may not be resolved by the visual system, the resulting moiré bands become visible, thus indicating illegal copying.

Consequently around the 1970s, the security printers extended the array of security features with

---

[22]An explanation of Moiré will be given in the next section.

[23]The French mathematician Moiré was fascinated by the interference of the spokes of the wheels of a carriage. The story goes that he watched the patterns as a young boy by hanging out of the carriage window with his head down.

[24]Rudolf L. van Renesse supplied the patterns.

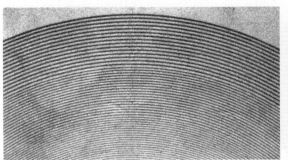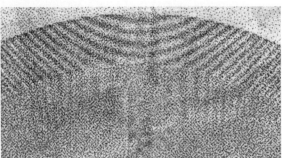

Figure 8.14. The Dutch NLG 100 note was one of the first notes equipped with a screen trap of concentric circles.[25] Left the original, right an inkjet copy.

moiré generating patterns. Next to exclusive types, difficult to reproduce colours and fine line structures, whose spatial frequency interferes with the scanning frequency has been added.

The majority of modern designed banknotes have incorporated some type of screen trap. The patterns work in theory, but their deterrent value is meagre because of the tricks[26] that experienced scanner operators have used, especially in the beginning, to avoid moiré effects. But today developments in copier technology, the rising quality of colour copiers, scanners, and printers and, last but not least, the appearance of the frequency modulated screen technology[27] have all reduced the effectiveness of moiré generating patterns.

8.4.5 Gualtiero Giori en Orlof intaglio

Along with the time line detailed above, there is the development of the security printing technology, and in this area Gualtiero Giori deserves an honoured place in security printing history. Before World War II, Gualtiero (Rino) Giori (1913–1992) experienced the joys and the problems of being an intaglio printer. After the war he presented a multi-colour intaglio concept together with a liquid wiping system. He founded Giori S.A. in 1951 and entered a collaborative agreement with Koenig

& Bauer in 1952 on banknote and security printing. The unique combination of salesman, printer, and inventor in one person was responsible for the transformation of Giori S.A. into the multinational business it is today, that is KBA-Giori. Giori's foresight industrialised [4; 5], the intaglio process and currently the majority of the world's banknotes are printed on Giori presses.

The many intaglio based security features have fully matured because of the performance of the Giori intaglio presses and new features, which are still being developed. These new features require digitally engraved intaglio images and it is beyond the scope of this section to digress on that subject, but suffice it to say that banknote engraving by hand is becoming a thing of the past.

Gualtiero Giori also experienced the added value of printing multi-colour images in close register. In Sections 8.3.2 and 8.3.4, security features such as colour and fine lines are elucidated. At least several printing forms and high quality optical systems are needed, respectively, for colour and fine line reproduction. The individual printing forms have to be positioned exactly on the press and the paper has to stay dimensionally stable between the successive print workings. The more printing forms the more difficult it is to stay in register. In the 1950s, Giori also launched the concept of a Simultan offset press which enables the security printer to print a build up of line images on the front and reverse side of the note in three colours, in close register, and in one pass. The mechanical accuracy of the machine and its printing forms is

[25]The idea of concentric circles stems from Dr. Ir.P. Koeze of the Dutch Central Bank. Rudolf L. van Renesse supplied the images.
[26]Tricks like scanning slightly out of focus and/or putting a transparent foil between a banknote and a scanning head.
[27]More on frequency-modulated screens can be found in Section 8.5.4.

less then 0.1 mm. The Simultan printing technology popularised the see-through or look-through[28] security feature.

Printing in close register as a feature against counterfeiting was known long before the arrival of the Simultan. Sir William Congreve (1772–1825) patented an ingenious construction for letterpress printing. Individual letterpress printing forms were first mechanically composed, before the multicolour image was transferred to paper and Congreve's printing idea had been applied for many security printings in the 19th century and during the first decades of the 20th century in India and other parts of the British Empire. The printing house Enschedé was also known since the 1920s for its two-colour close register work produced on patented RZ offset[29] presses.

As explained above, printing of multi-colour images in close register was one of the favoured innovations of Gualtiero Giori, and in his last years he also accomplished the printing of intaglio images in close register by exploiting the Orlof concept. From the past, the Orlof printing principle is defined as indirect transfer printing. In this process different inks from several printing plates are first collected on a blanket or rubber cylinder and then transferred to a single plate and printed onto paper.

Congreve printing as mentioned above is Orlof in letterpress and in Orlof intaglio the images of the inking rollers are first collected on the Orlof cylinder that is, a roller of large diameter with a relatively soft covering before being transferred to the intaglio plate cylinder. The revival of Orlof began in China[30] where the brilliant chief engineer Li Genxu of the Banknote Printing Corporation of the Peoples Bank of China designed Orlof offset machines for the third series of Ren Min Bi. Gualtiero

Giori and Albrecht Germann, the engineering director of Koenig & Bauer, transferred and modified the ideas of Li to intaglio printing and with the success of the now KBA-Giori, Orlof intaglio machines, the reinstatement of the Orlof principle for printing became a fact. An increasing amount of today's currency is printed on Orlof intaglio machines and the success of the Orlof intaglio technology is caused by the reduced consumption of intaglio ink, production speeds up to ten thousand sheets per hour, and the heavy tangible intaglio relief.

8.5 FROM 1980 UNTIL THE BEGINNING OF THE 21ST CENTURY. THE INFLUENCE OF THE DIGITAL REVOLUTION

8.5.1 The digital revolution

The introduction by Hewlett–Packard of the Think jet printer in 1984 marked a turning point in inkjet printing. A technology, which had primarily been implemented in industrial applications, embarked upon a development course of the home user as consumer of digital photography and printing. Unexpected low prices for a desktop printer capable of photographic image quality domesticated inkjet printing. Today installed machines of inkjet printers are in the hundreds of millions. These numbers have not arisen simply from a predictable improvement in the inkjet performance parameters. Parallel innovations in ink, colourants, digital halftone algorithms and colour reproduction have played an even more important role in this revolution. Adding fuel to the fire, the ever increasing computational power of PCs along with the plunging cost of the PC's bigger memory capacity have combined to energise this domestic digital imaging revolution. A parallel driving force has been the more or less synchronous development of PC operating systems and software. Microsoft presented its first version of the PC, Windows 1.0, in 1983. The next version appeared as Windows 2.0 in 1987 and in less than fifteen years, six new versions have been launched with the 2001 Windows XP as the most recent one.

In short at the turn of the century as society enters the 21st century, security printers and document issuers have been confronted with an enormous proliferation of reproduction centres. These

---

[28]The see-through or look-through feature consist basically of two complementary images that are printed on the front and the reverse side of a note. By looking at the note in transparency these two images merge.

[29]The sublime technical performance the Giori intaglio and Simultan press has been due to a great extent to the responsible KBA chief engineer Albrecht Germann. He started his career after WW II at the Roland factory and one of his first assignments was to reengineer the RZ press. I had the dubious honour of having to destroy the last RZ by turning it into scrap iron.

[30]Private communication from Albrecht Germann.

low cost devices produce and reproduce colour images with an astonishingly high quality and have completely changed the nature of counterfeiting into digifeiting.

8.5.2 Xerography

The threat of counterfeiting by copiers has its origin in 1938, when Chester Carlson invented a copying process based on electrostatics. He called it xerography from the Greek for dry writing. It was not until 1959, twenty-one years after this invention, that the first office copier using xerography was launched. Today xerography is synonymous with a gigantic worldwide copying industry. Xerox and other corporations make and market copiers that produce billions and billions of copies a year.

A typical xerographic copier uses a large metal drum, which is coated with a light sensitive me-

dium. A strip light illuminates the original document and a lens focuses the strip image on the drum; the image moves in synchronism with the rotation of the drum.

The copying process has [38: 157–158] seven steps (see Fig. 8.15).

(1) The drum is charged in the dark.
(2) The drum passes under the exposing slit and the projected image discharges the electrostatic charge in proportion to the luminance.
(3) Electrostatic charged toner is introduced on the drum surface where it accumulates on the areas that have retained their charge.
(4) The receiving paper is charged and brought into contact with the drum surface.
(5) Toner particles are pressed into the surface and fixed by heat.
(6) The drum is cleaned for the next cycle.

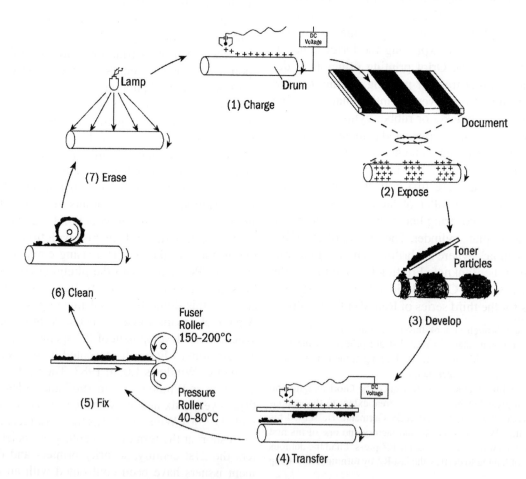

Figure 8.15. The cycle of operations of xerograph [38: 158].

(7) The drum is then exposed in order to dissipate any remaining charge.

From the 1960s until the late 1970s, the copiers reproduced in black and white and were neither threatening graphic arts and printers nor its *niche* of security printers. That threat changed when colour copying and laser printing arrived.

8.5.3 Colour copy machines

Additive and subtractive colour mixing were elucidated in Section 8.3.2. Additive is based on the colours Red, Green and Blue (RGB), and subtractive is based on the complementary colours CMY (cyan, magenta and yellow). Colour reproduction on a television screen is based on additive mixing while reproduction in print is based on subtractive mixing. The reader might recall that the entire range of colours cannot be faithfully reproduced which is illustrated with the aid of the *Commission Internationale de l'Eclairage* (CIE) diagram of Figure 8.16.

In Figure 8.16 all colours of light as well as of material objects are represented by a set of two numbers, x and y, and each is called the other's chromaticity coordinate.

From all the colours located within the bottom of the CIE diagram, only those colours within the triangle sRGB can be faithful reproduced on a colour television monitor. The colours within the irregular hexagon are the colours that can be reproduced with today's colour copying machines. A blue, red and green laser reproduces the gamut within the large triangle. So in order to optimise the effectiveness of the use of colour against colour reproduction, a person has to choose a colour outside of the RGB gamut and outside the colour copier hexagon, which is why the US dollar is green, the Euro ten is stone red, while the Euro twenty is blue-violet, etcetera.

Colour copying is a subtractive colour reproduction process and the copying machine needs three or four toners to reproduce a CMY or CMYK (CMY + black) image. The exposing light can be

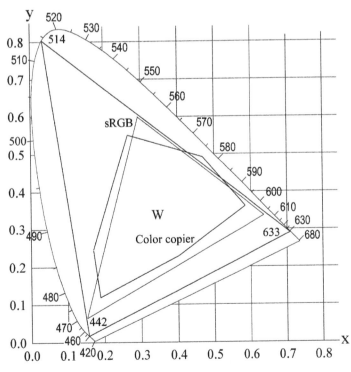

Figure 8.16. The CIE 1931, 2° observer chromaticity diagram showing the colour gamut (Courtesy by van Renesse [33]) of a colour copier, a standard RGB television screen and the gamut of three laser beam colours: HeNe (633 nm), Argon (514 nm) and HeCd (442 nm).

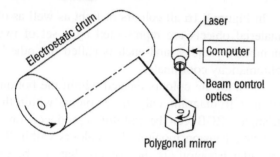

Figure 8.17. Laser scanner/printer [38: 160].

three or four successive strips of light with the appropriate filters or a row of RGB light emitting diodes. In the simplest version there are three or four separate passes, one for each toner. The size of the toner particles ranges from five-to-twenty microns, which originally limited the resolution of the first colour copying machines. Additionally, the colour reproducibility of the first colour copying machines was rather poor. Of the proven armoury against copying that is, fine line structures and colours that are difficult to reproduce were particularly effective against this first generation but that is no longer the case.

Around 1980 when the first Canon laser colour copying machines came on the market, colour reproduction advanced significantly. Instead of exposing the drum to an optical image, the drum would be scanned across its width [38: 160] by a tightly focused laser beam. That laser is fixed and a rotating polygonal mirror accomplishes the scanning. The beam intensity is steered by a computer and is varied appropriate to each pixel during the scan.

This development turned the copier into an intelligent reproduction machine. The next developments were focused on an increase in quality of the spot resolution on the drum and a reduction of the size of the toner particles. It finally brought the quality level of the colour copy closer to the quality level of four-colour offset printing. This advancement, together with the fact that the colour copier did not need to be operated by a printing professional, resulted in a very real threat for the security printing community. And worse is yet to come.

8.5.4 AM and FM screen technology

In the early 1990s, digital prepress technology was rapidly evolving. The fast growing power of computers enabled production and reproduction of colour with effective speed and so the prepress output became less dependent on its operators. But the moiré problem remained a challenge and there were some quite complicated and periodically ingenious solutions to diminish the influence of the so-called colour moiré rosette.[31]

Many years earlier in Germany, the head of the printing school at the technical university in Darmstadt Fischner believed that the detail of an image was dependent on the number of dots [16] used to reproduce the image. By the late 1970s two of his students began studies on small dot images, which were created digitally and thus called Frequency Modulated screens that used a small dot size. Their PhD thesis appeared in 1988.

The conventional method of producing continuous tone images, as explained in Section 8.4.3, is called Amplitude Modulation (AM) screening because the size of the dot or its amplitude varies. In areas where there are darker tints, the dots are larger and vice versa. Each dot always has the same distance to the next one.

With FM screening the dot sizes are the same, but their mutual distance varies and thus their number per unit area or frequency. In the early 1990s, when computers got powerful[32] enough to generate these dot structures, FM screens (also called stochastic or Monet screens) became available for the graphic arts. Printers started to experiment and soon discovered that with FM screening they could reproduce detailed images and enhanced colour and at the same time eliminate the rosette pattern or moiré.

FM screening [32: 147–148] also threatened some of the existing security features which is next illustrated. The minimal line width or resolution for the three major printing techniques ranges approximately from forty microns in dry offset to

---

[31] There are many textbooks on moiré patterns in four-colour printing. See for instance [20: 534] and [2: 191–201].
[32] A beautiful and early example of an FM screen application is the 1981 design of a Dutch definitive stamp of Queen Beatrix by Peter Struycken.

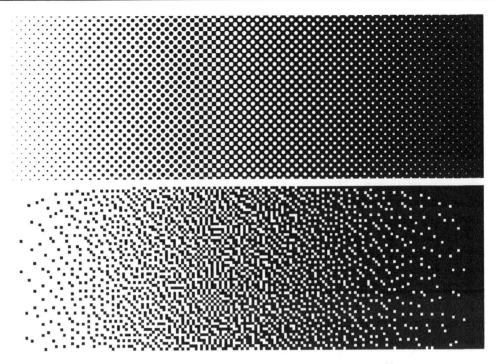

Figure 8.18. Halftone grids, the upper one with AM screening and the lower one[33] with FM screening.

fifteen microns in wet offset and intaglio. Copier manufacturers use dots per inch (dpi) in their specification and the copiers of today have a resolution of six hundred dpi (converted $\approx$ forty micron). The difference between AM and FM (see [16: 134–135]) is displayed in Figure 8.19 by presenting the information content in bit/cm^2 versus the dots size or minimal line width. The horizontal axis displays the minimal dot size. The vertical axis displays the information density. The hatched area covers the information density in halftone photographs.

Figure 8.19 tells us that a person can reproduce an information density, or read image, of slightly more than 10^5 bits/cm^2 for an AM screen and a minimal dot size of ten microns. But with an FM screen, a thirty-micron dot size is sufficient for the same information density. What a shock for the security printers to discover in the early 1990s that their fine line and moiré patterns, originally designed as a counterfeit deterrent, had become significantly less effective against FM screening.

---

[33]Rudolf L. van Renesse supplied Figure 8.18.

8.5.5 Expansion of the armoury

Recall that around 1970 the main printed security features were exclusive types, colours outside the reproduction gamut, fine line structures and patterns whose spatial frequency interfered with the scanning frequency.

These features are classified as *first-line of defence features*. *First-line* comprises tools for the general public to establish in a blink the authenticity of a banknote.

The *second-line of defence* is focused on professional money handlers such as cashiers and central banks. The devices are covert or rather a person would need a tool to detect alterations. A magnifying glass is such a tool, but the tool can also be a currency-sorting machine, which can detect certain paper or ink properties.

The *third-line of defence* is dedicated to experts and forensic laboratories. When the results of the first- and second-line inspections are inconclusive, a laboratory test then brings the solution. For this purpose dedicated markers or agents have been added in paper or ink to many of today's currencies.

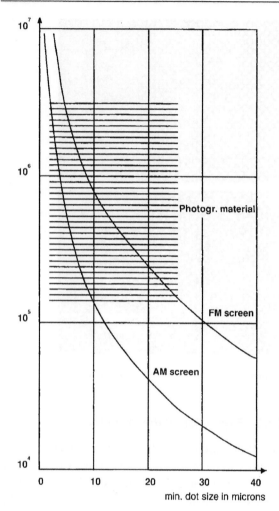

Figure 8.19. The information density in bits/cm^2 as a function of the minimal dot size in microns.

In the last two decades of the 20th century, security printers and document issuers were confronted with colour copiers combined with laser printers, FM screens, and home reproduction units consisting of a PC linked with a scanner and an inkjet printer. The existing armoury obviously became less effective if not obsolete and at least new *first-line* features were urgently requested. The threat was tackled this time by introducing:

- 3D devices,
- optically variable printed images and regarding the *second-line of defence,*
- printed covert images, tracers, and/or taggants.

8.5.6   3D devices

The idea speaks for itself. A printed image is two-dimensional; adding a third dimension creates a hurdle for the counterfeiter. The third dimension can be created by embossing the (paper) substrate or by adding a foil with a virtual (holographic) 3D image on the paper. The latter is coined DOVID, an acronym derived from Diffractive Optically Variable Image Device and in a later section there is more about DOVIDs.

Embossing is known from earlier times as seen with the dry seal cachet shown in Brion and Morreau [7: 3] of the first banknote in 1456. Paper is pressed into an engraved image, which results in plastic deformation (embossing) after the pressure is released. If a person puts ink on the engraved image, embossing results. This roughly explains why intaglio-printing shows a distinct relief, the tactility of which makes it fairly easy to confirm the presence of intaglio. One way to render printed images three-dimensionally is to make use of this relief [32: 203, 205].

The first people to file a patent on this type of latent or transitory image were Trevor Merry[34] and Robert Gordon Hutton in 1975 [21]. The underlying principle of their invention is displayed in Figure 8.20.

The latent image[35] of Figure 8.20(a) combines a background and a foreground pattern of fine parallel intaglio lines, the background pattern being aligned perpendicular to the foreground pattern. If the patterns are fine enough ($\approx$six lines per mm) they are not readily observed by the naked eye, so that under normal observation both foreground and background merge. When observed obliquely as shown in Figure 8.20(a$'$), foreground and background patterns will separate because the lines of one pattern will shield the (paper) substrate from view while between the lines of the other pattern the paper remains visible.

Today there are many banknotes equipped with a latent image security device. Elegant examples

---

[34]Trevor Merry told me that the commemorating stamps and lottery tickets for the Olympic games of 1976 were the first products featuring a latent image.

[35]Rudolf L. van Renesse uses in [33] transitory images, which is the general term for latent and transient images.

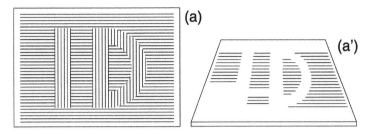

Figure 8.20. Latent image, from [33].

Figure 8.21. Latent image on Swiss fifty franc note with horizontal foreground line elements and vertical background line elements. Top left: normal illumination and observation. Bottom left and bottom right: oblique illumination from the top. (Courtesy Rudolf L. van Renesse [33].)

can be found in the current series of Swiss banknotes, as shown in Figure 8.21.

8.5.7 Optically variable printed images

Again, the idea is simple. Any reproduction starts with an image, and for an image a person would need illumination and an angle of observation. But a true reproduction becomes difficult if not impossible, if the colour or colour contrast changes with the angle of observation. This change of colour or

colour contrast is called geometrical metamerism[36] and can be brought about by thin-film interference colours. The BEP[37] ought to be credited for being

---

[36]Metamerism refers to a situation where two colour samples appear to match in less than one condition but not under another condition. Metamerism is usually discussed in terms of two illuminants whereby two samples may match less than one illuminant but not under another one. Other types of metamerism include geometrical metamerism and observer metamerism.

[37]Bureau of Engraving and Printing in Washington, USA.

Figure 8.22. OVI on US$ 100 note (Courtesy Rudolf L. van Renesse [33]) with a green-to-violet colour shift. Left the normal observation and above oblique observation.

the first to suggest optical variability as a counterfeit deterrent in one of the early meetings of the Four Nations group in the 1970s.

Printing inks are pigments in a vehicle. The vehicle is a liquid in which the pigments are dispersed giving them the mobility they need to flow through the printing press and onto the substrate. Inks, in general, are a careful compromise in technology, which implies that a change to optically variable pigments requests a change of the compromise. Inks with optically variable pigments were introduced in the late 1980s. Haim Bretler [8], director R&D of Sicpa, the leading supplier of security printing inks, announced the development of pigment flakes with interference properties at the SPIE conference in San José in 1990. The first banknote with an image printed with OVI (Optically Variable Ink) appeared in 1987 and was the commemorative sixty Baht note for the 60th birthday of King Rama IX of Thailand.

The latest US dollar versions as shown in Figure 8.22, the Euro notes, and many other currencies today are exploiting the OVI concept as a counterfeit deterrent.

8.5.8 Printed covert images, tracers and taggants

Printed security features, in general, ought to be difficult to reproduce and after reproduction something of the feature should be missing, preferably something striking. But in turning around this idea,

incorporate features that are not visible to the human eye but which show up after reproduction might also be a deterrent to counterfeiting. Another idea might be to incorporate features that locate copier type and serial number or that hamper the copier mechanism itself. These ideas are the main ones behind covert images, tracers and taggants, and they gained momentum in the 1990s. Their developments are still going on and only some early examples of each category will be highlighted here.

Covert images. This development in particular is based on the fundamental differences between human perception and digital sampling. The human visual system and digital scanners as employed by the latest generation of copying machines do not perceive the same image.

Today's colour copiers print structures with a resolution of 600 dpi and this is far outside the range of human vision. If the symmetry of the conventional dot screen of 600 dpi is 'broken' by replacing the screen dots by minimal lines [32: 173–193], a new degree of freedom is introduced. The result is that every small screen line element has an orientation angle that can be modulated. Hence it is called screen angle modulation with the acronym SAM. Now the screen angle can be linked to the local density of the image and since the spatial frequency of these rotated lines lies outside human visual perception, the orientation can be affected without affecting this perception. Covert information is then added to the image. But the variation in

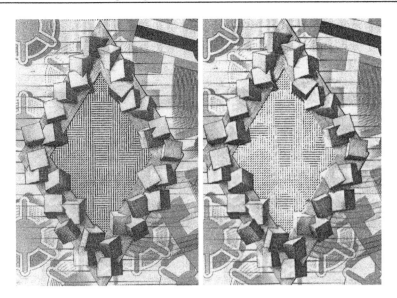

Figure 8.23. The first SAM structure as displayed inside the diamond. Left – the original image and on the right a colour copy [34].

the angle of the minimal lines does affect the output of digital scanners, which then overtly displays the intended visual image disturbances.

The SAM feature was incorporated for the first time in the Dutch NLG 100 design of 1992, see Figure 8.23.

Tracers. Modern colour copiers are individualised by their manufacturing brand and serial number and the question is raised if it is possible to trace the copier identity by putting a tracer in a printed output. The answer is affirmative and the print engine of the large and expensive colour copiers can generate a dot code linked to a type and to a serial number. In that way, a forensic investigation of counterfeits can use the tracer code to determine the copier type, the serial number and the eventual location.

Taggants. Quite another angle of approach is that of taggants. Their prime goal is prevention. Modern colour copying machines refuse to copy many of the more recent banknotes by detecting [25], a geometric pattern on the banknote. The taggants are unofficially called EURion constellation and have been the tangible result of SSG 1. This Special Study Group has been set up in the late 1980s as an international working group of experts of

central banks and note printers. Together with the Japanese producers of office machines, the threat was recognised and resulted in the successful development of the EURion image and its detector.

The EURion image consists of a geometric array of small circles and the abundance of many small circles in a well-defined colour, which also happens to be a design and a production constraint. Issuing authorities have to consider a new design or, even in the case of the US dollar, a multi-colour design.[38] The latter was one of the arguments to expand the mission of the SSG 1. It got actually a new name and a new task. In 1993 the Central Bank Counterfeit Deterrence Group (CBCDG) was established and is a working group of twenty-seven central banks and note printing authorities. The CBCDG [35], came up with a Counterfeit Deterrence System (CDS) which has been incorporated into PC image processing software and PC hardware such as HP colour printers. CDS does not track the use of a personal computer or digital imaging equipment but does refer users to the website *www.rulesforuse.com*. This site provides links

---

[38]Multi-colour colour use has been particular difficult for the US. The US dollar is printed on machines with a limited number of colours. For the recent upgrades the Bureau of Engraving and Printing acquired new offset machines that is, for EURion and background printing.

to the guidelines for the reproduction of banknotes for a large number of countries.

The underlying principle of the CDS is the embedding of imperceptible identifying codes. Such codes require machine authentication and dedicated software to be detected and/or decoded. The technology is known as Digital Watermarking. Its potential in the battle against digifeiting is obvious. Rather than adding features that are difficult to reproduce, codes can be embedded in the printing of documents that allow image-processing software to identify an attempt to capture or process the image and then abort it.

8.5.9 The armoury at the beginning of the 21st century

Nearly two ages of continuous improvement in printing and reproduction technology led to the following list of printed security devices at the beginning of the 21st century.

- Exclusive types
- Colours outside the reproduction gamut
- Fine line structures
- Patterns whose spatial frequency interferes with the scanning frequency
- 3D devices
- Optically variable printed images, and
- Printed covert tracers, taggants and/or images.

The continuous battle for the improvement of document security has undergone a fundamental change in the last decades. Instead of adding features that are difficult to reproduce, codes are embedded that thwart reproduction. Furthermore, the issuing authorities are searching more than ever for cooperation with the imaging industry also guided by the saying: "If you cannot beat them you better join them".

8.6 THE HISTORY OF SUBSTRATE-BASED SECURITY FEATURES

Paper has continuously been used as substrate for security documents, and this section focuses primarily on substrate-based security features. The principal features are either a modulation of paper thickness which leads to differences in transparency and to showing of watermarks or an incorporation of a foreign body, such as a thread within the paper mass.

These developments emphasise an increase of security, which is a response to the improvement and proliferation of reproduction technology. However from 1992 until 1996 the Reserve Bank of Australia (RBA) replaced[39] all of its paper currency by polymer currency. The RBA claims a significant increase in circulation life plus an increase in security. From that event, the substrate development efforts have focused on the increase of security and the increase of circulation life.

8.6.1 Watermarks

The use of watermarks to protect valuable documents printed on paper dates from the mid-13th century when several papermakers in the Ancona valley in Italy began using watermarks to differentiate their own paper production runs from those of other manufacturers. One paper maker from almost 750 years ago that used this early form of tracking and protection was known as Cartiere Miliani Fabriano and this company is still an active supplier for security paper today.

Watermarks are one of the most widely used security features with an estimated ninety percent of all banknotes being printed with them, and high quality watermarks are one of the most difficult features for a counterfeiter to simulate. Interpol reported 130 million counterfeits in 2000 and in only twelve percent of the cases had any attempt been made to simulate the watermark. There is a delicate balance between paper endurance and quality of the watermarks. The importance is in the fibre length. Short fibres are better for the production of fine details and render watermarks with the highest image quality that is, the best definition and the most tonal variations. However, short fibres have a negative influence on paper endurance. In contrast, long fibres can easily knit together and thus provide a stronger paper with a longer endurance but at the expense of fine watermark details.

---

[39] A director of one of the leading European currency printers used to say that the Australians now have plastic currency besides kangaroos.

Watermarks are normally produced in one of three following ways.

Mold made. By forming fibres onto a profiled porous mold, fewer fibres are deposited at elevated areas of the mold and more fibres at the indented areas. As a result, translucency, which is the paper's ability to transmit light, decreases with the amount of fibres deposited. Currency substrates are composed of cotton or cotton and flax, which is in contrast to most other papers, which are made of wood pulp. This process known as 'cylinder mold paper making' is characterised by multi-tone watermarks, which are also registered in the sheet of paper. Mold-made watermarks have the additional benefit of being visible in reflected light when viewed at an acute angle.

John Dickinson invented the mold-made process in England [1] in 1809. All Euro notes made today have a mold-made watermark.

Dandy roll. This process is an embossing method whereby an image is created by reorienting fibres in a sheet of paper shortly after formation. Pushing fibres from one area to the next using an embossed cylinder or roll creates light areas and dark areas when these surplus fibres build up around the edge of the lighter area.

While these watermarks are not as multi-tonal as mold-made versions, the paper has long cotton fibres thus giving it more strength. The watermark quality is traded for durability, a compromise considered suitable for the lower denominations. The US currency exploits dandy roll watermarks.

Electrotype. This method is formed using the cylinder mold papermaking process whereby water drainage is stopped in a limited area by the use of an electrotype. This technique creates a sharp light area, which is often the numeral of the denomination in the finished paper. All Euro notes today are equipped with mold-made + electrotype watermarks.

8.6.2 History of watermarks

There is a lot of literature [18: 35–44] available on the history of watermarks, its application in

Figure 8.24. Mold-made watermark + the denomination 100 as electrotype watermark in the former German DM 100 bank notes in transmission [33].

currency substrates, and watermark design [23: 116–149] in relation to its effectiveness [14] as a counterfeit deterrent. The following brief outline presents a few highlights.

The *Bank of England* introduced watermarks as counterfeit deterrents at the end of the 17th century. When some counterfeit notes were discovered in 1695, it was decided to introduce a special watermark and thereafter watermarks were to be an outstanding feature of all the Bank of England notes. From that time period, preventing people from forging banknotes relied principally on the watermark. At the beginning of the 18th century a French refugee, Henri de Portal started a paper mill and was awarded a contract in 1724 for producing the Bank of England paper; the firm Portals has held this contract ever since.

The watermark deterrent does work, but Hills [18: 37] reports that at the beginning of the 19th century (between 1812 and 1818) a total of 131.331 counterfeited pound notes were discovered. The Bank of England then again endeavoured to find ways to stop forgers and one notable person who came up with an improvement of security features in paper was Sir William Congreve. This remarkable inventor deserves a place in the history of security printing since his first objective regarding security features, as displayed in 1819 in one of his many patents, is still valid.

> Paper has to have clear distinguishing marks that would be as simple as they can be so that they may be immediately recognised and understood by the most unlearned persons, that all orders of people may be equally able to distinguish a bad note from a good one, by observing whether the said tests exists or not in any note offered.

The Bank of England considered some of his ideas as too radical for those days and embarked upon design changes of the watermark. More complex patterns were introduced to make counterfeiting more difficult. The history then successively goes further with problems and solutions associated with the production of watermark molds.

The development of the watermark in *France* is associated more closely with the centralised government. At the end of the 14th century, papermakers in France were obliged to guarantee size and quality of their products by using watermarks. Later during the reign of King Louis XIV watermark development was linked with revenues by an act that required official documents to have a special watermark thus being a revenue source for the government.

In 1800 of the Napoleonic period, the Banque de France was founded. The first notes were issued in 1808 and were protected with a watermark. In the second half of the 19th century, France adopted the invention of the cylinder mold paper machine. For over a century and until the appearance of the Euro, mold-made watermarks protected the French currency against being counterfeited.

The first *Dutch* banknotes were made in 1814 because De Nederlandsche Bank (DNB) wanted to issue paper currency as soon as possible after its foundation. As already explained in Section 8.6.1, the notes were secured with a border of music types and were printed on a paper substrate containing a watermark, which depicted a French eagle and the portrait of Napoleon. In retrospection, these two images used are a surprise since the French occupation of the low lands had just come to an end (1813) when Napoleon had been defeated and exiled.

The first forged Dutch banknotes appeared in 1836. DNB responded immediately and ordered a new series. DNB did not choose a new printed security feature but attained a higher security level by replacing the watermark for a more complex one.

Issuers, printers and paper makers did not yet realise that the counterfeit threat required a coordinated effort. That awareness came only in the last decades of the 20th century when, as explained in Section 8.5, reproduction technology became available for the laymen and was domesticated. Before that time, the watermark and the printed image developed in technique and design independently from one another.

8.6.3 Other first-line features

A multi-tone watermark of high quality is an outstanding feature for a first-line of defence against counterfeiting banknotes. The modulation of the mass of the paper substrate causes differences in transparency. But the paper substrate has other first-line features to offer by introducing a foreign body in the paper mass. Well-known examples are:

- Security threads
- Fibres
- Planchettes.

As an aside, there is another first-line feature [29], visible in transparency, called the Microperf®. The Microperf optical security feature is produced by using a laser that perforates banknote substrates with a series of circular or elliptical shaped holes between 85–130 microns and which are also only visible in transmitted light in the same manner as a watermark. The perforations created by a laser can be arranged in such a way as to produce an image of the denomination of the note. Microperf appeared for the first time in 1977 in the new Swiss currency. In 2004, the Lithuanian and Romanian currency adopted the Microperf security feature and it features from 2005 in the Russian rouble banknotes.

Security threads appeared officially in a patent issued by Crane USA in 1867, although John Dickinson (mentioned earlier) had already patented a process in 1829 in England for putting silk thread into paper to help prevent forgery. This silk-thread security paper was made famous by the legendary Penny Black stamps. There are also indications

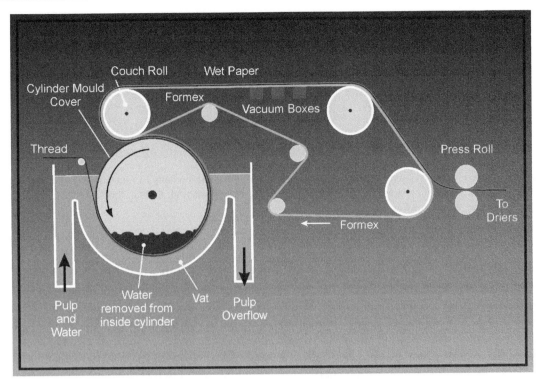

Figure 8.25. Mold-made[40] process flow.

[14] that the Japanese developed a metallic insert (gold or silver) in the surface of the paper to guarantee the authenticity of paper money as early as the 14th century. In the beginning of the 20th century there have been attempts to embed a platinum [13] wire into the paper substrate. But with the arrival of polymers just before WW II, the security thread started to gain in use. The first notorious example is the embedded thread in pound notes. The Bank of England protected the English currency from a large-scale counterfeiting attack by Hitler's Germany and with good cause as explained in Section 8.4.1.

As can be gathered from Figure 8.25 in which the method of incorporating a thread into a paper mass is displayed, threads in paper require a lot of mold-made paper technology. Today more than eighty percent of the world's currency incorporates a security thread. The threads enable verification in the first-line but are also important for second-line verification when high-speed currency sorter machines detect its presence and/or properties.

The most basic property of a thread for verification and security is that it be invisible in reflection yet visible in transmission. Circa 1960, threads used in this industry became far more sophisticated by using threads wider than 0.5 mm and when the polymer threads were doped with coloured pigments or pigments reacting on illumination with ultra violet (UV) light. Threads were also printed with micro text or even partially metallised. In 1980, Portals introduced Stardust® which is a thread that emerges[41] intermittently as a woven thread through the topside of the paper.

A highly-visible thread gained in effectiveness as part of the first-line defence by the end of the 20th century when threads were coated to show optical interference in reflection. Also changes in papermaking technology have allowed the incorporation of wider threads, the most recent developments [30] and [19] of which are Optiks® and Varifeye®. Optiks® is a 13-mm-wide thread, visible on the reverse side and 'windowed' on the front

---

[40]Courtesy Malcolm Knight, R&D director de la Rue.

[41]This thread concept is also known by professionals as Loch Ness.

Figure 8.26. *Stardust*®. In the Czech Republic 20 Korun note in specular reflection (left). In transmission (right), the coinciding barcode type watermark is inherent with the production technology of the windowed thread. Thread width 1.3 mm; negative lettering 1.0 mm high, as displayed in [33].

side of a banknote. Varifeye® is a see-through window with a genuine deckle edge combined with state-of-the-art security film elements.

The first appearance of *fibres and planchettes*[42] occurred in the 19th century. The first application of these features in currency substrates is vague, however. Hills mentions in *Papermaking in Britain 1488–1988* [18: 38–39] Congreve's experiments with multi-layer paper with different colours in each layer: "At Tumba in Sweden, one mold was dipped first into a yellow pulp and placed under a machine which scattered coloured fibres over part of it . . . gave a paper with a white surface on either side, yellow in the middle[43] and a band of security fibres in one part too".

No date is given for this statement, but it relates to the period of the Congreve experiments of 1819. Congreve's idea of introducing substrate colour as a first-line security feature finds support even today since the substrate of all euro denominations differs in colour.

In 1882 the paper mill of Smidt van Gelder in the Netherlands suggested a way of securing paper against counterfeiting by applying coloured fi-

bre tracks. They supplied samples[44] to the printing house Enschedé; but in the end, the watermark was preferred as a counterfeit deterrent.

In 1891, the American Bank Note Company introduced 'planchette paper' which contained coloured paper discs. These discs had various characteristics, but all of them were intended to be part of a first-line of defence against counterfeiting.

The fibre development moved away from visible fibres as a first-line feature to invisible fibres as a second-line feature. Nowadays most of the currency substrates incorporate fibres that are only visible under UV illumination. Planchettes are still in use as a first-line of defence feature against counterfeiting.

8.6.4 Second-line features

Under UV light, currency paper does not show a fluorescence, which is a built into the property of cotton fibres, used in paper. Fluorescence, then, is a second-line of defence feature but less effective today because genuine notes, which are washed, are also fluorescing under UV light. Modern fluorescent whitener powders for washing machines are responsible for this confusion.

By the 19th century, second-line features had already been suggested. In 1857, the Dutch Central Bank[45] proposed covering notes with a transparent coating in which a secret preparation was incorporated.

---

[42]Information provided by Dr. William Knight, Director R&D of the de la Rue Group in England.

[43]Congreve also tried to produce a laminate paper with a coloured inner layer. By the end of the 20th century, Domtar in Canada developed a sandwich paper Luminus which consists of a polymer body on front with a reverse side of a thin layer of paper. The polymer core carried a printed image which gave Luminus the feature of a coloured watermark; in addition, the sandwich effect improved the circulation properties. Unfortunately the sandwich construction resulted in a curl which prevents mass application.

[44]Museum Enschedé. HBA 02154. Letter of 26-09-1882 by P. Smidt van Gelder to the Enschedé board.

[45]Museum Enschedé, HBA 03150. Correspondence between DNB and the Enschedé board in 1857.

Of all other possible second-line features certainly the so-called MM-feature [22] deserves its place in the history of document security. The M's refer to its chemical structure and this MM-feature stems from the late 1970s and is a chemical agent with a peculiar luminescent property added to the paper mass. It is known that some of the lanthanides, which are embedded in a metal oxide lattice, are emitting radiation with a shorter wavelength after excitation with radiation with a longer wavelength. This kind of luminescence is known as anti-Stokes[46] or "up converting" since the emitted radiation has a higher frequency and thus a higher energy level than the exciting radiation.

This M-feature triggered an extensive search for other lanthanides to be embedded in the paper mass and there are many other up-converters available today. They all feature in the second-line of defence against counterfeiting. Many of the banknotes of today feature an up-converter and consequently many of the modern banknote-sorting machines are equipped with a detection device for these up-converters.

A few times in history even watermarks have been used as a feature in the second line. A typical example can be found in the Dutch twenty-five-guilder note, issued in 1990. The watermark of this note is extended with a special barcode [40] coined AQUS (from aqua system). This note was the first of a new series of which each denomination has its specific binary barcode. The barcode was over-printed, therefore not visible for the general public and the sorting equipment at the central bank had to be equipped with special detectors for this second-line feature. The overprinting inks have to be transparent for infrared radiation because the barcode is detected with infrared. The bits of the binary code are defined by detection in transmission of the positive or the negative gradient; respectively going from thicker paper to thinner or from thinner to thicker. This elegant detection method has the advantage of being independent of the substrate transmission and the occupied real estate is relatively small. The AQUS concept has been adopted

by a few other central banks and is incorporated in a less elegant form in the present euro notes.

8.6.5 Polymer substrate and the 'fight for life'

Banknotes at the end of their circulation life look fairly tattered. They are creased, dirty, torn and taped together, have dog-ears and holes, smell and may even transfer germs. All of these dirt-factors are especially true for the lower denominations since their exchange frequency in circulation is high.

Dirty circulation notes are more prone to counterfeit attacks than are clean[47] circulation notes and there are unfortunately many variables, such as the note coin boundary, the climatic conditions, the habits of the general public, etcetera that apply when deciding to take notes out of circulation by the issuing central banks. The logical question then arose about more how to find more durable substrates and in the 1940s the first trials started with a substrate that provides increased resistance to folding, tearing and dirtying. The American Bank Note Company, together with the American Cyanamid Company and the paper manufacturer Crane & Co,[48] led the way with a product they developed that was called Melamine. In 1955 Dupont launched Tyvek®, a non-woven product of strong polyethylene yarns.

Experimental areas were selected with severe climatic conditions and a small and controllable circulation, which resulted in field trials on the Caribbean and the Isle of Man. Security printers tell each other many anecdotic stories about what happened and what caused the 'short life' of the trials with the 'long life' substrates. But the main point comes down to poor ink adherence and poor dimensional stability of the substrate.

It was not and is not in the interest of the paper suppliers and security printers to extend the circulation life of the banknotes. So the search for extended currency life slowly[49] faded. Until the

---

[46]It is called Stokes when the wavelength of the excitation is shorter. Obviously there is no perpetual mobile involved because the emitted intensity is considerably lower than the excitation intensity.

[47]For this reason the Bank of Japan supervises a clean circulation.

[48]Crane & Co has been the supplier for the US dollar substrate from the beginning.

[49]At that time, a few currency printers (Belgium, The Netherlands and Switzerland) extended the circulation life by covering the notes with a protective coating.

1980s, when the Reserve Bank of Australia (RBA) was confronted with a circulation life of its five dollar note that lasted from six to eight months which is an extreme example to be compared with the eighteen to twenty-four month life span of US dollar notes.

The RBA then took the courageous decision to fund the research for a currency substrate with better circulation properties which resulted in the polymer substrate[50] Guardian® and in 1988 the first polymer note was issued. It was a commemorative note of AUD10, which marked Australia's bicentenary. From 1992–1996, the substrate of all the Australian notes was transferred from paper to polymer.[51] The established league of paper suppliers watched the development and saw to their amusement that in the beginning there were problems with ink adherence and dimensional stability. The league of suppliers was then inclined to conclude that the days of Tyvek and Melamine were being repeated. But the RBA put a lot of energy into the development and promotion of Guardian® and slowly but steadily convinced issuing institutions of neighbouring countries of this method's validity with the slogan "double price but four times longer in circulation".

The revolutionary polymer notes threatened to oust the paper substrate for the lower denominations. The currency paper makers realised the threat and responded with long life paper, coated paper, reinforced paper with threads and other improvements. This fight for life, literally, started by the end of the 20th century.

The real 'push' for marketing the RBA's "Plastic Technology" came from the Governors of Australia. Their initiative attempted to generate income to offset the huge development costs for the polymer substrate, which coincidently was attracting widespread criticism of the RBA in both the media and Parliament. Initially the task was assigned to Bob Larkin, the director of the printing works. But

Les Coventry, who was chief cashier at the time, finally picked up the promotion with remarkable energy. He certainly made the Guardian Project his own, so if any one person at the RBA can take credit for Guardian's success, it was Les Coventry.[52]

In addition to circulation life, there are other valuable ideas operating with the polymer substrate. It first has to be coated in order to obtain ink adherence; and in doing this partial or local coating, the RBA researchers[53] managed to generate a transparent window. The polymer substrate thus presented a new first-line security feature. The RBA did not hesitate and added 'increased security' to its longer life campaign. Also the transparent window leads to the new concept of self-authentication. Within this concept, tools are incorporated into the notes, which enable the general public to perform a simple check on a note to see if it is genuine, or not.

An example is the application of a coloured window on one side of the note and an image printed with metameric colours on the other side of the note. By folding[54] the coloured window over the metameric image, a colour contrast becomes visible.

Another example of self-authentication is applied in the New Zealand ten-dollar note issued in 2000. It possesses a so-called μSAM self-authentication feature. The message "Y2K" is modulated according to the SAM principle and becomes visible when the decoding line screen in the transparent window is folded over it.

The two examples need an expert's eye and as such the self-authentication was probably launched too early into circulation. On the other hand the paper makers were quick to discern the potential of the transparent window and in 2004 launched the 'windowed paper' concept.

The substrate development is certainly not over and goes forward with rigorous ideas, stimulated

---

[50]The polymer is BOPP, the acronym of Bi axial Oriented Poly Propylene.

[51]One of the European currency printers used to say that besides having kangaroos, the Australians also have plastic money.

[52]Private communication by Alex Jarvis, the former director of the Bank of England printing works.

[53]The Australian Research Institute CSIRO and the Australian note printing works NPA were contracted by the RBA.

[54]This principle is applied in 1999 in the Romanian 2000 Lei solar eclipse note.

on the one hand by the usual counterfeit threat and on the other hand, and for the first time in the history of document security, by the technological progress instigated by a new substrate.

8.7 THE HISTORY OF SECURITY FEATURES ADDED TO THE SUBSTRATE

The dramatic shifts in the reproduction technology since the 1970s, as indicated in the Sections 8.5.1–8.5.4, brought issuers and producers of security documents to the point of strengthening their innovation potential. They expanded in-house research and development groups; outsourced dedicated research assignments; and united to share efforts. The history of the development of holography runs more or less parallel with this growing awareness.

Returning for a moment to 1947, Denis Gabor first recognised the physical process of holography. Since the quality of his images was inferior to that of ordinary photography, he did not arouse must interest in his work. The main reason for the discouraging quality of his images was the absence of a light source with the combined intensity and coherence that was needed. That shortcoming changed with the arrival of the laser in 1960 and in 1962 when Emmet Leith and Juris Upathnieks produced the first three-dimensional image.

These excellent images astonished the scientific community, but security printers were barely aware of the potential of the holographic, three-dimensional image and did not immediately foresee any application. The legitimate reason for that lack of foresight is that holograms could be viewed in those days only with the aid of a coherent laser beam and so direct application in the area of visual security at that time was not possible.

In 1962 Yuri Denisyuk came up with white light reflection holography, based on the Lippmann–Bragg effect. For a floating image between the hologram machine and a viewer, laser light was not yet required and display holography finally gained momentum in 1968 when Stephen A. Benton invented white-light transmission holography while researching holographic television at Polaroid Research Laboratories. This type of hologram can be viewed in ordinary white light. The depth and brilliance of the image brought holography further into popular awareness, and the availability of the laser technology together with the appearance of the first holograms finally started directing research efforts towards holography based security features. In 1983 MasterCard International, Inc. launched the first credit cards carrying embossed holograms that were produced by the American Bank Note Company in New York. The MC holograms were the widest distribution of holography in the world at that time.

8.7.1 The early search for 3D devices

As suggested in Section 8.5.5, a printed image is two-dimensional, and adding a third dimension creates a greater hurdle for the counterfeiter. In the late 1970s, the first 3D experiments were focused on the embossing of the paper substrate. The limits of embossing by intaglio printing were stretched but it became clear that an intaglio image was always limited in its resolution due to the pure mechanical process of embossing and plastic deformation of paper. The latent image, as described in Section 8.5.6, was an elegant refinement, but its visibility degraded during circulation, besides which the latent image only shows under oblique illumination,[55] which is a handicap for detection by the general public.

An earlier 3D intaglio refinement occurred around the end of the 1960s and resulted in the printing of a small square filled with parallel lines. The embossed relief of the parallel lines can then be detected in the currency-sorting machine at the central bank. This device [32: 206] is called ISARD from the acronym Intaglio Scanning and Recognition Device. It is a second-line feature and was applied for the first time in 1968 on the Dutch ten-guilder note. In the euro notes of today the ISARD pattern is incorporated.

So by the end of the 1970s, it was generally recognised that the paper substrate had limitations and one way or another the surface of the paper

---

[55]There is a famous story about a currency printer showing notes with a latent image to his central bank customer. The customer looked at the notes in transmission and said that he was enthusiastic about the latent image!

substrate had to be upgraded. In 1978, Orell Füssli, the well-known security printer in Switzerland together with the optical research group of Landis & Gyr, gets the credit of being the first to present a diffractive optical element on a paper substrate at a Machinery Committee of the Banknote Printers Conference in Copenhagen. First the surface roughness of the paper substrate was levelled by the application of a coating. Then a die with a diffractive relief pattern embossed the coated paper surface and for the first time a diffractive pattern was observed from a paper surface. The die was fixed in a so-called Numerota, a letterpress-numbering machine of KBA-Giori. However the researchers of Landis & Gyr soon discovered that the paper surface proved to be unruly for optical effects and the coating did not withstand the extensive circulation tests. In 1981 under the research venture of the Swiss National Bank, Orell Füssli and Landis & Gyr decided to leave out the special coatings and to embark on foils added on the paper substrate. This method is the same as the course taken by the American Banknote Company; however, the latter had its focus on the credit card substrate.

8.7.2 The race of being the first

The major breakthrough for the application of 3D security devices took place in 1988 when the first hologram on a banknote appeared. The Austrian five thousand shilling note was issued with Mozart depicted in the hologram.[56] It was an embossed hologram in a foil and the Swiss company Landis & Gyr developed the embossing die while Kurz in Germany developed the foil technology. The Austrian printing works were bold enough to solve many start-up problems. The Kinegram for instance should stick to the paper during circulation, but attempts to remove the Kinegram should result in a destroyed kinegram in order to prevent the counterfeiter from reproducing or reusing it. The chosen hot melt technology proved to be reliable at last but the hot melt glue had to be modified during the production runs. The delicate balance between foil and paper will be elucidated in the next section.

Figure 8.27. The first kinegram[57] in circulation on the Austrian 5000 schilling note.

Across the ocean, the Canadians developed quite a different approach. Their starting point was not diffraction but optical interference. They developed a foil coated with several thin optical layers deposited under vacuum. The application of the foil on paper was original and ingenious. The interference image is printed in glue on the finished sheet of banknotes with a hybrid offset-gravure printing process. The glue is then activated by UV light and almost simultaneously the foil is brought into contact with the activated glue. The bond between activated glue and paper is stronger than the bond between the foil and its carrier[58] and the foil with the interference properties thus get attached to the paper.

The Canadians issued their twenty-dollar note with the interference foil patch three months after the issue of the Austrian five thousand shilling note. But neither the general public nor the holography or banking community had been aware that Orell Füssli had unofficially won the race. In1985, the Swiss National Bank commissioned Orell Füssli[59] to produce seventy-two million re-

---

[56]The Mozart feature had kinematics and was called a kinegram.

[57]Rudolf L. van Renesse was kind enough to supply this photograph of a kinegram.

[58]The first step in the vacuum deposition process is the deposition of a special release coating.

[59]Private communication of Adolf Kuhl in 2004.

Figure 8.28. Optical Security Device (OSD) on the Canadian $20 note [33] under near normal observation (left) and under oblique observation (right).

serve notes of one hundred Swiss franc note which were to have machine-readable optical tracks. Orell Füssli produced these notes from 1986–1988; however, the life of these reserve notes is contingent upon rules and in general appears only in emergencies.

A few pioneers of these first attached optical security features, such as Doug Overhill of the Bank of Canada, Werner Reinhart of Kurz and Jean François Moser of Landis & Gyr ought to be mentioned at this point. In addition, document security history has to credit Adolf Kuhl, manager director of Orell Füssli, for starting the cooperative effort with Landis & Gyr and with adding 3D holographic features to optical security devices.

8.7.3 The delicate bond between foil and paper

The development and fundamental technology as well as the properties of holograms are presented and described in many books and recently in *Optical Document Security* by van Renesse [33]. There are many studies regarding the perception of holographic security devices by general public users. However the bonding between foil and paper is responsible for trouble free application and circulation of the holographic patch on the substrate and this has remained rather underexposed. From 1985 onward and in a bit less then a decade, the bonding technology has overcome its start-up problems and has proven to be reliable and mature. A hot melt technique is used to bond the foil to the paper and the complex and delicate structure of the foil is shown in Figure 8.29.

The holographic grating in the metallised embossing layer is sandwiched between two protective layers. The hot melt ensures [41] that the foil adheres to the substrate. The total thickness ranges between three to five microns once it is bonded to the paper but excluded from the polyester film and release layer. The polyester carrier film has a thick-

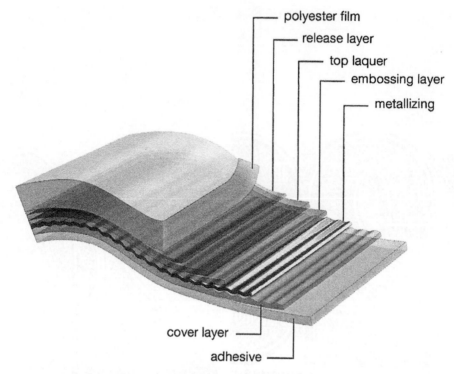

Figure 8.29. Exploded view[60], showing the layered structure of a security foil.

ness of nineteen microns, while the adhesive layer measures six microns. The paper roughness, or the peaks-to-valleys roughness, generally amounts to five microns so that the six microns of adhesive material are barely enough to compensate for the paper roughness of five microns. The option by pressing foil and paper together seems logical but confronts us with another problem.

The techniques to bond foil to paper all have pressure and temperature in common and the process is more or less similar to ironing. Put the foil on top of the paper, soften the hot melt by applying heat, and subsequently apply pressure. It is as simple as that, but mass production requests doing it online which occasioned process engineers much problem solving since both paper and foil have to move at identical speeds when the hot melt softens and gets sticky enough to adhere.

For this reason the first hot stamp foil printing machines were modified flatbed letterpress printing machines running at maximum speeds of four

thousand sheets per hour. The speed is understandably limited by the stop-and-go transport of foil and paper. In order to cope with the growing demand for an output increase up to the thousand sheets per hour, a rotary transport method had to be considered. But in that case paper and foil would be transported by friction, which in turn is a function of roughness and pressure. Due to the obvious difference in roughness, the friction between the foil and its driving roller is less than the friction between the paper and its driving roller. To compensate for this difference, the transport speed of the rollers had to be slightly adjusted when the paper and foil were not connected. As a consequence strain is exerted when the glue is activated and the foil is bonded to the substrate. To some extent, this strain can be absorbed by the foils elasticity. But if pressure increases, then simultaneously friction increases because of the increase difference between the friction and the strain.

The development of a foil application machine with an output of ten thousand sheets per hour that could accommodate this delicate bonding process

---

[60]Courtesy of Leonard Kurz GmbH.

was initiated by Albrecht Germann of Koenig & Bauer in Germany in the late 1980s and successfully manufactured in the early 1990s by his successor Johannes Schaede. The Optinota machine line of KBA-Giori today applies the holographic patches on the euro and other currencies.

Kurz in Germany more or less simultaneously developed machines for foil application in a continuous process as occurs in the paper mill industry. In particular Werner Reinhart of Kurz has been the driving force behind this development.

8.7.4   The prospects

At the start of the 21st century more than eighty issuing authorities have secured more than 150 denominations with a 3D device. These devices are called Diffractive Optically Variable Image Device (DOVIDs) and the estimated amount of DOVIDs in circulation is close to fifty billion (Fig. 8.30).

Even as DOVIDs have not been fully adopted by large banknote producers and consumers such as China and India, it can easily be forecasted that the exponential growth over the last fifteen years has not yet ended.

In general, it is, on one hand, difficult to reproduce or copy accurately and, on the other hand, just as difficult for the general public to remember and to identify a DOVID as being genuine. For that rea-son a return to the simple colour change as an identifying feature is foreseen, either as a stand-alone feature or in combination with a DOVID structure. The first signal of this tendency can be found in a review article [36], about printable holograms. And there is definitely more to come in the future.

8.8   A FEW COMMENTS ON THE ROLE OF PRINTING INK WITHIN THE HISTORY OF DOCUMENT SECURITY

It is normal to take ink's unparalleled ability to transfer essential information onto a chosen substrate for granted. Even though humans can create a huge variety of visible and recognisable images by applying a very thin film of ink and very fine dots, it is still normal to forget that ink has several unique properties, which must be carefully controlled throughout the printing process (both on contact with a press and with a substrate). Moreover, once applied, the ink must dry, or set, while subsequent prints made must be identical. To achieve all this, ink 'uses' just two components: *pigments* and a *vehicle* [42] and [43].

A *pigment* scatters and absorbs light, thus determining the final colour and opacity of a printed film. As a consequence, pigments should have sufficient colour strength, excellent light fastness, stability, adequate dispersion properties, and so on.

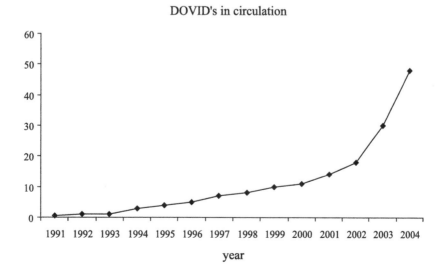

DOVID's in circulation

Figure 8.30. The increase [27] of DOVIDs in billion pieces over the years.

A *vehicle* is a liquid in which the pigments are dispersed, giving them the ability to flow through the printing press and onto the substrate. The vehicle, in general, is a simple solvent that is combined with a suitable binding agent. The latter is used to bind the pigment particles both to each other and to the substrate and is a non-volatile resin, which solidifies once the ink has been applied to the substrate. In other words, the solvent should provide mobility for as long as it takes to apply the ink film, whereupon, it must then dissipate.

Inks used in security printing are slightly more complicated, reflecting different printing process parameters, different drying mechanisms, and different requirements. However all the pigments found in security inks must contain the ingredients of what will be a first or second-line security feature. An example of a pigment based first-line feature is the Optically Variable Ink (OVI) as described in Section 8.5.7. The first OVI application[61] was the 60 Baht note of Thailand, [4] issued in 1987. Since then OVI has been applied either by intaglio or silkscreen in many currencies. However the majority of ink based security features are to be found as features in the second-line of defence.

Magnetic inks for the letterpress numbering printing technology appeared at the end of the 1960s. The Belgian national bank was among the first users and Sicpa supplied the inks.[62] Since magnetic pigments are generally brown or black, magnetic inks tend to be dark in colour; moreover, there is a direct correlation between the thickness of the ink layer and the strength of the signal. Most probably Thomas de la Rue in England developed in early days the magnetic numbering inks in parallel with Sicpa.

Up-converting inks are based on the anti-Stokes emission or MM feature described in Section 8.6.4. Up-converting occurs when a substance generates visible radiation once irradiated with a longer wavelength, for example infrared light. Up-converting pigments are based on combinations of the chemical elements of the lanthanide series of the periodic table that is, cerium to lutetium. In 1977, the up-converting feature first appeared in the paper substrate of German notes; while, up-converting inks stem from a later date.

Fluorescence or luminescent inks where the visible radiation results from invisible radiation with a shorter wavelength as for example UV light has been introduced as a covert feature in the early 1960s. When UV lamps became a common attraction in dance halls, the covert feature lost its anonymity.

Today the images purposely printed with luminescent inks are essential for the quick UV scan on genuineness by retailers.

Metamerism occurs when two substances display an identical colour when exposed to light source A (incandescent light, for example), but markedly different colours when exposed to light source B (daylight, for example). An image printed with metameric inks usually contains information that is practically invisible under normal lighting conditions; the information can only be seen if the image is exposed to a different light source. The NLG 250, issued by the Dutch central bank in 1985 was the first note in circulation with metamerism. The feature aimed to be a provisional solution for the professional money handlers.

In phosphorescence, as opposed to luminescence, radiation continues for a certain period once an irradiation source has been switched off or removed. Well-known applications of phosphorescence are the hands and faces of watches glowing in the dark. Since the 1990s, phosphorescence[63] has become a popular feature for the detection of the genuineness of banknotes by automatic banknote sorting machines.

The vehicle associated with the above-mentioned pigments, with its up-converting, luminescence, metameric, and phosphorescence properties is mainly tuned and dedicated towards the offset printing process. The relative thin layers of up to five microns contain enough pigment particles to display the desired effects.

Given the thickness of the ink layers as applied by the silkscreen or intaglio printing process, respectively up to and about twenty microns, the silk

---

[61]Confirmed in a private communication by Haim Bretler, former R&D director of Sicpa, Switzerland.

[62]Private communication from Haim Bretler, former R&D director of Sicpa Switzerland.

[63]The feature is marketed under the brand name 'Blink' by De La Rue of the UK.

screen ink and intaglio ink are able to accommodate a greater variety of pigments and sizes. The earlier mentioned OVI is a typical example. In the 1970s Sicpa developed infrared transparent and infrared absorbent intaglio inks and the first systematic incorporation[64] of this second-line feature happened in the 1976 issue of Swiss banknotes.

The tremendous progress of today's digital world is stunning. Much has been achieved and the cost of the effort almost lost. People are inclined to forget the importance of the amazingly smooth conversion from analogue to digital equipment; and still more, society often does not realise how marvellous it is that eyes can cope with pixels that are digitally generated and filled. Certainly, people, in general, are unaware of how much effort has been needed before this analogue to digital conversion could become reliable and reproducible.

Whereas society usually does not recognise the importance of ink in the security printing process, this injustice is not corrected either by this brief summary or by stressing the importance of the intricate combination of pigments and solvents tailor made for the several printing processes. Even so and within the framework of the history of document security, detailing the progress of printing security issues must include the first appearance of ink and its development if a true understanding and appreciation of first and second-line security features are to be understood.

8.9 RETROSPECTION

This report has covered quite a distance since the early dry seal cachet of the Stockholm's banco notes of 1666 to the intaglio printing and DOVID structures of currency in the 21st century. Staying ahead of counterfeiters is the driving force behind this increasingly complex technology that focuses on security features and which also is congruent to the growing influence of technology itself on human society. But the aim is still in accordance with the statement in 1819 of Sir William Congreve to develop features "as they may be immediately recognised and understood by the most unlearned persons".

Intaglio is the prime security feature of many current notes, and the USA one-dollar note as shown in Figure 8.31 is a prominent example of how the portrait and the border are printed in intaglio and dominate the design.

The evolution of the music types, as shown in Figure 8.3, resulted finally in intricate line patterns printed in accurate registered multiple colours. Figure 8.32 shows the offset background of a Chinese Yuan note.

The early *watermark* as shown in Figure 8.5 has since been perfected, and, in particular the watermarks of the Japanese currency, as shown in Figure 8.32, are of such exceptional quality and high reputation as to be a respectable counterfeit deterrent.

Finally the different *security features of today* are displayed in the specimen banknote of KBA-Giori, featuring Jules Verne the celebrated French author and scientific visionary. With more than fifty different security features, it is an overengineered banknote but nevertheless may be a glimpse some of the combined security features expected on future banknotes.

By the start of the 21st century, there is yet another element added to the continuous effort of thwarting counterfeiters – the competition between the suppliers of paper substrates and polymer substrates. Questions about the effectiveness of individual security features add momentum to this entire technological field. Rudolf L. van Renesse [31] is known for making the first attempt at the SPIE conference in January 1998, by ordering devices according to their spatial frequency along a horizontal axis and Schell later added [39] a subjective value for the counterfeit resistance on the vertical axis.

The final result is shown in Figure 8.35. The high value, with its spatial frequency at around 10 mm, on the left belongs to the watermark with its proven track record of more than two hundred years. The value on the right, with its spatial fre-

---

[64]Private communication of Haim Bretler, former R&D director of Sicpa Switzerland.

Figure 8.31. 'Classic' intaglio printing dominates the design.

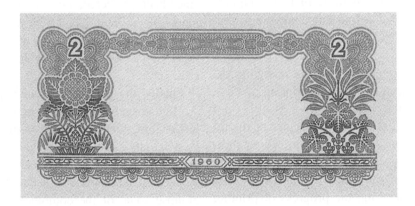

Figure 8.32. Multi-colour colour offset motive for the background of the 1960 2-yuan note of China [24: 45].

Figure 8.33. An example of a watermark in Japanese currency.

Figure 8.34. The 'roadmap' for security features at the beginning of the 21st century. Complex latent images appear from the black ornament and the globe in the left part of the note (above) as well as many other comparable 'nearly hidden' features, which appear on the sample note.

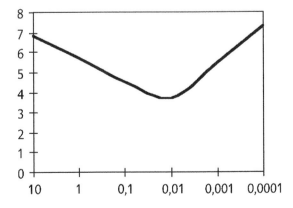

Figure 8.35. Evaluation of security devices. The spatial frequency of the device is displayed on the x-axis in mm. The y-axis displays on a subjective scale the resistance against counterfeiting.

quency around 0.1th of a micron, belongs to the colour shifting diffractive devices (DOVIDs) and interference devices such as optically variable ink. The x scale thus indicates the historic time line along which the development of security devices and features took place. More then two hundred years have passed before an equivalent feature to the watermark appeared and through this history of document security, it is reflected that any advancements are the same as that of the technological progress from the multi-tone watermark to the 'holographic watermark'.[65]

---

[65] Sometimes the DOVID type security devices are referred to as holographic watermarks or diffractive watermarks.

8.10 EPILOGUE

The occupation of the Dutch East Indies by the Japanese started in 1942 on the 8th of March. My father became a POW and was displaced after a few months to the notorious Burma railroad. My mother and I were confined in Japanese camps for more then three years.[66]

My mother cashed the family's savings at the bank before we actually went into the Japanese camps. It was, for those days, the considerable sum of eight hundred guilders. She decided to sew the eight notes of one hundred guilders into her belt. Doing this effort and with me, her two year old baby in the chair next to her, she noticed after the sixth note that two were missing. Searching for the missing two notes, she then discovered that I had them and, due to my swollen cheeks, apparently had put them into my mouth. More than thirty-five years later, I joined the Joh. Enschedé company as a security printer where I might then proceed to satisfy my Freudian appetite for currency.

I encountered many people during my involvement in the security printing industry of nearly thirty years and admire their professionalism and the lessons that I learned from them. These experiences are a direct and indirect part upon which I have based the history of document security. Together, these experiences are too numerous to thank individually. However allow me a few exceptions. I thank Karl de Leeuw for inviting me for the chapter on the History of Document Security and Rudolf van Renesse for his moral and actual support. Finally I owe an apology to my wife, the author Simone Schell. My support in the past to her thirty-five publications has been meagre in comparison to the great demands I made on her time for this publication only.

REFERENCES

[1] AD 1829 N° 5754 Manufacture of Paper. Dickinson's specification.

[2] J.M. Adams, P.A. Dolin, *Printing Technology*, Delmar, Albany (2002).

[3] G.A. Agoston, *Color Theory and Its Application in Art and Design*, Springer Series in Optical Sciences, Vol. 19, Springer-Verlag, Berlin (1979).

[4] K.W. Bender, *Geldmacher*, Wiley, Weinheim (2004).

[5] M.T. Bloom, *The Brotherhood of Money*, BNR Press, Port Clinton, OH (1983).

[6] M.T. Bloom, *Money of Their Own*, BNR Press, Port Clinton, OH (1982).

[7] R. Brion, J.L. Moreau, *Het bankbiljet in alle staten*, Mercatorfonds, Antwerpen (2001).

[8] H. Bretler, *Thin film devices in security printing inks*, SPIE Proceedings 1210 (1990), 78–83.

[9] D. Budd, G. Wyszecki, *Color in Business, Science, and Industry*, Wiley, New York (1975).

[10] D. Byatt, *Promises to Pay*, Spink & Son Ltd., London (1994).

[11] B. Coe, *The Birth of Photography*, Ash & Grant, London (1976).

[12] G. Davies, *History of Paper Money*, University of Wales Press, Cardiff (2002).

[13] *Document security – hanging by a thread*, Currency News 1 (1) (2003), 3.

[14] G. Detersannes, Publications du Musée de l'affiche et du tract, Paris (1997).

[15] Enz 5, *Brochure of Joh. Enschedé*, Joh. Enschedé, Haarlem (2001).

[16] G. Fisher, *Der frequenzmodulierte Bildaufbau – ein Beitrag zum Optimieren der Druckqualität*, Polygraph Verlag, Frankfurt am Main (1988).

[17] W. Hagen, *De grote geldvervalsing van het derde rijk*, Publiboek, Baart (1981).

[18] R.L. Hills, *Papermaking in Britain in 1488–1988*, The Athlone Press, London (1988).

[19] *Holier than thou – louisenthal introduces new transparent feature*, Currency News 3 (3) (2005), 4–5.

[20] R.W. Hunt, *The Reproduction of Color*, Fountain Press, Tolworth, England (1987).

[21] R.G. Hutton, T. Merry, US patent 4,033,059, July 5 (1977).

[22] W. Kaule, G. Schwenck, G. Stenzel, Patent WO8103507.

[23] Dr. Ir.P. Koeze, *Papier en water*, Gentenaar & Torley, Rijswijk (2000).

[24] W. Kranister, *The Moneymakers*, Black Bear Press, Cambridge (1989).

[25] M. Kuhn, available at www.cl.cam.ac.uk/~mgk25/eurion.pdf.

[26] O.M. Lilien, *History of Industrial Gravure Printing up to 1920*, Lund Humphries, London (1972).

[27] I.M. Lancaster, A. Mitchell, *The growth of optically variable features on banknotes*, SPIE Proceedings 5310 (2004), 34–45.

[28] D.L. MacAdam, *Colour Measurement, Theme and Variations*, Springer Series in Optical Sciences, Vol. 27, Springer-Verlag, Berlin (1981).

[29] *Microperf – simple yet secure*, Currency News 2(4) (2004), 3.

---

[66]The three of us survived and were reunited in September 1945.

[30] *Optiks – the future of wide thread technology*, Currency News 2 (5) (2004), 8.

[31] R.L. van Renesse, *Verifying versus falsifying banknotes*, SPIE Proceedings 3314 (1998), 71–85.

[32] R.L. van Renesse, **Optical Document Security**, Artech House, London (1998).

[33] R.L. van Renesse, **Optical Document Security**, Artech House, London (2005).

[34] R.L. van Renesse, *The secrets that lie within*, Currency News 2 (10) (2004), 4.

[35] R.L. van Renesse, *Technology profile*, Currency News 2 (10) (2004), 4–5.

[36] R.L. van Renesse, *Optically variable Features in Print*, Currency News 3 (1) (2005), 4–5.

[37] E.C. Rochette, **Making Money**, Jende Hagan Inc., Frederick, CO (1986).

[38] G. Saxby, **The Science of Imaging**, IOP Publishing Ltd, Bristol (2002).

[39] K. Schell, *Euro fysica*, NTvN 68 (1) (2002), 7–9.

[40] K.J. Schell, *A security printers application of lasers*, SPIE Proceedings 396 (1983), 135–140.

[41] K. Schell, *The delicate bond between paper and foil*, Keesing Journal of Documents 1 (3) (2003), 22–23.

[42] K. Schell, *All about ink?*, Keesing Journal of Documents 1 (5) (2003), 20–23.

[43] K. Schell, *All about ink, part 2*, Keesing Journal of Documents 1 (6) (2003), 14–17.

[44] K. Sierman, J. de Zoete, **Voor Stad en Staat**, Joh. Enschedé, Haarlem (2003).

[45] D. Standish, **The Art of Money**, Chronicle Books LLC, San Francisco (2000).

[46] W. Vanthoor, **De Nederlandsche Bank. Van Amsterdamse krediet instelling naar Europese stelselbank**, Boom, Meppel (2004).

[47] W. Vissering, **On Chinese Currency Coin and Paper Money**, Brill, Leiden (1877).

[48] J. Weatherford, **The History of Money**, Three Rivers Press, New York (1997).

9

FROM FRANKPLEDGE TO CHIP AND PIN: IDENTIFICATION AND IDENTITY IN ENGLAND, 1475–2005

Edward Higgs

University of Essex, United Kingdom

Contents

Abstract

This chapter is an attempt to look at the history of changing forms of personal identification in one country – Britain – over the last 500 years. It argues that what has been identified has changed over time, and this has significance. Identification practices have moved from identifying lineages and communities, via identifying citizens, to identifying bodies. In the process the concept of identification as the recognition of the citizen, has come to be confounded with the identification of the "anti-citizen". Identification has also increasingly become a technology, rather than a set of techniques. Part of the reason for this has been the increasing scale of interactions within society. When the vast majority of people lived, worked, received welfare and died within the same small community, forms of identification based on personal knowledge were adequate. However, when these social processes became dissociated during the nineteenth and twentieth centuries, and state systems began to function at the level of the Nation State, and beyond, such techniques were insufficient. Technologies that allowed data for identification to be stored remotely, and accessed over considerable time-spans, have become necessary. However, this story of identification has not necessarily been one of linear progression. DNA profiling leads one back to the identification of lineages rather than of individuals, and the credit card can be seen as a physical means of identification analogous to the medieval seal. Personal identification is thus not simply a contemporary problem, and placing it within a historical framework helps elucidate the nature of some modern dilemmas.

Keywords: seals, signatures, lineanges, forgeries, genealogy, clothing, impostor, frankpledge, presentment, coroner's court, parish registration, badging of the poor, branded, disguise, civil registration, birth certificate, census, places of record, signature, Poor Law, photography, tattoo, anthropometric, fingerprints, driving licences, immigration control, passports, MI5, old age pensions, National Insurance, income tax, credit cards, identity theft, welfare fraud, identity card, PIN, card, chip, identification cards, conscription, National Registration, terrorism, identity fraud, theft, DNA typing, iris recognition, face recognition, CCTV.

9.1 INTRODUCTION

This chapter is an attempt to look at the history of changing forms of identification in one country. This may appear somewhat parochial but can be justified on several grounds. First, historians are concerned with the unfolding of specific events and structures through time, and have to focus, therefore, on things that happened, or which at least might have happened. Being a British historian, that means writing about what happened in the historical community on which my own research is based. Also, I want to look at that community over a considerable period of time because I am interested in changes over what Ferdnand Braudel called the *longue durée* (Braudel [11]). Looking at more than one historic community over more than half a millennium would be very difficult for one person, if not impossible. However, since the development of techniques of identification has been an international project, looking at it in one country is merely to look at this global phenomenon from a particular perspective.

In addition, looking at this development from the English, or British, point of view is particularly fruitful because of the important role that country has played in the elaboration and dissemination of such techniques. After all, it was the British who introduced the first general 'scientific' system of fingerprinting into Western police work in 1901, and who first developed DNA profiling in the mid-1980s. England, as the centre of a world empire was also the nodal point through which, and from which, new techniques of identification flowed. Much of the groundwork for fingerprinting was laid by the British Raj in India, before being transplanted to the metropole and elsewhere. Similarly, the development of secret 'terrorist' identification registries in England grew out of the experience of holding down Ireland. Many of the forms of identification developed in England were subsequently exported to the other parts of the United Kingdom (i.e. to Scotland and Ireland), and then onto the colonies and dominions of the Empire.

The reason for looking at techniques of identification over a considerable period is that what has actually been identified is historically contingent – it has changed over time. Looking at these changes helps us to place some of our own concerns over identification into perspective, and helps to explain why some forms of identification might seem inappropriate in our own, networked world. Why, for example, do many English men and women bitterly resent today the idea of being issued with identity cards, and having to be fingerprinted, but see no problems in registering the birth of their children? Why is acquiring a driver's license an eagerly sought right of passage but DNA databases a sign of the development of George Orwell's Big Brother State?

These issues raise the question of what "identification" should mean in this context, and for once the *Oxford English Dictionary* [50] is actually rather helpful. It defines "identification" as the act or process of "establishing the identity of, [or] recognising", and also "the treating of a thing as identical with another". In turn, "identical" is defined as "Agreeing entirely in material, constitution, properties, qualities, or meaning: said of two or more things, which are equal parts of a uniform whole, individual examples of one species, or copies of one type, so that any one of them may, for all purposes, or for the purposes contemplated, be substituted for any other". We can thus understand identification in two, inter-related, senses: first, as the act of saying that person X is indeed the unique person X, and not someone else; and, secondly, that X has certain characteristics, which place him or her in a certain category with others. Another way of saying this is that X has either some unique characteristics, or some characteristics that he or she shares with some others but not all others. These definitions appear to be quite distinct, and in modern identification systems we tend to be thinking in terms of the former. But for practical purposes the two are often inextricably linked – a passport is a means of identification that identifies its holder as a particular person, but also as the citizen of a particular state.

This two-fold definition is perhaps more appropriate than the simple procedural one suggested by Roger Clarke in a seminal essay on human identification in information systems (Clarke [15]), in which the former "is the association of data with a particular human being". The problem with this definition is, of course, that the data associated with a particular human being may be incorrect, in which case the process fails to identify the individual. Also, the definition assumes that one is able to say that someone is a "particular" human being, which is the process of identification, so the definition is circular. Indeed, the association of data with a particular individual may efface that person's individuality, as was the case in Germany in the 1930s with respect to those described by the State as "Jews". In addition, it is what they share with others – nation, gender, or religion – which often gives individuals their strongest sense of personal identity.

The two-fold definition given above is also useful in that it helps us grasp what I have described as the contingency of what was being identified in England at various times over the past 500 years. We tend to think today that this must have been the unique characteristics of a person, especially of their body, but this has not always been the case. Indeed, I will argue here that the history of identification in England can be seen as a three-stage development:

> The early-modern period, from 1475 to, say, 1800, during which identification was by, and even of, the community, or of the lineage.
> The nineteenth and early twentieth century, when identification became increasingly of the idealised "citizen" through the latter's participation in the State.
> The period from about 1914 onwards, when identification increasingly came to mean identification of the body.

These stages overlapped, and in some ways we can now be seen as returning to some aspects of the first, but they help to characterise certain differences over time. The shifting relationship between the identification of the unique individual and the group into which he or she falls, is an important theme in the discussion that follows. Given the vast period being considered here, there will, inevitably, be gaps in the coverage of history of identification techniques in England. However, it is to be hoped that enough ground can be covered to give at least some idea of broad patterns and trends.

9.2 IDENTIFYING THE GREAT AND THE GOOD, AND THE INSIGNIFICANT AND THE BAD, IN EARLY-MODERN ENGLAND

Early-modern England has often been seen as the birthplace of what C.B. Macpherson described as "possessive individualism" (Macpherson [47]). This can certainly be seen in the development of political theories in the contract tradition, such as those of Thomas Hobbes [36] and John Locke [46]. But when it came to the techniques of identification used in this period, people were as much members of a neighbourhood community, or of a biological family, as unique individuals. This was even the case amongst the rich and powerful, whom one might imagine to have the greatest means of autonomous action.

It is certainly true that the wealthy had long had a technique for the individual authentication of documents in terms of the wax seal, although its use could be delegated to household officers (Clanchy [14: 308–317]). In the case of the monarch, for example, the staff of the keeper of the Great Seal had ceased to be part of the royal household in the reign of Henry III (1216–1272), and in the fourteenth century it ceased to follow the King in his progress around the country, and became permanently based in London [29: 7]. The dangers of this dissociation between seals and their "owners" may explain, in part, why their use was superseded by the personal signature. However, it should be noted that in the medieval period the signature on a property document was not necessarily that of the parties involved. The first signatures were notorial signs, the signatures of known scribes who could attest documents. In addition, early signatures were often drawn emblems, more like the impressions of seals than names spelt out in the modern sense (Clanchy [14: 304–308]; Fraenkel [23: 125–146]). However, the change, from identification as the possession of an object to the innate ability to perform a (fairly) unique act spread with literacy, and was given statutory force in England by the 1677 Statute of Frauds. This Act laid down that in future "for the prevention of many fraudulent practices which are commonly endeavoured to be upheld by perjury", agreements relating to property would only have the force of law if "putt in writeing and signed by the parties soe making or creating the same".

However, the aristocracy and gentry still tended to see themselves as the scions of lineages. Identification was bound up with proving descent through genealogies, and involved the use of antiquarians and heralds to prove that one had the right to a name, a title, to land, or to carry certain heraldic symbols. This involved the collection of documents and charters, or recording grants found in various series of royal enrolments (the patent and close rolls for example), and depositing them in the muniment rooms of aristocratic country houses (Wagner [71: 324–407]). Proving one's right to title in this way was fraught with difficulties and many families resorted to forgeries. In the reign of Charles II (1660–1685), for example, Henry Mordaunt, Earl of Peterborough published a charter of William the Conqueror in the eleventh century which purported to show his supposed ancestor, Osbert "le Mordaunt", being granted the manor of Radwell. This was despite the fact that Domesday Book, William's great land survey of 1087, showed that Radwell was in fact held by someone else. This was but one of the suspect charters produced by Peterborough and accepted by antiquaries. Such forgeries and uncertainties led to numerous legal cases respecting claims to title, as in the famous dispute over the barony of Abergavenny, which was heard by the House of Lords in 1605 (Round [61: 75–89, 290–302]). These lines of descent became publicly available in the eighteenth century in the printed volumes of the *Baronetage of England*, and in *Burke's Peerage* and *Burke's Landed Gentry* in the nineteenth.

This excessive preoccupation with identifying oneself in terms of one's genealogy was already being criticised in the early nineteenth century. Jane Austen satirised such pretensions in the opening pages of her novel *Persuasion*, poking fun at Sir Walter Elliot of Kellynch Hall, in Somersetshire:

> ... a man who, for his own amusement, never took up any book but the Baronetage; there he found occupation for an idle hour, and consolation in a distressed one; there his faculties were roused into admiration and respect, by contemplating the limited remnant of the earliest

> patents; there any unwelcome sensations, arising from domestic affairs changed naturally into pity and contempt as he turned over the almost endless creations of the last century; and there, if every other leaf were powerless, he could read his own history with an interest which never failed (Austen [4: 9]).

Visual identification through the wearing of certain colours and liveries, both by oneself and one's servants, was also of great importance. Indeed, so important were such external forms of identification that attempts were made to restrict the sorts of clothing that people could wear lest the "due subordination" of the social ranks was undermined. A series of Acts were passed from 1463 onwards, until by 1600 "excessive apparel" was controlled by seven statutes and ten royal proclamations. In a proclamation of the reign of Elizabeth I (1558–1603) it was laid down that:

> None shall wear ... cloth of gold, silver, or tinsel; satin, silk, or cloth mixed with gold or silver, nor any sables; except earls and all of superior degrees, and viscounts and barons in their doublets and sleeveless coats;
> Woollen cloth made out of the realm; velvet, crimson, scarlet or blue; furs, black genets, lucerns; except dukes, marquises, earls or their children, barons, and knights of the order;
> Velvet in gowns, coats, or outermost garments; fur of leopards; embroidery, pricking or printing with gold, silver, or silk; except baron's sons, knights, or men that may dispend £200 by year ... (Hunt [39: 119]).

To modern eyes such concerns over appearances and membership of particular lineages may seem somewhat trivial but it was anything but in the early-modern period. This can be seen in the career of the royal impostor Perkin Warbeck, who was born in Tournai in Flanders in the late fifteenth century. Warbeck arrived in Dublin, Ireland, in the employ of a silk merchant in 1491. Encouraged to dress up by his master in the latter's wears, he was taken by the townspeople to be Edward, Earl of Warwick, the Duke of Clarence's son, who they recognized because they had seen him in Dublin. Perkin vehemently denied all this and called for a crucifix and Gospel to swear an oath of denial. Subsequently, adherents of the Yorkist party persuaded him to impersonate Richard, Duke of York, the younger brother of Edward V of England. The House of York was eager to reclaim the throne from the Tudor Henry VII, who had killed the last Yorkist king, Richard III, at Bosworth Field in 1485 (Arthurson [3: 1–59]).

Perkin's claim was supported by the Holy Roman Emperor Maximilian I, by James IV of Scotland, and by Margaret of Burgundy, sister of Edward IV (and thus Richard's aunt) and the chief supporter of the Yorkist exiles. All these key international figures received the youth in their courts. Margaret had no children of her own, showed charitable generosity towards orphaned children, and may have seen her 'nephew' as a substitute son. She, and others, claimed to recognise Richard/Warbeck's looks and manners. The impostor's attempt to invade England in 1495 failed, and he went to Scotland where he married Catherine Gordon, a cousin of James IV. In 1497 Warbeck landed in Cornwall, proclaimed himself Richard IV, and raised a rebel army. His forces were defeated at Exeter by those of Henry VII, and the pretender fled. He was captured, admitted the whole story of his adventure, and was imprisoned. In 1499 he and the Earl of Warwick (son of the duke of Clarence) were hanged for plotting against the king. The fate of kingdoms could thus hang on personal identification understood in terms of appearances and family recognition of individuals as members of the lineage.

For the lower orders in early-modern England identification was a question of recognition in, or of, the local community. The State in this period was not the modern central state, rather it was distinct from the locality not by being central but by being more extensive than the locality – it was constituted by the active linkage of leading elements in local manors, parishes, and towns for the purposes of exercising authority. As Steve Hindle puts it [35: 20–23], one should, "think less of *government* as an institution or as an event, than of *governance* as a process, a series of multilateral initiatives to be negotiated across space and through the social order ...". The state formed a national community of those in every locality who were willing and able to involve themselves in the exercise of authority. The State as such was a composite affair,

with differing components having different orientations but with enough common interests to bind it together for most of the time. Mark Goldie has dubbed this composite state a "monarchical republic" (Goldie [27: 154]). That the "chief inhabitants" of the parish who undertook some of the functions of government at the micro-level had an understanding of the larger state is, of course, doubtful. But the amateur justices of the peace, who ran the counties certainly did (French [24]). This was not a Weberian central State, as modern sociologists understand it, but one that was still recognisable as a State, with officers wielding authority claimed as legitimate over defined territorial units (Braddick [9: 17–18]).

Because the State was local, identification within it was also local. Indeed, in the medieval period villagers were often identified as communities, rather than as individuals. Under the frankpledge system every householder in the country formed part of a group of 10 or 12 people, known as a "tithing". "Tithingmen" were responsible for the good behaviour of one another, and for bringing individual members to a manorial court to face charges. If a member of the tithing failed to appear, the rest of the tithingmen could be fined (Morris [48]). In the early-modern period the system fell into disuse but the link between the community and identification continued through the conventions of legal presentment. The jury originally was not a body of men chosen to adjudicate the likelihood of evidence presented in court but a group of local men who had first hand knowledge of what happened and could personally identify the guilty parties. But by the early-modern period juries had come to have the former function (Green [28]). However, in lower courts, such as the sheriff's tourn or court leet, the responsibility for accusing someone of a crime in court, or "presentment", might still lie with the parish community, often through the locally elected constable. In local ecclesiastical courts, for example, people could accuse their neighbours of fornication and adultery (Ingram [42]). Accusations of witchcraft could also be made within communities, as in the case of the famous witch trials in Salem, Massachusetts at the end of the seventeenth century. Identification was immediate and unmediated, and via the "common repute" of the locality.

Much the same could be said of the identification of the dead, which was done formally in local coroners' courts. These developed in the Middle Ages, and depended on personal identification by the first person who found the body (the "first finder"), usually a neighbour, rather than on any form of forensic science (Hunnisett [38: 1–36]). The vast majority of bodies were thus identified within the community, the usual exception being abandoned babies, or corpses that were washed up on beaches, or river banks. Thus, in the Liberty of St Etheldreda in Suffolk (the area covered by the hundreds of Carlford, Colneis, Loes, Plomesgate, Thredling and Wilford) as late as the years 1767 to 1858, "unknown" bodies brought before the local coroner's court represented 7.5 percent of the total. Of these, 153 in total, no fewer than 116 were 'cast up' from the sea, or found drowned in the River Orwell (Smith [64: 23–60]).

A coroner's inquest had to be held before the body of the deceased could be buried, and it was at this point that another local, State official – the local clergyman of the Church of England – played a role in registering identity. Attendance at the State Church was, of course, compulsory, and was a means of regulating the beliefs and manners of the population. In 1538, Thomas Cromwell, acting as Henry VIII's vicar-general, sent a series of injunctions to bishops of the Church in England. These included specific instructions to parish priests on teaching the people the rudiments of the faith; preaching Scripture; guarding against superstitious ceremonies; the placing of an English Bible in each church; and on the keeping of registers of the baptisms, weddings and funerals at which the clergy officiated. This was part of the process by which the medieval Catholic Church in England was brought under the control of the Tudor monarchs at the Reformation. In December of that year Thomas Cromwell issued a circular to justices of the peace denying rumours that parish registration was for the purpose of imposing taxes. He stated that the true reason for its introduction was "for the avoiding of sundry strifes, processes and contentions rising upon age, lineal descent, title of inheritance, legitimation of bastardy, and for

knowledge whether any person is our subject or no". The injunction respecting registration was repeated, and penalties for neglect prescribed, by an Act of 1597, which also directed that transcripts of the registers should be sent annually to a registrar in the local diocese (Higgs [32: 1–2]).

Such arrangements were plainly part of the early-modern concern with identifying lineages, and the descent of property, but the system of parish registration soon became enmeshed in identifying individuals for the purposes of Poor Law administration. The Poor Laws were a system developed from the sixteenth century onwards under which the people of the parish were to raise a local tax, the poor rate, for the relief of the indigent poor who "belonged" to the parish. By the 1662 Act of Settlement such a legal "settlement" was declared to be gained by birth, inhabitancy for a set period, apprenticeship, or service for forty days. Within this period all "intruders" could be removed from any parish by two justices of the peace, unless they settled in a tenement of the annual value of £10. Thus, identification depended upon parish registers, or the testimony of employers and neighbours (Slack [63: 194–195]). In 1697, an Act *For supplying some Defects in the Laws for the Relief of the Poor* required the "badging of the poor" – those in receipt of poor relief were required to wear, in red or blue cloth on their right shoulder, the letter "P" preceded by the initial letter of their parish (Hindle [34]).

The Poor Law system was partly intended to encourage and rationalise charity, but it was also a means of social control. If identification was through the stable community, then those who had no settlement were regarded with anxiety and fear by local and central elites as a possible source of insubordination. Hence, under the 1572 Vagabonds Act unlicensed "sturdy beggars" above 14 years of age were to be severely flogged and branded on the left ear unless some one would take them into service for two years. In case of a repetition of the offence, if they are over 18, they were to be executed, unless some one would take them into service. But for the third offence they were to be executed without mercy as felons. Under legislation passed in the reign of James I (1603–1625), anyone wandering about and begging was declared a rogue and a vagabond. Justices of the peace in petty sessions were authorised to have them publicly whipped, and for the first offence to imprison them for six months, for the second for two years. Incorrigible and dangerous rogues were to be branded with an R on the left shoulder and set to hard labour, and if they were caught begging again, to be executed without mercy. Justices could give needy individuals, such as ship-wrecked sailors, certificates to beg, but these became the subject of a lively trade in counterfeits (Slack [63: 94, 100, 118, 169]). Physical identification via the body was thus associated with criminals and the dregs of society, an association that can be traced back into classical antiquity (Gustafson [30]).

Since identification in the early-modern period was all about external show, and the recognition of individuals in the community, attempts to circumvent such identification involved parodying the dress of superiors, or donning disguise. "Lords of Misrule" paraded the streets at carnival time in mock liveries, carrying bogus heraldic symbols. At Christmas people disguised themselves and went to parties as "mummers", "maskers" or "guisers", a practice that was forbidden by law in a number of towns because of the opportunities for crime this offered (Hutton [40: 8–10, 116–117]). Rioters might dress up to disguise themselves, a tradition that was continued by the American colonists who dressed as red Indians during the Boston Tea Party of 1773. As late as the early Victorian period, Welshmen were dressing as women to attack tollgates during the Rebecca Riots. Similarly, poachers blacked up their faces to avoid being recognised by gamekeepers. According to E.P. Thompson [69] blacking was also taken up by members of local communities objecting to the corruption and abuse of power inherent in the use of royal forest law by corrupt political oligarchs. Deer parks were being carved out of local communities, denying the latter access to common lands, and were the target of armed encroachments. This led to the "Black Act" of 1723 "for the more effectual punishing wicked and evil-disposed persons going armed in disguise", which laid down the death penalty for those appearing in forests with their faces blacked up. Once again, the politics of personal identity were a matter of life and death.

9.3 CREATING THE MODERN CITIZEN AND ANTI-CITIZEN, 1830–1920

Many forms of identification found in the early-modern period continued into the industrial age. For example, personal identification continued to be the basis of the identification of bodies in coroners' courts well into the twentieth century (Purchase [54: 94]). However, in the nineteenth century such traditional methods were supplemented by new, centralised forms of identification. In part this reflected changes in the nature of the State. Local, communal forms of governance proved unable to cope with the creation of vast urban conglomerations, whilst local elites ceased to be interested in the locality as their wealth increasingly came from positions or shares in national, or international, companies, rather than land. Property came to be national rather than local, and authority passed out of the hands of parish officials and local justices of the peace, especially as policing and welfare came to be centralised. This led to the creation of systems of identification at the level of the Nation State. Identification could not be based on the "co-presence" of the identifier and the person being identified, and systems had to be developed in which means of identification could be stored and processed at a later date, and probably many miles from the local community. Rather than identifying members of such communities, these systems helped to create the identity of the citizen of the Nation State (Higgs [33: 106–108]).

This shift is clearly seen in the replacement of the ecclesiastical registration of baptisms, marriages and burials, with the civil registration of births, marriages and deaths. The 1836 Marriages and Registration Acts divided the country into registration districts, and appointed a local registrar of births, marriages and deaths for each. These officers, numbering 2,193 in 1839, were to be responsible for registering such events in their districts, and forwarding copies of the registrations of births, marriages and deaths to the General Register Office (GRO) in London. The latter was to be responsible for the overall administration of the system, and the maintenance of a central database of the registered information. Both the local registrars and the GRO could issue birth, marriage and death certificates. This new system was established because the old parish registration was seen as inefficient, with incomplete and damaged registers spread over thousands of parishes. There were also large numbers of "Nonconformists" – Roman Catholics or Protestants who did not belong to the Church of England – who did not register their baptisms, marriages and burials with the State Church. If genealogies were to be established, and thus the descent of property rights in an increasingly complex market economy, an exhaustive, and centralised, system for recording vital events was required. However, the possession of a birth certificate became in time a crucial means of personal identification, and the proof of age for numerous official purposes, including the right to work and to vote (Higgs [32: 1–21]).

However, it is important to recognise what did not happen at this date – there was no creation of a population register as was to happen in Scandinavia and continental Europe. In a population register each individual has to be registered with the state authorities, and notify those authorities of a change of address. The GRO collected data on births, marriages and deaths, and from 1841 organised a census of the entire population every 10 years. But no attempt was made to convert this system into a register to pin people down, or to issue internal passports. The records in the civil registration system were a resource upon which citizens could draw, if they so wished, and the census data was merely used for statistical purposes. Indeed, under the original 1836 Registration Act registration was in practice voluntary, since there were no fines for non-registration. This absence of a population register partly reflected the absence of military conscription, and an ingrained assumption that Liberty was liberty from the State, rather than liberty through the State (Higgs [33: 74–91]).

The GRO was situated in Somerset House in London, adjacent to the civil courts of law, and the "inns of court" where lawyers resided. The area rapidly became the location for the establishment of numerous other places of record. The Patent Office, set up under the Patent Law Amendment Act of 1852, also moved into nearby Chancery Lane, and budded off a Trade Marks Registry there

by 1875. A close neighbour here was the Public Record Office, the national archives, which had been set up as a central repository of title deeds. A Land Registry was created round the corner in Lincoln's Inn Fields under the provisions of the 1862 Land Registry Act. This was next door to the Design Registry, which had been set up in 1839 by the Board of Trade to protect title to property in textile and other manufactured designs.

In addition, a Principal Probate Registry for holding wills and other testamentary documents was attached to the new Court of Probate. The latter had been established in 1857 to replace the various ecclesiastical courts that had dealt so inefficiently with wills and administrations until that date. The wills repository was originally sited in Great Knightrider Street next to Doctors Commons, where lawyers specialising in probate cases resided, but moved to Somerset House by 1875. Here it lodged cheek by jowl with the GRO. The creation of this complex of places of record represented the creation of those generalised civil rights that helped to create civil society itself, although a civil society bounded by the institution of private property. The individual had come to be captured in a nexus of centralised state registration, which identified him (and it was usually a him) as a property-holding citizen (Higgs [33: 80–82]). This would explain the emphasis placed on the ability to sign one's name in the Victorian period. The signature was not just a measure of literacy but evidence of the ability to participate in the life of the polity as a rational agent. The citizen was expected to be an active participant, rather than someone to whom things were done.

But when these forms of identification ceased to convey rights they could be quickly challenged. This can be seen in the way in which birth registration came to be undermined by its association with smallpox vaccination. In 1853 the Vaccination Extension Act took the unprecedented step of making infant vaccination compulsory. The imposition of what was at that date an unpleasant and dangerous invasion into the bodies of helpless infants was bitterly contested throughout the nineteenth century by organisations such as the Anti-Vaccination League, which drew support from all social classes. The registration system was fully implicated in the vaccination system since, under the 1853 Act, medical practitioners were to transmit duplicate certificates of vaccination to the local registrars, who were to keep them in searchable registers. The central GRO was to provide the necessary books and forms for this purpose. Under the 1871 Vaccination Act (1867) Amendment Act, every registrar was to transmit a return of all births and deaths of children under 12 months, with addresses, to the local vaccination officer at least once a month. The Anti-Vaccination League encouraged parents not to register the births of their children, and as a result registration had to be made compulsory under the 1874 Births and Deaths Registration Act. The conflict between this form of state identification and individual conscience was removed when vaccination was made voluntary in the 1890s. Whereas anonymity was gained in the early-modern period by disguise, it was achieved in the nineteenth century through avoiding interaction with the central State (Higgs [33: 85–87]).

The means of identifying the "deserving poor" was also revolutionised in this period by the passing of the 1834 Poor Law Amendment Act. Under the "New Poor Law" those who could not support themselves had to apply to a Poor Law union workhouse for support or "relief". In theory there was no relief outside the workhouse, and "paupers" had to submit to imprisonment in all but name within its walls. The problem of identifying the poor in the community was thus circumvented via the erection of a system of physical surveillance within a carceral institution. In practice, there were various forms of out-relief practiced by the local Poor Law authorities but these were frowned upon by the central Poor Law Commission that policed the system (Crowther [19: 11–53]). But this theoretical circumvention of the problem of identification was undermined by the development of various forms of centrally paid welfare benefits, such as old age pensions, national insurance and soldiers' pensions, in the early twentieth century. The State could check, at least to some extent, that benefit applicants were who they were, and what they were, through official records such as birth certificates and census documents. To a considerable extent,

however, it had to fall back on having applications signed by "responsible persons" in the community – teachers, doctors, police constables – who could vouch for the details supplied. These details could then be checked in a sample of cases [17].

This crisis of identification led in 1918 to the suggestion by the Ministry of National Service that all servicemen should be fingerprinted in order to prevent fraud in the payment of war pensions. But this was opposed by the Ministry of Pensions and the War Office due to their concern over the public opposition that fingerprinting would arouse. Identification of the body was associated with the identification of criminals, rather than of free citizens, who were rational agents and identified through forms of state registration that conveyed rights, or protected them [66; 33: 139]. It was not the citizen whose body was identified but the "anti-citizen", the irrational, criminal, or foreign "other". In part, this represented the long-standing association of marking the body with the identification of "undesirables" already noted. However, it also reflected the development of systems of criminal identification in the late Victorian period.

Policing in the Victorian period was taken out of the hands of local village constables, and became the responsibility of the large, organised police forces of today. Given the scale of the areas covered by such forces (the whole of London in the case of the Metropolitan Police), local, communal knowledge could no longer be relied upon for the purposes of identification. 'Moral panics' in the 1860s, such as the London 'garrotting scare' of 1862, and general concern over crime rates, led to the passing of the Habitual Criminals Act in 1869. This set up a Habitual Criminals Register at the Metropolitan Police headquarters at Scotland Yard to record all persons convicted of crime in England. If they re-offended they could be placed under police supervision, in addition to any other sentence passed on them. A series of alphabetical registers of habitual criminals were produced giving names, "distinctive marks" (mainly tattoos), and descriptions. However, the Register was hardly ever used by the police, since it could not be accessed very easily. One needed a name to use it properly, which rather negated its use as a means of identification.

The published registers were only for prisoners released in one year and did not, therefore, contain all habitual criminals; few marks were really distinctive; and the registers were published up to 20 months after a convict's release – the period when he or she was most likely to re-offend [56: 215–216; 55: 8; 59: 4].

The new technology of photography was rapidly incorporated into such police work but with mixed results. Under the 1871 Prevention of Crimes Act, photographs of convicts were sent by the prison authorities to the Habitual Criminals Register, which held 34,000 such portraits by 1888. But sepia photographs could not be easily classified and indexed, especially by colour of eyes or hair, and the whole collection might have to be searched to find a particular individual. In addition, people could look alike, easily disguise their appearance, or change as they grew older. Experienced convicts were also adept at 'mugging' for the camera, distorting their faces in odd expressions [53; 56: 12–19]. The use of photography was also being introduced into French and German policing at about the same time, and proved equally problematic (Kaluszynski [44: 124]; Becker [6: 153–163]).

Some local forces began keeping their own records of criminals. At Birmingham there was a register with coloured drawings that showed the tattoo marks with which so many criminals ornamented their bodies. Tattooing appears to have enjoyed a renaissance in Europe after the discovery of the custom amongst Pacific islanders by Captain Cook in the late eighteenth century, and it appears to have become the symbol of the counter-culture of convicts, soldiers and sailors (Bradley [10: 136–155]). In Liverpool on the other hand, special registers were kept of the maiden names of the wives and mothers of criminals, because it was believed that in a large proportion of cases an offender, when he changed his name, took either his wife's or his mother's. But these were not attempts to maintain such an elaborately classified register of descriptions as at Scotland Yard. Where offenders were to be traced by an index of personal descriptions, it was, in the case of Liverpool at least, the Habitual Criminals Register that was used for the purpose [56: 219].

Subsequently, the need to supervise habitual criminals in London led to the creation of a separate Convict Supervision Office by the Metropolitan Police in 1880, with which the Habitual Criminals Register is often confused. Visiting officers verified the addresses and workplaces of those being supervised, and by 1886 its records covered 32,000 ex-convicts. The Office acted as a deterrent to re-offending but was also seen as leading to the "reformation, or restoration to honest labour, of old offenders, thereby preventing fresh crime". This bureau gradually expanded its remit as the movement of criminals into and out of London led to the compilation of records on men and women from across the whole country. The Supervision Office also compiled books of distinctive marks and descriptions of convicts, and kept albums of photographs but its small permanent staff – one chief inspector, one inspector, four sergeants and four constables in 1887 – could not keep pace with the work required. Nor was the speed of data retrieval very great. A parliamentary committee of 1893 discovered that in March of that year when 21 officers searched the Mark Registers at Scotland Yard for 27 prisoners, the total time spent doing so equalled 57 and a half hours, and yet they made only seven identifications. This was on average of more than two hours for each prisoner, and more than eight hours for each identification. This may explain why the Office failed to become the centre of criminal investigation in the Metropolitan area [51; 56: 217, 226; 59: 8–17].

Following the publication in 1894 of the report of a Home Office committee of inquiry into the best means for identifying habitual criminals, the Habitual Criminals Registry was amalgamated with a new anthropometric registry also being set up as a result of the committee's enquiries. The anthropometric registry was based on the system of identification invented by the chief of the police identification service in Paris, Alphonse Bertillon. The Bertillon system entailed photographing a subject looking directly at a camera, and then in profile, with the camera centred upon the right ear. Besides the two photographs, the subject's height was recorded, together with the length of one foot, an arm, and an index finger. This information was then archived as, or so it was believed, an infallible means of identifying the body. How far Bertillion's methods were linked to the contemporary speculations of those who thought that one could identify a specific criminal physiognomy, such as the Italian criminologist Cesare Lombroso, is difficult to determine. In 1882, after some years of experimentation, Bertillon began using his system of criminal identification on offenders detained at the Palais de Justice in Paris, although it was not until 1893 that a Criminal Identification Department was established there ([56: 5–35]; Joseph [43: 164–171]; Kaluszynski [44: 123-8]; Becker [6: 141–153]; Rhodes [60]).

However, not all of Bertillon's contemporaries were convinced of the accuracy of his scheme of measurement. Sir Francis Galton, the British father of eugenics and of modern statistics, was critical of Bertillon's system. In his work on human heredity, Galton had tried to establish the transmission of characteristics between generations via statistical correlations. This involved, in part, taking measurements from the bodies of differing generations in the same family. On the basis of these observations, Galton believed that Bertillon's system would lead to an unacceptably high rate of false identification because no account was taken of the relationships between different bodily characteristics. Instead, Galton was an enthusiast for fingerprinting, which he advocated in his book *Finger Prints*, published in 1892 (Galton [26]).

Fingerprints had been scientifically described in seventeenth-century Europe, and first classified in 1823 by the Czech anatomist, Jan Evangelista Purkinje. However, the first extensive use of such prints for official identification was made by the British Raj in Victorian India. A colonial civil servant, Sir William Herschel, began to use fingerprints, or palm prints, to identify Indians in Bengal in the mid-nineteenth century for the purpose of ensuring that native contracts could be made binding. Subsequently, Sir Edward Henry collected fingerprints as a means of identifying criminals whilst Inspector General of the Bengal Police at the end of the century. He also invented a practical way of classifying them for easy access, or at least took the credit for the work of his Indian subordinates

in doing so. The British favoured such methods because they claimed that all "natives" looked the same to the Western eye. Fingerprinting, eugenics and the belief in the inevitable inferiority of the "lower" races and classes, were thus inextricably intertwined (Bevan [8]; Sengoopta [62]; Cole [16]). It is understandable, therefore, that the fingerprinting of British servicemen after the First World War should have raised such objections. Criminals and native peoples, and the two were often linked in the British mind, might be identified via the body but Englishmen were to be identified as active citizens.

The new anthropometric registry at Scotland Yard began to store such prints in conjunction with measurements taken on the Bertillon system by prison warders. It was soon discovered, however, that the anthropometric system was cumbersome to implement and required costly equipment and training ([52]; Joseph [43: 170–171]). A further "Committee to Inquire into the Method of Identification of Criminals" was established in 1900. This recommended that there should be one central office, called the Criminal Registry, in which the records of English, Scottish and Irish criminals could be collected, and in which all the work of identifying and registering criminals should be carried out. This led to the amalgamation of the Habitual Criminals Registry and the Convict Supervision Office. At the same time the Bertillon system was dropped, and superseded completely by the archiving of fingerprints. The change was no doubt facilitated by the appointment of Sir Edward Henry as Assistant Commissioner of the Metropolitan Police in 1901, and then Commissioner, or head of the Met, in 1903. This represented the emergence of the modern system of centralised criminal identification in England, and an example of a transfer of techniques of identification from the Empire into the metropole [43: 171–174; 57].

The increasing speed of transport in the nineteenth century also added to the pressures for more national forms of identification. Local knowledge of the identity of criminals could not be relied upon when steam trains, and later the motor car, could carry people over great distances. This was especially serious in the case of the latter, which could cross several counties in a day, and invade the peace of small local communities that had been untouched by the railway. In 1903 the Motor Car Act laid down that all drivers of cars should have driving licences that could be revoked or suspended for traffic offences; that cars and their owners should be registered with local authorities; and that motor vehicles had to carry number plates that would allow the police to trace their owners in such registries. This piece of legislation needs to be seen in terms of the threat to the concept of community presented by the individualism and sheer anonymity of the motorist. The motorist was seen as an urban dweller on a spree, running down pedestrians, and invading the countryside in a cloud of dust. Members of Parliament were often moved to complain of the latter, alluding to the soiling of ladies' underwear on washing lines, and the virtual imprisonment of villagers by the noxious fumes and grit expelled from the cars of thoughtless motorists. The anonymity of motorists was heightened by the protective garb they wore. According to the Earl of Wemyss, speaking in the debates on the Motor Car Bill in the House of Lords, "Men went about in goggles and in a ghastly sort of headgear too horrible to look at, and it is clear that when they put on that dress they meant to break the law (Higgs [33: 115–118])".

At the same time as the internal "other" was being identified, and so constituted, so the external "other", the alien, was being delineated. Millions of people entered Britain from abroad in the nineteenth century, many on route to the USA, and had had to register themselves, but very little was done to keep them under surveillance. In 1905, after popular agitation against Jewish migrants into the East End of London, a new Aliens Act was introduced by the Conservative government. This set up a revamped system of immigration control and registration and placed responsibility for all matters of immigration and nationality with the Home Secretary, who had the power to deport immigrants considered to be criminals or potential paupers (Gainer [25]). As the First World War broke out in Europe, Parliament passed the 1914 Aliens Restriction Act, which gave the government greater powers to restrict movement into and out of the country. This increasing sensitivity to perceived threats

from abroad was shared by states across the developed world, and led to the creation of the modern system of international passports (Torpey [70: 111–117]).

The first modern UK passport was issued in 1915 when the 1914 Status of Aliens Act came into force. The passport contained the personal description of the holder, giving information on the shape of the person's face, their features (nose: large, forehead: broad, eyes: small), and their complexion, as well as a photograph and their signature. As such the passport combined a mixture of biometrics, photography, and old-fashioned chirography. The wartime restrictions were extended after the period of conflict by the Aliens Order of 1920, that laid down that anyone wanting to enter or leave the country had to carry a passport indicating his or her nationality and identity. In a world in which citizens were being mobilised for total warfare, all foreign nationals became a potential threat (Higgs [33: 140]).

Within Britain during the First World War, the entire German-born population, perhaps 80,000 in number, was seen as a potential fifth column for the Kaiser, and thus worthy of surveillance. In order to do this, MI5, the newly formed internal security service, compiled a register of all Germans and Austro-Hungarians in the country, possibly drawing upon the manuscript records of the 1911 census. According to MI5's own official history of the period

> the Registrar-General considered that the information in census returns had been obtained confidentially and that the police must not let it be known that they were being used for the purpose of [aliens] registration.

In time the Central Registry was to contain a card index in which enemy aliens were classified, in order of support for the British cause, as "AA" (Absolutely Anglicized or Allied), "A" (Anglicized/Allied), "AB" (Anglo-Boche), "BA" (Boche-Anglo), "B" (Boche) or "BB" (Bad Boche). This was the start of the creation of secret population listings, and by 1917 MI5 was receiving information on suspects from government departments, chief constables, and private individuals, from which it created registered files. The MI5

Registry (section H.2) distributed files to other divisions of MI5, and kept track of their movements. H.2 also produced various name, ship, subject and place indexes to the information in its records with geographical indexes organised down to the level of streets in London and large towns (Higgs [33: 110–111, 145]).

9.4 TWENTIETH-CENTURY IDENTIFICATION IN TOTAL WAR, TOTAL WELFARE AND TOTAL SHOPPING

The history of identification in the later twentieth-century Britain can be understood as a process of deepening and widening, driven by the needs of warfare, state welfare, and the increased penetration of financial systems into the lives of ordinary people. In the process, forms of identification, especially bodily identification, confined to non-citizens in the Victorian period were increasingly applied to everyone. This was accompanied by an increase in the role of technology in identification systems, and in the needs of identification within technology, especially IT.

The integration of the population into central state and national commercial identification systems was unprecedented. When old age pensions were introduced into Britain in 1908, 490,000 individuals qualified for the new transfer income, which was to be paid out of central taxes (Thane [67: 83–84]). In 1911 the National Insurance Act introduced some unemployment and medical benefits for about 2,250,000 skilled workers (Harris [31: 163]). By 1978, however, the Department of Health and Social Security had a database of 9,000,000 records covering the payment of retirement pensions and widows' benefits, and at the same date the National Insurance Contribution database contained 45,000,000 records [58: Appendix 6]. Similarly, the numbers of taxpayers increased enormously. In 1901 there were probably 1,000,000 active payers of income tax out of a total population of 41 million. Inflation and relatively fixed exemption limits meant that by the fiscal year 1989/90 there were 21,900,000 income tax payers

(Higgs [33: 144–145]). But how could one be certain that these records related to specific individuals, or that the information relating to them was correct?

At the same time, the reach of banks and other financial institutions into society was expanding, and interactions between them and their customers were increasingly being automated. In 1975 only about 9 million, or 45 percent, of all households had a bank account. These were mostly salaried people, or those with unearned income. By 1996, 94 percent of the adult population in the UK had some form of bank account, with 83 percent having a current account. Barclays Bank issued the first credit card in the UK in 1966, and installed the first cash point, a forerunner of the automated teller machine (ATM), a year later. By 1996 there were 560 credit cards per 1000 people in the UK and 550 debit cards, and 42 percent of the population held at least one card. In 2003 there were 46,461 ATMs in the UK, and 88,000,000 cards that could be used to get cash from them. Also in that year purchases on all plastic cards in the UK equalled £5,317,000,000, over triple the figure for 1993 (APACS [2]; Carrington [13: 18, 114–116]; Consoli [18: 8–9]). But how could one ensure that the people using such cards were the people they had been issued to, or who had the bank accounts that were associated with them?

These heightened requirements for mass forms of identification changed the nature of identification, certainly with respect to the State. As already noted, the identification of the citizen in the Victorian period was as much to do with recognizing rights, as with social control. William Beveridge, the architect of the Welfare State in Britain in the 1940s, saw social benefits in a similar light. They were to be given to everyone, so that no stigma would be attached to receiving them. However, this proved too expensive, and increasingly benefits were 'means tested', that is, they were given only to people whose income fell below a certain level. The result was similar to the Poor Laws – welfare recipients came to be seen as a burden, and as potential malingerers who needed to be watched (Higgs [33: 150]). Identification became punitive rather than a matter of recognition. Much the same could be said of the commercial sphere, where the number, scale and speed of electronic transactions opened up a whole new area of crime, and of uncertainty, through identity theft.

By the late twentieth century, "welfare fraud" had become an obsession of the political classes, and was calculated as possibly reaching as much as £7 billion per annum. Much of this was due, of course, to people giving false information about their financial circumstances but there were also numerous cases of false identities being used. In 2000, for example, the Department for Social Security noted the case of an organised gang that adopted the identities of 171 people from the Irish Republic. Documents relating to these people were then used them to make a string of false claims in the UK, leading to fraudulent payments in excess of £2 million. In another case, a woman stole nearly £60,000 by fraudulently continuing to claim a pension for her dead mother. This only came to light when enquiries were made ahead of the sending of a celebratory telegram for her mother's 100th birthday [1; 7; 21].

The benefits system tried to pin down the identity of claimants, much as it had done at the beginning of the century, through their production of traditional forms of documentation – birth certificates, marriage certificates, driving licences, and the like. All claimants had, in theory, to be issued with a registered National Insurance number (NINO), which increasingly became the "gateway" to all benefits. But there were all sorts of problems with these arrangements. Documents relating to other individuals could be copied, or forged. They could even be obtained quite legally, since anyone could purchase a birth certificate for any named person. Numerous people within the social security system could issue NINOs, and this opened the way to corruption and fraud. There were also so many different benefits, which were administered at so many different points, especially the local offices of the Department of Social Security and by local authorities, that it was difficult to keep track of who had claimed what. Ensuring that the information supplied was accurate, especially that relating to income, was also problematic [1; 7; 21].

The State attempted to get round these problems in a number of ways. Personal data on applicants was increasingly stored on computer, and shared across government departments and local authorities, allowing details to be crosschecked and verified. Taxation information from the Inland Revenue, for example, was supplied to the Department of Social Security to provide proof of income levels This was facilitated by the creation of a Personal Details Computer System. Only original documents could be presented when making claims, and the system for issuing NINOs was tightened up. Claimants were quizzed by "personal advisers" when they applied for benefits, and spot checks were made by inspectors making unannounced visits to their homes [1; 7; 21]. Increasingly the debate on combating welfare fraud came to focus on the benefits of introducing an "entitlement card" which could carry information on benefit claimants and be linked to a central register. This would have the advantage of creating a single "passport" which would relate databases across the State [37]. However, the entitlement card quickly became embroiled with arguments about the advisability of introducing a national identity card system.

For the State the fraudulent use of identity had numerous other aspects. A cross-Government study group led by the Cabinet Office was set up to establish the nature and extent of the problem, and estimated that in 2002 the minimum cost to the economy of identity fraud was £1.3 billion per annum. It listed some examples of the extent of such fraud in 2000/01, when:

- 3,231 driving tests were terminated prematurely because of doubts over the driver's identity;
- 1,484 fraudulent passport applications were detected;
- approximately 50 cases of fraudulent documentation were detected every month at Terminal 3, Heathrow Airport, London;
- in the course of a two week exercise targeted at Portuguese documents in June 2001, 59 fraudulent documents were detected at selected UK ports and the Benefits Agency National Identity Fraud Unit (NIFU). The majority were counterfeit identity cards, detected at NIFU;
- a Home Office study estimated that 1,500 women a year were trafficked for prostitution;

- in 1999 over 21,000 illegal immigrants were detected; during the same period 5,230 were removed or left voluntarily;
- 18,500 referrals were made to the Financial Services Authority under the money laundering regulations;
- 564 cases involving identity fraud were identified by the Benefits Agency's Security Investigation Service, whose specialist teams investigate organised fraud cases across the country [41].

There were also serious problems of "identity theft" within the commercial sector. APACS, the association of institutions delivering payment services to the British public, confirmed that total plastic card fraud rose by 18 percent to £478.8 million in the 12 months to June 2004. Of this, cash machine crime accounted for the fastest growing category during that year, resulting in losses of £61.1 million. One of the main reasons for this increase was the rise in the use of skimming devices, which copied card details, and miniature cameras, which recorded the personal identification numbers (PINs) used by cardholders at cash machines. Other types of cash machine fraud included:

"shoulder surfing" – where criminals looked over a cash machine user's shoulder to watch them enter their PIN, then stole the card by distracting them and picking their pockets; and card-trapping devices – where a device retained the card inside the cash machine, whilst the criminal tricked the victim into re-entering their PIN, which they recorded.

Over the same 12 month period, counterfeit card fraud equaled £123.0 million; and fraud involving lost and stolen cards rose by 11 percent to £118.8 million [12].

Many plastic card transactions at first depended for security on the use of the old identification technique of personal signatures. These had to be written on the back of the card on a special strip, and then cardholders using the card to purchase goods had to sign a receipt which the sales assistant could check against the card. From 1995 the banking industry began to develop a new technical solution to the problem of plastic card fraud, and joined

forces with retailers to implement a PIN verification system using secure chip cards. The UK's first "chip and PIN" transaction took place in 2003 in a public trial in Northampton, and from 2005 onwards the vast majority of face-to-face UK credit and debit card transactions will be authorised by the customer keying in their PIN on a card reader rather than by signing a receipt. APACS believes that when fully implemented, the chip and PIN verification system will more than halve predicted UK fraud losses [2], although how this can be squared with the rise of cash point fraud is less clear. Yet again, the solution to the modern problems of identification seems to involve some sort of identity card system linked to a register of identities.

This general trend in the UK towards identification cards of one form or other is a marked departure from the past. Unlike many countries in continental Europe, such as France and Germany, Britain did not establish a permanent identity card system in the twentieth century, except during wartime. As already noted, this partly reflected the lack of conscription, since Britain, as an island, did not require a large standing army. A system of National Registration was, however, introduced during the First World War to facilitate military recruitment and rationing, and a Civil Service Committee on National Registration recommended that the system should continue after 1918. This would have involved giving every citizen a certificate of registration that could act as a type of identity card. However, the proposal was rejected on the grounds of cost and because of the system's infringement of civil liberties, and National Registration lapsed. As noted above, liberty for the British meant freedom to participate in civil and political society, rather than to be regulated by the State (Higgs [33: 137–139]).

National Registration and the issuing of identity cards returned with the renewed threat of conflict with Germany. As early as 1935 the GRO was working on systems to establish a national register based on a census of the entire population, and these plans were put into operation on the outbreak of hostilities by the 1939 National Registration Act. The Register comprised "all persons in the United Kingdom at the appointed time" and "all persons entering or born in the United Kingdom after that time". A Schedule to the Act listed "matters with respect to which particulars are to be entered in Register". These were:

1. Names
2. Sex
3. Age
4. Occupation, profession, trade or employment
5. Residence
6. Condition as to marriage
7. Membership of Naval, Military or Air Force Reserves or Auxiliary Forces or of Civil Defence Services or Reserves.

Identity cards were also issued, which had to be produced to police when required. These gave the names and addresses of the individuals, and their National Registration numbers, as well as their signatures. As such, they were rather traditional documents, depending of chirography as the means of personal identification [49].

However, whereas National Registration lapsed after the end of hostilities in 1918, it continued after the end of the Second World War until the early 1950s, as did the issuing of identity cards. The Labour governments of this period retained the system because of the continuation of conscription during the early Cold War, the needs of the police and security forces, and the desire to plan society centrally. National Registration was, for example, the basis of patient registration for the National Health Service that was set up in 1947. National Registration was popularly regarded as intrusive, especially the need to show the card to police when requested. This was seen as treating the honest citizen as if he or she was a criminal, and undermining his or her willingness to co-operate with the police. The boundary between the criminal to be kept under surveillance, and the citizen to be allowed "to go about his business", was being effaced. Identity cards were abolished in 1952 by a Conservative government pledged to roll back the State (Higgs [33: 140–143]).

It was perhaps inevitable, however, that at the end of the twentieth century, in a period of heightened concern over crime, immigration, and terrorism, that the whole issue of identity cards should be resurrected. In a reversal of earlier patterns, calls for the introduction of such cards emanated as much from the Right of the political spectrum as from the Left, and were concerned predominantly with policing issues. In 1989 two private members bills were introduced into Parliament for the setting up a national identification system. Although the then Home Secretary made it clear that the government did not favour a compulsory national registration system, he also stated that he would examine the feasibility of a voluntary system. In the early 1990s concerns were expressed that the introduction of photographs and personal identity numbers on driving licenses might turn the latter into a de facto identity card. The government again made it clear that although it planned to make all drivers' details available to the police, drivers would not have to carry their licenses at all times. Demands for the introduction of identification cards surfaced in Parliament once more in 1994, and the then Prime Minister, John Major, stated that he believed that there was a strong case for such a means of checking identity (Higgs [33: 184–185]).

A change of government in 1997 did not lessen the pressure for identification cards to prevent "bogus" asylum seekers entering the country, or to combat international terrorism in the wake of the bombing of New York and Washington, DC in 2001. The new Labour administration was certainly willing to introduce cards for specific target groups. For example, in reaction to the case of Harold Shipman, a NHS general practitioner who may have murdered hundreds of his patients, the government announced that doctors were to be issued with a "smart card". This would hold a record of pre-employment checks and of the doctor's suitability for work with children; their police records; and any disciplinary proceeding against them by the General Medical Council (Higgs [33: 185–186]). In 2004 the government introduced an Identity Cards Bill, with the aim of "protecting people from identity fraud and theft and providing them with a convenient means of verifying their identity in everyday transactions". The system was also intended to:

> disrupt the use of false and multiple identities which are used by organised criminals and in a third of terrorist-related activity;

tackle illegal working and immigration abuse; ensure free public services are used only by those who are properly entitled to them [i.e. prevent welfare fraud]; and ensure British citizens are able to travel freely as international requirements for secure biometric identity documents developed [65].

Under the Identity Cards Bill the identification system would have a number of components: a national identification register; a national identity registration number; the national identity card; the establishment of legal obligations to disclose personal data; and the creation of new crimes and penalties to enforce compliance with the legislation. Biometrics would also be taken upon application for a card and for entry on the national identification register, and would be used thereafter for major 'events' such as obtaining a driving license, a passport, a bank account, benefits or employment. The iris and fingers of the applicant would be scanned, and then compared both with the biometric on the identity card (which would contain the biometrics in electronic form), and against a national database (which would also contain such biometrics) [68: 11–13].

These proposals indicate the importance of the development of a number of new means of bodily identification dependent on computer technology. These include DNA, and iris and face recognition. DNA typing was developed in the early 1980s by Professor Alec Jeffreys at the University of Leicester's Department of Genetics, who used a chemical process to force long DNA strands to separate into shorter pieces at predictable but variable points. These could then be arranged by size to create a set of horizontal bars resembling a bar code. This means of identification has subsequently been used to create national DNA databases for the purpose of criminal identification. Criminals have even been identified by matching samples taken at the scene of a crime with those of their kin (Higgs [33: 182–183]). In an analogous manner, iris recognition technology allows the identification of an individual by the mathematical analysis of the random patterns that are visible within the iris of the eye. The algorithms for iris recognition

were developed at Cambridge University's Computing Laboratory by Dr. John Daugman, and the system is currently being used to record foreigners entering the United Arab Emirates (Daugman [20]). In May 2003 the International Civil Aviation Organisation (ICAO) selected face recognition as the globally acceptable biometric for identification via machine-readable travel documents. The Australian Customs Service subsequently developed the "Smartgate" system to allow the identification of aircrew, and later all air passengers. Smartgate scans the photographs on passports and electronically compares them to the face of the passport holder. The system aims to catch criminals, terrorists, and illegal entrants that might use false travel documents [5].

What is intriguing about these technologies is the manner in which Anglo-Saxon countries, especially Britain, have been at the forefront of developing and implementing systems for the identification of persons as bodies. Why should polities that prided themselves on recognising citizens as active but voluntary participants in civil and political society, rather than as the criminal body, suddenly develop mass systems of identification dependent on recording physical attributes? In the case of Britain, it should be noted that the country also now 'boasts' the highest density of public CCTV cameras in the world [45]. This seems to point towards a fundamental reappraisal of the relationship between the State and its members. The boundary between the citizen and the "anti-citizen" is being eroded, with fundamental implications for civil liberties.

9.5 CONCLUSION

This has been a somewhat breathless race through some half a millennium of the history of identification in Britain. However, it is still possible to sketch out some general conclusions. First, what has been identified has changed over time, and this has significance. We have moved from identifying lineages and communities, via identifying citizens, to identifying bodies. In the process the concept of identification as the recognition of the citizen, has come to be confounded with the identification of the "anti-citizen". This is partly why identification has become a contentious issue in Britain.

Identification has increasingly become a technology, rather than a set of techniques. Part of the reason for this has been the increasing scale of interactions within society. When the vast majority of people lived, worked, received welfare, and died within the same small community, forms of identification based on personal knowledge were adequate. However, when these social forms became dissociated during the nineteenth and twentieth centuries, and state systems began to function at the level of the Nation State, such techniques were insufficient. When I can at the same time live in Oxford, work at the University of Essex, have a bank account in my home town of Lancaster in Lancashire, and speak at a conference in Canada, relying consistently upon someone who knows me as a means of identification is a not an option. Technologies that allowed data for identification to be stored remotely, and accessed over considerable time-spans, have become necessary.

However, this story of identification has not necessarily been one of linear progression. In some odd ways we have returned in contemporary Britain to certain aspects of the past. DNA profiling leads us back to the identification of lineages rather than of individuals, and the credit card can be seen, after all, as a physical means of identification analogous to the medieval seal. This is seen most clearly in the introduction of the eID card in Belgium, which allows citizens to use a card to electronically 'sign' documents [22]. Personal identification is thus not simply a contemporary problem, and placing it within a historical framework helps elucidate the nature of some of our modern dilemmas.

REFERENCES

[1] *A new contract for welfare: safeguarding social security*, British Parliamentary Papers, Cm 4276, 1997–1998.

[2] APACS website, available at http://www.apacs.org.uk/about_apacs/htm_files/figures.htm (15/3/2005).

[3] I. Arthurson, *The Perkin Warbeck Conspiracy 1491–1499*, Alan Sutton Publishing, Stroud (1997).

[4] J. Austen, *Persuasion*, Oxford University Press, Oxford (2004).

[5] Australian Customs website, available at http://www.customs.gov.au/site/ (23/3/2005).

[6] P. Becker, *The standardised gaze: the standardization of the search warrant in nineteenth-century Germany*, **Documenting Individual Identity. The Development of State Practices in the Modern World**, J. Caplan, J. Torpey, eds, Princeton University Press, Princeton (2001), pp. 139–163.

[7] *Beating fraud is everyone's business: securing the future*, British Parliamentary Papers, Cm 4012. 1997–1998.

[8] C. Bevan, **Fingerprints. Murder and the Race to Uncover the Science of Identity**, Fourth Estate, London (2002).

[9] M.J. Braddick, **State Formation in Early Modern England c. 1550–1700**, Cambridge University Press, Cambridge (2000).

[10] J. Bradley, *Body commodification? Class and tattoos in Victorian Britain*, **Written on the Body. The Tattoo in European and American History**, J. Caplan, ed., Reaktion Books, London (2000), pp. 136–155.

[11] F. Braudel, *History and the Social Sciences: The Longue Duree [1958]*, **Histories: French Constructions of the Past**, J. Ravel, L. Hunt, eds, New Press, New York (1995), pp. 115–145.

[12] Card Watch website, available at http://www.cardwatch.org.uk (22/3/2005).

[13] M.St.J. Carrington, Ph.W. Langguth, Th.D. Steiner, **The Banking Revolution: Salvation or Slaughter? How Technology is Creating Winners and Losers**, Pitman, London (1997).

[14] M.T. Clanchy, **From Memory to Written Record: England 1066–1307**, Basil Blackwell, Oxford (1993), pp. 308–317.

[15] R. Clarke, *Human identification in information systems: management challenges and public policy issues*, Information Technology and People 7 (1994), 6–37.

[16] S. Cole, **Suspect Identities: A History of Fingerprinting and Criminal Identification**, Harvard University Press, Cambridge, MA (2001).

[17] Committee on Periodical Identification of Government Pensioners. Selection of departmental delegates. Report, National Archives, London, Treasury Board Papers, T 1/11665/17893.

[18] D. Consoli, **The Evolution of Retail Banking Services in the United Kingdom: A Retrospective Analysis**, Centre for Research on Innovation and Competition, University of Manchester, Manchester (2003).

[19] M.A. Crowther, **The Workhouse System 1834–1929: The History of an English Social Institution**, Methuen, London (1983).

[20] J. Daugman, *Iris recognition for personal identification*, available at http://www.cl.cam.ac.uk/users/jgd1000/iris_recognition. html (23/3/2005).

[21] Department of Social Security, **Safeguarding Social Security: Getting the Information We Need**, Department of Social Security, London (2000).

[22] eID website, available at http://eid.belgium.be (29/4/2005).

[23] B. Fraenkel, *La signature. Genèse d'un signe*, Gallimard, Paris (1992).

[24] H.R. French, *Social status, localism and the 'middle sort of people' in England 1620–1750*, Past and Present 166 (2000), 66–99.

[25] B. Gainer, *The Alien Invasion: the Origins of the Aliens Act of 1905*, Heinemann, London (1972).

[26] Sir F. Galton, *Finger Prints*, Macmillan, London (1892).

[27] M. Goldie, *The unacknowledged republic: officeholding in early modern England*, **The Politics of the Excluded, c. 1500–1850**, T. Harris, ed., Palgrave, Basingstoke (2001), pp. 153–194.

[28] Th.A. Green, *A retrospective on the criminal trial jury, 1200–1800*, **Twelve Good Men and True: the Criminal Trial Jury in England, 1200–1800**, J.S. Cockburn, T.A. Green, eds, Princeton University Press, Princeton (1988), pp. 358–361.

[29] *Guide to the Contents of the Public Record Office*, Vol. 1, Legal Records, etc., HMSO, London (1963).

[30] M. Gustafson, *The tattoo in the later Roman Empire and beyond*, **Written on the Body. The Tattoo in European and American History**, J. Caplan, ed., Reaktion Books, London (2000), pp. 17–31.

[31] B. Harris, *The Origins of the British Welfare State: Social Welfare in England and Wales, 1800–1945*, Palgrave, London (2004).

[32] E. Higgs, *Life, Death and Statistics: Civil Registration, Censuses and the work of the General Register Office, 1837–1952*, Local Population Studies, Hatfield (2004).

[33] E. Higgs, *The Information State in England: The Central Collection of Information on Citizens, 1500–2000*, Palgrave, London (2004).

[34] S. Hindle, *Dependency, shame and belonging: badging the deserving poor, c. 1550–1750*, Cultural and Social History 1 (2004), 6–35.

[35] S. Hindle, *The State and Social Change in Early Modern England, c. 1550–1640*, Macmillan, London (2000).

[36] Th. Hobbes, *Leviathan, or, the Matter, Forme & Power of a Common-wealth Ecclesiasticall and Civill*, London (1651).

[37] Home Office website, available at http://www.homeoffice.gov.uk/docs2/fraudconsultation.html (21/3/2005).

[38] R.F. Hunnisett, *The Medieval Coroner*, Cambridge University Press, Cambridge (1961).

[39] A. Hunt, *Governance of the Consuming Passions. A History of Sumptuary Law*, Macmillan, London (1996).

[40] R. Hutton, *The Rise and Fall of Merry England: The Ritual Year 1400–1700*, Oxford University Press, Oxford (2001).

[41] *Identity fraud: A study*, Cabinet Office, London (2002).

[42] M. Ingram, *Church Courts, Sex and Marriages in England, 1570–1640*, Cambridge University Press, Cambridge (1990).

[43] A.M. Joseph, *Anthropometry, the police expert, and the Deptford Murders: the contested introduction of fingerprinting for the identification of criminals in late Victorian and Edwardian Britain*, **Documenting Individual Identity. The Development of State Practices in the Modern World**, J. Caplan, J. Torpey, eds, Princeton University Press, Princeton (2001), pp. 164–183.

[44] M. Kaluszynski, *Republican identity: Bertillonage as government technique*, **Documenting Individual Identity. The Development of State Practices in the Modern World**, J. Caplan, J. Torpey, eds, Princeton University Press, Princeton (2001), pp. 123–138.

[45] Liberty website, available at http://www.liberty-human-rights.org.uk/privacy/cctv.shtml (23/3/2005).

[46] J. Locke, *Two Treatises of Government [by J. Locke]. In the Former, the False Principles, and Foundation of Sir Robert Filmer are Detected and Over Thrown. The Latter is an Essay Concerning the True Original, Extent, and End of Civil Government*, London (1690).

[47] C.B. Macpherson, *The Political Theory of Possessive Individualism: Hobbes to Locke*, Clarendon Press, Oxford (1962).

[48] W.A. Morris, *The Frankpledge System*, Longmans, London (1910).

[49] National Identity Card: Wellcome Library, Medical Ephemera, CPH 37:15.

[50] *Oxford English Dictionary*, Oxford University Press, Oxford (1989).

[51] Police – Metropolitan: Augmentation of staff to meet increase of work in Convict Supervision Office: Request for increase in staff of Convict Supervision Office, National Archive, London, HO 45/9675/A46826.

[52] Police of England and Wales, National Archives, London, Treasury Blue Notes, T 165/21 (1904–1905).

[53] Prisons and Prisoners: (4) Other: Prevention of Crimes Act, 1871. Regulations for photographing prisoners: National Archives, London, HO 45/9320/16629C.

[54] W.B. Purchase, *Sir John Jervis on the Office and Duties of Coroners: with Forms and Precedents*, 8th edition, Sweet & Maxwell, Ltd, London (1946).

[55] Registry of criminals: National Archives, London, Metropolitan Police; Criminal Record Office: Habitual Criminals Registers and Miscellaneous Papers, MEPO 6/90 Pt 2.

[56] Report of a Committee appointed by the Secretary of State to inquire into the best means available for Identifying Habitual Criminals ..., British Parliamentary Papers (1893–1894), LXXII.

[57] Report of the Committee to Inquire into the Method of Identification of Criminals 1900: CRIMINAL: Report of Identification of Criminals Committee, 1900. Classification and use of fingerprints. Augmentation of staff in

Commissioners Office, National Archives, London, HO 144/566/A62042.

[58] Report of the (Lindop) Committee on Data Protection, British Parliamentary Papers (1978–1879), V.

[59] Report on the working of the Prevention of Crimes Acts by the Convict Supervision Office: National Archives, London, Home Office Registered Papers, Supplementary, HO 144/184/A45507.

[60] H.T.F. Rhodes, *Alphonse Bertillion: Father of Scientific Detection*, Harrap, London (1956).

[61] J.H. Round, *Peerage and Pedigree: Studies in Peerage Law and Family History*, Tabard Press, London (1910).

[62] C. Sengoopta, *Imprint of the Raj: How Fingerprinting Was Born in Colonial India*, Macmillan, London (2003).

[63] P. Slack, *Poverty and Policy in Tudor and Stuart England*, Longman, London (1988).

[64] L. Smith, D. Smith, *Sudden Deaths in Suffolk 1767–1858: A Survey of Coroners' Records in the Liberty of St Etheldreda*, Suffolk Family History Society, Ipswich (1995).

[65] Strengthening Security, Protecting Identity: Home Office Publishes Identity Cards Bill: Home Office Press Release 374/2004. Home Office website, available at http://press.homeoffice.gov.uk/press-releases/Strengthering_Security_Protecti.

[66] System of identifying discharged soldiers etc. by means of fingerprints: National Archives, London, Treasury Board Papers, T 1/12181/15387.

[67] P. Thane, *The Foundations of the Welfare State*, Longman, London (1982).

[68] *The Identity Project. An Assessment of the UK Identity Cards Bill & Its Implications. Interim Report*, Department of Information Systems, London School of Economics, London (2005).

[69] E.P. Thompson, *Whigs and Hunters: The Origins of the Black Act*, Peregrine, London (1977).

[70] J. Torpey, *The Invention of the Passport: Surveillance, Citizenship and the State*, Cambridge University Press, Cambridge (2000).

[71] A.R. Wagner, *English Genealogy*, Oxford University Press, Oxford (1972).

10

THE SCIENTIFIC DEVELOPMENT OF BIOMETRICS OVER THE LAST 40 YEARS

James L. Wayman

Office of Graduate Studies and Research, San Jose State University
San Jose, CA, USA

Contents

Abstract

This paper looks at the last 40 years of development of biometrics from a US perspective. We trace the major accomplishments of each decade from the 1960s to the 2000s. Particularly attention is paid to the 1970s explosion of interest in testing, multimodal decision making, vulnerability and consumer applications.

Keywords: automated personal identification, biometrics, facial recognition, fingerprinting, iris recognition, hand geometry, speaker recognition.

10.1 INTRODUCTION

Although the development of fully automatic systems for human recognition dates to the early 1960s, the use of the word 'biometrics' in this sense only began in the early 1980s. In the 1970s the field was known as "automated personal identification" (API), but before 1970, there seems to have been little self-consciousness that such a general field existed, beyond the individual, emerging technologies of fingerprint, signature, speaker and facial recognition. These individual technologies came out of very different scientific traditions: speaker recognition from acoustics and facial recognition from computer pattern recognition, for example. To this day, no single scientific publica-tion for automated personal identification exists; technical papers being published internationally in a wide variety of scientific journals. Consequently, a discussion of the history of biometrics is compli-cated by both the evolution of the term itself and the diversity of its sources.

As of this writing, the international standards committee on biometrics (ISO/IEC JTC1 SC37) is still debating the definition of 'biometrics'. In this chapter, I will define 'biometrics' very narrowly as "the automated recognition of living individuals based on biological and behavioral traits". Specif-ically excluded from consideration here are non-automated and forensic methods for human recog-nition. Thus, my use of 'biometrics' makes it a lim-ited subset of the larger field of human recognition science. The debate over whether these are 'recognition' or 'identification' technologies dates to the 1960s (Young [75]) and has not yet been fully resolved. In this chapter, we will ignore the distinctions and use the terms 'recognition' and 'identification' interchangeably.

My research methodology was to trace backwards the references in the open literature in the English language. Inevitably, the chain would be stopped with references to unpublished, internal reports. The U.S. Department of Defense, often via the Department of the Air Force, funded much of the early, published scientific work in the United States, so important primary sources were often in internal reports of defense contractors and National Laboratories. I am indebted to many friends, historians and archivists within those groups who were able to obtain release many of these internal documents, reviewed here for the first time. It was also possible to interview the surviving principals in many cases.

Clearly, this approach leads to many errors of omission and grossly under-reports contributions from non-English language sources. Historical overviews of the development of biometrics in all regions would be greatly welcome by the biometrics community, so that the major contributions internationally could be more fully acknowledged in subsequent historical overviews of the field.

Foregoing any discussion of the 19th century non-automated techniques of Bertillion, Faulds, Hershel and Galton, or the semi-automated 'voice-print' techniques of the mid-20th century (Potter, Kopp and Green [51]; Kersta [28]) we can begin our discussion with the first fully automated technologies of the 1960s.

10.2 THE TECHNOLOGY PIONEERS OF THE 1960s

One of the very first journal papers on biometrics, as defined above, appeared in *Nature* in 1963 (Trauring [61]), although the paper refers only to automatic fingerprint recognition. Even more interestingly however, the author, Mitchell Trauring, published an internal report for his employer speculating on the future of automatic fingerprint

recognition technologies (Trauring [62]). In this paper, he states that automated fingerprint recognition systems could provide "personal identity verification with the following desirable features: Speed – operation in seconds; Decentralization ... relatively inexpensive and requires no trained or skilled operating personnel; Ultravalidity ... not susceptible to forgery or theft; Convenience ... no need to carry identity cards or similar items".

Trauring predicts "A large, diverse market – The four features above promise products having a large, diverse market in nonmilitary, as well as some military applications; In credit systems ... retail stores, airline counters, etc.; With industrial and military security systems ... restricted access areas in industrial plants, offices; Personal lock ... it could replace key and combination locks".

Credit cards date only to the earliest years of the 1950s. Therefore, we can state that recognition of the potential for use of biometrics with credit cards is nearly as old as credit cards themselves.

Trauring's insight into the potential of biometrics was astounding, but technologically, he was only just slightly ahead of his competitors. In 1965, the U.S. Patent Office issued to IBM a patent for a *Personal Security System Having Personally Carried Card with Fingerprint Identification* (Classen [10]).

The first work on automatic facial recognition was that of Woodrow W. Bledsoe at the Panoramic Research Institute, Palo Alto, CA (Bledsoe [4]) around 1964. Bledsoe later became a professor at the University of Texas, Austin, which has archived his works. According to his University of Texas colleagues, "He was proud of this work, but because the funding was provided by an unnamed intelligence agency that did not allow much publicity, little of the work was published" (Faculty Council [14]). One of the few journal references to Bledsoe's work is in a 1971 paper on human/machine facial recognition (Goldstein, Harmon and Lesk [19]).

Bledsoe's initial approach was to manually mark various landmarks on the face (eye centers, mouth, etc.). These locations could be mathematically rotated by computer to compensate for pose variation. Distances between landmarks and distance ratios could be automatically computed and compared between images to determine goodness-of-fit. In further work in the 1965 time period, he experimented with automating the process of finding facial landmarks.

Bledsoe mentioned and quantatively explored the difficulties in facial recognition caused by changes in "head rotation and tilt, lighting intensity and angle, facial expression, aging, etc." and noted that the "correlation is very low between two pictures of the same person with two different head rotations". He acknowledged "other attempts at facial recognition by machine" but does not supply references to any earlier work by other researchers. In this research, he performed about 40,000 comparisons using a dataset of 2,000 images containing at least 2 separate images of each test subject. When Bledsoe left Panoramic Research, Inc. in 1966, the work was continued by Peter Hart at the Stanford Research Institute (Faculty Council [14]).

Automatic speaker recognition took off strongly in the 1960s. 'Visible speech', which used human examiners to recognize speakers from acoustic spectrograms output by analog computers, had been around since the 1940s (Potter, Kopp and Green [51]) and continued as a research area to the end of the 1960s. But visible speech was not an automated method for recognition, so is outside the scope of this paper. The *Journal of the Acoustical Society of America* carried most of the early work in both visible speech and 'automatic talker recognition'. One of earliest was papers on a completely automated method was that of Pruzansky in 1963 [52], followed by work in 1966 (Li, Dammann and Chapman [30]). The 1969 paper by Luck [32] was the first to use cepstral coefficients[1] as fundamental 'features' in speaker recognition. This is highly significant because cepstral coefficients are still used today as the basis of almost all speaker recognition systems. The point to be

made here is that, at the most fundamental technical level, biometrics is evolving very slowly from its origins.

The earliest study on automatic signature recognition appears to have been performed by US defense contractor North American Aviation in the early 1960s under funding from the U.S. Air Force (Maurceri [39]). Both pen acceleration and pen–paper contact was measured. An accelerometer was placed above the end of a ballpoint pen, mounted to the pen with a clear plastic strut. Current flow when the pen-to-conductive paper circuit was completed indicated pen–paper contact. The computation of distinctive features of the signature involved both analog and digital computers – the analog computers required to compute power spectral information of the two signals. The study reported correct recognition of 226 of 250 signatures, taken from a datasbase of 2,000 signatures collected at a rate of 5 signatures a day for 10 days from 40 subjects. The report references no earlier reports on automatic signature recognition, although the RAND Corporation was experimenting with digital tablets for the input of written information to computers at about the same time [8].

We cannot leave our discussion of the 1960s without some reference to the development of the Fast Fourier Transform (FFT) (Cooley and Tuckey [11]), allowing digital computers to efficiently compute frequency-domain measures, such as power spectra, which had previously required analog computers in speaker and signature recognition. Although this represented the most profound signal processing breakthrough of the entire 20th century, it had little practical impact on most of us until the 1980s. When FFT software became available in the 1960s, computers were main frames using punch cards as input. Output was rows and columns of numbers on large sheets of paper (graphical output was unheard of). Analog-to-digital (A/D) converters were only available to a few researchers in well-funded laboratories. Only when computers, graph machines and A/D converters became available on desktops around 1981 could more general researchers (like myself) begin to take advantage of the breakthrough that the FFT represented.

---

[1]The word 'cepstrum' is derived inverting the first four letters of 'spectrum'. Cepstral coefficients are computed by treating the log of the energy in the frequency representation of a speech signal as yet another signal, taking the Discrete Fourier Transfom of that spectral energy signal. The coefficients of this second Fourier transform are the 'cepstral coefficients'.

10.3 AUTOMATIC PERSONAL IDENTIFICATION OF THE 1970s

The 1960s was characterized by the independent development of fingerprint, voice, signature and facial recognition technologies. The 1970s marked a real "Coming of Age" for automatic human recognition because of the development of a self-consciousness of the existence of this field. Government was becoming aware of the technologies. The U.S. National Bureau of Standards (NBS), later to become the National Institute of Standards and Technology (NIST), started the formal testing of fingerprint systems in 1970 (Wegstein [73; 74]). By 1971, the US government was actively researching the use of automated fingerprint identification techniques, in particular holographic optical techniques, for forensic applications (McAlvey [45]).

Takeo Kanade demonstrated a rudimentary facial matching system in the Japanese pavilion at the 1970 Osaka World Exposition (Kanade [27]). In an automated approach similar to that of Bledsoe, this device attempted to find anatomical facial landmarks, such as eyes and chin, and determine distance ratios. These ratios were compared to those of famous people (John F. Kennedy and Marilyn Monroe among them) and exhibit visitors were told which person they most resembled. In the absence of any ground truth, this system could be considered a great success, although later analysis of the data revealed that the system had not reliably found the facial landmarks in question. To my knowledge, Kanade's 1977 book [26] on automatic facial recognition was the first to look closely at this technology.

The U.S. National Academy of Science organized a working group on speaker verification in 1971 (NAS [47]) and the Acoustical Society of America held a session at its annual meeting devoted to spoofing speaker recognition systems through mimicking behaviors (Lummis and Rosenberg [33]; Rekieta and Hair [55]; Rosenberg [57]). The real milestone of the 1970s for speaker recognition was the publication of two survey articles in the same issue of the *Journal of the Acoustical Society of America* in 1976 (Atal [1]; Rosen-

berg [58]). These articles demonstrate that speaker recognition was the most scientifically mature human recognition technology by the mid-1970s.

IBM contributed a fundamental work in 1970 [24] that appears to be the first to outline the three basic methods of automatic human identification: "1. By something he knows or memorizes. 2. By something he carries. 3. By a personal physical characteristic.". This idea was popularized through a widely quoted 1972 paper (Beardsley [2]). Future papers would rephrase this as: "what you are; what you have; what you know".

NBS sponsored a "Controlled Accessibility Workshop" in December of 1972, which considered all forms of API and concluded (Reed and Branstad [54]) that "there is a wide scope for further research. To develop methods of measuring the effectiveness of hardware/software against deliberate attack; To develop the methodology of a security measurement metric; To define precisely the measurements which determine the overall security of a system; To establish procedures for applying the security metric". In 2006, we still struggle with these same issues.

The year 1974 was a breakthrough year for what was now being referred to as "Automated Personal Identification", which had come to include hand geometry and signature recognition. The University of Georgia (US) had begun using hand geometry in their dormitory food service areas (Floyd [17]). Both the Stanford Research Institute in the US and the National Physical Laboratory in the UK had begun working on signature recognition systems.

The Calspan Corporation had begun to actively market fingerprint identification systems to both government and corporate customers, promising "truly reliable personal identification for use in: Airport maintenance and freight areas; Information storage areas; Hospital closed areas, drug storage areas; Apartment houses, office buildings, coed dorms; Prison areas; Computer terminal entry and access to information" (Calspan [6]). About this same time, the Rockwell Corportation founded a fingerprint technology development group, later to become Printrak, and untimately a division of Motorola.

In 1974, the U.S. Air Force announced an ambitious program for a military-wide biometric identification system for base access control called "Base and Installation Security System" (BISS) (Messner, Cleciwa, Kibbler and Parlee [46]). A test bed for evaluation of fingerprint, handwriting and voice recognition systems was created and testing was begun. The Air Force considered methods for hardening the system against forgery attacks and the possibility of combining biometric methods to reduce error rates. The system was intended for production at the end of 1981. This work is credited to both U.S. Air Force Headquarters, Electronic Systems Division and the Mitre Corporation. Ultimately, the world-wide system was never deployed, but the US military does to this day use biometrics for access control on a local basis at facilities throughout the world.

A most amazing short book was written in 1974 by D. Raphael and J. Young at the Stanford Research Institute, International (Raphael and Young [53]). They state:

The technology is available today to provide semiautomated and fully automated personal identification systems for use in business transaction and access control. How fast the transition will occur from manual to semiautomated systems in the latter 1970s to fully automated products in the 1980s will depend largely upon developments in the product features, multiple measurement systems, and technical and economic feasibility.

This interest in 'multiple measurement systems' and the BISS work on combining methods presage by 30 years what we now refer to as 'multimodal' biometrics.

The profusion of API systems without common standards also concerned Raphael and Young.

A standards agreement would have to be a precursor to acceptance by merchants since few could afford to install the multiplicity of API products corresponding to the multiplicity of charge cards they accept.

And concern over privacy issues with the use of API was developing.

A recent study indicates that one-third of the American public is already uneasy about the proliferation of databanks and computer accessible files that contain more or less personal information. Many of these people will voice a fear that API systems will contribute to the mammoth cross-indexing capability – the capability of gathering more and more facts from previously prepared, but separate files.

They further noted that

[i]n one study in a retail outlet, some established customers resented the need to 'prove' their identity by submitting to a thumbprint process each time a purchase was made. (In order to improve user acceptance, one marketeer has begun promoting thumbprints as 'thumb signatures'.)

The potential for biometric forgeries was also well understood by 1974:

Depending upon the system design, most forms of API will be vulnerable to defeat by mimicry (forgery), file fraud, or both. The threat of forgery will be greatest for systems that depending partly or wholly on behavioral measurements, such as handwriting or voice identification. Mimics who learn to duplicate or copy the learned behavior of another person may be authorized certain privileges by the system.

The Mitre Corporation began a formal test program in 1975, gathering over a thousand signature, fingerprint and voice samples on over 200 volunteers (Fejfar and Myers [16]; Haberman and Fejfar [20]). Error rates and transaction times are given in Table 10.1.

In a subsequent paper (Fejfar [15]), they 'fused' the data to look at error rate improvement under both 'AND' and 'OR' fusion logic. These are the first multimodal biometric test data I have found.

The annual International Conference on Acoustics, Speech and Signal Processing (ICASSP) was started in 1976 under the Institute of Electronic and Electrical Engineers (IEEE) and the *Transactions on Acoustics, Speech and Signal Processing* (later to split into the *Transactions on Signal Processing* and the *Transactions on Speech and Audio Processing*) was started in 1974. These provided communications vehicles for the increasing activities within the speaker recognition community.

By 1977, API had advanced to the point that NBS saw fit to publish formal *Guidelines on the Evaluation of Techniques for Automatic Personal Identification* as a *Federal Information Processing*

Table 10.1. Mitre test results (1975)

	Speaker verification		Signature		Fingerprint	
	FRR (%)	FAR(%)	FRR(%)	FAR(%)	FRR(%)	FAR(%)
Lab	0.2	4.4	3.2	1.7	4.6	2.2
Field	1.1	3.3	1.9	5.6	6.5	2.3
Seconds	6.2		13.5		8.9	

Standard (NBS [48]). This paper was absolutely remarkable in its sophistication and scope. The following API technologies are listed: faces, signatures, fingerprints, hand geometry, voice-prints, muscular–skeletal response patterns, ear features, dental characteristics, foot prints, patterns of the retina. The NBS paper gives details on both 'AND' and 'OR' logic for fusing multiple measures and discusses the possibility of spoofing. Twelve criteria for device evaluation are listed:

1. Resistance to deceit
2. Ease of counterfeiting an artifact
3. Susceptibility to circumvention
4. Time to achieve recognition
5. Convenience to the user
6. Cost of recognition device and of its use
7. Interfacing of device for intended purpose
8. Time and effort involved in updating
9. Processing required in computer system to support identification process
10. Reliability and maintainability
11. Cost of protecting the device
12. Cost of distribution and logistical support

Although retinal recognition was listed in the NBS *Guidelines*, the first US patent was not issued until 1978 (Hill [22]). Signature recognition was advanced in the 1970s by IBM (Herbst and Liu [21]; Liu, Herbst and Anthony [31]) and SRI (as referenced in Crane and Ostrem [12]).

The 1970s concluded with a nice book on *Identification Technologies: Computer, Optical and Chemical Aids to Personal ID* by George Warfel [70].

10.4 'BIOMETRICS' OF THE 1980s

None of the previously discussed works used the term 'biometrics'. This term and the discovery of these technologies by the press came in the 1980s. The first reference I have found to 'biometrics' in the sense of human identification is a 1981 article in the New York Times (Pollack [50]). The article references the use of fingerprint recognition for access control to the computer center at the First National Bank of Chicago and claims that access time has been reduced to 3 seconds.

But the banking world, in general, was not enamored with biometrics. A 1984 article in the *Journal of Financial Service Strategy* (Kuttler [29]) states, "Biometrics promise is long term. Biometric systems that would be cost effective in high-volume transaction environments are still in the early stages of R&D. Entrepreneurs active in such fields as hand geometry and voice print recognition can go only so far with the venture capital resources that are available. Some technology savvy bankers are therefore resigned to continuing relying on the less dependable, memorized personal identification numbers".

In France, the fingerprinting company, Morpho, was founded about 1981. The company, since acquired by Sagem and again by Safran, has become one of the biggest producers of fingerprint systems in the world.

In 1983, the U.S. Department of Energy began formal testing of biometrics at Sandia National Lab (SNL) and the U.S. Department of Defense began testing at the Naval Postgraduate School (NPS). Russ Maxwell and Gary Poock headed up these efforts, respectively. A 1984 report by SNL (Maxwell [40]) gave the prices for biometric equipment, shown here as Table 10.2.

Sandia wrote and published reports throughout the mid- and late-1980s on retina, signature, voice, fingerprint, palm crease, and hand geometry systems (Maxwell [41]; Maxwell and Wright [42];

Table 10.2. 1984 retail prices for biometric technologies

Speaker recognition	US$7,000
Retina	US$50,000
Fingerprint	US$10,000
Finger length	US$7,000

Maxwell [43]), motivated in part by a strong feeling of competition with the continuing biometrics work of Mitre (Maxwell [44]). In about 1985, NPS deployed retinal scanning to control access to their War Gaming Laboratory, one of the first uses of biometrics for access control to a secure U.S. Defense Department facility NPS also gave substantial support in this time period to Recognition Systems, Inc., for the refinement of hand geometry recognition.

In the mid-1980s, the State of California began collecting fingerprints as a requirement for all driver's license applicants. The US Congress became concerned with the problems of enforcing the 'one-driver, one-license, one-record' provision of a 1986 law (USPL [63]) affecting commercial truck drivers and passed a provision in 1988 (USPL [64]) requiring the development of standards for the biometric identification of commercial drivers. Despite this early interest, biometric identification is still not widely used with driver's licensing systems within the United States.

In 1985, George Warfel and Ben Miller joined together to begin publishing the *Personal Identification Newsletter* that continued into 1997, serving as the industry's first source of information on itself. The newsletter also carried regular reader surveys on the future of biometrics. Miller and Warfel also published an annual *PIN Biometrics Industry Sourcebook* that contained listings of the companies in the biometrics business. The *Sourcebook* indicates that by 1989 fingerprint recognition systems were being marketed as peripherals to control access to personal computers (Warfel and Miller [71]).

The first industry organization, the International Biometrics Association, was begun in about the 1986–1987 time period.

In 1987, Sirovich and Kirby published the first paper on the use of principal component analysis for human facial recognition, terming the principal components 'eigenpictures' (Sirovich and Kirby [59]). There was soon an explosion of papers dedicated to the spectral decomposition of faces for the purpose of recognition.

10.5 ORGANIZED ACTIVITIES OF THE 1990s

The 1990s saw the creation of organizations and conferences targeted specifically at the biometrics community. With encouragement from the UK Department of Trade and Industry, the Association for Biometrics (AfB) was founded in about 1991. Interest in the association quickly spread outside of the UK to the rest of Europe. I believe the AfB was the first European biometrics organization in existence.

The first "CardTech/SecurTech" meeting was held in 1991, established by Ben Miller, co-publisher of the *Personal Identification Newsletter*. Although attendance at the first meeting was about 850, by 2000 the annual meeting was attracting close to 10,000 participants.

In 1992, the first immigration system using biometrics (fingerprinting), the Schiphol Airport Travel Pass, debuted in Amsterdam. The system was quickly withdrawn, however, because of concerns over the potential for spoofing with artificial fingerprints. By 1994, the US followed by installing the INSPASS border crossing system based on hand geometry. This system ultimately was used at 9 US airports, until suspended in 2004.

The Research Division of the U.S. National Security Agency (NSA) organized and held the first "Biometric Consortium" (BC) meeting in October of 1992 at Ft. Meade, MD. The original impetus for the creation of the group is credited to Phil Galucci. This was an invitation only meeting open only to US government employees. In the early years, meetings were held about every 6 months. By the March, 1995 meeting, membership in the Consortium had broadened to include government contractors and a few industry representatives and the NSA had started a list serve service for Consortium members over the Internet. Attendance at the 1995 meeting was under 100, but interest in the practical application of biometrics to government problems, particularly in the area of secure physical access control, was clearly broadening. In 1996, the BC was officially commissioned by the US government's Executive Branch (through the Security Policy Board) as the focal point for activities in biometrics within the US government. With the realization within the NSA that biometrics had by now broadened beyond research into widely commercialized technologies, the NSA decided to share control of the Consortium with the Department of Commerce (through NIST) in 1998. NIST opened membership in the Consortium to the general public. By 2002, the annual meetings were attracting well over 1,000 interested individuals from both government and the private sector.

In 1994, the Consortium announced that they would establish a "National Biometric Test Center" (NBTC) and invited proposals for presentation at the 1995 meeting. The NBTC was ultimately established by the BC at San Jose State University in April of 1997. During this period, the test center ran the first carefully-controlled fingerprint algorithm test program, the International AFIS Benchmark, comparing about 4,000 electronically scanned 'file' prints of about 500 individuals to about 4,000 search prints collected about 6 weeks later. Both comparison algorithm and 'Level 1' binning algorithm performance was reported.

Upon conclusion of the contract with the BC in October of 2000, the Center published a "Collected Works" document online, summarizing the Center's contributions (Wayman [72]). The International AFIS Benchmark was picked up by the University of Bologna and renamed the Fingerprint Verification Competition (FVC) (Maio, Maltoni, Wayman and Jain [34]).

At approximately the same time as the US was starting NBTC, the European Commission founded the BIOTEST program at the U.K. National Physical Laboratory (NPL) under Dr. Tony Mansfield. The NPL biometrics lab went on to publish the extremely well received *Biometrics Product Testing Final Report* in 2001 (Mansfield, Kelly, Chandler and Kane [35]).

Through the mid-1990s, Sandia National Lab continued to publish performance test reports on biometric access control devices. The 1991 report (Maxwell and Wright [42]), comparing performance of hand geometry, signature, retinal, speaker and fingerprint devices, is considered a classic and was widely referenced into the next decade. But increasing market competitiveness was leading to increasing pressure on SNL to defend test results both prior to and after publication. One SNL report (Rodrigrez, Bouchier and Rudhle [56]) remains unpublished because of vendor concerns regarding the validity of the test data supplied by SNL test subjects. The 1996 report (Bouchier, Ahrens and Wells [5]), the last openly published, included a response by the tested vendor. Today, SNL continues to test biometrics under contract to the US government, but reports are no longer publicly released.

The SNL tests used 'live' human subjects directly interacting with the tested devices, but in the mid-1990s a new test genre was developing – the technology test – in which pre-collected, stored images from a database were run against competing algorithms of the same technology. By 1994, the Army Research Laboratory was testing facial recognition algorithms in the FERET program (Phillips, Moon, Rauss and Rizvi [49]) and NIST/NSA had begun formal testing of text-independent speaker recognition algorithms (Campbell [7]). FERET has evolved into the recurrent NIST Facial Recognition Vendor Tests (Blackburn, Bone, Grother and Phillips [3]) and the NIST/NSA program continues on an annual basis (Martin, Przybocki and Campbell [37]).

One of the biggest technical advances in the 1990s was the development of iris recognition (Daugman [13]). By around 1997, iris recognition had captured the entire small, but emerging, eye recognition market and retinal recognition systems were no longer commercially available. Other new technologies included facial thermography, which was commercialized briefly in the mid-1990s in the United States, and the first commercially available facial recognition systems, which hit the market about 1997. The access control market for biometric devices, however, continued to be dominated by hand geometry. The hand geometry devices of the 1990s were direct successors to the systems first fielded in the 1970s.

In September 1998 the International Biometrics Association of the 1980s was reborn as the International Biometric Industry Association, a non-profit trade association registered in the United States.

More US legislation was calling for the use of biometrics in government applications. The Illegal Immigration Reform and Immigrant Responsibility Act (USPL [65]) specified the creation of an Alien Border Crossing Card which shall "include a biometric identifier (such as the fingerprint or handprint of the alien) that is machine readable". The Transportation Equity Act for 21st Century (USPL [66]) stated that each Commercial Driver's License issued after January 1, 2001 must "include a unique identifier (which may include biometric identifiers) to minimize fraud and duplication", although the biometric option was not ultimately chosen.

The Thompson Media Publication *IDWorld*, in the Nov/Dec 1999 issue contained an analysis done by Eric Bowman of growth of the biometrics industry throughout the decade of the 1990s. That analysis is displayed as Table 10.3.

According to this analysis, hardware revenues increased 10-fold during the decade while hardware sales (in units) increased 100 fold. Clearly, the 1990s was an amazing decade in the growth of biometric applications.

As previously noted, the vulnerabilities of biometric systems to forgery attempts had been well recognized in the 1970s – the first artifical latex-based fingerprints can actually be traced to 1907

Table 10.3. Biometric industry growth in the 1990s

	Units	Change (%)	Hardware revenue ($M)	Change (%)	Ave. price ($)
1990	1,288		6.6		5,124
1991	1,675	30	7.3	11	4,358
1992	1,998	19	8.3	14	4,154
1993	3,073	54	10.1	22	3,287
1994	4,829	57	12.2	21	2,256
1995	6,450	34	14.7	21	2,279
1996	8,550	33	21.2	44	2,479
1997	28,391	232	33.0	56	1,162
1998	55,000	94	39.5	20	718
1999*	115,000	110	63.2	60	547

*Indicates estimate.

(Geller, Almog, Margot and Springer [18]). Public understanding of spoofing threats advanced rapidly in the late 1990s and early 2000s. Several imaginative studies showed how to trick biometric systems into accepting artificial images or body parts as genuine (Blackburn, Bone, Grother and Phillips [3]; Mastumoto and Mastumoto [38]; Thalheim, Krissler and Ziegler [60]; van der Putte and Keuning [69]). The NSA went on international television to demonstrate research on biometric spoofing countermeasures (CBS [9]).

10.6 THE 21ST CENTURY

Spurred by US legislation calling specifically for use of biometrics in border crossing applications (USPL [67; 68]), the US requested the formation of an international standards committee for biometrics. The committee (ISO/IEC JTC1 SC37) held its first meeting in November, 2002 and has begun drafting standards for biometric vocabulary, interfaces, application profiles, data formats, testing and social concerns.

Internationally, there has been great interest in biometrics for passenger applications. The International Civil Aviation Organization, a United Nations speciality agency, has issued recommendations for use of facial, fingerprint and iris images for border crossing, further enhancing interest in these technologies (ICAO [25]). The Privium program, using iris recognition for frequent travelers at Schiphol airport (Netherlands) has been instituted, and several biometric pilot projects for air travelers in the US, UK, Germany, Malaysia and Australia are now underway or have been recently concluded. Several countries have announced their intention to issue passports with embedded computer chips holding biometric images.

New commercially available technologies have included hand vein geometry and multi-spectral fingerprint imagery systems.

10.7 CONCLUSIONS

Any one of the broadly available technologies (face, speaker, fingerprint) considered in this chapter could have its own technical history written from the contributions of hundreds of talented and engaging scientists. In fact, so many researchers and administrators have participated in the development of biometric technologies over the last 4 decades that it is impossible to pick any single person as the 'grandparent' of biometrics or any single laboratory as its birthplace.

A quick overview of biometric history shows that much of what we consider to be 'new' in biometrics was really considered decades ago. But even today, there is much left to be done. The most efficient route will be to learn from our history, focusing our current efforts on that which is really yet unknown, not unknowingly rediscovering the developments of the past. At this writing in 2006, it is much too early to speculate on what the first decade of the new millennium will ultimately hold for biometrics. It seems clear, however, that the industry will continue to grow and that technical and human improvements to the systems will be made.

ACKNOWLEDGEMENTS

The author would like to thank Bob Allen, Rod Beatson, Joe Campbell, Adolph Fejfar, Ed German, Takeo Kanade, Tony Mansfield, Russ Maxwell, Donald Meagher, Ben Miller, Ken Moses, Nick Orlans, Ton van der Putte, Dawn Stanford, Mitchell Trauring, George Warfel, Larry Wright and James Young for their help in researching this paper.

REFERENCES

[1] B. Atal, *Automatic recognition of speakers from their voices*, Proc. IEEE 64 (1976), 460–475.

[2] C.T. Beardsley, *Is your computer insecure?*, IEEE Spectrum 9 (1) (1972), 67–78.

[3] D. Blackburn, M. Bone, P. Grother, J. Phillips, *Facial recognition vendor test 2000: evaluation report*, U.S. Department of Defense (2001), Available at www.frvt.org.

[4] W.W. Bledsoe, Man–machine facial recognition: Report on a large-scale experiment, Technical Report PRI:22, Panoramic Research Inc., Palo Alto, CA (1966).

[5] F. Bouchier, J. Ahrens, G. Wells, Laboratory evaluation of the IriScan prototype biometric identifier, Sandia National Laboratories, SAND96-1033 (1996). Available at http://infoserve.library.sandia.gov/sand_doc/1996/961033.pdf.

[6] Calspan Corp., Marketing documents (1974).

[7] J.P. Campbell, *Testing with the YOHO CD-ROM voice verification corpus*, **IEEE Proc. International Conference on Acoustics, Speech, and Signal Processing**, 8–12 May (1995), pp. 341–344.

[8] V. Campbell, *Why Johnniac can read*, American Heritage of Invention and Technology 21 (1) (2005), 55–56.

[9] CBS, Inc., National security nightmare, Sixty Minutes II, broadcast Feb. 13 (2001), Available for purchase on video tape at http://store.cbs.com.

[10] C.H. Classen, L.D. Green, *Personal security system having personally carried card with fingerprint identification*, IBM, U.S. Patent 3,383,657 (May 14, 1965).

[11] J.W. Cooley, J.W. Tukey, *An algorithm for the machine calculation of complex Fourier series*, Mathematics of Computation 19 (90) (1965), 297–301.

[12] H.D. Crane, J.S. Ostrem, *Automatic signature verification using a three-axis force-sensitive pen*, IEEE Trans. on Systems, Man and Cybernetics SMC-13 (3) (1983), 329–337.

[13] J. Daugman, *High confidence visual recognition of persons by a test of statistical independence*, IEEE Transactions on Pattern Analysis and Machine Intelligence 15 (11) (1993), 1148–1161.

[14] Faculty Council, University of Texas at Austin, In memoriam – Woodrow W. Bledsoe (2000), Available at http://www.utexas.edu/faculty/council/1998-1999/memorials/Bledsoe/bledsoe.html.

[15] A. Fejfar, *Combining techniques to improve security in automated entry control*, **Proc. 1978 Carnahan Conf. on Crime Countermeasures**, Mitre Corp. MTP-191, May 1978 (1978).

[16] A. Fejfar, J.W. Myers, *The testing of three automatic identity verification techniques*, **Proc. Int. Conf. on Crime Countermeasures**, Oxford, July, 1977 (1977).

[17] J.M. Floyd, *Biometrics at the University of Georgia*, **Proc. CardTech/SecurTech'96** (1996), pp. 429–230.

[18] B. Geller, J. Almog, P. Margot, E. Springer, *A chronological review of fingerprint forgery*, Journal of Forensic Science 44 (5) (1999), 963–968.

[19] A.J. Goldstein, L.D. Harmon, A.B. Lesk, *Identification of human faces*, Proc. IEEE 59 (5) (1971), 748–760.

[20] W. Haberman, A. Fejfar, *Automatic identification of personnel through speaker and signature verification – system description and testing*, **Proc. 1976 Carnahan Conference on Crime Countermeasures**, Univ. of KY, May (1976).

[21] N.M. Herbst, C.N. Liu, *Automatic signature verificaiton based on accelerometry*, IBM Journal of Research and Development 21 (1977), 245–253.

[22] R. Hill, *Apparatus and method for identifying individuals through their retinal vasculature patterns*, US Patent 4,109,237, August 22 (1978).

[23] J. Holmes, L. Wright, R. Maxwell, A performance evaluation of biometric identification devices, Sandia National Laboratories, SAND91-0276 (1991).

[24] IBM Corp., The consideration of data security in a computer environment, Report G520-2169, White Plains, NY (1970).

[25] International Civil Aviation Organization, Machine readable travel documents, technical report: Selection of a globally interoperable biometric for machine-assisted identity confirmation with MRTDs First Edition (2001).

[26] T. Kanade, *Computer Recognition of Human Faces*, Interdisciplinary Systems Research, 47 Birkhäuser, Basel, Stuttgart (1977).

[27] T. Kanade, personal communication (2003).

[28] L.G. Kersta, *Voiceprint identification*, Nature 196 (1962), 1253–1257.

[29] J. Kutler, *Biometric conversion*, Transitions: The Journal of Financial Service Strategy 4 (9) (1984), 16–20.

[30] K.P. Li, J.E. Dammann, W.D. Chapman, *Experimental studies in SV using an adaptive system*, Journal of the Acoustical Society of America 40 (5) (1966), 966–978.

[31] C.N. Liu, N.M. Herbst, N.J. Anthony, *Automatic signature verification system description and field test results*, IEEE Trans. on Systems, Man and Cybernetics SMC-9 (1) (1979), 35–38.

[32] J.E. Luck, *Automatic speaker verification using cepstral measurements*, Journal of the Acoustical Society of America 46 (4) (1969), 1026–1031.

[33] R.C. Lummis, A. Rosenberg, *Test of an automatic speaker verification method with intensively trained professional mimics*, Journal of the Acoustical Society of America 51 (1972), 131(A).

[34] D. Maio, D. Maltoni, J. Wayman, A.K. Jain, *FVC2000: Fingerprint verification competition 2000*, **Proc. 15th International Conference on Pattern Recognition**, Barcelona, September 2000, Available at www.csr.unibo.it/research/biolab.

[35] A.J. Mansfield, G. Kelly, D. Chandler, J. Kane, Biometric product testing final report, National Physical Laboratory, London, March 19 (2001), Available at www.cesg.gov.uk/technology/biometrics.

[36] A. Martin, M. Przybocki, G. Doddington, D. Reynolds, *The NIST speaker recognition evaluation – overview, methodology, systems, results, perspectives*, Speech Communications 31 (2000), 225–254.

[37] A. Martin, M. Przybocki, J. Campbell, *The NIST speaker recognition evaluation program*, **Biometric Systems: Technology, Design and Performance Evaluation**, J. Wayman, A. Jain, D. Maltoni and D. Maio, eds, Springer-Verlag, London (2005).

[38] T. Matsumoto, H. Matsumoto, *Impact of artificial 'Gummy' fingers on fingerprint systems*, Proc. SPIE 4677 (2002).

[39] A.J. Mauceri, Technical documentary report for feasibility study of personnel identification by signature verification, North American Aviation, Inc., Space and Information Systems Division, SID65-24, accession No. 00464-65 (1965).

[40] R.L. Maxwell, General comparison of six different personnel identity verifiers, Sandia National Laboratory (1984).

[41] R.L. Maxwell, *The status of personnel identity verifiers*, **Proceedings of the INMM 16th Annual Meeting** (1985).

[42] R.L. Maxwell, L. Wright, *A performance evaluation of personnel identity verifiers*, **INMM Proceedings**, Sandia National Laboratories, SAND87-0977C (1987).

[43] R.L. Maxwell, *Identity verifier performance*, Sandia National Laboratories, SAND88-0441c (1988). Presented at the Smart Card Applications and Technologies Conference in Atlantic City October 14 (1987).

[44] R.L. Maxwell, personal communication (2005).

[45] G. McAlvey, *A review of project SEARCH activities in identification 1969–1974*, Identification News, International Association of Identification (1974).

[46] W.K. Messner, G.A. Cleciwa, G.O. Kibbler, W.L. Parlee, *Research and Development of Personal Identify Verification Systems*, **Proc. 1974 Carnahan and International Crime Countermeasures Conference**, University of Kentucky, April 16–19 (1974).

[47] National Academy of Sciences, *Report of working group 53* (James L. Flanagan, chairman) National Research Council Committee on Hearing, Bioacoustics and Biomechanics, Research on Speaker Verification (1971).

[48] National Bureau of Standards, *Guidelines on the evaluation of techniques for automated personal identification*, Federal Information Processing Standard Publication, Vol. 48 (1977).

[49] P.J. Phillips, H. Moon, P. Rauss, S. Rizvi, *The FERET evaluation methodology for face-recognition algorithms*, **Proc. IEEE Conference on Computer Vision and Pattern Recognition**, San Juan, Puerto Rico (1997).

[50] A. Pollack, *Technology: Recognizing the real you*, New York Times, Financial Desk, Late City Final Edition, Section D, Page 2, Column 1, September 24 (1981).

[51] R. Potter, G. Kopp, H. Green, *Visible Speech*, Van Nostrand, New York (1947).

[52] S. Pruzansky, *Pattern-matching procedure for automatic talker recognition*, Journal of the Acoustical Society of America 35 (3) (1963), 354–358.

[53] D.E. Raphael, J.R. Young, **Automated Personal Identification**, SRI, International, Menlo Park, CA (1974).

[54] S. Reed, D. Branstad, Controlled accessibility workshop report, National Bureau of Standards, Tech. Note 827 (1974).

[55] T.W. Rekieta and G.D. Hair, *Mimic resistance of speaker verification using phoneme spectra*, Journal of the Acoustical Society of America 51 (1972), 131(A).

[56] J.R. Rodrigrez, F. Bouchier, M. Rudhle, A performance evaluation of biometric identification devices, Preliminary Draft (Unlimited Release), SAND93-1830, Sandia National Laboratory, Albuquerque, NM (1993).

[57] A. Rosenberg, *Listener performance in a speaker-verification task with deliberate impostors*, Journal of the Acoustical Society of America 51 (1972), 131(A).

[58] A. Rosenberg, *Automatic speaker verification*, Proc. IEEE 64 (1976), 475–487.

[59] L. Sirovich, M. Kirby, *Low-dimensional procedure for the characterization of human faces*, Journal of Optical Society of America 4 (3) (1987), 519–524.

[60] L. Thalheim, J. Krissler, P. Ziegler, *Biometric access protection devices and their programs put to the test*, C'T Magazine 11 (2002), 114, Available at www.heise.de/ct/english/02/11/114.

[61] M. Trauring, *On the automatic comparison of finger ridge patterns*, Nature 197 (1963), 938–940.

[62] M. Trauring, Automatic comparison of finger ridge patterns, Report No. 190, Hughes Research Laboratories (1961) (Rev. April 1963).

[63] U.S. Public Law 99-570, Commercial Motor Vehicle Safety Act (1986).

[64] U.S. Public Law 100-690, Truck and Bus Safety and Regulatory Reform Act (1988).

[65] U.S. Public Law 104-208, The Illegal Immigration Reform and Immigrant Responsibility Act (1996).

[66] U.S. Public Law 105-178, The Transportation Equity Act for 21st Century (1998).

[67] U.S. Public Law 107-56, The USA-PATRIOT Act (2001).

[68] U.S. Public Law 107-173, The Enhanced Border Security and Visa Entry Reform Act (2002).

[69] T. van der Putte, J. Keuning, *Biometrical fingerprint recognition: Don't get your fingers burned*, **Proc. IFIP TC8/WG8.8, Fourth Working Conference on Smart Card Research and Advanced Applications**, Kluwer Academic Publishers (2000), pp. 289–303.

[70] G. Warfel, **Identification Technologies: Computer, Optical and Chemical Aids to Personal ID**, Charles C. Thomas, Springfield, IL (1979).

[71] G. Warfel, B. Miller, **Biometric Industry Sourcebook**, Warfel and Miller Publishing, Washington, DC (1989).

[72] J.L. Wayman (ed.), **National Biometric Test Center Collected Works 1997–2000**, San Jose State University, San Jose, CA (2000), Available at www.engr.sjsu.edu/biometrics/nbtccw.pdf.

[73] J. Wegstein, Automated fingerprint identification, U.S. National Bureau of Standards, Tech. Note 538 (1970).

[74] J. Wegstein, Manual and automated fingerprint registration, U.S. National Bureau of Standards, Tech. Note 730 (1972).

[75] J. Young, personal communication, February 15 (2006).

PART 3

COMMUNICATION SECURITY

11

THE RISE OF CRYPTOLOGY IN THE EUROPEAN RENAISSANCE

Gerhard F. Strasser[*]

Dept. of Germanic and Slavic Languages and Literatures
The Pennsylvania State University
University Park, PA 16802, USA

Contents

Abstract

After a brief overview of some early cryptological inventions and the high quality of Arab cryptology, this chapter analyses the growing need for secure communication in Europe from about 1400 to 1650. It focuses on the important innovations in Italy, Germany and France in the 1500s, considers England and Spain in the 1600s, and presents the interconnection of mid-17th-century universal languages based on mathematical–combinatorial principles with cryptology, perhaps the most interesting development in this period.

Keywords: Arabic cryptology, Renaissance cryptology, nomenclators, Alberti, cipher disk, Trithemius, square table, Silvestri, codes, Bellaso, Cardano, Porta, Vigenère, "tableau", "Gustavus Selenus" (=Duke August the Younger of Brunswick-Lüneburg), Schwenter, Roger Bacon, Chaucer, Francis Bacon, biliteral cipher, Wilkins, musical cipher, Wallis, Morland, Falconer, Viète, Rossignol, Great Cipher, Comiers, Spanish cryptography, *Despacho Universal*, Bermudo, arithmetical nomenclator, universal languages, Becher, Kircher, Schott.

---
[*] Address for correspondence: Piflaser Weg 10 B,
D-84034 Landshut, Germany, E-mail: gfs1@psu.edu

11.1 THE ANCESTRY

11.1.1 Classical and medieval origins

Any discussion of the rise of cryptology in the Europe of the Renaissance period cannot consider this

15th-century development in a vacuum.[1] It is true that the increasing power of the Italian city states and a growing papal authority may have hastened the perceived need for modern cryptographic methods, but this phenomenon has to be seen from an historical perspective. It is also true that the various cryptographic methods developed in classical antiquity had been mostly lost during the Middle Ages. For this reason, it is interesting to see how gradually and over a period of perhaps 150 years early modern cryptologists became increasingly aware of classical western materials even though the entire body of advanced Arabic-Islamic cryptology remained inaccessible, as we shall see.

11.1.2 Some classical methods slowly rediscovered in the Renaissance

Most contemporary handbooks of cryptology will include a brief overview of the history of this discipline and will trace its origins back to Egyptian hieroglyphic encipherments, to India, Mesopotamia, and Biblical times. The Old Testament contained instances of what David Kahn terms "protocryptography" as the element of secrecy may well have been lacking in the three methods used [56: 76–79]: In the simplest substitution system, the *atbash* version, the last letter of the Hebrew alphabet replaced the first, and vice versa; the more sophisticated *albam* and *atbah* systems refined these substitutions and, in the case of *atbah*, involved Hebrew numbers. Nonetheless, these systems – while known among medieval Rabbinical scholars – did not have an impact on western cryptology before 1500.

This is also true of Greek contributions in the field that were reported by various classical historians, most of whom were not then available to early western cryptologists. Although recent research has not found any contemporary Greek sources [58], Roman authors – beginning with Cicero (106–43 BC) – considered the Spartan *skytale* a cryptographic device. Their – apparently erroneous – assumption was that such a "stick" or

"rod" was used by the warlike Spartans as far back as the fifth century BC, which would make the *skytale* the first cryptographic tool. It was supposed to be an ingeniously simple system, a wooden staff around which a strip of leather, cloth, or parchment was tightly wound in a spiral. The sender then wrote the secret message on the rounds of material along the length of the staff and unwound the strip. The letters, now disconnected, were scrambled or transposed and made no sense. The strip of material was then entrusted to a messenger, who might disguise it as a belt, with the letters turned to the inside. It was important that the recipient of the message rewrap the strip around a skytale of the very same diameter as the one used by the sender, in which case the words reappeared in their original order and could be read off. As convincing as this technique would seem, such a description cannot be found in the writings of contemporary Greek historians; the *skytale* may not have been anything "more than a piece of leather or parchment attached to a stick. The plaintext message was written on the leather or parchment which was then wound around the stick for easy transport", Thomas Kelly concludes after an exhaustive examination of the available Greek source materials [58: 259–260].

And while there was hardly any knowledge or use of the skytale in medieval or early modern cryptology, a dim notion of the two main categories of classical ciphers survived: Transposition ciphers (which rearrange the order of letters in a message, as was the case of the skytale) were still virtually unknown although a vague recollection of substitution ciphers (which replace letters or groups of letters with other letters or groups) survived.[2] Even the occasional monk or scribe who played with

---

[1] For the purposes of this analysis, the period under consideration – the Renaissance in central Europe – ranges from about 1400 (at least in Italy) to the 1660s. Interesting later developments are sometimes included.

[2] The term "cryptology" is used to designate the science of both making ciphers (namely "cryptography") and breaking them, "cryptanalysis". While the term cryptography has been used over the centuries, a related one, steganography – the art and science of writing hidden messages in a way that no one other than the intended recipient knows of the existence of such communication (for instance, a message written in invisible ink) – can primarily be found in the 16th and 17th centuries. Bauer [14: 9–13] differentiates between two forms of linguistic steganography, namely (1) the composing of a secret message and sending it as a piece of open communication under the assumption that nobody would consider it an encrypted message ("an open code") and (2) the preparation of a written or printed message where minute differences in writing, letter

a simple substitution system such as the one invented by and named for Caesar (who used it when he wrote to Cicero and other friends), a three-step shifted alphabet (D for *a*, E for *b*),[3] would have to be considered an exception. To add one more example from this period: The actual purpose of the so-called Tironian notes – originally devised in Roman times as a method for taking shorthand – became lost in the early middle ages; they were sometimes used in medieval royal documents as handy methods to identify or disguise signatures. Together with other signs, Pope Silvester II (999–1003) created a syllabary for his personal secret communication that included such notes [38: 54, 55]. By 1500, the true origin and meaning of these Tironian notes was completely forgotten, as the German abbot Trithemius reported. The early disappearance of this entire, important body of knowledge has been well researched and does not require further discussion here [56: 71–93].

11.1.3 Arabic-Islamic origins

While awareness of classical cryptology had thus been lost in central Europe during the Middle

Ages, manuscripts newly discovered in Turkey and the Middle East have demonstrated the advances in cryptology during the first three centuries of the Islamic civilisation, in other words, between 700 and 1000 AD [5]. In a recent article, Ibrahim A. Al-Kadi has highlighted areas that were particularly relevant to the creation of such means of concealed communication, among them the fields of translation and linguistic studies; the need of an efficient administrative organisation in the ever-growing Islamic empire; a high level of public literacy that increasingly necessitated the protection of sensitive materials; and the great advances in mathematics leading to, among other things, the foundation of algebra. One of the earliest writers on cryptology was Ismaìl al-Kindì (801–873), an encyclopedic author who composed the oldest existing treatise in this field. In his *Treatise on Cryptanalysis* [6], al-Kindì dealt with techniques of cryptanalysis culminating in the first description and use of statistical techniques needed to solve cryptographic materials. He also described the major types of cipher systems and analysed components of the Arabic language, such as phonetics, syntax and letter frequency, which led him to propose the earliest statistical examination of an Arabic text [6: 57, 58, 124–126, 166–170]. Later writers introduced further important concepts in an analysis of this kind, such as the minimum length of a text required to unambiguously solve a cryptogram. Some time before the decline of the Arab-Islamic civilisation set in after 1400, Ibn ad-Durayhim (1312–1359/62), perhaps the last great Arab cryptologist, gave a detailed description of eight cipher systems with numerous variations that culminated in a discussion of substitution ciphers leading to the earliest suggestion of a "tableau" of the kind that two centuries later became known as the "Vigenère table".[4] Ad-Durayhim's material [54] formed the basis for the section on cryptology in an enormous, 14-volume Arabic encyclopedia that was completed in 1412, where it appeared under the heading, "Concerning the concealment of secret messages within letters", which was part of the section "On the forms of correspondence" [56: 95–99].

---

font, or the insertion of dots or pin pricks would contain the code ("semagram"). The second method, while popular, can result in rather suspicious looking pieces and is much less secure.

The term steganography fell into disuse, but was resurrected in the electronic age as computer-generated secret communication is again embedded in data that appear perfectly harmless. Steganography must not be confused with stenography, namely shorthand. – In the August 11, 2005, *Manchester Guardian*, Ian Sample reported how the Christmas Eve, 2003, "high risk" alert issued by the US Department of Homeland Security was based on erroneous interpretation of the Arab al-Jazeera TV network's broadcasts. The alert that caused the grounding of nearly thirty international flights was based on steganographic analysis of "the equivalent of faces in clouds – random patterns" that the CIA analysts thought contained hidden terrorist messages interpreted as targeted flights and buildings [84].

[3]I am using capital letters for PLAINTEXT, lower-case italics for the *ciphertext*, and normal type for any key. Exceptions being the example taken from the Vatican archives (see Section 11.2.1) and the section on Leon Battista Alberti's cipher disk (see Section 11.2.2.1). The Vatican example is inconsistent; Alberti used capital letters for the ciphertext, lower-case letters for plaintext – which have been retained in this case. In some other instances, an author's differing use of such lettering has been retained in the reproduction of source materials.

[4]For a discussion of the Vigenère table see Section 11.2.2.7.

This valuable body of information lay buried in Islamic archives for centuries and is only now coming to light. It is clear that this loss of material, along with that of much of the classical knowledge of cryptology, was a major setback to this field, which began to re-establish itself in central Europe in the fourteenth century. It took Europe several hundred years to reach the sophistication that some of the Arab-Islamic authors had displayed as early as 850 AD.

11.2 EARLY INVENTIONS – 14TH AND 15TH CENTURIES

11.2.1 Preliminaries

The increasingly sophisticated communication in the northern Italian city states led to the development of simple methods of concealment. As early as 1226, Venetian clerks (there was no cipher secretary yet) substituted dots or horizontal crosses for vowels in letters to ambassadors, and by the end of the 14th century, other cities devised passable substitution ciphers. The earliest modern code can be found in correspondence between the Vatican and its nuncios, composed some time after 1330. It consists of a list of name-equivalents that fulfil the requirement of secrecy and have the added advantages of abbreviation [7; 68: 11–12]:

A	*Rex*	B	*ecclesia*
C	*charissimus*	D	*papa*
⋮	... ...		
I	*Petrus Duez*	q	*marescalcus papae*
e	*dominus de insula*	f	*dispensier*
x	*rex Boemie*		

Over time, individual syllables were identified with code groups, and a cipher alphabet was added allowing for the encryption of words or names not found in the code list. The structure of such nomenclators – half code list, half cipher alphabet[5] – remained the same from the Italian Renaissance all the way to about 1850.

Not only did Venice and other city states develop a certain cryptologic mastery and began employing special cipher secretaries, but so did other countries: France, in particular, established a tradition of high-level cipher security while other courts were sorely lacking in this regard. As late as 1605, Matteo Argenti, the papal cipher secretary, wrote in his elaborate manual of cryptology that he considered the danger of the solution of the Vatican's secret communication minimal at northern courts such as the emperor's, Poland's, Sweden's or in Switzerland – while he suggested great care when devising ciphers for use in France, England, Venice, or Florence [68: 54–65, 65–113].

11.2.2 Major 15th and 16th-century specific inventions

11.2.2.1 *Leon Battista Alberti.* One of the earliest cryptographic inventions in the West owes its development to a suggestion made around 1465 by Leonardo Dato, the papal secretary, who was concerned that he had to have other members of the curia become involved in cryptologic matters as he himself did not sufficiently master such communication. For this reason, he asked his friend, the Florentine architect, artist and art historian Leon Battista Alberti (1404–1472) – one of the towering figures of the Italian Renaissance [Grafton] – to devise a cryptographic method that he, Dato, could use without the help of others. And indeed, Alberti wrote *De componendis cyfris* in 1466/67 [1: 65–159; 68: 24–30, 125–141], a 25-page manuscript that constitutes the oldest surviving text on cryptanalysis in the West [69: 51].[6] David Kahn does

---

[5] In its earliest form, a nomenclator consisted of a relatively short list of names that could appear in the secret communi-

cation of a limited group of correspondents, along with letters or – some time later – code words representing the names. Thus, in the early example from the Vatican cited above, A would stand for *Rex,* x for *rex Boemie.* Some time later, in newer nomenclators, code words were introduced; TULIP could mean *Rex,* ROSE *rex Boemie.* An even later development was the introduction of a monoalphabetic cipher in addition to the list of names in order to encipher names or words not included in the original list. Ultimately, the list of code words could grow from a few dozen to several thousand, and the substitution alphabet might employ multiple substitutes, or homophones [73: 196–197].

[6] The Latin text – collated from all known manuscripts – has recently been translated into Italian and English [2; 3].

not think that Alberti developed the surprisingly detailed information on frequency analysis completely on his own [56: 126–130], and in fact some of these ideas are tantalisingly close to the Arab-Islamic treatises mentioned earlier – but Alberti could hardly have had any detailed knowledge of them, after all.

Alberti is perhaps not as well known for his cryptanalytic suggestions as he is for the famous cipher disk that he devised to make this very process much more secure (see Fig. 11.1).

After having provided an overview of various systems of encipherment, he proposed an invention that he deemed "worthy of kings": He devised a mechanism consisting of two round copper plates of which the larger – outer – is stationary, with the smaller one being a movable disk.[7] Each was divided into 24 equal parts or cells. He then wrote 20 capital letters of the alphabet on the outer circle, omitting H, J, K, U, W and Y as he deemed them unnecessary.[8] The remaining four cells carried the numbers 1 to 4. The 24 cells of the inner disk were inscribed with the 23 letters of the Latin alphabet in lower case (with j, u and w omitted), and the 24th reserved for the "et" symbol (&). The inner disk is then placed on the outer one and fixed with a needle in the centre so that the movable disk can rotate inside the outer circle.

To send a secret message to a correspondent, both the sender and the recipient of the cryptogram had to possess identical disks, as Alberti was sure to point out. To reach polyalphabeticity, the sender would place a given letter of the inner disk (Alberti selected "k") against any letter of the outer circle,

in his example "B". He needed to communicate to the recipient this first letter of the ciphertext and the corresponding letter, thus identifying the setting of the movable disk in relation to the stationary one. From this point on, an entire message could be enciphered with both disks remaining in place.

Yet Alberti went one major step further: He suggested that this correspondence between the two disks (B–k) could be changed a few words into the message, thus creating true polyalphabeticity: By moving the index letter "k" to point at "D" (to use Alberti's example again), all the other fixed letters on the outer disk received a new, different meaning. The plaintext word "no" that originally was enciphered as "FC" in one setting could become "ZE" in another. For each new setting of Alberti's disk, a new cipher alphabet came into play; with 24 settings on a disk, the limit, indeed, was 24 cipher alphabets in the Italian cryptologist's polyalphabetic system [45: 208–211].

But what were the four numbers on the outer disk used for? Meister stressed that Alberti further complicated his invention by introducing yet another first: enciphered code. The Florentine determined that the numbers 1 to 4 of the outer disk yielded 336 permutations in two-, three- and four-digit groups; his *Tabula numerorum* begins with the groups

11　21　31　41　111　121　131　141　211 ...
12　22　32　42　112　122　132　141　212 ...

and ends with

1441	2441	3441	4441
1442	2442	3442	4442
1443	2443	3443	4443
1444	2444	3444	4444	[1: 30–33].

Unlike the letters, the numerical code agreed upon between two correspondents did not and could not change; it could include entire phrases, such as the one used by Alberti for the number 12: *Naves quas polliciti sumus militum frumentoque refertas paravimus* (We have readied the promised ships and provided the troops and grain) but also refer to individuals – 132 might mean "rex" (king), 322 "papa" (pope). In a second step, Alberti enciphered these numbers according to whatever cipher

---

[7] The idea of using a disk for cryptologic purposes may have come from the *Ars inventiva veritatis* of the medieval Catalan philosopher Ramon Lull (ca. 1232–1315), where Lull proposed an "art of finding the truth" through the symbolic use of letters which stand for philosophical concepts [75: 680; 100: 58–60]. (Lull, in turn, may have had the idea from the *Sefir Yezirah* [The Book of Creation], one of the most important books of the cabbala.) Manuscripts of Lull's treatise were available in the 15th century. David Kahn saw in Lull the most plausible source for Alberti's idea of polyalphabetic substitution [55: 122–127].

[8] Since Alberti's cipher disk uses CAPITAL letters for *plaintext* on the outer wheel and lower case for he *ciphertext* on the inner disk, the convention as stated in footnote 3 above will not apply in this instance.

XVII
Formula

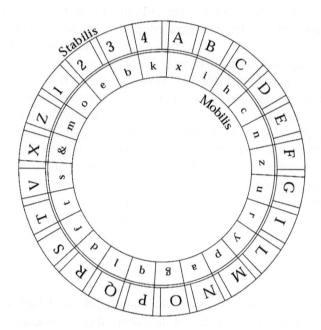

1   Forma seu exemplum Rotae et
tabellarum scilicet stabilis et mobilis,
quae supra à nobis descripta et
declarata fuit hic sequitur. *2ª m.*
exhibit tit. tantum Ch ·
figura: *om.* Co x; *tacent* La Pa;
*in ultima chart*a y*; *2ª m.* Ch

2   Mobilis: Stabilis - Mobilis *tantum* Fe

3   *Circulus stabili*s
     A B C D E F G I   L M N O P Q R S   T V X Z 1 2 3 4   *codd.*; *deleta* H Ch;
     A B C D E F G H I   K L M N O P Q R S   T V X Y Z **2** Va;
     A B C D E F G H I   K L M N O P Q R S   T V X Y Z - Mr

     *Circulus mobili*s
     om. Ba;
     u s q o m k h f   d b A c e g i l   n p r t   x z & y Ar Fe Va Lt;
     V S Q O M K H F   D B A C E G I L   N P - R T X Z 7 Mr;
     X I H C N Z V R   Y P A G Q L D F T S & M O E B K R^2;
     z y x u r o n m i   1 h g e d c b a & q t   p s f k Ch

4   Tit.: Tabulae numeralium *tantum* x Va;
Sequitur tabula numerorum
superius descriptorum *1ª m.* Ch ·
12-Naves quas polliciti sumus
militi frumentoque paravimus *in*
marg. Ar Fe; *recepit* Ba

Figure 11.1. Leon Battista Alberti's cipher disk.

formula would apply, which means that 12 might become "ZP" in one position and "DE" in another.

While Alberti's short treatise was not published for another one hundred years,[9] manuscript copies existed, and it seems to be clear that Matteo Argenti – who in his own, unpublished 1605 manual [9] suggested that he knew how to solve ciphers similar to Alberti's – was aware of these innovations. Indeed, the cipher disk was in use as late as the early 20th century.[10] Alberti, this intellectual giant of the Italian Renaissance, must be credited with "three firsts", to quote David Kahn [56: 130] – "the earliest Western exposition of cryptanalysis, the invention of polyalphabetic substitution, and the invention of enciphered code", making him "the Father of Western Cryptology". Thus the Florentine anticipated the polyalphabetical encipherment by means of a square table as proposed around 1500 by the German abbot Johannes Trithemius and anew in 1585 by the Frenchman Blaise de Vigenère.

11.2.2.2 *Johannes Trithemius.* While cryptology continued to be refined in 16th-century Italy, where polyalphabetic ciphers became increasingly complicated, the inventions of one cryptologist north of the Alps may have had an even greater – and arguably more immediate – impact than Alberti's. At a time when secret communication in the German lands still relied heavily on nomenclators or monoalphabetic substitution systems – at best using freely invented signs or symbols – Johannes Trithemius (1462–1516) took a major step forward in the field of polyalphabeticity. Since at least his second book dealing with secret communication, the *Polygraphiæ libri sex* of 1518, influenced Italian and French cryptologists soon after its publication, it is necessary to analyse the work of this German abbot and scholar at this point.

Trithemius excelled as a student at the University of Heidelberg, became co-founder of the Rhenish Literary Society, and after having entered the Benedictine monastery at Sponheim in 1483, he was elected abbot some eighteen months later, at age twenty-two. During the next decade, he published important works, hosted some of Germany's first-generation humanist scholars, and was consulted by Holy Roman Emperor Maximilian I. He also became deeply involved in hermetic thinking and esoterism and couched a number of his books in what appeared to be occult language [11: 187–200; 102: 32–63].

Trithemius' two cryptologic works were a case in point: In 1499, he boasted in a letter to a Carmelite friend of his plans for such a publication, the *Steganographia*, and outlined the contents of the first four of a planned eight books. Unfortunately, this summary became public. By 1500, he had completed only two books and parts of the third one before he stopped writing as word spread that this latest project was filled with black magic: It was rumoured to be replete with the names of the seven planetary angels and of spirits taken from the cabbala which would facilitate the transmission of secret communication. In actuality, the systems Trithemius proposed relied on non-mystical vowel-consonant substitutions and/or numerous variations on a relatively simple method in which only certain letters of apparent nonsense words signify the meaning, with other letters being nulls [56: 131]. The 31 chapters in Book I, for example, all follow a two-part set-up: The first part, the (enciphered) "conjuration" of spirits from the cabbala would yield the key to the analysis of the second part, the actual cryptogram, a secret message couched in a purported prayer (see Fig. 11.2).[11]

---

[9] Galland's reference to a 1568 first printing of this *Trattato* [47: 3] was verified by Sacco [83: 7–8]. The treatise was published in an Italian translation in Alberti's *Opusculi morali* [4].

[10] The Royal Dutch Navy used cipher disks in World War I.

[11] One of the cryptographic systems taken from Book I may serve as an illustration [100: 107–110; 102: 40–42]: Each of the 31 chapters stands under the sign and protection of the name of a spirit taken from the cabbala. Associated with this spirit is a *coniuratio*, a cipher that purports to conjure up the resident spirit and is indeed similar to a conjuration. For the decipherment of each of these 31 ciphers, Trithemius set up one overall rule: The first and last words are insignificant; the second, fourth, sixth (etc.) words have to be read in sequence, but only their second, fourth, sixth (etc.) letters are significant. The *coniuratio* of one of the simpler methods – under the protection of *Aseliel* – will elucidate the system. The conjuration reads as follows (with the significant letters capitalised, which is *not* found in the original, where every word is capitalised. The actual text is taken from the corrected version of an auto-

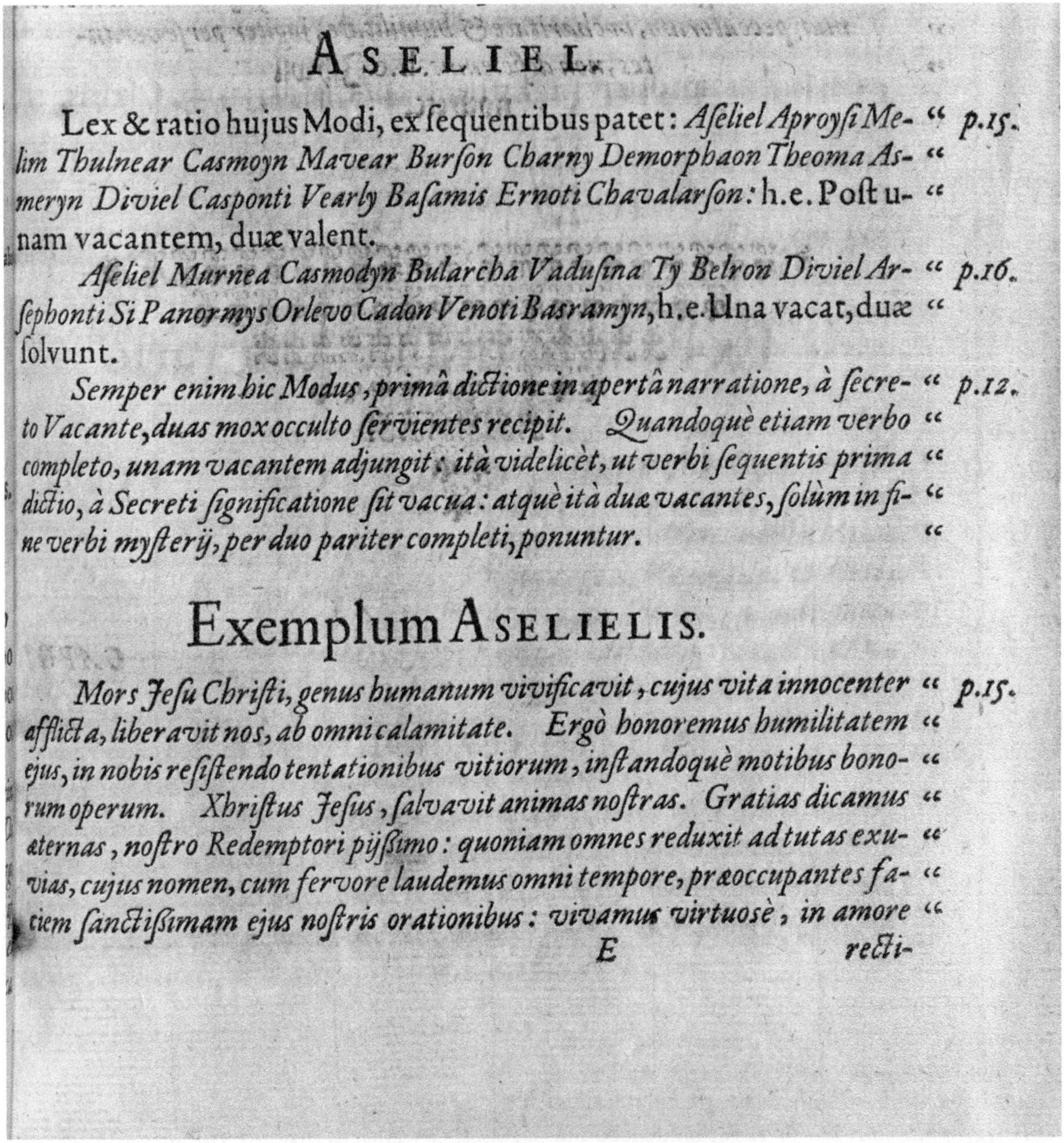

ASELIEL.

Lex & ratio hujus Modi, ex sequentibus patet: *Aseliel Aproysi Me-* " *p.15.*
lim Thulnear Casmoyn Mavear Burson Charny Demorphaon Theoma As- "
meryn Diviel Casponti Vearly Basamis Ernoti Chavalarson: h.e. Post u- "
nam vacantem, duæ valent.

   Aseliel Murnea Casmodyn Bularcha Vadusina Ty Belron Diviel Ar- " *p.16.*
sephonti Si Panormys Orlevo Cadon Venoti Basramyn, h.e. Una vacat, duæ "
solvunt.

   Semper enim hic Modus, primâ dictione in apertâ narratione, à secre- " *p.12.*
to Vacante, duas mox occulto servientes recipit.   *Quandoquè etiam verbo* "
completo, unam vacantem adjungit: ità videlicèt, ut verbi sequentis prima "
dictio, à Secreti significatione sit vacua: atquè ità duæ vacantes, solùm in fi- "
ne verbi mysterij, per duo pariter completi, ponuntur. "

Exemplum ASELIELIS.

   Mors Jesu Christi, genus humanum vivificavit, cujus vita innocenter " *p.15.*
afflicta, liberavit nos, ab omni calamitate.   *Ergò honoremus humilitatem* "
ejus, in nobis resistendo tentationibus vitiorum, instandoquè motibus bono- "
rum operum.   *Xhristus Jesus, salvavit animas nostras.* *Gratias dicamus* "
æternas, nostro Redemptori pijßimo: quoniam omnes reduxit ad tutas exu- "
vias, cujus nomen, cum fervore laudemus omni tempore, præoccupantes fa- "
ciem sanctißimam ejus nostris orationibus: vivamus virtuosè, in amore "
                      E   *recti-*

50   LIBER TERTIUS.   Cap.5.

  " *rectitudinis, abjicientes tumultum mundalium negociorum, justiciæ normam*
  " *sequamur, vicijs noxijs resistamus devotißimè, lachrymisquè negligentias*
  " *abluamus, in maximâ solicitudine, memores futuri judicij, cujus inæstima-*
  " *bilis horror, inferni pœnis, nequaquàm inferior:* h. e. Jch wil noch hint
umb XI/an der Porten clopfen: wart mir/und lais mich in.

Figure 11.2. *Coniuratio* and "Prayer" from the *Polygraphia* of Trithemius, taken from the corrected version in Duke August's 1624 handbook, *Cryptomenytices* [...]. Courtesy Bayerische Staatsbibliothek, Munich.

Such a dual encipherment process under the guise of seemingly occult language was beyond anyone's comprehension and could not but expose Trithemius to the accusation that he invoked black magic. This onus prevented publication of the *Steganographia* for over a century but meant that highly coveted manuscripts circulated among hermetic scholars such as Giordano Bruno or John Dee. It also caused the Catholic Church to place the first imprint of 1606 [108] on the Index of Prohibited Books some three years later. Books I and II

---

graph manuscript printed in the 1624 cryptologic compendium of Duke August the Younger of Brunswick-Lüneburg, who was the first to attempt an interpretation of the *Steganographia* [93: 49–50]):

> Aseliel aPrOySi melim ThUlNeAr casmoyn MaVeAr burson ChArNy demorphaon ThEoMa asmeryn DiViEl casponti VeArLy basamis ErNoTi chavalarson.

Stringing on the significant letters yields the solution to the purported plaintext that is associated with this conjuration: *Post unam vacantem, dvæ* [= duae] *valent* (after one non-significant [word, not letter], two are significant). This rule is to be applied to the second part of the conjuration, which means in this case that the first word of the innocent-looking, prayer- or letter-like text is to be skipped. The second and third, fifth and sixth (etc.) words (to be precise, only their first letters!) are significant. The text – in this example disguised as a prayer – is so convincing that it could be taken at face value and would not be considered a cryptogram:

> *Mors Jesu Christi, genus humanum vivificavit, cujus vita innocenter afflicata, liberavit nos, ab omni calamitate. Ergò honoremus humilitatem ejus, in nobis resistendo tentationibus vitiorum, instandòque motibus bonorum operum. Xhristus Jesus, salvavit animas nostras. Gratias dicamus æternas, nostro Redemptori piißimo*: [. . .]

Stringing along the significant first letters from the second, fourth, sixth (etc.) words (*j* from *J*esu, *c* from *C*hristi, *h* from *h*umanum, etc.) yields the actual cryptogram:

JCH VVIL NOCH HINT VMB XJ / AN DAER P[FORTEN CLOPFEN: VVART MJN / VND LAIS MJCH IN]; or, translated from Trithemius' 16th-century German, "I shall knock at the gate today at 11: wait for me and let me in".

It is quite obvious that nobody in the 16th century could conceive of this combination of conjuration and subsequent letter or prayer as what it really was, a two-part enciphered cryptographic message whose first part provided the key for the actual, innocent-looking cryptogram.

of the *Steganographia* were more or less correctly solved in the years following publication, but the notorious *Liber tertius*, the incomplete third book with its numerical–astrological ciphers, was indeed thought to contain "no cryptography at all", as generations of scholars and cryptologists had theorised [56: 131–132; 102: 41–44]. Finally, and almost 500 years after it had been written, the mathematical analysis independently undertaken by two Americans solved this centuries-old cryptological conundrum [39; 40; 80].

In 1508, when Trithemius drafted the second of his cryptologic works, the *Polygraphiae libri sex*, he was keenly aware of the allegations and accusations that had surfaced after the completed sections of the *Steganographia* had become known. This second book was published in Basel in 1518, two years after his death, and thus became the earliest printed book on cryptology. Trithemius now took a totally different approach: As the title implies, the "polygraphic" (or multiple writing) system basically provided hundreds of (ciphertext) columns of 24 Latin words. Based on Trithemius' alphabet of 24 letters, each ciphertext word corresponded to one of the 24 plaintext letters he used although these words themselves were not always alphabetised. The columns were so ingeniously arranged that they contained lists of nouns, adjectives, verbs, adverbs, participles or conjunctions and allowed for the creation of a syntactically and semantically acceptable "prayer" when the words corresponding to the plaintext letters were selected. The opening few rows of Book I of this "polygraphic" system – dubbed by cryptologists as the "Ave Maria cipher" – will illustrate this second method of Trithemius [107: fols A r°–A v°] (see Fig. 11.3).

If we want to encipher the word ABBA[S], we would select *Deus* from the first list, *clementissimus* from the second, followed by *regens* and *celos*. These words would form the beginning of a Latin prayer: *Deus clementissimus regens celos* [*manifestet optantibus lucem Seraphicam* / cum omnibus / *dilectis* / suis in / *perpetuum* / amen].[12]

---

[12] To the right of some columns, prepositions, pronouns or adverbs were printed vertically in order to provide syntactic continuity. In this sample sentence, they are marked off by slashes and not set in italics.

a	Deus	a	clemens	a	creans	a	celos
b	Creator	b	clementissimus	b	regens	b	celestia
c	Conditor	c	pius	c	consecrans	c	supercelestia
d	Opifex	d	pijssimus	d	moderans	d	mundum

Figure 11.3.

Book II produces 308 similar Latin lists, while in Books III and IV Trithemius once again goes back to the creation of columns of artificial words. Nonetheless, the justification for such a procedure – using "incantatory words" all over, David Kahn feels [56: 135] – is the abbot's intended expansion of his system into any language whatsoever [107: fol. g_1 r°]; this is why he provides linguistically "neutral" lists. In Book IV, in particular, Trithemius proposes to combine such universality with a cryptographic system in which not the first, but the second – or even further – letters of each word of the cover text would carry the secret message. Since these words are once again totally artificial, he suggests that anyone could devise similar lists; correspondents could thus prepare their own, identical sets (of code books, one is tempted to add), thereby obviating the need for the printed *Polygraphia*.[13] Taken as cryptologic devices, however, these systems might not be adequate as their underlying principle could easily be analysed, the abbot felt.

Having pointed out the relative insecurity of the "polygraphic" systems himself, Trithemius proceeded in Book V to introduce one of the fundamental inventions in the field of polyalphabeticity that he himself termed the *Tabula recta* or *Tabula transpositionis* although the system put forth is basically a polyalphabetic substitution.[14] In its simplest form, this "tableau" or square table, as it is now known, prints his normal 24-letter alphabet as the cipher alphabet, but not just once – the same sequence of letters is repeated in twenty-four lines, with each line shifting the first letter by one, thus producing the characteristic diagonal pattern of the original Trithemian square table (see Fig. 11.4).

Each cipher alphabet thus yields a Caesar substitution, but while monoalphabetic systems à la Caesar or Augustus were beginning to be insecure as frequency analysis slowly made its inroads, such a polyalphabetic device with its constant change of the cipher alphabet tremendously enhanced cryptologic security. Trithemius' own example, printed in the *Clavis* and using the first line also as the plaintext alphabet since no separate alphabet was provided, enciphers the Latin plaintext HUNC CAUETO UIRUM, QUIA MALUS

---

[13] It is in the Prefaces and separately paginated Keys (*Claves*) to Books III and IV that the idea first surfaces to use such cryptographic lists for the very opposite purpose for which they were originally intended, namely for open, worldwide communication. This usage forms the basis for Athanasius Kircher's re-interpretation of the cryptologic systems in Trithemius' *Polygraphia* almost 150 years later in his 1663 *Polygraphia nova et vniversalis* [102: 51–53].

In his *Traicté des chiffres* of 1586 [109: fol. 183 v°, fol. 146 v°], the French cryptologist Blaise de Vigenère suggested that Trithemius may have borrowed the idea for the cryptological system of the first two books of the *Polygraphia* from Johannes Reuchlin, his teacher at Heidelberg, later part of his circle of learned friends and the leading Hebrew scholar of his day. Reuchlin reported in Book III of his 1517 *De arte cabalistica* [81: fol. lxv r°, fol. lxviii v°] that he himself had made use of similar systems, which were based on the *notariacon*, the third part of the Jewish cabbala, a system in which a single letter can stand for an entire word, or new words are constructed by joining together the first letter (sometimes, other letters) of various words from a verse of the Bible. In this manner, new words can be extracted from a biblical text [102: 60, 61].

In one of the rare contemporary published documentations of the successful use of a cipher, Vigenère also reported [109: fols 183 r°–183 v°] that he was in the diplomatic service in Venice in 1569 when Sultan Selim made secret preparations to attack Cyprus. Since the Sultan feared that the Venetian ambassador might leak this information he required his communication to be unencrypted. At that point, a medical doctor offered to cloak the secret in a "prayer" with the help of the Ave Maria cipher – which was successfully sent to Venice [68: 24, 25].

[14] The final Book VI reproduces a number of authentic or spurious alphabets – including some taken from Beda Venerabilis or Otfried of Weißenburg – but also for the first time lists some thirty Tironian notes that Trithemius had discovered in an incorrectly identified medieval manuscript and recognised as such. He suggested the cryptological use of these letters or symbols in substitution systems.

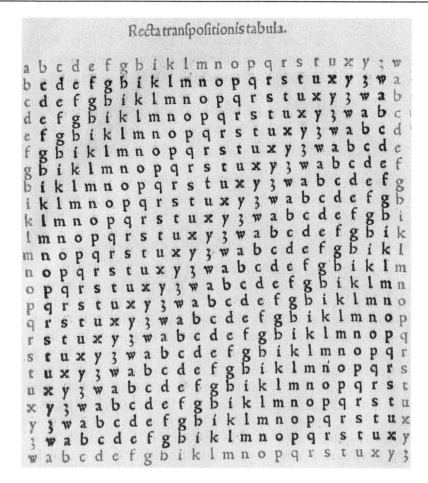

Figure 11.4. Tableau or square table of Johannes Trithemius. Courtesy Bayerische Staatsbibliothek, Munich.

EST [...] (Watch out for this man, because he is evil ...) as *hxpf gfbmcz fueib gmbet gexhsr ege* [...] [107: fols B4 r°–B4 v°].[15]

A comparison with the only other polyalphabetic system invented so far shows that Alberti's cipher disk did not bring into play a new alphabet with each letter; Kahn points out that Alberti's ciphertext would "mirror the obvious pattern of repeated letters of a word like *Papa* ("Pope"), or in English, *attack*, and the cryptanalyst would seize upon this reflection to break into the cryptogram" [56: 136]. In addition, Trithemius' system was the first to make use of a progressive key in which all the available cipher alphabets are exhausted before

any are repeated. And his method was particularly handy under adverse conditions; any soldier in the field could write up such a tableau (and destroy it after use) while it was more difficult to either carry a set of metal cipher disks or cut them out of a piece of paper or cardboard. This polyalphabetic method had an almost immediate and lasting effect on 16th-century cryptology, as we will see in the following. The *Polygraphia* was available much more easily than Alberti's treatise, which only appeared once in 1568, buried in a collection of his works, while Trithemius' cryptologic book was reprinted in 1550, 1564, 1571, 1600 and 1613, was translated into French (though heavily edited) in 1561 (with reprints in 1625 and 1626) [31], and saw this translation boldly plagiarised by a Frisian monk in 1621 [52].

---

[15]Since there is no separate plaintext alphabet, the H of HUNC thus remains an *h*; U becomes *x*, N is *p*, and so forth. Trithemius uses the letter "u" for "v", which his alphabet does not contain.

[Nr.] Signa	i. vel. Bd.	ij. vel. Af.	iij. vel. Dl.	iiij. vel. Cl.	v. vel. Ac.	vi. vel. Ba.
i.	Audio.	Bonum.	Cedo.	Diligo.	Expello.	Fallo.
ij.	Amo.	Bellum.	Confero.	Dormio.	Explico.	Falsum.
iij.	Aspicio.	Benefacio.	Concludo.	Dono.	Extollo.	Fallacie. [!]
...						
viij.	Amitto.	Bene.	Comparo.	Dominus.	Exemptio.	Forenses.
...						
x.	Actio.	Benedicio.	Colo.	Desidero.	Effectus.	Facio.
...						
xv.	Adiuuo.	Beatus.	Compello.	Dux.	Educo.	Fidus.
...						
xix.	Augustus.	Bamna. [!]	Consilia.	Damna.	Executor.	Flores.

Figure 11.5.

11.2.2.3 *Jacopo de Silvestri.* Returning to Italy, one early development frequently overlooked is the invention of Jacopo de Silvestri, a Florentine like Alberti. In 1526, he received the papal privilege to publish a slim book, *Opvs novvm* [95; 112: 5–9] which may well have been influenced by Alberti's work. Silvestri was the first cryptologist to include explanatory sections in Italian in his Latin treatise; he was also the earliest to suggest the use of (what he considered) hieroglyphic symbols and illustrations in cryptography [95: fol. 37 r°]. And he was one of the first authors to give an overview of antique methods of cryptography, beginning with the Greek skytale and then enhancing plain substitution methods by printing a cipher disk very similar to Alberti's without awareness of the quality of encipherment that it could provide [95: fol. 6 v°]. At the end of this overview, he demonstrated the inherent weaknesses of the methods described by showing how they could be deciphered; he even printed lists of vowel-consonant digrams and trigrams for use in cryptanalysis [95: fols 9 v°–11 v°].

Silvestri's most important contribution, though – made very probably without the knowledge of Trithemius' *Polygraphia* that was first published in Basel in 1518 – was the *Regvla componendi cipharam commvnem omnibus*, a cryptologic method designed to include classes of words, the gender and inflexions of nouns but also verb modes, tenses, and conjugations in the encipherment process. In his first example [95: fols 24 r°–25 v°], he suggested using early printed handbooks of orthography such as Tortelli's [105] and

attached the additional designations – his innovation – to the sequential numbers of each word of his sample sentence. In the second, more detailed example, he printed his own sample word lists and identified each column with a Roman numeral. Alternatively, and to increase security, he proposed the use of a capital letter together with one in lower case. The beginning of his own lists appears like this [95: fols 26 r°–27 v°] (see Fig. 11.5).

There is a certain similarity with earlier code lists – even the one going back to 1330 – but Silvestri's second sample sentence, BONA CONSILIA FACIUNT DOMINOS BEATOS [95: fols 27 v°–28 r°], shows the new direction he is taking: When BONUM is identified in the word lists with the signet *ij.i.* and/or *Af.i.*, the author proposes to add the designations for the form of the declination and the gender, namely *g* for nominative plural, *dd* for neuter. CONSILIUM can be found in column *iij./Dl.*, line xix. Since the qualifiers for BONUM also apply to CONSILIA, he attaches them to both. For the form of the verb FACIUNT we have *vi.x.* and/or *Ba.x.* plus the marks *n* for indicative, *s* for the present tense, and ÞÞ for the third person plural. The nominative form DOMINUS appears on line *viij.*, yielding the designation *iiij.viij.* or *Cl.viij.*; BEATUS (*ij/Af*) is on line xv. Both need to be qualified as accusative plural (*k*), masculine, *bb*. The entire sentence, enciphered (using the second columnar identification, namely the letter combination) appears as follows:

Af.i.Dl.xix.g.dd.Ba.x.n.s.ÞÞ.Cl.viij.Af.xv.k.bb.

Broken into word groups for easier reading (something that would not be done in actual cryptologic communication, though), this cryptogram is slightly clearer:

Af.i. Dl.xix. g.dd. Ba.x.n.s.ÞÞ. Cl.viij. Af.xv.k.bb.

Silvestri's method has thus reached a differentiation that was unheard of in the early 16th century; his basic idea – having undergone major modifications – was still found in 19th-century diplomatic communication. As it was, Silvestri's method increased the difficulty of his code lists to such a degree that the "explorers, defenders of the fatherland, pilgrims, merchants, soldiers, architects", and other illustrious persons for whom this work should be of great use – as Silvestri wrote on his title page – did not consider it as such. The very differentiation and specification reached with the help of his additional designators hindered cryptologic communication; moreover, the redundancy of the qualifiers provided unwanted cryptanalytic help. Yet they became an essential element in the universal language systems that were devised some 150 years later by authors and cryptologists such as Athanasius Kircher or Johann Joachim Becher, who will be discussed later.[16]

11.2.2.4 *Giovanni Battista Bel(l)aso.* While Silvestri's invention thus did not have any immediate impact on cryptology, a polyalphabetic device proposed by another Italian associated with the curia, Giovanni Battista Bel(l)aso (1505 to about 1565), became known quite soon. A native of Brescia, he was in the services of Cardinal Carpi. In 1553, Bellaso published his own enhancement of the polyalphabetic cipher in a small and now extremely rare book, *La Cifra del Sig. Giovan Battista Belaso* [sic] [20]. His system – which he apparently devised without the knowledge of Alberti's yet unpublished inventions – was reworked in two later editions: The second, 1555, book [21] "was a refinement of the basic system" [26: 47], while the third and most readily available edition of 1564 – re-using much of the preceding material of keyed reciprocal alphabets – added numerous

variations as far as usage is concerned [19]. The basic invention of Bellaso's, the introduction of the first literal key – which Bellaso named a "countersign" – was part of the earliest, 1553, book. It calls for the use of a keyword or phrase that could be easily remembered or changed; the key was to be placed above the plaintext message [70: 167]. The key letter matched with the plaintext letter determined the alphabet of a (Trithemian-type) tableau or square table that was to be used [26: 40–42; 68: 36–38]:

```
AB  a b c d e f g h i l  m
    n o p q r s t u x y z
CD  a b c d e f g h i l  m
    t u x y z n o p q r  s
EF  a b c d e f g h i l  m
    z n o p q r s t u x  y
GH  a b c d e f g h i l  m
    s t u x y z n o p q  r
IL  a b c d e f g h i l  m
    y z n o p q r s t u  x
MN  a b c d e f g h i l  m
    r s t u x y z n o p  q
OP  a b c d e f g h i l  m
    x y z n o p q r s t  u
QR  a b c d e f g h i l  m
    q r s t u x y z n o  p
ST  a b c d e f g h i l  m
    p q r s t u x y z n  o
VX  a b c d e f g h i l  m
    u x y z n o p q r s  t
YZ  a b c d e f g h i l  m
    o p q r s t u x y z  n
```

The example given with the message L[']ARMATA TURCHESCA PARTIRA A CINQUE DI LUGLIO to be enciphered and the countersign or key "virtuti omnia parent" is as follows, with the cyphertext written over the plaintext (see Fig. 11.6).

The first key letter (v) indicates which alphabet is to be used to encipher the plaintext letter L, namely "S". Plaintext A is to be enciphered by the "I" alphabet, and so on. This system, in particular, affords great flexibility as a cipher secretary could communicate with different correspondents in different keys that they, and they alone, were issued. Some of the additional ciphers included in

---

[16]See Sections 11.3.6.1.3 and 11.3.6.2.1.

S y b o v e y l d a n v o f s z l p i i n c v p n s h m l r n x o i z n r d
v i r t u t i o m n i a p a r e n t v i r t u t i o m n i a p a r e n t v i
L A R M A T A T U R C H E S C A P A R T I R A A C I N Q U E D I L U G L I O

Figure 11.6.

the 1555 and 1564 editions – among them a cipher with nomenclator and homophones, or diagrams with nomenclator – are extremely versatile and secure [26: 44–47]. Contrary to Silvestri's invention, Bellaso's basic cipher of 1553 was used right away and remained an important cryptological system for centuries. It anticipated modern polyalphabetic methods in which several keys are employed and varied at different intervals.[17]

---

[17]One of the other polyalphabetic cipher methods that Bellaso proposed in 1564 was later refined by Porta and reworked quite often: In this example, Bellaso operated without countersign (there are other methods with countersign or index letters) [26: 44–45]. He split a 20-letter alphabet in two half-lines, and – using, for example, the keyword "Saturno" – put its first syllable at the beginning of the first line, with the second syllable opening the second line of the alphabet and showing the remaining letters afterwards:

s a b c d e f g h i
t u r n o l m p q x

A second alphabet is created by moving *turno* one space to the right,

s a b c d e f g h i
x t u r n o l m p q

which means that after six further slides *turno* has arrived at the end of the line:

l m p q x t u r n o

Since each half-line has ten letters and the key word selected, Saturno, only seven, the letters of *turno* keep moving to the right, pushing the remaining letters to the beginning of the line – so that the last, tenth line looks like this:

s a b c d e f g h i
u r n o l m p q x t

During encryption, the first word of a letter will be enciphered with the first alphabet, the second with the second, and so on. If the key is longer than ten letters, the eleventh word is again enciphered by the first alphabet. Bellaso suggests to encipher an *x* by the alphabet of the preceding word in order to mark its end; if the plaintext word contained an X, the enciphered letter should be marked with a dot above.

To further enhance security, Bellaso proposes to change the alphabet after each letter of each plaintext word; at each new word, however, the selection of the following alphabet should be determined by means of an additional key that could either

11.2.2.5 *Girolamo Cardano*. At about the same time as Bellaso's first publication, the Milanese physician and mathematician Girolamo Cardano (1501–1576) wrote a plethora of books that appealed to a broad audience. They ranged from medical subjects all the way to gambling (and the earliest probability analysis), astrology, astronomy and also cryptology. A first set of materials related to cryptology appeared in his 1550 collection of scientific explanations, *De subtilitate libri XXI* [29], which saw numerous – frequently unauthorized – re-editions. The enormous success of this popular work caused the author to follow up with a 1557 expansion, *De rerum varietate libri XVII* [28]. David Kahn has discussed Cardano's most important cryptological invention, the autokey, a system that used the plaintext as a key to encipher itself and then started the key all over from the beginning with each next plaintext word [56: 143–145]. Unfortunately, this ingenious device was flawed in its original form and unusable as it created the possibility of multiple deciphered answers,[18] therefore resting Cardano's fame on the much simpler, so-

---

have been agreed upon in advance or signalled through the first letter of the new word.

[18]In the 1564 revision of his 1553 work, Bellaso described a form of autokey without Cardano's flaws; his autokey involved a mixed alphabet as a prerequisite [26: 50]. Sacco [83: 16–17] lists an ingenious polyphonic cipher of Cardano's that may be an expansion of a system used at the papal court around 1539; it shows Cardano well versed in various cryptographic methods: Using an alphabet reduced to 18 letters (substituting v for u, s for x and z, and marking the letter h [rare in Italian] with an accent), he devised a substitution cipher in which (1) each plaintext letter in the three columns on the right is enciphered with the corresponding capital letter (A, B, C) in the left column. (2) If the plaintext begins with a letter from the three right columns, a null is to be placed to the left of the cipher letter (in this example, the lower-case "b").

called Cardano grille [70: 163–165][19] – in itself not a cryptographic but a steganographic invention that actually was introduced by Silvestri, whom Cardano does not acknowledge [10: 100–101].[20]

This grille consists of a piece of cardboard or metal in which rectangular holes are cut at varying intervals, their height corresponding to that of a line of writing. The encipherer would then write individual letters, syllables or entire words of a secret message through these openings on a sheet of paper placed under the mask. After the removal of this grille, he would fill in the remaining blank spaces with a nondescript cover message. The decipherer has to possess the identical grille, which he would place on the complete text and read the hidden message through the windows. As with other steganographic systems – the prayer in Trithemius' "Ave Maria cipher" that is to hide the enciphered message, or the later musical ciphers being a case in point – the sender of such a message had to devise an innocent-looking cover text that would not automatically betray the existence of a hidden message. Nevertheless, the Cardano grille – which is anything but easy to use – was employed from the 16th well into the 18th centuries in the diplomatic correspondence of a number of countries [56: 145].[21]

The cipher and the example given are as follows:

	a	r	c
A			
	e	n	b
	i	d	g
B			
	o	l	q
	u	m	p
C			
	s	f	t

Plaintext: A T T E N D O   O R D I N I
Corresponding cipher: bA C C A A B B   B A B B A B
Cryptogram: bA–C–C A–A–BBB–A–BB–AB

[19]The Cardano grille is not to be confused with the so-called turning grille.

[20]In modern mechanics, Cardano's fame has survived to this day – he is the inventor of the Cardan gear or shaft . . . .

[21]Karl de Leeuw and Hans van der Meer [37] have analysed a variant of this grille in Dutch archives and determined 1745 as the probable time of its use.

11.2.2.6 *Giovanni Battista (della) Porta.* One generation younger than Bellaso and Cardano, Giovanni Battista della Porta (1535–1615) was a highly gifted and educated Neapolitan who not only founded the first association of scientists, the Accademia Secretorum Naturae, but also wrote poetry and more than a dozen well-received plays. Published at age twenty-three, his first book, the *Magia natvralis* of 1558 [79], dealt with "natural magic" in contrast with the purported black magic of Trithemius' *Steganographia*. It contains an early section on cryptology, *De Ziferis*. This material was expanded in 1563 when *De fvrtivis literarvm notis, vvlgò De Ziferis libri IIII* [78] appeared. In its four books, Porta gave an exemplary overview of the whole field of cryptology as it was known in his day [56: 137–143; 83: 18–20]. Since he was familiar with the entire relevant literature, he made valuable suggestions for cryptanalysts. The "probable word" becomes an important tool in the process when sender, recipient or subject of a message could be surmised, which was feasible in communication possibly issued by the curia or even by merchants where words such as *bank, money, treasure, or coins* could be expected.

Porta's second major addition to the field was the invention of the digraphic cipher, a means by which two letters could be replaced by one symbol: A tableau filled with 20 lines of artificial signs, 20 cells across, and with the alphabet written in the top row and down on the right side would enable a cryptologist to encipher the two plaintext letters P and L, for instance, with the symbol ▸☐◂, C and R with ▷▸, and so on (Fig. 11.7).[22]

In a second step that shows Porta's mastery of the subject he suggested that such digraphic ciphers could be arrived not only at with the help of a square table but also with its conversion to a pair of cipher disks (Fig. 11.8).

The twenty symbols on the inner disk and the Roman numerals on the stationary third circle are an expansion of Alberti's system, thus providing even more random choice and increasing the number of polyalphabets. David Kahn considers this Porta's lasting contribution to cryptology, namely

----

[22]For a mathematical analysis of Porta's tableau, see Bauer [14: 60 ff.].

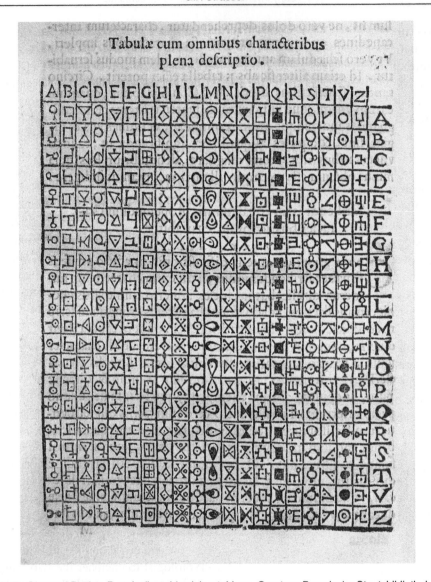

Figure 11.7. Giovanni Battista Porta's digraphic cipher tableau. Courtesy Bayerische Staatsbibliothek, Munich.

his "lamination of existing elements – the letter-by-letter encipherment of Trithemius, the easily changed key of Belaso, and the mixed alphabet of Alberti – into a modern system of polyalphabetic substitution" [56: 141].[23]

11.2.2.7 *Blaise de Vigenère.* Although it would appear that advances in modern cryptology were mostly devised in Renaissance Italy – with the exception of Trithemius – the French diplomat and mathematician Blaise de Vigenère (1523–1596) must be included in this group. While on a two-year mission in Rome around 1550 and on a second sojourn there in 1566 he became familiar with the publications of Trithemius, Bellaso, Cardano,

---

[23]Sacco [83: 18–20] reproduces Porta's enhanced version of the polyalphabetic cipher method devised by Bellaso (see footnote 17) that allows for an even greater personalisation – and security – of the cipher. (Porta nowhere acknowledged Bellaso although Bellaso's basic inventions of 1553 were included in Porta's 1563 work. Vigenère [109: fol. 36 v°] chided Porta for this appropriation of Bellaso's materials.) Sacco considers

Porta's multiple movable disks an ancestor of modern cipher machines of the "Enigma" type. See also Bauer [14: 135–136].

Figure 11.8. One of Giovanni Battista Porta's cipher disks. Courtesy Bayerische Staatsbibliothek, Munich.

Porta and a manuscript of Alberti's treatise.[24] After early retirement from court he returned to his country estate, where he wrote some twenty books – among them the 1586 *Traicté des Chiffres* [109]. In this 600-page tome, Vigenère presented a plethora of material ranging from an overview of cryptology to excursions into the cabbala and alchemy. It did introduce several important ciphers, though, in particular polyalphabetic systems with a variety of

key methods – words, phrases, poetry, or the progressive use of all the alphabets of the Trithemian-type square table chosen, with the Frenchman suggesting mixed alphabets at the top and side [56: 145–150; 96: 46–51].

Vigenère also devised his own autokey system, which – like Cardano's – used the plaintext as a key. But he made it usable with a so-called priming key, a single letter known to both encipherer and decipherer.[25] Unlike Cardano, he did not recom-

---

[24]Vigenère also saw a manuscript of Trithemius' *Steganographia* but could not make sense of it [14: 135–136; 109: fol. 182 r°].

[25]It enabled the recipient to decipher the first letter of the cryptogram, which allowed him to begin decryption. With the

mence his key with each plaintext word (a weakness of Cardano's method) but kept it running on. In a second system the cryptogram itself served as the key after a priming key, which had the advantage that this cryptogram-as-key was incoherent but the serious drawback that the key was actually right in front of the eyes of a cryptanalyst.

It is ironic that these methods – clearly described in the *Traicté* – went virtually unnoticed until the 19th century, when they were reinvented. Vigenère's fame, however, rests on a much simpler system than his autokey methods. The "Vigenère tableau" – as it is known – is yet another enhancement of the Trithemian square table. In modern handbooks, this tableau – further simplified from Vigenère's original, more sophisticated system[26] – is represented by a 26-letter alphabet (for the plaintext) that is written above a square of 26 horizontal alphabets, with each successive alphabet moving one space to the left of the previous one. On the left side, the key alphabet runs down along the 26 lines, basically repeating the first letter of each of these lines (see Fig. 11.10).

In order to bring the full potential of this system into play, a keyword has to be agreed upon between the persons exchanging a secret message. The encipherer would write this keyword above the message and repeat it again and again until each letter in the message is associated with the letter from the keyword. To generate the ciphertext, the key letter above the first plaintext letter would determine the particular row in the tableau which is to be used to find the substitute letter for the plaintext one, and so on.[27]

David Kahn criticises this Vigenère system as "clearly more susceptible to solution than Vigenère's original" [56: 148] and marvels how it could have been considered the "chiffre indéchiffrable" that it has come to be known. And while it is impregnable to classic frequency analysis and can employ a limitless number of keys, even this simplified form was hardly used for the next three centuries – before being re-invented in 1854 by a Bristol dentist and promptly broken by Charles Babbage, one of the mechanical geniuses of the 19th century.[28]

11.2.2.8 *Early cryptanalysts.* While many of the cryptologists discussed so far have included sections on cryptanalysis[29] and have attempted to show how their own systems and innovations added elements of cipher security that cryptanalysts would find hard to break, a number of cryptologists have come down to us primarily as specialists in decryption. Several of them were employed at Italian courts and held functions that would later become known as cipher secretaries. One of them was Cicco Simonetta, a Neapolitan in the service

---

first plaintext letter gained, he possessed the key to decipher the second cryptogram letter – and so on.

[26]Vigenère's least sophisticated full illustration of such a tableau (fold-out after fol. 185) uses a 20-letter (plaintext) alphabet (in red), beginning with "A", under which a second alphabet is printed in black, beginning with the letter "E". The first line of the tableau begins with the digraphs "*m* (red) *b* (black)", which continue in this line in dual alphabetic order ("*n c*", "*o d*", etc.). The second line of the square has the red letter slide to the left by two places (in other words, move from behind the black letter on its left to a position in front of it) and begins with "*n b*", "*o c*" and "*p d*"; the small red letters continue the 20-letter alphabet vertically while the black letters remain the same in each column.
There are two key alphabets written down to the left of the tableau, the rubricated 20 letters from the plaintext alphabet beginning again with "a", while the second (black) alphabet does not open with the same letter "e" as the horizontal line, but with "d". This second alphabet continues down with "e", "f", etc. (see Fig. 11.9).
It is quite clear that this tableau is by far more involved and secure than the square for which Vigenère has been known to cryptologists for centuries – unfairly so, David Kahn suggests [56: 147]. The 1624 cryptologic compendium published by the Duke of Wolfenbüttel, the standard work in the field for the rest of the century [93], prints at least half a dozen variants of Vigenère's tableaus, each one more sophisticated than what is now considered his major innovation.

[27]Simon Singh's example [96: 49–50] uses the following Vigenère square (see Fig. 11.11). He proposes the plaintext DIVERT TROOPS TO EAST RIDGE and uses the keyword "white". Encryption then looks as follows (see Fig. 11.12).

[28]Since Babbage never published this feat (he may have refrained from doing so in the national interest as it seems that the British military put his invention to good use in the Crimean War of 1855–1856 [97: 75]), it was Friedrich Wilhelm Kasiski, a retired officer in the Prussian army, who until the discovery of Babbage's notes in the 20th century became known as the person to have cracked the Vigenère tableau [96: 63–78; 57].

[29]Both Trithemius and Vigenère did not consider cryptanalysis very highly; the latter derided it [109: fol. 12 r°].

```
        A   B   C   D   E   F   G   H   I   L   M   N   O   P   Q   R   S   T   V   X
        E   F   G   H   I   L   M   N   O   P   Q   R   S   T   V   X   A   B   C   D
a   d   mb  nc  od  pc  qf  rg  sh  ti  ul  xm  an  bo  cp  dq  er  fs  gt  hu  ix  la
b   e   nb  oc  pd  qc  rf  sg  th  ui  xl  am  bn  co  dp  eq  fr  gs  ht  iu  lx  ma
c   f   ob  pc  qd  rc  sf  ...
```

Figure 11.9.

```
      A  B  C  D  E  F  G  H  I  J  K  L  M  N  O  P  Q  R  S  T  U  V  W  X  Y  Z
a     a  b  c  d  e  f  g  h  i  j  k  l  m  n  o  p  q  r  s  t  u  v  w  x  y  z
b     b  c  d  e  f  g  h  i  j  k  l  m  n  o  p  q  r  s  t  u  v  w  x  y  z  a
c     c  d  e  f  g  h  i  j  k  l  m  n  o  p  q  r  s  t  u  v  w  x  y  z  a  b
⋮     ⋮
z     z  a  b  c  d  e  f  g  h  i  j  k  l  m  n  o  p  q  r  s  t  u  v  w  x  y
```

Figure 11.10.

```
Plain    A  B  C  D  E  F  G  H  I  J  K  L  M  N  O  P  Q  R  S  T  U  V  W  X  Y  Z
Cipher   b  c  d  e  f  g  h  i  j  k  l  m  n  o  p  q  r  s  t  u  v  w  x  y  z  a
         c  d  e  f  g  h  i  j  k  l  m  n  o  p  q  r  s  t  u  v  w  x  y  z  a  b
         d  e  f  g  h  i  j  k  l  m  n  o  p  q  r  s  t  u  v  w  x  y  z  a  b  c
         e  f  g  h  i  j  k  l  m  n  o  p  q  r  s  t  u  v  w  x  y  z  a  b  c  d
         f  g  h  i  j  k  l  m  n  o  p  q  r  s  t  u  v  w  x  y  z  a  b  c  d  e
         g  h  i  j  k  l  m  n  o  p  q  r  s  t  u  v  w  x  y  z  a  b  c  d  e  f
         h  i  j  k  l  m  n  o  p  q  r  s  t  ...
```

Figure 11.11.

```
Keyword     w h i t e w h i t e w h i t e w h i t e w h i
Plaintext   D I V E R T T R O O P S T O E A S T R I D G E
Ciphertext  z p d x v p a z h s l z b h i w z b k m z n m
```

Figure 11.12.

of the Sforza at Milan who in 1474 wrote the earliest known manuscript devoted exclusively to cryptanalysis. In a two-page Latin tract he outlined thirteen rules for solving monoalphabetic substitution ciphers.[30] They were only applicable to ciphers in which word divisions were preserved, and which did not include homophones or nulls – thus relating to ciphers that were actually quite dated even in the Italy of his day [83: 12]. Nonetheless, they were practical and clear and complemented the more abstract analysis that Alberti had provided.[31]

The beheading of Simonetta in 1480 (he had assumed a leading rôle in the Milanese city government and was accused of treason) may have been the only violent death that met the group of cryptanalysts to be mentioned here. A generation later, Giovanni Soro (d. 1544) was appointed cipher secretary by the Venetian Council of Ten in 1506, and for almost forty years he raised the standards

---

[30]The material is accessible in an Italian translation in Sacco [83: 37–38].

[31]Kahn – who prints the opening statement of the manuscript suggesting ways to ascertain whether the enciphered material is in Latin or the vernacular – mentions that Milanese cryptology was so advanced that early in the 15th century the secretaries in Modena provided a more elaborate nomenclator for their envoys to Milan than for any other [56: 110–111].

of Venetian cryptology to a level that was considered even higher than that of the Roman curia. His own treatise, a *Liber zifrarum* – written in the early 1500s on the solution of Latin, Spanish and French ciphers and turned over to the Council toward the end of his career – is now lost, but his fame rests on the spectacular success he had in solving the ciphers of numerous Italian city states and foreign powers [56: 109–110]. The *Trattati di cifre* of a couple of his successors have survived; Francesco Marin's dates from 1578, Agostino Amadi's was written ten years later [67: 21–23]. Both of these treatises were highly practical in nature and intended for the eyes of the Venetian cryptologists only.

Cryptography at the Holy See reached a high level of perfection early on. Around 1540, Pope Paul III delegated all enciphered communication to Antonio Elio (1506–1576), who excelled as a cryptologist and devised a cipher that was so secure that it was used in papal diplomacy for another fifty years [83: 15].[32] In 1555, the title of Cipher Secretary was created; for twenty years – beginning with Giovanni Battista Argenti in 1585, who turned over the position to his nephew, Matteo, in 1595 – this position was held by the Argenti family until 1605. Relieved of his duties, Matteo Argenti[33] began to compile an important, 135-page manual of cryptology "that summarises the best in Renaissance cryptology", as David Kahn remarked [56: 112–114].

As could be expected, the level of cryptology at the court of the Medici in Florence was very high, earning this craft a place in the seventh book of Niccolò Machiavelli's 1521 work, *Il Libro dell'Arte della Guerra* [64]. In view of the advances in the field made in Italy by that time, the Florentine author could assert that a courier bearing enciphered dispatches "cannot know their secret contents, nor can there be any danger of his being discovered by the enemy" [75: 675]. By the mid-1500s, Pirro Musefili and, after him, Camillo Guisti devised complicated ciphers for the ducal

family and excelled at solving intercepted messages.

The last person to be mentioned in this section is a Frenchman, François Viète [Vieta] (1540–1603). A consummate mathematician, he not only became the "Father of Algebra" but also rose from parliamentary counsellor to privy counsellor to King Henry IV. Between 1588 and 1594 Viète solved several Spanish and Italian ciphers and nomenclators that helped the king in his conflicts with the Catholic League.[34] His success was due to his linking of his mathematical work to cryptanalysis, and on his deathbed, he wrote a memoir on decryption addressed to Henry IV's chief minister.[35] Its recent re-discovery affords a look at Viète's methods and allows us to compare his analytic and cryptanalytic skills that culminated in what he called the "infallible rule" for solving ciphers[36] [75: 682; 74].

11.2.2.9 *Summary of developments in the fifteenth and sixteenth centuries.* It is fair to say that most of the important cryptological inventions

---

[32]Elio (whose nickname was *Antonio delle cifre*) was held in such high esteem that in 1559 he was elevated to Bishop of Pola and Vicar of St. Peter, and finally to Patriarch of Jerusalem.

[33]On Matteo Argenti, see Sections 11.2.1 and 11.2.2.1.

[34]There is a well-documented story of Viète's working from October 28, 1589, to March 15, 1590, to decipher a new nomenclator (the usual alphabet with homophonic substitutions plus a code list of 413 terms) that Philip II of Spain used in his correspondence with Juan de Moreo, his ambassador at the French court. And while the solution came about one day after Henry IV had defeated the forces of the Catholic League and thus did not contribute to this victory, Philip II was so incensed when he heard about the deciphering of his new nomenclator that he complained to the pope and accused the French of having used black magic to find this solution. But the pope's cryptanalyst was none other than Argenti – who was well aware of Viète's qualifications.

[35]Peter Pesic [74: 18–19] suggests that Viète may actually have solved Matteo Argenti's ciphers, who came to France in 1600 in the service of Cardinal Cinzio Aldobdrandini, the papal nuncio. The Frenchman – who interacted with Aldobrandini at Lyon in 1600 – expressly mentioned details of the ciphers of Aldobrandini at the end of his 1603 memoir. Pesic speculates that the two cryptologists might have met in France in 1600 "and even compared notes".

[36]Pesic [75: 682] summarises these "Règles de Viète" as follows: His "infallible rule" is "based on systematic examinations of the relative frequencies of triads of ciphered symbols; the implicit precedence of such systematic rules over more haphazard (if very useful) habits of reliance on guessing probable words; and the recognition of what Viète calls *chiffres essentiels*, cardinal numbers left unenciphered, which shows the double significance of *chiffre* as cipher or as number".

and innovations in the period under consideration, in other words, the two centuries between about 1450 and 1650, were made by the end of the sixteenth century. The Italian, German and French systems that were available by 1600 were highly reliable and, as we have seen in the case of Vigenère, would remain so for another 250 years. An overview may be helpful in outlining the accomplishments achieved so far. Thus the development from monoalphabetic to polyalphabetic substitution might have progressed as follows:[37]

- monoalphabetic encipherment with permutation table[38]
- cipher disk as a substitution for the permutation table
- easy switching of alphabets by changing the cipher disk settings
- change of alphabets within a message (Alberti)
- progressive change of alphabets after each letter (Trithemius)
- key-dependent alphabet change after each letter (Bellaso)
- general polyalphabetic substitution (Porta)
- polyalphabetic substitution with autokey system and priming key (Vigenère)[39]

A look at the situation across central Europe in the seventeenth century will confirm how fundamental the cryptologic inventions and developments were that were made in the preceding period.

11.3 THE 17TH CENTURY

11.3.1 Preliminaries

The preceding overview has clearly shown how cryptological innovations in the European Renaissance began in Italy and received major inspiration from German and French contributors. An examination of these countries and several of the

neighbouring European states will indicate the increased awareness of cryptology in modern governance even though some of the cipher secretaries of these states – while excellent cryptanalysts – will not make major contributions to the field. In comparison with the earlier period, it will also become clear that the 17th century was a time of consolidation of cryptological knowledge and of indirect cross-connections, but that there were few lasting breakthroughs in the field.

11.3.2 Germany

It may be appropriate to begin with a look at cryptology in the Germany of the seventeenth century as the German lands were the principal location of the longest military conflict of the time, the Thirty Years' War (1618–1648). One might therefore expect further cryptologic stimuli from this area, but all indications are that the methods of encipherment used during these three decades of warfare were based on modifications of existing systems. A good indicator of the state of cryptology in Germany in the first quarter of the century is the 496-page compendium on secret communication that Duke August the Younger of Brunswick-Lüneburg (1579–1666) published in 1624. The young prince, who originally had no hopes of ruling the duchy, devoted his early life to scholarly pursuits, wrote a book on chess in 1616, and in the 1590s laid the foundation to what in 1666 had become one of the largest collections of books and manuscripts of his day. In 1624, he published *Cryptomenytices et Cryptographiæ Libri IX* [93], which was issued under the pseudonym of Gustavus Selenus. It began as a defence and explication of the two works of Trithemius and ended up being the earliest manual of cryptology. With almost two hundred pertinent primary books and manuscripts in his possession,[40] Duke August made use of just about all the authors discussed above and analysed the entire body of cryptologic material from ancient times to the 1621 plagiarism of Vigenère's *Traicté* [99: 83–86].

---

[37]This overview modifies a table prepared by Pommerening [77].

[38]A 17th-century enhancement of such monoalphabetic substitution ciphers is the *homophonic substitution cipher* (see footnote 59).

[39]Vigenère's *tableau* – for which he is unjustly known today – is actually a modification of the Trithemian square table, as we have seen (Section 11.2.2.7) and a less secure version than his very own inventions, which remained virtually unused until the 19th century.

[40]His library, which at the time of his death in 1666 numbered 135,000 titles, already comprised several thousand books and manuscripts at the time of the drafting of the *Cryptomenytices* in 1621/1622.

Duke August acquired the earliest cryptologic book in the German language too late to be included in this work: Samuel Zimmerman[n]'s *Etliche fürtreffentliche Gehaimnussen* of 1579 [119] was attached to a secretarial handbook. It discussed methods of secretly sending letters known from antiquity and the Middle Age ("attached to arrows or included in hollow canon balls"), dealt with papermaking [25: 53] to assure a high quality of the writing material on which sympathetic inks could be used, and treated other methods of concealment. On the last fifteen pages, Zimmermann listed numerous alphabets that even included those of the "Persians" and "Indians and Moors" [119: 148–163]. The Wolfenbüttel duke included the first and second editions of the earliest major cryptologic treatise written in German, Daniel Schwenter's *Steganologia & Steganographia*, both editions published under the pseudonym of Resene Gibronte Runeclus Hanedi in the 1610s [91; 92]. Schwenter, a professor of mathematics and Oriental languages at the Altdorf (Nuremberg) Academy, deemed it advisable not to reveal his authorship. He drew heavily on Italian source materials, in particular on Porta; his books focus on optical and acoustic telegraphs and even on telepathy and list various ways of communication by means of sympathetic inks. The two editions included in the *Cryptomenytices* and a later third one are well structured but do not add much new cryptologic matter.[41]

For well over half a century, Duke August's encyclopedic compendium served as a cryptologic reference in the German area and beyond.[42] Apart

from the (mostly) successful elucidation of Books I and II of Trithemius' *Steganographia*, its primary value lay in the presentation of cryptologic source materials that simply were not accessible to most individuals, as we have seen in the case of Jacopo de Silvestri and some of the other less well-known writers – including the 1593 *Scotographia* by Abram(o) Colorni [32]. This "Jew of Mantua", as he called himself, was in the service of the Hapsburg emperor Rudolph II at Prague, and his elaborate polyalphabetic substitution systems appealed to Duke August so much that no other author except Trithemius occupies more space in his compendium [93: 185–245].[43]

In view of the inclusiveness of this handbook it should therefore not surprise that the only other important cryptologic overview to appear in Germany in the 17th century leaned heavily on the *Cryptomenytices*. Yet contrary to Duke August, who meticulously documented his sources, Johann Balthasar Friderici virtually plagiarised the Lüneburg publication along with Porta and other relevant authors in his 1684 *Cryptographia oder Geheime schrifft- münd- und würckliche Correspondentz* [...] [44]. Although Friderici's handbook – written in German, after all – covered just about all aspects of cryptology, was very useful and finally replaced the dated German publications of Schwenter that were hardly available by then, it was tainted almost from the outset with the

---

[41] The *Steganologia* [...] *aucta* was reissued in the 1620s, too late for inclusion in the Duke's compendium. In 1636, the year of Schwenter's death, his widow and sons published a curious book, *Deliciæ Physico-Mathematicæ. Oder Mathemat: vnd philosophische Erquickstunden* [90], in which the author dealt with the art and mechanics of (concealed) writing (sections XIV and XV), which rely heavily on his earlier cryptologic publications but also on Porta and Duke August. The work was so successful that the Nuremberg poet and alderman Georg Philipp Harsdörffer published two sequels in 1651 and 1653 [50] that again included cryptographic details and were reissued to the end of the century.

[42] A letter written in 1655 indicates how widely used and available Duke August's *Cryptomenytices* must have been (at least in "interested" circles): Étienne Polier, a Swiss informant

residing in Paris, offered Samuel Hartlib, a German émigré scholar in London closely associated with Oliver Cromwell's government, to communicate privileged information, "even from the Cabinet". Such letters would have to be encrypted, and Polier suggested to use one of Hartlib's ciphers or one of his own, "ou que vous m'en indiquiez quelqu'un de ceux qui vous agree le plus dans Selenus qui en a traitté en Maistre": As a further alternative, Polier would thus leave it up to Hartlib to select a cipher from among those in Selenus's work [99: 98, 110]. Needless to say that Polier assumed that Hartlib also had access to Duke August's handbook.

[43] Wagner [112: 19] reports that the Venetian ambassador to Prague had asked the senate of his republic as early as 1591 to forbid for 25 years the reprint of a "trattato pertinente allo scrivere oscuro sotto il titolo di Scotographia" as it contained several new and universally useable ciphers. If the date of the request (Dec. 10, 1591) is correct it would mean that the ambassador had had knowledge of Colorni's manuscript two years before its publication and found it important enough to alert the Venetian authorities well in advance.

odium of plagiarism, at least in Germany.[44] One interesting cryptologic development in Germany in 1661 will be discussed below in connection with the re-use of the Trithemian "polygraphic" method for the purposes of universal communication: Johann Joachim Becher's *Character, Pro Notitia Linguarum Universali: Inventum steganographicum* [...] [17] borrows the underlying idea from a 1659 manuscript version of Athanasius Kircher's 1663 *Polygraphia nova et vniversalis* [62]. This folio-size work, which has to be seen in its Italian context, revisits Trithemius' system and adapts it to both universal and secret communication.

11.3.3 England

11.3.3.1 *The ancestry.* While the simple use of some kind of ciphers can be found in many medieval manuscripts, the first treatise to actually describe the use of cryptography was written by the English scholar and monk Roger Bacon (1220–1292). Written in the middle of the thirteenth century, the brief *Epistle on the Secret Works of Art and of Nature and on the Nullity of Magic* lists seven deliberately vague methods of communicating in secret, among them exotic alphabets, invented characters, figurative expressions, shorthand and "magic figures and spells" [56: 90].

A century later, Geoffrey Chaucer (1340?–1400) – one of England's most important late medieval poets but also an avid astronomer – demonstrated the use of cryptography in his treatise, *The Equatorie of the Planetis* [30]. He included several encrypted paragraphs explaining the use of the equatorie, an early astronomical instrument. They are enciphered with a symbol alphabet that replaced his plaintext letters – A, for example, is represented by a sign similar to a capital V, B by a symbol resembling a script alpha. Since the cryptograms are in Chaucer's own hand, they have become one of the most famous encipherments in history.[45]

Thomas Morus' "Utopian alphabet" that appeared in the early editions of his 1518 *De optimo reip. Statu deque noua insula Vtopia libellus* [...] [72: 13] should be mentioned here[46] as it inspired a number of later writers of Utopian novels to devise their own, more sophisticated devices that ultimately found their way into cryptologic use, as John Wilkins' example will show.

11.3.3.2 *The 17th century.* One of the Englishmen who early on realised the importance of secure communication was Francis Bacon (1561–1626). He felt that cryptology was not beneath the dignity of scholars as it assured safe scientific transfer. As a student in Paris – long before he entered politics and became Lord Chancellor – he invented his own substitution cipher.[47] Basically, his "Bi-formed Alphabet" or biliteral cipher, as it is now known, expressed all the letters of the alphabet with the first two, in other words, A and B. On the first level, Bacon devised the following substitution alphabet [73: 23; 102: 86–89] (see Fig. 11.13).

According to his own calculations, "the transposition of two letters through five places will yield thirty-two differences" [13: IV, 445–446].[48] Bacon proposed to work this rather conspicuous cipher into an inconspicuous cover letter, thus making use of steganography: The significant letters

---

[44]As early as 1689, Fabricius – also from Hamburg – identified it as such [41: 12–13].

[45]The authorship of this treatise has been disputed, but the work seems to be authentic.

[46]Duke August included it among his scurrilous alphabets, right below an alphabet reportedly used by Charlemagne [93: 282].

[47]Pesic [76: 200–203] quotes James Spedding, the 19th century editor of Bacon's works, who felt that Bacon's cipher could have been influenced by similar biliteral ciphers described in Vigenère's *Traicté* of 1586, where a table with four different versions of a cipher alphabet can be found [109: fol. 241 r°]. If Bacon had indeed devised his cipher during his Paris years (1576–1579), Spedding's assumption would violate chronology, though [13: I, 841–844]. However, both Vigenère and Bacon might have drawn on earlier ciphers by Trithemius, Bellaso, Cardano and, in particular, Porta. It should be noted that cryptography at the court of Elizabeth I was so advanced that a linear version of a cipher disk, the cipher slide, was used around 1600 [14: 54].

[48]In the age of the computer, it becomes clear that Bacon's biliteral cipher (first published in 1605) is very closely related to the binary scale as it was devised by Leibniz in connection with his calculating machine of 1671. Bacon's *a* corresponds to Leibniz's "0", his *b* to the "1" of Leibniz: $0 = 00000$; $1 = 00001$; $2 = 00010$; $3 = 00011$, etc. And John Napier, the inventor of logarithms, had illustrated the use of the binary scale in his 1617 *Rabdologiæ, seu Numerationis per virgules libri duo* [...].

A	B	C	D	E	F	G	H
aaaaa	*aaaab*	*aaaba*	*aaabb*	*aabaa*	*aabab*	*aabba*	*aabbb*
I	K	L	M	N	O	P	Q
abaaa	*abaab*	*ababa*	*ababb*	*abbaa*	*abbab*	*abbba*	*abbbb*
R	S	T	V	W	X	Y	Z
baaaa	*baaab*	*baaba*	*baabb*	*babaa*	*babab*	*babba*	*babbb*

Figure 11.13.

Retre / *at* a*s* q / *uick*l / y as po / ssi*bl* / e *to* pr　[event discovery by the enemy].
A　　　T　　　T　　　A　　　C　　　K

Figure 11.14.

of the cryptogram (only *a*'s and *b*'s, by definition) could be typeset or written in a slightly different font. Bacon's own example [13: IV, 446], namely the plaintext command, "FLY", is disguised in the steganographic message, "Do not go till I come". An example given by David E. Newton [73: 23] illustrates the two-step encipherment in greater detail: If the word ATTACK has to be enciphered, each of the six letters yields the following biliteral cipher:

A　　　T　　　T　　　A　　　C　　　K
aaaaa baaba baaba aaaaa aaaba abaab

The second, steganographic encryption could disguise the cryptogram with the help of all the letters *a* in roman script, while all the *b*'s could be set in italics. The ciphertext could be incorporated in a covertext that might actually carry the exact opposite meaning of the plaintext message:

Retre*a*t a*s* q*u*ic*k*ly as possi*b*le *t*o pr[event

discovery by the enemy].

In decryption, the covertext should be separated into five-letter segments that correspond to each of the letters in the word ATTACK. (The bracketed part of the covertext is irrelevant.) (see Fig. 11.14).

It is interesting that the presentation of the biliteral cipher follows a discussion of "Characters Real" which Bacon saw in Chinese ideograms, "which express neither letters nor words in gross, but Things or Notions" [13: IV, 399–400].[49] In his

opinion, even cryptographic signs could fall in this category – thus suggesting the connection between cryptology and (universal) languages that would become so important by the middle of the century [24: 25–29].

This connection is nowhere more relevant than in England. A number of writers dealing both in linguistic matters or the philosophy of language and in cryptology had expert knowledge of secret communication. One of these was Francis Godwin (1562–1633), whose utopian novel, *The Man in the Moone* [48], was posthumously published in 1638 and contained a sample of a musical language that also lent itself to cryptologic use as a musical cipher [102: 117–120].[50] As early as 1620–1621, Godwin and his son Thomas had submitted a design to King James I in which they suggested

---

[49]This quote is part of a much longer, important passage that first appeared in Bacon's 1605 work *Of the proficience and*

aduancement of Learning. (The passage was quoted almost verbatim in John Wilkins' early cryptological book, *Mercury, or, The Secret and Swift Messenger* [117: 106–107].) In Bacon's expanded 1623 translation (*De dignitate et augmentis scientiarum*) of the 1605 work he added, "any book written in characters of this kind can be read off by each nation in their own language" [13: IV, 439–440].

[50]Similar to Bacon – who had to have used earlier sources – Godwin may have been influenced by "the account of the tonal language in a redaction of the journals of Matteo Ricci (1552–1610) [82], the founder of the Jesuit mission in Peking, and a musical cipher explained and demonstrated by Porta" in the 1602 edition of *De fvrtivis literarvm notis* [78; 35: 325–326]. Schwenter's editions of the *Steganologia* [92: 303] contained an example of a musical cipher similar to the one shown by Porta in 1602; in 1621, Duke August commissioned a Lüneburg musician to compose an elaborate cipher in four voices for his 1624 *Cryptomenytices* [93: 325–328; 103: cols 784–787].

possibilities "for conveying Intelligence into Besieged Towns and Fortresses, and Receiving Answers therefrom" – material that may have been derived from optical and acoustic telegraphs described in Porta's and Vigenère's works [34: 309]. Having met with no response, they recast this matter in 1629 in the larger though not clearer *Nuncius inanimatus* [48], where they focused again on distant communication, in particular with the help of homing pigeons or beacons.

The two examples given in Godwin's "lunatique language" in *The Man in the Moone* amounted to a few musical notes that represented individual letters, which challenged John Wilkins (1614–1672). During his years as a chaplain in the service of noble families, he had already published what we would now call two "popular science books". The third volume that attempted to inform a general readership of difficult subject matter but was also a timely gift to the diplomats and leaders of the Civil War was his 1641 *Mercury; or, The Secret and Swift Messenger* [117]. As in the two previous books, Wilkins showed a solid mastery of the material both in terms of classical cryptology and the knowledge of his day[51] [102: 120–124]. In the preface, he referred to one of Godwin's works: "That which first occasion'd this Discourse, was the reading of a little Pamphlet, stiled, *Nuntius* [*sic*] *Inanimatus* [. . .]". And he easily added the letters missing in Godwin's musical alphabet, which Wilkins clearly considered a musical cipher – but also saw that "The Utterance of these Musical Tunes may serve for the Universal *Language*, and

the Writing of them for the Universal *Character*" [117: 77] (see Fig. 11.15).[52]

The use of this alphabet for cryptographic purposes may have been suggested to Wilkins by a chapter in a well-known book – *De prima scribendi origine et vniversa rei literariae antiqvitate* [53] – written in 1617 by a Flemish Jesuit, Hermann(us) Hugo [102: 90–95]. Aware of the latest publications on the Chinese and Japanese writing systems, Hugo devoted four of his 35 chapters to cryptology, the last one titled, "De occultâ scriptione per Notas" (Ch. XVIII). Among other methods of concealment, Hugo referred to the Tironian notes as a means of drawing up parallel lists of such notes in several languages[53] – a system that anticipates the works of Athanasius Kircher and Johann Joachim Becher in the 1660s, where the dual use of such lists for universal and closed communication along with their corresponding mathematical–combinatorial codes will be implemented.[54]

Although less known for his work as England's most esteemed cryptanalyst of the century, John Wallis (1616–1703) left behind a manuscript that was published posthumously in 1737. His amazing mathematical skills more than equalled those of his French counterpart, Antoine Rossignol (1600–1682). Wallis laid the groundwork for Newton's development of the calculus [75: 688–691], but his prowess also enabled him to decipher some of the most difficult letters and dispatches for Parliament

---

[51]Kahn [56: 155], who considers this book of Wilkins one of "only two English works of the period (to) merit attention" (the second being John Falconer's *Cryptomenysis Patefacta* [42]), zeroes in on "three kinds of geometrical cipher, a mystifying system in which a message is represented by dots, lines, or triangles" as one of Wilkins' inventions, although the latter's explanation of the use of these systems is anything but clear [117: 45–50]. An overview of Wilkins' cryptologic systems and adaptations of previous materials can be found in the otherwise problematic article by Thomas Leary [63: 233–240]. Wilkins also was the first to use the terms *cryptologia* and *cryptographia* with the meaning of secret communication; four years later, in 1645, James Howell coined the English term when he wrote, "cryptology, or epistolizing in a clandestine way, is very ancient" [14: 25].

[52]In the alphabet, the letters J and F have been switched. Earlier in his small book, Wilkins had prepared the idea of a universal character to be used for worldwide communication in Chap. XIII, "Concerning an Universal Character, that may be legible to all Nations and Languages. The Benefit and Possibility of this" [117: 55–58]. This idea became his life-long concern and in 1668 culminated in the publication of his most important work, *An Essay Towards a Real Character, And a Philosophical Language* [116].

[53]Hugo learned of these notes from Trithemius' *Polygraphia* [107], where some thirty notes were reproduced (see Sections 11.1.2 and 11.2.2.2, footnote 14). Duke August refers to Hugo's material in his *Cryptomenytices* [93: 370–371], where the cryptologic use of the notes prevails over that as a means of open, universal communication. Wilkins, too, lists Hugo's examples of *Tiro's Notes* in his Ch. XIII, the section devoted to universal communication, but was also familiar with *Selenus de Cryptogra.* and Duke August's shift toward their cryptologic use.

[54]See Sections 11.3.6.1.3 and 11.3.6.2.1.

Figure 11.15. John Wilkins's musical alphabet and/or musical cipher. Courtesy Bayerische Staatsbibliothek, Munich.

and, later, King Charles II and his successors over a period of seventy years. During the Civil War, he decrypted Charles' letters that were intercepted by Cromwell's efficient secretary, John Thurloe, whose staff screened the mail at the post office for intelligence-gathering purposes. And although Charles II knew that Wallis had been in parliamentary service, he retained him upon his ascent to the throne. Wallis apprised the ruling parties of major foreign plots; "his solutions – nearly all nomenclators, a few monoalphabetics – had considerable impact on current events", David Kahn feels [56: 166–169]. Near the end of his life, he even managed to have his grandson appointed as his successor and have him paid for the training he gave him in his last years [98: 83–84].

A contemporary of Wallis', Samuel Morland (1625–1695) was a successful diplomat in the ser-

vice of Cromwell and also – though certainly less spectacularly than Wallis, who handled the most difficult materials – worked with him as a cryptanalyst in the secret room connected to the post office that was supervised by Thurloe.[55] As he did with Wallis, Charles II retained Morland after the Restoration.[56] An avid inventor of mechanical devices – among them a calculating machine – Mor-

[55] The secret room next to the post office had its own, private entrance. Wallis, a certain Isaac Dorislaus, and Morland "opened, copied, and resealed letters from eleven p.m. until three or four in the morning, when the mails left the office" [113: 5].

[56] While in the service of Secretary Thurloe, Morland discovered a plot to assassinate Charles II. In order to ease his conscience, he secretly informed the exiled king – which caused Charles II at the Restoration to reward him royally and publicly acknowledge his debt to Moreland [27: 253–255].

land drew on his cryptologic experience when in 1666 he published a 12-page folio entitled, *A New Method of Cryptography* [71]. It shows his familiarity with quite a few letter transpositions and substitutions, but beyond that it describes and illustrates a *Machina Cyclologica Cryptographica*, Morland's enhancement of Alberti's pair of cipher disks [71: 11–12]. The Englishman's version sports the use of a stylus to be inserted into one of 25 holes or "Foramina" in the second of the three concentric circles of the upper disk with the help of a small thimble-like cap that ended in a thin point. The upper disk also has a square opening cut out of the peripheral circle with its 24 capital letters (Fig. 11.16).

When the alphabets on both disks are mixed, this arrangement has to be agreed upon by the two correspondents. The resulting cipher, as Augusto Buonafalce suggests, who analysed this rare cryptological publication of Morland's, "is equivalent to an autokey making a Beaufort-like substitution with mixed alphabets, using the ciphertext as a non-repeating key. One agreed-upon letter is used as a primer" [27: 257–258]. While Morland's invention is not a breakthrough in the field, the system is markedly more secure when the alphabets on the disks are mixed; it also involves the gadgetry of movable disks compared with Vigenère-type tableaus that would otherwise have been used.

"The second English book on the subject [of cryptology] excelled", according to David Kahn [56: 155], and its high quality is all the more surprising when one considers that John Falconer had written *Cryptomenysis Patefacta* in France some thirty years before its first publication in 1685 [42] while serving the Duke of York, the future King James II, during his period of exile there (1652–1660). Falconer, about whose life little is known except that he apparently was entrusted with the Duke's private cipher until he died in France, was a master of the trade. He structured his book in five sections, dealing with the ways of "Secrecy in Writing", such as cryptography, steganography, and simple "Rules for *Decyphering* them without a key"; of the "Methods of *Secret* Information, by *Signs* and *Gestures*"; by the "*Secrecy* consisting in Speech"; "Of Secret MEANS of CONVEYING

Written Messages"; and lastly, with "the several astonishing Proposals for *Secret Information* [...] concerning [Trithemius'] eight Books of *Steganography*" [42: fols b 5 r°–v°]. Falconer's sources are limited; they include Trithemius, Machiavelli, Kaspar Schott,[57] Francis Bacon, and Bishop Wilkins, but they suffice to provide a solid overview, which includes "the earliest illustration of a keyed columnar transposition, a cipher that is today the primary and most widely used transposition cipher" [56: 155]. One can only speculate as to why in his last chapter [42: 146–180] Falconer (unsuccessfully) attempts to explicate Trithemius' *Steganographia*, which he does by referring to Schott's similarly unsatisfactory material – and that, in turn, leads him to further speculation. This fifth section may, however, have been the reason for the manuscript's long-delayed publication in the last years of the reign of a king who espoused the Catholic cause and may have been interested in clearing Trithemius' name.

At the end of the period under consideration in this overview, in the middle of the 17th-century, the first adequate English handbook of cryptology was finally written; the second edition in 1692 may be an indication that Falconer's work was in demand even forty years after its redaction.

11.3.4   France

The discussion of Blaise de Vigenère's inventions[58] has shown that even in a country with purportedly higher cryptologic standards such polyalphabetic substitution ciphers were considered too complicated for everyday use. Secret communication – especially with commanders in the field, where speed and simplicity counted, relied on monoalphabetic substitution ciphers or, over time, on enhanced versions of this system, such as homophonic substitution ciphers [96: 52–55].[59]

---

[57] On Schott, see Section 11.3.6.2.2.

[58] See Section 11.2.2.7. For Matteo Argenti's 1605 assessment of France (and England, for that matter) as one of the countries where the court took great care when devising ciphers, see Sections 11.2.1 and 11.2.2.1.

[59] Singh's example is based on the frequency of English, where the letter "a" accounts for about eight per cent of all letters, "b" for only two. In order to avoid the single sym-

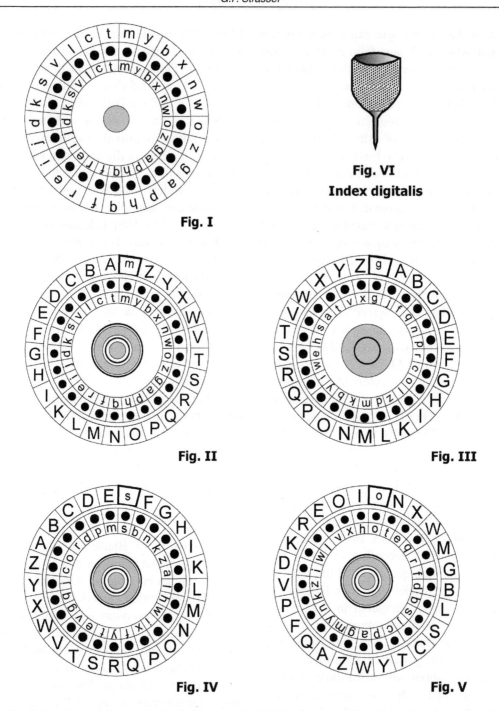

Fig. VI
Index digitalis

Fig. I

Fig. II

Fig. III

Fig. IV

Fig. V

Figure 11.16. Cipher disks of Samuel Morland's *Machina Cyclologica Cryptographica*. Courtesy Augusto Buonafalce, San Terenzo (Italy), and Brian Winkel, Cornwall, NY.

bol chosen to represent a plaintext A to occur eight times more than, for instance, the letter "j" (with only a one per cent frequency), eight symbols would have to be prepared for plaintext A. Such symbols could be two-digit numbers, which means that A would be enciphered with any of the "homophones" 09, 12, 33, 47, 53, 67, 78 or 92. This basically cancels out the frequency of each letter within a given language, something a cryptanalyst would go for first. (There are other

It is not known how sophisticated the enemy cipher was that was brought to Henry II, Prince of Condé, commander of the royal army, when he laid siege to the Huguenot city of Réalmont in April of 1628. The besieged forces refused to surrender, but for days nobody in Condé's camp could decipher the intercepted message, which was destined for other Huguenot troops in the area. When the prince learned that a 26-year-old mathematician in nearby Albi might be able to solve it, he sent him the cryptogram. Antoine Rossignol (1600–1682) deciphered it within the hour, and Condé learned that the Réalmont garrison desperately needed munitions. He sent the intercepted message and the clear text back into the city – and the Huguenot commander surrendered. Cardinal Richelieu, chancellor of King Louis XIII, attached Rossignol to the court when he learned of his qualities [56: 157–163]. Two years later, in 1628, when a Catholic army under Richelieu's command laid siege to La Rochelle, the last Huguenot stronghold, the young man similarly apprised the cardinal of the contents of intercepted messages to the English in which the desperate Huguenots requested their support from the sea. The Catholics sealed off the port, and one month later the Protestants capitulated.

These feats quickly established Antoine Rossignol at court. He not only served Louis XIII but was retained by the extravagant Louis XIV, next to whose study at Versailles he frequently worked. While his decades of service to the French rulers alone would assure Rossignol a place in the history of cryptology similar to that of his English counterpart, John Wallis, the Frenchman's fame nowadays is based more on the development of what is known as the Great Cipher. In the beginning of his career, Rossignol encountered nomenclators in diplomatic correspondence that listed both their plain and code elements in alphabetical order (or alphabetical and numerical order, if it was a number code). Basically, this was the parallel arrangement that had existed since the early Renaissance in Italy.[60]

Since Rossignol was not only called in occasionally to solve cryptograms or nomenclators, as was the custom at many neighbouring courts, but was permanently employed, he was interested in improving the security of the systems used by the French court for its dispatches. He thus proceeded to mix the code elements in relation to the plain ones. To accomplish this he drew up two lists, one for encoding with the plain elements in alphabetical order and the code elements mixed, and a second list with the code elements alphabetised or in numerical order while the equivalent plaintext words were out of order. The two lists of what now was a "two-part nomenclator" were called "tables à chiffrer" and "tables à déchiffrer"; the old, conventional system remained the "one-part" type. Within a short period, the two-part nomenclators were devised in other countries, too; by 1700, nomenclators had increased in size to perhaps 2000 or 3000 elements, which dramatically increased their security.

Like John Wallis, Antoine Rossignol trained a close relative – his son Bonaventure – to work with him and finally succeed him in what became known as the "Cabinet noir" that has given its name to the "Black Chambers" in other countries. The cryptologic invention for which Antoine and Bonaventure Rossignol are best known – and which defied cryptanalysis for two hundred years – is the Great Cipher of Louis XIV. This was a cipher that they devised for the king to encrypt his most secret messages. It continued to be used by Bonaventure's second son, whose

---

factors that still enable a cryptanalyst to attack such a homophonic substitution cipher, such as letter combinations common in a particular language ["th" in English], "q" only followed by one letter, namely "u", etc.) Yet while at first sight such a homophonic cipher is similar to a polyphonic one as each plaintext letter is enciphered in a number of ways, the fundamental difference between the two systems is that "once the cipher alphabet has been established, it remains constant throughout the process of encryption" [96: 54–55]. And even when further enhancements are worked into the homophonic cipher – such as by increasing the number of homophones beyond the letter frequency that was well established in the 16th or 17th centuries – it remains much more "user-friendly" than any polyalphabetic system.

[60]Rossignol himself must have taken advantage of this parallel arrangement in his own decryption efforts [56: 160–161]: If, for example, he realised that in an English message the number 137 meant *for* and 168 *in*, he immediately knew that 21 could not stand for *to* as its sequential number had to be much higher. He also knew that a word in between *for* and *in* – *hide*, for example – would fall in between the code numbers 137 and 168. See also Bauer [14: 73–76].

A. B. C. D. E. F. G. I. L. M. N. O. P. Q. R. S. T. V.
1. 2. 3. 4. 5. 6. 7. 8. 9. 10. 20. 30. 40. 50. 60. 70. 80. 90.

Figure 11.17. [30: 16]

death ended the long Rossignol family tradition. Antoine-Bonaventure did not leave behind any solutions, and the Great Cipher turned out to be so secure that it not only defied the efforts of all 17th-century enemy cryptanalysts but also of subsequent scholars and historians who could not access the enciphered papers of the reign of Louis XIV or XV in French archives for nearly two centuries. The cipher consisted of thousands of numbers, but only 587 different ones. Simon Singh [91: 55–58] has chronicled the painstaking analyses that Étienne Bazeries, an expert in the French Bureau du Chiffre of the Ministry of Foreign Affairs, undertook over a period of three years in the 1890s to finally solve this Great Cipher [15], in which the two Rossignols even had laid traps – one number represented neither a syllable nor a letter but was inserted to actually delete the previous number.

One last author should be mentioned here since he stands at the beginning of the algebraically oriented era of cryptology even though the date of his work almost reaches into the 18th century: Claude Comiers (d'Ambrun) (ca. 1600–1693), a canon at Embrun and later professor of mathematics at Paris, wrote about numerous scientific subjects, frequently in the *Journal des Sçavans*. He was well versed in the cabbala, whose mathematical resources may have influenced his cryptologic work. Around 1684, he turned blind and received a royal pension (hence his reference to the "aueugle [aveugle] royal" in the key he used from then on). At that time, he published his cryptographic system in three issues of the *Mercure Galant*. According to Joachim von zur Gathen, Comiers – who knew Vieta's work ("Mr. Viette, le Pere de nôtre Algebre specieuse, & le grand Déchifreur de son temps" [33: 62]) – presented "the first 'arithmetization' of cryptography" [110: 341–344]. Although Comiers never mentioned Trithemius or Vigenère and in his introductory pages went back to the general Caesar substitution that he attributed to the ancient

Hebrews, his point of departure is a Trithemius–Vigenère system with an 18-letter alphabet, which he writes as "ciphertext = cleartext + key', with the decryption rule 'cleartext = ciphertext − key" (Fig. 11.17).

And while Comiers does not have the notation or the mathematical terminology at his disposal, he describes several cryptosystems, such as "a monoalphabetic substitution given by a keyword followed by a letter-by-letter addition of another key word as superencipherment" [110: 343]. Central to his analysis is a large fold-out table with a Vigenère tableau in the middle, here taken from the 1690 book publication of the three articles, *L'Art d'Écrire et de parler occultement* [33: fold-out between pp. 12 and 13] (see Fig 11.18).

Right above the central Vigenère table, we find the illustration of Comiers' system (see Fig. 11.19).

The top line, the date in Latin – the 1690 date is used both in the 1684 articles and the 1690 book – is the actual plaintext, the bottom line Comiers' usual key ("Comier[s] aueugle royal"); the cryptogram is in the middle. To the left and right of the Vigenère table – and much better engraved in 1690 than in the magazine publication – are two banners with two alphabets. The top line on the left means that A is to be encrypted as 111, and B as the same number, but with a dot added, possibly 11·1 [110: 345]. The banner on the right follows the same system although it now uses geometric symbols instead of numbers. Below the Vigenère tableau, between the bells and the kettle drums, there is an 18-letter alphabet generated by the keyword *"Profetisandum"*. (The bells and the other "long range" instruments depicted could be used for the transmission of coded messages, as Comiers elaborates in several chapters of his publication.) The remaining letters *bcglq* complete the alphabet; once again, numerical equivalencies from 1 to 90 appear sequentially under the keyword-generated alphabet. Other cryptographic devices that are dealt with in the duodecimo-size book, such as a pigpen cipher and the line cipher under it, complete the fold-out.

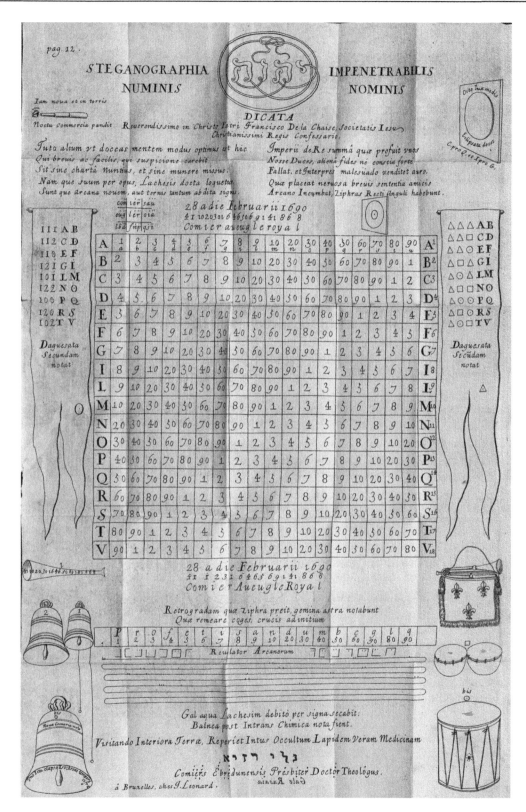

Figure 11.18. *Steganographia impenetrabilis* of Claude Comiers. Courtesy Bayerische Staatsbibliothek, Munich.

```
2 8 A D I  E F E B R  U A R I I  1 6 90
4 1 10 20 30 1 6 4 6 50 6 9 1 4 1 8 6 8
c o m  i  e  r a u e u  g l  e r o y a l
```

Figure 11.19.

Although Claude Comiers is not included in any handbooks of cryptology as his work appears to have been virtually overlooked,[61] he is a worthy successor to François Viète in his reliance on algebraic computation and the beginning "'arithmetization' of cryptography", the hallmark of 20th-century computerised cryptology.

11.3.5 Spain

11.3.5.1 *The early years of cryptology in Spain, the new world power.*   As in other central European countries – with the exception of Italy – cryptography on the Iberian Peninsula during the medieval period used substitutions, musical notation (neumes) or the reverse order of expressions to assure secure transmission [46: 89–93]. The beginnings of modern cryptology in Spain coincide with the expulsion of the Arab invaders toward the end of the fifteenth century. Under the reign of Ferdinand II (1452–1516) and Isabella I (the Catholic; 1451–1504), the kingdom became a world power, and secure communication soon was a major concern [56: 114–116]. Yet the new court lacked experienced cryptologists, and the first cipher system introduced in 1480 still used unwieldy Roman numerals in the nomenclators, thus rendering it highly unreliable [46: 94–95]. Simpler systems were introduced after Isabella's death in 1504, but major progress did not occur until the 1550s, when Philip II ascended to the throne (1527/1556–1598) and immediately changed the ciphers used during the reign of his father, Charles V (1500–1558). David Kahn considers Philip II's 1556 cipher "one of the best nomenclators of the day". It

comprised homophonic letter substitutions; equivalents for common digraphs and trigraphs; a small code in which words and titles were represented by two- and three-letter groups; and further refinements. It remained the basic cipher in Spanish cryptography well into the next century.

Similar to Giovanni Soro, the sought-after Venetian cryptanalyst of this period,[62] Philip II divided his systems into two groups: One, the *"cifra general"*, was used in communication among ambassadors and in exchanges with the king; the second, the *"cifra particular"*, was reserved for messages between the king and individual envoys. All of these originated in the command centre of Philip's administration, the *Despacho Universal*. Ciphers were periodically replaced [46: 96–101] or labelled up front to be used for a limited period only. Yet another group of nomenclators was reserved for the correspondence with the viceroys and governors of the new colonies. Philip's cipher service was so much up to date that Vigenère's tableau was known toward the end of the king's life.

Despite this seemingly streamlined operation, Spanish ciphers were read by cryptanalysts at several other courts, among them by François Viète in the service of Henri IV. In 1589, in the midst of the French king's combat with the Catholic League, Viète deciphered a dispatch that Philip II had sent to one of his commanders in the field, as we have seen.[63] In 1577, Philip's enciphered correspondence with Don Juan d'Austria, his half-brother and governor of the Spanish Netherlands, was intercepted by William of Orange. The Flemish cryptanalyst Philip van Marnix managed to read the dispatches, and William warned Sir Francis Walsingham, Queen Elizabeth's principal pri-

---

[61] Von zur Gathen [110: 348–349] refers to an anonymous 1770 publication – *Neueröffneter Schauplatz geheimer philosophischer Wissenschaften* [...] that proposes an even more general type of arithmetic cryptography in its Ch. XVI, "Von der logistischen Steganographia" [8: 382–392]. Since the unknown author does not list any sources, there is no way of telling whether Comiers' small booklet might have been known to him – or others after him.

---

[62] Soro (see Section 11.2.2.8) wrote a *Liber zifrarum* in the early 1500s on the solution of Latin, Spanish and French ciphers, which he turned over to the Council of Ten toward the end of his career (he died in 1544).

[63] See Section 11.2.2.8.

vate secretary, of an impending invasion [97: 36–43]. It was this decryption that convinced Walsingham of the value of cryptanalysis, which led to his hiring Thomas Phelippes as his cryptanalyst and to the establishment of a cipher school in London. Ultimately, Phelippes decrypted the 1586 correspondence in which Mary, Queen of Scots, condoned the planned assassination of her cousin Elizabeth – which led to Mary's execution.

11.3.5.2 *An inside look at the Despacho Universal in the seventeenth century.* An extremely rare *Traitté de l'Art de deschiffrer* [106], written in French between 1668 and 1714 by an unknown author close to the *Despacho Universal* at the Madrid court, sheds light at the level of sophistication of this agency one hundred years after Philip II's reign. After some introductory remarks on letter frequency and on the various types of ciphers, the anonymous author begins in Chapter 3 with the description of methods of decipherment of "simple ciphers" (monoalphabetic substitutions) both in Spanish and in French. He draws up eleven rules, which he then applies to ciphers both in Spanish and French. In Chapter 5, "composite ciphers" (digraphs) are analysed; three rules are established in the subsequent section.[64] Chapter 7 applies the newly proposed rules to a Spanish and a French "composite" cipher, the French one an intercepted original dispatch to Louis XIV from the ambassador at Nijmegen, sent in 1676. The step-by-step cryptanalysis of this second piece allows the reader to look at the workings of a seventeenth-century cipher bureau such as the Spanish *Despacho Universal*.

Chapter 9 briefly – and not very successfully – deals with transposition ciphers, while the tenth section is devoted to the description of a Vigenère tableau where the twenty-two letters are replaced with numbers.[65] Aware of the virtual impossibility of proposing any rules for the decryption of such a tableau, the author resorts to a truism in the conclusion of the chapter, "that this kind of cipher is not very heavily used in the cipher bureaus on account of their length and of the amount of time it would take to encipher and decipher a letter" [106: 95]. In Chapter 11 – the last one – the author presents *"Des Chiffres indéchiffrables"*, ciphers that in his opinion defy any decryption.

At least the first cipher deserves to be presented in brief [106: 96–97, xiii]: It is called an "encipherment through addition or subtraction" and lists the 22 letters of the alphabet used, with the numbers 1–22 under this alphabet (see Fig. 11.20).

The author then proposes to agree on an expression, a passage from the Holy Bible or a prayer formula such as the *Pater Noster* or the *Ave Maria,* which serve as a key. The key letters are written in a line above the plaintext to be enciphered, and the numbers listed above for each of the two respective columns are then added up together (see Fig. 11.21).

To decipher you would deduct from each of the cipher numbers the corresponding key number, which yields the number of the plaintext [46: 47].[66]

This anonymous treatise has afforded an inside look at the Spanish *Despacho Universal* and has once more shown that Vigenère's tableau was well known in cryptology but hardly used as it appeared unwieldy and presented as yet insurmountable problems in cryptanalysis.

---

[64]At the end of the description of the second rule, the anonymous author highlights the following as *"très important(es)"* and divulges that "two years past, the enciphered letter was sent to the Low Lands by the Secretary of State and by the order of His Majesty in order to test ['gain experience with'] the art that is the subject of this treatise" [106: xi, 55]. This is one of the rare instances which document that new ciphers were apparently sent out to be tested on cipher secretaries at other (Spanish) locations *before* being placed in general use – and the author adds that his treatise might well help in such a decryption process.

[65]In his introduction [106: xii], the editor of this treatise, Jérome Devos, refers to Kircher's *Polygraphia nova,* where such a combination of letters and numbers appears in an *"abacus numeralis"* (see Section 11.3.6.1.3).

[66]Jérome Devos, the editor of the treatise, considers this cipher a variant of the "Gronsfeld cipher" (its inventor, a certain Count Gronsfeld, is only known through a passage in the fourth part of Kaspar Schott's 1659 *Magia universalis* [89: IV, 65–69]. For Schott, see also Section 11.3.6.2.2). Newton explains the Gronsfeld cipher as follows [73: 127]: It "is an abbreviated form of the Vigenère cipher, in which the key consists of a series of numbers [...]. The number sequence indicates the alphabet to be used in enciphering each of the letters of the plaintext".

A B C D E F G H I L M N O P Q R S T V X Y Z
1 2 3 4 5 6 7 8 9 10 11 12 13 14 15 16 17 18 19 20 21 22

Figure 11.20.

Key   p a t e r n o s t e r q u i e s t
Plaintext L' A R M E E S T E N M A R C H E
Cipher   24 2 34 16 21 17 18 34 36 10 28 26 20 25 8 25 23

Figure 11.21.

11.3.6 Italy

11.3.6.1 *Seventeenth-century developments in Italy proper*

11.3.6.1.1 *The search for a universal language and its cryptological implications.* Luigi Sacco's overview of early cryptography in Italy [83] ended with the sixteenth century, which allowed him to include virtually all of the important earlier cryptologists of this country. There were indeed no major innovations in Italy before the 1650s, and when there was renewed interest in cryptology, it came about in connection with several linguistic publications. Although such materials are only relevant to this analysis as far as they relate to secret, closed communication, their interrelationship with the field of universal (written) communication requires a brief look at such inventions.

The publications we need to examine clearly profited from suggestions that had been made 150 years earlier, namely by Johannes Trithemius and, secondarily, Jacopo de Silvestri.[67] In the first four, in particular in Books III and IV of the 1518 *Polygraphia*, the German abbot gave clear indications as to how the rows of artificial words would lend themselves not only to secret but also to universal communication. And while the additional designators attached to the word lists in Silvestri's *Opvs novvm* were anything but desirable as far as cryptographic use was concerned since the redundancy of the qualifiers provided unwanted cryptanalytic help, they enhanced universal communication.

11.3.6.1.2 *Pedro Bermudo's 1653 universal language scheme – an "Arithmetical Nomenclator".* The publications of Trithemius and Silvestri were available and certainly known in mid-17th-century Rome, when first attempts were made by Catholic scholars to create means of universal written communication at a time when increased missionary efforts in South America and the Far East were hampered by a reliance on Latin as the language of interaction with the native peoples.[68] And while the earliest such universal language scheme only alluded to cryptographic use in its programmatic title, it clearly provided the impetus for further examinations: In 1653, the Spanish Jesuit Pedro Bermudo (1610–1664) published a large broadsheet during an extended stay in Rome, *Arithmeticus Nomenclator, Mundi omnes nationes ad linguarum & sermonis unitatem invitans, Auctore linguæ (quod mirêre) Hispano quodam, verè ut dicitur, muto* [22]. Bermudo's goal was to unite the nations of the world in one common language; he wanted to bring this about with the help of an "arithmetical nomenclator", thereby anticipating the mathematical–combinatorial methods upon which this type of universal (written) communication was built but also allowing for cryptographic use. The overall similarity of Bermudo's system with Silvestri's is evident, and the Spaniard could certainly have seen the 1526 publication in Rome.[69] His ephemeral broadsheet is only known

---

[67] See Sections 11.2.2.2 and 11.2.2.3.

[68] It is doubtful that suggestions such as the Englishman John Wilkins's proposal made in his 1641 *Mercury, or, The Secret and Swift Messenger* to interconnect closed and open communication and use cryptographic systems for universal interchange were known in Rome less than a decade later (see Section 11.3.3.2). For the following, see Strasser [102: 133–139].

[69] Similar to Silvestri, Bermudo uses a system of dots and accents to express tense, case and number relationships. He is

through a detailed analysis in one of the works of Kaspar Schott (1608–1666) [88: 478–505; 86], a German Jesuit who from 1652–1655 was Athanasius Kircher's assistant in Rome and brought the publication to his attention.[70] Schott confirmed that Bermudo was a deaf-mute (as the last part of the title indicates) and referred to one of the early manuals for their instruction,[71] which may have suggested to Bermudo the categorisation into 44 classes of the 1200 words and concepts that he included in his "nomenclator".[72] The beginning of the Confession of Faith (*Credo in Deum Patrem omnipotentem, etc.*) that Schott gives as an example may suffice to illustrate Bermudo's system and show the basic similarity with Silvestri's earlier version as well as with Kircher's printed material of 1663 and Johann Joachim Becher's variant of 1661 [88: 501]:

XXXIX.$_4$ (*Credo*)   XLII.8 (*in*)

III.1 ... (*Deum Patrem*)   XXXIII.47 (*omni-*)

XL.$_{23}$ ... (*potentem*),   XXXVI.17 ... (*creatorem*)

II.10 ..., (*cæli*)   XLI.$_{15}$ (*et*) I.21 ... | (*telluris*)   [etc.]

The text of the Creed that Bermudo reproduces shows one of the underlying shortcomings of any such system that is based on a limited vocabulary, which requires the frequent use of synonyms (*creatorem* instead of *factorem, telluris* for *terræ*).

Bermudo's broadsheet influenced Kircher's linguistic manuscript of 1659, which has to be seen as an attempt to placate Emperor Leopold I – whose predecessor, Ferdinand III (1608–1657), had asked the Jesuit almost a decade earlier "to propose a kind of *lingua universalis* that could serve as a universal means of written communication amongst all the peoples of this earth" [62: 6]. The late ruler also inquired how well the cryptographic system that Trithemius had proposed in the 1518 *Polygraphiæ libri sex* could be adapted for communication in several languages [62: 79; 102: 144–155]. Kircher sent copies of his manuscript to a few other rulers in addition to the imperial court in Vienna; one of the recipients was the Prince-Bishop of Mainz, Johann Philipp von Schönborn – whose medical doctor, Johann Joachim Becher, saw it. Becher profited from this information and immediately published his own, expanded universal language scheme in 1661; his *Character, Pro Notitia Linguarum Universali* [16] thus appeared two years before Kircher's *Polygraphia nova et vniversalis* [62].[73]

11.3.6.1.3 *Kircher's application of universal language schemes to cryptography.*

The fourteen-volume collection of letters to Athanasius Kircher (1602–1680), a German Jesuit who spent the second half of his life in Rome and soon became the clearing house for the correspondence with missionaries from all over the known world, shows his life-long interest in cryptological matters.[74] It

---

innovative when he subdivides his 1200 words into 44 categories.

[70] Schott's and Kircher's association is analysed in Paula Findlen's comprehensive collection of essays, in particular in her own introduction [43: 1–48].

[71] The instruction of deaf-mutes began in the Spanish nobility, where a high percentage of children were afflicted with this defect. In 1620, Juan Pablo Bonet (1579–1633) published the first such manual, with which Bermudo was probably taught. A British diplomat, Kenelm Digby (1603–1665), recast the material in an English version printed in Paris in 1644, and also in London; in 1651, a Latin translation appeared in Paris (*Demonstratio Immortalitatis Animæ Rationalis* [...]), to which Schott referred in his analysis of Bermudo's system [88: 483]. Bermudo, a high-ranking Spanish Jesuit, may also have known of the *Lingua Geral Brasilica* that two fellow Jesuits had devised at the end of the 16th century to facilitate communication between the missionaries and the South American Tupi Indians.

[72] Although Bermudo certainly would have known of the categorisation of the divine attributes that Ramon Lull had devised in the 13th century in his commonly known *Ars inventiva veritatis* (see Section 11.2.2.1, footnote 7), which clearly influenced the categorisation of (part of) Athanasius Kircher's 1659 language manuscript (see Section 11.3.6.2.1), the classification in his *Nomenclator* shows no such indebtedness and reflects more the grouping of the manuals for the deaf-mutes.

[73] Although there are clear differences between Kircher's 1659 manuscript and the 1663 publication of the *Polygraphia nova* (where the manuscript is primarily used in Book II), the structural similarities are close enough so that an analysis of the printed version will suffice for the purposes of this analysis.

[74] In 1640, he received a copy of an intercepted enciphered letter that the Swedish General Banér had sent; the "new" cipher for the Jesuit correspondence with India with the key word "Goa" issued in 1601 is also in his correspondence [102: 143–144]. Cryptologic topics are discussed more frequently after

may have taken Emperor Ferdinand III's request, however, for the Jesuit to analyse the linguistic aspects of the Trithemian *Polygraphia*, which he did in the first two books of his own *Polygraphia nova* of 1663.[75] An in-depth examination of the cryptological side of Kircher's folio-size publication, undertaken by George E. McCracken more than fifty years ago, provides the basis for the following analysis.

From a cryptological point of view, then, Kircher prepared "a polyglot code in five languages" in Book I, "a Trithemian cipher or open code" in Book II, and "a series of substitution ciphers based on a modified Vigenère table" in the third book [66: 218 ff]. Book I is made up of two polyglot dictionaries, of which the first – *Dictionarivm I* – is needed to encode a text in one of the five sample languages, Latin, Italian, French, Spanish, and German, with Latin being the base or reference language (see Fig. 11.22).

In cryptographic terms, then, this is the encoding dictionary; each of the five languages is alphabetised within itself, and for this reason, entries in a given line do not correspond. A message in any of the five languages can be transmitted in reference to the Latin base in the mathematical–combinatorial approach that Kircher may have gleaned from Silvestri and Bermudo; as in the two previous systems, there are numerous qualifiers

to specify case, number, tense, mode, etc.[76] Responding to the imperial charge, Kircher has thus prepared a means of encoding a text in one language, transmitting it numerically/mathematically, and decoding it with the help of a second listing. In this *Dictionarivm II*, the entries are arranged in 32 blocks of Roman numerals, each with between 32 and 40 lines numbered consecutively in Arabic numbers, thus enabling the recipient of a message to decode the numerical combination (e.g., XXII.17) in the language of his choosing. The last line of the first block may illustrate the semantic correspondence among the five words (but not an alphabetical one, of course) [62: 47] (see Fig. 11.23).

For cryptographic purposes, then, George McCracken has characterised Kircher's system in Book I as a "one-part code in Latin, with equivalent values (in the other languages) placed in adjoining columns" [66: 219].[77] And David Kahn credited the Jesuit's 1663 work with the creation of "a Marconi-like code of 1,048 words from each of five languages" [56: 845–846].[78] Some 250 years before Marconi's wireless transmission entailed

---

Kircher's publication of the 1646 *Ars magna Lvcis et Vmbræ*, where he proposed a number of projection systems that could also be used for secret communication in the section entitled, "*Cryptologia nova*" [59: 909–916]. As early as 1645, Kircher sent a "new steganography" manuscript (now lost) to the emperor in Vienna; at the end of the decade, he invented the first of several "*Cistae*", one of which would ultimately be adapted to cryptographic use in Book III of the *Polygraphia nova* [115: 265–287]. The most recent, specific discussions are Nick Wilding's book chapters [114; 115]; see also Strasser [102: 155–178]. The letters to Kircher are now available and searchable at the following URL: *http://www.lunaimaging.com* (registration [free] is required).

[75]The book consists of two separately paginated parts, the first half (pp. 1–148) containing the two sections responding to the Emperor's requests (*Syntagma I and II*) as well as Kircher's own cryptographic application of these purportedly linguistic sections (*System III*). The second half is a separate *Appendix Apologetica* (pp. 1–23); it has no particular cryptologic relevance.

[76]The signs representing the six Latin cases – to give only a few examples of the qualifiers Kircher devised – simply consist of the first letter of each case name, such as N for nominative, A with a small circle at the apex for accusative and A again for ablative, with a small circle at the apex of the former and at the lower right of the latter. Just as Kircher's entire vocabulary consisted of only 1046 words (not 1048), he reduced tenses to only three: present, perfect, and future. The first is marked by a U-shaped figure, the second by the same figure inverted; the future is shown by Roman numerals, I for the first person, II for the second, III for the third. – As has been mentioned in the case of Silvestri's and Bermudo's systems, such additional signs would constitute a grave defect if the primary purpose of the code were to create a means of safe, secret transmission. In Book I, however, this was not Kircher's purported first concern.

[77]McCracken clarifies that "a one-part code is that type which uses but a single list of [...] code groups, since when the latter are arranged in alphabetical or numerical order, their corresponding meanings will also be in alphabetical order. A two-part code, on the other hand, is one which is so constructed that the code groups have been assigned to their plain-text values in a mixed sequence, making necessary two lists". It appears that Kircher's system would thus be a two-part code as it needs both an encode and a decode – which it is not, though, the reason being the polyglot nature of the code.

[78]In *The Codebreakers*, Kahn does not consider Becher's 1661 publication nor Kircher's 1659 manuscript, which antici-

Latina.			Italica .			Gallica .			Hiſpanica .			Germanica.		
		A			A			A			A			A
Abalienare.	I.	1	aſtenere.	I.	4	abſtenir.	I.	4	ıbſtenir .	I.	4	ıbhalten .	I.	4
abdere.	I.	2	abbracciare,	II.	10	abayer.	XII.	35	ıbbraçar .	II.	10	ıbſchneiden .	I.	5
abire.	I.	3	abandonare .	VI.	23	abbaiſſer .	VII.	2	ıbrir.	II.	20	ıbeudvverden.	I.	21
abſtinere.	I.	4	abbaſſare .	VII.	2	abandonner.	VI.	26	ıbaxar .	VII.	2	abtragen.	III.	2
abſcindere.	I.	5	à che hora .	XIX.	5	acquerir .	I.	16	ıblaudar.	XIV.	21	abſteigen.	VII.	5
abſque te .	I.	6	acetoſa .	XVI.	11	accouſtumer .	V.	21	ıborrecer .	XV.	23	abhalten .	VII.	4
acuere.	I.	7	aſſendere .	XI.	21	accomoder .	X.	8	ıçotar.	XXIII.	19	abſchneyden.		
acetarium .	I.	8	accommodare .	X.	8	acheter.	VIII.	7	ıçucar.	XX.	2		XI.	33
accipiter.	I.	9	acquiſtare .	V.	35	acheuer.	XVII.	2	ıcordar.	XIX.	36	abbrechen .	XIV.	35
acipenſer.	I.	10	accoſtumare.	V.	21	aduehir.	I.	15	ıcabar.	IX.	7	abfagen.	XIX.	15

Figure 11.22. First few entries in *Dictionarivm I* of Book I of Kircher's *Polygraphia nova*. Courtesy Bayerische Staatsbibliothek, Munich.

Latina.	Italica.	Gallica.	Hispanica.	Germanica.
altitudo.40	*altezza.*40	hauteur.40	*altura.*40	höhe.40

Figure 11.23. [62: 47]

voluminous code books in several languages to cut down transmission cost, Kircher had anticipated this development.[79]

In Book II of the *Polygraphia nova*, Kircher reworked the material from his 1659 manuscript,[80] a Trithemian cipher that "is at once less original and less practical, since it is merely an adaptation of the form of open code invented a century and a half before" by the German abbot [66: 221–223]. *Syntagma II*, the "executive" summary, ends with the linguistically oriented statement, *VNIVS LINGVAE AD OMNES ALIAS TRADVCTIO* and thus purports to provide a means of translation of one language into all others (as far as polyglot lists were prepared, of course). Nonetheless, for the first time the *Syntagma* refers to the cryptographic use of this system. McCracken has pointed out that this Trithemian "open code" – as it is incorrectly referred to[81] – never provided the high amount of security that it could as its very existence is difficult to detect. Neither Trithemius nor Porta, who modified it in his 1563 *De fvrtivis literarvm notis* [78], nor Kircher in his adaptation of Porta or his own version managed to truly produce a cover letter that would not raise suspicion. Trithemius at least conceived of the cover letter (which does contain the secret message, after all) in the form of a prayer-like text: The "Ave Maria cipher" (in his example beginning with DEUS CLEMENTISSIMUS REGENS CELOS [...])[82] avoids the trap of presenting the cover letter itself as a letter, as Porta and Kircher then did. Thus even Kircher's own, painstakingly choreographed ciphertext would hardly be taken as normal corre-

___

pates his (printed) use of polyglot codes by four years – and thus his compliment of Kircher's earliest use of polyglot codes still applies.

[79] While reaction to Kircher's attempted universal communication – ideally to be used among Europe's ruling houses, where he sent numerous copies – was one of polite interest, there is one letter to Kircher using his polygraphic system that showed that such exchanges were workable: Juan Caramuel Lobkowitz, Bishop of Campagna and Satriano, a long-time scholarly friend of Kircher's and himself the author of mathematical and philosophical works, sent him a letter *"in lingua polygraphica"*. On August 4, 1663, he thanked Kircher for the copy of the *Polygraphia novà* and managed to draft a readable document that was limited to the 1046 terms in the dictionaries. He then provided translations into the four other languages that Kircher had listed [102: 164–165; 115: 292–294].

[80] See Section 11.3.6.1.2.

[81] In an open code such as the one devised by Trithemius (see Section 11.2.2.2), there are two messages: One that is to be seen by anyone who might intercept it and should be considered the real message – although it actually contained the secret communication that only the recipient would be able to read.

[82] See Section 11.2.2.2.

spondence – and while its secret meaning probably would not have been deciphered, it would have raised suspicion and never reached the intended recipient:

> Habui litteras tuas, magnificentissime Theophile, quas mihi tradidit Anastasius tabellarius tuus, & simul ac perlegi mandata tua, tibi opem ferre volui: nam tui gratia, omnibus posthabitis negotijs, Iosepho amantissimo amico tuo tredecim hungaros expendendas ex dotali pecunia curaui [. . .] [62: 88–106].[83]

Kircher's 40 tables with 22 letters each yielded this epistolary formula that was to encode the plaintext warning, CAVE A LATORE QUIA TIBI [INSIDIATUR].

While Kircher's reworking of the "Ave Maria cipher" for the purposes of universal communication may have been effective, its cryptographic adaptation simply was too involved. Nonetheless, the illustration of one of his famous *Cistae* or *Arcae* introduces yet another variant of Kircher's wooden combinatorial boxes that he had invented around 1650.[84] Just as the second version described in Book III of the *Polygraphia nova* – the *Arca Steganographica* – is primarily adapted to cryptological use,[85] the linguistic chest or *Arca Glottotactica* in Book II can be used for the mechanical decipherment of the sample letter. True to its name, though, it will allow for such a decryption not just in Latin but in the five languages shown in Book I; however, it will not accomplish more than what could have been done with the help of the forty printed tables of Book II alone. Nonetheless, the playful aspect of such a box with its tables written on the small wooden strips meant that the

little chests were highly coveted by the courts of his day.[86]

Book III of the *Polygraphia nova* is devoted almost exclusively to cryptological matters, as the chapter summary suggests, *De Technologia. siue De arcano Steganographico vniuersali combinatiuo rerum* [62: 128]. Kircher still relies on the Trithemian idea of a steganographic cover letter but also brings into play his mathematical-combinatorial experience, which is exemplified in the second, cryptological *Arca* that is similar to the earlier one primarily devoted to universal communication (Fig. 11.24).[87]

Instead of the 40 compartments of the earlier chest, this one only has 24, one for each of the

---

[83]McCracken [66: 222] introduced several errors (wrong words/expressions) in the text of his letter when he encoded the plaintext; even an experienced cryptologist apparently could easily pick the wrong word from the endless polyglot, alphabetic lists that Kircher had prepared.

[84]In 1650, Kircher published his major musicological compendium, *Mvsvrgia vniversalis, sive Ars magna consoni et dissoni*. In its second volume, he introduced for the first time an *Arca Musarithmica*, a composition tool which was to enable anyone to put together acceptable melodies following the author's combinatorial rules [61: II, 184].

[85]See Fig. 11.24.

[86]McCracken [66: 222–223] mentions that Kircher himself had a chest with five additional languages, and he possessed a set of lettered strips in a chest that had the properties of a Vigenère table: Instead of using his standard alphabet of 22 letters throughout, the second strip would begin with B and have A at the bottom, the third with C and A B at the end, etc. It is clear that the use of such tables "would have the effect of combining with the Trithemian cipher a polyalphabetic substitution providing the tables were used in the right way". Such a dual encryption would reach a spectacularly high level of security, of course.

[87]One such nicely crafted *Arca Steganographica* has survived in the Herzog Anton Ulrich-Museum in Brunswick [shelf number Kos 1457], which houses some of Duke August of Brunswick-Lüneburg's prized possessions. The duke – whose interest in matters cryptological (see Section 11.3.2) is reflected in his correspondence with Kircher from 1650 to his death in 1666 – was one of the few recipients of such a little chest; he left it to the one son of his (Ferdinand Albrecht, 1636–1687) who shared his interest in secret communication [102: 169–179].

Always taken by machines and gadgets, Kircher had already proposed such mechanical cryptologic devices in his 1643 investigation into magnetism, *Magnes, sive De arte magnetica*, where we find an illustration of a *Machina magnetica cryptologica* [60: between pp. 392 and 393]. Haun Saussy has likened it to "a sort of magnetic telegraph" [85: 265–267] more than to a cryptologic device. It consists of a series of spherical bottles which are positioned in line, close enough to each other so that the magnet contained in each of them can influence the next one. Once the magnet in the first bottle is turned to a letter marked on the outside of the bottle, the other magnets will spin "in sympathy", transmitting the letter all the way to the last, identically marked bottle. Saussy points out that this is "less a cryptographic than a semaphoric" device, though – and thus yet another one in a series of optical, acoustic, and mechanical cryptographic devices that Kircher proposed over the years in half a dozen of his works.

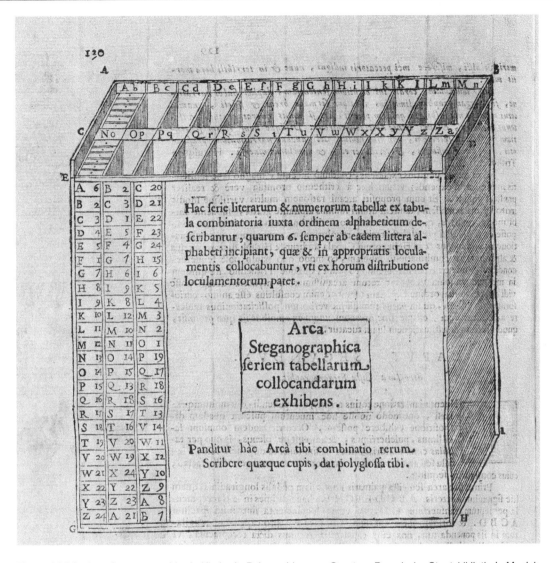

Figure 11.24. *Arca Steganographica* in Kircher's *Polygraphia nova*. Courtesy Bayerische Staatsbibliothek, Munich.

24-letter alphabets Kircher uses. The individual wooden strips have the same 24 letters written on them – but then each letter is combined with a number. Each column thus displays in a (secondary) column the numbers 1 through 24 that are mixed (though not thoroughly, as McCracken observed) [66: 223–224]. It is clear that this expanded system is a variation of a Vigenère table, as the illustration of the *Tabula Steganographica* will immediately show (Fig. 11.25).

The cryptologic systems in Book III, then, consist of four monoalphabetic and four polyalpha-betic substitution ciphers that are worked into a Trithemian-type cover letter.[88]

---

[88]The first two polyalphabetic types (after four monoalphabetic ciphers) may illustrate one of the more complicated encryption systems [66: 225–226]: In the fifth example, the first letter of the plaintext is enciphered by the first table, the second by the second, and so on. The sender of a message can also decide to begin with any table and continue in order, changing tables with each successive letter. In this case, the beginning table would have to be signalled to the recipient "with singular astuteness", as Kircher put it in his explanation [62: 135–136]. As long is this is successfully done, the added security would more than offset this disadvantage. As a further enhancement of this first of the four polyalphabetic encipherments, the au-

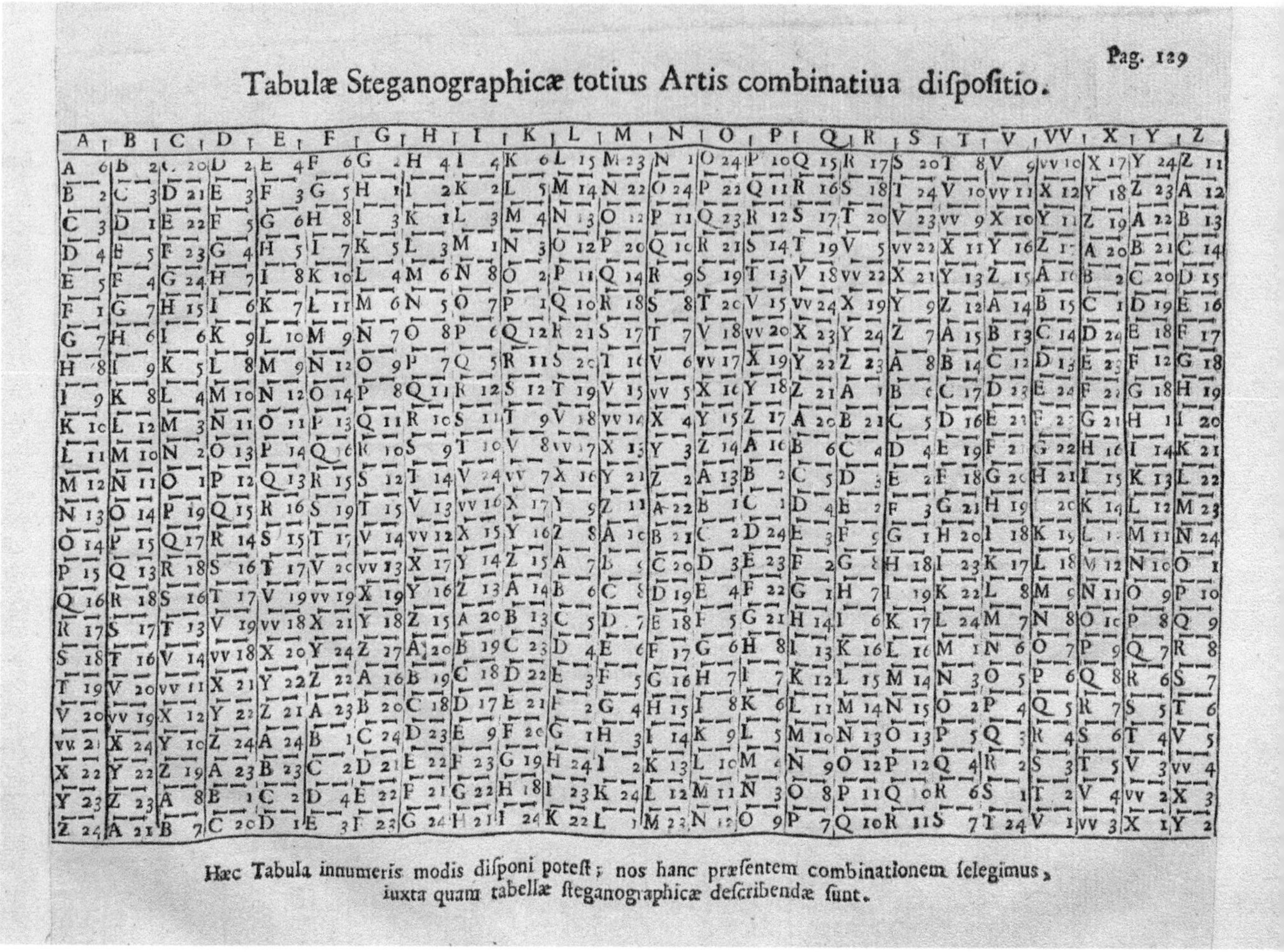

Tabulæ Steganographicæ totius Artis combinatiua dispositio.

A	B	C	D	E	F	G	H	I	I	K	L	M	N	O	P	Q	R	S	T	V	VV	X	Y	Z

(A complex combinatorial cipher table follows, with each row headed A–Z and each cell containing a letter paired with a number.)

Hæc Tabula innumeris modis disponi potest; nos hanc præsentem combinationem selegimus, iuxta quam tabellæ steganographicæ describendæ sunt.

Figure 11.25. *Tabula Steganographica* in Kircher's *Polygraphia nova*. Courtesy Bayerische Staatsbibliothek, Munich.

Key text: s a l v t e m i n e o q v i e s t v e r a s a l v s
("26 letters", Kircher notes)

Plaintext: C A V E A B E O Q V E M H A U D C O G N O V I S T I
("26 letters")

Figure 11.26. [62: 137–138].

thor proposes the use of a key word such as "Iesvs, Maria", etc. In this case, the first letter of the plaintext is enciphered by the table beginning with "I", the second by the table beginning with "E"; the key would have to be repeated if necessary.

The sixth example uses what is now called a "running key", which means that the key has the same length as the plaintext. The first of the following two lines is the key, the second the plaintext (see Fig. 11.26).

To encipher the "C", the "s"-table would be used; plaintext "A" would be enciphered with the "a" table, etc. Kircher proposes to transmit the numerical equivalents and place the pious key at the end of a cover letter as an acceptable salutation.

Duke Ferdinand Albrecht (whose castle was in Bevern), the youngest son of Duke August (see footnote 87), experimented with the eight cryptographic variants suggested by Kircher on a folio-size sheet that is attached to the inside cover of his own copy of the *Polygraphia nova* [102: 177]. He used "Fernando Alberto" as his key text and wanted to encipher the Low German "ICK BEEN SEER SIEK". The key is conveniently worked into an (Italian) cover letter: "*Mio caro Signore. Non so, si V. S. stá in cervello ó no, perche mai si ri-*

After having listed the eight different uses (or *"propositiones"*, as Kircher named them in reference to their mathematical operations) of the Vigenère table, he proceeded to once again display his linguistic interests: The stock letter addressed to "Carissime [Dearest] Theophylacte" that serves as a key (and precedes the secret letter) is repeated in Latin, Italian, French, Spanish, German, English, Dutch, Greek, Hebrew, and Arabic. The text of this template is so problematic that McCracken wondered whether the Jesuit author just wanted to display "his prowess in foreign tongues".[89] Nonetheless, the combination of a Vigenère-type polyalphabetic encipherment with steganographic means first suggested by Trithemius reaches an extremely high level of security.

11.3.6.2 *The impact of interconnections between linguistic and cryptological developments in Italy on similar mid-century publications in Germany.* The linguistic and cryptological materials of Silvestri and Bermudo were finally – and more easily – accessible north of the Alps through their treatment in Kircher's 1659 manuscript and his 1663 *Polygraphia nova* thanks to the numerous copies that the Jesuit dedicated to a number of German rulers.[90] For this reason, it may be defensible to look at several German publications in the framework of their Italian connections.

11.3.6.2.1 *Johann Joachim Becher's Assimilation of Kircher's 1659 Manuscript.* Within this context, a small octavo book put out in 1661 by Johann Joachim Becher and entitled, *Character, Pro Notitia Linguarum Universali, Inventum Steganographicum hactenus inauditum* [16], is most interesting – and controversial. There now is no doubt that Becher – at the time of the redaction of this work in the service of the prince-bishop of Mainz[91] – had seen Kircher's 1659 manuscript and drew inspiration from it. Just as in *Syntagma I of* Kircher's *Polygraphia nova* (and suggested in his earlier manuscript), Becher's method is based on two vocabulary lists, an encode and a decode. Only the encoding dictionary was completed, though, the *Lexicon Pro Resolutione primæ Characteris partis A. B. C. D.*[92] It is much more elaborate than Kircher's with its 9432 entries that are numbered sequentially from *A 1* to *Zythus 9432*. In addition, Becher lists hundreds of male and female first names, followed by place names from *Aachen 9715* to the last entry, *Zürich 10283*. Contrary to Kircher, who had thought about a simplified grammatical and morphological system in his qualifiers,

---

corda del nostro Guojan Battista. [...] *Stia bené. Bevere il 24 d'Aprile, l'anno 69. FERNANDO ALBERTO"*. The duke then proceeded to encipher the "I" (from "ICK") with the "f" table, which yielded the numerical value 7, and so on. The entire numerical cipher was: *7 2 16 21 5 18 11 19 5 3 18 15 6 19 9*. In order to safely transmit such a cipher, he proposed to include it in an astronomical calculation giving the "minutes" and "seconds" of various stellar positions – a rather common topic in the Renaissance that was often used for steganographic purposes, and one that Vigenère had considered reasonably safe [109: fol. 209 r°].

[89] The English translation of the Latin template begins as follows in Kircher's book [p. 143]: *"Knovv that I am very ill content vvith you because that you vvoulde not sende me your booke, I cannott imagine hovv I haue deserued that of you; novv I vvell perceaue you vvill doe very little for me vvhen you deny me soe small à matter. Your vvords and thoughts doe nott agree vvell one vvith another* [...]". Kircher proposed to mark in this passage each of the letters of his key with a small dot or a pin prick.

[90] While there was at least one English proposal for universal communication dating back to 1657 that could also be used "for secret Writing", Cave Beck's *Universal Character* [18] as well as other universal language schemes circulating in England at this time do not appear to have had any direct impact on the developments in Italy or Germany. This does not mean that the general interest on the British Isles in such forms of communication might not have been known among interested scholars on the continent [102: 127–132].

[91] See Section 11.3.6.1.2, and Strasser [101: 215–216]. Becher, who acknowledged a 1660 publication in his book, could thus have incorporated Kircher's concepts as the latter's manuscript arrived at Mainz no later than in the first half of that year. Further proof of this "inspiration" may lie in the fact that Archbishop Johann Philipp blocked the disbursement of the 100 ducats he had originally promised Becher for the writing of the *Character*, whereupon Becher left the Mainz court in a huff [51: 23; 102: 191–195]. In one of his later publications [17: 7–8], Becher refers to Kircher's and "a Spaniard's" works but misrepresents chronology when he states that they had published their systems "five years *after*" his own [my emphasis].

[92] Since the polyglot aspect of the work never came about during Becher's lifetime (sample lists in several languages were actually drawn up in 1676 and 1685 [102: 193]), the Latin encode list that is printed also serves as a decode with the help of the sequential numbering of each of the 10283 entries.

Becher sports no fewer than 173 such numerical suffixes – which rules out any memorisation as it might be possible in Kircher's case. Of course, both the large vocabulary and the generous amount of qualifiers would guarantee a more elegant means of communication. Contrary to what the author had discussed in his introduction, though, where he spoke of the necessarily different alphabetisation in the decoding lists in the six languages he proposed beyond Latin ("or even more" [16: fol. B 4 r°]), none of these was ever printed during his lifetime.[93]

Becher's one and only example of his universal character contained in the *Defensio Operis* that follows the Latin lists reproduces the dedication to Johann Philipp in his mathematical–combinatorial system [16: fol. N 5 v°]: "2770:169:3 / 6753:3 / 62 / 2614:3" stands for "*Eminentißimo Principi Electori ac Domino*", whereby 2770 is the place number (or "root" in Becher's terms) of *eminens* in his *Lexicon*, while 169 indicates the superlative and the last number, 3, the dative – thus *eminentissimo*. The following nouns – *Princeps* [Elector] and *Domino* – are again both in the dative case (3); the conjunction *ac* simply appears with its number, 62.

This is where the similarity with Silvestri, Bermudo and Kircher ends. Becher, who was also a commercial promoter at the Munich and Viennese courts, was among the earliest to envisage applications of such universal language schemes in international trade.[94] And he was the first to realise that the markets in the newly opened parts of the

world might not use Arabic numbers. To achieve total universality (or ethnic neutrality, one might say), he proposed a system of horizontal and vertical bars and dots that could serve as a non-biased means to transmit his numerical code[95] (see Fig. 11.27).

The cartouche-like frame indicates the number ("root") of each entry in the *Lexicon* in positions A to D (written from right to left, by the way). Single numbers are marked with dots in A (with a horizontal bar marking the number 5 in A, B, C, and D). Vertical bars in B signify tens, in C hundreds, in D thousands.[96] The graphic transposition of the 173 additional qualifiers into the universal character occurs in positions E to I. Thus Becher's system – as had already been discussed more than forty years ago, in the early days of the computer era [111: 47–57] – anticipates the binomial principle of this modern-day invention and lends itself to "machine translation".

After this look at one of the two pillars of Becher's theory of written communication, it should not surprise that the author considers any graphic transmission per se a form of secret writing, possibly of a steganographic nature – his own system being a case in point. Beyond that, various methods could increase transmission security, he adds in his introduction, referring to Trithemius and the cabbala [16: fols A 6 v°–A 7 v°]. At the end of the subsequent description of his book, he specifically proposes the cryptographic enhancement of his system by adding a numerical constant to both "root" and qualifiers in order to systematically render any transmission meaningless.

---

[93] Kaspar Schott, who compared Becher's and Kircher's systems in his 1664 *Technica curiosa* [88: 505–529], had access to a *Character Idiomate Germanico* (allegedly published in 1660) that cannot be found. Becher authorised this comparison and gave Schott a manuscript version of at least the German lists. Manuscripts of a glossary in 12 and in 34 languages have survived in Becher's files at Rostock University Library [101: 224–225].

[94] The very first suggestion of this kind was made by the Englishman Francis Lodwick (1619–1694), son of a Protestant Dutch immigrant and a French mother, who travelled all over Europe as a merchant. In several proposals (the earliest one dating back to 1647), he expressed his conviction that his *Common Writing* could be used in the expanding world trade. His materials remained virtually unknown until the twentieth century [102: 125–127].

[95] The attractive title page of Becher's *Character* shows an Egyptian obelisk inscribed with the "neutral", graphic version of his numerical code. In the preface to his readers, Becher discussed the fundamental importance of "graphics, that is the capacity for writing" for human society; in his own day, he posited, the art of writing could be subdivided into "hieroglyphics and steganography", thereby introducing the two pillars on which he wanted to build his *Character* [16: fols A 4 r°–A 5 r°; 101: 216–217].

[96] Although Becher's *Lexicon* comprises 10283 entries, he did not provide a place for more than four-digit numbers. Belatedly aware of this problem, he resorted to placing two horizontal bars (marking the number 5) above each other in D position (as seen in the lower right corner of Fig. 11.27).

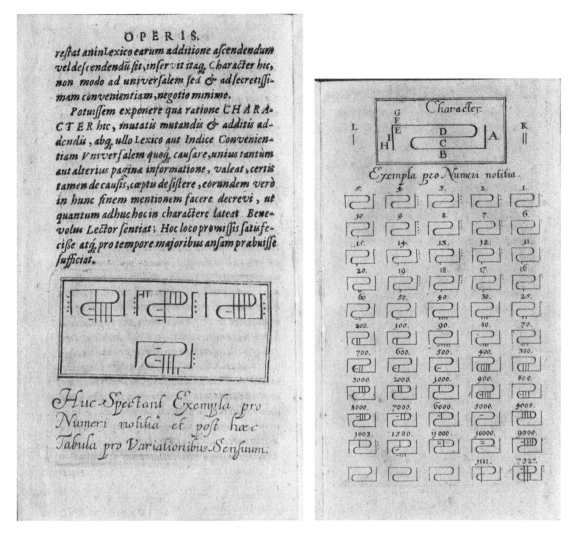

Figure 11.27. Becher's explanations and illustrations of his universal character. Courtesy Bayerische Staatsbibliothek, Munich.

The two correspondents would have to communicate this constant orally or in letters, the author suggests: stressing in the same breath, however, that he had not intended his invention for closed communication[97] [16: fols B 7 r°–v°]. Johann Christoph Sturm, the German scholar who actually prepared the missing polyglot listings fifteen years after Becher's death – and who only belatedly acknowledged the latter's authorship of the un-

derlying principle – also briefly discussed cryptographic enhancements of the combinatorial system and suggested the *multiplication* of the root number or code with an agreed-upon factor (instead of adding a constant, as Becher had proposed).[98]

---

[97]The inclusion of *inventum steganographicum* in the subtitle alone seems to contradict this assertion although Becher's interpretation of the art of writing ("hieroglyphics" and "steganography") is open to the assumption that any such system can be universally assimilated – in which case it would no longer be steganographic.

[98]In many ways, Becher's system caused a greater reaction in the scholarly community than Kircher's, not in the least on account of the detailed description of the version that Schott published in 1664 (see footnote 93): Sturm devoted more than twenty pages to Becher in 1676 (without ever mentioning his name) [104]; he did provide the missing polyglot lists, though. Not until 1685, when Sturm printed a brief clarification in his second volume, did he acknowledge Becher's authorship. Apart from drawing up these lists, Sturm's suggestion to *mul-*

11.3.6.2.2 *Kaspar Schott's cryptologic publications.* Kaspar Schott (1608–1666), a Jesuit from the Augsburg area, completed his studies in Sicily and then taught moral theology and mathematics in the Jesuit College at Palermo. Before being reassigned to the German province, he spent three years in Rome (1652–1655) as Kircher's editor and assistant. Since his mentor did not have time to publish all his findings, Schott considered it his duty to make them known in several of his own works, which all came out in Germany in the decade after his return to assume the post of professor of mathematics and ethics at the Jesuit College of Würzburg [86: 280–283].

We have already encountered Schott's most important work for this analysis, his 1664 *Technica curiosa*. The 1050-page tome is subdivided into twelve books that range in topics from an in-depth description of Otto von Guericke's famous "Magdeburg air-pump" to mechanical, hydrologic, chronometric, cabalistic and graphic "miracles". It is in this seventh section [88: 478–529] that Schott first described Kircher's intellectual priority following Emperor Leopold III's charge,[99] presented Pedro Bermudo's 1653 *Arithmeticus nomenclator*[100] and lastly analysed Becher's 1661 *Character*. The section presents the only sample lists beyond Latin[101] along with the only sample sentence in two languages and its transfer into the binomial graphic system Becher had invented. (Schott actually improved on Becher's method (see Fig. 11.29), allegedly because he could not print his graphics; his own version proceeds from left to right, and both the A–D and E–H groups are counted from the bottom up. The resulting binomial dot-and-bar sys-

tem would lend itself even more to computerised translation):[102]

23. 1.	15. 15.	35. 4.
Equus	comedit	gramen.
Das Pferdt	*isset*	*Graaß.*

In an overall assessment of the three systems, Schott refutes Becher's "universalist" call for a neutral character and suggests that those peoples which do not use Latin or Arabic numbers could easily learn them. Such numbers, Schott posits (thus indirectly favouring Kircher's system, where he also finds the best dictionaries), only have an auxiliary function and do not carry any intrinsic meaning – just as those in a nomenclator.

One more of Schott's works deserves a brief look: Only a year after the *Technica curiosa* with its cryptologic section, Schott published his most specific book, *Schola steganographica* [87].[103] To a much lesser degree than Duke August's or Schwenter's compendia, however, the book presents less of a systematic overview of the discipline than an in-depth analysis of Books II and III of Kircher's *Polygraphia nova* and an attempt at explicating Trithemius' cryptologic publications – unsuccessful as far as the problematic sections are concerned. Schott is well versed in the relevant literature, and while he seems to indulge in the construction of even further cryptologic boxes or tables, there is little new material [56: 154–155].[104]

---

tiply the root number instead of adding a constant was his only other original contribution [104; 110].

[99] Schott [88: 478–482]; also see Section 11.3.6.1.3.

[100] Schott [88: 483–505]; see Section 11.3.6.1.2.

[101] Schott [88: 508–510]; see Section 11.3.6.2.1. While the Latin *Lexicum* serves as an encode and decode for this reference language, two lists are needed for all other tongues as the German equivalent to the Latin reference word no longer retains the same, Latin, sequence, as the first few lines of Schott's sample list will show (which may have been provided by Becher himself) (see Fig. 11.28).

[102] For a comparison with Becher's own, much more "hieroglyphic" version of the universal character, see Fig. 11.27.

[103] The use of the adjective "*steganographica*" in the title is a misnomer (Schott had been referring to "*cryptographia*" or "*cryptologica*" up to this publication). He might have introduced "*steganographica*" since the term was in vogue at the time and could have been a selling factor. Schott's book clearly deals primarily with cryptological devices.

[104] The same holds true for Schott's various other publications that sport special sections on cryptology, such as his four volumes on *Magia universalis* of 1659, with the first book of Part IV devoted to *Magia cryptographica, et cryptologica* [89: IV, 1–90]. Similar sections can be found in several other of his works [99: 96–97; 109].

Lexicum [*sic*] Latinum A.		Lexicum Germanicum		Index Germanicum A.	
Ab	1	*Von*	1	*Abt*	4
Abacus	2	*Credentz Tisch*	2	*Angesicht*	28
Abavus	3	*Großvatter*	3	*Aufzug*	24
Abbas	4	*Abt*	4	(no more words provided in this column in Schott's sample	
Abbatia	5	*Abtey*	5	listing of the first five Latin words from A through H)	

Figure 11.28.

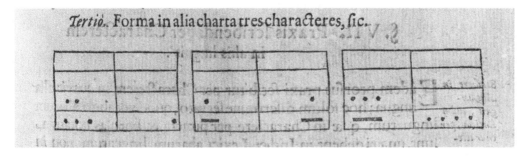

Figure 11.29. Schott's modification of Becher's universal character. Courtesy Bayerische Staatsbibliothek, Munich.

11.4 EUROPEAN CRYPTOLOGY IN THE RENAISSANCE – CONCLUSION AND OUTLOOK

The extensive view of cryptologic developments in major European centres[105] has shown how the discipline evolved along two lines that keep intersecting at various points. Contrary to the period before 1600, when so many of the major advances in the fields of cryptology and steganography occurred, the 17th century saw an increasing preoccupation of many cryptologists with linguistic matters, which had an impact on the systems they proposed: Francis Bacon, John Wilkins, but also the group around Bermudo, Kircher or Becher are primary proponents of this new interest. With clear exceptions – Rossignol's Great Cipher being one of them – the 17th century, then, is no longer the time of groundbreaking cryptologic inventions but frequently one of consolidation of earlier information.

While major advances in the field are thus rare in this later period, the increasing interest in universal communication (apart from Latin that saw its usefulness wane in the reaches beyond Europe) gave rise to a renewed preoccupation with the somewhat strange admixture between closed and open communication that had been suggested in the first quarter of the sixteenth century by Trithemius and Silvestri. Beginning with Pedro Bermudo's *Arithmeticus nomenclator* of 1653, scholars like Kircher, Becher and Schott developed what they called universal characters that also lend themselves to cryptographic use. From a mere cryptologic point of view, these systems appear to be un-

---

[105]This overview has not included states that are being discussed elsewhere in this volume, such as Holland. There are courts where cryptography certainly played an important rôle, such as in the Swedish army of the Thirty Years' War (see footnote 74). Nonetheless, after that cryptography does not seem to have been of too much concern in Sweden until the 20th century [56: 478]; in Russia, the beginnings of cryptological dealings coincide with the reign of Czar Peter the Great (1672–1725) and thus occur beyond the period under consideration here. – This analysis of cryptologic developments in the Renaissance also has not included the field of literature, where numerous rhetorical means were used during this period, in particular, for concealment. This could occur in the form of palindromes, acrostichons, chronostichons or chronograms, and similar techniques. Within the realm of literature, the vast areas of emblematics and hieroglyphics (as they were perceived in the Renaissance, in particular by Athanasius Kircher, and including Becher's similarly idiosyncratic view [see footnote 102]) invited cryptologic concealment and interpretation.

wieldy at first – though highly secure – but the numerical "interfaces" they employ actually are harbingers of a new technology, the commercial code books introduced in the 1850s in the exploding world of telegraph communication, in particular multilingual registers [56: 845–846]. The ingenious – and "value-neutral" – graphic rendering of such numerical interfaces as invented by Becher and refined by Schott, however, takes an even greater leap forward into the 20th century and its invention of the electronic computer.

ACKNOWLEDGEMENTS

I want to thank David Kahn for kindly perusing major portions of this material and for making valuable suggestions. Augusto Buonafalce has generously made available important materials, in particular an advance electronic copy of an article in press. I also want to express my appreciation to the Rare Books Departments of the Herzog August Bibliothek, Wolfenbüttel, Germany, and the Bavarian State Library in Munich for their assistance. Both institutions provided research carrels that greatly facilitated my work.

REFERENCES

[1] L.B. Alberti, *De componendis cyfris*, Available at http://www.apprendre-en-ligne.net/crypto/alberti/decifris.pdf.

[2] L.B. Alberti, *Dello scrivere in cifra* [= *De componendis cyfris*], ed. transl. from the Latin into Italian by A. Buonafalce, with a preface by David Kahn, Galimberti, Turin (1994).

[3] L.B. Alberti, *A Treatise on Ciphers* [= *De componendis cyfris*], ed. transl. from the Latin into English by A. Buonafalce, with a preface by David Kahn, Galimberti, Turin (1997).

[4] L.B. Alberti, *Opusculi morali*, transl. from the Latin into Italian by C. Bartoli, Franceschi, Venice (1568).

[5] I.A. Al-Kadi, *Origins of Cryptology: The Arab Contributions, Selections from CRYPTOLOGIA: History, People, and Technology*, Artech House, Boston, London (1998), pp. 93–122.

[6] M. Mrayati, Y.M. Alam, M.H. at-Tayyan (eds), **al-Kindi's Treatise on Cryptanalysis**, Series on Arabic Origins of Cryptology 1, King Faisal Center for Research and Islamic Studies, Riyadh, Saudi Arabia (2003).

[7] Anonymus, Manuscript, Vatican Archive, Arm. C. fasc. 78 nr. 5 (approx. 1330).

[8] Anonymus, *Neueröffneter Schauplatz geheimer philosophischer Wissenschaften, darinnen sowol zu der* [. . .] *Cryptologia, Cryptographia, Steganographia und Dechiffrirkunst Gehörige Anleitung Gegeben* [. . .] *Wird*, Montag, Regensburg (1770).

[9] M. Argenti, Unpublished Manual (1605), *Die Geheimschrift im Dienste der Päpstlichen Kurie* [. . .], A. Meister, ed., Quellen und Forschungen auf dem Gebiete der Geschichte 11, Schöningh, Paderborn (1906), pp. 54–65, 65–113.

[10] P.M. Arnold, *An apology for Jacopo Silvestri*, Cryptologia 4 (1980), 96–103.

[11] K. Arnold, *Johannes Trithemius (1462–1516)*, 2nd ed., Quellen und Forschungen zur Geschichte des Bistums und Hochstifts Würzburg 23, Schöningh, Würzburg (1991).

[12] F. Bacon, *Epistle on the secret works of art and the nullity of magic*, *The Cipher of Roger Bacon*, W.R. Newbold, ed., University of Pennsylvania Press, Philadelphia (1928), pp. 25–26.

[13] F. Bacon, *The Works of Francis Bacon*, in J.L. Spedding, R.L. Ellis and D.D. Heath, eds, 14 vols, Longman & Co, London (1858–1874).

[14] F.L. Bauer, *Decrypted Secrets: Methods and Maxims of Cryptology*, 3rd rev. edition, Springer, Berlin, London, New York (2002).

[15] Commandant [= Étienne] Bazeries, *Les Chiffres secrets dévoilés. Étude historique sur les chiffres*, Charpentier, Paris (1901).

[16] J.J. Becher, *Character, Pro Notitia Linguarum Universali. Inventum Steganographicum Hactenus Inauditum*, Ammon & Serlin, Frankfurt (1661).

[17] J.J. Becher, *Methodi Becherianæ didacticæ Praxis* [. . .], Zunner, Frankfurt (1669).

[18] C. Beck, *The Universal Character, By which all the Nations in the World may understand one anothers Conceptions, Reading out of one Common Writing their own Mother Tongues*, Maxey, London (1657).

[19] G.B. Bellaso, *Il Vero modo di scrivere in cifra con facilità, prestezza, et secvrezza* [. . .], Britanico, Brescia (1564).

[20] G.B. Bellaso, *La Cifra del Sig. Giovan Battista Belaso* [*sic*], Venice (1553).

[21] G.B. Bellaso, *Noui et singolari modi di cifrare* [. . .], Britanico, Brescia (1555).

[22] P. Bermudo, *Arithmeticus Nomenclator, Mundi omnes nationes ad linguarum & sermonis unitatem invitans, Auctore linguæ (quod mirêre) Hispano quodam, verè ut dicitur, muto*, Rome (1653), analysed in Kaspar Schott, *Technica Curiosa, sive, Mirabilia Artis libris XII. comprehensa*, Endter, Nuremberg (1664), pp. 478–505.

[23] N.L. Brann, *Trithemius and Magical Theology: A Chapter in the Controversy over Occult Studies in*

Early Modern Europe, SUNY Series in Western Esoteric Tradition, State University of Nyork Press, Albany, NY (1999).

[24] J.C. Briggs, *Francis Bacon and the Rhetoric of Nature*, Harvard University Press, Cambridge, MA (1989).

[25] I. Brückle, *The role of alum in historical papermaking*, Abbey Newsletter 17 (4) (1993), 53–57.

[26] A. Buonafalce, *Bellaso's reciprocal ciphers*, Cryptologia 30 (2006), 39–51.

[27] A. Buonafalce, *Sir Samuel Moreland's machina cyclologica cryptographica*, Cryptologia 28 (2004), 253–264.

[28] G. Cardano, *De rerum varietate libri XVII*, Petri, Basel (1557).

[29] G. Cardano, *De subtilitate libri XXI*, Petreius [= Petri], Nuremberg (1550).

[30] G. Chaucer, *The Equatorie of the Planetis*, D.J. Price, ed., Cambridge University Press, Cambridge (1955), Appendix I: Cipher Passages in the Manuscript.

[31] G. de Collange, *Polygraphie, et Vniverselle escriture Cabalistique de M.I. Tritheme Abbé*, Kerner, Paris (1561).

[32] A. Colorni, *Scotographia overo, scienza di scrivere oscvro, facilissima, & sicurissima, per qual si voglia lingua* [...], Sciuman, Prague (1593).

[33] C. Comiers (d'Ambrun), *L'Art d'Écrire et de parler occultement et sans soupçon*, Guerout, Paris (1690).

[34] H.N. Davies, *Bishop Godwin's 'Lunatique Language'*, Journal of the Warburg and Courtauld Institutes 30 (1967), 1068–1081.

[35] H.N. Davies, *The history of a cipher, 1602–1772*, Music and Letters 48 (1967), 325–329.

[36] J. Davys, J. Wallis, *An Essay on the Art of Decyphering: in which is inserted a discourse of Dr. Wallis, now first publish'd from his original manuscript* [...], L. Gilliver and J. Clarke, London (1737).

[37] K. de Leeuw and H. van der Meer, *A turning grille from the ancestral castle of the Dutch Stadtholders*, Cryptologia 19 (1995), 153–165.

[38] E. Dröscher, *Die Methoden der Geheimschriften (Zifferschriften) unter Berücksichtigung ihrer geschichtlichen Entwicklung*, Frankfurter Historische Forschungen, N.S. 3, Koehler, Leipzig (1921).

[39] T. Ernst, *Schwarzweiße Magie. Der Schlüssel zum dritten Buch der "Steganographia" des Trithemius*, Daphnis: Zeitschrift für Mittlere Deutsche Literatur 25 (1996), 1–205.

[40] T. Ernst, *The numerical-astrological ciphers in the third book of Trithemius' Steganographia*, Cryptologia 22 (1998), 318–341.

[41] J.A. Fabricius, *Decas decadum sive Centuria Plagiariorum & Pseudonymorum Centuria*, Lanckisch, Halle (1689).

[42] J. Falconer, *Cryptomenysis Patefacta: Or the Art of Secret Information Disclosed without a Key* [...], D. Brown, London (1685); 2nd edition, as *Rules for Explaining and Decyphering all Manner of Secret Writing* [...], D. Brown, London (1692).

[43] P. Findlen, *Introduction: the last man who knew everything ... or did he? Athanasius Kircher, S.J. (1602–80) and his world*, **Athanasius Kircher. The Last Man Who Knew Everything**, P. Findlen, ed., Routledge, New York and London (2004), pp. 1–48.

[44] J.B. Friderici, *Cryptographia oder Geheime schrifftmünd- und würckliche Correspondentz* [...] *zu machen und auffzulösen*, Rebenlein, Hamburg (1684 and 1685).

[45] J. Gadol, *Leon Battista Alberti: Universal Man of the Early Renaissance*, University of Chicago Press, Chicago/London (1969).

[46] J.C.G. Díaz, *Criptografía. Historia de la Escritura Cifrada*, Ed. Complutense, Madrid (1995).

[47] J.S. Galland, *An Historical and Analytical Bibliography of the Literature of Cryptology*, AMS Press, New York (1970), reprint of the edition published by Northwestern University Press, Evanston, IL (1945).

[48] F. Godwin, *The Man in the Moone: or a Discovrse of a Voyage thither* (1638), edited together with his 1629 work, *Nuncius Inanimatus*, by Grant McColley, Smith College Studies in Modern Languages XIX, No. 1, Smith College, Northampton, MA (1937).

[49] A. Grafton, *Leon Battista Alberti: Master Builder of the Italian Renaissance*, Hill & Wang, New York (2000).

[50] G.P. Harsdörffer, *Delitiæ* [later editions, *Deliciæ*] *Mathematicæ et Physicæ. Der mathematischen und philososphischen Erquickstunden Zweyter Theil*, Dümler, Nuremberg (1651 and later); *Dritter Theil*, Endter, Nuremberg (1653 and later).

[51] H. Hassinger, *Johann Joachim Becher 1635–1682: Ein Beitrag zur Geschichte des Merkantilismus*, Veröffentlichungen der Kommission für Neuere Geschichte Österreichs 38, Holzhausen, Vienna (1951).

[52] D. van Hottinga, *Polygraphie: ou Méthode universelle de l'escriture cachée* [...], Emden (1620); 2nd printing Groningen (United Provinces), no publisher (1621) [plagiarised version of Collange's 1561 translation of Trithemius' *Polygraphiae libri sex* of 1550].

[53] H. Hugo, *De prima scribendi origine et vniversa rei literariae antiqvitate* [...], Plantin, Antwerp (1617).

[54] M. Mrayati, Y.M. Alam, M.H. at-Tayyan (eds), *Ibn ad-Durayhim's Treatise on Cryptanalysis*, Series on Arabic Origins of Cryptology 3, King Faisal Center for Research and Islamic Studies, Riyadh, Saudi Arabia (2004).

[55] D. Kahn, *On the origin of polyalphabetic substitution*, Isis 71 (1980), 122–127.

[56] D. Kahn, *The Codebreakers: The Story of Secret Writing*, rev. edition, Scribner, New York (1996).

[57] F.W. Kasiski, *Die Geheimschriften und die Dechiffrir-Kunst*, E.S. Mittler, Berlin (1863).

[58] T. Kelly, *The myth of the Skytale*, Cryptologia 22 (1998), 244–260.

[59] A. Kircher, *Ars magna Lvcis et Vmbræ* [...], Scheus, Rome (1646).

[60] A. Kircher, *Magnes siue De arte magnetica opvs tripartitvm* [...], Grignani, Rome (1641).

[61] A. Kircher, *Mvsvrgia Vniversalis, sive Ars magna Consoni et dissoni*, vol. I: Corbelletti, vol. II: Grignani, Rome (1650).

[62] A. Kircher, *Polygraphia nova et vniversalis ex Combinatoria Arte Detecta*, Varesius, Rome (1663).

[63] T.(P.) Leary, *Cryptology in the 15th and 16th Century* [sic], Cryptologia 20 (1996), 223–242.

[64] N. Machiavelli, *Il Libro dell'Arte della Guerra*, [Heirs to P. di] Giunta, Florence (1521).

[65] G. Mancini, *Vita di Leon Battista Alberti*, Carnesecchi, Florence (1882).

[66] G. McCracken, *Athanasius Kircher's universal polgygraphy*, Isis 39 (1948), 215–228.

[67] A. Meister, *Die Anfänge der modernen diplomatischen Geheimschrift: Beiträge zur Geschichte der italienischen Kryptographie des XV. Jahrhunderts*, Schöningh, Paderborn (1902).

[68] A. Meister, *Die Geheimschrift im Dienste der Päpstlichen Kurie* [...], Quellen und Forschungen auf dem Gebiete der Geschichte 11, Schöningh, Paderborn (1906).

[69] C.J. Mendelsohn, *Bibliographical note on the "De Cifris" of Leone Battista Alberti*, Isis 32 (1940), 48–51.

[70] C.J. Mendelsohn, *Cardan on cryptography*, Scripta Mathematica 6 (1939), 157–168.

[71] S. Morland, *A New Method of Cryptography, Humbly presented to The Most Serene Majesty of Charles the II.* [...], n. p., London (1666).

[72] T. Morus, *De optimo reip. Statu deque noua insula Vtopia libellus uere aureus* [...], Froben, Basel (1518).

[73] D.E. Newton, *Encyclopedia of Cryptology*, ABC-CLIO, Santa Barbara, CA (1997).

[74] P. Pesic, *François Viète, father of modern cryptanalysis – two new manuscripts*, Cryptologia 21 (1997), 1–29.

[75] P. Pesic, *Secrets, symbols, and systems: parallels between cryptanalysis and algebra, 1580–1700*, Isis 88 (1997), 674–692.

[76] P. Pesic, *The clue to the labyrinth: Francis Bacon and the decryption of nature*, Cryptologia 24 (2000), 193–211.

[77] K. Pommerening, *Kryptologie: Die Erfindung der polyalphabetischen Verschlüsselung* (internet publication 1997, revised 2004), available at http://www.uni-mainz.de/~pommeren/Kryptologie/Klassisch/2_Polyalph/Renaissance.html (2004).

[78] G.B. (della) Porta, *De fvrtivis literarvm notis, vvlgò De Ziferis libri IIII*, Scotti, Naples (1563); 2nd expanded edition, Naples (1602).

[79] G.B. (della) Porta, *Magia natvralis sive de miraculis rerum naturalium libri IIII*, Cancer, Naples (1558).

[80] J. Reeds, *Solved: The ciphers in book III of Trithemius' Steganographia*, Cryptologia 22 (1998), 291–317.

[81] J. Reuchlin, *De arte cabalistica libri tres*, Anshelm, Hagenau (1517).

[82] M. Ricci, *De Christiana expeditione apud Sinas suscepta ab Societatis Jesu* [...], Mang, Augsburg (1615).

[83] L. Sacco, *Un primato italiano. La crittografia nei secoli XV e XVI*, Giacomaniello, Rome (1947).

[84] I. Sample, *Messages of fear in hi-tech invisible ink*, Manchester Guardian, August 11, 2005.

[85] H. Saussy, *Magnetic language: Athanasius Kircher and communication*, **Athanasius Kircher. The Last Man Who Knew Everything**, Paula Findlen, ed., Routledge, New York and London (2004), pp. 263–281.

[86] K. Schott, *La "Technica curiosa" di Kaspar Schott*, M. J. Gorman and N. Wilding, eds, M. Sonnino (transl.), Collana Tecnica curiosa 3, Edizioni dell'Elefante, Rome (2000).

[87] K. Schott, *Schola steganographica, in classes octo distributa* [...], Endter, Nuremberg (1665).

[88] K. Schott, *Technica Curiosa, sive, Mirabilia Artis libris XII. comprehensa*, Endter, Nuremberg (1664).

[89] K. Schott, *Thaumaturgus Physicus, sive Magiæ universalis naturæ et artis Pars IV.* [...], Schönwetter, Würzburg (1659).

[90] D. Schwenter, *Deliciæ Physico-Mathematicæ. Oder Mathemat: vnd philosophische Erquickstunden* [...], Dümler, Nuremberg (1636, reissued 1651 as *Theil I*).

[91] D. Schwenter, *Steganologia & Steganographia aucta, Geheime, Magische, Natürliche Red vnd Schreibkunst* [...], Halbmayer, Nuremberg (between 1615 and 1620); Another, revised and corrected edition, Dümler, Nuremberg (after 1620).

[92] D. Schwenter, *Steganologia & Steganographia nova, Geheime, Magische, Natürliche Red vnd Schreibkunst*, Halbmayer, Nuremberg (approx. 1610).

[93] G. Selenus [pseudonym for Duke August the Younger of Brunswick-Wolfenbüttel], *Cryptomenytices et Cryptographiæ Libri IX. In quibus & planißima Steganographiæà Johanne Trithemio* [...] *admirandi ingenij Viro, magicè & ænigmaticè olìm conscriptæ, Enodatio traditur*, Gebrüder Stern, Lüneburg (1624).

[94] W. Shumaker, *Renaissance Curiosa*, Medieval and Renaissance Texts and Studies 8, Center for Medieval and Early Renaissance Studies, Binghampton, NY (1982).

[95] J. de Silvestri, *Opvs novvm* [...], Rome (1521).

[96] S. Singh, *The Code Book: The Evolution of Secrecy from Mary Queen of Scots to Quantum Cryptography*, Doubleday, New York (1999).

[97] S. Singh, *The Science of Secrecy: The Secret History of Codes and Codebreaking*, Fourth Estate, London (2000).

[98] D.E. Smith, *John Wallis as a cryptographer*, Bulletin of the American Mathematical Society 24 (1918), 82–96.

[99] G.F. Strasser, *Die kryptographische Sammlung Herzog Augusts: vom Quellenmaterial für seine "Cryptomenytices" zu einem Schwerpunkt in seiner Bibliothek*, Wolfenbütteler Beiträge 5 (1982), 83–121.

[100] G.F. Strasser, *Herzog Augusts Handbuch der Kryptographie: Apologie des Trithemius und wissenschaftliches Sammelwerk*, Wolfenbütteler Beiträge 8 (1988), 99–120.

[101] G.F. Strasser, *Johann Joachim Bechers Universalsprachenentwurf im Kontext seiner Zeit*, **Johann Joachim Becher (1635–1682)**, G. Frühsorge and G.F. Strasser, eds, Wolfenbütteler Arbeiten zur Barockforschung 22, Harrassowitz, Wiesbaden (1993), pp. 215–232.

[102] G.F. Strasser, **Lingua Universalis: Kryptologie und Theorie der Universalsprachen im 16. und 17. Jahrhundert**, Wolfenbütteler Forschungen 38, Harrassowitz, Wiesbaden (1988).

[103] G.F. Strasser, *Musik und Kryptographie*, **Die Musik in Geschichte und Gegenwart: Allgemeine Enzyklopädie der Musik**, 2nd edition, L. Finscher, ed., vol. 6, Kassel [et al], Bärenreiter (1997), pp. 783–790.

[104] J.C. Sturm, [discussion of J.J. Becher's **Character**], **Collegium experimentale** [...], Part I Endter, Nuremberg (1676), pp. 74–99; Part II (1685), pp. 108–111.

[105] G. Tortelli, **Commentarium grammaticorum de orthographia dictionum e Graecis Tractarum proemium** [...], Jenson, Venice (1471, and later editions).

[106] Jé.P. Devos, H. Seligman (eds), **Traitté de l'Art de deschiffrer** [...], Université de Louvain, Travaux d'Histoire et de Philologie, 4e série, fasc. 36, Publications Universitaires de Louvain, Louvain (1967).

[107] J. Trithemius, **Polygraphiae libri sex**, Haselberg [of Aia], Basel (1518).

[108] J. Trithemius, **Steganographia. Hoc est: Ars per occvltam scriptvram animi svi volvntatem absentibvs aperiendi certa** [...]. **Praefixa est hvic operi sva clavis**, Berner, Frankfurt (1606).

[109] B. de Vigenère, **Traicté des chiffres, où secretes manieres d'escrire**, l'Angelier, Paris (1586 and 1587).

[110] J. von zur Gathen, *Claude Comiers: The first arithmetical cryptography*, Cryptologia 27 (2003), 339–349.

[111] W.G. Waffenschmidt, ed., **Allgemeine Verschlüsselung der Sprachen** [...]: **J. J. Becher. Zur mechanischen Sprachübersetzung. Ein Programmierungsversuch aus dem Jahre 1661**, Veröffentlichungen der Wirtschaftshochschule Mannheim, Reihe 1, 10, Kohlhammer, Stuttgart (1962).

[112] F. Wagner, *Studien zu einer Lehre von der Geheimschrift (Chiffernkunde)*, Archivalische Zeitschrift (Cologne) 12, Part II (1887), 1–29.

[113] S.E. Whyman, Postal Censorship in England 1635–1844, Available at http://www.postcomm.gov.uk/about-the-mail-market/uk-market-reviews/postalcensorship.pdf (2003).

[114] N. Wilding, *"If you have a secret, either keep it, or reveal it": cryptography and universal language*, **The Great Art of Knowing. The Baroque Encyclopedia of Athanasius Kircher**, D. Stolzenberg, ed., Stanford Univ. Press, Stanford (2001), pp. 93–103.

[115] N. Wilding, *Publishing the polygraphy: manuscript, instrument, and print in the work of Athanasius Kircher*, **Athanasius Kircher. The Last Man Who Knew Everything**, P. Findlen, ed., Routledge, New York and London (2004), pp. 283–296.

[116] J. Wilkins, **An Essay Towards a Real Character, And a Philosophical Language**, Gellibrand and Martin, for the Royal Society, London (1668).

[117] J. Wilkins, **Mercury; or, The Secret and Swift Messenger: Shewing, How a Man may with Privacy and Speed communicate his Thoughts to a Friend at any Distance**, I. Norton, London (1641).

[118] J. Wilkins, **The Mathematical and Philosophical Works** [...]. **Containing**, [...] **III. Mercury: Or, The Secret and Swift Messenger** [...]. J. Nicholson, London (1708).

[119] S. Zimmerman[n], **Etliche fürtreffentliche Gehaimnussen, verborgne, Mechanische, Apocryphische, vnd gleichsamb vbernatürliche Künsten, das Lesen vnd die Schreibery betreffendt**, author's press, Ingolstadt. It is attached to the author's **Titularbüech** (1579).

The History of Information Security: A Comprehensive Handbook
Karl de Leeuw and Jan Bergstra (Editors)
© Published by Elsevier B.V.

12

CRYPTOLOGY IN THE DUTCH REPUBLIC: A CASE-STUDY[1]

Karl de Leeuw

Informatics Institute
University of Amsterdam, The Netherlands
E-mail: karl.de.leeuw@xs4all.nl

Contents

Abstract

Mail interception and code-breaking played an important role in Dutch history. Already during the Revolt against Spain, the letters of King Philip to his governors and commanders were intercepted and decrypted on a regular basis. Later, Stadholder Frederick Henry's personal secretary, the playwright Constantijn Huygens, assumed such duties during military campaigns. After the Peace of Westphalia, the interest in code-breaking ceased. It was revived during the War of the Succession, when a Dutch code-breaker managed to monitor British diplomacy in the southern Netherlands. In 1751 a new Black Chamber saw the light to monitor Prussian and French diplomatic activity, which lasted until 1803. Until 1753 there was no effort to transfer code-breaking skills form one generation to another and neither were code-breakers involved in the construction of codes. Therefore the influence of cryptographic examples given in literature were quite marked.

Keywords: Dutch Revolt, War of the Spanish Succession, Seven Years' War, revolutionary wars, earlymodern diplomacy, mail interception, black chambers, earlymodern cryptography, book ciphers, Vigenere, turning grilles, two-part codes, empiricism, Pierre Lyonet, Samuel Egbert Croiset, Willem Jacob 's Gravesande, Abel Tassin d'Alonne, Constantijn Huygens the Elder, Marnix van Sint Aldegonde, Philip of Spain, Stadholder Frederick Henry, King-Stadholder William III, Grand Pensionary Anthony Heinsius, Stadholder William V, Princess Wilhelmina of Prussia, Grand Pensionary Laurens Pieter van de Spiegel, Anglo-Dutch relations.

12.1 INTRODUCTION

Contemporary warfare and espionage would be inconceivable without 'signals intelligence' or, abbreviated, 'sigint'; the analysis of intercepted messages of the enemy. Its role became paramount

---

[1]This contribution is an altered version of a chapter published previously in my dissertation "Cryptology and statecraft in the Dutch Republic" (Amsterdam 2000).

during the First World War mainly due to the introduction of the wireless as a means of communication between combat units and their military headquarters behind the frontlines and its importance has grown ever since, now involving permanently stationed intelligence units all over the world comprising of thousands of men [30: 1–16; 101: 298–299, 348–350]. In his famous *Codebreakers* David Kahn has presented us with several cases in which intelligence of this kind played a decisive role in the outcome of history. This happened for example in 1918 when French cryptanalysts managed to identify well in advance the exact spot where the Germans were to launch their final offensive and it happened again in 1941, at the eve of the outbreak of the war in the Pacific, when intercepted Japanese code material convinced Roosevelt that the Japanese had no intention whatsoever of settling the differences by negotiation [101: 1–5, 54–67, 340–347]. The latter also makes clear that, at least during the years immediately preceding the Second World War, sigint's role was not limited to warfare: it was just as important in the world of diplomacy, mainly in order to get a better assessment of the other countries goals and intentions. This may hardly seem surprising to us now but it constituted nonetheless a clear violation of diplomatic immunity and international law for which there seemed to exist no precedent whatsoever in times of peace. Particularly in Western democracies the monitoring of all communication lines of foreign embassies was re-introduced only reluctantly and relatively late because it seemed to terminate a long tradition of moral rectitude in public affairs. To put it in the words of the American secretary of State for Foreign Affairs during the early thirties, Henry L. Stimson: "Gentlemen don't read each other's mail".[2]

Kahn's book demonstrates, however, that the resistance and, especially, the moral indignation in the western world were partly caused by a short memory. During much of the early modern period it had been common practise to open and copy

diplomatic despatches as soon as they reached the central post office to be delivered or to be sent abroad, a habit that ceased to exist only as a consequence of 19th century liberalism. Most major powers sooner or later acquired signals units of their own, the so-called Black Chambers, solely responsible for the opening, copying and decoding of diplomatic despatches [101: 157–188]. In London the Black Chamber was located in a separate quarter of the Post Office that could only be entered from the Controller's House on Abchurch lane and it was attended by a staff permanently lodged in a nearby room. It was managed under direct supervision of the Director of the Post Office and strictly off-limits for ordinary employees. The salaries were paid, however, by the Foreign Office, and the despatches to be opened were identified through a general warrant, issued by one of the Secretaries of State [72: 62–65, 127–131, 138–142]. In France and Austria similar arrangements were made [79; 152: 92–107]. There is little doubt that these Black Chambers were part and parcel of the 18th century diplomacy, but in all other respects their role in history is still somewhat of an enigma. It is clear, however, that they first occurred in Renaissance Italy with the emergence of permanent diplomacy and that they subsequently found their way to other countries, with the conclusion of the Peace of Westphalia in 1648 and the beginning of permanent relations between Protestant and Catholic Courts probably acting as some sort of watershed [101: 125–156; 114; 115; 134]. From this time onwards, codes and ciphers were to be used regularly in diplomatic correspondence in as much as it was confidential and no messengers were available to carry them. The importance of this often-neglected aspect of foreign relations is still a matter of debate. In *The Western European Powers, 1500–1700* C.H. Carter states that the use of codes and ciphers during the early modern period was futile, because, generally speaking, due to the bad organisation of state bureaucracies, any document could be obtained at will by whoever was willing to pay.[3] Geoffrey Parker, however,

---

[2]See [101: 360]; for the emergence of the American Black Chamber see also [181]; about Stimson in particular [119: 639; 146: 188]; for a similar development in The Netherlands see [86: 57, 58].

[3][47: 233]; Carter acknowledges, however, that, occasionally, code-letters would be intercepted and broken and he even describes one of those cases himself in [48: 137–138].

only partly confirms this view. In his article about the Spanish Armada of 1588, he states that the British did, in fact, obtain much of their information directly from oral sources, but that this did not at all mean that the interception of letters was superfluous because they could not afford to remain solely dependent on hearsay. "What Elizabeth's ministers desperately needed", Parker writes, "was hard information from the pen of the king himself" and this they failed to intercept, because, "Philip II and his ministers studiously avoided writing down anything that might reveal the Armada plan to a third party" [123]. H.H. Rowen, writing about the middle of the 17th century, also states that the use of codes and ciphers was futile, albeit for different reasons than the previous authors. Rowen, taking for granted what the Dutch Grand Pensionary Johan de Witt says in one of his letters, believes that the Grand Pensionary's contemporaries would not have bothered at all, because any code or cipher to be devised could also be broken without much effort [129: 251]. What these authors all have in common is that they don't want to pay too much attention to the subject, because it seems irrelevant to them, at least for the period in history that they are dealing with. As a consequence, they are unable to account for the fact that, during the 18th century virtually all over Europe diplomatic despatches were intercepted on a daily basis and that it took highly trained professional cryptanalysts to make them accessible. This fact alone seems to indicate that there was no alternative and, secondly, that the codes and ciphers used, were sufficiently sophisticated to afford protection from almost anybody. This leaves us with a number of questions, not only with regard to the impact of these intelligence units on, broadly speaking, 18th century politics, but also with regard to the diffusion of cryptological knowledge from one country to another. The nature of the enterprise probably prohibited any direct exchange of information, but it seems unlikely that the development of Black Chambers in so many different countries in roughly the same period should have occurred only by chance. Which countries were included in this development and which were not? Was there any relation with the need to share information in

the context of alliances or was it rather obtained through bribery and espionage? Was there any relation at all with the rise of scientific thinking in various countries? And, last but not least, what role was played by the diffusion of cryptographic literature, most of it being written, as Kahn has noted, long before the rise of the Black Chambers outside Italy had begun [101: 156].

No effort has been made to answer all these related questions in a systematic fashion, but recently there has been some debate with regard to the diffusion of cryptographic literature that allows us to get to the depth of the problem. "These books", Kahn writes, "have a certain air of unreality about them (. . .). The literature of cryptology was all theory and no practise. The authors did not know the real cryptology that was being practised in locked rooms here and there throughout Europe, by uncommunicative men working stealthily to further the grand designs of state" [101: 156]. Gerhard F. Strasser, however, puts it differently. Like Kahn he signals a decline of cryptological literature during the 18th century, related to the rise of the Black Chambers. "It would have been counterproductive to communicate any new insight or development to the general public", Strasser says, "because the training of cryptologists had to remain exclusively an in-door matter" [147: 249]. According to Strasser, in the 16th and 17th centuries, this wasn't yet the case and therefore it might be possible to find links between the practise and development of cryptology and other bodies of knowledge, precisely through the examination of these books [147: 66]. Surprisingly, Strasser stresses, among other things, the relation with the rise of artificial languages [147: 13–14]. Strasser's chief protagonist is Athanasius Kircher who occasionally decoded intercepted Swedish letters on behalf of the German Emperor Ferdinand III. Kircher was interested in artificial languages as a means of facilitating written communication between scholars that spoke different languages. In a way, he was on the look out for a substitute for Latin but the whole enterprise also involved an effort to devise a language that was, unlike the existing ones, structured logically in the mathematical sense of the word [49: 49–60]. Originally there was no cryptological intention behind Kircher's language schemes but the third part of his *Polygraphia Nova* is solely devoted to cryptological aspects and possibilities of his system, integrating them with insights drawn from the classical authors in this field such as Trithemius, Porta, Alberti and Vigenère [49: 153–172].

It is important to note that this difference in approach (between Strasser and Kahn) really involves two sets of problems that only partially overlap. The first set relates to the way ciphers and codes were actually used. The second set has to do with supposed or possible links between the development of cryptology and other bodies of knowledge. Kahn's position can be roughly summarised as follows. All important works about cryptography that were written during the 16th and 17th centuries were of no consequence because they mainly deal with ciphers whereas, in real life, codes were being used almost on an exclusive basis [101: 125–126, 147, 160–161]. More often than not, these codes were devised by the same people that were engaged in breaking them; people who did not write books. Therefore, in as much as the use of codes during the early modern period shows any development at all, it was obviously caused by a dynamics of its own. Kahn explicitly rules out any exterior influence. "Before Friedman", Kahn writes, "cryptology eked out an existence as a study unto itself as an isolated phenomenon, neither borrowing from nor contributing to other bodies of knowledge. Frequency counts, linguistic characteristics, Kassiski-examinations – all were peculiar and particular to cryptology. It dwelt as a recluse in the world of science. Friedman led cryptology out of this lonely wilderness and into the broad, rich domain of statistics". This all happened only relatively recently, during the second quarter of this century [101: 383–384]. Obviously, Strasser who likes to think of the development of cryptology in terms of 'traditio et innovatio', basing his judgement, among other things, on the revival of the interest in kabbalah as manifested in the works of Trithemius and Vigenère, takes the opposite view [147: 66]. Although Kahn is the authority and Strasser is not, there are some elements in the older literature on the subject that indicate that he

may have a point. Aloys Meister states in his book on the papal ciphers that the use of codes and ciphers was changed dramatically during the 15th century as a consequence of the invention of the cipher wheel and that Vigenère was well aware of this development when he wrote his book almost one hundred years later [114: 25–26, 43]. Moreover, Charles J. Mendelsohn, carefully examining all influences at play in Vigenère's presentation of his famous 'Chiffre Carré', mentions a cabbalistic treatise by the name of 'Sepher Yetzira', that should have been of consequence, albeit not properly understood [116]. Of course, these examples do not prove that links like these still existed in a later period, but it seems obvious that this cannot be totally ruled out either. For the sake of argument it should be noted, however, that other positions could be taken as well. It is also possible to presume, for instance, that there was a link between cryptology and other fields of knowledge but that it didn't express itself in the classic highlights of cryptographic literature, always quoted in this respect. Or, to put it differently, there may be some connection between the advance in scientific thinking and the rise of the Black Chambers, independent of cryptographic literature. If there is, Strasser's argument in favour of a link with artificial languages is somewhat of a problem. The rise of artificial languages was a temporary, rather hybrid, phenomenon caused in part by a Baconian, fundamentally nominalistic, criticism of natural language and in part by a Aristotelian, that is a fundamentally realistic, belief that the taxonomies of language and nature were fundamentally identical and that absolute truth manifested itself in both. This attitude can be found with a great many scientists in the 17th and 18th centuries, among them Kircher and Leibniz, but it wasn't at all the direction Western thinking was about to take. This can best be demonstrated in the field of linguistics where the quest for artificial languages remained of no consequence at all. Far more important for the development of this field of knowledge were the contributions of John Wallis and the grammarians of Port Royal who stressed the rationality of natural languages as developed over the course of time.

The now almost forgotten Dutch linguist P.A. Verburg, writing about the history of linguistics during the early fifties, that is before Noam Chomsky revived the interest in the topic on a wider scale, coined the terms *scientalistisch rationalisme* to denote the first position and *practicaal rationalisme* to denote the latter [169: 248–250, 282, 322–323]. The latter was as much a result of the mechanistic-empirical approach of nature as developed by Newton, Boerhaave and Swammerdam as of a Cartesian quest for a logical order in all things. It was above all pragmatic; oriented towards the seeking of truth to be inferred indirectly through the comparing of analogies rather than to be revealed in one way or another [44: 39–44, 48, 49, 58–59, 100–108; 169; 139: 189, 194, 204, 218]. Or, to put it in other words, if there is to be any relation between the development of cryptology and other fields of knowledge at all, it is far more likely to be found elsewhere, either in the increased understanding of language due to the rise of linguistics or else, to the rise of empirical thinking in general. John Wallis, who was the first to develop an original and adequate model of English grammar, not dependent on any previous Latin example and who was also, so to speak, the founding father of the British Black Chamber, is a case in point [39: 309; 101: 166–169; 170: 20–29, 421–426].

This brings me to a second problem. The rise of the Black Chambers roughly coincides with what has been called by Ian Hacking *the emergence of probability* and it doesn't seem totally out of the question that there has been a connection of one kind or another. Kahn has made abundantly clear that the actual borrowing of a sophisticated, statistical instrumentarium by cryptanalysis is not to be expected before the rapid transformation of cryptography during the First World War had exhausted all traditional methods and this roughly coincides with the development of statistics as a science in its own right. The origin of probability theory, however, is much older, but it had to develop in the context of other disciplines. "In the 17th century", Hacking writes, "insurance and annuities were a focus of attention, in the 18th century a theory of measurement was needed, chiefly but not solely

for astronomy; in the latter part of the 19th century analysis of biological data demanded a mathematics that created biometrics" [84: 4].[4] Hacking stresses that probability wasn't the invention of one man: "In 1657 Huygens wrote the first probability textbook to be published; about that time Pascal made the first application of probability reasoning to problems other than games of chance, and thereby invented decision theory (...). Simultaneously, Leibniz thought of applying metrical probabilities to legal problems and annuities were being based on a sound actuarial footing by Hudde and De Witt (...). In short, around 1660 a lot of people independently hit on the basis of probability ideas (...)" [84: 11].

It is for our purpose also interesting to note how Hacking defines his subject: "(...) the probability that emerged so suddenly is Janus-faced. On the one side it is statistical, concerning itself with stochastic laws of chance processes. On the other it is epistemological, dedicated to assessing reasonable degrees of belief in propositions quite devoid of statistical background" [84: 12]. Surprisingly, this definition is equally applicable to cryptanalysis, being as it is, partly dependent on frequency counts and partly on what can be expected on the basis of syntax and what is known about the vocabulary likely to be used.

Obviously, one cannot hope to solve this problem by treating the history of cryptology as an isolated phenomenon. The contribution of men of science can not be investigated by restricting oneself solely to an examination of the cryptological literature, figuring so prominently in both Strasser's and, less marked and in a much different way, Kahn's reasoning. The research should, rather, focus on the people that were actually engaged in the field, either as the constructors of codes and ciphers or else as code-breakers, and, of course, on the nature of their commitment. This approach has the great advantage of covering a number of questions at the same time. First of all, it enables us to establish the exact nature of the conditions that gave rise

---

[4]I should like to thank professor H. Bos who pointed out to me that there may have been a connection between cryptology and probability theory already at this early stage and who gave me some relevant references.

to the application of codes and ciphers and to see how it was done. This will allow us also to investigate simultaneously the influence of cryptological literature and the contribution of men of science. Thirdly, it will show to what degree any development that occurred was due to an exchange of information with other countries, either through bribery or espionage or otherwise within the context of friendly collaboration. This last question clearly indicates that our research should also be concerned with the larger frame-work of international relations and the differences, temporary or otherwise, that may have occurred between various countries in the field of cryptology and on the impact this may have had on the competition for power. Of course, these questions have to be answered for every other country and the obvious way of doing so is to conduct a number of case studies. In a survey like this, the Dutch Republic merits our special attention, not only because of its paramount role in the advancement of science, but also due to its important role in the wars of the 17th and the early 18th centuries without being a great power in its own right an even lacking the ambition of becoming one. In this article, I shall first say something about Dutch society and Dutch politics in general, and show in what ways Dutch culture contributed to the rise of science. In the second part I shall come to the role of cryptography and code-breaking during the long conflict with Spain that lasted from 1568 until 1648. In the third part I shall treat the same topics for the period of standing diplomacy and coalition warfare, roughly between 1650 and 1750. In the fourth paragraph I shall deal with the changes that occurred after 1750, mainly in the use of codes by the Dutch government. In the fifth part I shall devote my attention to the increasing use of ciphers towards the end of the 18th century outside the state bureaucracy, mainly as a consequence of the political turmoil otherwise known as *the Age of the Democratic Revolution*.

12.2 THE CASE OF THE DUTCH REPUBLIC: GENERAL BACKGROUND

The Dutch Republic owed its independence to a revolt against its hereditary ruler, King Philip II

of Spain that was decided mainly during the last decade of the 16th century. The King of Spain had been engaged everywhere in his dominions in a relentless effort to centralise government at the expense of freedom of conscience and religion. Not surprisingly, the Dutch Revolt resulted in a political system that achieved the opposite: the full restoration of the ancient privileges of each province and town to mind its own business with the notable exceptions of foreign policy and the defence of the realm. This domain was left to the States-General, a body consisting of representatives of the provinces.[5] The matter of religion was solved by demanding that Calvinism should be the religion of everybody who held office and proclaiming full religious toleration for everybody else except for Roman-Catholics who were not allowed to practise their religion openly.[6]

The extreme wealth, due to a unprecedented economic expansion, and the confederate structure of the new Republic contributed to a climate that was highly favourable to the development of science, resulting in the founding of new universities in Leiden in 1575, in Franeker in 1585, in Groningen in 1614, and in Utrecht in 1636 [94: 569–575]. Another important tendency, however, was the rise of a new class of engineers, land surveyors and instrument builders outside the academic world that contributed to the development of a new category of applied sciences, oriented towards the solution of practical problems, such as the correct measurement of distances and weights and the calculation of movement [154: 41–44]. From the end of the 16th century onwards this manifested itself also, apart from the obvious expansion of shipping and overseas trade, in a Dutch dominance in such diverse fields as drainage and land reclamation, mining, harbour engineering and fortification building [94: 271–275]. In an ideological sense, the ground for this new development had been prepared by Calvinism, that showed a respect for manual labour wholly foreign to the Greco–Roman heritage that had dominated medieval thinking and

---

[5]For a recent, more extensive treatment of the political institutions of the Republic see [94: 276–306].

[6]The complicated matter of religion is also treated in this book. See [94: 361–398].

that was essentially biblical in inspiration [90: xii, xiii, 75]. Calvinism also fostered a new approach of nature that ultimately resulted in the triumph of the mechanistic–empiristic worldview over Aristoteleism and Cartesianism, associated with Newton and his followers [90: 29–30, 44, 51; 94: 581–591]. The Dutch Republic was an important battleground for the new science, but its influence was by no means exclusive or uncontested. Descartes had always exercised a considerable influence too, not only in the field of theology that gave rise to one of the few intellectual controversies that ever resulted in government intervention, but also in such diverse fields as mathematics and linguistics [94: 565–591, 889–993; 154: 38–45]. In fact, the discovery that Greek was totally different from Latin and merited a discipline of its own, was due to a student of Descartes: Tiberias Hemsterhuis whose pupils were instrumental in bringing about the Greek renaissance in Germany two generations later, associated with Winckelmann [81: 1–7, 107–110, 371]. In the field of mathematics proper, the Dutch Republic offered a less fertile ground however. At the beginning of the 18th century, the stimulating influence exercised by Cartesianism was still noticeable, but this did not mean that mathematics acquired recognition as a science in its own right. It was mainly seen as a necessity for the study of astronomy, physics, or military engineering. This development may partly be explained by the success of the empirical approach of the new physics, partly by an antiquated organisation of the university curricula that only allowed for some mathematical training during the initial years [35]. This did not preclude the Dutch scientific community, however, from participating in the debate about probability theory. The contribution of Christiaan Huygens has already been noted and clearly shows that the topic continued to be of interest to the Dutch from the very moment the concept evolved around 1660. Others could be mentioned too: Jacques Bernoulli for instance who formulated, during his stay at Groningen, the first limit theorem of probability [84: 125]. No doubt, this was motivated by the strong practical value of the issue, in such diverse fields as the organising of lotteries for the financial benefit of the state, insuring shipping, or calculating annuities.

Significantly, the booming of linguistics did *not* coincide with a rise in the interest in artificial languages or any efforts to devise one for that matter. Obviously, the utilitarian nature of Dutch science did not really encourage any efforts in this direction, at least not so during the latter part of the 17th century when French had already replaced Latin as a means of communication among nations. There may have been theological considerations at play too, however. Inasmuch as there was any debate on the subject at all, Dutch thinkers took the position that the quest for artificial languages was futile because mankind had already known a perfect language before the confusion of tongues. The remains of this original language were still around and they also believed firmly that it could be restored to its former glory. They only differed in opinion as to the question of whether it was Hebrew or Dutch that merited such high hopes [85: 1218–1219, 1222–1227].[7]

12.3 THE WAR WITH SPAIN

The Dutch Republic had been confronted with cryptography right from the very beginning of its existence. One of the first chroniclers of the Dutch Revolt, Everhardt van Reyd, tells how vulnerable Spanish communication-lines were because they always involved the passing of long distances over land and mainly across territory that wasn't under their control. The Revolt greatly and recurrently profited from the interception of letters exchanged between the Court of Spain and its civil authorities or armies based in The Netherlands. The opposite, Van Reyd reports, seldom happened, because the rebels took great care to send their letters either by boat or else over land only when short distances were involved. Apparently, the Spanish codes didn't pose much of a problem. Van Reyd mentions Charles Beaulieu from Valenciennes as the first one to know how to deal with intercepted, Spanish code-material. The first code to be broken was used in 1573 by the Spanish commander, Mondragon, in some letters written from Middelburg about the siege of Flushing and the fatigue

---

[7]I owe this reference to Dr. M.J. van der Wal.

and despair that it caused under the population of that city. Later, Van Reyd tells us, Beaulieu also decoded a number of letters on behalf of the English that revealed all the names of those who helped to prepare for an invasion of the British Isles, both in Ireland and in England proper. These letters were sent to him by the Queen of England which indicates firstly that his fame wasn't restricted to The Netherlands and, secondly, that the English had nobody who could deal with the problem at that particular moment. This incident also terminated Beaulieu's short-lived career as a code-breaker because he was never properly paid for his efforts, either by the Dutch or the English [163: 381–382].[8] From 1576 onwards the gap he left was filled by Philips van Marnix, heer van Sint Aldegonde, one of the leaders of the Dutch Revolt and, according to some, also the writer of what later became the Dutch national anthem.[9] At that time Marnix was already regularly supplied with code-books from the Spanish Court by the principal clerk of the King's secretary, Andreas Sapas, who was responsible for Dutch affairs. This clerk by the name of John of Castil, was born in Flanders as the son of a Spanish merchantman and continued to send fresh code-material for 10 years until he was caught and tried in 1581. In return, he received a pension of 300 crowns a year from the Prince of Orange. The contribution of Sapas' clerk had to remain a secret not only to the Spaniards but also to the rebels with the exception of some of the closest collaborators of the Prince of Orange. As a consequence, for some time Marnix had to keep up the appearance of a gifted code-breaker without being one [37: 44]. For all these years, the accuracy of Marnix's transcripts continued to amaze the public, however,

including the King of Spain who even suspected Marnix of using black magic [101: 118–119]. Surprisingly, Marnix continued to supply transcripts of intercepted code-material after Sapas' clerk had been caught. This is most remarkable because the Spanish had changed their code-books shortly after their treacherous clerk had been caught [67: 187]. In 1590 Marnix was also engaged in the solution of the so-called 'Escovedo-letters', a number of letters signed by King Philip himself and containing information about Spanish secret diplomacy towards England and France. These letters were intercepted by the English commander of Bergen op Zoom, colonel Morgan, and the information they contained was relevant not only to the Dutch, but also to the English and the French. Marnix was sent over personally to the British and the French Court to communicate their contents [166: 197–198]. It is tempting to speculate about contacts between Marnix, who stayed in Paris for three months, and the French principal code-breaker Viète who also had been engaged in the breaking of Spanish codes, or else between Marnix and the English code-breakers, first of all Leicester's secretary, Arthur Atey, who stayed with his master in the Republic [101: 116–117; 87: 15–16]. There is no doubt that Marnix had *some* skills as a code-breaker because he seems to have broken the Escovedo-letters all by himself, helped only by what he already knew about the way Spanish codes were built.[10] One may very well ask whether Marnix received some training by Viète during his long stay in Paris, possibly in return for the supply of the code-books that had been received from the bribed clerk at the Spanish Court.[11]

---

[8]This is remarkable, because Alan Haynes mentioned three names that were engaged in the deciphering of intercepted letters on behalf of Walshingham: John Somers, Lawrence Tomson and a Thomas Phelippes. Tomson and Somers were explicitly there to deal with difficult codes from Europe; Phelippes and Somers were monitoring the code-letters from and to Mary Queen of Scots. Perhaps these people gained their experience only after 1573. See [87: 10, 15–16, 26, 67].

[9]Although it has been generally believed that Marnix is he author of the 'Wilhelmus' for many years, the validity of this claim is still contested. See [112] and the review by A.Th. van Deursen [160].

[10]I disagree with Bor and Parker who both suggest that Marnix relied solely on espionage. Additional research in French and English archives, particularly related to the Earl of Leicester, may reveal more information, however; [43] does not give any reference to the contact that may have taken place between Marnix and Atey, or Leicester, for that matter, about the interception of letters. In fact, Atey seems to have been used by his master during these years mainly as a courier to travel back and forth between England and Holland.

[11]I haven't checked, however, whether Marnix and Viète were actually in Paris at the same time. Additional research in this direction is required.

It is also important to keep in mind that the apparent ease with which Spanish codes were broken during these years poses a problem. The Spanish codes weren't exactly simple. In some cases they consisted firstly of a substitution cipher yielding two or three homophones for most letters, in particular vowels and usually represented by single, imaginary characters, secondly of a list of roughly 200 syllables represented by one- or two-digit numbers, if necessary supplemented with a dot or a dash, and, thirdly, a nomenclature mostly containing roughly 500 items represented by syllables of two or three letters. These code-groups followed an alphabetical order that roughly, but not fully, coincided with the alphabetical order of the nomenclature itself. This pattern mostly prevailed, but the distribution of code-groups could still vary considerably. For instance the numbers or two or three letter-groups could be scattered randomly all over the code, to blur the connection between the alphabetical order of the nomenclature or list of syllables and the distribution of the code-groups [67: 60, 77]. To sum up, Spanish codes had a complicated structure and they weren't small. The guess-work that could always be used to break a simple homophonic substitution cipher, provided that enough code-material was available, simply wasn't sufficient. In a recent article about two hitherto unknown manuscripts by the French code-breaker François Viète, Peter Pesic stresses that his protagonist assembled masses of data and applied rigid mathematical rules, and that he departed from the principle that code-groups signifying syllables obeyed the same rules as code-groups signifying single letters inasmuch as frequency and their position in a word or a sentence were concerned. He was greatly helped by what he already knew, however, about the way Spanish codes were devised and used. The Spanish mostly sent various copies of one letter that weren't coded in an identical manner, so-called isologs [101: 335]. If one could get hold of two copies, a lot could be learned from the differences. Moreover, the Spanish indicated that a numeral was not to be read as a co-equivalent code-group by a stripe or some other mark and the value and the position of the numerals could be significant as well. For instance,

500 xx followed by 3000 yy was likely to mean 500 horseman and 3000 troops infantry, whilst 300.000 zz probably would mean 300.000 ducats. Additionally, the letters of the Spanish King often took the form of instructions, consisting of articles. The phrasing and naming of these articles were rather stereotyped, mostly beginning with a heading meaning 'memorial y instruction' and a beginning with the word 'que que', meaning 'moreover' [124]. Pesic believes that these manuscripts, or what was contained in them, must have been known to other cryptanalysts as well and that therefore Viète should be considered, as the title of Pesic's article indicates, the father of modern cryptanalysis. If he is correct, this would mean that Marnix must have had his support prior to his three months' stay in Paris in 1590 and in that case the same probably goes for that English code-breaker, Thomas Phelippes, in 1578 Walsingham's agent in France [101: 121]. It is also possible, however, that Marnix provided Viète with vital information about what was already known about Spanish codes through bribery.

Whatever may have been the case, Marnix's contribution marked the beginning of a continued Dutch effort to read intercepted Spanish code-material. In 1605, the Dutch Grand Pensionary Johan van Oldenbarnevelt, took on the engineer Jacques d'Alaume as a code-breaker at a salary of 1200 guilders a year. This happened at the recommendation of the new Stadholder, Prince Maurice.[12] Another important cryptanalyst during the long war against Spain was the famous Dutch poet and playwright Constantijn Huygens the Elder who, acting as private secretary to Maurice's successor and brother Frederick Henry, solved intercepted Spanish code-material literally on the battle-field [125].[13] There is a lot more to be known about this period in Dutch cryptanalysis.

---

[12][172: 89]. This code-breaker is first mentioned in the resolutions of the States-General on 20 December 1605, again on 3 January 1607 and for the last time on the first of July 1610. The way his name is spelled can vary, however, between Aléaume, Aleman and Aléaulme. This may indicate that his services were no longer required during the Twelve Years Truce. See [171].

[13]I owe this reference to J. van Meerwijk.

Did d'Alaume, whose activities probably were restricted to the period before the conclusion of the truce with Spain, train Huygens? Or did Huygens, who started to make cryptanalytical exercises already in 1616, possess some other source? [151].[14] Is there a connection between D'Alaume and Marnix? There has been little opportunity to find conclusive evidence to answer even one of these questions in Dutch sources, and it is also unclear whether Dutch code-breakers made any effort at all to improve the construction of Dutch codes and ciphers. The Dutch continued to use codes that were highly inferior to those of the Spaniards from the very beginning of the Revolt in 1572 onwards and right up until the end of 80-years' War in 1648. In fact, during this period, Dutch codes show remarkably little variation. During the initial years of the Revolt, there was little of a real foreign policy, except for the desperate efforts of William of Orange to get support for his case among the protestant Courts of Europe. During this period, Dutch codes mostly consisted of a monoalphabetic substitution cipher supplemented with a small nomenclature, roughly containing ten to fifteen codegroups signifying names and places. Sometimes, but not always, the monoalphabetic substitution cipher could contain two or three homophones for frequently occurring vowels or names.[15] It could also be supplemented by numerous empty codegroups, as for instance was the case in the letters written by William of Orange to his brother John in 1572.[16] They could either contain numerals or imaginary signs, occasionally resembling what later came to be known as the freemason's cipher [101: 772].[17] These examples all seem to

indicate that the leaders of the Revolt weren't too worried about their letters being intercepted and decoded. Probably they were delivered only by trusted friends or couriers who all knew very well how to avoid danger.[18]

This wasn't the case, however, with the first Dutch envoys that were sent abroad on behalf of the States General for a long time such as Gideon van Boetzelaar and Cornelis Haga. Gideon van Boetzelaar was sent to Paris in 1615 and was to stay there for at least three years. Boetzelaar had no couriers at his disposal of his own, but he had to report to Grand Pensionary Johan van Oldenbarnevelt on a regular, preferably a weekly, basis. Therefore he had to use the regular mail from Paris to Dieppe and Antwerp that was controlled by France and that had to cross enemy territory as soon as it was brought ashore. Obviously, Boetzelaar had every reason to be worried about a possible interception of his letters to the Grand Pensionary and he used a code-book that was more complicated than those used before. It consisted of a nomenclature roughly comprising of 80 items and a homophonic substitution cipher containing about 30 code-groups. The main thing was of course that the nomenclature was now substantially enlarged; most dignitaries at the French court, including the ambassadors of foreign nations, acquired a codegroup of their own and the same applied to those involved in the making of foreign policy at home. There was one other striking feature, however. The code-groups that were part of the nomenclature were clearly different from the code-groups that were part of the cipher because the first consisted entirely of two-digit-groups, occasionally supplemented with a dot or a dash, whereas the latter was composed exclusively of one-digit numbers and some Latin or Greek characters. Thus, the main principle of Boetzelaars code-book gave itself away [168]. The other Dutch envoy was Cornelis Haga who was sent to the Ottoman Empire

---

[14]In 1616 one finds the one-line entry: "Dechiffrere varie exercui". Unfortunately it doesn't say any more. In 1631, however, Huygens enquires about the fees received by Marnix and d'Alaume for their work as cryptanalists. Therefore it seems likely that he must have been active as a code-breaker at least from that year onwards. See [180: 331].

[15]William of Orange to Count William Landgrave of Hessen, Antorff, 9 April 1567. This letter is printed without the coded supplement in [83: 54–66].

[16]William of Orange to this brother John, Malines, 21 September, 1572 [83: 501–511]. The solution of the cipher was first published by C.M. van der Kemp in [1: 311–312].

[17]See for instance Junius to Counsillor Stöver, The Hague, 16/6 November 1609 [178]. Ref. [179] contains the key to this

cipher. It appears to be used for the first time in the correspondence between Willem Lodewijk and Jan de Oude in 1599. See [177; 97]. The preference for monoalphabetics and the use of imaginary signs shown in many of these ciphers seem to mirror the way ciphers were devised at the Court of Charles V. See [78].

[18]This is Van Reyd's explanation [163: 381].

to stay there for the rest of his life. Haga, too, had to report on a weekly basis and his letters had to travel an awfully long way (by way of Venice and over land to Paris) before they arrived at their destination. There was no question of a courier travelling back and forth, however, because this was considered much too expensive. Haga was provided with a similar code-book to the one given to Boetzelaar, but it was slightly improved. The nomenclature was somewhat enlarged, containing over 100 items, signified by numerals between 1 and 116. The order wasn't alphabetically but instead it was divided into groups related to certain topics, such as dignitaries at the Court, governors, friendly Christian princes and unfriendly ones and their agents, names of isles, cities and fleets. The cipher had no homophones and consisted wholly of imaginary characters.[19] As a result, Haga's code-book had the same flaw as the one belonging to Boetzelaar: it was only too easy to tell the cipher and the nomenclature apart. These early efforts to improve Dutch cryptography are somewhat puzzling because of the similarity they show with the codes used by the Spaniards. The whole idea of subdivision into various compartments for example, so typical of these early Dutch code-books, seems to have been borrowed from examples set by the Spaniards. This would indicate that Dutch code-breakers must have been involved somehow in the introduction of these new codes, as indeed the whole effort to ameliorate the standards of Dutch cryptography seems to suggest too. On the other hand, the level of Spanish cryptography, only too well-known to the Dutch code-breakers of the day, was never fully attained. Moreover, in a very rudimentary form, the use of separate compartments, one for the cipher and one for the nomenclature, can also be found in the small codes used by William the Silent and his brothers during the early days of the Revolt. The code-books for Haga and Boetzelaar may have simply been intended as enlargements of the monoalphabetic substitution ciphers that had been in use already for a long time. Therefore it seems hazardous to conclude from this

fact alone that either Huygens or D'Alaume were involved in the construction of these codes.

These cases do show, however, that, from the resumption of hostilities with Spain in 1621 onwards, the Dutch had more reason to worry about the interception of their mail than before but this did not result in any effort to apply sophisticated methods of encryption.[20] It is not entirely clear what this means, but there are several possibilities for explanation. First of all, it may indicate that the code-breakers who were around in Holland at that particular moment were not available to help to devise the ciphers and codes or that they simply weren't asked to do so. Secondly, it may indicate that the Dutch authorities simply had learned the wrong lesson from the successes of the early Dutch code-breakers. Some twenty yeas later Grand Pensionary Johan de Witt lightly professed his disregard for any efforts to improve encryption standards because it was impossible anyhow to devise a code or a cipher that could not be broken in the end [129: 251]. This statement may well have reflected a general attitude among the Dutch political elite that may have also existed much earlier. This would also imply, however, that the Dutch weren't sufficiently aware of the fact that there was still a lot to gain in this field. This may have been very well the case inasmuch as codes were involved. The discovery of the two-part code was a French affair and their is no indication that the principle was known in the Dutch Republic prior to 1684 [63: 148, 149].[21] It seems unlikely, however, that this also applies to the polyalphabetic ciphers that were described at length in the literature of the day. These books could be found in every respectable library in the world, including the Dutch Republic. In 1621 even a reprint appeared of Vigenère's book under a false name in Groningen, that was shamelessly dedicated to Stadholder Mau-

---

[19]Haga to the States-General, Constantinopel, 12 and 26 March 1620 and sheet with cipher and nomenclature without date or place. – ARA. Staten-General, inv. nr. 6894.

[20]The separation of the nomenclature and the cipher that was carefully retained in Dutch cryptography until the 1750's had one advantage of course: the breaking of the cipher would not give any clue as to the construction of the nomenclature.

[21]The Dutch authorities probably became familiar with the principle of the two-part code through the interception of a letter of the French ambassador Comte d'Avaux.

rice [58].[22] Therefore, the only explanation can be that the matter simply wasn't given sufficient attention because the construction of codes and ciphers was still considered or believed to be something merely accidental.

12.4 THE INTRODUCTION OF CODE-BOOKS FOR REGULAR USE BY THE STATES-GENERAL

This clearly changed after the Peace of Westphalia in 1648. The Peace Treaty marked the beginning of diplomatic relations between Protestant and Roman-Catholic powers and the opening of embassies all over Europe to facilitate contact between the nations. The Dutch Republic had little ambition in the field of foreign policy but the country had to go along with this new development.

On 14 October 1651 provisions were made for the delivery of all official mail. Basically, the federal government had three options. Firstly, it had the possibility of using 'ordinaris boden' who would simply hike or travel by public transport, such as stage-coach or towing barge. Of course, this was only suitable for communication in non-urgent matters that stayed inside the country. The second possibility was the use of one of the so-called 'postmeesters' at the disposal of the States-General. They took care of the mail delivery by their own means of transportation. Initially, there were three of them. After 1710, however, only one was left with four postboys at his disposal and five horses. This option was preferred whenever a matter was urgent or whenever a trip across the border was required. The regulations of the States-General on this matter also show, however, that only a small part of the so-called 'generaliteitspost' was delivered by servants of the States-General proper because of the enormous costs. The delivery of one letter by a postboy of the States-General could amount to 100 guilders whereas the delivery of the same letter by ordinary mail would cost only a few stivers [165: 71–72, 75–76]. This meant in effect that it was the exception rather than the

rule to make use of this possibility. Inasmuch as the weekly correspondence with Dutch ministers abroad was concerned, the Dutch authorities preferred by far to send their letters by ordinary mail, taking the risk of interception into the bargain.[23]

By 31 July 1651 the States-General had made an effort to set rules for the use of codes. The credentials of every envoy that was about to leave the country should always be supplemented with a code-book that was to be used whenever the diplomat saw fit [172: 87–94]. The code-letters should not be addressed to the plenum of the States-General but to the Greffier personally who would make a transcript and then submit it to the Secret Committee, a body wholly composed of senior members of the States General drawn from various provinces. This provision is most illuminating because it shows that the Dutch authorities knew only too well that the interception of mail abroad wasn't the only risk there. Due to the confederate structure of the Dutch State, the members of the States-General were not allowed to take any decisions at all without the prior consent of the governing bodies of the provinces they represented in The Hague. As a consequence, in the Dutch Republic more people were involved in the making of foreign policy than anywhere in the world. In fact, the whole idea of a Secret Committee was introduced under Grand Pensionary De Witt to facilitate the dealing with confidential matters by the States-General but this did not mean that its members could not be held accountable by the States-Provincial that had delegated them in the first place. During the 18th century matters got even worse because the

---

[22]On 23 May 1622 this book was presented by the author to the States Provincial of Utrecht who paid 12 pounds for it. See [126]. I owe this reference to J. van Meerwijk.

[23]Until 1751, the mail in the Dutch Republic was processed by local, non-governmental, post offices that possessed contracts with post offices abroad for the delivery of mail across the borders. Not surprisingly those of Rotterdam and Amsterdam were the most important, but there were also independent post offices at other cities such as Leiden and Utrecht, all exploiting only part of an intertwined network of postroutes. In 1750 however, in as much as the province of Holland was concerned, these post offices were forced to give up their independent existence and were brought directly under the control of the States-Provincial. This measure did not end the risk of interception, however, because outside the country the mail was still entrusted to postal organisations, controlled by foreign governments. See [122].

members of the Secret Committee lost the possibility to act without prior consent. This meant that the actual day-to-day conduct of foreign policy had to be restricted to the Greffier, the Grand Pensionary and, whenever present, the Stadholder, who could act after consultation with one or two members of the Secret Committee, whenever they thought necessary [28: 155, 160–165].[24] This all happened with the consent of the other members of the States-General and they all knew very well that it could only work when security at the registry was tight. On 20 April 1656, the States-General appointed a committee from within in order to investigate how security could be improved. The main questions were how the registry had to deal with the confidential letters received by the Greffier and what had to be done with the code-books that were used for the encryption of these letters [165: 66]. Probably as a consequence of the work of this committee two additional measures were taken. On 21 July 1657 the sending and receiving of the diplomatic despatches was laid into the hands of one official: the "Commissaris der publieke brieven van hare Hoog Mogenden Ministers buiten 's Lands resideerende" at a salary of 1000 guilders a year. He would deal with all matters related to the correspondence of Dutch ministers abroad, including the copying of incoming and outgoing letters and the paying of duties to the post office and the calling in of the federal postmaster [165: 76].[25] Secondly, by resolutions of 20 September and 4 October 1658, somebody was appointed specifically to make code-books, to keep them under lock and key and to watch over the encoding and decoding of confidential letters. This task was entrusted to Pieter van Peenen who was also responsible for the registry of marching orders to the federal army and all related expenses. Probably these tasks were

combined because both were of a highly confidential nature [165: 66].[26]

During the same period measures were taken to improve the standards of encryption. The Dutch ministers to be sent abroad were now furnished with an elaborate one-part nomenclature that comprised over 10.000 items or more, usually written on large folios with six columns showing 50 items each and folded into a booklet not exceeding 25 pages with a hard cover. It consisted of numerous homophones that were contiguously listed and it also contained code-groups for frequently occurring combinations of words that were inserted according to the initial letter of the initial word. For instance, the chapter headed by the character 'A' would also contain a code-group representing: 'aan zijne majesteit' (= to his majesty). The code-groups usually consisted of three-digit numbers, supplemented with a letter of the alphabet or an accent mark such as a slash through the first or the last digit of the code-group or a dash under or above it, or two or three dots. This, of course, hid the true size of the code and partly the way it was built, but it did not make up for the fact that it was basically a one-part-code.[27] Another remarkable feature of this new code-book was the fact that the value of the first code-group in the list always exceeded the number 200 or occasionally 250. The numbers below were reserved for a cipher consisting of 200 code-groups with 8 homophones or less for single letters, but also including code-groups for doubles, frequently occurring word-endings, articles and conjunctions. These ciphers were written on a separate double-sheet that could easily be inserted in or deleted from the code-book proper, usually containing two tables: one for enciphering and one for deciphering. Except for the fact that they did not exceed the number 200 or 250, the code-groups that were used in the cipher showed no internal order at all: they appeared to be jumbled as if in a tombola. In reality it seems more likely that a slide-rule was being used and that the

---

[24]For a detailed account of the complicated and frequently changing procedures at the Secret Committee see also [56: 252–281, 617–621].

[25]The first commissionary was Frank Bisdommer who was suceeded on 22 April 1688 by Jan Danckers, on 16 July 1722 to be succeeded by Otto Koetsch, on 11 January 1757 to be followed by Pieter le Clercq and on 24 December 1759 by Martinus Gousset and, finally, on 12 March 1788 by Johan Abraham Tinne.

[26]This combination lasted until 1672, when the marching orders were registered once again by the Stadholder, but apparently Van Peenen went on as a code-clerk. After his demise in 1691 he was succeeded by Johannes Adolphi who lived for only one year longer.

[27]For examples see [31].

13	a	b	c	d	e	f	g	h	j	k	l	m	n	o	p	q	r	s	t	u	w	x	y	z
	18	1	19	2	20	3	21	4	22	5	23	6	24	7	25	8	26	9	27	10	28	11	29	12
	47	30	48	31	49	32	50	33	51	34	52	35	53	36	54	37	55	38	56	39	57	40	58	41
	76	59	77	60	78	61	79	62	80	63	81	64	82	65	83	66	84	67	85	68	86	69	87	70
	104	87	105	88	106	89	107	90	108	91	109	92	110	93	111	94	112	95	113	96	114	97	115	98
	133	116	134	117	135	118	136	119	137	120	138	121	139	122	140	123	141	124	142	125	143	126	144	127
	157	145	158	146	159	147	160	148	161	149	162	150	163	151	164	152	165	153	166	154	167	155	168	156

Figure 12.1. Cipher table for ambassador Hochepied to Constantinople in 1747.

results were simply written down to avoid any mis-understanding. In fact, this slide-rule survived in the code-book that was handed to Ambassador Hop who was to leave for England in 1723 and there are also examples of ciphers written down in a way that still show a regular permutation of code-groups, achieved by moving a slice of paper containing a row filled with letters, while keeping a table with rows filled with numbers at its place.[28] Of course there were numerous different ways to rearrange these cipher tables and this was an important aspect of their attraction. A code-book could be supplemented with eight different tables that each received a identification number that could not be discerned from any ordinary code-group, intended to precede a coded period in order to indicate what table was being used at that particular time. Apparently, the singling out of numbers at will to identify the cipher-tables entailed the risk picking of one that was already being used as a ordinary code-group. Therefore, at some point or on certain occasions the principle of random organisation was abandoned in favour of a complicated structure that made sure that certain numbers within the range of the cipher were not being used. In the cipher for ambassador Hochepied to Constantinople, a small booklet consisting of 25 tables showing the letters of the alphabet in a row, heading columns of six numerical substitutes, with a permutation of one letter on every page while keeping the numerals in their place, the problem is solved by carefully setting apart a range of numbers to be used to identify the tables. For instance, the first five tables received the identification numbers 13 until 17, whereas the first number on the first row was 18 (= a), to be

followed by 1 (= b), to be followed by 19 (= a + 1) to be followed by 2 (= b + 1) until the numbers 29 and 12 were finally reached. The second row was constituted along the same principle, using 47 for a and 30 for b; the third row using 76 and 59; the fourth row using 104 and 87; the fifth row using 133 and 116; the sixth row using 157 and 145 and ending with 156 and 168. The other identification numbers were 42–46, 72–75, 99–103 and 128–132 (see Fig. 12.1).

The use of the slide-rule and the way identification numbers were inserted seem to point to Athanasius Kircher as a source of inspiration. In fact, it is a copy of his *Arca Stenographica* and an application of the principles described in his *Polygraphia Nova* [147: 169–174]. There is little doubt that whoever introduced this system was well aware of the cryptological literature of his day but this is all we know.[29] The only code-books that survived belong to the Fagel family archives. Therefore they cannot be traced back to a period prior to 1672 when the first member of this family assumed office at the States-General.[30] There is

---

[28]Cipher for the envoy Van Colst, leaving for Spain [25]; cipher for envoy Hochepied to leave for Constantinople in 1747 [70].

[29]Additional research in the elaborate correspondence of Kircher, kept at the Collegium Gregorianum in Rome may give a new perspective on this matter.

[30]The dynasty started in 1672 with Hendrik Fagel (1617–1690) who was the first one in his family to be appointed Greffier of the States-General on recommendation of Johan de Witt's successor, Grand Pensionary Gaspar Fagel. Remarkably, it seems that the Greffier's library did not contain a copy of Kircher's *Polygraphia Nova*. It did contain copies of other books by Kircher, however, such as *Oedipus Aegypticus* (1653) and *Obeliscus Pamphilius* (1650) and it also contained a number of classical works about cryptography, such as Porta's *De Occultis Literarum Notis* (1606), Gustavus Selenus' *Systema Integrum Cryptographica* (1624), various editions of Trithemius, Vigenère's *Traicté des Chiffres* (1586) and Gasparus Schotus' *Schola Steganographica* (1665). It is not entirely clear what this means. Of course the system was already introduced before the first Fagel became Greffier and

evidence, however, that the system, in broad outline at least, was already being used under Grand Pensionary Johan de Witt. The 18th century Dutch historian Johan Wagenaar presents, in his elaborate *History of the Fatherland*, a code-letter written in 1668 by the Dutch Ambassador in France, Coenraedt van Beuningen to Burgersdijk and Van der Togt that shows roughly the same construction as the codes used later.[31] The main resemblance is easily noted and lies in the use of two-digit code-groups not exceeding the value of 97 for single letters and three- and four-digit code-groups with a value not under 344 for the words derived from the nomenclature [174: Vol. XIII, 333]. This indicates that the system was already introduced under De Witt and this may point to Van Peenen as its originator. The evidence in this respect is rather inconclusive, however. The 19th century historian Vreede reports that the Dutch envoy Boreel who was sent to France in 1650 and again in 1664, received on his second trip a copy of the same code-book that he had used 13 years before [172: Vol. I, 90]. This would indicate that no change whatsoever occurred in the use of codes after Van Peenen had been appointed in 1658 and therefore the question about its origin remains.

The code-book determined Dutch cryptography well into the next century. Basically, the arrangement of the nomenclature never changed. There were small ones, consisting of 2000 or 3000 code-groups, medium-sized ones, consisting of roughly 5000–7000 code-groups and large ones, containing 8000–12000 groups, but there were no fundamental differences. The small nomenclatures were, in general, excerpts from the bigger ones and there seems to have been no awareness that any alternative existed. The size of the nomenclature varied with the importance of the mission. The large ones were used by Portland in France during his negotiations with Louis XIV in 1698 and again in France by Hop sent on a mission in 1718 and by Hamel Bruyninx in Vienna in 1721 [27]. The

small ones were used in Portugal by Houwens in 1718, by Keppel in Prussia in 1727 and by Hop in England in 1723 [26]. Van Colst, who was sent to Spain in 1719, only received 8 ciphers [25]. It is also clear that the system survived Van Peenen as a code-clerk. On 1 September 1691, his work as a code-clerk was taken over by Johan Adolphi who had only one more year to live. From 1692 onwards, the job was fulfilled by Greffier François Fagel. This implied probably that he encoded and decoded the letters personally but it seems unlikely that he also compiled the code-books [121].[32] This was done by one of the clerks at the Registry who was paid extra for it[33] (see Fig. 12.2).

It is not so easy to explain why so little changed in Dutch cryptography during all these years. The system was probably introduced because it offered – with numerous homophones in both nomenclature and cipher – many possibilities to suppress the frequency patterns. If used in a sophisticated way, it even may have been possible to write a coded letter without repeating a single code-group. There was no guarantee that it was used properly, however. In fact, Kircher advocated his system precisely because of the fact that it could be used just as well as a monoalphabetic substitution cipher, if the slides were fixed [147: 174]. Secondly, the system had the same flaw as those introduced earlier on behalf of Boetzelaar and Haga: it was relatively easy to tell apart the code-groups belonging to the cipher (with low values, usually not over 200) and the code-groups belonging to the nomenclature (with high values, starting with 200 or more) and this wasn't by accident. It was an unintended result of the effort to compensate for another, better-known, weakness of the traditional one-part code; the fact that a solution of single letters on the basis of frequency counts was just

he may not have been aware of its origin. It is also possible, however, that Kircher's book simply went astray. The library's catalogue was made only in 1802 for an auction at Christie's in London. See [11].

[31] Probably P. Burgersdijck (1623–1691), a republican politician from Leiden. See [20].

[32] These notes tell us that, in 1738, Greffier François Fagel transferred all his work related to the use of codes and ciphers to Lyonet because he had become to old to do it himself. This means that the Greffier had been doing this job for more than forty years! This statement is further corroborated by the fact that the Greffier's characteristic handwriting can be found in many code-books and coded letters either written above the lines to solve the code-groups as additions to, or changes of, certain entries.

[33] Application letter by Croiset to the Greffier for the vacancy left by Lyonet (draft) [121].

acoord	200	aan U Ho: Mo:	242	al de	284
accorderen	201	aan U L	243	al den	285
geaccordeert	202	aan U WelEd:	244	aldermeest	286
adverteren	203	aan UWelEdGes	245	aldewijle	287
geadverteert	204	aan welke	246	al die	288
advertentie	205	aan wien	247	alle die	289
aan	206	af	248	alle die geene	290
aan	207	afbreuck	249	aldaar	291
aan de	208	afdoen	250	aldoor	292
aan de	209	afgedaan	251	aldus	293
aan den	210	afdoeninge	252	alhier	294
aandenselven	211	afgegaan	253	alhoewel	295
aandewelke	212	afgaan	254	allemeest	296
verder tot	241	verder tot	283	verder tot	325

Figure 12.2. Part of a code-book of the States-General, late 17th or early 18th century.

as easy as in a monoalphabetic substitution cipher and that this solution also gave access to the structure of the nomenclature as a whole.[34] Thirdly, the nomenclature was never rearranged in any fundamental way. The smaller code-books would always roughly follow the order of the larger ones, only economising on the rendering of homophones and the listing of words or word-clusterings that almost meant the same. This of course meant that once a Dutch code was broken, it also could be used as a model for the solution of any other one.[35]

Initially this may have been caused by the fact that the Dutch simply didn't know any better but from 1684 onwards this was no longer the case. In his account of the wars and conflicts with France during the latter part of the 17th century, Wagenaar reports how Stadholder William III got hold of a letter written by the French ambassador D'Avaux

about his negotiations with the town council of Amsterdam that touched upon foreign policy, or to be more precise, about the delivery of troops to the Southern Netherlands to help Spain against the French invaders that had to be paid for by the town council. The letter was important to Prince William, not only because he wanted to know more about the intentions and goals of the French, but also as a way of showing that some members of the town council had no business of interfering with the foreign policy of the Republic: the exclusive domain of the federal authorities. D'Avaux's letter was intercepted just across the border by horsemen clearly belonging to the garrison of Maastricht who captured the ambassador's courier, robbing him of all his belongings except for his boots and his jacket. This incident caused much distress to the burgomasters of Amsterdam, who could only be reassured by D'Avaux's statement that the letter was fully coded and that nobody could read it without a key. On 16 February 1684, however, the Stadholder entered a meeting of the town council with a decoded copy of the letter, accusing the council's senior members Hooft and Hop of high treason. To avoid the impression of having violated diplomatic immunity, Prince William claimed that he had received the copy from the Governor of the Spanish Netherlands, De Grana. This, of course, only made D'Avaux laugh, but under the given circumstances there was little he could do to help his friends at the

---

[34] Apparently, even this wasn't widely known. During his captivity on the Isle of Wight in 1647–1648, the unfortunate King Charles I used a code like this to communicate with some of his followers in Scotland.

[35] It has not been possible to search the archives of Europe for intercepted Dutch code-material, but there are some examples to be found in literature. In his article on the Austrian Black Chamber, Franz Stix [79: 153–156] renders the names of several code-breakers able to deal with the Dutch language and he also mentions the amounts paid for the solution of Dutch codes: 600 guilders for larger ones (containing several thousands code-groups) and 200 guilders for small ones (not over 1000 code-groups).

town council. The only possibility was to pretend that William's code-breaker had interpreted the letter wrongly. To support his claim he brought forth a 'genuine' copy of the letter. This provoked a release of a copy on William's part too, at first leaving out many unsolved code-groups and somewhat later a version wholly in plaintext, accompanied by the solutions of other letters that were intercepted at the same time and that were even more compromising [174: Vol. XV, 180; 105: 95, 98, 107–109].

The identity of the code-breaker, however, has never been disclosed. Wagenaar only remarks that it was widely believed that the letter was decoded by a servant of the Spanish Governor De Grana. This may very well have been possible, because other sources seem to indicate too that De Grana had a code-breaker at his disposal in Brussels [68: vi]. There is nothing to prove, however, that De Grana was involved and it is also possible that the letter was decoded in the Dutch Republic proper. This may have been done by Constantijn Huygens the Elder who lived just long enough to do it, but another obvious candidate is Abel Tassin d'Alonne who even may have been trained by Huygens. D'Alonne acted as private secretary to Princess ary from the very moment onwards that she came to live in The Hague after she married Stadholder William III; and he stayed so after her return to England until her demise in 1695. It was generally believed that he was an illegitimate brother of the Stadholder-King but there is no conclusive proof of this. There is little doubt, however, that he was a man of influence at the Dutch Court and that he was a confidant of the royal couple, far more so than, for instance, Constantijn Huygens the younger, who served the Stadholder-King in the same capacity during his years in England [80: 442–443; 89: 97–98]. One of D'Alonne's tasks was the gathering of intelligence about the Court of the Stadholder-King's father-in-law, James II. This had already been the case in the Republic but the peak of his career came later, in England. There he was a key figure in the monitoring of all contacts between the exiled court and the Jacobites that had stayed behind, mainly through the interception of letters, but also by other means [98: Vol. II, 132; 99: Vol. IV, 721; 17: 3–7; 91: Vol. I, 98, 329]. There is no

evidence that he was also active as a code-breaker, but this seems highly probable. It is beyond doubt that he worked in this capacity during the War of the Spanish Succession when he was already in his sixties and it is extremely unlikely that such an exceptional talent as this could remain hidden until in later life [63: 147–155]. Therefore, D'Alonne seems to be a likely candidate to have been involved in the breaking of D'Avaux's code somehow, if only in the release of the second copy that may have shown a better understanding of Dutch politics.[36]

This incident shows that the Dutch, perhaps in collaboration with the Spaniards, knew how to deal with complicated French codes before the English did. This implies also that the Dutch, at least the Stadholder and some of his confidants, knew the principle of the two-part code from 1684 onwards – without applying it themselves.[37] This proves of course that there was no influence whatsoever exercised by the Dutch code-breakers on the development of Dutch cryptography, at least around the turn of the century, and it would be interesting to know why. It has not been so easy to find an explanation for this remarkable state of affairs. Part of it may be found in D'Alonne's activities as a code-breaker during the War of the Spanish Succession that have been described at length in my article about the Dutch Black Chamber in *The Historical Journal*. There is no need to incorporate the findings of this article in detail but one conclusion should be mentioned here. D'Alonne's efforts as a code-breaker were only required inasmuch as intercepts were concerned that could not be submitted to the Black Chamber in Hanover that also worked on behalf of the Dutch. In effect, this meant that it involved only intercepts that were made at the Post

---

[36] Apart from some references in contemporary literature, the only document that is left about this affair is a fully coded copy of the letter. See [130]. Additional research in the State Archives of Belgium or Spain may yield new information, however.

[37] There is no conclusive proof that the intercepted letter was coded with help of a two-part code, but it is extremely likely that it was. See [101: 160, 161; 107: 65–73]. The principle of the two-part code was applied in certain nomenclatures used by Heinsius, however. See two-part codes distributed to Grand Pensionary Heinsius and ambassador Schütz [19].

Office in Brussels, under the nose, but without the consent, of the British. D'Alonne was called in on these occasions because it could be done without informing anybody. Acting as private secretary of Grand Pensionary Heinsius, he worked in the same room as his master and they probably were seated at opposite desks. Therefore, D'Alonne's solutions did not even have to leave the Grand Pensionary's office. This was very different from the way other intercepts were treated. Usually made at the Post Office in Amsterdam, they had to be handed over to the Hanoverian envoy Bottmer who then sent them on, in all likelihood by courier to his capital. There the intercept was decoded and sent back, as soon as possible, to be subsequently read in the Secret Committee of the States General. This, unavoidably, meant that the enemy also knew about it, because the Secret Committee was notorious for its lack of efficiency in that respect. It is no accident that the collaboration between the Dutch and the Hanoverians in this field seems to have been terminated during the Summer of 1711 because of a leaking of information at the States General.[38]

---

[38]In [63: 151–152, 156] I suggest that the deterioration of the relations between the two countries had already started already in 1710, as a consequence of an intercepted letter that contained information about negotiations between the Dutch and the Bavarians related to the return of Max Emanuel to the Electoral Body. The chronology poses a problem, however. Georg Ludwig's letter to Bothmer is dated 14 March 1710; but the date given for the Dutch offer to Max Emanuel with respect to his return to the Electoral Body in [176] is July 1710. This probably indicates that Weber is wrong and that the contacts between the Bavarians and the Dutch started earlier on. Moreover, the end of the collaboration between the Dutch and the Hanoverians did not occur until after July 1711, because Neubourg's last letter to Heinsius dates from the 26th of that month. This implies that the collaboration was not terminated by the Dutch after all, but by the Hanoverians and for a different reason. In June 1711 Robethon reported that information contained in the intercepted letters was leaked by the Secret Committee of the States-General. See [21]. For the overall picture this doesn't seem very important, however. The understanding between the Dutch and the Hanoverians rapidly deteriorated because of the Dutch contacts with Max Emmanuel. The Hanoverians found out about this in an early stage as a result of an intercepted, Bavarian letter that was decoded by them. This marked the beginning of period of rising mutual distrust that ended with a total breach of confidence in the Summer of 1711. Moreover, Heinsius may have started sending only part of the intercepted letters after his unpleasant con-

This explains also why D'Alonne wasn't asked to improve Dutch codes. His efforts had to remain secret for everyone except the Grand Pensionary and a handful of people he could absolutely trust. The activities of the Dutch Black Chamber during the War of the Spanish Succession are also noteworthy from another perspective. D'Alonne's range as a code-breaker was much broader than the three cases described in detail in my article published in *The Historical Journal* and it is beyond doubt that other cryptanalysts were involved as well.[39] One of them was the famous Dutch natural scientist Willem Jacob 's Gravesande who worked in The Hague as a barrister from 1707 onwards and already at that time was well known for his contributions to the *Journal Littéraire* and who was occasionally asked to help with the decoding of documents, 'if', his first biographer Allamand writes, "nobody else was able to perform the task" [29: xi, xii, lvii]. In fact, in his first book on philosophy and mathematics, published in 1717, he devoted a single paragraph to the solution of an encrypted message to demonstrate what scientific thinking was all about: the formulation of a set of mutually exclusive hypotheses and the testing of them one by one until all except one could be rejected. Apparently, 's Gravesande considered the work of the code-breaker, or the cryptanalytic effort, to be a model of the scientific approach as he understood it to be: pragmatic, empirical and critical [132].[40] It is also hard to believe that he didn't use, or at least look for, mathematical–statistical devices to facilitate his work as a code-breaker. 's Gravesande had always been highly interested in probability theory. In 1712, he had already written a treatise about Arbuthnot's effort to prove the existence of Divine Providence on the basis of the equal distribution of birth rates of boys and girls in a city like London for a number of years. 's Gravesande's was not

---

versation with Bothmer in 1710, keeping those addressed to the Bavarian agent in Rotterdam back.

[39]See for instance a worksheet containing a one-part code and bearing the date '5 November 1713' that indicates that D'Alonne was also active during the peace-talks in Rastatt [106].

[40]This is a far cry indeed from Hacking's statement (25) that students of the physical sciences weren't interested in probability theory, because they were after *knowledge,* not *opinion.* For 's Gravesande's view of science see [131: 121–123].

interested in discarding the proof of Divine Providence – he firmly believed in it himself – but his focus lay elsewhere: in setting standards for the providing of mathematical–statistical evidence [138: 109]. Unfortunately, there are no documents left related to 's Gravesande's work as a code-breaker to substantiate any statement about a definite connection between the development of probability theory and the application of cryptanalysis but the question still remains. 's Gravesande's contribution is also puzzling for another reason, however. The Dutch scientist was a personal friend of the Oxford mathematician and astronomer John Keill (1671–1721) who had translated his book on the mathematical foundations of proof in the physical sciences in English, albeit not in a way that 's Gravesande approved of [29: liv; 131: 31]. From August 1712 onwards, after William Blencowe's suicide, Keill served as a code-breaker on behalf of the English Government [9: 310–311; 101: 169–170].[41] Of course there is very little reason to assume that Keill and 's Gravesande actually collaborated in this field. They probably first met in 1715, after 's Gravesande had already terminated his engagement as a cryptanalyst, but at least one may ask if they ever had the opportunity to talk about their experiences in this respect.[42] This matter is important because the involvement of both 's Gravesande and Keill seems to indicate that there may have been a connection between the rise of the Black Chambers and the development of scientific thinking after all. Both scientists were fervent adherents of the empiricism as first advocated by Newton. In fact, after 's Gravesande had been given a chair at Leiden University, he became the first and foremost defender of Newton's theories on

---

[41] Kahn characterises Keill as being totally incompetent, but I doubt this is true. There is some proof to the contrary in British archives, but to say that he wasn't a total failure doesn't imply that he was a great success! See for instance Keill to Tilson, 5 December 1715, reporting that he has *deciphered* the paper signed Nelson [3].

[42] There are no records of this at the Royal Society, neither correspondence nor minutes. This is hardly surprising, because the meetings were occasionally attended by likely victims such as the Spanish ambassador. Unfortunately neither 's Gravesande nor Keill left a personal archive worth mentioning.

Figure 12.3. Pierre Lyonet 1707–1789. Engraving by A. Schouman. Source: Rijksinstituut voor Kunsthistorische Documentatie, The Hague.

the continent [131: 115–118]. The least one can say is that experimental physics and the new empiricism in general, must have contributed to a mental attitude that also enhanced the understanding and development of methods of cryptanalysis, if so required.

12.5 LYONET'S CONTRIBUTION

This seems also to have been the case with Pierre Lyonet who entered the stage a quarter of a century later, ten years before the Stadholderate was restored and declared hereditary under Prince William IV as a result of a political crisis once again caused by a military conflict with France, better known as the War of the Austrian Succession. Lyonet was in the service of the Greffier as a translator, a so-called 'patent meester' or administrator of military expenses and as 'cipher clerk'. In this last capacity he was responsible for the encoding and decoding of all diplomatic despatches of the States General and for the supply of code-books to the Dutch embassies and ministers abroad. Lyonet had been asked to take this job in 1738 by the old

Figure 12.4. Samuel Egbert Croiset, 1734–1816. Oil painting by Taco Scheltema. Source: Museum voor Communicatie, Den Haag.

Greffier Hendrik Fagel's uncle François, who had shared Lyonet's broad and vivid interest in the arts and the sciences [167: 17–18].[43] At that time Lyonet was already somewhat of a celebrity. In his generation he was unsurpassed as a microscopist and from 1742 onwards he acquired fame as a particularly talented engraver of insects with a number of illustrations he made for books written by himself and by others.[44] He was a proud member of scientific academies all over Europe and he enjoyed being invited to the dinner table of the Russian and French ambassadors who were as equally fond of the minutiae of nature as he was [111]. His fame extended to Dutch government circles as well. In 1751 he sold an entire cabinet of butterflies to Anne of Hanover and it was Abraham Trembley, the private secretary of Willem Bentinck who had intro-

duced him to the Royal Society in London [57; 92: 27].

The way in which Lyonet got engaged as a codebreaker is described in full detail in my article about "the Black Chamber in The Dutch Republic and the Seven Years' War, 1751–1763" in *Diplomacy & Statecraft*. For the purpose of this article it is important to note, however, that it wasn't the result of a preconceived plan by the Dutch authorities who initially weren't ready to deal with matters like this at all. The decision to intercept the letters written by, or addressed to the Prussian envoy, taken in the autumn of 1751, was a novelty in Dutch politics that can only be explained by the deep insecurity of the Dutch authorities that resulted from the untimely death of Stadholder William IV and in particular their fear of Frederick the Great who might even have considered claiming the hereditary Stadholderate for himself. The Dutch authorities didn't expect that they needed the services of a Black Chamber, however. During the summer of the same year, they had managed, through bribery, to get hold of the code-book of the departing Prussian envoy D'Ammon and they were surprised to discover that it was of little use to them because the Prussians had supplanted the codebook on the arrival of the new envoy De Hellen. In the end the intercepted letters were decoded by the Black Chamber in London through mediation of the British Ambassador York and the Secretary of State for the Northern Department, Newcastle. The Dutch authorities never contemplated to open a Black Chamber in the Dutch Republic proper and Lyonet was only asked to make and store copies of the intercepted documents, because he was already serving as a code-clerk.[45]

Totally unaware of his predecessors in the Dutch Republic, Lyonet found out where the intercepted code-material went and what happened to it. Perhaps he was told from the outset, perhaps later, but it is beyond any doubt that he could not resist the challenge to see how far he could get on his own. He had always been interested in languages. As

[43] On François Fagel as a man of culture see [88].
[44] For Lyonet's work as an engraver and a microscopist see, in particular [167: 65–84].

[45] For a treatment in full detail see [60]. The most amazing thing of course is that the British were willing to help, a fatal mistake as we shall see.

a youngster, while at his parental home in Heusden, he had taught himself Italian without a grammar or a dictionary by 'deciphering' an edition of the adventures of Télémaque, using only his sense of logic [167: 10]. Of course, he didn't have to start from scratch. He already had in his possession the Prussian code-book that was used by the previous emissary D'Ammon. This told him what vocabulary was used and how the Prussian codes were organised in general. Moreover, he had received copies of the solutions by the Black Chamber in London. These solutions weren't complete; in every letter several words were left out and Lyonet started by filling in the blanks.[46] At the same time, he began working on the letters that were intercepted alongside those of De Hellen written by the Prussian envoy in London, Michel. In this case, there were no other solutions at hand but he did have all intercepted code-material at his disposal. It took him 18 months to solve both code-books.[47]

It is not entirely clear whether he acted with the consent of his masters, the Grand Pensionary and the Greffier. He was paid by Steyn 600 guilders a year for his efforts but this may have been paid only for his work as a copier and keeper of intercepted code-material. A better indicator is the fact that he was allowed to take on his cousin, Samuel Egbert Croiset, as an assistant; at first at his own expense, but from 31 March 1753 onwards at public expense, at 600 guilders a year [34: 245; 121].[48] This very moment suggests that there was some sort of connection with his work as a code-breaker because roughly at the same time he finished his work on the codes of De Hellen and Michel. It seems likely, therefore, that he acted on his own initiative but not without the consent of Steyn who

apparently wanted to wait and see what he would achieve. The advantages of Lyonet's contribution were of course undeniable. It made the Dutch less dependant upon the British for their information on Prussia then and for the future, and it opened the prospects of new initiatives in this field as well, with or without the approval of their British allies.

In August or September 1752 the decision was taken to extend the interception of mail to the despatches of the envoys of Cologne, Saxony–Poland and France as well.[49] The motives are not entirely clear but there is little doubt that the despatches of the new French ambassador Bonnac were a particular matter for concern from the moment he arrived in the Dutch Republic in December of that same year. He had orders to prepare the ground for the conclusion of an 'eternal peace-treaty' between Holland and France. This objective was to be reached by cultivating his contacts with the important members of the merchant elite in Amsterdam and promising them all sorts of trade-benefits if they would oppose the Republic's close alliance with England. This was more than his predecessors, Saint Contest and Durand, had tried to achieve and the conditions had improved greatly for such a policy because of the untimely death of Stadholder William IV and the resulting weakness of the Dutch Stadholderate now exercised by a *woman*. Therefore, the Dutch authorities requested that Bonnac's letters be investigated by the Black Chamber in London as well.[50] The matter was entrusted to bishop Edward Willes personally but it took him some time to succeed. In September 1753 he reached the conclusion that he needed more code-material, but it wasn't until the end of January 1754 that he could show a full compilation of Bonnac's principal code-book [42].

---

[46]The handwriting of Lyonet is clearly recognisable in the blanks that were left in the letters sent back by the Black Chamber in London [77].

[47]The opinion, stated here, that Lyonet got the idea himself in the Summer of 1751 and that he had to convince his superiors to go along with it and slavishly copied by Isings and Van Seters, is too ridiculous to be true.

[48]The view, expressed by Lyonet and Croiset, in their notes and memoirs, that Lyonet had to pay Croiset out of his own pocket until the Summer of 1756 is simply a distortion of the truth; for the fact that Croiset started to work for Lyonet already in April 1752; see application letter by S.E. Croiset to the Greffier for the vacancy of cipher secretary (draft).

[49]The decision to extend the interception of letters to those of the representatives of Cologne and Saxony–Poland was probably taken in August or early September 1752. The first intercepted letter by Kauderbach to the King of Saxony–Poland is dated 5 September 1752 and the first one by Cornet to the bishop of Cologne is dated 19 September 1752. Cornet also reported to a minister in the Palatinate: his first letter to Count Wagtendonk in Manheim is dated 12 September 1752. All items mentioned here are to be found in [76]. The first intercepted letter by the French ambassador Durand can be found here too and is dated 26 September 1752.

[50]This probably all happened during the first three months of 1753. See [16].

Initially, there was no intention of sending a copy of the code-book to Holland. The British preferred to keep the code-book and the intercepted code-material to themselves and they wanted to transmit only the contents of each letter that had been solved. This procedure, however, proved to be highly inefficient because of the loss of time caused by the sending of the intercepts overseas. Therefore, Lyonet would like to have received the code-material back from London as well in order to compile a copy of the code-book of his own. Apparently, Steyn and Bentinck shared his opinion. On 23 October 1753 Bentinck wrote to Newcastle about this matter trying to convince him that Steyn had taken sufficient precautions to prevent anybody finding out [22]. The British responded by sending a copy of their compilation of the entire code-book which was, of course, just as good.[51] From January 1754 onwards, all intercepted letters written by Bonnac were decoded in Holland by Lyonet himself [142]. The Black Chamber in the Dutch Republic had been able to broaden its scope only due to the crucial support of the British. This seemed to prove, however, that although Lyonet's activities as *code-clerk* were indispensable, his efforts as *code-breaker* were superfluous and there is no indication that Lyonet was ever asked to develop his expertise as a cryptanalyst.

This didn't bother Lyonet, however. The code-book from London provided him with the vocabulary and the general pattern of the codes used by the French at that particular time, just as had been the case with the Prussian codes a few years before. In the meantime Lyonet had learned what use he could make of this knowledge. Perhaps, he had this already in mind when he had Steyn ask for the code-material; perhaps it occurred to him later. Whatever the case may have been, in December 1755 the French code-books were changed once again and Lyonet tried to solve the new code, this time all on his own [4]. The political situation

had dramatically changed by then. The outbreak of war seemed only a matter of time. The Dutch had nothing to gain and favoured neutrality, but they were still obliged, by treaty, to send troops to the British. In fact, the change of code-books on the French side was a direct consequence of this development. The new code-books were brought by the special emissary Count d'Affry who was sent down specially in order to find out what side the Dutch Republic would take in case of war, if any at all. The answer was that they would stay out of the war, notwithstanding the fact that they would send troops to defend the British coast, if the British insisted.

In June 1756 Lyonet, after roughly six months of work, had broken d'Affry's code. He submitted a large number of neatly written transcripts of letters that had been written from the day of arrival to De Larrey, the private secretary of the Princess-Governess. On 16 June 1756 they were collated with some comments by De Larrey and brought to Princess Anne, at the time staying at 'Het Loo', the Stadholder's residence in the Eastern part of the country [8].

The results were bewildering. The French emissary d'Affry had an informer at the Court of the Princess-Governess, one of her oldest and most trusted friends from Frisia, her former private secretary De Back [46: 57–58, 157; 102: 106–108].

During the remaining and the following year Lyonet broke two more Prussian codes and two more French ones, belonging to Bonnac and D'Affry [121].

These repeated successes clearly showed that the Black Chamber was indispensable, although it took some time for the Dutch authorities to admit this [60: 1–30]. Thus, albeit with a slight exaggeration, it can be said that the Black Chamber had established its position only as a consequence of the tenacity of one man acting in the context of a growing political isolation the Dutch Government was reluctant to face.

The appointment of Lyonet as a code-clerk marked the beginning of a gradual improvement of Dutch codes and of the way they were used. At first, the code-book was only sized-down to an average length of roughly 4000 code-groups, but

---

[51]Yorke's letter to Hugh Jones on 25 January 1754 (see above) makes clear that at that particular time he hadn't received anything in return, notwithstanding the fact that in London the work was almost finished. It seems more likely therefore, that the code-book was sent, along with the transcripts, but not the original code-material, as was requested by Bentinck in his letter, dated 23 October 1753.

it now became a matter of procedure that every minister received a code-book of his own that he had to return whenever his mission ended.[52] Consequently, the legation's secretary also received a code-book that was used in the reports that were written on personal account, or whenever the Dutch emissary was absent and he also received a second code-book for use whenever one of the others seemed to be broken.[53] Occasionally, Dutch ministers would also receive a copy of the so-called 'chiffre commun', a separate code-book intended for the correspondence with Dutch ministers stationed in other capitals.[54] The fact alone that several code-books were being used at the same time and that they were changed every now and then indicates that Lyonet considered the code-book now to be a cryptographic device in its own right and no longer a mere addition to the cipher in order to avoid any unnecessary repetitions of code-groups and to limit the volume of code-material. From 1756 onwards, however, the very construction of the code-book changed too. This was mainly due to Lyonet's own experiences as a code-breaker that had made him increasingly aware of the vulnerability of the codes that were used by the Dutch themselves.[55] The new code-book seemed to have been intended as some sort of compromise between a one-part and a two-part code. The book still contained about 4000 items but each page was now di-

vided into five columns in stead of three and would have a box left blank at the top or the bottom of the page. The code-groups would follow the columns, their meanings would follow the rows. Thus there was no longer a direct correspondence between numerical and alphabetical value. The same book, however, could still be used for both decoding and encoding. Moreover, on each double page the square at the top or below, would be seen to contain a cipher and some code-groups for double letters, frequently occurring junctions and particles to make sure that it was always at hand and to avoid it being used in any stereotype manner. The book could consist either wholly of 4-digit code-groups or else of three digit code-groups with a mark of some sort, such as one or two dots, a dash above or under the number, or a slash through the first or the last digit of the number, and the nomenclature could vary considerably in composition or length.[56] Another remarkable feature was that it could be used both in Dutch and in French. Thus the Dutch word for sender ('afzender') could be followed by the French word 'age', meaning in Dutch either 'leeftijd' (= age) or 'tijd' (= era). In fact, a conscious effort was made to hide what language was being used in Dutch despatches. The beginning had always to be written *en clair* in Dutch, the actual code-letter could be written in either language, as long as it was preceded by some lines containing only nulls. In certain cases this would enhance the cryptological value of the code but apparently this result wasn't intended. For instance the code-group for 'altesse' (109⁻) would also signify, in a Dutch letter, 'hoogheid' or highness. This implies of course a slight breach of the one-part structure [74] (see Figure 12.5).

The 'modified' one-part structure, introduced by Lyonet, seemed to have been fairly common during late 18th century cryptography and the size of his codes wasn't unusual either, as indeed it would have been earlier in the century [101: 161]. This was caused by the fact that it was extremely troublesome to compile a new code-book.

---

[52]Each code-book was preceded by an annex or preamble, laying down the rules for its use, including the demand that it should be returned to Lyonet or Croiset by the minister personally, whenever his commission had ended. See for an example the code-book composed in 1791 by Croiset for Hogguer [71].

[53]The possession, and use, of additional code-books by the legation's secretary can be derived from examples of code-letters, written in the absence of an ambassador. See for instance: report signed by the legations'secretary De Swart, St Petersburg, 29 November 1787. This letter is coded, presumably by accident by two different code-books [145]. There are also examples that indicate that an ambassador could formally authorise the legation's secretary to use his code-book in his absence to deal with any current affairs. The absentee in this particular case was D.W. van Lynden [69].

[54]As was the case in 1751 with the newly appointed Dutch ministers at Berlin, Vienna, St Petersburg and Dresden who were to leave the Hague at the same time. See [13].

[55]This is Croiset's opinion and I see no reason whatsoever to doubt this. See the application by S.E. Croiset for the vacancy of cipher secretary (minute), 1789 [121].

---

[56]For an example of a 4-digit code see [70] for an example of 3-digit code see [23]. Croiset took great care to make an elaborate cipher for the Dutch envoy Hogguer who was about to leave for Russia in 1791, because there had been complaints about the opening of letters in that country. See [52].

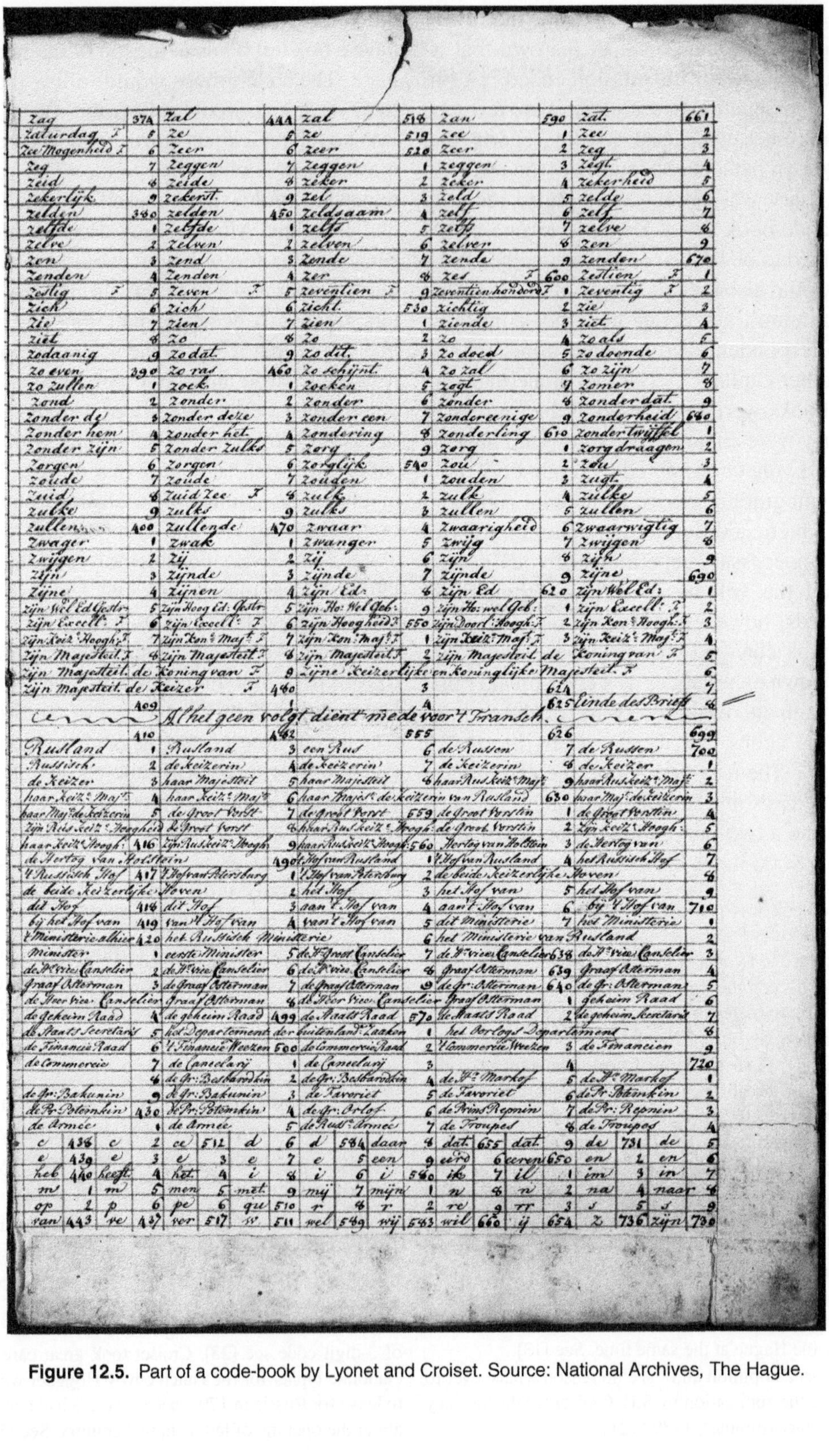

Figure 12.5. Part of a code-book by Lyonet and Croiset. Source: National Archives, The Hague.

It took Croiset, according to himself, several weeks to make a code-book, containing roughly 4000 groups that was strong enough to be used for years. Unfortunately the time taken for the construction of new code-books was at the expense of his activities as a code-breaker [52]. The rudimentary, one-part structure made it unnecessary to manufacture different books for decoding and encoding and the size alone seemed to promise that code would hold for some time. For the Dutch Black Chamber in particular it was important to economise. There were only two people involved and they had to take care of all incoming and outgoing code-letters, the making of the intercepts at the post office, the breaking of foreign codes and the manufacturing of code-books. In Britain it took roughly a dozen of these people to perform these tasks; of course the sheer volume of intercepted letters in Britain was much bigger but it seems beyond doubt that the Black Chamber in England was much better equipped. Therefore, it is hardly surprising that the British standards for government codes were superior. They used genuine two-part codes, comprising of roughly 3000 items signifying single letters, syllables, words and small wordclusters. There were also signs that annulled part of a cluster signified by a code-group. For instance, in code 1737 A code-group 469 signified 'prose, cute, in of, lyte, to'. If followed by code-group 1079, the last two words or syllables had to be struck out, leaving only 'prosecute in/of'. Another typical feature of British cryptography was that these code-books were always distributed in pairs. If code-book 1737 A was used to encode an outgoing letter, the answer had to be encoded with help of code-book 1738 B. This system was gradually introduced from 1737 onwards. In 1738 the code-book B mentioned above was issued; in 1742: code-book C followed; in 1750: code-books D and E; in 1755: F and G; in 1757: H and J; in 1759: K and L [2]. These code-books were used in various combinations by various British embassies; they may have been occasionally supplanted at a particular embassy, but they appear never to have been totally withdrawn from use [101: 173, 174].[57]

This was a big difference from French cryptography; French codes were changed every one or two years [60: 1–30].

Lyonet and Croiset were well aware that their codes could still be broken. In some countries, particularly in Sweden, in Russia and in Austria, Dutch embassy staff incidentally complained about the fact that their letters had been opened so clumsily that it could not be ignored, or that the contents of code-letters seemed to be already known before it was transmitted, or else, that a code-book had been in use for far too many years.[58] Hop, who had been representing the Republic at the British court for no less than thirty years, never used his code-book at all, because he wanted to avoid suspicion at any price and, apart from that, he knew for sure that the British would be able to break his code anyhow [14].

This leaves us with the question whether Lyonet's superiors showed any interest in his efforts to ameliorate Dutch codes at all and whether they properly understood what it was all about. There is little doubt that Dutch diplomats didn't always see the need of abiding by the rules set down by Lyonet and Croiset for the use of the codes that were supplied to them.[59] Moreover, both Greffier Hendrik Fagel and Grand Pensionary Laurens Pieter van de Spiegel stated at times that the improvements Lyonet added to Dutch cryptography were immaterial [121].

It should be noted, however, that the overtly stated 'contempt' for Lyonet's work was chiefly

---

[57]Kahn states that the Deciphering Branch devised the British code-books from 1745 onwards. This seems to be an error,

however, because the first code-book of this type was already introduced in 1737.

[58]For complaints from Stockholm see [158] and [175]; for a complaint from Turkey see [66]; for Austria see [12]; for Russia see [10; 15; 161]. The interception of diplomatic despatches in Russia took place, at the latest, from 1780 onwards, but, apart from this, too little is known. See [33]; see also A. Brikman's article about mail interception under Catharine II in *Russkaja starina*, VII (1873) 75–84. I should like to express my gratitude to Mrs. Natalia Chkalova, who made a summary for me in Dutch of this article which would otherwise not have been accessible to me.

[59]For an example of the latter see [164]. In this letter, the ambassador transgressed the important rule of not using single code-groups within a sentence, but always coding entire periods [164]; see also Lyonet to Fagel, without date or place, about transgressions and the improper use of a code-book by Van Landsberge [75].

motivated by political considerations. This had already been the case during the Seven Years' War when the Princess-Governess initially had refused to reward Lyonet properly for his exceptional achievement. Lyonet himself was convinced that this was solely due to his opinion that the Dutch Republic, at that particular moment, had more to fear from England than from France and therefore had to remain neutral. Perhaps unintentionally, Lyonet's political views brought him into the fold of the so-called 'republican party' that always had tried to curb the influence of the Stadholder and his Court on political matters; a movement that had gained strength among the merchant elite in Holland after the demise of Stadholder William IV in 1751, in particular among the town council of Amsterdam. Lyonet's convictions were probably already known to his superiors before he started to work as a code-breaker but the political climate rapidly deteriorated shortly before the outbreak of the war and Lyonet had even shown the audacity of ventilating his anti-British feelings in the presence of the Princess-Governess, who held the opposite view [121]. For the next twenty-five years, the political divergence between Lyonet and his masters seemed to have lost significance, mainly as a result of the repeated conflicts with England over transatlantic trade- and shipping interests and the increasingly anti-British mood in the country. This development culminated during a new, major conflict, caused by the outbreak of the American War of Independence in 1776 over the supply and shipment of goods for the American rebels and the repeated harassment of Dutch ships by British vessels that resulted from it. In 1780 the Dutch Republic seriously considered joining Spain, Sweden and Russia in the so-called 'League of Armed Neutrality' that promised mutual assistance in case any of their ships was to be confiscated again. This was followed by a proclamation of 'unlimited convoy' by the States General, meaning that the Dutch navy would fire if any Dutch merchantman was to be approached. The British reacted ferociously with a declaration of war on 23 December 1780 [59: 482–483; 136: 39–49]. Not surprisingly, this development roughly coincided with a peak in Lyonet's

career. From the beginning of the War of American Independence, the Dutch had to struggle to maintain their neutrality without sacrificing their interest in an American victory. Moreover, countries that used to play only a secondary role in Dutch foreign policy, at least as allies, such as Sweden, Russia and Spain, suddenly acquired a new importance, whereas an old enemy like France acquired the status of a friend. The increase and the sheer volume of Lyonet's work during this period seems to point in the direction of a high-tide of Dutch 'secret diplomacy' [34: 250–251].

12.6 THE RISE OF THE PATRIOT MOVEMENT

The political controversies in the Dutch Republic regained strength, however, after the defeat by England in 1784. During that year a humiliating peace treaty was concluded that clearly showed the military and political impotence of the Republic. The blame fell on Stadholder William V, now an adult although hardly mature and still under the supervision of Duke Louis of Brunswick, his former tutor. As a military commander the Stadholder could be held responsible for the bad performance of the Dutch defence. He became an easy target of a popular reform movement that was inspired by the example of the American Revolution and that wanted to change the whole frame-work of the Dutch State. In 1785 this reform movement, the so-called 'patriots', became strong enough to take over the control in the province of Holland. The Stadholder and his family had to leave The Hague in a hurry. Not surprisingly, Lyonet overtly supported the Patriots. In May 1784, he had even expressed his feelings in an anonymous letter to Prince William V, stating that he was a mere servant of the States-General and that he should behave accordingly [110]. This was all well remembered of course after the triumphant return of Stadholder to The Hague in September 1787. This event was basically a result of a counter-revolution, carefully prepared by the Stadholder's wife Princess Wilhelmina of Prussia, and executed in close collaboration with the British ambassador Harris and the Grand Pensionary of Zeeland, Laurens Pieter van de Spiegel, who assumed office as Grand Pensionary of Holland from that moment onwards. The Stadholders' most vehement opponents were either exiled or just removed from office and those remaining dared not speak. Apparently, Lyonet himself, already of old age, had little to fear. In September 1787, during an audience celebrating the return of the Stadholder to The Hague, he had shown the audacity of telling the Stadholder to his face, that he had not changed his heart, notwithstanding the defeat of his political friends [121]. But for his cousin Croiset, things looked different. Lyonet was already too ill to return to his old post and Croiset was already his successor in all but name. But Croiset had also shared the political

Frederica Sophia Wilhelmina
Prinses van Orange en Nassaü,
gebooren Prinses van Pruijsen enz. enz. enz.

Figure 12.7. Wilhelmina of Prussia 1751–1820. Engraving by Benjamin Samuel Bolomey. Source: Rijksinstituut voor Kunsthistorische Documentatie.

views of his uncle and this simply meant that he had been on the wrong side at the critical moment. In fact, he was lucky not to be sacked, like some of his 'fellow-patriots' who had held similar positions in the State Bureaucracy.[60] In January 1789, Lyonet's death provided the Orangists with the opportunity for revenge because Croiset's position had now to be formally reviewed. The obvious thing to do seemed to transfer to Croiset all responsibilities Lyonet had carried at the same pay.[61] The Stadholder and the Grand Pensionary, however, decided otherwise: Croiset was to be retained as 'cipher secretary' but not to hold the offices of translator and 'secretary of the foreign correspondence' that had fallen free as well and that could be so easily

M. L. P. VAN DE SPIEGEL
Raadpenſionaris van Holland.

Figure 12.6. L.P. van de Spiegel 1736–1800. Engraving by R. Vinkeles. Source: Rijksinstituut voor Kunsthistorische Documentatie.

[60][93]. Quarles and Slicher were sacked on 1 July 1788.

[61][121], draft for an application letter by S.E. Croiset to the Greffier for the vacancy of 'cipher secretary'.

combined [121]. There is no question that Croiset wasn't sacked, solely because he was the only one there to know how to deal with all the code-work for the States General and who was ready to do it.

His work as a *code-breaker* was increasingly a matter of concern. Croiset had not been able for considerable time to break the new Prussian and French codes that were introduced after the counter-revolution. This was mainly due to the fact that the newly introduced, Prussian codes were much more complicated than the old ones that had been introduced and also because the French suddenly started to change their code-books even more frequently than before. The fact that Croiset could not show any results for so long suggested that he didn't want to cooperate. In November 1788 he was even accused, in covert terms, of having revealed the secret of the Black Chamber to the French, out of political motives [121]. That this accusation had to be taken seriously can be derived from the fact the Grand Pensionary and the Stadholders' wife Wilhelmina took measures to exclude Croiset from dealing with confidential information altogether. They devised small codes of their own for their communication with the ambassadors in Prussia and England; the countries that were, in a manner of speaking, the 'life line' of the Restoration Regime. This was done for the sole reason to avoid the passing of these letters through 'certain' hands, as Van de Spiegel wrote to the Stadholder on 14 August 1788 [62: 107, 108].

In fact, the Grand Pensionary would have preferred to abolish the Black Chamber altogether. Through the interception of Prussian code-material, Croiset had access to everything that had been discussed between the Prussian ambassador and Van de Spiegel himself. If Croiset was as unreliable as he was suspected to be, the French had access to this information as well. That the Black Chamber was nevertheless retained was only due to the indecisiveness of William V who simply prevented the Grand Pensionary from taking firm action.[62] In the end no decision was taken at all. Croiset received a reward for every code he had broken but regarding the annual payment he felt entitled to,

[62]For this aspect of the relation between L.P. van de Spiegel and Stadholder Willem V see [36].

nothing definite was said, notwithstanding his recurrent efforts to press the matter with the Grand Pensionary and the Stadholder. This didn't mean that Grand Pensionary Van de Spiegel lacked understanding for Croiset's skills or achievements, as had been the case with the old Greffier who had even stated once that one didn't have to be a sorcerer to be a code-breaker. On the contrary, Van de Spiegel possessed several books on cryptography in his library, he devised codes and complicated ciphers himself and he had shown a lively interest in Croiset's worksheets and registers when they first met [121]. For Van de Spiegel's library see [150]; here the most important work on cryptography was [137]; for a Vigenère devised by the Grand Pensionary in captivity see [128].

This whole affair culminated during the final days of the old Republic, just before the breakdown of the Dutch defence in January 1795. During those last weeks the Dutch government had sent a delegation to Paris to negotiate an armistice or a truce of some sort. The delegation sent an elaborate report that was fully coded with the help of one of the regular code-books of the States-General. In this particular case, however, the Grand Pensionary felt unhappy in giving the report over to Croiset and he demanded to have the code-book instead, by way of one of Croiset's colleague's Van Lelyvelt. This was contrary to standard procedure and Croiset heavily protested, albeit giving in at the end. This incident constituted the pretext for the Grand Pensionary's arrest and his detainment after the French had occupied the country. He was accused of having tried to sell the country behind the back of the States-General, of high treason in other words, and the testimony of Croiset was part of the evidence to support this, as were a number of letters written by the Dutch ambassador in Prussia in a code that Croiset didn't recognise [127; 155; 153].

The Batavian Revolution solved much of Croiset's problems. Of course he favoured the new regime and he was, without much difficulty, incorporated in the new bureaucracy that was created swiftly by the revolutionaries who were inspired by French examples. His main task was, apart from his work for the Post Office, the construction of code-books and the encoding and decoding of letters for the new Department of Foreign Affairs [7; 148]. His work as a code-breaker continued until 1803, albeit restricted to the intercepted Prussian code-material. There was no real intention, however, to keep the Black Chamber open in the long run and to find a successor for Croiset. As long as Croiset remained in office as 'cipher secretary' this was of no consequence for the quality of Dutch codes. This changed however after his retirement. From 1806 or 1809 onwards the level of Dutch cryptography can be seen to deteriorate. This wasn't much of a disaster, of course, as long as Holland remained a satellite-state of France and ultimately ceased to exist, as was the case between 1810 and 1813. The bill was paid, however, after the Dutch had regained their independence. The return of the hereditary Prince William of Orange, who was shortly afterwards proclaimed King of a United Kingdom of The Netherlands, encompassing all of the Low Countries, including the former Austrian Netherlands, marked the beginning of a new era in Dutch foreign policy. For a short period in its history, the country resumed the role of a Great Power. In the field of cryptology it proved to be ill-prepared, however. Apparently, the construction of codes was now left to one, or more, civil servants at the Department without any cryptological training. As a result, the level of Dutch cryptography dropped dramatically.[63] The code-books that were handed to the Dutch ambassadors in France and in England were of surprisingly poor quality. As a result, between 1815 and 1840, many Dutch diplomatic despatches were decoded by the Black Chambers in both Paris and London, not to mention what happened in other capitals such as Berlin, Vienna and St Petersburg.[64]

12.7   THE USE OF CIPHERS DURING THE LATTER PART OF THE 18TH CENTURY

The last decades of the 18th century also saw a revival of the interest for, and the use of, ciphers.

This was partly, but not exclusively, caused by the political turmoil so characteristic of what has been justifiably called by Palmer, 'the Age of the Democratic Revolution'. The activities of Princess Wilhelmina and Grand Pensionary van de Spiegel have already been noted and can be directly linked to the crisis in the Dutch state. They were by no means the only ones, however. The political turmoil and the short civil war that resulted from it affected Orangists and Patriots alike and the increasing popularity of cryptography was not restricted to the Dutch Republic either. A number of new books were published, mainly in French, some in Dutch, that yielded recipes for encryption that were far from original and that were clearly derived from the preceding literature on the topic.[65] The increasing popularity of cryptography during this period did not go unnoticed and was even considered by some contemporaries as a topic worthy of satire.[66] The lack of originality exemplified so clearly in these little books does not mean, however, that the 18th century had lost all capacity for renewal [104: 30]. On the contrary, in my article written with Hans van der Meer about a turning grille used by Prince William IV, published in 1995 in *Cryptologia*, I am able to prove that a device that was described for the first time in a German mathematical magazine in 1795, was in fact already being used 50 years before [64: 153–165].[67] The occurrence of this device already at this early date indicates that the interest in sophisticated methods or devices for encipherment, as described by Alberti, Porta and Vigenère never totally waned but this didn't mean that they were slavishly copied. These devices seem rather to have been used for demonstration purposes and, probably also, to set standards for encipherment generally.[68] Apparently the design of ciphers was increasingly undertaken in an

[63] See for examples of codes used at the Department of Foreign Affairs between 1815 and 1830 [118] and [140]; for transcriptions of code-letters received from Paris between 1822–1833 see [24].

[64] Intercepted letters mainly written by baron Fagel during the years 1820 and 1821 [18]; intercepted Dutch diplomatic despatches, mainly written between 1823 and 1842 [41].

[65] For some examples in the Dutch language see [159: 188].

[66] See for an example of a mock-code the so-called Key-number of the political–satirical weekly magazine *Janus*, published on 14 June 1787. I owe this reference to Pieter van Wissing.

[67] For a similar devise used by the Hereditary Prince in 1800 see [159: 187–210].

[68] The treatment of methods of cryptanalysis, a topic omitted in Renaissance literature, points in the same direction. See for instance [38] that was also to be found in the library of the Dillenburg; [51] that contains examples in Dutch; as well as [54;

analytical way and it may even have figured as a pastime for those who felt the inclination, not unlike a game of chess. There is little reason to assume that the exercise of the art was ever limited to the professionals at the Black Chambers, however superior their understanding may have been. There *is* reason to assume, however, that the interest in sophisticated devices for encipherment such as the Vigenère, was motivated primarily by the *awareness* of the ease at which monoalphabetic substitution ciphers could be broken by professionally trained cryptanalysts. To put it in other words, only people who understood to a certain degree what the Black Chambers of the day were capable of, would see the significance of sophisticated methods of encipherment. This would mean of course that the interest in this field was restricted, in a broad manner of speaking, to the Courts of Europe and that it wasn't really to be found elsewhere.[69]

The knowledge of cryptology at the Dutch Court was first put to the test during the exile of the Stadholder and his family in Nijmegen between 1785 and 1787. The Stadholder, engaged in the organising of a counter-revolution, had to keep in touch with his adherents in the western part of the country, in particular L.P. van de Spiegel in Zeeland, the British ambassador Harris in The Hague and G.K. van Hogendorp in Rotterdam. These people all played an important role in the organising of a counter-revolutionary network and the British ambassador was also important for the support the Stadholder might or might not receive from abroad. There were couriers travelling back and forth almost on a daily basis between Nijmegen and The Hague and Nijmegen and Middelburg and this meant that they had to cross enemy territory and that their luggage might be checked by a Patriot patrol.[70] Therefore the letters had to be brought

in code. Apparently the task to make and distribute a cipher among the adherents of the Stadholder was entrusted to T.I. de Larrey, the secretary of the Stadholder who had stayed behind in The Hague. On 21 November 1786 he despatched a copy of the cipher to L.P. van de Spiegel, writing that other people who had to keep in touch with the Court in Nijmegen had also received one, such as the Viscount van Lynden, Mr. Van Teylingen and J.C. van der Hoop [156]. The cipher itself has been described in detail in my article written with Hans van der Meer about 'a homophonic substitution in the archives of the last Grand Pensionary of Holland', published in 1993 in *Cryptologia* [65: 225–236]. The simplicity of this cipher, and the fact that it was distributed to so many people, is remarkable because it afforded little protection and there was so much at stake during those years. The explanation need not to be sought, however, in any cryptological ignorance or a contempt of the analytical powers of the Patriot militias that stood the chance of intercepting them, but rather in the way they were used. All really important matters were communicated by word of mouth, the letters dealing only with small domestic arrangements or affairs.[71] The second test came in 1795 after the Stadholder and his family had been exiled once again from The Hague, this time under force of French arms that brought the patriots back in power. The Stadholder and his wife went to London to stay temporarily at Hampton Court as guests of the British Royal family [100: 176–184; 82: 257–285]. In Germany there were a number of Dutch refugees, mainly staying at Brunswick, and the remains of the Dutch army at Osnabrück under command of general Charles Bentinck and the youngest son of the Stadholder, Prince Frederick [55: 517–538]. In Holland, there were still the supporters of the Stadholder that had stayed behind, mostly members of noble or patrician families in The Hague. In the autumn of 1795, however, the hereditary Prince William went

---

149]. Ref. [135] reports how, during the 4th Anglo-Dutch War, a Dutch cipher was broken used in a document intercepted by the British navy aboard a Dutch ship.

[69] The West Indian Company, for instance, only had one cipher that was introduced in 1739 and was used over and over again during the 4th Anglo-Dutch War. See [61].

[70] As indeed happened to the servant of Mr. Boers who was, on his way from Amsterdam to Amersfoort with the luggage of his master, searched at Ankeveen. See [157]; for information about the courier between Nijmegen and Middelburg see [173].

---

[71] This can also be derived from the decoded message in the article mentioned above [65], stating that *young master Camper is to arrive with the next ferry carrying some important information*, almost certainly implying a message from Harris that England would declare war on France if it would interfere in the political strife in the Dutch Republic.

with his wife to Berlin to stay with his father-in-law, the King of Prussia. The Stadholder had to use the mail to keep in touch with his adherents on the continent and, inasmuch as nobody offered to carry any messages overseas, he had to resort to cryptography.[72] In London, the task to encode and decode letters, and probably also the construction and distribution of ciphers, was wholly left to Princess Wilhelmina who, as we have seen, had already gained experience in doing so during the exile of the Dutch Court at the Valkhof in Nijmegen, almost ten years before [162: 88–90; 120: 220]. She used a number of ciphers and codes that could vary according to circumstance.[73] For what was left of the Dutch army a variant of the Mirabeau was chosen because it was not very demanding [32: 19–20, 152].[74] For the friends in The Hague a very amusing code was devised that could make a letter dealing with politics pass for a letter dealing with family affairs. The Batavian Directorate would be called 'the ladies'; the Republic substituted by 'the farm'; the King of Prussia would become 'Charles', the revolution would become 'the commerce'; Prussian troops: 'a skein of grey wool' [108: 323–324]. This was done primarily because, during the first years of the Batavian Revolution all letters from or to England were opened at the Post Office in The Hague and superficially read, a security measure heavily protested by

Croiset as being of no use at all [148: 17]. The real danger, however, was with the Black Chamber in London that was ordered to break this code, most likely because the British Government was anxious to know whether or not the Stadholder was secretly negotiating his return to Holland and if so, on what conditions [40]. For the correspondence with the Hereditary Prince in Berlin, however, a totally different cipher was devised that merits our special attention. Princess Wilhelmina used a small two-part code, consisting of about 300 items in an apparently random alphabetical order, mostly referring to names of places or persons and, to a lesser degree, titles and concepts. The code-groups ranged from 10 to 334, the one-digit numbers being left to indicate the position of the letters in the words or names in the codelist. Thus words, not included in the nomenclature, could be formed by adding single digits between dots to code-groups that *were* included and indicating the end of a word by underlining the last digit. In this way the cipher provided for a great variety of means of representing the letters of the alphabet, while retaining all the advantages of a code that could be memorised.[75] This cipher is a modified form of the so-called 'bookcipher' that was already described by people such as Vigenère and Cardano, roughly based on the idea that two copies of the same book could be used as a code if a way was found to indicate the words required to constitute the message. Cardano proposed the use of his famous grille, Vigenère prescribed the use of transparent sheets of paper, of exactly the same size as the pages of the book with underlining of the relevant words of that page. Other authors would simply suggest to write down the page- and line-number, followed by a number to indicate the following-order of the relevant word on that line.[76] Clearly these 'original' variants of the bookcipher were quite laborious to use and therefore probably not much applied. The 18th century, however, looked for ways that made the bookcipher more convenient in its use, basically by limiting the volume of text on the one hand and numbering through the words of the entire text on the other.[77] This also happened in the

---

[72][108] states that these very important princely letters were carried generally by high-ranking officers that had remained loyal. Surely some of them were. The Stadholder had large masses of correspondence however, with various parties. He, or rather Princess Wilhelmina, would write twice a week with the members of her family alone, not to mention all other people loyal to her cause who had been exiled. Somehow I find it difficult to imagine that there were enough loyal, high-ranking officers travelling back and forth between Hamburg and London that could carry all these letters. Therefore, I take it, they frequently used the mail.

[73]Portfolio with ciphers and codes used by Her Royal Highness during the exile [144]. Only some of them are briefly described. The diversity and sophistication that manifests itself in this portfolio seems to have been the prerogative of the House of Orange.

[74]Typical for a Mirabeau is a subdivision of the alphabet into four of five numbered sequences of letters that could be memorised and the use of a horizontal line to indicate the number of the sequence above and the position of the letter within the sequence beneath, or vice versa; for examples see [5].

---

[75]For a more extensive description of this cipher and its background see [62].

[76]This small survey is mainly based on [113].

[77]For instance Breithaupt [38] and Thicknesse [149].

Hampton Court cipher, and in that respect it wasn't original, but it was original, however, in its application of the principle of 'local value' to transform the two-part nomenclature into a cipher.[78] This was totally unprecedented in the literature of the day it seems to reveal a typical 'mathematical' approach to secret writing.[79] There is nothing to link the construction of the cipher to a famous mathematician, however. In fact, there is sufficient reason to assume that Princess Wilhelmina constructed the cipher herself but she may have been inspired, apart from the example of the elaborate two-part code in the possession of the Stadholder, by a small poem that was used as a substratus for a cipher in a similar fashion for Prince Frederick during the years 1794–1795 [133: Vol. II, Suppl. 34, lxxii] (see Figures 12.8 and 12.9).

Here the same result was reached by truncating the words in the text through the indicating of the number of letters that had to be struck out behind a lunar sign added either at the left or the right side of the code-group that signified the relevant word. This small precursor or forerunner may have been devised by one of the mathematicians, present at that particular moment at the Dutch Court, probably the librarian Johann Friedrich Euler who had also been engaged in the education of the young princes.[80] In fact, there is evidence that Euler had already been devising codes and ciphers on behalf of the Stadholder for some years. In 1782, after the outbreak of the War with England, Stadholder William V was supplied by him with a genuine two-part code, consisting of roughly 4000 code-groups and also containing a subtraction- and addition-table for superencipherment, a device otherwise known from early 20th century secret writing[81] (see Figure 12.10).

The application of this table would have greatly improved the quality of Dutch cryptography but there is no evidence that it was much used.[82]

Therefore it is possible to say that the cipher reflected, in a very general way, a certain stage of mathematical thinking and for the sake of intellectual history it may be asked whether it could have occurred much earlier. Florian Cajori has stressed the importance of the 18th century and of Leibniz in particular for the development of mathematical notation, above all, because of his empirical-critical way of dealing with this topic [45: Vol. II, 180–185]. This would mean that this cipher cannot be expected to occur before Leibniz contributions had filtered through in the way mathematics was generally practised and understood, one or two generations later. This may seem, at first glance, a purely hypothetical matter, but there is more to it. The anatomy of the Hampton Court cipher closely resembles the anatomy of ancient Egyptian writing that was discovered some twenty years later by Alexander Young and Champollion. To cite E. Iversen: 'throughout their history each hieroglyphic sign had two independent functions. They could be used 'ideographically' to denote the object they represented (...). [In that case] a vertical stroke [was] placed after the hieroglyph itself to signify that it was to be understood as the actual material object it delineated. But each sign could also be used as a 'phonogram', as a purely phonetic element conveying the sound value expressed in the name of its prototype (...) [95: 14–15]. The similarity between the cipher of Hampton Court and the anatomy of the ancient Egyptian script is

---

[78]The principle of local value implies that the decimal order of a digit is denoted only through its position within a number. This is only possible because of the inclusion of the zero in the digital system. See [117].

[79]The first description is to be found in [103].

[80]For Johann Friedrich Euler (1741–1740), a cousin of the famous Leonhard in Saint Petersburg, see [73]. For his engagement at the Dutch Court see [6]; for Stamford, see [50] and [96]. The use of the lunar sign [for instance 8)24 or 24(8 also seems to point in the direction of Germany where it was used as symbolism for division. See [45: Vol. II, 269].

[81]This two-part code really consists of two one-part codes that only partially overlap with numerous additions that

seemed to have been made later. The handwriting suggests that we are dealing with Eulers's work here, but nothing definate can be said without asking the advice of an expert in this field. The strong mathematical approach, manifesting itself in the use of the substraction table, suggests that it is at least safe to say that we are dealing with the same person who devised the cipher for Prince Frederick that served as a source of inspiration for Princess Wilhelmina's bookcipher at Hampton Court [141].

[82]The Dutch emissary Van Rheede wrote to the Stadholder's secretary T.I. de Larrey, from Berlin, on 4 February 1792 that it was too much trouble for him to use the code-book and that he preferred writing 'en citron'. The code was used nontheless in a letter by Van Rheede written from Warszawa on 9 March 1793 [143].

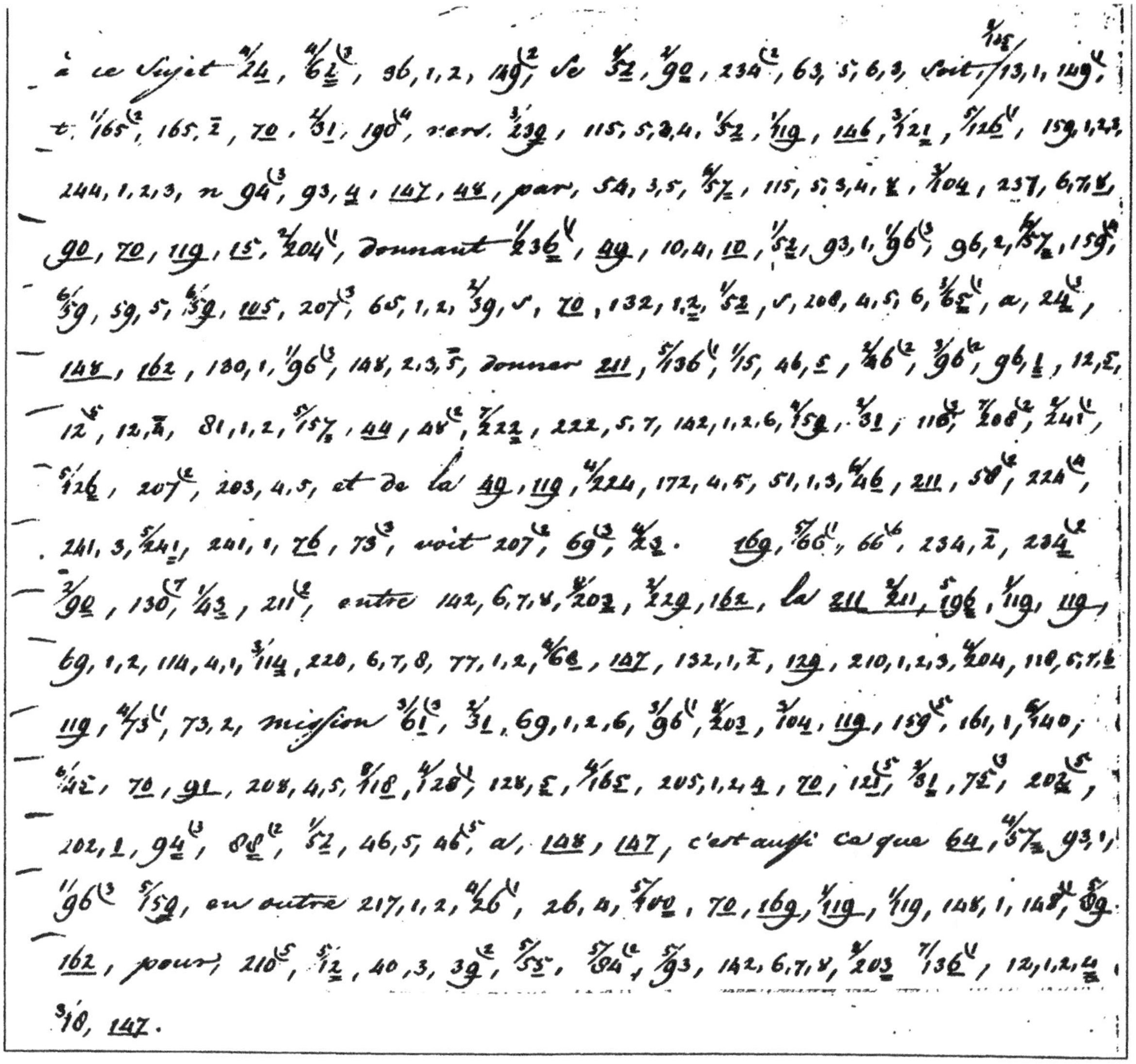

Figure 12.8. Leave written in cipher by Hereditary Prince William to Princess Wilhelmina of Prussia during their exile in Berlin and London. With kind permission of H.M. the Queen of The Netherlands.

of course purely accidental, but it seems highly unlikely that the fact that both phenomena occurred at roughly the same time was purely accidental as well. The breakthrough in the decipherment of hieroglyphic writing has usually been connected to the diminishing influence of the Neoplatonic conviction that hieroglyphs really did not constitute a script at all but rather an symbolic way to transmit occult ideas to the innate, of which Athanasius Kircher was the foremost exponent [53: 17–23]. In the field of symbolic logic, the field that investigates the general principles of rational procedure, a similar development took place. In his book on the development of this field, C.I. Lewis, mentions three important characteristics: (1) the use of ideograms in stead of the phonograms of ordinary language; (2) the deductive method – which may mean here simply that the greater portion of the subject matter is derived from relatively few principles by operations which are 'exact' and (3) the use of variables having a definite range of significance [109: 2]. Leibniz was the first to formu-

Figure 12.9. Cipher between Princess Wilhelmina and Hereditary Prince William during the their exile in London and Berlin. With kind permission of H.M. the Queen of The Netherlands.

late some sort of programme in this direction but he still thought of it as a means to achieve complete generality in all the sciences. In a way, this led Leibniz to reify the concepts he apply to both logic and nature [109: 8]. It was precisely this last element that was dropped by the next generation of mathematicians of whom Johann Heinrich Lambert (1728–1777) was a chief exponent and who chose a more modest approach, limiting its ambitions to the field of mathematics proper. To cite Lewis once more: 'Lambert [tried to] develop the calculus entirely from the point of view of intention; the letters represent concepts, not classes [109: 19]. In other words, only after Leibniz the realistic, or platonic, element, so typical of the quest for artificial languages of much of 17th century thinking was finally erased. The cipher of Hampton Court, with its sophisticated play of ideographic and phonographic elements, seems to have been a result of this development too. In that respect it is difficult to accept that artificial languages should have had a stimulating influence on the development of cryptography at all, as Strasser has stated on several occasions. Rather the opposite seems to be have been the case. Only after the quest

for artificial language had died out, the anatomy of writing could be approached in a new, more fruitful, way.

12.8 CONCLUSION

In Dutch history, code-breakers can be seen as active from the very beginning of the Revolt against Spain onwards. This clearly shows that Carter was wrong in his assertion that during much of the early modern period code-breaking only played a marginal part in political history, because there were so many other ways to obtain state-secrets. The opposite was true. During the Dutch Revolt against Spain the interception of enemy letters was an important item in the struggle for liberty and already at that time the art of code-breaking was practised in a professional way. There is no clear evidence, however, that these code-breakers also occupied themselves with the construction of codes and ciphers. It is true, during the long war against Spain, Dutch cryptography showed some similarities with contemporary Spanish cryptography but Dutch codes and ciphers were much smaller and

Tableau.

	A	B	C	D	E	F	G	H	I	K	L	M	N	O	P	Q	R	S
T		9	11	13	15	17	19	21	23	25	27	29	31	33	35	37	39	41
U		75	73	71	69	67	65	63	61	59	57	55	53	51	49	47	45	43
V		77	79	81	83	85	87	89	91	93	95	97	99	100	98	96	94	92
W		90	88	86	84	82	80	78	76	74	72	70	68	66	64	62	60	58
X		24	26	28	30	32	34	36	38	40	42	44	46	48	50	52	54	56
Y		22	20	18	16	14	12	10	8	6	4	2	1	3	5	7	8	9
Z		12	11	10	9	8	7	6	5	4	3	2	1	3	5	7	8	9
Aa		13	14	15	16	17	18	19	20	21	22	23	24	25	26	27	28	29
Bb		31	33	35	37	39	41	43	45	47	49	50	48	46	44	42	41	40
Cc		38	34	30	26	22	18	14	10	6	2	1	3	5	7	9	11	13
Dd		36	32	28	24	20	16	12	8	4	0	2	4	6	8	10	12	9

Figure 12.10. Addition or Substraction Table by Johann Friedrich Euler for the Stadholder's Codebook (1782). With kind permission of H.M. the Queen of The Netherlands.

showed a much more modest approach, probably due to the fact that there was no intention to open embassies all over the world on a permanent basis. Dutch diplomats were only occasionally sent abroad. This all changed in 1648, due to the Peace of Westphalia and the exchange of diplomats on a permanent basis between the Protestant and the Catholic powers. Roughly around 1650 a one-part code-book was introduced containing over 10.000 entries. This was supplemented with a set of ciphers, apparently inspired by Kircher's 'Arca Steganographica', a polyalphabetic device that worked with slides. This is remarkable because it proves that Kahn was wrong in his assertion that

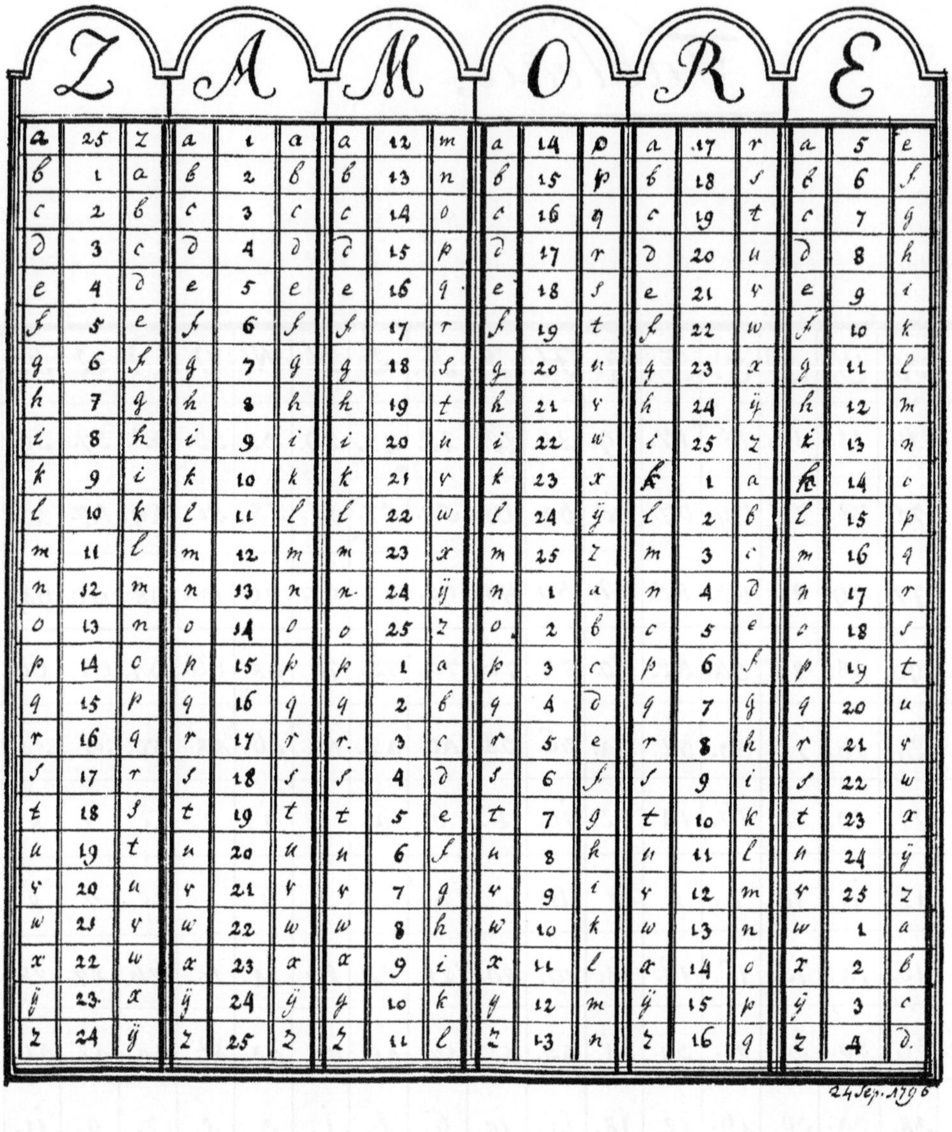

Z			A			M			O			R			E		
a	25	z	a	1	a	a	12	m	a	14	o	a	17	r	a	5	e
b	1	a	b	2	b	b	13	n	b	15	p	b	18	s	b	6	f
c	2	b	c	3	c	c	14	o	c	16	q	c	19	t	c	7	g
d	3	c	d	4	d	d	15	p	d	17	r	d	20	u	d	8	h
e	4	d	e	5	e	e	16	q	e	18	s	e	21	v	e	9	i
f	5	e	f	6	f	f	17	r	f	19	t	f	22	w	f	10	k
g	6	f	g	7	g	g	18	s	g	20	u	g	23	x	g	11	l
h	7	g	h	8	h	h	19	t	h	21	v	h	24	y	h	12	m
i	8	h	i	9	i	i	20	u	i	22	w	i	25	z	i	13	n
k	9	i	k	10	k	k	21	v	k	23	x	k	1	a	k	14	o
l	10	k	l	11	l	l	22	w	l	24	y	l	2	b	l	15	p
m	11	l	m	12	m	m	23	x	m	25	z	m	3	c	m	16	q
n	12	m	n	13	n	n	24	y	n	1	a	n	4	d	n	17	r
o	13	n	o	14	o	o	25	z	o	2	b	o	5	e	o	18	s
p	14	o	p	15	p	p	1	a	p	3	c	p	6	f	p	19	t
q	15	p	q	16	q	q	2	b	q	4	d	q	7	g	q	20	u
r	16	q	r	17	r	r	3	c	r	5	e	r	8	h	r	21	v
s	17	r	s	18	s	s	4	d	s	6	f	s	9	i	s	22	w
t	18	s	t	19	t	t	5	e	t	7	g	t	10	k	t	23	x
u	19	t	u	20	u	u	6	f	u	8	h	u	11	l	u	24	y
v	20	u	v	21	v	v	7	g	v	9	i	v	12	m	v	25	z
w	21	v	w	22	w	w	8	h	w	10	k	w	13	n	w	1	a
x	22	w	x	23	x	x	9	i	x	11	l	x	14	o	x	2	b
y	23	x	y	24	y	y	10	k	y	12	m	y	15	p	y	3	c
z	24	y	z	25	z	z	11	l	z	13	n	z	16	q	z	4	d

24 Sep. 1796

Figure 12.11. Vigenere by L.P. van de Spiegel (1799). Source: Rijksarchief in de Provincie Zuid Holland.

polyalphabetic ciphers were hardly ever used, at least for the Dutch Republic. This system was retained until the middle of the 18th century, without much alteration of the code-book, the ciphers being changed every now and then.

There is no proof, however, that this system was recommended by one of the Dutch cryptanalysts and it is beyond doubt that it was operated by code-clerks who had no experience in this field. This is also remarkable because the Dutch code-breakers had resumed their activities probably from 1684 onwards, focusing on French, and later also Bavar-

ian, intercepts. This situation can be explained, however, by the anomalous structure of the Dutch state that prevented the Dutch Black Chamber to develop into an institution, as was the case for instance in England and Austria.

Dutch codes were fundamentally changed only much later by the natural scientist Pierre Lyonet, who had gained experience in the breaking of Prussian and French codes from 1751 onwards. By then, Lyonet had been performing the job of code-clerk for many years already. This is, of course, a clear affirmation of Kahn's view that cryptology progressed as a consequence of the existence of the Black Chambers that acquired an expertise that remained exclusively theirs, and not, as Strasser has contended, as a result of the development of linguistics, in particular the effort to construct artificial languages. This does not mean, however, that the rise of the Dutch Black Chamber can be treated as an isolated phenomenon. Lyonet wasn't trained by other cryptanalysts; he was only helped by the result of the work undertaken by the Black Chamber in London on documents intercepted in the Dutch Republic. Lyonet had to reinvent the wheel, so to speak, all over again, with only his strong analytical capacities to rely on. In a way, this also holds true for Willem Jacob 's Gravesande, active 40 years earlier. 's Gravesande is also important because he saw a connection between code-breaking and probability theory, but there are no documents to corroborate the fact that he applied new statistical methods, although it is hard to believe that he didn't use them at all. It is beyond doubt, however, that these methods, if they ever existed, weren't communicated to Lyonet. Therefore it seems safe to say that the development of probability theory wasn't decisive for the emergence of the Black Chamber in the Dutch Republic but the existence of a critical–empirical tradition probably was. It should be noted, however, that this holds true for the 18th century in particular because it took almost a century for this tradition to mature. The Dutch code-breakers of the late 16th and 17th centuries were first and foremost humanists with a keen interest in both the arts and the sciences.

There is no doubt whatsoever that the development of cryptology in the Dutch Republic was unrelated to any efforts to construct artificial languages because in this country such efforts were never undertaken. There may be, however, in the field of the construction of ciphers, another influence that particularly manifested itself towards the close of the century. This had something to do with the advancement of mathematical notation and its application to the anatomy of writing in general. This seems to indicate that neither Kahn's nor Strasser's ideas about the development of cryptography are fully adequate. In Dutch cryptography, towards the end of the 18th century, innovations occurred that neither originated in a Black Chamber nor were dependent on an effort to construct an artificial language.

REFERENCES

[1] *Konst- en Letterbode* (1836).

[2] PRO. S.P. 106/15–39.

[3] Public Record Office (PRO). S.P. 35: 4/65.

[4] Bonnac to Rouillé, The Hague, 9 December 1755. – ARA. Fagel family, inv. nr. 5182.

[5] Charles Bentinck to William V, Bremen, 29 January 1796 and Nienburg 17 and 25 February 1796. – Royal Archives. Stadholder William V, inv. nr. 311 (62a).

[6] Copie de la lettre aux commissaires de Comitee dínstruction publique, The Hague, 15 March 1795. – Royal Archives. King William I, inv. nr. VIIIc-E18.

[7] Croiset to Dassevael, The Hague, 6 October 1812, presenting an outline of his career at the Ministry of Foreign Affairs. – ARA. Croiset family, inv. nr. 24.

[8] De Larrey to the Princess-Governess, The Hague, 16 June 1756. – ARA. Fagel family, inv. nr. 1401.

[9] *Dictionary of National Biography*, XXX, London (1892).

[10] Graaf van Rechteren tot Borgbeuningen to the States-General, St Petersburg, 23 September/4 October 1785. – ARA. Staten-Generaal, inv. nr. 7407.

[11] Hendrik Fagel the Oldest, *Nieuw Nederlandsch Biografisch Woordenboek (NNBW)*, III, p. 389.

[12] Hendrik Fagel to Burmania, The Hague, 19 February 1751. – ARA. Fagel family, inv. nr. 5135.

[13] Hendrik Fagel to Calkoen in Dresden, The Hague, 30 August 1751. – ARA. Fagel family, inv. nr. 5134.

[14] Hendrik Fagel to the new Dutch envoy in Great Britiain, Nagel, The Hague, 15 September 1788. – ARA. Fagel family, inv. nr. 5138.

[15] J.W. Hogguer to the States-General., St. Petersburg, 4/20 May 1791. – Ibidem, inv. nr. 7409.

[16] Joseph Yorke to Undersecretary Hugh Jones, The Hague, 27 April 1753. – British Library, Add. Mss. 35, 432.

[17] *Mémoirs du sieur John Macky (…), publiés sur le manuscrit original de l'auteur*, Amsterdam (1735).

[18] Ministere des Affaires Etrangers, Memoirs et Documents France, inv. nrs 732, 733 and 1931.

[19] Niedersächsisches Hauptstaatsarchiv, Hannover. Celle Br. 92: 124, ff. 25–29.

[20] *NNBW*, VII, 231.

[21] Robethon to Georg Ludwig, 16 June 1711. – Niedersächsisches Hauptstaatsarchiv, Hannover. Cal. Br. 24/3162, ff. 202–203.

[22] Willem Bentinck to Newcastle, The Hague, 23 October 1753. – Royal Archives, The Hague. Willem Bentinck van Rhoon.

[23] ARA, Fagel family, inv. nr. 1265.

[24] ARA, Frankrijk, inv. nrs 216–217.

[25] ARA, Fagel family, inv. nr. 1214.

[26] ARA, Fagel family, nrs 1212, 1218, 1223.

[27] ARA, Fagel family, inv. nrs 1210, 1213, 1217.

[28] J. Aalbers, *De Republiek en de vrede van Europa*, Groningen (1980), 155, 160–165.

[29] J.N.S. Allamand, *Histoire de la vie et des ouvrages de Mr. 's Gravesande*, **Oeuvres philosophiques et mathematiques de Mr. G.J. 's Gravesande**, Amsterdam (1774).

[30] C. Andrew, *The nature of military intelligence*, **Go, Spy the Land. Military Intelligence in History**, K. Nelson and B.J.C. McKercher, eds, Westport / London, (1992).

[31] ARA.- Fagel family, inv. nrs 1207–1223.

[32] C. Bazeries, *Les Chiffres Secret Dévoilés*, Paris (1901), pp. 19–20, 152.

[33] A. Beer, J. Ritter von Fiedler (eds), *Joseph II und Graf Ludwig Cobenzl. Ihr Briefwechsel*, **Fontes Rerum Austriacarum. Oesterreichische Geschichts-Quellen**, 2e Abth., Diplomataria et Acta, LIII, Wien (1901), p. 88.

[34] W.J.M. Benschop, *Secrete regeeringszorg met medewerking van het Haagsche postkantoor*, Bijdragen voor Vaderlandsche Geschiedenis en Oudheidkunde VIII-4 (1943), p. 245.

[35] P.P. Bockstaele, *Mathematics in the Netherlands from 1750 to 1830*, Janus. Revue internationale de l'histoire des sciences de la medicine, de la pharmacie et de la technique LXV (1978), 67–69, 80.

[36] J.C. Boogman, *Raadpensionaris L.P. van de Spiegel: een reformistisch-conservatieve pragmaticus en idealist*, **Mededelingen van de Koninklijke Akademie van Wetenschappen**, Afdeling Letterkunde, Nieuwe Reeks (1988) LI, p. 177.

[37] P. Bor, *Geschiedenis der Nederlandsche oorloghen*, 6 vols, Leiden (1621–1635) II.

[38] C. Breithaupt, *Ars decifratoria sive scientia occultas scripturas solvendi et legendi*, Helmstadii (1737).

[39] H.E. Brekle, *The Seventeenth Century*, **Current Trends in Linguistics**, Thomas E. Sebok, ed., The Hague/Paris (1975), p. 309.

[40] British Library, Add. Mss., 32,280, cipher keys Holland, fol. 61 (r.,v.).

[41] British Library, Add. Mss. 32279 and 32280.

[42] British Library, Add. Mss. 35, 432, f. 86, Yorke to Jones, 1 June 1753, f.152, Yorke to Jones, 28 September 1753, f. 156, Yorke to Jones, 3 October 1753, f. 162, Yorke to Jones, 16 October 1753, f. 172, Yorke to Jones, 30 October 1753, f. 182, Yorke to Jones, 16 November 1753, f. 202, Yorke to Jones, 25 December 1753, f. 210, Yorke to Jones, 8 January 1754, f. 218, Yorke to Jones, 18 January 1754, f. 222, Yorke to Jones, 25 January 1754, all from The Hague.

[43] H. Brugmans (ed.), *Correspondentie van Robert Dudley, graaf van Leicester en andere documenten betreffende zijn Gouvernement-Generaal in de Nederlanden, 1585–1589*, 3 vols, Utrecht (1931).

[44] P. Brunet, *Les physisiens hollandais et la méthode experimentale en France au XVIIIe Siècle*, Paris (1926), pp. 39–44, 48, 49, 58–59, 100–108.

[45] F. Cajori, *A History of Mathematical Notation*, 2 vols, La Salle, IL 3rd printing (1952) II, p. 269.

[46] A.C. Carter, *The Dutch Republic in Europe in the Seven Years' War*, London (1971), pp. 57–58, 157.

[47] C.H. Carter, *The Western European Powers, 1500–1700. The Sources of History*, London (1971), p. 233.

[48] C.H. Carter, *Secret Diplomacy of the Habsburgs*, New York/London (1964).

[49] J. Cohen, *On the project of a universal character*, Mind. A Quarterly Review of Psychology and Philosophy, LXIII (1954).

[50] H.T. Colenbrander, *Willem I, Koning der Nederlanden*, 2 vols, Amsterdam (1931) I, p. 20.

[51] D.A. Conrad, *Davides Arnoldi Conradi cryptographia denudata sive ars deciferandi (...)*, Lugundi Batavorum (1739).

[52] Additional notes by Croiset. – ARA. Croiset family, inv. nr. 17.

[53] M.V. David, *Le débat sur les écritures et l'hiéroglyphe aux XVIIe et XVIIIe siècles et l'application de la notion de déchiffrement aux écritures mortes*, Paris (1965).

[54] J. Davys, *An Essay on the Art of Decyphering*, London (1737).

[55] F. de Bas, *Prins Frederik der Nederlanden en zijn tijd*, 2 vols, Schiedam (1891) I, pp. 517–538.

[56] G. de Bruin, *Geheimhouding en verraad. De geheimhouding van staatszaken ten tijde van de Republiek*, Gravenhage (1991), pp. 252–281, 617–621.

[57] P. de Clercq, *Science at court: the eighteenth-century cabinet of scientific instruments and models of the Dutch Stadholders*, Annals of Science XXXXV (1988), p. 118.

[58] D. de Hottinga, *Polygraphie ou méthode universelle e léscriture cachée: avec les tables de figures concernantes l' éffect et l'intelligence d'elle: le tout compris en V livres*, Groningue (1621).

[59] J.C. de Jonge, *Geschiedenis van het Nederlandsche Zeewezen*, 5 vols, Amsterdam (1835–1843) V, pp. 482–483;

[60] K. de Leeuw, *The Black Chamber in the Dutch Republic and the Seven Years' War, 1751–1763*, Diplomacy and Statecraft X (1999).

[61] K. de Leeuw, *Geheimschrift op enkele kaarten en plattegronden van de verdedigingswerken rond de Surinamerivier, 1782*, Tijdschrift voor Zeegeschiedenis (1997), 160–178.

[62] K. de Leeuw, *Een lexicaal geheimschrift van Wilhelmina van Pruisen op Hampton Court*, De Achttiende Eeuw XXVII (1995), pp. 107–108.

[63] K. de Leeuw, *The Black Chamber in the Dutch Republic during the war of the Spanish Succession and its*

aftermath, 1707–1715, The Historical Journal XXXXII (1999).

[64] K. de Leeuw, H. van der Meer, *A Turning Grille from the Ancestral Castle of the Dutch Stadholders*, Cryptologia. A Quarterly Journal Devoted to Cryptology XIX (1995), 153–165.

[65] K. de Leeuw, H. van der Meer, *A homophonic substitution in the archives of the last Grand Pensionary of Holland*, Cryptologia, A Quarterly Journal Devoted to Cryptology XVII (1993), 225–236.

[66] F.G. Dedem van de Gelder to the Agent van Staat voor de Buitenlandse Betrekkingen, Constantinople, 10 July 1799. – ARA. Buitenlandse Zaken 1796–1810, inv. nr. 345.

[67] J.-P. Devos, *Les chiffres de Phillipe II et du Despacho Universal durant le XVIIe siècle*, Bruxelles (1950).

[68] J.-P. Devos, H. Seeligman (eds), *L'Art de deschiffrer. Traité de déchiffrement du XVIIe Siècle de la Secretairerie d' Etat et de Guerre Espagnol*, Louvain (1967).

[69] P.G. Duker to the States-General, Stockholm, 15 June 1781. – ARA. Staten-Generaal, inv. nr. 7236.

[70] ARA, Dutch legation in Turkey, inv. nr. 164.

[71] ARA, Dutch legations until 1813, Russia, inv. nr. 178.

[72] K. Ellis, *The Post Office in the Eighteenth Century*, London (1958).

[73] K. Euler, *Das Geschlecht Euler-Scholpi*, Gießen (1955).

[74] ARA, Fagel family, inv. nr. 1213.

[75] ARA, Fagel family, inv. nr. 5136.

[76] ARA, Fagel family, inv. nr. 5177.

[77] ARA, Fagel family, No. 5177.

[78] Franz Stix, *Die Geheimschriftenschlüssel der Kabinettskanzlei des Kaisers*, Nachrichten von der Gesellschaft der Wissenschaften zu Göttingen, Mittlere und Neuere Geschichte, Neue Folge, I (1936), 207–226; II (1937), 61–70.

[79] Franz Stix, *Zur Geschichte und Organisation der Wiener Geheimen Ziffernkanzlei von ihren Anfängen bis zum Jahre 1848*, Mitteilungen des Österreichischen Instituts für Geschichtsforschung I, Innsbruck (1937).

[80] R. Fruin (ed.), *Overblyfsels van geheugcenis der besonderster voorvallen uit het leeven van den heere Coenrat Droste*, Leiden (1879).

[81] J.G. Gerretzen, *Schola Hemsterhusiana*, Nijmegen/Utrecht (1940).

[82] P. Geyl, *Een Oranje in ballingscap, Studies en Strijdschriften*, Groningen (1958), pp. 257–285.

[83] Guillaume Groen van Prinsterer (ed.), *Archives où correspondance inédite de la Maison d'Orange-Nassau*, first series, 8 vols, Leiden (1835–1847), III, pp. 54–66.

[84] I. Hacking, *The Emergence of Probability*, Cambridge (1975).

[85] C. Hamans, *Universal language and the Netherlands, Transactions of the Fifth International Congress on the Enlightenment*, 4 vols, Oxford (1980) III.

[86] Robert D. Haslach, *Nishi no kaze, hare*, Weesp (1985).

[87] A. Haynes, *Invisible Power. The Elizabethan secret services, 1570–1603*, Wolfborofalls (1992).

[88] J. Heeringa, *François Fagel. Portret van een honnête homme*, Die Haghe LXXX, 43–127.

[89] K. Heeringa (ed.), *Gedenkschriften van Adriaan van Borssele van der Hooghe, heer van Geldermalsen, Archief. Vroegere en latere Meededelingen voornamelijk in betrekking tot Zeeland, uitgegeven door het Zeeuwsch Genootschap der Wetenschappen*, Middelburg (1916).

[90] R. Hooykaas, *Religion and the Rise of Modern Science*, Edinburgh / London (1972).

[91] J.H. Hora Siccama (ed.), *Journalen van Constantijn Huygens den zoon*, 2 vols, Utrecht (1881–1888) I.

[92] E. Hublard, *Le naturaliste hollandais Pierre Lyonet, sa vie et ses oeuvres, 1706–1789*, Bruxelles (1910).

[93] W.R. Hugenholtz, H. Boels, *De griffie van de Staten-Generaal en van de Nationale Vergadering, 1780-1798*, Tijdschrift voor Geschiedenis LXXXX (1977), 401–402.

[94] J.I. Israel, *The Dutch Republic. Its Rise, Greatness, and Fall, 1477–1806*, Oxford (1995).

[95] E. Iversen, *The Myth of Egypt and Its Hieroglyphs in European Tradition*, Copenhagen (1961), pp. 14–15.

[96] M. Jähns, *Geschichte der Kriegswissenschaften vornehmlich in Deutschland*, 3 vols, München (1891), III, pp. 1818, 2636.

[97] Jan de Oude to Willem Lodewijk (minutes). Royal Archives, Nassau-Beilstein, Jan de Oude, inv. nr. 895-2 (I).

[98] N. Japikse, *Prins Willem III. De Stadhouder-Koning*, 2 vols, Amsterdam (1930–1933) II.

[99] N. Japikse (ed.), *Correspondentie van Willem III en van Hans Willem Bentinck, eersten graaf van Portland*, 5 vols, 's-Gravenhage (1927–1928) IV, 721.

[100] N. Japikse, *De Geschiedenis van het Huis van Oranje-Nassau*, 2 vols, Den Haag (1938) II, pp. 176–184.

[101] D. Kahn, *The Codebreakers*, New York (1967).

[102] A. Kalshoven, *De diplomatieke verhouding tusschen Engeland en de Republiek (...) 1747–1756*, Gravenhage (1915), pp. 106–108.

[103] J.P. Klüber, *Kryptographik. Lehrbuch der Geheimschriftkunst*, Tübingen (1809), pp. 217–230.

[104] H. Koot, *Geheimschrift, Winkler Prins Encyclopedia* VII, Amsterdam (1935), p. 30.

[105] G.E. Kurtz, *Willem III en Amsterdam, 1683–1685*, Utrecht (1928), pp. 95, 98, 107–109.

[106] ARA, Legatiearchief Barbije, inv. nr. 17.

[107] E. Lerville, *Les cahiers secrets de la cryptographie*, Monaco (1972), pp. 65–73.

[108] C.L. Levoir, *Vorstelijk geheimschrift in vorige eeuwen*, F.J.L. Krämer, E.W. Moes and P. Wagner, eds, *Je maintiendrai. Een boek over Nassau en Oranje*, Leiden (s.a.) 227.

[109] C.I. Lewis, *A Survey of Symbolic Logic*, Berkeley (1918).

[110] Lyonet to the Stadholder on 10 May 1784 (minute). – Boerhaave Museum Leiden, arch 162-u.

[111] Lyonet to Hendrik Fagel, The Hague, 18 April 1753. – ARA. Fagel family, inv. nr. 2230.

[112] A. Maljaars, *Het Wilhelmus. Auteurschap, datering en strekking. Een kritische toetsing en nieuwe interpretatie*, Dissertatie Universiteit van Amsterdam (1996).

[113] S.M. Matyas, *The History of the Book Cypher*, Lecture Notes in Computer Science LXXXXVI (1985), 100–113.

[114] A. Meister, *Die Anfänge der modernen diplomatischen Geheimschrift*, Paderborn (1902).

[115] A. Meister, *Die Geheimschrift im Dienste der päpstlichen Kurie*, Paderborn (1906).

[116] C.J. Mendelsohn, *Blaise de Vigenère and the "Chiffre Carré*, Proceedings of the American Philosophical Society LXXXII (1940).

[117] K. Menninger, *Zahlwort und Ziffer*, 2 vols, Göttingen (1958) II, p. 205.

[118] ARA, Ministerie van Buitenlandse Zaken, diverse legaties: Denemarken, inv. nr. 17.

[119] E.E. Morrisson, *Turmoil and Tradition: a Study of the Life and Times of Henry L. Stimson*, Boston (1960).

[120] J.W.A. Naber, *Prinses Wilhelmina, gemalin van Willem V, Prins van Oranje*, Amsterdam (1908), p. 220.

[121] Notes and Memoirs by Lyonet and Croiset about their work as code-breakers – Boerhaave Museum Leiden, arch 162-r.

[122] J.C. Overvoorde, *De centralisatie van het Hollandsche postwezen in het midden der 18de eeuw*, Bijdragen voor Vaderlandsche Geschiedenis en Oudheidkunde, 4e reeks (1900) I, 208–214.

[123] G. Parker, *The worst kept secret in Europe? The European intelligence community and the Spanish Armada of 1588*, *Go, Spy the Land*, p. 54, footnote 23.

[124] P. Pesic, *François Viète, father of modern cryptanalysis – two new manuscripts*, Cryptologia XXI (1997), 21–26.

[125] W. Ploeg, *Constantijn Huygens en de Natuurwetenschappen*, Rotterdam (1934), pp. 88–89.

[126] Rijksarchief Utrecht, Resoluties Staten van Utrecht, 264-27.

[127] Rijksarchief Zuid-Holland, Provisioneele Representanten van het Volk van Holland 1795–1796, inv. nr. 113, rapport van de commissie in de zaak van Raadpensionaris van de Spiegel van 14 October 1795.

[128] Rijksarchief Zuid-Holland, Laurens Pieter van de Spiegel, inv. nr. 676.

[129] H.H. Rowen, *John de Witt, Grand Pensionary of Holland, 1625–1672*, Princeton (1978).

[130] Royal Archives, The Hague, Stadholder William III, inv. nr. XIII-1.

[131] E.G. Ruestow, *Physics at 17th and 18th Century Leiden: Philosophy and the New Science in the University*, Den Haag (1973).

[132] W.J. 's Gravesande, *Inleiding tot de wijsgeerte*, Amsterdam (1747), pp. 242–257.

[133] F.H.A. Sabron, *De oorlog van 1794-1795 op het grondgebied van de Republiek der Vereenigde Nederlanden*, 2 vols, Breda (1892–1893) II, Supplement 34, lxxii.

[134] L. Sacco, *Un Primato Italiano: la Crittografia nei Secooli XV e XVI*, Roma (1958).

[135] J.C.F. Scherber, *Der Dechiffrier-Schlüssel oder Entzifferung deutcher unter unbekannten Characteren versteckter Schrift*, Zelle (1798).

[136] J.W. Schulte Nordholt, *Voorbeeld in de verte. De invloed van de Amerikaanse revolutie in Nederland*, Baarn (1979), pp. 39–49.

[137] G. Selenus, *Systema Integrum Cryptographiae*, Lunenburg (1624).

[138] E. Shoesmith, *The continental controversy over Arbuthnot's argument for divine providence*, Historia Mathematica, XIV (1987), 109-passim.

[139] M.M. Slaughter, *Universal Languages and Scientific Taxonomy in the 17th Century*, Cambridge (1982), pp. 189, 194, 204, 218.

[140] ARA, Spanje, inv. nr. 187.

[141] Royal Archives, Stadholder William IV, inv. nr. 301 B.

[142] Royal Archives, Stadholder William V, inv. nr. 193. Lyonet's handwriting is clearly recognisable above the lines of the intercepted letters.

[143] Royal Archives, Stadholder William V, inv. nr. 205.

[144] Royal Archives, Stadholder William V, inv. nr. 339 (box 7).

[145] ARA, Staten-Generaal, inv. nr. 7408.

[146] H.L. Stimson, McG. Bundy, *On Active Service in Peace and War*, New York (1947).

[147] G.F. Strasser, *Lingua universalis. Kryptologie und Theorie der Universalsprachen im 16. und 17. Jahrhundert*, Wiesbaden (1988).

[148] E.A.B.J. ten Brink, *Geschiedenis van het Nederlandse Postwezen, 1795–1810*, Gravenhage (1950), 45 footnote.

[149] P. Thicknesse, *A Treatise on the Art of Deciphering and of Writing in Cipher*, London (1772).

[150] UB Amsterdam, Catalogus van (...) uitmuntende en welgeconditioneerde Boeken in allerlei Weetenschappen en Taalen waaronder een groot aantal van kostbaare en zeldaame Werken uitmunten, nagelaten bij wijlen Mr. L.P. van de Spiegel, al het welk publicq verkocht zal worden op maandag 12 October 1801 en volgende dagen binnen Utrecht door de Boekverkoopers J. Visch en C. van der Aa.

[151] J.H.W. Unger (ed.), *Dagboek van Constantijn Huygens*, published as a supplement to *Oud-Holland* (1885).

[152] E. Vaillé, *Le Cabinet Noir*, Paris (1950).

[153] J. Valckenaer, Advis in de zake van Mr. LP. van de Spiegel uitgebracht aan de Provisioneele Representanten van het Volk van Holland op 15 Februari 1796.

[154] K. van Berkel, *In het voetspoor van Stevin. Geschiedenis van de natuurwetenschap in Nederland, 1580–1940*, Meppel/Amsterdam (1985).

[155] L.P. van de Spiegel, *Journaal van mijne detentie in de Castelany van Hove, op de Oranje-zaal, anders het Huis in 't Bosch en op het Kasteel te Woerden*, Bijdragen en Mededeelingen van het Historisch Genootschap te Utrecht (BMGN), XV (1894).

[156] T.I. de Larrey to L.P. van de Spiegel, The Hague, 21 November 1786. – Rijksarchief Zuid-Holland, L.P. van de Spiegel, inv. nr. 23.

[157] W.A. van Citters to L.P. van de Spiegel, 12 August 1787. – Rijksarchief Zuid-Holland, L.P. van de Spiegel, inv. nr. 24.

[158] J.C. van der Borch to the States-General, Stockholm, 28 January 1794. – ARA, Staten-Generaal, inv. nr. 7237.

[159] K. de Leeuw, H. van der Meer, *Een roostergeheimschrift door Alexander baron van Spaen*, De Achttiende Eeuw XXV (1993), p. 188, footnote 6.

[160] A.Th. van Deursen, *Bijdragen en Mededelingen betreffende de Geschiedenis der Nederlanden, CXIII* (1998), pp. 390–392.

[161] D. van Hogendorp to Maarten van der Goes, St Petesburg, 7/19 June 1804. – ARA, Buitenlandse Zaken, 1796–1810, inv. nr. 318.

[162] G.K. van Hogendorp, *Brieven en Gedenkschriften*, 7 vols, s'-Gravenhage (1866–1903) II, pp. 88–90;

[163] E. van Reyd, *Historie der Nederlantsche Oorlogen, begin ende voortganck tot den Jaere 1601*, Amsterdam (1644).

[164] A.G. van Rheede to the Greffier, Berlin 9 January 1790. – ARA. Staten-Generaal, inv. nr. 6693.

[165] Th. van Riemsdijk, *De griffie van hare hoog mogenden: bijdrage tot de kennis van het archief van de Staten-Generaal der Vereenigde Nederlanden*, Gravenhage (1885).

[166] A.A. van Schelven, *Marnix van Sint Aldegonde*, Utrecht (1939), pp. 197–198.

[167] W.H. van Seters, *Pierre Lyonet*, Den Haag (1962), pp. 17–18.

[168] A.J. Veenendaal (ed.), *Johan van Oldenbarnevelt. Bescheiden betreffende zijn staatkundig beleid en betreffende zijn familie*, 3 vols, 's Gravenhage (1934–1967), III, ix–xiii.

[169] P.A. Verburg, *Taal en functionaliteit. Een historisch-critische studie over de opvattingen aangaande de functies der taal*, Wageningen (1952).

[170] E. Vorlat, *The Development of English Grammatical Theory, 1586–1737*, Leuven (1975), pp. 20–29, 421–426.

[171] G.W. Vreede, *Resoluties der Staten-Generaal 1576–1625*, 21 vols, 's Gravenhage (1915–1994) XIII, 412, res. 20 December 1605; ibidem, XIV, 9, res. 3 January 1607; ibidem, XV, 164, res. 1 July 1610.

[172] G.W. Vreede, *Inleiding tot eene geschiedenis der nederlandsche diplomatie*, 3 vols, Utrecht (1856–1861) I.

[173] G.W. Vreede, *Mr. Laurens Pieter van den Spiegel en zijne tijdgenoten (1737–1800)*, 4 vols, Middelburg (1874–1877) III, p. 534.

[174] J. Wagenaar, *Vaderlandsche historie*, 21 vols, Amsterdam (1749–1759) XIII.

[175] Secretary E. Wasmuth to the States-General, Stockholm, 15 February 1774. – ARA, Staten-Generaal, inv. nr. 7235.

[176] O. Weber, *Der Friede von Urecht. Verhandlungen zwischen England, Frankreich und den Generalstaaten, 1710–1713*, Gotha (1891), pp. 86–88.

[177] Willem Lodewijk van Nassau to Jan de Oude, 30 June 1599. Royal Archives, inv. nr. IX A 2-1.

[178] Willem Lodewijk van Nassau, inv. nr. IX P-4; Royal Archives.

[179] Willem Lodewijk van Nassau, Royal Archives, inv. nr. IX O.

[180] J.G. Worp (ed.), *Constantijn Huygens. Briefwisseling, 1608–1687*, 6 vols, 's Gravenhage (1911–1917) I, 331, letter by A. Ploos, The Hague, 4 October 1631.

[181] H.O. Yardley, *The American Black Chamber*, New York (1931).

13

INTELLIGENCE AND THE EMERGENCE OF THE INFORMATION SOCIETY IN EIGHTEENTH-CENTURY BRITAIN

Jeremy Black

University of Exeter
Exeter, United Kingdom

Contents

Abstract

Drawing together intelligence and information, this paper considers deciphering dispatches and news management as linked issues of analysis and policy.

Keywords: postal interception, deciphering, newspapers, news management, Parliament, diplomats.

13.1 INTRODUCTION

This essay brings together two topics that are generally treated separately: the secret world of espionage and counter-espionage and the so-called public sphere: the public world of coffee houses, newspapers and urban bustle which played an important and growing role in the politics of the period. The common theme was information and, more specifically, the concern of the state with the flow and content of information and its attempt to influence both [19].

Rather than taking the state as the prime mover, however, it is appropriate to note the extent to which government was responding to a wider social, cultural and intellectual current. In particular, there was a fundamental intellectual shift: to, or at least towards, a situation in which information, rather than received wisdom, had greater currency [18], both as a source of authority and as a practical guide to policy making. This shift is generally treated as an aspect of the history of science but in fact was far wider in its sources and applications. In the British case, what was termed political arithmetic was particularly important. It served as a valuable guide to policy formulation, execution *and* explication in the world of public politics that followed the 'Glorious Revolution' of 1688–1689 and the subsequent creation of a politico-governmental system in which annual sessions of Parliament were a central aspect of the political process, while accountability to Parliament had become more significant for government.

The social dimension of the information flow was a broadening of the section of society that comprised the 'political nation': those who expected to be aware of, and to have a view on politics. In part, this was due to the political transformation resulting from the Glorious Revolution, but, again, far more was at issue, not least a greater public curiosity that, in part, reflected a different attitude to news that affected the demand for information. In particular, although they remained very important, the role of community agencies – families, kindred, localities, confessional and economic groups – in formulating and disseminating news became less significant than external agencies. This was a long-term process, to which the role of state-directed religious change during and after the Reformation contributed greatly, but key elements were the increasing literacy that facilitated information demands, and the extent of socioeconomic change and opportunity that encouraged interest in the world outside the locality.

The state had far greater authority than its modern European counterparts but, in many respects, its power was less. In particular, the small size of the bureaucracy ensured that governance depended on co-operation as well as a measure of consent [3], and, in the field of information, the government lacked the means and what in military history is known as doctrine to permit an adequate engagement with the new information world in the form of government dominance of the provision of news and opinion. Pro-government newspapers and pamphlets were funded, but there were no opportunities for the totalitarian news-management to be seen in the twentieth century.

Instead, government attempts to affect the nexus of news and opinion reflected not its strength, but rather its weakness, a perspective that needs underlining as it is at variance with the commonplace assumption. For much of the period, this weakness was accentuated by and focused on particular challenges: specifically to the legitimacy of the government and to the security of the state from domestic and external threats. In the first case, the key problem stemmed from the Glorious Revolution, for the expulsion of the male line of the Stuarts in the person of James II and VII was far from consensual, which created particular problems for government news-management, and indeed, led to a fundamental dynastic challenge that remained militarily significant within Britain until the defeat of Charles Edward Stuart at Culloden in 1746 and in international terms until the naval victories off Lagos and Quiberon Bay in 1759 smashed French invasion plans. Issues of legitimacy thereafter were not dynastic in character, but their seriousness was readily apparent, particularly with the civil war within the empire known as the American War of Independence (1775–1783) and subsequently with the crisis caused by radicalism in the shadow of the French Revolution, especially with the Irish rising of 1798.

Issues of legitimacy were made more serious in security terms when, as with Jacobite plotting, they interacted with specific domestic and/or international crises, and here the interaction of intelligence and information becomes more pressing. If the government was to lessen this interaction or to cope with its consequences, then it was necessary to affect the flow of information. This was true in a double sense, first, the particular movement of items that might be part of, or contribute directly, to plots and, secondly, the more general flow that created a sense of weakness. In particular, it was necessary to ensure a positive foreign perception of developments within Britain, and, as far as the latter was concerned, to try to limit accounts of developments abroad that would encourage disaffection within Britain or, at least, challenge the sense of confident power on which the strength, and therefore purpose, of government greatly depended.

13.2   POSTAL INTERCEPTION

The first task was one in which postal interception played the key role, while the second was a more varied one in which news management was crucial. This news management was sometimes part of the intelligence world, particularly when the post of hostile newspaper figures was intercepted, but on the whole these methods were not employed. Instead, the government preferred to use judicial means to deal with publications judged seditious.

For both postal interception and news management, there is a certain amount of information, although there are no comprehensive records of a central agency permitting an evaluation of how the situation appeared to the government and enabling us to follow policy formulation in this field. As a consequence, the sources largely provide us with answers on how, not on why, and this may lead us to overplay or to underplay the significance of the issue to government.

The opening of the mail was not the most important step in the intelligence war, and the related attempt to thwart hostile information flows and, more generally, to gain knowledge of what was being written. The key element in the system was the deciphering [12], and this was a skill the British possessed in abundance in this period. Particularly after the appointment of the Reverend Edward Willes, later Bishop of Bath and Wells, as Decipherer in 1716, a post he held until his death in 1773, the British deciphering service was very successful. From Willes' appointment, the decipherer's office was continually staffed by members of his family until its abolition in 1844, and from 1762 onwards the entire office were members of the family [24]. From 1701, when the office was placed on a regular basis, being held in 1701–1703, in an important link with the world of science, by John Wallis, an eminent mathematician, until 1722, the salaries were paid at the Exchequer, thereafter by the Secretary of the Post Office, out of the Secret Service money until 1782, when the office was placed under the authority of the newly-created Foreign Office [15].

The success of the office can be gauged from its ability in the 1730s to decipher the codes of most European states, including Austria, Bavaria, Denmark, France, Naples, the Palatinate, Portugal, Prussia, Russia, Sardinia, Saxony, Spain, Sweden and the United Provinces. Evidence of a failure to decipher – either intercepts which were incompletely deciphered or not deciphered at all, or references to the problem in correspondence – is rare, although the office was constantly tested by new cipher keys.

Much of the strength of the British intelligence system (which was impacted in the more general information world) derived from allied co-operation, and the efforts made by ministers and diplomats in sustaining that co-operation were very important to Britain's foreign policy capability. Co-operation from the Dutch and Hanoverian governments was key and this in part reflected dynastic links. Co-operation from Hanover began before the Hanoverian accession, for William III was supplied with correspondence intercepted at Celle between Paris, Copenhagen and Stockholm [22]. The accession of George I in 1714 led to closer co-operation, with German experts being sent to London and Hanoverian intercepts being made available regularly by the king to his British ministers. The quality of Hanoverian interception remained high throughout the century, and co-operation with Britain continued [11]. Furthermore, in the early decades of the century, the Dutch maintained the tradition of assistance that had developed during William's reign.

The absence of a systematic survey both of surviving interceptions and of references to them makes it difficult to assess how far the effectiveness of the system changed during the century, a serious failing as it makes it difficult to establish the full range of the information at the government's disposal, and thus their sense of completeness in this respect. This was significant because interception and deciphering provided the principal means by which the government obtained information concerning foreign views of British policy and politics. They also offered the most reliable guide to the intentions and attitudes of hostile foreign powers. Intercepts therefore were part of a system of multiple inputs to ministers, both a kind of cat and mouse game played by all European governments and also a part of the diplomatic and political process. As far as the diplomatic community was concerned, intercepts gave the government a way to keep track of the gap between real and professed views.

This returns our attention to the information available to modern scholars. State Papers Foreign Confidential (SP 107) in the National Archives contains, bound at the start of SP 107/1A, a letter of 1 May 1838 from Francis Thomas of the Public Record Office giving an account of the collection including the passage:

> It is much to be regretted that so valuable a series of papers have so much suffered by damp, I understand from Mr. Lemon that he remembers these papers at the Old State Paper Office Scotland Yard, where they were thrown in a closet as papers of no consequence, and where an immense quantity perished. It will be manifest how much they have suffered from the separate statement enclosed, shewing the quantity or number

of volumes in each year. This will account for sometimes finding copies of the originals and no translations – sometimes translations without the copies of the originals – sometimes a postscript without the former part of the letter.

The Robert Lemon (1800–1867) consulted had been appointed Senior Clerk of the State Paper Office in 1835, and his father, also Robert (1779–1835), had been Deputy Keeper of State Papers from 1818 until his death, and had played a role in the removal of the State Papers from Scotland Yard and Great George Street.

The patchy survival referred to in the letter is very noticeable today in working in the collection, although it compares favourably with the deliberate destruction of material in 1837 when, on the death of William IV, King of Hanover as well as Britain, and the succession of his niece Victoria to the British but not the Hanoverian throne, many of the papers of the German Chancery in London were destroyed rather than being transported to Hanover. This might have provided very valuable material for the Anglo-Hanoverian intelligence link.

The statement referred to by Thomas is a listing of the number of documents per year:

1726 7 or 8 letters
1727 Ditto
1728 3 letters
1729 12 or 13 letters
1730 1 and a half volumes
1731 2 volumes
1732 3 volumes – none in July – only one in August and 5 in September
1733 11 volumes
1734 2 volumes – only 1 letter in May and none after June
1735 None
1736 Only 6 letters
1737 2 letters in February – 2 in March – then none till November in which there is but one
1738 About a volume and a half
1739 13 volumes
1740 12 volumes – but June, July, August and September are all contained in one of these volumes

1741 None in February, March, April, May, July – 1 letter in July – 1 in August – about 10 in September–October, November and December contain the most part of the letters of this year
1742 3 volumes
1743 6 volumes
1744 Not 20 letters
1745 2 volumes. None in January – only one letter in February – 3 in March – half a dozen in April – and none after September
1746
1747
1748 } None
1749
1750
1751 Only 2 letters, 1 in November – 1 in December
1752 Only 1 letter
1753 None
1754 About 25 or 30 letters – none till September in which month there is but one letter
1755 5 volumes many letters damaged by damp
1756 8 volumes part much damaged by damp
1757 4 volumes partly damaged by damp
1758 3 volumes
1759 1 volume. None in January, February, March, April. 1 in May – 2 in November
1760 5 volumes
1761 2 volumes. None in May – 4 letters in June – 2 in July – none after July
1762 None
1763 4 volumes. None till 29 August
1764 8 volumes
1765 6 volumes
1766 About 10 letters – the last is 2 May

This in fact is a less than complete account of SP 107. It omits Volumes 1B–D, which are substantial boxes, B containing a large quantity of intercepted and deciphered Swedish material from 1716–1717, a period of great significance due to Swedish intrigues with the Jacobites, and 1/C–D containing Dutch material from 1716 to 1732. Volume 110 is also omitted, although the covers it includes are endorsed with the initials of their readers, providing valuable guidance to how far information about intelligence circulated within government. It is clear that the majority of intercepts were read by the king. One cover is endorsed "Not read by the K. not worth sending His Majesty", while others reveal limitations in the deciphering branch and the importance of the Hanoverian connection. One cover is endorsed "These letters were written in February, March, and April, but the cypher was not discovered till the 29 June (1765)", and another "The reason that the enclosed are of so old a date is because they were sent to Hanover to be decyphered and translated from whence they have been rendered into English after their return".

Elsewhere in the National Archives, there are also important holdings. The miscellaneous supplementary volumes at the end of each series of State Papers Foreign contain material of value. SP 92/85-7 contains the intercepted correspondence of Giuseppe Ossorio, the well-informed Sardinian envoy, for 1733, and compares well to his reports in volume 40 of the *Lettere Ministri Inghilterra in the Archivio di Stato in Turin*. SP 94/246 contains intercepts of Spanish material for 1737–1738, in fact copies of material held in SP 107/21. SP 89/92 contains Portuguese material for 1762–1780, and 78/325 French material of 1777–1780.

In addition, copies of intercepted dispatches can be found in State Papers Foreign among the ordinary diplomatic correspondence, because sometimes intercepts were sent to diplomats whom it was felt would find them useful. Another series that contains much of value, and which has hitherto been comparatively neglected, is State Papers Regencies. This contains ministerial correspondence resulting from royal trips to Hanover, which were frequent from 1716 until 1755, and contains references to intercepts that are not otherwise known. State Papers Domestic contains copies of warrants from the Secretaries of State to the Postmasters General instructing them to open letters, as well as intercepts themselves, and also references to them that cast light on ministerial success in intercepting and deciphering correspondence.

The surviving private papers of ministers and diplomats are also significant. Modern regulations governing the ownership and scrutiny of the public papers of government officials are comparatively recent, and it was common for ministers and officials to retain possession of office papers. In some cases these are very considerable. Thus, the National Library of Australia in Canberra holds a collection of the papers of Charles, 2nd Viscount Townshend that includes 222 intercepts for the period 1724–1726 when Townshend was Secretary of State for the Northern Department [20]. The papers of an earlier holder of the office, Charles 3rd Earl of Sunderland, contain a large number of intercepts for the period 1716–1718, including French, Holstein, Jacobite, Savoyard, Spanish and Swedish material [8]. Similarly, the papers of Lord Carteret, holder of the post from 1742 until 1744, contain intercepts [5].

If it were only the papers of Secretaries of State that contain such material, it would be possible to argue that the information world of intelligence was clearly defined and thus, at least by modern standards, coherent and readily comprehensible. The location of surviving intercepts, however, suggests otherwise, for they can be found in the papers of ministers removed from formal competence in the field. This includes those of Sir Robert Walpole, First Lord of the Treasury from 1721 to 1742 [10]. The collections of other first ministers also contain important material. In 1766, a letter from George III to Charles, 2nd Marquess of Rockingham, preserved in the latter's papers held in Sheffield Archives, referred to an intercepted dispatch from the Prussian envoy, "You will laugh when you read the deciphered letter I have just seen of Baudouin, wherein he talks of a fresh change in the ministry. I should rather hope it is from want of sense than ill intention that he writes such gross falsehoods" [14]. The same year, a letter in the Chatham papers indicated success in intercepting French material: "There has been a large correspondence of the Duke of Choiseul's deciphered, which shows his sentiments very much at large" [26], the emphasis thus being on the information as to general attitudes acquired. In 1771, a letter from Thomas Bradshaw MP, a former senior civil servant, to his patron, Augustus, 3rd Duke of Grafton, until recently first minister, revealed that intercepts threw light on the crucial question of French policy during the Falkland Islands dispute, in

[A]n intercepted letter from Carraccioli makes it certain that the Duke de Choiseul meant war, and to amuse this Court with the hopes of peace – in this letter he recommends it to the King of Spain by all means to make up matters with England for *the present*, and to wait for a more favourable opportunity [4].

The material from the collections of ministers and diplomats suggests that an assessment of intelligence effectiveness available based simply on the holdings in SP 107 is insufficient. In combination with SP 107/1B–C, the Sunderland material suggests that the system was operating well in the late 1710s, while the range of intercepted material in the Townshend material – Austrian, Dutch, French, Hesse-Cassel, Modenese, Parmese, Sardinian, Saxon, Spanish, Swedish, Venetian – intercepted indicates considerable effectiveness in the mid-1720s, and creates a very different impression to that of SP 107 for those years.

The situation for the early 1750s is less happy. There is no significant holding among the papers of ministers and diplomats for these years. Furthermore the absence of intercepts in SP 107 between September 1745 and November 1751 is clarified by SP 95/136, a large volume of Swedish material from 1748 to 1755 that, however, is undeciphered. As Swedish material was deciphered earlier in the century, this underlines the suggestion that the efficiency of the British system had declined. George II certainly complained on this head, as Thomas, Duke of Newcastle, Secretary of State for the Northern Department, noted in 1752: "he run into one of the usual, but strongest declarations ... against our information from the intercepted letters, which told us nothing, for without *his* [27], we should know nothing" [21]. Nevertheless, there are indications of continued capability. Intercepts revealed Choiseul's plans for an invasion in 1759, French correspondence with Sweden was regularly deciphered in the mid-1760s and early 1770s, and French and Prussian codes in the late 1760s were broken [9;13;29].

The interception of Jacobite material was more difficult as the Jacobite information system was far more diffuse than those of foreign states which concentrated on diplomatic links. The government

had both to intercept messages and to seek to insert agents, in order to understand the nature as well as the content of the Jacobite information system [13].

The extent of Jacobite dependence on foreign assistance was unclear. In so far as it was large, then the intelligence task was the more familiar one of state-from-state intelligence gathering. This was seen with the crisis created by the Jacobite rising in 1745, as the government proved very weak in evaluating Jacobite plans and likely support within Britain, but far more successful in probing the possibility of foreign support. In July 1745, before the landing of Charles Edward Stuart (Bonnie Prince Charlie) in Scotland, William, Duke of Cumberland, the commander of the British forces fighting the French in the Austrian Netherlands, was ordered to be ready to send troops to England immediately. Later the same month, William, 1st Earl of Harrington, Secretary of State for the Northern Department, wrote to Cumberland from Hanover where he had accompanied George II:

I prepared your Royal Highness in my last for the possibility of the French court's being induced by their late successes [in the Austrian Netherlands] to mediate some attempt on his Majesty's British dominions. What we then only foresaw as a thing that might happen is now but too much verified, for the king has certain and infallible intelligence, and I am ordered to acquaint you in the utmost confidence that the resolution is actually formed at the court of France to execute immediately such an invasion.

Cumberland and Newcastle were instructed to make preparations, Newcastle being informed that "the scheme of a new invasion by France upon England ... does now appear to be a fixed and settled design for which everything is in earnest preparing". In explanation of these instructions, Harrington sent a copy of a letter from Frederick the Great of Prussia to Chambrier, his envoy in Paris, and another from the Danish envoy to Christian VI. Frederick's letter referred to information Chambrier had received from the pro-Jacobite French minister Cardinal Tencin that France was thinking of such an invasion [23].

Interceptions were of value not only for the light they threw on the policies of other states, but also on the relationship between these states and domestic politics. Indeed, this helps explain part of the pattern of surviving intercepts. They are scattered through the papers of Philip, 1st Earl of Hardwicke, Lord Chancellor from 1737 and 1756. These include intercepts of opposition correspondence, for example letters of Henry, 1st Viscount Bolingbroke in 1740 [7]. There are also opposition intercepts among the volume entitled *Letters and Papers of Sir Robert Walpole* that was part of the collection of his brother Horatio, now purchased by the British Library.

Opposition politicians indeed were aware that their letters were opened by the Post Office. William Pulteney, the leader of the opposition Whigs in the House of Commons, in a letter to George Berkeley in 1734 followed the statement that he was idle and intent on trifles by writing "Should the Post Office open my letter and read this frank confession from a man that has the appearance of so bustling a patriot, what would they say". The following year, he wrote "If you hear any news send it to me without reserve, never mind their opening letters at the Post Office, but bid them kiss your . . . whatever discovery they make" [6].

It was scarcely surprising that, once resigned, Townshend himself was cautious:

> I never write anything, but what I desire the ministry may see. There is no great skill nor dexterity in opening of letters, but such is the fate of this administration that they have managed this affair in such a manner as to lose all the advantage of it. Sir Robert [Walpole] having complained to a friend when he was last in these parts [Norfolk] that they had lost all the intelligence they used to get by the Post Office which was owning plainly that people were grown wiser than to write anything of consequence by the post [28].

13.3 NEWS MANAGEMENT

This shift of attention to domestic surveillance would not have represented a break for the ministry, because such surveillance was as one with that of foreign envoys and governments, as the common link was that of real or potential conspiracy. Intercepts could serve to compromise opposition politicians, as well as to provide vital informa-tion on their hopes and/or plans. For the government, there was not only these specific values, but also a more general desire to be in the dominant position as far as the flow of information was concerned. This had great value not only in terms of political self-confidence, but also because the flow, and thus availability, of information helped shape the political world. Parliamentary debates on foreign policy made this clear both in general terms and in specifics. As far as the former was concerned, the need to face Parliament at the beginning of each session ensured that the government both sought information on the international situation in order to put the best spin on its own position, and also needed to have better information than the opposition.

The best specific instance of what could happen if the battle for information dominance was lost was the Dunkirk debacle of 1730. Despite specific prohibitions in the treaties of Utrecht (1713) and The Hague (1717), the French had restored the harbour facilities at Dunkirk. The government had been well aware of this restoration and had complained to the French in 1727, 1728 and 1729, while the issue had been exploited by the opposition press over the previous three years. The ministry was aware that Dunkirk might be used as an issue by the opposition in the session of 1730, but they failed to gather adequate information to enable them to refute the attack in the Common or to prepare for a major press defence. Instead, after the Commons formed itself into a Committee of the Whole House to consider a motion from Sir William Wyndham, Tory leader in the House, for an examination of the state of the nation, Wyndham was able to use the restoration of Dunkirk's harbour as a proof of the government's failure to protect national interests, in part because he had witnesses ready. Government speakers tried to block this by arguing that it would be irregular to examine witnesses when the House had not been given prior notice, but this did not convince the Commons [25].

Information indeed became both the issue at stake and the one that determined the timing of the crisis. After the opposition had had its evidence about the restoration heard, on the following day it was resolved to address George II for the laying of all correspondence about Dunkirk before the House, but, in order to give time for this to be prepared, the debate on the state of the nation was adjourned a fortnight, which gave time for the government to trump fresh evidence of the French works then brought forward by the opposition by providing an official French assurance that the works had been performed without the authorisation of Louis XV, and that he had now ordered their demolition.

This issue was also fought out in the press, and this provided another aspect of the intelligence/information world. Diplomats were expected to provide information not only for the private consumption of government, but also for its more public face. There was an expectation that diplomats would contribute to the general pool of news that could be deployed by the government. Material for the official newspaper, the *Gazette*, was especially important in this light. Although the attempt to maintain a monopoly in printed news for the *London Gazette* (or *Gazette* for short) had collapsed with the Glorious Revolution, the *Gazette* was still the most widely circulated newspaper in 1705–1707 and, thereafter, even when it had declined in relative importance, successive governments still expected diplomats to provide regular reports for the benefit of the *Gazette*, and they obliged. The newspaper was frequently used as a source for items in other papers.

Demands for information were linked to another task for British diplomats, attempts to control the flow of unwelcome material. Diplomats had both to confront frequent complaints about reports concerning foreign powers in the British press, and also to seek to suppress foreign items unwelcome to the British government. Their unwelcome character stemmed not only from their influence on foreign audiences but also from the extent to which British newspapers were part of an international press world. In this, information moved freely, in part because foreign news was generally produced by a 'scissors-and-paste' technique, rather than by foreign correspondents. Thus, suppressing unwelcome foreign reports was an important aspect of the attempt to control the parameters within which the British press operated.

On 13 January 1731, the *Craftsman* printed a 'Hague letter' revealing and condemning the government's secret attempts to create an Anglo-Austrian alliance. This led to a sustained debate about the legality of such articles, and also to discussion of the correct foundations for the accuracy of newspaper reports. According to the *Free Briton* of 4 February 1731, "as to secretaries or ministers of state being the only evidence in such cases I think differently. I think concurrent and repeated advices, previous to such insinuations are an honest justification in the eye of reason and conscience". The *Craftsman* suffered legal action from the government, but was able to claim in its issue of 24 April that the alliance with Austria negotiated that March was a consequence of the paper's call for such an alliance.

The legal situation was difficult for the press publication of contentious material. In 1716, the postal interception system revealed that newspapers, especially the Tory *Post Boy*, did not dare print material offered by the Swedish envoy Count Gyllenborg, who sought to use them to stir up opposition to Britain's Baltic policy [17]. The following February, John Morphew of the *Post Boy* was indeed arrested for publishing an item on Anglo-Swedish relations [15].

The *Craftsman* of 27 February 1731 noted that it felt unable to print a critical Spanish declaration "as we are not yet furnished with sufficient proof to satisfy a court of justice that it was actually delivered", as in fact it had been.

The credit attached to foreign news led newspapers to emphasise the value of their foreign sources. Announcing that it was to become a daily, the *Post Boy* of 29 August 1728 stated that this new paper would include "as usual the Original Hague letter, confessedly superior to anything extant to that kind; which has never failed to give general satisfaction, and which can be procured by none but the proprietors of this paper". The reputation of such news led to issues of authenticity, with claims, as in the *London Evening Post* of 26 January 1740, that a major source of foreign news was information fabricated in London and inserted into the newspapers as coming from abroad, usually from The Hague.

13.4 NEWS AND ANALYSIS

The need for information, by both government and political commentators, varied greatly. It was pronounced when Britain was taking an active, interventionist stance in Continental power politics, as in the late 1720s and early 1730s. Conversely, an isolationist stance, as in the late 1760s and early 1770s, ensured that the need was less, although then there was the requirement for government to demonstrate that its failure to act, when for example the French purchased Corsica in 1768 or the First Partition of Poland occurred in 1772, did not betoken an inability to discern and defend national interests.

Interventionism required information if the government was to adopt the appropriate policy and to be able to defend it. This was particularly so as such interventionism depended on a successful interaction with other states and an adroit operation of the mechanisms available for power politics. Both were made more complex for British ministers (and for commentators) because in most states there was scant role for public politics. Instead, foreign policy was a mystery of state, formulated in conditions of secrecy in court contexts riven by factionalism. Issues of *gloire* and dynastic interest played a major role in policy. These and their application were difficult to discern, not least given the kaleidoscopic character of alliances [1].

This created major problems for processes of description and assessment, as information was generally unreliable and random, both of which contributed to uncertainty in analysis and debate. Far from recognising clearly the intentions and actions of others in an international system, ministers and commentators operated in a context of opacity. Furthermore, it proved difficult to apply the systematic ideas then available: the balance of power, and the relaxed concept of natural interests.

The ambiguity of these concepts interacted with the difficulty of acquiring accurate information about the plans of other states in order to produce a situation of great uncertainty. Looked at differently, this resulted in a wide field for speculation. The flexibility with which information and analysis was employed in political debate helped ensure

their value – they could serve all sides – but also created difficulties, as there was a desire for precision. Indeed, the extensive use in British discussion of the balance of power reflected in part the fascination with Newtonian physics and its mechanistic structures and forces, and, more generally, with reason or rationality. There was, indeed, an awareness of change and of the role of the vicissitudes of dynastic arrangements, the hazards of war, and all sorts of contingent fashions, all of which made international relations less calculable, but ministers and commentators tried to reduce those factors to the extent that such relations could be perceived as calculable. Thus, intelligence and information were designed to provide the necessary level of predictability that would make analysis possible and thus validate the contemporary conceptualisation of international relations. Information thus helped make the system seem rational, indeed gave it its systematic character, and, in turn, made it possible to analyse and predict developments rationally; rationality understood in terms of predictable cause and effect relations. If information could provide evidence of passion, greed, inertia and other 'irrational' forces at work, it could also make them comprehensible and thus manageable.

Knowableness was the key, as any emphasis on uncertainty challenged the possibility of ensuring order, peace, and, in British eyes, the national interests bound up with these. William, Lord Grenville, the Foreign Secretary, noted in 1798 that international relations in Eastern Europe and British policy toward the region had changed dramatically from the Ochakov Crisis of even years earlier:

> We are concluding an alliance with the Porte [Turks], in conjunction with Russia, which I certainly never thought I should live to see, much less to be an active instrument in, when in 1791 I came where I am. We are silly creatures with all our deep speculations, and our reasonings, and our foresights.

He then quoted from Horace's *Odes* (III, 29–30): "God buries in misty night the outcome of future time", adding that it

> is better philosophy than one thinks when one first reads it, full of confidence in all that one is to

do, and to see, and to foresee – But then I doubt whether because this sentiment is just, the right conclusion from it is that we have nothing to do in this world but to drink and to – [sic] [16].

Grenville's letter captured the tension between the attempt to shape events and the extent to which they defied shaping, in short the triumph of a realism about the limitations of calculation over confidence in the predictive nature of developments and policy initiatives. Yet, as he pointed out, as the response could not be to do nothing, so it remained important to use the calculation that information permitted to lessen the odds, or, at least, to counter the sense of uncertainty and helplessness. In terms of the different philosophies of foreign policy, intelligence-gathering therefore played a major role in both Whig and Tory assessments.

13.5 CHANGING INFORMATION NEEDS

If this situation did not present enough of a pressure on the provision of intelligence and the flow of information, it was accentuated by the spreading concerns of the British state. This was a dual process. First, the European political system changed both to incorporate a relatively distant state (Russia) and to ensure that conflict or the risk of conflict with other European powers arose from transoceanic disputes (the Ohio River country with France in 1754–1755, the Falkland Isles with Spain in 1770, Vancouver Island with Spain in 1790) and, secondly, that it was necessary to consider more non-European powers, as British interests expanded, first, in India, and, then, in the Malayan Peninsula (1780s), with, subsequently, in the 1790s, an attempt to improve relations with China.

The resulting pressure on information was acute and in the case of North America the system failed as the British and French governments found it difficult to agree on the cartography of the area in dispute. Similarly, in the Ochakov Crisis with Russia in 1791, there was a lack of clarity about the strategic importance of the fortress of Ochakov, and, in particular, about the extent to which it dominated the estuaries of the Bug and the Dnieper. This led to an attempt to secure expert advice, and to a controversy between diplomats. Given this dissension, it was scarcely surprising that parliamentarians and newspapers also disagreed about the importance of the fortress, as well as about the relationship between Russian gains and the balance of power. The problem facing parliamentarians discussing remote districts was mentioned by Richard Brinsley Sheridan who referred to the pro-government Sir William Young as having "expatiated with as much familiarity concerning the Dnieper and the Danube, as if he had been talking of the Worcestershire canal" [2].

This looked toward the nineteenth century, as did the Nootka Sound Crisis over Vancouver Island of the previous year. Ministers, politicians and commentators struggled to make comprehensible a wider world, and sought to understand its interactions. Existing institutional and intellectual strategies proved to have only limited value. Government responded by leaving a large margin of discretion in policy (and thus policy analysis) to local agents, and this was to be characteristic of much of the imperialist age, before the technology of telegraph, railway and steamship encouraged a sense of immediacy that corresponded to Victorian system-building and application of moral principles, ensuring new ethos and practises of interventionism and a drastically revised demand for, and provision of, intelligence and information.

REFERENCES

[1] J.M. Black, *European International Relations 1648–1815*, Macmillan, Basingstoke, UK (2002), pp. 5–49.

[2] J.M. Black, *British Foreign Policy in an Age of Revolutions, 1783–1793*, Cambridge University Press (1994), pp. 287, 298, 304, 311, 313.

[3] J.M. Black, *Kings, Nobles and Commoners. States and Societies in Early Modern Europe. A Revisionist History*, Macmillan, London (2004).

[4] Bradshaw to Grafton, 9 January 1771, Bury St. Edmunds, West Suffolk Record Office, Grafton papers no. 628.

[5] British Library, Department of Manuscripts, Additional Manuscripts (hereafter British Library Add.) 22539, 22542.

[6] British Library, Add. 22628, fols 71, 74.

[7] British Library, Add. 35586, fols 295–298.

[8] British Library, Add. 61567–61576.

[9] British Library, Add. 32309; Sir J. Fortescue (ed.), *The Correspondence of King George the Third from 1760 to December 1783*, 6 vols, London (1928), III, pp. 39–40.

[10] W. Coxe, *Memoirs of the Life and Administration of Sir Robert Walpole, Earl of Orford*, 3 vols, London (1798).

[11] K. Ellis, *The administrative connections between Britain and Hanover*, Journal of the Society of Archivists 3 (1969), 556–566.

[12] K. Ellis, *The Post Office in the Eighteenth Century: A Study in Administrative History*, Oxford University Press, London (1958).

[13] P. Fritz, *The anti-Jacobite intelligence system of the English ministers, 1715–1745*, Historical Journal 16 (1973), 265–289.

[14] George to Rockingham, April 1766, Sheffield Archives, Wentworth Woodhouse Muniments, R1-605.

[15] W. Gibson, *A eighteenth-century paradox: The career of the Decipherer-Bishop, Edward Willes*, British Journal for Eighteenth-Century Studies 12 (1989), 69–76.

[16] Grenville to Lord Mornington, 5 October 1798, British Library, Add. 70927, fols 24–25.

[17] Gyllenborg to the Swedish minister Baron Görtz, 23 October, 4, 10 November 1716, PRO. 107/1B, fols 172, 173, 210, 211, 218.

[18] D.R. Headrick, *When Information Came of Age. Technologies of Knowledge in the Age of Reason and Revolution, 1700–1850*, Oxford University Press (2001).

[19] E. Higgs, *The Information State in England*, Macmillan, Basingstoke, UK (2004).

[20] MS 1458 Folders 9/1–10, 10/1–4.

[21] Newcastle to his brother Henry Pelham, 26 July 1752, British Library, Add. 35412, fol. 209.

[22] S.P. Oakley, *The interception of posts in Celle, 1694–1700*, *William III and Louis XIV*, R.M. Hatton and J.S. Bromley, eds, Liverpool University Press (1968), pp. 95–116.

[23] PRO. SP. 43/36-7, Windsor Castle, Royal Archives, Cumberland Papers 3/270.

[24] J.C. Sainty, *Officials of the Secretaries of State 1660–1782*, London (1973), pp. 51–52.

[25] Stephen Fox to Henry Fox, 14 February 1730, British Library, Add. 51417, fols 38–39.

[26] W.S. Taylor, J.H. Pringle (eds), *Correspondence of William Pitt, Earl of Chatham*, 4 vols, London (1838–1840), III, pp. 94–95.

[27] Those from Hanover.

[28] Townshend to Samuel Buckley, Writer of the Gazette, 16 Sept. [no year given], Oxford, Bodleian Library, Department of Western Manuscripts, Ms. Eng. Lett. c. 144, fol. 267.

[29] B. Williams, *The Life of William Pitt, Earl of Chatham*, 2 vols, Longmans, Green and Co., New York (1913) I, p. 399.

14

ROTOR MACHINES AND BOMBES

Friedrich L. Bauer

Institute for Informatics
Technical University Munich
Germany

Contents

Abstract

    Section 1 covers the origin of the rotor idea, polyalphabetic substitution, its mechanisation and a historical review. Section 2 shows the Scherbius line that is, key generation and the reflector. Section 3 treats the rotor machines of the German armed forces with plugboard and its copies and competitors: TYPEX, NEMA, SIGABA, ... PURPLE. While Section 4 discusses the pioneering Polish cryptanalysis of the *Wehrmacht* ENIGMA and the works of Rejewski, Zygalski, Rozicki. Section 5 demonstrates the cryptanalytic activities of the British and United States against Germany that is, Turing, Welchman, Tiltman and Knox and against Japan: Friedman, Kullback, Rowlett. Section 6 concludes with the correct use of one-time random keys as the only stratagem for preserving secrecy.

14.1 THE ORIGIN OF THE ROTOR IDEA

14.1.1 Polyalphabetic substitution

Classical cryptology distinguishes two basic forms of encryption of a *plaintext* which is by transposition that is by jumbling the symbols (letters and numbers) and by substitution of the symbols, that is by a mapping of the plaintext symbol alphabet into a second, perhaps the same, alphabet, the *ciphertext* alphabet.

The rotor idea deals with substitutions and, in its pure form, with nothing else. A substitution may be denoted by a pairwise juxtaposition of plaintext symbols and ciphertext symbols. To give an example, for the plaintext alphabet {a, b, c, d, e} and the ciphertext alphabet {0, 1}, a substitution may be denoted by the pairing

$$a \mapsto 0, b \mapsto 1, c \mapsto 1, d \mapsto 0, e \mapsto 1$$

and thus there is the encryption abe $\mapsto$ 011.

The converse substitution reads

$$0 \mapsto a, 1 \mapsto b, 1 \mapsto c, 0 \mapsto d, 1 \mapsto e$$

and it gives the $2 \times 6 = 12$ possible decryptions

$$011 \mapsto abc, 011 \mapsto abe, 011 \mapsto acb,$$

$$011 \mapsto ace, 011 \mapsto aeb, 011 \mapsto aec,$$

$$011 \mapsto dbc, 011 \mapsto dbe, 011 \mapsto dcb,$$

$$011 \mapsto dce, 011 \mapsto deb, 011 \mapsto dec.$$

In other words, the decryption is not necessarily unique. Uniqueness of decryption means that the pairing is unambiguous from right to left that is *injective*. Furthermore, an encryption should be *definal*, that is defined for every plaintext letter as is the case here.[1]

Frequently, more is required than just these two conditions, namely that the encryption is *functional* that is uniquely defined and *surjective* that is the pairing is defined for every ciphertext letter. In this case, the plaintext alphabet and the ciphertext alphabet have the same cardinal number of elements,

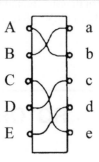

Figure 14.1. A permutation of five elements.

and if plaintext alphabet and ciphertext alphabet coincide, then the encryption is a permutation. The decryption is the inverse of this permutation. In the history of the rotor idea, it was taken for granted that permutations would be studied.

Example. [2] For the plaintext alphabet {a, b, c, d, e} and ciphertext alphabet {A, B, C, D, E}, the permutation may be a $\mapsto$ B, b $\mapsto$ A, c $\mapsto$ D, d $\mapsto$ E, e $\mapsto$ C or, as we shall write from now on

a	b	c	d	e
B	A	D	E	C

This can be visualised as in Fig. 14.1.[3]

Polyalphabetic substitution means that more than one substitution alphabet is provided for, for example 8 out of 120, and that a selection of the alphabet to be used is made according to a key made up from digits or slanted letters:

i	a	b	c	d	e
0	B	A	D	E	C
1	C	B	E	A	D
2	E	D	A	C	B
3	E	D	B	C	A
4	A	E	C	D	B
5	C	E	D	A	B
6	B	C	E	D	A
7	A	B	C	D	E

---

[1] If for some symbol no substitute is specified, it is frequently assumed that the substitution comprises a replacement of this symbol by itself.

[2] Following common cryptological usage, small letters are used for plaintext and capital letters for ciphertext.

[3] The direction from right to left corresponds to the layout in the Enigma, if seen from the front.

14.1.2 Mechanisation of substitutions and of their composition

The history of rotor devices started with the problem of mechanising a permutation. Mechanical devices seemed to have been too cumbersome for this purpose, and therefore fruitful attempts did not start before the technology of electrical relay circuits had come to some point of effectiveness. In the second decade of the 20th century, the time seemed to be ripe and the First World War pushed the development. Cryptography and cryptanalysis had arrived at a point where monoalphabetic encryption was totally insecure and only polyalphabetic encryption was likely to give security. Thus, the problem was to mechanise by electrical circuits a great number of different encryption steps such that the change from one to another could be made easily and quickly. Using plugboards, as was done around 1900 in telephone exchanges, was prohibitive. Using a row of contact switches with a number of possible positions seemed to be a solution. This shifting of alphabets was already present in the earliest ciphering tables such as the *tabula recta* of Trithemius [2: 115].

Figure 14.2 shows for an alphabet of five letters that are a realisation of five different permutations, obtained by shifts of the basic permutation of Fig. 14.1. The contact bar can be switched into five positions $i = 1, 2, 3, 4, 5$.

Compressed into a table, it shows the vertically shifted cyclic alphabets:

i	a	b	c	d	e
0	B	A	D	E	C
1	C	B	E	A	D
2	D	C	A	B	E
3	E	D	B	C	A
4	A	E	C	D	B

In actual cases, there can be an alphabet of 10 decimal digits as in the prototype ENIGMA Z 30 of 1931 recently discovered by Arturo Quirantes [31] in a Spanish Foreign Office archive. Scherbius had originally proposed 10-contact rotors, suited for the superencryption of numeral codes, of 25 (omitting j) or 26 letters, or even of 36 letters and decimal digits. But the number of contacts is limited by practical considerations, and so the number of positions that is the number of different permutations of this form may not be large enough to provide reasonable security.

14.1.2.1 *Cipher drums.* Apart from this drawback, there is another difficulty. The contact row is to be duplicated although this can be avoided if the cyclic nature of the shifted alphabets is taken into account by using a circular row instead of a linear row that is a rotatable contact drum called a cipher drum, German *Chiffrierwalze*, or French *tambour chiffrant*. Moreover, this adaptation allows connecting wires to be replaced by the safer use of sliding contacts on slip-rings. The resulting half-rotor is shown in Fig. 14.3.

14.1.2.2 *Rotors.* However, wire connections as well as slip-rings can be avoided by using an additional row of contacts on the input side of the contact bar that is the contact bar sandwiched between

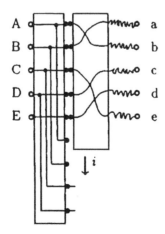

Figure 14.3. 'Half-rotor': input 5 slip-rings, output 10 positions of a contact drum.

Figure 14.2. Switch with five positions of a contact bar.

input and output contacts as shown in Fig. 14.4. The resulting five permutations compress into a table

i	a	b	c	d	e
0	B	A	D	E	C
1	D	C	B	E	A
2	B	E	D	C	A
3	B	C	A	E	D
4	E	C	D	B	A

and show diagonally shifted cyclic alphabets.

Note that in contrast to the vertically shifted alphabets, the diagonally shifted alphabets have columns where some letters are lacking and others occur more than one time. However, the letter distribution is $\{B^3DE\}$, $\{C^3EA\}$, $\{D^3AB\}$, $\{E^3BC\}$ and $\{A^3CD\}$ which is the same pattern in all five columns. This repetition of patterns is a general property of all rotor encryptions and follows immediately from the fact that the cycle decompositions[4] of the five permutations are

(ab) (cde),   (bc) (dea),   (cd) (eab),

(de) (abc),   (ea) (bcd).

The contact bar can again be replaced by a contact drum or a rotor. A complete circuitry for such a ro-

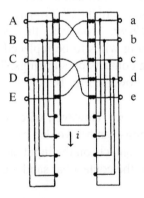

Figure 14.4. Switch with five positions of a sandwiched contact bar.

---

[4] If the basic permutation ($i = 0$) is denoted by P, the rotated permutations are of the form $\rho^{-i} P \rho^i$ ($i = 0, \ldots, 4$) and therefore have the same cycle decomposition ('characteristic'). This theorem (see Section 14.4.1.5) was the basis of the Polish break-through on Enigma and has at times been called "The theorem that won World War II".

tor with 10 contacts can be found in the Scherbius patent of 1918 (Fig. 14.5).

14.1.2.3 *Rotors in succession.* Although nothing prevented rotors from having many contacts for example the half-rotor of the Japanese *angō kikai taipu A* (Fig. 14.37) had 60; it seemed natural in the case of Z_N (the natural numbers modulo N) to have N contacts and thus only N alphabets. To achieve a high period for a progressive encryption where every available permutation is used before any permutation is repeated, the weak solution both Edward Hugh Hebern and Arthur Scherbius found was to place several rotors successively as in a counter, that is a regular or cyclometric rotor movement.

This idea was the essential step to rotor encryption. A battery of k rotors with N positions, with joint output–input, and the entire sandwiched between an entry drum and an exit drum will generate $\theta = N^k$ different permutations, provided that the basic permutations of the rotors are selected carefully enough and that no cancellation happens.

With $N = 26$, batteries of three, four or five rotors came into use, maximally leading to $\theta = 17{,}576$ and $456{,}976$ or $11{,}881{,}376$ different permutations and practically in almost progressive encryption not much smaller than these impressive numbers. It means that for four rotors, the period would not necessarily be exhausted for a message the length of a typical novel. And for five rotors, θ is about 12 million, which is many more letters than there are in the entire bible. Scherbius, in 1918, described [10] the use of 10 rotors with $N = 25$, leading up to $9.537\ldots \times 10^{13}$ permutations which is still much smaller than the total number of $25! = 1.551\ldots \times 10^{25}$ possible permutations. But 3 (Fig. 14.6), 4 (Fig. 14.3) or 5 (Fig. 14.30) moving rotors seemed for quite a time sufficient to prevent any attempt of an adversary to break an encryption that is properly done. Hebern thought in 1918 of using five rotors two of which had fixed positions.

14.1.3 Historical review

The rotor was invented or patented more or less at the same time by four different people: Hebern,

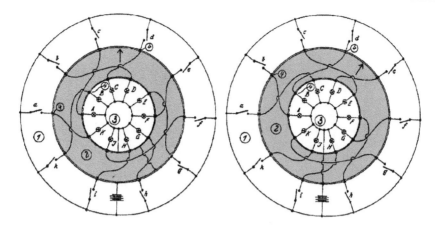

Figure 14.5. Switch with ten positions of a sandwiched contact drum, complete rotor apparatus. From the Scherbius patent No. 416,219 of 1918. Two positions of the middle ring.

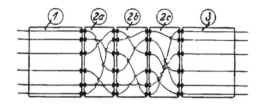

Figure 14.6. Arrangement of three rotors sandwiched between an entry drum and an exit drum. From the Scherbius patent No. 416,219 of 1918.

Scherbius, Koch, Damm (Fig. 14.7). None of them found fortune. Hebern was treated badly by the US Navy in 1934 and later by the US Government; in 1941 he lost a patent interference case against International Business Machines. He had little income when he died in 1952 from a heart attack. Scherbius suffered a fatal accident in 1929; his company's name, formerly Gewerkschaft Securitas, then Chiffriermaschinen Aktiengesellschaft, was changed in 1934 to Heimsoeth & Rinke under the lead of Dr. Rudolf Heimsoeth and Elsbeth Rinke, Berlin-Wilmersdorf (Uhlandstr. 136) which lasted at least until 1945. Koch, whose dubious role will be discussed later, died in 1928. Damm, an *homme galant*, died in 1927 before his work had been successfully recognised.

14.1.3.1 *Hebern.* There is no question that Edward Hugh Hebern (4/23/1869–2/10/1952) was an early inventor of a rotor encryption machine and for which he made drawings as early as 1917 of

a rotor connecting two typewriters electrically and thus having 26 alphabets available. The machine is on display at the National Cryptological Museum, Fort George G. Meade, MD (Fig. 14.8). Hebern started construction in 1918 of a rotor machine with 5 rotors. He called the drums codewheels. Due to circumstances, however, he filed for a US patent only after he came into contact with the Navy's Code and Signal Section and which was on March 31, 1921, later than Scherbius, Koch, and Damm. He finally received the patent (No. 1,510,441) on September 30, 1924, followed by another one (No. 1,683,072) on September 4, 1928. But in 1924, William Frederick Friedman (1891–1969) solved ten test messages, all approximately 320 characters in length, after Agnes Meyer (b. 1889), a cryptanalyst in the Navy's Code and Signal Section, in 1921 had solved a sample message [20: 415]. Starting 1928, the Navy purchased a few Hebern Electric Code machines named CSP 903 before replacing them with the ECM in the mid-1930s.

In 1926, the Hebern Electric Code, Inc. went bankrupt, but Hebern refused to give up. He hoped for further contracts with the navy and founded the International Code Machine Company in Reno, Nevada. His situation improved in 1928 when he could sell four five-rotor machines to the navy which was to initiate field tests. In 1931, the navy bought 31 machines for $54,480 to be used on flag ships. In 1932, Hebern finally designed the

Figure 14.7. Edward H. Hebern, Arthur Scherbius, Arvid G. Damm.

Figure 14.8. 'Hebern Electric Code' machine of ca. 1918 and commercial model of 1921.

HCM, a five-rotor machine with reasonably irregular rotor movement that satisfied the demands of the chief cryptologists of the navy (Lawrence F. Safford, 1893–1973) and of the army (Friedman). Unfortunately a defective construction that he sold in 1934 ended his sales to the navy although Hebern machines remained in use until 1942. The US Armed Forces switched to machines developed under Friedman since the late 1930s or early 1940s.

14.1.3.2 *Arthur Scherbius.* In a recent publication of the National Security Agency, a leaflet with the attractive title *The Bombe: Prelude to Modern Cryptanalysis* a historical error on the Enigma is repeated that has been copied for decades. It states that "Hugo Alexander Koch, a Dutchman, conceived the machine in 1919. Arthur Scherbius first produced it in 1923 commercially". This statement is historically incorrect.

The error is concerned with a true and a false inventor of the rotor principle: Arthur Scherbius

(10/30/1878–5/13/1929) and Hugo Alexander Koch (3/9/1870–3/3/1928). In 1967, David Kahn, in his famous book *The Codebreakers* [20: 420], correctly stated that on October 7, 1919, Koch, a Dutchman, filed for a Netherlands patent (No. 10,700) for a *Geheimschrijfmachine*. The patent contains a drawing showing a 4-rotor apparatus (Fig. 14.10). Kahn neglected to mention the German patent filed by Arthur Scherbius and gave only a facsimile of the US patent, filed February 6, 1923 (No. 1,657,411), that Scherbius was granted in 1926. In his 1991 book *Seizing the Enigma* [23: 31–38], Kahn gave more information, in particular that Scherbius on April 15, 1918 had approached the German Imperial Navy about his cipher machine patent for which he had applied on February 23, 1918 under file number Sch 52638 IX/42n.

The German patent (Fig. 14.9) that Kahn did not specify was issued July 8, 1925, as No. 416,219 with priority of February 23, 1918 [10]. Thus, Scherbius got a rotor invention patented more than

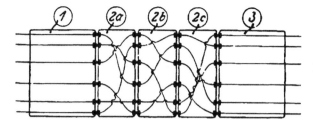

Figure 14.9. First page of the Scherbius patent of 1918.

a year earlier than Koch and also earlier than another rotor inventor, the Swede Arvid Gerhard Damm with his Swedish patent application of October 10, 1919. Scherbius, in 1918, described 10-contact rotors (Fig. 14.5) and 25-contact rotors, as well as combinations of 3 (Fig. 14.6) or more up to 10, such rotors which use electric or pneumatic switching. He also described a way to obtain irregular movement of rotors.

The patent numbers may be found in the book of 1927 on cipher machines by the Austrian criminologist Dr. Siegfried Türkel [37]. However, Türkel gave details only of the Koch patents and thus invited confusion. A second patent by Scherbius, No. 416,833 issued July 27, 1925 with priority of June 2, 1918, completed Scherbius' rotor invention.

Somehow, writers mangled this chronology into the following form: Koch invented the Enigma, and Scherbius, in 1923, bought the patents. It would go too far to list all the occurrences of this error found in the popular literature, it is found in the Antony Cave Brown's, *Bodyguard of Lies* (New York, 1975), on page 15: "Scherbius did build a machine from Koch's plans". Also in the Brian Johnson's, *The Secret War* (London, 1978) on page 309 "Koch does not appear to have constructed his invention; at all events a German engineer, Arthur Scherbius, bought the patent rights and made the machine, renaming it Enigma". The error also crept into Wladislaw Kozaczuk's *Enigma* [26: xiii]: "Later, the German engineer would develop the Enigma machine that had been patented in Holland in the fall of 1919 by the inventor Hugo Koch, who subsequently sold the patent to Scherbius".

The error may have been spread by the book of Cipher A. Deavours and Louis Kruh [9]. That book

explicitly mentions the Koch patent, filed on October 7, 1919, and furthermore listed are only the US patent, No. 1,584,660 by Scherbius, received May 11, 1926, for a 10-contact rotor, and the aforementioned No. 1,657,411, received in 1928, "for a machine he named 'Enigma' but which was not similar to the commercial model sold afterwards". The erroneous wording in the book reads (p. 7) "... and Arthur Scherbius, a German engineer acquainted with Koch who held patent rights to a *later* [my emphasis] design". In fact, Scherbius much later (Kahn, in *The Codebreakers* [20: 420], gives 1927), did indeed buy the Koch patents, but obviously not because he did not own patents before; presumably he wanted to protect his own patents.

The error can now be found on the Internet and thus it is propagated worldwide. The correct order in which the two inventors should be listed, is 'Scherbius–Koch'. (See Section 14.1.3.3.)

It should also be noted that Scherbius' company, Gewerkschaft Securitas, was founded in Berlin in July 1923 as the Chiffriermaschinen Aktiengesellschaft. Koch's patents were held by Naamlooze Vennootschap Ingenieursbureau Securitas in Amsterdam (see again Section 14.1.3.3).

Koch's patent-rights were transferred on May 5, 1922 to *N.V. Ingenieursbureau Securitas*; in 1927, Scherbius' company absorbed the patents.

14.1.3.3 *Koch and Securitas.* Until recently, the connection between Koch and Scherbius was believed to be purely commercial, deduced from the incorrect remarks "Koch invented the Enigma, and Scherbius, in 1923, bought the patents", with the variations from Antony Cave Brown, Brian Johnson and Wladislaw Kozaczuk. There was also a striking similarity between the names Gewerkschaft Securitas of the company that held the patents Scherbius filed in 1918 and the name Naamlooze Vennootschap Ingenieursbureau Securitas of the company that held Koch's patent rights after it was registered on May 4, 1922 in Amsterdam. The Naamloze Vennootschap Securitas was registered in the business of manufacturing and trading cipher machines. The transfer of the patents took place on May 5, 1922.

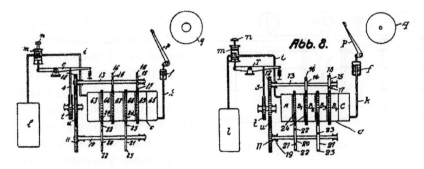

Figure 14.10. Drawings in Koch's Dutch patent 10,700 (left) and in German patent 425,147 of *Chiffriermaschinen A.G.* (right).

As found by a careful investigation Karl de Leeuw made recently [7], there is an even deeper connection. The German patent Nr. 425,147, filed on September 26, 1920 and owned by Chiffriermaschinen Aktiengesellschaft, since 1923 controlling Gewerkschaft Securitas, is an almost literal translation (see Fig. 14.10) of the Dutch patent No. 10,700 filed by Koch on October 7, 1919 the very patent that had been the reason for considering Koch as the inventor of rotor encryption. "This in itself indicates that Scherbius or one of his business friends was present at the very beginning of the development eventually resulting in Dutch *Securitas*" (Karl de Leeuw). Some duplicities in other patents corroborate this statement. "Therefore, it is likely that Naamlooze Vennootschap Ingenieursbureau Securitas was, or soon became, a Dutch front firm for Chiffriermaschinen Aktiengesellschaft in Berlin, founded with the sole purpose of transferring the production of rotor machines to Holland, whenever the continuation in Germany should prove difficult" (Karl de Leeuw). It was no longer possible to tell whether the rotor machine was either German or Dutch. That the machine could pass as either was a tremendous advantage in the light of the Treaty of Versailles which prohibited the production of war material in Germany.

But was Hugo Alexander Koch really an independent inventor, as many people had assumed? Because of de Leeuw's investigations, he may have been a contributor as Paul Bernstein and Willi Korn or Koch may have been merely a dummy. He owned only a minority of eight of the twenty shares

in the Ingenieursbureau Securitas company but the right to appoint the director and the board was held by Damaraland, a voting trust whose director was Adolf Hermkes, of German origin. It can be guessed "that Ingenieursbureau Securitas was simply a German subsidiary company in disguise" (Karl de Leeuw). It is also true that Koch, whose patent application was prepared by his brother-in-law Huybrecht Verhagen and the patent attorney J. Knoop Pathuis from the law firm Naamlooze Vennootschap Vereenigde Octrooibureaux" never built a machine for himself. Hugo Alexander Koch possibly could be dropped from the list of rotor inventors.

14.1.3.4 *Koch and Vereenigde Oetrooibureaux.* Koch turns up in quite another connection. In the beginning of September 1919, two Dutch navy officers, R.P.C. Spengler (1/14/1875–3/10/1955) and Th.H. van Hengel (10/27/1875–1/25/1939), contacted, without the navy's permission, the patent attorney A.E. Jurriaanse. Spengler and Hengel deposited their machine, a cipher machine, and drawings with the request to prepare a patent application that could be filed as soon as the navy had expressed its interest. On October 28, the navy gave permission to go ahead and on November 29, 1919, Jurriaanse filed for a patent on behalf of the navy officers; the application was registered under number 13 461 and entitled Application for a cipher machine. It happened that Jurriaanse was partner in Naamlooze Vennootschap Vereenigde Octrooibureaux, a law firm accustomed to doing business with the Dutch navy. But, Naamlooze Vennootschap Vereenigde Octrooibureaux was also involved in the handling of Koch's patent application. Koch filed his application on October 7, 1919 roughly one month after Spengler and van Hengel had demonstrated their invention to Jurriaanse, but three weeks before the navy gave the inventors the signal to go ahead.

Karl de Leeuw disclosed in 2003 that Koch was only a figurehead for his brother-in-law, the patent attorney Huybrecht Verhagen, another partner in Naamlooze Vennootschap Vereenigde Octrooibureaux, who had pinched the rotor idea from van Hengel and Spengler. They had Sea Lieutenant W.K. Maurits, a mechanical engineer, build a model for the Dutch navy following their instructions as early as 1915, two years before Hebern and three years before Scherbius.

Spengler and van Hengel came too late. The fact that the law firm was dealing with colliding claims came out in the open in June 1920, and law suits continued for some years, until on December 1923 the Committee of Appeal decided questionably against the navy officers. Consequently Spengler and van Hengel did not get their patents and until 2003 [7] were never even credited for the invention of the rotor machine.

14.1.3.5 *Spengler and van Hengel.*

In fact, R.P.C. Spengler and Th.A. van Hengel are the very first inventors of the rotor cipher machine. Their invention goes back to 1915, two years before Hebern made his first drawings and three years before Scherbius filed his patent.

Spengler and van Hengel were lieutenant commanders in the Dutch navy squadron in the Dutch East Indies, which was surrounded by dozens of warships of belligerent forces and of the USA. To avoid any conflict was a strategic and political goal which needed better communication security than the outdated Navy code books for tactical warfare afforded. Thus, the two gentlemen who had previously shown their engineering skills were ordered in January 1915 to do something to help improve the situation, and they invented a cipher machine. Very little is known about the actual way by which they arrived at this conclusion, and, unfortunately, neither the original machine nor their drawings have survived. But Spengler and van Hengel opposed Koch's patent for his rotor machine

by claiming that his drawings were identical with theirs. If they did not lie, then their machine was a 4-rotor machine without reflectors, and Koch had stolen their intellectual property. Indeed there is no other logical explanation for its sudden appearance out of the blue.

14.1.3.6 *Damm.*

Three days after Koch, the Swede Arvid Gerhard Damm filed October 10, 1919 for a Swedish patent and obtained it for his company Aktiebolaget Cryptograph under No. 52,279, US Patent No. 1,502,376. In contrast to all other patents mentioned so far, the patent was concerned with the half-rotor. Actually, Damm is out of place in this context in that he used his 5-contact half-rotors (Fig. 14.10) in pairs for a POLYBIUS-type cipher with substitution by bigrams.

The half-rotors of Damm showed up [14: 12] in the *ango kikai taipu A* (Cipher Machine A), called RED in American jargon. Apart from a fixed permutation by a plugboard, it had two half-rotors with 26 slip-rings (Fig. 14.35). The wiring permuted the six vowels onto them and for that reason also the 20 consonants; the half-rotors had 60 exit contacts since 60 is the least common multiple of 6 and 20. The reason for this disadvantageous separation may have been the tariff regulations of the international telegraph union requiring pronounceable words.

When Damm died in 1927, his company was taken over by Boris Hagelin (7/2/1892–9/7/1983), who abandoned the half-rotor in 1935, and in 1939 renamed the company Aktiebolaget Ingenjörsfirman Cryptoteknik. In the sequel, Hagelin, in 1935, replaced the half-rotors by a bar drum, also called lug cage or German Stangenkorb, for performing BEAUFORT encryption steps. Nevertheless, in later models he used irregular movements produced by step figures of the key wheels. For the machines C-35/C-36, the number of key wheels was increased to five to give a period of $17 \cdot 19 \cdot 21 \cdot 23 \cdot 25 = 3,900,225$; in a later model C-38, improved circa 1938 on the advice [9: 182] of the famous Swedish cryptologist Yves Gylden, six key wheels were used, for a period of $17 \cdot 19 \cdot 21 \cdot 23 \cdot 25 \cdot 26 = 101,405,850$. Hagelin received an order from France for 5000 machines to be fabricated under license by Ericsson–Colombes. In the Second World War, 140,000 machines were built in the USA under license to the typewriter company L.C. Smith & Corona and named C-38 by Hagelin, M-209 by the US Army, and CSP 1500 by the US navy. Late Hagelins also known as hags had a period of $29 \cdot 31 \cdot 37 \cdot 41 \cdot 43 \cdot 47 = 2,756,205,443$ that is over 2 billion.

14.2 THE SCHERBIUS LINE OF COMMERCIAL ROTOR MACHINES

14.2.1 Scherbius 1918–1924

14.2.1.1 *The patent.* On February 23, 1918, the engineer and inventor Dr. Arthur Scherbius (10/ 30/1878–5/13/1929) who was living at Berlin-Wilmersdorf, Hildegardstrasse 17 filed under the sign Sch 52638 IX/42n at the Reichspatentamt a patent application for a *Chiffrierapparat*, an electric cipher machine, einfacher und leistungsfähiger (simpler and more efficient) than the ones hitherto known. The German patent, supplemented by an application under the sign Sch 53189 IX/42n on June 21, 1918, was granted rather late on July 8, 1925 under the number DRP 416,219 for Scherbius' company Gewerkschaft Securitas in Berlin, which later founded Chiffriermaschinen A.G. (Berlin W35, Steglitzer Str. 2). The patent application in the USA was made on February 6, 1923 and the US Patent No. 1,657,411 was eventually granted in 1926. A British patent No. 231,502 was applied for on March 25, 1925 with German priority on March 26, 1924.

On April 15, 1918, Scherbius wrote to the Office of the Imperial Navy (*Reichs-Marineamt*) and offered his invention for examination, mentioning that "the commercial exploitation is at present assigned to Certified Engineer E. Ritter & Co." and pointing out that for the multi-rotor machine the key variation was so great that without knowledge of the key even with an available plaintext and its corresponding ciphertext and with the possession of the machine, the key could not be found since it is impossible to run through 6 billion rotor starting positions generated by 7 rotors with 25 positions each. Scherbius noted correctly that it would

only make sense to search for a key in this way if the same key would be maintained for a long time. A confidential position paper of July 16, 1918 of the *Marineamt*, Cipher Department DH came to the conclusion that the cipher machine would electrically transform the plaintext consisting of letters or numbers from a code book into a cipher script simply by means of a keyboard. The price for a single machine, including a coupled typewriter, was given at about 4000–5000 Mark approximately 1000$ and the time for delivery was 8 weeks. The Imperial Navy stated that the device offered high security even if it fell into the hands of an enemy. But it did not buy the machines because the prevailing opinion about existing ciphering by hand that it was sufficient and the use of machines was not worthwhile. Instead, the navy suggested to the German *Auswärtiges Amt* (Foreign Office) that they should examine the use of the machine for diplomatic correspondence. The Foreign Office also declined the offer.

14.2.1.2 *Commercial exploitation.* In the early 1920s, Scherbius tried to exploit his invention commercially. In the first models called typing machines ENIGMA A (Fig. 14.11) and ENIGMA B (Fig. 14.12), the typewriter was an integral part. The ENIGMA A of 1923, using a type-wheel, was presented in 1923 at the International Postal Congress held in Bern, and in 1924 at the International Postal Congress held in Stockholm. The ENIGMA B of 1924 used type-bars for capital letters and minuscules.

Both machines were 15 inch high and weighed more than 100 pounds. The ENIGMA A used the regular alphabet enriched with two additional characters for spaces and together there where $28^4 = 614,656$ different rotor positions and almost as many alphabets provided the rotors were suitably chosen.

The nucleus of Scherbius' invention was what is nowadays called rotors in the form of wheels or codewheels or drums called in German Walze although Scherbius used the expression *Leitungszwischenträger* in 1918 and later he called them *Durchgangsräder*. Figure 14.13 shows an arrangement of four rotors with an input keyboard for 10 figures and an output lamp field.

Figure 14.11. Scherbius' ENIGMA A of 1923.

Figure 14.12. Scherbius' ENIGMA B of 1924 with type-bar printing.

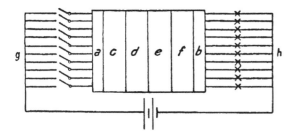

Figure 14.13. Battery of four rotors sandwiched between an input keyboard and an output lamp field.

In 1918, A. Scherbius and E. Ritter founded the firm Scherbius & Ritter, which transferred the cipher patent to Gewerkschaft Securitas. On July 9, 1923 Gewerkschaft Securitas founded the Chiffriermaschinen AG.

14.2.2 General remarks on key generation

Crypto machines of sophistication frequently have a double function: They perform polyalphabetic encryptions, and they generate their own key sequence for the selection of these encryptions. If keytext generation is included, it is because of the crucial issue of mechanisation. Progressive encryption means in the ideal case that all of the available alphabets are used before a particular alphabet is used a second time. Then, the period d of a progressive encryption coincides with the cardinal number θ of the set of encryption steps.

The abbot Johannes Trithemius (1452–1516) used the shifted standard alphabets straightforwardly one by one, and this was still done in the Cryptograph of Wheatstone, 1867 involving the use of a fixed key of period 24. The use of keys by Giovan Battista Bellaso (1555) allowed the experienced encryptor enough irregularity in the selection of the alphabets.

Arvid Gerhard Damm, one of the inventors of the rotor principle, in his Swedish patent application of October 10, 1919 made a first attempt at irregularity. Where four gears or key wheels' move each of the four half-rotors a varying number of positions after each encryption step. This irregularity was rather superficial and more likely to impress a naive person was the period $d = 17 \cdot 19 \cdot 21 \cdot 23$ of the half-rotor movement; at more than 150,000 it was about one third of $\theta = 26^4 = 456,976$. It was almost a progressive encryption.

Irregular movements through gap–tooth cog wheels with varying numbers of teeth and gaps were also used by the cipher machine of Alexander von Kryha (patent filed January 16, 1925, German patent 434,642), but with a period of between 260 and 520 the machine was cryptologically very weak [9: 10–11, 151–170].

14.2.2.1 *The Bernstein patent for key generation by irregularly dispersed cams.* In his patent application of 1918, Scherbius mentioned the possibility of a very regular form of rotor movement called Transport der Zwischenleitungsträger that is the cyclometric movement similar to counters. In this case each rotor, after turning a full round, moves the next higher rotor by just one position. A different kind of key generation was used in the ENIGMA A and ENIGMA B. In a patent application filed September 26, 1920 (German patent No. 425,147), Scherbius mentioned geared drive wheels with irregularly dispersed cams. On March 26, 1924, Paul Bernstein filed for a patent (German Patent No. 429,122) and gave an example taken from the ENIGMA A and ENIGMA B where the four gears had gaps such that one wheel with 11 positions had 5 teeth and 6 gaps, one wheel with 15 positions had 9 teeth and 6 gaps; one wheel with 17 positions had 11 teeth and 6 gaps; and one wheel with 19 positions had 11 teeth and 8 gaps. Thus, a period of $d = 11 \cdot 15 \cdot 17 \cdot 19 = 53,295$ was obtained for the rotor movement, which was about one ninth of $26^4 = 456,976$, and could provide almost progressive encryption.

The geared drive wheels with notches, however, turned up again in the 1928 ENIGMA G (*Enigma Schlüsselmaschine mit 4 Walzen und Zählwerk*).

14.2.2.2 *The Korn patent for key generation by multiply-notched rings.* In the ENIGMA C of 1926 and in all of its followers, except the ENIGMA G, the movement of the rotors is similar to the movement of counters. The ENIGMA G of 1928 had a modified form of key generation by a pinion/cog-wheel movement of the rotors with 11, 15 and 17 ratchet notches on the alphabet rings (11-15-17 ENIGMA) and, naturally, like the ENIGMA C and ENIGMA D without a plugboard – following rather closely Willi Korn's German patents No. 534,947 and No. 524,754 filed on November 9, 1928 and January 30, 1929 respectively; British patent No. 231,502; US patent No. 1,938,028 of 1933.

While in later versions of the ENIGMA whenever two ratchet notches where used on an alphabet ring, the period of the alphabets was halved, Korn's patent gave the full period of $26^4 = 456\,976$ for the ENIGMA G with its four moving rotors. Moreover, its cyclometric rotor movement allowed, in connection with a revolution counter ('counter' ENIGMA), backward and forward movement by means of a crank. It is hard to see why this method was not followed. However, the alphabet rings, sitting on the rim of the rotor core and fixed with a pin, were used in all the models following the ENIGMA G. Starting with the ENIGMA I, the ratchet notch was rigidly affixed to the alphabet ring which gave increased security since the turnover position would vary.

14.2.3 Reflector and exchangeable rotors

Given a rotor battery for encryption, the decryption required an interchange with the input keyboard and the output lamp field. The obvious way to achieve this objective is to use exchangeable multiple plug-ins for the input keyboard and for the output lamp field. A solution which did not include any additional equipment was to turn around the packet of rotors, provided the contacts are symmetric. More convenient, however, is a commutator, as shown in Fig. 14.14 for 6 letters. Such

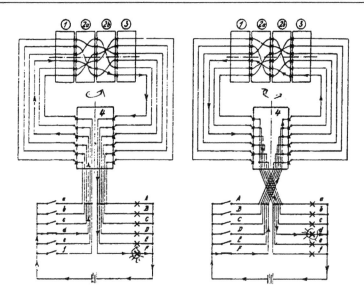

Figure 14.14. Multi-switch with two positions for encryption and decryption (from the Scherbius patent No. 416,219 of 1918).

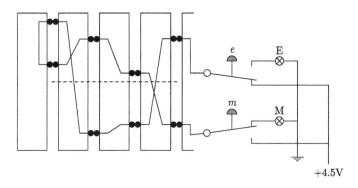

Figure 14.15. Electric current in a 3-rotor ENIGMA C with reflector U and stator S, for push-button *e* and lamp M (or for push-button *m* and lamp E).

a contrivance was used in the ENIGMA A and ENIGMA B.

14.2.3.1 *The Korn patent for the reflector.* Scherbius' colleague Willi Korn from Berlin-Friedenau filed March 21, 1926 a German patent No. 452,194 for the seemingly clever idea of introducing a reflector in German called *Umkehrwalze*, in Bletchley Park (see Section 14.5.1) sometimes misspelled Umkerwaltz and pronounced Uncle Walter. It made the encryption an involution and thus encryption and decryption were identical and no extra commutator was necessary. The number $n = 2\nu$ of contacts is now necessarily even, and the reflector U connects ν pairs of contacts thus

$U^{-1} = U$. Let S denote the substitution effect of the battery of rotors without the reflector; then, the entire arrangement performs the encryption $P = S^{-1}US$ and thus $P^{-1} = P$. Since each rotor came into play twice, the misleading idea was that the resulting permutation was mixed better and therefore the number of rotors proper could be reduced from four in the ENIGMA A and ENIGMA B to three in the commercial ENIGMA C of 1925/1926 and ENIGMA D of 1927 which was called in the British jargon the index machine (Hugh Foss) and the Glowlamp machine (Gordon Welchman). All members of this family of KORN encryption steps are now properly self-reciprocal permutations and $N/2$ pairs of letters are swapped (Fig. 14.15).

DEUTSCHES REICH

AUSGEGEBEN AM
14. NOVEMBER 1928

REICHSPATENTAMT

PATENTSCHRIFT

№ 452 194

KLASSE **42**n GRUPPE 14

C 38024 IX|42n

Tag der Bekanntmachung über die Erteilung des Patents: 20. Oktober 1927.

Chiffriermaschinen Akt.-Ges. in Berlin*).

Elektrische Vorrichtung zum Chiffrieren und Dechiffrieren.

Patentiert im Deutschen Reiche vom 21. März 1926 ab.

Elektrische Chiffriervorrichtungen oder -maschinen sind bekannt, bei denen Kontakttastenreihen als Gebestellen, eine Vertauschungsvorrichtung in Gestalt von gegen-
5 einander verdrehbaren Chiffrierwalzen und eine Schreibmaschine mit Typenhebeln oder umlaufendem Typenrad als Anzeigestellen verwendet werden. Bei solchen Chiffriermaschinen waren die verdrehbaren Chiffrier-
10 walzen zwischen festen Endwalzen angeordnet, und die Stromführung war derart, daß der elektrische Strom am Ende des Chiffrierwalzensatzes durch die eine feste Endwalze eintrat und denselben durch die andere
15 feste Endwalze wieder verließ. Für die Umstellung der Chiffriermaschine von Chiffrieren auf Dechiffrieren und umgekehrt war ein besonderer Umschalter vorgesehen. welcher. bewirkte. daß der elektrische Strom
20 den Chiffrierwalzensatz beim Dechiffrieren in umgekehrter Richtung durchwanderte wie beim Chiffrieren. Die Erfindung überwindet diese Nachteile, indem zur Vermeidung eines Umschalters
25 zum Umstellen der Vorrichtung von Chiffrieren auf Dechiffrieren und umgekehrt eine der Endwalzen des Chiffrierwalzensatzes als Umkehrwalze für die Rückführung des elektrischen Stromes durch den Chiffrierwalzen-
30 satz ausgebildet ist. Durch diesen Rückgang des Stromes durch den Chiffrierwalzensatz

findet eine weitere Verwürfelung statt. Infolge dieser Anordnung ist es möglich, mit verhältnismäßig wenig Chiffrierwalzen aus-
35 zukommen und trotzdem eine große Chiffriersicherheit aufrechtzuerhalten. Dadurch ist eine Chiffriervorrichtung nach der Erfindung geeignet. weitgehendste Verwendung nicht nur für kaufmännische oder diplo-
40 matische Zwecke, sondern auch für Armee und Marine zu finden.
Die Erfindung ist in den beiliegenden Zeichnungen beispielsweise veranschaulicht, und es stellt dar:
45 Abb. 1 ein teilweiser Grundriß der Vorrichtung bei teilweise abgenommenem Deckel,
Abb. 2 ein Schaltungsschema für den Stromverlauf zweier Tasten und Glühlampen.
50 Abb. 3 eine Seitenansicht und einen Teilschnitt durch einen Einzelteil nach einer besonderen Ausführungsform,
Abb. 4 eine Seitenansicht der Vorrichtung von rechts gesehen mit dem die Vorrichtung
55 abdeckenden Kasten,
Abb. 5 eine Seitenansicht der Vorrichtung von links gesehen ohne den die Vorrichtung abdeckenden Kasten.
Die Vorrichtung besteht aus einem Chiffrierwalzensatz *A*. einem Tastensatz *B* als
60 Gebestellen und einem Glühlampensatz *C* als Anzeigestellen.

*) Von dem Patentsucher ist als der Erfinder angegeben worden:
Willi Korn in Berlin-Friedenau.

Figure 14.16. First page of the Korn patent of 1926. Line 30 reads: *Durch diesen Rückgang des Stromes durch den Chiffrierwalzensatz findet eine weitere Verwürfelung statt.*

It was considered to be an advantage that encryption and decryption coincided and that a switch was no longer needed. The introduction of the reflector, however, had the consequence that no letter could be encrypted as itself. Ironically, this detail would turn out to be a cryptologic disaster.[5] Like-

wise, the fact that the electric current went through six rotors was interpreted wrongly by the inventor (Fig. 14.16) as additional cryptanalytical security.

14.2.3.2 *The Korn patent for exchangeable rotors.* March 11, 1926 Korn filed another patent No. 460,457 providing for exchangeable rotors

---

[5][2], Sections 11.2.4, 14.1, 14.5.1, 19.7.2.

(Fig. 14.17) which meant that for a p-rotor machine, there are $\binom{n}{p}$ possibilities to choose p rotors from a set of n, where $p \leqslant n$; giving $n!/(p!(n-p)!)$ choices. For each choice, there are $p!$ rotor orders possible, leading to a total of $n!/((n-p)!)$ rotor combinations.

For the first time the commercial ENIGMA D was furnished with $n > 3$ exchangeable rotors and allowed $n!/(n-3)! = n \cdot (n-1) \cdot (n-2)$ possibilities of the rotor selection and arrangement, for example, 60 for $n = 5$, 336 for $n = 8$.

Figure 14.17. Rotor with 26 contacts from the ENIGMA D of 1927.

14.2.3.3 *The settable reflector.* In the commercial ENIGMA C (Fig. 14.18) the reflector U could be inserted in two fixed positions and along with the six possible permutations of the rotors could provide $26^3 \cdot 2 \cdot 6 = 210{,}912$ initial settings with hopefully the same number of resulting *different* substitution alphabets. The number of all different properly self-reciprocal substitution alphabets that is of all possible KORN encryption steps was $26!/(13! \cdot 2^{13}) = 7.91 \cdot 10^{12}$. In the commercial ENIGMA D (Fig. 14.19) the reflector could be set in 26 positions just as the three rotors; from the outside, it looked like another that is a fourth rotor the reflecting rotor U which did not move during enciphering. Now the number of all initial settings was $26^3 \cdot 26 \cdot 6 = 2{,}741{,}856$.

14.2.3.4 *The anomaly in the cyclometric movement.* In the ENIGMA C, ENIGMA D, and in the 3-rotor *Wehrmacht* models, the movement of the rotors was accomplished by using one notch at the alphabet ring of each rotor. However, it was not a regular movement (like that of a counter) due to an anomaly in the construction of the cam mech-

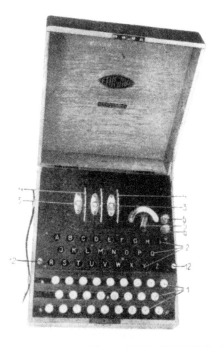

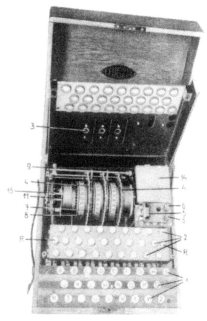

Figure 14.18. ENIGMA C of Scherbius and Korn, 1925.

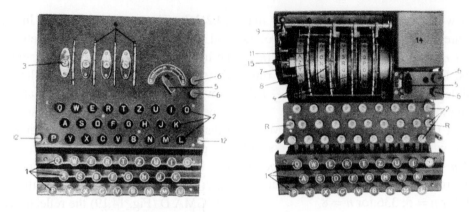

Figure 14.19. ENIGMA D of Scherbius and Korn, 1927.

anism. Whenever the leftmost or slow rotor R_L stepped, the medium rotor R_M made an extra step.

$$
\begin{array}{ccc}
\vdots & \vdots & \vdots \\
A & D & P \\
A & D & Q \\
A & E & R \\
\rightarrow \quad B & F & S \\
B & F & T \\
B & F & U \\
\vdots & \vdots & \vdots \\
R_L & R_M & R_N
\end{array}
$$

Therefore, the period was a bit less than the maximal one, $\theta = 26^3$, namely, $26 \cdot 25 \cdot 26 = 16,900$. Except for this stepping anomaly, there was almost cyclometric rotor movement.

14.2.3.5 *The commercial ENIGMA line.* With the ENIGMA C, introduced in 1925/1926, Scherbius left the provision of a writing device and used, as already mentioned in the patent application, glow-lamps in a display; the ENIGMA C was battery-operated (4.5 volt) and was called *Glühlampentype*. The cipher text was to be read off character by character and written down by hand; it was then usually sent by Morse wireless. The keyboard and the lamp field were alphabetic.

The commercial ENIGMA D machine, introduced in 1927, was widely used and was used in Sweden, The Netherlands, Japan, Italy, Spain, UK and the USA, and was bought legally by the Polish *Biuro Szyfrów*. The keyboard was essentially the same as the standard German typewriter. Its successor, ENIGMA K, was delivered 1938–1940 to the Swiss Army with a US code-name INDIGO. Because of the lax use of stereotypes on the Swiss side, ENIGMA K was routinely broken by both the German and the Allied cryptanalytic services in particular by the Polish Cadix group working October 1940–March 1941 in the Château des Fouzes near Uzès in the unoccupied part of southern France [26: 125].

The ENIGMA Z 30 of 1931 [31], with an alphabet of 10 decimal digits, presumably was a prototype only and destined for enciphering numerical code.

The price of a single ENIGMA D in 1928 was about 600 Reichsmark or US$140. About a hundred ENIGMA Ks were sold in 1941 for 760 Reichsmark a piece. An ENIGMA C was said to have been purchased in 1925 by Dillwyn Knox in Vienna. When working on the ENIGMA of the SD, Keith Batey found out that some of its wheels were identical to a wheel of the ENIGMA C.

14.3 ROTOR MACHINES FOR THE GERMAN ARMED FORCES

The German *Reichsmarine* had started experiments in 1925 with a 28-contact 3-rotor ENIGMA version (*Funkschlüssel* C, Fig. 14.20). It appeared in February 1926 with an alphabetically ordered keyboard comprising additional characters Ä, Ö and Ü (X bypassed) called the O-Bar machine in Britain.

Figure 14.20. Funkschlüssel C of the *Reichsmarine*, 1926; with additional vowels Ä, Ö and Ü (X bypassing the rotors). Keyboard with alphabetic ordering of the keys. The three rotors can be chosen out of five available ones.

The reflector could now be inserted in four fixed positions, denoted by $\alpha, \beta, \gamma, \delta$. In 1933, minor modifications were made to the *Funkschlüssel C*; a 28-contact version including Ä, Ü only was in test use. In the *Reichswehr* models ENIGMA G, introduced on July 1, 1928 under Major Rudolf Schmidt[6] and ENIGMA I (1930) there were again 26 contacts and a keyboard similar to that of the standard German typewriter up to the position of the letter P (Fig. 14.40). The reflector had one fixed position. There was also an ENIGMA II with a typewriter used only by the highest commands; it was considered unpractical and unreliable, and found little use.

14.3.1 Plugboard

14.3.1.1 *The Polish Biuro Szyfrów.* The Poles were known for the high standard of their cryptanalytic abilities. Under the guidance of the later Colonel Captain Jan Kowalewski (1892–1965) they won their war against Russia in 1920 with the help of cryptanalysis [25: 239]. Consequently

---

[6]Major Rudolph Schmidt headed the army *Chiffrier-Stelle* during 1925–1928 and he became a *Generaloberst* later. His brother Hans-Thilo Schmidt was found to be a spy for France, since 1931 and who stole important ENIGMA information while working in the *Chiffrier-Stelle*. He was discovered in March 1943 and committed suicide on September 19, 1943 [2: 390]. See Section 14.4.1.2.

after July 15, 1928 with so many radio signals of German origin encrypted with a new method appearing in the ether, the Polish Bureau under the leadership of Major Franciszek Pokorny became alarmed and recalled the commercial ENIGMA, which had been on the open market since 1926. Initially the Polish Bureau was not able to break it. The internal wiring of the rotors in the military ENIGMA G of 1928 and ENIGMA I of 1930 was different from that in the commercial version, of course. What the Bureau could not know, however, was that ENIGMA I performed a puzzling fixed superencryption by way of a plugboard.

The Polish *specialists* in B.S.-4 under Captain Maksymilian Ciężki (1899–1951) even found that the first six letters of every message had a peculiarity and based on recommendations provided with the commercial ENIGMA, the Poles successfully guessed, that the first six letters were an encrypted message key. However, in 1931 they had almost given up any hope of success.

14.3.1.2 *The need for a plugboard.* The commercial ENIGMA despite of claims of being secure was repeatedly cryptanalysed in the 1920s and 1930s, by British, French, Swedish, German and presumably other services, in particular during the Spanish Civil War 1936–1939. A method that nearly everyone seems to have discovered was called *méthode des bâtons*; the methods was called cliques on the rods by British cryptanalysts. On April 24, 1937, Dilly Knox, at that time the leading British cryptologist, broke into enciphering by an ENIGMA K, given by the Germans to the Italians and Spanish [26: 48, 146]. In 1946 Rosario Candela, described the method in 1946 for the first time in the open literature [5] calling it the method of isomorphs.

To explain this expression the plugboardless ENIGMA equation

$$c_i = p_i S_i U S_i^{-1},$$

where p_i is the ith plaintext character, S_i is the ith substitution performed by the rotors R_N, R_M, R_L, U is the reflector and c_i is the ith ciphertext character is rewritten in the form

$$c_i S_i = p_i S_i \cdot U.$$

Ring setting	k	a	b	c	d	e	f	g	h	i	j	k	l	m	n	o	p	q	r	s	t	u	v	w	x	y	z
A	0	E	K	M	F	L	G	D	Q	V	Z	N	T	O	W	Y	H	X	U	S	P	A	I	B	R	C	J
B	1	J	L	E	K	F	C	P	U	Y	M	S	N	V	X	G	W	T	R	O	Z	H	A	Q	B	I	D
C	2	K	D	J	E	B	O	T	X	L	R	M	U	W	F	V	S	Q	N	Y	G	Z	P	A	H	C	I
D	3	C	I	D	A	N	S	W	K	Q	L	T	V	E	U	R	P	M	X	F	Y	O	Z	G	B	H	J
E	4	H	C	Z	M	R	V	J	P	K	S	U	D	T	Q	O	L	W	E	X	N	Y	F	A	G	I	B
F	5	B	Y	L	Q	U	I	O	J	R	T	C	S	P	N	K	V	D	W	M	X	E	Z	F	H	A	G
G	6	X	K	P	T	H	N	I	Q	S	B	R	O	M	J	U	C	V	L	W	D	Y	E	G	Z	F	A
H	7	J	O	S	G	M	H	P	R	A	Q	N	L	I	T	B	U	K	V	C	X	D	F	Y	E	Z	W
I	8	N	R	F	L	G	O	Q	Z	P	M	K	H	S	A	T	J	U	B	W	C	E	X	D	Y	V	I
J	9	Q	E	K	F	N	P	Y	O	L	J	G	R	Z	S	I	T	A	V	B	D	W	C	X	U	H	M
K	10	D	J	E	M	O	X	N	K	I	F	Q	Y	R	H	S	Z	U	A	C	V	B	W	T	G	L	P
L	11	I	D	L	N	W	M	J	H	E	P	X	Q	G	R	Y	T	Z	B	U	A	V	S	F	K	O	C
M	12	C	K	M	V	L	I	G	D	O	W	P	F	Q	X	S	Y	A	T	Z	U	R	E	J	N	B	H
N	13	J	L	U	K	H	F	C	N	V	O	E	P	W	R	X	Z	S	Y	T	Q	D	I	M	A	G	B
O	14	K	T	J	G	E	B	M	U	N	D	O	V	Q	W	Y	R	X	S	P	C	H	L	Z	F	A	I
P	15	S	I	F	D	A	L	T	M	C	N	U	P	V	X	Q	W	R	O	B	G	K	Y	E	Z	H	J
Q	16	H	E	C	Z	K	S	L	B	M	Z	O	U	W	P	V	Q	N	A	F	J	X	D	Y	G	I	R
R	17	D	B	Y	J	R	K	A	L	S	N	T	V	O	U	P	M	Z	E	I	W	C	X	F	H	Q	G
S	18	A	X	I	Q	J	Z	K	R	M	S	U	N	T	O	L	Y	D	H	V	B	W	E	G	P	F	C
T	19	W	H	P	I	Y	J	Q	L	R	T	M	S	N	K	X	C	G	U	A	V	D	F	O	E	B	Z
U	20	G	O	H	X	I	P	K	Q	S	L	R	M	J	W	B	F	T	Z	U	C	E	N	D	A	Y	V
V	21	N	G	W	H	O	J	P	R	K	Q	L	I	V	A	E	S	Y	T	B	D	M	C	Z	X	U	F
W	22	F	V	G	N	I	O	Q	J	P	K	H	U	Z	D	R	X	S	A	C	L	B	Y	W	T	E	M
X	23	U	F	M	H	N	P	I	O	J	G	T	Y	C	Q	W	R	Z	B	K	A	X	V	S	D	L	E
Y	24	E	L	G	M	O	H	N	I	F	S	X	B	P	V	Q	Y	A	J	Z	W	U	R	C	K	D	T
Z	25	K	F	L	N	G	M	H	E	R	W	A	O	U	P	X	Z	I	Y	V	T	Q	B	J	C	S	D

Figure 14.21. *Wehrmacht* ENIGMA rotor I, 26 alphabets.

$c_i S_i$ is a *monoalphabetic* image of $p_i S_i$ under U, $c_i S_i$ and $p_i S_i$ are said to be *isomorphic*. In principle, U would not need to be self-reciprocal.

The method uses a plaintext–cryptotext compromise: searching for the position of a short probable plaintext word that is short enough that the middle rotor is kept at rest during it's enciphering, such that no contradiction to the isomorphism would occur or in jargon called a non-scritching position. Actually, U is self-reciprocal in the ENIGMA case which leads to further possibilities of exclusion of mis-hits by scritching as well as to a positive confirmation of a suitable position by the appearance of a self-reciprocal substitution. The possibility of mishits is reduced; the self-reciprocal character of the ENIGMA helps the unauthorised decryptor. Now, the concrete isomorphism reads

$$c_i \rho^{k-i} R_N \rho^{-k+i} = p_i \rho^{k-i} R_N \rho^{-k+i} \cdot U',$$

R_N being the fast rotor, U' the virtual reflector which combines the actions of R_M, R_L, U, R_L^{-1},

R_M^{-1} where k denotes the shift in the position of the core R_N which is looked for in order to accomplish isomorphism.

To give an example, assume R_N is the known rotor I of the *Wehrmacht* ENIGMA fallen into the hands of an enemy, generating with ring setting A for core position $k = 0$, 26 rotated alphabets (Fig. 14.21).

Assume that with rotor I the following ENIGMA encryption of a probable word was obtained:

```
r e c o n n a i s s a n c e
U P Y T E Z O J Z E G B O T
```

Among the maximal 26 shifts to be tested, let $k = 0$:

```
    r  e  c  o  n  n  a  i  s  s  a  n  c  e
↓   0  1  2  3  4  5  6  7  8  9  10 11 12 13
    U  F  J  R  Q  N  X  A  W  B  D  R  M  H
    A  W  C  Y  R  G  U  Q  I  N  N  D  S  Q
↑   0  1  2  3  4  5  6  7  8  9  10 11 12 13
    U  P  Y  T  E  Z  O  J  Z  E  G  B  O  T
```

and run into a scritching, such that (RY) and (RD) violate uniqueness, (AQ) and (HQ) as well as (BN) and (DN) violate injectivity, and (AQ) and (QR) as well as (QR) and (RD) violate self-reciprocity.

Next, test $k = 1$:

r	e	c	o	n	n	a	i	s	s	a	n	c	e
↓ 1	2	3	4	5	6	7	8	9	10	11	12	13	14
R	B	D	O	N	J	J	P	B	C	I	X	U	E
H	S	H	N	U	A	B	M	M	O	J	K	X	C
↑ 1	2	3	4	5	6	7	8	9	10	11	12	13	14
U	P	Y	T	E	Z	O	J	Z	E	G	B	O	T

and run again into scritchings. (JA) and (JB) as well as (BS) and (BM) violate uniqueness, (RH) and (DH) as well as (PM) and (BM) violate injectivity, and (CO) and (ON) as well as (NU) and (UX), (JB) and (BM), (IJ) and (JA) violate self-reciprocity.

The next shifts give scritchings again and the example shows almost the worst case that is not earlier than for $k = 24$ is successful: there are no scritchings,

r	e	c	o	n	n	a	i	s	s	a	n	c	e
↓ 24	25	0	1	2	3	4	5	6	7	8	9	10	11
J	G	M	G	F	U	H	R	W	C	N	S	E	W
U	Z	C	Z	B	J	O	T	A	M	Q	E	S	A
↑ 24	25	0	1	2	3	4	5	6	7	8	9	10	11
U	P	Y	T	E	Z	O	J	Z	E	G	B	O	T

and the self-reciprocal virtual reflector U' is revealed: it reads

(AW), (BF), (CM), (ES), (GZ),

(HO), (JU), (NQ), (RT).

Thus, the entrance for a break is made and Reciprocal ENIGMAs', as *Hugh Foss* calls ENIGMAs with reflectors, without a plugboard are unsafe (*Foss* 1927 [12: 45]). It took the Germans longer to discover this defect and they invented the plugboard which now prevents this attack. The probable word had to be mapped at first with the plugboard substitution which presumably is not known to the enemy.

Therefore, the ENIGMA I of the *Heer* introduced June 1, 1930 which later became the common *Wehrmacht* ENIGMA was protected by adding a variable super-encryption by way of a plugboard that provided an additional entry substitution T, called the *Steckerverbindung* or cross-plugging which was conveniently though unnecessarily self-reciprocal and correspondingly an exit substitution T^{-1}. This addition resulted in the ENIGMA equation with plugboard

$$c_i = p_i T S_i U S_i^{-1} T^{-1},$$

where T is the plugboard substitution in the form

$$c_i T S_i = p_i T S_i \cdot U.$$

Now, $c_i T S_i$ and $p_i T S_i$ are *isomorphic*. But since T is not known to the enemy, the method of isomorphs does not work.

Although the introduction of the plugboard did thwart the method of isomorphs attack, it did not mean that there would not exist other means methods insensitive to *steckering*. That the Germans did not think of such methods was stubborn and haughty. The Poles were successful, the *Wehrmacht* ENIGMA continued to be unsafe for almost as long as the *Wehrmacht* existed.

14.3.1.3 *Cross-plugging.* A plugboard can implement the most general fixed substitution of the 26 characters. The ENIGMA plugboard of 1930 allowed cross-plugging only, i.e., self-reciprocal substitutions by crossing wires by swapping pairs of wires by using double-ended connectors (Fig. 14.22). Such reflections were used in the ENIGMA plugboard (German *Steckerbrett*).

The number $d(k, N)$ of plugboard reflections depends on N, the number of rotor contacts and the number k of cinch plugs used.

$$d(k, N) = \frac{N!}{2^k \cdot (N - 2k)! \cdot k!} = \binom{N}{2k} \cdot \frac{(2k)!}{2^k k!}$$

$$= \binom{N}{2k} \cdot (2k - 1)!!,$$

where

$$(2k - 1)!! = (2k - 1) \cdot (2k - 3) \cdots \cdot 5 \cdot 3 \cdot 1$$

$$= \frac{(2k)!}{2^k k!}.$$

Properly self-reciprocal permutations (genuine reflections) require N to be even and with $N = 2\nu$

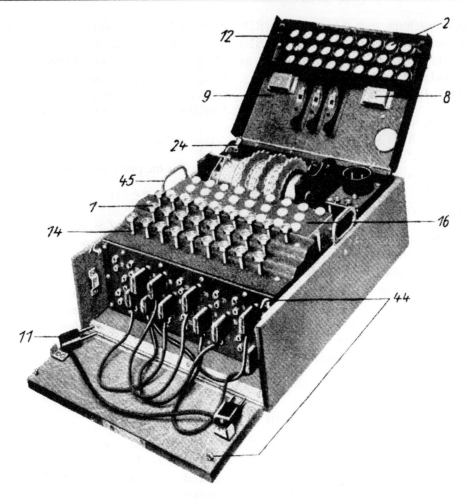

Figure 14.22. Funkschluessel M (1934) with Plugboard.

such permutations consist of ν 2-cycles. The number $d(\frac{N}{2}, N)$ of all genuine reflections is then

$$d\left(\frac{N}{2}, N\right) = (N-1)!!$$
$$= (N-1) \cdot (N-3) \cdot \cdots \cdot 5 \cdot 3 \cdot 1$$
$$= \frac{(2\nu)!}{\nu! 2^\nu} \approx \frac{\sqrt{(2\nu)!}}{\sqrt[4]{\pi \cdot (\nu + \frac{1}{4})}}.$$

The approximate value with a relative error $< 10^{-3}$ for $N \geqslant 6$ is a rather good upper bound for $(N-1)!!$.

For fixed N, however, $d(k, N)$ is maximal for $k = \nu - \lfloor \sqrt{(\nu+1)/2} \rfloor$ and $d(5, 26) \approx 5.02 \cdot 10^9$, $d(6, 26) \approx 1.00 \cdot 10^{11}$, $d(7, 26) \approx 1.31 \cdot 10^{12}$,

$d(8, 26) \approx 1.08 \cdot 10^{13}$, $d(9, 26) \approx 5.38 \cdot 10^{13}$, $d(10, 26) \approx 1.51 \cdot 10^{14}$, $d(11, 26) \approx 2.06 \cdot 10^{14}$, $d(12, 26) \approx 1.03 \cdot 10^{14}$, $d(13, 26) \approx 7.91 \cdot 10^{12}$, $d(3, 10) = 3150$, $d(4, 10) = 4725$, $d(5, 10) = 945$. Note that $^2\log d(10, 26) \approx 47.1$ [bit], $^2\log d(11, 26) \approx 47.5$ [bit], $^2\log d(12, 26) \approx 46.5$ [bit] and for all reflections, $^2\log \sum_{k=1}^{13} d(k, 26) \approx {}^2\log 5.33 \cdot 10^{14} \approx 48.9$ [bit].

The ENIGMA I of the *Reichswehr* (1930) and the *Wehrmacht* ENIGMA originally used six double-ended two-line connectors; later, beginning on October 1, 1936, five to eight; from January 1, 1939, seven to ten; and from August 19, 1939 ten prevailed but key net BROWN of the Luftwaffe, used only six or seven double-ended two-line connectors for cross-plugging.

Figure 14.23. 3-rotor *Wehrmacht Heer* ENIGMA (1937).

14.3.2 The *Wehrmacht* ENIGMA

In 1934 when the *Reichsmarine* and the *Heer* agreed on a common version called the *Wehrmacht* ENIGMA or Service Enigma (Fig. 14.23), this agreement was achieved under pressure from Colonel Erich Fellgiebel (1886–1944).[7]

14.3.2.1 *German navy, army and air force.* The three rotors of the *Reichsmarine* (whose name was changed in 1935 to *Kriegsmarine*) ENIGMA called

Funkschlüssel M, introduced in October 1934, could now be selected respectively from a set of five (1934), seven (1938), or eight (1939) rotors and moreover, they could be permuted. They were marked with the roman numerals I, . . . , VIII. Before December 15, 1938, the army (*Heer*) released only three of the five rotors provided for their ENIGMA. The air force, too, introduced on August 1, 1935 the *Wehrmacht* ENIGMA for its new *Luftnachrichtentruppe*. Removing the exchangeable three rotors is shown in Fig. 14.24. The reflector presumably marked A of the *Wehrmacht* ENIGMA was replaced on November 1, 1937 by

[7]Fellgiebel was later (from 1939) Major-General and Chief, OKW Signal Communications.

Figure 14.24. The three removable rotors of the *Wehrmacht Heer* ENIGMA (1937).

Umkehrwalze B. In mid-1940, C turned up, was rarely used, and was later withdrawn. Thereafter a pluggable reflector D a reflector which could be simply rewired, was first observed on January 2, 1944 in traffic with Norway. Figure 14.25 shows a cipher document of the *Luftwaffe* ENIGMA from 1944, which indicated that in 1944 the pluggable reflector was changed every 10 days and that some *Steckerverbindungen* were changed every 8 hours.

A special device, the *Uhr* box was introduced in 1944 to replace the steckering of the *Wehrmacht* ENIGMA plugboard by a non-reciprocal substitution, which also could be changed easily by turning the knob. Despite the extra security it added, it was not widely used. However, it did convincingly demonstrate that the restriction of the plugboard to self-reciprocal substitutions was not a cogent consequence of the use of the self-reciprocal reflector, since the encryption was still self-reciprocal.

The maximal length of a message for the Army ENIGMA was limited to 180 characters and after January 13, 1940 to 250 characters, and to 320 characters for the Navy ENIGMA. Longer messages had to be divided into parts.

The rotors could be inserted in arbitrary order into the ENIGMA. Until the end of 1935, this wheel order called the *Walzenlage* and the cross-plugging (*Steckerverbindung*) were fixed for three months. Beginning January 1, 1936, they changed every month; and from October 1, 1936, every day. Later, during the Second World War, they were changed every eight hours. For the Germans, it needed the increasing pressure the development of

the war brought about to do what they should have done from the very beginning.

Another invention Willi Korn had made as early as 1928 did not develop into a cryptological fiasco. Instead the ring which allowed the rotor position that is the alphabet ring or index ring (German *Sperr-Ring*) to be read was made movable, like a tire mounted around the rim of the rotor core; and with its position with respect to the core, the ring setting (German *Ringstellung*), could be fixed with a pin. Starting with the ENIGMA I, the ratchet notch was rigidly affixed to the alphabet ring which gave increased security by multiplying the key space size.

14.3.2.2 *Multi-notched rotors.* Notches were located for the *Wehrmacht* ENIGMA rotors I–VIII as follows.

Rotor	I II III IV V VI, VII, VIII
Notch(es) at letter	Y M D R H H U

The notches were at different positions of the alphabet ring[8] and were to meant to provide some irregularity which was only a *complication illusoire*. Worse yet, it was a complication that defeats itself as Kahn ironically said: if all rotors had the notches cut at the same letter, then the cryptanalysts would not have been able to find which rotor was used as the fast rotor by finding out for known rotors which letter caused the turnover. The *Kriegsmarine* seemingly found out about this problem and cut the notches on the new rotors VI and VII (1938) and VIII (1939) at the same positions. Moreover, these rotors had *two* notches: one at the letter H, one at U. Unlike in the commercial ENIGMA D, the notches were on the alphabet ring and the turnover positions depended on the ring setting. Although by using two notches the period was halved and the danger of a superimposition [2: 375] increased. As a countermeasure, the permissible length of any one message had been drastically limited and this change made cryptanalysis much more difficult:

---

[8]For rotors I to V, the turnover happened when with a difference of 19 letters the transitions $Q \to R$, $E \to F$, $V \to W$, $J \to K$, $Z \to A$ occurred in the window. At Bletchley Park there was a corresponding and a rather silly mnemonic line *Royal Flags Wave Kings Above*.

Geheime Kommandosache! Jeder einzelne Tagesschlüssel ist geheim! ... im Flugzeug verboten!

Luftwaffen-Maschinen-Schlüssel Nr. 2744

Achtung! Schlüsselmittel dürfen nicht unversehrt in Feindeshand fallen. Bei Gefahr restlos und frühzeitig vernichten.

Nᵒ 000082

Tag	Walzenlage	Ringstellung	an der Umkehrwalze	Steckerverbindungen 1–10	Zusatzsteckerverbindungen 1900 / 2300	Kenngruppen
31	III V I IV	17 11 04	TS IM	TW BI UY GP CK JQ DL RV EM AH	NS FO	kim pwh · sbx csw
30	I IV V	08 17 21	UX IB	LS DH MT EO AP UZ PQ WY BK GR	CI JN	uaq omn · ume duf
29	V II III	11 14 05	SV CO	DO JW CN IV PZ BM HU AL FR KX	EQ GT	don cqo · xum bpg
28	II IV V	02 20 16	NQ BP	NT HK BW EP LQ AU OY PJ CX GI	DE MR	lui pyg · sby dtq
27	III V IV	18 13 22	GL AD	HM GV KZ AI DQ NR ES BL OU FT	CF JY	cmy fqr · acl bur
26	I III II	24 10 01	HK BW	GW AQ MO PV PS DI RU JZ BN EH	KT CL	kbj ysq · udm cns
25	IV I III	04 25 23	DZ NE	LT DR QX AG IN EU BJ KP FW CM	SS HO	kqs yar · vdb com
24	V III I	09 19 06	CS MP	GL MY CR HN JX DT AF PU IQ BO	EW KS	cms aoj · sod auh
23	IV I V	15 03 19	BH LX	IT DV HQ AJ MU EX KO OS FY LN	BP GZ	kra yas · xun cob
22	I V III	12 26 07	IR	EY JL AK NV FZ CT BP MX BQ GS	DW IO	jdm uhf · xuo bph
21	III IV II	15 09 12	AG FK	JF DT QS HL AE NW CU IK FX BR	MV GO	jpf aok · iys btx
20	IV II I	02 22 05	QV	HT NP AM DX GJ KQ BS OV ER CW	IU PL	boy wac · uow cse
19	V I III	08 19 17	EW NE	GM OX BT QU DP HJ FK SW AN EL	CY IR	xjc wad · unj ctd
18	III IV I	11 21 01	DZ NE	KW IP DM SV JR CX EN AZ QT BU	FH GY	kpn rsi · vcm bpo
17	I V II	18 23 14	BH MP	BV HW AR NX DS PT CZ PI LY EJ	GK MQ	kdx crq · vcn cod
16	III IV V	16 04 07	CS LX	LU CV FM KR BY GN QW DJ PS AO	EI HX	lgx jri · uob sur
15	V III IV	24 13 10	BO CE	HZ NQ AD TV IX AM BG LO CE RY	JU PP	wpt vhy · zos aus
14	I IV II	06 20 25	AG FK	FN UY CJ IW LP AS DK GQ MO BZ	ET HR	wog hxi · xxi bpi
13	III II I	03 26 18	IR LX	KR IZ AT NV BH MP CG OY ES DP	UW LQ	lqv iqb · ssy coe
12	IV II III	04 11 15	FK	DT JV HS OI AY KU EN PQ LR BW	MF GO	zic myt · zof dtr
11	V I IV	16 07 02	BG RU	JS PW AV QX DN IZ KM CO EG PL	HY BR	inf xbm · krs dug
10	IV III II	20 12 14	TV	PS CQ JO PR AW HV EZ KN DU GT	IL BY	ink acu · zxj cnu
9	III II V	06 18 10	BG RU	HX TZ MX LW GQ AD NY BE CS JP	RV IO	efm pmi · snw cof
8	V I III	01 21 17	AG RU	GU SW BF RX EV OT LQ CH IP KY	JM NZ	imy rjw · tjm cog
7	II IV I	25 08 23	EI QX	CX AZ DV KT HU LW GP EY MR PQ	IN OS	inv rkc · snx bpj
6	IV II V	13 26 03	PZ	DV LP NQ GZ OS FK BW MR IT HX	UY BJ	yvu hsb · swq aut
5	III I II	24 19 22	KN PM	SY EK NZ OR CG JM QU PV BI LW	TX DP	seu iqe · swr auv
4	II IV I	17 05 09	QX OW	BD GV AX KP EM FN CW RU HO JT	IL GS	sfj hxj · sxk dpt
3	V II IV	20 16 11	GP CR	JT NW DU EO KV BY PS HQ IM LX	GP CR	clx sbn · xxa buk
2	II III V	14 03 19	HL DS	RW OQ GI AZ EJ MS CU DH PY BF	LV TX	ljs jre · spq coh
1	III I IV	18 24 15	DS OW	NP JV LY IX KQ AO DZ CR FT EM	GS HW	plf dgw · tjn cnv

Figure 14.25. Cipher document No. 2744 of the Luftwaffe ENIGMA meant presumably for September 1944 with a column Steckerverbindungen an der Umkehrwalze.

"We would have had great trouble if each wheel had had two or three turnover positions instead of one" (Welchman).

The rotors of the ENIGMA T (Tirpitz) (see Section 14.3.2.3), had five notches:

Rotor	I, III	II, IV	V, VII	VI, VIII
Notches at letter	EHMSY	EHNTZ	GKNSZ	FMQUY

14.3.2.3 *Other organisations.*

The railroad company *Deutsche Reichsbahn* and the Post Office *Deutsche Reichspost*, as well as the *Sicherheits-dienst* (SD for security service) which ran from September 1, 1937 used less-secure older ENIGMA models without a *Steckerbrett* or plugboard. This was careless, however, because messages concerning railroad transports in Russia for example were liable to give clues to the enemy.

Knox, Tiltman, and other old-timers in Bletchley park enjoyed reading the signals picked up. Lieutenant-Colonel John Tiltman (1894–1982), from 1940 onward, solved the German railway traffic, later codenamed Rocket [12: 72; 18: 116].

Knox, in 1941, even broke into the *Abwehr* traffic [18: 123]. The Italian navy used a commercial ENIGMA D with rewired rotors which led to the Italian defeat in the battle of Matapán on March 28, 1941 [23: 139; 18: 3].

The German *Abwehr*, the Intelligence Service of the German armed forces, played an exceptional role as far as ENIGMA goes. It used a version (Fig. 14.26) with rewired rotors of the old 3-rotor plus reflector ENIGMA G of 1928 with a pinion/cog-wheel movement of the rotors, with 11, 15 and 17 notches on the index rings, 11-15-17 ENIGMA, and without a plugboard (see Section 14.2.2.2). A mechanical revolution counter ("counter enigma") allowed moving the rotor positions backward and forward by means of a small crank [18: 123] and [17: 41].

A few specimens of a 3-rotor ENIGMA T (Tirpitz) likewise without a plugboard but with five-notched rotors, were destined for the Japanese navy, but did not leave the harbour and were captured by the allies in Brittany near Lorient.

14.3.2.4 *The QWERTZU riddle.*

In the commercial ENIGMA, the keyboard used the standard

Figure 14.26. The *Abwehr* ENIGMA G-312.

German typewriter (Fig. 14.40) and the input keys were connected correspondingly to the sequence of contacts of the stator:

q w e r t z u i o a s d f g h j k p y x c v b n m l
A B C D E F G H I J K L M N O P Q R S T U V W X Y Z

Reordered, the following mapping *H* of the input alphabet to the stator was obtained:

a b c d e f g h i j k l m n o p q r s t u v w x y z
J W U L C M N O H P Q Z Y X I R A D K E G V B T S F

This method provided a fixed superencryption, and since the plugboard gave a variable superencryption anyhow was considered unnecessary for the *Wehrmacht* ENIGMA to use furthermore an mapping in addition. Therefore, a change was made to the identical mapping where the connections from the keys to the contacts of the stator were in alphabetic order which turned out to be a double-cross the Germans had unintentionally played on the British. When the British had found out that the *QWERTZU* mapping no longer worked, they

tried to find another useful mapping with the exclusion of the trivial identical one which they thought would not be used by the clever Germans. The end of this story will be told later in connection with Rejewski's achievements (footnote 16).

14.3.2.5 *The German navy: M4.* The *Kriegsmarine*, as it was renamed in 1935, was always suspicious that its *Funkschlüssel* M3 could be compromised, and introduced on February 1, 1942 for the key net *Triton* a new version of the *Funkschlüssel* M4 with a fourth rotor marked β and therefore called *Griechenwalze*. The extra rotor could be set but was not moved during encryption. The 4-rotor ENIGMA was first used only by the U-boats in the Atlantic. By July 1, 1943, an additional rotor γ came into use. To ensure compatibility of the new 4-rotor ENIGMA with the old 3-rotor ENIGMA, the old reflector B or C was split into a fixed thin reflecting disk called a B-thin or C-thin and the turnable additional rotor β or γ, respectively.[9]

---

[9]Rohwer presumed in 1978 a further rotor α [32: 173]; Deavours and Kruh in 1985 followed him [9: 140]. To this David

VIII. Beiſpiel.

17. Gültiger Tagesſchlüſſel:

(Ausſchnitt aus der für die Verſchlüſſelung des Klartextes
in Betracht kommenden Schlüſſeltafel, z. B. »...........«
Maſchinenſchlüſſel für Monat Mai«)

Datum	Walzenlage	Ringſtellung	Grundſtellung
4.	I III II	16 11 13	01 12 22

Steckerverbindung	Kenngruppen-Einſatzſtelle ...... Gruppe	Kenngruppen
CO DI FR HU JW LS TX	2	adq nuz opw vxz

Figure 14.27. From the *Wehrmacht* ENIGMA operating manual dated June 8, 1937. Setting the ring position and rotor order was a navy officer's prerogative. The basic wheel setting was later expressed by letters, $A = 01$, $B = 02$, etc.

A mixed combination B-thin and γ or C-thin and β occurred occasionally. The Greek rotors β and γ could be set, but could not step during operating the ENIGMA M4.

14.3.3   ENIGMA operating manual

The German *Heer* and *Luftwaffe* used ENIGMAs from the level of regiments upwards, the navy on large ships. An estimated total of 200 000 machines is certainly far too high while 30 000 is a low estimate. A total of 40,000–50,000 may be more accurate.

It has been reported, naturely by inofficial sources, that after the Second World War victorious countries sold captured ENIGMA machines which were still widely thought to be secure, to developing countries. David Kahn, in a review of Winterbothams book, wrote on 29.12.1974 in the New York Times "that the British gathered up all the Enigma machines they could find and sold them to developing countries, confident that they could continue reading the messages of the machines's new owners".

14.3.3.1 *Wehrmacht 1934–1938.* Figure 14.27 shows a section from the ENIGMA operating manual of 1937, with the daily changing rotor order, ring settings, and cross-pluggings. The basic wheel

Kahn [23: 230] "No α rotor was ever recovered". In fact, splitting the reflector A, which had disappeared in 1937, did not make sense.

setting (German *Grundstellung*) characterises the situation of the three rotors when enciphering is started. The *Kenngruppen* had no proper cryptological meaning.

With 6 different rotor orders (*Heer* from 1934), 60 orders (navy from 1934, *Heer* and *Luftwaffe* from 1938), or 336 different rotor orders (Navy from 1939) and with $26^3 = 17,576$ different ring settings and moreover with 1.00×10^{11} *Steckerverbindungen* when using 6 *stecker* and 1.08×10^{13} when using 8 *stecker* and even 1.51×10^{14} *Steckerverbindungen* when using 10 *stecker*, the number of variations of the external key tokens was so big that naive people in the *Wehrmacht* staff had been deeply impressed.

The *Grundstellung* was set out in a key list and was to be used for every single message, but the message setting (also called indicator or text setting and *Spruchschlüssel* in German) was not determined beforehand but rather dealt with by a key negotiation, as discussed below. To use the ENIGMA for this purpose was understandable since the ENIGMA machine was considered invincible; and it was possibly considered particularly clever by the authorities although an enciphering of higher security would have been necessary. But neither *Heer* nor *Luftwaffe* provided for additional measures to protect the enciphered *Spruchschlüssel*; only the navy used a bigram superencipherment on the basis of bigram tables from 1937 (see Section 14.4.1.3).

Until 1938 the following procedure held: With a general daily key, comprising the rotor order, the cross-plugging, the ring setting, and the general basic wheel setting of the three rotors, the sender chose at random a 3-letter group, enciphered it with the general basic wheel setting, and sent this enciphered indicator; subsequently the recipient deciphered it and it served on both sides as the *Spruchschlüssel* or plain indicator for the message that followed.

But to prevent garbling in the case of noisy wireless transmissions, the *Spruchschlüssel* first was doubled, and the resulting 6-letter group was enciphered with the general basic wheel setting, a precaution that had already been recommended by Chiffriermaschinen A.G. for the commercial ENIGMA C.

14.3.3.2 *Wehrmacht 1938–1940.* The hints obtained by the intelligence service that this key negotiation was flawed or was no longer safe precipitated a new procedure, introduced on September 15, 1938. No longer was a general basic wheel setting used for the entire day, but rather every message preamble contained a freely chosen 3-letter group basic wheel setting transmitted in plain,[10] followed by the *Spruchschlüssel* still doubled and enciphered with this basic wheel setting.

To give an example: If a transmission starts with RTJWA HWIK. ..., then /rtj/ is the plain basic wheel setting and WAH WIK is the doubled indicator enciphered with this basic wheel setting. The recipient determines with /rtj/ from WAH WIK the doubled plain indicator (which is to have the pattern *123 123*); the message is deciphered with the first three letters as the indicator.

14.3.3.3 *The end of the indicator doubling.* Unfortunately, the new procedure did not remove the weakness of the old one because the doubling was continued which meant that a dependence among the first and the fourth, the second and the fifth, and the third and the sixth letter, and offered a possibility for a break. Not until May 1, 1940 was

the weakness eliminated.[11] That the indicator doubling had been unnecessary was confirmed when the wireless traffic still ran smoothly. The damage, however, had already occurred and was irreparable. Misgivings that the ENIGMA enciphering was no longer secure arose within the German leadership, in particular that of the navy during the course of the war. "Zweifel an der Sicherheit der Schlüsselmittel [sind] bei der U-Boot-Führung immer wieder einmal aufgetreten, vor allem dann, wenn Geleitzüge offensichtlich die Vorpostenstreifen der U-Boote umgingen oder U-Boote auf Versorgungstreffpunkten überrascht wurden. ...Stets waren – wenn auch manchmal etwas gewaltsam – andere Erklärungen möglich" (Hans Meckel) [32: 130].[12] The warnings were repeatedly diverted but halfhearted steps were taken, such as the introduction of the variable wiring of the plugged reflector or the *Uhr*, and this alteration occurred rather late.

14.3.4 ENIGMA copies and competitors

14.3.4.1 *TYPEX.* Britain also used rotor machines in the Second World War such as TYPEX or Type-X developed 1934 by Wing Commander O.G.W. Lywood and three other members of the Royal Air Force (RAF) as a private venture and which was an improved version of the commercial ENIGMA. After mid-1941 some TYPEX models had a plugged reflector, and instead of the plugboard there was an entrance substitution performed by two fixed rotors which wisely were not self-reciprocal (see Section 14.2.3.1) but the disastrous reflector was not abolished.

Figure 14.28 shows the TYPEX, ready in April 1935 and improved by May 1938 but not yet commercially available. The TYPEX in some respects was similar to the 3-rotor ENIGMA since among its five rotors the two next to the entry were

---

[10]The daily changing of the *Ringstellung* made this transmission as secure or as insecure as the previous use of the dailychanging general basic wheel setting.

[11]The enciphering of the doubled indicator was continued for the 4-rotor *Abwehr* ENIGMA which was a version of the ENIGMA G with rewired rotors until January 1, 1944 when some improvement was made.

[12]"Doubts concerning the security of encryption techniques occurred time and again with the submarine control in particular when convoys avoided the spearheads of submarine packs or if submarines were taken by surprise at refueling points ...Alternative explanations although sometimes contrived were always possible.".

Figure 14.28. TYPEX Mark II.

settable, but wisely did not move during operation. In this respect, TYPEX was cryptanalytically equivalent to a 3-rotor ENIGMA with a non-self-reciprocal plugboard. The light-bulb output of the ENIGMA was replaced by a Creed tape printer. Essential differences existed in the rotor movement: It was regular, too, but basically multi-notched whereas the ENIGMA had only the rotors VI, VII and VIII equipped with two notches. The notched rim was rigidly fixed to the rotor rim as in the commercial ENIGMA and the rotor core was a wiring slug sitting in a receptacle carrying the rotor rim and the alphabet. In some later models, the wiring slugs could be inserted in two orientations, P or inversely P^{-1}. In a typical version, five slugs could be selected out of ten. There were rims with five, seven, and nine notches; in the last case the notches were arranged so that a turnover occurred when in the window one of the letters B G J M O R T V X was shown. All rotors when used together had identical notching patterns. TYPEXs were still being used by the British until at least 1956 and have been supplied to NATO and some Commonwealth countries.

14.3.4.2 *NEMA.* From 1947, the Swiss army and diplomatic service used an ENIGMA variant NEMA *Neue Maschine* Modell 45, developed by the mathematicians Hugo Hadwiger (1908–1981), Heinrich Weber (1908–1997) and more recently the army cryptologist Paul Glur, built by Zellweger A.G., Uster (Fig. 14.29). It had ten rotors: four (out of six) enciphering ones and a reflector and the other ones served for rotor movement only. The treacherous use of a reflector was kept [35]. The NEMA replaced the commercial ENIGMA K.

14.3.4.3 *SIGABA.* In the early 1930s the USA, under the early influence of William Friedman (1891–1969) and on the basis of the Hebern development, there was a more independent line of rotor machines leading in 1933 to the M-134-T2 and then to the M-134-A (SIGMYC). Since 1933, under pressure from the navy, the Electric Cipher Machine ECM Mark I was designed which is a one-rotor-machine with rotor movement controlled by a tape and which finally fulfilled the highest requirements.

But Friedman's colleague Frank Rowlett (1908–1998) succeeded in inventing further improvements, leading in 1936 to the ECM Mark II, often simply called ECM, with the US army also called the M-134-C and the SIGABA, with the Navy CSP 889 (Fig. 14.30). The SIGABA had 15 rotors; apart from 5 turning cipher rotors, 5 rotors for irregular movement were sitting in a basket, another five were equivalent to a plugboard. With these additional rotors and without using a reflector, it proved in the 1940s to be the best rotor machine in the western world. It never fell into the enemy's hands and it was the most expensive machine. The system was in use until 1959. The Germans obviously did not succeed in breaking the SIGABA.

A machine with 3 rotors and a reflector, based on the ENIGMA, the Converter M-325 and dubbed SIGFOY, was designed by Friedman and built after 1944. Because of some practical drawbacks, it was not generally introduced. Friedman had applied for a patent on August 11, 1944; he received it on March 17, 1959 under No. 2,877,565. The

Figure 14.29. Swiss rotor machine NEMA Model 45 (1947).

Figure 14.30. Rotor machine ECM Mark II (M-134-C SIGABA, CSP 889).

Combined Cipher Machine (CCM) also named CSP-1700 used two adaptors for a connection between SIGABA and TYPEX. It turned out to be insecure.

"The SIGABA was far better than the ENIGMA" (David Kahn) [21: 28].

14.3.4.4 *KL-7.* The rotor machine KL-7 of the North Atlantic Treaty Organization (NATO) was based on US experiences. The NATO used the KL-7 machine from the late 1940s until the early 1980s, although by 1963 it was operationally obsolete. The KL-7 (Fig. 14.31) had seven cipher rotors; it was in its mechanical aspects faintly similar to the British TYPEX by using interchangeable coding cylinders and rings. Plastic slip rings which controlled the irregular rotor movement could be permuted among the coding cylinders. Each rotor had 36 contacts, providing for letters and digits. The KL-7 was one of the last genuine rotor machines ever produced.

The security of these machines had to be very good since they were available even to some non-NATO countries and were sure to fall into the other side's hands. This openness shows that cryptologists in the 1960s had fully accepted Kerckhoffs' maxim that a cryptosystem must be safe even if the device is in the hands of enemies. On the other hand, only superpowers can afford the effort necessary for the cryptanalysis of good encryptions which gave the US, with the support of NSA, a marked lead. Indeed, in 1962 the US Officer Joseph G. Helmich sold to the Sovjets technical information about rotors and key lists (see Section 14.3.4.8); he was arrested in 1982 by the FBI. Use of the KL-7 ended at the latest by 1985 after the Walker espionage case but by that time it was outdated anyway.

14.3.4.5 *OMI.* An interesting postwar variant of the ENIGMA with seven rotors and a fixed reflector was the OMI Cryptograph-CR, built and marketed by the Italian company *Ottico Meccanica Italiana* in Rome (Fig. 14.32). The rotors could be assembled from a receptacle that contained notches, and a pair of rotor cores; in this way, one could speak of 14 rotors because stepping was coupled in pairs or of 7 rotors with a choice from $\binom{14}{2}$ which equals 91 rotors.

14.3.4.6 *FIALKA.* The Russian ENIGMA counterpart M-125, a 10-rotor machine with a fixed reflector, was named фиалка Fialka, violet. Alternatingly, five rotors moved in one and five in the other direction. The plugboard substitution was controlled by a punched card and could be varied easily and quickly. Figure 14.33 shows a version with a Russian–German keyboard of 30 keys used by the *Nationale Volksarmee* of the former German Democratic Republic. For Fialka 30 cyrillic letters are placed along the rims of the rotors in the order

А Б В Г Д Е Ж З И К Л М Н О П Р С Т У
Ф Х Ц Ч Ш Щ Ы Ь Ю Я Й

Figure 14.31. Rotor machine KL-7 (cover name ADONIS).

Figure 14.32. Rotor machine of the Ottica Meccanica Italiana.

The rotor movement is controlled by seven pegs corresponding to the notches of a multi-notched ENIGMA, forming patterns such as

○ ○ ○ ● ○ ○ ○ ○ ○ ● ● ● ○ ○ ○ ○ ○ ○ ○ ○ ● ○ ○ ● ○ ○ ○ ○ ●

Ten rotors can be selected out of twenty, and the rotor wiring can be easily changed.

14.3.4.7 *Green, red and purple: the Japanese way.* Following a familiar pattern, the Japanese studied the machines of other countries, in particular those that were accessible through the patent literature: the ENIGMA, the machines of Damm and Hagelin, and those of the US. For machines with the Latin alphabet, the common Hepburn transliteration of *kana* into the Latin alphabet was used.

The Japanese imitation of the ENIGMA D, denoted GREEN by American cryptanalysts, was a strange construction with four vertically mounted rotors (Fig. 14.34) and did not achieve great importance.

Next, the half-rotors of Damm showed up in the *Angōki Taipu A* (Cipher Machine A), called RED in American jargon. Apart from a fixed permutation by a plugboard, it had a half-rotor with 26 slip

rings and 60 contacts (Fig. 14.35). The wiring permuted the 6 vowels onto them and therefore also the 20 consonants. Twice 60 exit contacts were used; note that 60 is the least common multiple of 6 and 20. The reason for this rather disadvantageous separation into 20 plus 6 may have been in the tariff regulations of the international telegraph union, requiring pronounceable words [33: 117].

The Japanese RED machine was a very poor cryptosystem, not much better than Kryha's machine. Its predecessor M-2, used since mid-1933, had already two half-rotors, one with 6 and one with 20 slip-rings, and followed Damm's patent 1,502,376 of July 22, 1924. Lawrence Safford claimed that it was completely reconstructed in 1936 by Agnes Meyer Driscoll, the former Miss Agnes Meyer, Navy Principal civilian cryptanalyst 1924–1959. In the RED machine, rotor movement was accomplished by a gear with 47 positions, with 4, 5 or 6 gaps. Cryptologically it performed two separate ALBERTI encryptions of the vowel and the consonant group. It is not at all surprising that the RED machine (Fig. 14.36) was attacked in 1935 by Kullback and Rowlett of the US Army and solved in 1936 [9: 218].

Figure 14.33. Soviet rotor machine Fialka (courtesy R.F. Staritz).

Figure 14.34. Japanese ENIGMA imitation, GREEN machine.

Figure 14.35. Half-rotor with 26 slip-rings in the Japanese machine *Angōki Taipu A*.

In the spring of 1936, Werner Kunze at *Pers Z* of the German *Auswärtiges Amt* directed his interest to M-1 called ORANGE in American jargon, working with the *kana* alphabet. Jack S. Holtwick from the US navy had a similar goal. They both succeeded, Kunze finally with RED by August 1938 [20: 437]. In 1934, RED was also broken by the British, presumably by Hugh Foss and Oliver Strachey at Government Communications Headquarters [12: 131].

In 1937, Japan started the development of a much more secure encryption machine. It replaced the RED machine in the diplomatic service and was put in operation in February 1939. The first unreadable messages went in March 1939 from Warsaw to Tokyo. The *Angōki Taipu B* (Cipher Machine B, also 97-*shiki obun injiki*, alphabetic typewriter '97), called PURPLE in American jargon, included a new feature and was used for the first time by the Japanese, namely, stepping switches or uniselectors, known from telephone exchanges. The separation into two groups of 6 and 20 characters was kept although later the six characters no longer had to be vowels. It turned out that the number of available alphabets was cut to 25, and the mapping was quite irregular and determined by the internal wiring. To find this fact needed the concentrated, months-long work of a whole group of people, not only William Friedman [15] and Frank Rowlett but also Robert O. Ferner, Albert W.

Figure 14.36. American reconstruction RED of the *Angōki Taipu A* with two half-rotors.

Small, Samuel Snyder, Genevieve Feinstein (née Grotjan) and Mary Jo Dunning. They first found the mapping of the 6-vowel group, and they had indications that the number of alphabets was 25. But as for the 20-consonant group, they were in the dark, and no one could identify a known electromechanical encryption step that would produce the observed effects. When the situation seemed almost hopeless by midsummer 1940, a newly arrived recruit from MIT, Leo Rosen [9: 238], was initiated, and he hit upon the idea that the Japanese may have used stepping switches (Fig. 14.37). That discovery gave the work a fresh incentive, and the mystery was soon solved. There were three banks of stepping switches, and the wiring connections could be established. On September 20, 1940, an important discovery was made by Genevieve Grotjan who was looking for letter intervals [22: 284] that would reveal the advancement of PURPLE's stepping switches. Only a week later and after 18 months of work, the first complete PURPLE solution was achieved. A working reconstruction of the machine was built; on February 7, 1941, the British in Bletchley Park were given a copy. The RED machine had paved the way, and many weaknesses in the encryption discipline of the Japanese gave clues, hints, and cribs, but it was a victory

Figure 14.37. Stepping switch bank of the Japanese PURPLE machine.

of the US army cryptanalytic bureau "that has not been duplicated elsewhere...the British cryptanalytic service and the German cryptanalytic service were baffled in their attempts" (Friedman).

Strategically, the PURPLE break was of highest importance; the Americans spoke of MAGIC. But the British had their victory, too, over the German ENIGMA; they called it ULTRA. However, David Kahn reported that the Russians, Jürgen Rohwer and Otto Leiberich that the Germans also solved PURPLE which had a theoretical security significantly greater (Stephen J. Kelley [24: 178]) than the 3-rotor ENIGMA and comparable to the 4-rotor ENIGMA.

Once the internal wiring of the PURPLE machine was understood, it proved to possess only mediocre security comparable to RED. It seems that the Japanese underrated the cleverness of the Americans; also they believed that their language would protect them and would not be understood fully elsewhere. The follow-up machines they built used stepping switches too and were only slightly more complicated: one, called CORAL in the American jargon, gave up the separation into $20 + 6$ and it was finally broken by OP-20-GY, the cryptanalysis department of the cryptological branch of the navy OP-20-G with the help of Conel Hugh O'Donel Alexander from GC&CS in March 1944 [12: 148]. Another one, called JADE, was unique because it printed in *kana* symbols. Otherwise it had only minor added complications and was broken in due course [9: 213].

The Japanese also had a transparent system in their daily plugboard arrangements and the bad habit of sending changes to the keying as encrypted messages thus keeping the foe up to date once encryption was broken. Even the key to the keys that is the rule for constructing keys, was discovered following the Jade break in 1944 by Lieutenant-Commander Frank Raven, US Navy [9: 249, 12: 149].

14.3.4.8 *KW7, KW26.* KW-7 was a one-time key on-line rotor machine developed in the USA for multinational communications, SIGABA was considered too good to be shared. A later model KW-26 was presumably the very target of Helmich's endeavour.

14.4 POLISH CRYPTANALYSIS OF THE *Wehrmacht* ENIGMA

14.4.1 The Polish original break

As Władysław Kozaczuk disclosed[13] in 1967, a typical case of cryptotext–cryptotext compromise allowed that from 1932 onward the Polish *Biuro Szyfrów* under Major Gwido Langer ('Luc') with its cryptanalytic service B.S.-4 under Maksymilian Ciężki (1899–1951) and the young mathematicians Marian Rejewski, Jerzy Różycki, Henryk Zygalski penetrated the encryption of the German *Wehrmacht* whose practise of transmission of ENIGMA-encrypted radio signals in the Eastern provinces of Prussia gave a copious supply of cryptotexts.

It was a typical problem of machine encryption with a key sequence generated by the machine itself. If every message was started with its own initial setting, immediate superimposition was then inhibited. But it was widely considered too difficult and prone to mistakes to prearrange such new initial settings for every message. This complication did not hold for ENIGMA only but is a general problem for key negotiation and key administration. Therefore, indicators are used for transmitting the message setting which of course, should not give the setting plainly but encrypted.

Thus, it was thought to be clever to use the encryption machine itself for this purpose. One specific weakness with both the army and air force ENIGMAs was that the encryption of the indicator was based *exclusively* on the ENIGMA itself. However, for the 4-rotor navy ENIGMA at least a bigram superencryption was done starting May 1, 1937.

---

[13]In his book *Bitwa o tajemnice: Służby wywiadowcze Polski i Rzeszy Niemieckiej 1922–1939*, Warsaw 1967, which was largely overlooked in the West although a review was published in a Göttingen journal in 1967. However, in a foreword to *Breach of Security* by David Irving, London 1968, in 1968, Donald Cameron Watt revealed that in 1939 Britain "received from Polish Military Intelligence keys and machines for decoding German official military and diplomatic ciphers". This seemed unbelievable, until in 1973 the French General Gustave Bertrand, who had been involved in the deal, confirmed it. A second book *W kręgu Enigmy* was published in 1979, and an English-language translation by Christopher Kasparek appeared in 1984 [27].

Another weakness was that, as described below, the plain indicator was doubled before encryption, the old treacherous trick, thus introducing a tiny cryptotext–cryptotext compromise at a fixed place at the very beginning of the message. This compromise was the fault not of the machine but of the rules for operating it. Bletchley Park called the indicator doubling system boxing or throw-on.

The fancy idea of doubling the indicator seems to go back to recommendations of Scherbius' Chiffriermaschinen A.G. for the commercial ENIGMA of 1924. It is the doubling of the indicator in conjunction with the self-reciprocal encryption that the ENIGMA performed, and some help from outside gave the entrance to the break. Until 1940 the Germans did not think of such a possibility.

14.4.1.1 *Indicator doubling helped Rejewski.* Weakening the security of the cryptotext proper by using an indicator was accepted without scruples by the Germans since the indicator was thought to be well protected by the ENIGMA which was judged to be *indéchiffrable*. No-one observed the vicious circle. Anyhow, the ENIGMA seemed to have enough combinatory complexity.

The reason for doubling the plain indicator that is the double encipherment of each message setting[14] was that radio signals were frequently disturbed by noise. An encrypted indicator corrupted during transmission would cause nonsense in the authorised decryption and increase risks if repetition had to occur. To transmit the whole cryptomessage twice for the sake of error-detection was considered out of the question. Thus, the error-detecting possibility was restricted to the plain indicator. But this restriction led to a much more dangerous compromise which could have been avoided. The doubling was not necessary at all because when it was discontinued on May 1, 1940 ENIGMA traffic did run smoothly.

The Polish bureau found out about the indicator around 1930. It became clear to them that two signals which started with the same 6-letter group showed a higher character coincidence near κ_d,

the index of coincidence for the German language, thus the two signals invited superimposition. As a consequence, the initial setting of the rotors was determined by the first 6-letter group, it was an indicator. Since it was clear that it was not plain, it could only be encrypted. Next the Polish bureau had to discover how was this done?

14.4.1.2 *Help from France.* It is not known whether the Poles had reasons to believe the Germans would be careless and use the ENIGMA itself for the indicator encryption. In any case in 1931 the Poles obtained help from their French friends. The spy Hans-Thilo Schmidt (1888–1943) with code name *ASCHE* (the French pronunciation of *H*) was working until 1938 in the *Chiffrier-Stelle* of the *Reichswehrministerium*. Since October 1931 he had forwarded via the secret agent 'Rex' (Rodolphe Lemoine, alias Rudolf Stallmann) some manuals on the use of the ENIGMA on the encryption procedure, and even *Tagesschlüssel* for September and October 1932 including the ring settings and cross-pluggings for these two months to the French *grand chef*, then Major, later *Général* Gustave Bertrand (1896–1976) with codenames Bolek and Godefroy who forwarded them in turn to the Polish bureau. Hans-Thilo was arrested on March 23, 1943 and allegedly shot in July 1943; actually he committed suicide on September 19, 1943.[15] Ciężki's young aide, the highly gifted[16] Marian Rejewski (1905–1980) who he knew since 1929 and whom he had hired permanently in 1932

---

[14]Indicator doubling, as recommended for the commercial ENIGMAs, was not questioned for the military use.

[15]His brother, *Generaloberst* Rudolf Schmidt, commander of the 2. *Panzerarmee*, was dismissed on April 1, 1943.

[16]Marian Rejewski's intuitive abilities are illustrated by the following episode. In the commercial ENIGMA D, the contacts on the entry ring belonged to letters in the order QWERTZU . . . of the letters on the keyboard which seemed to be different with the military ENIGMA I. It was vital for a Polish success to find the right mapping. Around New Year 1933 Rejewski argued that the Germans rely upon order and tried the alphabetic order (see Section 14.3.2.3) and that was correct. Knox had long racked his brain about this question, was told the solution in July 1939 by Rejewski. Penelope Fitzgerald, Knox's niece reported that Knox was furious when he learned how simple it was [13]. Later, he chanted "*Nous avons le QWERTZU, nous marchons ensemble*" (Peter Twinn [18: 127]). The story proves that in 1939 the British were behind the Polish standard.

1.	AUQ	AMN	9.	HNO	THD	17.	NXD	QTU	25.	SJM	SPO
2.	BNH	CHL	10.	HXV	TTI	18.	NLU	QFZ	26.	SUG	SMF
3.	BCT	CGJ	11.	IKG	JKF	19.	OBU	DLZ	27.	TMN	EBY
4.	CIK	BZT	12.	IND	JHU	20.	PVJ	FEG	28.	TAA	EXB
5.	DDB	VDV	13.	JWF	MIC	21.	QGA	LYB	29.	USE	NWH
6.	EJP	IPS	14.	KHB	XJV	22.	RJL	WPX	30.	VII	PZK
7.	FBR	KLE	15.	LDR	HDE	23.	RFC	WQQ	31.	VQZ	PVR
8.	GPB	ZSV	16.	MAW	UXP	24.	SYX	SCW	32.	WTM	RAO

33.	WKI	RKK
34.	XRS	GNM
35.	XOI	GUK
36.	XYW	GCP
37.	YPC	OSQ
38.	ZZY	YRA
39.	ZEF	YOC
40.	ZSJ	YWG

Figure 14.38. Different observed encrypted doubled indicators to the same *Tagesschlüssel*.

had first to find out how the French gift could be made useful. According to an appendix by Tadeusz Lisicki in a book by Garliński [16: 192], he proceeded as follows:

He had already guessed or learned from the ASCHE material that each signal began with the 6 letters of the encrypted doubled indicator.[17]

Let $A_1, A_2, A_3, A_4, A_5, A_6$ (Rejewski used the letters A, B, C, D, E, F) denote the permutations performed upon the 1st, 2nd, 3rd, ..., 6th plaintext letter, starting with some basic wheel setting.

The properly self-reciprocal character of the ENIGMA was known. Therefore, from $aA_i = X$ and $aA_{i+3} = Y$ ($i = 1, 2, 3$) it follows that $XA_iA_{i+3} = Y$. The known characters X, Y standing in the 1st and 4th, or the 2nd and 5th, or the 3rd and 6th positions of the cryptotext thus impose conditions on the three products A_iA_{i+3} of the unknown reflections $A_1, A_2, A_3, A_4, A_5, A_6$.

Figure 14.38 presents 40 different out of 65 encrypted doubled indicators for one and the same day presumably in 1933 which shows that for A_1A_4

the character a goes into itself (1.);

likewise,

the character s goes into itself (24.),

while

the character b goes into c and vice versa (2., 4.)

and

the character r goes into w and vice versa (22., 32.).

For the other characters it turns out that they belong under A_1A_4 to the cycles

(d v p f k x g z y o)

(5., 30., 20., 7., 14., 34., 8., 38., 37., 19.)

and

(e i j m u n q l h t)

(6., 11., 13., 16., 29., 17., 21., 15., 9., 27.).

In short, A_1A_4 has two cycles of ten characters, two 2-cycles (b c), (r w) of length two and two 1-cycles (a), (s) of length one, also called fixpoints; and since this embraces all 26 characters, A_1A_4 is fully determined. The cycle determination is complete if every character occurs at least one time in the first, the second, and the third position; as a rule this determination requires fifty to a hundred messages and this much was certainly the result of a busy manoeuvre day.

For A_2A_5 and A_3A_6 the work is similar, altogether there is the lucky result

$$A_1A_4 = \text{(a) (s) (b c) (r w)}$$
$$\text{(d v p f k x g z y o) (e i j m u n q l h t)}$$
$$A_2A_5 = \text{(d) (k) (a x t) (c g y)}$$
$$\text{(b l f q v e o u m) (h j p s w i z r n)}$$
$$A_3A_6 = \text{(a b v i k t j g f c q n y)}$$
$$\text{(d u z r e h l x w p s m o)}.$$

---

[17]Rejewski observed that signals starting with the same first letters always showed identical fourth letters, likewise, for the second and the fifth letters and for the third and the sixth (see Fig. 14.38). This repetition was a clear hint at indicator doubling.

The 1-cycles or females[18] play a particular role. Since each one of the permutations A_1, A_2, A_3, A_4, A_5, A_6 because of its self-reciprocal character consists of 2-cycles or swappings only, $A_i A_{i+3} x = x$ implies that there exists a character y such that $A_i x = y$ and $A_{i+3} y = x$ that is both A_i and A_{i+3} contain the 2-cycle $(x\ y)$. In the given example, both A_1 and A_4 contain the 2-cycle (a s).

A theorem of group theory about products of properly self-reciprocal permutations states that the cycles of $A_i A_{i+3}$ occur in pairs of equal length: if

A_i contains the 2-cycles

$$(x_1 y_1), (x_2 y_2), \ldots, (x_\mu y_\mu),$$

and

A_{i+3} contains the 2-cycles

$$(y_1 x_2), (y_2 x_3), \ldots, (y_\mu x_1),$$

then

$A_i A_{i+3}$ contains the μ-cycles

$$(x_1 x_2 \ldots x_\mu), (y_\mu y_{\mu-1} \ldots y_1)$$

and vice versa.

Thus, if one of the cycles of $A_i A_{i+3}$ is written in reversed order ($\leftarrow$) below the other one, then the 2-cycles of A_i can be read vertically, provided that the cycles are in phase. To find this phase is the problem, and it could be solved by exhaustion, for $A_1 A_4$ above in $2 \cdot 10$ trials and for $A_2 A_5$ above in $3 \cdot 9$ trials.

14.4.1.3 *Rejewski's shortcut.* However, Marian Rejewski found a shortcut. By observing that the encrypted indicators used actually showed deviations from equal distribution, which probably meant that the German crypto clerks, like most people playing in the lottery, were unable to choose

the message settings truly at random. Thus, Rejewski directed his interest primarily to conspicuous patterns, and he was right to do so. In fact, the German security regulations were not too clear on this point, and a German officer who had given the order to take as message setting the end position of the rotors in the previous message could argue that he had made sure the message setting was changed after every message.

Thus, it was common practise to use even stereotyped 3-letter groups like /aaa/, /bbb/, /sss/. When in the spring of 1933 the mere repetition of letters was explicitly forbidden, it was already too late. The Poles had made their entry into the obscurity of the ENIGMA. Later, the bad habit developed of using horizontally or vertically adjacent letters on the keyboard: /qwe/, /asd/ (horizontally); /qay/, /cde/ (vertically), and so on.

Rejewski's frequency argument was that the most frequently occurring encrypted indicator 24. SYX SCW, which had occurred five times, should correspond to a most conspicuous pattern. There were still a number of those to be tested. Assume we test with the plain indicator aaa. This test indicator fits in A_1 with the 2-cycle (a s), in A_2 it gives the 2-cycle (a y), in A_3 the 2-cycle (a x); it fits in A_4 with the 2-cycle (a s), in A_5 it gives the 2-cycle (a c), in A_6 it gives the 2-cycle (a w). Thus, for A_3 and A_6 the phase of the two cycles of $A_3 A_6$

$$\rightarrow (\overset{\downarrow}{a}\ b\ v\ i\ k\ t\ j\ g\ f\ c\ q\ n\ y)$$

$$\leftarrow (x\ l\ h\ e\ r\ z\ u\ d\ o\ m\ s\ p\ w)$$

is already determined. In a zig-zag the 2-cycles of A_3 and A_6 and beginning with (a x) it can be calculated that:

$$A_3 = (a\ x)\ (b\ l)\ (v\ h)\ (i\ e)\ (k\ r)\ (t\ z)\ (j\ u)$$
$$(g\ d)\ (f\ o)\ (c\ m)\ (q\ s)\ (n\ p)\ (y\ w)$$
$$A_6 = (x\ b)\ (l\ v)\ (h\ i)\ (e\ k)\ (r\ t)\ (z\ j)\ (u\ g)$$
$$(d\ f)\ (o\ c)\ (m\ q)\ (s\ n)\ (p\ y)\ (w\ a).$$

A_3 contains among others the 2-cycle (q s). Thus, the plain indicator to 1. AUQ AMN has the pattern $**s$ and since A_1 contains among others the 2-cycle (a s), it has the pattern $s*s$. A guess is made

---

[18] In the jargon of Bletchley Park the term "female" was used, originating from the Polish pun *te same* ("the same") $\leftrightarrow$ *samiczka* ("female"). Most people in Bletchley Park did not know and in fact did not need to know the Polish origin and found their own explanations such as *female*: a screw for a threaded hole.

sss:	AUQ	AMN	ddd:	IKG	JKF	xxx:	QGA	LYB	ert:	VQZ	PVR
rfv:	BNH	CHL	dfg:	IND	JHU	bbb:	RJL	WPX	ccc:	WTM	RAO
rtz:	BCT	CGJ	ooo:	JWF	MIC	bnm:	RFC	WQQ	cde:	WKI	RKK
wer:	CIK	BZT	lll:	KHB	XJV	aaa:	SYX	SCW	qqq:	XRS	GNM
ikl:	DDB	VDV	kkk:	LDR	HDE	abc:	SJM	SPO	qwe:	XOI	GUK
vbn:	EJP	IPS	yyy:	MAW	UXP	asd:	SUG	SMF	qay:	XYW	GCP
hjk:	NXD	FBR	ggg:	KLE	QTU	ppp:	TMN	EBY	mmm:	YPC	OSQ
nml:	GPB	ZSV	ghj:	NLU	QFZ	pyx:	TAA	EXB	uvw:	ZZY	YRA
fff:	HNO	THD	jjj:	OBU	DLZ	zui:	USE	NWH	uio:	ZEF	YOC
fgh:	HXV	TTI	tzu:	PVJ	FEG	eee:	VII	PZK	uuu:	ZSJ	YWG

Figure 14.39. 40 different indicators decrypted.

that the plain indicator to AUQ AMN reads sss, then in A_2 and apart from (a y) also (s u) can be determined. Thus, the phase for the cycles of A_2 and A_5 is also fixed.

$$\rightarrow (\overset{\downarrow}{a} \, x \, t) \, (b \, l \, f \, q \, v \, e \, o \, \overset{\downarrow}{u} \, m) \, (d)$$

$$\leftarrow (y \, g \, c) \, (j \, n \, h \, r \, z \, i \, w \, s \, p) \, (k).$$

In a zig-zag the 2-cycles of A_2 and A_5, beginning with (a y), can be calculated:

$$A_2 = (a \, y) \, (x \, g) \, (t \, c) \, (b \, j) \, (l \, n) \, (f \, h) \, (q \, r) \, (v \, z)$$
$$(e \, i) \, (o \, w) \, (u \, s) \, (m \, p) \, (d \, k)$$
$$A_5 = (y \, x) \, (g \, t) \, (c \, a) \, (j \, l) \, (n \, f) \, (h \, q) \, (r \, v) \, (z \, e)$$
$$(i \, o) \, (w \, u) \, (s \, m) \, (p \, b) \, (k \, d).$$

Another frequently occurring encrypted indicator was 22. RJL WPX, which occurred four times. The corresponding plain indicator has the pattern *bb. A_1 can only have the 2-cycles (r b) or (r c). In the first case with the more likely plain indicator bbb, A_1 contains the 2-cycle (b r) and A_4 the 2-cycle (r c). For a pairing of the 10-cycles, another encrypted indicator may be used, say 15. LDR HDE. Since A_3 and A_6 contain (r k) and (k e), A_2 and A_5 contain (d k) and (k d), and the pattern of the plain indicator is *kk. This observation suggests again the stereotype kkk with the result that A_1 and A_4 contain the 2-cycles (l k) and (k h). Thus, the phase for the cycles of A_1 and A_4 is also completely fixed.

$$\rightarrow (a) \, (\overset{\downarrow}{b} \, c) \, (d \, v \, p \, f \, \overset{\downarrow}{k} \, x \, g \, z \, y \, o)$$

$$\leftarrow (s) \, (r \, w) \, (i \, e \, t \, h \, l \, q \, n \, u \, m \, j)$$

and thus

$$A_1 = (a \, s) \, (b \, r) \, (c \, w) \, (d \, i) \, (v \, e) \, (p \, t) \, (f \, h)$$
$$(k \, l) \, (x \, q) \, (g \, n) \, (z \, u) \, (y \, m) \, (o \, j)$$
$$A_4 = (s \, a) \, (r \, c) \, (w \, b) \, (i \, v) \, (e \, p) \, (t \, f) \, (h \, k)$$
$$(l \, x) \, (q \, g) \, (n \, z) \, (u \, y) \, (m \, o) \, (j \, d).$$

Altogether the first six permutations read in ordered form:

$$A_1 = (a \, s) \, (b \, r) \, (c \, w) \, (d \, i) \, (e \, v) \, (f \, h) \, (g \, n)$$
$$(j \, o) \, (k \, l) \, (m \, y) \, (p \, t) \, (q \, x) \, (u \, z)$$
$$A_2 = (a \, y) \, (b \, j) \, (c \, t) \, (d \, k) \, (e \, i) \, (f \, h) \, (g \, x)$$
$$(l \, n) \, (m \, p) \, (o \, w) \, (q \, r) \, (s \, u) \, (v \, z)$$
$$A_3 = (a \, x) \, (b \, l) \, (c \, m) \, (d \, g) \, (e \, i) \, (f \, o) \, (h \, v)$$
$$(j \, u) \, (k \, r) \, (n \, p) \, (q \, s) \, (t \, z) \, (w \, y)$$
$$A_4 = (a \, s) \, (b \, w) \, (c \, r) \, (d \, j) \, (e \, p) \, (f \, t) \, (g \, q)$$
$$(h \, k) \, (i \, v) \, (l \, x) \, (m \, o) \, (n \, z) \, (u \, y)$$
$$A_5 = (a \, c) \, (b \, p) \, (d \, k) \, (e \, z) \, (f \, h) \, (g \, t) \, (i \, o)$$
$$(j \, l) \, (m \, s) \, (n \, q) \, (r \, v) \, (u \, w) \, (x \, y)$$
$$A_6 = (a \, w) \, (b \, x) \, (c \, o) \, (d \, f) \, (e \, k) \, (g \, u) \, (h \, i)$$
$$(j \, z) \, (l \, v) \, (m \, q) \, (n \, s) \, (p \, y) \, (r \, t).$$

The reconstruction of all plain indicators used on this busy manoeuvre day is now possible (Fig. 14.39), as reported by Tadeusz Lisicki (1910–1991). The bad habits of the German ENIGMA operators are evident. First, the use of stereotypes has

Figure 14.40. Keyboard of the Service ENIGMA.

led to multiple use of identical indicators, something that should not happen. Second, a look on the keyboard of the ENIGMA (Fig. 14.40) is disconcerting since only two out of forty triples, namely abc and uvw, are not keyboard stereotypes; instead, they are alphabet stereotypes. Neither the crypto clerks nor their signal officers would have dreamed that peaceful practise transmissions with an innocently invented combat scenario would give away so much of the secret of the ENIGMA that is the message setting (*Spruchschlüssel*).[19]

14.4.1.4 *Finding the rotor wiring.* The Polish Biuro Szyfrów certainly learned something from the content of the decrypted signals. But more importantly the compromise had endangered the wiring of the rotors. Since the indicator analysis involved only the first six letters, it was mostly the core of the fast rotor R_N that is the rightmost one which was moved, and the two other ENIGMA rotors remained in 20 out of 26 cases at rest. This fact together with the material from ASCHE, was sufficient for Rejewski in December 1932 to reconstruct the wiring of the fast rotor core, and since the rotor order at that time was changed every quarter and from 1936 every month, only later every day each rotor came finally under examination by the Polish Bureau. Once all the plain indicators of a day were decrypted, all the signals of that day could be decrypted with the help of Polish ENIGMA replicas, the first of which were ready by 1933. By mid-1934, over a dozen were built in the factory AVA in Stepinska Street, Warsaw. Nonetheless, by the next day if the sequence of the drums could be changed or the number of plug connections could be increased.

---

[19]Dilly Knox also found, as indicated by Patrick Mahon [29: 12], the method which was called by Turing [36: 24] a Saga he was told when he came to Bletchley Park, but Knox was unable to apply it by not knowing (see footnote 16) the correct QWERTZU which also prevented him from solving Enigma's rotor wiring.

14.4.1.5 *A catalogue for the cycles.* Reconstruction of the basic wheel setting (German *Grundstellung*) was accomplished with the help of a theorem from group theory which was highlighted by Deavours as the theorem that won World War II. It says that:

> if A and ρ are permutations, then the permutation $\rho^{-1}A\rho$ has the same cycle decomposition (or characteristic) as the permutation A.

Therefore, Marian Rejewski and his co-workers made use of the fact that the cycle lengths in the three observable $A_i A_{i+3}$ are independent of the choice of the cross-plugging and of the ring setting. The number of essentially different cycle arrangements is the number of partitions of $26/2 = 13$, which is 101. Three such partitions in general uniquely characterise the $6 \cdot 26 \cdot 26 \cdot 26 \approx 10^5$ basic wheel settings. In the example above the partitions or characteristics are $10 + 2 + 1$, $9 + 3 + 1$, and 13. Rejewski, supported by Różycki and Zygalski, was now able to produce in over one year with the help of the ENIGMA replica a catalogue for every rotor order containing the partitions of the cycles for all basic wheel settings. For this purpose, an electromechanical device called the 'cyclometer' was built by AVA. The *Biuro Szyfrów* finished the catalogue in 1937; to find the *Tagesschlüssel* then took no longer than ten to twenty minutes. Unfortunately for the Poles, on November 1, 1937 the Germans changed the reflecting rotor. To prepare the catalogue with the new reflector took "somewhat less than a year's time" (Rejewski).

14.4.1.6 *Finding the right ring setting: /anx/.* There remained the problem of finding the correct ring setting on the rotor core. The exhaustive treatment could be simplified by an observation Rejewski had made in 1932, due to the material of ASCHE that is: most plaintexts started with /anx/, where /x/ was the common replacement for word space and /an/ means 'to'. According to Kerckhoffs' admonition, it had to be expected that the machine was in the wrong hands, so it was unproductive to use a stereotyped beginning. This trivial case of a plaintext–cryptotext compromise (Section 14.3.1.2) will be covered in Section 14.4.1.8.

14.4.1.7 *Grill method.* In earlier times, Rejewski temporarily used a method he called the grill method (*metoda rosztu*), which was manual and tedious as Rejewski had commented. It was usable only as long as the number of cross-pluggings was small. It was six up to October 1, 1936. The grill method served to determine the ring setting of the fast rotor. It was abandoned after Zygalski sheets and *bomby* were available. The procedure used the matrix of the 26 alphabets generated by the fast rotor (see for example Fig. 14.21) and the first six permutations $A_1, A_2, A_3, A_4, A_5, A_6$ reconstructed from the enciphered doubled indicator. Some details of the procedure have been published in 1980 by Tadeusz Lisicki as an appendix in Józef Garlińskis book [16: 192].

14.4.1.8 *The 4-rotor Navy Enigma.* Introduction of additional bigram tables for the superencryption of the indicators (see Section 14.5.1.2) in April 1937 stopped the Poles from further reading the *Marine* ENIGMA and this was a very sad blow for them. However, by a combination of luck and skill they recovered slightly on May 8, 1937 when they found a probable word or crib. They had observed that sometimes very long messages were broken into parts, the continuation being marked by a prefix /fort/ as in *Fortsetzung* and also by a reference number, which had before mid-1937 its figures encoded by the letters in the top line of a German typewriter keyboard

1	2	3	4	5	6	7	8	9	0
q	w	e	r	t	z	u	i	o	p

and enclosed by /y/.../y/ and then doubled. Thus, /forty/ was a crib for such messages. They now had a guess and tried it successfully since the continuation of the plaintext obtained read /fortyweepyyweepy/, *Fortsetzung 2330*. After this entrance into the decryption, the Poles had no difficulty in finding the rotor order, *Ringstellung* and steckering of this particular message, and since these were changed not too frequently, they had good reasons to hope for a complete break once they had found the basic wheel setting (*Grundstellung*). Luckily for Poland, a German torpedo boat with the call sign AFÄ had not been provided in time with the

instructions for the new indicating system and sent on May 1, 1937 a message in the old system that the Poles were familiar with and since more messages were exchanged on May 2 and May 3 there was enough information for the basic wheel setting to be discovered since a guess for the rotor order and steckering was already known from the /fortyweepyyweepy/ message. It turned out that indeed a message of April 30, broken from the old system, had the same rotor order and steckering. Thus, the Poles tried and were successful in breaking individual messages from the intermediate days May 2, 3, 4, 5 altogether about fifteen per day with a total of almost 100 not knowing how the new indicator system worked.

Two and one half years later, Alan Turing in England found that out. But the Poles expressed already the conjecture that some kind of a bigram substitution was involved. With the next change in the rotor order and steckering the Polish were no longer capable of deciphering German codes. Nonetheless in 1939 their results were admired by a the few people in Bletchley Park who were allowed to know about this achievement rightfully. Forty Weepy was for a long time a magic formula for insiders.

This success was only possible because of the *properly* self-reciprocal character[20] of the ENIGMA rotor encryption and the reflecting rotor of Scherbius and Korn turned out to be a grandiose illusory complication.

14.4.2 September 1938: new encryption procedure, new additional rotors

In 1938, the situation became aggravated. The Germans changed the encryption procedure on September 15, and introduced on December 15 a fourth and a fifth rotor, giving $60 = 5 \cdot 4 \cdot 3$ instead of $6 = 3 \cdot 2 \cdot 1$ possible rotor orders.

14.4.2.1 *Finding the wiring of the new rotors.* The Poles had to find out the wiring of the new rotors quickly, and they were lucky. Among the

---

[20]The *simply* self-reciprocal encryption machines of Boris Hagelin, allowing 2-cycles *and* 1-cycles, did not suffer from *this* defect. Nevertheless from 1942 the M-209 was broken by the Germans in North Africa [20: 460].

traffic they regularly decrypted were signals from the S.D. that is the *Sicherheitsdienst*, the intelligence service of the Nazi Party. The S.D. did not change their encryption procedure before July 1, 1939 even though they introduced the new rotors in December 1938. These rotors came from time to time into the position of the fast rotor and their wiring could be reconstructed the in same way as previously was done with the first three rotors.

The use of two methods, one of which possibly was compromised, was a grave error.

As an aside, there is a story of how B.S.-4 came to read the S.D. signals. The *Sicherheitsdienst* officers were distrustful of everybody and encoded their messages by hand before giving them to an ENIGMA operator for superencryption. The Poles, decrypting all ENIGMA traffic, obtained meaningless text and thought at first the cryptotext was encrypted in a different system. Then, one day in 1937, the three letter word /ein/ was read which could only mean that a plaintext group by was mistakenly mixed with a code; probably the numeral 1 had not been transcribed and the ENIGMA operator knew no better than to send /ein/ instead. The Poles then found it easy to break the simple hand-encrypted code.

14.4.2.2 *The new procedure suggests mechanised pattern search.* The new encryption procedure valid until the end of April 1940 did not use the same basic wheel setting for all messages of the day, but each message had an arbitrary basic wheel setting, which should precede the signal in plain. With the basic wheel setting (B.P.: indicator setting), as before, a randomly chosen and doubled plain indicator was encrypted and also used as the message setting for the encryption of the text. To give an example, for a signal following the plain preamble and beginning with RTJWA HWIK..., /rtj/ is the basic wheel setting. WAH WIK would be the encrypted doubled indicator, encrypted with /rtj/. In the sequel his will be written as rtj | WAH WIK. The authorised recipient would use the basic wheel setting /rtj/ to find given WAH WIK the plain indicator doublet which has the pattern *123123* and then decrypt the cryptotext with the first three letters, which constitute the true indicator, as the message setting.

Thus

$$\text{rtj} \quad \updownarrow \quad \begin{array}{cccccc} 1 & 2 & 3 & 1 & 2 & 3 \\ W & A & H & W & I & K. \end{array}$$

As long as the ring setting and rotor order had not fallen into the wrong hands, the foe could do nothing with the openly displayed basic wheel setting. The search space still contained respectively 105,456 or 1,054,560 possibilities, that is 26^3 ring settings, and 6 rotor orders, which were increased by December 1938 to $60 = 5 \cdot 4 \cdot 3$ rotor orders.

The methods Rejewski and his friends used until this time no longer worked since they were based on the multiple use of the same basic wheel setting for a full day. But the Germans, almost incredibly, kept the doubling of the plain indicator[21] and thus allowed the attack of searching for a pattern that is for the pattern *123123*, at a known position. This method would have worked before, too, but needed much more work. In the fall of 1938, however, there was no choice but to find alternatives. The Poles therefore thought of mechanisation.

14.4.3 The Polish bomba, Zygalski sheets and the clock method

14.4.3.1 *Mechanised pattern search.* Rejewski ordered in October 1938 six machines from the factory Wytórnia Radiotechniza AVA, each one simulating one of the six rotor orders used them to test in parallel the 17 576 positions of the rotor cores in B.P. called rod positions. This process needed at most 110 minutes.

The correct ring setting on the rotor cores was found using the 1-cycles in the following way. The machine was built from three pairs of ENIGMA rotor sets. In each pair the rod-positions of all rotors were shifted by three. The position of the rotor sets of the first pair shifted by one against the position of the rotor sets of the second pair which in turn was shifted by one against the position of the rotor sets of the third pair.

---

[21] While the double encipherment of each message setting was discontinued by May 1, 1940 for the service ENIGMAs, it was reintroduced for the navy key net *Süd* around August 1941 and was still employed in January 1944 which was an astonishing blunder (Ralph Erskine [11: 468]).

As soon as there was enough material to provide three encrypted doubled indicators such that the same character appeared once in the first and the fourth positions, once in the second and the fifth, and once in the third and the sixth position such as the letter W in the following example which goes back to Rejewski.

rtj	WAH	WIK
dqx	DWJ	MWR
hpl	RAW	KTW

Now as in Section 14.4.1.2, the data exhibited a 1-cycle or fixpoint and a promising attack was enabled (Fig. 14.41). The machine was started with the three initial settings /rtj/, /dqx/ and /hpl/, and the common character W was input repeatedly as a test character until in each one of the three pairs the same character triple occurred twice that is the pattern *123123* was found. Such a coincidence triggered a simple relay circuit to stop the whole machine, whose appearance led the Poles to call it the *bomba*.[22] Of course, sometimes there were mishits.

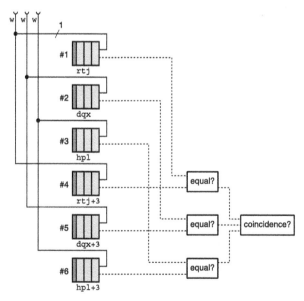

Figure 14.41. Abstract function of a Polish *bomba* (October 1938).

---

[22]According to Tadeusz Lisicki [26: 65], it was originally named by Jerzy Różycki after an ice-cream bombe. While they were eating it, the idea for the machine came to him and his friends.

If in this way the core position was revealed, a comparison of the encrypted doubled indicators with the encryptions produced by the ENIGMA replica reconstructed the ring setting and the cross-plugging. In this way, all signals of the day and later of the eight-hour shift from one and the same key net could be decrypted. During the war there were up to 120 such key nets.

The *bomba* was sensitive to cross-pluggings, and the method worked only if the test character as in the example above the letter W was unplugged. The likelihood for this to happen was around 50%, as long as five to eight plugs were used. If three different fixpoints had been allowed, as the British originally had in mind, the likelihood would have dropped to 12.5%.

14.4.3.2 *Zygalsky sheets.* For overcoming the difficulties with the new encryption procedure, another kind of mechanising was developed in autumn 1938, just at the right moment, by Henryk Zygalski (1908–1978). This new method involved a 'punch card catalog' for determining the *Tagesschlüssel* from about ten to twelve arbitrary fixpoints. For the six rotor orders it was calculated for A_1A_4, A_2A_5 or A_3A_6 whether or not for a wheel setting $\langle R_L \rangle \langle R_M \rangle \langle R_N \rangle$ of the rotors R_L, R_M, R_N some fixpoint is possible at all. For each one of the 26 letters $\langle R_L \rangle$ this result was recorded in a $\langle R_M \rangle \times \langle R_N \rangle$ matrix by (Fig. 14.42) a punched hole or "female". Roughly 40% of the squares on a sheet contained holes. In fact, to allow full overlay, sheets of 51×51 fields that were made of horizontal and vertical duplicates, were used. By superimposing the sheets aligned according to their wheel setting $\langle R_M \rangle \langle R_N \rangle$ the core position was determined, as a rule uniquely as soon as ten to twelve fixpoints were available. Most importantly, the method was *insensitive to the cross-plugging used* and was still useful when ten plugs were used after August 19, 1939 just as long as the double encipherment lasted. The Zygalski sheets had the drawback that a different sheet was needed for each rotor order and their number grew rapidly that is from $3! = 6$ as long as three wheels were in use and to $5!/2! = 60$ as soon as five wheels came into use even up to $8!/5! = 336$ once the navy ENIGMA

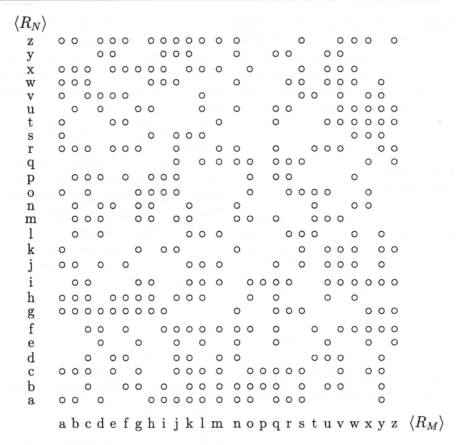

Figure 14.42. Zygalski sheet K_{14}^{413} for rotor order IV–I–III, $\langle R_L \rangle = k$. The holes show possible 1-cycles or fixpoints of $A_1 A_4$ for basic wheel setting $\langle R_L \rangle \langle R_M \rangle \langle R_N \rangle$.

had eight wheels available. Therefore, when the Germans introduced on December 15, 1938 the fourth and fifth rotor, the Poles were helpless for quite a while.

14.4.3.3 *Różycki's clock method for finding the fast rotor: early Banburism.* Tadeusz Lisicki has reported that the method called Banburismus by the British (Section 14.5.3) was independently discovered by Jerzy Różycki in the mid-1930s, he called it clock method thus comparing the fast rotor with the minute-hand and the middle rotor with the hour-hand of a clock. Różycki's method was particularly useful when the new rotors were introduced.

14.4.3.4 *A general remark.* The German side had always tried by suitable encryption security not to become victims of superimposition of in-phase

cryptotext segments with the same key as described by François Kerckhoffs van Nieuwenhoff (1835–1903) but rather had fallen into a more trivial trap the doubling of the indicator.

14.5 BRITISH AND US CRYPTANALYSIS OF THE *Wehrmacht* ENIGMA

14.5.1 British–Polish cooperation

In a meeting on January 9–10, 1939 in Paris, the Polish Lieutenant Colonel Gwido Langer (1894–1948) supplemented his French connections by contacts with his British colleagues. Sensing an increasing danger of war, a closer cooperation was desired but the Polish side did not yet reveal their successes. This meeting was before the British–Polish treaty of May 1939.

However, this attitude changed and Alfred Dillwyn Knox, the leading British cryptanalyst in the Foreign Office[23] arranged a meeting on July 24–25, 1939 in Pyry near Warsaw. On the British side in addition to Knox, participants were Alastair Denniston (1881–1961), the head of the Government Code and Cypher School, who believed in the importance of mathematics for cryptanalysis and Commander Humphrey Sandwith, the mysterious Mr. Sandwich in the Polish documents. The French were represented by Commandant Gustave Bertrand and Capitaine Henri Braquenié while the Polish side was by Ciężki, Langer and the Grand Chef Colonel Stefan Mayer. Rejewski, Różycki, and Zygalski proudly presented all their results in the Pyry forest south of Warsaw. On this occasion, the French as well as the British were promised Polish replicas of the ENIGMA with all of its five rotors, and the one for GC&CS was handed over in the diplomatic a pouch by Major Bertrand to General Menzies in London on August 16, 1939. The British were also informed about the Cyclometer, the Zygalski sheets, and the *bomba*, and were given decoded German messages. "They were speechless, and maybe for the first time the British experts dropped their arrogance" (Penelope Fitzgerald [13]).

The crisis that led to the Munich Conference of September 1938 had made clear that the outbreak of war no longer could be ruled out. The British Government decided to expand GC&CS. As a consequence, the existing housing as part of the Foreign Office at Whitehall, hitherto known as Room 47 was no longer sufficient for the upcoming tasks. A new location was found in Bletchley Park, B.P. for short, radio codename and also cover name Station X, located some fifty miles north of London where it was less likely to be bombed. Before war broke out in 1939, the GC&CS was established there and reinforced. Among its many duties was decryption of the ENIGMA.

Oliver Strachey, a veteran of the army's WWI codebreaking operations[24] had on several occasions arranged working contacts between the young

Alan Mathison Turing (1912–1954), who already had a reputation as a logician and had been interested since childhood in cryptology, and the GC&CS. Knox was a classics scholar, who in 1915 had preferred Room 40 of the Admiralty to a Fellowship at King's College in Cambridge, and who already had experience with the unsteckered ENIGMA used by the Italians; however he lacked the mathematical training necessary to make further progress. Strachey may have sensed this. On September 4, 1939 [23: 94], the second day of the Second World War, Turing reported to B.P. He worked on breaking the naval ENIGMA and on a further development of the Polish *bomba* and was later joined in this by Gordon Welchman (1906–1985), who also had arrived September 4. Turing had experience with relay circuitry and thus was not merely theoretically interested in cryptology. His contacts with GC&CS may have reached back to 1936 [19: 148].

14.5.1.1 *Zygalski sheets made in Bletchley Park.* The group around Knox and Turing made good use of the Polish proposal of punched sheets that is the Zygalski sheets which they called canvasses or Jeffreys sheets after John Jeffreys who supervised their preparation. However, some work was necessary in B.P. in order to cope with now sixty instead of formerly six rotor orders, the number of sheets being ten times bigger. The first complete sets were ready in January 1940. Around January 24, 1940, using Zygalski sheets, B.P. for the first time broke an ENIGMA key, the key Red for January 6 of the carelessly transmitting *Luftwaffe*, and B.P. continued to do so using the sheets made in the Cottage in Bletchley Park and later in Banbury.

14.5.1.2 *Turing's analysis of the Navy indicator system.* Informed after Pyry about the Forty Weepy results, Alan Turing started in September 1939 where the Polish had left off in May 1937. Turing has described this in his typewritten treatise entitled the Prof's book probably written in 1940.

---

[23] Knox died February 27, 1943 from stomach cancer.

[24] Oliver Strachey (1874–1960), husband of the feminist Ray Strachey, father of the computer scientist Christopher Strachey, and brother of the writer Lytton Strachey, replaced in

1941 in the Canadian cryptanalytical services called the Examination Unit the former US Major Herbert Osborne Yardley who had fallen into disgrace in the United States.

He starts with four messages from May 5 with 8-letter indicators and 3-letter message settings

K F J X	E W T W	P C V
S Y L G	E W U F	B Z V
J M H O	U V Q G	M E M
J M F E	F E V C	M Y K

and says: "The repetition of the E W [in the first two lines] combined with the repetition of the V suggests that the fifth and sixth letters describe the third letter of the window position [message setting], and similarly one is led to believe that the first two letters of the indicator represent the first letter of the window position, and that the third and fourth represent the second. Presumably this effect is somehow produced by means of a table of bigramme equivalents of letters, but it cannot be done simply by replacing the letters of the window position with one of their bigramme equivalents, and then putting in a dummy bigramme, for in this case the window position corresponding to J M F E F E V C would have to be say M Y Y instead of M Y K. Probably some encipherment is involved somewhere".

Therefore, Turing expects a more complicated procedure, the two most natural alternatives being

(1) encoding the letters of the message setting by bigrams and enciphering the result at the *Grundstellung* or
(2) enciphering the message setting at the *Grundstellung* and encoding the letters of the result by bigrams.

Turing thinks that "The second of these alternatives was made far more probable by the following indicators [and message settings] occurring on the 2nd May 1937 traffic

E X D P	I V J O	V O P
X X E X	J X J Y	V U E
R O X X	J L W A	N U M

With this second alternative we can deduce from the first two indicators that the bigrammes E X and X X have the same value, and this is confirmed from the second and third [indicators] where E X and X X occur in the second position instead of the first".

Turing's guesses were supported when the message settings V O P, V U E, N U M were decoded using the basic wheel setting obtained from the AFÄ messages. The problem of the navy indicator system was fundamentally solved by the end of 1939. However, without knowing the bigram tables that were used, no further progress could be expected. Turing decided early in 1940 to analyse German navy signals that had been intercepted in November 1938, when only 6 steckers were in use and hand methods attacking cribs were possible. In particular, interrogation in November 1939 of a prisoner of war, Funkmaat Meyer, disclosed that the German navy now used spelling for numerals, so, for example, Forty Weepy was now /fortzwodreidreinul/. Thus, the Forty Weepy method gave cribs. Alan Turing, Peter Twinn, and two girls, as Turing calls them, started an attack using an EINSing catalogue (see Section 14.5.4.2) on the November 28, 1938 traffic. After a fortnight of work the code of that day was broken and five others between November 24 and 29 came out on the same rotor order. The rotor order and *Ringstellung* seemed to remain constant for about a week; the number of steckers was still 6, moreover, the same letter was never steckered on two consecutive days which was a serious mistake. Reconstructing the bigram table was at that moment only partly possible and the only hope of success was oriented towards a pinch. It actually happened April 26, 1940, when a German Q-boat, the trawler *Polares* or *Schiff 26*, was seized off Ålesund [23: 117], giving the steckering and message setting for April 23 and 24, operator logs giving cribs for April 25 and 26 traffic and, most important, exact details of the method of working of the indicator system which confirmed Turing's discoveries.

The method as such is outlined in [2 (2007): 61, 444]. It uses two trigrams picked from a book that is the K-book or *Kenngruppenbuch*. While the first trigram *Schlüsselkenngruppe* or key net indicator, say C I V, had no immediate cryptological importance, the second one (in German *Verfahrenkenngruppe*) say T O D, when decrypted with the *Grundstellung*, gave the message setting to be used by the sender and by the recipient. Thus, the *Verfahrenkenngruppe* can be called an encrypted message setting. The basis for the bigram substitution was now, along with two dummies, the grouping

```
* C I V    Q C I V
T O D *    T O D X
```

Using the cribs that the Ålesund event had produced, further entries in the bigram table were possible. In small steps, some progress was made during 1940, using suspicions that in June the bigram tables had changed. The signals of May 8 turned out to be particularly obstinate, and only in November an old hand, Hugh Foss, solved them together with its paired day, May 7. Later still, the signals of June 27 were broken and no change in the bigram tables had happened. Messages continued to be broken until July 1, 1940 when the bigram tables were finally changed. In February 1941, the traffic of April 28 was broken on a crib, using a BOMBE for 336 rotor orders in succession. Shortly after a planned raid, the taking of the trawler *Krebs* [23: 130] near the Lofoten Islands ended the difficult period and started a rich flow of decrypts of the naval ENIGMAs produced by a choir of crib runs on BOMBEs, EINSing (Section 14.5.4.2), and especially the Banburismus procedure (Section 14.5.3).

14.5.2 BOMBEs

14.5.2.1 *Turings motivation for the BOMBE.* When in mid-January 1940 Turing met Rejewski, who had fled to France, in Gretz-Armainvilliers, north of Paris, he was, according to Rejewski, very interested in the Polish ideas for defeating the cross-plugging. By then, he had arrived at his own ideas (see Section 14.5.2.2), but of course he could not mention how far he had advanced. It was only natural that Turing tried to improve the Polish *bombe* to make them insensitive to cross-plugging like the Zygalski sheets. The British, like the Poles, were afraid that with fewer and fewer unsteckered or self-steckered letters their methods would soon become useless. Thus, Turing wanted to get rid of the restriction of self-steckered letters.

In the *bomba* construction, the Poles had followed essentially the electrical design of the ENIGMA, providing through the reflector a single

path between the key that was pressed for a plaintext test letter input and the light bulb to indicate the corresponding cryptotext letter output. For the purpose of encryption, this design was quite practical.

In late 1939 Turing wanted in contrast, as Joan Murray née Clarke remembers the argument [18: 113], to test all 26 letters in parallel to see what output they would have which would allow a simultaneous scanning of all 26 possibilities of the test letter at each position of the wheels. Thus, Turing thought of replacing the Scherbius rotors by different ones (Turing rotors), each one having on both the entry and the exit sides two concentric rings of contacts, one for the journey towards the reflector and one for the return journey, both mimicking *the same* ENIGMA rotor. Likewise, the reflector would have two rings of contacts. This modification would have input and output of 26 wires in parallel, and result in a double-ended scrambler (as it was called by Welchman), see Fig. 14.43 or in US parlance the commutator and which strictly represents a classical ENIGMA substitution $A_i = S_i U S_i^{-1}$ for $i = 1, \ldots, 26^3$. In accordance with its self-reciprocal character, the battery of scramblers had to be input–output symmetrical which was provided for by a symmetric wiring between the contacts of the inner and outer rings of the reflector.

In this hypothetical Turing version, the Polish *bomba* amounted to three closed cycles, each built from two double-ended scramblers; one such cycle is shown in Fig. 14.44. Turing had thus managed to strip off the superencryption by the cross-plugging mechanically. And he recognised that the 1-cycles, the said *females* of the Zygalski sheets were natural fixpoints of a mapping and could be determined by

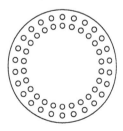

Figure 14.43. The two contact rings of the double-ended scrambler.

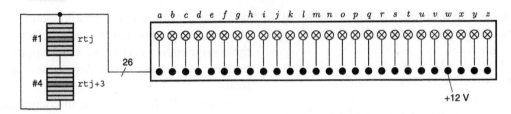

Figure 14.44. Hypothetical Turing version of the Polish *bomba* with simultaneous scanning. Diagram of one of the three feedback cycles of length 2.

an iterative feedback process, which normally diverged and thus indicated that the rotor positions in question did not allow a fixpoint. If it did not diverge, it gave the fixpoint. The logician Turing, familiar with the *reductio ad absurdum*, thus turned to the general principle of feedback.

Technically, the distinction between the divergent and the non-divergent case was made by a test register attached to the 26-line bus of the feedback which can be seen in Fig. 14.44 from #4 to #1. Voltage is applied to the wire belonging to the test letter which is W in the shown example. In the divergent case all light bulbs of the test register light. In the non-divergent case the feedback cycle of the fixpoint is electrically isolated from the remaining wiring; correspondingly, either exactly one light bulb that is the one belonging to W is lit or all light bulbs but this case depends on whether the cross-plugging was correctly chosen or not.

A battery of double-ended scramblers was to be moved simultaneously. Thus Turing could have actually simulated a Polish *bomba*. But he wanted more and the actual development took a different, much more general path (see Section 14.5.2.2).

Turing may have earlier thought of making use of probable words to break into ENIGMA. After the Pyry meeting and having heard about the *bomba*, he turned his thoughts to mechanising his own method. The major advantage with the device Turing had in mind was that it not only found the rotor order, compare the Zygalski sheet, but it also found at least one *stecker*. "It was he who first formulated the principle of mechanising a search for logical consistency based on a probable word" wrote Andrew Hodges, Turing's biographer [19: 179].

Late in 1939 when he was still a novice Welchman, by the way, arrived independently at similar conclusions although at first he was not involved in the ENIGMA decryption by machines. He also reinvented the Zygalski sheets, not knowing that John Jeffreys in another building already had a production line going. Likewise, Welchman did not know of Turing's ideas but the mutual unawareness of these two men's efforts was intentional Turing's ideas and precautions were guided by Knox's remark at Pyry that the Germans could again change their encryption procedure and give up the indicator doubling. Then, having learned in the meantime from decryptions many of the habits and styles of the Germans, the British hoped to be able to produce the necessary feedbacks efficiently with probable words the Germans used so plentifully such as /wettervorhersage biskaya/, /wettervorhersage deutsche bucht/ or /obersturmbannführer/, obergruppenführer/ or /keine besonderen ereignisse/. Thus, B.P. was prepared on May 1, 1940 shortly before the campaign in France, for when the next change occurred that is when *Heer* and *Luftwaffe* dropped the indicator doubling and put the Zygalski sheets out of action. The Polish *bomba* would have been useless. Moreover, Turing had designed the BOMBE in a universal and most flexible way.

If, however, the probable word method would not work or if not enough BOMBEs were available, then there was still the possibility of a deep adjustment of the messages for superimposition [2: 61], and the Banburismus procedure (see Section 14.5.3.1) could be performed to obtain a reduction of the number of rotor orders to be tested.

14.5.2.2 *The Turing BOMBE: feedback cycle attack.* In the last quarter of 1939, Turing's design had progressed far enough that Bletchley Park was allowed to ask the British Tabulating Machine Company in Letchworth to build a machine which was also called BOMBE. Harold 'Doc' Keen, with a crew of twelve people, finished it by March 1940. Later, he was equally successful in building the 4-rotor-BOMBE MAMMOTH.

The Turing BOMBE prototype Victory [3: 154] entered service on March 14, 1940 [3: 340]. On May 1, 1940, when the prototype BOMBE had made its first steps and the double encipherment of each message setting was discontinued the ENIGMA system was essentially open to the British. A probable word attack, a plaintext–cryptotext compromise, profited from the astuteness of the Turing (Welchman) feedback idea pursued since 1939. And Turing had designed the British BOMBE in a way that allowed this attack. With the algebraic methods they had learnt from the Poles since 1939, a method was resumed that Rejewski had started with in 1932 (see Section 14.4.1.6).

Now, the British had to prepare the menu day by day, and they found enough cribs to do so. Meanwhile, they were helped by continuing violations of even the simplest rules of cryptosecurity on the German side. John Herivel observed in May 1940 that for the first signal of the day the wheel setting was frequently very close to (if it did not actually coincide with) the position of the wheels for the ring setting of the day (see Herivel tip Section 14.5.6.3). Moreover, the use of stereotyped indicators continued which the British placed under the heading Cillies (see Section 14.5.6.1).[25] There was also the abuse of taking as the indicator the basic wheel setting, referred to as JABJAB by Dennis Babbage (see Section 14.5.6.2). When the German supervisors finally reacted, the damage was already irreparable.

---

[25]Singular: Cilli. Sometimes interpreted sillies. Welchman in 1982 said: "I have no idea how the term [sillies] arose" [38: 99]. Budiansky in 2000 mentioned the abbreviated name of the girlfriend of a German wireless operator [3: 143]. Sebag-Montefiore in 2000 commented that the word comes from CIL, which was the first message setting worked out in this way [26:81]. None of these explanations is convincing.

Cryptosecurity discipline was lowest in the Air Force of the pompous parvenu Hermann Göring. From May 26, 1940 onward before the Turing–Welchman BOMBE was working, mathematicians and linguists in B.P. regularly managed to read the ENIGMA signals of the *Luftwaffe* referred to as key net Red, while for the signals of the *Kriegsmarine* referred to as key net Dolphin, *Heimische Gewässer*, and later *Hydra* they had to wait until June 1941 before they had mastered the bigram superencryption of the message keys. In December 1940, they succeeded in breaking into the radio signals of the SS key net Orange. From September 1942 onward, Field Marshal Rommel's ENIGMA traffic with Berlin (key net Chaffinch) was no longer secure, and from mid-1942 onward the British achieved serious and lasting breaks especially in the heavy *Luftwaffe* traffic in key net Wasp of *Fliegerkorps* IX, Gadfly of *Fliegerkorps* X, Hornet of *Fliegerkorps* IV, Scorpion of *Fliegerführer Afrika*. Most obstinate, to British judgement, was the radio communication of the German *Heer*, which was a consequence of the solid training of the operators. Before the spring of 1942, no ENIGMA traffic line of the *Heer* except one, Vulture I in Russia (June 1941), was broken.

In the general probable word attack that Turing and in parallel Welchman used instead of the three isolated, two-fold cycles of the Polish *bomba* an entire system of feedback cycles formed by a battery of first up to 10 and later up to 12 double-ended scramblers. Such feedback cycle systems, which are *independent of the steckering*, are obtained from a juxtaposition of a probable word and a fragment of the cryptotext. Fortunately, for sufficiently long probable words, the non-coincidence exhaustion method that is no letter can be mapped into itself allows exclusion of many juxtapositions; furthermore, conspicuous probable words are often at the beginning or the end of the message unless Russian copulation that is cutting in two parts and joining with the first part after the second has been used. Therefore, it is not unrealistic to establish a new feedback cycle system for every new juxtaposition; there are not too many.

#		1		4	5		7	8	9		11	12	13	14	15		17		19				24
		o	b	e	r	k	o	m	m	a	n	d	o	d	e	r	w	e	h	r	m	a	c	h	t

O V R L J B Z M G E R F E W M L K M T A W X T S W V U I N Z

Figure 14.45.

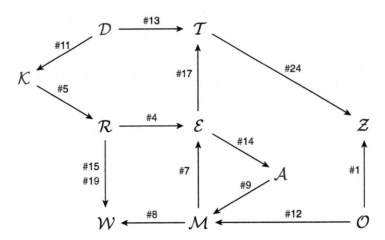

Figure 14.46. Plaintext letter/cryptotext letter pairings for juxtaposition or crib.

The following example[26] goes back to C.A. Deavours and L. Kruh. Let

OVRLJ BZMGE RFEWM LKMTA

WXTSW VUINZ GYOLY FMKMS GOFTU

EIUQR GKJOC GUZBN VGXFY BNLRR

NNNHO LKKSZ DQATZ JKUYP . . .

be the cryptotext and /oberkommandoderwehrmacht/ the probable word. The third leftmost position not excluded by scritching (Section 14.3.1.2) gives a correct position of the crib (see Fig. 14.45) with a core of 10 letters $A\ D\ E\ K\ M\ O\ R\ S\ W\ Z$ connected by 14 transitions. (Note: calligraphic letters like $\mathcal{E}$ stand for both e and E.)

The 13 different pairings of plaintext and cryptotext letter can be compressed into a directed graph with 10 nodes and 13 edges, shown in Fig. 14.46. The graph contains one true cycle ($\mathcal{MEA}$):

#		7	9	14
		m	a	e
		E	M	A

[26]Rotor order IV I II, reflector B, cross-plugging $(VO)(WN)(CR)(TY)(PJ)(QI)$. Ring setting AAA, message setting tgb.

The self-reciprocal character of the double-ended scramblers, reflected in the symmetric electrical connection of their inputs and outputs, means a transition to an undirected graph. From this graph, a subgraph may be selected, in jargon called a menu, for our example the graph with 8 nodes shown in Fig. 14.47 at the upper right corner. Each cycle called closure by Turing in this subgraph establishes a feedback in the Turing BOMBE setup. A menu with 6 letters and 4 cycles is of course more lucrative than one with 12 letters and one cycle since it reduces the danger of mis-hits.

Corresponding to such a subgraph with 10 edges, 10 double-ended scramblers are now connected by means of 26-line buses and a test register is attached, say at $\mathcal{E}$ (Fig. 14.47). To some entry, say e, voltage is applied.

As mentioned above, the positions (14, 9, 7) form a cycle or closure. Denoting the internal contacts with $a, b, c, \ldots, y, z$, the cross-plugging with T, and the substitution performed by scrambler #i with A_i; we obtain the following relations.

$$eT = mT P_7, \quad mT = aT P_9,$$
$$aT = eT P_{14} \quad \text{or} \quad eT = eT P_{14} P_9 P_7.$$

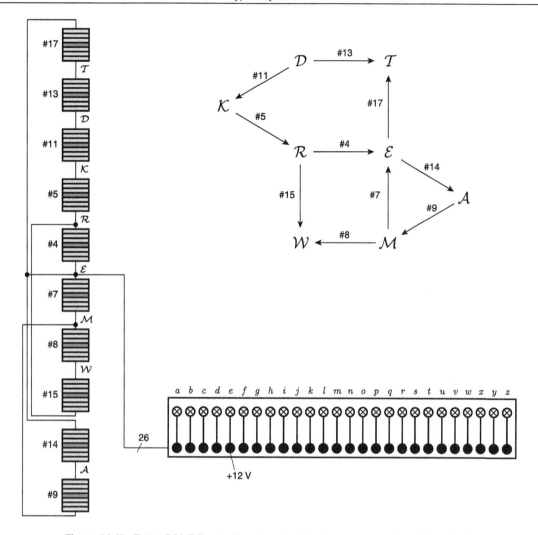

Figure 14.47. Turing BOMBE setup for subgraph of feedback cycle system of Fig. 14.46.

Thus, eT is a fixpoint of $P_{14}P_9P_7$.

But the positions $(4, 15, 8, 7)$ form also a cycle and we have the relations

$$eT = rT\,P_4, \quad wT = rT\,P_{15},$$
$$wT = mT\,P_8, \quad eT = mT\,P_7;$$

since the scrambler substitutions are self-reciprocal, we obtain

$$eT = mT\,P_7, \quad mT = wT\,P_8, \quad wT = rT\,P_{15},$$
$$rT = eT\,P_4 \quad \text{or} \quad eT = eT\,P_4P_{15}P_8P_7.$$

Thus, eT is also a fixpoint of $P_4P_{15}P_8P_7$.

Moreover, the positions $(4, 5, 11, 13, 17)$ form a cycle: using again that the scrambler substitutions are self-reciprocal, we obtain the relations

$$eT = rT\,P_4, \quad rT = kT\,P_5, \quad kT = dT\,P_{11},$$
$$dT = tT\,P_{13}, \quad tT = eT\,P_{17}.$$

Thus, eT is even a fixpoint of $P_{17}P_{13}P_{11}P_5P_4$.

Assume that the position of the scramblers is not the correct one. Then normally that is if enough cycles exist the voltage spreads over the total system and all the test register light bulbs are lit. A relay circuit discovers this divergent case and moves the scramblers on to the next position.

Now assume that the position of the scramblers is the correct one that is the one used for encryption such that the scrambler #4 maps $rT = $ /r/ into $eT = $ E. Then there are two subcases. If the cross-plugging is correctly chosen that is the entry e to which voltage is applied is indeed /e/, then the voltage does not spread, and apart from the light bulb belonging to e no lamp is lit. If, however, the cross-plugging is not correctly chosen, then normally that is if enough cycles exist the voltage spreads the whole remaining system and all lamps are lit except one, which indicates the cross-plugging. In both of these convergent subcases, the machine setting and the light bulb indication can be noted down. The scrambler position determines the correct position of the rotor core. It can be a mishit. This fact can be quickly decided by using the resulting setting to try to decrypt the surrounding text.

The possibility of Turing's feedback cycle attack was totally overlooked by the newcomer Gisbert Hasenjäger (1919–2006), formerly a gifted student of the German mathematician Heinrich Behnke and now responsible for the security of the ENIGMA in the Cipher Branch OKW, *Referat* IVa, named Security of own ciphers. In 1942 section IVa was founded, rather late, in 1942 and was headed by Karl Stein (1913–2000), who was about to become a famous topologist.

This attack, as was shown above, is strongly supported by the properly self-reciprocal character of the ENIGMA using KORN encryption steps; however, it would also work in principle for non-self-reciprocal double-ended scramblers although such cycles occur much less frequently. For example, there is only one true cycle in Fig. 14.46 and very long probable words would be needed to make the attack succeed, or a larger menu would be needed. For example, in the feedback cycle system of Fig. 14.46, the crib would allow the adjunction of a node $\mathcal{U}$ connected with $\mathcal{A}$, a node $\mathcal{V}$ connected with $\mathcal{M}$, or a few more. There is, compared to Fig. 14.47 one more cycle $(\mathcal{TZOME})$ in the example of Fig. 14.47 that is from $\mathcal{T}$ over $\mathcal{Z}, \mathcal{O}, \mathcal{M}, \mathcal{E}$ to $\mathcal{T}$. But its analysis would increase from 10 to 13 (#24, #1, #12) the number of scramblers needed.

14.5.2.3 *The Welchman diagonal board.* Gordon Welchman (1906–1985) improved the Turing feedback cycle attack quite decisively by taking all relations affected by the self-reciprocal character of the typical ENIGMA cross-plugging explicitly into account. Whenever Turing's BOMBE stopped, the nodes like $\mathcal{A}, \mathcal{D}, \mathcal{E}, \mathcal{K}$ and so on were assigned certain internal contacts.

Figure 14.48 shows such a halting configuration with two interpretations $\mathcal{A}$-a, $\mathcal{E}$-e which are self-steckered; the two interpretations $\mathcal{D}$-t and $\mathcal{T}$-d would indicate a cross-plugging $(\mathcal{TD})$. But the two interpretations $\mathcal{W}$-m and $\mathcal{M}$-x contradict the self-reciprocal character of the cross-plugging. The BOMBE should not have halted in such a configuration and the contradiction found by this reasoning should have caused divergence inside the BOMBE's electrical wiring.

Welchman found in November 1939 a surprisingly simple electrical realisation of such formation of the reflexive hull with respect to cross-plugging, the diagonal board shown in Fig. 14.49. Its functioning is explained in Fig. 14.50. Assume $eT = dT\ P_{11}P_5P_4$. The bold overlay in the wiring shows how the electrical connection made by the scramblers from e in bus $\mathcal{E}$ to d in bus $\mathcal{D}$ is supplemented by the diagonal board with a fixed electrical connection from d in bus $\mathcal{E}$ to e in bus $\mathcal{D}$.

With Welchman's improvement, Turing's feedback cycle attack attained its full power and the efficiency of the BOMBE increased dramatically so that fewer cycles were needed to fill the test

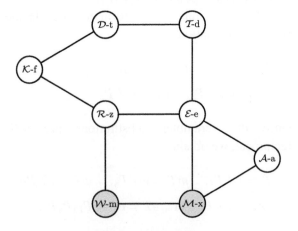

Figure 14.48. Contradictory halting configuration.

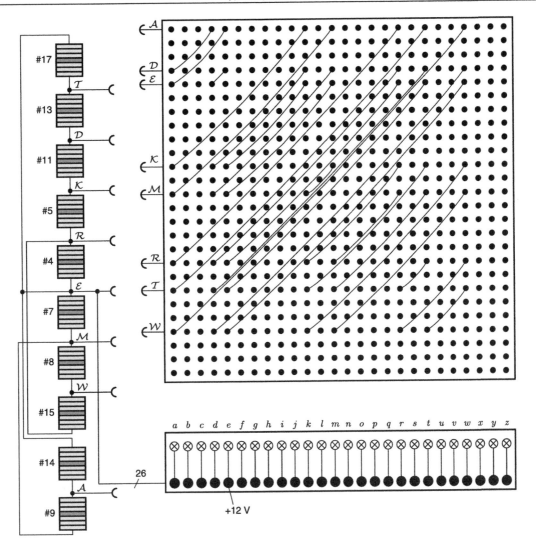

Figure 14.49. Welchman BOMBE setup for feedback cycle system of Fig. 14.46.

register. This improvement not only helped to save scramblers, it also allowed shorter cribs and thus increased the chance that the middle rotor remained at rest. So Welchman is the true hero of the BOMBE story in Bletchley Park. Devours and Kruh [9: 123] formulated it in a way that may have consoled Hasenjäger:

> It is doubtful that anyone else would have thought of Welchman's idea because most persons, including Turing, were initially incredulous when Welchman explained his concept.

14.5.2.4 *BOMBEs in Bletchley Park.* The Turing BOMBE prototype Victory had originally no

diagonal board. Starting August 8, 1940 [3: 341], improved versions with a diagonal board called Turing–Welchman BOMBEs followed [38: 147], nicknamed Agnes, Jumbo, Funf. Ming was ready by May 26, 1941. People in Bletchley Park who called the aggregate of scramblers together with the test register and the diagonal board a bomb were not made aware of the Polish origin of the name and idea. Agnes,[27] the first Turing–Welchman production BOMBE after the prototype Victory, was ready by mid-August 1940; a bomb run needed 17 minutes for a complete exhaustion

---

[27] Turing had dubbed it originally 'Agnus Dei'.

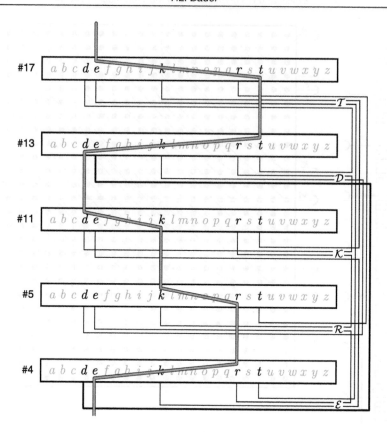

Figure 14.50. Functioning of the 'diagonal board' of Welchman for the example of Fig. 14.46.

of one rotor order. In the spring of 1941 there were 8 BOMBEs at work including Ming (end of May 1941) and 12 towards the end of the year built by the British Tabulating Machine Company in Letchworth. The number increased rapidly to 30 in August 1942, 60 in March 1943, and about 200 at the end of the Second World War. B.P. cracked tens of thousands of German military messages a month.

By mid-1943 the United States helped with an effort called Project Beechnut. At the Eastcote outstation [3: 300, 398], operated since February 1, 1944 by the 6812th Signal Security Detachment, US Army, there were 10 British BOMBEs in operation: Atlanta (Fig. 14.51), Boston, Chicago, Houston, Minneapolis, New York, Omaha, Philadelphia, Rochester and San Francisco. Other outstations were located in Adstock, Gayhurst and Stanmore. The BOMBEs had 36 scramblers, a scrambler being formed by three vertically adjacent rotors.

Fourteen BOMBEs, called JUMBO, had an attachment called a machine gun to resume work when a stop occurred and a contradiction of the steckers was found.

However, running the bombe on naval ENIGMA needed in the worst case 336 wheel orders to be checked, against 60 for the air force and army ENIGMA. Such a complete run would take at least about 96 hours (4 days), and Hut 8, working against the naval ENIGMA, was short of bombe time in 1942.

14.5.2.5   *The US BOMBEs.*   After the BRUSA pact on Cooperation in Code/Cipher Matters was settled in May 1943, with British help American groups developed high-speed versions of the BOMBE which continued to do service well after the war had ended. To give just a different example, there was the Russian Problem on Deciphering diplomatic messages presumably encrypted with a one-time-pad. Repeated use of the alleged one-time-pad had first to be discovered and then to be used for decryption. This issue gave an immense

Figure 14.51. British BOMBE Atlanta standard model in Eastcote with 36 scramblers.

amount of work for suitable machines and led to the top secret Venona breaks [3: 307–310, 330].

Both army and navy developed high-speed versions of the BOMBE. A first step was the X-68 003 of the US Army (SIS) which was a genuine relay machine constructed by Samuel B. Williams from Western Electric/Bell Labs, in operation since October 1943 and equipped with 144 double-ended scramblers. It became known[28] as Madame X which used stepping switches to allow for a quick change of the crib. The simulation of scramblers by relays was slow but avoided rotating masses. Developed with the help of Bell Laboratories, it was directed against 3-rotor ENIGMAs, and was unwieldy against the 4-rotor ENIGMAs of the *Kriegsmarine*. Only one Madame X was actually built, and it corresponded in power to six or eight British BOMBEs. Design and construction cost a million dollars which was not thought cost-effective compared with the navy's BOMBEs.

---

[28]It is unclear whether the name is an allusion to Agnes Driscoll née Meyer, the brave fighter for pure cryptanalysis, who was referred to in Op-20-G as Madam X.

14.5.2.6 *BOMBEs against M4.*   For the navy cipher branch Op-20-G, Joseph Desch of National Cash Register Company in Dayton, Ohio [12: 191], who had experience with rapid circuitry for elementary particle counters and thus a reputation in electronics, accepted in September 1942 the ambitious commission to build 350 BOMBEs, each one several times larger than the Turing–Welchman BOMBE. Moreover, they expected to have these machines operating by the spring of 1943. Desch and his group "thought that American technology and mass production methods could work miracles" (Colin Burke [4: 284]). Desch, however, rejected the request by Joseph N. Wenger to build an electronic version by stating that "An electronic BOMBE was an impossibility" [4: 283]. He was wise because for a "super-BOMBE" he had calculated 20,000 tubes, while the British needed only about 2000 tubes for COLOSSUS.

The British could do no more than help the navy of their ally. Howard Engstrom, a mathematics professor at Yale University and naval reserve officer [4: 224; 11: 174], had sent Lieutenant Joseph Eachus from OP-20-G in July 1942 to Bletchley Park, where he was given, after some delays and

Figure 14.52. US Navy BOMBE for 4-rotor ENIGMA.

excuses, complete wiring diagrams and blueprints of the actual BOMBEs [3: 238]. Turing travelled November 7–13, 1942 on the *Queen Elizabeth* over the Atlantic Ocean and departed again from New York harbour on the night of March 23, 1943 on the *Empress of Scotland* to arrive safely back in England on March 29. He made good use of the four months in the USA, meeting also Claude Elwood Shannon at Bell Labs while doing work on voice scrambling. Turing found Op-20-G well budgeted for money and with the most able people, but security measures were tighter than those for the atomic bomb project and prevented a more serious contact. NCR in Dayton, Ohio provided the setting for the BOMBEs. Despite the efforts of Eachus, it took longer than expected, and by the spring of 1943 only two prototypes, Adam and Eve, were halfway ready. Franklin Delano Roosevelt himself gave the project support and impetus. Meanwhile, the situation in the Atlantic improved for the Allies, mainly thanks to B.P. decryptions.

Desch had particular problems with fast spinning electromechanical scramblers with brush contacts and by mid-June 1943 it was hoped that the difficulties would soon be overcome. On July 26,

1943, 13 production models did not function at all and it looked like the whole project would be killed.

But Desch did not give up. The mechanical difficulties were surmounted step by step and reliability was increased. In September 1943 the first machines built by NCR were sent from Dayton to Washington, where they were put into operation. By mid-November 50 BOMBEs were in use and 125 were altogether built. In 1944, success was certain although it took slightly longer than optimistically projected and the cost was almost three times as much as planned. A Desch BOMBE N-530, N-1530 would cost $45 000. This project was so secret that the machine did not even have a name.

Britain did not have the first of its very few 4-wheel bombes until early summer 1943. The US Navy 4-rotor BOMBE (Fig. 14.52) Desch had developed comprised 16 4-rotor scramblers, a scrambler being formed by four vertically adjacent rotors and a Welchman diagonal board, and it was 200 times faster than the Polish *bomba* and 20 times faster than the Turing–Welchman BOMBE even though the specification had said 26 times faster. It was 30% faster than the 1943 British Bombe attachment WALRUS (Cobra) directed against the 4-rotor ENIGMAs of the *Kriegsmarine*. Op-20-G had caught up. By December 1943 the decryption of a *Triton* ENIGMA signal took on average only 18 hours, compared to 600 in June 1943. In contrast to their British cousins, they did the localising of the scrambler positions and the whole job control by digital electronics with 1500 thyratrons which were gas-filled tubes. The Desch BOMBEs proved to be so reliable that by the end of 1943 all work on the decryption of the *Triton* key net of the German U-boats was assigned to the US navy which was a great step forward from the rivalries of mid-1942.

The BRUSA pact *Cooperation in Code/Cipher Matters* of May 1943 between the USA and the UK, and for their navies in particular the Holden[29] Agreement of October 1942 had made this development possible. It "began to move the two nations towards a level of unprecedented cooperation" (Burke [4: 272]) in cryptanalysis. But some frictions and tensions remained. "It was not until the UKUSA agreement of 1946 that the two nations forged that unique relationship of trust that was maintained throughout the Cold War" (Burke [4: 272]).

However, "by the time of the BRUSA pact the army's SIS, much more than [the navy's] Op-20-G, had become a subsection of GC&CS" – writes Burke [4: 308] with some exaggeration. In the transition to electronic machines, SIS built in 1945 a successor to the relay-based AUTOSCRITCHER [8] which was correspondingly called the SUPERSCRITCHER (see Section 14.5.4.3). Op-20-G also built DUENNA, which did not enter service until November 1944, and the British built GIANT, names that did not exist until recently in the open literature [4: 309, 435]. All these machines were directed against cross-plugging and reflector plugging.[30]

---

[29]Carl F. Holden, Capt. US Navy, Director of Naval Communications.

[30]The expression scritchmus comes from the jargon of Bletchley Park. "I cannot now recall what technique was nicknamed a scritchmus", wrote Derek Taunt and continued that "the method was used to telling effects by Dennis [Babbage]" [18: 110]. For the meaning see also Section 14.3.1.2. Ralph Erskine thinks scritching comes from the phrase scratching out contradictions.

Another subject Alan Turing approached in 1943, after his return from the USA, was speech encipherment. The design of DELILAH [19: 273] started in September 1943, construction in June 1944; it was just finished by May 6, 1945.

14.5.2.7 *BOMBEs against Japanese rotor machines.* VIPER and PYTHON were American machines directed against the Japanese rotor cipher machines. They were built from relays and stepping switches, and little by little, equipped with electronic additions. Finally, towards the end of the war, the inevitable transition to fully electronic machines was made: Op-20-G built RATTLER against Japan's JN-157.

14.5.3 The Polish clock method and Banburismus at B.P.

Direct adjustment of the messages was impossible for the indicator systems used with the ENIGMA. Thus something like a search for repetitions or a coincidence count was necessary. Which leads to three stages of work.

14.5.3.1 *Stage 1.* The adjustment of the messages was Stage 1 of both the clock method of Jerzy Różycki (1909–1942) and its elaboration by Alan Turing and Jack Good [18: 155] in Bletchley Park. There, the mechanism used for performing this adjustment was called Banburismus because the long overlay sheets of paper or banburies containing the messages in a 1-out-of-26 code (Fig. 14.53 which shows that a hole was punched in the marked letter). They were produced in Banbury, a little town near Oxford. For the purpose of in-depth adjustments Turing and Good[31] developed a particular method of scoring the repeats also known as ROMSing which means weighing the repetitions or in Good's words using the weight of evidence. Intuitively one would expect that, for example, two bigram repetitions favoured adjustment more than four monogram repetitions. Turing used a logarithmic unit called ban which was

---

[31]Good, who joined B.P. on May 27, 1941, considered himself as the main statistical assistant [18: 155] of Turing. Using the Sequential Bayes rule Good gave a sound probabilistic basis to the more intuitive results of Turing on multigram repetitions.

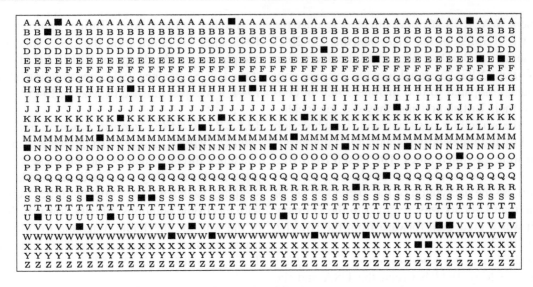

Figure 14.53. Perforated sheet with the cryptotext segment.

N U B A I V S M U K H S S P W N V L W K A G H G N U M K W D L N R W E Q J N X X V V O A E G E U

a decimal counterpart to Shannon's binary unit of information called bit; 1 [ban] $\hat{=} 1/^{10} \log 2$ [bit]. 1 [deciban] ≈ 0.332 [bit], the practical unit used in Bletchley Park, corresponds to 1 [dB] (deciban). Good introduced the unit halfban.

Turing and Shannon obviously developed their ideas independently of each other and only met by the end of 1942. "Turing and I never talked about cryptography" said Claude Shannon. It could well be, as Ralph Erskine thinks, that before July 1941 the cryptanalysts in Bletchley Park were unaware of the *Kappa* test of Friedman or the *Phi* test of Kullback. Despite Good's explanations by way of a 'Sequential Bayes Rule', it remains unclear how Turing found the importance of what he called repetition frequency in his Treatise on the Enigma, the Prof's Book written in late summer or early autumn 1940. Repetition frequency is what Hugh Alexander[32] called repeat rate and also what Jack Good later called weight of evidence.

Anyhow, Turing was ahead of Friedman in one respect. Friedman's philosophy of the index of coincidence was based primarily on investigating the coincidence of single characters [2: 301], and even in Kullback's work of 1935 [28] only the coincidence of bigrams was given attention and

the coincidence of trigrams was only marginally mentioned. However, in the Prof's Book and in Alexander's summary thereof, the probability for coincidence of *n*-grams with rather high *n* is discussed and the method of scoring is sketched. Clearly, one *n*-gram repeat, for example, scores more points than *n* single repeats, and for tetragrams and hexagrams Alexander mentions their repeat rate which is about $100 \cdot (\frac{1}{26})^4$ and $1500 \cdot (\frac{1}{26})^6$. This value, however, is much larger than $(\kappa_d)^4 = 16 \cdot (\frac{1}{26})^4$ or $(\kappa_d)^6 = 64 \cdot (\frac{1}{26})^6$ (with $\kappa_d \approx 2 \cdot \frac{1}{26}$) which one would have if four or six independent coincidences occurred. Heptagrams ($n = 7$), octagrams ($n = 8$), and sometimes even enneagrams ($n = 9$) turning up in examples are considered to give practical certainty about the adjustment differences.

14.5.3.2 *Stage 2.* We now give two examples of the Stage 2 of the method. The clock method of Jerzy Różycki was intended for dealing with the indicator doubling system of the 3-rotor *Wehrmacht* ENIGMA. The adjustment of the cryptotexts results in a difference in the plain indicators. If two messages are found with encrypted 3-letter indicators that coincide in the first two letters, say A U Q and A U T, then the plain indicators also coincide in the first two letters and the difference of

---

[32]C.H.O'D. Alexander, in 1942 successor of Alan Turing as head of Hut 8 of B.P.

their third letters that is the shift for adjustment, equals the difference of Q and T. Thus, at least the fast rotor R_N whose initial setting was indicated by the third letter of the message setting was under observation. Deavours and Kruh [9: 142] give the following example. From several pairs of in depth messages, the third position of the encrypted indicator showed the following observed values:

first message	R F N B D T N M K M
second message	F K K Y Y Y Q Q O C
adjustment	07 12 03 11 04 02 14 21 06 06.
difference	
modulo 26	

These data form two disconnected chains of 8 and 4 letters.

```
1 2345 678 9 10 11 12 13 14 15 16 17 18 19 20 21 22 23 24 25 26
R...Q..F.M.....CN..K.....O
B.....D.T.Y...........
```

Sliding the plaintext alphabet along the first chain, a non-crashing situation with a reciprocal pair, namely (k n), is found for the following shift mod 26:

```
u v w x y z a b c d  e f g h i j  k l m n o p q r s t
R . . . Q . . F. M.....CN.. K.....O
```

Using all possibilities for reciprocal pairs, the following hull is obtained

```
u v w x y z a b c d  e f  g h i j  k l m n o p q r  s t
R . . . Q . . F J M . B . . . C N . D  K T . Y U . O
```

where, among the letters J B D T Y U that appear additionally, B D T Y have the correct relative distance that is found in the second chain. There are the seven 2-cycles (b f), (c j), (d m), (k n), (q y), (r u). Thus, fourteen possible starting positions of the fast rotor are revealed by the fourteen cases B C D F J K M N O Q R T U Y of the third letter of the encrypted message setting.

We shall see how for the rotors I...V, usually with some luck, the position of one notch (Section 14.2.3.5) could be determined, and thus which rotor was used as the fast rotor. This simplification meant in the Polish *bomba* (Section 14.4.3.1) and

with the Zygalski sheets (Section 14.4.3.2) a reduction in the number of rotor orders to be tested from 60 to $12 = 4 \cdot 3$.

The situation is not very different with the navy indicator system, which was the object of Turing's elaboration of Różycki's clock method. Again, if a message pair is found with *Verfahrenkenngruppen* that is encrypted message settings that coincide in the first two letters, say B B C and B B E, then the message settings also coincide in the first two letters, and the difference of their third letters that is the shift for adjustment equals the difference of C and E.

The following example goes back to a 1945 report [29: 18] by A.P. Mahon[33] that has recently been released: From several pairs of messages adjusted with the Banbury sheets, with the following pairs of nearby encrypted message settings and corresponding adjustment differences mod 26

B B C	B B E	02	E N F	E P Q	07
R W C	R W L	13	I U S	I U Y	03
Z D R	Z I X	05	S U D	S W I	23
P I C	P N X	21			

the third position of the encrypted indicator shows the following values.

first message setting	C C R C	F S D
second message setting	E L X X	Q Y I
adjustment difference		
modulo 26	02 13 05 21	07 03 23

The first four of these data form a cluster of the letters R X C E L

```
1 23456 789 10 11 12 13 14 15 16 17 18 19 20 21 22 23 24 25 26
R....X....C.E.........L..
```

Sliding the plaintext alphabet along the chain until a non-crashing situation is found, the following shift *modulo* 26 is possible.

```
t u v w x y z a b c d e f  g h i j k l m n o p q r s
R . . . X . . . C. E.........L..
```

---

[33] Patrick Mahon, a linguist, became head of Hut 8 when Alexander left in September 1944. With him was an excellent team: Rolf Noskwith, Joan Clark who once had been close to Alan Turing, and Richard Pendered.

Forming the hull of reciprocal pairs and using in addition the remaining three data, a non-crashing solution is obtained:

```
t u v w x y z a b c d e f g h i j k l  m n o p q r s
R S . .  Y X I . . D C F E . . Z . Q .  . . . L T U
```

There are seven 2-cycles: (cd)(ef)(iz)(lq)(rt) (su)(xy). Thus, fourteen possible starting positions of the fast rotor are revealed by the fourteen cases C D E F I L Q R S T U X Y Z of the third letter of the encrypted message settings.

14.5.3.3 *Stage 3.* In Stage 3 and using the second example for the pair RWC and RWL) of encrypted message settings, the plain message settings are of the following form **d, **q. Stephen Budiansky writes [3: 348] "The brilliant part of Turing's method[34] was the next leap" and Mahon describes it as follows: The turnover notch was located at a different position on the various rotors. On rotor IV, for example, it occurred between J and K (see Section 14.2.3.5). He continues to show that a contradiction arises since the transition at J → K lies between d and q and would mean a turnover of the middle wheel. Thus, rotor IV is ruled out as the fast rotor; likewise, rotor II with a transition at E → F and rotors VI, VII and VIII with a transition at M → N.

Indeed, the clock method and Banburismus at B.P. served to rule out rotors in the rightmost position and thus to reduce drastically the number of rotor orders to be tested. Usually for the Navy ENIGMA with luck from 336 to 42 = 7 · 6, otherwise that is in the case of rotors VI, VII, VIII to 126 = 3 · 42 and for the army and air force ENIGMA from 60 to 12 = 4 · 3. By more sophisticated reasoning rotors could sometimes be excluded as a middle rotor.

Until September 1943 Banburismus was an essential help at B.P., when enough Turing–Welchman BOMBEs were available. Joan Murray

---

[34]Budiansky forgets to mention that this idea was also developed by Różycki in about 1935: "...the clock, devised by Jerzy Różycki ...made it possible in certain cases to determine which rotor was at the far right side on a given day in a given Enigma net" (Władisław Kozaczuk, in 1984 [26: 290, 263]).

née Clarke (1917–1996) is said to have been one of the best Banburists in B.P. (Hugh Alexander [1: 52], Rolf Noskwith [12: 209]).

As mentioned at the end of Section 14.5.1.2, it was November 1940 before the Banburismus procedure helped Hugh Foss break a naval Enigma setting for the first time.

14.5.4 Other methods

14.5.4.1 *Depth cribbing and in-depth adjusting by double cribbing.* A pair of isologs in an ENIGMA cipher, forming a depth of two, allows one to make a particular use of a crib for one of the ciphertexts to obtain a fragment of the plaintext belonging to the other ciphertext. Not only lead identical ciphertext letters in some position to identical plaintext letters in this position but, moreover due to the self-reciprocal property this also holds crosswise which is demonstrated in the following example (Patrick Mahon [29: 32]) by way of the bold letters:

```
1 2 3 4  5 6  7 8 9  10 11 12 13 14 15 16 17 18 19 20
B H N W  S M  S A W  M  N  T  C  K  N  N  P  Z
w e t t  e r  f u e  r  d  i  e  n  a  c  h  t
C N N J  T R  Q N W  S  T  T  C  X  R  C  D  S  L  D
. . t .  . m  . e .  .  i  e  .  .  n  .  .  .  .
```

In the circumstances of this example, mine sweepers played a role and the message might contain a boat number; moreover, a tripling of the m designating the boat class could be guessed. This leads to conjecture

m i t m m m d r e i s i e b e n e i n s, in clear M 371.

Since the crib, as usual was to be tested only in the few non-scritching positions (see Section 14.5.2.2), the method is quite effective provided that good insight into the enemy's usage has already been built up.

Cribbing can also be used for the in-depth adjusting of two ciphertexts. A hit is

```
1 2 3 4 5 6  7 8  9 10 11 12 13  14 15 16 17 18 19 20 21 22 23 24 25
Q W A W S U  H D W M T N C N K H P Z F H Y
  v o r h e r s a g e b e r e i c h d r e i
F D Q R L T  U L  E W G D Q  P  B O X N Z R N C I  O Z
z u s t a n d o s t  w a e r t i g e r k a n a l x
```

This method, which helped the Banburists in particular during the blackout caused by the introduction of the M 4 in 1942, died with Banburismus in autumn 1943 after enough BOMBEs were available.

14.5.4.2 *EINSing: a clever brute-force method.*

For progressive polyalphabetic encryption with a known sequence of known alphabets, all that is needed is to bring the alphabets in phase with the cipher text. For not too long periods, this correlation can be easily mechanised by prefabricating a catalogue with the cryptotext equivalents of a very frequent short word, such as in German /der/, /und/, /die/ or a word often used in certain circumstances, such as the German numeral /eins/. For example, the 26 positions of the single rotor I of the *Wehrmacht* ENIGMA (see Fig. 14.54) encipher /eins/ as follows (and lead to the following catalogue in alphabetic order).

A	L	Y	F	F		A	M	U	V	P
B	F	L	U	X		B	Q	Q	M	C
C	B	Q	Q	M		E	C	P	I	O
D	N	K	N	W		F	L	U	X	B
E	R	R	J	C		G	L	H	U	I
F	U	S	T	W		G	V	X	Y	Z
G	H	A	A	B		H	A	A	B	G
H	M	P	S	C		H	N	X	F	N
I	G	L	H	U		I	J	V	V	W
J	N	I	R	Z		I	K	D	K	U
K	O	E	X	T		J	R	W	B	S
L	W	O	R	P		K	S	O	A	Q
M	L	V	W	B	→	L	V	W	B	M
N	H	N	X	F		L	Y	F	F	A
O	E	C	P	I		M	P	S	C	H
P	A	M	U	V		N	F	P	S	X
Q	K	S	O	A		N	I	R	Z	J
R	R	M	K	U		N	K	N	W	D
S	J	R	W	B		O	E	X	T	K
T	Y	S	A	C		O	P	Q	Z	V
U	I	K	D	K		O	R	W	O	Y
V	O	P	Q	Z		R	M	K	U	R
W	I	J	V	V		R	R	J	C	E
X	N	F	P	S		U	S	T	W	F
Y	O	R	W	O		W	O	R	P	L
Z	G	V	X	Y		Y	S	A	C	T

For the cryptotext fragment UYPMMBMDYL-VWBHZB, four-letter groups are to be looked up successively in the catalogue and at the tenth trial there is the hit <u>LVWB</u> with setting $i = 14$, key letter *M*. Entering the complete cryptotext fragment UYPMMBMDY<u>LVWB</u>HZB into the table of rotated alphabets (Fig. 14.54) reveals the complete plaintext which turns out to read /nummerzwo**eins**aqt/; the surrounding of /eins/ makes sense and thus gives confirmation.

The method was applied by Turing for the 17,576 rotor positions of the 3-rotor *Wehrmacht* ENIGMA to decipher the remaining signals of a single day after one signal was broken and thus the wheel order and steckering of this day were known.

Turing had chosen /eins/ as a probable word when he found in his deciphered naval traffic of five days in November 1938 that /eins/ occured in about ninety percent of all messages. Another test word used was /krkr/ which indicated a priority message. In 1940 the British started using this method, called EINSing (Fig. 14.54), and finally even dedicated machines called drag grenades used for mechanising were built by the US navy, allowing arbitrary cribs of four letters or less for example frequent words such as /der/ or /und/ and also camouflaged priority tokens such as /bine/ or /muke/.

14.5.4.3 *AUTOSCRITCHER and SUPERSCRITCHER.*

A variant of the method of isomorphs (see Section 14.3.1.2) enables the determination of all 2-cycles of an unknown reflector *U*. This method became necessary for the Allies, when in early 1944 the Germans from time to time used a plugged reflector in the *Luftwaffe*-ENIGMA (see Section 14.3.2.1). Now, the method of isomorphs was carried through for all $26^3 = 17,576$ initial indexes, with suitable probable words. This procedure was not manually feasible and therefore required special machines. The relay machine AUTOSCRITCHER (workable 1944) and the electronic machine SUPERSCRITCHER (workable 1946) were built in the USA by the F Branch of the Army Signal Security Agency under the command of Colonel Leo Rosen.[35]

---

[35]In the open literature, the words *scritch*, *scritchmus* were used without detailed explanation, e.g., in 1993 by Derek

	a	b	c	d	e	f	g	h	i	j	k	l	m	n	o	p	q	r	s	t	u	v	w	x	y	z	
A	E	K	M	F	L	G	D	Q	V	Z	N	T	O	W	Y	H	X	U	S	P	A	I	B	R	C	J	
B	J	L	E	K	F	C	P	U	Y	M	S	N	V	X	G	W	T	R	O	Z	H	A	Q	B	I	D	
C	K	D	J	E	B	O	T	X	L	R	M	U	W	F	V	S	Q	N	Y	G	Z	P	A	H	C	I	
D	C	I	D	A	N	S	W	K	Q	L	T	V	E	Ⓤ	R	P	M	X	F	Y	O	Z	G	B	H	J	n
E	H	C	Z	M	R	V	J	P	K	S	U	D	T	Q	O	L	W	E	X	Ⓨ	F	A	G	I	B		u
F	B	Y	L	Q	U	I	O	J	R	T	C	S	Ⓟ	N	K	V	D	W	M	X	E	Z	F	H	A	G	m
G	X	K	P	T	H	N	I	Q	S	B	R	O	Ⓜ	J	U	C	V	L	W	D	Y	E	G	Z	F	A	m
H	J	O	S	G	Ⓜ	H	P	R	A	Q	N	L	I	T	B	U	K	V	C	X	D	F	Y	E	Z	W	e
I	N	R	F	L	G	O	Q	Z	P	M	K	H	S	A	T	J	U	Ⓑ	W	C	E	X	D	Y	V	I	r
J	Q	E	K	F	N	P	Y	O	L	J	G	R	Z	S	I	T	A	V	B	D	W	C	X	U	H	Ⓜ	z
K	Ⓓ	J	E	M	O	X	N	K	I	F	Q	Y	R	H	S	Z	U	A	C	V	B	W	T	G	L	P	w
L	I	D	L	N	W	M	J	H	E	P	X	Q	G	R	Ⓨ	T	Z	B	U	A	V	S	F	K	O	C	o
M	C	K	M	V	Ⓛ	I	G	D	O	W	P	F	Q	X	S	Y	A	T	Z	U	R	E	J	N	B	H	**e**
N	J	L	U	K	H	F	C	N	Ⓥ	O	E	P	W	R	X	Z	S	Y	T	Q	D	I	M	A	G	B	**i**
O	K	T	J	G	E	B	M	U	N	D	O	V	Q	Ⓦ	Y	R	X	S	P	C	H	L	Z	F	A	I	**n**
P	S	I	F	D	A	L	T	M	C	N	U	P	V	X	Q	W	R	O	Ⓑ	G	K	Y	E	Z	H	J	**s**
Q	Ⓗ	E	C	Z	K	S	L	B	M	Z	O	U	W	P	V	Q	N	A	F	J	X	D	Y	G	I	R	a
R	D	B	Y	J	R	K	A	L	S	N	T	V	O	U	P	M	Ⓩ	E	I	W	C	X	F	H	Q	G	q
S	A	X	I	Q	J	Z	K	R	M	S	U	N	T	O	L	Y	D	H	V	Ⓑ	W	E	G	P	F	C	t
T	W	H	P	I	Y	J	Q	L	R	T	M	S	N	K	X	C	G	U	A	V	D	F	O	E	B	Z	
U	G	O	H	X	I	P	K	Q	S	L	R	M	J	W	B	F	T	Z	U	C	E	N	D	A	Y	V	
V	N	G	W	H	O	J	P	R	K	Q	L	I	V	A	E	S	Y	T	B	D	M	C	Z	X	U	F	
W	F	V	G	N	I	O	Q	J	P	K	H	U	Z	D	R	X	S	A	C	L	B	Y	W	T	E	M	
X	U	F	M	H	N	P	I	O	J	G	T	Y	C	Q	W	R	Z	B	K	A	X	V	S	D	L	E	
Y	E	L	G	M	O	H	N	I	F	S	X	B	P	V	Q	Y	A	J	Z	W	U	R	C	K	D	T	
Z	K	F	L	N	G	M	H	E	R	W	A	O	U	P	X	Z	I	Y	V	T	Q	B	J	C	S	D	

Figure 14.54. Cipher table for rotor I of the *Wehrmacht* ENIGMA and Method of EINSing for decrypting UYPMMBMDY<u>LVWB</u>HZB.

As far as steckering is concerned, the plugboard ruins the method of isomorphs because the unknown plugboard connection or *steckering* hides the probable plaintext word. Now, it is necessary to find repeated pairs of plaintext and corresponding cryptotext characters. Each group of such pairs is mapped under the plugboard substitution into a group of corresponding characters from two isomorphs. Isomorphism requires that the groups are

kept together when all rotor positions are tested. The machines AUTOSCRITCHER and SUPER-SCRITCHER were designed to carry out this task, too. The US navy's DUENNA and the British GIANT were related.

A manual procedure called the Hand-Duenna was described in 1944 by Hugh Alexander and now contained in the Fried Reports of the U.S. Army liaison, National Archives and Records Administration (NARA), Record Group 457, NSA Historical Collection, Box 880, Nr. 2612.

14.5.5 Help from characteristics of German cryptographic procedure

On a number of occasions, prescribed rules of the cryptographic procedure with the intent to avoid

---

Taunt when describing the atmosphere of the work at Bletchley Park with reference to the duties of Dennis Babbage (see Section 14.5.2.6), mentioning also the pluggable reflector [18: 108]. The origin of the term is therefore to be sought in Britain. Recent work by Cipher A. Deavours ([8: 137], 1995) has established the connection with the method of isomorphs. (*Scritch* is, according to Merriam–Webster, a dialect variant of *screech*.)

regularity turned out to be *complications illusion-aires* that actually helped the British. "German methodicalness is here again at the bottom of a very foolish cryptographic habit" as stated in documents of the Signal Security Agency, 1944.

In exceptional cases, even German top leaders gave the enemy support through odd orders. For example Karl Dönitz in his famous Order of the Day (March 14, 1942) to his '*U-bootsmänner*' announced his promotion to Großadmiral. Many re-encodements in different keys gave excellent long cribs.

14.5.5.1 *Wheel order and steckering rules.*

The German air force [23: 113] and the German navy [1: 68] never used the same rotor in the same position two successive days of the same keying period. Clearly, such a thing should not occur regularly, but to prevent it strictly meant that the number of rotor orders that the bombes had to test was reduced considerably. For the navy ENIGMA, instead of the $8 \cdot 7 \cdot 6 = 336$ cases of selecting tree rotors out of eight, the number was only $7 \cdot 6 \cdot 5 = 210$, a considerable reduction to 62.5%.

In addition the German navy always had a rotor VI, VII or VIII in their wheel order. Alexander stated "as the three latest introduced they had a special sanctity for the keymaker". This rule alone reduced the number of 336 wheel orders to be tested by 60 (Alexander [1: 68]).

Likewise, a letter was never steckered for two days in succession. This rule was a most encouraging discovery. In 1939 when still only six steckers were used, "it meant that having broken one day twelve unsteckered letters for the next day were known so that methods applicable to the easier unsteckered machine that is the commercial ENIGMA could be used" (Alexander [1: 23]).

14.5.5.2 *Paired days.*

The complete key of the German naval ENIGMA procedure did not change every day. On the second day of a pair of odd-numbered and even numbered days of the month the wheel order and the ring setting would be the same as for the first day, only the plugboard steckering and the basic wheel setting changed. In months with 31 days there was a triplet of days

with the same wheel order and the ring setting. With known wheel order, only one BOMBE run was necessary (Alexander [1: 13]).

When on, say the first and second day the wheel order was determined to be say II VI VIII and on the 5th VIII VII IV, then on the 3rd and 4th neither II nor VIII would be in the first position and neither VI nor VII in the second nor VIII nor IV in the third. This convention reduced the number of wheel orders to be tested considerably (Alexander [1: 32]).

14.5.5.3 *Consecutive stecker knock-out.*

On the plugboard, the German air Force [23: 113] never connected a letter to the next in alphabetic order. This rule became so certain that the British introduced a special modification in their BOMBES called Consecutive Stecker Knock-Out (C.S.K.O.). which takes advantage of this fact in BOMBE operations. If a stop possible solution is reached which would require the steckering of two alphabetically consecutive steckers, the BOMBE automatically eliminates such a stop. Thus the number of false stops is materially reduced.

Mahon [29: 87] as well as Alexander [1: 61] report on a process called Stecker Knockout (S.K.O.) which according to Mahon was a lengthy and laborious hand method requiring rather long cribs of 400 letters or more which was successfully used on May 27, 1943 on messages in the keynet Shark when the wiring of the new rotor C Caesar and reflector Gamma was to be found out.

It seems that there was no connection between C.S.K.O. and S.K.O.

14.5.5.4 *Parkerism.*

In late 1941, the Germans decided to organise their ENIGMA traffic, avoiding passing too much of it on any one key by introducing a greater number of key nets. Most of the keys were issued for a month at a time and were changed every day. The production of keys grew into a business. It was natural to keep lists of the keys issued, and for saving of time and effort keys from previous periods could be reused. Of course, the British also kept records of the keys they had broken, and Reg Parker, who kept the catalogue of broken keys in Hut 6, one day discovered a regularity in the repeating of keys. After a few days of a new month, the observed regularities gave first hints about wheel orders, ring setting, steckering and so on [38: 131, 167]. Parkerism, as the effect was called, blossomed in 1942. Later it came less relevant possibly because the Germans had started to monitor some of their procedures.

14.5.5.5 *METEO code.* The Meteo code was a three-letter code used by the German air force for communicating weather information between an airfield and aircraft in flight. It first appeared at the turn of 1939/1940, but was not given much attention then. It was finally found out that a self-reciprocal permutation was being applied to the letters of each code group in the message. Then the astonishing discovery was made that this permutation was the stecker permutation of the ENIGMA key for the day.

So the first step of the Knox method, as it was called [38: 222], was to break the Meteo code for each day, which was not difficult. This result gave the steckering, and together with a Herivel tip (see Section 14.5.6.3) for finding the ring setting, and the use of cillies (see Section 14.5.6.1) for finding the wheel order, and the rest was easy. The Knox method was used by the Polish at Bruno and Cadix (Section 14.5.7).

14.5.5.6 *'pronounceables' and 'guessables' – continued use of indicator doubling.* It may seem to be incredible, since the German navy was always more worried about security than *Heer* and *Luftwaffe*, but the following is true. The German navy continued after May 1940, as Bletchley Park discovered, the dangerous use of double encipherment for message setting in at least five key nets: Grampus (*Poseidon*), Trumpeter (*Uranus*), Porpoise (*Hermes*), Sunfish (*Tibet*), and Seahorse

(*Bertok*), Bletchley Park even missed the key net *Süd*, the precursor of *Poseidon*, *Uranus* and *Hermes*. As one would expect, although the use of keyboard sequences for message keys had disappeared, there were from time to time pronounceable pords and guessable words in use, such as /eva/, /amt/ [1: 40].

14.5.6 Help from bad habits of German crypto operators

14.5.6.1 *Cillies.* "When cillies were discovered by Dilly Knox in late January 1940, they enormously reduced the work involved in using the Zygalski sheets. ... After 1 May, when the Zygalski sheets became useless following the dropping of double enciphered message keys, cillies became a vital part of breaking ENIGMA by hand, even when the first usable BOMBE entered service in August. Although more BOMBEs became available in 1941, cillies greatly eased the burden on them, thereby making bombes available for work against keys which could not otherwise have been attacked. Cillies were still valuable in 1943, and probably until a German procedural change in mid-1944" (Ralph Erskine [12: 453]).

Cillies resulted from an interplay between two different errors some unskilled ENIGMA operators loved to make in a multi-part message. One was their practise of leaving the rotors untouched when they reached the end of some part of a multi-part message, which meant that the three letters then showing in the rotor windows, say BDH (Fig. 14.55), formed the new basic wheel setting, which was therefore transmitted in unenciphered form in front of the next part. Since a letter count of each message part was included in the preamble, the unenciphered message key of the preceding

Part	Basic wheel setting	Plain message key	Enciphered message key	Length
1	GXO	/asd/	QPG	238
2	BDH	/fgh/	RZL	241
3	FPO	/qay/	PRB	244
4	QKI	/paw/	FML	141
5	PGH		MOT	95

Figure 14.55. Multi-part message (plain message keys are deduced).

	-11	-10	-9	-8	-7	-6	-5	-4	-3	-2	-1	0
								d	e	f	g	h
	s̲	t̲̄	ū	v	w	x	y	z	a	b	c	d
											a̲	b̄

Figure 14.56. b̄ūd, b̄t̄d, a̲t̲d̲, or a̲s̲d̲ are $9 \cdot 26 + 4$ steps before BDH.

R_L	R_M	R_N
I	II	III
IV	II	III
V	II	III
I	II	V
III	II	V
IV	II	V

Figure 14.57. 6 possible wheel orders under knowledge of N: III or V, M: II.

part could be calculated. Assume the openly transmitted next basic wheel setting was BDH and the length of the preceding part was $238 = 9 \cdot 26 + 4$. Then counting back 4 steps from H gives d as the third letter. Counting back 9 steps from D gives overlined u or u as the second letter, depending on whether a turnover happened in the middle rotor, provided that no turnover happened in the slow rotor, which leaves b as the first letter; otherwise this procedure gave underlined s or s as the second letter and a as the first letter if a turnover happened in the slow rotor (Fig. 14.56).

The other mistake was to use pronounceable words such as /ada/, or guessable words such as /ler/ (from Hitler) or /rom/ (from Rommel) or even keyboard sequences as message keys after enciphering in the preamble. This mistake frequently made it possible to decide among the four candidates, in our case among b u d, b t d, a t d, or a s d: /asd/ is a keyboard sequence (see Fig. 14.40) and a good guess for the unenciphered message key of part 1. Likewise, $241 = 9 \cdot 26 + 7$ steps back from FPO gives a choice among the four candidates f g h, f f h, e f h and e e h, among which /fgh/ is the favourite. Next, QKI and $244 = 9 \cdot 26 + 10$ delivers q b y, q a y, p a y and p z y with the keyboard sequence /qay/; finally, PGH and $141 = 5 \cdot 26 + 11$ gives p b w, p a w, o a w and o z w; and /paw/ is a keyboard sequence. The cilli has given four cribs of three letters each, which is a very good start.

Apart from finding the message key, there was connected finding whether a turnover occurred or not and thus information was obtained about the rotors used as the fast and the middle rotor. In the case used here the fast rotor is II, the only one that has a turnover between D and H; the middle rotor is III or V and there remain 3 choices for the slow rotor and altogether 6 rotor orders instead of 60 (Fig. 14.57).

August 1944 the *Heer* changed the procedure to require operators to use fully random lists of message keys. This improvement ended the use of the cillies.

A related oddity was reported in 1982 by Gordon Welchman [38: 104]. He gave the following example of what he also called sillies. In a three-part message the basic settings and encrypted message settings were

First part: QAY MPR

Second part: EDC LIY

Third part: TGB VEA.

The basic settings QAY, EDC and TGB are obviously keyboard diagonals. It can be guessed that the alternating diagonals /wsx/, /rfv/, /zhn/ are the plain message settings that is

at the setting QAY, MPR is the encrypted /wsx/,

at the setting EDC, LIY is the encrypted /rfv/,

at the setting TGB, VEA is the encrypted /zhn/.

This gives three cribs of three letters each, a good start.

14.5.6.2 *JABJAB.* Dennis Babbage, in mid-1940, found another incredible narrow-mindedness of German ENIGMA operators. To select as message setting the same 3-letter-group as chosen for a basic setting simply by moving back the right-hand wheel three places before encoding the message. This error happened particularly often in connection with the use of keyboard stereotypes. According to Welchman [38: 102], JABJAB was coined from the letters /jab/ used in its first appearance. Sebag-Montefiore [34: 338] gives a similar explanation for cilli: the letters /cil/.

14.5.6.3 *Herivel tips.* John Herivel observed in May 1940 [38: 222] that for the first signal of the day the wheel setting was frequently close to if not identical with the position of the wheels for the ring setting of the day that is the Herivel tip. "To set an alphabet ring on a wheel, he [the ENIGMA operator] would probably hold the wheel in one hand so that the clip position was facing him, and then rotate the ring until the correct letter was opposite the clip position" (Welchman). Welchman gives an example where among 30 message settings of a specific day four formed a cluster in particular H S K, G T K, G R I, F R J. The cryptanalysts would guess that the ring setting is within the cube of 27 triplets

L: F or G or H
M: R or S or T
N: I or J or K

and would start with the most probable six triplets

L: G
M: R or S
N: I or J or K,

which a tremendous reduction of the 17,576 possible ones.

As mentioned in Section 14.5.5.5 in connection with METEO code, by a combination of sillies for determining the message setting and Herivel tips for determining the ring setting, and many breaks were made, the first one was on May 22, 1940.

Since the key net Red of the *Luftwaffe* was particularly interesting for the British after the 1 May 1940 blackout and was one of the first broken before the BOMBEs were usable, it seems likely that the arrogant *Luftnachrichtentruppe* was the main provider in sillies, Herivel tips, and JABJABs. Sebag-Montefiore states that they were explicitly forbidden in the German manuals. But who would have dared to supervise Görings braintrust?

14.5.6.4 *Sundries.* There are many minor details of cunning cryptanalytical tricks, examples being the utilisation of the *Werftschlüssel* used in the German Navy for home-based communication which was not too resistant to cryptanalysis and gave many cribs [12: 189], or the gardening of cribs

by provoking certain actions of the enemy [2: 378, 433]. It would go too far trying to list them all.

14.5.7 1940: The Polish cryptanalysts in France and in Britain

Rejewski, Zygalski and Różycki, escaping the Polish disaster, fled via Roumania to France. End of September 1939, they joined the French radio intelligence group under Commandant Bertrand (Barsac) in the *Château de Vignolles* near Gretz-Armainvilliers with cover name *Poste de Commandement Bruno*, 30 miles southeast of Paris. Starting January 3, 1939, when the British liaison officer Capt. MacFarlane brought the first set and carried on until the German attack (*Fall Gelb*) on France in May 1940, Group Z worked with Zygalski sheets provided by Bletchley Park, and solved mainly German Army administration signals from key net Green, nearly the same number as B.P. On January 17, 1940 the messages of October 28, 1939 were solved and on January 28, 1940 those of September 3, 1939). Later and more importantly the *Luftwaffe* signals from key net Red, and finally, after the Norway invasion the signals of *Fliegerführer* Trondheim (key net Yellow starting April 10, 1940).

On the British side, the first success concerned the Green key of October 25, 1939, broken around January 17, 1940 and the Red key of January 6, 1940 which was broken around January 25, 1940. Communication between B.P. and *Bruno* was well established and by April 1940 even a direct teletype line was operating.

After the collapse of France, *P.C. Bruno* was first transferred on June 24, 1940 to Oran in Algeria then in October 1940 transferred back to the *Château des Fouzes* near Uzès with Cover name Cadix in the unoccupied part of France. Since the Zygalski sheets had become useless by May 1, 1940, the Poles and British had to rely for a while on Cillies and Herivel tips (Sections 14.5.6.1, 14.5.6.3). The Polish unit under Lieutenant-Colonel Gwido Langer, named 'Expositur 3000' by the British, was evacuated November 9, 1942 after the landing of the Allies in North Africa and the German occupation of the rest of France. Rejewski and Zygalski were imprisoned for a while in Spain and reached London by way of Gibraltar on August 3, 1943. They continued with cryptanalytic work on hand ciphers; however, they were kept away from Bletchley Park where the Turing–Welchman BOMBEs and the COLOSSUS machines were located.

14.6 CONCLUSIONS

Rotor machines are historical artifacts. No new rotor machines have had a chance to find a market since 1965, and thus machines for deciphering rotor encryptions such as BOMBEs now only hold the interest of historians.

There are several reasons for this development. Although rotor encryptions, like many other encryption techniques, can very easily, cheaply, and quickly be simulated by a modern computer, there is no need to program an available computer which can range from a small laptop to a big supercomputer with this simulation. The real need is to run a safe, unbreakable cryptosystem with an infinite, genuinely random key, in short a holocryptic key on a computer. Thus, the task is reduced to the generation of a holocryptic key. The first to give a stringent definition of a holocryptic key was Gregory J. Chaitin in circa 1975, based on the work of A.N. Kolmogorov.

We prefer the algorithmic definition Chaitin has given in [6: 47]:

> A holocryptic key is an infinite index sequence of key elements such that for every finite subsequence there does not exist a shorter algorithmic characterisation than the listing of the subsequence (in short: 'randomness by algorithmic incompressability').

From this definition, it follows immediately that no holocryptic key can be generated by a finite machine which leads to two interpretations.

Theoretically, no machine-generated holocryptic key exists.

And practically, approximations are to be constructed that come arbitrarily close to the theoretical holocryptic key.

One theoretical possibility in the latter case is to overcome the finiteness of the machine by allowing quantum effects; in fact one of the early practical solutions was to generate alleged holocryptic

key sequences by electronic noise. This idea leads into the field of quantum cryptology where security relies on physical laws and no longer relies on limited machine power. This is a fashionable topic but "in any case, there will be many years or decades before the quantum computer will be in our daily life" (D. Meinrup [30: 128]).

REFERENCES

[1] H. Alexander, *Cryptographic History of Work on the German Naval Enigma* (PRO HW 25/1).

[2] F.L. Bauer, *Decrypted Secrets, Methods and Maxims of Cryptology*, 3rd edition, Springer-Verlag, Berlin (1997). 4th edition 2007.

[3] S. Budiansky, *Battle of Wits*, The Free Press, New York (2000).

[4] C. Burke, *Information and Secrecy*, The Scarecrow Press, Metuchen, NJ (1994).

[5] R. Candela, *Isomorphism and its Application in Cryptanalysis*, New York (1946).

[6] G.J. Chaitin, *Randomness and mathematical proof*, Scientific American 232 (5), 47–52.

[7] K. de Leeuw, *The Dutch invention of the rotor machine, 1915–1923*, Cryptologia 27 (2003), 73–94.

[8] C.A. Deavours, *The Autoscritcher*, Cryptologia 18 (1994), 137–148.

[9] C.A. Deavours, L. Kruh, *Machine Cryptography and Modern Cryptanalysis*, Dedham, MA (1985).

[10] Deutsches Reich, Reichspatentamt. Patentschrift Nr. 416 219 Klasse 42n Gruppe 14 (Sch 52638 IX/42n) Gewerkschaft Securitas in Berlin, *Chiffrierapparat*, Patentiert im Deutschen Reich vom 23 Februar 1918 ab Issued July 8 (1925).

[11] R. Erskine, *Naval Enigma: An astonishing blunder*, Intelligence and National Security 11 (1996), 468–473.

[12] R. Erskine, M. Smith (eds), *Action This Day*, Bantam Press, London (2001).

[13] P. Fitzgerald, *The Knox Brothers*, Macmillan, New York (1977).

[14] W. Freeman, G. Sullivan, F. Weierud, *Purple revealed: Simulations and computer-aided cryptanalysis of Angooki Taipu B*, Cryptologia 27 (2003), 1–43.

[15] W.F. Friedman, Preliminary Historical Report on the Solution of the B-machine, 14 October (1940).

[16] J. Garliński, *The Enigma War*, Charles Scribner's Son, New York (1980).

[17] D.H. Hamer, *G-312: An Abwehr ENIGMA*, Cryptologia 24 (2000), 41–54.

[18] F.H. Hinsley, A. Stripp (eds), *Codebreakers*, Oxford University Press, Oxford (1993).

[19] A. Hodges, *Alan Turing: The Enigma*, McGraw-Hill, New York (1982).

[20] D. Kahn, *The Codebreakers*, Macmillan, New York (1967).

[21] D. Kahn, *Why Germany lost the Code War*, Cryptologia 6 (1982), 26–31.

[22] D. Kahn, *Pearl Harbour and the inadequacy of cryptanalysis*, Cryptologia 15 (1991), 273–294.

[23] D. Kahn, *Seizing the Enigma*, Houghton-Mifflin, Boston (1991).

[24] S.J. Kelley, *Big Machines*, Aegean Park Press, Walnut Creek, CA (2001).

[25] W. Kozaczuk, *Geheimoperation WICHER*, Bernhard & Graefe Verlag, Koblenz (1889).

[26] W. Kozaczuk, *ENIGMA*, University Publications of America, Frederick, MD (1985).

[27] W. Kozaczuk, *ENIGMA – How the German Machine Cipher Was Broken, and How It Was Read by the Allies in World War Two*, Arms and Armour Press, London (1984).

[28] S. Kullback, *Statistical Methods in Cryptanalysis*, Aegean Park Press, Laguna Hills, CA (1976).

[29] P. Mahon, *History of Hut Eight*, (PRO HW 25/2) (1945).

[30] D. Meinrup, *Zur Mathematik von Quantencomputern*, Math. Semesterberichte 53 (1976), 109–128.

[31] A. Quirantes, *Model Z: A numbers-only Enigma version*, Cryptologia 28 (2004), 153–156.

[32] J. Rohwer, E. Jäckel (eds), *Die Funkaufklärung und ihre Rolle im Zweiten Weltkrieg*, Motorbuch Verlag, Stuttgart (1970).

[33] F. Rowlett, *The Story of Magic*, Aegean Park Press, Laguna Hills, CA (1998).

[34] H. Sebag-Montefiore, *Enigma: The Battle for the Code*, Weidenfeld & Nicolson, London (2000).

[35] G. Sullivan, F. Weierud, *The Swiss NEMA cipher machine*, Cryptologia 23 (1999), 310–328.

[36] A. Turing, *Mathematical Theory of ENIGMA Machine* (PRO HW 25/3).

[37] S. Türkel, *Chiffrieren mit Geräten und Maschinen*, Verlag von Ulrich Mosers Buchhandlung, Graz (1927).

[38] G. Welchman, *The Hut Six Story*, Simon and Schuster, New York (1983).

15

TUNNY AND COLOSSUS: BREAKING THE LORENZ *Schlüsselzusatz* TRAFFIC

B. Jack Copeland

The Turing Archive for the History of Computing
University of Canterbury
New Zealand

Contents

Abstract

More sophisticated than Enigma, the Lorenz SZ40 *Schlüsselzusatz* (codenamed 'Tunny' by the British) was used to protect the highest levels of German Army traffic. The attack on Tunny spawned the world's first large-scale electronic computer, Colossus. Built by the brilliant engineer T.H. Flowers, Colossus was in use by the Bletchley Park codebreakers at the beginning of 1944 (two years before the American ENIAC first operated, although the ENIAC is often said to have been the first electronic computer).

Keywords: Automatic Computing Engine (ACE), Bletchley Park, Colossus, first electronic computers, Fish, T.H. Flowers, Heath Robinson, Lorenz SZ40/42 *Schlüsselzusatz*, Manchester 'Baby' computer, M.H.A. Newman, Statistical Method, Tunny, A.M. Turing, Turing machine, W.T. Tutte.

15.1 INTRODUCTION

Colossus, the first large-scale electronic computer, was used against the German system of teleprinter encryption known at Bletchley Park as 'Tunny'. Technologically more sophisticated than Enigma, Tunny carried the highest grade of intelligence. From 1941 Hitler and the German High Command relied increasingly on Tunny to protect their communications with Army Group commanders across Europe. Tunny messages sent by radio were first intercepted by the British in June 1941. After a year-long struggle with the new cipher, Bletchley Park first read current Tunny traffic in July 1942. Tunny decrypts contained intelligence that changed the course of the war in Europe, saving an incalculable number of lives.

The Tunny machine was manufactured by the German Lorenz company.[1] The first model bore the designation SZ40, 'SZ' standing for 'Schlüsselzusatz' ('cipher attachment'). A later version, the SZ42A, was introduced in February 1943, followed by the SZ42B in June 1944 ('40' and '42' appear to refer to years, as in 'Windows 97'). Tunny was one of three types of teleprinter cipher machine used by the Germans. (The North American term for 'teleprinter' is 'teletypewriter'.) At Bletchley Park (B.P.) these were given the general cover name 'Fish'. The other members of the Fish family were *Sturgeon*, the Siemens and Halske T52 *Schlüsselfernschreibmaschine* ('Cipher Teleprinter Machine'), and the unbreakable *Thrasher*.[2] Thrasher was probably the Siemens T43, a one-time-tape machine. It was upon Tunny that B.P. chiefly focused. (Bletchley's work on Sturgeon is described in [34].)

15.2 THE TUNNY MACHINE

The Tunny machine, which measured $19''$ by $15\frac{1}{2}''$ by $17''$ high, was a cipher *attachment*. Attached to a teleprinter, it automatically encrypted the outgoing stream of pulses produced by the teleprinter, or automatically decrypted incoming messages before they were printed. (Sturgeon, on the other hand, was not an attachment but a combined teleprinter and cipher machine.) At the sending end of a Tunny link, the operator typed plain language (the 'plaintext' of the message) at the teleprinter keyboard, and at the receiving end the plaintext was printed out automatically by another teleprinter (usually

onto paper strip, resembling a telegram). The transmitted 'ciphertext' (the encrypted form of the message) was not seen by the German operators. With the machine in 'auto' mode, many long messages could be sent one after another – the plaintext was fed into the teleprinter equipment on pre-punched paper tape and was encrypted and broadcast at high speed. Enigma was clumsy by comparison. A cipher clerk typed the plaintext at the keyboard of an Enigma machine while an assistant painstakingly noted down the letters of the ciphertext as they appeared one by one at the machine's lamp-board. A radio operator then transmitted the ciphertext in the form of Morse code. Morse code was not used with Tunny: the output of the Tunny machine, encrypted teleprinter code, went directly to air (see Copeland [10: ch. 7]).

International teleprinter code assigns a pattern of five pulses and pauses to each character. Using the Bletchley convention of representing a pulse by a cross and no pulse by a dot, the letter C, for example, is •xxx•: no-pulse, pulse, pulse, pulse, no-pulse. More examples: O is ••xx, L is •x••x, U is xxx•• and S is x•x••. (The complete teleprinter alphabet is shown in Appendix 1.) When a message in teleprinter code is placed on paper tape, each letter (or other keyboard character) takes the form of a pattern of holes punched across the width of the tape. A hole corresponds to a pulse (cross).

The first Tunny radio link, between Berlin and Athens/Salonika, went into operation on an experimental basis in June 1941.[3] In October 1942 this experimental link closed down, and for a short time it was thought that the Germans had abandoned the Tunny machine.[4] Later that same month Tunny reappeared in an altered form, on a link between Berlin and Salonika and on a new link between Königsberg and South Russia.[5] At the time of the allied invasion in 1944, when the Tunny system had reached its most stable and widespread state,[6] there were 26 different links known to the British.[7]

---

[1] The physical Tunny machine is described in Section 11 of *General Report on Tunny*, and in [11]. The machine's function and use is described in Sections 11 and 94 of *General Report on Tunny*. *General Report on Tunny* was written at Bletchley Park in 1945 by Tunny-breakers Jack Good, Donald Michie and Geoffrey Timms; it was released by the British government in 2000 to the National Archives/Public Record Office (PRO) at Kew (document reference HW 25/4 (Vol. 1), HW 25/5 (Vol. 2)). A digital facsimile is available in The Turing Archive for the History of Computing, http://www.AlanTuring.net/tunny_report.

[2] On Thrasher, see Section 93 of *General Report on Tunny*.

[3] *General Report on Tunny*, p. 14. *General Report on Tunny* mentions that the first messages on the experimental link passed between Vienna and Athens (p. 297).

[4] *General Report on Tunny*, p. 320.

[5] *General Report on Tunny*, pp. 14, 320, 458.

[6] *General Report on Tunny*, p. 14.

[7] *General Report on Tunny*, p. 14.

Figure 15.1. The Tunny machine with its twelve encoding wheels exposed. *Source*: *General Report on Tunny*. Crown copyright; reproduced by permission of the National Archives Image Library.

B.P. gave each link a piscine name: Berlin–Paris was Jellyfish, Berlin–Rome was Bream, Berlin–Copenhagen Turbot. The two central exchanges for Tunny traffic were Strausberg near Berlin for the Western links, and Königsberg for the Eastern links into Russia.[8] In July 1944, the Königsberg exchange closed and a new hub was established for the Eastern links at Golssen, about 20 miles from the Wehrmacht's underground command headquarters south of Berlin. During the final stages of the war, the Tunny network became increasingly disorganised.[9] By the time of the German surrender, the central exchange had been transported from Berlin to Salzburg in Austria.[10]

There were also fixed exchanges at some other large centres, such as Paris.[11] Otherwise, the distant ends of the links were mobile. Each mobile Tunny unit consisted of two trucks.[12] One carried the radio equipment, which had to be kept well

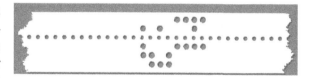

Figure 15.2. Punched paper tape containing the letters c o l o s s u s in teleprint code. The smaller holes in the centre (called 'sprocket holes') engage a toothed wheel that drives the tape.

away from teleprinters for fear of interference. The other carried the teleprinter equipment and two Tunny machines, one for sending and one for receiving. This truck also carried a device for punching tapes for auto transmission. Sometimes a land line was used in preference to radio.[13] In this case, the truck carrying the Tunnies was connected up directly to the telephone system. (Only Tunny traffic sent by radio was intercepted by the British.)

As with the Enigma, the heart of the Tunny machine was a system of wheels. Some or all of the wheels moved each time the operator typed a character at the teleprinter keyboard (or in the case of

---

[8] *General Report on Tunny*, p. 395.

[9] *General Report on Tunny*, p. 15.

[10] *General Report on Tunny*, p. 15.

[11] *General Report on Tunny*, p. 5.

[12] *General Report on Tunny*, p. 4.

[13] *General Report on Tunny*, p. 5.

an 'auto' transmission from a pre-punched tape, each time a new letter was read in from the tape). There were twelve wheels in all. They stood side by side in a single row, like plates in a dish rack. As in the case of Enigma, the rim of each wheel was marked with numbers, visible to the operator through a window, and somewhat like the numbers on the rotating parts of a combination lock.

From October 1942 the operating procedure was this. Before starting to send a message, the operator would use his thumb to turn the wheels to a combination that he looked up in a codebook containing one hundred or more combinations (known as the QEP book). At B.P. this combination was called the *setting* for that particular message. The wheels were supposed to be turned to a new setting at the start of each new message (although because of operator error this did not always occur). The operator at the receiving end, who had the same QEP book, set the wheels of his Tunny machine to the same combination, enabling his machine to decrypt the message automatically as it was received. Once all the combinations in a QEP book had been used it was replaced by a new one.

The Tunny machine encrypted each letter of the message by *adding* another letter to it. (The process of adding letters together is explained in the next paragraph.) The internal mechanism of the Tunny machine produced its own stream of letters, known at B.P. as the 'key-stream', or simply *key*. Each letter of the ciphertext was produced by adding a letter from the key-stream to the corresponding letter of the plaintext.

The Tunny machine adds letters by adding the individual dots and crosses that compose them. The rules that the makers of the machine selected for dot-and-cross addition are simple. Dot plus dot is dot. Cross plus cross is dot. Dot plus cross is cross. Cross plus dot is cross. In short, adding two sames produces dot, and adding a mixed pair produces cross. (Computer literati will recognise Tunny addition as boolean XOR.)

For example, if the first letter of the plaintext happens to be M, and the first letter of the key-stream happens to be N, then the first letter of the ciphertext is T: adding M (●●xxx) and N (●●xx●) produces T (●●●●x).

M		N		T
●	+	●	=	●
●	+	●	=	●
x	+	x	=	●
x	+	x	=	●
x	+	●	=	x

The German engineers selected these rules for dot-and-cross addition so that the following is always true (no matter which letters, or other keyboard characters, are involved): adding one letter (or other character) to another and then *adding it again a second time* leaves you where you started. In symbols, $(x + y) + x = y$, for every pair of keyboard characters x and y. For example, adding N to M produces T, as we have just seen, and then adding N to T leads back to M:

T		N		M
●	+	●	=	●
●	+	●	=	●
●	+	x	=	x
●	+	x	=	x
x	+	●	=	x

This explains how the receiver's Tunny decrypted the ciphertext. The ciphertext was produced by adding a stream of key to the plaintext, so by means of adding exactly the same letters of key to the ciphertext, the receiver's machine wiped away the encryption, exposing the plaintext again.

For example, suppose the plaintext is the single word 'COLOSSUS'. The stream of key added to the plaintext by the sender's Tunny might be: WZHI/NR9. These characters are added serially to the letters of 'COLOSSUS':

C + W   O + Z   L + H   O + I   S + /   S + N   U + R   S + 9.

This produces

XDIVSDFE

(as can be checked by using the table in Appendix 1). 'XDIVSDFE' is transmitted over the link. The Tunny at the receiving end adds the same letters of key to the encrypted message:

X + W   D + Z   I + H   V + I   S + /   D + N   F + R   E + 9.

This uncovers the letters

COLOSSUS.

The Tunny machine in fact produces the key-stream by adding together two other letter streams, called at B.P. the *psi*-stream and the *chi*-stream (from the Greek letters psi (ψ) and chi (χ)). The psi-stream and the chi-stream are produced by the wheels of the Tunny machine. Let us consider the wheels in more detail.

The twelve wheels form three groups: five psi-wheels, five chi-wheels and two motor wheels. Each wheel has different numbers of cams (sometimes called 'pins') arranged evenly around its circumference (the numbers varying from 23 to 61). The function of the cam is to push a switch as it passes it, so that as the wheel rotates a stream of electrical pulses is generated. The operator can adjust the cams, sliding any that he selects sideways, so that they become inoperative and no longer push the switch when they pass it. The wheel now causes not a uniform stream of pulses as it turns, but a pattern of pulses and non-pulses – crosses and dots. The arrangement of the cams around the wheel, operative or inoperative, is called the *wheel pattern*.

Prior to the summer of 1944 the Germans changed the cam patterns of the chi-wheels once every month and the cam patterns of the psi-wheels at first quarterly, then monthly from October 1942. After 1 August 1944 wheel patterns changed daily. The changes were made according to books of wheel patterns issued to Tunny units (different links used different books).

It is the patterns of the cams around the wheels that produces the chi-stream and the psi-stream. Whenever a key is pressed at the keyboard (or a letter read in from the tape in 'auto' mode), it causes the five chi-wheels to turn in unison, just far enough for one cam on each wheel to pass its switch. Depending on whether or not that cam is operative, a pulse may or may not be produced. Suppose, for example, that the cam at the first chi-wheel's switch produces no pulse and the cam on the second likewise produces no pulse at its switch, but the cams on the third and fourth both produce a pulse, and the cam on the fifth produces no pulse. Then the pattern that the chi-wheels produce at this

point in their rotation is ●●xx●. In other words, the chi-stream at this point contains the letter N. The five psi-wheels also contribute a letter (or other keyboard character) and this is added to N to produce a character of the key-stream.

A complication in the motion of the wheels is that, although the chi-wheels move forward by one cam *every* time a key is pressed at the keyboard (or a letter arrives from the tape in auto mode, or from the radio receiver), the psi-wheels move irregularly. The psis might all move forward with the chis, or they might all stand still, missing an opportunity to move. This irregular motion of the psi-wheels was described as 'staggering' at B.P. Designed to enhance the security of the machine, it turned out to be the crucial weakness.

Whether the psi-wheels move or not is determined by the motor wheels (or in some versions of the machine, by the motor wheels in conjunction with yet other complicating factors). While the psis remain stationary, they continue to contribute the same letter to the key. So the chis might contribute

…KDUGRYMC…

and the psis might contribute

…GGZZZWDD….

Here the chis have moved eight times and the psis only four.

15.3   A SAMPLE DECRYPT

Pages 452–453 display a rare survivor – a word-for-word translation of an intercepted Tunny message.[14] Dated 25 April 1943 and signed by von Weichs, Commander-in-Chief of German Army Group South, this message was sent from the Russian front to the German Army High Command ('OKH' – *Oberkommando des Heeres*). It

---

[14]British message reference number CX/MSS/2499/T14; PRO reference HW1/1648. Words enclosed in square brackets do not appear in the original. (Thanks to Ralph Erskine for assistance in locating this document. An inaccurate version of the intercept appears in (Hinsley [19: 764, 765]).)

To OKH/OP. ABT. and to OKH/Foreign Armies East,
from Army Group South IA/01, No. 411/43, signed von
Weichs, General Feldmarschall, dated 25/4:-

Comprehensive appreciation of the enemy for "Zitadelle"

In the main the appreciation of the enemy remains the
same as reported in Army Group South (Roman) IIA, No.
0477/43 of 29/3 and in the supplementary appreciation
of 15/4. *[In Tunny transmissions the word 'Roman' was
used to indicate a Roman numeral; '29/3' and '15/4' are
dates.]*

The main concentration, which was already then
apparent on the north flank of the Army Group in the
general area Kursk-Ssudsha-Volchansk-Ostrogoshsk, can
now be clearly recognised: a further intensification of
this concentration is to be expected as a result of the
continuous heavy transport movements on the lines
Yelets-Kastornoye-Kursk, and Povorino-Svoboda and
Gryazi-Svoboda, with a probable (B% increase) *['B%'
indicated an uncertain word]* in the area Valuiki-Novy
Oskol-Kupyansk. At present however it is not apparent
whether the object of this concentration is offensive
or defensive. At present, (B% still) in anticipation of
a German offensive on both the Kursk and Mius Donetz
fronts, the armoured and mobile formations are still
evenly distributed in various groups behind the front
as strategic reserves.

There are no signs as yet of a merging of these
formations or a transfer to the forward area (except
for (Roman) II GDS *[Guards]* Armoured Corps) but this
could take place rapidly at any time.

According to information from a sure source the
existence of the following groups of the strategic
reserve can be presumed:- A) 2 cavalry corps (III GDS
and V GDS in the area north of Novocherkassk). It can
also be presumed that 1 mech *[mechanised]* corps (V GDS)
is being brought up to strength here. B) 1 mech corps
(III GDS) in the area (B% north) of Rowenki. C) 1
armoured corps, 1 cavalry corps and probably 2 mech
corps ((Roman) I GD Armoured, IV Cavalry, probably (B%
(Roman) I) GDS Mech and V Mech Corps) in the area north
of Voroshilovgrad. D) 2 cavalry corps ((B% IV) GDS and
VII GDS) in the area west of Starobyelsk. E) 1 mech
corps, 1 cavalry corps and 2 armoured corps ((Roman) I

GDS (B% Mech), (Roman) I GDS Cavalry, (Roman) II and XXIII Armoured) in the area of Kupyansk-Svatovo. F) 3 armoured corps, 1 mech corps ((Roman) II Armoured, V GDS Armoured, (B% XXIX) Armoured and V GDS Mech under the command of an army (perhaps 5 Armoured Army)) in the area of Ostrogoshsk. G) 2 armoured and 1 cavalry corps ((Roman) II GDS Armoured, III GDS Armoured and VI GDS Cavalry) under the command of an unidentified H.Q., in the area north of Novy Oskol.

In the event of "Zitadelle", there are at present approximately 90 enemy formations west of the line Belgorod-Kursk-Maloarkhangelsk. The attack of the Army Group will encounter stubborn enemy resistance in a deeply echeloned and well developed main defence zone, (with numerous dug in tanks, strong artillery and local reserves) the main effort of the defence being in the key sector Belgorod-Tamarovka.

In addition strong counter-attacks by strategic reserves from east and southeast are to be expected. It is impossible to forecast whether the enemy will attempt to withdraw from a threatened encirclement by retiring eastwards, as soon as the key sectors *[literally, 'corner-pillars']* of the bulge in the frontline at Kursk, Belgorod and Maloarkhangelsk, have been broken through. If the enemy throws in all strategic reserves on the Army Group front into the Kursk battle, the following may appear on the battle field:- On day 1 and day 2, 2 armoured divisions and 1 cavalry corps. On day 3, 2 mech and 4 armoured corps. On day 4, 1 armoured and 1 cavalry corps. On day 5, 3 mech corps. On day 6, 3 cavalry corps. On day 6 and/or day 7, 2 cavalry corps.

Summarizing, it can be stated that the balance of evidence still points to a defensive attitude on the part of the enemy: and this is in fact unmistakable in the frontal sectors of the 6 Army and 1 Panzer Army. If the bringing up of further forces in the area before the north wing of the Army Group persists and if a transfer forward and merging of the mobile and armoured formations then takes place, offensive intentions become more probable. In that case it is improbable that the enemy can even then forestall our execution of Zitadelle in the required conditions. Probably on the other hand we must assume complete enemy preparations for defence, including the counter attacks of his strong mot *[motorised]* and armoured forces, which must be expected.

gives an idea of the nature and quality of the intelligence that Tunny yielded. The enciphered message was intercepted during transmission on the 'Squid' radio link[15] between the headquarters of Army Group South and Königsberg.

The message concerns plans for a major German offensive in the Kursk area codenamed '*Zitadelle*'. Operation *Zitadelle* was Hitler's attempt to regain the initiative on the Eastern Front following the Russian victory at Stalingrad in February 1943. *Zitadelle* would turn out to be one of the crucial battles of the war. Von Weichs' message gives a detailed appreciation of Russian strengths and weaknesses in the Kursk area. His appreciation reveals a considerable amount about the intentions of the German Army. British analysts deduced from the decrypt that *Zitadelle* would consist of a pincer attack on the north and south flanks ('corner-pillars') of a bulge in the Russian defensive line at Kursk (a line which stretched from the Gulf of Finland in the north to the Black Sea in the south).[16] The attacking German forces would then attempt to encircle the Russian troops situated within the bulge.

Highly important messages such as this were conveyed directly to Churchill, usually with a covering note by 'C', Chief of the Secret Intelligence Service.[17] On 30 April an intelligence report based on the content of the message, but revealing nothing about its origin, was sent to Churchill's ally, Stalin.[18] (Ironically, however, Stalin had a spy inside Bletchley Park: John Cairncross was sending raw Tunny decrypts directly to Moscow by clandestine means.[19])

The Germans finally launched operation *Zitadelle* on 4 July 1943 (Hinsley [19: 626]). Naturally the German offensive came as no surprise to the Russians – who, with over two months warning of the pincer attack, had amassed formidable defences. The Germans threw practically every panzer division on the Russian front into *Zitadelle* (Hinsley [19: 625]), but to no avail, and on 13 July Hitler called off the attack (Hinsley [19: 627]). A few days later Stalin announced in public that Hitler's plan for a summer offensive against the Soviet Union had been "completely frustrated" (Hinsley [19: 627]). *Zitadelle* – the Battle of Kursk – was a decisive turning point on the Eastern front. The counter-attack launched by the Russians during *Zitadelle* developed into an advance which moved steadily westwards, ultimately reaching Berlin in April 1945.

15.4   CENTRAL FIGURES IN THE ATTACK ON TUNNY

Colossus was the brainchild of Thomas H. Flowers (1905–1998). Flowers joined the Telephone Branch of the Post Office in 1926, after an apprenticeship at the Royal Arsenal in Woolwich (well known for its precision engineering). Flowers entered the Research Branch of the Post Office at Dollis Hill in North London in 1930, achieving rapid promotion and establishing his reputation as a brilliant and innovative engineer. At Dollis Hill Flowers pioneered the use of large-scale electronics, designing equipment containing more than 3000 electronic valves ('vacuum tubes' in the US). First summoned to Bletchley Park to assist Turing in the attack on Enigma, Flowers soon became involved in Tunny. After the war Flowers pursued his dream of an all-electronic telephone exchange, and was closely involved with the groundbreaking Highgate Wood exchange in London (the first all-electronic exchange in Europe).

Max H.A. Newman (1897–1984) was a leading topologist as well as a pioneer of electronic digital computing. A Fellow of St John's College, Cambridge from 1923, Newman lectured Turing on mathematical logic in 1935, launching Turing[20]

---

[15]Copy of message CX/MSS/2499/T14, PRO document reference HW5/242, p. 4.

[16]'A Postponed German Offensive (Operations ZITADELLE and EULE)' (anon., Government Code and Cypher School, 7 June 1943; PRO reference HW13/53), p. 2.

[17]Documents from G.C. & C.S. to Churchill, 30 April 1943 (PRO reference HW1/1648). An earlier decrypt concerning *Zitadelle* (13 April 1943), and an accompanying note from 'C' to Churchill, are at HW1/1606.

[18]Tape-recorded interview with Harry Hinsley (Sound Archive, Imperial War Museum, London, reference number 13523).

[19]Cairncross [2: 98], Hinsley [21: 322, 323], interview with Hinsley (see above).

[20]Newman in interview with Christopher Evans (The Pioneers of Computing: An Oral History of Computing, Science Museum, London).

on the research that led to the 'universal Turing machine', the abstract universal stored-program computer described in Turing's 1936 paper 'On Computable Numbers'. At the end of August 1942 Newman left Cambridge for Bletchley Park, joining the Research Section and entering the fight against Tunny. In 1943 Newman became head of a new Tunny-breaking section known simply as the Newmanry, home first to the experimental 'Heath Robinson' machine and subsequently to Colossus. By April 1945 there were ten Colossi working round the clock in the Newmanry. The war over, Newman took up the Fielden Chair of Mathematics at the University of Manchester and – inspired both by Colossus and by Turing's abstract 'universal machine' – lost no time in establishing a facility to build an electronic stored-program computer. On 21 June 1948, in Newman's Computing Machine Laboratory, the world's first electronic stored-program digital computer, the Manchester 'Baby', ran its first program.

John Tiltman (1894–1982) was seconded to the Government Code and Cypher School (GC & CS) from the British army in 1920, in order to assist with Russian diplomatic traffic (Erskine and Freeman [15]). An instant success as a codebreaker, Tiltman never returned to ordinary army duties. From 1933 onward he made a series of major breakthroughs against Japanese military ciphers, and in the early years of the war he also broke a number of German ciphers, including the army's double Playfair system, and the version of Enigma used by the German railway authorities. In 1941 Tiltman made the first significant break into Tunny. Promoted to Brigadier in 1944, he went on to become a leading member of GCHQ, GC & CS's peacetime successor. Following his retirement from GCHQ in 1964, Tiltman joined the National Security Agency, where he worked until 1980.

Alan M. Turing (1912–1954) was elected a Fellow of King's College, Cambridge in 1935, at the age of only 22. 'On Computable Numbers', published the following year, was his most important theoretical work. It is often said that all modern computers are Turing machines in hardware: in a single article, Turing ushered in

both the modern computer and the mathematical study of the *un*computable. During the early stages of the war, Turing broke German Naval Enigma and produced the logical design of the 'Bombe', an electro-mechanical code-breaking machine (Copeland [8]). Hundreds of Bombes formed the basis of Bletchley Park's factory-style attack on Enigma. Turing briefly joined the attack on Tunny in 1942, contributing a fundamentally important cryptanalytical method known simply as 'Turingery'. In 1945, inspired by his knowledge of Colossus, Turing designed an electronic stored-program digital computer, the Automatic Computing Engine (ACE). At Bletchley Park, and subsequently, Turing pioneered Artificial Intelligence: while the rest of the post-war world was just waking up to the idea that electronics was the new way to do binary arithmetic, Turing was talking very seriously about programming digital computers to think. He also pioneered the discipline now known as Artificial Life, using the Ferranti Mark I computer at Manchester University to model biological growth. (For further information on Turing see Copeland [6; 9].)

William T. Tutte (1917–2002) specialised in chemistry in his undergraduate work at Trinity College, Cambridge, but was soon attracted to mathematics. He was recruited to Bletchley Park early in 1941, joining the Research Section. Tutte worked first on the Hagelin cipher machine and in October 1941 was introduced to Tunny. Tutte's work on Tunny, which included deducing the structure of the Tunny machine, can be likened in importance to Turing's earlier work on Enigma. At the end of the war, Tutte was elected to a Research Fellowship in mathematics at Trinity; he went on to found the area of mathematics now called graph theory.

15.5 BREAKING THE TUNNY MACHINE

From time to time German operators used the same wheel settings for two different messages, a circumstance called a *depth*. It was thanks to the interception of depths, in the summer of 1941, that the Research Section at B.P. first found its way into Tunny.

Prior to October 1942, when QEP books were introduced, the sending operator informed the receiver of the starting positions of the 12 wheels by transmitting an unenciphered group of 12 letters. The first letter of the 12 gave the starting position of the first psi-wheel, and so on for the rest of the wheels. For example, if the first letter was 'M' then the receiver would know from the standing instructions for the month to set his first psi-wheel to position 31, say. At B.P. this group of letters was referred to as the message's *indicator*. (Sometimes the sending operator would expand the 12 letters of the indicator into 12 unenciphered names: Martha Gustav Ludwig Otto ..., for example, instead of MGLO ....) The occurrence of two messages with the same indicator was the tell-tale sign of a depth.

So when on 30 August 1941 two messages with the same indicator were intercepted, B.P. suspected that they had found a depth. As it turned out, the first transmission had been corrupted by atmospheric noise, and the message was resent at the request of the receiving operator. Had the sender repeated the message identically, the use of the same wheel settings would have left B.P. none the wiser. However, in the course of the second transmission the sender introduced abbreviations and other minor deviations (the message was approximately 4000 characters long). So the depth consisted of two not-quite-identical plaintexts each encrypted by means of exactly the same sequence of key – a codebreaker's dream.

On the hypothesis that the machine had produced the ciphertext by adding a stream of key to the plaintext, Tiltman added the two ciphertexts. If the hypothesis were correct, this would have the effect of cancelling out the key (since, as previously mentioned, $(x + y) + x = y$). The resulting string of approximately 4000 characters would consist of the two plaintexts summed together character by character. (This is because $(K + P) + (K + P) = ((K + P) + K) + P = P + P$, where K is the key, P is the plaintext, and $K + P$ is the ciphertext.)

Tiltman managed to prise the two individual plaintexts out of this string (it took him ten days). He guessed at words of each message, and Tiltman was a very good guesser. Each time he guessed a word from one message, he added it to the characters at the right place in the string, and if the guess

was correct an intelligible fragment of the second message would pop out. For example, adding the probable word 'geheim' (secret) to characters 83–88 of the string revealed the plausible fragment 'eratta' (Bauer [1: 372]). This short break can then be extended to the left and right. More letters of the second message are obtained by guessing that 'eratta' is part of 'militaerattache' (military attache), and if these letters are added to their counterparts in the string, further letters of the first message are revealed. And so on. Eventually Tiltman achieved enough of these local breaks to realise that long stretches of each message were the same, and so was able to decrypt the whole thing.

Adding the plaintext deduced by Tiltman to its corresponding ciphertext revealed the sequence of key used to encrypt the messages. These 4000 characters of key were passed to Tutte and, in January 1942, Tutte single-handedly deduced the fundamental structure of the Tunny machine. He focused on just one of the five 'slices' of the key-stream, the top-most row were the key-stream to be punched on tape. Each of these five slices was called an 'impulse' at B.P. (In the 'Colossus' punched tape shown earlier, the first impulse is ●●●●xxxx, the second is x●x●●●x●, and so on.)

The top-most impulse of the key-stream, Tutte managed to deduce, was the result of adding two streams of dots and crosses. The two streams were produced by a pair of wheels, which he called 'chi' and 'psi'. The chi-wheel, he determined, always moved forward one place from one letter of text to the next, and the psi-wheel sometimes moved forwards and sometimes stayed still. It was a remarkable feat of cryptanalysis. At this stage the rest of the Research Section joined in and soon the whole machine was laid bare, without any of them ever having set eyes on one.

15.6 TURINGERY

Now that Bletchley knew the nature of the machine, the next step was to devise methods for breaking the daily traffic. A message could be read if the wheel settings and the wheel patterns were known. The German operators themselves were

Figure 15.3. Alan M. Turing. *Source*: Beryl Turing and King's College Library, Cambridge.

revealing each message's setting via the 12-letter indicator. Thanks to Tutte's feat of reverse-engineering, the wheel patterns were known for August 1941. The codebreaker's problem was to keep on top of the German's regular changes of wheel-pattern.

In July 1942 Turing invented a method for finding wheel-patterns from depths – 'Turingery'. Turing was at that time on loan to the Research Section from Hut 8 and the struggle against Naval Enigma (Tutte [33: 359, 360]). Turingery was the third of the three strokes of genius that Turing contributed to the attack on the German codes, along with his design for the Bombe and his unravelling of the form of Enigma used by the Atlantic U-boats (see Copeland [6]). As fellow codebreaker Jack Good observed, "I won't say that what Turing did made us win the war, but I daresay we might have lost it without him".[21]

Turingery was a hand method, involving paper, pencil and eraser. Beginning with a stretch

of key obtained from a depth, Turingery enabled the breaker to prize out from the key the contribution that the chi-wheels had made. The cam-patterns of the individual chi-wheels could be inferred from this. Further deductions led to the cam-patterns of the psi- and motor-wheels. Once gained via Turingery, this information remained current over the course of many messages. Eventually the patterns were changed too frequently for any hand method to be able to cope (there were daily changes of all patterns from August 1944), but by that time Colossus, not Turingery, was being used for breaking the wheel patterns.

Basic to Turingery was the idea of forming the *delta* of a stream of characters. (Delta-ing a character-stream was also called 'differencing' the stream.) The delta of a character-stream is the stream that results from adding together each pair of adjacent letters in the original stream. For example, the delta of the short stream MNT (sometimes written ΔMNT) is produced by adding M to N and N to T (using the rules of dot-and-cross addition given previously). The delta of MNT is in fact TM, as the following table shows (the shaded columns contain the delta):

M	N	T	M + N	N + T
•	•	•	•	•
•	•	•	•	•
x	x	•	•	x
x	x	•	•	x
x	•	x	x	x

The idea of the delta is that it tracks *changes* in the original stream. If a dot follows a dot or a cross follows a cross at a particular point in the original stream, then the corresponding point in the delta has a dot (see the table). A dot in the delta means 'no change'. When, on the other hand, there is a cross followed by a dot or a dot followed by a cross in the original stream, then the corresponding point in the delta has a cross. A cross in the delta means 'change'. Turing introduced the concept of delta in July 1942, observing that by delta-ing a stretch of key he was able to make deductions which could not be made from the key in its un-deltaed form.[22]

---

[21]Good in interview with Pamela McCorduck (McCorduck [28: 53]).

[22]*General Report on Tunny*, p. 313.

Turingery worked on deltaed key to produce the deltaed contribution of the chi-wheels. Turing's discovery that delta-ing would reveal information otherwise hidden was essential to the developments that followed. The algorithms implemented in Colossus (and in its precursor Heath Robinson) depended on this simple but brilliant observation. In that sense, the entire machine-based attack on Tunny flowed from this fundamental insight of Turing's.

How did Turingery work? The method exploited the fact that each impulse of the chi-stream (and also its delta-ed form) consists of a pattern that repeats after a fixed number of steps. Since the number of cams on the 1st chi-wheel is 41, the pattern in the first impulse of the chi-stream repeats every 41 steps. In the 2nd impulse the pattern repeats every 31 steps – the number of cams on the 2nd chi-wheel – and for the 3rd, 4th and 5th impulses, the wheels have 29, 26 and 23 cams respectively. Therefore a hypothesis about the identity, dot or cross, of a particular bit in, say, the first impulse of the chi will, if correct, also produce the correct bit 41 steps further on, and another 41 steps beyond that, and so on. Given 500 letters of key, a hypothesis about the identity of a single letter of the chi (or delta-ed chi) will yield approximately 500/41 bits of the first impulse, 500/31 bits of the second impulse, 500/29 bits of the third, and so on – a total of about 85 bits.

In outline, Turing's method is this. The first step is to make a guess: the breaker guesses a point in the delta-ed key at which the psi-wheels stayed still in the course of their 'staggering' motion. Whatever guess is made, it has a 50% chance of being right. Positions where the psis did not move are of great interest to the breaker, since at these positions the deltaed key and the deltaed chi are identical. (The reason for this is that the deltaed contribution of the psis at such positions is ●●●●●, and adding ●●●●● to a letter does not alter the letter.) Because the key is known, the letter of the deltaed chi at the guessed position is also known – assuming, of course, that the guess about the psis not having moved is correct. Given this single letter of the deltaed chi, a number of bits can then be filled in throughout the five impulses, by propagating to the left and right at the appropriate periods.

Now that various bits of the delta chi are filled in, guesses can be made as to the identity of other letters. For example, if one letter of the delta chi is ●???● and the corresponding letter of the delta key is ●xxx● (C), the breaker may guess that this is another point at which the psis stood still, and replace ●???● in the delta chi by ●xxx●. This gives three new bits to propagate left and right. And so the process continues, with more and more bits of the delta chi being written in.

Naturally the breaker's guesses are not always correct, and as the process of filling in bits goes on, any incorrect guesses will tend to produce clashes – places where both a cross and a dot are assigned to the same position in the impulse. Guesses that are swamped by clashes have to be revised. With patience, luck, a lot of rubbing out, and a lot of cycling back and forth between putative fragments of delta chi and delta psi, a correct and complete stretch of delta chi eventually emerges.

15.7 TUTTE'S STATISTICAL METHOD

Tunny could now be tackled operationally, and a Tunny-breaking section was immediately set up under Major Ralph Tester.[23] Several members of the Research Section moved over to the 'Testery'. Armed with Turingery and other hand methods, the Testery read nearly every message from July to October 1942[24] – thanks to the insecure 12-letter indicator system, by means of which the German operator obligingly conveyed the wheel setting to the codebreakers. In October, however, the indicators were replaced by numbers from the QEP books, and the Testery, now completely reliant on depths, fell on leaner times. With the tightening up of German security, depths were becoming increasingly scarce. The Research Section renewed its efforts against Tunny, looking for a means of finding wheel settings that did not depend on depths.[25]

In November 1942 Tutte invented a way of discovering the settings of messages not in depth. This became known as the 'Statistical Method'. The rub

---

[23] *General Report on Tunny*, p. 28.

[24] *General Report on Tunny*, p. 28.

[25] *General Report on Tunny*, pp. 28, 320–322.

Figure 15.4. William T. Tutte.

was that at first Tutte's method seemed impractical. It involved calculations which, if done by hand, would consume a vast amount of time – probably as much as several hundred years for a single, long message, Newman once estimated.[26]

The necessary calculations were straightforward enough, consisting basically of comparing two streams made up of dots and crosses, and counting the number of times that each had a dot, or cross, in the same position. Today, of course, we turn such work over to electronic computers. When Tutte shyly explained his method to Newman, Newman suggested using high-speed electronic counters to mechanise the process. It was a brilliant idea. Within a surprisingly short time a factory of monstrous electronic computers dedicated to breaking Tunny was affording a glimpse of the future.

Electronic counters had been developed in Cambridge before the war. Used for counting emissions of sub-atomic particles, these had been designed by C.E. Wynn-Williams, a Cambridge don (Wynn-Williams [35; 36]; see also Hull [22], de Bruyne and Webster [13]). Newman knew of Wynn-Williams' work, and in a moment of inspiration he saw that the same idea could be applied to the Tunny problem. Within a month of Tutte's

inventing his statistical method Newman began developing the necessary machine. He worked out the cryptanalytical requirements for the machine and called in Wynn-Williams to design the electronic counters. Construction of Newman's machine started in January 1943 and a prototype began operating in June of that year, in the newly formed Tunny-breaking section called the 'Newmanry'. The prototype machine was soon dubbed 'Heath Robinson', after the famous cartoonist who drew overly-ingenious mechanical contrivances.

Tutte's method delivered the settings of the chi wheels. Once the Newmanry had discovered the settings of the chis by machine, the contribution that the chis had made to the ciphertext was stripped away, producing what was called the 'de-chi' of the message. The de-chi was made by a replica of the Tunny machine, designed by Flowers' Post Office engineers at Dollis Hill. The de-chi was then passed to the Testery, where a cryptanalyst would break into it by 'ordinary' pencil-and-paper methods requiring only (as a wartime document described it) "the power of instantaneous mental addition of letters of the Teleprint alphabet".[27]

The reason it was feasible to break the de-chi by hand was that the staggering motion of the psi-wheels introduced local regularities. Once the contribution of the chis had been stripped out of the key, what remained of the key contained distinctive patterns of repeated letters, e.g. ...GGZZZWDD..., since while the psis stood still they continued to contribute the same letter. By latching onto these repetitions, the cryptanalyst could uncover some stretches of this residual key, and this in turn enabled the settings of the psi-wheels and the motor-wheels to be deduced. For example, adding the guessed word 'dringend' ('urgent') to the de-chi near the beginning of the message might produce 888EE00WW – pure gold, confirming the guess. With luck, once a break was achieved it could be extended to the left or right, in this case perhaps by trying on the left 'sehr9' ('very' followed by a space), and on the right ++M88, the code for a full stop (see Appendix 1). Once the codebreaker had a short stretch

---

[26]Newman in interview with Evans.

[27]*General Report on Tunny*, p. 22.

of the key that the psi-wheels had contributed, the wheel settings could usually be obtained by comparing the key to the known wheel patterns. When all the wheel-settings were known, the ciphertext was keyed into one of the Testery's replica Tunny machines, and the German plaintext would emerge.

In order to illustrate the basic ideas of Tutte's method for finding the settings of the chi wheels, let us assume that we have an intercepted ciphertext 10,000 characters long. This ciphertext is punched on a tape (we call this the 'message-tape'). An assistant, who knows the chi-wheel patterns, provides us with a second tape (the 'chi-tape'). This assistant has worked out the machine's entire chi-stream, beginning at an arbitrarily selected point in the revolution of the chi-wheels, and stepping through all their possible joint combinations. (Once the wheels have moved through all the possible combinations, their capacity for novelty is exhausted, and should the wheels continue to turn they merely duplicate what has gone before.) The complete chi-stream is, of course, rather long, but eventually the assistant does produce a roll of tape with the stream punched on it. The sequence of 10,000 consecutive characters of chi-stream that was used to encrypt our message is on this tape somewhere – our problem is to find it. This sequence is called simply 'the chi' of the message. Tutte's method exploits a fatal weakness in the design of the Tunny machine, a weakness again stemming from the staggering motion of the psi-wheels. The central idea of the method is this: *The chi is recognisable on the basis of the ciphertext, provided the wheel patterns are known*. Tutte showed by a clever mathematical deduction that the delta of the ciphertext and the delta of the chi would usually correspond slightly. That *slightly* is the key to the whole business – any degree of regularity, no matter how weak, is the cryptanalyst's friend. The slight regularity that Tutte discovered could be used as a touchstone for finding the chi. (Readers interested in Tutte's mathematical reasoning will find the details in Appendix 2; at present we will concentrate on how the method is carried out.)

We select the first 10,000 characters of the chi-tape; we will compare this stretch of the chi-tape with the message-tape. Tutte showed that in fact we need examine only the *first* and the *second* of the five horizontal rows punched along the chi-tape, the first and second impulses (these two rows are the contributions of the first and second chi-wheels respectively). Accordingly we need consider only the first and second impulses of the message-tape. This simplifies considerably the task of comparing the two tapes. Because Tutte's method focused on the first and second chi-wheels it was dubbed the '1 + 2 break in'.[28]

Here is the procedure for comparing the message-tape with the stretch of chi-tape we have picked. First, we add the first and second impulses of the message-tape and form the delta of the resulting sequence of dots and crosses. (For example, if the sequence produced by adding the two impulses begins **x•x...**, the delta begins **xx....**) Second, we add the first and second impulses of the 10,000-character piece of chi-tape, and again form the delta of the result. Next we lay these two deltas side by side and count how many times they have dots in the same places and how many times crosses. We add the two tallies to produce a total score for this particular piece of the chi-tape. We are looking for a match between the two deltas of around 55%. Tutte showed that this is the order of correspondence that can be expected when the piece of chi-tape under examination contains the first and second impulses of the actual chi.

The first score we obtain probably will not be anything special – for we would be extremely lucky if the first 10,000 characters of chi-stream that we examined were the chi of the message. So next we shift along one character in the chi-stream and focus on a new candidate for the message's chi, the 2nd through to the 10,001st characters on the chi-tape (see the diagram). We add, delta, and count once again. Then we shift along another character, repeating the process until we have examined all candidates for the chi. A buoyant score reveals the first and second impulses of the actual chi (we hope).

Once a winning segment of the chi-tape has been located, its place within the complete chi-stream tells us the positions of the first and second chi-wheels at the start of the message. With

---

[28] *General Report on Tunny*, p. 20.

these settings in hand, a similar procedure is used to chase the settings of the other chi-wheels.

As mentioned previously, the cause of the slight regularity that Tutte latched onto is at bottom the staggering movement of the psi-wheels – the great weakness of the Tunny machine. While the psis remained stationary, they continued to contribute the same letter to the key; and so, since delta-ing tracks change, the delta of the stream of characters contributed by the psis contained more dots than crosses (recall that a cross in the delta indicates a change). Tutte calculated that there would usually be about 70% dot in the delta of the sum of the contributions of the first two psi-wheels.

The delta of the plaintext also contained more dots than crosses (for reasons explained in Appendix 2, which included the fact that Tunny operators habitually repeated certain characters). Tutte investigated a number of broken messages and discovered to his delight that the delta of the sum of the first two impulses was as a rule about 60% dot. Since these statistical regularities in the delta of the psi and the delta of the plain both involved a predominance of dot over cross, they tended to reinforce one another. Tutte deduced that their net effect, in favourable cases, would be the agreement, noted above, of about 55% between the processed ciphertext and the processed chi.

Tunny's security depended on the appearance of randomness, and here was a crack in the appearance. The British seized on it. If, instead of the psi-wheels either all moving together or all standing still, the designers had arranged for them to move independently – or even to move regularly like the chis – then the chink that let Tutte in would not have existed.

15.8 HEATH ROBINSON

Smoke rose from Newman's prototype machine the first time it was switched on (a large resistor overloaded). Around a vast frame made of angle-iron wound two long loops of teleprinter tape. Resembling an old-fashioned bed standing on end, the frame quickly became known as the 'bedstead'. The tapes were supported by a system of pulleys and wooden wheels of diameter about ten inches.

Each tape was driven by a toothed sprocket-wheel which engaged a continuous row of sprocket-holes along the centre of the tape (see Fig. 15.2). The tapes were driven by the same drive-shaft and moved in synchronisation with each other at a maximum speed of 2000 characters per second. To the amusement and annoyance of Heath Robinson's operators, tapes would sometimes tear or come unglued, flying off the bedstead at high speed and breaking into fragments which festooned the Newmanry.

One tape was the message-tape and the other the chi-tape. In practice the chi-tape might contain, for example, only the first and second impulses of the complete chi-stream, resulting in a shorter tape. The drive mechanism was arranged so that as the tapes ran on the bedstead, the message-tape stepped through the chi-tape one character at a time (see Fig. 15.5). Photo-electric readers mounted on the bedstead converted the hole/no-hole patterns punched on the tapes into streams of electrical pulses, and these were routed to a 'combining unit' – a logic unit, in modern terminology. The combining unit did the adding and the delta-ing, and Wynn-Williams' electronic counters produced the scores. The way the combining was done could be varied by means of replugging cables, a primitive form of programming. (The combining unit, the bedstead and the photo-electric readers were made by Post Office engineers at Dollis Hill[29] and the

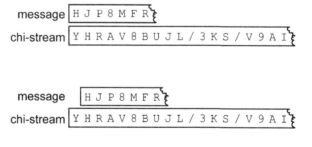

Figure 15.5. Stepping the ciphertext through the chi-stream, looking for the starting position of the chi-wheels.

---

[29]Letter from Harry Fensom to Copeland (4 May 2001).

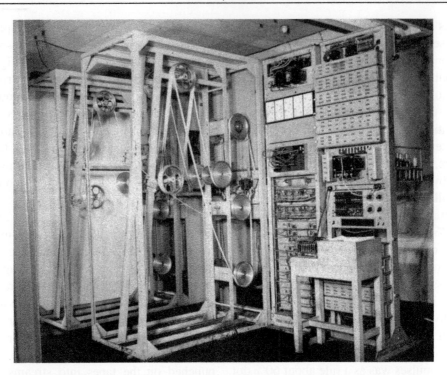

Figure 15.6. This machine, eventually called 'Old Robinson', replaced the original Heath Robinson (the two were of similar appearance). To the left are the two large metal frames called 'bedsteads', which held the tape-drive mechanism, the photo-electric readers, and the two tapes supported by pulleys. One tape contained the ciphertext and the other held impulses from the chi-wheels of the Tunny machine. To the right are the 'combining unit' and the electronic counters. *Source*: *General Report on Tunny*. Crown copyright; reproduced by permission of the National Archives Image Library.

counters by Wynn-Williams' unit at the Telecommunications Research Establishment (TRE).)

Heath Robinson worked, proving in a single stroke that Newman's idea of attacking Tunny by machine was worth its salt and that Tutte's method succeeded in practice. However, Heath Robinson suffered from 'intolerable handicaps'.[30] Despite the high speed of the electronic counters, Heath Robinson was not really fast enough for the codebreakers' requirements, taking several hours to elucidate a single message.[31] Moreover, the counters were not fully reliable – Heath Robinson was prone to deliver different results if set the same problem twice. Mistakes made in hand-punching the two tapes were another fertile source of error, the long chi-tape being especially difficult to prepare. At first, undetected tape errors prevented Heath Robinson from obtaining any results at all.[32]

And paramount among the difficulties was that the two tapes would get out of synchronisation with each other as they span, throwing the calculations out completely. The loss of synchronisation was caused by the tapes stretching, and also by uneven wear around the sprocket holes.

The question was how to build a better machine – a question for an engineer. In a stroke of genius, the electronics expert Thomas Flowers solved all these problems.

15.9 FLOWERS, THE NEGLECTED PIONEER OF COMPUTING

During the 1930s Flowers pioneered the large-scale use of electronic valves to control the making and breaking of telephone connections.[33] He was swimming against the current. Many regarded

---

[30]*General Report on Tunny*, p. 328.

[31]Newman in interview with Evans.

[32]*General Report on Tunny*, p. 328.

[33]Unless indicated otherwise, material in this chapter relating directly to Flowers derives from (1) Flowers in interviews with

Figure 15.7. Thomas H. Flowers.

the idea of large-scale electronic equipment with scepticism. The common wisdom was that valves – which, like light bulbs, contained a hot glowing filament – could never be used satisfactorily in large numbers, for they were unreliable, and in a large installation too many would fail in too short a time. However, this opinion was based on experience with equipment that was switched on and off frequently – radio receivers, radar and the like. What Flowers discovered was that, so long as valves were switched on and left on, they could operate reliably for very long periods, especially if their 'heaters' were run on a reduced current.

At that time, telephone switchboard equipment was based on the *relay*. A relay is a small, au-

Copeland, 1996–1998 (2) Flowers in interview with Christopher Evans in 1977 (The Pioneers of Computing: an Oral History of Computing, Science Museum, London.).

tomatic switch. It contains a mechanical contact-breaker – a moving metal rod that opens and closes an electrical circuit. The rod is moved from the 'off' position to the 'on' position by a magnetic field. A current in a coil is used to produce the magnetic field; as soon as the current flows, the field moves the rod. When the current ceases, a spring pushes the rod back to the 'off' position. Flowers recognised that equipment based instead on the electronic valve – whose only moving part is a beam of electrons – not only had the potential to operate very much faster than relay-based equipment, but was in fact potentially more reliable, since valves are not prone to mechanical wear.

In 1934 Flowers wired together an experimental installation containing three to four thousand valves (by contrast, Wynn-Williams' electronic counters of 1931 contained only three or four valves). This equipment was for controlling connections between telephone exchanges by means of tones, like today's touch-tones (a thousand telephone lines were controlled, each line having 3–4 valves attached to its end). Flowers' design was accepted by the Post Office and the equipment went into limited operation in 1939. Flowers had proved that an installation containing thousands of valves would operate very reliably – but this equipment was a far cry from Colossus. The handful of valves attached to each telephone line formed a simple unit, operating independently of the other valves in the installation, whereas in Colossus large numbers of valves worked in concert.

During the same period before the war Flowers explored the idea of using valves as high-speed switches. Valves were used originally for purposes such as amplifying radio signals. The output would vary continuously in proportion to a continuously varying input, for example a signal representing speech. Digital computation imposes different requirements. What is needed for the purpose of representing the two binary digits, 1 and 0, is not a continuously varying signal but plain 'on' and 'off' (or 'high' and 'low'). It was the novel idea of using the valve as a very fast switch, producing pulses of current (pulse for 1, no pulse for 0) that was the route to high-speed digital computation. During 1938–1939 Flowers worked on an experimental high-speed electronic data store embodying this idea. The store was intended to replace relay-based data stores in telephone exchanges. Flowers' long-term goal was that electronic equipment should replace all the relay-based systems in telephone exchanges.

By the time of the outbreak of war with Germany, only a small number of electrical engineers were familiar with the use of valves as high-speed digital switches. Thanks to his pre-war research, Flowers was (as he himself remarked) possibly the only person in Britain who realised that valves could be used reliably on a large scale for high-speed digital computing.[34] When Flowers was summoned to Bletchley Park – ironically, because of his knowledge of relays – he turned out to be the right man in the right place at the right time.

Turing, working on Enigma, had approached Dollis Hill to build a relay-based decoding machine to operate in conjunction with the Bombe (the Bombe itself was also relay-based). Once the Bombe had uncovered the Enigma settings, the machine requisitioned by Turing would be used to decipher the message automatically and print out the German plaintext.[35] Dollis Hill sent Flowers to Bletchley Park. He would soon become one of the great figures of World War II codebreaking. In the end, the machine Flowers built for Turing was not used, but Turing was impressed by Flowers, who began thinking about an electronic Bombe, although he did not get far. When the teleprinter group at Dollis Hill ran into difficulties with the design of the Heath Robinson's combining unit, Turing suggested that Flowers be called in. (Flowers was head of the switching group at Dollis Hill, located in the same building as the teleprinter group.) Flowers and his switching group improved the design of the combining unit and manufactured it.[36]

Flowers did not think much of the Robinson, however. The basic design had been settled before he was called in and he was sceptical as soon as Morrell, head of the teleprinter group, first told him about it. The difficulty of keeping two paper tapes in synchronisation at high speed was a conspicuous weakness. So was the use of a mixture of valves and relays in the counters, because the relays slowed everything down. Heath Robinson was built mainly from relays and contained no more than a couple of dozen valves. Flowers doubted that the Robinson would work properly and in February 1943 he presented Newman with the alternative of a fully electronic machine able to generate the chi-stream (and psi- and motor-streams) internally.[37]

Flowers' suggestion was received with 'incredulity' at TRE and Bletchley Park.[38] It was thought that a machine containing the number of valves that Flowers was proposing (between one and two thousand) "would be too unreliable to do useful work".[39] In any case, there was the question of how long the development process would take – it was felt that the war might be over before Flowers' machine was finished. Newman pressed ahead with the two-tape machine. He offered Flowers some encouragement but effectively left him to do as he wished with his proposal for an all-electronic machine. Once Heath Robinson was a going concern, Newman placed an order with the Post Office for a dozen more relay-based two-tape machines (it being clear, given the quantity and very high importance of Tunny traffic, that one or two machines would not be anywhere near enough). Meanwhile Flowers, on his own initiative and working independently at Dollis Hill, began building the fully electronic machine that he could see was necessary. He embarked on Colossus, he said, "in the face of scepticism"[40] from Bletchley Park and "without the concurrence of BP".[41] "BP weren't interested until they saw it [Colossus] working", he recollected.[42] Fortunately, the Director of the Dollis Hill Research Station, Radley, had greater faith in

---

[34]Flowers in interview with Copeland (July 1996).

[35]Flowers in interview with Copeland (July 1998).

[36]Flowers in interview with Copeland (July 1996); *General Report on Tunny*, p. 33.

[37]Flowers in interview with Copeland (July 1996); Flowers [16: 244].

[38]Flowers, T.H. Colossus – Origin and Principles, typescript, no date, p. 3; Coombs in interview with Christopher Evans in 1976 (The Pioneers of Computing: An Oral History of Computing, Science Museum, London). 'Incredulity' is Flowers' word.

[39]Flowers, Colossus – Origin and Principles, p. 3.

[40]Flowers in interview with Copeland (July 1996).

[41]Flowers in interview with Copeland (July 1996).

[42]Flowers in interview with Copeland (July 1996).

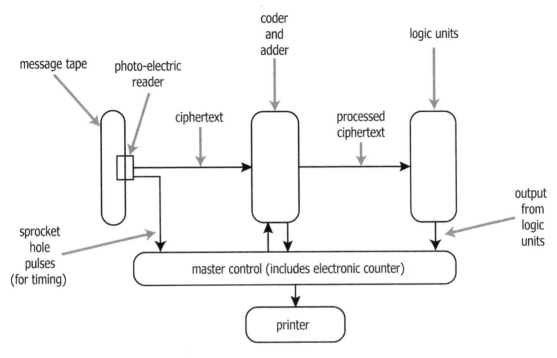

Figure 15.8. Colossus (from a sketch by Flowers).

Flowers and his ideas, and placed "the whole resources of the laboratories" at Flowers' disposal.[43]

15.10 COLOSSUS

The prototype Colossus was brought to Bletchley Park in lorries and reassembled by Flowers' engineers.[44] It had approximately 1600 electronic valves and operated at 5000 characters per second. Later models, containing approximately 2400 valves, processed five streams of dot-and-cross simultaneously, in parallel. This boosted the speed to 25,000 characters per second. Colossus generated the chi-stream electronically. Only one tape was required, containing the ciphertext – the synchronisation problem vanished. (Flowers' original plan was to dispense with the message tape as well and set up the ciphertext, as well as the wheels, on valves; but he abandoned this idea when

it became clear that messages of 5000 or more characters would have to be processed.[45])

The arrival of the prototype Colossus caused quite a stir. Flowers said:

> I don't think they [Newman et al.] really understood what I was saying in detail – I am sure they didn't – because when the first machine was constructed and working, they obviously were taken aback. They just couldn't believe it! ...I don't think they understood very clearly what I was proposing until they actually had the machine.[46]

On what date did Colossus first come alive? In his written and verbal recollections Flowers was always definite that Colossus was working at Bletchley Park in the early part of December 1943.[47] In three separate interviews he recalled a key date quite specifically, saying that Colossus carried out its first trial run at Bletchley Park on 8 December 1943.[48] However, Flowers' personal diary for

---

[43]Flowers, Colossus – Origin and Principles, p. 3.

[44]K. Myers, Dollis Hill and Station X, in The Turing Archive for the History of Computing, http://www.AlanTuring.net/myers.

[45]*General Report on Tunny*, p. 35.

[46]Flowers in interview with Evans.

[47]Flowers [16: 245]; Flowers in interview with Evans.

[48]Flowers in interview with Copeland (July 1996); Flowers in interview with Darlow Smithson (no date); Flowers in interview with staff of the Imperial War Museum, London (1998).

1944 – not discovered until after his death – in fact records that Colossus did not make the journey from Dollis Hill to Bletchley Park until January 1944. On Sunday 16 January Colossus was still in Flowers' lab at Dollis Hill. His diary entry shows that Colossus was certainly working on that day. Flowers was busy with the machine from the morning until late in the evening and he slept at the lab.

Flowers' entry for 18 January reads simply: "Colossus delivered to B.P.". This is confirmed by a memo dated 18 January from Newman to Travis (declassified only in 2004). Newman wrote "Colossus arrives to-day".[49] Colossus cannot therefore have carried out its first trial run at Bletchley Park in early December. What did happen on 8 December 1943, the date that stuck so firmly in Flowers' mind? Perhaps this was indeed the day that Colossus processed its first test tape at Dollis Hill. "I seem to recall it *was* in December", says Harry Fensom, one of Flowers' engineers.[50]

By February 1944 the engineers had got Colossus ready to begin serious work for the Newmanry. Tutte's statistical method could now be used at electronic speed. The computer attacked its first message on Saturday 5 February. Flowers was present. He noted laconically in his diary, "Colossus did its first job. Car broke down on way home".

Colossus immediately doubled the codebreakers' output.[51] The advantages of Colossus over Robinson were not only its greatly superior speed and the absence of synchronised tapes, but also its greater reliability, resulting from Flowers' redesigned counters and the use of valves in place of relays throughout. It was clear to the Bletchley Park authorities – whose scepticism was now completely cured – that more Colossi were required urgently.

Indeed, a crisis had developed, making the work of Newman's section even more important than before. Since the German introduction of the QEP system in October 1942, the codebreakers using hand-methods to crack Tunny messages had been reliant upon depths, and as depths became rarer during 1943, the number of broken messages reduced to a trickle.[52] Then things went from bad to worse. In December 1943 the Germans started to make widespread use of an additional device in the Tunny machine, whose effect was to make depth-reading impossible (by allowing letters of the plaintext itself to play a role in the generation of the key). The hand breakers had been prone to scoff at the weird contraptions in the Newmanry, but suddenly Newman's machines were essential[53] to all Tunny work.

In March 1944 the authorities demanded four more Colossi. By April they were demanding twelve.[54] Great pressure was put on Flowers to deliver the new machines quickly. The instructions he received came "from the highest level" – the War Cabinet – and he caused consternation when he said flatly that it was impossible to produce more than one new machine by 1 June 1944 (Flowers [16: 246]).

Flowers had managed to produce the prototype Colossus at Dollis Hill only because many of his laboratory staff "did nothing but work, eat, and sleep for weeks and months on end" (Flowers [16: 245]). He needed greater production capacity, and proposed to take over a Post Office factory in Birmingham. Final assembly and testing of the computers would be done at his Dollis Hill laboratory. Flowers estimated that once the factory was in operation he would be able to produce additional Colossi at the rate of about one per month (Flowers [16: 246]). He recalled how one day some Bletchley people came to inspect the work, thinking that Flowers might be "dilly-dallying": they returned "staggered at the scale of the effort".[55] Churchill for his part gave Flowers top priority for everything he needed.[56]

By means of repluggable cables and panels of switches, Flowers deliberately built more flexibility than was strictly necessary into the logic units

---

[49]M.H.A. Newman, Report on Progress (Newmanry, 18 January 1944; PRO document reference HW14/96), p. 4.

[50]Letter from Fensom to Copeland (18 August 2005).

[51]*General Report on Tunny*, p. 35.

[52]*General Report on Tunny*, p. 34.

[53]*General Report on Tunny*, p. 28.

[54]*General Report on Tunny*, p. 35.

[55]Flowers in interview with Copeland (July 1996).

[56]Note from Donald Michie to Copeland (27 May 2002), reporting a disclosure by Coombs in the 1960s.

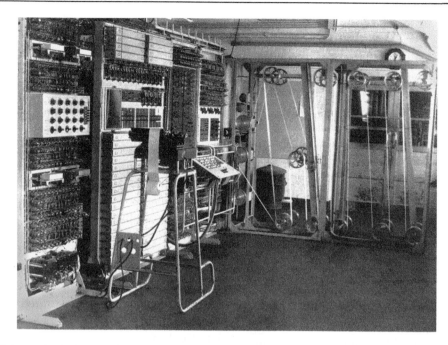

Figure 15.9. Colossus. In the foreground is the automatic typewriter for output. The large frames to the right held two message tapes. As one job was being run, the tape for the next job would be loaded onto the pulleys, so saving time. Using a switch on the selection panel, the operator chose to run either the 'near' or the 'far' tape. *Source*: *General Report on Tunny*. Crown copyright; reproduced by permission of the National Archives Image Library.

of the prototype Colossus. As a result, new methods could be implemented on Colossus as they were discovered. In February 1944 two members of the Newmanry, Donald Michie and Jack Good, had quickly found a way of using Colossus to discover the Tunny wheel patterns.[57] Flowers was told to incorporate a special panel for breaking wheel patterns in Colossus II.

Colossus II – the first of what Flowers referred to as the 'Mark 2' Colossi[58] – was shipped from Dollis Hill to Bletchley Park on 4 May 1944.[59] The plan was to assemble and test Colossus II at Bletchley Park rather than Dollis Hill, so saving some precious time (Chandler [4: 261]). Promised by the first of June, Colossus II was still not working properly as the final hours of May ticked past. The computer was plagued by intermittent and mysterious faults (Flowers [16: 246]). Flowers struggled to find the problem, but midnight came and went. Exhausted, Flowers and his team dispersed at 1 AM

to snatch a few hours sleep.[60] They left Chandler to work on, since the problem appeared to be in a part of the computer that he had designed. It was a tough night: around 3 AM Chandler noticed that his feet were getting wet.[61] A radiator pipe along the wall had sprung a leak, sending a dangerous pool of water toward Colossus.

Flowers returned to find the computer running perfectly. "Colossus 2 in operation", he noted in his diary.[62] The puddle remained, however, and the women operators had to don gumboots to insulate themselves.[63] During the small hours Chandler had finally tracked down the fault in Colossus (parasitic oscillations in some of the valves) and had fixed it by wiring in a few extra resistors (Flowers [16: 247]). Flowers and his 'band of brothers' (Coombs [5: 259]) had met BP's deadline – a deadline whose significance Flowers can only have guessed at.

[57] *General Report on Tunny*, p. 461.

[58] Flowers' personal diary, 4 May 1944.

[59] Flowers' personal diary, 4 May 1944.

[60] Flowers' personal diary, 31 May 1944.

[61] Letter from Chandler to Brian Randell, 24 January 1976; unpublished manuscript by Gil Hayward *1944–1946* (2002).

[62] Flowers' personal diary, 1 June 1944.

[63] Hayward, *1944–1946*.

Less than a week later the Allied invasion of France began. The D-day landings of June 6 placed huge quantities of men and equipment on the beaches of Normandy. From the beachheads the Allies pushed their way into France through the heavy German defences. By mid-July the front had advanced only 20 or so miles inland, but by September Allied troops had swept across France and Belgium and were gathering close to the borders of Germany, on a front extending from Holland in the north to Switzerland in the south.[64]

Since the early months of 1944, Colossus I had been providing an unparalleled window on German preparations for the Allied invasion.[65] Decrypts also revealed German appreciations of Allied intentions. Tunny messages supplied vital confirmation that the German planners were being taken in by Operation *Fortitude*, the extensive programme of deceptive measures designed to suggest that the invasion would come further north, in the Pas de Calais (Hinsley [20: 47–65]). In the weeks following the start of the invasion the Germans tightened Tunny security, instructing operators to change the patterns of the chi- and psi-wheels daily instead of monthly. Hand methods for discovering the new patterns were overwhelmed. With impeccable timing Colossus II's device for breaking wheel patterns came to the rescue.

Once Flowers' factory in Birmingham was properly up and running, new Colossi began arriving in the Newmanry at roughly six week intervals. Eventually three were dedicated to breaking wheel patterns.[66] Flowers was a regular visitor at B.P. throughout the rest of 1944, overseeing the installation programme for the Mark 2 Colossi.[67] By the end of the year seven Colossi were in operation. They provided the codebreakers with the capacity to find all twelve wheel settings by machine, and this was done in the case of a large proportion of decrypted messages.[68] There were ten Colossi in operation by the time of the German surrender in 1945, and an eleventh was almost ready.

[64]Hinsley [20]: maps "OVERLORD" (frontispiece) and "September position 1944" (facing p. 365).

[65]Some crucial decrypts are listed by Hinsley [20: ch. 44 and appendix 10].

[66]*General Report on Tunny*, p. 36.

[67]Flowers' personal diary for 1944.

[68]*General Report on Tunny*, p. 35.

15.11 MISCONCEPTIONS ABOUT COLOSSUS

One of the most common misconceptions in the secondary literature is that Colossus was used against Enigma. Another is that Colossus was used against not Tunny but Sturgeon – an error promulgated by Brian Johnson's influential television series and accompanying book *The Secret War* (Johnson [25: 339–347]). There are in fact many wild tales about Colossus in the history books. Georges Ifrah even states that Colossus produced *English* plaintext from the German ciphertext (Ifrah [24: 218])! As already explained, the output of Colossus was a series of counts indicating the correct wheel settings (or, later, the wheel patterns). Not even the de-chi was produced by Colossus itself, let alone the plaintext – and there was certainly no facility for the automatic translation of German into English.

An insidious misconception concerns ownership of the inspiration for Colossus. Many accounts identify Turing as the key figure in the designing of Colossus. In a biographical article on Turing, the computer historian J.A.N. Lee said that Turing's "influence on the development of Colossus is well known" [26: 671], and in an article on Flowers, Lee referred to Colossus as "the cryptanalytical machine designed by Alan Turing and others" [26: 306]. Lee asserted: "Newman fully appreciated the significance of Turing's ideas for the design of high-speed electronic machines for searching for wheel patterns and placings on the highest-grade German enciphering machines, and the result was the invention of the 'Colossus'" [26: 492]. Even a book on sale at the Bletchley Park Museum states that at Bletchley Park "Turing worked … on what we now know was computer research" which led to "the world's first electronic, programmable computer, 'Colossus'" (Enever [14: 36–37]). *Time* magazine says, "At Bletchley Park, Alan Turing built a succession of vacuum-tube machines called Colossus that made mincemeat of Hitler's Enigma codes" [18: 82].

The view that Turing's interest in electronics contributed to the inspiration for Colossus is indeed common. This claim is enshrined in codebreaking exhibits in leading museums; and in the

Annals of the History of Computing Lee and Holtzman state that Turing "conceived of the construction and usage of high-speed electronic devices; these ideas were implemented as the 'Colossus' machines" (Lee and Holtzman [27: 33]). However, the definitive 1945 *General Report on Tunny* makes matters perfectly clear: "Colossus was entirely the idea of Mr. Flowers" (p. 35). By 1943 electronics had been Flowers' driving passion for more than a decade and he needed no help from Turing. Turing was, in any case, away in the United States during the critical period at the beginning of 1943 when Flowers proposed his idea to Newman and worked out the design of Colossus on paper. Flowers emphasised in an interview that Turing "made no contribution" to the design of Colossus.[69] Flowers said: "I invented the Colossus. No one else was capable of doing it".[70]

In a recent book on the history of computing Martin Davis offers the following garbled account of Colossus:

> Some of the methods ... used were playfully called *turingismus* indicating their source. But turingismus required the processing of lots of data and for the decryption be [sic] of any use, the processing had to be done very quickly. ... In March 1943, Alan Turing sailed home from a visit of several months in the United States ... He whiled away the time during his Atlantic passage by studying [an] RCA catalogue, for it had been found that vacuum tubes could carry out the kind of logical switching previously done by electric relays. And the tubes were fast ... Vacuum tube circuits had in fact been used experimentally for telephone switching, and Turing had made contact with the gifted engineer, T. Flowers, who had spearheaded this research. Under the direction of Flowers and Newman, a machine, essentially a physical embodiment of turingismus, was rapidly brought into being. Dubbed the Colossus and an engineering marvel, this machine contained 1500 vacuum tubes (Davis [12: 174–175]).

Here Davis conflates Turingery, which he calls 'turingismus', with Tutte's statistical method. (*ismus* is a German suffix equivalent to the English *ism*. Newmanry codebreaker Michie explains

the origin of Turingery's slang name 'Turingismus': "three of us (Peter Ericsson, Peter Hilton and I) coined and used in playful style various fake-German slang terms for everything under the sun, including occasionally something encountered in the working environment. Turingismus was a case of the latter."[71]) Turing's method of wheel breaking from depths and Tutte's method of wheel setting from non-depths were distant relatives, in that both used delta-ing. But there the similarity ended. Turingery, Tutte said, seemed to him "more artistic than mathematical"; in applying the method you had to rely on what "you felt in your bones" (Tutte [33: 360]). Conflating the two methods, Davis erroneously concludes that Colossus was Turingery in hardware. But as explained above, Turingery was a hand method – it was Tutte's method that "required the processing of lots of data". Tutte's method, not Turingery, was implemented in Heath Robinson and Colossus. "Turingery was not used in either breaking or setting by any valve machine of any kind", Michie underlined.[72]

15.12 POSTWAR

If Flowers could have patented the inventions that he contributed to the assault on Tunny, he would probably have become a very rich man. As it was, the personal costs that he incurred in the course of building the Colossi left his bank account overdrawn at the end of the war. Newman was offered an OBE for his contribution to the defeat of Germany, but he turned it down, remarking to ex-colleagues from Bletchley Park that he considered the offer derisory.[73] Tutte received no public recognition for his vital work. Turing accepted an OBE, which he kept in his toolbox.

At the end of hostilities, orders were received from Churchill to break up the Colossi, and all involved with Colossus and the cracking of Tunny were gagged by the Official Secrets Act. The very existence of Colossus was to be classified indefinitely. Flowers described his reactions:

---

[69] Flowers in interview with Copeland (July 1996).
[70] Flowers in interview with Copeland (July 1996).

[71] Letter from Michie to Copeland (29 July 2001).
[72] Letter from Michie to Copeland (28 November 2001).
[73] Peter Hilton in interview with Copeland (May 2001).

When after the war ended I was told that the secret of Colossus was to be kept indefinitely I was naturally disappointed. I was in no doubt, once it was a proven success, that Colossus was an historic breakthrough, and that publication would have made my name in scientific and engineering circles – a conviction confirmed by the reception accorded to ENIAC, the U.S. equivalent made public just after the war ended. I had to endure all the acclaim given to that enterprise without being able to disclose that I had anticipated it. What I lost in personal prestige, and the benefits which commonly accrue in such circumstances, can now only be imagined. But at the time I accepted the situation philosophically and, in the euphoria of a war that was won, lost any concern about what might happen in the future (Flowers [17: 82–83]).

ENIAC, commissioned by the US army in 1943, was designed to calculate trajectories of artillery shells. Although not operational until the end of 1945 – two years after Colossus first ran – ENIAC is standardly described as the first electronic digital computer. Flowers' view of the ENIAC? It was just a number cruncher – Colossus, with its elaborate facilities for logical operations, was "much more of a computer than ENIAC".[74]

The Newmanry's Colossi might have passed into the public domain at the end of the fighting, to become, like ENIAC, the electronic muscle of a scientific research facility. The Newmanry's engineers would quickly have adapted the equipment for peacetime applications. The story of computing might have unfolded rather differently with such a momentous push right at the beginning. Churchill's order to destroy the Colossi was an almighty blow in the face for science – and for British industry.

In April 1946, codebreaking operations were transferred from Bletchley Park to buildings in Eastcote in suburban London.[75] At the time of the move, the old name of the organisation, 'Government Code and Cypher School', was formally changed to 'Government Communications Headquarters' (GCHQ).[76] Six years later another move commenced, and during 1952–1954 GCHQ shifted its personnel and equipment, including its codebreaking machinery, away from the London area to a large site in Cheltenham.[77] Some machines did survive the dissolution of the Newmanry. Two Colossi made the move from Bletchley Park to Eastcote, and then eventually on to Cheltenham.[78] They were accompanied by two of the replica Tunny machines manufactured at Dollis Hill.[79] One of the Colossi, known as 'Colossus Blue' at GCHQ, was dismantled in 1959 after fourteen years of postwar service. The remaining Colossus is believed to have stopped running in 1960.

During their later years the two Colossi were used extensively for training. Details of what they were used for prior to this remain classified. There is a hint of the importance of one new role for these Newmanry survivors in a letter written by Jack Good:

I heard that Churchill requested that all Colossi be destroyed after the war, but GCHQ decided to keep at least one of them. I know of that one because I used it myself. That was the first time it was used after the war. I used it for a purpose for which NSA [National Security Agency] were planning to build a new special-purpose machine. When I showed that the job could be carried out on Colossus, NSA decided not to go ahead with their plan. That presumably is one reason I am still held in high regard in NSA. Golde told me that one of his friends who visits NSA told Golde that I am 'regarded as God' there.[80]

After Bletchley's own spectacular successes against the German machines, GCHQ was – not unnaturally – reluctant to use key-generating cipher machines to protect British high-grade diplomatic traffic. Instead GCHQ turned to one-time pad. Sender and receiver were issued with identical key in the form of a roll of teleprinter tape. This would be used for one message only. One-time pad is highly secure. The disadvantage is that a complex and highly efficient distribution network is required to supply users with key. It is probably

---

[74]Flowers in interview with Copeland (July 1996).

[75]P. Freeman, How GCHQ Came to Cheltenham (undated, GCHQ), p. 8.

[76]P. Freeman, How GCHQ Came to Cheltenham, p. 8.

[77]P. Freeman, How GCHQ Came to Cheltenham, p. 30.

[78]Unpublished manuscript by Gil Hayward, Dollis Hill at War (2002).

[79]G. Hayward, Dollis Hill at War.

[80]Letter from Jack Good to Henry H. Bauer (2 January 2005).

true that GCHQ initially underestimated the difficulties of distributing key.

The GCHQ Colossi assisted in the production of one-time pad. Ex-Newmanry engineers used some of Flowers' circuitry from Colossus to build a random noise generator able to produce random teleprinter characters on a punched tape. This device, code-named 'Donald Duck', exploited the random way in which electrons are emitted from a hot cathode. The tapes produced by Donald Duck were potential one-time pad. The tapes were checked by Colossus, and those that were not flat-random were weeded out. Newmanry-type tape-copying machines were used to make copies of tapes that passed the tests, and these were distributed to GCHQ's clients.

Probably the Colossi had additional postwar applications. They may have been used to make character counts of enemy cipher traffic, searching for features that might give the cryptanalysts a purchase. Perhaps the GCHQ Colossi were even used against reconditioned German Tunny machines. Many Tunnies were captured by the invading British armies during the last stages of the war. If the National interest so dictated, Tunny machines may have been sold to commercial organisations or foreign powers, and the resulting traffic read by GCHQ.

Until the 1970s few had any idea that electronic computation had been used successfully during the Second World War. In 1975, the British government released a set of captioned photographs of the Colossi.[81] By 1983, Flowers had received clearance to publish an account of the hardware of the first Colossus (Flowers [16]). Details of the later Colossi remained secret. So, even more importantly, did all information about how Flowers' computing machinery was actually used by the codebreakers. Flowers was told by the British authorities that "the technical description of machines such as COLOSSUS may be disclosed", but that he must not disclose any information about "the functions which they performed".[82] It was

rather like being told that he could give a detailed technical description of the insides of a radar receiver, but must not say anything about what the equipment did (in the case of radar, reveal the location of planes, submarines, etc., by picking up radio waves bouncing off them). He was also allowed to describe some aspects of Tunny, but there was a blanket prohibition on saying anything at all relating to "the weaknesses which led to our successes". In fact, a clandestine censor objected to parts of the account that Flowers wrote, and he was instructed to remove these prior to publication.[83]

There matters more or less stood until 1996, when the US government declassified some wartime documents describing the function of Colossus. These had been sent to Washington during the war by US liaison officers stationed at Bletchley Park. The most important document remained classified, however: the 500 page *General Report on Tunny* written at Bletchley Park in 1945 by Jack Good, Donald Michie and Geoffrey Timms. Thanks largely to Michie's tireless campaigning, the report was declassified by the British Government in June 2000, finally ending the secrecy.

15.13 COLOSSUS AND THE MODERN COMPUTER

As everyone who can operate a personal computer knows, the way to make the machine perform the task you want – word-processing, say – is to open the appropriate program stored in the computer's memory. Life was not always so simple. Colossus did not store programs in its memory. To set up Colossus for a different job, it was necessary to modify some of the machine's wiring by hand, using switches and plugs. The larger ENIAC was also programmed by re-routing cables and setting switches. The process was a nightmare: it could take the ENIAC's operators up to three weeks to set up and debug a program (Campbell-Kelly [3: 151]). Colossus, ENIAC, and their like are called 'program-controlled' computers, in order to distinguish them from the modern 'stored-program' computer.

---

[81] The photographs were released to the Public Record Office (PRO reference FO 850/234).

[82] Personal files of T.H. Flowers (24 May 1976, 3 September 1981).

[83] Personal files of T.H. Flowers (3 September 1981).

This basic principle of the modern computer, that is, controlling the machine's operations by means of a program of coded instructions stored in the computer's memory, was thought of by Turing in 1936. At the time, Turing was a shy, eccentric student at Cambridge University. His 'universal computing machine', as he called it – it would soon be known simply as the universal Turing machine – emerged from research that no-one would have guessed could have any practical application. Turing was working on a problem in mathematical logic, the so-called 'decision problem', which he learned of from lectures given by Newman. (For a description of the decision problem and Turing's approach to it, see (Copeland [7: 45–53]).) In the course of his attack on this problem, Turing thought up an abstract digital computing machine which, as he said, could compute "all numbers which could naturally be regarded as computable" (Turing [31: 249]). The universal Turing machine consists of a limitless memory in which both data and instructions are stored, in symbolically encoded form, and a scanner that moves back and forth through the memory, symbol by symbol, reading what it finds and writing further symbols. By inserting different programs into the memory, the machine can be made to carry out any algorithmic task. That is why Turing called the machine *universal*.

Turing's fabulous idea was just this: a single machine of fixed structure that, by making use of coded instructions stored in memory, could change itself, chameleon-like, from a machine dedicated to one task into a machine dedicated to a completely different task – from calculator to word processor, for example. Nowadays, when many have a physical realisation of a universal Turing machine in their living room, this idea of a one-stop-shop computing machine is apt to seem as obvious as the wheel. But in 1936, when engineers thought in terms of building different machines for different purposes, the concept of the stored-program universal computer was revolutionary.

In 1936 the universal Turing machine existed only as an idea. Right from the start Turing was interested in the possibility of building such a machine,[84] as to some extent was Newman, but before

the war they knew of no practical way to construct a stored-program computer. It was not until the advent of Colossus that the dream of building an all-purpose electronic computing machine took hold of them. Flowers had established decisively and for the first time that large-scale electronic computing machinery was practicable, and soon after the end of the war Turing and Newman both embarked on separate projects to create a universal Turing machine in hardware. Racks of electronic components from the dismantled Colossi were shipped from Bletchley Park to Newman's Computing Machine Laboratory at Manchester. Historians who did not know of Colossus tended to assume quite wrongly that Turing and Newman inherited their vision of an electronic computer from the ENIAC group in the US.

Even in the midst of the attack on Tunny, Newman was thinking about the universal Turing machine. He showed Flowers Turing's 1936 paper about the universal machine, 'On Computable Numbers', with its key idea of storing symbolically encoded instructions in memory, but Flowers, not being a mathematical logician, "didn't really understand much of it".[85] There is little doubt that by 1944 Newman had firmly in mind the possibility of building a universal Turing machine using electronic technology. It was just a question of waiting until he "got out".[86] In February 1946, a few months after his appointment to the University of Manchester, Newman wrote to the Hungarian-American mathematician von Neumann (like Newman considerably influenced by Turing's 1936 paper, and himself playing a leading role in the post-ENIAC developments taking place in the US):

> I am … hoping to embark on a computing machine section here, having got very interested in electronic devices of this kind during the last two or three years. By about eighteen months ago

---

[84]Newman in interview with Evans.

[85]Flowers in interview with Copeland (July 1996).

[86]Letter from Newman to von Neumann (8 February 1946) (in the von Neumann Archive at the Library of Congress, Washington, DC; a digital facsimile is in The Turing Archive for the History of Computing, http://www.AlanTuring.net/newman_vonneumann_8feb46).

Figure 15.10. The first stored-program electronic computer, built in Newman's Computing Machine Laboratory at the University of Manchester. *Source*: Department of Computer Science, University of Manchester.

Figure 15.11. The pilot model of Turing's Automatic Computing Engine, the fastest of the early machines and precursor of the DEUCE computers. *Source*: National Physical Laboratory. Crown copyright.

I had decided to try my hand at starting up a machine unit when I got out. ... I am of course in close touch with Turing.[87]

The implication of Flowers' racks of electronic equipment was obvious to Turing too. Flowers said that once Colossus was in operation, it was just a matter of Turing's waiting to see what opportunity might arise to put the idea of his universal computing machine into practice. (By the end of the war, Turing had educated himself thoroughly in electronic engineering: during the later part of the war he gave a series of evening lectures "on valve theory" (Turing [32: 74]).) Turing's opportunity came along in 1945, when John Womersley, head of the Mathematics Division of the National Physical Laboratory (NPL) in London, invited him to design and develop an electronic stored-program digital computer. Turing's technical report 'Proposed Electronic Calculator' (in [9]), dating from the end of 1945 and containing his design for the ACE, was the first relatively complete specification of an electronic stored-program digital computer.[88] The slightly earlier 'First Draft of a Report on the EDVAC' (in [30]), produced in about May 1945 by von Neumann, was much more abstract, saying little about programming, hardware details, or electronics. (The EDVAC, proposed successor to the ENIAC, was to be a stored-program machine. It was not fully working until 1952 (Huskey [23: 702]).) Harry Huskey, the electronic engineer who subsequently drew up the first detailed hardware designs for the EDVAC, stated that the "information in the 'First Draft' was of no help".[89] Turing, in contrast, supplied detailed circuit designs, full specifications of hardware units, specimen programs in machine code, and even an estimate of the cost of building the machine.

Turing asked Flowers to build the ACE, and in March 1946 Flowers said that a "minimal ACE" would be ready by August or September of that year.[90] Unfortunately, however, Dollis Hill was overwhelmed by a backlog of urgent work on the national telephone system, and it proved impossible to keep to Flowers' timetable. In the end it was Newman's team who, in June 1948, won the race to build the first stored-program computer. The first program, stored on the face of a cathode ray tube as a pattern of dots, was inserted manually, digit by digit, using a panel of switches. The news that the Manchester machine had run what was only a tiny program – just 17 instructions long – for a mathematically trivial task was "greeted with hilarity" by Turing's team working on the much more sophisticated ACE.[91]

A pilot model of the ACE ran its first program in May 1950. With an operating speed of 1 MHz, the pilot model ACE was for some time the fastest computer in the world. The pilot model was the basis for the very successful DEUCE computers, which became a cornerstone of the fledgling British computer industry – confounding the suggestion, made in 1946 by Sir Charles Darwin, Director of the NPL and grandson of the great Darwin, that "it is very possible that ... one machine would suffice to solve all the problems that are demanded of it from the whole country".[92]

REFERENCES

[1] F.L. Bauer, *The Tiltman break*, **Colossus: The Secrets of Bletchley Park's Codebreaking Computers**, B.J. Copeland et al., Oxford University Press, Oxford (2006).

[2] J. Cairncross, **The Enigma Spy: The Story of the Man who Changed the Course of World War Two**, Century, London (1997).

[3] M. Campbell-Kelly, *The ACE and the shaping of British computing*, **Alan Turing's Automatic Computing Engine: The Master Codebreaker's Struggle to Build the**

[87]Letter from Newman to von Neumann (8 February 1946).

[88]A digital facsimile of the original typewritten report is in The Turing Archive for the History of Computing, http://www.AlanTuring.net/proposed_electronic_calculator.

[89]Letter from Huskey to Copeland (4 February 2002).

[90]Status of the Delay Line Computing Machine at the P.O. Research Station (anon., National Physical Laboratory, 7 March 1946; in the Woodger Papers (catalogue reference M12/105); a digital facsimile is in The Turing Archive for the History of Computing, http://www.AlanTuring.net/delay_line_status).

[91]Michael Woodger in interview with Copeland (June 1998).

[92]C. Darwin, Automatic Computing Engine (ACE) (National Physical Laboratory, 17 April 1946; PRO document reference DSIR 10/275); a digital facsimile is in The Turing Archive for the History of Computing, http://www.AlanTuring.net/darwin_ace.

Modern Computer, B.J. Copeland, ed., Oxford University Press, Oxford (2005).

[4] W.W. Chandler, *The maintenance and installation of Colossus*, Annals of the History of Computing 5 (1983), 260–262.

[5] A.W.M. Coombs, *The making of Colossus*, Annals of the History of Computing 5 (1983), 253–259.

[6] B.J. Copeland (ed.), **The Essential Turing**, Oxford University Press, Oxford (2004).

[7] B.J. Copeland, *Computable numbers: A guide*, **The Essential Turing**, B.J. Copeland, ed., Oxford University Press, Oxford (2004).

[8] B.J. Copeland, *Enigma*, **The Essential Turing**, B.J. Copeland, ed., Oxford University Press, Oxford (2004).

[9] B.J. Copeland (ed.), **Alan Turing's Automatic Computing Engine: The Master Codebreaker's Struggle to Build the Modern Computer**, Oxford University Press, Oxford (2005).

[10] B.J. Copeland et al., **Colossus: The Secrets of Bletchley Park's Codebreaking Computers**, Oxford University Press, Oxford (2006).

[11] D. Davies, *The Lorenz cipher machine SZ42*, Cryptologia 19 (1995), 517–539.

[12] M. Davis, **The Universal Computer: The Road from Leibniz to Turing**, Norton, New York (2000).

[13] N.A. de Bruyne, H.C. Webster, *Note on the use of a thyratron with a Geiger counter*, Proceedings of the Cambridge Philosophical Society 27 (1931), 113–115.

[14] E. Enever, **Britain's Best Kept Secret: Ultra's Base at Bletchley Park**, 2nd edition, Alan Sutton, Stroud (1994).

[15] R. Erskine, P. Freeman, *Brigadier John Tiltman: One of Britain's finest cryptologists*, Cryptologia 27 (2003), 289–318.

[16] T.H. Flowers, *The design of Colossus*, Annals of the History of Computing 5 (1983), 239–252.

[17] T.H. Flowers, *D-day at Bletchley Park*, **Colossus: The Secrets of Bletchley Park's Codebreaking Computers**, B.J. Copeland et al., Oxford University Press, Oxford (2006).

[18] F. Golden, *Who built the first computer?*, Time, March 29, no. 13 (1999), 82.

[19] F.H. Hinsley et al., **British Intelligence in the Second World War**, Vol. 2, Her Majesty's Stationery Office, London (1981).

[20] F.H. Hinsley et al., **British Intelligence in the Second World War**, Vol. 3, Part 2, Her Majesty's Stationery Office, London (1988).

[21] H. Hinsley, *The counterfactual history of no ultra*, Cryptologia 20 (1996), 308–324.

[22] A.W. Hull, *Hot-cathode thyratrons*, General Electric Review 32, (1929), 390–399.

[23] H.D. Huskey, *The development of automatic computing*, **Proceedings of the First USA–JAPAN Computer Conference**, Tokyo (1972).

[24] G. Ifrah, **The Universal History of Computing: From the Abacus to the Quantum Computer**, Wiley, New York (2001).

[25] B. Johnson, **The Secret War**, British Broadcasting Corporation, London (1978).

[26] J.A.N. Lee, **Computer Pioneers**, IEEE Computer Society Press, Los Alamitos (1995).

[27] J.A.N. Lee, G. Holtzman, *50 years after breaking the codes*, Annals of the History of Computing 17 (1999), 32–43.

[28] P. McCorduck, **Machines Who Think**, W.H. Freeman, New York (1979).

[29] D. Murray, **Murray Multiplex: Technical Instructions, Manual No. 1, General Theory**, Creed and Coy, Croydon (no date).

[30] N. Stern, **From ENIAC to UNIVAC: An Appraisal of the Eckert–Mauchly Computers**, Digital Press, Bedford, MA (1981).

[31] A.M. Turing, *On computable numbers, with an application to the Entscheidungsproblem*, Proceedings of the London Mathematical Society 2(42) (1936–1937), 230–265. Reprinted in B.J. Copeland (ed.) *The Essential Turing*, Oxford University Press, Oxford (2004).

[32] S. Turing, **Alan M. Turing**, W. Heffer, Cambridge (1959).

[33] W.T. Tutte, *My work at Bletchley Park*, **Colossus: The Secrets of Bletchley Park's Codebreaking Computers**, B.J. Copeland et al., Oxford University Press, Oxford (2006).

[34] F. Weierud, *Bletchley Park's Sturgeon – The fish that laid no eggs*, **Colossus: The Secrets of Bletchley Park's Codebreaking Computers**, B.J. Copeland et al., Oxford University Press, Oxford (2006).

[35] C.E. Wynn-Williams, *The use of thyratrons for high speed automatic counting of physical phenomena*, Proceedings of the Royal Society, Series A 132 (1931), 295–310.

[36] C.E. Wynn-Williams, *A thyratron scale of two automatic counter*, Proceedings of the Royal Society of London, Series A 136 (1932), 312–324.

APPENDIX 1. THE TELEPRINTER ALPHABET

In teleprinter code (Murray [29]) the letters most frequently used are represented by the fewest holes in the tape, which is to say by the fewest crosses (in B.P. notation). For instance, E, the commonest letter of English, is **x●●●●**, and T, the next most frequent, is **●●●●x**. The table opposite gives the 5-bit teleprinter code for each character of the teleprint alphabet.

The left-hand column of the table shows the characters of the teleprint alphabet as they would

Conventional name	Impulse 1	2	3	4	5	Meaning In letter shift	In figure shift
/	•	•	•	•	•	(no meaning)	
9	•	•	x	•	•	space	space
H	•	•	x	•	x	H	£
T	•	•	•	•	x	T	5
O	•	•	•	x	x	O	9
M	•	•	x	x	x	M	full stop
N	•	•	x	x	•	N	comma
3	•	•	•	x	•	carriage return	carriage return
R	•	x	•	x	•	R	4
C	•	x	x	x	•	C	colon
V	•	x	x	x	x	V	equals
G	•	x	•	x	x	G	@
L	•	x	•	•	x	L	close bracket
P	•	x	x	•	x	P	0 (zero)
I	•	x	x	•	x	I	8
4	•	x	•	•	•	line feed	line feed
A	x	x	•	•	•	A	dash
U	x	x	x	•	•	U	7
Q	x	x	x	•	x	Q	1
W	x	x	•	•	x	W	2
+ or 5	x	x	•	x	x	move to figure shift	(none)
− or 8	x	x	x	x	x	(none)	move to letter shift
K	x	x	x	x	•	K	open bracket
J	x	x	•	x	•	J	ring bell
D	x	•	•	x	•	D	who are you?
F	x	•	x	x	•	F	per cent
X	x	•	x	x	x	X	/
B	x	•	•	x	x	B	?
Z	x	•	•	•	x	Z	+
Y	x	•	x	•	x	Y	6
S	x	•	x	•	•	S	apostrophe
E	x	•	•	•	•	E	3

Source: *General Report on Tunny*, p. 3.

have been written down by the Bletchley codebreakers. For example, the codebreakers wrote '9' to indicate a space (as in 'to9indicate') and '3' to indicate a carriage return.

The 'move to figure shift' character (which some at Bletchley wrote as '+' and some as '5') told the teleprinter to shift from printing letters to printing figures; and the 'move to letter shift' character (written '−' or '8') told the machine to shift from printing figures to printing letters. With the teleprinter in letter mode, the keys along the top row of the keyboard would print QWERTYUIOP,

and in figure mode the same keys would print 1234567890.

Most of the keyboard characters had different meanings in letter mode and figure mode. In figure mode the M-key printed a full stop, the N-key a comma, the C-key a colon, and the A-key a dash, for example. (Unlike a modern keyboard, the teleprinter did not have separate keys for punctuation.) The meanings of the other keys in figure mode are given at the right of the table.

To cause the teleprinter to print 123 WHO, ME? the operator must first press figure shift and key Q W E to produce the numbers. He or she then drops into letter mode and keys a space (or vice versa), followed by W H O. To produce the comma it is necessary to press figure shift then N. This is followed by letter shift, space, and M E. A final figure shift followed by B produces the question mark.

$$+\mathrm{QWE}-9\mathrm{WHO}+\mathrm{N}-9\mathrm{ME}+\mathrm{B}.$$

Often Tunny operators would repeat the figure-shift and letter-shift characters, sending a comma as ++N−− and a full stop as ++M−−, for example. In this case the operator would key

$$++\mathrm{QWE}--9\mathrm{WHO}++\mathrm{N}--9\mathrm{ME}++\mathrm{B}.$$

Presumably the shift characters were repeated to ensure that the shift had 'taken'. These repetitions were very helpful to the British, since a correct guess at a punctuation mark could yield six characters of text (including the trailing 9).

APPENDIX 2. THE TUNNY ENCIPHERMENT EQUATION AND TUTTE'S 1 + 2 BREAK IN

First, some notation. P is the plaintext, C is the cipher text, χ is the stream of letters contributed to the message's key by the chi-wheels and Ψ is the stream contributed by the psi-wheels. $\chi + \Psi$ is the result of adding χ and Ψ using the rules of Tunny-addition. ΔC is the result of delta-ing the ciphertext, $\Delta(\chi + \Psi)$ is the result of delta-ing the stream of characters that results from adding χ and Ψ, and so forth.

Since C is produced by adding the key to P, and the key is produced by adding χ and Ψ, the fundamental encipherment equation for the Tunny machine is:

$$C = P + \chi + \Psi.$$

C_1 is written for the first impulse of C (i.e. the first of the five streams in the teleprint representation of the ciphertext); and similarly P_1, χ_1 and Ψ_1 are the first impulses of P, χ and Ψ respectively. The encipherment equation for the first impulse is:

$$C_1 = P_1 + \chi_1 + \Psi_1.$$

Delta-ing each side of this equation gives

$$\Delta C_1 = \Delta(P_1 + \chi_1 + \Psi_1).$$

Delta-ing the sum of two or more impulses produces the same result as first delta-ing each impulse and then summing. So

$$\Delta C_1 = \Delta P_1 + \Delta\chi_1 + \Delta\Psi_1.$$

Likewise for the second impulse:

$$\Delta C_2 = \Delta P_2 + \Delta\chi_2 + \Delta\Psi_2.$$

Adding the equations for the first and second impulses gives

$$\Delta C_1 + \Delta C_2 = \Delta P_1 + \Delta P_2 + \Delta\chi_1$$
$$+ \Delta\chi_2 + \Delta\Psi_1 + \Delta\Psi_2$$

which is the same as

$$\Delta(C_1 + C_2) = \Delta(P_1 + P_2)$$
$$+ \Delta(\chi_1 + \chi_2) + \Delta(\Psi_1 + \Psi_2).$$

Because of the staggering motion of the psi-wheels, $\Delta(\Psi_1 + \Psi_2)$ turns out to be about 70% dot. But adding dot leaves you where you started: cross plus dot is cross and dot plus dot is dot. It follows that the addition of $\Delta(\Psi_1 + \Psi_2)$ more often than not has no effect. So

$$\Delta(C_1 + C_2) = \Delta(P_1 + P_2) + \Delta(\chi_1 + \chi_2)$$

is true more often than not.

Tutte also discovered that $\Delta(P_1 + P_2)$ is approximately 60% dot. This effect is the result of various factors, for instance the Tunny operators' habit of repeating certain characters (see Appendix 1), and contingencies of the way the individual letters are represented in the underlying teleprinter code – for example, the delta of the sum of the first and second impulses of the common bigram (or letter pair) DE is dot, as it is for other common bigrams such as BE, ZE, ES. So it is true more often than not that

$$\Delta(C_1 + C_2) = \Delta(\chi_1 + \chi_2).$$

Tutte's '1 + 2 break in' is this. $\Delta(C_1 + C_2)$ is stepped through the delta-ed sum of the first and second impulses of the entire stream of characters from the chi-wheels. Generally the correspondence between $\Delta(C_1 + C_2)$ and a strip from the delta-ed chi of the same length will be no better than chance. If, however, $\Delta(C_1 + C_2)$ and a strip of delta-ed chi correspond more often than not, then a candidate has been found for $\Delta(\chi_1 + \chi_2)$, and so for the first two impulses of χ. The greater the correspondence, the likelier the candidate.

The History of Information Security: A Comprehensive Handbook
Karl de Leeuw and Jan Bergstra (Editors)

16

BORIS HAGELIN AND CRYPTO AG: PIONEERS OF ENCRYPTION

Silvan Frik

Crypto AG
Zug, Switzerland

Contents

Abstract

The trend towards integrated electronics has inevitably changed the demands made on the production structures and the staff of a cipher equipment supplier. Sophisticated production processes are now completely automated. Today, the amount of mechanical work is very small, e.g. the production of the housing. By contrast, the use of brain power is constantly increasing. The Swiss-located company Crypto AG, founded by Swedish cryptography-pioneer Boris Hagelin, has always been on the top of these developments. Crypto AG provides in its 55th year of succesful history more than 130 countries the best-possible services and solutions on the spot.

Keywords: B-211, chain, electric random generator, Enigma, The Hag, M-209, One Time Pads, PEB unit.

16.1 VIA THE NOBEL FAMILY TO A.B. CRYPTOGRAPH

Boris Caesar Wilhelm Hagelin was born on 2 July 1892 in Adjikent near Baku (Azerbaijan), the son of Karl Wassiljewitj Hagelin and Hilda Hagelin, née Kiander.[1] Due to his father's work, he spent his childhood in various different places and travelled a lot. As a result, he became multilingual at an early age. Hagelin began his professional career in 1915, after completing his studies and military service.

He was taken on originally for a year's work experience by the electrical engineering group ASEA. Since his training was in mechanical engineering, he needed to get experience in electrical engineering and worked in different departments at ASEA. When he stayed on at the company, he was placed in the international department dealing with tenders. Since Boris Hagelin was fluent in five languages, he was given the job of looking after foreign customers.

Due to a happy combination of circumstances and events, Boris Hagelin is linked directly to the history of the famous Nobel family. Hagelin's grandfather found work in the engineering factory of Immanuel Nobel, the father of the inventor of

---

[1] A detailed biography of Boris Hagelin can be found in a book by Stadlin [4]. The present account is based on that biography. A further recommended source is Kahn [1].

dynamite and founder of the Nobel Prize, Alfred Nobel. His father received financial support from Emanuel, the nephew of the Nobel Prize founder, towards the end of his studies. It was also the beginning of a life-long, close friendship between the two, a friendship which would, many years later, have a decisive impact on the direction of Boris Hagelin's life.

In the autumn of 1919, Boris' father wanted his son to have a chance to get to know the American oil industry. Boris made his first trip to America, taking with him some orders for ASEA in the railways sector. However, after being refused a salary rise, Hagelin resigned from ASEA and took on a temporary job with Standard Oil. After returning to Sweden, he was faced with the question of his future professional career. He opened a small engineering office, with a contribution of 50,000 Swedish Crowns from Emanuel Nobel. Once he had established himself in Stockholm, his services were soon required again by the Nobels. One of the tasks entrusted to him by Emanuel Nobel was of a very special nature and was to have a decisive influence on his life: it was the business of cipher machines.

A man, who had some acquaintances in Nobel's St. Petersburg office visited Nobel in Stockholm and told him about a company, A.B. Cryptograph, whose products were bound to become very important. Hagelin's father and Emanuel Nobel realised that the possibility of mastering a simple, mechanically generated secret language would undoubtedly bring great advantages – especially in a country where it had always been difficult to keep business secrets. So the financially failing A.B. Cryptograph, with the inventions of the engineer Arvid Damm, was supported by the Nobels.[2] This is how, in 1922, Hagelin established a link with Damm: At

first he was a kind of supervisor, then he became involved in the technical work of the Stockholm office, and eventually after Damm's death, he became the (unpaid) manager of A.B. Cryptograph.[3]

16.2 HAGELIN'S LIFE'S WORK BEGINS . . .

Employed at A.B. Cryptograph as Dr. Nobel's trustee, Boris Hagelin was now officially a member of the board of directors. On his first visit to the factory, a piece of equipment had been put on a table for the purpose of demonstration. It consisted of a keyboard equipped with contacts, a so-called electrical cryptograph, and two typewriters controlled by an electric magnet. During the encryption process, the plaintext was typed on the keyboard, and the plaintext was then printed on one side and the ciphertext on the other side. In the decryption process, the drums were simply reversed.

By the time Hagelin joined the company, Damm had already done quite a lot of work, though mainly in the form of prototypes – which is what had used up all the company's capital and was the reason for Nobel's help. Damm's basic invention was his 'chain'. This did indeed consist of a chain which could be separated into individual links and in which the links were either flat or had an indentation (so they were inactive or active links). The operator could assemble chains of different lengths – within certain limits – with any chosen sequence of inactive and active links. The chains were inserted in the machines designed by him and thus controlled the movements of the encryption mechanism.

The chain was used for the following machines: A pocket device called A-22, which was intended for military purposes.[4] A few of these machines

---

[2]In the history of encryption, Arvid Damm appears in 1915 with his first patent. He succeeded in finding backers from the Swedish financial and industrial world in Stockholm who were interested in his invention, and A.B. Cryptograph was founded in 1916. At the end of 1927, Damm fell ill, and died in 1928. Damm had a brother, who was a mathematician, and who had probably given him the necessary understanding of the mathematical aspect of cryptology. But it is said that he was not a good designer. For more information about Damm's machines and how they functioned, see also Kahn [1: 208ff].

[3]Damm is reputed to have been quite an unconventional person – probably also a reason why Hagelin was placed there by his father and Nobel as a sort of 'supervisor'. To what extent Damm was aware of this is not known, but it could explain later critical comments he made about further developments of his inventions.

[4]The use of the chain was to play another important role here: It was self-destructing if the machine cover was opened with the press of a button – the chain would then fall out and come apart into its individual links. This prevented anyone from reconstructing the function of one of the most important parts of the ciphering mechanism if a machine were captured.

Figure 16.1. The pocket device A-22.

were produced, but they were never put to use. The next product was the development of an office machine. Like the A-22, it was equipped with a drum which contained a number of alphabet-strips with different cipher sequences which could be changed. This drum was covered and on pressing a key, a shutter opened and the corresponding secret or plaintext cipher appeared and could be noted down. The chain's effect was that, depending on its combination, the drum was moved forward or backward by one step.

Damm then designed a simplified electrical crypto-machine, the B-18, which was to play an important role later on. This machine, too, was made up of several components. A keyboard, a motor-driven encryption unit and an output unit which could either be a typewriter or a punch for telegraph operation. On Hagelin's suggestion, a typewriter model which was quite new then, the Woodstock-Electric (the first electrically driven typewriter), was used.

In 1925, Boris Hagelin heard the news that the Swedish General Staff intended to procure cipher machines. It first acquired a German Enigma machine for trial purposes. Very quickly, Hagelin arranged a visit to the responsible procurement officer and told him about the products of the Swedish company A.B. Cryptograph. However, the company could not offer suitable products; the Enigma machine with a keyboard and light panel seemed to be what the General Staff desired. At least he got a concession that the procurement of the Enigma machine would be delayed by six months to give him an opportunity to come up with his own machine.

With an additional 500 Swedish crowns from the company Nobel, Hagelin now had to design a new machine. Since Hagelin could no longer rely on Damm, who was by now in Paris, he saw the solution in adapting the ciphering mechanism of Damm's B-18 machine to suit his purpose. He constructed a device that included the corresponding ciphering component. The device was then equipped with a keyboard and a light panel like the Enigma, although the functions were quite different. The prototype was ready in time and was thoroughly tested, with an actuary approving the ciphering reliability.

Hagelin got the order and although it was not very large, the machine with the model designation B-21 was the first commercial success of A.B. Cryptograph. A description of the machine was of course sent to Arvid Damm to his new abode in Paris. Damm is said to have criticised this machine, for reasons that are no longer known, but the Swedish Army liked it and placed a large order in 1926.[5] It was probably true that the machine was not really that good, but it was good enough to be of service to the customers for many years. Repeat orders still came in after World War II.[6] And at that time Hagelin didn't even know what cryptology was. The small-scale production at the time had to take place in very poorly equipped premises, first in a rubber factory, then a neon factory.

Hagelin's travels in connection with cryptography only started in the early 1930s. By chance, he met a businessman who was a successful representative of the American company Dictaphone and with whom a cooperation agreement was signed. However, it was years before he achieved a breakthrough. This came about because he approached an office manager whose department needed new equipment. The office manager apparently expressed an interest in Hagelin's design, but wanted a machine that printed. So once again, Hagelin had to improvise: The light panel in the B-21 was removed and Hagelin designed a printer in which

---

[5]Kahn [2: 425].

[6]Some improvements were made over the years by, e.g., linking the B-21 to an electric typewriter, which made it much easier to use when stationary.

Figure 16.2. The famous B-211.

the ciphertext – as required – only contained letters, while the plaintext could contain all the normal letters and symbols. A mechanical drive was added which made it possible to achieve an operating speed of approximately three ciphers per second, which was a considerable achievement for that time. The ciphering mechanism was driven by a torch battery and for emergencies, a hand crank was also provided. This is how the famous B-211 model was born.

For its time, the B-211 was by far the most practicable cipher machine in the world. One has to assume now, however, that it did not offer adequate security. Apparently, experts have developed methods that enable a message encrypted by the B-211 to be deciphered, provided the analyst detects a letter sequence of five ciphers, which occurs in any place in the message.[7] There were repeated claims that the British and Americans were able, even before World War II, to decipher French messages. However, this is difficult to imagine, since the French army purchased another 100 B-211 machines after the war.[8] But it was indeed the case after the war. Joseph Petersen provided the Dutch with information about it, although the French were not aware of it.[9]

After Emanuel Nobel's death in 1932, his heirs did not want to increase the investment in A.B. Cryptograph and dissolved the company. Hagelin's

activities were integrated into A.B. Ingeniörsfirman Teknik. Later, Hagelin and his staff founded A.B. Cryptoteknik, since growing numbers of orders for the B-211 were coming from France and they needed additional capital from Hagelin's father. The profit made until the outbreak of the war in 1939 was enough to set up a workshop in Stockholm. This became Hagelin's first factory in which modern equipment could be installed and efficient production be started.

It is interesting that at the time, a half-finished design would later be of vital importance for the entire activity. It is worth briefly introducing this machine: It was an automatic coin changing machine. The purpose of the design was to enable people to insert any size and number of coins (up to a specified maximum) into the machine and in return get a receipt or travel ticket of a specified value by entering it on a keyboard, plus, of course, any change due. The most important component was the counter: This consisted of a drum fitted with moveable bars (27 bars for up to three-digit amounts). The bars had so-called riders which were fitted in such a way that one bar was moved for a number one, two bars for a number two, etc. This functioned because the keys were like levers which moved against the drum when a key was pressed, causing the drum to turn once.

After suffering a nervous breakdown in the spring of 1934, Hagelin received an inquiry whether his factory was able to design a pocket-size cipher machine that printed. At first, he had no idea how to go about this, because the design he had used for the B-21 and B-211 could not be reduced to the required degree. Later, it occurred to him to adapt the counter of the automatic coin changing machine for the cipher machine.[10] After all, the ciphering process involves calculations just like in a counting machine, the only difference being that the counting always had to be 'wrong'. In this incredible way, the new machine was born, the Type C-machine, which would gradually achieve more sales than all others.

---

[7] Schmeh [3: 14ff].
[8] Kahn [1: 216].
[9] Kahn [2: 690f].

[10] It is not possible to determine how much Hagelin actually contributed to this invention. He undoubtedly recognised the connections and significantly used and developed the invention.

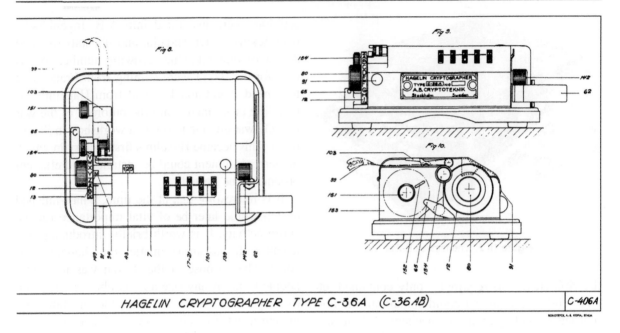

HAGELIN CRYPTOGRAPHER TYPE C-36A (C-36AB) | C-406A |

Figure 16.3. A drawing of the development of the C-36.

The idea used in the counter drums of the coin dispenser, with the shiftable bars and levers, was now used and changed as follows for the new cipher machine: The machine was (initially) fitted with 25 bars on a drum. Five levers were arranged, each of which was operated by a pinwheel. The bars were divided into five groups, the first had one bar with rider, or lug, the second had two bars with a lug each set opposite the second lever, the third one four, the fourth eight and the fifth ten bars. The pinwheels had different divisions: 17, 19, 21, 23 and 25 – so there was no common factor and had the same number of adjustable pins as their divisor. The pins could be set in either an active or inactive position, whereby the active pins pressed the associated levers against the drum. With each activation, a typewheel connected to an alphabet wheel was put in the desired position (the plaintext letter to be encrypted, or the ciphertext letter to be decrypted) and then the drum was rotated once. Where levers were pressed against the drum, the corresponding bars were moved and acted on the typewheel, which was then rotated by the same number of steps as the number of the moved bars, with the resulting letter being printed.

Finally, the pinwheels were rotated forward by one step, and since no common factor was present for the steps, the pinwheels could only return to their common starting position after $17 \times 19 \times 21 \times 23 \times 25$ activations. This meant that a length of 3,900,225 ciphers could be achieved with this new type of cipher machine, an enormously long period for that time. It should be noted that it was possible, with the groups of moveable bars of 1–2–4–8–10, to achieve all steps from 0 ($= 26$ steps) to 25. A prototype was built, since Hagelin first had to ascertain that it was indeed possible to make a 'pocket-size machine'. It is said that a licensee later expressed the criticism that because of its small dimensions, it was very difficult to manufacture the machine. But in fact, the small dimensions played a major part in the machine's growing popularity. It was accepted by the French and in 1934; Hagelin was awarded the Légion d'Honneur for this invention.

The new machine was initially called C-34 after the year it was built. However, it soon lost any claim to be a 'pocket-size machine' – the customer expressed the wish to have a protective hood added. Moreover, a small amusing detail was added, which allowed strapping the machine to the operator's thigh. It allowed him, in the worst case in a field operation, to beat a quick retreat with the machine attached to him.

In the late 1930s, Hagelin also started to be interested in a possible market in the USA. He made his first trip there in 1937, though it was mostly a pleasure trip. But he did meet the world-famous cryptologist William Friedman and began a friendship which lasted until Friedman's death in 1969. During a visit to Washington, interest was expressed in a C-device with a keyboard. In 1939, the first prototype was finished and Hagelin made his second trip to the USA. But the demonstration was flawed and Boris Hagelin was forced to return to Stockholm without having achieved any results at a time when World War II had already started.

16.3 THE CRUCIAL JOURNEY

The winter of 1939/1940 was extremely stressful, both for the company and for Hagelin himself. He was preoccupied with making another trip to the USA in a last attempt to win over the Americans for his C-device. When he arrived there, he met up with his employees Paulding and Hedden, who had already arranged a visit to the US Signal Corps. On his arrival, he first met his old friend Friedman and some of his army colleagues. The devices were demonstrated and he was told to make contact again a week later. When he did so, he was told in Mr. Friedman's office that the decision had been taken to order another 50 machines so as to test them very thoroughly. Hagelin ordered the machines by telegraph and within a very short time, they arrived in Washington by American air courier. They were paid for promptly which, given the limited possibilities of importing currency, was urgently needed. In the spring of 1941, the Hagelins returned to Washington to visit Friedman and wait for the final answer from the responsible authorities of the Signal Corps. But on each visit, Boris Hagelin was brusquely dismissed and told that he would be called if he was needed.

In June 1941, he was finally informed that the machines had been tested and that he should prepare for sizable orders. Immediately after his arrival in the USA, Hagelin's representatives there had looked for a suitable manufacturer and eventually found the right one with L.C. Smith & Co. This company, controlled by the Smith family, owned a large typewriter factory which had just recently been converted to produce guns. They also had a factory to produce travel typewriters, located in a small town in New York State – they made the well-known 'Corona' typewriter and employed approximately 900 people in its production. The plan was to convert this factory into the production of war material and was thus perfectly suited to Hagelin's purpose.

The technical manager of the factory was provided with a machine and drawings. After the drawings had been examined and the machine tried out, the price was agreed. But it still took more than a year before the machine was approved and the first order was placed, despite the fact that the first machines were due to be delivered just five months later. These were extremely tough terms that demanded an extraordinarily good performance from the Corona factory, since all the drawings still had to be converted from the metric system into feet and inches. This also required certain compromises. Furthermore, Friedman was keen to simplify the handling of the machine as much as possible. In the first place, the possibility of adjusting the input and output alphabet in relation to each other, as with the normal machines, was abandoned. Secondly, the locks used to lock the protective hood and the machine lid were also left out. The machines were intended for use at the frontline.

The preparations for starting the production made very rapid progress and were completed 60 days later – even though it had involved making all the tools. There was one snag during start-up, however, which was to be expected: There was one single component for which the dimensions had been specified, but whose final shape still hade to be determined in trials. It was a dimension that allowed only the smallest tolerances, otherwise the machine would constantly have been blocked during use.

After production had started properly, problems arose once the USA had been drawn into the war, and the demands on the industry's performance capability were constantly intensified. It became more and more difficult not just to get the necessary raw materials on time, but to procure them at all.

The first order was placed by the Signal Corps, i.e. the Army, followed straight away by orders from the Navy and the Air Force. The number of machines ordered quickly reached 50,000, which occasioned a great celebration.

16.4 THE LONG STAY IN AMERICA

In spring 1942 Hagelin received a large order for a machine which he had been unable to sell in 1939, because it wasn't functioning properly: the BC-543 (a variety of the C-38) with a keyboard and electric drive. This machine was intended for use by the OSS (Office of Strategic Services) at American stations abroad. Although the machines were manufactured in Stockholm, the distribution was to be based in the USA. In addition, operators had to be trained and service had to be ensured. For this purpose, Hagelin bought a house in Greenwich, Connecticut, about an hour by train from New York. A small workshop was set up in the cellar to service the BC-machines and Hagelin found a Swedish-born American who was able to provide excellent support in servicing them.

In the summer of 1942, Boris Hagelin and his friends Paulding and Hedden were asked to come to Washington. The Navy was about to place a fairly large order with them, but it demanded the full disclosure of the cost calculations. At the time, a new system was being introduced for the procurement of war material, which allowed only a specified profit margin. Renegotiations were introduced for orders that had already been placed, which meant that new, reduced prices were negotiated for many products. This also applied to material already delivered. The Navy considered Hagelin's prices excessive and the Army was also in favour of renegotiation. Hedden proved to be a cunning negotiator and managed to avoid renegotiation by selling the patent for the C-machines to the War Department. The price for the patent was equal to the total of the commitments already entered into by the customer. Of course, Hagelin had not the least inclination to pay 80% tax on such a profit in Sweden, so that after a strict interrogation by the Treasury Department, he paid 25 % capital gains tax in the USA. However, to fulfil all the formalities, Hagelin had to leave the country and apply from 'abroad' for a new residential visa (he only had a visitor's visa) to be allowed back.

The way in which the cryptography business was organised was that Hagelin's representatives, Paulding and Hedden, had founded a company called Hagelin Cryptograph. The L.C. Smith & Corona company became a subcontractor for the Hagelin company and Hagelin would personally get a fixed amount for each machine sold to the American army. The preliminary negotiations, in which Corona was also included, were conducted mainly by Hedden. Hagelin was asked to come to Washington for the final deal. The contract, that was finally signed, stipulated that 26,500 machines would be delivered between January and September 1942 and another 46,500 between January and September 1943! A total of 140,000 units of the slightly modified C-36 machines[11] were eventually built, a lot more than originally planned. The Americans called it M-209[12] or simply 'The Hag'. Ironically, it was also used by the Italian Navy.[13] Since all the business problems had been dealt with, the Hagelins returned to Sweden in September 1944. Boris Hagelin had made a lot of money in the USA. Some of it was invested in real estate in Sweden; the rest remained in the USA in the form of stocks and shares.

A.B. Cryptoteknik was very busy during the war. New models were also developed and built according to Hagelin's instructions. First, the machine with a keyboard, which he had presented in the USA in 1939, was completed. At the same time, the pocket-size machine was refined by integrating a double printer, so that both the input text and the processed text were printed simultaneously. In the first C-446 model, there were two paper tapes, in the second, final C-448 model one single, wider tape was needed, on which both texts were printed alongside each other. A cutting device made it easy to separate the two tape segments after the two texts had been compared to check them (this was the main purpose of the double printer).

---

[11]The C-36 and the C-38 also function according to the pinwheel principle and are more developed versions of the C-34.

[12]A detailed description of how the M-209 functioned can be found in Kahn [2: 427ff].

[13]Kahn [2: 427].

Figure 16.4. An M-209 in operation.

The C-38 came to be the real bestseller for A.B. Cryptoteknik. The main customers were government offices in France, Italy and Finland. From 1941, the Germans built a similar machine, called 'Schlüsselgerät 41', which often led it to be called C-41 – but Hagelin had nothing to do with it.[14]

However, after the war business was slack at A.B. Cryptoteknik. Even Hagelin was convinced at the time that with 'permanent peace' having been achieved, future interest in cipher machines would remain minimal. So he prepared himself to spend the rest of his life in the country and busy himself with the production of roof tiles and forest management. But the cipher machine business could not be neglected, since the company still had important customers – but not enough to cover costs. At first, the production of oil heating installations provided the company with good extra income, while it also searched for other potential products to manufacture. Then, against Hagelin's wish, a large project was accepted: the design and construction of an adding machine. The work took several years, soaked up large sums of money and was eventually abandoned. The company had more luck with smaller items: paging machines and pencil sharpeners. The pencil sharpener was later taken over by another manufacturer, when A.B. Cryptoteknik was dissolved in 1958, and it is still produced to-

day. They also made record players, but that did not grow into a worthwhile business.

Hagelin himself was very keen to revive the cryptography business if possible. The C-machine, which had sold in such huge quantities during the war, now had to be considered as somewhat outdated. Hagelin decided to improve it so that the security of encryption was increased. He remained convinced to the end that this old C-design, dating back to 1934, was still a secure machine, provided it was used properly. Just when Hagelin wanted to get going with this project, the following happened: A new law came into force, which required that inventions in the military field first had to be offered to the government, before the inventor was allowed to use it for other purposes. This was a delicate situation for Hagelin, since he was specifically targeting foreign markets with new developments.

It was a coincidence that supplied the solution, which would turn out to be a complete success: One day, Boris Hagelin received a phone call from the company L.M. Ericsson, which had had a visit from Dr. Edgar Gretener from Zurich. Gretener had built a small teletype machine and was looking for a representative in Sweden. He was given Hagelin's name and so they started talking. Since he had often thought about ciphering arrangements for teletype machines, Hagelin suggested to Gretener that they should initiate collaboration for the design and manufacture of so-called telecipher machines. They came to an agreement, and this opened up the opportunity for him to move to Switzerland and work on the new inventions without interruptions. A.B. Cryptoteknik could then make use of them without coming into conflict with the law. So first he had to design a machine that could function with Gretener's teletype machine.

Thanks to Gretener acting as intermediary, Hagelin obtained a residence permit for Switzerland for one year and on 1 October 1948 he and his wife left Sweden. Friends suggested that they should settle in Zug in central Switzerland. The Hagelins found a villa in Weinbergstrasse, above the town of Zug, which had very few buildings at that time. There were severe restrictions on currency exports in force in Sweden at the time, so that

---

[14]Schmeh [3: 147f].

Hagelin was only permitted to take 500 Swedish crowns out of the country and could only pay for the shipment of his furniture in crowns as far as the Swedish border.

But fortunately, Hagelin had concluded an agreement with his employee, Hedden, before leaving America. A provision in the agreement stated that, if there was risk that his funds in the USA could end up in foreign possession, he had the right to place these funds into the hands of a trustee. When concluding the agreement, he had been more concerned with the possibility that the Russians might occupy Sweden. This agreement now became the lifeline for Hagelin when he was asked by the Swedish national bank to change his funds deposited in the USA into Swedish currency. Hagelin referred the foreign currency department to Hedden, who obviously refused to comply with this demand. After this incident, he clearly became persona non grata with the Swedish authorities. However, his considerable US funds enabled him to live without financial worries in Switzerland, where the tax system was more favourable.

The attractive tax situation in Switzerland was undoubtedly one of the main reasons why Hagelin moved to Switzerland, but it was not the only one. Having been largely spared the horrors of World War II, and having remained neutral, Switzerland was a perfect location. The market had since prospered and become internationally integrated. The high level of skills and the existing human resources in the field of precision engineering (mainly because of the watch industry) also influenced Hagelin's decision. Switzerland enjoys a reputation for excellent quality and complete reliability.

In addition, over the years, the government has created further incentives, such as the numerous opportunities for export financing and export guarantees. Another aspect to be highlighted is the unique, liberal legislation regarding exports – Switzerland is probably the only country that allows any type of algorithm to be exported. Imports and exports of industrial goods enjoy *total freedom from customs charges and quotas,* as a matter of principle.

For the customers of Crypto AG this means, among other things, that its products are not subject to any governmental or other influences. It means that the customers can acquire security technology of the highest standards and also benefit from a simple and uncomplicated export procedure. Switzerland has no export restrictions for algorithms.[15] This enables Crypto AG to export the highest-possible security level globally to any country that is not subject to UN sanctions. It places customers on the same security level as a major power.

After Hagelin had settled into the villa, he visited Gretener's office every day to help with the construction of the telecrypto machine specially designed for him. The machine was finished in quite a short time and a number were sold together with Gretener's teletype machine. However, it seems that cooperation with Gretener soon became difficult, because Hagelin had to bear all the costs while, in his view, Gretener reaped the benefits. They parted after less than a year, which was tantamount to a notification from the Aliens Registration Office asking him to leave the country after a given time. However, Hagelin succeeded in getting a new residence permit as a private individual.

16.5   THE NEW START IN SWITZERLAND

Hagelin was now able to devote himself completely to the task of developing new machines for A.B. Cryptoteknik. He equipped two rooms in his villa for the purpose, an administrative and a design office. For a while, he even managed to employ his most experienced designer, Lindmark,[16]

---

[15]It is difficult to judge to what extent Crypto AG benefits from the fact that other industrial countries, in particular the USA, but also Germany and France, have some severe export restrictions and so the competitor companies from those countries were, and are, not in a position to supply the strongest algorithms. After all, 'soft' factors such as Swiss neutrality are also important in purchasing decisions, particularly with regards to the other countries referred to, but also arguments such as the sophisticated security architecture offered by the product.

[16]He had been recruited by engineer Damm as early as 1916 and was instrumental in ensuring that the new type C device had very good and functional equipment.

in Zug. He also employed two Swiss nationals: An engineer, Oskar Stürzinger, whom he had met at Dr. Gretener's company and whose special skills in telegraph engineering were of great benefit later on, and his wife who was an excellent secretary. Since, as a foreigner, Hagelin was not allowed to employ people for business activities, the Swiss company Crypto AG was founded in 1952.

By a lucky coincidence, he was able to build a small laboratory and workshop opposite his villa. His wife had noticed some excavations taking place there, and after inquiring what was being built, the Hagelins were told that the plan was to build a two-storey garage. They were not very keen on having a garage next door and managed to buy the plot from the owner. This is how the first workshop was established in Zug.

His first job in Switzerland was to update the C-machine, now over 25 years old. Although the C-machine had a long key period, it consisted of six short, interlocking periods. Gradually, a method was developed by which the ciphertexts that were produced with the aid of this machine could be successfully decrypted. Strict rules had to be followed during encryption, to avoid giving one's opponent an opportunity to find weaknesses and turn them to his advantage. There have been such cases, especially in critical situations.[17]

Hagelin came up with a design, which made the movement of the pinwheels irregular. This machine, called the CX-52, was produced as a prototype in Stockholm and received very different responses from potential customers. In this machine, the number of divisions for the pinwheels was increased by using more divisions. At the time, some experts insisted on having as long a key series as possible, which was extremely complicated, even

Figure 16.5. Boris Hagelin with his famous M-209 in the factory in Zug.

at the expense of the key length. Since it was found after a while that it was possible to unintentionally produce extremely short periods with the CX-52, methods were increasingly devised that would lead to irregular series sequences, but always with a guaranteed minimum length. The result of further development was that it became possible to offer customers from different countries machines with different switching systems (but with many possible variations). This is how the different types were gradually produced.

Instead of building one autonomous machine with a keyboard and electrical drive that was compatible with the CX types, a sort of base was developed which contained the keyboard and the drive. This made it possible for customers who had bought the manual machine to later buy bases, once their operation had reached a level where a larger volume made faster and easier processing of the secret correspondence desirable. This arrangement also allowed customers in head offices to operate the manual devices with settings for several different recipients with one single base, if necessary. Based on the earlier trials with direct teletype cipher machines, which had led to a temporary co-

---

[17] Hagelin repeatedly recounted what the former boss of the 'Deuxième Bureau' in France had told him after the war: He was present at the American landing in North Africa and knew the people in charge of the American encryption service. They made a bet. The French claimed that they would be able to decrypt the coded messages produced with the M-209 – of course the Americans claimed the opposite. The result was: As long as the American troops were stationary, it was impossible to decipher their messages, but when they were involved in battles, the rules were neglected and it was possible to decipher the messages.

operation with Dr. Gretener, a new model was built in Stockholm in 1952 under the direction of Boris Hagelin Jr., which was largely based on the components of the C-machine. It was very compact and above all very modern in the cryptology it used. It was called the TX-52.

Hagelin also began to devote himself to a very ambitious project. He wanted to design a sort of universal machine. He intended it to be capable of operating both as a line-bound machine (teletype machine) and as an autonomous machine and include the following functional parts: a keyboard, a printer (with text tape), telegraphic sending and receiving units, a detector and a punch. The input units could be the keyboard or the detector, which would be controlled by a punched text tape, as well as the telegraphic receiver unit. The output could be in the form of a printed tape, a punched tape or the telegraphic dispatch unit. It included everything that could possibly be demanded of a cipher machine. A prototype was produced and given the type designation TMX-53. But there were difficulties with the sending and receiving contacts, which would not be overcome, although this should not have been that difficult. But only the one prototype was ever produced, which now graces the museum of Crypto AG in Switzerland.

In the same year, the attention of Hagelin and his staff was drawn to another area of encryption, the use of so-called 'One Time Pads'. These involved tables with random key character sequences of which at least two identical pads were always produced; one for the sender, the other one for the recipient. Such pads were only used once. If the key character sequences were really completely random, then totally secure secret messages would be produced. But initially, the keys would only be reasonably random if they were produced by means of dice, which was very time-consuming. Attempts were also made to produce such tables with the aid of typewriters, where totally random keys were depressed. However, it was found that it was impossible to avoid certain character sequences because people always develop particular work patterns which are repeated.[18]

Hagelin decided to design a device that would be as simple as possible. A number of type wheels, corresponding to a typewriter line, were mounted on a spindle with some friction. This spindle was given a rotating pulse and the wheels turned with it. Since they did not all weigh the same and the friction was not the same in each of them either, they came to a halt in a fairly irregular sequence. A sheet placed above the line of characters was printed, but shifted by one line and the process was repeated. However, this first attempt proved to be too simple.

Hagelin then had a slightly crazy idea: The position of every single type wheel was to be determined by a process where, for each operation, a number of small balls had to be admitted into a channel and detected. The respective number of the detected balls was determined by making one ball slightly bigger than the others – so it could not enter the channel. For each operation, all balls would be vigorously shaken, the channel would be opened and after the detection a character would be printed. There were 40 such 'cells' placed next to each other, so that 40 characters could be printed simultaneously. The system worked very well, and ten machines of this design, with the type designation CBI-53, were built. As a printing unit, it operated at a total capacity of 400 characters a minute, i.e. at the speed of a normal teletype machine, with always one line with eight character groups of five characters being printed in one stroke.

This machine was assembled by an outside design engineer, because there were not yet enough design engineers employed by him. The manufacture of the resulting machine proved to be unusually expensive. Since at that time, devices for the same purpose were being designed with electronic random distribution of the characters, Crypto AG gradually tried to develop similar equipment. For Hagelin, this first step into electronics was quite daunting. His training as an engineer was focused on mechanical engineering, and he had only a little training as an electrical engineer. For a long time he was resistant to embarking on this electronic adventure.

---

[18]The process has two challenges, however: Firstly, it is not that easy to generate a 'genuinely random key', and secondly, the handling of the key is very exacting. For more information about this, see Wobst [5: 60–63].

First of all, Hagelin devoted himself to developing Crypto AG and its production of mechanical and electro-mechanical cipher machines and devices. Having perfected the CX-machine with its different versions, additional refinements were introduced. One of these was the possibility, designed for customers from the Middle East, of getting messages printed in Arabic, which was essential for diplomatic traffic across public telegraph networks. This involved quite a simple trick: During encryption, a double type wheel was used with an alphabet set of Arabic symbols for the plaintext and an alphabet set of Latin symbols for the ciphertext. Since Arabic is written from left to right, one alphabet was turned upside down. The process required double-sided type wheels, with one side for the encryption and the other side for the decryption, with Latin script for the input and Arab script for the output.

A further improvement was made by using type wheels with changeable types. This was initially practiced in a primitive way in France, with lose characters being used. However, the design produced by Crypto AG enabled the desired substitutions to be made without having to remove the characters from the type wheel. The use of such type wheels was a great cryptological step forward, particularly if compared to the original version where the type sequence on the wheels was assumed to be known to the enemy.

Another improvement was an arrangement that was used fairly early on: This allowed changing the position of the 'riders' or lugs, on the drum bars, which made it possible to achieve the most diverse patterns. This was the starting point for the different switching systems mentioned above. To set the pins on the pinwheels in cases where this was done in a central office, a tool was designed which allowed the desired pin combination to be set in a single operation. This meant that the operator could have a number of pinwheels for his machine, and use different ones in accordance with his instructions. It took several years to conceive and implement the different systems of movement in particular.

It is also worth mentioning a machine that was developed later, the so-called PEB unit.[19] Quite early on, Hagelin was thinking about introducing an efficiency measure for those cases where secret messages could be transmitted by telegraph across the customers' own lines. The PEB unit included a punch. This meant that during encryption the ciphertext was produced in the form of a punched tape, by means of a BC- or BCX-machine connected with a cable. During deciphering, when the received text was printed out as a punched tape, it was inserted in the detector in the PEB, and the original plaintext was automatically printed on the cipher machine. This device, which was launched on the market in 1963, received a very favourable response, becoming extremely popular as a time and work-saving device.

The last, purely mechanical cipher machine designed by Hagelin was registered as a patent in 1955. It is interesting that the C-machine was originally designed as a pocket-size machine, but was later fitted with a base plate and a protective hood and could be converted into an easily portable machine. The French police now expressed the desire for a genuine pocket-size device to be built. That is how the CD-55 was created, of which more than 12,000 units were supplied. The CD-55 offered all the possibilities of corresponding with the C-machines with regard to substitution organs and alphabet ring. But it was not compatible with the CX-machines.[20] Since it was very small, it was not possible to integrate a printer and the texts read from it had to be written down.

Crypto AG sold a large number of the CD-machines. For a long time, Hagelin regretted not having followed up the new developments he had devised. A prototype was built which had a CD-machine with a detachable addition. This had an

---

[19]This was a tape perforator/tape reader. This machine was easier to handle for telegraphic traffic. Kahn [1: 235] explained: "During encipherment a punched tape with ciphertext was made on the PEB and was dispatched directly over the punched tape transmitter of the teleprinter station. At the receiving end the cipher was punched on the tape which could then be deciphered automatically with the PEB".

[20]The exact functioning and the differences between the C-, the CX- and the CXM-machines can be found in Kahn [1: 228].

Figure 16.6. A CD-55.

alphabet tape with letters in the normal sequence. This tape could adopt 26 different positions and was equipped with a linear motion instead of a rotation of the alphabet disk in the CD. This was the input alphabet. A thin disk was provided for the output, printed with 26 different alphabet sequences. During operations, the alphabet tape could take up 26 different positions and one of the 26 disk alphabets would then come into the reading position. Now, the letter was identified on the disk which was opposite the letter on the tape and this was noted down.

In the prototype, both sides of the disk were printed in such a way that one side could be used during encryption and the other during decryption. Since the alphabet sequence on the input tape was normal, it could quickly be read off. The idea was to equip each machine with a number of alphabet disks and to use them in turn as agreed. Hagelin also thought about replacing the pinwheels by movable pins with a larger number of disks with high and low positions, a type of pinwheel with fixed pins. Such pinwheels and alphabet disks were easy and cheap to produce and would take up very little space (to carry). They would produce a huge number of variations. Such a device would have been a very easy and simple means of

encryption for travelling diplomats, for example. Throughout his life, Hagelin was convinced that such CD-machines had advantages over electronic pocket devices. One should remember that electronic devices need a supply of electricity, while a mechanical device can always be used.

The CX- and CD-machines were later also designed to generate ciphertext based on the 'One Time Pad' system, i.e. where a random key sequence was in principle only used for one message. This was made possible by the fact that, after the attempt to manufacture devices for the random key series (the CBI devices), engineer Stürzinger designed an electric random generator which produced the key sequences in the form of a punched telegraph tape. It was now possible to build accessories for the C- and CD-machines, which used punched tape instead of pinwheels to activate the substitution mechanism.

The use of random key series in the form of punched tape gave total security, but it also had important drawbacks. On the one hand, the supply system had to function faultlessly which, especially at times of war or crisis, was not always possible, and on the other hand it was vital to prevent access by outsiders. The tapes had to remain totally secret and could only be used by authorised staff.

Since great advances were also being made in the art of deciphering (later thanks to the use of computers), it was necessary to build machines that were even more secure than the CX-machines. This meant a change to using 4 rotors, which had already been done in the German 'Enigma' machine in the early 1920s, and was patented in the USA by the American, Hebern.[21] Machines of this type had a number of permutators, the so-called rotors, in which 26 contacts, which were assigned to 26 character keys, were scrambled and at the exit led to output sequences marked with letters.

At first, Hagelin did not seem to have great admiration for the Enigma machine, since it was easy to decipher. However, this failed to take into account that it was greatly improved during World

---

[21]For information about Hebern's rotor encryption machine or the functioning of a rotor machine, see particularly Schmeh [3: 47ff].

Figure 16.7. An HX-63.

War II and Hagelin apparently found out that a similar machine (originally built for the Navy) was used in the United States. There were therefore strong supporters at Crypto AG in favour of a machine with 4 rotors and a small number of such machines was built. A switching system developed for the CX-machine was used for this, which provided the rotors with regular movement with a guaranteed series length and was also equipped with quite a new type of rotors.

A lot of effort was required before a good solution was found and two important new designs were incorporated: Firstly, the wires in the individual rotors could be changed at will, and secondly, the so-called loopback system was used. Instead of the normal 26 contacts, 41 were used and the surplus ones were 'looped back'. Thus it could happen that a closed electric circuit went several times through the rotors between input (keyboard) and output (printer). All switching circuits could also be changed at will via a so-called modifier.

This rotor machine, which was launched on the market under the name HX-63 after almost a decade of development work, was a mechanical masterpiece. That it was only sold in limited volumes is due on the one hand to the high manufacturing costs, and on the other hand to the arrival of the electronic age, which had already begun.

But first, an improved version of the telecipher machine was built, model T-55. Although this still had the C-drum as the ciphering mechanism, it was used in a different way than the T-52, which was still based on the principles of the C-36. Now, two detectors were used: one served as the punched tape sender and the other one could be used, when switching off the mechanical ciphering mechanism, for the deployment of a key tape. So it was possible to use arbitrary keys.

On 1 June 1953, Hagelin's son Boris Jr., who had been working at A.B. Cryptoteknik, also came to Zug. While Oskar Stürzinger was in charge of the factory and the development section, Hagelin junior looked after the business side. He was later sent to Washington, where he controlled sales in North and South America until his death in 1970 in a car accident. The business continued to grow. At first, the company was mainly involved in development work and also took on the sale of A.B. Cryptoteknik's machines abroad. When it became clear that Crypto AG was also better placed to manufacture products – there were much greater opportunities to procure components more cheaply from specialist companies than in Sweden – the company started its own production. This had, as a by-product, a positive influence on the unique security philosophy of Crypto AG: To this day, all security-relevant processes are produced exclusively in-house, which means that no third party can gain access. So the workshop premises were extended.

In 1958, Hagelin decided to take an important step to consolidate the company. It had proved to be inefficient to operate a production facility in Sweden as well as in Switzerland, given the fairly limited volume of business. So A.B. Cryptoteknik was dissolved, and the entire production was moved to Zug. The former manager of A.B. Cryptoteknik, Sture Nyberg, also moved to Zug. The business was thriving and it was time to think about finding a location for a new building. Crypto AG found its definite location in Steinhausen, a small town close to Zug. The new building was formally opened in 1966 and the properties in Zug were sold. For Hagelin, this new factory, which was extended again in 1975, can be seen as a final milestone of his work which spanned more than 50 years. The workforce, which consisted of around 60 employees in Stockholm, had now grown to approximately 350.

The production of the Swedish machines, particularly the CX-machine, was reorganised. But even during the time in Zug the development trend was towards electronics. Hagelin himself was not able to contribute much in that field. The random generator designed by Stürzinger was the first product in this field. After the first units were built, the device was studied in detail. This enabled cryptologically faultless units to be built in production runs. It became clear that an important accessory was missing: Not only was it necessary to punch two identical tapes, but it should also be possible to identify these. For this purpose, Hagelin succeeded in designing a numbering machine which printed both tapes with a mark and a serial number after 10 or 200 characters, as required. An American punch was used for the perforation, which was capable of 120 punches per second. It was therefore not easy to build an equally fast numbering machine, since the tapes had to be stopped during printing.

As demand for telecipher machines rose, it became clear that there was also a market for radio transmitters. This market was controlled by the big manufacturers, but the radio systems that were available were expensive and needed a high power supply and very large, cumbersome antenna systems. These two features could present a problem particularly for embassies of countries who wanted to have their own, wireless connections with their ministries when opening up an embassy.

Once again by chance, Crypto AG came into contact with a Professor Vierling.[22] He had acquired an invention that made it possible to build radio devices for teletypewriters that were not only relatively cheap, but also used little electricity and only needed small antennas. Vierling sold his rights and Crypto AG made contact with the inventor, an engineer called Schindler, who then joined Crypto AG and remained there until his retirement. Initially, he was given the task of developing the device for factory production. Subsequently, he was always involved when sold equipment had to

be assembled and put into operation. There were also occasions when only a demonstration of the equipment was required. Originally, the machine was heavy and expensive at the development stage, but it became a very worthwhile accessory for the telecrypto devices. The French foreign ministry became the first major customer and used these units not only in the Middle East but also in the Far East.

There was one curious occasion that achieved fame. When the 1967 war broke out, Amman was without any telegraphic links with the outside world for several days. The installation of Crypto AG at the embassy there was the only one that functioned – precisely because of the small antenna and its own power supply.

Another development was a ciphering machine which Vierling had worked on and considered as his own invention. Through him, Crypto AG came into contact with one of his former employees. Vierling lived in the 'Fränkische Schweiz', a region of Bavaria in Germany, and during the war he had his headquarters in Feuerstein castle high up in the mountains, where he concerned himself with radio messages. He had a small workshop and a service agreement for telephone equipment with the German postal services. At that time, his hobby was to develop an electronic organ. Hagelin was interested in this not just for musical reasons, but also because he saw a business opportunity. Although it did not in the end materialise, the radio transmission system became a great commercial success.

After the success of Stürzinger's electronic random generator, greater emphasis was gradually put on electronics at the Zug offices. The first result was the so-called 'simulator'. This attracted a very good customer: the foreign ministry of a major country, which used the BC machines with key tape control for its foreign communications. But these machines turned out to be too slow for the operations, which led to the development of the 'simulator', the monster of 'mechanics and electromechanics', given its large size and its complexity. Several units were put into service, but were later replaced by modern, fully electronic devices. Hagelin was now reaching a point where his personal input in the new developments had to cease. Electronics were not his field and he was getting too old to get to grips with this discipline.

---

[22]Crypto AG established contact with Vierling via other Germans. The peace treaty after World War II did not allow any activities in Germany that were connected to cipher machines, among other things. For this reason, a 'study group' was founded which initially concerned itself with random generators.

Many improvements and adaptations were developed in Hagelin's field of mechanical cipher machines over the next few years. He continued to monitor and control the work in the design office and decided which inventions would be manufactured and which one not. However, he devoted most of his time to managing the business, which involved him in a lot of travelling, though almost always in Europe. The only trips outside Europe during his time at Crypto AG were across the Atlantic, where his son needed his support a few times, to North Africa in 1958 and to Iran. He travelled to Iran to be present at the Shah's coronation. Hagelin had the honour of presenting him with a gift at the reception and exchange a few words with him.

But most of his trips were to France, where Crypto AG continued to do good business for several years, until the French developed their own machines. However, this only happened once they had bought the French Crypto patents for the CX-machine. During his visits, Hagelin got to know the members of the NATO headquarters. His dinner invitations were accepted with alacrity by the Belgians, Dutch, Danes and French. However, it transpired that it was not possible to transact business directly with NATO – the manufacturer had to be from a NATO country. But those convivial evenings were not without benefits, because the machines were regularly bought by NATO members for their own, national use. Many years later, in 2007, Crypto AG is still supplying 19 of the 26 NATO countries who place great store by the independence of Crypto AG for their national networks.[23]

Hagelin also visited Holland, Belgium and Denmark and above all Sweden, where he always spent the summer with his wife. Germany also became an important customer. Here, Hagelin met the famous Dr. Rudolf Hell in 1951,[24] who had heard about him. He received a proposal from the former 'study group' to acquire a patent for a cipher machine. They decided to cooperate in this transaction. Hell built an initial prototype, but it transpired that the functional principle could not achieve a satisfactory operating speed. When analysing the ciphering principle, Hagelin found that they could adapt their CX-machine in such a way that it would operate with the working mechanism of this invention. Eventually, this machine was accepted by the German army and a large number of these units were manufactured in Hell's factory in Kiel.

With advancing age and with the changeover of new designs to electronics, Hagelin's personal involvement gradually diminished. He used to say that he didn't understand electronics, but that he knew exactly what could be done with it.[25] When the new factory was built in Steinhausen, Hagelin decided that he would no longer work at the factory on a daily basis. He had an incredibly gifted employee in Sture Nyberg,[26] who had already worked in his Swedish company A.B. Cryptoteknik. He was appointed director of Crypto AG Zug, while Hagelin took the position of chairman of the board of directors. He had now reached the point where his life's work had drawn to a close. He kept contact with Crypto AG, but the actual management had passed to Sture Nyberg. In 1970, Boris Hagelin withdrew from active involvement in Crypto AG and was appointed honorary president. His time in the shadow of Nobel was at an end. Boris Hagelin died in 1983 in Zug.

16.6 THE ADVENT OF NEW TECHNOLOGY AT CRYPTO AG

The era of electro-mechanics was slowly coming to an end in the 1970s. A new technology, electronics, was increasingly displacing electro-mechanics. The fundamental innovation was the possibility of digitising numbers and characters. This made it possible to alter information in computers with the

---

[23]For more detail, see Wobst [5: 385ff].

[24]Rudolf Hell was one of the most ingenious individuals and most important inventors of the 20th century. His inventions accelerated worldwide communication and transformed the working day. He is considered to be the original inventor of the fax machine.

[25]Kahn [2: 434].

[26]Sture Nyberg was the first manager and then director of Crypto AG, until his retirement in 1976. He proved to be the perfect choice. He made a major contribution to analysing the security of the manufactured machines (Kahn [1: 206]) and was probably pivotal in ensuring the company's successful leap into electrotechnics.

help of mathematical algorithms, instead of just substituting them. The security and speed of encryption acquired a completely new dimension. At Crypto AG, the huge potential of digitisation was quickly recognised and large sums were invested in this new technology at a very early stage. Electronics also opened up completely new possibilities for automating the security processes. Boris Hagelin's goal of ensuring security without placing demands on the user was becoming ever more realistic.

As early as the end of the 1960s, a new text encryption system was developed, a so-called offline device with integrated printer for plain and encrypted information. This was the first device in which Crypto AG used a key generator based on ICs. A punched tape reader/puncher was also developed for it, since the encrypted messages were frequently transmitted via teletype machine and telex. This product was mostly used in diplomatic services.

Shortly afterwards, the first fully electronic online text encryption unit was launched on the market: The T-450 could be used in asynchronous and synchronous mode at speeds of 50, 75 and 100 baud. Thanks to a highly stable clock generator, the T-450 units would remain synchronous in a shortwave radio network, even if there were no radio transmissions for more than a day. They were used worldwide in civilian and military government offices, as well as international organisations. Crypto AG was also a pioneer in the field of digital voice encryption, with the launch of the famous CRYPTOVOX CSE-280, the VHF transmitter/receiver, at a speed of 16 kBps. A large range of accessories enabled it to be used as a fixed, mobile or relay station. This system, too, was a great commercial success. Another Crypto AG innovation at that time was a telephone voice encryption unit for civilian and military use, with permutation of the voice packages on a time and frequency axis.

Between 1976 and 1983, an entire text encryption family was developed. This consisted of a manual text encryption unit, a portable text encryption unit (available in a civilian and military version), a radiation-protected teletype terminal with integrated encryption, a powerful offline text encryption device, as well as an online text

encryption unit linked to computer-based message transmission. All the units in this range were cryptologically compatible, making them very flexible to use. The range continues to be updated and supplemented to this day, for example with a radiation-proof off/online terminal with integrated encryption and screen, which is still successfully used in many countries. These devices are now being replaced with computer-based messaging solutions.

From around the mid-1980s, more than 10,000 units of a digital voice encryption device for tactical army and navy applications, the famous CVX-396 with a speed of 9.6 or 16 kBps, were produced. In 1985, a duplex voice encryption unit, mainly aimed at telephone lines, came on the market, which permutates voice packets on a time and frequency axis. In addition, a very successful military voice encryption unit has been available since the end of the 1980s, which has very good voice quality and channel robustness on poor short wave radio communications. A new model has become available recently. The rediscovered qualities of HF and VHF have resulted in a growing demand among users for encryption units that can be used in a wide range of applications. The policy pursued by Crypto AG in recent years of continually developing its security technology for these applications is now paying off: Several products are available that are suited to different operating conditions and existing infrastructures. The product range is divided into three lines: voice encryption, data transmission and messaging.

Crypto AG is also very successfully in marketing solutions in the field of PSTN and GSM telephony that are compatible with each other. As early as the beginning of the 1980s, with the introduction of the G3 fax generation, Crypto AG developed a fax encryption unit. At the time, the G3 fax was as big as a tall refrigerator and extremely expensive. Crypto AG then produced fax machines with integrated encryption. In many regions, for example in the Gulf states, the company not only introduced fax encryption in the government departments with great success, but fax transmissions as such. Because the fax manufacturers were changing their models over very short periods, Crypto AG decided to build 'stand-alone' fax encryption

Figure 16.8. The Ethernet Encryption HC-8440 1 G.

units and continues to market these with great success to this day. Today, even a colour fax option is available.

The first data encryption unit was developed in 1976. Moreover, in the first half of the 1980s, a device for low bit rates was developed. After that, Crypto AG launched a link encryption device on the market. Recently, it presented a complete world novelty in the form of the Broadband Encryption Gigabit Ethernet units: These devices transmit up to 10 gigabit wirespeed of data per second via fibre optics. Crypto AG achieved a real breakthrough as the market leader in the field of data encryption with the HC-7000 system, which now consists of single and multi-channel encryption units for voice and data that can be used both in the VHF/UHF and HF fields.

The communication society continuously offers new opportunities for civilian and military users. By now, system providers and operators offer virtually unlimited transmission capacities. However, the security of the information that is transmitted does not always keep pace with the network capabilities. This means that the user of these links and systems must ensure the security of the transmitted and stored information. Today, cryptography is a highly mathematical science integrated in modern IT infrastructures. The encryption process takes place in specifically tamper-protected hardware. The modern key management system guarantees the total secrecy of the communication keys and supports uninterrupted operation. Encryption takes place in the background and is available at the press of a key. The design allows the user to control and profile the vital parts of the algorithms

used by him, which places him in the position of performing the encryption process completely independently.

The trend towards integrated electronics inevitably also changed the demands made on the production structures and the staff. It led to a rise in productivity with fewer employees. Sophisticated production processes, such as the fitting of circuit boards, were now completely automated. Today, the amount of mechanical work is very small, e.g. the production of the housing. By contrast, the use of *'brain power'* is constantly increasing. Today, Crypto AG employs software specialists, cryptologists, mathematicians, scientists and circuit technicians. At present, it has a staff of around 230. In addition, the company operates Regional Offices in Africa, South America, as well as several in the Middle East and Far East, to enable to provide its customers in more than 130 countries the best-possible service on the spot.

REFERENCES

[1] D. Kahn (ed.), *The story of the Hagelin cryptos*, Cryptologia XVIII (3) (1994).

[2] D. Kahn, **The Codebreakers. The Comprehensive History of Secret Communication from Ancient Times to the Internet**, Scribner, New York (1996).

[3] K. Schmeh, **Die Welt der geheimen Zeichen. Die faszinierende Geschichte der Verschlüsselung**, W3L-Verlag Herdecke, Dortmund (2004).

[4] H. Stadlin, *100 Jahre Boris Hagelin: die G eschichte des Firmagründers,* Hauszeitung der Crypto AG, No. 11, Zug (1992).

[5] R. Wobst, **Abenteuer Kryptologie. Methoden, Risiken und Nutzen der Datenverschlüsselung**, Addison-Wesley, München (2001).

The History of Information Security: A Comprehensive Handbook
Karl de Leeuw and Jan Bergstra (Editors)

17

EAVESDROPPERS OF THE KREMLIN: KGB SIGINT DURING THE COLD WAR

Mathew M. Aid

Washington, DC, USA

Contents

Abstract

Since the end of the Cold War and the collapse of the Soviet Union in 1991, a great deal of new information has become available concerning Soviet Signals Intelligence (SIGINT) operations during the Cold War from declassified documents, books and articles published in Russia, and from interviews with former Soviet intelligence officers. This chapter attempts to set out what can definitively be established concerning the SIGINT collection activities of the former Committee of State Security (KGB). In aggregate, this new material indicates that the KGB's SIGINT activities were much larger and more successful than previously believed. Former Soviet intelligence officials have indicated that while SIGINT supplanted Human Intelligence (HUMINT) as the principal intelligence source for the KGB by the early 1970s, internecine warfare within the Soviet intelligence community and the Soviet Union's inability to develop high-speed computers and other critically important telecommunication technologies, prevented Soviet SIGINT from fully achieving its potential.

Keywords: Signals Intelligence (SIGINT), Communications Intelligence (COMINT), Electronic Intelligence (ELINT), Committee for State Security (KGB), Chief Intelligence Directorate (GRU), Soviet Union (USSR), counterintelligence, cryptanalysis, radio intercept.

17.1 INTRODUCTION

When compared with the National Security Agency (NSA) and other Western intelligence organizations, relatively little is known publicly about the Signals Intelligence (SIGINT) operations of the Soviet Union during the Cold War. In part, this is due to the lingering heavy mantle of secrecy that continues to cover this subject inside Russia despite the end of the Cold War and the collapse of the Soviet Union. It is also true that much of what was written in the public realm in the West, such as it was, about Soviet SIGINT was flat-out wrong or inaccurate. Moreover, a review of declassified US intelligence reporting on the Soviet intelligence services revealed that these documents, which were based largely on defector accounts, were not surprisingly highly inaccurate in their description of Soviet SIGINT. It is now clear that US intelligence, in general, knew relatively little about Soviet cryptanalytic work until the mid-1970s, when a series of defectors from within the Soviet SIGINT system opened our eyes to the danger posed by this highly secretive foe. And although the Anglo-American intelligence services knew more about Soviet radio intelligence work, including radio intercept, direction-finding and traffic analysis, even in this area our knowledge of Soviet work in this field was grossly deficient [38].

It was not until after the end of the Cold War and the collapse of the Soviet Union in the early 1990s that the US intelligence community finally learned the full nature and extent of Soviet SIGINT successes and failures during the Cold War, thanks in part to formerly secret KGB documents obtained from the former Soviet republics, as well as from the fruits of an NSA-sponsored special interrogation program of former KGB and GRU intelligence officers who had left Russia (confidential interviews).

Because of space constraints, this paper sets out in summary fashion what is now known about the SIGINT effort of the Soviet Committee for State Security (KGB) during the Cold War on the basis of newly released Russian historical materials, declassified American intelligence reports, and information obtained from a series of confidential interviews conducted over the past 20 years with former American, European, and Russian intelligence officials who were knowledgeable about the Soviet Cold War SIGINT effort.

First, it is important for the reader to understand that Soviet SIGINT collection efforts during the Cold War were far more fragmented than the comparable cryptologic programs conducted in the United States or Great Britain. There were not one, but rather three Soviet intelligence organizations engaged to varying degrees in SIGINT collection and processing – the Committee for State Security (*Komitet Gosudarstvenoi Bezopastnosti*, or

KGB), the Chief Intelligence Directorate of the Soviet General Staff (*Glavnoye Razvedatelnoye Upravleniye*, or GRU) and the Intelligence Directorate of the Soviet Navy (*Razvedatelnoye Upravleniye Voennoj Morskoj Flot*, or RU VMF) [12; 14]. The Soviet Air Defense Forces (PVO) also had their own independent network of strategic SIGINT collection sites situated around the Soviet periphery. For instance, during the Cold War the PVO's 141st Independent Radiotechnical Brigade OSNAZ (whose cover designation was Military Unit 86621) was assigned to the Central Asian Military District. The Brigade was located near the town of Chingel'dy, 40 kilometers from the city of Kapchagay in southern Kazakhstan. This brigade copied air-related communications targets in China, Pakistan, Iran, Iraq, Afghanistan, South Korea, and even the US [66].

But by far, the two largest and most important Soviet SIGINT organizations belonged to the KGB and the GRU. The KGB focused its SIGINT efforts on non-military signals traffic. The two most important targets were foreign diplomatic communications and foreign clandestine intelligence radio traffic, although the KGB also devoted significant cryptologic resources to collection of commercial communications. The KGB's foreign intelligence and counterintelligence directorates also planted bugs in foreign embassies, recruited foreign cipher clerks, and stole foreign cryptographic materials to assist the service's SIGINT collection efforts. For its part, the GRU focused its attention on intercepting and decrypting the communications of foreign military forces, especially those of the US [36: 47].

17.2 THE GENESIS OF KGB SIGINT

The KGB's SIGINT organization underwent numerous changes in name and organization since its beginnings shortly after the Russian Revolution of 1918, growing slowly larger and more important over time until it finally became a behemoth in the early 1970s. A summary of the organizational development of the KGB SIGINT organization since its inception in 1921 is contained in Table 17.1.

It took Feliks Dzerzhinsky's feared Secret Police, the *Cheka*, almost three years after the Russian

Table 17.1. Chronological development of the KGB SIGINT organization

Spetsial'niy Otdel (Special Department) VCHK: January 1921–May 1921
Spetsotdel pri VCHK: May 5, 1921–February 6, 1922
Spetsotdel pri GPU: February 6, 1922–November 2, 1923
Spetsotdel pri OGPU: November 2, 1923–July 10, 1934
Spetsotdel pri GUGB NKVD SSSR: July 10, 1934–December 25, 1936
9th Otdel GUGB NKVD SSSR: December 25, 1936–March 28, 1938
3rd Spetsotdel NKVD SSR: March 28, 1938–September 29, 1938
7th Otdel GUGB NKVD SSSR: September 29, 1938–February 26, 1941
5th Otdel NKGB SSSR: February 26, 1941–July 31, 1941
5th Spetsotdel NKVD SSSR: July 31, 1941–November 3, 1942
5th Upravleniye NKVD: November 3, 1942–April 14, 1943
5th Directorate NKGB SSSR: April 14, 1943–May 4, 1946
6th Directorate MGB: May 4, 1946–November 19, 1949
Main Directorate of Special Services (GUSS) TsK VKP: November 19, 1949–March 14, 1953
Main Directorate of Special Services (GUSS) of TsK CPSU: March 1953–March 1954
8th Directorate MVD SSSR: March 14, 1953–March 18, 1954
8th Main Directorate KGB: March 18, 1954–July 1978
16th Directorate KGB: created in 1973.
8th Main Directorate KGB: July 1978–1991
FAPSI: Created in 1991

Revolution of October 1918 to get into the code-breaking business. By decree of January 28, 1921, a small Cheka cryptologic service called the *Spetsial'niy Otdel* (Special Department) was formed. This unit was usually referred to simply as the 'Spetsotdel'. The Spetsotdel commenced operations on May 5, 1921. From 1921 until 1935, the offices of the Spetsotdel were located on the top floor of a modest office building owned by the People's Commissariat of Foreign Affairs on Kuznetskiy Most Street in downtown Moscow. In 1935, the department was moved to a specially-guarded suite of offices within the Lubyanka office complex along with the rest of the KGB headquarters staff [7: 173–174].

The first head of the Spetsotdel was one of the Soviet secret police's most feared officers, Gleb Ivanovich Bokiy [47: 21]. Born in Tbilisi, Georgia in 1879, Bokiy was a longtime Bolshevik revolutionary and a personal friend of Lenin. After the 1917 Revolution, he joined the Cheka and was named the head of the all-important Petrograd (Leningrad) Cheka, where he developed a well-deserved reputation for fierceness and cruelty against the enemies, real or imagined, of the nascent Soviet regime. During the Russian Civil

War, he was the head of the Cheka Special Purpose Detachment on the Eastern Front that fought the White forces of Admiral Kolchak until September 1919, where his duties included rooting out and liquidating White Guard forces and so-called 'counter-revolutionaries'. He then was the chief of the Cheka in Turkmenistan from September 1919 until August 1920, where he enhanced his already formidable reputation for frightening brutality while crushing Muslim separatists in the region. Some consider Bokiy to be one of the prime architects of the GULAG, the vast Soviet prison system created to house Russia's political prisoners. Bokiy ran the Spetsotdel for the next 16 years, from 1921 until he was purged in 1937 at the height of the "Great Terror" [67: 114; 75: 412–413].

Because of the severe shortage of trained cryptologists in Russia, the ever resourceful Bokiy brought into the Special Department a small number of older but experienced cryptologists, most of whom had worked as codebreakers for the Black Chambers (*Cherniy Kabinet*) of the Tsarist government before and during World War I [37]. Among the men brought in by Bokiy were the talented cryptanalysts Professor Vladimir Ivanovich Krivosh-Nemanich, who had worked as a codebreaker for the Black Chamber of the Tsarist Ministry of Foreign Affairs since 1905; Ivan Aleksandrovich Zybin, the Tsarist Ministry of the Interior's star cryptanalyst since 1901; and I.M. Yamchenko [75: 417]. Also recruited were former Tsarist cryptologists such as G.F. Bulat, E.S. Gorshkov, E.E. Kartaly, Yevgeniy E. Morits, and the former Tsarist cryptanalysts Boris Alekseevich Aronskiy and F.A. Blokh-Khatskelevich. These men were to form the core of Bokiy's Spetsotdel for rest of the 1920s and much of the 1930s [74: 325, 327; 75: 419].

Day-to-day supervision and management of the work of these elderly former Tsarist codebreakers was given to a small group of Bokiy's most trusted deputies, all of whom were veterans of the GULAG prison camp system or who had been involved with Bokiy in suppressing "bandit and counter-revolutionary activities" in the early 1920s, including Spetsotdel section chiefs A.G. Gusev, A.M. Pluzhnikov and Fedor Eichmans [75: 419].

The Tsarist codebreakers, however, were not sufficiently numerous, nor nearly well enough versed in modern cryptanalytic techniques, forcing Bokiy to once again go out and recruit personnel for his department from outside the Cheka. In 1923, Bokiy convinced the Red Army's intelligence staff to lend him one of their most experienced intelligence officers, Major (later Colonel) Pavel Khrisanfovich Kharkevich. Born in 1896 in the village of Pisarevka near the Volga River city of Voronezh, Kharkevich went to high school in Orel, then obtained his officer's commission in the Russian Army following graduation in 1916 from the Alekseevskiy Military Academy. During World War I, Kharkevich had commanded the intelligence staff of one of the Russian Army's elite units, the 1st Guards Rifle Regiment. After the Russian Revolution, he joined the Red Army in 1918, serving as commandant of the Eastern Department of the Red Army Military Academy from 1919 until he joined the Spetsotdel in 1923. Kharkevich was to serve as Gleb Bokiy's military deputy for the next seven years, from 1923 to 1930. After leaving the Special Department in 1930, Kharkevich spent the next nine years trying to build up the GRU's own decryption service until he was eventually purged in 1939 [52: 486].

But Colonel Kharkevich's attempts to escape the clutches of Bokiy's Spetsotdel were unsuccessful.

In the early 1930s, at Bokiy's suggestion, a joint OGPU/Fourth Department (GRU) cryptanalytic unit was established within the Spetsotdel to handle both civilian and military SIGINT processing, analysis and reporting. The head of the unit was Gleb Bokiy. His deputy was Colonel Kharkevich, who also concurrently headed the Military Section of the Special Department. The addition of Colonel Kharkevich's GRU decryption unit roughly doubled the size of the Spetsotdel. By 1933, Gleb Bokiy's little unit had grown from the 20 people it started with in 1921 to some 100 military and civilian personnel [74: 328].

The earliest efforts of Bokiy's Special Department were initially focused on those countries which bordered the Soviet Union, such as Turkey, Persia (Iran) and Japan, as well as more distant strategic targets such as the US, Germany, France, China and Bulgaria. The first cryptographic system solved by the Spetsotdel was a German five-digit cipher system used to carry German diplomatic traffic between Berlin and the German embassy in Moscow, which was solved in June 1921. Two months later, a high-level Turkish diplomatic cipher system was solved in August 1921. As of 1925, the Special Department was reading in part or in whole the diplomatic cipher systems of 15 nations. In 1927, the Special Department had its first break into Japanese cryptographic systems, followed in 1930 by initiation of its first work against American cryptographic systems [74: 329–334]. During the 1930s, coded Japanese diplomatic and military radio and telegraph traffic was the Special Department's primary target, with much of the unit's material coming from compromised cryptographic documents provided to the unit by OGPU spies in Japan and elsewhere [7: 174].

The spies of the Foreign Department (INO) of the OGPU provided the Special Department's codebreakers with reams of stolen cryptographic material. One of the most important hauls were copies of more than ten British diplomatic codes, which were forwarded to Bokiy's unit from the OGPU *Rezidentura* in London in the early 1920s. With these ciphers in hand, the Special Department was able to read British diplomatic traffic coming in and out of the British embassy in Moscow, including telegrams about British-Russian relations, British military activities, and even information about British intelligence activities in Central Asia in the 1920s. With the help of the INO, in 1924 two codes used by the Polish military intelligence service were solved, allowing the Russians to read the encrypted cable traffic between the Polish General Staff in Warsaw and the Polish military attaches in Moscow, Paris, London, Revel (now called Tallinn), Washington and Tokyo [74: 333–334].

The *Spetsotdel* in the late 1930s sent "Special Operational Groups" of trained cryptologists to Spain, China and Mongolia to perform intelligence collection missions. Shortly after the Spanish Civil War began in July 1936, a KGB Special Operational Group comprised of cryptanalysts and linguists was sent to Spain together with a radio intercept group to provide intelligence support to Republican forces during the Spanish Civil War. The group established its operations center outside the port city of Barcelona, and promptly went to work trying to solve the ciphers of Francisco Franco's Spanish Nationalist forces, as well as the traffic of the German and Italian forces operating in Spain. Some of their attention, however, was spent spying on the communications of their erstwhile Republican allies. The group pulled out of Spain shortly before the collapse of the Republican military forces in 1939 [74: 347–348].

A second *Spetsotdel* "Special Operational Group" was sent to China in early 1938 to support Mao Tse-tung's Chinese Communist forces in their fight against the Japanese military. In the one and one-half years (19 months) the group spent in China, the group reportedly solved ten different Japanese military cipher systems [74: 348]. In May 1939, a third Special Operations Group was sent to Mongolia shortly after the Battle of Khalkin-Gol. Establishing their operations center in the Mongolian capital of Ulan Bator, the Special Operations Group immediately began work on the cryptographic systems of the Japanese Kwantung Army in northern China and Manchuria. Cryptanalytic solutions of Japanese military communications traffic were forwarded directly to the commander of Soviet forces in the region, General G.K. Zhukov [74: 348–349].

The so-called "Great Terror" of the mid- to late-1930s decimated the OGPU-Fourth Department SIGINT unit. Between 1937 and 1938, more than 40 Spetsotdel officials and staff members were purged and executed on a variety of trumped-up charges [75: 475]. The department's founder and chief for 16 years, Gleb Bokiy, was arrested in May 1937 on charges of spying for a foreign power after being denounced by a former subordinate. After a trial whose outcome had already been predetermined, Bokiy was executed in the basement of the Lubyanka on November 15, 1937 by a bullet in the back of the head fired by the dreaded Kommandant of the NKVD, Major Vasilii Mikhailovich Blokhin. His body was cremated in secret later that night at the Moscow crematorium next to the famous Donskoy Monastery outside Moscow, and his ashes dumped in a mass grave on the grounds of the monastery [75: 463, confidential interview].

Two months later, on July 22, 1937, Bokiy's longtime deputy, Fedor Ivanovich Eichmans, was arrested, convicted on September 3, 1938 in a secret trial of spying for Great Britain, then shot at the Kommunarka execution grounds south of Moscow the same day [54]. Another of Bokiy's chief deputies, Aleksandr Georgiyevich Gusev, was also shot at the Kommunarka execution ground outside Moscow on April 19, 1938 [55]. Virtually all of the Spetsotdel section heads were executed in 1937–1938, including the head of the Spetsotdel Laboratory, E.E. Goppius (shot in 1937); the chief of the Spetsotdel Technical Section, A.D. Churgan (shot in 1938); section chief G.K. Kramfus (shot in 1938); section chief N.Ya. Klimenkov (shot in 1938); and the head of the Spetsotdel's intercept unit, V.D. Tsibizov (shot in 1938) [47: 28, 45]. And finally, almost all of the Tsarist-era linguists and cryptanalysts that Bokiy had gone to such great lengths of recruit were also executed, including linguist Professor Vladimir Ivanovich Krivosh-Nemanich; and arguably Bokiy's best codebreaker, Ivan Alekseevich Zybin, who was shot on December 13, 1937. Needless to say, the loss of so many of the Spetsotdel's best managers and technical personnel in such a short period of time had a severe impact on the organization's ability to perform its mission in the years that followed.

In the years following Bokiy's removal, the KGB's SIGINT organization went through a period of what can only be described as turmoil and upheaval. Bokiy's successor as head of the Spetsotdel, Major Isaak Ilyich Shapiro, lasted only from May 1937 until April 1938 [7: 225]. In 1938–1939, the *Spetsotdel* was redesignated at the 7th Department (*7 Otdel*) of the GUGB. The head of the 7th Department from March 28, 1938 until April 8, 1939 was an OGPU officer named Captain Aleksandr Dmitriyevich Balamutov, who had no previous experience with SIGINT or cryptanalysis whatsoever. There apparently were few tears shed within the 7th Department when he was transferred to another assignment that was not as taxing [56].

Balamutov's successor as head of the 7th Department was Major (later Major General) Aleksei Ivanovich Kopitsev, a 27 year old professional cryptologist, who ran the unit with distinction from April 9, 1939 until July 31, 1941, a little more than a month after the German invasion of Russia began on June 22, 1941. Born in the Smolensk Oblast' in 1912, Kopitsev had been a member of the Komsomol from 1930 until 1938, when he was finally admitted to full membership in the CPSU [8: 38].

17.3   WORLD WAR II AND THE REBIRTH OF SOVIET SIGINT

The German invasion of Russia on June 22, 1941, codename *Operation Barbarossa*, changed everything in how Soviet SIGINT was run. On August 1, 1941, a new NKVD SIGINT organization, the 5th Special Department (*5-ogo Spetsotdel*), was created and placed under the command of a longtime NKVD counterintelligence officer named Major Ivan Grigoryevich Shevelev. Shevelev had no prior experience in the field of cryptology. Rather, his entire career up until that time had been spent in the field of internal security, where he specialized in crushing anti-Soviet dissent and "counter-revolutionary" activities inside the Soviet Union [8: 38]. A tough, no-nonsense officer, but a surprisingly able administrator with a talent for empire-building, Shevelev was to run the State Security SIGINT organization in its various guises and forms for the next 13 years. Shevelev left the

technical and operational management of the 5th Special Department to his deputy, Major Kopitsev [67: 242].

During the first full year of the Great Patriotic War with Nazi Germany, Shevelev and Kopitsev rapidly expanded the size and capabilities of the NKVD 5th Special Department, recruiting a sizable corps of talented personnel, many from academia, to fill out the organization, including some of the best mathematicians, engineers and linguists then available in the Soviet Union. Many of these recruits were to form the professional core of the KGB SIGINT program for the next forty years. By the end of World War II, the 5th Directorate controlled the single largest concentration of mathematicians and linguists in the Soviet Union (confidential interview).

But it should be noted that the NKVD was not the only game in town. The Soviet military intelligence service, the Chief Intelligence Directorate (*Glavnoye Razvedatelnoye Upravleniye*, or GRU), has been in the SIGINT business even longer than the NKVD. The difference between the two organizations was that, in general, the GRU devoted a greater percentage of its resources to SIGINT collection than the NKVD, and as a result, the GRU's SIGINT organization was always significantly larger and more diverse than that of its civilian State Security counterpart (confidential interview).

Unlike the NKVD, the GRU historically kept its SIGINT collection and processing organizations separate from each other. The GRU had its own cryptanalytic organization that was significantly larger than that of the NKVD. In August 1940, the GRU decryption unit was redesignated as the 10th Department of the General Staff Intelligence Directorate of the Red Army (*10-y Otdel, RU GSh KA*) [9: 34; 75: 479]. The commander of the 10th Department was Colonel N.A. Filatov. As of 1941, the GRU's Decryption Intelligence Service (DRS) was commanded by a veteran GRU signals officer named Colonel Aleksei Aleksandrovich Tyumenev [9: 34; 52: 478].

A separate organization within the GRU handled SIGINT intercept. In May 1939, the GRU radio intercept unit was redesignated as 7th Department

(Radio Intelligence) of the 5th Directorate NKO, under which name it existed from May 1939 to June 1940 [2]. In June 1940, the 6th Department (Radio Intelligence) was merged with the GRU's 8th Department (Radio Communications) to form a new unit designated the 8th Department (Radio Intelligence and Radio Communications) (*8-y Otdel (radiorazvedki i radiosvyazi)*), under which designation it existed from June 1940 to February 1942. The chief of the new 8th Department was Engineer 1st Rank I.N. Artem'ev [2]. As of 1941, the GRU also managed a large-scale field radio intercept organization that provide COMINT direct support to the Red Army. During the 1930s, a GRU Radio Reconnaissance Department was organized, subordinate to which were the first of the newly formed "radio battalions of special designation" (*radiodivizioni osobogo naznacheniya*, or *ORD OSNAZ*). At the time that World War II began, there were 16 OSNAZ radio battalions active throughout the Soviet Union [3]. According to another source, the head of the GRU Radio Intelligence Department (*Otdel radiorazvedki i perekhvata*) from 1939 to 1946 was Major General Roman Samuilovich Pekurin, a specialist in long-range radio communications, who also headed the GRU department responsible for radio communications from 1936 to 1946 [77].

Former Soviet intelligence officers in interviews have stated that, by and large, the NKVD and GRU SIGINT organizations, did not perform particularly well during the first year of World War II, which is still referred to in Russian literature as "The Great Patriotic War". The badly understrength and underequipped NKVD and GRU SIGINT organizations tripped over themselves in their effort to outdo each other, which oftentimes resulted in massive duplication of effort and wasted resources at a time when the Soviet military was struggling for survival against the onslaught of the German Wehrmacht (confidential interviews).

By special decree issued by the Peoples Comisariat of Defense on October 23, 1942, Major Shevelev's NKVD 5th Special Department was merged with the GRU decryption service to form a single integrated wartime cryptologic organization that was designated the NKVD 5th Directorate, which was formally activated on November 3, 1942. Pursuant to this order, all GRU SIGINT units, including the GRU decryption service, as well as all Red Army 'special purpose' radio battalions (*radiodivizioni OSNAZ*) and independent radio reconnaissance units, were placed under the operational control of Shevelev's NKVD 5th Directorate in Moscow for the duration of the war [47: 78; 9: 3, 10]. When the NKVD was renamed the Peoples Commissariat of State Security (NKGB) in May 1943, the directorate was renamed the 5th Directorate of the NKGB. As the size and importance of his unit grew during the war, so did Shevelev's standing. By the end of the war, Shevelev had been promoted to the rank of Lieutenant General [8: 38].

As time went by, the combined NKVD and GRU SIGINT organizations became progressively better organized and far more proficient in performing their assigned intelligence collection and processing missions. Frontline Red Army *OSNAZ* tactical SIGINT units became increasingly efficient in monitoring German troop movements and identifying German troop concentrations along the full length of the Eastern Front. Radio counterintelligence units of the NKVD successfully monitored German intelligence activities, including locating a number of clandestine radio transmitters used by German agents operating within the Soviet Union. Soviet Naval SIGINT was able to monitor the movements and activities of German fleet units in the Baltic and Black Seas, while the COMINT units assigned to the Soviet Air Force as time went by experienced considerable success monitoring the operational and tactical radio traffic of German Air Force (the *Luftwaffe*) commands operating on the Eastern Front [1; 29].

17.4 SOVIET COMINT REORGANIZES IN POST-WORLD WAR II ERA

After the end of the Second World War, Soviet SIGINT operations were reorganized to reflect dramatically reduced postwar fiscal and manpower resources, as well as the broad range of new global geostrategic intelligence targets. On May 4, 1946, the Peoples Commisariat of State Security

Table 17.2. Organizational structure of NKGB 5th directorate – 1945

Command Staff
Directorate Secretariat
1st Section (Germany)
2nd Section (Japan-Manchukuo)
3rd Section (Britain, United States)
4th Section (Italy, Spain)
5th Section (France, Belgium)
6th Section (Balkans, Scandinavia, Finland)
7th Section (Turkey, Iran, Iraq, Afghanistan)
8th Section (China)
9th Section (code preparation for Russian government)
10th Section (cipher book compilation for Russian government)
11th Section (encrypted communications for NKVD operational components)
12th Section (encrypted communications for NKVD internal components)
13th Section (encrypted communications for NKVD 1st Foreign Intelligence Directorate)
14th Section (encrypted communications for NARKOM and other services)
15th Section (unknown)
16th Section (agent-operational)

Source: Assotsiatsiya issledovateley istorii spetssluzhb im. A.X. Artuzova, *Shpionami ne rozh-dayutsya: Istoricheskiy spravochnik* (Moscow: Artuzovskaya Assotsiatsiya, 2001), pp. 5–6.

(NKGB) was disbanded, and all existing Soviet intelligence and security missions were transferred to the newly created Ministry of State Security (*Ministerstvo Gosudarstvennoy Bezopasnosti*, or MGB) [15: 10]. The MGB took over the assets of General Shevelev's wartime NKGB 5th Directorate as well as the NKGB's independent decryption unit, the 2nd Special Department (*2-e spetsotdel NKVD SSSR*). Both units were merged together to create a new unified MGB SIGINT collection and processing organization designated the 6th Directorate (*6-e Upravleniye MGB SSSR*). Under the new organizational scheme, the GRU's SIGINT components were released from the control of the MGB, and allowed once again to develop on their own [47: 139–140]. Command of the MGB 6th Directorate remained with Lt. General Shevelev for the next three and one-half years, from May 1946 to October 1949 [8: 38].

Former Soviet intelligence officials confirmed in interviews that the relative importance of General Shevelev's 6th Directrorate declined somewhat after the end of World War II. Like its American and British counterparts, General Shevelev's 6th Directorate lost many of its best personnel once the war was over as the unit's wartime civilian and military recruits were demobilized and allowed to return to their prewar civilian occupations. The loss of so many of the organization's best personnel naturally had a deleterious effect on the directorate's performance and production. The 6th Directorate also fought a series of losing battles over money and resources with other components of the MGB. The principal winner of these battles was the MGB's human intelligence (HUMINT) arm, which reasserted its primacy as the principal source for foreign intelligence within the Soviet intelligence community. Moreover, the MGB's internal security directorate, which was the largest component of the MGB, took the best manpower and the majority of the MGB's fiscal resources in the years immediately after the end of World War II, which was a reflection of the overall sense of paranoia which was so prevalent in Soviet security policy in the postwar era. The result was that the productivity of General Shevelev's directorate declined in the years immediately after the end of World War II, and research and development work on much-needed analytic machinery and computer equipment for cryptologic application, which was rapidly progressing in the US and elsewhere in the West, came to a virtual standstill in the USSR (confidential interviews).

17.5 THE MGB RADIO COUNTERINTELLIGENCE SERVICE

To make matters worse, General Shevelev's SIGINT directorate had to compete against a larger and better-funded cryptologic competitor within the MGB. Following the end of World War II, the MGB formed a large and well-funded SIGINT counterintelligence organization that was deliberately kept separate from General Shevelev's 6th Directorate. The origins of this highly specialized unit can be traced back to May 1942, when a Radio Counterintelligence Section was created within the NKVD's 2nd Special (Operational Technical) Department. The formal designation of the unit was the 5th Section of the NKVD 2nd Special Department (Operational Technical), which was headed throughout World War II by a veteran NKVD counterintelligence officer named E.P. Lapshin [9: 7].

Lapshin's SIGINT counterintelligence service was treated as a dedicated adjunct to the massive NKGB counterintelligence directorate, which used this unit to locate the radio transmitters of foreign agents operating inside the Soviet Union, as well as monitor around-the-clock the radio stations situated outside the USSR that were communicating with these agents [46: 461–462].

Shortly after the end of World War II, on May 4, 1946 there was formed a new unit called Department "R" (*Otdel "R"*) of the MGB, which was responsible solely for radio counterintelligence [47: 232]. The headquarters of Department "R" was located in the Kommunarka suburb southwest of Moscow near the present day headquarters in Yasevevo of Russia's foreign intelligence service, the SVR (confidential interview). The chief of this department from May 1946 to January 15, 1952 was V.M. Blinderman [47: 140]. His successor was Colonel L.N. Nikitin, who headed Department "R" from January 1952 until his death on November 3, 1953 [47: 116, 211–212].

The KGB SIGINT counter-espionage service, which by the early 1950s consisted of several thousand personnel in Moscow and throughout the USSR, proved to be an extremely important and productive tool for detecting and locating the radio transmitters of American and British agents parachuted into Russia during the late 1940s and early

1950s [46: 476, 478–479]. For example, the first Swedish agent sent into Estonia was Ernst Mumm (codename "*Nixi*"), who was secretly landed in Estonia in 1947. He was followed on October 15, 1948 a three-man agent team, who were landed on an isolated beach near the town of Lohusalu in northwestern Estonia. The Soviets detected their radio signals shortly after they came ashore, leading to an intensive manhunt which resulted in the team being captured within weeks of landing [42: 251–252; 49: 237].

17.6 THE FAILED SOVIET EXPERIMENT WITH CRYPTOLOGIC UNIFICATION

By the late 1940s, Soviet government officials responsible for managing the Soviet intelligence community came to the conclusion that the disparate and competitive cryptologic efforts of the MGB and the GRU were wasteful, and not achieving the operational efficiencies needed to make them effective intelligence contributors in the fiscally austere postwar era. According to former Soviet intelligence officials, the resource-strapped MGB and GRU SIGINT components were found to be "stepping on each other's shoes", copying each other's targets and occasionally stealing each other's personnel. Moreover, there were complaints from both the MGB and the GRU that they were not receiving sufficiently clear tasking guidance from the Kremlin or the Soviet Ministry of Defense about which SIGINT targets they should be focusing on, or what priority to assign to these targets. It was hoped that a consolidated MGB-GRU SIGINT organization, modeled on General Shevelev's successful wartime SIGINT organization, would solve all of these problems (confidential interviews).

On October 19, 1949, by special decree of the Politburo of the Communist Party of the Soviet Union (CPSU), General Shevelev's MGB 6th Directorate was deactivated, and its personnel and resources were merged with the GRU's Decryption Intelligence Service (DRS) to form on November 15, 1949 a new joint KGB–GRU SIGINT collection and processing organization called the Chief Directorate of Special Services (*Glavnoye*

Table 17.3. Organization of the chief directorate of special services (GUSS) – 1949

1st Decryption–Intelligence (*Deshifroval'noye–Razvedyvatel'noye*) Directorate
2nd Radio Intelligence (*Radio Razvedki*) Directorate
3rd Encryption (*Shifroval'noye*) Directorate
Operational Scientific Research Institute No. 1 (NII-1)
Technical Scientific Research Institute No. 2 (NII-2)
Higher School of Cryptography (*Vyshaya shkola kriptografii*, VshK)
Central Radio Establishment (*Tsentral'ny Radiouzel*) of Special Services (TsRSS)

Source: Assotsiatsiya issledovateley istorii spetssluzhb im. A.X. Artuzova, *Shpionami ne rozhdayutsya: Istoricheskiy spravochnik* (Moscow: Artuzovskaya Assotsiatsiya, 2001), p. 6.

Upravleniye Spetsial'noy Sluzhby, or GUSS) of the Communist Party of the Soviet Union (CPSU) Central Committee, which as it's name implies, was directly subordinate to the Central Committee in the Kremlin [75: 486; 47: 142; 9: 3, 10].

The first commander of GUSS was the ubiquitous General Shevelev, who remained on at the helm of GUSS from November 1949 to June 1952, when he was promoted and transferred to non-SIGINT related intelligence duties within the MGB [8: 38].

The man chosen to succeed General Shevelev at the helm of GUSS was an intelligence novice with high-level political patrons within the Kremlin, Colonel (later Major General) Ivan Tikhonovich Savchenko. A former political officer in the Red Army in the Ukraine during World War II, Savchenko was a protégé of the two leaders of the Ukrainian Communist Party, Nikita Khrushchev and Leonid Brezhnev, both of whom would dominate the Russian political scene for the next 25 years. Before transferring to the KGB in 1951, Savchenko had served for four years as the deputy chief of the Directorate of Communist Party Cadres in the Ukraine from 1947 to 1951. Before that, Savchenko, then the senior NKVD official in the western Ukraine, was responsible for combating Ukrainian nationalist forces operating in the western Ukraine from 1944 to 1947. In the MGB, Savchenko served as the deputy chairman of the MGB from 1951 to 1952, then head of the Chief Directorate of Special Services (GUSS) from June 1952 to March 1953. After the dissolution of GUSS, General Savchenko briefly served as chief of it's KGB successor, the 8th Chief Directorate,

from March 1953 to March 1954 before being promoted to the position of one of the KGB's deputy chiefs, which also came with a seat on the KGB Presidium [47: 286–287; 8: 38].

Former Soviet intelligence officers consider the GUSS an experiment that was doomed to failure given the bitter internecine warfare that marked KGB–GRU relations in the post-World War II era. The first chief of GUSS, General Shevelev, tried his best to build the foundation for a consolidated peacetime cryptologic organization, but ultimately could not overcome the lack of fiscal and manpower resources which his organization had to live with. Lack of analytic machinery severely limited the ability of even the best of the Soviet cryptanalysts at GUSS to solve the increasingly complex machine cipher systems then being introduced into use by the US government and military. Moreover, Shevelev's efforts to keep the GRU's military SIGINT components working harmoniously within the GUSS organizational framework were repeatedly sabotaged by senior GRU and Russian Navy intelligence officials, who incessantly lobbied within the Kremlin for greater independence for their organizations from what they perceived as the dictatorial rule of the MGB officers who ran the most important components of the GUSS, who they argued had little appreciation for the urgent need for greater SIGINT collection emphasis on military-oriented intelligence targets. Moreover, Shevelev's successor as head of GUSS, General Savchenko, proved to be a major disappointment given his lack of understanding of the technical aspects of the GUSS cryptologic mission, as well as his well-advertised ambition to get promoted out of GUSS as quickly as possible to a more senior position within the Soviet intelligence establishment (confidential interviews).

This is not to say that GUSS was a complete failure. Many of the MGB's SIGINT successes during the late 1940s and early 1950s can be directly attributed to its clandestine human intelligence (HUMINT) effort. For example, from 1945 to 1952, the MGB was able to listen in on conversations taking place inside the office of the US ambassador to Moscow because MGB's operatives had planted a bugging device inside the Great Seal of the United States that was mounted on the wall of the American ambassador's office [6: 338; 36: 46].

17.7   THE KGB AND GRU GO THEIR OWN SEPARATE WAYS

In April 1953, less than a month after the death of Josef Stalin, the Ministry of State Security (MGB) was merged with the Ministry of the Interior (MVD) to form a "super intelligence and security service" under Laventi Beria. But after Beria's downfall and subsequent execution, on March 14, 1954, the MGB was dissolved and was replaced by a new intelligence and security organization called the Committee for State Security (*Komitet Gosudarstvenoi Bezopastnosti*, or KGB). At the same time, the CPSU Politburo ended Stalin's experiment with a unified SIGINT service. By order of the Soviet Council of Ministers, dated April 24, 1953, the Chief Directorate of Special Service (GUSS) was disbanded and its assets divided into three separate SIGINT collection and processing organizations: a KGB Special Service; a General Staff Intelligence Directorate (GRU) Special Service; and a Russian Navy Special Service [9: 3].

The newly created KGB "Special Service" cryptologic organization was designated the Eighth Chief Directorate, which absorbed the non-military MGB SIGINT and COMSEC units that had previously been subordinate to GUSS [7: 450]. With its headquarters located inside the KGB's Lubyanka headquarters complex in downtown Moscow, the Eighth Chief Directorate was responsible for virtually all aspects of KGB communications, communications security, and radio intelligence work. The

Eighth Chief Directorate operated and maintained the KGB's extensive telecommunications networks both inside the USSR and overseas, including support of clandestine agent communications controlled by the First Chief Directorate; developed, designed, and produced communications equipment for all other branches of the KGB; and operated and maintained all special radio and telephone systems for use by the Soviet leadership, which were operated by the KGB Directorate of Government Communications Troops (*Upravleniye vojsk pravitel'stvennoy svyazi*). Cryptanalysis was the responsibility of the directorate's Department D (*Otdel "D"*). As of the 1970s, a special research laboratory was maintained by the KGB at Kuntsevo, west Moscow to specifically handle SIGINT research and development problems. In 1959, the Eighth Chief Directorate absorbed the KGB's 4th Special Department (Operational–Technical Directorate, or OTU), which had been responsible for radio development, radio operator training, and radio monitoring work [16: 27–28; 43: 20; confidential interviews].

Savchenko's replacement as head of the Eighth Chief Directorate in March 1954 was Major General Vasilii Andreevich Lukshin, a 42-year old technocrat and like his predecessor, a Ukrainian political protégé of Nikita Khrushchev and Leonid Brezhnev. And like his predecessor, General Lukshin had no prior intelligence experience, having previously served as the deputy head of the powerful CPSU Central Committee's Administration and Finance Department in the Kremlin. A capable administrator with strong political ties to the Soviet leadership in the Kremlin, Lukshin was to run the Eighth Chief Directorate for the next seven years, from March 1954 until March 1961 [47: 272; 8: 38].

The GRU also rebuilt its own independent SIGINT collection and processing organization following the breakup of GUSS. In 1953, all GRU decryption personnel and equipment resources formerly subordinate to the GUSS were split off and redesignated as the GRU Decryption Service (*Deshifroval'naya Sluzhba GRU*). In 1954, all Soviet military SIGINT collection units were brought under the control of a newly created GRU staff organization called the 2nd Department (Radio Intelligence), including the dozens of independent "Special Designation" radio battalions (*radiodivizion OSNAZ*) deployed throughout the USSR and Eastern Europe, which were redesignated Radiotechnical Detachments (*radiotekhnicheskiy otryad*) [9: 11].

In May 1955, the year-old GRU 2nd Department (Radio Intelligence) was expanded into the newly created GRU 6th Directorate (*6-e Upravleniye*), which is sometimes referred to today as the Radio and Radio-Technical Intelligence Directorate of the GRU [9: 11]. The 6th Directorate was assigned the responsibility for managing all GRU COMINT and ELINT collection, processing, analysis and reporting functions. The 6th Directorate, however, had no management control over the SIGINT components of the Russian Navy or the National Air Defense Forces (PVO), nor did it have any power over any of the operational SIGINT units assigned to Russian Armies, or for the several "Special Designation" (*SPETSNAZ*) units whose function was strategic and/or tactical electronic warfare and jamming [16: 72]. The GRU maintained a separate cryptanalytic organization called the Decryption Intelligence Service (DRS), which from its headquarters on the Komsomol'skiy Prospekt in downtown Moscow worked on solving the various foreign military cryptographic systems that were then being intercepted [16: 77]. The 6th Directorate received its intercepts from three separate sources: a nationwide complex of large COMINT collection facilities run from the GRU central SIGINT processing center at Klimovsk, located 20 miles south of Moscow; from Russian Army SIGINT collection units assigned to the 15 Military Districts within the USSR and the four Groups of Soviet Forces in Eastern Europe; and finally, from a small number of clandestine GRU listening posts (called Technical Service Groups) hidden within GRU "*Rezidenturas*" at selected Russian diplomatic establishments overseas [16: 72].

During the 1950s, both the KGB and the GRU invested heavily in building "strategic" collection systems to intercept high-frequency (HF) radio traffic. The Soviets began developing and fielding a new generation of domestically produced HF intercept systems shortly after the end of World War II.

One of the first of these systems was a circularly-disposed antenna array (CDAA) called *Krug*, an omni-directional radio intercept and HFDF antenna system which was based on the captured German *Wullenweber* antenna design. Each *Krug* antenna array was 130-feet in diameter, and was capable of covering the frequency range of 1.5–20 MHz. The system's range was estimated at 5,000 to 6,000 miles. The Soviets began development of the *Krug* antenna system in 1946 using a number of captured German scientists and engineers, who worked on the program from January 1947 to June 1952 at a closed electronics research facility called MGB 568 at Novogorsk near the town of Fenino. Three prototype *Krug* antenna systems were built, one being constructed at near the town of Zheleznodorozhnyy north of Moscow, and another situated 30 kilometers north of Moscow [10: 3]. Mass production of the new antenna system began in 1951 at a state-owned radio manufacturing plant in Leningrad [17: 33; 33: 147; 59: 1–3; 65: 14]. Field deployment of the *Krug* antenna system began in 1952, and continued at a rapid rate throughout the 1950s [33: 342, fn 43]. By the summer of 1965, CIA KH-4 *Keyhole* reconnaissance satellites had located a total of 31 *Krug* antenna complexes situated around the periphery of the Soviet Union [60].

Some of these sites were run independently by either the KGB, GRU or Naval GRU, or in a few cases, manned jointly by all three of the SIGINT organizations. The KGB SIGINT operators were uniformed KGB personnel assigned to KGB Government Communications Regiments (*polk pravitel'stvennoy svyazi*). These regiments also operated most of the high-level Russian government national communications networks. For example, the large Rustavi listening post outside Tbilisi, the capital of Soviet Georgia, which copied Turkish and other Middle Eastern radio traffic, was operated by a KGB Government Communications Regiment (confidential interview). Military-run strategic listening posts inside the USSR and Eastern Europe were operated by Independent Special Purpose Radio Regiments (*otdel'niy radiopolk OSNAZ*), each of which was managed by the Intelligence Directorates of the local Military District or Group of

Table 17.4. Location of KGB/GRU KRUG strategic intercept/HFDF sites

Verolantsy/Gatchina	5934N 2949E	8.0 nm W of Gatchina Airfield, Leningrad Oblast'
Odessa #1	4626N 3030E	9.5 nm SW of Odessa
Odessa #2	4631N 3033E	7.5 nm NW of Odessa
Murmansk/Murmashi	6849N 3249E	3.3 nm SE of Murmashi
Podolsk/Vlasyevo	5528N 3722E	7.7 nm WNW of Podolsk, 24.0 nm SW Moscow
Moscow/Vnukovo	5559N 3726E	Krug and Thick Eight HFDF antennas
Podolsk/Klimovsk	5523N 3728E	4.5 nm SW Podolsk, GRU Central SIGINT Facility
Khimki	5555N 3737E	a/k/a Moscow/Veshki, 11.0 nm North of Moscow
Lyubertsy	5542N 3759E	a/k/a Moscow/Temnikovo, 12.3 nm SE Moscow
Krasnodar	4509N 3847E	10.8 nm NW of Krasnodar
Arkhangel'sk #1	6425N 4040E	1.5 nm NW of Isakogorka
Arkhangel'sk #2	6430N 4045E	6.0 nm SE of Arkhangel'sk
Rustavi	4124N 4507E	8.4 nm SE of Rustavi, southeast of Tbilisi
Sumgait/Shuraabad	4046N 4928E	12.5 nm NW of Sumgait, outside Baku
Izgant	3810N 5805E	1.0 nm SW of Izgant, 20.0 miles NW of Ashkabad
Sverdlovsk	5647N 6055E	a/k/a Aramil', 10.5 nm SE of Sverdlovsk
Vorkuta	6739N 6350E	9.5 nm NNW of Vorkuta
Tashkent #1	4119N 6926E	4.0 nm E of Tashkent
Tashkent #2	4109N 6925E	12.6 nm SSE of Tashkent
Ili	4358N 7731E	15.7 nm NE of Ili
Novosibirsk	5515N 8319E	
Oyek/Koty	5234N 10432E	a/k/a Kuda, 19.0 nm NE of Irkutsk
Chita #1	5208N 11327E	6.5 nm NW of Chita
Chita #2	5210N 11330E	8.5 nm N of Chita
Tiksi	7138N 12841E	
Yakutsk	6155N 12937E	7.5 nm SW of Yakutsk
Khabarovsk #1	4830N 13518E	8.0 nm E of Khabarovsk
Khabarovsk #2	4825N 13521E	10.5 nm SE of Khabarovsk
Khabarovsk #3	4823N 13516E	8.0 nm SE of Khabarovsk
Petropavlovsk #1	5302N 15849E	5.5 nm E of Petropavlovsk
Petropavlovsk #2	5305N 15822E	6.0 nm South of Petropavlovsk
Beringovskiy	6304N 17906E	a/k/a Ugol'nyy, 6.0 nm W of Beringovskiy

Soviet Forces headquarters, but which forwarded their intercepts to the GRU 6th Directorate's headquarters in Moscow [16: 84].

Both the KGB and GRU also operated a small number of clandestine SIGINT collection facilities overseas. For instance, until July 1951 the Soviets maintained a clandestine listening post at Whetstone in North London under the cover of being a TASS News Agency radio broadcast monitoring facility. The station had been opened during World War II with the approval by the British government. Only after British defense and security officials vehemently complained about the station's activities did the British government force its closure [4: 407]. The GRU was the earliest practitioner of placing radio intercept sites inside Soviet em-

bassies around the world. During the 1950s, GRU listening posts inside the Soviet embassy and consulate in London intercepted the radio transmissions of MI-5 surveillance teams that were engaged in monitoring the movements and activities of Soviet diplomatic and intelligence personnel [83: 52–53, 91–97].

Defectors have revealed that the Soviets experienced some significant cryptanalytic successes during the 1950s. The KGB's 8th Chief Directorate solved some of the diplomatic systems used by France, West Germany, Italy, Belgium and Japan, although apparently little success was experienced against high-level American or British diplomatic machine cipher systems [7: 455]. For its part, the GRU 6th Directorate continued to focus its attention on intercepting and analyzing the encrypted and plaintext radio traffic of the military forces of the US, the Western European nations comprising NATO, and the Peoples Republic of China [16: 72]. For instance, in October 1958 the Soviets publicly announced that their intelligence services had detected 32 recent US nuclear weapons tests. The US intelligence community concluded that the Soviets could only have found out about some of the extremely low-yield tests through SIGINT, since the Russians were not believed to have been capable at the time of having detected these atomic tests through conventional seismic monitoring [19].

By the late 1950s, it became clear to senior KGB officials that the 8th Chief Directorate was experiencing significant successes in the SIGINT field, in large part because of the foreign cryptographic materials being obtained by the KGB's clandestine operatives around the world. In order to further enhance these efforts, in 1959 the chairman of the KGB, Aleksandr Shelepin, ordered the creation of a Special Section within the KGB First Chief Directorate to coordinate HUMINT collection efforts outside of the Soviet Union in support of the KGB's SIGINT collection program. The first head of the KGB Special Section was Aleksandr Sakharovsky. Much of the Special Section's efforts were directed against US codes and ciphers [7: 457]. At the same time that the Special Section was created, a 1959 KGB directive ordered all of its overseas stations (*Rezidentura*) to emphasize the recruitment of foreign nationals with access to cryptographic materials, such as code clerks, secretaries, typists, stenographers, and cipher machine operators [16: 43].

17.8   RADIO COUNTERINTELLIGENCE IN THE 1950s

On March 14, 1953, the State Security radio counterintelligence organization, Department "R", was redesignated as the 4th Special Department of the MVD (*4 Spetsotdel*, or *4 S/O MVD SSSR*), then after the creation of the KGB, renamed once again on March 18, 1954 as the 4th Special Department of the KGB (*4 S/O KGB SSSR*) [47: 232]. The first acting chief of the 4th Special Department as

of January 1954 was S.V. Kanishev [47: 204]. He was replaced later that year by P.F. Kuznetsov, who headed the 4th Special Department for the remainder of its existence from 1954 until July 1959 [47: 150, 212–214]. On July 2, 1959, the KGB 4th Special Department was disbanded, and its SIGINT counterintelligence mission was taken over by the Operational–Technical Directorate (*Operativno–Tekhnicheskoe Upravleniye*) of the KGB [47: 157, 232].

The Radio Counterintelligence Service continued to provide the KGB 2nd Main Directorate (Counterintelligence) with important intelligence information about the activities of American, English, French and Swedish agents operating inside the Soviet Union. For instance, on the night of May 6–7, 1954, the CIA parachuted two agents, Kalju N. Kukk (*Karl*) and Hans A. Toomla (*Artur*) into southern Estonia near the village of Auksaar. The KGB Radio Counterintelligence Service quickly homed in on their radio transmissions and located their source. On July 11, 1954, KGB security forces struck, killing Toomla and capturing Kukk. Kukk refused to cooperate with the KGB and was subsequently executed by firing squad [46: 516; 49: 239–240; 78: 54, 56].

17.9   SOVIET SIGINT IN THE 1960s

The KGB Eighth Chief Directorate grew relatively slowly during the 1960s, which was a reflection of the continued dominance of HUMINT as the predominate intelligence discipline within the KGB (confidential interview). In March 1961, Lt. General Lukshin was replaced as head of the Eighth Chief Directorate by Lt. General Serafim Nikolaevich Lyalin [47: 273]. Born in 1908, Lyalin had previously served as the deputy chief of the KGB's 2nd Chief Directorate, which was responsible for counterintelligence inside the Soviet Union, then chief of the Operational–Technical Directorate (OTU) since its creation on June 2, 1959. Lyalin's tenure as chief of the OTU from 1959 to 1961 gave him the much-needed grounding in the technical aspects of SIGINT collection and processing that were to serve him well when he was named head of the Eigth Chief Directorate.

General Lyalin held the title of chief of the Eighth Chief Directorate for six years from March 1961 to October 18, 1967. He would go on to serve as the chief of the KGB Directorate responsible for the city and province of Moscow [57; 58]. General Lyalin's career in the KGB, however, came to an abrupt end in 1973 after his nephew, Oleg Adolfovich Lyalin, then a junior KGB intelligence officer stationed in London, defected to England in 1971 after having spied for MI-6 for many years [81: 136].

Following Lyalin's departure, the KGB leadership decided that a more technically-oriented individual was needed to manage the KGB's still growing SIGINT collection and processing operations. From July 2, 1968 to August 1971, the head of the 8th Chief Directorate was a civilian engineer recruited from outside the KGB named Nikolai Pavlovich Yemokhonov. Born in 1921 in the Ukrainian coal mining town of Kuznetsk, Yemokhanov had been a member of the CPSU since 1947. After serving in the Russian Army during World War II, Yemokhanov graduated from the Military Communications Academy in Leningrad in 1952 with a degree in radio engineering. Following graduation, Yemokhanov worked in a secret military-run research and design institute outside Moscow on sensitive communications and communications security projects, rising to the position of department head by 1958. Before being named head of the Eighth Chief Directorate, Yemokhanov had served as the director of the Central Scientific Industrial Radiotechnical Institute in Moscow from 1964 to 1968 [47: 168, 263].

The size and scope of the KGB's SIGINT effort continued to grow during the 1960s, but continued to lag well behind in terms of money and personnel resources which the KGB devoted to HUMINT collection. Still, the Eighth Chief Directorate managed to achieve some impressive results, testifying to the growing importance of SIGINT within the KGB. In 1960, the KGB's Eighth Chief Directorate deciphered 209,0000 diplomatic messages from 51 countries [84]. In 1962, the KGB Eighth Chief Directorate monitored 1,170 radio stations in 118 countries around the world, including encrypted diplomatic communications traffic belonging to 87

"capitalist governments". Most of the KGB's SIGINT focus was against the diplomatic and clandestine communications traffic of the US. According to a declassified KGB report to the CPSU Presidium, in 1962 74.7% of all KGB intercepts were American diplomatic, military or clandestine communications targets (the previous year 72.2% of all intercepts came from US targets) [44: 17–18]. In 1967, the Eighth Chief Directorate's SIGINT organization was actively monitoring 2,002 radio stations in 115 countries around the world. In 1967, the KGB's codebreakers read traffic in 152 encryption systems belonging to 72 "capitalist countries", solved 11 new encryption systems, and deciphered and translated 188,400 messages [45: 9].

The Cuban Missile Crisis of 1962 revealed to the Soviet senior political and intelligence leadership just how bare KGB and GRU HUMINT assets in the US were. At the time of the crisis, the KGB's best source of "political intelligence" in Washington was Johnny Prokov, the lead bartender at the Tap Room of the National Press Club in Washington, DC, who passed on to the KGB *Rezidentura* gossip picked up from American newspapermen. The KGB *Rezidentura* then dressed up the information and passed it on to Moscow as intelligence information received from a reliable Washington insider [35: 257–258]. The lack of Soviet HUMINT assets in the US during the Missile Crisis also highlighted how important SIGINT had become within the Soviet intelligence community. GRU SIGINT played an important, albeit unheralded role during the Cuban Missile Crisis. The Soviet Navy's spy trawler (AGI) *Shkval* was deployed in the South Atlantic during the crisis, where its movements and communications were closely monitored by the U.S. Navy at the height of the crisis [63]. On October 25, 1962, at the height of the crisis, the GRU listening post inside the Russian embassy in Washington intercepted an order from the Joint Chiefs of Staff placing the Strategic Air Command (SAC) on fullscale nuclear alert [35: 258; 71: 197–198]. In the 1960s, the GRU *Rezidentura* in New York City monitored with great success the unencrypted radio transmissions of FBI agents following GRU intelligence operatives around town [82: 61].

Following the Cuban Missile Crisis, the Soviets rapidly built up a sizable SIGINT collection infrastructure in Cuba. NSA detected the establishment on November 27, 1962 of a Russian COMINT intercept and HFDF network in Cuba. This Soviet SIGINT unit remained in Cuba long after Cuban combat troops left the island [18]. The CIA and NSA first became concerned about the significant threat posed to US government and military communications from Soviet SIGINT sites in Cuba shortly after the end of the Cuban Missile Crisis [20; 21]. By 1964, six small Soviet HF intercept and direction-finding facilities had been identified by US intelligence in Cuba at Holguin, Remedios, Guayabal, Artemisa, Bejucal and Caraballo [61]. In the mid-1960s, the GRU opened a radio intercept station at Lourdes, Cuba, whose mission was to intercept U.S. Navy and other American military HF communications traffic [70: 723–725].

The KGB ventured into the field of clandestine SIGINT collection during the 1960s. The first KGB embassy listening post, codename *RADAR*, was established in Mexico City in 1963. The station monitored radio and telephone communications coming from the US embassy in Mexico City as well as from the CIA station. The first KGB clandestine intercept station in the US, codenamed *POCHIN*, was set up inside the Soviet embassy in Washington, DC in 1966. In 1967, a station called *PROBA* was established inside the Soviet consulate in New York City [6: 343]. Both stations monitored a wide range of communications targets, but amongst the most important were the radio traffic of police and FBI surveillance teams [16: 29].

The KGB radio counterintelligence service also continued to actively try to find foreign agents operating inside the USSR and Eastern Europe. In 1962, the KGB radio counterintelligence service was still actively monitoring the clandestine radio transmissions of the US, British, French and West German foreign intelligence services. The major accomplishment of the radio counterintelligence service in 1962 was the identification of clandestine radio links between the US embassies in Moscow, Warsaw and Prague with CIA-operated communications centers in Frankfurt and Cyprus

[44]. In 1967, the KGB radio counterintelligence service was actively monitoring all clandestine radio traffic emanating from or destined for 24 communications centers operated by US and other hostile foreign intelligence services, and monitored clandestine radio traffic from 108 clandestine radios operating around the periphery of the Soviet Union. Three new clandestine radios were identified by the KGB listeners as operating inside North Vietnam, but none were identified inside the USSR itself [45].

KGB HUMINT operations designed to steal or compromise foreign cryptographic systems and/or materials were also accelerated in the 1960s. On March 11, 1961, the Central Committee of the CPSU and the Soviet Council of Ministers issued a classified directive ordering the KGB and GRU to intensify their efforts to steal foreign cryptographic materials. In 1962, for example, KGB agents succeeded in stealing the plans for two new foreign-made cipher machines, although the declassified document provides no further details of this theft, such as the make and model of these machines or where they were stolen from [44]. In April 1964, State Department security officials discovered 17 active audio-surveillance microphones planted at strategic locations inside the walls of the US embassy in Moscow, which ultimately grew to 39 clandestine listening devices. US security officials concluded that these microphones allowed the KGB to listen to what was going on inside the US embassy, including the embassy's code room, for eleven years between 1953 and 1964 [34]. In 1967, KGB operatives surreptitiously obtained 36 "objects of cryptologic interest" to help further the progress of Soviet intelligence [45].

17.10 SOVIET SIGINT DURING THE 1970s

According to former Soviet intelligence officers, the 1970s was the era when SIGINT finally surpassed HUMINT as the premier producer of intelligence information for Soviet political and military intelligence consumers (confidential interviews). In 1977, a senior U.S. Defense Department official, Dr. Gerald P. Dineen, testified before Congress that the Soviets "... maintain the largest signals intelligence establishment in the world ... operating hundreds of intercept, processing and analysis facilities with heavy exploitation of unsecured voice communications" [39]. A 1976 secret study done by the CIA concluded that the USSR was spending approximately $6.3 billion on intelligence collection, the largest component of which was the $3.1 billion that the CIA believed the Soviets were spending on satellite reconnaissance systems, which included research and development and intelligence processing costs. The next largest component was the $1.4 billion that the Soviets were believed to be spending on land-based SIGINT collection systems. Airborne and seaborne reconnaissance systems, most of which were involved in SIGINT collection, accounted for another $700 million in spending. Finally, Soviet clandestine HUMINT collection spending amounted to an estimated $700 million, while intelligence research, analysis, and production accounted for only $400 million in spending [22].

The growing importance of SIGINT within the Soviet intelligence community was marked by the dramatic growth of both the KGB and GRU SIGINT organizations. The biggest winner was the KGB, which by the mid-1970s was estimated to consist of some 410,000 personnel, including 175,000 border troops and 65,000 uniformed troops performing internal security functions [16: 18]. Much like its parent organization, the fortunes of the KGB's SIGINT organization prospered greatly during the 1970s. The KGB SIGINT organization, the Eighth Chief Directorate, changed dramatically during the 1970s. In 1971, Nikolai Pavlovich Yemokhonov was replaced as head of the Eighth Chief Directorate by a career KGB officer with experience in technical intelligence, Lt. General G.A. Usikov (1917–1990), who ran the Directorate for four years from August 1971 until August 1975 [47: 171–172].

On June 21, 1973, the KGB Presidium issued Directive No. 0056, which stripped the COMINT intercept and processing missions from KGB Eighth Chief Directorate, and instead gave these responsibilities (with all personnel and equipment assets formerly part of the Eighth Directorate) to a newly established unit called the KGB 16th Directorate, which was created for the sole purpose of managing the KGB's larger and increasingly important SIGINT operations [47: 171]. The 16th Directorate had three primary missions: intercepting foreign communications; deciphering foreign encrypted communications; and the technical penetration of foreign installations located inside the USSR. The CIA estimated that by the early 1980s, the 16th Directorate consisted of some 1500 personnel in Moscow, not including intercept personnel stationed elsewhere in the USSR and overseas [23: 37]. In fact, Soviet sources indicate that by the end of the 1970s, the 16th Directorate's headquarters staff consisted of more than 2000 officers [68]. Following the divorce, the Eighth Chief Directorate became a purely communications security (COMSEC) organization, consisting of approximately 5000 personnel [6: 346; 7: 528; 23: 37; 43: 20; 73: 143].

The first chief of the 16th Directorate was General Nikolai Nikolaevich Andreyev, who ran the organization from July 1973 until he became the head of the Eighth Chief Directorate in August 1975 [47: 171–172]. Born in 1929 and a geologist by training, General Andreyev had joined the KGB in 1959 at the age of 30, and worked his way up through the ranks of the Eighth Chief Directorate by managing a series of highly successful SIGINT operations during the 1960s and early 1970s. In August 1975, Andreyev was laterally transferred to replace General Usikov as the head of the Eighth (COMSEC) Chief Directorate. General Andreyev ran the Eighth Chief Directorate for an astounding 16 years, from August 1975 until the collapse of the Soviet Union and the dissolution of the KGB in the fall of 1991. His replacement as head of the KGB 16th Directorate was his former deputy, Major General Igor Vasiliyevich Maslov, a highly-respected veteran SIGINT officer, who would run the 16th Directorate from 1975 until the dissolution of the KGB in the fall of 1991 [6: 346; 7: 528; 43: 20; 47: 172; 73: 143].

The 16th Directorate worked closely with the 30-man 16th Department of the KGB's foreign intelligence organ, the First Chief Directorate. Established in 1969, the 16th Department was responsible for managing the KGB's espionage programs to obtain foreign cryptologic materials by clandestine means [23: 53; 7: 529]. In the early 1980s, the 16th Department was headed by A.V. Krasavin [6: 352].

Many, but not all, of the KGB's SIGINT successes during the 1970s were directly connected with the successes of the KGB's HUMINT collection efforts. For example, between 1976 and 1984, the KGB was able to monitor communications and correspondence being generated or received inside the US embassy in Moscow and the consulate in Leningrad by virtue of dozens of clandestine listening devices planted in the embassy and consulate typewriters. The listening devices were not discovered by NSA technical security specialists until 1984 as part of a Top Secret operation designated *Project Gunman* (Interview with Walter G. Deeley, 1989). It was not until mid-1984 that the U.S. State Department discovered that the new US embassy building in Moscow, then nearing completion, was riddled with bugs and eavesdropping devices planted by the KGB during the construction process [50: A1].

As had been the case since the end of World War II, the vast majority of the Soviet's SIGINT resources during the 1970s were dedicated to monitoring US government, military and commercial communications traffic. In 1975, NSA estimated that well over 50% of the Soviet and Warsaw Pact SIGINT collection resources were targeted against the United States, which was referred to within the Soviet intelligence community as "Special Target No. 1" (*Spetsial'naya tsel' No. 1*) [62: 2]. The Soviets experienced great success collecting crucially important commercial intelligence using the KGB listening posts in Cuba and the US. For example, in 1972 the Soviets purchased 25% of the US grain harvest using intercepts of commercial telephone calls to keep Soviet negotiators informed as to changes in market prices for grain throughout the negotiations [80: 122].

During the mid- to late-1970s, due to a lack of significant HUMINT successes in the US, SIGINT coming from the KGB's clandestine listening posts inside the Russian embassy in Washington, DC had become the most important source of information about what was going on inside the Ford

and Carter administrations [6: 348]. Much of the KGB's SIGINT was obtained from clandestine listening posts established inside Soviet diplomatic establishments overseas. By the 1970s, there were five *POCHIN* clandestine SIGINT intercept sites in various Soviet diplomatic establishments in the Washington, DC area (including *POCHIN-1* at the Soviet embassy, and *POCHIN-2* at the Soviet residential complex), and four *PROBA* intercept stations in the greater New York City area, including *PROBA-1* inside the Soviet mission to the UN and *PROBA-2* at the Soviet diplomatic dacha in Glen Cove, Long Island [6: 344].

Because of the success of the embassy intercept operations, in the early 1970s the scope of KGB embassy clandestine SIGINT operations were dramatically increased. On May 15, 1970, KGB chairman Yuri Andropov approved the creation of 15 new embassy listening posts in Washington, New York, Montreal, Mexico City, Tokyo, Peking, Teheran, Athens, Rome, Paris, Bonn, Salzburg, London, Reykjavik and Belgrade [6: 346]. In the early 1970s, the KGB established separate radio intercept sites in their diplomatic establishments in New York City, Washington, DC and San Francisco (codenamed *RAKETA*, *ZEFIR* and *RUBIN* respectively) to monitor FBI counterintelligence communications traffic [6: 348–349]. In 1976, another station, codenamed *VESNA*, was established inside the Russian consulate in San Francisco [6: 348]. By 1979, there were 34 KGB embassy listening posts in operation in 27 countries around the world [6: 634–635, fn 63].

One of the National Security Agency's most sensitive operations conducted in the mid-1970s was a program of intercepting the radio traffic of Russian KGB and GRU listening posts in the US and elsewhere overseas in order to determine which US government telephone calls the Soviets were intercepting [51]. As of the late 1970s, this NSA program helped the FBI determine which U.S. Government telephone calls the KGB was intercepting from inside the Soviet diplomatic establishments in Washington, New York and San Francisco. The program's name was *Project Aquarium*. The *Aquarium* intercepts revealed which communications circuits the Russians were monitoring,

Table 17.5. KGB embassy listening posts – 1979

Washington, DC, USA (POCHIN)
New York City, USA (PROBA)
San Francisco, USA (VESNA)
Ottawa, Canada
Montreal, Canada (VENERA)
Lourdes, Cuba (TERMIT-S)
Brasilia, Brazil (KLEN)
Mexico City, Mexico (RADAR)
Reykjavik, Iceland (OSTROV)
London, England (MERCURY)
Oslo, Norway (SEVER)
The Hague, The Netherlands (TULIP)
Brussels, Belgium (VEGA)
Paris, France (JUPITER)
Bonn, Germany (TSENTAVR-1)
Cologne, Germany (TSENTAVR-2)
Salzburg, Austria (TYROL-1)
Vienna, Austria (TYROL-2)
Berne, Switzerland (ELBRUS)
Geneva, Switzerland (KAVKAZ)
Lisbon, Portugal (ALTAY)
Rome, Italy (START)
Belgrade, Yugoslavia (PARUS)
Athens, Greece (RADUGA)
Ankara, Turkey (RADUGA-T)
Istanbul, Turkey (SIRIUS)
Nairobi, Kenya (KRYM)
Cairo, Egypt (ORION)
Damascus, Syria (SIGMA)
Teheran, Iran (MARS)
Hanoi, Vietnam (AMUR)
Jakarta, Indonesia (DELFIN)
Beijing, China (KRAB)
Tokyo, Japan (ZARYA)

Source: Christopher Andrew [6: 634–635, fn 63].

such as the fact that the Russians were listening to the telephone calls of US Attorney General Griffin Bell, who the intercepts caught discussing classified information on an insecure telephone line [5: B15].

Like the KGB, the GRU's SIGINT capabilities grew dramatically during the 1970s. During the mid-1970s, the GRU 6th Directorate was commanded by Lt. General Aleksandr Ignat'evich Paliy [48: 97]. In 1971, the GRU 6th Directorate activated the so-called Special Center (*Spetsialniy Tsentr*) near the Sokolovskiy train station, 25 miles from downtown Moscow, which used banks of Russian-made mainframe computers to process the

rising volume of radio intercepts being collected by GRU intercept facilities around the world [16: 73]. By the end of the 1970s, the GRU's conventional COMINT intercept infrastructure had grown to massive proportions. A list of GRU and Soviet military listening posts, other than the aforementioned *Krug* strategic radio intercept and HFDF sites, is contained in Table 17.6. The GRU's global network of clandestine embassy listening posts was also significantly larger than that of the KGB. As of the early 1970s, the larger GRU SIGINT organization was operating clandestine listening posts with Russian embassies and trade missions in 48 countries [16: 72].

By the mid-1970s, CIA intelligence assessments held that the Soviets depended heavily on the huge GRU SIGINT network for the bulk of their warning intelligence reporting. The distinguished intelligence historian David Kahn has written that: "Determining whether the United States and its allies were planning a nuclear war against the Soviet Union was the primary task of Soviet COMINT during the Cold War" [43: 14]. It seems clear that the GRU did a reasonably good job of providing Soviet intelligence consumers with this kind of intelligence information throughout the Cold War. Even though they could not read the traffic, GRU SIGINT closely monitored all of the communications traffic associated with US strategic forces looking for changes in tempo or volume that might be indicative of US nuclear forces being placed on alert [24: 59; 36: 50]. A 1973 US European Command intelligence summary reported that GRU listening posts in the Soviet Union and Eastern Europe, supplemented by SIGINT coverage from 14 clandestine SIGINT stations inside Soviet embassies in Western Europe, focused on the activities of US and NATO strategic and tactical air and ground forces in Europe. Among the traffic being monitored were combat readiness checks (such as monitoring nuclear Emergency Action Messages), the progress of major NATO exercises, the realtime status of NATO forces, the flight activity of US airborne command posts, and the travel of NATO VIPs [32]. By the mid-1970s, NATO intelligence officials were beginning to wonder how the Soviets knew so precisely when NATO released nuclear

Table 17.6. Soviet non-Krug HF intercept/HFDF sites

Arsenyev	4410N 13319E	3 NM NE Arsenyev, FIX-24
Artagla	4136N 4454E	THICK EIGHT
Ashkhabad	3758N 5817E	FIX-24
Astrakhan Airfield	4611N 4804E	THICK EIGHT
Baku	4021N 5016E	18 nm east of Baku, FIX-24
Bataysk	4706N 3941E	Unkown CDAA antenna type
Brest	5208N 2342E	THICK EIGHT
Dikson	7333N 8035E	Unknown CDAA antenna type
Kabalovka	6003N 3014E	THICK EIGHT
Kazan	5545N 4915E	FIX FOUR HFDF
Kiev	5022N 3025E	THICK EIGHT
Kiev	5035N 3046E	15 nm NE of Kiev, FIX-24
Krasnoye Selo	5942N 3003E	THICK EIGHT
Krokhalevo	5857N 5525E	nr Tomsk, FIX-24
L'vov	4955N 2404E	THICK EIGHT
Magadan	5937N 14948E	THICK EIGHT
Magadan	5937N 15048E	3 nm north of Magadan, FIX-24
Minsk	5400N 2739E	THICK EIGHT
Moscow/Ostafyevo	5533N 3728E	THICK EIGHT
Murmansk	6851N 3809E	6 nm south of Murmansk, FIX-24
Narva	5920N 2814E	Suspected THICK EIGHT
Nomme	5920N 2437E	THICK EIGHT
Petrozavodsk	6154N 3413E	THICK EIGHT
Petrozavodsk	6152N 3412E	Unknown CDAA type
Riga	5637N 2401E	12 nm south of Riga, FIX-24
Sevastopol	4433N 3333E	4 nm SE Sevastopol, FIX-24
Simferopol	4458N 3402E	THICK EIGHT
Sovetskaya Gavan	4902N 14017E	30 nm NE of Sovetskaya Gavan, FIX-24
Tallinn	5929N 2451E	3.5 nm NE Tallinn, FIX-24
Uglovoye	4323N 13158E	FIX EIGHT
Ussuriysk	4357N 13148E	FIX-24
Ventspils	5622N 2134E	1 nm south of Ventspils, FIX-24
Ventspils	5622N 2134E	0.5 nm south of Ventspils, FIX EIGHT
Yuzhno Sakhalinsk	4657N 14241E	THICK EIGHT
Yuzhnyy	4952N 3608E	THICK EIGHT

weapons to its military commanders during military exercises in Europe. The answer was SIGINT [40; 31: 23].

During the 1970s, both the KGB and the GRU moved heavily into the field of intercepting foreign communications traffic being carried by communications satellites beginning in the mid-1970s. According to declassified CIA documents, by the early 1980s the KGB and GRU were jointly operating ten SATCOM intercept stations inside the USSR and overseas. Among the communications traffic these stations were intercepting was foreign commercial communications traffic, including scientific and technological data, that was being trans-

mitted over the INTELSAT communications satellite network [25: 21]. The largest of these SATCOM intercept stations was located in Cuba. On April 25, 1975, the KGB was authorized to establish a new SATCOM intercept post at Lourdes in Cuba, whose codename was *TERMIT-P*. The station was initially equipped with a single 12-meter fixed parabolic antenna and a mobile 7-meter parabolic antenna mounted on a lorry, both of which were used for intercepting American satellite communications (SATCOM) traffic. The Lourdes SATCOM intercept station, run jointly by the KGB's 16th Directorate and the GRU 6th Directorate, commenced operations in December 1976 [6: 349].

By the 1980s, the Lourdes SATCOM intercept station had become the KGB's principal sources of high-quality economic and political intelligence concerning the US [70: 734].

17.11 SOVIET SIGINT DURING THE 1980s

As SIGINT became increasingly important to both the KGB and GRU during the 1980s, the size and importance of the SIGINT components of both Soviet intelligence services grew commensurately. By 1989, the combined KGB-GRU SIGINT organizations reportedly consisted of over 350,000 personnel operating 500 listening posts in the USSR and Eastern Europe, as well as covert listening posts inside Soviet embassies and consulates in 62 countries. This compared with the 75,000 personnel comprising the NSA-managed US SIGINT System. This naturally made the Soviet SIGINT system the largest in the world [11: 73; 13: 22].

By the mid-1980s, the KGB 16th Directorate's headquarters staff in Moscow consisted of an estimated 1500 personnel, which did not include the thousands of military and uniformed KGB communications personnel performing the radio intercept function for the KGB (confidential interview). There are no hard figures available for how much the Soviet Union and its allies spent on SIGINT, but according to a former KGB official, by the late 1980s SIGINT collection was reportedly eating up 25% of the KGB's annual operating budget [70: 726].

The same held true for the GRU, which was even more dependent on SIGINT for its supply of hard intelligence information about US and NATO military activities around the world. According to a former GRU intelligence officer, by the early 1980s the GRU had become increasingly dependent on COMINT because it consistently proved to be a more productive and reliable source for strategic intelligence than the more traditional HUMINT sources (confidential interview). A spy ring within the US Navy led by former Navy warrant officer John A. Walker, Jr., spied for the GRU and KGB for twenty years before Walker was arrested in May 1985 after his former wife turned him in to the FBI. From December 1967 until his arrest,

Walker's spy network provided the Soviets with near continuous access to the Navy's most sensitive ciphers, causing immense damage to US national security. Using the cipher keys and other highly classified cryptographic materials provided by the Walker spy ring, the KGB's SIGINT organization was reportedly able to exploit some of the US Navy's most sensitive communications. According to former KGB defector Vitaly Yurchenko, "It was the greatest case in KGB history. We deciphered millions of your messages. If there had been a war, we would have won it" [43: 16; 53: D1].

During the 1980s, the KGB and GRU jointly operated a small number of conventional SIGINT collection facilities outside the Soviet Union. The largest of these intercept stations was the huge 28 square-mile Lourdes Central SIGINT Complex at Torrens, Cuba, which was manned by 2,100 KGB and GRU personnel, making it the largest Soviet listening post outside the USSR. A reinforced motorized rifle battalion from the Soviet combat brigade, based nearby at the Santiago de las Vegas Camp, was responsible for providing security for the Lourdes station [26: 53, 60; 79]. The relative importance of the SIGINT product coming from the Lourdes station cannot be understated. According to a 1993 statement by Cuban Defense Minister Raul Castro, Russia got about 75% of its strategic military intelligence information from the Lourdes listening post [30]. During the *Achille Lauro* crisis in 1985, NSA was able to decrypt the secure telephone calls between senior KGB officials at the KGB–GRU listening post at Lourdes, Cuba and the KGB *Rezidentura* in Washington, DC about who was to intercept which US government communications circuits during the crisis [64].

The Lourdes station was not the only Soviet SIGINT collection facility located outside the Soviet Union. Declassified CIA documents reveal that as of the mid-1980s, there were three Soviet SIGINT collection sites in Vietnam, with the largest being a multi-hundred man joint KGB–GRU station located just outside the Soviet naval base at Cam Ranh Bay [26: 91]. The Cam Ranh Bay listening post was used to track the movements of US Navy warships in the South China Sea, monitor Chinese naval and naval air activity on Hainan

Island, and monitor the radio traffic of US Navy forces in the Far East, including the big Navy bases in the Philippines [41]. There was also a large joint KGB–GRU listening post in South Yemen, located adjacent to the Soviet military's Salah ad Din Communications Facility, situated just west of the Yemeni capital of Aden [26: 107]. From 1987 to 1997, the GRU operated a listening post in North Korea, codenamed *Ramona*, which was situated near the town of Ansan, located south of the North Korean city of Sariwon. The station's main target was Japanese and South Korean communications traffic, as well as US military radio traffic emanating from the Far East [76].

The KGB and GRU also continued their program of expanding the size and capabilities of their dozens of embassy listening posts around the world. In January 1987, Soviet diplomats and intelligence officials prevented Canadian firefighters from battling a blaze which consumed the Soviet consulate in Montreal so as to prevent Canadian security officials from gaining access to the sensitive KGB–GRU listening post located on the top floor of the consulate [72].

17.12 THE STRENGTHS AND WEAKNESSES OF SOVIET SIGINT

Available evidence suggests that SIGINT was one of the most reliable and productive sources of information available to the Soviet intelligence community. Former Soviet and Western intelligence officials confirmed in interviews that the Soviet SIGINT effort was much larger, more extensive, and far more successful than the US intelligence community had previously been aware of during the Cold War. Unfortunately, most of the details concerning these cryptologic successes remain highly classified in both the West and Russia (confidential interviews).

The KGB's SIGINT effort alone produced a healthy cornucopia on a daily basis of foreign diplomatic, military, and commercial communications traffic, which was highly prized inside the Kremlin. Every day the KGB sent a selection of key intercepts in a bound volume called

the *Red Book* to the top six members of the Politburo, although the KGB officials typically excised any materials from the decrypts that ran contrary to the prevailing political trends of the time within the Kremlin [6: 352–353]. Sadly, because of the paucity of declassified policymaking materials available from the Russian archives concerning the Soviet era, it is impossible at this date to accurately posit an opinion about how important SIGINT was to Soviet civilian officials and military commanders during the Cold War.

Sources generally tend to agree that Soviet SIGINT accomplishments were considerable during the Cold War, although it is not yet possible to empirically measure this level of accomplishment. And while it is generally recognized that despite their technological inferiority to their Western counterparts, both the KGB and the GRU SIGINT organizations had two major advantages. First, they could press-gang just about any personnel that they wanted into their service. This meant that the KGB and GRU SIGINT organizations employed, or had direct access to many of the best mathematicians and telecommunications engineers available to the Soviet Union. And second, the KGB and GRU both closely integrated their HUMINT and SIGINT efforts, using operatives to steal foreign cryptographic material to great effect [7: 451].

But according to published reports, declassified documents and discussions with American intelligence officials and Soviet defectors, there were numerous problems with the Soviet SIGINT effort. Throughout the Cold War, the Soviet SIGINT organizations consistently suffered from one major deficiency: they lacked the technological wherewithal of their Western SIGINT counterparts. Of particular concern was the fact that the KGB and GRU lacked high-powered computers and sophisticated processing equipment, which severely limited their intercept processing and cryptanalytic capabilities [7: 451]. For instance, in the late 1970s the KGB's Eighth Chief Directorate possessed only one Russian-made mainframe computer for use in enciphering and deciphering work, which was manufactured by the KGB computer factory, called "Polin's Enterprise" [73: 164]. As of the mid-1980s, Russia still had not been able to match their American and Japanese counterparts by developing a working supercomputer. A 1985 CIA assessment found that: "The best Soviet high-performance computers are 30 to 60 times slower that US or Japanese supercomputers and lack good mass storage devices and systems software. The Soviet's lag in supercomputer development is estimated to be 10 to 15 years" [27]. This lack of high-tech equipment severely impeded the ability of Soviet cryptanalysts to solve sophisticated machine cipher systems, and impaired even traffic analysis by Soviet intelligence analysts during the Cold War. This meant that much of the SIGINT that the Soviets were able to exploit were low-level cryptographic systems as well as the unencrypted voice communications of NATO forces [6: 338; 39; 43: 7; 69: 89].

Other technical problems dogged the Soviet SIGINT effort throughout the Cold War. The Soviets perpetually lagged far behind the US and Great Britain in the highly technical field of ELINT collection, again largely because of Soviet technological deficiencies. The Soviets first airborne ELINT collection effort was noted by US intelligence in 1953 (Great Britain and the US began flying ferret missions more than ten years earlier during World War II); the first Soviet ground-based ELINT equipment was observed in East Germany in 1954; and the first Soviet ELINT collection trawlers (AGI) were not spotted on the high seas until 1956 [65]. The first Russian dedicated ELINT satellite was launched into space in 1967, four years after the first successful American launch, and this and subsequent generations of Soviet ferret satellites never came close to the capabilities of their American counterparts [28]. The Soviets were never able to put a COMINT satellite into orbit, severely inhibiting the ability of the KGB and GRU to monitor what the American military was doing. Nor were the Soviets ever able to field a particularly effective airborne SIGINT collection system, nor used their fleet of attack submarines for intelligence collection purposes anywhere near as well as the Americans and British. Much of the equipment used by the Soviet SIGINT system, such as intercept receivers and direction finding equipment, was crude and ineffective compared with that used by their Western counterparts. Poor communications facilities and equipment meant that intercepts oftentimes took days to reach Moscow. A lack of computer processing equipment meant that the KGB and the GRU had to employ tens of thousands of personnel to manually sort and process intercepts which NSA and its partners accomplished with computers (confidential interviews).

Other problems dogged the SIGINT efforts of the KGB and the GRU throughout the Cold War. First, for a variety of reasons, most of them bureaucratic in nature, centralization and rationalization of the Soviet SIGINT effort never took hold during the Cold War. As a result, because of the intense inter-service rivalry that had always existed between the KGB and the GRU, there was little cooperation between the two services in their respective SIGINT endeavors. The KGB and GRU SIGINT organizations were constantly stepping on each other's shoes. Turf battles between the two agencies were frequent as the KGB and GRU strove constantly to 'scoop' their competition by obtaining the hottest intelligence. The KGB frequently intercepted foreign military radio traffic in violation of the division-of-labor agreements between the two intelligence agencies. The GRU also frequently violated the KGB's 'turf' by copying foreign diplomatic and intelligence traffic as well (confidential interviews). It was not uncommon for both the KGB and GRU to operate clandestine intelligence radio intercept stations inside the same Russian embassies around the world, in many cases monitoring the same targets without any coordination of effort whatsoever (confidential interview). For example, both the KGB and the GRU maintained their own clandestine SIGINT stations inside the Russian embassies in London, Washington and other major capitals.

Second, duplication of the SIGINT effort within the Soviet military forces was equally bad. during the Cold War the GRU was never able to fully control the SIGINT activities of the Soviet armed services, which in fact paid only nominal lip service to GRU control but pretty much whatever they wanted. For example, the Russian Navy and the Strategic Air Defense Forces (PVO) maintained their own sizable independent SIGINT collection and processing resources that were separate and distinct from those of the Army-dominated GRU (confidential interview). Third, the manpower-intensive SIGINT organizations of the KGB and the GRU, like so many other Soviet-era bureaucracies, were bloated and wasteful. Fourth, analysis and reporting of Soviet SIGINT was made to confirm to existing political and ideological norms in the USSR, with KGB and GRU officials oftentimes deleting information from intercepts that did not fit the existing political beliefs held inside the Politburo [6: 352–353; confidential interviews]. And fifth, SIGINT collaboration between the KGB and GRU and their Eastern European allies was strictly limited and oftentimes strained because Moscow did not trust them [6: 351]. As the Eastern European services became more independent beginning in the 1970s, they began to monitor the activities of their Soviet counterparts. For example, in the late 1970s the KGB discovered that the Polish intelligence service was eavesdropping on Russian political communications [73: 16–17].

In the end, because of continued secrecy surrounding the subject, we are left with more questions than answers when it comes to trying to properly place into context the role of SIGINT in Soviet foreign intelligence gathering during the Cold War. From the information now available, we can posit that SIGINT was probably a more important intelligence source for the Soviet intelligence community during the Cold War, especially during the period from the early 1970s until the demise of the USSR in 1991. More difficult to answer because of a lack of publicly-available documentation is the all-important question of how important the intelligence information derived from SIGINT was in the formulation and execution of Soviet foreign policymaking and the conduct of military operations. To answer this question, historians must await the declassification of further documentary historical materials in Russia, or the publication of more substantive memoirs by former Soviet intelligence officials than have been released to date. In the interim, however, we shall have to depend on documents released by the CIA and other western intelligence services for much of what we can discern about the nature and extent of Soviet SIGINT during the Cold War.

REFERENCES

[1] Armed Forces Security Agency, *Russian Signal Intelligence*, February 15, 1952, RG-457, Historic Cryptologic Collection, Box 1134, File: ACC 17912 Russian SIGINT 1941–1945, National Archives, College Park, Maryland.

[2] Agentura, available at http://www.agentura.ru/dosie/gru/imperia/prilog/

[3] Agentura, available at http://www.agentura.ru/dosie/gru/structure/osnaz/

[4] R.J. Aldrich, *The Hidden Hand: Britain, America and Cold War Secret Intelligence*, John Murray, London (2001).

[5] J. Anderson, *Project Aquarium: tapping the tappers*, Washington Post, December 2 (1980).

[6] C. Andrew, *The Sword and Shield: The Mitrokhin Archive and the Secret History of the KGB*, Basic Books, New York (1999).

[7] C. Andrew, O. Gordievsky, *KGB: The Inside Story*, Harper Collins Publishers, New York (1990).

[8] A.X. Artuzova, *Assotsiatsiya issledovatelej istorii spetssluzhb im. A.X. Artuzova*, **Rukovoditeli spetssluzhb SSSR i Rossii: Biograficheskiy slovar'**, Artuzovskaya Assotsiatsiya, Moscow (2001).

[9] A.X. Artuzova, Assotsiatsiya issledovateley istorii spetssluzhb im. A.X. Artuzova, **Shpionami ne rozhdayutsya: Istoricheskiy spravochnik**, Artuzovskaya Assotsiatsiya, Moscow (2001).

[10] AIN/SRR/ATIC-4-55, ATIC, Air Intelligence Information Report, Radio Direction Finder Equipment for Intercept Purposes, March 30, 1955, RG-457, Historic Cryptologic Collection, Box 1134, File: ACC 17916N, National Archives, College Park, Maryland.

[11] D. Ball, *Soviet Signals Intelligence*, **International Countermeasures Handbook**, 12th edition, EW Communications, Palo Alto, CA (1986).

[12] D. Ball, *Soviet Signals Intelligence (SIGINT)*, Strategic and Defence Studies Centre, Canberra (1989), pp. 3–15.

[13] D. Ball, *Signals Intelligence in the Post-Cold War Era*, Institute of Southeast Asian Studies, Singapore (1992).

[14] D. Ball, R. Windrem, *Soviet Signals Intelligence (SIGINT): Organization and management*, Intelligence and National Security (1989), 621–659.

[15] ST-21, Central Intelligence Agency, Soviet Intelligence: Organization and Functions of the Ministry of State Security (MGB), June 1948, RG-263, CIA Reference Collection, Document No. CIA-RDP78-02546R000100130001-3, National Archives, College Park, Maryland.

[16] CIA, Directorate of Intelligence, Foreign Intelligence and Security Services: USSR, undated but circa 1975. Copy in author's collection.

[17] Central Intelligence Agency, Foreign Radar Recognition Guide, September 1, 1959, RG-263, CIA

Reference Collection, Document No. CIA-RDP78-02646R000400160001-6, National Archives, College Park, Maryland.

[18] CIA, Memorandum, Soviet Presence in Cuba, December 7, 1962, Annex I, p. 1-2, CIA Electronic FOIA Reading Room, Document No. 0000878001, http://www.foia.cia.gov

[19] Letter, Dulles to Gray, October 24, 1958, enclosure: Comments on Baldwin Article, p. 1, CIA Electronic FOIA Reading Room, Document No. 0000483739, available at http://www.foia.cia.gov

[20] Memorandum, McClelland to Executive Director, Security of Voice Communications Circuitry Between the White House and Nassau, January 7, 1963, RG-263, CIA Reference Collection, Document No. CIA-RDP80B01676R003100240017-6, NA, CP.

[21] Memorandum, McCone to Members of the National Security Council, March 20, 1963, RG-263, CIA Reference Collection, Document No. CIA-RDP80B01676R003100240017-6, NA, CP.

[22] ER 76-4480, Memorandum, Graybeal to Deputy Director of Central Intelligence, Cost of the Soviet Intelligence Effort, December 7, 1976, RG-263, CIA Reference Collection, Document No. CIA-RDP79M00467A002400050002-9, NA, CP.

[23] CIA, Directorate of Intelligence, Soviet Intelligence: KGB and GRU, January 1984. A copy in author's collection.

[24] NIE 11-3/8-76, National Intelligence Estimate, Soviet Forces for Intercontinental Conflict Through the Mid-1980s, December 21, 1976, CIA Electronic FOIA Reading Room, Document No. 000028136, available at http://www.foia.cia.gov

[25] NI IIM 82-10006, Interagency Intelligence Memorandum, The Technology Acquisition Effort of the Soviet Intelligence Services, June 1982, CIA Electronic FOIA Reading Room, Document No. 0000261337, available at http://www.foia.cia.gov.

[26] NIE 11-6-84, National Intelligence Estimate, Soviet Global Military Reach, November 1984, CIA Electronic FOIA Reading Room, Document No. 0000278544, available at http://www.foia.cia.gov

[27] CIA, Selected Items from the National Intelligence Daily, November 13, 1985, p. 5, CIA FOIA Electronic Reading Room, available at http://www/foia.cia.gov.

[28] Interagency Intelligence Memorandum, Soviet Dependence on Space Systems, November 1975, p. 10, declassified and on file at the National Security Archives, Washington, DC.

[29] CSAW-RUN-C-5-49, Department of the Navy, US Naval Communications Supplementary Activity, Radio Intelligence in the Red Army (April 1943), May 20, 1949, RG-457, Historic Cryptologic Collection, Box 1437, File: ACC 45020 Radio Intelligence in Red Army 1935-1943, National Archives, College Park, Maryland.

[30] U.S. Defense Intelligence Agency Testimony to U.S. Senate Select Committee on Intelligence, Worldwide Threat to U.S. National Security, August 1996, available at http://www.securitymanagement.com/library/000255.html

[31] Improving Defense Concepts for the NATO Central Region, European-American Institute for Security Research, April (1978).

[32] US European Command, COMSEC Assessment During October 1973 Mid-East Conflict, December 11 (1973), p. A–2.

[33] N. Friedman, *Seapower and Space*, Naval Institute Press, Annapolis, MD (2000).

[34] Telegram, Embassy in the Soviet Union to the Department of State, April 29, 1964; Telegram, Embassy in the Soviet Union to Department of State, May 24, 1964; and USIB-D-9.7/2, Memorandum, Chairman of USIB Security Committee to United States Intelligence Board, Preliminary Damage Assessment of Technical Surveillance Penetration of the U.S. Embassy, Moscow, June 1, 1964, all in U.S. Department of State, *Foreign Relations of the United States 1964–1968: Vol. XIV*. Office of the Historian, Washington, DC (2001).

[35] A. Fursenko, T. Naftali, *One Hell of a Gamble: Khrushchev, Castro & Kennedy, 1958–1964*, W.W. Norton & Co., New York (1997).

[36] R.D. Glasser, *Signals intelligence and nuclear preemption*, Parameters, June (1989).

[37] T.R. Hammant, *Russian and soviet cryptology: some communications intelligence in Tsarist Russia*, Cryptologia, July (2000).

[38] R.J. Hanyok, Letter to the Editors by Robert J. Hanyok, *Soviet COMINT during the Cold War*, Cryptologia, XXIII (2) (1999), 167–168.

[39] U.S. House of Representatives, Committee on Appropriations, Department of Defense Appropriations for 1978, (1977), Part 3, p. 639.

[40] U.S. House of Representatives, Committee on Armed Services, Department of Defense Authorization for Appropriations for Fiscal Year 1981, 96th Congress, 2nd Session (1980), p. 1946.

[41] *Russia's History at Cam Ranh Bay*, Jane's Intelligence Review, December (2001).

[42] I. Jürjo, *Operations of Western intelligence services and Estonian refugees in post-war Estonia and the tactics of KGB counterintelligence*, **The Anti-Soviet Resistance in the Baltic States**, Du Ka, Vilnius (2000).

[43] D. Kahn, *Soviet COMINT in the Cold War*, Cryptologia, January (1998).

[44] Semichastniy to Khrushchev, Report, O rezul'tatakh raboty Komiteta gosbezopasnosti pri Sovete Ministrov SSSR i ego organov na mestakh za 1962 god, February 1, 1963, Dmitrii Antonovich Volkogonov Papers, Reel 18, KGB, September 1932–January 1988 (1 of 3), Manuscript Division, Library of Congress, Washington, DC (1963).

[45] Andropov to Brezhnev, Report, O rezul'tatakh raboty Komiteta gosbezopasnosti pri Sovete Ministrov SSSR i ego organov na mestakh za 1967 god, May 6, 1968, Dmitrii Antonovich Volkogonov Papers, Reel 18, KGB, September 1932–January 1988 (1 of 3), Manuscript Division, Library of Congress, Washington, DC (1968).

[46] Vyshaya Krasnoznamennaya Shkola, Komiteta Gosudastvennoj Bezopasnosti pri Sovete Ministrov SSSR imeni F.E. Dzerzhinskogo, Istoriya Sovetskikh Organov Gosudarstvennoj Bezopasnosti, KGB Higher Red Banner School, Moscow (1977). Copy in author's collection.

[47] A.I. Kokurin, N.V. Petrov (eds), *Lubyanka Organy: VChK-OGPU-NKVD-NKGB-MGB-MVD-KGB: 1917–1991 Spravochnik*. Mezhdunarodnyy Fond "Demokratiya", Moscow (2003).

[48] A.I. Kolpakidi, ***Entsiklopediya Voennoy Razvedki Rossii***, ACT – Astrel, Moscow (2004).

[49] M. Laar, *The armed resistance movement in Estonia from 1944 to 1956*, ***The Anti-Soviet Resistance in the Baltic States***, Du Ka, Vilnius (2000).

[50] G. Lardner Jr., *Unbeatable bugs: The Moscow embassy fiasco: KGB defector says he warned U.S. 10 years ago of built-in sensors*, Washington Post, June 18 (1990).

[51] Levi to The President, Memorandum, no subject, January 6 (1976), p. 1, Declassified Documents Retrieval Service.

[52] V.M. Lur'ye, B.Ya. Kochik, ***GRU: Dela i Lyudi***, Izdatel'stvo "Olma Press", Moscow (2002).

[53] W.S. Malone, W. Cran, *Code name catastrophe: how Moscow cracked our secret cipher systems*, Washington Post, January 22 (1989).

[54] http://www.memo.ru/memory/communarka/Chapt11.htm#_KMi_2909

[55] http://stalin.memo.ru/names/p103.htm

[56] http://www.memo.ru/history/NKVD/STRU/by_year.htm

[57] http://www.memo.ru/history/NKVD/STRU/54-60.htm

[58] http://www.memo.ru/history/NKVD/STRU/61-67.htm

[59] National Photographic Interpretation Center, Photographic Interpretation Report, Introduction to the Krug Installation Series, October 1965, RG-263, CIA Reference Collection, Document No. CIA-RDP78T04759A001900010018-8, National Archives, College Park, MD.

[60] CIA, NPIC, Imagery Analysis Division, TCS 9794/65, CIA/PIR 61018, Photographic Intelligence Report: Krug Sites, USSR, August 1965, RG-263, CIA Reference Collection, NA, CP.

[61] SC-01395/64, NPIC/R-61/64, *Photographic Interpretation Report: SIGINT and Selected Other Electronic Sites in Cuba*, January 1964, pp. 1–4, RG-263, CIA Reference Collection, Document No. CIA-RDP78B04560A002100010004-1, NA, CP.

[62] NSA/CSS, Communications Security – The Warsaw Pact COMINT Threat, January 1975, p. 2, NSA FOIA.

[63] http://www.nsa.gov/docs/cuba/images/19621023_1.pdf

[64] O. North, *The world according to Oliver North*, Washington Post, December 21 (1986), D1.

[65] *Soviet ELINT Capabilities*, ONI Review 15 (1) (1960), Operational Archives, Naval Historical Center, Washington, DC.

[66] Posbyashchaetsya 141 OrtBr OSNAZ!, available at http://brigada141.boom.ru

[67] N.V. Petrov, K.V. Skorkin, ***Kto Rukovodil NKVD: 1934–1941***, Zven'ya, Moscow (1999).

[68] D. Prokhorov, *The electronic ears of Moscow*, Sekretnyy Materialy, June 7 (2004).

[69] J.W. Rawles, *Soviet SIGINT platforms range from trawlers to consulates*, Defense Electronics, February (1988), 89.

[70] W. Rosenau, *A deafening silence: US policy and the sigint facility at Lourdes*, Intelligence and National Security, October (1994).

[71] H. Rositzke, ***The KGB***, Doubleday, New York (1981).

[72] F. Shalom, *Soviets, firefighters argue as consulate blaze spreads*, The Globe and Mail, January 15 (1987), A1.

[73] V. Sheymov, ***Tower of Secrets***, Naval Institute Press, Annapolis (1999).

[74] T.A. Soboleva, ***Taynopis' v istorii Rossii***, Mezhdunarodniye otnosheniya, Moscow (1994).

[75] T.A. Soboleva, ***Istoriya Shifroval'nogo Dela v Rossii***, Olma Press, Moscow (2002).

[76] A. Soldatov, *Posledniy sekret imperii*, Versiya, January 28 (2002), available at http://www.agentura.ru/dosie/countries/inobases/ramona/

[77] Evrei v sovetskoy razvedke (okonchaniye), July 5, (2002), available at http://www.sem40.ru/magendavid/015.shtml

[78] Soviet Information Bureau, *Caught in the act: Facts about U.S. espionage and subversion against the USSR*, Soviet Information Bureau, Moscow (1960).

[79] Memorandum, ARA/CCA – Robert B. Morley to ARA – Mr. Aronson, Soviet Assets in Cuba and Potential Trade-offs, May 29 (1990), p. 2, U.S. Department of State Electronic FOIA Reading Room, http://www.state.gov

[80] G.W. Weiss, *Duping the soviets: the farewell dossier*, Studies in Intelligence 39 (5) (1996), unclassified edition.

[81] N. West, ***Games of Intelligence***, Crown Publishers, New York (1990) p. 136.

[82] D. Wise, ***Nightmover***, Harper Collins Publishers, New York (1995).

[83] P. Wright, ***Spycatcher***, Viking, New York (1987).

[84] V.M. Zubok, *Spy vs. Spy: The KGB vs. the CIA, 1960–1962*, Cold War International History Project Bulletin 4 (1994), 22–33.

The History of Information Security: A Comprehensive Handbook
Karl de Leeuw and Jan Bergstra (Editors)

18

NATIONAL SECURITY AGENCY: THE HISTORIOGRAPHY OF CONCEALMENT

Joseph Fitsanakis

King College
Bristol, TN, USA

Contents

Abstract

The National Security Agency (NSA) is often described as America's most secretive intelligence institution, operating with a "near-pathological passion for security". Nevertheless, an overly intimidated academic community ought to share the blame for NSA's obscurity in modern historiography. For, despite its innumerable gaps, the source material currently available on the NSA allows us to provide a general outline of the Agency's development, enriched with more operational detail than ever before. This chapter aspires to fill an astonishing gap in the relevant literature by tracing NSA's history, from the end of the Korean War until the crisis days of September 11, 2001 and the invasion of Iraq. In doing so, it highlights the Agency's operational successes and failures and discusses its controversial role in some of the most significant global events in modern history.

Keywords: AFSA, Armed Forces Security Agency, Cipher Bureau, COINTELPRO, COMPUSEC, COMSEC, cryptanalysis, cryptography, cryptology, Cuban Missile Crisis, Data Encryption Standard, DES, ECHELON, ENIGMA, espionage, Gulf of Tonkin incident, HACKINT, Huston Project, INFOSEC, INFOSEC, intelligence, John Anthony Walker Jr., key escrow, MAGIC, MEGAHUT, MI8, MID, Military Information Division, National Computer Security Center, National Reconnaissance Office, National Security Agency, NCSC, NRO, NSA, Office of Naval Intelligence, ONI, operation ANADYR, operation DESERT SHIELD, operation DESOTO, operation IVY BELLS, operation KEYHOLE, operation MINARET, operation MONGOOSE, operation SHAMROCK, Pearl Harbor, project GUARDRAIL, project HOMERUN, PURPLE, RADINT, SCS, SIGINT, Signal Security Agency, Signals Intelligence Service, SIS, Special Collection Service, spies, spying, SSA, surveillance, telecommunications security, TELINT, UKUSA, 9-11, September 11, 2001, USS Liberty, USS Pueblo, VENONA, Walker spy ring, Washington Conference on the Limitation of Armaments, Watergate, weapons of mass destruction, wiretapping, WMD.

18.1 INTRODUCTION

It is said that no individual better personifies the institutional character of the National Security Agency (NSA) than its first Director, Lieutenant-General Ralph J. Canine (Fig. 18.1). So inclined was he toward operational covertness, that he never physically occupied the Director's office. In his five years at that post, few at NSA's Fort George G. Meade, Maryland, headquarters (see Fig. 18.6) knew what he looked like. Equally rare were his appearances before the US Congress – usually for the odd budget appropriations hearing. His few trusted colleagues say that his favorite response to Congressional inquisitiveness was "Congressman, you don't really want to know the answer to that. You wouldn't be able to sleep at night" [24: 585].

Canine's attachment to secrecy is still appreciated by an agency whose *Security Guidelines Handbook* describes concealment as "the most important individual responsibility" [11: 5]. Today, NSA's official Hall of Honor calls Canine "a figure of mythic stature" and grants him with securing NSA's survival, transforming it from a struggling bureaucracy to a "crucible for cryptologic centralization".

While this centralization has historically sheltered NSA from scholarly intrusion [291: 2], an overly intimidated research community ought to share the blame for the Agency's contemporary obscurity. Indeed, the scant literature on NSA is saturated with self-fulfilling prophecies of mystery and concealment, to the degree that intelligence observers are often restrained by their own writings. The recurring dictum that NSA is America's "most secret intelligence agency" [22: title page; 224: 15] is a central component of this mythology. In reality, there are numerous dark corners in America's intelligence leviathan that enjoy a seclusion far greater than NSA's – not least is the National Reconnaissance Office (NRO), which arguably deserves the crown of America's most concealed spying institution [108: 136; 208: 99].

NSA's operational remoteness is undisputed [247: ix], as is the Agency's "near-pathological passion for security" [61: 249] that has prevented broader understanding of the historical role of

Figure 18.1. The NSA's first Director Ralph J. Canine. Courtesy of the National Security Agency.

American signals intelligence (SIGINT) [2: 60]. Yet our ignorance also reflects a profound deficit in scholarly imagination [5: 2], for even when important SIGINT-related material is available, researchers are often slow to acknowledge it. Thus, information on the selling of US cryptologic secrets to the Japanese by Herbert O. Yardley – the prewar icon of American cryptology – has been in the National Archives since 1940, though it took almost 30 years for American historians to notice [132: 100]. Similarly, intelligence scholars familiar with the 1921 Washington Conference on the Limitation of Armaments, may think that a comprehensive analysis of the event is unattainable without considering the American insight into Japanese diplomatic messages. Yet this parameter is ignored in the relevant literature [72; 143], which perhaps aptly illustrates the description of secret intelligence by two historians as "the missing dimension of most diplomatic history" [7: 1].

Today, the knowledge base that survives both NSA's silence and the researchers' deafness is relatively minuscule. We probably know more about Mycenaean bureaucracy or 9th century Chinese warfare, than about NSA's institutional changes following '9-11'. Nevertheless, despite its innumerable gaps, the source material currently available allows us to provide a general outline of NSA's development, enriched with more operational detail than ever before.

18.2 NSA'S INSTITUTIONAL LINEAGE

NSA and its predecessors emerged in fulfillment of an historical intelligence requirement of governing structures – namely to maintain the privacy of their communications while undermining everyone else's. Its primary operational mission, which over the years has remained remarkably constant[1] [288: 40], is to safeguard the information and communications security of components of the US government (INFOSEC and COMSEC), while enabling the latter to challenge the communications security of alien entities through SIGINT. The development of this two-fold cryptologic assignment is guided by the requests of NSA's clients, which include the CIA, the Department of Defense, and the White House. It is also guided by technological developments in communications. These have, at various times, enriched the Agency's operational burden with such tasks as telemetry intelligence (TELINT), involving surveillance of spacecraft emissions, acoustical intelligence (ACOUSTINT), relating to underwater sonar tracking of vessels, and electronic intelligence (ELINT), including radar intelligence (RADINT) [51: 18; 208: 102, 103, 177].

Stripped of all its technological enhancements, NSA's basic operational goal has been a feature of American government ever since its revolutionary birth. There are indications of limited communications security awareness by US officials dating as far back as 1789, as evidenced by – among others – the Dumas and Lovell ciphers [267: 3, 22ff, 40]. Over the following century, the expansion of America's diplomatic and consular presence [144:

---

[1]Compared to that of other US intelligence agencies, particularly the metamorphosis-prone Central Intelligence Agency (CIA) [204: 209; 288: 151, 163ff; 291: 40].

19, 41] increased official cryptographic demand, leading to the utilization of variable cipher systems, and the establishment of the first regular government code room in 1867 [267: 151; 215; 237; 240; 87: 10].

NSA's direct institutional roots are often traced back to 1885, the year when the US Army's intelligence organization, the Military Information Division (MID), was established. Operating out of a single room in the War Department's Miscellaneous Branch, the MID possessed no precise SIGINT mission. Its task was to amass and disseminate throughout the military, foreign tactical and strategic information [240; 150: 116]. Even though institutional genealogy links the MID with NSA, the latter displays equal operational similarities with the Bureau of Navigation's Office of Naval Intelligence (ONI) – America's first formal, permanent intelligence institution, established in 1882 [68: 197, 204; 75: 12]. By 1914, when the MID had lost most of its funding and eclipsed as a distinct body within the War Department, ONI had managed to strengthen its Congressional funding, explore ways to embed secret agents in Japan, and pursue wireless cryptology to the extent of laying the foundations of the US Naval Radio Service [75: 101; 76: 39; 240: 3, 5].

ONI's bureaucratic achievements, however, were limited. It spent nearly twenty years desperately trying to overcome systematic under-funding and be acknowledged as an equal member of naval establishment. It achieved its occasional impact on decision-making primarily by hammering its data through the closed office doors of disinclined navy or government bureaucrats; even its first communications code was contracted from Europe [75: 26, 27, 70, 141, 142; 240: 3]. Such tribulations were indicative of a broader "state of cryptographic innocence" [6: 33] that marked successive US administrations. Thus, in April 1917, when the country declared war on Germany and the Central Powers, American military institutions were "as ill prepared for signals warfare as [they were] for trench warfare" [5: 7]. With its intelligence functions atrophied, the government possessed virtually no official cryptanalytic ability [5: 6, 7; 33: 99; 61: 42; 76: 7; 150: 147; 240: 8].

Attempts by alarmed US Army officials to overcome this immense cryptologic deficit were twofold. Five days after America's entry into war, they intensified their contacts with the Department of Ciphers at Riverbank Laboratories, a private research institute in Geneva, Illinois, where they appealed to the patriotism of its eccentric millionaire owner [22: 28; 61: 16–39, 43; 153: 17ff]. A month later, a directive was issued by the chief of staff for creating a Military Intelligence Section (MIS) tasked with performing espionage and counterintelligence in the US and abroad [33: 113]. The Section's head, major Ralph Van Deman, a perceptive administrator who had lobbied for the establishment of MIS since 1915 [150: 147, 148], utilized his newfound administrative status to launch a Cable and Telegraph Section. Also known as MI8,[2] the Section was in essence America's "first civilian codebreaking organization" [24: 20]. To lead MI8, Van Deman selected Herbert O. Yardley, a former cipher clerk in the State Department. By 1918, Yardley had overseen MI8's transformation from a total of three clerks to a fully functioning cryptologic unit of over 150 staff members. He had also coordinated MI8's compartmentalization into Code Compilation, Communications, Shorthand, Secret Ink, and Code and Cipher Solution subsections and, by war's end, his office was producing quasi-regular estimative intelligence bulletins [5: 8; 22: 7; 92: 46, 47; 150: 148; 220: 4]. Despite its somewhat rushed setup, MI8 managed to quickly obtain daily access to foreign and domestic cable traffic though deals with US cable companies. Having secured its supply channels, it eventually solved Central and South American diplomatic codes, and read a number of diplomatic messages from Germany [5: 9; 142: 353; 288: 224]. In nearly thirty months, MI8 deciphered almost 11,000 foreign government messages – a volume described as impressive or simply respectable, depending on the source [23: ix; 220: 53]. The Section's cryptanalytic work also aided the arrest by US authorities of German spy Lothar Witzke,[3] responsible for the 1916 sabotage explosion at New York's Black Tom

island [150: 145, 149; 241: 353, 354; 279: 12, 240, 245].

It is often asserted that between MI8 and NSA the "line of succession [. . .] is continuous" [157: 20]; see also [23: ix]. If this is so, then NSA ought to assume a hereditary stake to MI8's logistical and operational shortcomings, as well as achievements. MI8 bears at least partial responsibility for failing to instill COMSEC principles among US Army officers engaged in the field of battle – an oversight that cost numerous US lives in Europe [142: 331ff]. Additionally, the Section's obduracy in refusing to collaborate with elements of US Navy intelligence in the war effort was reckless by any standards of evaluation [6: 33; 75: 116; 150; 290: 137]. ONI, NSA's other distant ancestor, demonstrated inadequacies of similar magnitude. It is true that, throughout World War I, the organization was systematically hindered by lack of governmental funding, which – with the exception of some wireless frequency direction-finding accomplishments [76: 54; 220: 4] – restrained its SIGINT product far below its true potential [50: 18; 75: 138]. But even assuming unlimited funds were available, it is doubtful that ONI's politicization would have allowed it to fulfill its wartime mission. The organization's increasing obsession with internal security, and the racist prejudices of its executives, directed its espionage activities inward [76: 7]. Gradually, most of ONI's wartime activities concentrated on the surveillance of African Americans, as well as Irish, Mexican and Indian immigrants to the US, which were considered "easy prey for German agents" [240: 18, 71]. What is more, the organization's sinister partnership with the American Protective League, "World War I's most extensive and unprincipled" citizen vigilante group [76: 95], eventually unleashed a form of domestic counterintelligence hardly distinguishable from government-sponsored terrorism, which ruthlessly targeted American Jews and labor union members, among others [75: 119, 142; 76: 012; 150: 149ff; 158: 98; 203: 161; 240: 33ff].

These shortcomings were instrumental in MI8's and ONI's limited impact on the war effort. Essentially, there is currently no proof that their intelligence output affected the war in any way [150:

---

[2] MI8 was later renamed the Cipher Branch. Throughout its existence, its unofficial title was the Black Chamber.

[3] Alias Pablo Waberski.

149], or "had the slightest impact on American diplomacy during the war or during the negotiations at the [1919] Versailles Peace Conference [[4]]" [5: 9].

Following the war, internal security remained ONI's pivotal preoccupation – a seemingly spiraling surveillance project to which even military intelligence eventually lent a hand. The Army was instrumental in singling out labor activists during the notorious 1920 Palmer Raids, and spying on the activities of dozens of legal organizations, including the Industrial Workers of the World (IWW) and the American Federation of Labor (AFL) [72: 291; 240: 206, 216]. Conversely, ONI showed a preference for African American activism, Christian groups, such as the National Federation of Churches, suffragists, and the Communist Party, whose headquarters it subjected to an "orgy of vandalism" during a raid in 1929 [74: 291; 76: 38, 41, 45, 53]. In an ominous precursor to the Watergate affair of the 1970s, the Navy agency was also utilized in 1930 by US President Herbert C. Hoover to illegally burglarize offices of the Democratic Party [76: 3].

Evidently, such ultra-patriotic feats had little effect on US Secretary of State Henry L. Stimson, who, in 1929, decided to liquidate Yardley's organization on the ethical grounds that "[g]entlemen do nor read each other's mail" [5: 15; 22: 17; 61: 114; 129: 45; 234: 188]. Much has been written on Stimson's decision, particularly since his reasoning appeared to introduce a moral concept rarely observed in intelligence targeting. As he explained later,

> [I] adopted as [my] guide in foreign policy a principle [I] always tried to follow in personal relations – the principle that the way to make men trustworthy is to trust them [. . .]. In 1929 the world was striving with good will for lasting peace, and in this effort all the nations were parties [. . . A]s Secretary of State, [I] was dealing as a gentleman with the gentlemen sent as ambassadors and ministers from friendly nations [234: 188], see also [183: 639].

[4]This is not to imply that US cryptanalysts were encouraged to target the Conference. The participating US Secretary of State objected to a large SIGINT section descending on Paris [266: 10].

Stimson's outlook was neither as rare nor as isolated as it seems. State Department Undersecretary Joseph Cotton agreed with his superior, and President Hoover raised no objection to the Department's decision [127: 203]. Additionally, Stimson's moralistic attitude was not alien to Yardley's group. In 1918, a staunch Catholic State Department official had tried to shut down MI8 for breaking the official diplomatic code of the Vatican [87: 32n, 33n].[5] Furthermore, Stimson's decision should be considered within a broader context of developments. Firstly, since 1922, the Cipher Bureau – as MI8 was known after 1919 – had succumbed to a gradual operational and budgetary demise [23: x; 132: 106, 109]. Its peculiar cryptanalytic insistence for diplomatic, rather than military, codes had alienated the Army, which rarely utilized the Bureau's SIGINT product [5: 29]. Secondly, following the end of World War I, private cable companies refused to illegally supply the codebreakers with copies of international cables, causing the Bureau's intercept channels to run almost completely dry [5: 12, 13, 31; 6: 50; 22: 11, 12]. The organization's idleness was coupled with severe budgetary setbacks imposed by postwar realities, which forced a remarkable staff shrinkage from over 300 in 1918, to less than a dozen in 1924 [22: 15, 16; 142: 359; 150: 164]. Even prior to this fatal crisis, the Bureau's output had been at best unexceptional [5: 12; 31: 31, 32; 150: 148], with the exclusion of Latin America.[6] Its neglect for the code systems of major world powers was unjustified. Even in the case of Japan, the solution of its diplomatic codes, allowing the reading of Japanese messages from 1920 onward [87: 22, 30n], amounted to very little in practice. Specifically, the manner in which this secret knowledge

[5]Stimson's intervention was not the last time moral codes directly impacted US SIGINT policy. In 1933, the State Department aired moral objections to a plan for radio intelligence proposed by the War Department. A year later, US Army chief of staff General Malin Craig reduced the Army's codebreaking budget for ethical reasons [5: 28; 77: 9].

[6]The Bureau's concentration on Latin America has been erroneously described as of marginal importance [5: 11; 87: 55], though in reality it was justified in the wider political context of the 1823 Monroe doctrine [211: 397ff; 214: 5], and America's 66 armed interventions in Latin America from 1806 to 1917 [186: 37–78, 137–284; 257: 85–87].

was utilized by US diplomats, namely to humiliate Japan in the 1921 Washington Conference on the Limitation of Armaments [149: 249; 157: 21, 22], has been hailed as a success by some US intelligence historians [65: 11; 23: x; 129: 51]. In reality, Tokyo's acute embarrassment awakened a nationalist backlash that led to Japan's aggressive imperialist campaigns in Asia and, eventually, its military attachment to the Axis [1: 52; 72: 215, 216; 87: 26; 131: 156].

Last but not least, Stimson may have known that Yardley had systematically embezzled Cipher Bureau's budget funds, and had passed on substantial American cryptographic secrets to the Japanese [6: 51; 87: 56; 150: 165; 157: 33]. This latest parameter in the equation may be significant in more ways than one, particularly if NSA's reluctance until 1988 to publicize Yardley's disloyalty [132: 100] represents a systemic policy of concealment. Could it be that the Agency still considers other, more recent, cases of betrayal as too damaging to reveal?

Unbeknownst to Stimson, however, the Army did not share his sanguine view of global affairs. Instead of terminating its cryptological functions as instructed, it clandestinely transferred the Cipher Bureau's duties to a new cryptologic unit under the Signal Corps, named Signals Intelligence Service (SIS) [22: 026, 027; 150: 165]. Ironically, therefore, the genesis of NSA's direct forerunner occurred through an alarming disregard by the military establishment of directives by its civilian commanders – including the President himself [5: 28; 216: ff37]. Selected for heading the clandestine organization was former chief cryptanalyst to the War Department, William F. Friedman (Fig. 18.2), whose cryptologic contacts with the government dated back to his days at Riverbank Laboratories.

Friedman's place in the pantheon of American cryptology is well deserved, though not for the reasons to which many historians point. For instance, although Friedman is often accredited with solving the Japanese PURPLE cipher[7] [61: passim; 168: 120], the honor actually belongs to the cryptanalytic team led by his civilian assistant Frank B.

Figure 18.2. William F. Friedman. Courtesy of the National Security Agency.

Rowlett [100: 95; 157: 39; 121: 117; 202: 163]. Furthermore, while Friedman is usually identified as NSA's "intellectual founder" [71: 50], his tenure at SIS was not marked by any extraordinary institutional breakthroughs. For nearly two decades, the organization remained chronically under-funded, severely under-staffed, technologically emaciated [22: 31; 50: 19; 5: 44, 233; 87: 95], and isolated from most components of naval intelligence, with whom it was reportedly "not on speaking terms" [6: 52] for most of the 1930s [150: 166; 49: 56; 220: 6; 87: 33, 80ff; 92: 50]. Additionally, its operational disposition was essentially bureaucratic and too cumbersome to detect the rapidly changing military and diplomatic environment prior to the outbreak of World War II [5: 233]. Thus, it was only in 1938 that SIS decided to target Italian communications, while intercept coverage of German message traffic was not intensified until 1939 [5: 47, 61]. It is therefore not surprising that, even as late as 1936, contact between SIS and the White House was virtually non-existent – a term that also

---

[7]Even though PURPLE is often referred to as a "code" (see for instance [168: 120]), it was actually a machine-generated cipher [287: 86].

describes SIGINT's impact on US war diplomacy up to 1945 [5: 41, 229, 243].

What *was* a breakthrough during Friedman's leadership of the US Army's COMSEC and SIGINT functions was his apposite concentration on cryptological education, as opposed to operations [157: 39; 5: 15, 29]. This reversal in emphasis laid the practical foundation for the first-ever systematic production of cryptologic expertise in the US military [220: 6], as well as the theoretical backdrop to America's first cipher machine, the ECM Mark II/Sigaba M-134 [100: 95; 121: 112]. SIS was also more adept at securing intercept channels, both though illegal negotiations with the commercial cable sector [5: 9, 50; 220: 6], and by means of the US Army's expanding radio interception units [5: 34, 48; 22: 32; 61: 121].

SIS's methodical training was at the root of the unit's greatest cryptanalytic success, namely the solving of the Japanese Foreign Office's Alphabetical Typewriter'97, dubbed PURPLE by the Americans [141: 108]. Intelligence gained through PURPLE, along with contributions by ONI[8] – known collectively as MAGIC [87: 100] – constituted the bulk of America's SIGINT advantage during World War II [141: 109; 157: 057; 168: 205]. MAGIC, particularly its PURPLE component, denotes a remarkable cryptanalytic aptitude, which flourished despite SIS's striking technological backwardness [50: 20; 61: 138, 139]. Furthermore, MAGIC intelligence resulted in significant symbolic and tactical American victories, including the Battle of Coral Sea, the Battle of Midway, and the assassination of Japanese Admiral Isoroku Yamamoto [22: 43; 61: 188ff; 129: 91, 171; 131: 296–306; 141: 107ff; 142: 573; 150: 185ff; 156: 104–118; 168: 139; 165: 17–41; 202: 46; 220: 10; 202: 317ff, 459–463, 527; 22: 171ff, 291–343].[9] Ironically, these

military achievements have overshadowed PURPLE's most valuable contribution to US diplomacy, namely the elevation of America's bargaining status in its dealings with the cryptological superpower that was then Britain [5: 82]. In hindsight, it is apparent that the sharing of PURPLE's information with Britain allowed the US access to Britain's German-related intelligence – in the form of ENIGMA decrypts. More substantially, it also marked the American foundations of "the modern era of cryptanalytic collaboration" [111: 143] – culminating today in the shape of the UKUSA Security Agreement [208: 1; 49: 272; 68: 173].

Contemporary MAGIC accounts frequently overlook dissenting voices calling for re-examining the misleading interpretations of Japan's intentions prior to the attack on Pearl Harbor, to which MAGIC appears to have contributed. In his detailed study of MAGIC, Japanese historian Keiichiro Komatsu uncovered numerous critical distortions and mistranslations in MAGIC reports, which Komatsu charges with sabotaging "efforts made by the participants on both sides to achieve a successful outcome and avert the conflict" [149: 289, 247ff]. Articulate dissenters have been further marginalized in the historiography of the Pearl Harbor attack. Given the considerable extent of US intercept coverage of Japanese communications at the time [61: 169; 76: 172, 176; 77: 13; 92: vii; 150: 170], most scholars consider the attack a US intelligence *failure* blamed on either (a) lack of relevant information in decrypted diplomatic communications [101: 94; 142: 1–67; 156: 51, 52], or (b) structural delays in intercepting, decrypting, translating, or alerting the proper authorities of Japanese communications indicating a pending attack [22: 36; 61: 168, 169; 87: 333; 121: 122–128; 129: 51; 150: 172ff; 168: 120; 230: 2].[10] Few, however,

---

[8]Notably in breaking contemporary editions of the Japanese JN-series Imperial Navy code, which ONI had systematically targeted since 1920 [77: 13; 87: 37; 149: 249; 156: 55; 157: 19–48; 202: 76; 272: 5].

[9]MAGIC intelligence also divulged Japan's decision to join Germany in its attack on the USSR, and assisted in the destruction of a number of Japanese fleet units and components of the German U-boat fleet [22: 43; 68: 118, 126, 168; 129: 171; 202: 045, 527].

---

[10]A series of US intelligence catastrophes of equal magnitude to the Pearl Harbor attack – given the reaction time available to US forces – took place on the same day. Extremely damaging Japanese attacks on the Clark Field US military base in the Philippines, and subsequent attacks on Lunzon, Khota Baru, Guam, Hong Kong and Wake island, among other places, wiped out nearly half of General Douglas MacArthur's Far East Air Force, "including half his fleet of B-17s" [101: 267, 268]; see also [150: 176]. Yet, these incidents are almost completely ignored in the relevant intelligence literature.

have seriously examined traffic analysis output by US Navy radio surveillance, which, according to one of the most technically substantiated studies on the subject, "ensured foreknowledge of Japan's intentions and actions in the North Pacific. [Thus,] Japan's impending attack could not have been a complete surprise to every single member of the [US Navy]" [272: 116]; see also [132: 267; 235: passim]. Intelligence scholars have indeed been too slow in shifting their attention from structural deficiencies in prewar US SIGINT to "gross neglect or careful design as the explanation for Hawaii's lack of preparation" [272: 116] on December 7, 1941.

18.3 NSA'S GENESIS IN CONTEXT

NSA's genesis is commonly viewed as a "logical outcome" [61: 231] of the inevitable political backlash caused by the Pearl Harbor attack [142: 674]. Although this causal link is genuine, a more comprehensive conceptualization of the Agency's birth must embrace a contemporary series of equally momentous intelligence shortcomings. These include the failure to anticipate the detonation of the USSR's first atomic device in 1949, several years ahead of American estimates [128: 23, 24; 200: 53], grave misconceptions about the ideological nature and tactical planning of Mao Zedong's People's Liberation Army (PLA) in revolutionary China [3: 68; 174: 7ff; 239: 90, 96] and, perhaps most of all, the Korean War debacle.

In 1943, SIS was renamed to Signal Security Agency (SSA). Six years later, it changed its title to Armed Forces Security Agency (AFSA), and assumed the coordination of the strategic activities of all US SIGINT organizations [142: 675; 113: 106; 245: 28, 29]. By the morning of June 25, 1950, when 70,000 North Korean troops crossed into South Korea, AFSA was floating in operational limbo, struggling to assert its authority over aggressively competitive agencies with SIGINT components, including ONI, MID, Federal Bureau of Investigation (FBI), CIA, Air Force Security Service (AFSS), and the State Department [22: 51; 120; 208: 250; 288: 164]. The Agency did not have its own logistical facilities, and its token budget

did not allow for sufficient allocation of staff resources to the Korean peninsula until well after the breakout of hostilities [12; 120]. Even at that time, AFSA's North Korean department consisted of two part-time cryptanalysts and a single linguist working in the absence of a Korean language typewriter or even a dictionary [24: 25; 120]. It is thus not surprising that the Agency was startled by the North Korean invasion – as was the CIA[11] and President Harry S. Truman: his administration was notified of the attack by a United Press Agency news reporter, causing one observer to describe the intelligence fiasco as "more disgraceful than Pearl Harbor" [116: 166]; see also [22: 49; 150: 228ff; 200: 21].

Crucial files on relevant US SIGINT output during the Korean War are still classified, and are expected to remain so for the foreseeable future [213: ix]. NSA maintains that few of the hundreds of North Korean messages intercepted shortly before the war were read, none of which gave any advanced warning of an invasion [120]. The Agency's position is less clear on its failure to accurately detect plans of the People's Republic of China to engage in the war on the side of North Korea. It claims that AFSA repeatedly pointed to major movement of Chinese military divisions along Manchuria and toward Korea, and even identified the name of the PLA commander assigned to lead the Chinese intervention. Yet the Agency appears to suggest that, even if further information had been produced on China's determination to enter the war, it would have fallen on the deaf ears of Truman administration officials and General Douglas MacArthur, who was notorious for ignoring intelligence that contradicted his plans [77: 234]; see also [3: 111ff; 24: 29; 44: 108; 120; 156: 23; 213: 122; 200: 115–116; 237: 163, 164].

An even blurrier aspect of the SIGINT background to the Korean War involves the role of British informers for the Soviets Guy Francis de Moncy Burgess, Donald Duart Maclean and

---

[11]It is true that the CIA had actively speculated that Kim Il Sung had the capability and "was in a position to launch an invasion at any time of his choosing" [44: 40]; see also [200: 21]. But its reports had consistently discouraged the possibility that such an invasion might occur in 1950 [92: 71]. Eventually the CIA was "the last to find out about the invasion" [150: 228].

Harold Adrian Russell (Kim) Philby. It is known that, following the North Korean invasion, Maclean was promoted to head the British Foreign Office's American Department. At the same time, Philby and Burgess, in their positions as first and second secretary of Britain's Washington embassy, would have been privy to most high-grade US SIGINT, through the UKUSA SIGINT collaboration agreement formalized three years earlier [22: 315; 208: 5ff]. Through their Soviet contacts, all three would have been able to extensively advise the Chinese and North Koreans on relevant US intelligence estimates [44: 100ff; 61: 208; 100: 216; 194: 128; 197: 110; 213: 86, 407ff].[12]

All detail aside, the disastrous outcome of America's military participation in Korea was largely blamed on AFSA's underperformance [120; 213: x]. Yet the political momentum leading to AFSA's transformation into NSA was not related solely to the Korean debacle. The Agency's reorganization was part of a grander postwar modernization and redeployment of the American diplomatic, military and intelligence machinery, in response to the country's new global status as a superpower. NSA was essentially designed to provide SIGINT support to a host of new American and American-inspired policies and institutions, including the Department of Defense, the National Security Council (NSC), the Joint Chiefs of Staff (JCS), the North Atlantic Treaty Organization (NATO), the Truman Doctrine, the Marshall Plan, and the policy of communist containment [179: 1, 38; 201: 30ff; 217: 1; 288: 78]. NSA's institutional function, meant to consolidate federal SIGINT logistics and operations, was an essential element in what an analyst of the 1947 Hoover Commission[13] had termed "[a strategy] fully tuned to the intelligence needs of the atomic era" that was "imperative not only for

---

[12]After defecting to Moscow, Philby, who held an official pass to the National Security Agency [195: 245], admitted that his position in Washington was "even more advantageous than my earlier post as head of the counter-espionage section in London" [195: 254].

[13]The Commission on Organization of the Executive Branch of the Government, popularly known by the name of its chairman, former US President Herbert Hoover, laid the foundations of the intelligence reorganization that eventually generated NSA.

the proper discharge of [America's] obligations to its own people at home, but also for the fulfillment of its new responsibilities as the greatest power on earth" [109: 7, 279].

18.4   1952–1960: ATTAINING AN INSTITUTIONAL IDENTITY

Truman's secret executive directive establishing NSA and its charter was a product of his frustration with the lack of centralized SIGINT coordination. Perhaps because of this frustration, the document was extremely short and vague, and unaccompanied by relevant legislation – which can be interpreted as Truman's unwillingness to seriously engage in reorganizing communications intelligence [288: 192, 211]. None of this was much help to NSA, which, according to one of its former historians, "opened its doors in 1952 under siege conditions" [24: 355]. In its desperate efforts to acquire operational authority over the SIGINT components of other agencies, NSA received little encouragement from the executive.[14] Furthermore, in decreeing that the Agency should assist the Department of Defense "in certain circumstances" requiring "close [communications intelligence] support" [244: 7], the directive set no precise limits on the Pentagon's authority over NSA. The Agency has since had to learn to preserve its civilian identity by resisting constant attempts to militarize it [22: 3, 75–77; 24: 341; 113: 106; 208: 100, 249].

During its formative years, NSA's primary means of delineating its institutional identity were logistical in nature. It won its first big bureaucratic battle by becoming the first US cryptologic agency in history to retain its personnel numbers in a postwar period – approximately 9,000 in 1957 and over 10,000 in 1960 [22: 33; 137; 142: 677; 288: 192]. It achieved this by sustaining a high annual budget – reportedly $480 million in 1960 [142: 684] – particularly after an influential 1956 report by a White House commission[15] recom-

---

[14]The 1958 report by the President's Science Advisory Committee, entitled *Improving the Availability of Scientific and Technical Information in the United States*, commonly known as the Baker Report, was a rare exception.

[15]This was the report by the Commission on Organization of the Executive Branch of the Government, also known as the Second Hoover Commission.

Figure 18.4. Atlas computer. Courtesy of the National Security Agency.

Figure 18.3. Deputy Director Louis W. Tordella. Courtesy of the National Security Agency.

mended that "[m]onetary considerations [regarding NSA] should be waved" (cited in [24: 356]). The completion in 1957 of the Agency's Fort George G. Meade headquarters was yet another important element in this logistical equation. The $30 million headquarters also hosted NSA's brand new 24-hour SIGINT Command Center [24: 10; 277: 206].

The Agency's computer power also accelerated, spurred by two successive computer-friendly Directors, General Canine (1952–1956) and Lieutenant-General John A. Samford (1956–1960), as well as by the Agency's influential Deputy Director, Louis W. Tordella (1958–1974) [17: 14; 22: 87] (Fig. 18.3). NSA's Atlas (Fig. 18.4) and Atlas II machines, ordered by AFSA, served the Agency for a number of years, along with the Bogart general-purpose and the PACE-10 analog desktop computers, as well as the SOLO – a smaller, transistorized version of the Atlas II [22: 99; 24: 583; 41: 17, 110, 112, 113, 128]. Soon however,

the Agency assembled considerable contracting capability and, encouraged by the aforementioned Baker Report, implemented project LIGHTNING in 1957. At that time, LIGHTNING was history's largest and costliest governmental computer research program, bringing together a consortium of industry and academic computer centers tasked with enhancing NSA's computer potency. It eventually produced NSA's extremely powerful Stretch supercomputer, delivered to the Agency in 1962 [22: 10; 24: 585].

Conversely, little is known about NSA's fulfillment of its COMSEC mission during the 1950s. It is presumed that the Agency must have scrutinized its internal communications at an early stage, along with reexamining the security of governmental communications. This effort is evidenced by its technical contributions to COMSEC, which include the KL-7 rotor machine and the very successful KW-26 cryptographic device for teletypewriter communications, which remained in use until the early 1980s [17: 14, 16; 24: 101; 71: 28; 147: 11].

Perhaps most importantly, during the 1950s NSA accomplished a spectacular augmentation of its intercept capabilities. The postwar US occupation of the Federal Republic of Germany, Austria and Japan allowed for the establishment of American military bases on European and Asian soil with strong NSA presence [22: 159; 48: 62; 239: 78ff]. More SIGINT field sites rapidly mushroomed through formal agreements with a host of foreign governments. They included facilities in Behshahr and Kabkan (Iran), Habbaniyah (Iraq, "of immense value to the intelligence community" [154: 32, 33]), Karamürsel (Turkey), Asmara (Eritrea), Sidi Yahia (Morocco) and, probably the most valuable, the Peshawar listening station in Pakistan – described by one US official as "a box seat that we'd be fools to relinquish" [113: 107ff]; see also [48: 62]. Additionally, the aforementioned UKUSA SIGINT arrangement with the governments of Britain, Canada, Australia and New Zealand allowed the US "access to intelligence information it would not otherwise be able to acquire" [208: 6ff], and stretched its geographical presence from the south Pacific to the arctic [24: 142]. By 1956, NSA had more than doubled its field sites from 42 to 90 [17: 14].

The early 1950s also marked the beginning of the Agency's first known application of underwater acoustic surveillance, known as SOSUS (for SOund SUrveillance System). The scheme constituted a primary component of the US Navy's Integrated Undersea Surveillance System (IUSS), employed as a monitoring mechanism of Soviet submarine traffic. SOSUS consisted of an interconnected complex of bottom-mounted extra-sensitive hydrophones – codenamed CEASAR – able to detect acoustic radiation emitted by traveling submarines. SOSUS, which incorporated NSA's subcontracted JEZEBEL and MICHAEL projects, was not the only such undertaking by NSA, but rather the only one currently identified in relative detail [51: 179ff; 65: 24ff; 113: 118; 208: 200].

NSA's ambitious collection programs were almost totally directed on the USSR. This "myopic" [155: 130] intelligence policy bore the mark of America's broader Cold War strategy. What is more, it was implemented at the expense of the world's remaining intelligence targets, which the Agency characteristically bundled together in a single subsection named ALLO – ALL Others [22: 90, 212]. Yet, despite NSA's insistence, the USSR remained a denied area, wherein US intelligence collection efforts through conventional methods were consistently ineffective [19]. The US intelligence community's frustration over the Soviet problem gave rise to the U-2 air reconnaissance project, known as project HOMERUN

[24: 35, 36]. Unlike the AO-1 Mohawk, which the Agency acquired in 1959, the U-2 was not a dedicated SIGINT aircraft. It was a civilian project, directed by the CIA, initially without NSA's participation [31: 31, 121; 241: 107, 147, 148]. Following test flights, and a daring first espionage mission over Moscow, Leningrad and the Baltic coast on July 4, 1956 [31: 121; 51: 80], the U-2 was fitted with NSA's System V intercept equipment, which recorded a variety of ground and microwave communications. More importantly, by engaging in repeated unauthorized intrusions into Soviet airspace, the airplane used System V to detect Soviet radar emissions while electronically stimulating Soviet early warning systems [24: 44; 31: 158; 113: 110; 159: 27].

The U-2 has been hailed as "one of the greatest intelligence achievements in history" [31: 365]; see also [79: 67]. Indeed, combined with NSA's other SIGINT assets, the U-2's collection capabilities permitted effective intelligence monitoring of Soviet missile development and space programs[16] [24: 360; 31: 148; 48: 3, 61, 62, 148, 149; 51: 95; 150: 245; 276: 56, 57]. Furthermore, they significantly contributed to a decade of relatively accurate US estimates of the USSR's defensive capabilities [32: 5; 159: 27]. The then Deputy Director of Central Intelligence revealed that, upon assuming his post in 1959, U-2 missions were collecting nine-tenths of Washington's reliable intelligence on the USSR (cited in [31: 5]). Ultimately, U-2 operations significantly contributed to the increase of identified Soviet targets by the Pentagon, from a handful in 1956 to around 20,000 in 1960 [31: 156].

This success, however, was accompanied by considerable setbacks. The project's concealment from US government officials and Congress members [22: 138, 139; 31: 56] did not prevent the Soviets from gathering precise information on most aspects of the U-2 [31: 236]. Soviet radar systems detected every single U-2 espionage encroachment on USSR airspace [51: 95], and on Mayday 1960, the Soviets actually managed to shoot down the twentieth such mission over their territory. Their

---

[16]Including accurate prediction reports of the launching of *Sputnik 1*, history's first artificial satellite, launched by the USSR in 1957 [51: 95; 70: 99, 100; 79: 168; 119: 29].

eventual revelation that they had captured the airplane's CIA-employed pilot effectively disproved the US government's previous denials that the downed U-2 had been conducting espionage authorized by President Dwight D. Eisenhower. The latter's public confirmation of the Soviet charges, on May 11, did little to alleviate the embarrassment of his subordinates who had already lied on record about the U-2 operation [24: 58, 61; 31: 52, 58; 276: 82]. The U-2 episode was neither the first nor the last deliberate downing of a US aircraft during the Cold War. Ever since November 1951, when Soviet fighters brought down a US Navy P2V Neptune airplane over Siberia, dozens of US aircraft met the same fate, resulting in the demise of at least 64 NSA staff[17] [19; 31: 78, 159, 321; 113: 104; 142: 720; 159: 27]. But the U-2 revelations marked the first-ever disclosure of a major US espionage operation in peace time, and gave the American public a bitter taste of the conspiratorial mindset instilled by the Cold War on the nation's governing officials [31: 114, 247, 249; 276: 108]. It was "the first time many [Americans] learned that their leaders did not always tell them the truth" [31: xi].

Another window to the USSR was provided by operation BRIDE, inherited by NSA from AFSA, and later dubbed VENONA by the Agency [29: xxiin; 222: 96]. VENONA represented a rare success by American cryptanalysts in their attack on Soviet cryptosystems [24: 355, 358]. Bolshevik and Soviet communications had attracted the US intelligence community's interest since 1917 [133: 151, 152]. Some Russian messages were solved by MI8, and SIS's small Russian subsection managed to read a number of Red Army codes [5: 204, 209, 219, 269n]. In general, however, Soviet encrypted communications became totally impenetrable[18] by US attacks after 1927 [5: 195, 196; 157:

44]. A duplication mistake of additive pages made by Soviet cipher clerks in 1942 is usually accredited with enabling the US intelligence generated by VENONA [29: vii; 196: xv]. In reality, what facilitated VENONA was the apparently independent 1943 decision by SIS to form a Soviet cryptanalytic section [5: 228; 29: xiii; 100: 239]. Without this systematic effort to exploit previously intercepted diplomatic communications of the USSR – a US ally at the time – the Soviet duplication mistake of 1942 would probably have gone unnoticed.

Between 1945 and 1954, VENONA cryptanalysts deciphered, to various degrees, nearly 3,000 Soviet messages. It was a tiny fraction of the approximately two million that had been intercepted [222: 96], but enough to uncover a number of Soviet wartime espionage activities in the US and Britain – such as operation CORRIDOR and the compromising of America's MANHATTAN project – implicating over 400 identified individuals – most of them American citizens or Soviet diplomats [100: 237]. Dozens of others whose aliases are mentioned in VENONA decrypts remain unidentified, while some were identified but never charged by the US Justice Department for reasons that remain unknown [2: 76; 197: 38]. Nevertheless, the decrypts helped expose some of the most proliferate wartime Soviet spies in the West, such as David Greenglass, Harry Dexter White, Alger Hiss, Donald Duart Maclean, Emil Klaus Julius Fuchs, Theodore Alvin Hall, and Ethel and Julius Rosenberg, to name a few[19] [29: xxi, xxv, xxvi; 100: 237; 197: 260; 222: xxvi].

VENONA revealed the extent of America's penetration by Soviet intelligence and caused NSA to probe the political background of members of its staff, often to illegal degrees. Soon after its formation, the Agency's Office of Security Services initiated file compilations containing the political backgrounds and personality traits of 'suspicious' NSA employees. By 1974, when this practice was terminated, the Office's External Collection Program had utilized unwarranted electronic and physical surveillance against several present or former NSA workers [182: 76, 77]; see also [118: 176; 208: 103;

---

[17]Some observers dispute official US casualty disclosures, pointing to unofficial Soviet allegations that 730 American air personnel were captured, and probably many more killed, by the Soviet Union during the Cold War [37: 114].

[18]NSA's successful interception of Soviet internal Teletype communications in the 1950s does not fall into this category, as the intercepted messages were unencoded [222: 262]. Furthermore, claims that NSA, in association with Britain's Government Communications Headquarters (GCHQ), reconstructed a Soviet cipher machine dubbed ALBATROSS remain unsubstantiated [286: 148].

---

[19]Some, such as Schecter and Schecter [222: xxvi], would add J. Robert Oppenheimer to this list.

242: 196]. Yet despite the Agency's precautions, Soviet and other cryptological agencies repeatedly penetrated its operational security. The Soviets had been warned about VENONA's progress as early as 1944 by White House economist Lauchlin Currie [29: xiv]. Four years later, AFSA linguist and Soviet agent William W. Weisband probably confirmed VENONA's success to the Soviets, prompting them to change all Warsaw Pact encryption systems on October 29, 1948 – NSA's 'Black Friday', a term denoting what two of its historians called "the most significant intelligence loss in US history" [120]; see also [24: 23, 24, 355; 29: xxvii; 196: xv]. Other Soviet moles soon inflicted further injuries on NSA's mission. US Army sergeant Roy A. Rhodes and officer Robert Lee Johnson both supplied the Soviets with detailed cryptographic information between 1952 and 1965 [142: 685; 271: 170–174]. In 1954, NSA cryptographer Joseph S. Petersen was arrested and convicted for delivering similar information to Dutch cryptologists. Nine years later, NSA clerk-messenger Jack Edward Dunlap committed suicide after Agency investigators closed in on him for passing NSA secrets to the Komitet Gosudarstvennoy Bezopasnosti (KGB) for nearly six years[20] [22: 131, 150–152; 82: 14; 83: 197ff; 84: 73ff; 142: 690, 691, 696; 197: 20, 85; 208: 261; 245: 214, 218; 277: 207ff; 278: 209]. It was probably with Dunlap's assistance that Soviet SIGINT intercepted and deciphered US presidential messages in the late 1950s [245: 78; 277: 208]. By November 1960, when two NSA analysts, William H. Martin and Bernon F. Mitchell – both clandestine Communist Party USA (CPUSA) members – defected to the USSR [22: 141ff; 24: 92, 093, 543; 142: 693; 245: 211], it was evident that known instances of Soviet penetration of NSA operations were simply the tip of the iceberg [250: passim].

There is little information in the public record about NSA's assessment of two defining events of the 1950s: the Suez crisis and the Cuban Revolution. The joint attack by Britain, France and Israel on Egypt, following the latter's nationalization of the Suez Canal, was the Agency's first encounter of a major military crisis [24: 39]. Scholarship is divided in its assessment of NSA's performance. Most researchers assert that the crisis occurred too soon for the new Agency to respond effectively. Its severe lack of Middle Eastern linguists, and its inability to decrypt radio communications between Britain and France, is said to have significantly hindered its intelligence capability. It follows that this must have contributed to the failure of America's intelligence apparatus to enlighten the government's policy maneuvers [24: 39, 40; 95: 184; 146: 81, 84; 166: 258]. But others point to informal British allegations that British, French and Israeli communications were readable by NSA's cryptanalysts [142: 729; 208: 261]. Hence, although NSA's success in decrypting these encoded exchanges was far from absolute, the Agency was probably privy to the intentions of its allies [61: 4, 238; 79: 168]. If these contentions are accurate, then NSA's intelligence must have been immeasurably useful to US diplomats engaged in the Anglo-American mediation project ALPHA, and the Agency deserves at least partial recognition for America's emergence as the Middle East's major Western powerbroker following the crisis [95: 186; 166: 34].

In 1958, two years after the Suez crisis, NSA was instructed by US Defense Secretary Neil H. McElroy to "establish a SIGINT collection capability in [. . .] Cuba [. . .] as a matter of the highest intelligence priority" (cited in [24: 99]). Additionally, it should be presumed that NSA enjoyed near-complete access to Cuban telephonic communications until Compañía Cubana de Teléfonos was nationalized, in late 1959.[21] Yet President Eisenhower is reported to have "heard no alarm bells

---

[20]NSA should be faulted for neglecting to protect information transported through their courier system. Around the time Dunlap was copying messages contained in his message pouch, the KGB courier system consisted of "a metal container not much larger than a shoe box [secured by] a minute pyrotechnic charge that would explode if the box was improperly opened or even suddenly shaken" [36: 114].

[21]Compañía Cubana de Teléfonos (CCT) was owned by US-based International Telephone and Telegraph (ITT) and managed by a US Naval Academy graduate [193: 44, 45]. Considering ITT's institutional overlap with American national security goals at the time [221: 41–56, 183ff, 269], access by US intelligence to CCT's networks was almost certainly extensive, if not total.

from his advisers" [193: 92] signaling the impending overthrow of Cuban dictator Fulgencio Batista y Zaldívar. It is unknown whether, on the day before Batista's escape from Havana, NSA was in agreement with CIA's assessment that Fidel Castro Ruz's guerillas had failed to amass broad popular support [193: 246]. Nor is there any knowledge of assistance by NSA to the CIA's sabotage and subversion on the island. It is known that CIA's project CUBA included interception operations to which NSA staff could have contributed, such as the aborted 1960 attempts to bug the Soviet consulate in Miramar and the Xinhua News Agency offices in Vedado [85: 52, 53; 125: 13]. Finally, the possibility should be entertained that, after 1960, agents of Cuba's Dirección General de Inteligencia may have infiltrated NSA, as they did the CIA [190: 57].

18.5 1961–1973: ACCELERATION AND CONTINUITY

In the thirteen years following 1960, five different generals and vice-admirals occupied NSA's head office.[22] Remarkably, their rapid replacements do not appear to have restrained NSA's organizational continuity. During their tenure, the Agency managed to enhance its research and instructive capabilities [22: 114, 344], and to further consolidate its SIGINT authority over US armed forces, through the establishment of its sister agency, the Central Security Service [22: 155, 156; 260: 1]. This consolidation came with unprecedented budgetary and staff increases: by 1969, NSA's expanded headquarters administered nearly 100,000 employees scattered throughout the globe. The Agency was funded by an estimated two billion dollars annually, making it the largest, costliest American intelligence agency [22: 60; 24: 339; 113: 105; 142: 677; 245: 77]. Budgetary acceleration helped expand the Agency's computer power: HARVEST, at the time the "world's largest computing system" according to Tordella (cited in [22: 87]), remained

NSA's electronic backbone from 1962 until the mid-1970s [17: 20]. Shortly prior to HARVEST's decommission, NSA began systematic utilization of computer-assisted communications interception, based on the EXPLORER and GUARDRAIL computerized systems [24: 343]. The latter helped upgrade the Agency's airborne reconnaissance program – particularly the BOXTOP component of the Peacetime Airborne Reconnaissance Program (PARPRO) – incorporating U-2, RC-135/U/V, and SR-71 manned airplanes [24: 246, 320ff; 51: 162ff, 320ff; 150: 272; 228: 1; 245: 139]. These airplanes were largely responsible for NSA's accurate intelligence on the Chinese nuclear program of the 1960s [31: 392; 113: 114].

The Agency's underwater surveillance project, initiated in the 1950s, persisted [24: 104; 245: 130]. In 1961 it was coupled with an equally ambitious naval SIGINT undertaking in – not always smooth [162: 27; 155: 8; 24: 255, 316] – cooperation with the US Navy. The scheme, which emulated the Soviets' modification of old naval vessels into floating SIGINT stations,[23] aimed to increase intercept coverage of primarily non-Soviet targets around the world. The collection success of the project's first two ships, the USS *Private Jose F. Valdez* and the USS *Joseph E. Muller*, both commanded solely by NSA, prompted the construction of second-generation SIGINT vessels. As a result, by 1970 NSA co-directed an extensive SIGINT fleet [22: 215–233; 24: 185ff; 81: 8; 155: 8]. NSA's dry land listening bases also increased, both in terms of facilities and staff. In 1965, more than 1,000 NSA employees were stationed on Japan's Kamiseya naval support base, with thousands more elsewhere around the world [22: 205; 24: 147, 156; 113: 125; 134: 52]. Despite the critical loss of its best-performing station in Pakistan's Peshawar, in 1968 [113: 107, 108], the Agency still collected SIGINT from over 2,500 intercept positions worldwide – aircraft and vessels included [113: 107; 142: 719].

---

[22]They were Lieutenant-Generals Gordon A. Blake (1962–1965), Marshall S. Carter (1965–1969) and Samuel C. Phillips (1972–1973), and Vice-Admirals Laurence H. Frost (1960–1962) and Noel A. M. Gaylor (1969–1972).

[23]Ever since the mid-1950s, most of Soviet SIGINT's continental US collection was facilitated through ships named Naval Auxiliary Intelligence Collectors [155: 8], which included such vessels as the *Barometr*, *Arban*, *Gidrolog* and *Teodolit*.

By 1973, the performance of these stations was significantly aided by NSA's increasing utilization of satellite surveillance systems. NSA had maintained a stake in the US satellite program as early as 1954, with the WS-117L – a project by the US Air Force reconnaissance [73: 11]. In the summer of 1960, a US Navy orbital launch vehicle carried Solrad 1, the first satellite incorporating an NSA ELINT component.[24] Solrad 3, a second Navy satellite carrying a SIGINT apparatus, was launched from Cape Canaveral a year later. Discerning the specifics of NSA's satellite reconnaissance program is a challenging task, particularly since it secretly overlaps with the satellite ventures of other organizations, including the CIA, the National Aeronautics and Space Administration (NASA), and branches of the US military. But the Agency's satellite expansion should not necessarily be viewed as a natural outgrowth of its mission. NSA battled against image-oriented reconnaissance agencies – including the CIA and Air Force elements [51: 222, 223] – to engrain its SIGINT requirements in the satellites' technical features. Eventually, NSA's close collaboration with the NRO, founded to consolidate the construction and management of all relevant US satellite ventures, was achieved in opposition to the CIA's wishes. It was out of these bureaucratic quarrels that NSA's participation in some of the most prolific spy satellite programs, including POPPY, RHYOLITE/AQUACADE, BYEMAN, CANYON, JUMPSEAT/TRUMPET and KEY-HOLE, emerged [22: 193–202; 24: 369; 51: 190ff; 54: 5–8; 102: 61–62; 160: 61–62; 208: 177ff].

NSA's investment in space reconnaissance advanced its SIGINT collection and assisted the Agency in substantiating its dominance in the UKUSA Security Agreement [7: 10; 208: 7, 8]. Although the Agreement was "a direct extension" [208: 4, 135] of Anglo-Saxon intelligence cooperation during World War II, its functional features were standardized in the years following its March 1946 formalization [168: 278]. Through its possession of advanced computerized satellite platforms during the UKUSA's formative years, NSA

overturned Britain's World War II SIGINT dominance and fashioned the treaty after its own intelligence requirements [51: 189, 190]. The 1971 standardization of UKUSA's technical collection features, codenamed ECHELON,[25] is an example of NSA's institutional dominance. ECHELON is an automated interception and relay system of global proportions, comprising a complex network of wireline and wireless interception facilities, consolidating the collection capabilities of the US, Britain, Canada, Australia and New Zealand. Through ECHELON, these countries share their SIGINT capacities as exercised by each other in their corresponding geographic sectors of responsibility [86: 48; 208: 5]. Although no party to the UKUSA Agreement has ever confirmed ECHELON's existence, its function has been abundantly established in the literature since 1996, when New Zealand's role in the system was exposed [86: 67–70; 117: passim]. Qualified testimonies in US and European parliamentary and governmental reports have ascertained the SIGINT status and technical features of ECHELON to the extent that they are "no longer in doubt" [131: 133], see also [86: 55–59, 71–76].

Throughout the Cold War, SIGINT satellites were directed primarily, though not entirely, against the USSR and its political allies. Although they assisted NSA's TELINT and RADINT missions [24: 153ff; 245: 18], they did not augment the Agency's record of solving high-grade Soviet codes and ciphers [24: 110]. This resulted in NSA's failure to foresee a number of Soviet political and military decisions, including the 1968 offensive in Czechoslovakia [24: 153, 475]. Behind NSA's sterility lay the USSR's unparalleled supply of intelligence staff resources, allowing the widespread utilization of one-time 'gamma' encryption pads [100: 95; 242: 71, 72]. To overcome this difficulty, NSA was forced to employ creative means of collecting useful unencrypted or pre-encrypted Soviet communications. This effort sparked a series of relatively successful projects such as GAMMA

---

[24]This marked a considerable advantage over the USSR, whose first known SIGINT satellite, COSMOS 148, was not successfully launched until March 1967 [51: 267].

[25]Considering NSA's practice of changing formal control terms once they are publicly identified, it is unlikely that the ECHELON system is still known by that name within the Agency.

GUPY, wherein US Army personnel stationed at the US embassy in Moscow acquired and easily unscrambled telephonic communications of Soviet government officials made from their Chaika executive limousines [22: 283; 155: 8]. Other noteworthy projects of the period include HOLYSTONE, BARNACLE and, particularly, IVY BELLS – the softwired interception of Soviet underwater military cables – and a number of tempest surveillance projects, whereby NSA SIGINT vessels approached known communist bloc navy facilities to record radiation emissions from Soviet cipher machines [24: 110, 370ff; 51: 139; 283: 449, 450].

Other unorthodox means of acquiring cryptological intelligence involved bribing foreign embassy employees [142: 686] or physically breaking into foreign diplomatic facilities in search of cryptographic manuals or to bug cryptographic appliances. In the late 1960s and early 1970s, under SCOPE, a joint NSA/FBI project, a number of embassies located in Washington, DC, were clandestinely entered, including those of the Republic of Vietnam, Israel and Chile [274: 153, 175, 175n, 294]. Reportedly, members of the group arrested in 1972 in the foiled burglary at the Watergate complex had earlier been employed under SCOPE [274: 178; 269]. Arguably the most ambitious known such operation was MONOPOLY, another joint NSA/FBI plan to compromise the communications security of the Soviet embassy in Washington, DC, while the latter was under construction. A spacious tunnel was clandestinely dug under the embassy grounds, and equipped and staffed by NSA at the cost of "several hundred million dollars" [212]; see also [275: 98, 104, 105].

MONOPOLY was among a number of NSA operations eventually compromised by Soviet-funded agents who continued to penetrate the Agency despite its general reorganization of security practices in 1963. In July of that year, Victor Norris Hamilton, a former cryptanalyst at NSA, defected to the USSR and publicly denounced NSA practices in a letter to the newspaper of the Supreme Soviet, *Izvestia* [142: 696; 277: 208; 245: 211; 208: 261]. Further damage to the Agency was caused by Robert S. Lipka, an Agency analyst who, between 1965 and 1967, supplied the Soviets with microfilms of photographed NSA documents [145: 341,

343]. But NSA's most destructive mole was not an Agency employee, but US Navy Chief Petty Officer John Anthony Walker, Jr., who began spying for the Soviets in 1967. From 1970, through his position as a Registered Publications System custodian, he had unlimited access to nearly 100 percent of NSA's cryptographic code systems and manuals designed for the US armed forces [36: 140; 283: 479]. By 1984, when he was arrested by the FBI, Walker had recruited his brother Arthur, his son Michael Lance, and his best friend Jerry Alfred Whitworth – all employed in various posts by the US Navy [27: passim]. In seventeen years, the four members of the Walker spy ring provided Soviet intelligence with keylist and circuit diagrams and frequencies for numerous NSA cipher systems including the KW-7, KY-8, KW-47, KWR-37, KG-26 and the KL-47, which NSA had considered unbreakable [71: 31, 116, 124, 153; 224: 35; 24: 245, 277]. The compromised information allowed the KGB to decrypt at least one million US messages [36: 412; 155: 86], prompting the Soviet intelligence agency to consider the Walker case as its paramount espionage achievement of the Cold War, and probably the greatest and longest security breach in the annals of spying [24: 277; 36: 412; 155: 86; 197: 77].

NSA's cryptographic endeavors were further compromised in 1968 by the unprecedented capture of the USS *Pueblo* (Fig. 18.5) SIGINT vessel by the Democratic People's Republic of Korea (DPRK). The *Pueblo* was not the first NSA ship to be targeted by a foreign nation: in 1967, Israeli naval and air forces engaged in the Six-Day War had attacked the USS *Liberty* off the Gaza coast, allegedly mistaking it for an Egyptian freighter. The US government's weak diplomatic response, despite the high American casualty rate of 34 dead and over 170 wounded, was almost certainly not agreed to by NSA,[26] and did not help dispel accusations that Israeli explanations for the attack were all too readily accepted on the political level [24:

_____

[26] An outraged Louis Tordella, NSA's Deputy Director at the time, authored a memorandum detailing a US government plan to sink the bombed, yet still floating, *Liberty* so as not "to inflame [US] public opinion against the Israelis" (cited in [24: 223]).

Figure 18.5. The USS Pueblo SIGINT vessel. Courtesy of the National Security Agency.

228, 231; 67: 31; 81: 151ff].[27] Despite the popular awareness of the USS *Liberty* incident, flamed by the surrounding diplomatic controversy, its outcome caused little cryptological damage to NSA.

The contrary can be said about the seizure of the USS *Pueblo*. The ship, bearing NSA's "euphemistic designation" [24: 94] of Auxiliary General Research (AGER) vessel number two, was executing a joint US Navy/NSA SIGINT mission

against North Korean and Soviet targets, codenamed PINKROOT ONE [63; 155: 69, 70]. On January 23, while sailing 25 miles off Wonsan, DPRK, the vessel was chased by four DPRK patrol boats and eventually boarded by armed troops. After becoming the first US Navy vessel since 1807 to be captured without resisting, and the first American vessel to be seized in international waters in over 150 years, the *Pueblo* was towed to Wonsan, where it remains today [24: 281; 155: 82; 163: 2]. The ship's surviving crew was released a year later, after the US government confessed to spying on the DPRK and issued an apology [245: 127; 162: 3]. It is true that, in pursuit of its SIGINT mission which required geographic proximity to its target area, the *Pueblo* had been instructed to violate DPRK ter-

---

[27]Widespread speculation about the reasons behind the Israeli attack on the *Liberty* – namely that the NSA ship was targeting Israeli and French communications and potentially sharing its findings with the Egyptian government [88: 44; 150: 270] – is destined to remain conjectural unless the US government decides to declassify an accurate description of the "mission and assigned tasks of the *Liberty* in the eastern Mediterranean on June 8, 1967" [67: 29].

ritorial waters in the days leading up to its capture [155: 91]. Furthermore, citing DPRK's numerous official protests to the increasing presence of spy ships in its waters [155: 61; 162: 8, 63; 228: 7; 245: 139], the Agency had advised the US Navy and JCS to reconsider the aggressive character of PINKROOT ONE. But no agency briefed the *Pueblo* on the North Korean warnings [150: 271; 155: 62ff; 162: 61; 228: 4, 5].

Ultimately, it was NSA who suffered the operational cost of *Pueblo's* capture. The ship's crew was immobilized before completing the destruction of cryptological documents and equipment onboard. Consequently, only ten percent of the 600 pounds of documents onboard were eliminated. The remaining volume, confiscated by DPRK officials, included cipher machine operation manuals, NSA keylists and, equally importantly, intelligence collection requirement sheets, spelling out in detail NSA's SIGINT targets in North Korea and beyond [24: 263–266, 275; 155: 84; 162: 157]. Additionally, the North Koreans acquired intact at least five NSA cryptographic devices and radio interceptors – the KW-7, R390A, WLR-1, KG-14, KW-37 and KL-47 – as well as numerous TELINT and RADINT receivers, teletypes and SIGINT recorders [24: 276, 277; 36: 90; 155: 83ff; 162: 29]. The damage to NSA's secrets was so extensive that a number of knowledgeable US intelligence officials expressed the desire "to destroy the ship before it reached Wonsan, even if it meant killing American crew members" [155: 84]. Most of all, the Agency dreaded the certainty that the North Koreans would eventually share the confiscated material with the Soviets – which would be the first time the latter would have the opportunity to inspect NSA's cryptological handiwork in such detail [36: 90]. To lessen the damage, NSA altered the keylists and a number of vital technical specifications to the compromised devices, which allowed their continued use. However, what the Agency failed to anticipate was the role of Soviet moles Walker and Whitworth, who soon supplied the updated NSA keylists and technical alterations to the KGB [24: 276, 277; 36: 90].

Among NSA's operation arenas impacted by the *Pueblo* compromise was Vietnam. Historically, the Agency's activity in Indochina is almost as old as America's postwar military interest in the area [4: 43–67; 227: 7]. NSA units were operative against the Viet Minh as early as 1952, before the termination of French colonial rule [30: 20; 24: 287]. After initially conducting SIGINT collection from listening posts in neighboring countries, a small US SIGINT presence was established in the Republic of Vietnam (South Vietnam) in April 1961, a few days after US President John F. Kennedy authorized the move [24: 288–290; 30: 21, 388]. In the same year, a US Army SIGINT officer became America's first military fatality in the country [24: 290–291; 110: 10].

NSA's role was also central in the incident that sparked America's involvement in the Vietnam War: the US government accused the Democratic Republic of Vietnam (North Vietnam) of attacking an NSA SIGINT ship, the USS *Maddox*, and an accompanying destroyer, the *USS Turner*, on international waters on August 2 and August 4, 1964 [112: 34]. On August 5, US President Lyndon B. Johnson submitted a request to Congress for a joint resolution authorizing the government to repel attacks against US forces in Vietnam. The resolution, later cited by the Johnson administration as explicit authorization for the Vietnam War, was approved by both houses of Congress, with only two Senators voting against it [4: 183]. However, with NSA's implicit cooperation, the Johnson administration failed to inform Congress that the USS *Maddox* had been deployed in the Gulf of Tonkin as part of operation DESOTO – a joint US Navy/NSA program of coastal intelligence collection against North Vietnam, China and DPRK [64: 118; 112: 34; 227: 15]. Furthermore, during the time of the attack, *Maddox's* DESOTO mission was to provide support to Operations Plan 34A (OPLAN34A), a paramilitary sabotage engagement against North Vietnamese coastal targets, authorized in 1963 by a joint CIA/JCS/US Navy task force by the name of Special Operations Group (SOG)[28] [64: 91, 92, 118–121; 113: 96, 125; 180: 5, 50; 227: 15]. The first reported attack on the vessel was in fact a response to a

---

[28] After July 1964, the Special Operations Group was known as the Studies and Observations Group [64: 296n9; 180: 6].

direct assault by SOG paramilitary forces against North Vietnamese coastal radar facilities the night before [227: 13; 93: 64n; 155: 118]. The second reported attack on the *Maddox*, which, according to the US government, caused no damage to the ship, never actually occurred. As the then US Secretary of Defense Robert S. McNamara, and NSA Deputy Director Tordella later admitted, the US allegation was based on a series of NSA misinterpretations of intercepted North Vietnamese messages [24: 299; 155: 17; 187: 34].

Following the beginning of large-scale American military involvement in Vietnam, NSA's overall intelligence performance can be characterized as ineffectual. As its predecessor agencies had done in World War I, NSA underestimated the cryptanalytical resourcefulness of the enemy, and failed to instill adequate COMSEC principles among US troops engaged in the field of battle [30: 396, 402, 403, 405, 406; 93: 11; 110: 44, 46, 74; 177: 132ff]. What is more, the "virtually nonexistent" [177: 21] voice radio COMSEC was not seriously combated by the Agency until 1969 when its PRC-77, PRC-25 and NESTOR security devices were distributed in sufficient numbers to US ground forces [30: 407; 110: 126; 210: 36, 40, 92, 93, 122, 123]. NSA's reluctance was apparently caused by high research and development costs, and the fear that cryptographic devices introduced in the field of battle would inevitably be seized by communist troops [30: 398–400].

Despite the Agency's immense technological SIGINT superiority over the North Vietnamese [22: 104, 165; 24: 227, 330; 30: 111, 304, 391; 150: 255; 177: 116; 210: 3, 18; 282: 065], its operational successes – such as operation STARLIGHT, the EB-66 tactical electronic reconnaissance/warfare aircraft, and project TEABALL – were limited [110: 35; 112: 67, 88; 262: 12ff]. NSA's concentration on Soviet linguistics had impeded investment in other languages, including Vietnamese, Chinese, Khmer and Cham. This critical shortage actually worsened as the war progressed [24: 302, 303; 176: 27]. Coupled with the near-impenetrable security of North Vietnamese codes and ciphers [162: 220, 221; 177: 116], NSA's linguistic blindfold seriously disabled the Agency's intelligence functions, to the extent that US military planners often

avoided consulting it prior to military operations [99: 39; 138: 116]. To make matters worse, the cryptographic compromise of the *Pueblo* seizure, coupled with the information supplied to the Soviets by the Walker spy ring, and the losses of US cryptographic material to the North Vietnamese forces during the war, significantly aided the North Vietnamese Army and the Viet Cong [27: 210, 211; 30: 202; 36: 140; 155: 86; 177: 98–100; 245: 73].

Throughout the Vietnam War, the absence of any US diplomatic or other human intelligence (HUMINT) in the Democratic Republic of Vietnam [4: 190], propelled NSA into becoming, in the Agency's own words, the most critical "single element in the United States government [. . .] in national decisions [. . .]" [NSA report cited in 24: 292]. Yet the inevitable over-reliance on SIGINT significantly hampered US intelligence capabilities, notably in the 1968 Tet Offensive, a 'shock-and-awe' undertaking by the Viet Cong credited with signifying the beginning of the end of US involvement in the war [122: 188; 138: ix, 279n; 177: 97; 192: 167; 273: 274]. Although SIGINT traffic analysis allowed NSA to predict imminent attacks against South Vietnam, the absence of radio traffic by Viet Cong rebels operating within South Vietnam caused the Agency's warnings to mistakenly identify rural border, instead of interior urban, areas as the target of the offensive [4: 256, 257; 93: 128; 138: 144; 273: 1, 202, 203, 214, 215, 273, 257].

The Agency proved somewhat more useful to American strategic interests closer to home, particularly in the case of Cuba. Much has been made of John F. Kennedy's 1961 decision to intensify intelligence monitoring of the island, which involved an escalation of NSA activities [24: 75, 76; 22: 249; 125: 118; 137]. In reality, far from conveying hawkish intentions, Kennedy's decision was aimed at rectifying an intelligence oversight by the previous administration: in January 1961, Castro's government demanded that the US embassy reduce its staff to eleven, the number of diplomats Cuba maintained in the US. In response, Washington announced the termination of all diplomatic and consular relations with the island, and promptly recalled all of its 300 diplomats [190: 8; 270: 8, 59].

Since most of the recalled officers fulfilled intelligence functions, the US was left without an organized espionage presence on Cuba. NSA was one of the few US government agencies capable of filling that void [125: 104, 105].

The Agency's precise role in CIA-led sabotage and subversion schemes against the Cuban government – such as operation MONGOOSE/JMWAVE – is not known. NSA was kept informed of at least parts of MONGOOSE, because it is reported as having disagreed with CIA's plans to sabotage facilities of the Compañía Cubana de Teléfonos, which served the Agency as a useful intelligence platform [24: 118]. Additionally, it is possible that NSA collaborated in MONGOOSE operation's "intelligence phase [of what were] referred to as study flights" [125: 118]. It is also asserted that NSA had no supportive role in the Bay of Pigs invasion [24: 77], though that remains unverified.

NSA is correctly credited as being the first US intelligence agency to detect signs of operations ANADYR and KAMA – Soviet plans to establish a military base in Cuba "with intermediate-range and medium-range ballistic missiles, nuclear-armed diesel submarines, bombers and over 50,000 Soviet troops" [98: 64]; see also [137; 155: 239n4; 164; 2: 77]. NSA's information, later corroborated by numerous air reconnaissance flights, led to what became known as the Cuban Missile Crisis of 1962 [189: 19; 190: 81ff]. The Agency's intelligence contribution during the days of the crisis remains largely unknown, but, despite its shortage of Spanish linguists at the time, NSA is said to have "added materially to all other intelligence" (Director of Central Intelligence John A. McCone cited in [102: 44]; see also [35: 174]). Yet the important issue, largely overlooked in the relevant literature, is NSA's delay in discovering ANADYR/KAMA until after significant numbers of Soviet troops and armaments had reached Cuba – even at a time when the island was supposedly among the Agency's intelligence priorities [34: 4; 148: passim; 280: 692]. Another issue in need of examination is the alleged Soviet control of Cuban defense maneuvers during the crisis. In the absence of evidence concerning the USSR's control of vital military bases and armaments, US intelligence – NSA included

– assumed that it was the Soviets, not the Cubans, that controlled both the missiles and antiaircraft defense systems on the island [24: 123; 102: 33]. That latter proved to have been a dangerous mistake. Soviet communications intercepted during the crisis, and decrypted by the Agency two years later, reveal that Cuban troops had stormed the Soviet military facilities in Los Angeles, Cuba, killing and wounding at least eighteen Soviet troops in the process. It was these Cuban troops, and not Soviets, who used the facility's surface-to-air missile system to shoot down an American U-2 aircraft on October 27, 1962 [123].

18.6 1974–1990: FROM WATERGATE TO INFOSEC

The conclusion of the Vietnam War resulted in unspecified, though undoubtedly considerable, budgetary setbacks for NSA. Nevertheless, the sheer volume of hard intelligence produced by the Agency [22: 65] allowed it to maintain its institutional stature. In the late 1970s and early 1980s, the organization had approximately 70,000 employees, and commanded a truly impressive array of financial resources [22: 2–4; 24: 549]. In 1986, as the Cold War drew to a close, NSA fought intensely to retain funding for its SIGINT ventures, which by that time exceeded imagery intelligence by a margin of four to one [51: 201, 221; 243].

During that period, NSA lost control of a number of its most important listening posts – in Behshahr and Kabkan, Iran, after the Iranian Revolution, and in Samsun and Karamürsel, Turkey, for a few years following the Turkish invasion of the Republic of Cyprus [130: 67; 132: 175; 159: 148; 232: passim]. Despite of these losses, the Agency's technical collection and analysis capacity reached its prime, mainly due to ECHELON's expansion and increased computer processing power, automation, and storage capabilities [17: 20; 22: 101, 104; 24: 394, 591–593; 54: 13; 141: 38; 182: 73; 208: 96]. The subsequent freeing of cryptological resources facilitated by digital automation, allowed NSA to gradually address security issues arising from the proliferation of network computing. Thus,

Figure 18.6. The NSA's head quarter Ford Meade. Courtesy of the National Security Agency.

along with its conventional COMSEC duties – notably the development of the HW-28 electronic cipher device and the KG-84 general-purpose cryptographic device – the Agency incorporated computer security (COMPUSEC) – later renamed information systems security (INFOSEC) – into its duties. Within this context NSA's Computer Security Center – later renamed National Computer Security Center (NCSC)[29] – was founded in 1981, reaching an annual budget of approximately $40 million by 1987 [71: 70; 80] (see Fig. 18.7).

Despite the Agency's efforts, many of its SIGINT operations and resources were compromised

by individuals disloyal to its mission. In late 1985, a former NSA intelligence analyst named Ronald William Pelton was arrested for causing the Agency "inestimable damage [by providing] Soviet agents [with] an incredibly detailed account of US electronic espionage capabilities" [261: 19]. It is conjectured that among NSA's operations terminally compromised by Pelton were the aforementioned IVY BELLS, as well as microwave communications interception and COMSEC systems [2: 70; 24: 370, 374; 197: 81, 138; 224: 35; 283: 448, 451]. The Agency suffered further compromises of its activities by the seizure of various COMSEC material by North Vietnamese forces during their capture of Saigon, as well as by a number of other known Soviet moles, including cipher

---

[29]In parts of the literature, the NCSC appears misnamed Computer Security Technical Evaluation Center or Computer Security Institute [22: 362; 80].

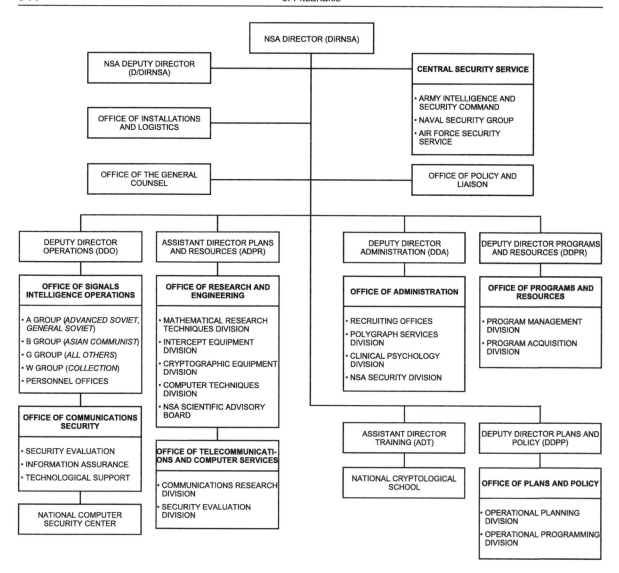

Figure 18.7. Organizational chart, National Security Agency, c. 1983 (abbreviated).

clerks Christopher Boyce and Daulton Lee, former NSA cryptologist David Sheldon Boone, and CIA employee William Kampiles, who is believed to have divulged NSA's KEYHOLE spy satellite project [24: 352, 353; 85: 159, 160; 181; 185; 197: 53; 207: 203–205].

In the long run, the espionage accomplishments of the USSR were outspent by NSA's "gargantuan" [24: 383]; see also [22: 204] satellite proliferation. The Soviets were ultimately unable to compete with the overwhelming space reconnaissance encirclement facilitated by the NSA/NRO/NASA/US Navy partnership. The MAGNUM SIGINT satellite alone, launched into orbit by NASA's *Discovery* space shuttle in 1985, is reported to have cost $300 million [51: 192, 193; 208: 181]. Other new and ongoing satellite projects, such as the JUMPSEAT/TRUMPET RADINT ferret and the CHALET/VORTEX/MERCURY SIGINT collection platform [51: 182, 192, 223, 224; 54: 6; 208: 180; 283: 196], proved as important as the Agency's ground command centers. These facilities – RUNWAY, SILKWORTH, STEEPLEBUSH and MOONPENNY, among others – enabled the efficient management of the satellites, and secure downlinking and distribution of acquired data throughout NSA's dispersed facilities [54: 6]. It is possible that the increased collection resulting from these technological developments facilitated NSA's first-ever decryption of high grade Soviet communications around 1981, a cryptanalytic accomplishment that one observer – who places the breakthrough in 1979 – describes as "one of the most important since World War II" [24: 370].

Yet while NSA accelerated its cryptanalytic supremacy over international targets, revelations about the Agency's record in the domestic field stigmatized its reputation at home. On May 14, 1973, former presidential counsel John Wesley Dean III gave the US Senate's Watergate investigative committee[30] access to what the *New York Times* described a few days later as "the 1970 intelligence plan" [66: 1]. The documents referred to what became known as "the unused spy plan" [8: 770] and is today termed the Huston plan – after its chief proponent, Tom C. Huston, former US Army intelligence officer and White House counsel to US President Richard M. Nixon [182: 49]. Although Huston was indeed the plan's chief designer, the scheme's informal title is incorrect: a more appropriate designation would be the Nixon–Sullivan–Huston plan. In drafting it, Huston was acting under "heavy pressure" [230: 72] from Nixon to direct the entirety of America's intelligence apparatus against political dissidents. Huston was further prompted by FBI assistant Director William C. Sullivan, with whom he shared an "ideologically compatible" [242: 17] political viewpoint. Sullivan urged Huston to set up an Interagency Committee on Intelligence to examine the possible "secret removal of legal restraints on the CIA, the FBI, NSA, and other US intelligence agencies so they could spy on American citizens" [135: 134]; see also [242: 25; 258: 924–986]. In July 1970, Nixon authorized the plan as laid out in the Committee's 43-page *Special Report*, only to change his mind five days after US Attorney – General John N. Mitchell warned him of its illegality [24: 431].

Nixon's eventual cancellation of the plan did not appease the Senate Watergate committee, whose

---

[30]Known by its official name as the Select Committee on Presidential Activities, or by the name of its chairman, Samuel James Ervin, Jr., the Ervin Committee.

chair, Sam Ervin, described the Huston papers as reflecting the administration's "Gestapo mentality" (cited in [8: 771]). Furthermore, memoranda exchanged between members of the Interagency Committee on Intelligence revealed that NSA had apparently tried to conceal the Committee's tracks by circulating its communications under its own TOP SECRET: HANDLE VIA COMINT CHANNELS ONLY classification [242: 13, 16, 25]. NSA's critical role in the Huston affair dragged the unwilling Agency into the political limelight, as three more Senate and House committees took turns subjecting it to unprecedented degrees of Congressional inquisitiveness. Before 1976 was over, the Church [258], Pike [251] and Abzug [252] committees had uncovered NSA domestic collection activities much older than even Huston himself could have suspected. In 1945, NSA's predecessor, SSA, had secretly managed to secure the intercept cooperation of the nation's largest telecommunications carriers. The latter had agreed to continue to illegally provide the Agency with copies of all telegrams crossing US borders, despite the official termination of wartime censorship restrictions [22: 237–241; 242: 120]. Codenamed operation SHAMROCK, the scheme continued uninterrupted until May 1975, when Congressional disclosure of the operation appeared imminent [133: 112; 182: 76; 242: 121]. SHAMROCK survived communications digitization, and was practiced by NSA without regard to legal constraints or governmental approval, which the Agency never sought after receiving Truman's tacit consent [22: 244, 245; 118: 174; 133: 112; 233; 252: 256; 274: 400]). Congress estimated that, in the last years of SHAMROCK's operation, and with CIA's illegal assistance [274: 400], NSA analysts reviewed about 150,000 out of millions of copied telegrams they received each month – making SHAMROCK history's "largest governmental interception program affecting Americans" [258: 740].

Congress found that NSA had recently terminated another secret domestic interception program codenamed MINARET. Under MINARET, NSA systematically spied on the foreign communications of approximately 75,000 US-based individuals and organizations between 1962 and 1973, according to watch lists mainly supplied to the Agency by the CIA, FBI, the Defense Intelligence Agency (DIA), the Bureau of Narcotics and Dangerous Drugs (BNDD), and the Secret Service [118: 176; 133: 105, 106; 182: 74; 208: 103]. MINARET's main operational mission was to spy on those

> involved in domestic antiwar and civil rights activities in an attempt to discover if there was "foreign influence" on them [...]. The program applied not only to alleged foreign influence on domestic dissent, but also to American groups and individuals whose activities "may result in civil disturbances or otherwise subvert the national security of the US" [258: 739]; see also [24: 429; 242: 121ff].

It should be recognized that, despite their illegality, both MINARET and SHAMROCK were not far removed from the political context in which they were conceived and applied. They were not isolated ideological paraphernalia of NSA executives, nor were they exceptional specimens of executive absolutism in the years that sparked FBI's COINTELPRO[31] and COMINFIL,[32] CIA's CHAOS[33] and MERRIMAC,[34] and the HUAC[35] hearings. Rather they were intelligence encapsulations of "the searing impact of the Cold War on national institutions and values" [242: 13]. SHAMROCK, in particular, was a natural operational outgrowth of the "intimate institutional interface" [91: 405] between intelligence agencies and telecommunications service providers – a relationship that largely facilitated domestic espionage during the Cold War and remains active today [91: passim]. The Huston proposal, on the other hand,

was potentially far more significant, in the operational sense, as its application would have certainly augmented MINARET's – and thus CHAOS' – activities to unprecedented degrees [133: 105].

The revealing paradox of the Huston plan is that NSA had secretly conducted extensive domestic communications collection – through MINARET and SHAMROCK – for decades prior to seeking official presidential sanctioning for Huston. What is more, Nixon's eventual refusal to authorize the plan did not result in the Agency's termination of similar programs already underway [22: 275–277; 134: 147, 149; 191: 95; 242: 35; 274: 156]. Thus, the accomplishments of the Congressional investigations of NSA between 1974 and 1976 were numerous: for a while, they penetrated the Agency's venerable secrecy protections that shielded it from public scrutiny ever since its birth [2: 64, 65, 70, 71; 4: 43; 22: 72, 88; 24: 136; 32: 16; 155: 6; 230: 4ff]. Both NSA and the government fought hard against this momentary and momentous exposure [22: 30, 304, 305; 133: 92; 132: 218; 191: 104, 105; 230: 63; 252: 6], perhaps failing to recognize that it symbolized American society's "historic [...] breakdown of the Cold War consensus" [191: 4]; see also [230: 10] that had previously afforded considerable operational permissiveness to US intelligence agencies. It was indeed this fundamental social shift that forced the first-ever public testimony by an NSA Director [32: 17, 18; 230: 72]. It also laid the foundations for permanent Congressional intelligence oversight committees, the Foreign Intelligence Surveillance Act (FISA) and the US Signals Intelligence Directive 18 (USSID 18), which today govern – albeit with limitations [43: passim; 175: passim] – NSA's activities against US persons [24: 440ff; 28: 13; 32: 19, 22; 182: 75; 191: 177; 230: 10; 233].

From an institutional viewpoint, the Congressional investigations also revealed the complex layers of NSA's operational detachment from other components of the US government [242: 38, 39]. They established, for instance, that NSA's motivation to participate in the Huston plan was due not so much to its ideological compliance with the Nixon administration, as to its desire to resume project SCOPE [24: 432, 433; 208: 251; 209:

---

[31]COunter INTELligence PROgram (1956–1971): Operation against mostly legal US-based dissident political organizations.

[32]COMmunist INFILtration (1962–1968): A sub-project of COINTELPRO targeting primarily Martin Luther King, Jr., and the Southern Christian Leadership Conference.

[33]CHAOS (1967–1974): An extensive operation to spy on legal peace and leftwing political groups in the US, linked to NSA's MINARET.

[34]MERRIMAC (1967–1974): A sub-project of CHAOS, geographically limited to the Washington, DC, area.

[35]House Un-American Activities Committee (1938–1975): the main Congressional vehicle during the years 1947–1954 of the American phenomenon later known as McCarthyism.

283, 284; 242: 15–19; 274: 150, 155ff, 271ff]. This embassy break-in scheme had facilitated some of NSA's greatest cryptanalytic achievements until 1967, when FBI Director J. Edgar Hoover, fearing SCOPE's exposure, had refused to continue authorizing it. Three years later, in cooperation with FBI elements sympathetic to its mission, NSA saw the Huston plan as "a heaven-sent opportunity" (Tordella cited in [242: 19]) to circumvent Hoover's objections to SCOPE [133: 84; 242: 35; 248: 304, 305; 22: 278; 274: 155, 176n].[36]

Similar degrees of functional autonomy were displayed by NSA's activities in regards to project X and the Soviet brigade crisis, in 1977 and 1979 respectively. Project X was NSA's alliance with ONI in sharing SIGINT technology with the Republic of South Africa, in return for access to intercepted data of Soviet maritime activities in the area. Because exporting the equipment was in violation of US laws, NSA and ONI set up a front company, Gamma Systems, and recruited weapons dealer James H. Guerin to run the shipments. The operation was terminated in 1977, after the Carter administration decided to halt all US intelligence relations with the racist state. Guerin was later convicted of weapons smuggling to South Africa, and of a money-laundering scheme that one US judge described as "the largest fraud ever perpetrated in North America" [178]; see also [24: 375, 382; 96: 58–60; 37: 253]. Two years later, NSA sparked the Soviet brigade crisis, after it mistakenly issued a report claiming that military activities were organized in Cuba under a newly arrived Soviet combat brigade [189: 19; 247: vii]. After causing considerable uproar upon the US government, the Agency reexamined its data and concluded that the Soviet brigade was not combat-ready. It also discovered that the brigade had been in Cuba for at least 17 years, ever since the Kennedy administration had consented to a limited Soviet military presence on the island for training purposes [102: 42, 43; 189: 19, 50; 190: 223]. NSA's confusion was caused by its habitual tendency to forward intelligence to governmental decision makers without

first corroborating it with analytical agencies, thus hoping to get "credit for the scoop" [247: x]; see also [134: 55, 56; 246].

These instances are not meant to illustrate that NSA achieved – at any point in its institutional lifetime – an operational independence verging on executive uncontrollability. On the contrary, the Agency proved particularly compliant in providing SIGINT and COMSEC support to Ronald W. Reagan's administration's ventures in El Salvador, Nicaragua, and Honduras [40: 262, 263; 209: 377; 229: 153]. It also dutifully assisted the illegal Iran-Contra operation in the 1980s [2: 65; 135: 131; 134: 53; 152; 155: 7; 229: 274; 265: 8, 76, 104, 176–181; 283: 413, 486]. Yet it would be erroneous to distinguish NSA's closed door attitude from the mentality of any other secretive institution which – as a former CIA Director remarked with reference to NSA's Soviet brigade mishap – has learned to utilize mandate ambiguities "in taking advantage [...] to do what it wants" [247: x].

18.7 1991–2001: BREAKDOWN

On December 25, 1991, Mikhail S. Gorbachev resigned as president of the USSR, and the US formally recognized the independence of former Soviet republics. But the liquidation of Soviet authority came too abruptly to cause waves of excitement in America's intelligence community. For NSA, it entailed a radical reallocation of its budget, most of which was assigned to Soviet targets until as late as 1980 [164]. The Agency clearly recognized the imminence of such a reallocation [21: 37]. Yet its motivation to implement it was largely enforced by world developments, such as America's increasing military presence in the Middle East. In 1990, NSA had failed to perceive the Iraqi military's intention to invade Kuwait, until shortly before the attack [9]. In the days leading to America's armed intervention, the Agency had to struggle to penetrate the US-developed Iraqi COMSEC systems, prompting one commentator to exclaim that "Saddam knows more about American battlefield intelligence than all but a handful of other foreign leaders, thanks to Ronald Reagan, William Casey and aides" [126]; see also [106; 268: A10]. NSA finally overcame

---

[36]NSA and FBI did in fact resume SCOPE, possibly under a different name, after Hoover's death in 1972 [24: 433; 274: 143n].

this obstacle by saturating the battlefield with an unprecedented host of airborne SIGINT and electronic warfare platforms [238; 238]. Eventually, NSA's SIGINT support was credited with ensuring DESERT SHIELD's tactical efficiency, including in the Battle of Khafji, Iraq's only organized military offensive encountered in the campaign [2: 62; 10; 62; 107]. The "transparency of the battlefield" [136: 112] achieved by NSA, prompted the US military to dub Gulf War I as history's "first space war" [94: 85; 188: 409].

Other developments forcing NSA's allocation of resources to non-Soviet targets were the US interventions in Haiti, Somalia and the former Yugoslavia, which impelled the Agency to augment its Creole, Somali, Arabic, Serbo-Croatian, Albanian and Macedonian linguistic components [24: 554; 172: 39; 281: 35]. Nuclear weapons control has also been a major point of concentration for NSA since 1991. It is estimated that the Agency was behind 90 percent of nonproliferation-related intelligence acquired by the US in the first half of the 1990s [24: 418]. In later years, most of this effort has focused on perceived potential nuclear technology suppliers, such as China and Pakistan, as well as perceived seekers, such as North Korea, Iran, Iraq and, until recently, Libya [13; 16; 24: 407; 185].

Despite these changes, and although by 1993 less than 15 percent of NSA's annual budget was devoted to the former USSR [164], it must be assumed that the Agency directed much of its attention to its own internal emergencies. In 2001, FBI arrested its Supervisory Special Agent Robert Phillip Hanssen on charges of sporadic espionage for the USSR and Russia that spanned nearly 22 years [275: 8]. Even though Hanssen was working for the FBI, his espionage activities caused incalculable damage to NSA: most of the information he compromised related to NSA, and included precise data on the Agency's penetration of Soviet satellite systems, its cryptanalytic shortcomings, and its MONOPOLY operation against the Soviet embassy in Washington, DC [172: 38; 261: 9, 10; 275: 284]. Ironically, despite Hanssen's damage, which one governmental report described as "possibly the worst intelligence disaster in US history" [261: 1], he was the least of NSA's troubles.

In the decade following the Soviet Union's meltdown, NSA's budget was reduced by approximately 30 percent [164; 198]. Despite large budgetary increases awarded to the Agency in 1998 [69], its income remained far below Cold War levels. Consequently, by early 2001, NSA staff had been reduced by between 24 and 30 percent [24: 549, 576; 32: 5; 108: 137; 164; 260: 6]. The rate of workforce cuts was endorsed by Congressional overseers, who, in 1997, reproached the Agency for allocating nearly 40 percent of its budget to staff payrolls, instead of to "investments in new technologies" [249: 96]. The criticism was not groundless: NSA had been forced to discontinue nearly half of its spy satellite projects by 1994 [24: 549]. Still, new generations of SIGINT satellites, such as the INTELSAT 8 and the IOSA and RUTLEY series, actually increased the Agency's collection capability though simultaneous multichannel monitoring [24: 408, 466; 54: 6]. This intensification, along with the rapidly increasing infrastructure and traffic volume of global telecommunications, overwhelmed NSA's analytical and processing elements. Thus, by 2001, NSA was unable to process more than a tiny fraction of the estimated two million text and oral messages intercepted each hour by each of its listening posts [55: 35; 108: 134; 164; 172: 38]. The Agency was frustratingly slow to address this problem: advanced SIGINT automation devices, including text retrieval, machine translation and speech recognition systems, were gradually introduced in the late 1990s, accompanied by the inevitable assortment of limitations, software bugs and glitches [24: 555ff; 54: v].

NSA's severe technological inadequacies were further highlighted in January of 2000, when the Agency's entire information processing system unexpectedly shut down for 72 hours, grinding related functions to a halt [132: 282; 185]. Even worse for the Agency, the ensuing investigation into the incident detected organizational mismanagement at the root of computer inadequacies [104: 98]. This was not the first time NSA's executives had been blamed for functional shortcomings: throughout the 1990s, the Agency's Congressional and Defense overseers had consistently described NSA as an institution "in desperate need of organizational restructuring and modernization of its information technology infrastructure" (cited in [199]), and "lacking the [...] resources needed even to maintain the status quo, much less meet emerging challenges" [259: 39]; see also [254: 9, 10; 255: 12; 260: 2, 3]. In response to constant criticism, NSA launched at least two major reorganizations. In 1992, Director Vice-Admiral John M. McConnell authorized the purging of 40 percent of deputy director and nearly 30 percent of middle managerial positions, among other structural changes [108: 135; 260: 6]. Seven years later, Director Lieutenant-General Michael V. Hayden initiated his "100 Days of Change" scheme, under which whole operational committees and groups deeply embedded in NSA's structure were liquidated [24: 470ff; 32: 9; 259: 39].

Such adjustments did moderate the bureaucratic authority of "unwieldy senior management groups that held much of [NSA's] power" [24: 472]. But the technological problems persisted. By the time Director Hayden launched his "100 Days of Change", it was becoming apparent that the Agency was losing the commercial encryption dispute [54: 19; 24: 463; 136: 7]. This contentious issue had its roots in 1971, when IBM developed Lucifer, the first known American cipher intended for commercial distribution in civilian telecommunications. NSA pressure convinced IBM to downgrade Lucifer's encryption key to 56 bits, in return for its widespread distribution under NSA-sanctioned Data Encryption Standard (DES). NSA thus hoped to achieve the dual aim of promoting the relative security of American civilian telecommunications, without losing access to intercepted messages, or compromising its own algorithms [71: 59; 140: 151]. But complications emerged when three academics at the University of California at Berkeley and the University of Stanford discovered public-key cryptography. Some at NSA considered the invention a threat to the Agency's cryptanalytic mission. In October 1977, an NSA employee tried without success to prevent the inventors from officially presenting public-key cryptography at a conference, citing potential arms trafficking violations [71: 60ff; 141: 199]. A year later,

when NSA attempted, again without success, to introduce patent restrictions on cryptographic research, it became clear that the Agency's intent was to monopolize the release and distribution of cryptographic material in the US [22: 351; 24: 381; 141: 199]. In a number of public statements, NSA Director, Vice-Admiral Bobby Ray Inman, argued for imposing governmental controls over cryptography on national security grounds, as was the case with atomic energy [71: 63; 226: passim]. The dispute carried over into the 1980s and 1990s, with NSA marketing two tamper-resistant integrated-circuit chips, the Clipper series (telephony) and the Capstone (networked computers). Under the Agency's 'key escrow/key recovery' schemes, separate decrypting keys were to be held in secure databases accessible by government officials for legitimate law enforcement purposes [54: 15; 71: 210, 215–218]. NSA also tried, with little success, to restrict the export of strong encryption products, citing national security requirements [26: 26; 71: 208].

Ultimately, the issue of whether NSA intends to systematically sabotage American civilian cryptology is not as interesting as the issue of why it appears unable to do so. There is little doubt that the William J. Clinton administration's 1998 decision to relax export restrictions on commercial encryption favored commercial interests, while being potential anathema to NSA's ability to read intercepted messages [169]. Yet the decision is well within the character of American capitalism, particularly when examined in the context of the relationship between national security and neo-liberal economic doctrines. Indeed, the recent history of US communications liberalization shows that market interests often tend to dominate intelligence interests. The neo-liberal restructuring of the US telecommunications industry – initiated with the breakup of American Telephone & Telegraph (AT&T) in the early 1980s – has consistently threatened the ability of intelligence agencies to police the pattern and content of intercepted communications. The breakup of AT&T literally crippled law enforcement and security agencies by greatly increasing the numbers of actors involved in administering commercial telecommunications

[90: 230–235]. For this reason, the US Department of Defense, representing NSA among others, lobbied against AT&T's divestiture, warning the Reagan administration of "severe problems [that will] confront the Department of Defense if this network is broken up" (US Deputy Secretary of Defence, Frank C. Carlucci cited in [59: 225]; see also [39: 53, 54]). Admittedly, therefore, the government's subsequent decision to implement AT&T's breakup was a direct affront to the interests of the nation's national security apparatus.

Telecommunications deregulation continued to preoccupy NSA throughout the 1990s. By 1991, the rapid emergence of fiber optic cables as the principal channel of communications was seriously challenging the Agency's worldwide SIGINT collection ability. Instead of microwave modes of transmission, which NSA effortlessly accessed through its satellite network, interception of fiber optic systems required physical proximity to the cables, thus threatening to leave the Agency "with an increasingly irrelevant constellation of SIGINT satellites" [136: 6, 7]; see also [32: 41; 206: 22]. Despite this predicament, which one Congressional committee described as "a crisis" [136: 6], NSA had to battle hard in convincing the US government to prevent a consortium of American corporations from installing fiber optic communication systems in the Soviet Union [173: 8; 184]. Four years later, the Clinton administration actively promoted a contract by US companies to enhance China's military command and control systems (including the TigerSong air defense system), despite NSA objections centered on its inability to intercept the proposed networks [105; 231: passim].

The spread of fiber optic systems, combined with the increasing prevalence of strong encryption, may prove critical to NSA's future. The Agency calculates that by 2008 the vast majority of all communications will be encrypted [24: 464] – a trend that could totally neutralize its surveillance capabilities [32: 4; 82; 132: 282; 263]. More importantly, NSA's inability to influence these critical technological developments comes with the realization that its own interests are far from dominant in the overall debate. Indeed, critical political decisions appear to be primarily steered by financially-oriented state agencies whose priority is to safeguard and promote the competitive nature of the market, rather than to enhance, or even maintain, national security mandates.

18.8 2001 AND AFTER: REORGANIZING UNCERTAINTY

Soon after the dramatic events of September 11, 2001, some rushed to link NSA's structural and technological problems of the 1990s with the Agency's failure to detect the hijackers' plans [108: 134]. But such analyses were, for the most part, politically motivated, produced by observers predisposed to holding the Clinton administration's security policy responsible for the intelligence failure that was '9-11'. In reality, neither budget cuts, workforce reductions, nor even the export of strong encryption, had anything to do with NSA's failure. The '9-11' perpetrators stationed in the US communicated with their comrades abroad via Internet chat links and prepaid cellular telephones, none of which relied on unbreakable encryption, or indeed any encryption at all [25: 248; 42]. Even if NSA surveillance had detected those channels of communication, it is unlikely that its SIGINT analysts would have gained any insight to the planned attacks. NSA had monitored Osama bin Laden's cellular telephone from 1995 until 1998 [25: 163, 168; 108: 129], but even though that line was used to coordinate al-Qaeda's bombings of US embassies in Kenya and Tanzania, at no point did NSA analysts extract any specific warnings [25: 168]. The same applies to telephone messages intercepted by NSA satellites on September 10, 2001, and not translated until two days later. Even though much has been made of the two-day lapse, the messages – "the match begins tomorrow", "tomorrow is zero day" – were too vague and could not have been expected to trigger any specific alerts [108: 128, 129; 113: 75, 147, 249]. Nor were NSA's budget and staff reductions directly to blame for '9-11'. Despite the Agency's relative contraction, its budget was retained at extremely generous levels, exceeding $17 billion between 1995 and 1999 – with billions more going toward NSA/NRO satellite programs. Furthermore, the size of NSA's workforce exceeded that of the CIA and FBI combined – without including an estimated 25,000 employees operating the Agency's listening posts worldwide [22: 481; 24: 481].

In reality "questionable management decisions" [255: i] were chiefly instrumental in NSA's performance prior to '9-11'. The allocation of vast budgetary portions to sustain the Agency's enormous arcane workforce had probably less to do with legitimate workload requirements, and more with the gigantism inherent in the American cultural context [132: 6]. This imbalance reinforced NSA's bureaucratic dispositions and marginalized technological and logistical innovation at a time when the rate of the digital telecommunications revolution peaked worldwide [256: vi]. Characteristically, NSA's total workforce *decreased* in the years following '9-11', after it was decided that large segments of its staff had to be driven to retirement for a radical workforce reorganization to flourish [25: 356; 256: vi]. Equally critical was the Agency's inability to trail the decentralized, non-hierarchical al-Qaeda network [32: i]. Despite a broad realization that its post-Cold War SIGINT targeting ought to diversify, NSA appeared unwilling to radically restructure its surveillance philosophy. Consequently, despite the experience of the 1993 World Trade Center bombing in New York city, NSA's conventional regime of SIGINT collection left al-Qaeda's unpredictable non-state actors essentially untouched [32: 3, 4; 188: 87; 255: v]. Finally – not in order of importance – NSA displayed little initiative in combating the culture of animosity and estrangement among America's intelligence organizations. The Agency's systematic reluctance to reveal its sources, methods and weaknesses to external agencies caused a pattern of "deliberate withholding of information" [151] often resembling a feud between NSA and agencies it considered unworthy adversaries, rather than partakers in its mission [2: 65; 24: 420; 25: 226; 114: 39; 188: 88; 208: 240; 283: 49].

If the extent of intelligence failures is weighed by the resulting loss of life, then a cynic could be excused for advocating a shift in attention away from NSA's performance in '9-11', and toward its role in America's subsequent war on Iraq. In February 2003, NSA audio intercepts supplied the crucial exhibit in a United Nations (UN) presentation in which US Secretary of State Colin L. Pow-

ell requested the Security Council's authorization for Iraq's invasion. His speech argued for the harboring by Iraq of weapons of mass destruction (WMD), and marked the first ever disclosure of audio intercepts at a UN event [206]. NSA Director Hayden later admitted that the excerpts used in Powell's presentation were "ambiguous" [25: 373]. Judging by America's subsequent failure to discover traces of WMD in occupied Iraq, these excerpts were also misleading. Recent assessments by qualified observers suggest that the intercepts were construed by the George W. Bush administration to justify its war on Iraq [25: passim]. This possibility, however, does not absolve NSA from its role in the controversy. Although the Agency cannot be expected to interpret its SIGINT product, it is required to substantiate it in the context of corroborating data, prior to legitimating it by imparting it to its clients. In the case of Iraq's purported WMD, NSA either relinquished this responsibility, or withheld its objections to the distortion of its intelligence by its customers. Either way, its posture was instrumental in facilitating apparently misleading premises for war – for the second time in its history.

Nevertheless, it is plausible that the Agency's preoccupation with its own restructuring barred it from concerning itself with the government's use of its intelligence. Since '9-11', NSA has undergone the most radical organizational and logistical shake-up in its history. Some changes have been predictable: the Agency's budget has "skyrocketed" [108: 137], probably at an increase far in excess of the $1 billion Hayden requested for upgrading NSA's computer systems [114: 71]. Its hiring quota has also increased to nearly Cold War levels, reaching 1,500 recruits per year in 2004 [18; 25: 356; 219]. More importantly, the Agency's overall methodology has shifted toward increased flexibility. After hiring an advertising agency, NSA hosted its first-ever press conference and even a job fair [132: xii; 130]. Linguists and information technology experts make up the brunt of new recruits, while non-classified tasks are increasingly subcontracted to industry [60; 163]. By June 2004, 2,690 private sector enterprises constituted NSA's contracting base [52]. At the root of this transformation lie attempts for responsible budgetary management, ensuring, according to Hayden, "that our mission drives our budget decisions" (cited in [25: 111]).

Organizational progress aside, critical technological problems persist, as NSA's intelligence capabilities are constantly thwarted by widespread use of strong encryption and fiber optic transmission systems. The solution apparently points to what some call "probably the most profound change in [SIGINT] history" [24: 465; 25: 148; 215], namely the revolutionary transformation of SIGINT agencies from "passive gatherer[s] to [...] proactive hunter[s]" [256: v] of information. Just as the tapping of fiber optic facilities requires physical proximity [78], so communications surveillance will increasingly require interception prior to encipherment – that is, intercepting messages "before cryptographic software has a chance to scramble them" [24: 479]. Since 1978, this has been the task of the Special Collection Service (SCS), a joint CIA/NSA effort to acquire pre-encrypted data [24: 477; 215; 283: 313, 314]. Focusing on the endpoint collection of information at rest – as opposed to the conventional midpoint collection of information in motion – SCS utilizes a variety of means in pursuit of its mission – ranging from computer hacking (HACKINT)[37] and parabolic and tempest surveillance, to recruiting foreign diplomatic staff with access to desired information [24: 479, 480; 25: 146–148; 164]. It is true that SCS espionage has enjoyed considerable success in the past, particularly in operations by Special Collection Elements in Eastern Europe in the 1980s, and more recently through MEGAHUT, a joint SCS/FBI project in New York targeting foreign diplomatic representations [145: 222, 223; 275: 59, 59n; 283: 314]. But if the paradigm shift in SIGINT collection is as momentous as some predict, then SCS could be expected to gradually outgrow its organizational dependence on NSA and the CIA alike, and mutate into yet another separate institutional entity in

America's already fragmented intelligence constellation.

Questions remain as to NSA's target priorities following '9-11'. The unanticipated dismantling of the Agency's Bad Aibling listening station in Germany – a major Cold War SIGINT beehive – might offer some clues, particularly when juxtaposed with NSA's publicized hunt for primarily Arabic and Chinese linguists [218; 219]. But the Agency has been unable to dispel accusations that it systematically directs its SIGINT network on foreign companies competing against US corporations for international contracts [54: 17, 18; 57]. In 2000, following a European Parliament report alleging NSA's involvement in economic espionage against European firms, French cabinet ministers publicly urged French companies to utilize encryption as a means of thwarting UKUSA's ECHELON surveillance [15]. A few months later, NSA officials refused to attend scheduled meetings in Washington, DC, with a European Parliament delegation requesting the Agency's detailed response to the allegations [170]. This intensified accusations that, since the end of the Cold War, the Agency is acting on executive instructions to "aggressively support US bidders in global competitions where advocacy is in the national interest" [57]. NSA officials have repeatedly confronted these claims, arguing that economic espionage is precluded by Congressional oversight and the American "business ethic" (former NSA Director Vice-Admiral William O. Studeman cited in [224: 15]; see also [170]). Other qualified observers have noted that US corporations competing internationally are already dominant in their fields of operation and would rarely ask for SIGINT support, particularly from government [136: 33; 285].

Undoubtedly, NSA conducts operations against friendly countries. Operations by American cryptologic agencies against allied nations, including Britain, France, Canada, Italy, Turkey, Greece, Australia and Israel, date back to 1917 and are well established in the literature [5: 189ff; 6: 49; 22: 142, 144, 321; 53; 56: passim; 61: 4, 123, 238; 87: 027n; 88: 45; 142: 686, 728; 202: 45; 208: 239, 263; 245: 95; 289: 332]. The same applies to interception operations against international bodies,

---

[37] Also termed "offensive information operations" by NSA [236]. In 1999, one such operation targeted databases of foreign banks in Russia, Greece and Cyprus, suspected of handling investments belonging to Yugoslavia's former president Slobodan Miloševiç [264].

such as NATO and the UN [24: 023; 25: 358, 359; 61: 004; 141: 046; 205: 075; 208: 263; 223: 093, 094; 274: 153]. In 1963, an NSA defector revealed that the US State Department often read diplomatic messages intended for UN missions in New York before they even reached their final destination [139: 728; 245: 216]. In 2003, despite earlier assurances by NSA that its analysts were required to purge names of UN officials from their intelligence reports [24: 444], the Agency was revealed to have spied on at least six UN Security Council delegations while the US intensified its Iraq war diplomacy [45]. NSA's UKUSA British partner, GCHQ, was found to have tapped the telecommunications facilities of the UN Director General himself [167]. Also in recent years, NSA has targeted a number of high-profile UN agents, including members of the United Nations Special Commission (UNSCOM) and the International Atomic Energy Agency (IAEA) [255: 359; 103; 161]. Perhaps more substantially, NSA is reported to have conspired since 1957 with one of the world's leading cryptological equipment vendors in penetrating the enciphered communications of non-enemy nations that use that vendor's devices [22: 322; 61: 4]. A former NSA employee has claimed that, through its advanced knowledge of that vendor's products, NSA "routinely intercepted and deciphered [...] encrypted messages of 120 countries" [171] for several decades [225]. Similar allegations have emerged in relation to purported deals between NSA and computer software vendors Microsoft and Lotus. Despite the companies' dismissal of the allegations [14], in 1997 the Swedish government protested that it suffered privacy compromises by its use of Lotus software [13; 54: vii].

While NSA's 'no comment' policy has shielded it from having to confront such allegations, US government officials have suggested that the nature of intelligence work often commands the conduct of espionage against allied governments and corporations [136: 33, 37, 231n10; 284]. But they claim the motivation behind such measures is legitimate, and includes the supervision of economic and weapons sanctions, as well as the detection of money laundering and the prevention of economic espionage conducted by allied nations against US

companies [38; 58; 86: 101; 89: 96; 115; 124; 136: 162]. On the other hand, observers note that American law does not currently prevent NSA from assisting specific US companies whose financial success it may deem advantageous to America's national interests [24: 423].

Historically, NSA's espionage mission has always incorporated an economic element, particularly during the Cold War, when both the US and the USSR considered economic and industrial espionage a legitimate means of inflicting political wounds on their ideological rivals [89: 119; 46: 910, 911]. Today, the systematic acquisition of economic intelligence – namely insights into "the economic decisions and activities of foreign governments" [136: 33] – is still required from NSA by its customers [55: 34; 136: 32]. This is arguably distinct from economic and industrial espionage, which denotes delivering sensitive information to American corporations. But the demarcating line between the two is admittedly thin, particularly since economic intelligence shaping the formation of national policy can often assist the financial pursuits of US companies operating under that policy. Additionally, history has shown that NSA tends to abide by ethical standards of conduct only when forced by the implementation of binding legal frameworks. Thus, while repeated allegations of NSA's economic espionage remain unsubstantiated [24: 426; 86: 135], critics are essentially justified in questioning the unlegislated application of the Agency's economic intelligence activities.

18.9 EPILOGUE: THE INTRIGUE OF LIMITATIONS

Having recently stepped into the turbulent 21st century, NSA remains a rare example of an American institution confronting not only an uncertain future, but also an uncertain past. The modest unlocking of the nation's largest intelligence organization constitutes an indisputable trend over recent years. But its products have yet to enable historical scholarship to offer even an approximate evaluation of the Agency's overall performance [172: 38].

The discernable historical patterns in NSA's esoteric institutional development are limited. Clearly, the Agency's known history "is marred almost as much with mud as with medals" [113: 104]. During the past twenty years, its intelligence has probably helped safeguard innocent civilians in more instances than publicly acknowledged. Other times, its inaccurate information generated misleading *casi belli* for conflicts that ended countless lives. Its valuable insight has often been poorly utilized by its customers. Periodically, its intelligence product shared and perpetuated the "deeper intellectual misjudgment[s] of [...] central historical reality" [47: 397] that habitually marred America's Cold War policy. Although its cryptological resources were usually unequal to the demands placed upon them, its intelligence product remained extremely sought after, even during America's greatest intelligence setbacks [24: 457; 97: 54; 114: 136, 202]. Its organizational and technological transformations have been driven mostly by emergencies, which probably explains why its institutional memory appears vivid.

The darkest NSA enigmas concern its contributions – or lack thereof – to America's military and diplomatic record. What, if any, was the Agency's role in the series of coups and assassinations sponsored by the CIA in Syria in the late 1940s and 1950s? Did it assist in clandestine operations against Patrice Hemery Lumumba in the Congo in the 1960s? We know that in 1967 "NSA cryptanalysts [...] cracked the Greek army's code" [245: 95]. Does that imply that NSA was aware of a pending April 1967 coup by the military group that later became know as 'the colonels', and ruled Greece until 1973? What was NSA's SIGINT presence in Angola between 1975 and 1980? In Afghanistan between 1979 and 1992? In Thailand, Laos and Cambodia between 1961 and 1975? In Chile in September 1973? Did its analysts foresee the 1974 invasion of Cyprus, or the Portuguese revolt of the same year? Did it suffer significant cryptologic loss during the 1979 siege of the US embassy in Tehran? Does it share responsibility for the failed mission to rescue US hostages held in the embassy? We know that the Agency was "a primary source of information during the war in Bosnia from 1991 to 1996" [2: 62]; see also [20]. But how accurate was the intelligence generated by its intercepts?

Answers to these and many other questions remain hostage to NSA's stringent declassification policy – a byproduct of some legitimate, and many illegitimate, assumptions about the nature of democratic societies. Even after the advancements in accountability achieved by the Church, Pike, and Abzug committees in the 1970s, the Agency still appears extremely reluctant to relinquish even basic information to its few Congressional overseers [1; 22: 1; 253: 21, 22]. This unfortunate attitude has caused one observer to wonder "whether the elected representatives of the American people may not be trusted with information handled daily by typists and technicians" [142: 701].

Ultimately, the future of historiography on American intelligence in general, and NSA in particular, depends principally on the instinctive curiosity of scholarly communities equipped with the methodological understanding of the subject's paramount importance. For better of for worse, NSA has been a prime element in the machinery of America's global dominance. Inevitably, the quest for a deeper comprehension of the nature and scope of America's status in the world cannot claim completion unless it intrudes into the cloaked universe of the National Security Agency.

REFERENCES

[1] D. Acheson, *This Vast External Realm*, Norton & Company, New York, NY (1973).

[2] M.M. Aid, *Not so anonymous: Parting the veil of secrecy about the National Security Agency, A Culture of Secrecy: The Government Versus the People's Right to Know*, A.G. Theoharis, ed., University Press of Kansas, Lawrence, KS (1998), pp. 60–82.

[3] B. Alexander, *The Strange Connection: US Intervention in China, 1944–1972*, Greenwood Press, Westport, CT (1992).

[4] G.W. Allen, *None So Blind: A Personal Account of the Intelligence Failure in Vietnam*, Ivan R. Dee, Chicago, IL (2001).

[5] D. Alvarez, *Secret Messages: Codebreaking and American Diplomacy, 1930–1945*, University Press of Kansas, Lawrence, KS (2000).

[6] C. Andrew, *Codebreakers and foreign offices: The French, British and American experience*, **The Missing Dimension: Governments and Intelligence Communities in the Twentieth Century**, C. Andrew and D. Dilks, eds, University of Illinois Press, Urbana, IL (1984), pp. 33–53.

[7] C. Andrew, D. Dilks, *Introduction*, **The Missing Dimension: Governments and Intelligence Communities in the Twentieth Century**, C. Andrew and D. Dilks, eds, University of Illinois Press, Urbana, IL (1984), pp. 1–16.

[8] Anonymous, *The traps in Watergate*, The Nation 216 (25) (1973), 770–771.

[9] Anonymous, *Invasion tip*, Aviation Week and Space Technology 133 (6) (1990), 15.

[10] Anonymous, *The battle damage assessment challenge*, Aviation Week and Space Technology 134 (5) (1991), 9.

[11] Anonymous, **Security Guidelines Handbook**, National Security Agency/Central Security Service, Fort Meade, MD [copy on file with author] (1993).

[12] Anonymous, **The Origins of NSA**, Center for Cryptologic History, Fort George G. Meade, MD, http://www.nsa.gov/publications/publi00015.cfm, last accessed 17 January 2004 (1996).

[13] Anonymous, *China assists North Korea in space launches*, Kyodo News Agency 23 February, http://www.findarticles.com/p/articles/mi_m0WDQ/is_1999_March_1/ai_54069918, last accessed 15 August 2003 (1999).

[14] Anonymous, *Microsoft denies spying charges*, Computergram International, 07 September, http://www.findarticles.com/p/articles/mi_m0CGN/is_3741/ai_55695355, last accessed 23 June 2003 (1999).

[15] Anonymous, *France accuses US of spying*, BBC News Online, 23 February, http://news.bbc.co.uk/1/hi/world/654210.stm, last accessed 23 February 2000 (2000).

[16] Anonymous, *Pentagon says Libya trying to get missile know-how from China*, Kyodo News Agency, 17 April, http://www.findarticles.com/p/articles/mi_m0WDQ/is_2000_April_17/ai_61968334, last accessed 15 August 2003 (2000).

[17] Anonymous, **Cryptologic Excellence: Yesterday, Today, and Tomorrow**, Center for Cryptologic History, Fort George G. Meade, MD (2002).

[18] Anonymous, *National Security Agency To Hire 1,500 People by September 2004*, Office of Public and Media Affairs, National Security Agency, Fort George G. Meade, MD, 07 April, http://www.nsa.gov/releases/relea00076.cfm, last accessed 11 April 2004 (2004).

[19] Anonymous, **Dedication and Sacrifice: National Aerial Reconnaissance in the Cold War**, Center for Cryptologic History, Fort George G. Meade, MD

http://www.nsa.gov/publications/publi00003.cfm, last accessed 17 January 2004 (undated).

[20] R. Atkinson, *GIs signal Bosnians: Yes, we're listening*, The Washington Post, 18 March (1996), A14.

[21] S. Baker, *Should Spies Be Cops?*, Foreign Policy 97, (1994), 36–52.

[22] J. Bamford, **The Puzzle Palace: A Report on America's Most Secret Agency**, Houghton Mifflin, Boston, MA (1982).

[23] J. Bamford, *Introduction*, **The Chinese Black Chamber: An Adventure in Espionage**, H.O. Yardley, ed., Houghton Mifflin, Boston, MA (1983), pp. vii–xxiv.

[24] J. Bamford, **Body of Secrets: Anatomy of the Ultra-Secret National Security Agency**, Doubleday, New York, NY (2001).

[25] J. Bamford, **A Pretext for War: 9/11, Iraq, and the Abuse of America's Intelligence Agencies**, Doubleday, New York, NY (2004).

[26] J.P. Barlow, *Decrypting the puzzle palace*, Communications of the ACM 35 (7) (1992), 25–31.

[27] J. Barron, **Breaking the Ring: The Spy Family That Imperiled America**, Houghton Mifflin, Boston, MA (1987).

[28] E.B. Bazan, **The Foreign Intelligence Surveillance Act**, Novinka Books, New York, NY (2002).

[29] R.L. Benson, M. Warner, *Preface*, **Venona: Soviet Espionage and the American Response, 1939–1957**, R.L. Benson and M. Warner, eds, National Security Agency and Central Intelligence Agency, Washington, DC (1996), pp. vii–xxxiii.

[30] J.D. Bergen, **Military Communications, A Test for Technology: The United States Army in Vietnam**, United States Army, US Government Printing Office, Washington, DC (1986).

[31] M.R. Beschloss, **Mayday: Eisenhower, Khrushchev and the U-2 Affair**, Harper & Row, New York, NY (1986).

[32] R.A. Best, Jr., **The National Security Agency: Issues for Congress**, Congressional Research Service, The Library of Congress, Washington, DC, 16 January (2001).

[33] B.W. Bidwell, **History of the Military Intelligence Division, Department of the Army General Staff, 1775–1941**, University Publications of America, Frederick, MD (1986).

[34] J.G. Blight, D.A. Welch, *What can intelligence tell us about the Cuban Missile Crisis, and what can the Cuban Missile Crisis tell us about intelligence?*, **Intelligence and the Cuban Missile Crisis**, J.G. Blight and D.A. Welch, eds, Frank Cass, London, UK (1998), pp. 1–17.

[35] J.G. Blight, D.A. Welch, *The Cuban Missile Crisis and intelligence performance*, **Intelligence and the Cuban Missile Crisis**, J.G. Blight and D.A. Welch, eds, Frank Cass, London, UK (1998), pp. 173–217.

[36] H. Blum, *I Pledge Allegiance: The True Story of the Walkers, An American Spy Family*, Simon & Schuster, New York, NY (1987).

[37] W. Blum, *Killing Hope: US Military and CIA Interventions Since World War II*, Common Courage Press, Monroe, ME (2004).

[38] P. Blustein, M. Jordan, *US eavesdropped on talks, sources say*, The Washington Post, 17 October (1995), 12A.

[39] G.H. Bolling, *AT&T, Aftermath of Antitrust: Preserving Positive Command and Control*, National Defense University, Washington, DC (1983).

[40] R. Bonner, *Weakness and Deceit: US Policy and El Salvador*, Times Books, New York, NY (1984).

[41] D.L. Boslaugh, *When Computers Went to Sea: The Digitization of the United States Navy*, IEEE Computer Society Press, Los Alamitos, CA (1999).

[42] F. Bowers, *Via eavesdropping, terror suspects nabbed*, The Christian Science Monitor, 02 June (2004), 2.

[43] N.K. Breglio, *Leaving FISA behind: The need to return to warrantless foreign intelligence surveillance*, The Yale Law Journal 113 (1) (2003), 179–218.

[44] W.B. Breuer, *Shadow Warriors: The Covert War in Korea*, Wiley, New York, NY (1996).

[45] M. Bright, E. Vulliamy, P. Beaumont, *UN launches inquiry into American spying*, The Observer, 09 March (2003), 4.

[46] W.J. Broad, *Evading the Soviet ear at Glen Cove*, Science 217 (3) (1982), 910–911.

[47] Z. Brzezinski, *Power and Principle: Memoirs of the National Security Adviser*, Farrar, Straus & Giroux, New York, NY (1983).

[48] R. Bulkeley, *The Sputniks Crisis and Early United States Space Policy: A Critique of the Historiography of Space*, Indiana University Press, Bloomington, IN (1991).

[49] C. Burke, *Information and Secrecy: Vannevar Bush, Ultra, and the Other Memex*, The Scarecrow Press, Inc., Metuchen, NJ (1994).

[50] C. Burke, *Automating American cryptanalysis 1930–1945: Marvelous machines, a bit too late*, **Allied and Axis Signals Intelligence in World War II**, D. Alvarez, ed., Frank Cass Publishers, London, UK (1999), pp. 18–39.

[51] W.E. Burrows, *Deep Black: Space Espionage and National Security*, Random House, New York, NY (1986).

[52] G. Cahlink, *Security agency doubled procurement spending in four years*, Government Executive, 01 June, http://govexec.com/dailyfed/0604/060104g1.htm, last accessed 03 June 2004 (2004).

[53] D. Campbell, *British MP accuses US of electronic spying*, The New Scientist 71(1012) (1976), 268.

[54] D. Campbell, *The State of the Art in Communications Intelligence of Automated Processing for Intelligence Purposes of Intercepted Broadband Multi-Language Leased or Common Carrier Systems and Its Applicability to COMINT Targeting and Selection, Including Speech Recognition*, European Parliament, Directorate General for Research, Luxembourg (1999).

[55] D. Campbell, *Shhh...they're listening*, The UNESCO Courier 54 (3) (2001), 34–35.

[56] D. Campbell, L. Malvern, *1980: America's big ear on Europe*, The New Statesman 12(584) (1999), 47.

[57] D. Campbell, P. Lashmar, *The new Cold War: How America spies on us for its oldest friend, the dollar*, The Independent, 02 July (2000), 5.

[58] W.M. Carley, *As Cold War fades, some nations' spies seek industrial secrets*, The Wall Street Journal, 17 June (1991), A1, A5.

[59] A.B. Carter, *Telecommunications policy and US national security*, **Changing the Rules: Technological Change, International Competition and Regulation in Communications**, R.W. Crandall and K. Flamm, eds, The Brookings Institution, Washington, DC (1989), pp. 221–256.

[60] J. Chaffin, *US turns to private sector for spies*, The Financial Times, 17 May (2004), 7.

[61] R.W. Clark, *The Man Who Broke Purple: The Life of Colonel William F. Friedman, Who Deciphered the Japanese Code in World War II*, Little, Brown and Company, Boston, MA (1977).

[62] E. Cody, *Allies test joint land, sea, air assaults*, The Washington Post, 17 February (1991), A29.

[63] Commander, US Naval Forces, Japan, *USS Pueblo (AGER-2) Operational Orders*, United States Navy, 18 December, http://www.usspueblo.org/v2f/attack/missionorders.html, last accessed 09 October 2004 (1968).

[64] K. Conboy, D. Andradé, *Spies and Commandos: How America Lost the Secret War in North Vietnam*, University Press of Kansas, Lawrence, KS (2000).

[65] O.R. Cote, Jr., *The Third Battle: Innovation in the US Navy's Silent Cold War Struggle With Soviet Submarines*, Naval War College Newport Papers 16, Naval War College Press, Newport, RI (2003).

[66] J.M. Crewdson, *Documents show Nixon approved partly 'illegal' '70 security plan*, The New York Times, 07 June (1973), 1, 36–37.

[67] A.J. Cristol, *The Liberty Incident: The 1967 Israeli Attack on the US Navy Spy Ship*, Brassey's, Inc., Washington, DC (2002).

[68] J. Debrosse, C. Burke, *The Secret in Building 26: The Untold Story of America's Ultra War Against the U-Boat Enigma Codes*, Random House, New York, NY (2004).

[69] J. Diamond, *Spy satellites take spending cut in intelligence budget*, Associated Press, 09 October, http://www.floridatoday.com/space/explore/stories/1998b/100998r.htm, last accessed 10 October 1998 (1998).

[70] P. Dickson, *Sputnik: The Shock of the Century*, Walker & Company, New York, NY (2001).

[71] W. Diffie, S. Landau, *Privacy on the Line: The Politics of Wiretapping and Encryption*, MIT Press, Cambridge, MA (1998).

[72] R. Dingman, *Power in the Pacific: The Origins of Naval Arms Limitation, 1914–1922*, The University of Chicago Press, Chicago, IL (1976).

[73] R.A. Divine, *The Sputnik Challenge*, Oxford University Press, New York, NY (1993).

[74] F.J. Donner, *The Age of Surveillance: The Aims and Methods of America's Political Intelligence System*, Knopf, New York, NY (1980).

[75] J.M. Dorward, *The Office of Naval Intelligence: The Birth of America's First Intelligence Agency, 1865–1918*, Naval Institute Press, Washington, DC (1979).

[76] J.M. Dorward, *Conflict of Duty: The US Navy's Intelligence Dilemma, 1919–1945*, Naval Institute Press, Annapolis, MD (1983).

[77] E.J. Drea, *MacArthur's ULTRA: Codebreaking and the War Against Japan, 1942–1945*, University Press of Kansas, Lawrence, KS (1992).

[78] B. Drogin, *NSA blackout reveals downside of secrecy*, The Los Angeles Times, 13 March (2000), 1.

[79] A.W. Dulles, *The Craft of Intelligence*, Harper & Row, New York, NY (1963).

[80] M. Edwards, *Concerns over data security stimulate countermeasures*, Communications News, March, http://www.findarticles.com/p/articles/mi_m0CMN/is_n3_v21/ai_557091, last accessed 14 August 2003 (1984).

[81] J.M. Ennes, Jr., *Assault on the Liberty: The True Story of the Israeli Attack on an American Intelligence Ship*, Random House, New York, NY (1979).

[82] D. Ensor, *Biggest US spy agency choking on too much information*, Cable News Network, 25 November, http://www.cnn.com/US/9911/25/nsa.woes/, last accessed 26 November 1999 (1999).

[83] J.E. Epstein, *The spy war*, The New York Times Magazine, 28 September (1980), 11–14.

[84] J.E. Epstein, *Deception: The Invisible War Between the KGB and the CIA*, Simon & Schuster, New York, NY (1989).

[85] F. Escalante, *The Secret War: CIA Operations Against Cuba, 1959–1962*, Ocean Press, Melbourne, Australia (1995).

[86] European Parliament, Temporary Committee on the ECHELON Interception System, Report on the Existence of A Global System for the Interception of Private and Commercial Communications (ECHELON Interception System), The European Parliament, Luxembourg, July (2001).

[87] L. Farago, *The Broken Seal: The Story of Operation Magic and the Pearl Harbor Disaster*, Random House, New York, NY (1967).

[88] P. Fellwock, *US electronic espionage: A memoir*, Ramparts 11 (2) (1972), 35–50.

[89] J.J. Fialka, *War by Other Means: Economic Espionage in America*, W.W. Norton & Company, New York, NY (1997).

[90] J. Fitsanakis, The nerves of government: Electronic networking and social control in the Information Society, PhD dissertation, Department of Politics, University of Edinburgh, UK (2002).

[91] J. Fitsanakis, *State-sponsored communications interception: Facilitating illegality*, Information, Communication and Society 6 (3) (2003), 404–429.

[92] H.P. Ford, *Estimative Intelligence: The Purposes and Problems of National Intelligence Estimating*, University Press of America, Lanham, MD (1993).

[93] H.P. Ford, *CIA and the Vietnam Policymakers: Three Episodes, 1962–1968*, Center for the Study of Intelligence, Central Intelligence Agency, Washington, DC (1998).

[94] B.E. Fredriksson, *Space power in joint operations: Evolving concepts*, Air and Space Power Journal 28 (2) (2004), 85–95.

[95] S.Z. Freiberger, *Dawn Over Suez: The Rise of American Power in the Middle East, 1953–1957*, Ivan R. Dee, Chicago, IL (1992).

[96] A. Friedman, *Spider's Web: The Secret History of How the White House Illegally Armed Iraq*, Bantam Books, New York, NY (1993).

[97] D.A. Fulghum, *SIGINT aircraft may face obsolescence in five years*, Aviation Week and Space Technology 145 (17) (1996), 54.

[98] A. Fursenko, T. Naftali, *Soviet intelligence and the Cuban Missile Crisis*, Intelligence and the Cuban Missile Crisis, J.G. Blight and D.A. Welch, eds, Frank Cass, London, UK (1998), pp. 64–87.

[99] R.A. Gabriel, *Military Incompetence: Why the American Military Doesn't Win*, Hill & Wang, New York, NY (1985).

[100] J. Gannon, *Stealing Secrets, Telling Lies: How Spies and Codebreakers Helped Shape the Twentieth Century*, Brassey's, Washington, DC (2001).

[101] M. Gannon, *Pearl Harbor Betrayed: The True Story of A Man and A Nation Under Attack*, Henry Holt & Co., New York, NY (2001).

[102] R.L. Garthoff, *US intelligence in the Cuban Missile Crisis*, Intelligence and the Cuban Missile Crisis, J.G. Blight and D.A. Welch, eds, Frank Cass, London, UK (1998), pp. 18–63.

[103] B. Gellman, *US spied on Iraqi military via UN*, The Washington Post, 02 March (1999), A1.

[104] J.A. Gentry, *Doomed to fail: America's blind faith in military technology*, Parameters 32 (4) (2002), 88–103.

[105] J. Gerth, *US rethinking a satellite deal over links to Chinese military*, The New York Times, 18 June (1998), A1.

[106] B. Gertz, *US breathes easier as it spots Iraq's jamming gear*, The Washington Times, 09 October (1990), A8.

[107] B. Gertz, *Russian voices directing Iraqis*, The Washington Times, 13 February (1991), A1.

[108] B. Gertz, ***Breakdown: How America's Intelligence Failures Led to September 11***, Regnery Publishing, Inc., Washington, DC (2002).

[109] F. Gervasi, ***Big Government: The Meaning and Purpose of the Hoover Commission Report***, Whittlesey House, New York, NY (1949).

[110] J.L. Gilbert, ***The Most Secret War: Army Signals Intelligence in Vietnam***, Military History Office, US Army Intelligence and Security Command, Fort Belvoir, VA (2003).

[111] L.A. Gladwin, *Cautious collaborators: The struggle for Anglo–American cryptanalytic co-operation, 1940–43*, ***Allied and Axis Signals Intelligence in World War II***, D. Alvarez, ed., Frank Cass Publishers, London, UK (1999), pp. 119–145.

[112] J.D. Glasser, ***The Secret Vietnam War: The United States Air Force in Thailand, 1961–1975***, McFarland & Co., Inc., Jefferson, NC (1995).

[113] J.C. Goulden, ***Truth is the First Casualty: The Gulf of Tonkin Affair, Illusion and Reality***, James B. Adler, Inc., Chicago, IL (1969).

[114] B. Graham, J. Nussbaum, ***Intelligence Matters: The CIA, the FBI, Saudi Arabia, and the Failure of America's War on Terror***, Random House, New York, NY (2004).

[115] S. Gregory, Economic intelligence in the post-Cold War era: Issues for reform, paper delivered at the Intelligence Reform in the Post-Cold War Era conference, The Woodrow Wilson School of Public and International Affairs, Princeton University, January, http://www.fas.org/irp/eprint/snyder/economic.htm, last accessed 12 January 2003 (1997).

[116] J. Gunther, ***The Riddle of MacArthur: Japan, Korea, and the Far East***, Harper & Brothers, New York, NY (1951).

[117] N. Hager, ***Secret Power: New Zealand's Role in the International Spy Network***, Craig Poton, Nelson, New Zealand (1996).

[118] M.H. Halperin, J.J. Berman, R.L. Borosage, C.M. Marwick, ***The Lawless State***, Penguin, New York, NY (1976).

[119] B. Harvey, ***Race Into Space: The Soviet Space Programme***, Ellis Horwood, Ltd., Chichester, UK (1988).

[120] D.A. Hatch, R.L. Benson, ***The Korean War: The SIGINT Background***, National Security Agency, Fort George G. Meade, MD, http://purl.access.gpo.gov/GPO/LPS23352, last accessed 08 June 2004 (2000).

[121] H. Haufler, ***Codebreakers' Victory: How the Allied Cryptographers Won World War II***, New American Library, New York, NY (2003).

[122] G.C. Herring, ***America's Longest War: The United States and Vietnam, 1950–1975***, Wiley, New York, NY (1979).

[123] S.M. Hersh, *Was Castro out of control in 1962?*, The Washington Post, 11 October (1987), H1.

[124] J. Hillkirk, *Trade secrets: Next frontier for US spies*, USA Today, 05 June (1990), A7.

[125] W. Hinckle, W.M. Turner, ***The Fish is Red: The Story of the Secret War Against Castro***, Harper & Row, New York, NY (1981).

[126] J. Hoagland, *America's Frankenstein's monster*, The Washington Post, 07 February (1991), A19.

[127] G. Hodgson, ***The Colonel: The Life and Wars of Henry Stimson, 1867–1950***, Alfred A. Knopf, New York, NY (1990).

[128] D. Holloway, ***The Soviet Union and the Arms Race***, Yale University Press, New Haven, CT (1983).

[129] W.J. Holmes, ***Double-Edged Secrets: US Naval Intelligence Operations in the Pacific During World War II***, Naval Institute Press, Annapolis, MD (1979).

[130] P.M. Holt, ***Secret Intelligence and Public Policy: A Dilemma of Democracy***, CQ Press, Washington, DC (1995).

[131] J.S. Hopkins, *Secretive NSA is open for business*, The Baltimore Sun, 02 June (2004), 1C.

[132] S. Howarth, ***Morning Glory: A History of the Imperial Japanese Navy***, Hamish Hamilton, London, UK (1983).

[133] L.K. Johnson, *A Season of Inquiry: Congress and Intelligence*, Dorsey Press, Chicago, IL (1988).

[134] L.K. Johnson, ***America's Secret Power: The CIA in a Democratic Society***, Oxford University Press, New York, NY (1989).

[135] L.K. Johnson, ***Secret Agencies: US Intelligence in a Hostile World***, Yale University Press, New York, NY (1996).

[136] L.K. Johnson, ***Bombs, Bugs, Drugs, and Thugs: Intelligence and America's Quest for Security***, New York University Press, New York, NY (2000).

[137] T.R. Johnson, D.A. Hatch, ***NSA and the Cuban Missile Crisis***, Center for Cryptologic History, Fort George G. Meade, MD, http://www.nsa.gov/cuba/index.cfm, last accessed 19 January 2004 (1998).

[138] B.E. Jones, ***War Without Windows: A True Account of a Young Army Officer Trapped in an Intelligence Cover-Up in Vietnam***, The Vanguard Press, New York, NY (1987).

[139] D. Kahn, ***The Codebreakers: The Story of Secret Writing***, Macmillan, New York, NY (1967).

[140] D. Kahn, *Cryptology goes public*, Foreign Affairs, 58 (1) (1979), 141–159.

[141] D. Kahn, ***Kahn on Codes: Secrets of the New Cryptology***, Macmillan Publishing Company, New York, NY (1983).

[142] D. Kahn, ***The Codebreakers: The Story of Secret Writing***, Scribner, New York, NY (1996).

[143] N. Kawamura, ***Turbulence in the Pacific: Japanese-US Relations During World War I***, Praeger, Westport, CT (2000).

[144] C.S. Kennedy, *The American Consul: A History of the United States Consular Service, 1776–1914*, Greenwood Press, New York, NY (1990).

[145] R. Kessler, *The Bureau: The Secret History of the FBI*, St. Martin's Press, New York, NY (2002).

[146] C.C. Kingseed, *Eisenhower and the Suez Crisis of 1956*, Louisiana University Press, Baton Rouge, LA (1995).

[147] M. Klein, *Securing Record Communications: The TSEC/KW-26*, Center for Cryptologic History, Fort George G. Meade, MD (2003).

[148] K. Knorr, *Failures in National Intelligence Estimates: The case of the Cuban missiles*, World Politics 16 (3) (1964), 455–467.

[149] K. Komatsu, *Origins of the Pacific War and the Importance of Magic*, Curzon Press, Ltd, Richmond, UK (1999).

[150] M.L. Lanning, *Senseless Secrets: The Failures of US Military Intelligence from George Washington to the Present*, Birch Lane Press, New York, NY (1996).

[151] G. Lardner, Jr., *Agency is reluctant to share information*, The Washington Post, 19 March (1980), 4.

[152] G. Lardner, Jr., *Uncompromising NSA frustrated North prosecutors*, The Washington Post, 19 March (1990), A1, A4.

[153] P. Leary, *The Second Cryptographic Shakespeare*, Westchester House Publishers, Omaha, NE (1990).

[154] M. Ledeen, W. Lewis, *Debacle: The American Failure in Iran*, Alfred A. Knopf, New York, NY (1981).

[155] M.B. Lerner, *The Pueblo Incident: A Spy Ship and the Failure of American Foreign Policy*, University Press of Kansas, Lawrence, KS (2002).

[156] A. Levite, *Intelligence and Strategic Surprises*, Columbia University Press, New York, NY (1987).

[157] R. Lewin, *The American Magic: Codes, Ciphers and the Defeat of Japan*, Penguin Books, New York, NY (1983).

[158] E. Lewis, *Public Entrepreneurship: Toward A Theory of Bureaucratic Political Power*, Indiana University Press, Bloomington, IN (1980).

[159] D.T. Lindgren, *Trust But Verify: Imagery Analysis in the Cold War*, Naval Institute Press, Annapolis, MD (2000).

[160] R. Lindsey, *The Falcon and the Snowman: A True Story of Friendship and Espionage*, Vintage Books, New York, NY (1980).

[161] D. Linzer, *IAEA leader's phone tapped: US pores over transcripts to try to oust nuclear chief*, The Washington Post, 12 December (2004), A01.

[162] R.A. Liston, *The Pueblo Surrender: A Covert Action by the National Security Agency*, M. Evans and Co., Inc., New York, NY (1988).

[163] V. Loeb, *NSA to turn over non-spy technology to private industry*, The Washington Post, 07 June (2000), A29.

[164] V. Loeb, *Test of strength*, The Washington Post, 29 July (2001), W08.

[165] W. Lord, *Incredible Victory*, Harper & Row, New York, NY (1967).

[166] W.S. Lucas, *Divided We Stand: Britain, the US and the Suez Crisis*, Hodder & Stoughton, London, UK (1991).

[167] E. MacAskill, P. Wintour, R. Norton-Taylor, *Did we bug Kofi Annan?*, The Guardian, 27 February (2004), 1.

[168] K. Macksey, *The Searchers: How Radio Interception Changed the Course of Both World Wars*, Cassell, London, UK (2003).

[169] R. MacMillan, *US releases new encryption export rules*, Newsbytes News Network, 31 December, http://www.findarticles.com/p/articles/mi_m0NEW/is_1998_Dec_31/ai_53505647, last accessed 12 September 2003 (1998).

[170] R. Macmillan, *European officials leave US in huff over spy network*, Newsbytes News Network, 10 May, http://www.findarticles.com/p/articles/mi_m0NEW/is_2001_May_10/ai_74437581, last accessed 11 January 2003 (2001).

[171] W. Madsen, *The NSA's Trojan whore?*, Covert Action Quarterly 63 http://www.caq.com/CAQ/caq63/caq63madsen.html, last accessed 25 February 1999 (1998).

[172] T.W. Maier, *Uncovering cloak-and-dagger conspiracies*, Insight on the News 17(28) (2001), 36–39.

[173] C. Mason, *US opposes US West's plans in Soviet Union: Plan to build fiber optic*, Telephony 218 (24) (1990), 8, 9, 11.

[174] E.R. May, *The Truman Administration and China, 1945–1949*, J.B. Lippincott Company, Philadelphia, PA (1975).

[175] J.D. Mayer, *9-11 and the secret FISA court: From watchdog to lapdog?*, Case Western Reserve Journal of International Law 34 (2) (2002), 249–252.

[176] J.A. McChristian, *Vietnam Studies: The Role of Military Intelligence, 1965–1967*, United States Army, US Government Printing Office, Washington, DC (1974).

[177] J.W. McCoy, *Secrets of the Viet Cong*, Hippocrene Books, New York, NY (1992).

[178] J. Mintz, *Inman may be queried on arms dealer*, The Washington Post, 20 December (1993), A12.

[179] R.C. Moe, *The Hoover Commissions Revisited*, Westview Press, Boulder, CO (1982).

[180] E.E. Moïse, *Tonkin Gulf and the Escalation of the Vietnam War*, The University of North Carolina Press, Chapel Hill, NC (1996).

[181] S. Moreno, V. Loeb, *Ex-army cryptologist accused of spying*, The Washington Post, 14 October (1998), B01.

[182] R.E. Morgan, *Domestic Intelligence: Monitoring Dissent in America*, University of Texas Press, Austin, TX (1980).

[183] E.E. Morrison, *Turmoil and Tradition: A Study of the Life and Times of Henry L. Stimson*, Houghton Mifflin, Boston, MA (1960).

[184] B.E. Mullen, *NSA said to oppose Soviet fiber optic cable plan*, Newsbytes News Network, 20 March,

http://www.findarticles.com/p/articles/mi_m0NEW/is_1990_March_20/ai_8288966, last accessed 15 August 2003 (1990).

[185] B. Murray, *National Security Agency recovers from 3-day computer outage*, Newsbytes News Network, 31 January, http://www.findarticles.com/p/articles/mi_m0HDN/is_2000_Jan_31/ai_59119231, last accessed 16 August 2003 (2000).

[186] Musicant, *The Banana Wars: A History of United States Military Intervention in Latin America from the Spanish-American War to the Invasion of Panama*, MacMillan Publishing Company, New York, NY (1990).

[187] J.A. Nathan, *Robert McNamara's Vietnam deception*, USA Today (magazine) 2604 (1995), 32–35.

[188] National Commission on Terrorist Attacks, *The 9/11 Commission Report: Final Report of the National Commission on Terrorist Attacks Upon the United States*, W.W. Norton & Co., New York, NY (2004).

[189] D.D. Newsom, *The Soviet Brigade in Cuba: A Study in Political Diplomacy*, Indiana University Press, Bloomington, IN (1987).

[190] C. Nieto, *Masters of War: Latin America and United States Aggression from the Cuban Revolution Through the Clinton Years*, Seven Stories Press, New York, NY (2003).

[191] K.S. Olmsted, *Challenging the Secret Government: The Post-Watergate Investigations of the CIA and FBI*, The University of North Carolina Press, Chapel Hill, NC (1996).

[192] B. Palmer, Jr., *The 25-Year War: America's Military Role in Vietnam*, University Press of Kentucky, Lexington, KY (1984).

[193] T.G. Paterson, *Contesting Castro: The United States and the Triumph of the Cuban Revolution*, Oxford University Press, New York, NY (1994).

[194] H.A.R. Philby, *My Silent War*, MacGibbon & Kee, London, UK (1968).

[195] H.A.R. Philby, *Lecture to the KGB, July 1977*, *The Private Live of Kim Philby: The Moscow Years*, R. Philby, M. Lyubimov and H. Peake, eds, St Ermin's Press, London, UK (1999).

[196] C.J. Phillips, *What made Venona possible?*, *Venona: Soviet Espionage and The American Response, 1939–1957*, R.L. Benson and M. Warner, eds, National Security Agency and Central Intelligence Agency, Washington, DC (1996), p. xv.

[197] C. Pincher, *Traitors: The Labyrinths of Treason*, Sidgwick & Jackson, London, UK (1987).

[198] W. Pincus, *Military espionage cuts eyed*, The Washington Post, 17 March (1995), A6.

[199] W. Pincus, *NSA system crash raises hill worries: Agency computers termed out of date*, The Washington Post, 02 February (2000), A19.

[200] R.E. Powaski, *March to Armageddon: The United States and the Nuclear Arms Race, 1939 to the Present*, Oxford University Press, New York, NY (1987).

[201] J. Prados, *Keepers of the Keys: A History of the National Security Council from Truman to Bush*, William Morrow and Company, Inc., New York, NY (1991).

[202] J. Prados, *Combined Fleet Decoded: The Secret History of American Intelligence and the Japanese Navy in World War II*, Random House, New York, NY (1995).

[203] W. Preston, Jr., *Aliens and Dissenters: Federal Suppression of Radicals, 1903–1933*, Harper & Row, New York, NY (1963).

[204] H.H. Ransom, *Secret intelligence in the United States, 1947–1982: The CIA's search for legitimacy*, *The Missing Dimension: Governments and Intelligence Communities in the Twentieth Century*, C. Andrew and D. Dilks, eds, University of Illinois Press, Urbana, IL (1984), pp. 199–226.

[205] E. Reid, *Time of Fear and Hope: The Making of the North Atlantic Treaty, 1947–1949*, McClelland and Stewart, Toronto, Canada (1977).

[206] T. Reid, *No place to hide from hi-tech spies in the sky*, The London Times, 07 February (2003), 22.

[207] J.T. Richelson, *The Wizards of Langley: Inside the CIA's Directorate of Science and Technology*, Westview Press, Boulder, CO (2001).

[208] J.T. Richelson, D. Ball, *The Ties That Bind: Intelligence Cooperation Between the UKUSA Countries – the United Kingdom, the United States of America, Canada, Australia and New Zealand*, Allen & Unwin, Boston, MA (1985).

[209] M. Riebling, *Wedge: The Secret War Between the FBI and the CIA*, Alfred A. Knopf, New York, NY (1994).

[210] T.M. Rienzi, *Vietnam Studies: Communications Electronics, 1962–1970*, United States Army, US Government Printing Office, Washington, DC (1972).

[211] F.J. Rippy, *Latin America: A Modern History*, University of Michigan Press, Ann Arbor, MI (1958).

[212] J. Risen, L. Bergman, *US thinks agent revealed tunnel at Soviet embassy*, The New York Times, 04 March (2001), 1.1.

[213] P.C. Roe, *The Dragon Strikes: China and the Korean War, June–December 1950*, Presidio Press, Inc., Novato, CA (2000).

[214] N.C. Ronning, *Introduction*, *Intervention in Latin America*, N.C. Ronning, ed., Alfred A. Knopf, New York, NY (1970), pp. 3–23.

[215] A. Roslin, *Cyberspies and saboteurs: Hackers on the payroll of US security agencies*, The Montreal Gazette, 06 October (2001), A1.

[216] F.B. Rowlett, *The Story of Magic: Memoirs of an American Cryptologic Pioneer*, Aegean Park Press, Laguna Hills, CA (1998).

[217] D.F. Rudgers, *Creating the Secret State: The Origins of the Central Intelligence Agency, 1943–1947*, University Press of Kansas, Lawrence, KS (2000).

[218] D. Ruppe, *US to close eavesdropping post*, ABC News, 02 June, http://www.globalsecurity.org/org/news/2001/010601-echelon.htm, last accessed 11 July 2003 (2002).

[219] A. Sabar, *Want to be a spy? NSA is hiring*, The Baltimore Sun, 10 April (2004), 1A.

[220] L.S. Safford, *A brief history of communications intelligence in the United States, **Listening to the Enemy: Key Documents on the Role of Communications Intelligence in the War with Japan***, R.H. Spector, ed., Scholarly Resources, Inc., Wilmington, DE (1988), pp. 3–12.

[221] A. Sampson, *The Sovereign State: The Secret History of ITT*, Hodder & Stoughton, London (1973).

[222] J. Schecter, L. Schecter, *Sacred Secrets: How Soviet Intelligence Operations Changed American History*, Brassey's, Inc., Washington, DC (2002).

[223] S.C. Schlesinger, *Act of Creation: The Founding of the United Nations*, Westview Press, Boulder, CO (2003).

[224] P. Schweizer, *Friendly Spies: How America's Allies Are Using Economic Espionage to Steal Our Secrets*, The Atlantic Monthly Press, New York, NY (1993).

[225] S. Shane, T. Bowman, *US Secret Agency Scored World Coup: NSA Rigged Machines for Eavesdropping*, The Baltimore Sun, 03 January (1996), 1A.

[226] D. Sharpley, *Intelligence agency chief seeks dialogue with academics*, Science 202 (4366) (1978), 407–410.

[227] E.Y. Siff, *Why the Senate Slept: The Gulf of Tonkin Resolution and the Beginning of America's Vietnam War*, Praeger, Westport, CT (1999).

[228] R.R. Simmons, *The Pueblo, EC-121, and Maya-Guez Incidents: Some Continuities and Changes*, Occasional Papers/Reprints Series in Contemporary Asian Studies, School of Law, The University of Maryland, College Park, MD (1978).

[229] H. Sklar, *Washington's War on Nicaragua*, South End Press, Boston, MA (1988).

[230] F.J. Smist, Jr., *Congress Oversees the United States Intelligence Community, 1947–1994*, The University of Tennessee Press, Knoxville, TN (1994).

[231] A. Smith, *A sale to Red China we will one day regret*, Insight on the News 15(20) (1999), 28–30.

[232] H. Smith, *US aides say loss of post in Iran impairs missile monitoring ability*, The New York Times, 02 March (1979), 1, 8.

[233] L.B. Snider, *Unlucky shamrock: recollections from the Church Committee's investigation of NSA*, Studies In Intelligence, winter, http://www.cia.gov/csi/studies/winter99-00/art4.html, last accessed 20 June 2004 (1999).

[234] H.L. Stimson, M. Bundy, *On Active Service in Peace and War*, Harper and Brothers, New York, NY (1946).

[235] R. Stinnett, *Day of Deceit: The Truth About FDR and Pearl Harbor*, The Free Press, New York, NY (1999).

[236] I. Stokell, *Cyberspace is potential battleground: NSA chief*, Newsbytes News Network, 17 October, http://www.findarticles.com/p/articles/mi_m0NEW/is_2000_Oct_17/ai_66162678, last accessed 20 August 2003 (2000).

[237] I.F. Stone, *The Hidden History of the Korean War*, Monthly Press Review, New York, NY (1969).

[238] M. Streetly, *Allies take advantage of electronic support*, Flight International, 30 January (1991), 10.

[239] Y. Sugita, *Pitfall or Panacea: The Irony of US Power in Occupied Japan, 1945–1952*, Routledge, New York, NY (2003).

[240] R. Talbert, Jr., *Negative Intelligence: The Army and the American Left, 1917–1941*, University Press of Mississippi, Jackson, MS (1991).

[241] P. Taubman, *Secret Empire: Eisenhower, the CIA, and the Hidden Story of America's Space Espionage*, Simon & Schuster, New York, NY (2003).

[242] A. Theoharis, *Spying on Americans: Political Surveillance from Hoover to the Huston Plan*, Temple University Press, Philadelphia, PA (1978).

[243] R.C. Toth, *Head of NSA is dismissed for opposing budget cuts*, The Los Angeles Times, 19 April (1985), 1.

[244] H.S. Truman, Communications Intelligence Activities [memorandum to the secretaries of state and defense outlining the charter of the National Security Agency], 24 October, http://jya.com/nsa102452.htm, last accessed 15 November 2002 (1952).

[245] A. Tully, *The Super Spies: More Secret, More Powerful Than the CIA*, William Morrow & Company, Inc., New York, NY (1969).

[246] S. Turner, *The stupidity of intelligence: The real story behind that Soviet combat brigade in Cuba*, The Washington Monthly, February, http://www.findarticles.com/p/articles/mi_m1316/is_v18/ai_4118413, last accessed 14 August 2003 (1986).

[247] S. Turner, Foreword, *The Soviet Brigade in Cuba: A Study in Political Diplomacy*, D.D. Newsom, Indiana University Press, Bloomington, IN (1987), pp. vii–xiii.

[248] S.J. Ungar, *FBI*, Little, Brown & Co., Boston, MA (1976).

[249] US Congress, Commission on the Roles and Capabilities of the United States Intelligence Community, *Preparing for the 21st Century: An Appraisal of US Intelligence*, Government Printing Office, Washington, DC (1996).

[250] US Congress, [House] Committee on Un-American Activities, *Security Practices in the National Security Agency: Defection of Bernon F. Mitchell and William H. Martin*, Committee on Un-American Activities, Washington, DC, 13 August (1962).

[251] US Congress, [House] Select Committee on Intelligence, *United States Intelligence Agencies and Ac-*

tivities, Government Printing Office, Washington, DC (1976).

[252] US Congress, [House] Subcommittee of the Committee on Government Operations, *Interception of Nonverbal Communications by Federal Intelligence Agencies*, Government Printing Office, Washington, DC (1976).

[253] US Congress, [House] Committee on Government Operations, *Report on the Computer Security Act of 1987*, Government Printing Office, Washington, DC, 17 March (1987).

[254] US Congress, [House] Permanent Select Committee on Intelligence, *Intelligence Authorization Act for Fiscal Year 1999*, Government Printing Office, Washington, DC (1998).

[255] US Congress, [House] Permanent Select Committee on Intelligence, *Intelligence Authorization Act for Fiscal Year 2000*, Government Printing Office, Washington, DC (1999).

[256] US Congress, [House] Subcommittee on Terrorism and Homeland Security of the Permanent Select Committee on Intelligence, *Counterterrorism Intelligence Capabilities and Performance Prior to 9–11*, Government Printing Office, Washington, DC, July (2002).

[257] US Congress, [Senate] Committee on Foreign Relations and Committee on Armed Services, *The Situation in Cuba*, Government Printing Office, Washington, DC (1962).

[258] US Congress, [Senate] *Final Report, Book III: Supplementary Detailed Staff Reports on Intelligence Activities and the Rights of Americans*, Select Committee to Study Governmental Operations With Respect to Intelligence Activities, Government Printing Office, Washington, DC (1976).

[259] US Congress, [Senate] Select Committee on Intelligence, *Special Report*, Government Printing Office, Washington, DC (2001).

[260] US Department of Defense, Inspector General, Intelligence Review Directorate, Policy and Oversight, *Final Report on the Verification Inspection of the National Security Agency*, Government Pringing Office, Washington, DC (1996).

[261] US Department of Justice, Commission for the Review of FBI Security Programs, *A Review of FBI Security Programs*, Government Printing Office, Washington, DC (2002).

[262] G. Van Nederveen, *Signals intelligence support to the cockpit*, paper delivered at the 6th International Command and Control Research and Technology Symposium, Command and Control Research Program, United States Department of Defense, Annapolis, MD, 19–21 June (2001).

[263] G.L. Vistica, E. Thomas, *Hard of hearing*, Newsweek, 13 December (1999), 78.

[264] G.L. Vistica, Cyberwar, sabotage, Newsweek, 31 May (1999), 38.

[265] L.E. Walsh, *Firewall: The Iran-Contra Conspiracy and Cover-Up*, W.W. Norton & Company, New York, NY (1997).

[266] A. Walworth, *Wilson and His Peacemakers: American Diplomacy at the Paris Peace Conference, 1919*, W.W. Norton & Company, New York, NY (1986).

[267] R.E. Weber, *United States Diplomatic Codes and Ciphers, 1775–1938*, Precedent Publishing, Inc., Chicago, IL (1979).

[268] T. Weiner, *Iraq uses techniques in spying against its former tutor, the US*, The Philadelphia Inquirer, 25 January (1991), A1, A10.

[269] T. Weiner, *Tapes reveal Nixon's ideas on assassination, burglaries*, The Oregonian, 26 February (1999), A09.

[270] R. Welch, Jr., *Response to Revolution: The United States and the Cuban Revolution, 1959–1961*, The University of North Carolina Press, Chapel Hill, NC (1985).

[271] N. West, *The Circus: MI5 Operations, 1945–1972*, Stein and Day, New York, NY (1984).

[272] T. Wilford, *Pearl Harbor Redefined: USN Radio Intelligence in 1941*, University Press of America, Lanham, MD (2001).

[273] J.J. Wirtz, *The Tet Offensive: Intelligence Failure in War*, Cornell University Press, Ithaca, New York (1991).

[274] A. Wise, *The American Police State: The Government Against the People*, Random House, New York, NY (1976).

[275] A. Wise, *Spy: The Inside Story of How the FBI's Robert Hanssen Betrayed America*, Random House, New York, NY (2002).

[276] A. Wise, T.B. Ross, *The U-2 Affair*, Random House, New York, NY (1962).

[277] D. Wise, T.B. Ross, *The Invisible Government*, Random House, New York, NY (1964).

[278] D. Wise, T.B. Ross, *The Invisible Government*, Vintage Books, New York, NY (1974).

[279] J. Witcover, *Sabotage at Black Tom: Imperial Germany's Secret War in America, 1914–1917*, Algonquin Books, Chapel Hill, NC (1989).

[280] R. Wohlstetter, *Cuba and Pearl Harbor: Hindsight and foresight*, Foreign Affairs 43(4) (1965), 691–707.

[281] D.K. Wood, J.B. Mercier, *Building the ACE in Kosovo: Analysis and control element*, Military Intelligence Professional Bulletin 27 (1) (2001), 33–36.

[282] R. Wood, *Call Sign Rustic: The Secret Air War Over Cambodia, 1970–1973*, Smithsonian Institution Press, Washington, DC (2002).

[283] B. Woodward, *Veil: The Secret Wars of the CIA, 1981–1987*, Simon and Schuster, New York, NY (1987).

[284] J.R. Woolsey, *Why we spy on our allies*, The Wall Street Journal, 17 March (2000), A18.

[285] J.R. Woolsey, Transcript of Briefing Delivered at The Foreign Press Center, Washington, DC, 07 March,

http://cryptome.org/echelon-cia.htm, last accessed 12 March 2000 (2000).

[286] P. Wright, *Spy Catcher: The Candid Autobiography of A Senior Intelligence Officer*, Viking, New York, NY (1987).

[287] F.B. Wrixon, *Codes, Ciphers and Other Cryptic and Clandestine Communication: Making and Breaking Secret Messages from Hieroglyphs to the Internet*, Black Dog & Leventhal Publishers, Inc., New York, NY (1998).

[288] E.R. Yardley, *Memories of the American Black Chamber*, *The Chinese Black Chamber: An Adventure in Espionage*, H.O. Yardley, ed., Houghton Mifflin, Boston, MA (1983), pp. 221–225.

[289] H.O. Yardley, *The American Black Chamber*, Aegean Park Press, Laguna Hills, CA (1931).

[290] H.O. Yardley, *The American Black Chamber*, Naval Institute Press, Annapolis, MD (2004).

[291] A.B. Zegart, *Flawed by Design: The Evolution of the CIA, JCS, and NSC*, Stanford University Press, Stanford, CA (1999).

19

AN INTRODUCTION TO MODERN CRYPTOLOGY

Bart Preneel

Dept. Electrical Engineering-ESAT/COSIC
Katholieke Universiteit Leuven
Kasteelpark Arenberg 10, B–3001 Leuven-Heverlee, Belgium
E-mail: bart.preneel@esat.kuleuven.be

Contents

Abstract

This paper provides an overview of cryptographic algorithms; it discusses the role of cryptology in information security. Next it presents the different types of algorithms for encryption and data authentication and explains the principles of stream ciphers, block ciphers, hash functions, MAC algorithms, public-key encryption algorithms, and digital signature schemes. Subsequently the design and evaluation procedures for cryptographic algorithms are discussed and a perspective is offered on research challenges.

Keywords: cryptology, encryption, data authentication, symmetric cryptography, public-key cryptography, hash functions, MAC algorithms, cryptanalysis.

19.1 INTRODUCTION

Our society is more and more dependent on information systems. An ever increasing number of interactions is moving on-line, supported by a growing infrastructure of networks (wired and wireless), servers and databases. End users, organisations, and governments are exchanging information and setting up e-services; after the end of the dot-com hype, it is easy to see a steady growth of e-commerce and e-business, and even the first successes in e-government. In the next years we can expect the growth of mobile services, and in five to ten years there will be an emergence of pervasive computing and ambient intelligence.

However, information systems and communication networks are increasingly subjected to a broad range of abuses. The risk for misuse has increased considerably since potential attackers can operate from all over the globe. Moreover, if someone gains access to an electronic information system, the scale and impact of the abuse can be much larger than in a brick-and-mortar or paper-based context. Hackers break into computer systems and networks with many goals: just for fun, to obtain sensitive data, for financial gain, to corrupt critical data, to bring down the system, which is called a denial of service attack, or to use them as a launching platform for other attacks, such as a distributed denial of service attack. Malware, such as computer viruses and worms, spreads at a growing speed through our networks. For example, the Sapphire/Slammer worm in January 2003 needed only 8.5 minutes to infect more than seventy thousand machines. While the current anti-virus software is effective for known malware, new malware keeps frequently appears and keeps causing damage in the time window between the detection and the effective update of the anti-virus software. Eavesdropping information over open networks is becoming a larger threat; all forms of wireless communications, for example mobile phones, Wireless LANs (WLAN) and Personal Area Networks (PAN) are particularly vulnerable. Privacy of users is compromised in various ways: sometimes data collection is pushed to the limit for commercial reasons; search and storage technologies become more and more effective; and standardised data formats, such as XML make data collection easier. SPAM is a growing problem: in 2006, forty to eighty percent of all email traffic consisted of unsolicited bulk email, but these figures vary depending upon the source consulted.

There is no doubt that in the past twenty-five years society's security technology has become more sophisticated, and people have realised that security problems cannot be solved only by technology and that securing a system requires an integrated and process-oriented approach that takes into account the complete context of the development process, user education, secure deployment, management of security systems which include audit and intrusion detection. Nevertheless, for the time being the security of the on-line environment is getting worse in spite of technologies efforts and insights. The following reasons can be given for this:

- as mentioned above, the growth of information systems and their increased connectivity implies that the target and motivation for attackers are growing as well; simultaneously, the time between the discovery of a security weakness and its exploitation is decreasing;
- systems become increasingly complex, open, and interconnected; however, security and complexity do not mix very well. There seems to be a lack the know-how and the tools to make a large system bug-free; a substantial part of the bugs has security implications; hence, the deployment of new systems keeps creating new problems;
- the threshold for attacking systems has become lower because of the on-line availability of a large number of hacker tools, such as password sniffers, password crackers, and virus authoring tools, which implies that unsophisticated attackers can launch smart attacks, that is write a clever virus using a few mouse clicks;
- the entities who need to invest in security may not be the ones who benefit from it, which can create a market failure that can only be resolved with regulation or self-regulation, such as a gradual increase of liability for software.

In the 1990s, protocols have been developed to secure communications: the most successful ones for the Internet are Transport Layer Security (TLS) and Secure Shell (SSH) at the transport layer and IPsec at the network layer [75]. S/MIME, the IETF standard to protect email, seems to be slow to catch on. The trend for the coming years is the improvement of the security of the end systems. Important initiatives are the Trusted Platform Module (TPM) from the Trusted Computing Group (TCG) [77], formerly TCPA and Microsoft's Next Generation Secure Computing Base (NGSCB), formerly known as Palladium [56]. In both approaches, the goal is to add a simple trusted subsystem to the computer system, which can verify the configuration, remotely attest the configuration, and store data securely. NGSCB adds additional trusted functionality in a secure subsystem.

Within all of these security mechanisms, cryptology plays an essential role. Without cryptology, securing the infrastructure which consists of communication networks and computer systems would not be possible. Cryptology also plays an essential role in protecting applications such as e-banking, financial transactions, electronic elections, and storage of sensitive data such as health information. While cryptology as an art is very old, it has developed as a science in the last sixty years; most of the open research dates from the last thirty years. A significant number of successes has been obtained, and it is clear that cryptology should no longer be the weakest link in our modern security systems. Nevertheless, as a science and engineering discipline, cryptology is still facing some challenging problems.

This article intends to present the state of the art and to offer a perspective on open research issues in the area of cryptographic algorithms. The paper will cover the principles underlying the design of cryptographic algorithms and will distinguish between algorithms for confidentiality protection that is, the protection against passive eavesdroppers and for authentication that is, the protection against active eavesdroppers, who try to modify information. The paper will also review the different approaches taken in cryptography and discuss the issues and research problems that arise when selecting, designing, and evaluating a cryptographic algorithm. Finally some concluding remarks will be presented.

19.2 ENCRYPTION FOR SECRECY PROTECTION

The use of cryptography for protecting the secrecy of information is as old as writing itself. For an excellent historical overview, see Kahn [44] and Sing [74]. Some people might say that cryptography is the second oldest profession. The basic idea in cryptography is to apply a complex transformation to the information to be protected. When the sender (usually called Alice) wants to convey a message to a recipient (Bob), the sender will apply to the *plaintext* P the mathematical transformation $E(\cdot)$. This transformation $E(\cdot)$ is called the encryption algorithm; the result of this transformation is called the *ciphertext* or $C = E(P)$. Bob, the recipient, will decrypt C by applying the inverse

transformation $D = E^{-1}$, and in this way he recovers P or $P = D(C)$. For a secure algorithm E, the ciphertext C does not make sense to an outsider, such as Eve, who is tapping the connection and who can obtain C, but not obtain any partial information on the corresponding plaintext P.

This approach only works when Bob can keep the transformation D secret. While this secrecy is acceptable in a person-to-person exchange, it is not feasible for large scale use. Bob needs a software or hardware implementation of D: either he has to program it himself or he has to trust someone to write the program for him. Moreover, he will need a different transformation and thus a different program for each correspondent, which is not very practical. Bob and Alice always have to face the risk that somehow Eve will obtain D (or E), for example, by breaking into the computer system or by bribing the author of the software or the system manager.

This problem can be solved by introducing into the encryption algorithm $E(\cdot)$ a secret parameter, the key K. Typically such a key is a binary string of forty to a few thousand bits. A corresponding key K^* is used for the decryption algorithm D. One has thus $C = E_K(P)$ and $P = D_{K^*}(C)$. (See also Fig. 19.1 which assumes that $K^* = K$.) The transformation strongly depends on the keys: if one uses a wrong key $K^{*\prime} \neq K^*$, then a random plaintext P' and not the plaintext P is obtained. Now it is possible to publish the encryption algorithm $E(\cdot)$ and the decryption algorithm $D(\cdot)$; the security of the system relies only on the secrecy of two short keys, which implies that $E(\cdot)$ and $D(\cdot)$ can be evaluated publicly and distributed on a commercial basis. Think of the analogy of a mechanical lock: everyone knows how a lock works, but in order to open a particular lock, one needs to have a particular key or know the secret combination.[1] In cryptography the assumption that the algorithm should remain secure even if it is known to the opponent is known as 'Kerckhoffs' principle'. Kerckhoffs was a 19th century Dutch cryptographer who formulated this principle.

---

[1] Note however that Matt Blaze has demonstrated in [13] that many modern locks are easy to attack and that their security relies to a large extent on *security through obscurity*, that is, the security of locks relies on the fact that the methods to design and attack locks are not published.

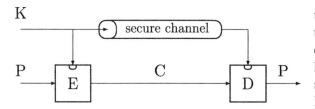

Figure 19.1. Model for conventional or symmetric encryption.

A simple example of an encryption algorithm is the Caesar cipher after the Roman emperor who used it. The plaintext is encrypted letter by letter; the ciphertext is obtained by shifting the letters over a fixed number of positions in the alphabet. The secret key indicates the number of positions. It is claimed that Caesar always used the value of three, such that AN EXAMPLE would be encrypted to DQ HADPSOH. Another example is the name of the computer HAL from S. Kubrick's *A Space Odyssey* (2001), which was obtained by replacing the letters of IBM by their predecessor in the alphabet. This transformation corresponds to a shift over twenty-five positions or minus one position. It is clear that such a system is not secure, since it is easy to try the twenty-six values of the key and to identify the correct plaintext based on the redundancy of the plaintext.

The *simple substitution* cipher replaces a letter by any other letter in the alphabet. For example, the key could be

ABCDEFGHIJKLMNOPQRSTUVWXYZ

MZNJSOAXFQGYKHLUCTDVWBIPER

which means that an A is mapped to an M, a B to a Z, and so on; hence, THEEVENING would be encrypted as VXSSBSHFHA. For an alphabet of n letters (in English $n = 26$), there are $n!$ substitutions, which implies that there are $n!$ values for the secret key. Note that even for $n = 26$ trying all keys is not possible since $26! = 403291461126605635584000000 = 4 \cdot 10^{26}$. Even if a fast computer could try one billion (10^9) keys per second, it would take one billion years to try all the keys. However, it is easy to break this scheme by frequency analysis: in a standard English text, the character E accounts for 12 out of every 100 characters, if spaces are omitted from the plaintext. Hence it is straightforward to deduce that the most common ciphertext character, in this example S corresponds to an E. Consequently, the key space has been reduced by a factor of twenty-six. It is easy to continue this analysis based on lower frequency letters and based on frequent combinations of two (e.g., TH) and three (e.g., THE) letters which are called digrams and trigrams respectively. In spite of the large key length, simple substitution is a very weak cipher, even if the cryptanalyst only has access to the ciphertext. In practice, the cryptanalyst may also know part of the plaintext, for example a standard opening such as DEARSIR.

A second technique applied in cryptology is a *transposition cipher* in which symbols are moved around. For example, the following mapping could be obtained:

TRANS		ORI S
POSIT	$\longrightarrow$	NOTIT
IONS		OSANP

Here the key would indicate where the letters are moved. If letters are grouped in blocks of n (in this example $n = 15$), there are $n!$ different transpositions. Again solving this cipher is rather easy, for example by exploiting digrams and trigrams or by fragments of known plaintexts. In spite of these weaknesses, modern ciphers designed for electronic computers are still based on a combination of several transpositions and substitutions (cf. Section 19.2.1.3).

A large number of improved ciphers has been invented. In the 15th and 16th century, *polyalphabetic substitution* was introduced, which uses t different alphabets. For each letter one of these alphabets is selected based on some simple rule. The complexity of these ciphers was limited by the number of operations that an operator could carry out by hand. None of these manual ciphers is considered to be secure today. With the invention of telegraph and radio communications, more complex ciphers have been developed. The most advanced schemes were based on mechanical or electromechanical systems with rotors, such as the famous Enigma machine used by Germany in World War II and the Hagelin machines. Rotor machines were used between the 1920s and the 1950s. In spite of the increased complexity, most of these schemes were not sufficiently secure in their times One of the weak points was users who did not follow the correct procedures. The analysis of the Lorenz cipher resulted in the development of Colossus, one of the first electronic computers.

A problem not yet been addressed is how Alice and Bob can exchange the secret key. The easy answer is that cryptography does not solve this problem; cryptography only moves the problem and at the same time simplifies the problem. In this case the secrecy of a (large) plaintext has been reduced to that of a *short* key, which can be exchanged beforehand. The problem of exchanging keys is studied in more detail in an area of cryptography that is called *key management*, which will not be discussed in detail in this paper. See [55] for an overview of key management techniques.

The branch of science which studies the encryption of information is called *cryptography*. A related branch tries to break encryption algorithms by recovering the plaintext without knowing the key or by deriving the key from the ciphertext and parts of the plaintext; it is called *cryptanalysis*. The term *cryptology* covers both aspects. For more extensive introductions to cryptography, the reader is referred to [20; 47; 55; 61; 73; 76].

So far this paper has assumed that the key for decryption K_D is equal to the encryption key K_E, or that it is easy to derive K_D from K_E. These types of algorithms are called *conventional* or *symmetric* ciphers. In *public-key* or *asymmetric* ciphers, K_D and K_E are always different; moreover, it should be difficult to compute K_D from K_E, which has the advantage that one can make K_E public which has important implications for the key management problem. The remainder of this section discusses symmetric algorithms and public-key algorithms.

19.2.1 Symmetric encryption

This section introduces three types of symmetric encryption algorithms: the one-time pad also known as the Vernam scheme, additive stream ciphers, and block ciphers.

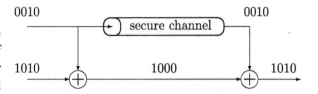

Figure 19.2. The Vernam scheme or one-time pad.

19.2.1.1 *The one-time pad or the Vernam scheme.* In 1917 G.S. Vernam invented a simple encryption algorithm for telegraphic messages [79]. The encryption operation consists of adding a random key bit by bit to the plaintext. The decryption operation subtracts the same key from the ciphertext to recover the plaintext (see Fig. 19.2). In practice, Vernam stored the keys on paper tapes.

The Vernam scheme can be formally described as follows. The ith bit of the plaintext, ciphertext and key stream are denoted with p_i, c_i and k_i, respectively. The encryption operation can then be written as $c_i = p_i \oplus k_i$. Here $\oplus$ denotes addition modulo 2 or exclusive or. The decryption operation is identical to the encryption or the cipher is an involution. Indeed, $p_i = c_i \oplus k_i = (p_i \oplus k_i) \oplus k_i = p_i \oplus (k_i \oplus k_i) = p_i \oplus 0 = p_i$. Vernam proposed use of a perfectly random key sequence, that is, the bit sequence k_i, $i = 1, 2, \ldots$, should consist of a uniformly and identically distributed sequence of bits.

In 1949 C. Shannon, the father of information theory, published his mathematical proof that shows that from observing the ciphertext, the opponent cannot obtain any new information on the plaintext, no matter how much computing power he has [70]. Shannon called this property *perfect secrecy*. The main disadvantage of the Vernam scheme is that the secret key is exactly as long as the message. C. Shannon also showed that the secret key cannot be shorter if one wants perfect secrecy. Until the late 1980s, the Vernam algorithm was used by diplomats and spies and even for the red-telephone system between Washington and Moscow.

Spies used to carry key pads with random characters: in this case p_i, c_i and k_i are elements from $\mathbb{Z}_{26}$ representing the letters A through Z. The encryption operation is $c_i = (p_i + k_i) \bmod 26$ and the decryption operation is $p_i = (c_i - k_i) \bmod 26$. The keys were written on sheets of paper contained on a pad. The security of the scheme relies on the fact that every page of the pad is used only once, which explains the name one-time pad. Note that during World War II the possession of pads with random characters was sufficient to be convicted as a spy.

The one-time pad was also used for Soviet diplomatic communications. Under the codeword Venona, US cryptologists attempted to break this system in 1943, which seems ridiculous as it offers perfect secrecy. However, after two years, it was discovered that the Soviets used their pads twice that is, two messages were encrypted using the same key stream. This error was due to time pressure in the production of the pads. If c and c' are ciphertexts generated with the same pad, one finds that

$$c \oplus c' = (p \oplus k) \oplus (p' \oplus k)$$
$$= (p \oplus p') \oplus (k \oplus k) = p \oplus p'.$$

This analysis implies that one can deduce from the ciphertext the sum of the corresponding plaintexts. Note that if the plaintexts are written in natural language, their sum $p \oplus p'$ is not uniformly distributed, so it is possible to detect that the correct c and c' have been matched. Indeed, if c and c' would have been encrypted with different keys k and k', the sum $c \oplus c'$ would be equal to $(p \oplus p') \oplus (k \oplus k')$ which is uniformly distributed. By guessing or predicting parts of plaintexts, a clever cryptanalyst can derive most of p and p' from the sum $p \oplus p'$. In practice, the problem was more complex since the plaintexts were encoded with a secret method before encryption and cover names were used to denote individuals. Between 1943 and 1980, approximately three thousand decryptions out of twenty-five thousand messages were obtained. Some of these plaintexts contained highly sensitive information on Soviet spies. The Venona successes were only made public in 1995 and teach an important lesson: in using a cryptosystem errors can be fatal even if the cryptosystem itself is perfectly secure. More details on Venona can be found in [37]. Section 19.4.3 discusses another weakness that can occur when implementing the Vernam scheme.

19.2.1.2 *Additive stream ciphers.* Additive stream ciphers are ciphers for which the encryption consists of a modulo 2 addition of a key stream to the plaintext. These ciphers try to mimic the Vernam scheme by replacing the perfectly random key stream by a pseudo-random key stream which is generated from a short key. Here pseudo-random means that the key stream looks random to an observer who has limited computing power. In practice one generates the bit sequence k_i with a keyed finite state machine (see Fig. 19.3). Such a machine stretches a short secret key K into a much longer key stream sequence k_i. The sequence k_i is eventually periodic. One important but insufficient design criterion for the finite state machine is that the period has to be large (2^{80} is a typical lower bound) because a repeating key stream leads to a very weak scheme (cf. the Venona project). The values k_i should have a distribution that is close to uniform; another condition is that there should be no correlations between output strings. Note that cryptanalytic attacks may exploit correlations of less than 1 in 1 million.

Formally, the sequence k_i can be parameterised with a security parameter. For the security of the stream cipher one requires that the sequence satisfies every polynomial time statistical test for randomness. In this definition polynomial time means that the complexity of these tests can be described as a polynomial function of the security parameter. Another desirable property is that no polynomial time machine can predict the next bit of the sequence based on the previous outputs with a probability that is significantly better than one-half. An important and perhaps surprising result in theoretical cryptology by A. Yao shows that these two conditions are in fact equivalent [84].

Stream ciphers have been popular in the 20th century: they operate on the plaintext character by character, which is convenient and allows for a simple and thus inexpensive implementation. Most of the rotor machines are additive stream ciphers. Between 1960 and 1990, stream ciphers based on Linear Feedback Shift Registers (LFSRs) were very popular. (See for example the book by Rueppel [67].) An example of a simple LFSR-based stream cipher is presented in Fig. 19.4. However, most of these algorithms were trade secrets;

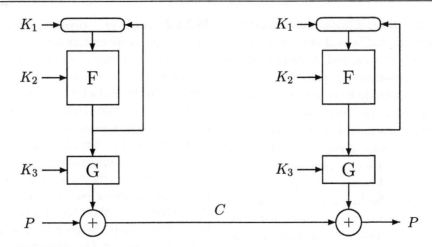

Figure 19.3. An additive stream cipher consisting of a keyed finite state machine. The initial state depends on K_1, the next state function F depends on K_2 and the output function G depends on K_3. The three keys K_1, K_2 and K_3 are derived from the user key K; this operation is not shown.

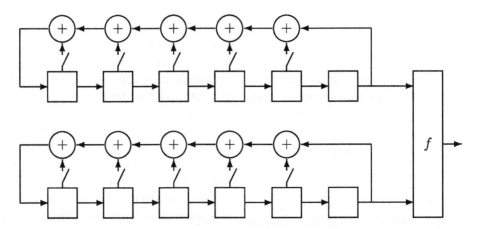

Figure 19.4. Toy example of a non-linear combiner: this additive stream cipher consists of two LFSRs (the squares represents single bit memories or D flip-flops) and a non-linear combining function f. Not all the feedback elements will be active in the LFSR: the key can consist of the feedback functions and the initial states of the LFSRs.

every organisation used its own cipher, and no standards were published. The most widely used LFSR-based stream ciphers are A5/1 and A5/2, which are implemented in hardware in the GSM phones. There are currently more than one billion GSM subscribers. The GSM algorithms were kept secret, but they leaked out and were shown to be rather weak [10]. In the last 15 years it has become clear that most LFSR-based stream ciphers are much less secure than expected. (See for example [17; 53].)

RC4, designed in 1987 by Ron Rivest, is based on completely different principles. RC4 is de-

signed for eight-bit microprocessors and was initially kept as a trade secret. It was posted on the Internet in 1994 and is currently widely used in browsers (TLS protocol). While several statistical weaknesses have been identified in RC4 [29; 59] the algorithm still seems to resist attacks that recover the key.

In the last fifteen years, a large number of very fast stream ciphers has been proposed that are software oriented, suited for 32-bit processors, and that intend to offer a high level of security. However, for the time being weaknesses have been identified in most proposals and no single scheme has emerged as a standard or a *de facto* standard. Nevertheless, stream ciphers can be very valuable for encryption with very few hardware gates or for high speed encryption. Developing strong stream ciphers is clearly an important research topic for the years ahead.

19.2.1.3 *Block ciphers.* Block ciphers take a different approach to encryption: the plaintext is divided into larger words of n bits, called *blocks*; typical values for n are 64 and 128. Every block is enciphered in the same way, using a keyed one-way permutation that is, a permutation on the set of n-bit strings controlled by a secret key. The simplest way to encrypt a plaintext using a block cipher is as follows: divide the plaintext into n-bit blocks P_i, and encrypt these block by block. The decryption also operates on individual blocks:

$$C_i = E_K(P_i) \quad \text{and} \quad P_i = D_K(C_i).$$

This way of using a block cipher is called the Electronic CodeBook (ECB) mode. Note that the encryption operation does *not* depend on the location in the ciphertext as is the case for additive stream ciphers.

Consider the following attack on a block cipher, the so-called tabulation attack: the cryptanalyst collects ciphertext blocks and their corresponding plaintext blocks which is possible as part of the plaintext is often predictable; these blocks are used to build a large table. With such a table, one can deduce information on other plaintexts encrypted under the same key. In order to preclude this attack, the value of n has to be quite large (e.g., 64 or 128). Moreover, the plaintext should not contain any repetitions or other patterns, as these will be leaked to the ciphertext.

This last problem shows that even if n is large, the ECB mode is not suited to encrypt structured plaintexts, such as text and images. This mode should only be used in exceptional cases where the plaintext is random, such for as the encryption of cryptographic keys. There is however an easy way to randomise the plaintext by using the block cipher in a different way. The most popular mode of operation for a block cipher is the Cipher Block Chaining (CBC) mode (see Fig. 19.5). In this mode

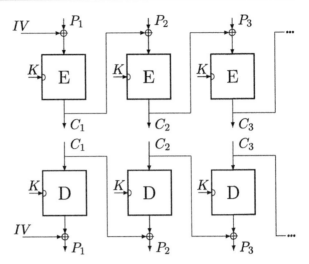

Figure 19.5. The CBC mode of a block cipher.

the different blocks are coupled by adding modulo 2 to a plaintext block the previous ciphertext block:

$$C_i = E_K(P_i \oplus C_{i-1}) \quad \text{and}$$

$$P_i = D_K(C_i) \oplus C_{i-1}.$$

Note that this randomises the plaintext and hides patterns. To enable the encryption of the first plaintext block ($i = 1$), one defines C_0 as the initial value *IV*, which should be randomly chosen and transmitted securely to the recipient. By varying this value, one can ensure that the same plaintext is encrypted into a different ciphertext under the same key. The CBC mode allows for random access on decryption: if necessary, one can decrypt only a small part of the ciphertext. A security proof the CBC mode has been provided by Bellare et al. [4]; it holds as long as the opponent can only obtain ciphertext corresponding to chosen plaintexts. Note that the CBC mode is insecure if the opponent can choose ciphertexts and obtain the corresponding plaintexts.

One can also use a block cipher to generate a key stream that can be used in an additive stream cipher: in the Output FeedBack (OFB) mode the block cipher is applied iteratively by feeding back the n-bit output to the input; in the CounTeR (CTR) mode one encrypts successive values of a counter. An alternative stream cipher mode is the Cipher FeedBack (CFB) mode; this mode is slower but it has better synchronisation properties. The modes of operation have been standardised in FIPS 81 [25] (see also [58] which adds the CTR mode) and ISO/IEC 10116 [40].

Block ciphers form a very flexible building block. They have played an important role in the past twenty-five years because of the publication of two US government Federal Information Processing Standards (FIPS) for the protection of sensitive but unclassified government information. An important aspect of these standards is that they can be used without paying a license fee.

The first standardised block cipher is the Data Encryption Standard (or DES) of FIPS 46 [24], which was published in 1977. This block cipher was developed by IBM together with National Security Agency (NSA) in response to a call by the US government. DES represents a remarkable effort to provide a standard for government and commercial use; its impact on both practice and research can hardly be overestimated. For example, DES is widely used in the financial industry. DES has a block size of 64 bits and a key length of fifty-six-bits. (More precisely, the key length is 64 bits, but eight of these are parity bits.) The fifty-six-bit key length was a compromise: it would allow the US government to find the key by brute force that is, by trying all $2^{56} \approx 7 \cdot 10^{16}$ keys one by one but would put a key search beyond limits for an average opponent. However, as hardware got faster, this key length was sufficient only for ten to fifteen years; hence, DES reached the end of its lifetime in 1987–1992 (see Section 19.4.3 for details). The block length of 64 bits is no longer adequate either because there exist matching ciphertext attacks on the modes of operation of an n-bit block cipher which require about $2^{n/2}$ ciphertext blocks [45]. For $n = 64$, these attack require four billion ciphertexts and with a high speed encryption device this number is reached in less than a minute. The DES design was oriented towards mid-1970s hardware. For example, it uses a sixteen-round Feistel structure (see Fig. 19.6) which implies that the hardware for encryption and decryption is identical. Each round consists of non-linear substitutions from six bits to four bits followed by some bit permutations or transpositions. The performance of DES in software is suboptimal; for example DES runs at 40

cycles/byte on a Pentium III, which corresponds to 200 Mbit/s for a clock frequency of 1 GHz.

In 1978, one year after the publication of the DES standard, an improved variant of DES was proposed: triple-DES consists of three iterations of DES: $E_{K_1}(D_{K_2}(E_{K_3}(x)))$. Only in 1999 this variant has been included into the third revision of FIPS 46 [24]. The choice of a decryption for the middle operation is motivated by backward compatibility: indeed, choosing $K_1 = K_2 = K_3$ results in single DES. Three-key triple DES has a 168-bit key, but the security level corresponds to a key of approximately one hundred bits. Initially two-key triple DES was proposed, with $K_3 = K_2$: its security level is about eighty to ninety bits. On first examination, the double-DES key length of 112 bits appears sufficient; however, it has been shown that the security level of double-DES is approximately seventy bits. For an overview of these attacks see [55]. The migration from DES to triple-DES in the financial sector was started in 1986, but it is progressing slowly and has taken twenty years to complete. Triple-DES has the disadvantage that it is rather slow (115 cycles/byte on a Pentium III) and that the block length is still limited to 64 bits.

In 1997 the US government decided to replace DES by the Advanced Encryption Standard (AES). AES is a block cipher with a 128-bit block length and key lengths of 128, 192, and 256 bits. An open call for algorithms was issued; fifteen candidates were submitted by the deadline of June 1998. After the first round, five finalists remained and in October 2000 it was announced that the Rijndael algorithm, designed by the Belgian cryptographers Vincent Rijmen and Joan Daemen, was the winner. The FIPS standard was published in November 2001 [28]. It may not be a coincidence that the U.S. Department of Commerce Bureau of Export Administration (BXA) relaxed export restrictions for US companies in September 2000. Note that otherwise it would have been illegal to export AES software from the US to Belgium. In 2003, the US government announced that it would also allow the use of AES for secret data, and even for top secret data; the latter application requires key lengths of 192 or 256 bits. Rijndael is a rather elegant and mathematical design. Among the five finalists

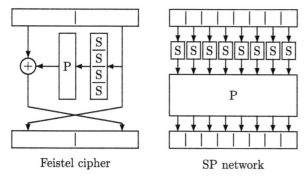

Feistel cipher   SP network

Figure 19.6. One round of a Feistel cipher (left) and of a Substitution–Permutation (SP) network. Here S represents a substitution and P a permutation, which can be a bit permutation or an affine mapping. In a Feistel cipher a complex operation on the right part is added to the left part; it has the advantage that the decryption operation is equal to the encryption operation, which simplifies hardware implementations. The substitution operation does not need to be invertible here. In an SP network all the bits are updated in every round, which guarantees faster diffusion of information; the substitution and permutation operations need to be invertible.

it offered the best combination of security, performance, efficiency, implementability, and flexibility. AES allows for efficient implementations on 32-bit architectures (15 cycles/byte on a Pentium III), but also for compact implementations on eight-bit smart cards. Moreover, hardware implementations of AES offer good trade-offs between size and speed. AES consists of a Substitution–Permutation (SP) network with ten rounds for a 128-bit key, twelve rounds for a 192-bit key and fourteen rounds for a 256-bit key (see Fig. 19.6). Each round consists of non-linear substitutions (from eight bits to eight bits) followed by some affine transformations, which move around the information. Note that Rijndael also supports 192-bit and 256-bit block lengths, but these have not been included in the AES standard. For a complete description of the AES and its design, see [19].

There exist many other block ciphers; a limited number of these has been included in products, such as Camellia, members of the CAST family, FEAL, Gost, IDEA, Kasumi, and Skipjack. For more details the reader is referred to the cryptographic literature.

19.2.2 Public-key encryption

The main problem left unsolved by symmetric cryptography is the key distribution problem. Especially in a large network it is not feasible to distribute keys between all user pairs. In a network with t users, there are $t(t-1)/2$ such pairs; hence even for one thousand users approximately half a million keys are needed. An alternative is to manage all keys in a central entity that shares a secret key with every user. However, this entity then becomes a single point of failure and an attractive target of attack. A much more elegant solution to the key management problem is offered by public-key cryptography invented in 1976, independently by Diffie and Hellman [21] and by Merkle [54].

19.2.2.1 *Public-key agreement.* A public-key agreement protocol allows two parties who have never met to agree on a secret key by way of a public conversation. Diffie and Hellman showed how to achieve this goal using the concept of *commutative one-way functions*. A *one-way function* is a function that is easy to compute but hard to invert. For example, in a block cipher, the ciphertext has to be a one-way function of the plaintext and the key: it is easy to compute the ciphertext from the plaintext and the key, but given the plaintext and the ciphertext it should be hard to recover the key otherwise the block cipher would not be secure. Similarly it can be shown that the existence of pseudo-random string generators, as used in additive stream ciphers, implies the existence of one-way functions. A *commutative one-way function* is a one-way function for which the result is the same independent of the order of the evaluation: for a function $f(\cdot, \cdot)$ with two arguments $f(f(z, x), y) = f(f(z, y), x)$.

The candidate commutative one-way function proposed by Diffie and Hellman is $f(\alpha, x) = \alpha^x \bmod p$; here p is a large prime number (large means 1024 bits or more), $x \in [1, p-1]$, and α is a generator mod p, which means that $\alpha^0, \alpha, \alpha^2, \alpha^3, \ldots, \alpha^{p-2} \bmod p$ run through all values between 1 and $p-1$. For technical reasons, it is required that p is a safe prime, which means that $(p-1)/2$ is a prime number as well. The Diffie–Hellman protocol works as follows (see also Fig. 19.7).

$$\begin{array}{ccc}
\textit{Alice} & & \textit{Bob} \\[2em]
x_A \in_R [1, p-1], \alpha^{x_A} \bmod p & \xrightarrow{\quad \alpha^{x_A} \bmod p \quad} & x_B \in_R [1, p-1], \alpha^{x_B} \bmod p \\[2em]
k_{AB} = (\alpha^{x_B})^{x_A} \bmod p & \xleftarrow{\quad \alpha^{x_B} \bmod p \quad} & k_{AB} = (\alpha^{x_A})^{x_B} \bmod p
\end{array}$$

Figure 19.7. The Diffie–Hellman protocol.

- Alice and Bob agree on a prime number p and a generator $\alpha \bmod p$;
- Alice picks a value x_A uniformly at random in the interval $[1, p-1]$, computes $y_A = \alpha^{x_A} \bmod p$ and sends this to Bob;
- Bob picks a value x_B uniformly at random in the interval $[1, p-1]$, computes $y_B = \alpha^{x_B} \bmod p$ and sends this to Alice;
- On receipt of y_B, Alice checks that $1 \leqslant y_B \leqslant p-2$ and computes $k_{AB} = y_B^{x_A} \bmod p = \alpha^{x_A x_B} \bmod p$;
- On receipt of y_A, Bob checks that $1 \leqslant y_A \leqslant p-2$ and computes $k_{BA} = y_A^{x_B} \bmod p = \alpha^{x_B x_A} \bmod p$;
- Alice and Bob compute the secret key as $h(k_{AB}) = h(k_{BA})$, with $h(\cdot)$ a hash function or MDC (see Section 19.3.1).

It is easy to see that the commutativity implies that $k_{AB} = k_{BA}$, hence Alice and Bob obtain a common value. Eve, who is eavesdropping the communication, only observes $y_A = \alpha^{x_A} \bmod p$ and $y_B = \alpha^{x_B} \bmod p$; there is no obvious way for her to obtain $k_{AB} = (\alpha^{x_B})^{x_A} \bmod p$. If Eve could compute discrete logarithms, that is, derive x_A from y_A and/or x_B from y_B, she could of course also derive k_{AB}. However, if p is large, this problem is believed to be difficult (cf. infra). Eve could try to find another way to compute k_{AB} from y_A and y_B. So far, no efficient algorithm has been found to solve this problem which is stated as the Diffie–Hellman assumption: it is hard to solve the Diffie–Hellman problem, that is, to deduce $(\alpha^{x_B})^{x_A} \bmod p$ from $\alpha^{x_A} \bmod p$ and $\alpha^{x_B} \bmod p$. If the Diffie–Hellman assumption holds, the Diffie–Hellman protocol results in a common secret between Alice and Bob after a public conversation. It is clear from the above discussion that the Diffie–Hellman problem cannot be harder than the dis-

crete logarithm problem. It is known that for some prime numbers the two problems are equivalent.

It is very important to check that $y_A, y_B \notin \{0, 1, p-1\}$: if not, Eve could modify y_A and y_B to one of these values and ensure in this way that $k_{AB} \in \{0, 1, p-1\}$. However, the Diffie–Hellman protocol has another problem: how does Alice know that she is talking to Bob or vice versa? In the famous person-in-the-middle-attack, Eve sets up a conversation with Alice which results in the key k_{AE} and with Bob which results in the key k_{BE}. Eve now shares a key with both Alice and Bob; she can decrypt all messages received from Alice, read them, and re-encrypt them for Bob and vice versa. Alice or Bob are unable to detect this attack; they believe that they share a common secret only known to the two of them. This attack shows that the common secret can only be established between two parties if there is an authentic channel, that is, a channel on which the information can be linked to the sender and the information cannot be modified. The conclusion is that the authenticity of the values y_A and y_B has to be established, by linking them to Alice and Bob respectively. One way to achieve this goal is to read these values or hash values of these values (see Section 19.3.1) over the phone; this solution works if Alice and Bob know each other's voices or if they trust the phone system to connect them to the right person.

The Diffie–Hellman problem has another limitation: Alice and Bob can only agree on a secret key, but Alice cannot use this protocol directly to tell Bob to meet her tonight at 9 PM. Alice can of course use the common key k_{AB} to encrypt this message using the AES algorithm in the CBC-mode. We will explain in the next section how public-key encryption can overcome this limitation.

19.2.2.2 *Public-key encryption.* The key idea behind public-key encryption is the concept of *trapdoor one-way functions* [21]: Trapdoor one-way functions are one-way functions with an additional property: given some extra information, the trapdoor, it becomes possible to invert the one-way function.

With such functions Bob can send a secret message to Alice without the need for prior arrangement of a secret key. Alice chooses a trapdoor one-way function with public parameter P_A, that is, Alice's public key and with secret parameter S_A, that is, Alice's secret key. Alice makes her public key widely available. For example, she can put it on her home page, but it can also be included in special directories. Anyone who wants to send some confidential information to Alice computes the ciphertext as the image of the plaintext under the trapdoor one-way function using the parameter P_A. Upon receipt of this ciphertext, Alice recovers the plaintext by using her trapdoor information S_A (see Fig. 19.8). An attacker, who does not know S_A, sees only the image of the plaintext under a one-way function and will not be able to recover the plaintext. The conditions which a public-key encryption algorithm has to satisfy are:

- the generation of a key pair (P_A, S_A) has to be easy;
- encryption and decryption have to be easy operations;
- it should be hard to compute the public key P_A from the corresponding secret key S_A;
- $D_{S_A}(E_{P_A}(P)) = P$.

Note that if a person wants to send a message to Alice, that individual has to know Alice's public key P_A and has to be sure that this key really belongs to Alice and not to Eve, since only the owner of the corresponding secret key will be able to decrypt the ciphertext. Public keys do not need a secure channel for their distribution, but they do need an authentic channel. As the keys for encryption and decryption are different, and Alice and Bob have different information, public-key algorithms are also known as *asymmetric algorithms.*

Designing a secure public-key encryption algorithm is apparently a difficult problem. From the large number of proposals, only a few have survived. The most popular algorithm is the RSA algorithm [66], which was named after its inventors R.L. Rivest, A. Shamir and L. Adleman. RSA was published in 1978; the patent on RSA has expired in 2000. The security of RSA is based on the fact that it is relatively easy to find two large prime numbers and to multiply these while factoring their product is not feasible with the current algorithms and computers. The RSA algorithm can be described as follows:

key generation: Find two prime numbers p and q with at least one hundred fifty digits and compute their product, the modulus $n = p \cdot q$. Compute the Carmichael function $\lambda(n)$, which is defined as the least common multiple of $p - 1$ and $q - 1$. In other words, $\lambda(n)$ is the smallest integer which is a multiple of both $p - 1$ and $q - 1$. Choose an encryption exponent e, which is at least 32 to 64 bits long and which is relatively prime to $\lambda(n)$, that is, has no common divisors with $\lambda(n)$. Compute the decryption exponent as $d = e^{-1} \bmod \lambda(n)$ using Euclid's algorithm. The public key consists of the pair (e, n), and the secret key consists of the decryption exponent d or the pair (p, q);

encryption: represent the plaintext as an integer in the interval $[0, n - 1]$ and compute the ciphertext as $C = P^e \bmod n$;

decryption: $P = C^d \bmod n$.

The prime factors p and q or the secret decryption exponent d are the trapdoor that allows the inversion of the function $f(x) = x^e \bmod n$. Indeed, it can be shown that $f(x)^d \bmod n = x^{ed} \bmod n = x$. Note that the RSA function is the dual of the Diffie–Hellman function $f(x) =$

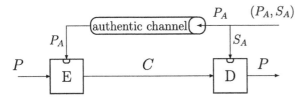

Figure 19.8. Model for public-key or asymmetric encryption.

$\alpha^x \bmod p$, which has a fixed base and a variable exponent.

Without explaining the mathematical background of the algorithm, it can be seen that the security of the RSA algorithm depends on the factoring problem. Indeed, if an attacker can factor n, he can find $\lambda(n)$, derive d from e and decrypt any message. However, in order to decrypt a ciphertext it is sufficient to extract modular eth roots. Note that it is not known whether it is possible to extract eth roots without knowing p and q. The RSA problem is the extraction of random modular eth roots since this corresponds to the decryption of arbitrary ciphertexts. Cryptographers believe that the RSA problem is hard; this assumption is known as the RSA assumption. It is easy to see that the RSA problem cannot be harder than factoring the modulus. Some indication has been found that the two problems may not be equivalent.

For special arguments, the RSA problem is easy. For example -1, 0 and 1 are always fixed points for the RSA encryption function and for small arguments, $P^e < n$ and extracting modular eth root simplifies to extracting natural eth roots, which is an easy problem. However, the RSA assumption states that extracting *random* modular eth roots is hard, which means that the challenge ciphertext needs to be uniformly distributed. Such a uniform distribution can be achieved by transforming the plaintext with a randomising transform. A large number of such transforms is known and many of these are ad hoc so there is no reason to believe that they should be effective. In 1993, Bellare and Rogaway published a new transform under the name Optimal Asymmetric Encryption (OAEP) together with a security proof [5]. This proof shows that an algorithm that can decrypt a challenge ciphertext without knowing the secret key, can be transformed into an algorithm that computes a random modular eth root. The proof is in the random oracle model, which means that the hash functions used in the OAEP construction are assumed to be perfectly random. However, seven years later Shoup pointed out that the proof was wrong [72]; the error has been corrected for by Fujisaki et al. in [30], but the resulting reduction is not meaningful that is, the coupling between the two problems is not very

tight in this new proof (except when e is small). Currently the cryptographic community believes that the best way of using RSA is the RSA-KEM mode [57; 63], which is a *hybrid* mode in which RSA is only used to transfer a session key while the plaintext is encrypted using a symmetric algorithm with this key. It is interesting to note that it has taken more than 20 years before cryptographers have understood how RSA should be used properly for encryption.

19.2.2.3 *The factoring and the discrete logarithm problem.* The more complex properties of public-key cryptography seem to require some 'high level' mathematical structure; most public-key algorithms are based on problems from algebraic number theory. While these number theoretic problems are believed to be difficult, it should be noted that there is no mathematical proof that shows that these problems are hard. Moreover, since the invention of public-key cryptography, significant progress has been made in solving concrete instances. This evolution is due to a combination of more sophisticated algorithms with progress in hardware and parallel processing. Table 19.1 summarises the progress made in factoring over the past forty years. It is believed that the discrete logarithm problem mod p is about as difficult as the factoring problem for the same size of modulus. This equivalence only holds if p satisfies certain conditions; a sufficient condition is that p is a safe prime as defined above.

The best known algorithm to factor an RSA modulus N is the General Number Field Sieve. It has been used in all factoring records since 1996 and has a heuristic asymptotic complexity

$$O\big(\exp\big[(1.923 + o(1)) \times (\ln N)^{1/3} \cdot (\ln \ln N)^{2/3}\big]\big).$$

Note that this asymptotic expression should be used with care; extrapolations can only be made in a relatively small range due to the $o(1)$ term. Lenstra and Verheul provide an interesting study on the selection of RSA key sizes [51]. Currently it is believed that factoring a 1024-bit RSA modulus (308 digit) modulus requires 2^{80} steps. With special hardware proposed by Shamir and

Table 19.1. Progress of factorisation records for products of two random prime numbers. One MIPS year (MY) is the equivalent of a computation during one full year at a sustained speed of one Million Instructions Per Second, which corresponds roughly to the speed of a VAX 11/780

Year	# digits	# bits	Computation (MY)
1964	20	66	
1974	45	150	0.001
1983	50	166	
1984	71	236	0.1
1991	100	332	7
1992	110	365	75
1993	120	399	835
1994	129	429	5000
1996	130	432	1000
1999	140	465	2000
1999	155	512	8400
2003	174	578	
2005	200	663	

Tromer [69], the following cost estimates have been provided: with an investment of US$ ten million, a 1024-bit modulus can be factored in one year (the initial R&D cost is US$ twenty million). A 768-bit modulus can be factored for US$ five thousand in ninety-five days, and a 512-bit modulus can be factored with a US$ ten thousand device in ten minutes. Note that these cost estimates do not include the linear algebra step at the end; while this step takes additional time and effort, it should not pose any unsurmountable problems. Nevertheless, these estimates show that for long-term security, an RSA modulus of 2048 bits is recommended.

19.2.2.4 *Basing public-key cryptology on other problems.*

There has been a large number of proposals for other public key encryption algorithms. Many of these have been broken, the most notable example being the class of knapsack systems. The most important alternative to RSA is the ElGamal scheme, which extends the Diffie–Hellman scheme to public-key encryption. In particular, the group of integers mod p can also be replaced by a group defined by an elliptic curve over a finite field, as proposed by Miller and Koblitz in the mid-eighties. Elliptic curve cryptosystems allow for shorter key sizes that is, a 1024-bit RSA key corresponds to a 170-bit elliptic curve key, but the operations on an elliptic curve are more complex (see [2; 12; 36]). Other alternatives are schemes based on hyperelliptic curves, multivariate polynomials over finite fields, lattice-based systems such as NTRU and systems based on braid groups;[2] while these systems have particular advantages, it is believed that they are not yet mature enough for deployment. It is a little worrying that our digital economy relies to a large extent on the claimed difficulty of a few problems in algebraic number theory.

19.2.2.5 *Applying public-key cryptology.*

The main advantage of public-key algorithms is the simplified key management; deployment of cryptology on the Internet largely relies on public-key mechanisms (e.g., TLS and SSH [75]). An important question is how authentic copies of the public keys can be distributed; this problem will be briefly discussed in Section 19.3.2. The main disadvantages are the larger keys (typically 64 to 512 bytes) and the slow performance: both in software and hardware public-key encryption algorithms are two to three orders of magnitude slower than symmetric algorithms. For example, a 1024-bit exponentiation with a thirty-two-bit exponent takes 360 µs on a 1 GHz Pentium III; this corresponds to 2800 cycles/byte; a decryption with a 1024-bit exponent takes 9.8 ms or 76 000 cycles/byte. This speed should be compared to 15 cycles/byte for AES. Because of the large difference in performance, the large block length which influences error propagation and the security reasons indicated above, one always employs *hybrid* systems: the public-key encryption scheme is used to establish a secret key, which is then used in a fast symmetric algorithm.

19.3 HASHING AND SIGNATURES FOR AUTHENTICATION

Information authentication includes two main aspects:

- *data origin authentication*, or who has originated the information;

---

[2]Braid groups are non-commutative groups derived from geometric arrangements of strands; in a non-commutative group $a \cdot b$ is in general not equal to $b \cdot a$.

- *data integrity*, or has the information been modified.

Other aspects which can be important are the timeliness of the information, the sequence of messages, and the destination of information. These aspects can be accounted for by using sequence numbers and time stamps in the messages and by including addressing information in the data. In data communications, the implicit authentication created by recognition of the handwriting, signature, or voice disappears. Electronic information becomes much more vulnerable to falsification as the physical coupling between information and its bearer is lost.

Until the mid-1980s, it was widely believed that encryption with a symmetric algorithm of a plaintext was sufficient for protecting its authenticity. If a certain ciphertext resulted after decryption in a *meaningful* plaintext, it had to be created by someone who knew the key, and therefore it must be authentic. However, a few counterexamples are sufficient to refute this claim. If a block cipher is used in ECB mode, an attacker can easily reorder the blocks. For any additive stream cipher, including the Vernam scheme, an opponent can always modify a plaintext bit even without knowing whether a zero has been changed to a one or vice versa. The concept of meaningful information implicitly assumes that the information contains redundancy, which allows a distinction of genuine information from arbitrary plaintext. However, one can envisage applications where the plaintext contains very little or no redundancy, for example, the encryption of keys. The separation between secrecy and authentication has also been clarified by public-key cryptography: anyone who knows Alice's public key can send her a confidential message, and therefore Alice has no idea who has actually sent this message.

Two different levels of information authentication can be distinguished. If two parties trust each other and want to protect themselves against malicious outsiders, the term *conventional message authentication* is used. In this setting, both parties are at equal footing; for example, they share the same secret key. If however a dispute arises between them, a third party will not be able to resolve it. For example a judge may not be able to tell whether a message has been created by Alice or by Bob. If protection between two mutually distrustful parties is required, which is often the case in a commercial relationships, an electronic equivalent of a manual signature is needed. In cryptographic terms this is called a *digital signature*.

19.3.1 Symmetric authentication

The underlying idea is similar to that for encryption, where the secrecy of a large amount of information is replaced by the secrecy of a short key. In the case of authentication, one replaces the authenticity of the information by the protection of a short string, which is a unique fingerprint of the information. Such a fingerprint is computed as a hash result, which can also be interpreted as adding a special form of redundancy to the information. This process consists of two components. First the information is compressed into a string of fixed length, with a cryptographic hash function. Then the resulting string, the hash result, is protected as follows:

- either the hash result is communicated over an authentic channel (e.g., it can be read over the phone). It is then sufficient to use a hash function without a secret parameter, which is also known as a Manipulation Detection Code or MDC;
- or the hash function uses a secret parameter (the key) and is then called a Message Authentication Code or MAC algorithm.

19.3.1.1 *MDCs.* If an additional authentic channel is available, MDCs can provide authenticity without requiring secret keys. Moreover an MDC is a flexible primitive, which can be used for a variety of other cryptographic applications. An MDC has to satisfy the following conditions:

- preimage resistance: it should be hard to find an input with a given hash result;
- second preimage resistance: it should be hard to find a second input with the same hash result as a given input;
- collision resistance: it should be hard to find two different inputs with the same hash result.

An MDC satisfying these three conditions is called a *collision resistant* hash function.

For a strong hash function with an n-bit result, solving the first two problems requires about 2^n evaluations of the hash function. This implies that $n = 90, \ldots, 100$ is sufficient (cf. Section 19.4.3); larger values of n are required if one can attack multiple targets in parallel. However, finding collisions is substantially easier that finding preimages or second preimages. With high probability a set of hash results corresponding to $2^{n/2}$ inputs contains a collision, which implies that collision resistant hash functions need a hash result of 160 to 256 bits. This last property is also known as the *birthday paradox* based on the following observation: within a group of twenty-three persons the probability that there are two persons with the same birthday is about fifty percent. The reason is that a group of this size contains 253 different pairs of persons, which is rather larger compared to the 365 days in a year. The birthday paradox plays an essential role in the security of many cryptographic primitives. It is important to note that not all applications need collision resistant hash functions; sometimes preimage resistance or second preimage resistance is sufficient.

The most efficient hash functions are dedicated hash function designs. The hash functions MD4 and MD5 with a 128-bit hash result are no longer recommended. For the devastating attacks by Wang et al., see [80]. The most popular hash function today is SHA-1, but it has been pointed out by Wang et al. [81] that a shortcut collision attack exists which requires effort 2^{63} rather than 2^{80}. RIPEMD-160 is an alternative; both SHA-1 and RIPEMD-160 offer a 160-bit result. Recent additions to the SHA-family include SHA-256, SHA-384 and SHA-512 (see FIPS 180-2 [26]). The ISO standard on dedicated hash functions (ISO/IEC 10118-3) contains RIPEMD-128, RIPEMD-160, SHA-1, SHA-256, SHA-384, SHA-512 and Whirlpool [41]. Part 2 of this standard specifies hash functions based on a block cipher while Part 4 specifies hash functions based on modular arithmetic.

19.3.1.2 *MAC algorithms.* MAC algorithms have been used since the 1970s for electronic transactions in the banking environment. They require

the establishment of a secret key between the communicating parties. The MAC value corresponding to a message is a complex function of every bit of the message and every bit of the key; it should be infeasible to derive the key from observing a number of text/MAC pairs or to compute or predict a MAC without knowing the secret key.

A MAC algorithm is used as follows (cf. Fig. 19.9): Alice computes for her message P the value $\text{MAC}_K(P)$ and appends this MAC to the message (here MAC is an abbreviation of MAC result). Bob recomputes the value of $\text{MAC}_K(P)$ based on the received message P, and verifies whether it matches the received MAC result. If the answer is positive, he accepts the message as authentic that is, as a genuine message from Alice. Eve, the active eavesdropper, can modify the message P to P', but she is not able to compute the corresponding value $\text{MAC}_K(P')$ since she is not privy to the secret key K. For a secure MAC algorithm, the best Eve can do is guess the MAC result. In that case, Bob can detect the modification with high probability: for an n-bit MAC result Eve's probability of success is only $1/2^n$. The value of n lies typically between 32 and 96. Note that if separate techniques for encryption and authentication are combined, the keys for encryption and authentication need to be different. Moreover, the preferred option is to apply the MAC algorithm to the ciphertext since this order of operations protects the encryption algorithm against chosen ciphertext attacks.

A popular way to compute a MAC is to encrypt the message with a block cipher using the CBC mode which is another use of a block cipher and to keep only part of the bits of the last block as the MAC. However, Knudsen has shown that this approach is less secure than previously believed [46]. The recommended approach to use CBC-MAC consists of super-encrypting the final block with a different key, that may be derived from the first key. This scheme is known as EMAC; a security proof for EMAC has been provided by Petrank and Rackoff in [60]. Almost all CBC-MAC variants are vulnerable to a birthday type attack which requires only $2^{n/2}$ known text-MAC pairs [62]. Another popular MAC algorithm is HMAC, which derives a MAC algorithm from a hash function such

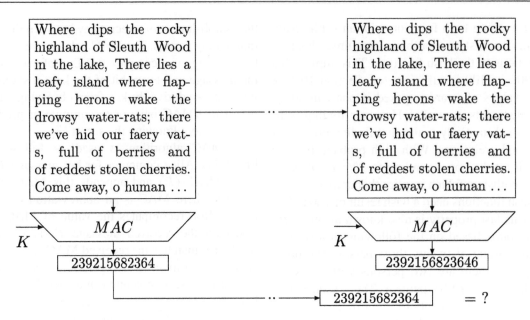

Figure 19.9. Using a Message Authentication Code for data authentication.

as SHA-1 [26]; an alternative for HMAC is MDx-MAC [62]. A large number of MAC algorithms has been standardised in ISO/IEC 9797 [39].

For data authentication, the equivalent of the Vernam scheme exists which implies that a MAC algorithm can be designed that is unconditionally secure in the sense that the security of the MAC algorithm is independent of the computing power of the opponent. The requirement is again that the secret key is used only once. The basic idea of this approach is due to Simmons [73] who defined authentication codes and Carter and Wegman [16; 82] who used the term universal hash functions. The first ideas date back to the 1970s. It turns out that these algorithms can be computationally extremely efficient since the properties required from this primitive are combinatorial rather than cryptographic. The combinatorial property that is required is that if one takes the average over a key, the function values or pairs of function values need to be distributed almost uniformly. This property is much easier to achieve than cryptographic properties which require that it should be hard to recover the key from input–output pairs which is a much stronger requirement than a close to uniform distribution. Recent constructions are therefore one order of magnitude faster than other cryptographic primitives, such as encryption algorithms

and hash functions, and achieve speeds up to 1–2 cycles/byte on a Pentium III for messages longer than 256 bytes (e.g., UMAC [11] and Poly1305-AES [8]). A simple example is the polynomial hash function (see [43]). The key consists of two n-bit words denoted with K_1 and K_2. The plaintext P is divided into t n-bit words, denoted with P_1 through P_t. The MAC value which consists of a single n-bit word is computed based on a simple polynomial evaluation:

$$\text{MAC}_{K_1, K_2}(P) = K_1 + \sum_{i=1}^{t} P_i \cdot (K_2)^i,$$

where addition and multiplication are to be computed in the finite field with 2^n elements. It can be proved that the probability of creating another valid message/MAC pair is upper bounded by $t/2^n$. A practical choice is $n = 64$ which results in a 128-bit key. For messages up to one megabyte, the success probability of a forgery is less than $1/2^{47}$. Note that K_2 can be reused; however, for every message a new key K_1 is required. This key could be generated from a short initial key using an additive stream cipher but then the unconditional security is lost. However, it can be argued that it is easier to understand the security of this scheme than that of a computationally secure MAC algorithm. An even better way to use universal hash functions is to apply a pseudo-random function to its output concatenated with a value that occurs only once, for example, a serial number or a large random number.

19.3.2 Digital signatures

A digital signature is the electronic equivalent of a manual signature on a document. It provides a strong binding between the document and a person, and in case of a dispute, a third party can decide whether or not the signature is valid based on public information. Of course a digital signature will not bind a person and a document, but will bind a public key and a document. Additional measures are then required to bind the person to his or her key. Note that for a MAC algorithm, both Alice and Bob can compute the MAC result; hence, a third party cannot distinguish between them. Block ciphers and even one-way functions can be used to construct digital signatures but the most elegant and efficient constructions for digital signatures rely on public-key cryptography.

We now explain how the RSA algorithm can be used to create digital signatures with message recovery. The RSA mapping is a bijection, more specifically a trapdoor one-way permutation. If Alice wants to sign some information P intended for Bob, she adds some redundancy to the information, resulting in $\widetilde{P}$ and decrypts the resulting text with her secret key. This operation can only be carried out by Alice. Upon receipt of the signature, Bob encrypts it using Alice's public key and verifies that the information $\widetilde{P}$ has the prescribed redundancy. If so, he accepts the signature on P as valid. Such a digital signature which is a signature with message recovery requires the following condition on the public-key system: $E_{P_A}(D_{S_A}(\widetilde{P})) = \widetilde{P}$. Anyone who knows Alice's public key can verify the signature and recover the message from the signature.

Note that if the redundancy is left out, any person can pick a random ciphertext C^* and claim that Alice has signed $P^* = C^{*e} \bmod n$. It is not clear that P^* is a meaningful message which will require some extra tricks, but it shows why redundancy is essential. A provably secure way to add

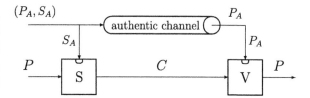

Figure 19.10. A digital signature scheme with message recovery based on a trapdoor one-way permutation; S and V denote the signing and verification operation respectively.

the redundancy is PSS-R [6]; however, in practice other constructions are widely used, and most of them combine a hash function with a digital signature scheme (Fig. 19.10).

If Alice wants to sign very long messages, digital signature schemes with message recovery result in signatures that are as long as the message. Moreover, signing with a public-key system is a relatively slow operation. In order to solve these problems, Alice does not sign the information itself but rather the hash result of the information computed with an MDC (see also Fig. 19.11). This approach corresponds to the use of an MDC to replace the authenticity of a large text by that of a short hash value (cf. Section 19.3.1). The signature now consists of a single block which is appended to the information. This type of signature scheme is sometimes called a digital signature with appendix. In order to verify such a signature, Bob recomputes the MDC of the message and encrypts the signature with Alice's public key. If both operations give the same result, Bob accepts the signature as valid. MDCs used in this way need to be collision resistant: if Alice can find two different messages (P, P') with the same hash result, she can sign P and later claim to have signed P' (P and P' will have the same signature!).

Note that there exist other signature schemes with appendix, such as the DSA from FIPS 186 [27] which are not derived immediately from a public-key encryption scheme. For these schemes it is possible to define a signing operation using the secret key and a verification operation using the public key without referring to decryption and encryption operations. The security of the DSA is based on the discrete logarithm problem as the Diffie–Hellman scheme. There also exists an

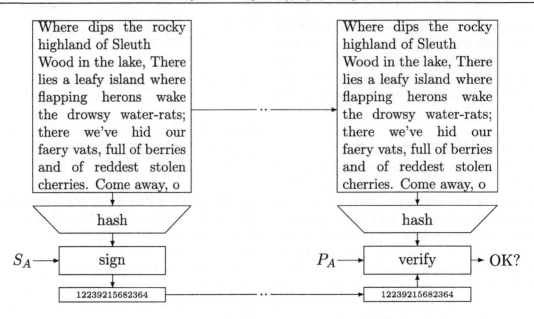

Figure 19.11. A digital signature scheme with appendix.

elliptic-curve based variant of DSA called ECDSA. There are many more digital signature schemes than public-key encryption schemes. Other digital signature schemes include ESIGN, Fiat-Shamir, Guillou-Quisquater and Schnorr.

19.3.2.1 *Certificates.* Digital signatures move the problem of authenticating data to the problem of authenticating a link between an entity and its public key. This problem can be simplified using digital *certificates.* A certificate is a digital signature of a third party on an entity's name, its public key, and additional data, such algorithms and parameters, key usage, begin and end of validity period. It corresponds roughly to an electronic identity. The verification of a certificate requires the public key of the trusted third party, and in this way the problem of authenticating data will be replaced by authenticating a single public key. In principle authenticating this key is very easy: it can be published in a newspaper, listed on a webpage, or added in a browser. The infrastructure which provides public keys is called a Public Key Infrastructure (PKI). Deploying such an infrastructure in practice is rather complex since it is necessary to consider revocation of certificates, upgrades of keys, multiple trusted parties, and integration of the public key management with the application (see for example [1]).

19.4 ANALYSIS AND DESIGN OF CRYPTOGRAPHIC ALGORITHMS

In this section we compare several approaches in cryptography. Next we describe the typical phases in the life of a cryptographic algorithm. Finally we discuss why, in spite of the progress in cryptology, weak cryptographic algorithms are still in use and which problems cryptographers are facing in the next decades.

19.4.1 Different approaches in cryptography

Modern cryptology follows several approaches: the information theoretic approach, the complexity theoretic approach, the bounded-storage approach, the quantum cryptology approach and the system based approach. These approaches differ in the assumptions about the capabilities of an opponent, in the definition of a cryptanalytic success, and in the notion of security.

From the viewpoint of the cryptographer, the most desirable are unconditionally secure algorithms. This approach is also known as the *information theoretic* approach. It was developed in the seminal work of C. Shannon in 1943 and published a few years later [70]. This approach offers a perfect solution since the security can be proved independent of the computational power of the opponent, and the security will not erode over time. However, few such schemes exist that are secure in this model; examples are the Vernam scheme (Section 19.2.1) and the polynomial MAC algorithm (Section 19.3.1.2). While they are computationally extremely efficient, the cost in terms of key material may be prohibitively large. For most applications users have to live with schemes which offer only conditional security.

A second approach is to reduce the security of a cryptographic algorithm to that of other well-known difficult problems or to that of other cryptographic primitives. The *complexity theoretic* approach starts from an abstract model for computation and assumes that the opponent has limited computing power within this model [32]. The most common models used are Turing machines and RAM models. This approach was started in cryptology by Yao [84] in 1982. It has many positive sides and has certainly contributed towards moving cryptology from an art to a science:

- It forces the formulation of exact definitions and the clear statement of security properties and assumptions, which may seem trivial, but it has taken the cryptographic community a long time to define what secure encryption, secure message authentication, secure digital signatures, and secure authenticated encryption are. It turns out that there are many variants for these definitions depending on the power of the opponent and the goals, and it takes a substantial effort to establish the relationship between them. For more complex primitives, such as e-payment, e-voting, interactions with more than two parties, and in general multi-party computation establishing correct definitions is even more complex.
- The complexity theoretic approach results in formal reductions. It can be formally proven that if a particular object exists, another object exists as well. For example, one-way functions imply digital signature schemes, or the existence

of pseudo-random permutations, which corresponds to a secure block cipher, is sufficient to prove that the CBC mode with secret and random initial value is a 'secure' mode for encryption assuming that the opponent can choose plaintexts and observe the corresponding ciphertexts. The terms secure have very specific definitions as explained above. These reductions often work by contradiction. For example, it can be shown that if an opponent can find a weakness in CBC-encryption using a chosen plaintext attack, the implication is that there is a weakness in the underlying block cipher which allows a distinction of this block cipher from a pseudo-random permutation. Once the proofs are written down, any person can verify them which is very important, as it turns out that some of these proofs have very subtle errors.

The complexity theoretic approach also has some limitations:

- Many cryptographic applications need building blocks, such are one-way functions, one-way permutations, collision resistant compression functions, and pseudo-random functions, which cannot be reduced to other primitives. In terms of the existence of such primitives, complexity theory has only very weak results: it is not known whether one-way functions exist. In non-uniform complexity, which corresponds to Boolean circuits, the best result proved thus far is that there exist functions which are twice as hard to invert as to compute, which is far too weak to be of any use in cryptography [38]. To some extent cryptographers need to rely on number theoretic assumptions, such as the assumptions underlying the security of the Diffie–Hellman protocol and RSA. For the others, we rely on unproven assumptions on functions, such as RC4, DES, AES, RIPEMD-160 and SHA-1 which means that even when there are strong reductions, the foundations of the reductions are still weak.
- Sometimes the resulting scheme is much more expensive in terms of computation or memory than a scheme without security proof; however, in the last decade a substantial effort has been

made to improve the efficiency of the constructions for which there exists a security proof.

• The security proof may not be very efficient. Some reductions are only asymptotic or are very weak. For example, for RSA-OAEP, if the security property is violated that is, a random ciphertext can be decrypted, the security assumption, in this case the computation of a modular eth root is infeasible, is only violated with a very small probability. Many reductions require the random oracle assumption, which states that the hash function used in the scheme behaves as a perfectly random function. It has been shown that this approach can result in security proofs for insecure schemes; however, these reductions still have some value as a heuristic in the sense that these schemes are typically better than schemes for which no reduction in the random oracle model is known.

The term concrete complexity has been coined by Bellare and Rogaway to denote security proofs with concrete reductions focusing on efficiency that is, without asymptotics and hidden constants. For an overview of complexity theoretic results in cryptology, the reader is referred to the work of Goldwasser and Bellare [34] and Goldreich [33].

There has also been some interest in the bounded-storage model where it is assumed that the storage capacity of an adversary is limited (see for example [22]). This model can be considered to be part of the information theoretic approach if one imposes that only limited information is available to the adversary or part of the complexity theoretic approach if space rather than time is considered to be limited.

Quantum cryptology does not work based on any computational assumptions, but rather it starts from the assumption that quantum physics provides a complete model of the physical world. It relies on concepts such as the Heisenberg uncertainty principle and quantum entanglement. Quantum cryptography provides a means for two parties to exchange a common secret key over an authenticated channel. The first experimental results have been obtained in the early 1990s and ten years later several companies offer encryption products based on quantum cryptography. While this offers a fascinating approach, quantum cryptology needs authentic channels which seems to make them not very useful to secure large open networks such as the Internet. Indeed, without public-key cryptography, the only way to achieve an authentic channel is by establishing prior secrets manually, which seems rather impractical for a large network. Moreover, most implementations today have practical limitations. They are not yet compatible with packet switched systems; they typically allow for low speeds only although high speed links have been demonstrated recently; and the communication distances are limited. For high-speed communication, one still needs to stretch the short keys with a classical stream cipher; of course, the resulting scheme then depends on standard cryptographic assumptions which to some extent defeats the purpose of using quantum cryptology in the first place. Moreover, it can be expected that the current implementations are vulnerable to side channel attacks as discussed below.

In view of the limitations of the above approaches, modern cryptology still has to rely on the *system-based* or *practical approach*. This approach tries to produce practical solutions for basic building blocks such as one-way functions, pseudo-random bit generators or stream ciphers, pseudo-random functions, and pseudo-random permutations. The security estimates are based on the best algorithm known to break the system and on realistic estimates of the necessary computing power or dedicated hardware to carry out the algorithm. By trial and error procedures, several *cryptanalytic principles* have emerged, and it is the goal of the designer to avoid attacks based on these principles. The second aspect is to design *building blocks with provable properties*, and to assemble such basic building blocks to design cryptographic primitives. The complexity theoretic approach has also improved our understanding of the requirements to be imposed on these building blocks. In this way, this approach is also evolving towards a more scientific approach. Nevertheless, it seems likely that for the next decades cryptographers still need to rely the security of some concrete functions such as AES, SHA-512, and their successors.

19.4.2 Life cycle of a cryptographic algorithm

This section discusses the life cycle of a cryptographic algorithm, evaluates the impact of open competitions, and compares the use of public versus secret algorithms.

A cryptographic algorithm usually starts with a new idea of a cryptographer and a first step should be an evaluation of the security properties and of the cryptographic algorithm in which the cryptographer tries to determine whether or not the scheme is secure for the intended applications. If the scheme is unconditionally secure, he has to write the proofs and convince himself that the model is correct and matches the application. For computational security, it is again important to write down security proofs and check these for subtle flaws. Moreover, a cryptographer has to assess whether the assumptions behind the proofs are realistic. For the system-based approach, it is important to prove partial results and write down arguments which should convince others of the security of the algorithm. Often such cryptographic algorithms have security parameters, such as the number of steps, the size of the key. It is then important to give lower bounds for these parameters and to indicate the value of the parameters which corresponds to a certain security level.

The next step is the publication of the algorithm at a conference, in a journal, or in an Internet Request for Comment (RFC) which hopefully results in an *independent* evaluation of the algorithm. Often more or less subtle flaws are then discovered by other researchers. These flaws can vary from small errors in proofs to complete security breaks. Depending on the outcome, this evaluation can lead to a small fix of the scheme or to abandoning the idea altogether. Sometimes such weaknesses can be found in real-time when the author is presenting his ideas at a conference, but often evaluating a cryptographic algorithm is a time consuming task; for example, the design effort of the Data Encryption Standard (DES) has been more than seventeen man-years and the open academic evaluation since that time has taken a much larger effort. Cryptanalysis is quite destructive; in this respect it differs from usual scientific activities, even when proponents of competing theories criticise each other.

Few algorithms survive the evaluation stage; ideally, this stage should last for several years. The survivors can be integrated into products and find their way to the market. Sometimes they are standardised by organisations such as the National Institute of Standards and Technology, US (NIST), IEEE, IETF, ISO, or ISO/IEC.

As will be explained below, even if no new security weaknesses are found, the security of a cryptographic algorithm degrades over time; if the algorithm is not parameterised, the moment will come when it has to be taken out of service or when its parameters need to be upgraded and often such upgrades are not planned for.

19.4.2.1 *Open competitions.* In the past several open competitions have been launched to stimulate this open and independent evaluation process: such competitions were organised in the US for the selection of DES (1974) and AES (1997), in Europe by the RIPE (1988–1992) [65] and NESSIE (2000–2003) [57; 63] projects and in Japan by the CRYPTREC project (2000–2003) [18]. While each of these competitions has slightly different goals, it is clear that they have advanced the state of the art and have helped to develop better solutions. The main disadvantage of most of these competitions is that the results are not much better than the state of the art at the time that the competition started which is a direct consequence of the fact that the designs cannot be modified too much during the process. However, it turns out that the lessons learned during these competitions have been very valuable.

19.4.2.2 *Public versus secret algorithms.* The open and independent evaluation process described above offers a strong argument for publishing the details of a cryptographic algorithm. Publishing the algorithm opens it up for public scrutiny and is the best way to guarantee that it is as strong as claimed. Note that a public algorithm should not be confused with a public-key algorithm. Published algorithms can be standardised, and will be available from more than one source.

Nevertheless, certain governments and organisations prefer to keep their algorithms secret. These groups argue that obtaining the algorithm raises an additional barrier for the attacker. Moreover, governments or organisations may want to protect their know-how on the design of cryptographic algorithms. Note, however, that obtaining a description of the algorithm is often not harder than just bribing a single person. An unacceptable excuse for using a secret algorithm is the desire to hide the fact that an insecure algorithm is being used. The use of secret algorithms is acceptable in closed systems provided that sufficient experience and resources are available for *independent* evaluation and re-evaluation of the algorithm.

19.4.3 Insecure versus secure algorithms

Today secure cryptographic algorithms are publicly available that offer good performance at an acceptable cost. Nevertheless, one finds in many applications insecure cryptographic algorithms. This section explains some of the reasons for this occurrence.

19.4.3.1 *Incompetence.* Cryptography is a fascinating discipline which tends to attract do-it-yourself people who are not aware of the scientific developments of the last sixty years. Their home-made algorithms can typically be broken in a few minutes by an expert. For example, popular software sometimes encrypts data by adding a constant key word to all data words and it is of course very hard to beat the performance of this solution.

19.4.3.2 *Political reasons.* In a large part of the 20th century, governments have attempted to control know-how on cryptology and the use of cryptology. This situation results in the use of deliberately weak algorithms, such as A5/1 and A5/2 in the GSM phones or in short keys such as forty bit RC4 keys in browsers, and fifty-four-bit session keys in GSM phones.[3] General export from the US was only allowed for short key lengths: 40 bits or 56 bits for symmetric ciphers and 512 bits for factoring based systems (RSA) and discrete logarithm modulo a large prime (Diffie–Hellman). The US

export restrictions have been lifted to a large extent in January and October 2000 (see Koops [50]). In several countries, domestic controls were imposed; the best known example is France, where the domestic controls were lifted in January 1999. While the export restrictions are now of smaller concern, it still takes a long time before all legacy systems with short key lengths have been upgraded. Moreover, many countries still have legislation that restricts usage, importing and exporting of cryptographic algorithms.

19.4.3.3 *Economic reasons.* Even when it is obvious that an algorithm needs to be replaced or upgraded for technical reasons, it may not be economical to do so. Upgrades may be delayed because it is estimated that the risk of using weak cryptographic algorithms is acceptable and because the system has many other much weaker points. Economic reasons can probably explain why the financial sector has taken almost twenty years to upgrade DES to triple-DES.

19.4.3.4 *Brute force attacks become easier over time.* Brute force attacks are attacks which exist against any cryptographic algorithm that is conditionally secure, no matter how it works internally. These attacks only depend on the size of the external parameters of the algorithm, such as the block length of a block cipher or the key length of any encryption or MAC algorithm. It is the task of the designer to choose these parameters such that brute force attacks are infeasible.

A typical brute force attack against an encryption or MAC algorithm is an exhaustive key search which is equivalent to breaking into a safe by trying all the combinations of the lock. The lock should be designed such that a brute force attack is not feasible in a reasonable amount of time. This attack requires only a few known plaintext/ciphertext or plaintext/MAC pairs which can always be obtained in practice. It can be precluded by increasing the key length. Note that adding one bit to the key doubles the time for exhaustive key search. It should also be guaranteed that the key is selected uniformly at random in the key space.

---

[3]It was discovered that ten bits out of 64 bits are set to zero by the operators at the request of national authorities.

On a standard PC, trying a single key for a typical algorithm requires between 0.1 and 10 μs depending on the complexity of the algorithm. For example, a forty-bit key can be recovered in one-to-one hundred days. If a Local Area Network (LAN) with one hundred machines can be used, the key can be recovered in twenty minutes to a day. For a fifty-six-bit key such as DES, a key search requires a few months if several thousand machines are available as has been demonstrated in the first half of 1997. However, if dedicated hardware is used, a different picture emerges. M. Wiener has designed in 1993 a US$ one million hardware DES key search machine that can recover a fifty-six-bit DES key in about three hours [83]. If such a machine would be built in 2006, it would be about 300 times faster and thus recovering a fifty-six-bit key would take a little over half a minute on average. In 1998, a US$ two hundred fifty thousand dollar machine called Deep Crack was built that finds a fifty-six-bit DES key in about four and one-half days [23]; the design which required fifty percent of the cost has been made available for free.

These numbers makes a person wonder how the US government could publish in 1977 a block cipher with a fifty-six-bit key. However, it should be taken into account that a variant of Moore's law formulated in 1967 [68] states that computers will double their speed every eighteen months which implies that in 1977, recovering a fifty six-bit key with a US$ ten million hardware machine would take about twenty days; such a machine was clearly only feasible for very large organisations including the US government. This discussion also explains the initial controversy over the DES key length.

Experts believe that Moore's law will be holding for at least another fifteen years which means that if data needs to be protected for fifteen years against an opponent with a budget of US$ ten million, a key length of at least ninety bits is needed, which corresponds to the security level of three-key triple-DES (see Section 19.2.1.3). However, as the cost of increasing the key size is quite low, it is advisable to design new algorithms with variable key size from 128 to 256 bits. Indeed, searching for the smallest AES key (128 bits) is a factor 2^{72} times more expensive than finding a DES key.

Even with a key search machine of US$ 10^{12}, it would take in 2024 one million years to recover a 128-bit key. This calculation shows that if symmetric algorithms with 128-bit keys are used, it may no longer necessary to worry about exhaustive key search for the next decades.

19.4.3.5 *Shortcut attacks become more efficient.* Many algorithms are less secure than suggested by the size of their external parameters. It is often possible to find more effective attacks than trying all keys (see for example the discussion on two-key and three-key triple-DES in Section 19.2.1.3). Assessing the strength of an algorithm requires cryptanalytic skills and experience as well as hard work. During the last fifteen years, powerful new tools have been developed which includes differential cryptanalysis [9], which analyses the propagation of differences through cryptographic algorithms; linear cryptanalysis [52], which is based on the propagation of bit correlations; fast correlation attacks on stream ciphers [53]; and algebraic attacks on stream ciphers [17]. For example, the FEAL block cipher with eight rounds which was published in 1987, can now be broken with only ten chosen plaintexts. Similar progress has been made in the area of public key cryptology, for example, with attacks based on lattice reduction [42] and factoring methods discussed in Section 19.2.2.2.

19.4.3.6 *New attack models.* The largest threats however originate from new attack models. One of these models is a quantum computer. Another new threat is the exploitation of side channel attacks and attacks based on fault injection.

Feynman realised in 1982 that computers based on the principles of quantum physics would be able to achieve exponential parallelism: having n components in the quantum computer would allow it to perform 2^n computations in parallel. Note that in a classical computer, having n computers can speed up the computation with a factor up to n. A few years later, Deutsch realised that a general purpose quantum computer could be developed on which any physical process could be modelled, at least in principle. However, the first interesting application of quantum computers outside physics was proposed by Shor in 1994 [71] who showed that quantum computers were perfectly suitable to number theoretic problems, such as factoring and discrete logarithm problems. This result implies that if a large quantum computer could be built, the most popular public-key systems would be completely insecure. Building quantum computers however is a huge challenge. A quantum computer maintains state in a set of qubits. A qubit can hold a one, or a zero, or a superposition of one and zero; this superposition is the essence of the exponential parallelism, but it can be disturbed by outside influences, called decoherence. The first quantum computer with two qubits was built in 1998. Currently the record is a quantum computer with seven bits, three of these used for quantum error correction. This allowed the factorisation of the integer fifteen in 2002 [78]. Experts are divided on the question whether sufficiently powerful quantum computers can be built in the next fifteen to twenty years, but no one seems to expect a breakthrough in the next five to ten years. For symmetric cryptography, quantum computers are less of a threat since they can reduce the time to search a $2n$-bit key to the time to search an n-bit key, using Grover's algorithm [35]. Hence doubling the key length (from 128 to 256 bits) offers an adequate protection. Collision search on a quantum computer reduces from $2^{n/2}$ steps to $2^{n/3}$ steps [35], so it is sufficient to increase the number of bits of a hash result from 256 to 384.

Cryptographic algorithms have to be implemented in hardware or software, but it should not be overlooked that software runs on hardware. The opponent can try to make life easier by obtaining information from the hardware implementation, rather than trying to break the cryptographic algorithm using fast computers or clever mathematics. Side channel attacks have been known for a long time in the classified community; these attacks exploit information on the time to perform a computation [48], on the power consumption [49], or on the electromagnetic radiation [31; 64] to extract information on the plaintext or even the secrets used in the computation. A very simple side channel attack on the Vernam scheme (see Section 19.2.1)

exploits the fact that if there is a logical zero on the line, this can be the result of $0 \oplus 0$ or $1 \oplus 1$. If the device implementing the scheme is not properly designed, the two cases may result in different electrical signals which leaks immediately information on half of the plaintext bits.

Protecting implementations against side channel attacks is notoriously difficult and requires a combination of countermeasures at the hardware level such as adding noise, special logic, decoupling power source, at the algorithmic level, for example, blinding and randomisation and at the protocol level, for example, frequent key updates (see [15]). While many countermeasures have been published, many fielded systems are still vulnerable. This vulnerability is in part due to cost reasons and delays in upgrades and also due to the development of ever more sophisticated attacks. Developing efficient implementations which offer a high security level against side channel attacks is an important research challenge for the coming years.

The most powerful attacks induce errors in the computations, for example, by varying clock frequency or power level or by applying light flashes. Such attacks can be devastating because small changes in inputs during a cryptographic calculation typically reveal the secret key material [14]. Protecting against these attacks is non-trivial since it requires continuous verification of all calculations, which should also include a check on the verifications, and even this countermeasure may not be sufficient as has been pointed out in [85]. It is also clear that pure cryptographic measures will never be sufficient to protect against such attacks.

19.5 CONCLUDING REMARKS

This paper only covers part of the issues in modern cryptology since the discussion is restricted to an overview of cryptographic algorithms. Other problems solved in cryptography include identification, timestamping, sharing of secrets, and electronic cash. Many interesting problems are studied under the concept of secure multi-party computation, such as electronic elections and the generation and verification of digital signatures in a distributed way. An important aspect is the underlying key management infrastructure which ensures that secret and public keys can be established and maintained throughout the system in a secure way which is where cryptography meets the constraints of the real world.

This paper has demonstrated that while cryptology has made significant progress as a science, there are some interesting research challenges ahead both from a theoretic and from a practical viewpoint. Progress needs to be made in theoretical foundations, such as how to prove that a problem is hard, the development of new cryptographic algorithms, such as new public-key cryptosystems which do not depend on algebraic number theory and light-weight or low footprint secret-key cryptosystems, and secure and efficient implementations of cryptographic algorithms. Hopefully this will allow the cryptographic community to build schemes that offer long term security at a reasonable cost.

REFERENCES

[1] C. Adams, S. Lloyd, *Understanding PKI. Concepts, Standards, and Deployment Considerations*, 2nd edition, Addison-Wesley, Boston, HA (2003).

[2] R.M. Avanzi, H. Cohen, C. Doche, G. Frey, T. Lange, K. Nguyen, F. Vercauteren, *Handbook of Elliptic and Hyperelliptic Curve Cryptography*, H. Cohen, G. Frey, eds, Chapman & Hall/CRC, London (2005).

[3] M. Bellare, R. Canetti, H. Krawczyk, *Keying hash functions for message authentication*, *Advances in Cryptology, Proceedings Crypto'96*, Lecture Notes in Computer Science 1109, N. Koblitz, ed., Springer-Verlag, Berlin (1996), pp. 1–15. Full version http://www.cs.ucsd.edu/users/mihir/papers/hmac.html.

[4] M. Bellare, A. Desai, E. Jokipii, P. Rogaway, *A concrete security treatment of symmetric encryption*, *Proceedings 38th Annual Symposium on Foundations of Computer Science, FOCS'97*, IEEE Computer Society, Washington, DC (1997), pp. 394–403.

[5] M. Bellare, P. Rogaway, *Random oracles are practical: A paradigm for designing efficient protocols*, *Proceedings ACM Conference on Computer and Communications Security*, ACM Press, New York, NY (1993), pp. 62–73.

[6] M. Bellare, P. Rogaway, *The exact security of digital signatures. How to sign with RSA and Rabin*, *Advances in Cryptology, Proceedings Eurocrypt'96*, Lecture Notes in Computer Science 1070, U. Maurer, ed., Springer-Verlag, Berlin (1996), pp. 399–414.

[7] C.H. Bennett, G. Brassard, A.K. Ekert, *Quantum cryptography*, Scientific American 267 (4) (1992), 50–57.

[8] D.J. Bernstein, *The Poly1305-AES message-authentication code*, *Fast Software Encryption*, Lecture Notes in Computer Science 3557, H. Gilbert, H. Handschuh, eds, Springer-Verlag, Berlin (2005), pp. 32–49.

[9] E. Biham, A. Shamir, *Differential Cryptanalysis of the Data Encryption Standard*, Springer-Verlag, Berlin (1993).

[10] A. Biryukov, A. Shamir, D. Wagner, *Real time cryptanalysis of A5/1 on a PC*, *Fast Software Encryption*, Lecture Notes in Computer Science 1978, B. Schneier, ed., Springer-Verlag, Berlin (2002), pp. 1–18.

[11] J. Black, S. Halevi, H. Krawczyk, T. Krovetz, P. Rogaway, *UMAC: Fast and secure message authentication*, *Advances in Cryptology, Proceedings Crypto'99*, Lecture Notes in Computer Science 1666, M. Wiener, ed., Springer-Verlag, Berlin (1999), pp. 216–233.

[12] I.F. Blake, G. Seroussi, N.P. Smart, *Elliptic Curves in Cryptography*, Cambridge University Press, Cambridge (1999).

[13] M. Blaze, *Rights amplification in master-keyed mechanical locks*, IEEE Security & Privacy 1 (2) (2003), 24–32.

[14] D. Boneh, R.A. DeMillo, R.J. Lipton, *On the importance of eliminating errors in cryptographic computations*, Journal of Cryptology 14 (2001), 101–119.

[15] J. Borst, B. Preneel, V. Rijmen, *Cryptography on smart cards*, Journal of Computer Networks 36 (2001), 423–435.

[16] J.L. Carter, M.N. Wegman, *Universal classes of hash functions*, Journal of Computer and System Sciences 18 (1979), 143–154.

[17] N. Courtois, W. Meier, *Algebraic attacks on stream ciphers with linear feedback*, *Advances in Cryptology, Proceedings Eurocrypt'03*, Lecture Notes in Computer Science 2656, E. Biham, ed., Springer-Verlag, Berlin (2003), pp. 345–359.

[18] CRYPTREC project, http://www.ipa.gov.jp/security/enc/CRYPTREC/index-e.html.

[19] J. Daemen, V. Rijmen, *The Design of Rijndael. AES – The Advanced Encryption Standard*, Springer-Verlag, Berlin (2001).

[20] D.W. Davies, W.L. Price, *Security for Computer Networks. An Introduction to Data Security in Teleprocessing and Electronic Funds Transfer*, 2nd edition, Wiley, New York (1989).

[21] W. Diffie, M.E. Hellman, *New directions in cryptography*, IEEE Transactions on Information Theory IT–22 (6) (1976), 644–654.

[22] S. Dziembowski, U. Maurer, *Optimal randomizer efficiency in the bounded-storage model*, Journal of Cryptology 17 (2004), 5–26.

[23] EFF, *Cracking DES. Secrets of Encryption Research, Wiretap Politics & Chip Design*, O'Reilly, London (1998).

[24] FIPS 46, *Data Encryption Standard*, Federal Information Processing Standard, NBS, U.S. Dept. of Commerce (January 1977) (revised as FIPS 46-1 (1988); FIPS 46-2 (1993), FIPS 46-3 (1999)).

[25] FIPS 81, *DES Modes of Operation*, Federal Information Processing Standard, NBS, U.S. Dept. of Commerce (December 1980).

[26] FIPS 180, *Secure Hash Standard*, Federal Information Processing Standard (FIPS), Publication 180, National Institute of Standards and Technology, U.S. Department of Commerce, Washington, DC (May 11, 1993) (revised as FIPS 180-1 (1995); FIPS 180-2 (2003)).

[27] FIPS 186, *Digital Signature Standard*, Federal Information Processing Standard, NIST, U.S. Dept. of Commerce (May 1994) (revised as FIPS 186-1 (1998); FIPS 186-2 (2000); change notice published in 2001).

[28] FIPS 197, *Advanced Encryption Standard*, Federal Information Processing Standard, NIST, U.S. Dept. of Commerce (November 26, 2001).

[29] S. Fluhrer, I. Mantin, A. Shamir, *Weaknesses in the key scheduling algorithm of RC4*, *Selected Areas in Cryptography*, Lecture Notes in Computer Science 2259, S. Vaudenay, A. Youssef, eds, Springer-Verlag, Berlin (2001), pp. 1–24.

[30] E. Fujisaki, T. Okamoto, D. Pointcheval, J. Stern, *RSA-OAEP is secure under the RSA assumption*, *Advances in Cryptology, Proceedings Crypto'01*, Lecture Notes in Computer Science 2139, J. Kilian, ed., Springer-Verlag, Berlin (2001), pp. 260–274.

[31] K. Gandolfi, C. Mourtel, F. Olivier, *Electromagnetic analysis: Concrete results*, *Proceedings Cryptographic Hardware and Embedded Systems – CHES 2001*, Lecture Notes in Computer Science 2162, C.K. Koç, D. Naccache, C. Paar, eds, Springer-Verlag, Berlin (2001), pp. 251–261.

[32] M.R. Garey, D.S. Johnson, *Computers and Intractability: A Guide tot the Theory of NP-Completeness*, Freeman, New York (1979).

[33] O. Goldreich, *Foundations of Cryptography: Volume 1, Basic Tools*, Cambridge University Press (2001).

[34] S. Goldwasser, M. Bellare, *Lecture Notes on Cryptography*, http://www.cs.ucsd.edu/users/mihir/papers/gb.html.

[35] L.K. Grover, *A fast quantum mechanical algorithm for database search*, *Proceedings 28th Annual ACM Symposium on Theory of Computing*, ACM Press, New York, NY (1996), pp. 212–219.

[36] D. Hankerson, A. Menezes, S. Vanstone, *Guide to Elliptic Curve Cryptography*, Springer-Verlag, Berlin (2004).

[37] J.E. Haynes, H. Klehr, *Venona. Decoding Soviet Espionage in America*, Yale University Press, New Haven (1999).

[38] A.P.L. Hiltgen, *Construction of feebly-one-way families of permutations*, *Proceedings Auscrypt'92*, Lecture Notes in Computer Science 718, J. Seberry, Y. Zheng, eds, Springer-Verlag, Berlin (1993), pp. 422–434.

[39] ISO/IEC 9797, Information technology – Security techniques – Message Authentication Codes (MACs). Part 1: Mechanisms using a block cipher (1999). Part 2: Mechanisms using a dedicated hash-function (2002).

[40] ISO/IEC 10116, Information technology – Security techniques – Modes of operation of an n-bit block cipher algorithm (1997).

[41] ISO/IEC 10118, Information technology – Security techniques – Hash-functions, Part 1: General (2000). Part 2: Hash-functions using an n-bit block cipher algorithm (2000). Part 3: Dedicated hash-functions (2003). Part 4: Hash-functions using modular arithmetic (1998).

[42] A. Joux, J. Stern, *Lattice reduction: a toolbox for the cryptanalyst*, Journal of Cryptology 11 (1998), 161–185.

[43] G.A. Kabatianskii, T. Johansson, B. Smeets, *On the cardinality of systematic A-codes via error correcting codes*, IEEE Transactions on Information Theory IT–42(2) (1996), 566–578.

[44] D. Kahn, *The Codebreakers. The Story of Secret Writing*, MacMillan, New York (1967).

[45] L. Knudsen, Block Ciphers – Analysis, Design and Applications, PhD thesis, Aarhus University, Denmark (1994).

[46] L. Knudsen, *Chosen-text attack on CBC-MAC*, Electronics Letters 33 (1) (1997), 48–49.

[47] N. Koblitz, *A Course in Number Theory and Cryptography*, Springer-Verlag, Berlin (1987).

[48] P. Kocher, *Timing attacks on implementations of Diffie–Hellman, RSA, DSS, and other systems*, *Advances in Cryptology, Proceedings Crypto'96*, Lecture Notes in Computer Science 1109, N. Koblitz, ed., Springer-Verlag, Berlin (1996), pp. 104–113.

[49] P. Kocher, J. Jaffe, B. Jun, *Differential power analysis*, *Advances in Cryptology, Proceedings Crypto'99*, Lecture Notes in Computer Science 1666, M. Wiener, ed., Springer-Verlag, Berlin (1999), pp. 388–397.

[50] B.-J. Koops, *Crypto law survey*, http://rechten.kub.nl/koops/cryptolaw

[51] A.K. Lenstra, E.R. Verheul, *Selecting cryptographic key sizes*, Journal of Cryptology 14 (2001), 255–293.

[52] M. Matsui, *The first experimental cryptanalysis of the Data Encryption Standard*, *Proceedings Crypto'94*, Lecture Notes in Computer Science 839, Y. Desmedt, ed., Springer-Verlag, Berlin (1994), pp. 1–11.

[53] W. Meier, O. Staffelbach, *Fast correlation attacks on stream ciphers*, Journal of Cryptology 1 (1989), 159–176.

[54] R. Merkle, *Secrecy, Authentication, and Public Key Systems*, UMI Research Press, Ann Arbor (1979).

[55] A.J. Menezes, P.C. van Oorschot, S. Vanstone, *Handbook of Applied Cryptography*, CRC Press, London (1997).

[56] Microsoft NGSCB, http://www.microsoft.com/resources/ngscb/

[57] NESSIE, http://www.cryptonessie.org

[58] NIST, *SP 800-38A Recommendation for Block Cipher Modes of Operation – Methods and Techniques* (December 2001).

[59] S. Paul, B. Preneel, *Analysis of non-fortuitous predictive states of the RC4 key stream generator*, **Progress in Cryptology**, Lecture Notes in Computer Science 2904, T. Johansson, S. Maitra, eds, Springer-Verlag, Berlin (2003), pp. 30–47.

[60] E. Petrank, C. Rackoff, *CBC MAC for real-time data sources*, Journal of Cryptology 13 (2000), 315–338.

[61] B. Preneel, V. Rijmen (eds), *State of the Art in Applied Cryptography*, Lecture Notes in Computer Science 1528, Springer-Verlag, Berlin (1998).

[62] B. Preneel, P.C. van Oorschot, *MDx-MAC and building fast MACs from hash functions*, **Proceedings Crypto'95**, Lecture Notes in Computer Science 963, D. Coppersmith, ed., Springer-Verlag, Berlin (1995), pp. 1–14.

[63] B. Preneel, A. Biryukov, C. De Cannière, S.B. Ors, E. Oswald, B. Van Rompay, L. Granboulan, E. Dottax, G. Martinet, S. Murphy, A. Dent, R. Shipsey, C. Swart, J. White, M. Dichtl, S. Pyka, M. Schafheutle, P. Serf, E. Biham, E. Barkan, Y. Braziler, O. Dunkelman, V. Furman, D. Kenigsberg, J. Stolin, J.-J. Quisquater, M. Ciet, F. Sica, H. Raddum, L. Knudsen, M. Parker, *Final Report of NESSIE, New European Schemes for Signatures, Integrity, and Encryption*, Lecture Notes in Computer Science, Springer-Verlag, in print.

[64] J.-J. Quisquater, D. Samide, *ElectroMagnetic Analysis (EMA): Measures and countermeasures for smart cards*, **Smart Card Programming and Security, International Conference on Research in Smart Cards, E-smart 2001**, Lecture Notes in Computer Science 2140, I. Attali, T. Jensen, eds, Springer-Verlag, Berlin (2001), pp. 200–210.

[65] RIPE, *Integrity Primitives for Secure Information Systems. Final Report of RACE Integrity Primitives Evaluation (RIPE-RACE 1040)*, Lecture Notes in Computer Science 1007, A. Bosselaers, B. Preneel, eds, Springer-Verlag, Berlin (1995).

[66] R.L. Rivest, A. Shamir, L. Adleman, *A method for obtaining digital signatures and public-key cryptosystems*, Communications ACM 21(2) (1978), 120–126.

[67] R.A. Rueppel, *Analysis and Design of Stream Ciphers*, Springer-Verlag, Berlin (1986).

[68] R.R. Schaller, *Moore's law: past, present, and future*, IEEE Spectrum 34(6) (1997), 53–59.

[69] A. Shamir, E. Tromer, *Factoring large numbers with the TWIRL device*, **Advances in Cryptology, Proceedings Crypto'03**, Lecture Notes in Computer Science 2729, D. Boneh, ed., Springer-Verlag, Berlin (2003), pp. 1–26.

[70] C.E. Shannon, *Communication theory of secrecy systems*, Bell System Technical Journal 28 (4) (1949), 656–715.

[71] P.W. Shor, *Algorithms for quantum computation: discrete logarithms and factoring*, **Proceedings 35th Annual Symposium on Foundations of Computer Science**, S. Goldwasser, ed., IEEE Computer Society, Washington, DC (1994), pp. 124–134.

[72] V. Shoup, *OAEP reconsidered*, **Advances in Cryptology, Proceedings Crypto'01**, Lecture Notes in Computer Science 2139, J. Kilian, ed., Springer-Verlag, Berlin (2001), pp. 239–259.

[73] G.J. Simmons (ed.), *Contemporary Cryptology: The Science of Information Integrity*, IEEE Press, New York, NY (1991).

[74] S. Singh, *The Code Book. The Science of Secrecy from Ancient Egypt to Quantum Cryptography*, Anchor, Wilington (2000).

[75] W. Stallings, *Cryptography and Network Security*, 3rd edition, Prentice Hall, New York (2003).

[76] D. Stinson, *Cryptography. Theory and Practice*, 3rd edition, CRC Press, London (2005).

[77] TCG, http://www.trustedcomputinggroup.org.

[78] L.M.K. Vandersypen, M. Steffen, G. Breyta, C.S. Yannoni, M.H. Sherwood, I.L. Chuang, *Experimental realization of Shor's quantum factoring algorithm using nuclear magnetic resonance*, Nature 414 (2001), 883–887.

[79] G.S. Vernam, *Cipher printing telegraph system for secret wire and radio telegraph communications*, Journal American Institute of Electrical Engineers, Vol. XLV (1926), 109–115.

[80] X. Wang, H. Yu, *How to break MD5 and other hash functions*, **Advances in Cryptology, Proceedings Eurocrypt'05**, Lecture Notes in Computer Science 3494, R. Cramer, ed., Springer-Verlag, Berlin (2005), pp. 19–35.

[81] X. Wang, Y.L. Lin, H. Yu, *Finding collisions in the full SHA-1*, **Advances in Cryptology, Proceedings Crypto'05**, Lecture Notes in Computer Science 3621, V. Shoup, ed., Springer-Verlag, Berlin (2005), pp. 17–36.

[82] M.N. Wegman, J.L. Carter, *New hash functions and their use in authentication and set equality*, Journal of Computer and System Sciences 22 (3) (1981), 265–279.

[83] M.J. Wiener, *Efficient DES key search*, Presented at the Rump Session of Crypto'93, reprinted in **Practical Cryptography for Data Internetworks**, W. Stallings, ed., IEEE Computer Society, Washington, DC (1996), pp. 31–79.

[84] A.C. Yao, *Theory and applications of trapdoor functions*, **Proceedings 23rd IEEE Symposium on Foundations of Computer Science**, IEEE Computer Society, Washington, DC (1982), pp. 80–91.

[85] S.-M. Yen, M. Joye, *Checking before output may not be enough against fault-based cryptanalysis*, IEEE Transactions on Computers 49 (9) (2000), 967–970.

PART 4

COMPUTER SECURITY

20

A HISTORY OF COMPUTER SECURITY STANDARDS

Jeffrey R. Yost

Charles Babbage Institute, University of Minnesota
Minnesota, USA

Contents

Abstract

While computer security has only achieved widespread attention in recent years, the topic has a long and complex history. This chapter begins by briefly Surveying Computer Security Standards within the broader context of technical standard setting. It then details early work to address vulnerabilities resulting from electronic emanation, heightened Concerns with the proliferation of computer networking, the ongoing influence of the Department of Defense (DoD) with standards work, and the establishment of formalized national (TCSEC in the US) and international (Common Criteria) computer security standards. Early cryptographic research in the academy and the emergence of the software security industry are also explored, along with the relationship between security and privacy. Ultimately, formalized government standards have been limited, and the global commercial sector has been increasingly influential in seeking to mitigate accelerating threats.

Keywords: computer security, computer security standards, public key, Clipper Chip, RSA Data Security, computer security industry, Orange Book, TCSEC, Common Criteria, Bell–LaPadula.

In the 1980s, early forms of malware, or malicious software, primarily computer viruses that were passed by exchanging infected disks from computer to computer, posed substantial threats to the integrity and operations of computer software and systems. More recently, computer crimes, including denial of service attacks and computer-based identity theft, along with new categories of malware, such as worms and Trojan horses, have exacerbated the problem. The renaissance of the Internet and the ubiquity of its use with the advent and growth of the World Wide Web and browser software during the early to mid-1990s further fueled threats of increasingly destructive network-spread malware, cyberterrorism and other forms of computer crime. Concomitantly, these threats have bolstered computer security as: a growing area of study within computer science departments in the academy; a fundamental concern and procurement expense for government, corporations and other organizations; a rapidly expanding sector of the software industry; and an increasingly common topic explored by mass media.

Computer security, however, has a much longer history than most of the literature on the issue implies. It, like digital computing more generally, is an area that first developed from the needs of and funding by the U.S. Department of Defense (DoD). As digital computing technology evolved and proliferated, and the earliest networks were formed, a number of DoD-funded researchers led projects to address potential computer security threats, and to establish common or standard, security protocols. The following chapter briefly touches upon the early history of technical standard setting work in other fields, and then goes on to survey the origin of computer security standards, changes in computer security with advancing technology, and the appearance and activities of a broad set of institutional, corporate, and individual actors that became involved in defining and shaping computer security standards in government and industry.

While the history of computer security standards holds some commonalities with what we know of the broader range of technical standard setting work, computer security standards have also developed within some unique contexts and arguably have been less influential and useful than technical standards in many domains.

20.1 SETTING TECHNICAL STANDARDS: A BRIEF OVERVIEW

Technical standard setting is a rich historical topic that has been understudied. To the extent that historians have explored certain aspects of technical standard setting, it typically has been episodic and as a bit part to larger story of economic and technological change, such as in David S. Landes' *The Unbound Prometheus: Technological Change and Industrial Development in Western Europe from 1750 to the Present* or David F. Noble's *America By Design: Science, Technology, and the Rise of Corporate Capitalism*. In a small number of works, the history of technical standards has figured more prominently, including David A. Hounshell's *From the American System to Mass Production, 1800–1932*; Janet Abbate's *Inventing the Internet*; and Merritt Roe Smith's *Harpers Ferry Armory and the New Technology: The Challenge of Change*.

The US government, and in particular the military, has long had a hand in setting many types of technical standards, primarily for military, national security and trade purposes. In 1824 the US Congress formed the Office of Standard Weights and Measures. Throughout most of the 19th century, its work was generally limited to setting and regulating measurement for fair commerce. Also during the first half of the 19th century, the US military helped set standards and specifications with armory practice as the nation transitioned from craft to factory methods in manufacturing rifles [54]. This was done with the aim of achieving interchangeability of parts to ease the task of battlefield repairs, as well as to increase efficiency of production [54].[1] Later, the need for efficient rail transportation during the Civil War led to heightened recognition of the need for standard track gauges (distance between rails) [40].

---

[1]As a result of this book and the broader recognition of the impact of armory practice on factory methods, this is one of the more thoroughly explored episodes of standard setting work in the historical literature. The work highlights the difficulties with instituting standards that conflict with established cultural practices. It also indicates that interchangeability came at great expense rather than leading to lower cost production in its early years.

Figure 20.1. National Bureau of Standards' Standards Eastern Automatic Computer (SEAC), c. early 1950s. (Courtesy of the Charles Babbage Institute.)

Many agencies and departments of the federal government have played a role in setting standards to facilitate trade, better ensure public safety (standards of air travel and air traffic control), and for other purposes. Additionally, industry trade associations and non-government technical organizations have had a major role in standard setting work [52]. More recently, international standards organizations have become ever more influential in an era increasingly defined by global commerce, perspectives, and policies.

At the end of the nineteenth century, the Office of Standard Weights and Measures purview expanded into conducting science and engineering research for the government in physics, chemistry, engineering and metallurgy. In recognition of this broader program, the organization's name was changed to the National Bureau of Standards (NBS) in 1901. The work of the NBS broadened into digital computing in the early 1950s with the Standards Eastern Automatic Computer (SEAC) and Standards Western Automatic Computer (SWAC).

The NBS became the primary computer security standards setting organization for the federal government in 1965, and though its relative importance in this area has waned to a certain extent, it (NBS changed its name to National Institute of Standards and Technology, or NIST, in 1990) has remained fundamental to this field to the present day [40].

20.2 DIGITAL BEGINNINGS, PHYSICAL SECURITY AND ELECTRONIC RADIATION

Before exploring the role of the NBS/NIST in computer security standards, and then introducing and outlining the growing role of the National Security Agency (NSA) in this field, the chapter will first introduce the earliest appearance of computer security standards. These standards related to physical security and were designated by the military.

In the early to mid-1940s the first meaningful digital computer, the Electronic Numerical Integrator and Computer (ENIAC), was designed and de-

Figure 20.2. Electronic Numerical Integrator and Computer (ENIAC) at the University of Pennsylvania's Moore School of Electrical Engineering, c. 1946. (Courtesy of the Charles Babbage Institute.)

veloped at the University of Pennsylvania's Moore School of Electrical Engineering. It had been financed by the Army Ballistic Research Laboratory (BRL) to provide a tool to quickly and accurately solve mathematical equations for ballistic firing tables during World War II. Though it was not completed until several months after the end of the war, the United States almost immediately became engaged in the Cold War with the Soviet Union, and the DoD continued to finance a number of computer development and design projects at universities, national laboratories, and a few select firms. By the early to mid-1950s, each of the four longtime leaders of the American office machine industry, Burroughs Corporation (accounting machines), National Cash Register (NCR – cash registers and accounting machines), Remington Rand (typewriters and tabulating machines), and International Business Machines (IBM – tabulating machines and punched cards), had entered the computer industry.

One of the first applications of the ENIAC was calculating equations for the DoD's classified "Super" project to develop a hydrogen bomb. Main-

frame computers of the early post-World War II era cost from several hundred thousand to millions of dollars, and were only installed at and controlled by elites in the military and to a lesser extent industry (defense contractors in the aerospace field, large insurance firms, etc.). Many of the calculations completed on these early digital computers were secret. Further revising and enhancing security procedures on national defense matters (which had begun with legislation in 1936) in 1944 the military began to use the classification "Top Secret" for documents and information that could lead to "grave damage to the nation" if compromised. In 1946 a number of Atomic scientists who had worked on the Manhattan Project successful lobbied congress to create a civilian nuclear research agency, the Atomic Energy Commission (AEC). This organization adopted the military's classification system for secret information and documentation, and coupled with related legislation, this formally placed restrictions on the export of nuclear information [50: 48–50].

Despite the secretive nature of the DoD, and to a lesser extent, industry, early computer security was fairly straightforward and posed few additional problems regarding the protection of confidential information. In the first decade after World War II computer security was typically limited to physical security. It was simply one element of the more general security of the installations where computers were housed. Security protocols were designed to focus upon and prevent or address theft, vandalism, sabotage and natural disasters (hurricanes, floods, tornadoes, volcanoes, earthquakes, etc.). Guards and alarms were used to prevent unauthorized personnel from entering and engaging in theft or destructive acts. One or a small number of skilled computer technicians usually handled the operation of the machines at an installation. Scientific researchers often had no physical contact with the computers that crunched their numbers. The complexity and specialized knowledge involved in operating early mainframes ensured and justified limited access and contributed a substantial degree of security. Input/output (I/O) systems of the time typically consisted of punched cards or paper tape containing encoded data, and jobs were processed in batches, where one job was done before the next one began. Computer operators were few in number, tended to go through strict screening, and rarely posed any risk to the security of computer systems or the data processed by them.

Electronic radiation or emanation, however, did present some potential computer security risks for early mainframe computers. This was a problem for secret communication (or opportunity for spying) that was also present with cipher machines. In his autobiography, one senior British Security officer in the early post-World War II era, Peter Wright, states:

> Any cipher machine, no matter how sophisticated, has to encipher the clear text of the message into a stream of random letters. In the 1950s, most advanced ciphers were produced by typing the clear text into a teleprinter, which connected into a separate cipher machine, and the enciphered text clattered out on the other side. The security of the whole system depended on thorough screening. If the cipher machine was not electromagnetically screened from the input machine carrying the clear text, echoes of the unencoded message might be carried along the output cables along with the enciphered message.

> With the right kind of amplifiers it was theoretically possible to separate the "ghosting" text out and read it off [60: 11].

In fact, it was more than just a "theoretical" possibility. Such techniques allowed the British security agents, in MI 5 and GBHQ, to frequently read the high-grade French cipher used by the French Embassy in London in the early 1960s [60].

All electronic equipment emits a level of electrical and electromechanical radiation. During the 1950s some government computer specialists became concerned that "electronic eavesdroppers" could capture and "decipher" emanations from mainframe computers, and could do so with little risk of detection. The level of electronic radiation and the distance of the potential eavesdropper were fundamental factors in determining whether or not emanations were decipherable. Given the secure surroundings of most early government computer installations processing classified data the risks were modest. Nevertheless, by the latter part of the 1950s the government set the first standard, called TEMPEST, for the level of emanations that remained acceptable when classified information was being processed. In the following decades there were refinements to the original TEMPEST standard on electromechanical radiation. TEMPEST became a wide-ranging term for technology that suppressed signal emanations from electronic equipment, as well as for the research and development underlying such efforts. TEMPEST products were larger, heavier, and more expensive than their non-TEMPEST counterparts, and generally worked by creating containers or barriers to shield electronic emanations [47].

Physical shields can take the form of specially designed computer rooms or even entire buildings (Secure Compartmented Information Facilities or SCIFs), but more commonly were containers that surrounded and became part of computing equipment. This generally did not pose any feasibility challenges given the remote locations and abundance of space for early military computers. For instance, Lawrence Livermore National Laboratory's Jerome A. Russell described general measures taken at the lab by the mid-1960s to prevent "someone outside the fence" picking up the noises:

With the teletype setup, we have a multiprogramming or multiprocessing system, which we call
Octopus. We have twisted par cables carrying the
teletype leads to the physicists' and mathematicians' offices. These cables are shielded according to a classified regulation which says you have
to have a shield on it of a certain nature ... We
don't share the telephone facility with the regular
voice-lined systems [48].

Such security, however, became more difficult
and less practical as mainframes, and later, minicomputers began to proliferate in number and location to more "open" environments by the late
1960s.

Under TEMPEST, techniques were also used
to protect information by adding additional radi-
ating equipment to generate "noise" and confuse
and suppress the ability of eavesdroppers to comprehend signals. During the 1960s, even though
considerable attention was paid to TEMPEST protection, it lacked uniformity, and the various DoD
agencies often had to establish criteria on a project-
by-project basis within Requests For Proposals
(RFPs). To achieve greater uniformity, in 1974, the
first Industrial Tempest Program (ITP) was established with three goals: to set commercial standards; to designate criteria for testing equipment
that met the set standards; and to certify vendor
equipment. This allowed and encouraged vendors
to develop "off-the-shelf" TEMPEST computers
and other communications equipment that the government could purchase without devoting time, attention, and expense to the issue of electronic emanations. Over the years, the DoD and other areas
of the federal government have remained heavily
involved with monitoring and scrutinizing the research, development, standard setting, testing and
sale of TEMPEST products [47: 255].

TEMPEST, however, only addressed one potential type of computer security vulnerability: electrical emanation. While this form of risk remained
and TEMPEST equipment has continued to be important, wired access to machines and transmission
of data over networks soon became the fundamental focus and driving force to devote significant
time and resources to the potential and real problem of computer security, and to identify and implement appropriate mechanisms or standards during the 1960s and 1970s [6; 18].

20.3 THE EARLY LEADERSHIP OF WILLIS WARE IN COMPUTER SECURITY RESEARCH

In the 1950s, the physical isolation of machines
and their data and programs gradually began to
wither alongside early digital computer networking
technology, forever altering the computer security
landscape. During the mid-1950s IBM received a
major contract to work in conjunction with Massachusetts Institute of Technology (MIT) to develop
computers for the Semi-Automatic Ground Environment (SAGE), a complex system of radar and
networked computers designed to provide early detection against an enemy air attack.[2] Preceding and
influencing the development of the SAGE project,
MIT's Project Whirlwind (originally a flight simulator project that evolved into a broader and more
influential digital computer development project
during the early 1950s) introduced the technology of time-sharing. MIT remained a leader in this
area, and the school's Multiple Access Computer
project, or Project MAC (started in 1963), focused
on the further development and extension of time-
sharing technology [45].

Time-sharing was one of the main areas of
research supported by ARPA's Information Processing Techniques Office (IPTO) and typically involved connecting multiple users at simple teletypewriter terminals to a central computer using
telephone lines. By 1963 MIT could connect two-
dozen users at once over a network. The purpose of
time-sharing was to more effectively utilize expensive processing and memory resources of computers. A time-sharing system allocated split-second
slices of time to individual users in rapid rotation
to seek to create the user-experience of a dedicated
stand alone, or individual system. In university settings, the system often had to be set to prevent overwriting, but access to reading and modifying data
was not perceived as a significant issue at many of
the early university computer centers [10].

A far different situation existed in the defense
community. Nevertheless, there were also close
ties between the military and a small number of

---

[2]For an impressive analysis of the work of MIT and MITRE
on Whirlwind and SAGE see Redmond and Smith [45].

elite universities in science and technology. With SAGE and other projects, there was a very close partnership between Massachusetts Institute of Technology and a number of areas of DoD research and development. The research and use of time-sharing at leading universities (like MIT), information technology firms (IBM and others), and in the military led to a new environment with regard to computer security. This was recognized by several individuals who became pioneers in the information technology security field.

In the mid-to-late 1960s personnel in or associated with the DoD increasingly came to recognize the importance of computer security in light of time-sharing and other computer networking activity.

Willis Ware was one of the early leaders in identifying and trying to better understand the nature of the emerging problem of computer security in an environment defined by the proliferation of time-shared computing systems. Ware had a strong background in computing and had deep connections to DoD research and development. He had worked with John von Neumann at the Institute for Advanced Study, and completed his PhD in electrical engineering at Princeton University in 1951. Von Neumann, one of the principle figures in early computer technology and a fundamental force behind the definition of the principle architecture in early digital computing (von Neumann Architecture), is examined with great skill and insight by leading computer historian William Aspray in *John von Neumann and the Origins of Modern Computing*.

Willis Ware went on to serve in the Mathematics Department, and later as the longtime Chair of the Computer Science Department of the RAND Corporation. In its first decades, The RAND Corporation, which was formed in Santa Monica by the Air Force in 1946, existed primarily as an Air Force-sponsored research entity.

The Air Force gave the RAND Corporation a great amount of leeway in the 1940s, 1950s and 1960s to decide how best to spend research money to address military issues and problems. While the Air Force was also conducting some internal research on computing systems and classified data

Figure 20.3. Computer security pioneer RAND Corporation's Willis Ware speaking at the 1979 National Computer Conference. (Courtesy of the Charles Babbage Institute.)

(including studies at the Air Force Rome Air Development Center), the combination of having a conglomeration of leading scientists and engineers in the computing field, substantial resources, past exposure to and work on military systems, and the interaction of these scientists and engineers with the broader technical community made RAND an ideal setting for recognizing and seeking to address the problem of computer security. As Ware put it:

> We would talk amongst ourselves in the hallways, at conferences, and gradually there emerged a concern that we really ought to do something about finding out how to protect computers and all the information in them, because the country had become too dependent on them [57].

In the 1960s the RAND Corporation, and its spin-off that did much of the programming and system integration work for SAGE, the System Development Corporation (SDC), engaged in some of the first of the so-called "penetration studies" to try to infiltrate time-sharing systems in order to test their vulnerability.

Simultaneously, the National Security Agency, a secretive intelligence gathering organization formed by President Harry S. Truman in 1952, became involved with computer security. In one of the rare public appearances by a NSA employee during the 1960s, the agency's Bernard Peters collaborated with Ware in leading a session on computer security at the 1967 Spring Joint Computer Conference in Fort Meade, Maryland. Both Peters and Ware emphasized the sea change in computer security that had occurred with time-sharing. They both saw how current control programs were entirely unable to maintain the military's classification protocol in comparison to that which existed with paper documents. More specifically, time-sharing systems were incapable of meeting the security needs involved with keeping data separate – providing read-and-write access as appropriate to individuals with designated clearances, while denying access to people without proper clearances [58].

The time-sharing system used at SDC in the mid-1960s for instance accommodated up to 53 users at one time and trying to partition the system so that one user did not have access to another's data was an ongoing challenge. When working on classified DoD material SDC had to "kick everyone out" who did not have clearance for the data being processed. New programs often provided the greatest hurdles, as errors could result in jumping from one part of the system to another. An additional issue was the residual information sometimes left behind after overwriting. To address these types of challenges, Peters concentrated on the importance of the monitor or operating system to ensure security, the certification of the monitor, and embedding critical security functions in small amounts of code [33: 43].

Within the Defense community the growing recognition of the security issue, brought about by the increased prominence and importance of resource-sharing systems (time-sharing, multiprogramming, etc.), resulted in the Defense Science Board establishing a task force in October 1967 to examine security problems with such computing systems. Willis Ware chaired this task force, which included members from the RAND Corporation, SDC, a number of major defense contractors, academic computer scientists from colleges and universities, NSA officials, and CIA officers. This group recognized the vast range of hardware and software systems already in existence and sought to provide a "general compilation of techniques and procedures" that could be flexible and broadly useful in protecting confidential military information. At the same time, the group was cognizant of the limitations involved given the variety of systems, the newness of the field, and continuous innovation. While understanding the advantage of broad standards, they stressed the inevitable need to solve many problems on a case-by-case basis. The group deliberated for two years, bringing in a host of additional experts to testify. On February 11, 1970 they completed their classified report, entitled, *Security Control for Computer Systems: Report of Defense Science Board Task Force on Computer Security* [58].[3]

The Defense Science Board report was by far the most important and thorough study on technical and operational issues regarding secure computing systems of its time period. In general, the report emphasized how technology had surpassed the knowledge and ability to implement appropriate controls to ensure security for confidential military information. The report focused on the distinction between *open* environments, which were characterized by access to non-cleared users, unprotected consoles, and unprotected communications; and *closed* environments, which consisted of only cleared users, protected consoles, and protected communications. Specifically, the report contained seven primary conclusions:

- Providing satisfactory security controls in a computer system is in itself a system design problem.
- Contemporary technology can provide a secure system acceptably resistant to external attack, accidental disclosures, internal subversion, and denial of use to legitimate users for a closed environment.

[3]The Report was declassified in 1975 and was republished by the RAND Corporation in 1979.

- Contemporary technology cannot provide a secure system in an open environment.
- It is unwise to incorporate classified or sensitive information in a system functioning in an open environment.
- Acceptable procedures and safeguards exist and can be implemented so that a system can function alternatively in a closed and open environment.
- Designers of secure systems are still on the steep part of the learning curve and much insight and operational experience with such systems is needed.
- Substantial improvement (in cost and performance) in security controlling systems can be expected if certain research areas are pursued [58: iv].

In addition to the conclusions, the report identified a fundamental problem with the current policy of the DoD on matters of computer security. Despite the fact that a number of systems were not entirely secure, the task force recognized that the situation might grow worse if all such systems lacking strong security were off limits, as was then stipulated. The group believed that exposure to and use of systems was critical to gaining understanding and allowing the development of design and operational guidelines. The sole action item of the report stated:

> The security policy directives presently in effect prohibit the operation of resource-sharing computer systems. This policy must be modified to permit contractors and military centers to acquire and operate such systems. This first step is essential in order that experience and insight with such systems be accumulated, and in order that technical solutions be tried [58].

Ware's task force also identified one "near action item", to establish a technical agent to define procedures and techniques for certifying security-controlling systems, especially computer software. The group believed that the need for such an agent was immediate, but it also posed a challenge in that it required advanced technical expertise in several disciplines. The group stated that the responsibility for finding and overseeing the work of such an agent could fall upon the NSA, the Defense Intelligence Agency (DIA), Joint Tactical SIGINT Architecture (JTSA), or a multi-service agency. Finally, the Task Force made one last recommendation, as well as provided a prediction for the near future. They recommended that their group be maintained to provide advice to the Directorate of Security Policy, the Technical Agent, and the designers, certifiers, and users of secure systems. They also emphasized that in the future it would be the computer industry that would have to provide systems with appropriate safeguards. Given this, the Task Force believed their report should be circulated relatively broadly to government agencies, industry, research groups and defense contractors [58].

20.4 CONTINUING CHALLENGES WITH COMPUTER SECURITY AT THE DoD

These last points were related to perhaps the greatest early challenge to establishing computer security standards in what amounted to a secret community within and tied to the DoD. The conundrum was that while there was a recognition of the need to initially classify reports and documents useful to understanding the technical side of security standards, at the same time there was a need for an ever broader community, particularly the computer industry, to effectively take advantage of the rapidly growing possibilities of information technology as the Cold War heightened. Greatly simplified, it played into a tradeoff of speed versus security that had long existed within secret scientific communities within and tied to the DoD. There were also significant costs involved. About ten to fifteen years prior to the widespread proliferation of time-sharing systems, in 1953 the University of California Radiation Laboratory reported their security costs at $503,079 and Los Alamos indicated theirs were approximately $383,000. Adding the complexity of time-sharing systems could only exacerbate the challenges and costs involved (into many millions of dollars), but at the same time this was fundamental to advancing possibilities for security upon which it was difficult to assign a monetary value [50: 54].

The Science Defense Board's 1970 report generally sought to establish policies and mechanisms that would allow the incorporation of existing security infrastructure for classified military information into different computer system environments. The seemingly simple policy used with documents, that individuals could not "read up", or see materials above their level of clearance (the four levels were "top secret", "secret", "confidential" and "unclassified" – an individual with a secret clearance, for instance, could read secret, confidential and unclassified documents, but not top secret ones) was not easy to implement in computing systems given demands of users for computing resources and the internal functioning of existing operating systems.

Operating systems are large and complex and it is difficult to know all the conditions and internal workings of these systems. In essence, multilevel computer security, or security adhering to different levels of classification, was a microcosm of a fundamental and daunting question underlying the young field of computer science, "how did systems actually operate and how did this change under different conditions and contexts?". Ware's committee had correctly identified the tremendous complexity of computer security both within, and potentially outside of, the military and classification environment. The group had offered a number of insights and recommended mechanisms for the future, but they were not able to provide solutions to the many problems that they raised – other than stressing the need for more openness and the involvement of a wider community to figure out remedies to the dilemmas at hand. Following the Ware-led study, a number of projects on computer security were funded by various agencies of the defense community, but most of these were similar to the Ware project in that they were long on problem identification, and short on solutions or workable practices.

20.5 JAMES P. ANDERSON AND THE AIR FORCE'S FOCUS ON COMPUTER SECURITY

In early 1972, after a couple years of relative uncertainty, Major Roger Schell of the Air Force

Electronic Systems Division initiated a new and influential effort. James P. Anderson, a computer consultant out of Fort Washington, Pennsylvania, headed the study. James Anderson's project was more tightly defined than the exploratory and broader Ware study. It was focused solely on the problem as it existed in the Air Force and was completed in a matter of months in October 1972. Anderson and his group had a dire assessment of the situation, indicating that there was no current system that could operate securely and have multiple or differential levels of access. He stressed that infiltration groups testing systems, or so-called "tiger teams", could usually break through security mechanisms currently in use, and that inefficiencies with regard to the Air Force's computer systems were resulting in costs of approximately $100 million annually. Additionally, on the rare occasions when tiger teams failed to infiltrate systems, this did not necessarily guarantee that adequate security was in place [33].

At the heart of the problem was the fact that resource-sharing systems relied on operating systems to maintain their separation, and users commonly program these operating systems to accomplish their work. In essence, users with different security clearances were accessing the same primary storage, and thus, had access to the same data. Additionally, various applications might contain a "trap door" inserted by programmers to secretly gain subsequent access to systems. Systems developed by programmers without clearances could be particularly vulnerable, or at the very least, uncertain with regard to security. Anderson's group, which consisted of members of NSA, Thompson–Romo–Wooldridge, Inc. (TRW), the SDC and MITRE (a MIT spin-off non-profit research corporation formed in July 1958), was not optimistic about the future, and emphasized how major and immediate changes were necessary. Chief among the recommendations was to allocate $8 million to a research and development computer program that consisted of discontinuing the integration of security functions within existing operating systems in favor of creating separate "security kernels", or smaller, simpler, peripheral operating system-like pieces of code to interact directly with system hardware. An implementation of the kernel concept was achieved during the early 1970s using a kernel called HYDRA in association with an operating system for the Carnegie Mellon Multi-Mini-Processor (C.mmp) at Carnegie Mellon University [12].

Most significantly, Anderson's group introduced the notion of a reference monitor to enforce authorized access relationships between subjects (generally perceived as people at the time) and objects (a passive entity that contains or receives information), and sought to implement a "no read up" system regarding classified data. The security kernel was the key tool to implement the resource monitor concept [2].

20.6   BELL AND LaPADULA: MODELING OF COMPUTER SECURITY AND THE FOUNDATION FOR STANDARDS

In the years following the work of James Anderson's group, a number of researchers within and outside of the DoD conducted research and wrote reports on multilevel security (MLS) systems. Of these, none would be more influential than the model developed by David Elliott Bell and Leonard J. LaPadula, or the Bell–LaPadula model. Bell and LaPadula recognized that "no read up" was not enough to provide security for a military system because individuals, processes, or programs (subjects) could (in the case of individuals inadvertently or intentionally) write confidential information from higher clearances or classifications to lower ones. Broadening the definition of subjects (to include processes and programs) recognized and attempted to address the possible existence of a Trojan horse (a program that in addition to its designated and perceptible operation or purpose also had a surreptitious or concealed function). Daniel Edwards of the NSA first used the term when he served on Anderson's group, and Trojan horse attacks have long been a major concern of those working to achieve more secure computer systems [33: 46]. Under the Bell–LaPadula model, subjects could not read up, nor could they write down – in theory, and possibly practice, they could write what they could not read. This was

a major development mathematically, but it presented problems with implementation. It also highlighted the inflexibility between the military needs and the functionality that was necessary for some non-military systems.

Technically, in considering the actual incorporation of the insights of the Bell–LaPadula system, the larger problem of computer security, understanding how a kernel, or an overall operating system served as a correct implementation of a mathematical model, moved to the forefront. Implementing models in working systems was something that was often easier said than done, and was at the heart of software engineering – of understanding how code conforms to a mathematical ideal. In the 1970s "program proof" became a major area of computer science. While studied in a variety of places, SRI (initially standing for Stanford Research Institute) became a leader in researching "program proof" or verification in software. In 1973 SRI began a project called Provably Secure Operating System (PSOS) to design an operating system to meet security requirements and verify security properties. During the decade SRI's Peter Neumann and Richard Feiertag also worked on developing a kernel model as opposed to an entire operating system. Progress was slow at SRI, as were efforts at other centers focusing on computer security such as the University of California at Los Angeles.

Besides the complexity and challenges involved with designing secure kernels and systems, and verifying the security achieved, other issues existed within the ever changing technical and organizational environment. For instance, classifications within the military were not static, but changed over time. This could pose a number of technical and logistical hurdles. Moreover, computer security needs outside the military did not necessarily correspond to the military model.

20.7   MOVING TOWARD A COMPUTER SECURITY ORGANIZATIONAL INFRASTRUCTURE

In 1965 the Brooks Act designated the National Bureau of Standards as the agency in charge of research and standard setting for federal procurement. During the 1970s the NBS concentrated its efforts on computer security in two fundamental areas: standards work for researching, building, testing, and procuring secure systems; and the development of national standards for cryptography. While cryptography provides tools for authentication, verification, and secure communications, and is thus related to the system security work discussed above, it is at the same time relatively distinct as an area of research and policy. Given this, NBS and other organizations' system security work will be discussed first, and later in the chapter the research, politics, and organizational efforts regarding cryptography standards will be addressed.

In the late 1960s and early 1970s the NBS engaged in a small number of research efforts and organized collaborations to investigate computer security. On some of these efforts NBS collaborated with other organizations, including the Association of Computing Machinery (ACM). In 1973 NBS established a formal research program in developing standards for computer security, and by 1977 it was running invitational workshops of experts from government, industry, and academia in the computer and software engineering field to define problems and attempt to find solutions to ensure secure computing systems. As a result of the workshops NBS sought to define policy for sensitive, but not classified, data stored in and communicated between computing systems. It attempted to establish evaluation criteria and approved lists of computing equipment/software systems for handling classified data. While the NBS was providing some leadership, there was still substantial research conducted by DoD entities that lacked central coordination. During the mid-1970s ARPA and the different branches of the military were still conducting research with minimal coordination between them. In 1979, as another outgrowth of the workshops, the MITRE Corporation was given the task of developing criteria for evaluating the trust that could be placed in computing systems for handling classified data. Around this time, the Deputy Secretary of Defense gave the Director of the NSA the duty of extending the use of trusted computer systems within the DoD.

With this responsibility, in 1981, the NSA established the DoD Computer Security Center (CSC). This formalized the NSA's computer security standard setting authority – previously it had been important in shaping the policy of the NBS in this area. Several years later the CSC expanded substantially and assumed control of computer security for the entire federal government, and was renamed the National Computer Security Center (NCSC). This organization would formalize the strong work achieved in the 1970s and first years of the 1980s to produce a landmark work on computer system security.

20.8 THE ORANGE BOOK (TCSEC)

During the 1970s, after the Anderson task force had developed some key principles, Bell and La-Padula provided a fundamental computer security model. Work on design and security verification was also underway at SRI, UCLA, and elsewhere. The military recognized a heightened need to have standard computer security criteria for evaluation in order to avoid independent verification and validation for each new system being considered. The NCSC tackled the problem of bringing key elements of the work of the 1970s in the computer security field into a working framework to standardize the evaluation of system security.

The work of the NCSC and MITRE resulted in the *Department of Defense Trusted Computer System Evaluation Criteria* (TCSEC) book, often referred to as "The Orange Book" because of the color of its cover. The Orange Book was by far the most influential publication to date (published originally in August 1983, and republished with minor revisions in 1985) on computer security, and remained the most significant document on the topic for many years to come [15].

Chief among the goals with *TCSEC* was to standardize government procurement requirements. It provided a structure for manufacturers to evaluate and measure the security of their systems. More specifically, it outlined a mechanism to designate different levels of security as well as a means to test if a particular system met a designated level. *TCSEC* recognized security as a continuum and not an absolute, as well as the great value of having a common language to understand different levels of security within computing systems. The book established standards for the degree of trust that could be placed in computer systems by designating the criteria or properties for achieving a particular level of security. The fundamental underlying premise of *TCSEC* is that it is possible to evaluate a system and accurately place it a category that is useful to understanding the degree of trust that is built in with regard to being able to process and maintain the security of a range of classified and unclassified data. At its heart, trust refers to the assurance an evaluator can have that a system does what it is supposed to do and does not do what it is not supposed to do.

Within the *TCSEC* there are four basic categories of security protection (listed in order from the strongest protection to the weakest):

A   Verified Protection
B   Mandatory Protection
C   Discretionary Protection
D   Minimal Security Protection

Each of these categories then have a couple or more subdivisions designated by numbers, with higher numbers representing greater security within a category than lower numbers. For instance, C3 is a level or rating of greater security than C2, which in turn, is greater than C1. Any level of B, of course, offers a higher security designation than any of the subcategories of level C (C1, C2 and C3) [15].

TCSEC refers to the totality of protection mechanisms of a computing system as the Trusted Computing Base (TCB). It recognizes that not all parts of the operating system need to be trusted, but rather that the computer architecture, assurance mechanisms, and elements of the TCB must be understood well enough to know that the base is protected against intentional and unintentional interference and tampering. *TCSEC* emphasizes that manufacturers should make the TCB as simple as possible for the functions it is to perform. For higher levels, the security elements should be centralized in a security kernel that can then be carefully scrutinized to assure the TCB offers superior security protection against tampering. Among

other operations, this kernel implements the reference monitor to enforce authorized access relationships between subjects and objects in the system.

The different categories, and subcategories contain a host of requirements in many different areas: discretionary access requirements; identification and verification of user requirements; testing requirements; auditing requirements; system architecture requirements; documentation requirements; and design specification and verification requirements. For summary purposes some basic criteria for the different categories and subcategories, from least secure to most, follows.

Minimal security systems, or D systems, offer little or no security features. It is a category for all systems that do not fit into any higher classifications. The types of systems (such as personal computers) that would fit in this category are not submitted for evaluation. Thus, it is a category that contains no systems, as it would offer no value to manufacturers to have such a designation [47: 155–159].

C1 Systems provide very minimal security functions. The level of security of such systems is only appropriate for cooperating users who are all at the same security level. The main requirements protect against users making accidental mistakes that might damage the system (the system is protected against user programs). While there is discretionary protection of files, passwords, and designated access, controls are far less than higher systems. The designation was of limited use in the 1980s and few systems have ever been designated in this subcategory.

C2 Systems, on the other hand, require accounting and tracking of individual users on the systems, more involved discretionary controls (designated read and write access), object reuse (to assure data left on a system does not become available to a subsequent – unauthorized – user), and far more extensive testing and documentation requirements than C1 Systems. C2 Systems have been far more common than C1 and included Digital Equipment Corporation's VAX/VMS 4.3 and Control Data Corporation's Network Operating System [11; 47: 155–159].

There is a significant leap between the requirements of C and B1 Systems. All B1 subcategories and higher (B2, B3 and A1) require a "Mandatory Access Control" which stipulates all files (and other significant objects) are given sensitivity labels to enforce the systems security. Protection of files is not discretionary, but instead is designated through a clearly defined classification and access system. The architecture of B1 and higher systems maintains concrete distinctions between the secure parts and non-secure parts of the systems. IBM's MVS/ESA and Unisys' OS 1100 were classified as B1 Systems. B2 primarily just extended rather than added to the security features of B1, but it did offer "least privilege", which limits users and program access to the shortest necessary time to complete a function. Testing requirements are significantly higher than B1. Honeywell's Information Systems' Multics System is one of the few to achieve a B2 classification. Likewise, B3 Systems are rare and require even tighter structure to the TCB to prevent penetration of the system. Honeywell Federated Systems XTS-200 had this classification.

A1 systems are the highest rated systems in the *TCSEC* classification, but the publication does discuss the possibility of higher-level systems that have even more stringent designations for architecture, testing, and formal verification (A2). A1 systems go beyond B3 systems in requirements for trusted distribution (in shipping the system to the customer). They require formal proof that the actual design of a system completely matches the specifications or mathematical model. During the 1980s just two systems received the A1 designation: Honeywell Information Systems' Secure Communications Processor (SCOMP) and Boeing Aerospace's Secure Network Server (SNS) Multilevel Secure (MLS) Local Area Network (LAN) system [15].

At the various levels, each of these systems achieved its authentication through extensive testing known as the Department of Defense Information Technology Security Classification and Accreditation Process (DITSCAP). This involves testing and evaluation of both the technical aspects of systems, but also the mechanisms and practices of its operators. Final declaration is typically made by a senior operational commander with specified authority to approve a computer system given the potential risks.

Prior to *TCSEC*, there was no common and efficient language for different people and organizations to communicate about the level of security of computer systems. The laborious establishment of detailed specifications of security for each system was time consuming, expensive, and inefficient. Even more problematic, however, was the association of different meanings or criteria for what specifications meant, how they were measured, and the degree to which they were verified. The aim of *TCSEC* was to ease this process, to make it more efficient, accurate, and reliable by providing clear levels and criteria for measurement that would in turn give manufacturers guidance in how to build their trusted systems for classified or sensitive applications. At the customer end, which was typically a government agency or entity, and often one within the DoD, the acquisition process could be quicker, easier and more precise.

20.9 COMMON CRITERIA AND THE GLOBALIZATION OF COMPUTER SYSTEM SECURITY STANDARDS

While substantial and influential work on computer security was achieved during the 1970s and 1980s in the United States by the DoD, it was not alone in this field, business and the academy, both in the US and internationally were becoming increasingly prevalent in developing and influencing standards. The networked world, as the ARPANET and other networks were transformed into the Internet by the early 1980s, were fundamentally changing the environment and the overarching models for standards. Of the critiques of the relevance of the DoD spawned Bell–LaPadula model and *TCSEC* to other environments and aims, none was stronger than a talk given by David Clark, a senior research scientist at MIT's Laboratory for Computer Science (LCS) and David Wilson, the director of information security services for Ernst & Whitney, at the 1987 Institute for Electrical and Electronics Engineers (IEEE) symposium on security and privacy held in Oakland, California. In this paper these scientists argued that most discussions of computer security have focused on a military-oriented model that privileges disclosure as the primary issue even though the "core of data processing [is] concerned with business operation and control of assets ... of data integrity". They claimed the military "lattice" model was simply insufficient to control disclosure and provide integrity in commercial settings [11: 184].

Internationally, a considerable amount of research and standard setting was achieved in European countries, Canada, and other nations. In Europe, during the 1980s, standards were established independently by the United Kingdom's Communications–Electronic Security Group (CESG), Germany's Bundesamt fur Sicherheit in der Informationstechnik (BSI), Frances's Central Service for Information System Security (SCSSI) and The Netherlands' National Communication Security Agency (NLNCSA). These nations pioneered with creating internationally recognized standards in a joint effort called Information Technology Security Evaluation Criteria (ITSEC).

Although the standards that developed in Europe were quite similar to those outlined in the United States in TCSEC, novel research and projects were undertaken in applying security standards to new types of systems. One such instance was the UK Ministry of Defense's project to develop a unique microprocessor called VIPER (Verifiable Integrated Processor for Enhanced Reliability) at the Royal Signals and Radar Establishment (RSRE) in 1987. Back in 1986 the UK Cabinet Office's Advisory Council for Applied Research and Development suggested that mathematical proof should become standard for any system in which a system failure could result in ten or more deaths. VIPER was an outgrowth of this, but whether the constructed 32-bit system achieved mathematical proof was controversial. British firm, Charter Technologies Ltd., was a licensee of the RSRE's VIPER and took legal action against the Ministry of Defense arguing that the claim of proof was a misrepresentation [34: 159–164]. Noted sociologist and historian of technology Donald MacKenzie has emphasized how the VIPER episode highlights that using the term "proof" without clear definition and precision can be dangerous. More broadly, he asserts that the reputation of mathematics for precision and certainty

have been rhetorically appropriated by those advocating moving from "an empirical to a deductive approach to computer system correctness" [34: 183]. To address this problem, MacKenzie calls for "a sociology of proof" to better understand the notion, understanding, and construction of this term in its mathematical and computer systems contexts [34: 159–164]. This challenge becomes even greater as the need for true international standards becomes ever more imperative in our increasingly global economy.

Recognizing the increasing benefits for efficiency in research, standard setting, and particularly procurement processes that are possible with broader international cooperation, an attempt was made in the first half of the 1990s to bring together a number of allied nations to have a common set of computer security standards. This resulted in a global set of standards that superseded the Orange Book or *TCSEC* in the United States; ITSEC, for member European nations; and Canadian Trusted Computer Product Evaluation Criteria (CTCPEC).

Various books with different colors, commonly referred to collectively as the Rainbow series, modified and extended *TCSEC* for greater specialization and sought to address security needs within the changing environment of computing, software, and networking technology. Like *TCSEC*, the later books aimed to provide a standard for manufacturers regarding security features and assurance levels in order to provide widely available systems meeting certain "trust" requirements for sensitive applications.

Of particular significance, the so-called Red Book was issued by the National Computer Security Center in 1987, and provided interpretations of *TCSEC* security features, assurance requirements, and rating structure to networks of computers, including local area networks (LANs), wide area networks (WANs) and internetwork systems. It also documented the different security services (such as communications integrity, transmission security, and denial of service) with regard to LANs and WANs.

This new set of international computer security standards was developed through a collaboration of the United States, Great Britain, France, Germany, The Netherlands and Canada in conjunction with the International Standards Organization (ISO). The system was developed and refined in the mid-1990s and the first official set of standards was completed in 1996. These standards were named the Common Criteria (CC), but upon their minor revision to create CC version 2.0 in 1998, they also became known as ISO 15408.

The Common Criteria defines seven security levels, called Evaluation Assurance Levels, which are labeled from EAL1 to EAL7. Higher number levels represent systems that have successfully undergone more stringent and extensive evaluations and offer greater trust. These levels are closely related to those set in ITSEC and *TCSEC*. For instance, EAL3 corresponds very closely *TCSEC* level C2.

Common Criteria certification is a rigorous process, especially at the higher number levels, and is achieved through testing by a third party laboratory that has been accredited by the National Voluntary Laboratory Accreditation Program (NVLAP). The National Information Assurance Partnership (NIAP) was created to oversee use of the Common Criteria as the standard for evaluation in the United States, and similar bodies exist in other member nations. In addition to the originators of the Common Criteria, other countries joined in the late 1990s under a sub-treaty level Common Criteria Mutual Recognition Agreement (MRA), which was established in 1998 by the United States, France, Germany and Canada. Australia, Great Britain and New Zealand joined in 1999, and the following year Norway, Spain, Italy, Greece, The Netherlands, Finland and Israel signed on and became members. Common recognition of standards between all of these countries exists at the lower levels (EAL1 to EAL4), but typically countries will only accept their own certification for levels EAL5 to EAL7.

While *TCSEC* and subsequent global standards of the Common Criteria have been of fundamental importance to computer security systems professionals in government and industry, they are entirely outside of most individuals' understanding of computer security. Computer security remained a relatively obscure topic to most people before the advent of personal computers and the widespread dissemination of these machines in peoples' offices and homes. Some became interested in computer security as a facet of their interest in insuring the protection of personal privacy. More individuals were exposed to issues of computer security through focusing events surrounding current and future threats and new forms of malware in the age of ubiquitous computer networking with the Internet and World Wide Web.

20.10 THE ORIGIN AND EARLY DEVELOPMENTS OF CRYPTOGRAPHIC RESEARCH IN THE ACADEMIC COMMUNITY

Research and use of cryptography in the military, for diplomacy, and in other realms of society has been used for many centuries. This included codes (substitution or exchange of words and phrases) as well as more complex cryptography using cipher text (substitution of individual letters) to conceal embedded information from all but the intended recipient. The first known use of a cipher text in the military is described in Julius Caesar's *Gallic Wars*. Caesar discussed how he sent Cicero a message replacing Roman letters with Greek ones. Caesar also developed the so-called "Caesar shift cipher", which shifted the alphabet forward three places in the cipher text [53: 9–11]. Between the ninth and thirteen centuries, Arab scholars pioneered new types of cryptographic systems, as well as cryptanalysis, or techniques to unearth patterns to decipher secret codes, such as the study of frequency and patterns of the cipher text. In one of the most famous historical uses of cryptography, Mary, Queen of Scots used cipher text to communicate with conspirators to assassinate Queen Elizabeth (the Babington Plot) and escape from her imprisonment. Thomas Phelippes, a cryptographer gifted in the art of frequency analysis, deciphered the messages between Mary Queen of Scots and her conspirators, exposed the plan to Queen Elizabeth, and Mary was beheaded [53: 42–44].

Each distinct cipher has a corresponding algorithm, or the general specification of the method of encryption, and a key, or the detailed rules for a particular encryption. Over the years these algorithms and keys have become increasingly sophisticated, to stay ahead of the many corresponding advances in the art of cryptanalysis. The highest level of this work has been conducted by secretive government agencies such as the GBHQ in Great Britain and the NSA in the United States. Top scholars in the area worked in anonymity developing systems and analytical techniques. In addition to this secret research, several pioneering scholars outside the secret community tackled a fundamental problem with the practical use of cryptography. In doing so, they made a critical discovery, and changed the face of the history of cryptography.

Whitfield Diffie, a recent graduate of MIT in the mid-1960s, began working at the school's non-profit spin-off corporation MITRE to avoid the draft. He worked under mathematician Roland Silver for MITRE and also had the opportunity to work at MIT's Artificial Intelligence (AI) Laboratory as a resident guest of esteemed AI expert Marvin Minsky. After his MITRE funding ran out, he moved to Stanford to work in the Stanford AI lab with John McCarthy. During this time, Diffie became increasingly interested in the possibility of applying mathematics and cryptography to achieve private communications.

Diffie was enthralled by David Kahn's book *Codebreakers*, a monumental and original work that detailed the development and use of secret communications from ancient times to the present and brought the NSA to public attention, an agency that prior to Kahn's book was so secretive and little known that some Washington insiders jokingly claimed the acronym stood for "No Such Agency" [32: 14]. With the influence of Kahn's book, his devouring anything he could find to read on cryptography, his interaction with various leading computer scientists, and his cross country travels to research the topic and meet with people, Diffie's interest in this area escalated. At the same time, his suspiciousness of the NSA also grew. On one of his road trips to do research and seek out people knowledgeable in cryptography, he visited and gave a talk at IBM's Thomas J. Watson Laboratory, where Alan Konheim suggested he look up someone interested in cryptography on the Electrical Engineering faculty at Stanford University: Martin Hellman.

Diffie and Hellman met in 1975 and immediately were rejuvenated by their common interest in cryptography, a field that Hellman had deep interest in but was frustrated by as a result of most of the high-level research in the field taking place behind the so-called triple fence of the NSA compound. One practical problem with cryptography at this time was that of key distribution. To use cryptography required a key distribution center. Individuals could not communicate prior to exchanging keys or having this function done through a trusted centralized authority. There were two primary issues with cryptography, privacy and authentication – making certain communications were not intercepted and deciphered by a third party and knowing that the person who claimed to send a message actually had sent a particular message. Diffie and Hellman, along with a third collaborator who came from Berkeley to join them at Stanford, Ralph Merkle, developed public key cryptography with the concept of a trap door one way function (a function easily solved in one direction but not the other). Diffie and Hellman reported on their ideas in their seminal article, *New Directions in Cryptography*, published in November 1976 in *IEEE Transactions on Information Theory*. The article began:

> We stand today on the brink of a revolution in cryptography. The development of cheap digital hardware has freed it from the design limitations of mechanical computing and brought the cost of high grade cryptographic devices to where they can be used in such commercial applications as remote as cash dispensers and computer terminals. In turn, such applications create a need for new types of cryptographic systems which minimize the necessity of secure key distribution channels and supply the equivalent of a written signature ... *Public key distribution systems* offer a different approach to eliminating the need for a secure distribution channel [16: 644].

Hellman and Diffie's public key system rested on the notion that the traditional single symmetric secret key would be split so there would be both a public key and a private key, and that these two keys would be mathematically linked (a trap door one way function). The public key would be associated with a particular individual and would be readily known and accessible – it could be published in a directory, just like someone's name and phone number in the phone book. The private key, however, would be kept secret. An individual would send a message to someone encrypting it with the recipient's public key and their own private key (creating a trap door). This idea of splitting the key was completely counter-intuitive with accepted notions of cryptography, and in this, laid the brilliance of Hellman and Diffie's idea.

In addition to Diffie and Hellman's great theoretical accomplishment with key distribution and privacy, the paper also addressed digital signatures for reliable authentication. This would have a profound impact down the road as the work of Diffie and Hellman was complemented and expanded upon by a group of researchers at MIT. First, however, it is important to address some background to the emergence of government encryption standards and the conflict that was developing between the government, particularly the NSA, and the small group of university-based scientists working on cryptography, led by Hellman and Diffie.

20.11 EARLY ATTENTION TO AND INVESTING IN COMPUTER SECURITY SYSTEMS DEVELOPMENT IN THE PRIVATE SECTOR

The computer security environment of the first half of the 1970s was beginning to change. While most investment in the field, including task forces and tiger teams, was funded by the DoD, one private sector company was becoming increasingly focused on the issue: IBM. Between the last half of the 1950s and first half of the 1960s IBM had built SAGE computers for the Air Force; the so-called Defense Calculator, commercialized as the IBM 701, for defense and industry; and Stretch, an unprecedented and ambitious computer for Los Alamos National Laboratory. The firm had also invested heavily in developing a backward compatible series of computers and software with the System/360 series and OS/360 operating system

that was announced in 1964. IBM's past experience drove home the importance of future military contracts, but leaders at the company also came to recognize that computer security would be of fundamental importance to all of their different markets in the future.[4] This led the company to commit to spend $40 million on computer security issues over a five-year period. By the late 1960s IBM, and its leader, Thomas Watson, Jr., believed it was critical for the computer industry to take the lead in addressing problems regarding computer security and privacy. In a 1968 address to fellow industrialists, Watson, Jr. stated:

> I believe we in industry must continue to improve existing technological safeguards which limit access to information stored in electronic systems; and we must devise new ones as the needs arise ... Moreover, I believe we in industry must offer to share every bit of specialized knowledge we have with the users of our machines – the men who set their purposes – in determination to help secure progress and privacy, both together [26].

Despite some existing knowledge concerning the commitment to and the expenditures by IBM on computer security work, the details of the firm's early activities in this field are still somewhat sketchy. The company produced literature for its customers outlining computer security broadly and indicated the wide scope of the firm's computer security agenda [26]. At the same time, much of the specific effort was devoted to developing and enhancing a single IBM product, Resource Security System (RSS), to address operating system design. Generally speaking, RSS was created to improve the existing level of security in resource-sharing systems [2: 5].

20.12 CRYPTOGRAPHIC RESEARCH AND THE EARLY DIGITAL COMPUTER INDUSTRY

IBM's computer security work, however, was not limited to focusing just on the design and workings

---

[4]Early time-sharing and computer networking work at IBM was funded by the government – IBM's contract for SAGE computers, funded by the Air Force, provided the computer giant with hundreds of millions of dollars in the second half of the 1950s.

of operating systems such as RSS, and the underlying issues involving access to memory. Cryptography was also a fundamental concern and area of IBM's research. In the NSA, this effort was continuous from its founding in 1952, and the work of this secretive agency was far more advanced than anything existing outside of it. For centuries ever more advanced ciphers and deciphering methods had been developed to try to ensure secret communications, while simultaneously uncovering the secret code used by enemies. The work of the allies in breaking the Enigma code, which used a sophisticated mechanical rotor machine and changed ciphers daily, and other encryption tools and techniques in the US, Europe, and throughout the world, have an extensive and growing scholarly and popular literature.[5]

In the late 1960s, IBM focused attention on cryptographic work, hiring Horst Feistel to lead a group that would come up with a data encryption system that was named Demon, and later renamed Lucifer. In later iterations, Lucifer evolved into what was established as the Data Encryption Standard (DES) in 1976 by the National Bureau of Standards. In 1973 Feistel published some details on Lucifer in *Scientific American*, the most detailed work to that date openly published on current cryptographic research. Feistel had come up with, or at least wrote about, substitution boxes (S-boxes). S-boxes were algorithms for nonlinear equations specifying how letters of cipher code could be shifted. Some believed that the NSA had provided this important development to Feistel in order to embed a controlled level of cryptography into the commercial sector [32: 42–43].[6]

The NSA, as critics in the academic cryptography community asserted at the time and most people now accept, had a strong hand in influencing the NBS on the structure and specifications of DES (reducing the key size to only 56 bits).[7] Diffie, Hellman, and other academic critics of DES believed that the standard was constructed to enable

the secret agency to continue to have the capacity to break it (through a brute force attack), or that DES might have even included a trap door, offering easy decryption by the NSA (the NSA would provide tamper-proof DES chips to approved companies). The critics and some negative press influenced some corporations such as Citibank to steer clear of DES and develop its own proprietary standard.

With DES, IBM management thought cryptography, a subset of and area contributing to the larger field of computer and communications security, had essentially been solved. Horst Feistel and the cryptographic group at IBM were somewhat letdown by the fact that IBM's top management thought the work in their area was done, and that the company could focus all its computer security research resources on the larger problem of operating system security [23].

The tension between Hellman, Diffie, and a growing group of academic researchers and the NSA and federal government would grow more intense and involve additional individuals as researchers at MIT extended the developments of Diffie, Hellman, and Merkle, and sought to commercialize encryption technology. Conflicts would also erupt over intellectual property between the leading cryptographic researchers at Stanford University and those at MIT.

20.13 RSA DATA SECURITY: PIONEERING A SECURITY SOFTWARE INDUSTRY

Diffie, Hellman and Merkle had created an outline or general explanation for using one-way functions as the basis for secret communication, but had not developed a workable system. This was accomplished on the heals of the Stanford researchers by a group of three scholars at MIT, Ronald Rivest, Adi Shamir and Leonard Adleman, who collectively would be awarded the Turing Award for their discovery. Rivest was intrigued by Diffie and Hellman's "New Directions" article and became increasingly focused on trying to provide the mathematical formulas that would take public key cryptography from theoretical construct to workable

---

[5]A sampling of important and influential works include: Kahn [28], Denning [13], Diffie and Landau [17], de Leeuw [30], Hodges [24], Singh [53], Levy [32].

[6]A colleague of Feistel, Alan Konheim, believed that the idea for S-boxes was given to Feistel by the NSA.

[7]DES was formally adopted by NBS in July 1977.

algorithms for a public key system. Rivest's system rested on a one way function created by taking the product of two very large randomly chosen prime numbers. The public key was the product of the prime numbers. The decryption key, however, could only be uncovered by knowing the primes – a factoring problem that was nearly impossible to do given the size of the randomly chosen prime numbers. While Rivest came up with the algorithm, he had interacted closely with Shamir and Adleman and insisted that all three names appear on the MIT/Laboratory for Computer Sciences Technical Memo where the findings were first published. Like Diffie and Hellman's article, it contained a prescient opening:

> The era of electronic mail may soon be upon us; we must insure that two important properties of the current "paper mail" system are preserved [privacy and signatures] [46].

Unlike, Diffie and Hellman's conceptualization, the so-called Rivest, Shamir and Aldeman Algorithm (RSA Algorithm) was something more concrete that could be tested to see if it stood up to scrutiny. Rivest worked with famed mathematician Martin Gardner to outline the algorithm in Gardner's column in *Scientific American*. Appearing in the August 1977 issue, the column offered a cash prize ($100) to anyone who could prove the system did not work.

By the mid-1970s an increasing number of large corporations in petroleum, banking, and other industries were beginning to encrypt their electronic correspondence [27: 23]. Despite the fact that Rivest, Shamir, and Adleman had no real business experience, they decided to form a company to commercialize the technology of the RSA Algorithm. They received some money from an outside investor (physician and businessman Jack Kelly) and purchased rights to the invention from MIT for $150,000. The company, RSA Data Security, was formed in the early 1980s. It received several other modest investments, but did not have a true product to sell, was burning through cash at a rapid rate, and was in deep financial trouble. In 1985, the company's marketer Bart O'Brien called upon an old friend he had worked with at Paradyne, Jim Bidzos, to come on board, originally just for some management consulting, but soon thereafter, to lead the firm [5].

Bidzos, a talented businessman who had done programming and networking systems work at IBM, and had experience in running his own international IT consulting venture, brought much needed management and discipline to the organization. During the mid-1980s things remained tough for RSA, and after a short while Bidzos was essentially the entire company, as the others departed. The firm had little money, and the founders, Rivest, Shamir and Adleman, were not active in the enterprise. Bidzos recognized the great opportunity for RSA was to license its encryption technology to be used in the products of other firms. The company rounded the corner to financial security and substantial profitability later in the decade when he orchestrated a deal with Lotus Development to include RSA's algorithm in Lotus Notes, and subsequent sales to Microsoft and Apple Computer [5].

Bidzos also initiated the annual RSA Data Security conference. In the late 1980s, this was a relatively small event, but Bidzos insightfully made it a broad and inclusive forum for issues of concern regarding computer security. By this time the battle between the NSA and government and the academic cryptographers at universities had become more intense. The government used export restrictions (laws prohibiting the export of weapons or technical know-how harmful to national security) to try to restrict the academic freedom of both the academics as well as to restrict encryption in products. The NSA wanted to be in sole control of the type and strength of cryptography available to individuals and organizations both in the United States and abroad. With an increasing amount of revenue coming from foreign markets, this created a major disincentive to produce products with embedded encryption technology. By bringing different groups together to air their opinions and differences, including individuals from the government, the RSA Data Security Conference grew to be the premier event each year in the field of computer security.

Critics of the government's effort to control encryption standards emphasized the right to privacy, as well as the importance of having access to foreign markets. Concerns mounted that the government, and especially the NSA, was fighting for standards that were not strong enough to withstand the massive brute force attacks (using extensive computing power to break encryption keys) possible given the computer infrastructure of the NSA, or that standards included a backdoor that gave the NSA ready access.

Bidzos wanted to establish RSA Data Security's technology as the standard, and the widespread attention and visibility of the RSA Data Security Conference helped achieve this. RSA Data Security received major deals with Netscape and other firms by the mid-1990s as the true value of RSA's technology became apparent in the era of ubiquitous computer networking ushered in by the World Wide Web. While RSA had become successful as a business with the Lotus and Apple contracts, and particularly the Netscape and other deals of the Web era, the greater opportunity for RSA lay with digital signatures and authentication. As Rivest, Adleman and Shamir had vaguely alluded to back in the late 1970s, signatures would be fundamental as computer networking grew.

A critical issue with having the digital signature standard, and later domain name standards, rested with the broader IT industry acceptance of the standard. Bidzos insightfully recognized that this would best be accomplished in collaboration outside the confines of RSA Data Security. Along with a group of partner corporations, including Microsoft, Netscape, American Online (AOL), IBM, and Visa, Bidzos and RSA Data Security spun-off and formed VeriSign in 1995 as an independent corporation. This company grew much larger than RSA Data Security, and Bidzos served as Chair of the Board in its early years (and is currently Vice Chair). In 1996 Bedford, Massachusetts-based Security Dynamics Technology, Inc., a firm specializing in authentication that also began in the early 1980s, acquired RSA Data Security for approximately $300 million. The following year it also purchased DynaSoft, a leader in access control solutions. Shortly thereafter Security Dynamics held roughly two-thirds of both the encryption and authentication markets, and sold the popular product line of SecurID software. While the security

products market was without industry-wide standards, the consolidation of the late 1990s and the substantial market share of Security Dynamics led to the emergence of some product standards, such as the SecurID line. These were used by many firms and individuals as electronic commerce began to grow rapidly – eventually reaching more than 15 million users. As SecurID took off and the company also developed Secursight enterprise security suite (providing enterprise-wide authentication solutions) in the late 1990s, the company also recognized the strategic importance of the RSA's brand and took advantage by changing its name to RSA Securities, Inc. in 1999.

By this time the security software industry had grown to include other firms that formed fronts outside of the encryption and authentication areas, most importantly, virus detection software. Symantec, a firm that was formed by Dr. Gary Hendrix and a few of his colleagues from the Stanford Research Institute (SRI), was involved in many areas of programming during the 1980s. The firm moved into the security software field after going public in 1990 and merging with Peter Norton Computing, a firm that behind its founder and leader Dr. Peter Norton had created powerful virus detection software. While Symantec continued an acquisition spree throughout the decade, and had a portfolio of software applications of varying types, its Norton brand and security software became the centerpiece of the corporation. Symantec emerged as the leading security software company in the world in terms of revenue and net profits.

Unlike RSA Data Security, Symantec was developing products in part for individual personal computer users rather than just for companies and organizations. The security area has also continued to grow as computer, software and service giants including IBM and Microsoft, have extended their presence in the field. Microsoft founder Bill Gates gave the keynote address at the 2004 RSA Data Security Conference and he emphasized the greatest portion of Microsoft's $6 Billion annual research budget is now targeted in the software security area [21].

The security software industry emerged because RSA Data Security, VeriSign, Symantec, and later other firms, were developing products outside of the context of government that increasingly came to be standards. They were superior to anything that the government was developing, or at very least, better than anything they were releasing outside the protected confines of the NSA. These products were also developed specifically for commerce rather than military purposes. Needs were different and information integrity played a much larger role in this arena than in the DoD, where the paradigm was more closely aligned with protecting classified information and integrating computing technology with existing protocols of command and control, disclosure, and the overall infrastructure for classified military information.

In the commercial world there was also growing corporate involvement between computer security firms and many of the largest firms in the world to address computer security/data integrity concerns. Many companies in the Fortune 100 belong to the International Information Integrity Institute (I4) – an organization that confidentially shares information on best practices and emerging technologies and techniques in computer security to better facilitate communication and commerce in a global economy where the World Wide Web has fundamentally transformed the importance of effective electronic commerce infrastructure, policies, and performance to survive and thrive.

20.14 COMPUTER SECURITY AND SOCIETY

Computer security and computer privacy are unique and separate concepts, but are deeply interrelated and have been in a long and continuing historical dialog. While governments have been at the forefront of defining computer security standards, individual citizens have viewed the growth of massive databases and networking as potential and real threats to unauthorized use of personal information.

Computer privacy emerged as a major public issue for the first time in the second half the 1960s and led some individuals to give initial thought and attention to the notion of computer security – long before networked computers became commonplace in offices and homes, but after they had

become fixtures within many government agencies and corporations.[8] The relationship between privacy and security, like that between privacy and technology more generally, has little meaning outside of defined contexts. Privacy is commonly framed as a right, which is defended fiercely by individuals and businesses. Invasions of privacy create a perceived or real need by the state to protect the greater good of society and to better ensure the security and well being of its people. Computer databanks can potentially reduce individual privacy, while computer encryption technology often facilitates private communications. Such private communications could have positive or negative impacts depending on the context. Private communications can protect individuals and businesses, allow law enforcement officials to communicate privately, but also can aid criminals or terrorists in planning crimes or attacks. Some companies want unregulated use of encryption technology to ensure privacy of communication and the protection of trade secrets, yet buy, sell, and trade databases of information on consumers to better apply market segmentation strategies. The individuals or groups and situational factors dictate perspectives. Furthermore, such perspectives are not static, and commonly are influenced by focusing events, such as the terrorists' attacks on the World Trade Center and the Pentagon on September 11, 2001.

While the history of the evolving and ever contextually based relationship between privacy and

---

[8]Privacy is both a cultural concept and a legal right that has long been debated in the United States regarding its constitutional foundation. While it is not formally delineated in the constitution, many have argued it is an implied constitutional right. The "right to privacy", first articulated in a *Harvard Law Review* article in 1890 in response to the advent of handheld cameras and an emerging paparazzi photographing the rich and famous has evolved to become the basis for fundamental and controversial legal rights in the United States regarding contraception (Griswold v. Connecticut (1967)), and abortion rights (Roe v. Wade (1974)). Alongside federal rulings and laws on a small number of privacy related issues, much legislation on privacy has been passed by individual states. Computer security, on the other hand, is a broad term referring to the design of systems, and procedures for using them; and additionally software or hardware, or other technologies or practices to safeguard the integrity and control access to information on, or transmitted over networks by, computers.

security has existed for many decades, and even centuries, computer security and privacy did not register on the radar of many individuals, groups, and institutions prior to the latter part of the 1960s. Growing concern about the protection of financial and other information on individuals on government computers, the consolidation of databanks, and use of social security numbers as a standard identifier, led to substantial controversy and public backlash in the second half the 1960s. The failure of Lyndon B. Johnson's 1967 proposal for a National Data Center, to greatly centralize government information maintained on individuals, was indicative of the public concerns for individual privacy in the era of computers. Despite the abandonment of the National Data Center the amount and concentration of information on individuals by government continued and privacy concerns persisted.

Computer databanks and privacy concerns were fueled by a number of influential books in the late 1960s and early 1970s. Most notably among these were legal scholars Alan Westin's *Privacy and Freedom* (1967) and Arthur L. Miller's *The Assault on Privacy* (1971). The former was a multifaceted look at privacy, technology, and law, inspired by and attentive to the heightening and changing uses of computers, while the latter concentrated almost exclusively on the new threats to individual privacy posed by the growing use and connectivity of computer databanks. Such works, however, tended to be more focused on real and potential abuses by government, as opposed to outside infiltration and the security of systems, or inappropriate uses in the private sector.

The ongoing controversy concerning government as Big Brother, with massive data banks of information on individuals, planted a seed, and the concern would grow as commercial computer networks came into more common use. Such concerns accelerated rapidly, as networked minicomputers, and particularly personal computers, became increasingly commonplace, and infiltration of corporate and government systems became increasingly common [55].[9]

---

[9]In 1976 the Government Accounting Office estimated that many of the federal government's 9,000 computers were in-

Growing public pressure for safeguarding and restricting the government's use of concentrated databases of information on individuals led Secretary of Health, Education, and Welfare Elliot Richardson to initiate a committee to study and report on government information systems and personal privacy, the Automated Personal Data System (APDS) Committee in 1972. As a result of Willis Ware's early involvement and expertise in computer security, coupled with his strong concern for the privacy of individuals, Richardson asked Ware to join, and later chair, the APDS. This committee's work led to the Privacy Act of 1974 and the establishment of the Privacy Protection Study Commission (PPSC). The work of these two groups represented the most important achievements during the 1970s to examine threats to individual privacy posed by government computer databases regarding financial, medical and other types of personal information [61]. While the Privacy Act of 1974 gave individuals in the United States the right to inspect records on themselves held by the federal government, this did not extend to state and local government databases, nor those held by corporations and non-government organizations. In general the US has favored a policy of responding to privacy problems that arise and gain attention, not to develop a broad and comprehensive policy on privacy, as has been the case throughout Western Europe and a number of other nations in the world during the past quarter century [61].

Overall, despite the Privacy Act and legislation at the state level, government data bases containing private information on individuals, such as the systems of the Social Security Administration, failed in tests to prevent infiltration. Such failures and abuses led President Jimmy Carter to seek to tighten security over the government's increasingly unwieldy networks of computers in order to protect individual privacy and prevent fraud and crime.

National privacy legislation has generally been more prominent in Western Europe than in the

---

sufficiently protected against sabotage, vandalism, terrorism, or natural disaster. In June 1976 a computer programmer who claimed to only be trying to prove the laxness of government computer security was convicted for tapping into classified federal energy information.

United States. In Great Britain, the government asked Dr. Ross Anderson of the Computer Laboratory of Cambridge to evaluate and help define security standards for patient clinical information within the country's National Health Service (NHS) network. With ethical conduct of patients rather than national security as the goal, protocols and standards for secure medical information systems have evolved differently, and have been influenced not only by national groups, but also international ones, such as the European Standardisation group for Security and Privacy of Medical Informatics (CEN TC 251/WG6) that has mandated the encryption of patient medical health data on large networks. In 1996 Anderson set out rules that could be used to uphold the principle of patient consent independently of the details of specific systems, and provide models and protocols that gave the medical profession an initial response to creeping infringement of patient privacy by NHS computer systems [3].

As privacy concerns have escalated, broad public concern with privacy has not been limited to government medical and social welfare databases. Concerns about the vulnerability of DoD systems heightened with the growth of networked personal computers and hackers and computer criminals in the 1980s. In 1983, as the film "War Games" was introducing a scary scenario of hacker infiltration of DoD computers controlling the nation's nuclear arsenal, a number of military experts expressed concerned about the penetrability of the department's computer systems. In the second half of that year, the DoD, in its biggest move ever to prevent illegal penetration of its computers, announced it would split its computer networks into separate parts for civilian and military users, and thus limit access to university-based researchers, trespassers, and possible terrorists or spies [10]. While this enhanced security, the notion of impregnable DoD computers was relegated to an idyllic and distant past. For instance, in 1988 the *New York Times* reported that a West German citizen had secretly gained access to more than 30 DoD and DoD contractor computers for more than two years [37].

On a different front, during the 1980s, the controversy over encryption standards between the government, particularly the NSA and Commerce Department, and academic cryptographers and industry leaders (essentially just RSA Data Security prior to the 1990s) received modest, but persistent attention from the *New York Times* and other media sources. This conflict, however, remained a fairly low level issue until the early 1990s when the Clinton Administration backed a plan for a key escrow phone encryption system using DES, and later a DES-like Slipjack algorithm (which used twice as many substitutions as DES, and thus provided much stronger security) that would allow government to have a trap door (Law Enforcement Access Field, or LEAF) to listen in on encrypted communications of this commercial cryptosystem, AT&T Telephone Security Device (TSD) 3600 – or the so-called Clipper Chip [49].[10]

The idea for the Clipper Chip originated with the assistant deputy director of the NSA, Clinton Brooks, and National Institute of Standards and Technology acting director Ray Krammer. The argument Brooks, Krammer, and other proponents made for the Clipper Chip was that it was critical for law enforcement and national security to ensure the future ability of the government to wiretap the phones to catch and prosecute criminals and terrorists. Furthermore, it could obviate the government's real or perceived need for export controls.

The Clipper Chip was heavily pushed by the agency, NIST, the FBI, and others in government, and rekindled the outcries of government abuse that was different in form, but similar in degree and tone to the public response two and a half decades earlier to President Lyndon B. Johnson's proposed National Data Center. While the Computer Security Act of 1987 had shifted authority in the field away from the NSA and to NBS (and later NIST), the influence of the secretive intelligence organization remained. This was in part because of some common beliefs among the leaders of NBS/NIST and NSA, but also a result of the far greater human and computing resources and capabilities of the NSA relative to the NBS/NIST.

---

[10]Ironically the original name for the government's trap door for Clipper was LEEF, or Law Enforcement Exploitation Field.

The annual RSA Data Security Conference event provided a perfect forum for computer scientists, government officials, and industrialists to debate the Clipper Chip, and the rapidly growing event that early on had drawn less than 100 grew exponentially. RSA's Jim Bidzos was the master of ceremonies for these conferences and became the leading spokesperson opposing the Clipper Chip. Bidzos, however, was far from alone on this issue. Opposition to the Clipper Chip included software security companies, advocacy groups and political pundits from the right and the left, many members of the public, and foreign governments (which strongly opposed the importation of an American technology that allowed the US government unfettered access to eavesdrop). A *New York Times*/CNN poll in at the beginning of 1994 showed more than 80 percent of the American public opposed the Clipper Chip [32: 261]. Vice President Albert Gore had originally been sold on the proposed chip by the NSA and FBI, and the Clinton Administration remained firmly in favor of the Clipper Chip throughout 1993 and the first part of 1994. However, soon thereafter the administration grew weary of this losing battle, gained greater appreciation for the drawbacks of the technology, and withdrew support.

The Clipper Chip controversy highlighted the fact that the privacy of individuals and organizations remained a significant factor in drawing peoples' attention to computer security issues. The media's embracement of the issue of computer security fit in with the growing tendency toward sensationalism that attracted attention by instilling fear and feeding the insecurities of their viewers and readers. While computer security was an important issue and threats were real, most reporting took a shallow approach to the underlying complexity of the issue in favor of personalized accounts of cyberpunks, malicious hackers, cyber criminals, and the dangers faced. This got many people to think about computer security on one level, but overall, in terms of secure systems and practices, most remained uninformed or oblivious. Paradoxically, computer security and computer crime, like crime more generally, simultaneously exists as a frequent mass media topic (indicating its popularity in a ratings driven media environment) and a topic that most people prefer not to think about. As longtime computer crime and computer security expert Donn Parker has argued:

> Everybody likes to be able to communicate in confidence and secrecy when they need to and everybody likes to have an assurance that they are free from attacks from criminals, but as far as security itself is concerned, the locked doors and filing cabinets and having to lock your computer, it is unpleasant and inconvenient and it subtracts in business from profitability, productivity, and growth. So nobody likes it. I have spent thirty-five years of my career working in a field nobody really likes. It has been kind of interesting. The journalists like it for the sensationalism. It sells newspapers; you can get people disturbed [42].

20.15 CONCLUSION

Computer security has been a fundamental aspect of digital computing for decades, with both getting their start and early funding from the DoD, but with developments in industry and the academy becoming all the more prevalent over the past two decades. While the DoD continues to be a force in both networked computing and computer security, its relative place has diminished to an extent as a large and vibrant computer industry has been the epicenter of many computer technology developments and the software security trade has become a fundamental sector of IT security. Likewise, the pre-eminent position of the US in computer security standards has declined as the importance of research throughout the world and international standards becomes more critical in an increasingly global environment and economy. Standards today are created by the marketplace and through corporate interactions to a greater extent than ever before relative to the military.

While security products abound and leading ones create some degree of standardization, the reality that no product or system is impenetrable becomes all the more clear. Increasing the dialogue about the historically subverted topic of computer security, both publicly, and when necessary, in closed settings such as leading international corporations becomes all the more important. Sharing useful knowledge and techniques, as is done confidentially by major global corporations as part of I4, can lead to a more secure environment with greater information integrity. On both an individual and institutional level, recognizing that products cannot provide magic bullets and there will always be threats, infiltration, and destruction is a critical first step. At the same time, unpredictability can be a very valuable tool, as computer criminals prey on knowing how both people and computing and software systems work, and the continuity and timing of processes. Developing, using, and frequently changing complex passwords, varying the routine of computing systems, and continuously altering the procedures we use to interact with computers can make computer crime far more difficult. In this respect, the notion of standards is altered from clearly defined processes and adhering to exact specifications as was done within the Bell–LaPadula and TCSEC models, to defining broader evaluation techniques, recognizing differences, embracing variance, and never abandoning efforts to ensure greater integrity of data. Overall, a multifaceted approach of products and processes will be increasingly important to best address an uncertain future with regard to the challenging issue of computer security as computer networking becomes increasingly ubiquitous.

REFERENCES

[1] J. Abbate, *Inventing the Internet*, MIT Press, Cambridge, MA (1999).

[2] J.P. Anderson, *Computer Security Technology Planning Study*, Deputy for Command and Management Systems HQ Electronic Systems Division, Bedford, MA (1972).

[3] R. Anderson, *Security Engineering: A Guide to Building Dependable Distributed Systems*, Wiley, New York (2001).

[4] W. Aspray, *John von Neumann and the Origins of Modern Computing*, MIT Press, Cambridge, MA (1990).

[5] J. Bidzos oral history interview by J.R. Yost, Mill Valley, California (December 11, 2004).

[6] H.W. Bingham, Security Techniques for EDP of Multilevel Classified Information, Technical report No. RADC-TR 65-415, Rome Air Development Center, Griffiss Air Force Base, New York (December 1965), National Bureau of Standards Collection, Charles Babbage Institute, University of Minnesota.

[7] W.J. Broad, *Every computer 'whispers' its secrets*, New York Times (April 5, 1983), C1.

[8] W.J. Broad, *Global computer network split as safeguard*, New York Times (October 5, 1983), A13.

[9] D. Burnham, *Computer safeguard is broken in U.S. test*, New York Times (January 31, 1977), 18.

[10] M. Campbell-Kelly, *From Airline Reservations to Sonic the Hedgehog: A History of the Software Industry*, MIT Press, Cambridge, MA (2003).

[11] D. Clark, D. Wilson, *A comparison of commercial and military security policies*, *Proceedings of the IEEE Symposium on Security and Privacy*, IEEE Computer Society Press, Los Alamitos, CA (1987), pp. 184–194.

[12] E. Cohen et al., *HYDRA: the kernel of a multiprocessor operating system*, Communications of the ACM 17: 6 (1974), 337–345.

[13] D.E. Denning, *Cryptography and Data Security*, Addison-Wesley, Reading, PA (1999).

[14] J.B. Dennis, *Segmentation and the design of multiprogrammed computer systems*, Journal of the Association of Computing Machinery 12 (1965), 589–602.

[15] Department of Defense Trusted Computer System Evaluation Criteria, DoD 5200.28 STD, Department of Defense, Washington, DC (1985).

[16] W. Diffie, M.E. Hellman, *New directions in cryptography*, IEEE Transactions on Information Theory IT-22 (1976), 644–654.

[17] W. Diffie, S. Landau, *Privacy on the Line*, MIT Press, Cambridge (1998).

[18] W. van Eck, *Electromagnetic radiation from video display units: an eavesdropping risk?*, Computers and Security 4(4) (1985), 269–286.

[19] L.D. Faurer, *Computer security goals of the Department of Defense*, Computer Security Journal (1984).

[20] *G.A.O. Finds Security is Lax at U.S. Computer Installations*, New York Times (May 11, 1976), 16.

[21] W. Gates, Remarks, RSA Data Security Conference (February 15, 2005), http://www.microsoft.com/billgates/speeches/2005/02-15RSA05.asp.

[22] L. Goff, *Worm Disables Net*, Computerworld 33(40) (1999), 78.

[23] M. Hellman, oral history interview by J.R. Yost, Palo Alto, California (November 11, 2004), Charles Babbage Institute, University of Minnesota.

[24] A. Hodges, *Turing: The Enigma*, Simon & Schuster, New York (1983).

[25] M. Hunter, *U.S. to tighten computer security to halt abuses*, New York Times (July 27, 1978), A14.

[26] International Business Machines, *The Considerations of Data Security in a Computer Environment*, National Bureau of Standards Collection, Charles Babbage Institute, University of Minnesota.

[27] D. Kahn, *Tapping computers*, New York Times (April 3, 1976), 23.

[28] D. Kahn, *The Codebreakers: The Story of Secret Writing*, Macmillan, New York (1967).

[29] D.S. Landes, *The Unbound Prometheus: Technological Change and Industrial Development in Western Europe from 1750 to the Present*, Cambridge University Press, London (1969).

[30] K. de Leeuw, *The Dutch invention of the rotor machine, 1915–1923*, Cryptologia 27(1) (2003), 73–94.

[31] E.C. Lesser, G.T. Thompson, *How a hacker tried to fool a security expert*, New York Times (February 22, 1995), D19.

[32] S. Levy, *Crypto: How the Code Rebels Beat the Government – Saving Privacy in the Digital Age*, Viking, New York (2001).

[33] D. MacKenzie, G. Pottinger, *Mathematics, technology, and trust: Formal verification, computer security, and the U.S. Military*, IEEE Annals of the History of Computing 19(3) (1997), 41–59.

[34] D. MacKenzie, *Knowing Machines: Essays on Technical Change*, MIT Press, Cambridge, MA (1996).

[35] J. Markoff, *Can computer viruses be domesticated to serve mankind?*, New York Times (October 6, 1991), E18.

[36] J. Markoff, *Computer intruder is put on probation and fined $10,000*, New York Times (May 5, 1990), 1–2.

[37] J. Markoff, *Top-Secret and Vulnerable*, New York Times (April 25, 1988), D1.

[38] J. Markoff, *Virus Barely Causes Sniffle in Computers; Feared Computer Plague Passes with Very Few Infections*, New York Times (March 7, 1992), 1–2.

[39] A. Miller, *Privacy: Computers, Databanks, and Dossiers*, University of Michigan Press, Ann Arbor, MI (1970).

[40] D.F. Noble, *America by Design: Science, Technology, and the Rise of Corporate Capitalism*, Knopf, New York (1977).

[41] D.B. Parker, *Fighting Computer Crime*, Charles Scribner & Sons, New York (1983).

[42] D.B. Parker, oral history interview by J.R. Yost, Los Altos, California (May 14, 2003), Charles Babbage Institute, University of Minnesota.

[43] E. Passaglia, K.A. Beal, *A Unique Institution: The National Bureau of Standards, 1950–1969*, National Institute of Standards and Technology, Gaithersburg, Maryland (1999).

[44] C. Pursell, *The Machine in America*, Johns Hopkins University Press, Baltimore, MD (1995).

[45] K. Redmond, T.M. Smith, *From Whirlwind to MITRE: The R&D Story of the SAGE Air Defense Computer*, MIT Press, Cambridge, MA (2000).

[46] R.L. Rivest, A. Shamir, L. Adleman, A Method for Obtaining Digital Signatures and Public Key Cryptosystems, MIT Laboratory for Computer Science Technical Memo 82 (MIT LCS TM-28).

[47] D. Russell, G.T. Gangemi Sr., *Computer Security Basics*, O'Reilly & Associates, Cambridge, MA (1991).

[48] J.A. Russell, Electromagnetic Radiation from Computers, Security in the Computing Environment: A Summary of the Quarterly Seminar, Research Security Administrators-June 17, 1965, System Development Corporation, Santa Monica, CA (1966), National Bureau of Standards Collection, Charles Babbage Institute, University of Minnesota.

[49] B. Schneier, *Secrets & Lies: Digital Security in a Networked World*, Wiley, New York (2000).

[50] R.W. Seidel, *Secret scientific communities: classification and scientific communication in the DOE and DoD*, **The History and Heritage of Scientific and Technological Information Systems**, M.E. Bowden, ed., Information Today, New Bedford, NJ (2000).

[51] B. Sinclair, *At the turn of a screw: William sellers, the Franklin Institute, and a standard American thread*, Technology and Culture 10 (1969), 20–34.

[52] B. Sinclair, *Philadelphia's Philosopher Mechanics: A History of the Franklin Institute, 1824–1865*, Johns Hopkins University Press, Baltimore (1974), pp. 170–194.

[53] S. Singh, *The Code Book: The Evolution of Secrecy From Mary, Queen of Scots to Quantum Cryptography*, Doubleday, New York (1999).

[54] M.R. Smith, *Harpers Ferry Armory and the New Technology: The Challenge of Change*, Cornell University Press, Ithica, New York (1977).

[55] *Theft by computer: Programmer taps U.S. data secrets*, New York Times (June 17, 1976), 58.

[56] F.M. Tuerkheimer, *The underpinnings of privacy protection*, Communications of the ACM 36(8) (1993), 69–74.

[57] W.H. Ware, oral history interview by J.R. Yost, Santa Monica, California, RAND Corporation (August 11, 2003), Charles Babbage Institute, University of Minnesota.

[58] W.H. Ware, Security controls for computer systems (U): Report of Defense Science Board Task Force on Computer Security, Office of the Director of Defense Research and Engineering, Washington DC (11 February 1970).

[59] A.F. Westin, *Privacy and Freedom*, Atheneum, New York (1967).

[60] P. Wright, *Spy Catcher: The Candid Autobiography of a Senior Intelligence Officer*, Viking, New York (1987).

[61] J.R. Yost, *Reprogramming the Hippocratic Oath: An Historical Examination of Medical Informatics and Privacy*, **Proceedings of the Second Conference on the History and Heritage of Science and Technological Information Systems**, W.B. Rayward and M.E. Bowden, eds, Information Today for Chemical Heritage Foundation, Medford, New Jersey (2004), pp. 48–60.

21

SECURITY MODELS

Dieter Gollmann

Hamburg University of Technology
D-21071 Hamburg, Germany

Contents

Abstract

Security models are formal descriptions of security policies or abstract models of access control systems. In the first case, they are a starting point when designing security systems. In the second case, they provide a framework for developing theories of access control. This chapter gives a survey of important security models and of fundamental theorems about access control.

Keywords: security models, multi-level security, information flow, Bell–LaPadula, Biba, Clark–Wilson, Chinese Wall, enforcement monitors.

21.1 INTRODUCTION

From early on, computer security has been looking towards formal models for developing security systems to a high level of assurance. The Anderson report (1972) summarising the findings of a planning study for the US Air Force (USAF) on security requirements already comments:

> In order to provide a base upon which a secure system can be designed and built, we recognise the need for a formal statement of what is meant by a secure system – that is a model or ideal design [2: 16].

The development plan put forward states further:

> The model development is considered to be the first step of the development, as it establishes the technical requirements of such a system [2: 23].

The development process proposed starts from a model of the security requirements that serves as a yardstick for analysing the top level specification of the system to be built. In turn, the top level specification is refined and a series of lower level specifications lead finally to the actual implementation of the system. Consistency between the different levels of specifications may be checked informally, or by formal means for the highest degree of assurance. In the context of such a design process, the security model is a formal description of the security policy the system should enforce. Indeed, this definition of security models is frequently found in the security literature.

However, two cautionary remarks are in place. The Anderson report actually takes a wider view of the aspects covered by a security model and not only includes the policy but also the environment the system will exist in. Secondly, there is another class of security models that are not intended to capture a specific policy but are abstract models of authorisation systems. Such models play an important part in the theoretical analysis of access control and will be the subject of Section 21.3.

This chapter is organised as follows. We conclude the Introduction with a few remarks on state machine models. Section 21.2 presents models for specific classes of security policies, starting from military multi-level security policies (see Section 21.2.1.1) and progressing to commercial application oriented policies. Section 21.3 gives a survey of generic authorisation models and of the theoretical results derived in these models. Section 21.4 comments on current developments in IT security that have taken the place of research on security models.

21.1.1 State machine models

State machines (automata) are a popular tool for modelling computing systems in general. They also play an important role in computer security. Their essential features are the concept of a *state* and of state changes occurring at discrete points in time. A state is an abstract representation of the system under investigation at any given moment in time. It should capture exactly those aspects relevant to the problem at hand. The possible *state transitions* are specified by a partial state transition function which defines the next state depending on current state and input. An output may also be produced and there may be a set of final *accepting* states.

To capture a specific security property of a system in a state machine model, one first characterises all the states that fulfil this property. Then

one checks whether all state transitions preserve this property. If this is the case, then one can prove by induction a "Basic Security Theorem" guaranteeing that the desired security property always will hold if the system had started in an initial state having this property. Such theorems are a feature of state machine modelling and do not follow from the particular security policy examined.

21.2 POLICY MODELS

The security models in this section capture specific (classes of) security policies. In each instance, we will first sketch the policy to be modelled, and then describe the model and assess its impact and relevance.

21.2.1 Bell–LaPadula model (1973)

Published in 1973, the *Bell–LaPadula model* (BLP) is probably the most famous of the security models [4]. It was developed by David E. Bell and Leonard J. LaPadula at MITRE in a project for the USAF at the time of the first concerted efforts to design secure multi-user operating systems initiated by the Anderson report. The concepts, and the terminology, of BLP had a noticeable influence on later work on security models. The original reports on BLP have been re-published in the *Journal of Computer Security* [6].

21.2.1.1 *Background.* In the 1970s and 1980s the main sponsor for research on computer security was the US defence sector, in particular the USAF. In this military environment, the data requiring protection is classified information, and there existed well-established policies regulating physical access to classified paper documents. Documents are assigned *security levels*, typically using labels like *unclassified*, *confidential*, *secret* and *top secret*. Users are assigned *clearance* levels. A user's clearance dictates which documents can be obtained.

Adapted to IT systems, these policies are known as *multi-level security* (MLS) policies. The Trusted Computer System Evaluation Criteria (*Orange Book*) [37] uses the term *mandatory access control (MAC)* when referring to policies that make use of security levels and to *discretionary access control (DAC)* for policies based on named users and named objects. For some time, the terms discretionary and mandatory access control were used in the security literature almost exclusively in this restricted MLS interpretation. Today, they are often used to distinguish between policies that are defined at the discretion of the user and policies that are mandated by the system.

21.2.1.2 *The model.* The BLP model translates the security policies for handling classified data to rules enforced by a *reference monitor* in a computer system. When a process requests access to a resource, the reference monitor decides whether to grant or deny access. The process (the subject making the request) runs with a security level typically established during logon from the user's clearance. The operating system also maintains security levels associated with resources (the objects of access requests) such as files, directories, printers, network ports, etc.

The BLP model is a state machine model of MLS policies that prevent information flowing downwards from documents with a high security level to documents with a low security level, i.e. unauthorised declassification of data. BLP only considers the information flow that occurs when a subject performs an access operation on an object. The access operations considered, `execute`, `read`, `append` and `write`, are typical for operating systems. To simplify our discussion, we will refer to operations that *observe* an object (`read`, `write`) and to operations that *alter* an object (`append` and `write`).

Access permissions are defined both through an access control matrix and through a partially ordered set of security levels. It is thus possible to compare security levels, and it is customary to say that a higher security level *dominates* a lower security level.

21.2.1.3 *Security policies.* The *simple security property* (ss-property) straightforwardly captures the policy that users may read objects only if they have the required clearance.

ss-property: A request to observe an object is granted only if the (logon) security level of the subject dominates the classification of the object.

In this traditional *no read-up* policy that applies when a person requests access to a classified document the subject acts as an observer. However, in a computer system subjects are processes and can act as channels by reading one memory object and transferring information to another memory object. In this way, a malicious subject could declassify data.

Transferring a policy for paper documents to IT systems thus raises an issue that did not exist before. There is thus a further need for a policy to control write access. For this purpose, BLP includes a *no write-down* policy, the so-called *∗-property* (star-property). The first version of BLP had not yet included this property, and the symbol ∗ was reputedly used as a placeholder until a proper name for the policy would be found.

∗-property: A request to alter an object is only granted if the classification of the object dominates the (current) security level of the subject. Furthermore, the security level of any object the subject may alter must dominate the security level of any object the subject may observe.

The ∗-property blocks illegal information flow downwards between objects via a subject but it creates a new problem. How can a high level user send legitimate messages to low level users? To do so, a high level subject could be temporarily downgraded. For this very reason, BLP maintains a logon security level and a current security level for subjects. Downgrading the current security level implies that a subject immediately forgets all it knew at the previous security level. This would be implausible when subjects are human beings. However, processes in a computer system do not have 'memory' like a person, but they have access to memory objects and only 'know' the contents of the objects (files) they are allowed to observe. In this situation, a temporary downgrade indeed solves the problem.

Alternatively, a set of subjects exempted from the ∗-property could be identified. These subjects are called *trusted subjects*. By definition, a *trusted subject* may violate the security policy. In the early years of computer security the adjective *trusted* was used precisely as an indicator for system components that can hurt you. In contrast, a component was called *trustworthy* if there were reasons to believe that it would work as expected.

In BLP, discretionary access control policies are expressed by an access control matrix and captured by the *discretionary security property* (ds-property):

ds-property: An access request is granted only if permitted by the access control matrix.

21.2.1.4   *The basic security theorem.*

A state in the BLP model is called *secure*, if the ss-, ∗- and ds-properties are satisfied. A state transition is called secure, if it maps a secure state to a secure state. The Basic Security Theorem of BLP then states that if all state transitions in a system are secure and if the initial state of the system is secure, then every subsequent state will also be secure, no matter which inputs occur. In practice, this theorem limits the effort needed to verify the security of a system. Each state transition can be checked individually to show that it preserves security, and a secure initial state has to be identified.

21.2.1.5   *Tranquility.*

In 1987, McLean [31] triggered a heated discussion about the value of BLP by putting forward a system that contained a state transition, which downgraded all subjects and all objects to the lowest security level, and entered all access rights in all positions of the access control matrix. The state reached by this transition is secure according to the definitions of BLP, but a system that can be brought into a state where everyone is allowed to read everything is not secure according to the common intuition of security. Hence, BLP should be regarded as a flawed security model.

Bell's counterargument stated that if the requirements call for such a state transition, then it should be possible to express it in the security model. If it is not required, then it should not be implemented. The existence of such a downgrade operation is thus not a problem of the model but a problem of correctly capturing the security requirements.

At the root of this disagreement is a state transition that changes access rights. Such changes would be possible within the general framework of BLP but the model was intended for systems that are running with fixed security levels. The property that security levels and access rights never change is called *tranquility*.

21.2.1.6   *Aspects and limitations of BLP.*

BLP is a very significant security model. It has played an important role in the design of secure operating systems, consider e.g. its contributions to the design of the Multics operating system [5], and to the design of MLS database management systems. Many other models build on its features, and often follow quite closely the terminology established for BLP. There was a time when it was treated almost as a universal security model, but it is neither the case that all meaningful security policies can be expressed as MLS policies, nor are all relevant modes of information flow captured in BLP.

21.2.2   MLS databases

The development of BLP was closely related to work on the design of Multics, and the BLP security properties treat objects as atoms without internal structure. This is quite appropriate when dealing with operating systems. To construct a model that captures MLS policies in relational databases, one must however take into account that the objects in a database management system (DBMS) are databases, tables, records, integrity constraints, etc., and that the relationships between the structural elements of database objects can give raise to new kinds of information flow not captured by the BLP model.

Consider a confidential table that contains a secret record. Records in the table are uniquely identified by their *primary key* (see e.g. [13]). A user cleared at level confidential will not see the secret record and may by accident try to enter a new record with the same primary key as the secret record. When the DBMS issues an error message to protect the integrity of the table the user would learn about the existence of the secret record. In general, telling a low-level subject that a requested operation is not permitted because of the previously unknown existence of a high-level object constitutes information flow from a high security level to a low security level. Transmission channels that can be used to violate a security policy are called *covert channels*. Eliminating covert channels caused by the internal structure of database objects was a major challenge in adapting BLP to relational databases.

21.2.2.1 *SeaView (1988).* The Secure Data Views (SeaView) model [15; 30] was developed by SRI International and Gemini Computers, Inc., in a project sponsored by the USAF in the mid-1980s. The designers of the SeaView model took the decision to hide from the users the security labels that are used internally by the DBMS. Databases, tables, records, and fields all can have their own security label. When a low table contains a high record, a low user would not see the record. When a low record contains a high field, a low user would see the record with a NULL value in the high field. When the low user tries to update this field, the DBMS has three options to react:

- refuse to perform the update.
- overwrite the old value.
- perform the update *and* keep the old value: this is called *polyinstantiation*.

The first option introduces a covert channel revealing that this location contains a data value at a higher security level. The second option does not give away information about the existence of a high data value but leaves high users in the unfortunate situation that low users can unwittingly destroy their data. Both options are unattractive and we are left with polyinstantiation. This term indicates that there may be several tuples for the same primary key.

To reconcile MLS security requirements with the necessities of relational data bases, the labels of all fields in a tuple become part of the primary key in the SeaView model. To address a tuple in a relation, one has to specify a vector of labels in addition to the original primary key. This extended primary key allows again a unique identification of tuples. While normal tables are two-dimensional structures, multi-level tables can be viewed as three-dimensional structures with the co-ordinates (original primary key, attribute, security label).

The original polyinstantiation integrity property made updates rather cumbersome and bloated the size of a polyinstantiated relation. This aspect of SeaView has been remodelled in the Sandhu–Jajodia model, see e.g. [32–34], that introduces improvements to the decomposition strategy for multi-level relations and to the polyinstantiation rules.

21.2.2.2 *Insert low (1992).* When a DBMS keeps security labels as internal information, it will not only withhold sensitive data from unauthorised users, it also hides their existence. On one hand, this offers more security. On the other hand, polyinstantiation was introduced to close a potential covert channel and had to be integrated with the relational database model.

A collision with the fundamental integrity properties of relational databases was avoided but the problem reappears when application specific integrity constraints are added to an MLS DBMS. To fit into the MLS model, every integrity constraint has to be labelled with an access class. The following consistency rule applies.

> The access class of an integrity constraint must dominate the access classes of the relations the constraint applies to.

A familiar problem emerges when an integrity constraint is labelled at a higher level than a data item it refers to. If a low level subject wanted to update a data item that is subject to a high level constraint, the DBMS could either allow the subject to make the change and potentially violate the integrity constraint or disclose the existence of the high level constraint to maintain consistency. This time, there is no technical trick to escape from this trap.

A database captures external facts. A polyinstantiated database creates ambiguity about these external facts as there exist different entries for the same external entity. The SWORD DBMS [38] developed at the Royal Signals and Radar Establishment in the UK avoided these problems by implementing MLS policies without hiding the existence of data items or constraints. Security labels are attached to each data item and relation (although not to tuples) and are visible to every user who is cleared to see the relation. Rejection of an access request does not constitute an illegal information flow because security levels can be seen by all subjects.

An *insert low* strategy is used to insert data into the database without creating an illegal information flow. A subject *creator* with access class 'system low' is the only entity entitled to create new items in the database. To create a high level data item, the *creator* first creates this data item and stores some placeholder information in it. The existence of this data item is thus known to everybody. Next, the *creator* sets the classification of the data item to 'high'. Thus, the classification of the data item is known to everybody. Now, high level users can write to the data item, which cannot be accessed by low level subjects. The same approach can be used with integrity constraints, hiding their contents but not their existence. A low level subject trying to update a data item that is subject to a high level constraint now can be told about the existence of the constraint and be prevented from performing the update.

With insert low, multi-level tuples can be created without polyinstantiation and the ensuing expansion of the key space and the relations. As a potential disadvantage, all data items have to be created and properly classified by low level subjects. Not all security policies will agree with such a classification procedure. In addition, high level subjects are somewhat impeded as they need a low level helper whenever they create a new data item.

21.2.3 Biba model (1977)

The Biba model was develop by Kenneth J. Biba at MITRE, again in a project for the USAF. A first version of the model was published in June 1975, a revision in April 1977 [7]. The Biba model captures multi-level integrity policies. Such policies are the dual of the BLP confidentiality properties. Subjects must not read-down, i.e. observe objects with lower integrity labels, and must not write-up, i.e. alter objects with higher integrity labels. The integrity labels form a *lattice* so the greatest lower

bound and the least upper bound of two labels are well defined.

In contrast to BLP, this model did not capture existing procedures for handling paper documents. The Biba model thus has more the flavour of an academic exercise exploring how MLS integrity policies might be defined. There are policies with static integrity levels that are the exact dual of the BLP policies, even following the language of BLP:

Simple integrity property: A subject may alter an object only if the subject's integrity level dominates the integrity level of the object.

Integrity ∗-property: A subject that can observe an object *o* is only permitted to alter objects with integrity levels dominated by the level of *o*.

Low watermark policies dynamically adjust the integrity level of an entity if it has come into contact with a low level entity.

Subject low watermark property: Subjects can observe objects at any integrity level. When a subject observes an object, the integrity level of the subject is downgraded to be the greatest lower bound of the integrity levels of the subject and the object before the operation.

Object low watermark property: Subjects can alter objects at any integrity level. When a subject alters an object, the integrity level of the object is downgraded to be the greatest lower bound of the integrity levels of the subject and the object before the operation.

The Biba model can be extended to include an access operation `invoke`. A subject can invoke another subject, e.g. a software tool, to access an object. If we want to make sure that invocation does not bypass the mandatory integrity policies we could add the

Invoke property: Subjects may only invoke other subjects at lower integrity levels.

Otherwise, a 'dirty' subject could use a 'clean' tool to access, and contaminate, a clean object. Alternatively, we may want to use tools exactly for this purpose. Dirty subjects should have access to a clean object, but only when they use clean tools.

This tool may perform consistency checks to ensure that the object remains clean. Integrity protection mechanisms in operating systems that use protection rings fall into this category. In a scenario where a clean subject should not use dirty tools we could adopt the

Ring property: A subject can observe objects at all integrity levels. It can only alter objects at lower integrity levels and it can invoke only subjects at higher integrity levels.

The Biba model never had the same impact as BLP. The most obvious cause is the lack of widely used label-based integrity policies that would be captured by (some of) the Biba properties.

21.2.4 Clark–Wilson model (1987)

In 1987, David D. Clark (MIT) and David R. Wilson (Ernst & Whinney[1]) published an influential paper on commercial security policies [12] that reflects the experience of accountancy and consultancy businesses, and also the trend away from multi-user operating systems to PC-based IT applications in the commercial world that had started in the early 1980s.

In business applications security requirements are predominantly (data) integrity concerns, addressing the unauthorised modification of data, fraud, and errors. *Internal consistency* requirements refer to the internal state of a system and can be enforced by the computing system. *External consistency* requirements deal with the relation of the internal state of a system to the real world and are enforced by means outside the computing system, e.g. by auditing.

Clark and Wilson discuss the difference between military and commercial security requirements. The relative importance of confidentiality and integrity may indeed differ between both worlds but the military also has integrity requirements and commerce has confidentiality requirements. For the historian, there is a much more relevant transition. The access operations in the Clark–Wilson model are programs performing complex

[1]Ernst & Whinney merged with Arthur Young in 1989 to form Ernst & Young.

application specific manipulations, not the simple, generic operating system functions of BLP. Security models have moved from general purpose multi-user operating systems (BLP) to application oriented IT systems (Clark–Wilson).

Generic mechanisms for enforcing integrity in commercial applications are *well-formed transactions*, i.e. data items can be manipulated only by a specific set of programs and users have access to programs rather than to data items, and *separation of duties*. Users have to collaborate to manipulate data and therefore to collude to penetrate the security system.

In the Clark–Wilson model programs are an intermediate layer between subjects and objects. Subjects are authorised to execute certain programs. Data items can be accessed through specific programs. It is testimony to the influence of BLP that Clark and Wilson write about "labelling subjects and objects with programs instead of security levels". Inputs to the system are captured as unconstrained data items (UDIs). The data items governed by the security policy are called constrained data items (CDIs). Conversion of UDIs to CDIs is a critical step that cannot be controlled solely by the security mechanisms in the system. CDIs can be manipulated only by transformation procedures (TPs). The integrity of a state is checked by integrity verification procedures (IVPs).

Five general *certification rules* suggest the checks that should be conducted so that the security policy is consistent with the application requirements.

- IVPs must ensure that all CDIs are in a valid state when the IVP is run.
- TPs must be certified to be valid, i.e. valid CDIs must always be transformed into valid CDIs. Each TP is certified to access a specific set of CDIs.
- The access rules must satisfy any separation of duties requirements.
- All TPs must write to an append-only log.
- Any TP that takes an UDI as input must either convert the UDI into a CDI or reject the UDI and perform no transformation at all.

Four general *enforcement rules* describe the security mechanisms within the computer system that should enforce the security policy. These rules have some similarity with discretionary access control in BLP.

- The system must maintain and protect for each TP a list of CDIs the TP is certified to access.
- The system must maintain and protect for each user a list of TPs the user can execute.
- The system must authenticate each user requesting to execute a TP.
- Only a subject that may certify an access rule for a TP may modify the respective entry in the list. This subject must not have execute rights on that TP.

The Clark–Wilson model is a framework and guideline ('model') for formalising security policies rather than a model of a specific security policy. Instantiating a policy is a more complex process than with BLP, entailing the definition of CDIs and TPs, and the labelling of subjects and objects with TPs, i.e. the assigning of access rights. An implementation of the Clark–Wilson model using capabilities is described in [26]. An extension of the Biba model providing mandatory integrity controls that can be used to implement the Clark–Wilson model is presented in [28].

21.2.5 Chinese Wall model (1989)

The Chinese Wall model [11] was proposed by David F.C. Brewer and Michael J. Nash of Gamma Secure Systems Ltd. Gamma was, and at the time of writing still is, an independent UK security consultancy business. The term Chinese Wall was introduced in the US after the stock exchange crash in 1929 to denote a conceptual communications barrier between different divisions of a financial institution installed to avoid conflict of interest.

Chinese Wall policies were thus a well established term in the financial sector when this model was proposed in the aftermath of the deregulation of UK financial markets that had culminated in the deregulation of the London Stock Exchange, Britain's 'big bang' on 26 October 1986. As companies in the financial sector were given permission to offer a wider range of services, and as those companies were increasingly using IT systems for conducting their business, regulators had to be convinced that there were sufficient barriers also within the IT systems of a company to prevent conflicts of interest, internal fraud, insider trading, and the like.

In financial services, analysts dealing with different clients have to make sure that no conflicts of interest arise in their work. Such conflicts would occur when an analyst works for clients that are direct competitors in the same market, or when the owners of the clients are competitors. When modelling a Chinese Wall policy, the subjects are the analysts and the objects are items of information concerning a single company. All objects concerning the same company are collected in a *company dataset*. *Conflict of interest classes* indicate which companies are in competition. The *security label* of an object is its company data set together with its conflict of interest class. *Sanitised information* has been purged of sensitive details and is not subject to access restrictions.

Initially, a subject is allowed to access any object. Once an object has been accessed, a conceptual wall has to be erected, with all objects that could be the cause of a conflict of interest at the other side of the wall. Thus, a subject's access rights my change dynamically in every state transition. Following BLP, the Chinese Wall model has a ss-property and a ∗-property.

ss-property: A subject is granted access to an object only if the object is in the same company dataset as an object already accessed by the subject, or if the object belongs to an entirely different conflict of interest class.

The ss-property does not prevent information flow where subjects figure as information channels. Consider, for example, two competitors A and B that have their accounts with the same bank. An analyst dealing with company A and the bank would be a conduit for information about A flowing to the bank. An analyst dealing with company B and the bank now has access to information about a competitor's business. Write access is thus regulated as follows:

∗-property: Write access is only permitted if no object containing unsanitised information can be read, which is in a different company dataset to the one for which write access is requested.

The Chinese Wall model had also been of particular interest to researchers as commercial policies with dynamically changing security labels added a new dimension to the design of security models.

21.3 SECURITY MODELS AND THE THEORY OF ACCESS CONTROL

In parallel with the development of models for security policies of existing military and commercial applications, there exists another line of work, focused on exploring the theoretical limitations of security models. The security models presented in this section are not formalisations of any given policy but are abstract models of enforcement mechanisms. Their purpose is the analysis of the intrinsic complexity of core problems in access control, or to provide generic frameworks for analysing information flow in computer systems (information flow and non-interference models).

To appreciate the results quoted in the following, the reader should be familiar with basic concepts from computational complexity theory such as decidability, see, e.g., [17] for a introduction to this topic. Informally, a problem is *undecidable* if no computable algorithm can exist that solves all instances of this problem.

21.3.1 Harrison–Ruzzo–Ullman model (1976)

Michael A. Harrison, Walter L. Ruzzo (UC Berkeley) and Jeffrey D. Ullman (Princeton University) derived fundamental results in a formal model of *authorisation systems* where access rights may be changed and where subjects and objects can be created and deleted [22]. An access matrix records the current rights subjects have on objects. There are primitive operations for creating and deleting subjects and objects, and for entering and deleting entries of the access matrix. From these primitive operations, commands can be constructed. For example, the following command creates a new file, defines its owner, and grants the owner read and write access to that file by adding entries into the access matrix M.

command *create_file(s, f)*
  create f

enter $\underline{o}$ into $M_{s,f}$
enter $\underline{r}$ into $M_{s,f}$
enter $\underline{w}$ into $M_{s,f}$
end

The effect of a command is recorded as a change to the access matrix. To verify that a system complies with a given policy on the allocation of access rights, one would have to check that there exists no way for undesirable access rights to be granted. In the security literature, this is described as the *leaking* of access rights.

> An access matrix is said to be *safe* with respect to a given access right, if no sequence of commands will enter the right into the matrix, i.e. into a state that leaks the right.

Verifying compliance with a security policy in the HRU model thus corresponds to verifying safety properties. Harrison, Ruzzo and Ullman gave the following general result.

Theorem 1. *Given an access matrix M and a right r, verifying the safety of M with respect to the right r is an undecidable problem.*

Thus, there is no single universal algorithm that solves all instances of the safety problem in HRU. This situation changes when some aspects of the HRU model are restricted. For example, *mono-operational* commands contain only a single operation. *Mono-conditional* commands contain only a single condition. A system is called *monotonic* if access rights, subjects, and objects can be created but not deleted.

The following theorem collects the main decidability results derived for a number of variations of the HRU model.

Theorem 2.

- *The safety property for authorisation systems without create operations is PSPACE-complete.*
- *The safety property for monotonic authorisation systems is undecidable [21].*
- *The safety property for monotonic mono-conditional authorisation systems is decidable [21].*
- *The safety property for mono-operational authorisation system is decidable.*

- *The safety problem in monotonic authorisation systems is undecidable if commands may have up to two conditions* [21].
- *The safety problem for arbitrary authorisation systems is decidable if the number of subjects is finite* [29].

21.3.2 Take–grant model (1976)

The take–grant model proposed by Anita K. Jones (Carnegie–Mellon University), Richard J. Lipton, and Lawrence Snyder (Yale University) describes capability-based systems [25], building on earlier work on authorisation systems [20; 23].

We sketch an elementary version of the model that uses just two access rights read, write. More elaborate versions of the take–grant model exist. The model is given as a directed graph. The nodes of the graph are subjects and objects. The edges of the graph are labelled with the access rights read, write and the combination read + write. When subject s_1 has read right on subject s_2 it can acquire capabilities from s_2. When subject s_1 has write right on subject s_2 it can grant its own capabilities to s_2. There are three transition rules, *take*, *grant* and *create*.

- Take: A subject s can acquire right r on y if s has read right on x and x has right r on y.
- Grant: x can acquire right r on y if there exists a subject s that has write right on x and right r on y.
- Create: A subject s may create a new subject to which it has read and write rights.

The create rule creates a new node in the graph. The take and grant rules can add new edges, or a new label to an existing edge. One can then ask security management questions of the kind: "Given two nodes of a graph and some label, is there a sequence of take and grant transitions so that an edge with the given label is added between those two nodes?". This is the safety problem for the take–grant model.

Theorem 3. *The safety problem for the take–grant system with subjects and objects is solvable in linear time.*

21.3.3 Information-flow model (1976)

Compared to BLP, the information-flow model developed by Dorothy E. Denning (Purdue University) [14] has a richer model of the processes that run in a computer system, thereby capturing a wider range of information flows than BLP. Processes execute programs that are written using three constructors,

- assignments, e.g. y:= x;
- conditional statements, e.g. IF x=0 THEN y:=1;
- sequential execution of statements.

Informally, an operation causes an information flow from an object x to an object y, if observing y tells us more about x. If we already knew x, no information can flow from x. An assignment y:= x; can cause an explicit information flow. Conditional statements cause an implicit information flow. Observing y after the conditional statement IF x=0 THEN y:=1; may tell something about x even if the assignment y:= 1; had not been executed. For example if $y = 2$, we would know that $x \neq 0$.

Information theory provides a precise and quantitative definition of information flow. The amount of information that can be derived from an observation is given by the entropy of the object (variable) observed. In the MLS information flow model all objects are labelled with elements from a lattice of security levels. Information flow from an object o_1 to an object o_2 is permitted only if the level of o_2 dominates the level of o_1. Any information flow that violates this rule is illegal. A system is called *secure* if there is no illegal information flow. A system is called *precise* if no legal information flow is blocked.

The advantage of such a model is that it covers a wider range of information flows. The disadvantage is that it can become more difficult to analyse the security of a given system. Checking whether a given information flow model is secure or precise is undecidable [24].

21.3.4 Non-interference model (1982)

The *non-interference* model by Joseph A. Goguen and Jose Meseguer (SRI International) presents an alternative to information-flow model [19]. It provides a different formalism for describing what a subject knows about the state of the system. Subject s_1 does not interfere with subject s_2 if the actions of s_1 have no influence on s_2's view of the system. Currently, information-flow and non-interference models are active areas of research rather then the basis of a practical methodology for the design of secure systems.

21.3.5 Execution monitors (2000)

The last contribution to the theory of access control to be discussed attempts to characterise the policies that can be enforced by the typical access control mechanisms in use today. The term *Execution Monitoring* (EM) was introduced by Fred B. Schneider (Cornell University) for enforcement mechanisms that monitor execution steps of a target system and terminate the target's execution if a violation of the security policy is about to occur [36]. Such mechanisms are deployed today in firewalls, operating systems, middleware architectures like CORBA, or in web services.

Execution monitors have two important limitations. They do not have a model of the target system so they cannot predict the outcomes of possible continuations of the execution they are observing, and they cannot modify the target system before executing it. Important access control mechanisms outside EM are therefore compilers and theorem-provers, as they analyse a static representation of the target and can deduce information about all of its possible executions, and in-line reference monitors and reflection in object-oriented systems, as they modify the target.

21.3.5.1 *Properties of executions.* Executions of the target system are sequences of steps. The precise nature of these steps will depend on the actual target. A security policy is defined as a predicate on the set of executions. A set of executions is called a *property* if membership of an element is determined by the element alone.[2] A security policy must therefore be a property to have an enforcement mechanism in EM.

However, not every security policy is a property. Some security policies cannot be defined as a predicate on individual executions, e.g. information flow policies demanding that a low user cannot distinguish an execution where a high user is active from some other execution where the high user is inactive. Moreover, not every property is EM enforceable. Enforcement mechanisms in EM cannot look into the future when making decisions on an execution. When a 'secure' execution has an 'insecure' prefix an execution monitor has to prohibit the insecure prefix and stop the execution. For such policies, EM would be a *conservative* approach that stops more executions than necessary.

21.3.5.2 *Safety and liveness.* Among the properties of executions there are two broad classes of particular significance.

- Safety properties: nothing bad can happen. (The safety property of access matrices in the HRU model meets this description.)
- Liveness properties: something good will happen eventually.

There exists a close relationship between safety and the type of policies that can be enforced by execution monitors. Informally, if an execution is unsafe there has to be some point in the execution after which it is no longer possible to revert to a safe continuation of the execution.[3]

If the set of executions for a security policy is not a safety property, then there exists an unsafe execution that could be extended by future steps into a safe execution. Such policies do not have an enforcement mechanism from EM. So, if a policy is not a safety property, it is not EM enforceable. Conversely, execution monitors enforce policies that are safety properties. However, not all safety properties have EM enforcement mechanisms. We summarise the findings in [36]:

Access control policies define restrictions on the operations principals can perform on objects. These policies define safety properties. Partial executions that end attempting an unacceptable operation will be prohibited.

---

[2]This observation provides a link to the literature on linear-time concurrent program verification [1].

[3]A formal definition of safety properties can be found in [27].

Information flow policies restrict what principals can infer about objects from observing system behaviour. These policies do not define sets of executions that are properties. Thus, information flow cannot be a safety property and in turn cannot be enforced by EM.

Availability policies restrict principals from denying others the use of a resource. These policies do not define safety properties. Any partial execution could be extended so that the principal would get access to the resource in the end. Availability policies that refer to a Maximum Waiting Time (MWT) [18] are safety properties. Once an execution has waited beyond the MWT, any extension will naturally also be in violation of the availability policy.

21.4 CURRENT WORK

Research on security models had been very active in the 1970s and 1980s. At that time, most of the fundamental questions about access control had been answered. Since then, research on access control has moved on driven by the changes in the security policies that had to be enforced. Today's policies are customised and application specific so a small set of generic policy models no longer meets the requirements.

Research on *trust management* (PolicyMaker [10]; KeyNote [9]) develops approaches to decentralised access control, defining flexible languages for expressing policies, but not prescribing the types of policies that can be enforced. Work on policy languages that have a formal execution model (e.g. Binder [16]) takes research in this direction further. The challenge is the design of languages that are sufficiently flexible so that meaningful policies can be expressed and at the same time have efficient policy evaluation algorithms. The policies encountered in today's applications are often too complex and too dynamic to be captured by a simple security model that can be completely described in a single research paper. The Cassandra project [3] modelling the policies for the UK Electronic Health Records system in constrained datalog may serve as an example for research linking concrete policies to the theory of access control.

Finally, customised security models that make use of ideas from the early security model continue to be used in the formal analysis of high-assurance systems, e.g. for smart cards [8; 35].

REFERENCES

[1] B. Alpern, S. Schneider, *Defining liveness*, Information Processing Letters 21 (4) (1985), 181–185.

[2] J. Anderson, Computer security technology planning study, Tech. Rep. 73-51, U.S. Air Force Electronic Systems Technical Report (October 1972).

[3] M.Y. Becker, P. Sewell, *Cassandra: Flexible trust management, applied to electronic health records*, **Proceedings of the 17th IEEE Computer Security Foundations Workshop**, Washington, DC (2004), pp. 139–154.

[4] D.E. Bell, L.J. LaPadula, Secure computer systems: Mathematical foundations and model, Tech. Rep. M74-244, The MITRE Corporation, Bedford, MA (May 1973).

[5] D. Bell, L. LaPadula, Secure computer system: Unified exposition and Multics implementation, Tech. Rep. ESD-TR-75-306, The MITRE Corporation, Bedford, MA (July 1975).

[6] D.E. Bell, L.J. LaPadula, *MITRE technical report 2547 (secure computer system): Volume II*, Journal of Computer Security 4 (2/3) (1996), 239–263.

[7] K. Biba, Integrity consideration for secure computer systems, Tech. Rep. ESD-TR-76-372, MTR-3153, The MITRE Corporation, Bedford, MA (April 1977).

[8] P. Bieber, J. Cazin, P. Girard, J.-L. Lanet, V. Wiels, G. Zanon, *Checking secure interactions of smart card applets*, **ESORICS 2000**, F. Cuppens et al., eds, LNCS 1895, Springer-Verlag (2000), pp. 1–16.

[9] M. Blaze, J. Feigenbaum, J. Ioannidis, A.D. Keromytis, The KeyNote Trust-Management System Version 2, Internet informational RFC 2704 (September 1999).

[10] M. Blaze, J. Feigenbaum, J. Lacy, *Decentralized trust management*, **Proceedings of the 1996 IEEE Symposium on Security and Privacy**, IEEE Computer Society, Washington, DC (1996), pp. 164–173.

[11] D. Brewer, M. Nash, *The Chinese Wall security policy*, **Proceedings of the 1989 IEEE Symposium on Security and Privacy**, IEEE Computer Society, Washington, DC (1989), pp. 206–214.

[12] D. Clark, D. Wilson, *A comparison of commercial and military computer security policies*, **Proceedings of the 1987 IEEE Symposium on Security and Privacy**, IEEE Computer Society, Washington, DC (1987), pp. 184–194.

[13] C.J. Date, **An Introduction to Database Systems, Volume I**, 8th edition, Addison Wesley, Reading, MA (2003).

[14] D.E. Denning, *A lattice model of secure information flow*, Commun. ACM 19 (5) (1976), 236–243.

[15] D. Denning, T. Lunt, R. Schell, W. Shockley, M. Heckman, *The SeaView security model*, **Proceedings of the 1988 IEEE Symposium on Security and Privacy**, IEEE Computer Society, Washington, DC (1988), pp. 218–233.

[16] J. DeTreville, *Binder, a logic-based security language*, **Proceedings of the 2002 IEEE Symposium on Security and Privacy**, IEEE Computer Society, Washington, DC (2002), pp. 105–113.

[17] M.R. Garey, D.S. Johnson, **Computers and Intractability**, Freeman, New York (1979).

[18] V.D. Gligor, *A note on denial of service in operating systems*, IEEE Transaction on Software Engineering 10 (3) (1984), 320–324.

[19] J. Goguen, J. Meseguer, *Security policies and security models*, **Proceedings of the 1982 IEEE Symposium on Security and Privacy**, IEEE Computer Society, Washington, DC (1982), pp. 11–20.

[20] G.S. Graham, P.J. Denning, *Protection, principles, and practice*, **Proceedings of the 1972 AFIPS Joint Computer Conference**, Vol. 40, AFIPS Press, Montvale, NJ (1972), pp. 417–429.

[21] M.A. Harrison, W.L. Ruzzo, *Monotonic protection systems*, **Foundations of Secure Computation**, R.D. Demillo et al., eds, Academic Press, New York (1978), pp. 337–365.

[22] M.A. Harrison, W.L. Ruzzo, J.D. Ullman, *Protection in operating systems*, Communications of the ACM 19 (8) (1976), 461–471.

[23] A.K. Jones, Protection in programmed systems, PhD. thesis, Carnegie–Mellon University (1973).

[24] A.K. Jones, R.J. Lipton, *The enforcment of security policies for computation*, ACM Operating Systems Reviews 9 (5) (1975), 197–206. (Proceedings 5th Symposium on Operating System Principles.)

[25] A.K. Jones, R.J. Lipton, L. Snyder, *A linear time algorithm for deciding security*, **Proceedings 17th Annual Symposium on Foundations of Computer Science**, IEEE Computer Society, Washington, DC (1976).

[26] P. Karger, *Implementing commercial data integrity with secure capabilities*, **Proceedings of the 1991 IEEE Symposium on Research in Security and Privacy**, IEEE Computer Society, Washington, DC (1991), pp. 130–139.

[27] L. Lamport, *Checking secure interactions of smart card applets*, **Distributed Systems: Methods and Tools for Specification: An Advanced Course**, M.W. Alford et al., eds, LNCS 190, Springer-Verlag (1985), pp. 119–130.

[28] T. Lee, *Using mandatory integrity to enforce "commercial" security*, **Proceedings of the 1991 IEEE Symposium on Research in Security and Privacy**, IEEE Computer Society, Washington, DC (1991), pp. 140–146.

[29] R.J. Lipton, L. Snyder, *On synchronization and security*, **Foundations of Secure Computation**, R.D. Demillo et al., eds, Academic Press, New York (1978), pp. 367–385.

[30] T. Lunt, D. Denning, R. Schell, M. Heckman, W. Shockley, *The SeaView security model*, IEEE Transactions on Software Engineering 16 (6) (1990), 593–607.

[31] J. McLean, *Reasoning about security models*, **Proceedings of the 1987 IEEE Symposium on Security and Privacy**, IEEE Computer Society, Washington, DC (1987), pp. 123–131.

[32] R.S. Sandhu, S. Jajodia, *Polyinstantiation integrity in multilevel relations*, **Proceedings of the 1990 IEEE Symposium on Research in Security and Privacy**, IEEE Computer Society, Washington, DC (1990), pp. 104–115.

[33] R.S. Sandhu, S. Jajodia, *Referential integrity in multilevel secure databases*, **Proceedings of the 16th National Computer Security Conference**, NIST (1993), pp. 39–52.

[34] R.S. Sandhu, S. Jajodia, T. Lunt, *A new polyinstantiation integrity constraint for multilevel relations*, **Proceedings of the 6th IEEE Computer Security Foundations Workshop**, IEEE Computer Society, Washington, DC (1990), pp. 159–165.

[35] G. Schellhorn, W. Reif, A. Schairer, P. Karger, V. Austel, D. Toll, *Verification of a formal security model for multiapplicative smart cards*, **ESORICS 2000**, F. Cuppens et al., eds, LNCS 1895, Springer-Verlag (2000), pp. 17–36.

[36] F.B. Schneider, *Enforceable security policies*, ACM Transactions on Information and System Security 3 (1) (2000), 30–50.

[37] US Department of Defense, DoD trusted computer system evaluation criteria, DOD 5200.28-STD (1985).

[38] A. Wood, S. Lewis, S. Wiseman, The SWORD multilevel secure DBMS, Tech. Rep. RSRE 92005, DRA, Malvern (1992).

22

COMPUTER SECURITY THROUGH CORRECTNESS AND TRANSPARENCY

Hans Meijer, Jaap-Henk Hoepman, Bart Jacobs and Erik Poll

Institute for Computing and Information Sciences (ICIS)
Radboud University Nijmegen
P.O. Box 9010, 6500 GL Nijmegen, The Netherlands
E-mails: hans.meijer.de.la.arcilla@gmail.com, jhh@cs.ru.nl, bart@cs.ru.nl, erikpoll@cs.ru.nl

Contents

Abstract

The computer security problem is nearly as old as the computer itself. The necessity of a proper hardware-based protection was already recognized in the 1950s. During the 1960s computer crime emerged together with the deployment of mainframe computers. The computer security problem is aggravated by software being buggy, making it vulnerable to malicious attacks. The ubiquity of software bugs was one of the main reasons for proclaiming the 'software crisis' in the late 1960s. The resulting software engineering research appeared to consider the security problem as just a program correctness issue until the outbreak of computer worms and viruses on the Internet in the 1980s. Since then, it has become clear that a multi-faceted approach is necessary, with a possibly central role for open transparent software.

Keywords: computer security, program correctness, open source.

22.1 INTRODUCTION AND PROBLEM STATEMENT

Security is about regulating access to assets. This generally involves some combination of technical, organizational and legal measures [32; 55].

A company's employee may have a key or smart card to access certain assets, e.g. the building or her own office, and may need to wear a badge. A visitor, on the other hand, must ring a doorbell and identify himself at the reception, whereupon he is collected by the employee who expects the visitor. If an intruder manages to sneak in and steal something, he or she may be caught and brought to court.

When access regulation is automated in computer programs, the threats to security are basically the same: bugs or holes, attacks, backdoors. However, organizational and legal measures are usually less powerful in this case, and technical measures become much more important.

A computer program should not only be *correct* (that is, do what it is supposed to do), but also be both *invulnerable* (resilient to attacks) and *honest* (not maliciously helping an attacker). In other words, a program should do *exactly* what it is supposed to do: it should not do too little (being vulnerable) or too much (helping an attacker).

Hence an important question is: how can we obtain trust in computer programs designed for security sensitive tasks? The historical development of the answers to this question forms the topic of this chapter.

Cryptography will not play an important role in this chapter. The focus is on the programs that carry out security sensitive operations, either with or without cryptography. The cryptography itself is seen as a black box.

22.1.1 Correctness and security

Somewhat surprisingly, correctness and security turn out to be different phenomena.

The correctness of a computer program is not an absolute notion, and only makes sense with respect to a specification [5]. Specifications typically describe desirable program properties such as "an account's balance is never less than n" where n is

some given limit. A program is called correct with respect to a particular specification if it satisfies the properties formulated in the specification. Specifications are ideally described in a formal language, as starting point for a formal correctness proof (see Section 22.3.2).

Security of programs is not an absolute notion either. A program which constitutes a security risk in one environment may be harmless in another. Security goals like, for example, the confidentiality or integrity of certain data that are handled by a program, are commonly formulated as *security policies*, which regulate the access to assets that are controlled by a program. Such policies are usually stated informally, for instance by stating which behaviour is not allowed. There are currently no well-established logics for formulating and proving security policies for concrete programs.[1]

It may be argued that security is an aspect of correctness, involving special program properties, and this view will become more prominent in the future. Currently, security properties and the way of establishing them seem sufficiently different from correctness to regard them as belonging to separate albeit related fields. For instance, Bell–LaPadula style requirements [6] about levels of confidentiality are usually not considered to be correctness requirements.

22.1.2 Leaks and attacks

In practice, correctness comes before security: it is unlikely that an incorrect program is secure.

For instance, most programming languages allow one to define a so-called array of n items, usually indexed from zero onward. It is obviously an error to read or to write nonexistent items with an index n or above, or less than zero. Indices will in general have to be computed at run-time. For each array access the programmer should therefore either *prove* that it is safe, or insert code to check that it is. An implementation might automatically insert such checks, but such checks slow down programs considerably. Newer programming languages, such as Java and C#, do insert such run-time checks.

---

[1] There is however work on security of protocols (e.g. [12]) at a higher level of abstraction.

Many programs suffer from an infamous instance of this flaw: a parameter of type 'string' (i.e., an array of characters) is copied into a local buffer without checking the string's length, which enables an attacker to pass an overlong string containing malicious code which will overwrite existing code. This is usually called a *buffer overflow* attack [2] which has been and continues to be one of the most frequently occurring methods used to break security [39]. Just as an illustration, for the application ssh, which has been specifically designed to provide security, CERT has given five security alerts in the period 2001–2003 (see http://www.cert.org/advisories), and all but one of these alerts concern buffer overflows.

Malicious programs typically exploit such flaws in operating systems or application software. Program correctness is therefore of vital importance for computer security. However, a program is not guaranteed to be invulnerable if it is free of bugs. Many programs do *more* than they are supposed to do. They may either

1. actively compromise the confidentiality or integrity of information, or
2. passively leak (access to) information, making them vulnerable to attacks.

Thus programs may be insecure in an aggressive sense or in a defensive sense, where the defensive insecurity seems to be the primary concern. For ideally, a well-defended program is invulnerable, and if all software would be absolutely invulnerable, malice would become futile.

One might conclude that software should just also be correct in the sense that it does *nothing* what it is *not* supposed to do, for instance leaking (access to) information. However, such negative requirements are usually difficult to specify and prove, and it is also difficult to know when the collection of requirements is complete.

22.2 COMPUTER SECURITY

Software security became a *serious* issue only when the Internet started to get widely used, around 1990. In other words, software security became a serious issue with *widely distributed*

processing. The problem however emerged as early as 1965 with the advent of multiprogramming. In the intervening period, software developers were primarily concerned with getting their programs free of bugs, leaving the security issues to the operating system.

22.2.1 The early days

Until around 1965, program correctness was not really an issue, let alone software security. The first programs were developed in a benign environment, where all programmers and users were trusted. Programs were developed and run in isolation with only the developer or an *operator* sitting in front of the computer's console. Results were printed on paper or stored on punched cards or magnetic tapes for later use in other programs. There was hardly any other interaction between programs.

It was seemingly considered inevitable that programs contained flaws, even if they were used to control hazardous chemical or nuclear plants or to process financial transactions. Nevertheless *computers* were not considered a security risk, not even with respect to the doom-scenario of an accidental world-wide nuclear war. Stanley Kubrick's film *Dr. Strangelove* (1964) illustrates this beautifully. John Badham's somewhat less reputed film *War Games* (1983) was one of the first to reach a broad audience and make explicit that a supposedly highly secure computer system could be broken into, albeit accidentally.

The idea was that even professionals make errors, and that it was their duty to make sure that these errors would happen as few as possible. Programmers were supposed to be able to verify that their programs were absolutely free of faults by critically inspecting their code and by thoroughly debugging and comprehensively testing it.

Moreover, the notion that the tools used to develop and run programs should be reliable as well was equally underdeveloped. In the very beginning, even the hardware would not always operate correctly for a longer period of time. Normally, the programmers would easily notice this, and just try again, for instance after some cooling off period. Also, errors in compilers were usually quite obvious, and corrected.

Precisely because programs inevitably contained flaws, protection was already a feature in many of the early computers. For instance, the IBM System/360, developed around 1960, had a well-designed protection facility, comprising a processor state bit, privileged instructions, and hardware interrupts, among other things. It could prevent programs from accessing certain parts of memory and from performing certain 'privileged' actions such as input or output. These safeguards could be used to enforce the regulation of the access to files at any level (`read`, `write`, `execute`, etc.).

Since hardware checks cannot easily be tampered with, a cleverly designed and implemented operating system could protect itself against buggy or malicious user programs and in principle reach a very high level of security. Hardware mechanisms are still the ultimate foundation of computer security techniques.

22.2.2 Early incidents

Even in those days there were reports of fraud where programmers included backdoors in their programs, for instance by letting them transfer tiny amounts of money from random bank accounts to one particular account. Computer personnel may have tried to make the computer pay them a higher salary. Programmers may have tried to obtain sensitive information by inspecting the data left behind in primary memory[2] or secondary memory such as hard disks.[3]

Such incidents were not really considered computer security risks but just 'normal' fraud, albeit with modern means.

One intriguing fact is the development, as early as August 1961, of the program DARWIN, "A Game of Survival and (Hopefully) Evolution", on an IBM 7090 at Bell Labs [60]. In this game, apparently conceived by Vyssotsky, programmed by McIlroy, and 'defeated' by Morris, self-replicating computer programs had to try to destroy each other in order to become the sole survivor. In other words, the game was unmistakeably based on worm- and virus-like behaviour. However, the game was created just for fun and was played for only a couple of weeks and subsequently lost when the IBM 7090 retired.

Vyssotsky, Morris and McIlroy casually refer to a contestant program to be written for DARWIN as a *virus* in a letter to the editor [60] sent in response to [1]. This might be the first time a malicious program was called a virus, contrary to hacker's lore (as mentioned in Section 22.2.5). DARWIN and other legendary programs also inspired A.K. Dewdney to create the game CORE-WAR [22] which is still played by *aficionados* today.

Curiously enough, Robert Morris Sr., one of the authors of [60], was the father of Robert Morris Jr. who wrote and released the Internet Worm in 1988, which was the first to make the general public aware of the perils of the internet. Thus it seems fair to say that it took computer worms and viruses only a generation to become hostile.

There have been more apparently isolated incidents of worm-creating programs. In 1974, halfway between DARWIN and the Internet Worm, John Walker [61], then working at UNIVAC, created a game ANIMAL which rapidly became popular among colleagues and friends. Tired of sending tapes back and forth, he created a little subroutine PERVADE which he attached to ANIMAL. Whenever the game was played, this subroutine would copy or update the game and itself into all directories in the host machine that it could access. Since the game was also played by personnel having privileged access, and users of different installations would regularly exchange tapes, after a few months the game could be found at virtually every installed UNIVAC computer.

22.2.3 Multiprogramming

From 1965 onwards, multiprogramming was rapidly becoming popular. It was a technical solution to the efficiency problem that programs running *stand alone* would leave the expensive computer idle during relatively long periods of input–output activity. The advantages of cooperating and communicating processes were soon recognized and gave rise to many new research issues.

---

[2]The leading technology was *core storage*, using small magnetic rings which retained their state when power was removed. Modern memory technology is usually *volatile*.

[3]The forerunner of hard disks being drum storage.

Computers became more and more powerful and more and more popular. Moreover, the *time sharing* variant of multiprogramming made it possible to let many users and programmers work interactively with the computer at the same time. The next logical step was to create a network by connecting two computers (being interactive users of each other), possibly by a telephone connection. Most of these ideas emerged long before 1965 [43], but they typically started to be common from 1965 onwards.

Multiprogramming (together with timesharing, multiprocessing, and distributed computing) caused an explosive increase in the complexity of software. The problems arising from correctly and efficiently cooperating processes turned out to be quite difficult which was the main reason for proclaiming the *software crisis* [44]. The need to professionalise *software engineering* became apparent.

Apart from the complexity issues, multiprogramming also implied that more than one program was running on the same computer at the same time. In retrospect, this started the computer security problem: with multiprogramming the confidentiality, integrity and availability of one program could be under attack by an arbitrary concurrently running program.[4]

Consequently, there appear to have been two rather separate developments during the 1970s and 1980s. On the one hand, there was the issue of program correctness and software engineering. In this field of research, software security, if it was discussed at all, was clearly considered to be just a software correctness problem. This field, which mostly ignored security issues, is discussed in Section 22.3. One the other hand, the research in (computer) system security focused on measures, methods, tools, and techniques to make computer systems secure. These are discussed in Sections 22.2.4 and 22.2.5.

22.2.4 Operating systems security

Early work by Bingham [8] may be one of the first to address the security *problem* in a multiprogramming environment. Unfortunately we have not

been able to recover the actual text. It seems to focus on data integrity, i.e. protection against accidental damage to data belonging to a concurrently running process, by what is commonly referred to as *memory protection*.

Security was also one of the many important novel issues in the development, in the years 1965–1970, of the influential multiprogramming ('multiplexing') operating system MULTICS.

Dennis and Van Horn [21] proposed *capabilities* as a more general protection mechanism in a multiprogramming environment. Essentially, these capabilities give subjects certain rights to access objects (or *assets*). Many variants of this idea of *access control* have since been proposed and implemented. As an aside, the paper above also proposed the `fork`- and `join`-mechanism well known from UNIX.

In 1970, a task force of the US Department of Defense clearly stated the possible security problems of 'resource sharing' systems [62]. The report can be found, together with other seminal papers and reports, on a website of the Computer Security Resource Center [16]. The report looks surprisingly modern, using many of today's terms: vulnerability, threat, trap-door, dependability, certification. It even already discusses (physical) attacks on networks, and suggests using encryption in 'open environments'.

The title of a book published in 1973 [42], which basically only discusses 'white collar computer crime', indicates that by that time three main aspects were distinguished: security (protection against attackers), accuracy (absence of error) and privacy (protection against sniffers). Unfortunately, since then many different terms have been introduced, partly as a result of changing views of the problem, but also just because one could not agree on the right terminology. We have seen terms like 'dependability', 'recoverability' and 'availability' for protection against attackers, 'integrity' and 'reliability' for absence of error, and 'secrecy' and 'confidentiality' for protection against sniffers. What these terms mean changes over time. Since the late 1980s, the trio Confidentiality, Integrity, Availability seems somewhat stable, perhaps because of its appealing acronym CIA.

---

[4]This was illustrated, perhaps for the first time, by the already mentioned game Darwin.

Around 1974, computer security appears to emerge as a research area of its own. The Computer Security Institute [17] is founded, issuing its first newsletter and organising its first conference. In 1976 several reports and books appear [27; 51; 54] which are still considered valuable by specialists in the field [50] today.

From these titles it is clear that computer *crime* had become a serious issue. The Ribicoff report led to the first legislation on computer crime [53].

Linden [40] gives a good impression of the developments from 1965 to 1975 and of the state of the art. From conversations with Robert Courtney Jr. he reports that for the year 1974 alone *one* source had identified 339 cases of computer-related crime with an *average* loss of $544,000, the median loss being close to the average. Details of these crimes were classified but most cases were fraud committed by employees.

Linden's paper proposes a 'principle of least privilege' and the use of abstract data types and modules which indicates an increasing interest in a more abstract and formal approach. In the ACM Computing Surveys Issue of September 1981, two survey articles illustrate this development [14; 38].

While this line of research continued in the 1980s, focusing on protection in operating systems,[5] the problems of confidentiality, integrity and availability reappeared overnight on a far larger scale by the end of the decade with the advent of the world-wide internet.

22.2.5 Internet security

In the 1970s, as a result of the development of computer chips, the personal computer appeared, initially in many varieties, and became very popular. These computers lacked many of the known security measures discussed in the previous section, such as memory protection or access control, since security was not considered to be an issue on these single-user systems. Personal computers used light-weight, inexpensive and high-volume removable storage media (such as floppy disks)

which intensified the exchange of software. The personal computer became the breeding ground for a new phenomenon, the computer virus.[6]

The development of networks, begun in the 1960s, intensified the exchange of software. Originally, networks were only private, within or between companies and institutions, but after 1980 the general public was given access to *bulletin boards*, the forerunners of internet service providers. As a result, security attacks became more anonymous, cunning and violent.

Viruses spread by attaching themselves to programs which are transported to other computers. Whenever these host programs are executed, the virus temporarily takes control with the current access rights and replicates by attaching a copy of itself to as many programs as possible. Apart from replicating, viruses typically do something annoying or destructive, such as erasing data files, or installing a 'Trojan horse' which may set up a connection with the creator of the virus who then can remotely take over the computer.

In 1988, Robert Morris Jr. released the infamous Internet Worm [33]. Unlike viruses, worms do not require a host program or file to exist or replicate. In itself this release was not a novel event. However, perhaps for the first time, it employed the vastness and the interconnectedness of the internet to spread very quickly and cause major inconveniences in many companies, universities, hospitals, etc. This Morris Worm was particularly important because it was the first event of its kind to generate a lot of publicity in the media and thus to make the general public aware of the computer security problem.

The Internet Worm resulted in a classic 'denial of service' attack. Nevertheless the world-wide spread of worms and viruses shifted the focus of computer security research. Good protection mechanisms are still important, but worms show that they can often be easily circumvented by using flaws in programs. As the internet connects virtually every computer to every other computer, program correctness becomes an urgent security issue.

---

[5]The fact that computer security remained an area of interest for specialists only is for instance illustrated by the observation that the 1983 printing of a popular textbook [41] still mentions the problem of computer security only very marginally.

[6]The term 'computer virus' was allegedly (re-)coined in 1983, by Len Adleman.

Ever more people communicate over the internet, often without even knowing it. They are ever longer on-line and ever less skilled. This situation is very attractive to all sorts of criminals, from unruly kids to vandals and thieves. The internet has become a dangerous place. Worms, viruses and Trojan horses are relatively new phenomena, but most threats are just old tricks by new means, albeit on a much larger scale. Guessing passwords is almost routine. It is easy to obtain illegal copies of copyrighted material. Communications can be tapped or even intercepted and modified, known as a man in the middle attack. Counter-measures include biometrics, information hiding, encryption and safe communication protocols.

Confidentiality, Integrity and Availability are still important security issues. However, the perils of the internet have emphasised the importance of authorisation and authenticity.

However, it seems obvious that one aspect of integrity, namely program correctness, towers over every other aspect. The majority of security breaches employ programming flaws. The next sections discuss how programming research has tackled this problem and to what extent it has helped to solve the computer security problem.

22.3 PROGRAM CORRECTNESS

The software crisis gave birth to the software engineering discipline. The principal issue of software engineering is program correctness, by managerial or technical means. A software system should do what it is supposed to do according to its requirements or specification. However, it should also do so with reasonable efficiency and even have other desirable properties such as usability, portability, and extendability. Moreover, last but not least, it should be delivered on time and at the estimated cost. A famous account of why this was (and still is) hard was given by Frederick P. Brooks [11].

A wide range of methods is employed to get programs more or less correct. With a posteriori methods, one first writes a program and then shows its correctness, either in an informal way, e.g. by *testing* it, or in a formal way, by *verifying* or *proving* it. Alternatively, one may try to write programs in

such a way that they are a priori correct. Here, too, there is a spectrum from informal *structured programming*, via *paradigmatic programming*, to formal *program derivation* methods.

22.3.1 Testing

Traditionally, programmers *test* their programs in order to establish that they do what they are supposed to do. They run them on test data and verify that the outcome is as expected. While being developed, programs usually run into all sorts of errors: syntactic errors, typing errors, run-time errors,[7] and erroneous results. Often, programmers find themselves caught in endless debugging sessions.

For many programmers the large amount of all these errors is most embarrassing, as is the time it can take to find out what caused them. However, as a result of this, when (at last!) all tests show the expected outcomes, the average programmer will subsequently have a lot of confidence in his or her program.

However, as Dijkstra appropriately remarked, testing may reveal the presence, but not the absence of errors [26].

Some errors, and often the more nasty ones, will only show up under rare circumstances.

- After years, some user may come along who feeds the program an unusual, maybe unintended, combination of inputs. Deeply nested conditional constructs like

```
if (...) {
  ...
} else if (...) {
  ...
} else {
  ...
  if (...) ...
  ...
}
```

may be in operation for years before a particular branch is executed for the first time.

---

[7]An example being incorrectly computed references to memory fields. Such errors are usually very difficult to detect.

- A *pointer* variable intended to contain references to memory locations normally holds the distinguished value `NIL` to indicate that it is not pointing to anything yet. Forgetting this is a common flaw. Since `NIL` is usually represented by zero, and unassigned memory normally contains zeroes, this bug may go undetected for years, until the program is run on a computer where unassigned memory contains, for example, arbitrary values, in which case the pointer almost certainly points to a 'forbidden' location.
- Normally the program communicates with peripheral hardware or other programs in more or less the same way until for some reason there is a unexpected delay.
- A program uses a very large buffer and does therefore not check for overflowing of this buffer, for example, for reasons of efficiency, but which some day nevertheless happens.

Although in recent years methods have been developed for more systematic testing, where in particular extreme boundary situations are taken into consideration, testing as it is done in practice is still insufficient for gaining trust, especially with respect to modern complex systems.

22.3.2 Verification

In 1967, Robert W. Floyd proposed to tag programs (actually, flow charts) with *assertions*, boolean expressions over the variables used in the program [29]. By writing appropriate assertions between any two successive statements, as well as before the first statement (the *precondition*) and after the last statement (the *postcondition*), a programmer could establish that the program variables *invariantly* satisfy the assertions during any execution of the program.

This approach allows a programmer to prove the statement "if the variables satisfy the precondition before execution, they satisfy the postcondition after execution". If one can find assertions such that this statement is equivalent to the specification of the program, one has proven the program to be partially[8] correct.

---

[8]Partially because one does not know yet whether the program will ever terminate.

With the related idea of 'runtime assertions' one actually checks the assertions during program execution to the effect that the program finishes with an error message whenever an assertion does not hold. However, this approach is a weak variant: (a) checking *all* assertions would slow down the program considerably, (b) if an assertion does not hold, the error has already occurred and the damage may have been done, and (c) in many cases users will not be happy with programs which just give up – they want programs to do their job. However, strategically and sparingly placed runtime assertions may be helpful as emergency exits in order to avoid tedious proofs of 'obviously correct' code.

Floyd's work opened the field of formal program verification. In order to prove that the assertions are maintained, one needs to know exactly what each computational unit, such as language statement, machine instruction, system call, does. This requires a detailed account in a precise language, preferably mathematics, of the semantics of the language, the properties of all processors involved, the behaviour of system calls. Several essentially equivalent styles of semantics were developed, for example

- operational semantics [30], where the meaning of a language element is defined as the effect it will have on a given (abstract) computer,
- axiomatic semantics [34], where that meaning is defined by the element's postcondition, given a precondition (or the other way around) and the meaning of its constituent elements, if any, and
- denotational semantics [57], where it is defined as a mathematical function, its 'denotation', for example mapping states to states of some abstract mathematical domain.

One might say that operational semantics was *en vogue* until Hoare formalised Floyd's assertions.

Axiomatic semantics is usually formalised as *Hoare logic* with axioms giving a pre- and postcondition relation for each individual language construct and with inference rules for composite constructs. This means that a standard deductive proof method can be used for formal program verification.

Unfortunately, even moderate-sized programs have specifications that are so complicated that just finding the intermediate assertions is a major task, let alone proving them correct. If ever, this is only possible with the use of automatic (interactive) proof tools like PVS [49], Coq [7] and Isabelle [46], which, thanks also to increasingly faster computer hardware, only now, in the first decade of the 21st century, begins to become useful in practice. For proving full-blown software packages correct, let alone secure, there is still a long way to go. However, the first tools capable of verifying real programs, albeit for programs of limited size and relatively simple properties, do exist now, for instance ESC/Java [28].

Since Floyd's ideas the theory of formal program correctness has been well developed while that of formal program security is still in its infancy. Meanwhile, one needed to find other ways to obtain trust in computer programs.

22.3.3 Structured programming

Around 1968 the computer community, that is manufacturers, programmers, scientists, began to realise that delivering well-behaving software according to plan was far more difficult than was thought. Computer scientists, then, began to emphasise the need to write good programs.

E.W. Dijkstra is the principal architect of this discipline [23]. To begin with, he advocated simplicity and rigour in the development of a program. He argued successfully that *jumps* were only needed for repetitions (loops) and choice (conditionals), as well as subroutine calls, in larger programs, and should be banned as independent language constructs [24]. This observation led to the ever since popular *structured programming* [19] paradigm: use as few language constructs as possible and use them in a structured, nested, modular way.

Although structured programming vastly improved the quality of the products of nearly all programmers, it was not the panacea sought for. One can still disguise an incomprehensible mess as a structured program and, consequently, still introduce any imaginable error. In particular, one can still write programs which are functionally incorrect, have other (un)intended effects, and/or are vulnerable to attacks by other programs.

22.3.4 Formal program derivation

Around 1974, Dijkstra developed the *weakest precondition* technique [25] for deriving programs from a given postcondition by successive applications of rules reflecting semantic properties of the building blocks of the programming language. This technique went immediately back to Floyd's and Hoare's work.

Another descendant of these developments was *transformational programming* [13], where one develops a program 'algebraically' from the specification by applying transformation rules which preserve correctness and are themselves proven correct or axiomatic. Well-known transformation rules are *unfolding* (of a function call) and *folding* (the reverse).

These methods for program design are normally used to arrive at classical 'imperative' programs. If all goes well, the resulting program does indeed obey its specification. However, this still does not rule out the possibility that the program does *more* than its specification requires, nor does it guarantee that the specification is strong enough, for instance to exclude leaks.

The program may have side effects which are not covered by the specification. The side effects may be unintentional but destructive to cooperating processes, or intentional and possibly malicious. For instance, a program may correctly and harmlessly compute the nth prime number and at the same time try to peek into another program's data.

It is quite hard to rule out such superfluous behaviour. The program may need a large number of auxiliary variables (temporary memory space) which one doesn't want to mention in the specification. Any of these variables might cause side effects.

In general, a given specification will have many implementations and most if not all of them will be much more detailed. It is the task of the tools which process the implementation (see Section 22.3.6) to make sure that these details are inaccessible to anyone else, including programmers, users, or system administrators.

22.3.5 Programming paradigms

The structured programming approach is applicable to any programming language, even machine language. A further step towards correctness and security was the popularisation, in the early 1970s, of special programming languages which were, and still are, advocated to facilitate, or ideally to enforce, writing good programs.

With object-oriented languages [4; 20; 35] one decomposes a program into modules, specifies these modules and then proves each module using the specifications of the other modules. Here mutual dependency requires induction. The approach is data-oriented, hence the name: a program is designed as a collection of interacting objects. Each module is a class of such objects. Object-oriented languages aim at encapsulating (i.e. hiding) implementation details, such as temporary memory space, in these classes.

In functional programming [9; 43; 59], the use of memory is made implicit or invisible. A program is constructed as a hierarchy of elements which have a pure input–output behaviour.[9]

Logic languages [3; 37] are similar in this sense. However, the elements are now predicates, and the result of a logic program is any combination of instances of logic variables which satisfy some 'root' predicate.

All these languages make it easier to avoid errors. Many errors are less easier to introduce, for instance, those caused by type-inconsistencies such as applying string-operations to integers. Nevertheless, it is still possible to make any conceivable error. After all, the restrictions the languages impose do not affect their expressive power. Thus, special programming paradigms will at best make it harder to introduce errors.

The concept of proof-carrying code [45] was an important step in recognising the potential of program verification as a means of establishing security. The idea here is not only to use program verification to enforce security properties but also to provide evidence – a proof – for this that can be independently verified, providing a way to establish trust in program code that comes from a potentially untrusted supplier.

---

[9]In other words, the elements are functions, hence the name.

22.3.6 The role of language processors

Programs written in some source language (which actually means *any* language except pure machine language) are processed before and while they are run. They are compiled, statically linked together, enriched with library or API routines, loaded into memory, and run in some environment – an operating system, in general. Some parts are linked to programs while they are running. Traditionally, programs are run directly by a hardware processor, but they may as well be run, albeit usually slower, by a software processor, such as an interpreter or *virtual machine*.

Each of these processors represents an opportunity both for enforcing *and* for breaking security. On the one hand, a compiler may insert malicious code into each program it compiles [58]. On the other hand, most compilers perform various kinds of static checks, i.e. on the basis of the program text, thus without executing it, such as type consistency. The compiler may also try to check statically certain semantic conditions, such as the out-of-boundary conditions for arrays (buffers) mentioned earlier, or insert dynamic checks (run-time assertions). Some of these checks may be delegated to linkers and loaders.

As has been discussed in Section 22.2.1, hardware processors routinely perform various checks. Some processors may detect that a program tries to compute unsound results such as an integer which is too large to be represented. Other processors support the integrity of the operating system and that of concurrently running processes. In principle, these hardware mechanisms may enforce a very high level of protection and security. Although originally designed to enable an operating system to protect itself against good-natured but faulty user programs, there is no reason to doubt that these mechanisms could equally well protect any program from other, malicious, programs. In practice, however, for reasons of efficiency, the mechanisms are not always sufficiently powerful or, as a result of programming errors, not employed to their full potential.

Similarly, virtual machines can check various sorts of complicated conditions, albeit at an efficiency penalty. As an example, the Java 'sandbox' provides a run-time environment where a program's outbound accesses are severely restricted. The Java class loader and the Java virtual machine perform many checks during run-time, especially with respect to overflow exceptions and pointer handling.[10] These checks provides a form of memory protection which is more fine-grained and flexible than the memory protection provided by operating systems discussed in Section 22.2.4. The protection is realised exclusively by means of software, whereas operating system memory protection ultimately relies on hardware.

Sun Microsystems, the developers of Java, have also developed the (sub)language and platform *JavaCard* for use in multi-application smart cards. Since the security of smart cards is paramount and smart card applications, also called 'applets', are relatively small, as is the platform itself, JavaCard is a good candidate to be completely verified.

22.4 PROGRAMMING TRANSPARENCY

A responsible programmer will use a good programming language, carefully design programs, continually inspect source code, and reason about it (or even prove many parts of it correct), choose trusted language processors, and extensively test the resulting executables.

However, even in a good language one can write faulty programs. Making good designs is very hard. It is very difficult and time-consuming to prove source code correct, while testing can offer only limited guarantees. Testing is particularly weak in discovering unintended malicious functionality or vulnerabilities, while such properties are also very difficult, if at all possible, to completely describe formally. Code inspection may be more successful at detecting such errors, especially when it is done by other people.

Altogether the software crisis seems still unaverted, despite enormous advances in design methods, languages, paradigms, implementation techniques, tools for testing and verification, and

---

[10]Nevertheless, in an early version of Java (JDK 1.1.1), its security manager, one of its program processors, was discovered to accidentally give its clients write (instead of only read) access to an object containing vital security data [10].

in understanding the issues. The explosive growth of computer capacity, operating systems, and applications made the situation even worse.

22.4.1 Auditing

It may be very helpful to explain one's code to someone else, or to ask someone to read and criticise it. Many software houses let their programmers inspect each other's code.

Code inspection may be greatly facilitated by explicitly formulating as many assertions as possible. One should at least *try* to formulate that the program will not respond to possible attacks and does not leak valuable information through parameters or results, global variables, communicated messages, input/output, or leave such information behind in parts of memory such as buffers or the system stack.

Inspection of another programmer's code can nevertheless be very time-consuming. Therefore, simpler *auditing* methods are often used. One can for instance make a list of checks which can be performed relatively easily and use the outcomes of the checks to estimate the quality of the code.

The checks may vary a lot. One can examine the style of design, coding, and documentation, observe the behaviour of the program-in-execution, perform certain standard tests which look for well-known pitfalls, consider how the program is used in an organisation, set up particular attacks, look for certain patterns in the code, etc.

Many institutions have formulated such criteria for evaluating the quality of developed or commissioned software, in particular with respect to security, before putting it into actual use. A well-known example is the Orange Book [47] of the United States Department of Defense. Several organisations have participated in developing the so-called Common Criteria [15], the international ISO standard for the evaluation of the security of IT products. The Common Criteria distinguishes certifications at several level, with seven being the highest. Like the Orange Book, the Common Criteria insists on the use of formal methods for the highest level of certification.

In such auditing systems either the purchaser of the software, or an independent certified, i.e.

trusted by purchaser *and* vendor, evaluating institution acts as the 'someone else' who inspects the code among other things. However one could also invite the general public to critically examine the code.

Programmers have always exchanged (sources of) programs they wrote for themselves and which they chose not to commercialise. When bulletin boards became popular around 1980, many programmers put their programs in the public domain by uploading them to these bulletin boards, often including their code code.

In order to profit from the benefits of code inspection on a larger scale programmers should publish their source code, and this is the core of the philosophy of the *Open Source Initiative* [48]. The basic idea is that if *anyone* can inspect the source code, then any remaining errors are more easily and quickly discovered.

22.4.2   An early example

The renowned computer scientist Donald E. Knuth was one of the first to apply this principle to a complex piece of software. Around 1980 he wrote a comprehensive typesetting system[11] which he called TEX. He decided to make TEX publicly available, together with the source. He challenged the users of TEX to discover any errors in it, under the restriction that only he and nobody else was allowed to change the source.

However, Knuth explicitly recognized that others might want to make money by selling readily installable distributions of TEX, for different platforms and printers as long as it would maintain full compatibility and not restrict the use of TEX in any way. Knuth did not want to force other people to make such software freely available as well.

22.4.3   Free software

Slightly later, Richard M. Stallman [56; 63] went a step further. For more than ten years, being a system developer at MIT, he had been part of a community of programmers who were used to sharing

the programs they wrote, freely adapting, extending, and improving them. As a user of such programs, one could participate in finding the cause of obscure flaws, correct them, replace algorithms by better ones, add new features, and so on.

A user of 'proprietary' software cannot even begin to look for the cause of errors, and under Knuth's restrictions one cannot actively help to develop the program further. A program sharing community tries to reach Knuth's implicit goal by always obeying the rule that adapting an existing shared program never leads to a *restriction* in features or quality.

In 1984, when the working environment at MIT changed, Stallman decided to try to create an independent program sharing community. He coined the term 'free software' as an analogue to 'free speech' to emphasise the independence of the community and the fact that the software it developed was not bound by legislation, copyright, patents, or licenses.

Stallman took personal consequences. He left his job and tried to earn his living by teaching, selling ready-to-go distributions, merchandising, and from donations.

He also tried to use only 'free' software whenever and as soon as it was available, for example TEX. He realised that he should first develop free software for programmers, that is an operating system (GNU), an editor (GNU Emacs), and a compiler (gcc). Together, these programs became the GNU project [31].

Based on his views, which were and still are tightly connected to his ethical, political and societal views, he created the 'Free Software Foundation' in 1985. The principles of the FSF are laid down in its 'General Public License'.

The principal idea of the GPL is that any software distributed under this license may freely be copied or modified, and redistributed with the same rights. In particular, one is not allowed to turn GPL software into proprietary software.

Nowadays, a lot of software, especially from academics, is developed and distributed under this GPL. Stallman's software sharing community is now a world-wide society. There are even signs that software companies are beginning to recognise the benefits of freely developed software.

---

[11]Because he found the currently available typesetting programs inadequate for typesetting his series of books *The Art of Computer Programming* [36], in particular with respect to mathematic formulae.

22.4.4 Open software

The choice of the term 'free' turned out to be somewhat unfortunate since people started to interpret it as just 'free of charge'. But free software in the sense of the FSF is not necessarily free of charge, rather free for anyone to copy, modify or distribute under the conditions of the GPL. In order to avoid this confusion in terms, most people now refer to *free software* as *open software* [48]. However, note that TEX is open but not free. Also, according to the FSF, large amounts of open software are distributed under weaker licenses than GPL, leading to proprietary software being referred to as open, and open software being patented and thus made proprietary. It is not the point of this chapter to take sides in this debate, but the term 'open' will be used in this chapter for software which can be inspected by anyone without restriction.

Many open source projects have evolved. One of the better known ones is Linux, which during the 1990s grew out from a self-chosen practical exercise of computer science student Linus Torvaldsi, who still plays the central role in the development of Linux. This development heavily uses the internet, allowing fast updates despite the large number of participants. Eric Raymond, one of the principals of the Open Source Movement, compared this way of working with a bazaar [52].

From the perspective of software security one may, perhaps somewhat disrespectfully, consider open software as the ultimate form of code inspection. Indeed for some of the most heavily used open programs such as TEX and gcc, this may have resulted in complex programs which are now virtually secure, that is correct, trustworthy and invulnerable.

22.4.5 Public software

In a sense, the free and open software movement advocates a way of working which is everyday practice at research institutes all over the world. The essence of scientific research is that a scientist makes his or her findings as public as possible. An obvious reason to do so is to let anybody know, but it is even more important to provoke comments, criticism, or even refutation. Other people may freely use the published results to find better or newer results.

This open and free method of research has proven so successful that modern society routinely supports it financially through universities, science foundations, and specialised research institutes. Nearly all fundamental research is financed from public funds.

Obviously, one cannot compare a twenty page scientific article with a one hundred thousand line computer program; while it is reasonable to expect that a significant number of scientists read that entire article, no one will read that entire program. But if we would restrict ourselves to the software which has become fundamental to society (admittedly as a result of private enterprise), it might well be wise to follow the track of the free and open software movement.

Governments (as well as many non-governmental and non-profit organisations like privately financed universities) might take the lead by (a) using open software on all its computer platforms whenever there is a choice, and (b) explicitly financially supporting the development of fundamental software in a world-wide open competition akin to how the international scientific community works. It is a disquieting thought that more and more societal issues are essentially dependent on buggy software which is developed only for profit.

To a large extent, computer science already works this way. It has produced many results which are open to criticism, free to apply, and ready to provoke new ideas: algorithms, methods, designs, computer architectures and programming languages, among other ideas. But it is the source code, the actual implementation which is not (yet) subject to such public scrutiny.[12]

Of course, private companies may still want to develop their own specialised software, just as some private companies do their own confidential research and development. However, fundamental software should be developed in the open, especially when security is involved.

---

[12]That is, not as such. Many participants of the open source movement already work for government and/or academia.

22.4.6 Fundamental software

We think that operating systems and internet software are fundamental, as well as the software to develop them, such as compilers and libraries, as was already recognized by the Free Software Foundation. Also many applications on which society depends (software for voting, financial software, to name only two) are fundamental. Note that hardware is an entirely different matter. Hardware bugs are an exception, while software bugs are the rule, and malicious hardware is a highly unlikely phenomenon.

Another metaphor may illustrate some of this. Governments are responsible for the quality of a country's infrastructure, such as roads, railways, the power network, and buildings. Depending on the economical and political insights of a country, part of the infrastructure is built and/or exploited by private companies, part by public companies, and part by public companies using a private company model, but all this takes place under strict, essentially 'open', supervision by the government. This was not always the case, for instance with the railroads. In contrast, cars, similar to applications, are built by private companies, and only marginally examined, for safety.

This comparison may also contribute to the debate on software patents and intellectual property which has emerged in recent years. We think that the arguments of the supporters of software patents are valid for software that is not 'fundamental' in the above sense and therefore belongs to the market of proprietary software. However we agree with opponents that fundamental software should not be patented and should remain free to be used within the limits of the law, just as the results of fundamental, publicly financed research are available to the public.

22.4.7 Discussion

It may seem that the open source movement is willing to present an empty goal to potential attackers, which would be true if existing closed software would be opened overnight. However, it is very likely that programmers would work much more carefully if they know that their software will be open from the start. Moreover, when software is successfully attacked it usually gets disclosed anyway.

Open software enables code inspection, but there is no *guarantee* that it will ever be inspected in an appropriate way [18; 64]. Usually, source code is very large, contains a surprising amount of details, and is written in a hard to read language. Nevertheless, if a user runs into a problem, he or she can immediately look into the code and try to solve the problem, ask a friend to do so, or else a certified evaluator of his or her own choice. With proprietary software one can only report the error to the manufacturer, or the retailer, and then wait and see.

Free software may be modified freely, which means also by badly trained programmers. However, if the open source community is large enough, the software is incrementally developed and constantly evaluated. If the community also has a minimum of organisation, bad effects will disappear. The same principle is used successfully in a different domain by the open internet encyclopedia Wikipedia. Linux is the prime example in the field of software.

22.5 CONCLUSION

Software security involves the question of how to remove or avoid vulnerabilities from software in order to avoid attacks by malicious software such as viruses, worms, or backdoors. Such vulnerabilities are the result of bugs, which is a euphemism for design or implementation errors. Bugs may also cause software to become unintentionally malicious.

During the 1960s the computer community became aware of the seriousness of buggy software. Two lines of research emerged, which developed into separate fields, which were only superficially aware of each other for many decades. One field, the field of program correctness discussed in Section 22.3, focused on techniques to get rid of bugs by either a priori avoiding them or a posteriori removing them, by formal or informal means. Here security was not explicitly considered, at least not until the advent of Proof Carrying Code [45]. The other field focused on theories, protocols, algorithms, techniques, tools, etc. for making systems secure. Broadly speaking, the first field concentrated on implementation, the second on design.

Around 1990, with the explosive growth of the internet, worms, viruses and Trojan horses began to have a world-wide impact. As a result, society at large became aware of the seriousness of the computer security problem. However, the solution is still far away. Operating system developers may pay ever more attention to Confidentiality, Integrity and Availability; many software companies may develop virus killers, firewalls, intrusion detectors; researchers world wide may investigate security protocols and cryptography, the threat of viruses, spies or intruders has not diminished. Formal methods become ever more powerful and popular but it is still not even clear how the notion of security should be formalised.

In our opinion, no single approach will suffice. The threats should be counter-attacked on many fronts. New security measures and protocols must be developed; the engineering of secure software must be strongly encouraged; the formalisation of security must be investigated; and formal program verification must be enhanced. But at the same time, fundamental software should be open, for everyone to inspect, test, or verify. Programmers of such open software should develop the habit of expressing their intentions not only by providing documentation and test cases but also by decorating their programs with provable assertions and using certified library software as much as possible.

Perhaps some day software will normally be open and freely developed in publicly financed but independent institutions in the style of academic research institutes. This might lead to an ever growing library of modules which are recognised as absolutely correct against a given specification, trustworthy and invulnerable and which are assessed by publicly assigned independent authorities.[13] Any program is then correct, trustworthy, and invulnerable if it is so in itself and provably only uses modules from this library.

---

[13]Comparable with aviation authorities, for instance.

ACKNOWLEDGEMENTS

We thank Karl de Leeuw for his continuing support and encouragement and Donn B. Parker for his input.

REFERENCES

[1] $\aleph_0$. *Computer Recreations 'Darwin'*, Software – Practice and Experience 1 (2) (1971), 93–96.

[2] $\aleph_1$, *Smashing the stack for fun and profit*, Phrack 7 (1996).

[3] K.R. Apt, V.W. Marek, M. Truszczynski, D.S. Warren, (eds), *The Logic Programming Paradigm: A 25-Year Perspective*, Springer-Verlag, New York (1999).

[4] K. Arnold, J. Gosling, *The Java Programming Language*, 2nd edition, Addison-Wesley Publishing Co. (1998).

[5] J.W. de Bakker, *Mathematical Theory of Program Correctness*, Prentice-Hall, Inc., London (1980).

[6] D.E. Bell, L.J. LaPadula, Secure computer systems: Unified exposition and multics interpretation, Technical Report NMTR-1997, The Mitre Corporation, March (1976). ESD-TR-75-306.

[7] Y. Bertot, P. Castéran, *Coq'Art: The Calculus of Inductive Constructions*, Volume XXV, Texts in Theoretical Computer Science. An EATCS Series, Springer-Verlag, New York (2004).

[8] H.W. Bingham, Security Techniques for EDP of Multilevel Classified Information. Technical Report RADC-TR-65-415, Rome Air Development Center, Griffiss Air Force Base, New York, December (1965). Available at http://hbingham.com/technical/radctr.htm.

[9] R. Bird, *Introduction to Functional Programming using Haskell*, 2nd edition, Prentice Hall Press (1998).

[10] B. Bokowski, J. Vitek, *Confined types*, *Object-Oriented Programming, Systems, Languages and Applications (OOPSLA)*, Volume 34, Denver, Colorado, ACM Press, New York (1999), pp. 82–96.

[11] F.P. Brooks, *The Mythical Man-Month*, Addison-Wesley, Reading, MA (1995).

[12] M. Burrows, M. Abadi, R. Needham, *A logic of authentication*, ACM Transactions on Computer Systems 8 (1) (1990), 18–36.

[13] R.M. Burstall, J. Darlington, *A Transformation System for Developing Recursive Programs*, Journal of the ACM 24(1) (1977), 44–67.

[14] M.H. Cheheyl, M. Gasser, G.A. Huff, J.K. Millen, *Verifying Security*, ACM Computing Surveys 13 (3) (1981), 279–339.

[15] See http://csrc.nist.gov/cc/.

[16] See http://csrc.nist.gov/publications/history/.

[17] See http://www.gocsi.com/.

[18] C. Cowan, *Software security for open-source systems*, IEEE Security & Privacy 1 (1) (2003), 38–45.

[19] O.-J. Dahl, E.W. Dijkstra, C.A.R. Hoare, **Structured Programming**, Academic Press, London, New York (1972).

[20] O.-J. Dahl, K. Nygaard, *SIMULA: an ALGOL-based simulation language*, Communications of the ACM 9(9) (1966), 671–678.

[21] J.B. Dennis, E.C. Van Horn, *Programming semantics for multiprogrammed computations*, Communications of the ACM 9(3) (1966), 143–155.

[22] A.K. Dewdney, *In the game called Core War hostile programs engage in a battle of bits*, Computer Recreations, Scientific American (1984), 14–22.

[23] See http://www.cs.utexas.edu/users/EWD/ewd02xx/EWD249.PDF.

[24] E.W. Dijkstra, *Go to statement considered harmful*, Communications of the ACM 11 (3) (1968), 147–148.

[25] E.W. Dijkstra, **A Discipline of Programming**, Prentice Hall, New York (1976).

[26] E.W. Dijkstra, *The humble programmer*, Communications of the ACM 15 (10) (1972), 859–866.

[27] **Encyclopedia of Computer Science**, Litton Educational Publishing, Inc. (1976).

[28] C. Flanagan, K.R.M. Leino, M. Lillibridge, G. Nelson, J.B. Saxe, R. Stata, *Extended static checking for Java*, **ACM SIGPLAN 2002 Conference on Programming Language Design and Implementation (PLDI'2002)**, (2002), pp. 234–245.

[29] R.W. Floyd, *Assigning meanings to programs*, **Mathematical Aspects of Computer Science, Proceedings of Symposia in Applied Mathematics 19**, J.T. Schwartz, ed., American Mathematical Society, Providence, RI (1967), pp. 19–32.

[30] J.V. Garwick, *The definition of programming languages by their compilers*, **Formal Language Description Languages for Computer Programming**, North-Holland, Amsterdam (1966), pp. 139–147.

[31] See http://www.gnu.org.

[32] D. Gollmann, **Computer Security**, 2nd edition, John Wiley & Sons, New York (2006).

[33] K. Hafner, J. Markoff, **Cyberpunk: Outlaws and Hackers on the Computer Frontier**, Simon & Schuster, New York (1991).

[34] C.A.R. Hoare, *An axiomatic basis for computer programming*, Comm. ACM 12 (10) (1969), 576–580, 583.

[35] D.H.H. Ingalls, *The Smalltalk-76 programming system design and implementation*, **Proceedings of the 5th ACM SIGACT-SIGPLAN Symposium on Principles of Programming Languages**, ACM Press (1978), pp. 9–16.

[36] D.E. Knuth, **The Art of Computer Programming**, Vols 1–3, Addison-Wesley Publishing Company, 1997–1998. Available at http://www-cs-faculty.stanford.edu/~knuth/taocp.html.

[37] R. Kowalski, *Algorithm = logic + control*, Communications of the ACM 22 (7) (1979), 424–436.

[38] C.E. Landwehr, *Formal models for computer security*, ACM Computing Surveys 13 (3) (1981), 247–278.

[39] K.-S. Lhee, S.J. Chapin, *Buffer overflow and format string overflow vulnerabilities*, Softw. Pract. Exper. 33(5) (2003), 423–460.

[40] T.A. Linden, *Operating system structures to support security and reliable software*, ACM Computing Surveys 8 (4) (1976), 409–445.

[41] S.E. Madnick, J.J. Donovan, **Operating Systems**, McGraw-Hill, Inc., New York (1974).

[42] J. Martin, **Security, Accuracy, and Privacy in Computer Systems**, Prentice Hall, London (1973).

[43] J. McCarthy, P.W. Abrahams, D.J. Edwards, T.P. Hart, M.I. Levin, **The LISP 1.5 Programmers' Manual**, MIT Press, Cambridge, MA (1962).

[44] P. Naur, B. Randell (eds), **Software Engineering, Report on a conference sponsored by the NATO Science Committee**, NATO Scientific Affairs Division, Brussels, Belgium (1969).

[45] G.C. Necula, *Proof-carrying code*, **Proceedings of the 24th ACM SIGPLAN-SIGACT Symposium on Principles of Programming Languages (POPL'97)**, Paris (1997), pp. 106–119.

[46] T. Nipkow, L.C. Paulson, M. Wenzel, **Isabelle/HOL – A Proof Assistant for Higher-Order Logic**, LNCS, Vol. 2283, Springer, Berlin (2002).

[47] Department of Defense, Trusted Computer Systen Evaluation Criteria, Technical Report 5200.28-STD, DoD, December (1885).

[48] See http://www.opensource.org.

[49] S. Owre, J.M. Rushby, N. Shankar, *PVS: A Prototype Verification System*, **11th International Conference on Automated Deduction (CADE)**, Saratoga, NY, D. Kapur, ed., Lecture Notes in Artificial Intelligence, Vol. 607, Springer-Verlag (1992), pp. 748–752, http://www.csl.sri.com/papers/cade92-pvs/.

[50] R. Johnston via D. Parker, personal communication.

[51] D.B. Parker, **Crime by Computer**, Charles Scribner's Sons, New York (1976).

[52] E.S. Raymond, **The Cathedral and the Bazaar: Musings on Linux and Open Source by an Accidental Revolutionary**, O'Reilly, New York (2001).

[53] See http://www.cybercrimelaw.net/tekster/background.html.

[54] US Senator Abe Ribicoff, Problems Associated with Computer Technology in Federal Programs and Private Industry, Prepared by the Committee on Government Operations, Technical report, United States Senate, (1976).

[55] B. Schneier, **Secrets and Lies**, John Wiley & Sons Inc., New York (2000).

[56] R. Stallman, *Why software should be free*, **Computers, Ethics, and Social Values**, D.G. Johnson, H. Nissenbaum, eds, Chapter 3, Prentice Hall (1995), pp. 190–200.

[57] R.D. Tennent, *The denotational semantics of programming languages*, Communications of the ACM 19 (8) (1976), 437–453.

[58] K. Thompson, *Reflections on trusting trust*, Communications of the ACM 27 (8) (1984), 761–763.

[59] D.A. Turner, *A new implementation technique for applicative languages*. Software – Practice and Experience 9 (1) (1979), 31–49.

[60] V.A. Vyssotsky, R. Morris Sr., M.D. McIlroy, *Letter to the Editor*, Software – Practice and Experience 2 (1) (1972), 91–96, Transcript available at http://www.cs.dartmouth.edu/~doug/darwin.pdf.

[61] See http://www.fourmilab.ch/documents/univac/animal.html.

[62] W.H. Ware, Security Controls for Computer Systems (U): Report of the Defense Science Board Task Force on Computer Security, Technical report, The RAND Corporation, Santa Monica, CA, February (1970).

[63] S. Williams, *Free as in Freedom. Richard Stallman's Crusade for Free Software*, O'Reilly, London (2002).

[64] B. Witten, C. Landwehr, M. Caloyannides, *Does open source improve system security?*, IEEE Software, September–October (2001), 57–61.

23

IT SECURITY AND IT AUDITING BETWEEN 1960 AND 2000

Margaret van Biene-Hershey

Vrije University, Amsterdam, The Netherlands

Contents

Abstract

This chapter presents the development of the IT security and the IT audit function from the time when information technologies started to have an impact on the administration of large corporations up to the start of the 21st century. Starting in the late 50s and taking time frames of about 10 years the reader is given an impression of the significant information systems and information technologies of that period and the developments in IT security and IT audit. Each timeframe is given a period name to emphasis the impact IT was having on business: Electronic Data Processing (60s), Automation (70s), Integration and Diversification (80s) and Contagion (90s). The evolution from the late 1950s through to the start of the second millennium is described for IT and IS, IT security and IT auditing.

Keywords: audit, automation period, contagion period, controls, effectiveness, efficiency, electronic data processing period, enterprise, expectation gap, information systems, information technologies, IT environment, integration and diversification period, management of IS and IT, security, security measures.

23.1 INTRODUCTION

This article presents the development of the IT security and the IT audit function. The foundations for this development are found in the growth of and changes in the use of Information Technologies (IT) for business purposes. In timeframes of about 10 years an impression is given of the attitudes and challenges for IT, IT security and IT auditing. Each timeframe is given a period name to emphasis the impact IT was having on business: Electronic Data Processing (60s), Automation (70s), Integration and Diversification (80s) and Contagion (90s). The evolution from the late 1950s through to the start of the second millennium is described for IT, IT security and IT auditing. The chosen characterization of IT is limited to a perspective relevant for the changes that occur in the IT security and IT audit functions. This implies that the automation of factories does not receive attention because the main emphasis on IT security and IT auditing has been primarily directed towards information systems for bookkeeping, marketing and general management information purposes along with the supporting IT. This treatment of the history of information technologies does not start with Babbage in the middle of the 19th century, but in the middle of the 20th century when IT auditing begins. IT security starts to receive attention in the late 1960s and early 1970s.

Because it takes about ten years for the repercussions from new IT to be truly visible in enterprises, most techniques are only discussed in the decade following their introduction onto the market. A survey of the literature will show publications about IT developments 10 years or more in advance of the moment when the use of the developments has any real impact on business, government or society in general.

It is easy to loose oneself in the buzzwords and the changing definitions of terms for information systems and information technologies. The terminology of the decade is referred to but the reader is expected to understand what was meant at that time without extensive definitions.

The term enterprise is used to represent any kind of large organization (government, profit or non-profit). Large organizations are where IT security and IT audit are most prevalent. These organizations are the initiators of the developments that occur in IT security and IT audit. The term business unit will be used to refer to a specific part of an enterprise with an autonomous responsibility for a set of goals of the enterprise.

IT security and IT auditing are in different stages of development in the same timeframe but their development is intertwined.

Security cannot be seen as a pure staff function completely structured within one unique organizational unit. But to attain a measure of security, action must be taken and investments made to protect the specific assets that are threatened. Security in an organization is everybody's concern and can not be adequately attained without close attention to making the whole organization security minded.

The important advisory role of security specialists is well recognized. IT security is, irrespective of the manner in which the responsibilities for IT security are structured, always part of a more general responsibility for the total security of an enterprise.

The definition of computer security according to NIST is:

The protection afforded to an automated information system in order to attain the applicable objectives of preserving the integrity, availability and confidentiality of information system resources (includes hardware, software, firmware, information/data and telecommunications) [25].

This definition came into common use in the 1990s.

IT auditing is, exclusively, a staff-function within an enterprise or it is a function that is carried out by independent auditing firms. When IT auditing is a staff-function within an enterprise, it is usually part of the general internal audit function. External IT auditors originally were only found working in accounting firms. Today many companies functioning in an IT advisory capacity also offer IT audit services. Auditing functions should not be confused with internal control functions. The auditor audits the (internal) control and the security measures of the enterprise.

IT auditing is a function that must not only form an opinion on the adequacy of the IT related control measures but also gives an opinion on the adequacy of the IT security in relation to the enterprise.

The definition of IT audit for this article is deemed to be: an independent and impartial assessment leading to a clear and concise opinion on the adequacy of part or all of the information systems and the related information technologies in development and in use in the enterprise [5].

These are the definitions of IT security and IT auditing used from the 1980s onward.

23.1.1 IT security in perspective

Security stems from the need to react to threats or prevent potential threats that could lead to unacceptable consequences. Cave men who felt threatened by their environment took all kinds of measures to help them feel secure in what was perceived to be a hostile environment. A number of essential aspects of security are visible in this example: the concept of a threat, the evaluation of the perceived threat (vulnerability) and the necessity to take action upon detection, and foremost the necessity to institute adequate measures to limit the risk. Security is definitely a primary human requirement.

Even IT security recognizes environmental threats as one of the types of threats that IT security must adequately deal with. Obviously there are many generic threats: natural, intentional and unintentional. These generic threats to IT security, as assessed during the growth in the use of IT, shape the role that IT security is expected to fulfil. When and why IT security came into existence is a subject for the sections to come in this article. It suffices in this introduction to say that security is the freedom from risk or danger. It is safety and the assurance of safety.

Unfortunately advancements in technology can and do go beyond the individual's ability to assess these threats and take adequate measures. Hence it becomes a necessity to rely upon IT security specialists. Specifically the weaknesses in information technologies force enterprises to make use of experts trained in IT security.

IT security can require a high degree of expertise that it is not expedient for an enterprise to maintain, for example: for testing the rigor of encryption techniques. This has led to independent firms that specialize in advising on IT security issues. These firms and their specialists are supporting enterprises from the 1970s onward.

An issue for the reader to assess after reading this article is the degree to which the supplier of the technologies should be held more responsible for adequate security measures than is common practise today.

23.1.2 Justification of auditing

Since IT auditing has its roots in the accountancy profession, an understanding of the importance of IT auditing requires looking back quickly at the development of the audit profession.

It all seems to have started in the Roman army. Since the soldiers could not read, they could not be sure that they received their due. So the Romans instituted the function of auditor. He who listened could also read and so he read and listened when the soldiers received their payments. Thus the auditor gave assurance to the soldiers that they were not being cheated.

The Scots in the seventeenth century were making use of professional bookkeepers who were not in their employment to make up financial statements as required for their joint ventures.

Already in the nineteenth century England and Scotland recognized their professional bookkeepers as fulfilling a necessary role in commerce. The first legal recognition of the independent professional bookkeeper was in 1844 with the Joint Stock Companies Act. This document stated that a profit and loss statement that was to be approved by the shareholders required certification by an auditor. The British are credited with the introduction of this concept in the United States and Canada in order to protect their large financial interests there.

The late nineteenth and most of the twentieth century see the development and strengthening of the financial auditing profession and the creation of other types of audits, for example: milieu audits, audits of social regulations, legal audits and IT audits. The reasons for this growth can be found in:

Importance of business for society,
Complexity of the financial interrelationships among businesses,
Business globalization,
Bureaucracy,
Government regulation of business,
Use of technologies in the broadest sense of the word [18].

This all brings back visions of the poor Roman soldier who couldn't read. The crux lies in the necessity to trust someone else, with adequate expertise, to assure you that things are being done as intended. Although this statement implies inherent shortcomings in the human condition, it also points out the ability of the individual to recognize these shortcomings and act upon them.

IT auditing came into existence in the early 1960s. This is when enterprises really begin to take advantage of the available technology and the first computer centres come into existence. Accountants are confronted for the first time with important financial information that is no longer being maintained on paper but digitally in the computer centres. EDP auditors, as they are called in those days, are needed to assist the accountant in obtaining the evidence needed to approve financial statements made by the enterprise. The accountants can read but they can't read digitalized data as such.

An auditor advises on actions to be taken to reduce the risks implied in their findings. An auditor does not initiate nor carry out the actions to alleviate the findings. An auditor must maintain an independent and impartial position in relation to the control and security measures of the enterprise.

23.2 THE ELECTRONIC DATA PROCESSING PERIOD (THE SIXTIES)

This period is called the EDP period because; during the 1950s and most of the 1960s the use of computers are introduced to take over simple processes done by people and do them much faster and more accurately. No major changes are made in the structure of the work processes in the organization. Computers allow enterprises to do a lot more work with the same number of people, and in some cases even with fewer people. Particularly in financial institutions, the computer is a major contributor to an improved efficiency at the operational level of the enterprise.

When things go wrong it is said to be the computer's fault. This gives business unit personnel some frustration but it was handy to have something else to blame mistakes on.

Management is very happy with the improved efficiency and integrity brought to the business by the use of computers.

23.2.1 Character of IT during the EDP period

The first computers being used in enterprises are perceived to be very expensive and their use was limited. Core memory, the memory in which the actual tasks and data are stored that are required during processing, is expensive and a determining factor for the computer infrastructure. Improvements in the memory technology occur quickly, allowing for faster and more reliable processing. The replacement of the vacuum tube with the transistor and printed circuit technology speed up computers dramatically.

The computer begins to take over the registration of the actual bookkeeping in enterprises.

23.2.2 IT environment during the EDP period

Initially applications are written in assembler languages which allow the use of mnemonics instead of machine code instructions when programming [36]. Application programs read a punch card (this entailed programming the successful reading of a punch card and the transport of that card through the reader); they perform the commands needed to process the data, produce a punched card (with the same sort of specific coding as required for reading a punch card), and finally write a line of data on the printer (again all the code to achieve this was in-line in the application program). Where magnetic tape units are available they are used to access data from one tape and write data to a new tape on a different tape unit. This process is repeated until all punch cards have been read into the computer, processed and the output produced. Initially, there are no buffering techniques because of the limited availability of memory and for that same reason; there are no operating systems to manage multitasking. This is the way the first generation computers worked [11; 21].

The second generation of computers is marked by the replacement of vacuum tube technology by transistors which are much faster and less error prone. This new technology is also cheaper than the first generation technology. The computer suppliers develop Input/Output Supervisors (IOS) that are resident in memory and make life a lot easier for the application programmers. Nothing much changes in the structure of the programs except that the programmer could call upon subroutines from the I/O supervisor to perform the input and output functions and the I/O supervisor does some error control reporting back to the application as to whether or not the command has been performed successfully. It becomes more common to retain data on magnetic tape; this makes the computer programs faster and eliminates most punch card output. The computers still run one program at a time and the application usually simply waits for the I/O to be performed. Already in the early 1960s the so-called higher level programming languages like COBOL (Common Business Oriented Language) and FORTRAN (FORmula TRANslator) are available.

In 1964, IBM announces what was branded as the third generation computer. It makes use of the 360 Operating System. This computer is boldly announced as a general purpose computer. The new operating system incorporates a PCP (Primary Control Program) with the I/O supervisor. The PCP is quickly replaced by the real third generation of operating systems (and computer infrastructure). MFT (Multiple Fixed Tasks), and then MVS (Multiple Variable Storage) making it possible to run many tasks 'at the same time' on the same computer [27].

Buffering techniques and this multitasking greatly improve the speed of the job runs (batch processing). The word 'fixed' versus the word 'variable' indicates the first evolution in the sophistication of the memory management allowing for a more efficient use of the available memory and the central processing unit. The first multiprocessing computers are introduced and attempts with some success are made with paging algorithms. Memory paging becomes at least theoretically possible.

Towards the end of the 1960s terminals are introduced chiefly to replace the punch card data input.

The first disk drives make it possible to store data that can perform updates in place and make data constantly available (on-line) to automated processes. Tape handling, which slowed down the processing throughput time of applications, is being eliminated.

IBM was the major manufacturer of IT for data processing purposes. This position was maintained for about three decades [17].

23.2.3 IT applications during the EDP period

After the introduction of multitasking, IT applications are divided into input processing, functional processing and output processing.

An explicit control on the completeness and correctness of the processing is finally done by the business unit responsible for the process. In this batch world, it means that the control is based on error detection and resulted in the business unit carrying out correction–booking procedures or even demanding reruns of entire batches.

Application programming moves away from the assembler programming language to the use of COBOL [12]. Programs can be developed more quickly than with the assembler languages. However there is some contention as to whether or not programming languages can produce runtime code that is as efficient as assembler. The business units receive their first operational support of the business processes through data processing.

23.2.4 IT operations during the EDP period

The IT processing is batch oriented. The IPO as it is called (Input–Process–Output) is organized into jobs that carry out the three activities consecutively in what are called job runs. In other words volumes of input are entered into the computer, then the required functions are carried out against the data and the results stored in files and for printing these results. The printing is all done by the so called print spooling program that prints any data placed on the spool file. Master files are updated by the application and saved as starting point for the next run of that application. This does result in minimal changes in the operational procedures in the business unit. Input must still be prepared, delivered to the computer centre to be punched onto punch cards and serve as input for an application. The application processes the input and produces paper output that is distributed to the predefined destinations in the enterprise and, after the necessary controls, output is also sent to clients. An important continuity measure involves making copies of the master files from disks to magnetic tapes. Sometimes the input data and the output data as printed are also copied to tapes. These tapes are put in safes at sites some distance from the computer centre. All these activities are computer centre responsibilities to administer. The IPO batch structures cause some changes to the operational procedures in the business units.

The computer centre usually checks the completeness of the processing before printing and sending the paper output to a business unit. The computer centre is also responsible for corrective action whenever the process presents an error code to the master console operator. Often a programmer is called in to correct application code. This usually meant a rerun of the entire batch. Batch runs could take many hours to complete and even days.

The main concern of the computer centre manager is about the continuity of processing and timely error recovery within the computer centre (availability issues). Tape backup procedures are devised and carried out by the computer centre personnel. Usually the continuity concept is based on 24 hour cycles, weekly cycles or even longer ones and generations of backup data sets.

By the end of the EDP period digitalized data is being exchanged between enterprises. Magnetic tapes are the chief medium used.

The documentation delivered with the computer installation about the computer and operating system, linkage editor and compilers are extensive. Revisions to the documentation were sent at least on a monthly basis. At that time there were claims that IBM was a major publisher printing more pages and distributing more documents worldwide per year than many publishers of books.

23.2.5 IT security in the EDP period

The continuity (actually the availability of the data along with the back up and recovery procedures for the data) of the computer installation is a major issue for computer centre management. The physical security of the tape safe on location at the computer centre and the procedures to get into that safe are soon considered important. Sometimes the risk caused by insufficient procedures is only recognized after some programmer has gone into the tape safe and taken a tape for use in some extensive testing of a new IT application. Preferably the programmer has taken one with some unique master file on it.

Physical security of safes and procedures to assure an adequate registration of the tape content are not typically considered to be IT security issues. Initially the only real IT security issue is the maintenance of alternate power supplies and this is an issue that the computer manufacturer gives advice on.

Environmental issues are of course very important for large computer installations. Physical security of the computer room is a topic but this does not require any IT expertise. The same is true for the security measures to ensure only authorized access to buildings, special rooms within buildings and the physical protection of tapes. Advice is obtained from external security specialists on these subjects.

So there is no real specialized IT security within the enterprise – yet. General security specialists are employed or called upon as advisers to advise and implement the necessary general security measures to protect the IT investments.

23.2.6 IT auditing in the EDP period

IT auditing comes into existence because the public accountant, forming an opinion on the completeness and correctness of the financial data, is confronted with master files on magnetic tape containing accounts and client information. Of course, the organization has copies of these files on paper. However the integrity of the processing and of the representation on paper is completely dependent upon the integrity of the master files and on top of that the hardcopies represent too much paper to work through. Accountants start employing programmers to write audit-software programs (now generally called CAATs: Computer Aided Audit Techniques) so that they can form an opinion on the integrity of the master files themselves.

Already in the early 1960s many accountants begin to recognize the necessity of understanding a lot more about computers and in particular computer centre organizations. Doubts arise about the quality of the digitalized data being produced by IT applications. After all if the data in the computer can not be trusted, even the audit-software programs may not reveal the errors in the master files.

The accountant had come to the conclusion that the computer centre organization, and the computers, presented inherent risks for the integrity of data. First, in the punch card era, there were machines for punching new cards in the computer room. Computers are expensive and are therefore used day and night. The console operators and other personnel can easily generate new unauthorized input for the computer. Tapes are, from an auditor's point of view, a pleasant medium. It is

rather difficult to do a rewrite (update in place) on magnetic tape. Hash totals across all, or important parts of the data, helped to ensure the integrity of the data. Control totals across the tape can be used to assess the completeness and correctness of some financial data. Unfortunately, data can be added to an existing tape and the control totals recalculated.

The financial auditor was not overly uncomfortable about the IT applications because the most important controls on the integrity of the processes were still carried out in the business units. A financial auditor with a feel for EDP could easily advise the business on the kinds of controls that needed to be implemented in the automated processes as well as in the business units.

When disks arrive in the computer centre they, by comparison, make much greater demands on the measures to ensure the integrity of the computer processing than tape or punch cards had. Updates in place are the big advantage but this update function can be carried out indiscriminately leading to a loss of data integrity.

Disk drives also complicate the backup and recovery procedures and the long term retention of data. The EDP auditor is called in to give advice to the financial auditor on the safeguards that need to be instituted in computer centres to prevent unauthorized updates to data on disks.

Disks are permanently available on the computer to all computer personnel with access to the computer and to all automated processes running on the computer. Continuity of data on disks is also a bit different from tapes. One disk can have many different datasets and, on the other hand, some datasets were soon so big that they couldn't be copied to one tape. When backups of disk drives are made by copying the whole disk drive, it is found to be very time consuming and expensive not just because of the time needed to make the backup but also because it is very complicated and time consuming to recover a specific data set that is necessary for a rerun of an IT application. Copying significant data sets to tape was the most common solution.

Once multitasking has come to fruition, IT auditing of computer centres becomes important, at least in large financial institutions.

IT auditing is primarily an audit of the computer centre's organizational procedures for planning and carrying out the production processes, the master console operations, the procedures for the punch card typists, the specific procedures for backup and recovery of data and processes, the long term retention of data, the tape registration, tape procedures, supplies of tapes and the physical security of the computer centre environment.

After rules begin to be applied for entrance to the computer room itself, these procedures are of course also audited.

Segregation of duties between the development of financial IT applications and the computer centre organizations are a principle issue for the financial auditors. It must be remembered that it was the convention to organize the technical system programming people within the computer centre organization. The poor troubleshooting procedures are major concerns. During troubleshooting it is surprisingly easy to make unchecked changes to data because detective controls are often insufficient and no stringent rules are followed for the performance of trouble shooting. The stringent controls carried out by the business units to assure the integrity of the processing does to some extent alleviate this threat. Formal rules and procedures for allowing system programmers access to the computers when production data is on-line are needed. Troubleshooting procedures needed to be instituted so that the troubleshooting is controllable and sometimes, for specific modules even auditable. Explicit automated controls across the data maintained on disk are necessary but not always implemented.

Some concern on the part of the financial auditor about the integrity of computer programs is justified. Due to the frequency with which programs in production fail and need repair (troubleshooting), there is an on-going integrity issue. Where the integrity of code is in doubt or simply needs verification, EDP auditors are asked to perform code inspections. Whether code inspection or extensive testing is most effective for assessing the integrity of code was a subject of much discussion [14]. The use of assembler language, especially on computers with variable length instruction codes, makes inspecting code next to impossible.

Assessing the measures taken to assure the continued proper segregation of duties becomes part of the computer centre audit procedure. However, the EDP auditor's most common activity is still the writing audit software applications for the financial auditor.

It must be mentioned that there are nearly no interpretive programming languages in use for developing large financial systems (COBOL is the chief programming language) and this gives considerable comfort to those trying to assure themselves that programs can only be changed by the people responsible for development and maintenance. This statement is only true if there are no COBOL-compilers available on the computer performing production processing and no unauthorized changes can be introduced into the production computer environment.

Another nice thing about COBOL was the clear separation of data definitions from the processing statements. This fact made the examination of COBOL programs comparatively easy.

The clear separation of data and processes in the most commonly used operating system infrastructure and programming language is another plus that the later generations of operating system architects neglected to appreciate. (Microsoft's operating systems are the most persistent example of this effect.)

It needs to be mentioned that the continuity issues, that the auditor assesses, relate mostly to the continued operations of the computer installations within one physical computer centre. In particular computer centres are expected to deliver output on time. "On time" means within the agreed upon production cycle.

The audits in this period are mainly audits after the fact to ascertain the degree to which the financial auditor can rely upon the electronic data that it is complete, correct and timely. Documentation of information systems and computer centre procedures are considered important but are usually assessed to be lacking.

In this period, large financial institutions start employing IT auditors.

The roots for ISACA (Information Systems Audit and Control Association) are in the late 1960s. The first name for the organization was EDP Auditors Association. It was later changed to

ISACA. ISACA becomes the only global association for professional information systems auditors today [19].

23.3 AUTOMATION PERIOD (THE SEVENTIES)

The end of the 1960s and the beginning of the 1970s see the initial use of technologies that lay the basis for the further use of IT in the twentieth century. These technologies had real consequences for the way the enterprise organized its operational processes, allowing not only for more *effective* operational processes in the enterprise, but eventually also for automated systems that support the management processes. These new applications are called information systems instead of applications. The use of the word *effective* instead of *efficient*, as discussed in the EDP period shows a conscious shift in management's expectations as to what the goals in this period are for the newly developed information systems.

Management is not as happy about the effectiveness of the IT investments as such as they were about the efficiency thereof in the EDP period. The business units do not receive their new IT applications as well as they had the initial IT support. Debit to this situation are the major changes that occur in the organization of new business processes and the changes to the way the business units are expected to work with these new systems. Personnel's own lack of self-confidence also leads to a distrust of the new systems.

The computer centres are getting bigger and bigger. The management of the CC organizations are primarily concerned with stabilizing their production processes, eliminating errors in and increasing the availability of their computer installations.

IBM gives a lot of attention to improving the continuity of their computers; while computer centre management concerns itself with improving the effectiveness of the back up, recovery and contingency measures.

Automation is understood to imply that major changes occur in the way the organization carries out its business processes in order to take advantage of the new IT. The business processes are reorganized to take better advantage of the available technologies.

There are new information technologies available that make it all possible.

23.3.1 Character of IT during the automation period

The reference to discernable new generations of computer technologies was short lived and ends in the EDP period. Already in the automation period the advances in hardware innovations are no longer as important for marketing IT as they were in the EDP Period. Data bases and networking are the motors behind the improved use of computers in the enterprise.

The development of tailor made information systems, chiefly using COBOL, become major projects. These projects often take more than five years to complete.

23.3.2 IT environment during the automation period

In the early 1970s IBM is seen taking over the marketing and support of IMS from the developers, North American Rockwell. IMS is an hierarchical data base management system and a data communications management system. IDMS, Adabas and Total are more examples of prominent Data Base Management Systems (DBMS). Data directory dictionaries and warehousing are being discussed in the literature and implementations attempted. Ad hoc management information is being generated through the use of RPGs (report program generators). In the EDP Period, the IBM computer's operating system recognized when it was accessing data and when it was accessing machine code to be run as computer programs. The introduction of data bases further extends the possibilities to control the use of data. There are three important areas of improvement for the information system architectures. First the concept of physical data as structured in the actual data base and logical data schemas used by the applications make it possible to organize and re-organize the data in the data bases without requiring changes to the existing processing programs that use the data (except if the lengths of existing fields are changed). This phenomenon is called data independence. The second improvement, or opportunity, is the use of the same data by many information systems. This makes it possible to write programs that access data hither to only available to one process at a time and to integrate that data into reports across the business unit boundaries. Thirdly access control of the use of data is implemented into the data base management system. This is a very important preventive control measure. It limits the use of data to those authorized to access the data. Given the extensive use of data on disks that are permanently on-line, this access control facility carried out by the DBMS is a necessary improvement.

The integrity of data receives a much broader significance. Integrity of data can no longer be managed exclusively within a business unit. Data consistency is required across the boundaries of the business units. Data base administrators and data base managers with enterprise wide responsibilities become part of the organization.

Many enterprises get themselves into trouble making changes to the way the operating system functions. The time and costs of this tailoring for the new releases become prohibitive. IBM responds to this problem by introducing its OS/VS operating system that requires a minimum of tailoring for use in the enterprise. System programmers (who tailor the operating system) are sparse and expensive. This new design objective for the operating systems promises to alleviate this situation.

The introduction by IBM of its SNA (Systems Network Architecture) is the real starting point for the use of telecommunications to support the business processes. IMS supports the use of SNA too. Other DBMSs do not. Integrating telecommunications and data base management in one package makes the access control on the use of data in the newly developed systems easier. The systems are called real time on-line systems but most of them only support some degree of real-time updating. The batch capabilities of IMS are more commonly used than most people are willing to admit.

In 1970 at the Spring Joint Computer Conference, Bell and Church [1] present their architecture that is to become the basis for the TCP/IP protocol. The PMS and ISP were made available in the ARPA network that was funded by the American

Department of Defence (DoD). This ARPA network evolves into what is now called the Internet.

IBM introduces RACF (Resource Allocation Control Facility) as an add-on to their MVS (Multiple Virtual Systems). RACF uses resource access profiles when requesting initial access to the resource to assure that the actual use of those resources is conform the intent as designated in the access profiles. OS/VS too is an improvement over MVS (Multiple Virtual Storage). Virtual memory proved possible (the initial paging algorithm problems are solved). The old physical limitations on the amount of memory available against the size of an individual program are surmounted.

Digital Equipment Corporation's (DEC) PDP has made its entrance into the enterprise. At that time in large enterprises their computers are referred to as mini computers.

23.3.3 IT applications during the automation period

There is a movement away from the term EDP to the use of IT when referring to the information technologies being used to support the business units.

The 1970s see the professional application system development becoming very important. Methods and techniques are developed to guide the developer through the development process. These System Development Methods eventually define all the procedures needed from the incipient strategic decision making through the development, introduction of applications into production, with the ensuing replacement of old applications, and the maintenance of the operational applications. The SDM is usually accompanied by a set of techniques to assist the organization in performing the many tasks such as documenting, interviewing and decision making. At least partly due to the embracement of SDMs, business units are introduced to the necessity of actively participating in the system development process in order to obtain systems that do what business considers to be important [29]. Even strategic management starts to concern itself with projects representing major investments for the enterprise. There are a lot of project failures. Projects are considered to be failures whenever:

the IT applications never go into production, the project does not achieve the implementation of key functions defined as necessary by the business unit, or the project's development costs become so exorbitant that the IT application can not be considered cost effective.

Many new programming languages are introduced but COBOL remains dominant.

The restructuring of business processes from batch oriented input, processing, output systems into transaction oriented systems is expensive, time consuming and more complicated than the IT-professionals initially expected it to be. Business consultants are hired and the accountant as advisor on matters of IT looses ground to experts specialized in the implementation of these new business processes.

The investment in data communication technologies in order to build networks is used almost exclusively for internal networks. There are only limited connections to other institutions.

Small businesses take advantage of computers. The 'other' computer companies, as they are called, are able to prosper in this market. The use of different computer infrastructures, of data base management systems and of data communication protocols begins to have an effect on the thinking as to which techniques best fit the business requirements. The decision making process with respect to IT becomes more complicated than in the EDP period.

Although Information management is a buzzword, it is not really achieved in the automation period.

23.3.4 IT operations during the automation period

The Computer Centre organization (CC) is the data processing centre in the enterprise. The CC goes through some major changes. Digitalizing data is no longer done within the scope of the CC organization. The preparation of digitalized input for the computer is restructured in the business units. The result of this reorganization is to make the business unit responsible for the integrity of the input and to relieve the CC organization of any direct responsibility for the correctness of the input.

On-line processing changes the thinking of the CC management on continuity issues. Now availability must include the availability of processes and not just of data to the business units. Response times become an issue when processing transactions. Check-pointing, back-outs and restarting are new processes added to the automated applications that a CC must manage. Procedures for backup and recovery after calamities and for the long term retention of data, which is specifically required by law, are issues requiring management decisions. Initially the management of the computer centre is expected to make all the right decisions to ensure the continuity of the processing. In the CC with many different transaction and data base oriented systems, contingency planning becomes very complicated.

The introduction of terminals and display units is the starting point for discussions on the scope of responsibility of the CC organization. The expectations of the business unit and the perception of the CC organization are not always well aligned.

Developing computer networks is a new issue that the IT organization is not yet equipped to manage. Much time is spent trying to prove the viability of investments in computer networks and of course in defining the kind of personnel necessary to build and manage them.

IT consultants have become a very common phenomenon in the IT organization.

23.3.5 IT security in the automation period

In May of 1972 IBM announces four studies for the purpose of gaining more information about data security and identifying user requirements for this vital subject. It is important to note that at least one manufacturer of IT (IBM) recognized its responsibility to advise and to support secure IT.

The three external study sites were: Massachusetts Institute of Technology, TRW Systems, Inc., and the Management Information Division of the State of Illinois. The internal site was IBM's Federal Systems Division. The findings were published in six volumes [15]. This initiative played an important part in the thinking about data security throughout this period. Although the term was data security, the studies did address more than just data security. In particular generic issues that needed to be addressed in defining a secure operating system were defined. A secure operating system is a sine qua non for data security.

Data security with all its ramifications for the enterprise soon became a major subject of discussion. Most security consultants rightly considered introducing data security into large organizations a project that required the involvement of top management and many years to complete. There was a lot of counselling of management on this issue. The project approach for the achievement of a secure IT environment is illustrated in the IBM publications Data Security Design Handbook written in Sweden [16]. This type of document is a guide for the: analysis of deficiencies in and threats to computer operations, education on data security, risk assessment and choice of measures, planning and control and finally the design of data protection measures. Most of these guides include, of course, an executive summary.

Data security had become a very real issue. Thanks to data bases there is some concern about the ownership of data. After all the data was all being stored in these large data bases and the data was being used by many different business units. Who could be held responsible for the data security?

Data security gets management's attention. The implementation of access control using an access control facility available from the market is the major task that leads to the big security projects of this period.

Secure at this time is generally taken to be synonymous with data security. Data security is defined as the results achieved by means of data protection which, in turn, is described as measures taken against undesirable events, whether intentional or unintentional, that cause data to be modified, destroyed, or disclosed [16].

Given the limited network domains within the enterprise and the inter-company networking, contingency planning for the network was limited to alternate routing of network connections to the computer centre (and sometimes many computer centres).

The very large financial institutions were already being confronted with insurmountable problems when trying to find a cost effective solution for the creation of alternate sites. On-line transaction processing was too voluminous to allow for alternate site recovery procedures for an entire computer centre. Smaller businesses were able to solve this problem because their automated support was less complex and one alternate site could serve many businesses. Channel speeds are the most important factor that limits how much data can be backed up for recovery purposes cost effectively. The technology for the big installed computer bases was just not available. The only alternative was to select important jobs from a business point of view and provide an alternate site for their processing. If the computer centre needed to make use of an alternate site, it could take more than 24 hours to update the databases at the alternate sites with the data that was not at the alternate site at the time of the catastrophe. Some information system architectures did allow for regional processing at different sites. This did make the takeover process easier. The investments in extra hardware were much higher because all sites required enough hardware to manage the total peak processing if a calamity at one of the sites should occur.

Of course strikes involving all enterprise CCs is and was management's real nightmare because no amount of IT can give the enterprise the required protection.

23.3.6 IT auditing in the automation period

The increased involvement of the business unit in the development of systems lays bare a lack of understanding of information technologies within the business units. The clove that is recognized to exist between business units and the IT development units widens beyond that of the former period to include disappointments with the new or improved IT applications, the response times of transaction based systems and ultimately even the quality of the communication from the CC organization concerning matters of IT support.

Top management becomes less secure about whether or not the investment decisions in IT are the right ones.

As seen with the early Romans, the Scots in the seventeenth century and in the 1960s with the financial auditors, management now looks to an independent and impartial expert to help fill in the blanks. The EDP auditor has another sponsor for audits in addition to the financial auditor: the management of the enterprise. Management is concerned with the continuity and integrity of all the data and processes of the enterprise. IT is now automating a lot more than just the financial data. Therefore, in the 1970s, the scope of the IT audit is broadened immensely.

The character of IT had changed drastically by the end of the automation period. At the business unit level, the use of on-line transaction processing makes processing of the business activities faster and more natural for the business unit to perform than it had been. At the technical level, the IT processing has seen its greatest change. The data base management systems are performing preventive controls on the integrity of data instead of the input processing applications of the EDP period. Some of the measures to ensure the authorized use of data are also part of the DBMS, others part of the access control facility and finally access control measures are also necessary in the IT application. Two kinds of output come out of the systems: the direct feedback on the individual transaction and the paper output still required by most organizations for control purposes. The business unit controls are now for ascertaining that the controls in the automated process work properly, and not for directly controlling the correctness of the individual transactions. New control processes are designed into the systems to give the business unit the information it still requires to explicitly ascertain that the automated business processes function properly. For example, when financial transactions are being processed in real time, two different control processes are considered necessary. One to ascertain that each processed transaction has been processed completely and correctly. The second process is to ascertain that all transactions introduced for processing have been processed and that no processing of unauthorized transactions has taken place.

The use of System Development Methods leads to better documentation of information systems for future maintenance, for operational support in the computer centre organization and for the business units. A lot of project documentation is produced during the development phases in order to facilitate management decisions, to control the project's progress and to define tasks necessary for the completion of the project.

The EDP auditor is expected to ascertain the continued working of the proper division of duties among the business units, the development units and the CC organization. This segregation of duties between development and use of IT is a fundamental preventive control measure necessary to prevent unauthorized changes to programs in production under the care of the CC.

A principle change in thinking about effective control measures occurs with the arrival of transaction processing. Preventive measures baked into the automated applications and the supporting technologies are rapidly becoming more important than detection measures. Detection measures are always necessary to ascertain that the preventive measures have worked properly. The main reason for this change in attitude is that some transaction processing creates business vulnerabilities that can not be corrected after the fact without still incurring unacceptable losses. Relying solely on detective measures is no longer an option. This in turn places heavier demands on the integrity of the automated processes, that the preventive measures are adequate and working.

Due to the importance of the preventive controls in automated business processes, it becomes necessary to institute and control a stringent segregation of responsibilities between the automated support of the development processes and the operational production processes. Getting compilers and other programming tools banned from the production process environment requires a lot of effort. CC organizations only capitulate, if ever, after the presence of programming tools in the production environment proves to have led to undesired production processes.

Even today in some UNIX environments, master console operators and other computer room personnel continue to demand the availability to them of functions that should be removed from the scope of their daily access profiles. Allowing them to maintain these extra functions only necessary in special situations, at the very least corrupts the trust that can be placed in the integrity of logging (an important detective control necessary to ascertain the proper working of the preventive controls).

When compilers are available in the production environment the control on unauthorized changes to application processes is totally detective.

Access control facilities are introduced into or on the operating systems to facilitate the definition of different operating environments for processing functions that could otherwise cause integrity threats if run in the same environment, for example the automated support of the development personnel versus the operational production applications. Access control facilities are also particularly useful to limit the functions and data made available to computer centre personnel.

The system programmer (responsible for maintaining the technical systems within the enterprise) is no longer structured within the CC. Technical issues requiring the attention of these specialists are no longer exclusively managed by computer centre organizations. The IT auditor is very happy with the move of these IT specialists out of the CC organization because it strengthens the degree to which the organization can rely upon its preventive controls.

The data input functions are laid in the hands of the business units. Any application development processes that are not as yet sufficiently segregated from the CC are now of necessity placed in business units or defined as independent IT development units serving all the business units of the enterprise.

The shortcomings in the use of IT and the responsibility of the financial auditor to approve the yearly profit and loss statement lead to discussions about auditing through the computer versus auditing around the computer. In a highly automation dependent organization, it becomes very clear that auditing around the computer is not cost effective even if it is theoretically possible. The crux of the issue hinges on the importance of the completeness of the yearly profit and loss statement. Different countries had different opinions as to the importance of this completeness issue. This problem received a great deal of attention after the infamous American Equity Funding fraud became public. The insurance company had an automated system in which bogus life insurance policies were being maintained. It took some years before the accounting firm discovered this fact [22].

The scope of IT auditing is broadened with the audit of the development of business systems and the audit of the business system controls in the operational automated systems. Because the integrity of the automated processes is now also an issue, the EDP auditor starts to participate, as auditor, more actively in the development process. The EDP auditor is expected to assure management that the systems as developed have all the necessary control measures built into them.

The information systems auditor (IS auditor) is born.

The computer centre audit is deepened. The CC audit now includes the way in which the CC manages its relations with the business units, development units, backup and recovery procedures for business units and the operational access control procedures and profiles.

The dependency on preventive control measures in the supporting technologies (DBMS, network protocols and operating systems), the role of access control facilities, the segregation of processing into different processing environments and the use of data communications to support the business processes all lead to the creation of auditors, called technical (EDP) auditors [4]. In the late 1970s, Technical auditors assess the security developments with respect to the access control facilities, effectiveness of the measures planned to protect the processing environments from each other, the rest risks or potential risks still present among different operational processing environments, the implementation of the access control facilities and the projects for the realization of the data communications networks.

The third IT audit specialty is discernable: technical auditing.

The financial auditor is still an important audit sponsor but management had by the end of the

1970s also discovered IT auditing. The internal audit departments doing general auditing also makes use of the IT auditors for CAATs.

Auditors in the 1970s make thankful use of the Computer Control Guidelines [8] published by the Canadian Institute of Chartered Accountants in 1970 and of the Computer Audit Guidelines [9] that followed in 1975.

Systems auditability and control receives a lot of attention. These reports called the SAC Study (Systems Auditability and Control Study) and were published in 1977 [34; 33; 32].

23.4 INTEGRATION AND DIVERSIFICATION PERIOD (THE EIGHTIES)

The 1980s see, thanks to the integration of data in data bases, real information systems take advantage of the integrated data now being managed by the data base management systems. The realization of the much sought after management information system becomes common place [35].

Enterprises are making use of packages (ready made information systems) to speed up the use of IT in the company, to prevent project development disasters and to lower the information system development costs (for example Hogan, an American banking package). From an information technology perspective, there is an enormous diversification in technologies available to the enterprise and there are even more suppliers of these technologies. Time to market becomes the key issue for these suppliers.

Enterprises concern themselves with the globalization of their IT support.

Technologies are getting cheaper and cheaper. Enterprises are making more and more use of information systems and information technologies; the net result is an enormous growth in the IT budgets.

IT security is accepted as a necessary function in the enterprise. Even the Dutch National Bank starts advising all banks, if they have not done it already, to institute an IT security function within their organizations.

23.4.1 Character of IT during the integration and diversification period

System software (OS, DBMS and telecommunications) is no longer the development issue of this period. The realization of the dreams for adequate management information is made possible by the data base technology of the previous period. Telecommunications makes the use of IT support for the business processes more natural and diversified. IT networks are used to integrate the enterprise with the computer centre organizations that supply services to them.

The mainframes get bigger and faster but there are no major changes to the infrastructure at the hardware and operating system level that directly affect the IT applications. Bigger and more integrated hierarchical data bases support most of the IT applications running on mainframes. Even global networks are implemented.

The computer centre organization no longer uses the term IT support but IT services. This change in terminology represents a change in attitude in the CC organization. The big computer centre organizations go through an important transition from system managers focused primarily on issues within the computer centre to managers of a service organization that are focused on delivering services to the business units. The focus moves from managing computer installations, relations with IT suppliers and their own internal organizations to service level management, and delivering cost effective services.

23.4.2 IT environment during the integration and diversification period

The big news is the arrival of the office computer, as it was called in enterprises. It is usually a stand alone computer that is considered absolutely necessary for the business unit because the big mainframe applications are not being built fast enough or are deemed to be too expensive.

Terminals are connected to the office computers which are, in turn, connected to the mainframes. Many terminals are connected directly to the mainframes with only very limited intelligence in concentrators. Unfortunately, there is no real standardization of system software across office computers from different suppliers. UNIX was not a real supplier independent standard.

By the late 1980s, stand alone personal computers as well as local area networks are in use.

Relational data base management systems and SQL are first introduced in the office computer environment. Data bases are maintained at the business unit level with no control over the consistency of data definitions across business units within the enterprise. This data soon is deemed to be a threat to the overall integrity of management information at the corporate level.

Back up and recovery of the office systems is easier to organize than for the mainframes but business units with no expertise or real interest in running office computers are confronted with this and other issues like authorized access to the office computers and the distribution of output from the IT applications. These applications were seen as off line systems when they had no connection to the central mainframes and not taken seriously by the professional IT organization. Soon demands to be connected to the mainframes are being made and the professional IT organization has difficulty refusing these requests.

IBM stops delivering documentation in printed form. System documentation is electronic and the enterprises must themselves print anything needed.

Global enterprises expand their networks to reach all the corners of the earth. The office computers in use in the smaller international business units are connected to the central mainframes at head office as are the mainframes of the larger business units around the world. TCP/IP begins to replace all other network protocols (Datanet/1, SNA, etc). Digital telephone exchanges make their appearance and companies begin to think about integrating voice and data in the company owned networks. The term telecommunications replaces the term data communications.

23.4.3   IT applications during the integration and diversification period

COBOL is loosing importance. Office computers bring their own high level programming language for developing applications. Credit, for example,

was the name of a programming language introduced by Philips for their office computers. Fortunately this attempt to bind enterprises into the use of hardware from a particular manufacturer of computers was short lived. But today Microsoft and Macintosh still do their best to lock customers into their hardware platforms. Diversification is not always cost effective.

More or less standardized nonprocedural languages are being introduced and are replacing the use of COBOL.

Most automated information systems are connected to other information systems either because they pass results of their processing to other information systems for further processing, or because the data maintained in the data base is used to generate ad hoc management reports. Mainframes no longer perform all the processing needed to maintain the integrity of data. Office computers do controls on the input before sending data to the mainframe. They perform parts of the actual bookkeeping procedures for some accounts or even do the complete bookkeeping process within the business unit.

Secretaries are being confronted with word-processing systems. Initially these are stand alone systems.

On-line processing for business applications is becoming commonplace. Real-time applications can be found, for example, in industry where production lines are being automated and in municipalities where it is common to have the traffic lights managed by computers.

The use of telecommunications particularly among banks, their corporate customers and business service support organizations is ubiquitous. The SWIFT network, started in the 1970s, is operational exclusively for international money payments among banks. Banks offer large corporations on-line cash management facilities.

Management, concerned with the costs incurred due to failed projects and due to projects realizing too little functionality too late, considers packaged applications as a possible alternative to the tailor made systems. Initially packages are used within business units for applications with little or no relationship to the primary enterprise activities. Smaller companies are using packages much more comprehensively. The documentation of the packages is of course an issue for evaluation when choosing the best suited package. Unfortunately larger business units usually find shortcomings in the functionality. This implies "improvements" to make the package acceptable. This tailoring of packages to specific needs leads to higher and even unacceptable costs for upgrading when new releases arrive. This situation is not unlike problems caused by the tailoring of the operating systems a decade before. Since package suppliers only support old releases for a limited period of time, either the organization considers the package as stable and functionality adequate and themselves capable of maintaining it, or, in order to achieve new functionalities, the organization is forced to replace the entire package. Many packages do not survive far beyond this decade because even the suppliers loose the capability to maintain the packages.

The development of applications for use in global enterprises was costly. Not just from acquisition perspective but also because of the complexities of getting the business systems up and running in the local business units. Packages were bought and installed internationally with varying degrees of success for the enterprise in is entirety.

23.4.4 IT operations during the integration and diversification period

IT operations needs to be treated from three different perspectives. First the arrival of office computers has organizational consequences. Secondly, the computer centre organization undergoes some significant changes in its own operational procedures. Finally, management recognizes that IT is a strategic issue requiring more attention than ever before.

Office computers, whatever their size, are not generally considered to be the responsibility of computer centres. The budgets for the office computers are not often part of the professional IT budget but part of the budget of the business unit. These office computers require personnel to manage the installations, advisers and consultants to assist with the installation and use of the office computers.

Too often the enterprise does not see these extra costs and complexities coming and fails to regulate their acquisition and use before incurring operational problems. Different business units buy the office computer that they like best. Business units make use of different computers that use different operating systems and have different operational support requirements. Since the business units choose their own office computers, the business unit is always very satisfied with what they get.

All is well until these business units decide that their systems need to be connected to the mainframes in order to take advantage of the data maintained on the mainframes.

Connecting office computers to the mainframe proves to be more complicated than expected. Data integrity problems due to the differences in the data definitions or even the absence of data entities or attributes mean changes to programs and data on the installations managed by the business unit and/or in the computer centre organization. The computer centre organization is faced with new telecommunication protocols that must be maintained for communication with the various office computer installations. Due to the unfortunate habit that systems have of failing, expertise is needed on all these different protocols that are in the enterprise. Business units are sometimes anxious to delegate the entire operational management of their office computers to the computer centre organization. Computer centres do not always have the tools or the experts to manage these installations. When too late or inadequate data comes from the office computer causing the mainframe application to be too late with output, the business units consider this tardiness as mainframe failures. The computer centre organization looses its complete control over the availability of their IT services to the business units.

Business units are getting into trouble because their systems are too difficult to manage and/or because the firm that has delivered the system does not give the continued support of that system as required by the business unit. Generally mainframe professionals have their hands full realizing the business support expected from the mainframe systems.

The use of office systems with IT specifically for the particular business unit does give assurance that only data from that office computer can be delivered to the mainframe. The reason for this lies in a combination of factors. Many of these systems use their own telecommunications protocols and/or the formatting of their data is unique for that process. For example a personnel sub-system running on an office computer within a central personnel department will be the only system that can generate data for updating personnel records on the mainframe. Other office computers in the factory or wherever not only do not have the programs to generate the data, they are not from the same manufacturer and the mainframe does not recognize them as being authorized to use the programs (possibly because of the difference in communication protocol) so the factory can not update the personnel database.

The dependency on the integrity of IT processes, the complexity of introducing new IT processes into the production environment, the number of changes to existing systems, the necessity of managing changes to the IT infrastructure without endangering the integrity of the IT processes, all lead to the introduction of what is called a change management organization. This organization becomes and still is responsible for managing all changes to the IT infrastructure and to the IT application architectures.

The globalization requires computer centres to be operational 24 hours a day 7 days a week. This has its consequences not only for computer centre personnel but also for the operating systems and the telecommunication protocols. In particular updating functionality and removing bugs requires stopping all network traffic to perform the maintenance. Eventually the suppliers rise to the occasion and make it possible to perform technical maintenance without stopping all systems. But the required strategies do not really receive attention until the mid-1990's [31].

In banking, telecommunications replaces the use of magnetic tapes for communicating bulk data.

The authorized use of hardware, system software, IT applications and data becomes very complicated to maintain. It requires the continued attention of the computer centre organization.

Computer operations processes are automated. Two such processes important enough to mention here are the production planning and control and the storage management. By the end of this period magnetic tapes as mass storage medium are replaced with cartridges and tape robots. The tape registrations and the many limitations of magnetic tape are solved either through the use of cartridges or through automated storage management systems for disk data sets. The complete disappearance of the magnetic tape unit in the computer centre is not entirely possible. Long term retention of data implies that some data must be available for as long as 30 years. Computer centres find ways to maintain the readability of data when the carriers of that data are no longer readily available.

Computer centre organizations start negotiating Service Level Agreements between themselves and their clients (the business units). The aim of these SLAs is to make explicit the degree of service (availability of IT applications, response times on transactions, helpdesk support etc.). Sometimes the SLAs include price agreements for that service.

From the corporate perspective, IT seems to be getting out of hand. The total enterprise wide expenditures on IT are very difficult to define accurately.

Strategic management has a lot more IT to make policies on than ever before. In the EDP period, IT budgets were the only issue. In the automation period, top management becomes more concerned with the effectiveness of these budgets. Finally, the alignment of the IT investments with the business strategies, the relationship between IT investments and other investments lead to organizations introducing formal procedures for controlling the IT investments. IT projects are only started after explicitly stated objectives for supporting the business unit with these investments have been approved. Security policy statements and IT security in particular, must be endorsed by top management. Management is confronted at the very least with the necessity to make policies for preferred suppliers, for telephone and telecommunication networks, data base management systems, management of global IT, the use of consultancy firms, outsourcing and the degree to which business units could make autonomous decisions on their IT support.

Finally large enterprises consolidate their computing into a limited number of global computer centres in order to lower the operational costs of the enterprise wide use of IT.

During the integration and diversification period the clove between the professional IT organization and the business units has a new element as a result of the IT investments initiated and managed by the business units. In order to formalize the communication, the business units institute information managers to liaison between them, strategic management and the professional IT organizations.

The ownership by the business units of IT applications and data is finally achieved.

23.4.5 IT security in the integration and diversification period

The IT security function becomes common in enterprises. Sometimes it appears for the first time within already existing security departments, even within the internal audit, and sometimes the function is placed in the computer centre organization. Most often it becomes a specialized staff function with responsibility for initiating IT security policies, advising on the IT security investments and assessing the threats and vulnerabilities of the enterprise with respect to IT security. IT security projects are specifically created to achieve the required overall security objectives. These projects are often carried out under the auspices of the head of IT security. Once an IT security policy has been put into place, defining and implementing the necessary security measures becomes part of projects to achieve new business objectives/products or services. The continued effective security of the operational IT support is a responsibility for business units and computer centre organizations to implement and maintain. IT security now addresses at least the following issues: privacy of data, confidentiality of data, protection of processes against misuse or unauthorized changes, continued availability of computer resources and the integrity of all IT security measures. It does not address the integrity controls for complete and correct data; the delegation of responsibilities within the organization (authorization structures) required to ensure the integrity of data nor does it address the consistent implementation of the authorization policies in the automated systems.

The IT security policy is usually based on a "need to know" principle. Privacy laws influence the measures chosen to ensure the privacy of data. IT security recognizes the necessity for stronger authentication techniques than passwords and the use of encryption techniques.

IT security officers manage contingency drills to examine the strengths and weaknesses of the procedures for diverting computer production to other computer centres. Obviously, contingency planning and procedures include procedures for key business units. Drills are necessary to assess the adequacy of the contingency measures.

Policy decisions on the use of personal computers and in particular the measures taken to connect a personal computer to the mainframes involved advice from IT security.

The policies on authentication techniques are the result of advice from the IT security.

In general, security policies are mandated only after Security Officers and IT Security Officers have rendered consistent advice on these policies.

23.4.6 IT auditing in the integration and diversification period

At the beginning of this period, IT auditors generally feel a lot more confident about the quality of internal control procedures in relation to the enterprise vulnerabilities than at the end of this period. The diversification, is considered to pose a threat to the integrity of data and processes.

IT security is a relatively new subject. Security does have management's attention.

The IT auditor must assess the adequacy of the implemented IT security.

Access control profiles tend to be implemented to allow too much functionality to those individuals using the access profiles. Assessments of the operational access control profiles are very popular among auditors because there are always critical remarks to be made. Even so this period can be characterized as a period of great improvement in the quality of IT security measures in the mainframe environment.

The computer centre organizations are complex, highly automated and the personnel professional. IT auditors must know their stuff to have any added value in the complex of computer installations that are being managed within enterprises.

The automated support of the change management unit is a new subject for audit as is the implementation and use of all the "automation of the automation" tools (for example: production planning and control for the automated information systems, storage management, network management, management of changes and distribution of infrastructure and information systems).

Access control is a subject of concern to the IT auditor. It is generally known who is responsible for which access control profiles in the enterprise and procedures are being instituted to ensure that the automated access controls conformed to the delegation of duties (authorizations) within the enterprise. Placing extra access control functions within the office computers has advantages and disadvantages that always need to be assessed. Since up to the early 1990s telecommunication with external organizations is chiefly computer to computer, the authentication techniques are often burnt into the computer hardware and part of the technical protocol for starting up the point to point connections. For the inter-enterprise communications, there are some uses of encryption techniques for communicating passwords. The internal telecommunication is usually from personnel to main frame or office computer. The authentication is too often just a simple small alphanumeric password entered into the system from the keyboard. Personnel do not take these passwords very seriously nor are the controls on the passwords rigorous. Some organizations are using TCP/IP and many large organizations are still using proprietary data communication protocols.

The first Local Area Networks appear on the business scene and without much thought on the part of the suppliers to the security issues. The password regimes in use are not rigorous enough to be able to protect LANs against unauthorized use.

The lack of segregation of data and processes in PCs as well as the impossibility of creating trustworthy detection measures lead to many discussions on where and how personal computers can and cannot be used to support business processes.

When personnel computers are used instead of the old terminal technologies, IT auditors usually advise getting rid of disk drives on the PCs, changing the standard implementation of the personnel computer and many other techniques to limit the risk of viruses, unauthorized applications and in general misuse of the added intelligence and flexibility that the personnel computer offers.

Computer centre management is thinking more strategically about their investments in information technologies and their relationships with their clients: the business units. Service level agreements with the departments being serviced are a new phenomenon [27]. The IT auditor needs to be able to address these issues as well as the assessment of the day to day operations.

Audit departments start creating systems to monitor the automated business activities and report exceptional situations so that auditors can assess the degree to which the internal controls are working. Initially, these tools were called "audit software" by the eighties they are being called CAATs (Computer Assisted Audit Techniques). This new term implicitly refers to a broader functional use than simply the one time creation of a report for the financial auditor to use when gathering evidence and making findings on the integrity of the bookkeeping process. For example CAATs may include applications to select exceptional transactions from the input and store them in a data base for further examination when the department responsible for the transaction is being audited.

As auditing becomes more important because of the overall view which the IT auditor has over the IT applications and IT infrastructure in development and in use, audit committees and boards of directors become interested in the findings of the IT auditors with respect to the control and security measures in the enterprise.

The first corporate governance reports [10] (COSO) are not concerned specifically with IT but the improved corporate governance has its effect on the way IT auditing is carried out and on the reporting procedures for the IT audit findings.

Privacy laws are introduced in many countries; these influence the thinking on the structuring of some information systems and the responsibilities of the computer centre organizations to protect all data concerning persons. The organization of the

IT security function and the effectiveness of the IT security measures is a new IT audit subject.

The office computers can be considered to form mini computer centres. Their importance for the integrity of the bookkeeping is one of the reasons why IT auditors are asked to audit them. When the office computers are connected to the mainframe, it becomes necessary to form an opinion on the IT security of the connection to the mainframe and the production environment on that office computer.

A steady improvement and diversification of the system development methods initially available from the early 1970s occurs in this period [2].

The increased accessibility of data across the traditional lines of business causes IT auditors to have major concerns about the overall consistency of the data being used for management information. This in turn meant some concerns about the integrity of corporate data being used for strategic decision making. The introduction of ERP software in the 1990s gives management the opportunity to truly integrate the use of data elements across business processes. Complete banking packages for smaller banks are available early in the eighties (for example: Midas from BIS, Atlas from Internet Software Services) but large businesses had to wait for applications like SAP.

The management policies concerning the use of IT in the enterprise are used as standards of conduct when auditing the way IT is managed and used (for example: IT security policy, supplier policy, project structures, the SDM).

23.5 CONTAGION PERIOD (THE NINETIES)

Client/Server (C/S) technology, network applications, ERPs (Enterprise Resource Programs) and of course the Internet are technologies that define the character of IT in the contagion period. Along with these technologies, the maintenance to systems dictated by the arrival of the second millennium had to be completed before the first of January 2000. This Y2K maintenance problem gave the 1990s a unique character.

Buzz words include knowledge systems, Client/Server (C/S), EDI (Electronic Data Interchange), E (Electronic)-commerce, GUIs (Graphic User Interfaces) and outsourcing [23].

The term contagion is used for this period because of the way the use of IT spread throughout governments, businesses of all sizes and the public. Whole new services and IT products become available.

23.5.1 Characteristics of IT during the contagion period

The introduction of new technologies combined with the high degree of integration of IT applications as well as network application technologies make characterizing the IT technologies separately from the IT applications difficult.

Internet is the motor that facilitates much of the expansion of the use of IT in enterprises and the acceptance of IT by the general population.

IT management is concerned with global IT infrastructures and managing highly integrated information system architectures that support the organization at all levels and across all business units.

The IT environment is ready for some radical changes based on client/server technology and the use of the Internet.

23.5.2 IT environment during the contagion period

If the distance between the computer centres is limited, mainframe infrastructures finally allow for the real-time backup of data among large computer centres. It is a step in the right direction where contingency is concerned but the big computer centres remain vulnerable.

Channel speeds remain from the 1960s onward too slow in relation to CPUs and the amount of digital data being maintained on the installations to allow an efficient dissemination of data across different computer centres.

Client/server technologies (C/S) with GUIs greatly increase the flexibility of the use of IT in the enterprise. But their introduction is not seamless. Servers are not fast enough in relation to the amount of data they maintain. Initially, there are no tools for managing infrastructures based on C/S technologies, for arranging backup and recovery or for distributing applications and upgrades for applications throughout the C/S infrastructure. A closer look at the Gartner European Symposium 93, shows much attention to strategic management issues and presentations on the possible information architectures and information technology infrastructures but nearly no attention to the tools needed and actually available for managing a large global client/server infrastructure [13]. Each enterprise in the early 1990s had to solve these problems for itself. In response to the vacuum left by the manufacturers of C/S technologies Computer Associates, among other software suppliers for businesses, developed many tools to support the operations of large distributed information systems and infrastructures. Unicenter and Tivoli are two often used products managing large decentralized computer infrastructures.

Towards the end of the integration and diversification period, the Internet is cut loose by the American Department of Defence which originally created and paid for its development. The Internet comes into its own.

While all personal computers in the enterprise are connected exclusively through the enterprise network, a secure and efficient communication throughout the enterprise is possible. However this infrastructure does not last long. The possibility to access the Internet directly from the personal computers in the enterprise proves impossible to contain.

Thanks to the Internet, home computing is born. Partly because of this, enterprises are embracing TCP/IP. TCP/IP also proves to be common in many smaller businesses. Enterprises turn to the Internet infrastructure to support their own internal communications. Unfortunately TCP/IP has proven to be more difficult to secure than the internal communication protocols of the seventies and eighties.

As in the IBM dominated EDP period, there is one dominate supplier. Microsoft's position of dominance over the (personal) computer market is not unlike IBM's dominance of the mainframe market.

One communication protocol, TCP/IP dominates communication infrastructures. This dominancy has many advantages for the developments of this period.

Unfortunately from a control and security point of view the weaknesses in each of these technologies result in a negative influence on the total enterprise security (Microsoft and TCP/IP).

23.5.3 IT applications during the contagion period

Artificial Intelligence systems are being developed and proving to have distinct advantages for supporting complicated decision making and for controlling automated processes.

Object oriented programming languages are widely used along with object oriented data bases for developing network applications.

IT is not just used for the inter-company communications through EDI (Electronic Data Interchange) but also for the so-called E-commerce. E-commerce implies the integration of commercial processes between businesses and between businesses and banks without any involvement of personnel for confirming these processes. Business applications make use of the Internet not just for the communication with other businesses but also with private customers.

The year 2000 proved to be a date with enormous repercussions for the character of IT applications development in the 1990s. Or was it maintenance? The convention of using two digits to represent the year in a date (YY) instead of four digits (19YY) could not be continued after 1999 without creating major integrity problems. There were two prevalent strategies to solve the problem. Some enterprises built shells around their existing application base so that the data in the data bases and the applications underwent minimal changes and the integrity of dates starting in 2000 in relation to the older dates is maintained. Smaller organizations often took the opportunity to replace their application base with packages that were millennium proof. A third technique was of course to make all dates recorded in data bases millennium proof by using four digit dates instead of two digit dates. Unfortunately the data independence between programs and data bases is usually not rigorous enough to allow for a change in the length of an item without at least recompiling the programs and redesigning and reprogramming the output forms to accommodate four digit dates. In organizations with a very large IT application base, this method was prohibitively expensive. One of the reasons for the prohibitive costs may well have been in some cases that the documentation of the applications was particularly out of date.

The available programming languages, the development of network applications, the desire to develop quick and dirty applications and the millennium, all contribute to the demise of COBOL as a major programming language for new uses of IT in the business environment. Most of the new languages and the supporting operational environments do not support the segregation of data and processes (C++ and Java are examples of two such languages).

The big news is the success of ERPs (Enterprise Resource Programs). ERPs are transaction oriented systems that automate all the main business areas of the enterprise on a real-time basis and make use of non-redundant data. As these systems replace the tailor made systems and the assortment of packages used to support diverse functions within the organization (from personnel departments, marketing and bookkeeping to inventory), the integrity of data and management information is greatly improved. Of course the ERPs were millennium proof far enough in advance of the year 2000 to allow for timely upgrades and implementations in enterprises.

23.5.4 IT operations during the contagion period

The three-tier IT infrastructure is here to stay. Although the buzzword in this period was two tier infrastructures, which refer to standalone client/ server applications; actually enterprises are using three tier infrastructures.

The computer centre organization is starting to loose its dominancy. The computer centre has a wide diversity of clients to deal with. E-commerce gives a whole new urgency to the necessity, availability and continuity of the computer centre organization. Discontinuity in service is now not only unacceptable for the enterprise itself but it becomes visible to other enterprises and private persons. Thus IT becomes a threat to the continuity of the enterprise.

Outsourcing of computer centre operations to lower the operational costs of IT is popular. Although the usual argument for outsourcing is overall cost reduction, there are at least two reasons for outsourcing that are more important than direct cost savings: improvement of the quality of the IT services to clients and personnel and more efficient management of IT services [24].

Mainly during the 1990s the British Central Communications and Telecommunications Agency (CCTA) publishes a series of booklets devoted to describing important tasks that are necessary for operating computer centres. These publications represent the first extensive treatment of what a manager of a computer centre must organize and manage [6].

Legislatures discover the importance of regulating IT. Computer criminality, ownership of programs (copyrights), telecommunications as well as privacy are subjects receiving attention not just at the national level but also among nations.

Strategic management is confronted with higher total IT costs mainly due to the speed with which the new technologies become outdated and the resulting pressure to see these costs as expenditures instead of investments. The growth in IT costs is largely caused by end-user computing (desktop automation) and by the maintenance costs necessary to prepare for the new millennium.

23.5.5 IT security in the contagion period

Because of the growth in the use of telecommunications in general (client/server technology) and Internet in particular, IT security becomes an important issue at all echelons of the enterprise.

Secure use of personal computers in financial applications causes many large companies to loose a lot of money trying to design and implement adequate measures to ensure the integrity of personal computers used as clients in client/server applications.

Eventually the so-called thin client [30] becomes the most chosen option because all important functionality required to support the distributed processing is placed on the server where it is securable.

The any to any concept of the Internet protocol causes another problem for enterprises heavily dependent upon network applications. Enterprises do not want just anybody able to access all of their systems – as is implied in the "any to any" concept. It is diametrically opposite to the "need to know" principle usually stated as a requirement in the IT security policy. Limiting the otherwise unbridled access to some applications to only those clients authorized to use them is a problem that the IT security is asked to solve. A plethora of encryption techniques with the accompanying key management procedures and different kinds of firewalls in conjunction with each other is the result. Enterprises look to Trusted Third Parties to help secure the authorized use by clients of network applications.

The information security standard BS7799 is an important contributor to a common practice when instituting security measures in enterprises [7].

The weaknesses in the TCP/IP protocol become liabilities. Virus control is a major concern.

Encryption techniques are being applied for confidentiality and integrity of telecommunications. The choice of techniques and key management are important new IT security responsibilities.

23.5.6   IT auditing in the contagion period

Projects to introduce an ERP into the enterprise are achieved thanks to the expertise of consultants versed in the implementation of the specific ERP being implemented. Large projects of strategic importance to the enterprise are usually subjected to IT audits and thus a new area of expertise for the IT auditor comes into existence. The IT auditor must know the ERP well enough to assure management that the strategic goals are (going to be) met by the project, that the control measures are well chosen, that the security measures are adequate and that the system to be made operational will be maintainable.

Client/server technology is sold to enterprises without any concern for tools for maintaining big client server infrastructures or for the security requirements some uses of client/server technology require. IT auditors attempt to make management conscious of these shortcomings.

Encryption and key management are new areas that management expects the auditor to audit.

Establishing the adherence of the organization to constraints on procedures as dictated by law is another new subject that the IT auditor is still expected to report on. Adequate privacy measures become a corporate concern [28]. Advise for businesses on measures to take to adequately protect the privacy of the individual are forthcoming. Child pornography becomes an issue for management and IT auditors are also asked to examine the adequacy of the detection techniques and procedures concerning pornography. Management is primarily concerned with their own liability should pornography be found within their enterprise infrastructures. The IT auditor is now expected to form an opinion on the quality of all measures taken in the enterprise to meet all legal requirements including the privacy law requirements [26].

The use of personal computers in the enterprise for personal purposes is generally frowned upon by management, not just because of pornography laws and other liability risks but also due to the hidden costs. Management needs to limit its liability where the use of illegal packages is concerned. Naturally the integrity of information may not be compromised when using personal computers for personal activities.

Business vulnerabilities due to the inherent weaknesses in the Internet protocol combined with the use of personal computers receives a lot of IT audit attention. The initiatives to develop adequate preventive measures in personal computers are met with very limited success. The very weak segregation of the user of the personal computer from the application code and the near impossibility of creating incorruptible independent logging functions on a PC form limiting factors for the functionality that can be responsibly placed on a PC.

The newer and better versions of COBIT [20] are published by ISACF. ISACF is the Information System Audit and Control Foundation. It is an independent organization affiliated with ISACA (Information System Audit and Control Association). COBIT is a comprehensive framework for IT auditors to use as a reference when substantiating opinions to management on IT control measures implemented or planned.

23.6 CONCLUSIONS

Through the decades starting with the 1970s, the effectiveness of the investments in IT is a source of concern.

A clove between IT and the business units it supports starts as early as the 1960s and continues to widen until, in the 1990s, when personal computers and Internet become commonplace.

The IT control and IT security measures within the enterprise are probably most effective in the 1980s.

IBM's announcement in 1964 of a general purpose computer proved, when viewed from within the scope of a machine for supporting bookkeeping processes, to have been no exaggeration. Unfortunately most of the information technologies poured on the market place since then have been designated by the suppliers as general purpose.

Before an enterprise buys IT products, obviously it will first decide what the intended responsible use of the IT product is and then scrutinize the products and assess the measures needed to tailor a product to the control and security requirements. When IT is being bought by informed customers, the customer is able to estimate the time and effort to create the management tools for the operational support of the IT product and to create the security measures needed to use the technology safely. Sadly, the 1990s prove that "informed customers" is a utopia in IT. Even if it weren't, the position is hardly tenable that it is much cheaper for each and every enterprise to decide upon their management tools and their own security measures. Remember the problems in the 1970s that tailoring of operating system led to or, in the 1980s and 1990s the problems incurred after too much tailoring of packaged applications were carried out?

More important is the inherent architectural weakness present whenever IT control and IT security measures are add-ons to IT applications and system software. The control and security is never as effective as when the measures are integrated into design of the IT application or system software.

The use of information technologies for building and exploiting automated information systems has undergone an enormous rationalization. ERPs have solved development challenges that went far beyond the development expertise that could be found even in large enterprises.

The vulnerability of the huge centrally managed data bases in enterprises remains an issue that will probably only finally be solved with processing strategies that are developed to safely exploit the computing capacity of the Internet. But this route will only be responsibly taken when telecommunications and operating systems software are a lot more controllable and securable (with preventive measures and measures for detection) than the present technologies.

Whatever the reasons may be, it is clear that not much was learnt from the successes and failures in the use of information technologies from 1960 through 1990. Today there are very basic improvements needed in the IT products.

The big mistakes that could have been avoided include:

- Implementing new operating system infrastructures that do not distinguish data from programs (machine code) in the structure of their interfaces and protocols (Microsoft operating system infrastructure);
- Creating programming languages that do not support a segregation of responsibility between the developer, business unit's use of the functionality and the operational system (Java and other object-oriented languages);
- Implementing a communication protocol that does not allow for (TCP/IP):
  - explicit routing of packages through predefined channels,
  - complete and correct end to end transmission as part of the network protocol,
  - functionality to implement closed user groups within the network protocol,
  - logging at certain nodes to allow for tracing packages and error recovery,
  - transport of packets containing data and those containing programs as distinguishable entities to the receiving computer installation and/or servers;
- Suppliers delivering information technologies with no regard for the impact of the technology

on the management of the IT support or the possible security threats inherent in the technology (personal computers, telecommunications protocols, mobile telephones etc.) [3].

The greatest shortcoming is the failure to develop standards for different control and security profiles. Standard profiles must become mandatory design criteria for the design of IT products and the design criteria used for a specific product made public for the benefit of consumers. Only then will the consumer be able to choose IT products that can be used responsibly within the enterprise.

REFERENCES

[1] American Federation of Information Processing Societies (IFIP), *AFIPS Conference Proceedings 1970*, Montvale, NJ, Library of Congress Catalog Card Number 55-44701 (1970), pp. 351–374.

[2] A. Anderson, Foundation – Method/1 Information Planning, Version 7.5 (1988).

[3] M.E. van Biene-Hershey, Audit van IT en het IT-auditberoep (2004).

[4] M.E. van Biene-Hershey, *Auditing the Technical EDP Organization*, NGI/SIC, Amsterdam, The Netherlands (1985).

[5] M.E. van Biene-Hershey, *IT Auditing, an Object Oriented Approach*, Delwel Publishers (1996).

[6] British Central Communications and Telecommunications Agency (CCTA), more than 40 booklets between 1990 and 1999, IT Infrastructure Library, Copyright Unit, Her Majesty's Stationery Office, St Clements House, 2-16 Colegate, Norwich NR3 1BQ.

[7] British Standard, Code of Practice BS7799, September (1993).

[8] Canadian Institute of Chartered Accountants (CICA), Computer Control Guidelines, Toronto, Canada (1970).

[9] Canadian Institute of Chartered Accountants (CICA), Computer Audit Guidelines, Toronto, Canada (1975).

[10] Committee of Sponsoring Organizations of the Treadway Commission (COSO), Internal Control – Integrated Framework (1992).

[11] H.J. Cozijnsen, R.J. van Biene, *Programmeren*, Wolters-Noorthoff nv Groningen, The Netherlands (1971).

[12] W.B.C. Ebbinkhuijsen, *COBOL*, Samson Uitgeverij, Alphen aan den Rijn, Brussel (1987), pp. 3ff.

[13] Gartner Group, European Symposium 93, Volumes I and II (1993).

[14] T. Gilb, *Reliable Data Systems*, Universitetsforlaget Oslo–Bergen–Trosø (1971).

[15] IBM Corporation, Data security and data processing volume 1–6, Technical Publications – Systems, White Plains, NY 10604; IBM code G320-1370 through G320-1376 (1974).

[16] IBM Svenska Aktiebolag, *Data Security Design Handbook*, Stockholm, Sweden (1978).

[17] IBM, System/network control center overview, processes and products, International Business Machines Corporation G226-3551-03 (1983).

[18] Intermediar, *De Accountant Verklaard*, Kluwer BV, Amsterdam (1977), pp. 23–25.

[19] ISACA, www.isaca.org, overview and history.

[20] ISACA, CobiT Framework, http://www.isaca.org (1996).

[21] M.C. Kelly, W. Aspray, *Computer, A History of the Information Machine*, Basic Books, New York (1996).

[22] G. Manne, *Cracking the Books II: Reliving Equity Funding, the Cal Ripken of Stock Frauds*, www.thestreet.com/stocks/accounting/789337.html (1999).

[23] Moret Ernst & Young Telecom Group, Trends in Telematica, Amsterdam (1995).

[24] Moret Ernst & Young, Trends in Uitbesteding (1995).

[25] National Institute of Standards and Technology (NIST), *An Introduction to Computer Security: The NIST Handbook*, Technology Administration, U.S. Department of Commerce, Special Publication 800-12.

[26] NOREA, IT-Recht in volgelvucht, Duthler Associates, Den Haag (2005).

[27] R. Paans, *A Close Look at MVS Systems: Mechanisms, Performance and Security*, Elsevier Science Publishers B.V. (1986), p. 14.

[28] Registratiekamer, Beveiliging van persoonsregistraties, Registratiekamer, November (1994).

[29] J.C. Shaw, W. Atkins, *Managing Computer System Projects*, McGraw-Hill Book Company (1970).

[30] P. Schay, *Client Server*, Annual Symposium on the Future of Information Technology, GartnerGroup, Paris, France (1993).

[31] J. Schulman, *Software Management Strategies*, Annual Symposium on the Future of Information Technology, Gartner Group 2–5, Paris, France, November 1993.

[32] Stanford Research Institute, Data Processing Audit Practices Report (1977).

[33] Stanford Research Institute, Data Processing Control Practices Report (1977).

[34] Stanford Research Institute, Systems Auditibility and Control Study, Executive Report (1977).

[35] R.J. Thierauf, *Effective Management Information Systems*, Charles E. Merril Publishing Company (1984).

[36] R. Yarmish, J. Yarmish, *Assembly Language Fundementals, 360/370, OS/VS DOS/VS*, Addison-Wesley Publishing Company, Inc., Philippines (1979).

The History of Information Security: A Comprehensive Handbook
Karl de Leeuw and Jan Bergstra (Editors)

24

A HISTORY OF INTERNET SECURITY

Laura DeNardis

Information Society Project, Yale Law School
Yale, USA

Contents

Abstract

The Internet and its predecessor networks evolved in an era devoid of home Internet access or personal computers and in a closed and trusted user environment predominantly in academic, research, and military contexts in the United States. Network security was important but did not have the same complexity it would later assume when the network expanded into business environments, across the globe, into homes, and over the open airwaves of wireless. A watershed event occurred in the fall of 1988, when a self-propagating computer program (the Morris worm) disrupted or crashed thousands of Internet-connected computers. Since this attack, security incidents and challenges such as worms, viruses, wireless vulnerabilities, denial of service attacks, spam, identity theft, and spyware have increased annually even while national economies and national security operations have become increasingly dependent on the Internet. This chapter traces the history of Internet vulnerabilities and solutions within the social milieu in which they materialized and raises questions about whether the Internet will ever be secure.

Keywords: Internet security, computer networking, history of the Internet, computer security, network protocols.

24.1 PROLOGUE

Since its inception in the late 1960s, the Internet has evolved from an experimental network allowing resource sharing among a few researchers into a global platform for personal communications and electronic commerce. The Internet materialized during an era devoid of personal computers or home Internet access. The United States Department of Defense (DoD) originally funded and inaugurated the Internet's precursor, ARPANET, through its Advanced Research Projects Agency (ARPA). Access was somewhat restricted and users predominantly existed in academic, research, and military contexts within the United States. In this relatively trusted and closed user context, security was important but did not have the same complexity it would later assume when the Internet expanded into business contexts, across the globe, into homes, and over the open airwaves of wireless. Some underlying security vulnerabilities were recognized even in the early 1980s, before the World Wide Web or home Internet access. But it was the notorious Morris Internet worm of 1988 that underscored the need for more potent Internet security. As national economies and national security operations became increasingly dependent on the Internet, both commercial and freely available security products sought to address a variety of needs including access control, information privacy and integrity, user authentication, and virus protection. Nevertheless, security incidents have increased annually. Costly worms, viruses, spam, and spyware have plagued users. Access mechanisms like broadband and wireless have created a spate of security challenges and cyberterrorism has emerged as a concern. Internet security history is infused with not only technical and economic complexities but numerous political issues: the United States government categorizing encryption technologies as firearms relative to export controls; concern about the threat of cyberterrorism; balancing law enforcement and national security with privacy and civil liberties; the use of Denial of Service (DoS) attacks to politically or economically chastise corporate entities, the media, or nations; global power struggles over wireless Internet standards; and debates about technical monocultures like Microsoft facilitating security vulnerabilities. This historical examination traces the progression of Internet security vulnerabilities and solutions within the social milieu in which they materialized and raises questions about whether the Internet will ever be secure.

24.2 CLOSED WORLD ORIGINS

To communicate, computers must adhere to standard rules, or protocols, specifying how to format, address, and exchange information. The development and standardization of the Internet protocols and infrastructure in the 1970s materialized in an environment inhabited by relatively trusted users and shaped by United States Department of Defense (DoD) requirements. Historians Janet Abbate

[1] and Thomas Hughes [35] emphasize the influence of US government funding on the inception of the Internet. The US DoD funded ARPA, the Advanced Research Projects Agency which would eventually develop the ARPANET, in 1958 as a Cold War response to the Soviet Union's launching of Sputnik and concern that the United States was trailing the Soviet Union in scientific research and development. Another US military concern, as Paul Edwards stresses in *The Closed World, Computers and the Politics of Discourse in Cold War America*, was the preservation of command and control communications capability in the even of a nuclear attack [24]. This requirement for survivable communications influenced the development and selection of packet switching as an underlying infrastructural approach for ARPANET, the predecessor to the Internet. RAND Corporation's Paul Baran, the inventor of packet switching in the United States, designed a system that used a "store and forward" technique breaking information into small segments and transmitting each segment to its destination over dynamically available paths. In response to the need for survivable communications and in contrast to the prevailing telecommunications approach, packet switching provided redundancy, distributed switching nodes, and alternative paths.

Vinton Cerf and Robert Kahn [13], the primary authors of what would become the universal Internet protocols, TCP/IP, spearheaded the emergence of TCP/IP in this environment when ARPANET was funded and shaped by the United States defense requirements. The DoD selected TCP/IP as the standard for military ARPANET use in the early 1980s, helping TCP/IP gain momentum. Some standards specifications for TCP/IP were expressly linked to DoD requirements. For example, the 1980 and 1981 Internet standards specifications for TCP were entitled "DoD Standard Transmission Control Protocol". The stated motivation for the TCP standard was primarily to provide reliable, host-to-host communications for military applications, "especially in the presence of communication unreliability and availability in the presence of congestion ..." [49; 50]. Reliability, availability and performance were the primary concerns for TCP, not security. Security requirements were important, but not paramount relative to other concerns. The Internet user context was relatively closed because, with the mid-seventies exception of the Altair home computer kit, personal computers were not available and public or home access seemed implausible. As Hafner and Lyon describe in *Where Wizards Stay up Late: The Origins of the Internet*, the term "hacking" described creative programming, and malicious network users were almost nonexistent in the early ARPANET environment [33].

Business ARPANET use was almost nonexistent and the networks supporting business communications and electronic commerce, which required robust security, were based on proprietary protocols like IBM's Systems Network Architecture (SNA) and Digital Equipment Corporations DECnet architecture rather than on TCP/IP. Proprietary protocols were developed and controlled by a single networking vendor and inherently closed to outside access. These protocols were not publicly available and even networks within a single institution lacked interoperability because one manufacturer's products were incompatible with other product lines. Institutions wanting corporate-wide networking completely depended upon a single manufacturer for all their network infrastructure products. The major drawback to these proprietary environments was the difficulty in interoperating with business partners and the economic disadvantages of being locked into a single vendor's product line. Lack of interoperability nevertheless provided an ancillary benefit of inherent security because outsiders could not easily access proprietary networks. In contrast, TCP/IP promised much needed interoperability between disparate computing networks, and openly available protocols enabled the emergence of competing, yet interoperable, product lines.

By the late 1980s, discussions occurred within the Internet community about security vulnerabilities extant in Internet software, hardware, and protocols but a major Internet security attack had yet to occur. Security weaknesses in UNIX, and especially BSD (Berkeley Software Distribution), were well understood by UNIX system users, but there existed an underlying assumption of "trust between scholars in the pursuit of knowledge, a trust upon which the users of the Internet have relied for many years" [26].

24.3 BLACK THURSDAY

A watershed event for Internet security occurred in the fall of 1988 when a self-propagating computer program rapidly spread across the Internet and disrupted or crashed thousands of Internet-connected computers.

At 11:28 p.m. pacific standard time on November 2, the following message appeared on the TCP/IP Usenet discussion board:

> *"We are currently under attack from an Internet VIRUS. It has hit UC Berkeley, UC San Diego, Lawrence Livermore, Stanford, and NASA Ames."* [90].
>
> Peter Yee
> NASA Ames
> comp.protocols.tcp-ip newsgroup

Individuals at Stanford University, MIT, RAND, Berkeley and other sites similarly reported virus attacks [25]. Launched late in the day on Wednesday, November 2, 1988, the self-replicating computer code, then alternately referred to as a worm or virus, created overwhelming and widespread outages that effectively halted computer usage for several days at major universities, military installations, and research facilities. The virus infected Sun Microsystems' Sun 3 computers and VAX systems running BSD Unix (Berkeley Software Distribution) and derivatives of BSD [63]. As historian Paul Ceruzzi addresses in *A History of Modern Computing*, the development of UNIX variants in academic environments promoted the ability of users to easily share files, but this same feature also rendered UNIX systems susceptible to unauthorized access and viruses [19]. The virus program, written in the C computer language, entered through holes in sendmail (an email routing utility), finger (a utility allowing users to obtain information about each other), and through password guessing [53].

The code was technically "benign" in that it did not erase data, read private messages, gain privileged access to information, or deposit Trojan horse programs. However, by replicating itself on infected machines, the virus had the devastating effect of consuming resources to the point of disabling systems. By Thursday morning, the attacks were widespread and Internet computer experts embarked on a two day coordinated effort, primarily working at academic sites, to identify and thwart the virus [60]. Individuals at geographically dispersed sites communicated with each other through mailing lists, telephone calls, and Usenet discussion board postings.

Exactly how many computers did the worm infect? The often cited number is 6000, derived in part because MIT Professor and Vice President of Information Systems, James Bruce, estimated to reporters from a Boston CBS affiliate that the attack affected 10% of MIT's computers [25]. Reporters extrapolated this 10% estimate to the entire 60,000 computers then connected to the Internet to derive the number 6000. The outages certainly were widespread, but the exact number of infected computers remains unknown.

Within a week of the attack, FBI investigators suspected Cornell University graduate student, Robert T. Morris, of releasing the self-replicating program. Ironically, Morris was the son of Robert H. Morris, a chief scientist at the National Security Agency's (NSA) National Computer Security Center (NCSC) located in Bethesda, Maryland. Founded in 1981 as the DoD's Computer Security Center and renamed in 1985, the NCSC established security requirements for computing technologies that supported classified information [44]. The elder Morris had previously testified before the United States Congress about computer security vulnerabilities. The younger Morris had also been acutely aware of network security problems and had participated in discussions with other graduate students about the many security vulnerabilities within various Internet systems and protocols. He had previously published the paper, "A Weakness in the 4.2 BSD Unix TCP/IP Software", in which he described specific security weaknesses in the BSD version of Unix. Morris, who had discovered BSD vulnerabilities while working at AT&T Bell Laboratories, suggested about TCP/IP connected

machines that, "perhaps steps should be taken to reduce their vulnerability to each other" [43].

Morris was convicted in the United States District Court, Northern District of New York, of unauthorized access to a federal government computer and unauthorized access resulting in monetary damages greater than $1000 [82], both violations of the Computer Fraud and Abuse Act (CFAA). The expenditures required to terminate the spread of the worm and restore systems totaled between $200 to more than $53,000 at some sites, in addition to the opportunity cost of computer outages. The courts sentenced Morris to 400 hours of community service, three years of probation, and a fine of $10,050. Morris claimed his motivation for unleashing the program was to illustrate inherent Internet security vulnerabilities and the inadequacy of prevailing network security measures [66]. The Cornell Commission investigating the incident noted that the vulnerabilities were well known and required no "act of genius or heroism" to attack [26].

Although the Internet was twenty years old in 1988, the public and the judiciary system were generally incognizant of its existence and function. Even the legal judgments about the Morris case required a definition of the Internet. As explained in *The United States of America v. Robert Tappan Morris* (1991), "Morris released the worm into INTERNET, which is a group of national networks that connect university, governmental, and military computers around the country".

24.4 INTO THE PUBLIC CONSCIOUSNESS

"US Computer Systems Attacked by 'Virus'", read the headline of the London Financial Times on the morning of November 5, 1988 [36]. News of the Morris worm reverberated across the globe, making front page news in the New York Times [40], the Toronto Star [52] the Washington Post [34], and the Sydney Morning Herald [10], among many others. In the context of 1988, the Internet was relatively obscure and offered minimal global connectivity. The general public had little first hand knowledge of the Internet. Yet, the security attack attracted considerable press, with reports of the "computer virus" appearing in the United States on National Public Radio, local evening news shows, and some national news programs including the Today show. An account by MIT researchers involved in thwarting the spread of the worm described how the media interest was so intense the MIT News Office had to arrange a press conference. The researchers added "We were amazed at the size of the press conference – there were approximately 10 TV camera crews and twenty-five reporters" [25]. The computing team at the University of California, Berkeley, reported that aggressive press "hounding" encumbered their efforts to handle the crisis [54]. Some media accounts of the incident did not directly refer to the Internet, instead describing the problem generally as a computer virus that infected researchers nationally. For example, a series of Boston Globe articles appeared in the immediate aftermath of the incident including, "Computer Virus Halts Research" [3] Computer Invasion Seen as a Warning" [4] and "Computer Intruder Found Doors Unlocked" [56]. Nevertheless, publicity about the Morris worm brought the Internet and associated security issues to some extent into the public consciousness.

The New York Times interviewed Robert Morris' father, the National Computer Security Center's chief scientist, about the worm and the elder Morris suggested, "It has raised the public awareness to a considerable degree. It is likely to make people more careful and more attentive to vulnerabilities in the future" [40]. At the time, a University of Utah computer scientist, Donn Seeley, called Morris' statement "one of the understatements of the year", and emphasized that "our community has never before been in the limelight in this way, and judging by the response, it has scared us" [60].

24.5   THE BIRTH OF CERT

In addition to raising public awareness about computer security and about the existence of computer networks, the Morris worm reverberated through the Internet user community and United States governmental agencies. Within a week of the attack, the National Computer Security Center convened a series of "postmortem" meetings to discuss the nature and possible repercussions of the security incident. The first meeting occurred on November 8, 1988, at the National Security Agency Headquarters in Maryland and included approximately fifty attendees. In addition to those from the NCSC, attendees came primarily from academics (e.g. the Massachusetts Institute of Technology (MIT), Harvard, and the University of California, Berkley) and government (the Federal Bureau of Investigation (FBI), the Defense Communications Agency (DCA), and the Defense Advanced Research Projects Agency (DARPA)) [25]. The purpose of the meeting was to discuss what had happened, how the attack was eventually thwarted, how to prevent future attacks, and what could be learned from the incident. One attendant, Eugene Stafford of the Purdue Department of Computer Sciences, called it a "hastily-convened workshop" and recounted that "the topic of discussion was the program and what it meant to the Internet community" [63].

The meeting proceedings attributed the termination of the Morris worm to an informal network of computer professionals it termed the UNIX "old-boy network". The group recommended formalizing this ad hoc network by launching a centralized group that could coordinate responses to security attacks and also act as a clearinghouse of information that might help avert future problems. Within a month of the NCSC meetings, and after another security incident, DARPA (formerly ARPA) announced the establishment of a new organization, the Computer Emergency Response Team (CERT) to spearhead responses to computer security problems [21]. In addition to coordinating responses to security attacks, part of the organization's responsibilities would include reporting incidents, conducting security research, and educating the computer user community about security issues. CERT was funded and launched by the U.S. Defense Department, which identified computer network security as an important national issue, but the nascent organization actually existed and operated at the Software Engineering Institute (SEI), a research center located at Carnegie Mellon University in Pittsburgh, Pennsylvania. The SEI was already federally funded by DARPA with the objective of adapting software innovations to national defense applications.

At the inception of CERT, the Internet was comprised of fairly distinct, though interconnected, network communities such as ARPANET, Milnet, NSFnet and Bitnet. The DARPA-sponsored CERT would focus on ARPANET and Milnet and DARPA's press release about CERT's formation indicated that "each major computer community may decide to establish its own CERT" [35]. This call for multiple CERT organizations seemed to directly contradict the objective of establishing a centralized coordination center for incident response and advisory provisioning. Nevertheless, the newly formed CERT reported six distinct security incidents in its brief 1988 inaugural year. Throughout the first years of CERT, most advisories announced vulnerabilities in the dominant systems of the time, Digital Equipment Corporation (DEC) and Sun Microsystems platforms, and involved the implementation of vendor-provided software upgrades. Other activities included issuing advisories about worms, dispelling unsubstantiated rumors of Internet attacks propagated by the media, and alerting users to the social engineering aspects of network security, such as remedying lax password strategies. One noticeable phenomenon was the emerging involvement of vendors in identifying vulnerabilities and providing associated software patches. Some of the CERT advisories were simply a recapitulation of vendor security bulletins. For example, one 1990 CERT advisory identified vulnerabilities in the SunOS sendmail feature, a vulnerability that Sun had already reported and addressed with a software solution [14]. As CERT described in its advisory, "This incident underscores the need for system administrators to maintain an awareness of the steps their vendors are taking to improve the security aspects of their products, and to seriously consider upgrading system configurations when solutions to security problems are made available" [35]. Maintaining awareness of multiple vendors' security advisories was not as straightforward as it would later become with the invention of the World Wide Web.

24.6 THE COMMERCIALIZATION OF SECURITY

The advent of the Web in the early 1990s and the associated introduction of Web browser software contributed to a deluge of Internet connectivity among corporations. Whereas researchers, academics, and military communities previously comprised the majority of Internet users, business applications of the Internet suddenly altered these demographics. Early business Internet connectivity emerged in an ad hoc manner whereby individuals or departments established connections usually in the absence of centralized Internet strategies. In other words, Internet connectivity and especially the development of internal web sites were usually grassroots phenomena in corporations rather than part of a uniform strategic decision. In many cases, disparate departments within the same company purchased Internet services and software from different vendors. The heterogeneous nature of the emerging business Internet connectivity, along with the sudden exposure of internal corporate networks to external Internet users, created an immediate security concern. An associated rise of commercial security products and services sought to meet emerging business requirements.

Connecting business networks to the Internet created several immediate security vulnerabilities: exposure to inherent security vulnerabilities in protocols and systems; increased exposure to computer viruses and worms; the possibility of unauthorized access to corporate data and computing resources; data interception and modification; and password theft. Internet connectivity potentially exposed corporate networks to hackers trying to gain unauthorized access for curiosity or for more nefarious purposes like stealing data or disrupting systems. Corporate espionage emerged as a concern, in that rivals might attempt to infiltrate systems for competitive advantage. Additionally, Internet access provided another venue for rogue or disgruntled employees to inflict damage. Because of exposure, requirements for greater access control, authentication, data privacy, and data integrity became critical.

Commercial firewall products became the first imperative and entered the market in the early

1990s to provide access control, the establishment and enforcement of rules about who may access various computing resources on a network. Historically, firewalls were brick walls built between structures to prevent the spread of fire. Analogously, an Internet firewall is software or hardware that regulates access between a private and a public network. The first firewalls were simply routers that filtered packets based on information such as source and destination IP (Internet Protocol) addresses and port numbers. These routers, called screening or packet filtering routers, operated under prespecified rules dictating which traffic to allow and which to prohibit. In the context of the early 1990s, widely available routers such as those by Cisco, 3COM and Wellfleet, were placed between a trusted internal network and the public Internet and adapted to perform packet filtering security services. Firewalls enabled users to specify which IP addresses to allow and which to prohibit for a specific port, with each communications port accommodating traffic of a specific application such as email (SMTP) or web access (HTTP). One of the limitations of early firewall implementations was that they screened traffic based on network information (e.g. IP addresses), not by application or user. An early advancement was the development of proxy-based firewall features, which operated at the application level rather than the network level and acted as a "proxy" between users and applications. The first commercially available firewall, DEC's Secure External Access Link, or DEC SEAL, employed a combination of packet filtering and application level proxy services. Between 1991 and 1994, the other commercially available firewall products introduced were Raptor Systems Eagle, Advanced Networks and Services' (ANS) Inter-Lock, Trusted Information System's (TIS) Gauntlet, and Checkpoint Software's Firewall-1 product [62]. TIS also offered a free product called the TIS Firewall Toolkit. These firewall products typically resided on a dedicated workstation such as a SUN SparcStation or an IBM AIX platform.

Early firewalls were sometimes challenging to use and predicated upon the oversimplified assumption that internal network users were trustworthy and external users were untrustworthy. In reality, some security problems originated within trusted, internal networks while some trusted users like mobile employees or business partners were physically extraneous to trusted networks. Implementing firewalls as a singular gatekeeper or choke point between private networks and the public Internet quickly became inadequate. Usage patterns simply did not comply with the external, untrusted user and internal, trusted user model. For example, business-to-business transactions over the Internet (called B2B) necessitated the extension of business networks to external users and applications. In the Internet parlance of the 1990s, "secure" uses of the Internet between businesses or between businesses and customers (B2C or Business-to-Customer) were termed "extranets", a modification of the term "intranet", which described applications of the Internet, and particularly web technologies, for internal business applications. The extension of internal business applications over the Internet to external trusted partners and customers was accompanied by the proliferation of mobile and remote employees requiring secure access. Firewall products added firewall-to-firewall encryption, user authentication, and virus scanning. Although firewalls quickly became a staple of corporations' Internet security strategies, they represented only one aspect of security.

Commercial enterprises using the Internet to connect geographically distributed employees or business partners were especially concerned about authenticating users. Were customers, vendors, or partners logging on from somewhere on the Internet who they claimed to be? Businesses understood that hackers could potentially intercept passwords transmitted over the public Internet or assume an authorized user's identity. Companies such as Security Dynamics introduced authentication techniques based not only on something a user knows (i.e., password), but something a user holds. Security Dynamics introduced the SecurID card, a small card containing a digital display of a single-use password that changes every few seconds in coordination with a security server. Requiring both a conventional password and a single-use number displayed on the SecurID card provided greater user authentication.

Another concern for enterprises using the Internet for internal business communications or external information exchange with partners was protecting the integrity and privacy of information while transmitted over the Internet. Businesses were accustomed to secure private networks and leased lines or third party, relatively proprietary Electronic Data Interchange (EDI) services for commerce with business partners. The Internet offered the potential for inexpensive and standardized business transactions with partners and customers and the possibility of reaching a more global customer base. This very ubiquity and ease of connectivity to the Internet made the protection of information traversing the network a significant security concern. A major commercial response to the need for secure business communications over the Internet was the introduction of a spate of Internet Virtual Private Network (VPN) services from traditional telecommunications carriers like MCI, AT&T and Sprint as well as newer Internet Service Providers (ISPs) including PSINet and ANS. VPNs offered network services over the public Internet with added inherent features like performance guarantees and security enhancements. To protect the privacy of business information traversing the public Internet, these services offered end-to-end encryption. Encryption techniques over the Internet offered the promise of protecting the privacy and integrity of information, but encountered obstacles. Multiple encryption standards for information exchange over the Internet initially impinged upon marketplace encryption decisions and legal impediments complicated the deployment of global electronic commerce services over the Internet.

24.7 ENCRYPTION AS MUNITIONS

The expansion of the public Internet presented the prospect of inexpensive global electronic commerce and communications and businesses recognized that cryptographic solutions were technically imperative to securely deploying the Internet for business transactions. Many individual users also wanted their personal email communications kept private. "Public key cryptography" responded to

these demands. Cryptographic techniques use a secret code key to encrypt information and, generally speaking, there are two encryption approaches, symmetric and asymmetric cryptography. In symmetric encryption approaches, both the sender and receiver use the same key to encode and decode information. This approach has obvious limitations in an electronic network environment. Transmitting information over an electronic network using symmetric encryption requires sending the key itself over the network. If a network is secure enough to transmit an encryption key in the open, why is encryption necessary in the first place? Exchanging messages with multiple parties using this approach also requires tracking a different key to use with each party. Asymmetric cryptography, such as RSA public key encryption, overcame these weaknesses. RSA was named after three MIT professors, Ronald Rivest, Adi Shamir and Len Adleman, who, in the late 1970s, devised a public key cryptographic system based on multiplying two lengthy prime numbers [55]. Public key systems were first introduced by Whitfield Diffie and Martin Hellman in 1976 [23]. Rather than transmitting a key over the public Internet, the sender and receiver each have two keys, a public one available to anyone and a private key. To transmit an encrypted message, the sender obtains the receiver's publicly available key and uses it to encrypt the message. Only the intended receiver's private key is able to decode the message encrypted with the receivers' public key.

Several Internet encryption approaches based on public key cryptography emerged including email encryption tools like Pretty Good Privacy (PGP), developed by Phil Zimmerman in 1991. Simson Garfinkel's history of PGP provides an account of how American Phil Zimmerman, an antinuclear activist and computer scientist, developed PGP and encountered significant patent and regulatory difficulties [30]. Other encryption approaches emerging in the early 1990s were primarily developed by software companies or business consortia seeking a common standard for secure business transactions over the web. For example, US based RSA Security introduced the S/MIME (Secure Multipurpose Internet Mail Extensions) encryption standard, which was subsequently enhanced by the Internet Engineering Task Force [57]. Secure HTTP

(S-HTTP) was promoted by CommerceNet, a consortium of businesses interested in promoting a standard for secure business transactions over the web to help commercialize the Internet. Netscape introduced the Secure Sockets Layer (SSL) encryption standard in 1994 with its first version of the Netscape Navigator browser. SSL-based Public Key Infrastructure (PKI) services emerged as a means for authenticating user identities, ensuring data integrity and protecting transmitted information. A web browser could obtain a commerce site's public key and encrypt data prior to transmitting private information such as a credit card. Public key encryption approaches not only protected data privacy but provided an option for greater authentication through digital certificates uniquely linking each party to its public key. Digital certificates linking a party and its public key are vouched for, or "signed", by an external party called a Certificate Authority (CA) or Trusted Third Party (TTP), whose own digital certificates are incorporated into browser software [38]. What makes a "third party" sufficiently trustworthy to vouch for the security and digital identity of a web commerce site, for example? In other words, who certifies that the CA is trustworthy enough to certify another party? Independent groups, government entities, and standards bodies have become involved in certifying Certification Authorities. For example, WebTrust is an independent organization that lists the CAs that have passed an audit by an independent practitioner such as Deloitte & Touche, Ernst & Young, or KPMG. The independent practitioner evaluates the CAs based on established criteria consistent with American National Standards Institute (ANSI) guidelines for PKI policies and procedures. VeriSign, headquartered in Mountain View, California, a large trust infrastructure provider, developed a global network of security services including EuroTrust, VeriSign Australia, and numerous worldwide offices. Other Certificate authorities have included Entrust, Comodo, RSA Security, and Digital Signature Trust, among others. An enormous industry developed around SSL certificates and many of the Certificate Authorities were established during the mid- to late 1990s.

The economic and social requirement for Internet encryption encountered the reality that cryptography is a politically charged technology, obviously playing a critical role in military, diplomatic, intelligence and crime fighting strategies. Commercial and freely available encryption solutions designed to address emerging business and personal requirements for secure Internet transmissions initially encountered regulatory obstacles. Impediments in the use of cryptography for international commerce initiatives included encryption export restrictions and a patchwork of distinct regulations varying from nation to nation. Some countries banned encryption outright, some required licenses, and some imposed export restrictions based on the strength of the encryption (i.e., key length). The United States encryption export policy during the explosive 1990s growth of the Internet prohibited the export of strong encryption products with the exception of financial institution transactions or if US law enforcement was provided legal recourse to obtain a decoding key.

Prior to 1996, the United States government grouped cryptography in the category of munitions, placing encryption under the requirements of the U.S. International Traffic in Arms Regulations (ITARS). Commensurate with firearm regulations, encryption products could not be exported without a license and could not be exported to certain designated countries including Syria, Cuba, Iraq and Iran. Businesses exporting Internet encryption products were, in essence, arms dealers under the prevailing regulatory code. The most powerful cryptography businesses could legally export was, for quite some time, 40-bit encryption, relatively easy to break. Business requirements for protecting the confidentiality of proprietary commerce existed in tension with national security and law enforcement interests. The rational for the restrictions addressed concerns about the potential erosion of intelligence gathering ability or criminal investigation facility.

As expressed by TCP/IP developer Vinton Cerf [12], encryption products considerably stronger than the products restricted by US export laws were already available internationally, thereby challenging the national security justification for restrictions which, he argued, only limited valid commerce. The concern was that the encryption restrictions hindered the pace of international electronic commerce over the Internet, failed to address national security concerns, and eroded the competitive position of nascent security businesses in the United States relative to security vendors from countries not bound by similar regulatory limitations. World Wide Web inventor Tim Berners-Lee suggested that US encryption export bans exasperated software manufacturers who have had to develop two different software versions, one with strong encryption, and one with weaker and exportable encryption. Berners-Lee also noted that encryption bans have frustrated the open source code software community "in which distribution of the source code (original written form) of programs is a basic tenet" [7].

To illustrate the seriousness with which the US government approached the Internet encryption export issue, it opened a federal investigation in 1993 of Phil Zimmerman, the developer of the 1991 encryption tool, PGP [30]. A version of PGP appeared as freeware on international FTP servers, a violation of United States encryption export law. By 1996, the Department of Justice announced it would close its investigation of Zimmerman or generically anyone allegedly involved in PGP distribution in 1991 [67]. Also in 1996, the United States modified its encryption policy to remove non-military encryption products from the U.S. Munitions List [20]. Additionally, the new policy allowed the export of encryption products of up to 56-bit key length but required licensing and plans for "key recovery", also called key escrow. Multinational businesses understood that key recovery would potentially enable governments to read their data. They also faced the challenges of concurrently attempting to adhere to encryption regulations that varied from country to country. Many other impediments to Internet commerce arose from encryption regulations. Would a business offering encrypted transactions to customers using, for example, strong 128-bit browser encryption, have to verify online that its customer was in the same country? Though later regulatory modifications eased export restrictions considerably,

Internet users faced a complicated and changing patchwork of regulations and security vendors believed they were competitively impeded by restrictions.

24.8 THE ERA OF CYBERTERRORISM

Throughout the 1990s, as national economies and national security operations became increasingly dependent upon information infrastructures, including the Internet, the possibility of cyberterrorism emerged as a concern. Industries such as finance, transportation, healthcare, energy, defense, retail, and agriculture, relied increasingly upon computer networks for day-to-day operations while, contemporaneously, the number of security incidents rose exponentially. The decade also experienced a series of terrorist attacks on US targets, including the 1993 World Trade Center bombing, the 1996 attack on the Khobar Towers military complex in Saudi Arabia, the 1998 US embassy bombings in Africa, and the bombing of the USS Cole in October, 2000. The confluence of high-profile terrorist attacks, the rise in Internet security threats, and corresponding economic dependencies on the Internet drew attention to the potential threat and consequences of cyberterrorism.

Government concern within the United States was expressed in a variety of ways. President Clinton, in 1996, issued Executive Order 13010 establishing the President's Commission on Critical Infrastructure Protection [51]. The commission identified the threat of cyberterrorism as an ancillary threat to physical terrorism, explaining that, "Today, the right command set over a network to a power generating station's control computer could be just as effective as a backpack full of explosives, and the perpetrator would be harder to identify and apprehend". Oft repeated descriptions of potential cyberterrorism such as an "electronic Pearl Harbor" reflected angst about the potential threat. By 1998, the US government established the National Infrastructure Protection Center (NIPC) as a mechanism to alert the Internet community about possible cyber attacks and to respond to attacks. It was well known that high profile terrorist organizations like Al Qaeda, Hezbollah and

Hamas were using the Internet to communicate. A February, 2000, statement to Congress by Director of the NIPC, Michael Vatis, warned that the increasing technical ability of terrorist groups raised the specter of possible cyberterrorism. In particular, Vatis warned that "Cyber terrorism, by which I mean the use of cyber tools to shut down critical national infrastructures (such as energy, transportation, or government operations) for the purpose of coercing or intimidating a government or civilian population – is thus a very real, though still largely potential threat" [85]. During this time, an Australian named Vitek Boden reportedly broke into the computer system controlling a Queensland, Australia sewage treatment plant and intentionally released millions of liters of raw sewage into rivers, parks, and a hotel grounds, serving as an anecdotal example of the potential for hackers to use computers to disrupt physical infrastructures. Boden, eventually sentenced to two years in prison for his actions, was reportedly a disgruntled former employee who had helped develop the sewage system he attacked [48].

While awareness of the threat of cyberterrorism already existed in the United States throughout the 1990s, concern about an assault on critical information infrastructure became more pronounced after the terrorist attacks of September 11, 2001. The US government's "National Strategy to Secure Cyberspace" (February, 2003) emphasized that the nation's security and economy were dependent upon information infrastructures, including the Internet, and that computer networks controlled real world systems like water distribution, electrical grids, financial markets, trains, and radar [79]. The document formally established a role for the US Department of Homeland Security (founded in November, 2002) in cyberspace security. By June, 2003, the Department of Homeland Security announced the creation of a National Cyber Security Division (NCSD) tasked with reducing vulnerabilities, issuing warnings, and responding to security incidents, responsibilities seeming to mirror the original objectives of Carnegie Mellon's CERT [72]. Tom Ridge, then Secretary of Homeland Security, appointed Amit Yoran, former Vice President for Managed Security Services at Symantec Corporation, as Director of the new NCSD.

What exactly would qualify as cyberterrorism? Internet worms had cost billions in financial devastation so were they considered cyberterrorist attacks? Dorothy Denning defined cyberterrorism as follows:

> "... a computer-based attack or threat of attack intended to intimidate or coerce governments or societies in pursuit of goals that are political, religious, or ideological. The attack should be sufficiently destructive or disruptive to generate fear comparable to that from physical acts of terrorism. Attacks that lead to death or bodily injury, extended power outages, plane crashes, water contamination, or major economic losses would be examples" [22].

The term, cyberterrorism, generally described potential destructive acts motivated by political objectives. Under this definition, even costly Internet worms would not necessarily qualify as cyberterrorism. The sewage control system attack by disgruntled employee, Vitek Boden, would not qualify. Though evidence of a cyberterrorist attack had yet to materialize, the threat of large scale cyberterrorism loomed large because of the potential consequences to any nation economically and functionally dependent upon the Internet and other critical information infrastructures.

Consequentially, a specific action taken by the U.S. Homeland Security Department's NCSD in September, 2003, was the inauguration of a new federally run Computer Emergency Response Team (CERT) that would supercede the longstanding, privately run CERT operating at Carnegie Mellon University. The Homeland Security Department anticipated that the new organization would involve, at least initially, close cooperation between the Homeland Security Department's National Cyber Security Division and Carnegie Mellon's CERT. The announcement of the US CERT included an anticipation that the group would coordinate with private sector security vendors and other CERT organizations around the world [73]. The establishment of US CERT raised the issue of how to appropriately and effectively structure institutional coordination among many CERTs in many nations such as Japan, The Netherlands, Great Britain and Australia. Was it possible to achieve rapid coordination between hundreds of distinct organizations? The establishment of a federal government run CERT illustrated the extent of federal concern about the cyberterrorism threat to national economic and political structures and also raised the issue of the relationship between the public and private sectors in coordinating Internet security prevention and remediation.

Despite federalization of the CERT function, much of the responsibility for implementing effective Internet security in the face of potential cyberterrorism has rested with the private sector. The private sector includes both institutional and individual users of the Internet and also private Internet vendors that manufacture the software and hardware that often contain the vulnerabilities exploited in attacks. Within the private user sector, there has been ongoing debate about the actual threat of cyberterrorism, with many companies placing strategic emphasis on the "what" rather than the "who". In other words, Internet user organizations have focused on securing vulnerabilities rather than distinguishing between multiple attack sources ranging from teenage hackers to corporate criminals to terrorist organizations. Private Internet vendors, such as software manufactures, have carried considerable responsibility for the product vulnerabilities that have enabled security attacks. However, in the United States, software is exempt from usual product liability laws, illustrating that private market forces, like public oversight, are not alone sufficient to combat potential cyberterrorism or other Internet security threats.

24.9 ONGOING PUBLIC PLAGUES: WORMS AND VIRUSES

While no instances of actual cyberterrorism had yet been recorded, costly Internet attacks in the form of viruses and worms incessantly plagued Internet users throughout the 1990s and into the 21st Century. Long after the Morris worm of 1988, Internet worms continued to exploit protocol and product vulnerabilities and have presented an inexorable security challenge. In the globally interconnected and ubiquitous Internet environment, worms, as well as viruses, can not only damage data, compromise resources, and debilitate networks, but consume billions of dollars in Internet resources and

human administrative capital. In Internet parlance, the term "virus" describes malicious code cloaked in a legitimate program that activates when a user performs some legitimate function. For example, a virus may be embedded in an email attachment or web link and is executed when the attachment is downloaded or the link is selected by a user. In other words, a virus requires activation by a user. In contrast, an Internet worm is self propagating, autonomously replicating itself from system to system without action by a user. Worms primarily exploit software security holes and usually are solved by installing vendor provided software patches or upgrades.

The potency of viruses became clear in 1999 when a particularly virulent virus known as Melissa propagated rapidly across the Internet as an infected electronic mail attachment in Microsoft Word 97 or Word 2000 [18]. The email subject line of a Melissa-containing message usually read, "Important Message from <name>" but opening the attachment would infect a user's computer and trigger the emailing of virus copies to the first 50 names found in the user's Microsoft Outlook address book. Although other similar viruses had existed, Melissa garnered great concern because it spread so rapidly. According to CERT's account of the virus dissemination, a single location apparently received more than 30,000 Melissa mail messages within 45 minutes. Because of its exponentially rapid replication, the virus overloaded electronic mail servers and shut down major computer networks. Melissa infected more than a million Internet-connected computers in North America and resulted in financial damages of at least $80 million [74]. The virus was launched by David Smith, 31, of New Jersey, who pled guilty to the crime and eventually served 20 months in prison. Smith used a stolen America Online account to launch the virus on the "Alt.Sex" Usenet group [76].

Demand for anti-virus software rose in response to the Melissa calamity but anti-virus programs have only protected against known threats. Within a year of Melissa, an even more destructive program, the "I Love You virus", spread globally using similar tactics but spreading even faster, using more destructive tactics like overriding files, and resulting in massive financial losses. The I Love You, or "Love Bug" virus originated in the Philippines and quickly spread throughout Asia, Europe and North America. According to Congressional testimony about the virus in the United States, an estimated 65% of North American businesses had been impacted, infecting approximately 10 million computers and costing, in North America alone, an estimated $950 million [80].

Even virus hoaxes have been problematic because they consume system and human resources. Hoaxes like the "Good Times" virus of 1994 have periodically circulated through Internet email and have emulated self replicating worms as they have been invariably passed from user to user. The Good Times virus hoax warned Internet email users that opening any email message with the text "Good Times" in the subject line would erase their hard drives. The contents of an email subject line were incapable of launching a virus, but the hoax propagated nevertheless and attributed the warning to the United States Federal Communications Commission (FCC). The Good Times virus behaved like a real virus because thousands of users forwarded the email warning, consuming network and system resources and diminishing user productivity. In a self fulfilling prophecy, when users forwarded the email, the term, "Good Times", automatically appeared in the subject line, the very offense the email forewarned. The hoax compelled the US government to release a public service announcement dispelling the rumor [71]. Similar hoaxes have warned users to avoid certain email lines like "Guts to Say Jesus", "Undelivered Mail", "Penpal Greetings" and countless others. Security software vendors like Symantec and McAfee have maintained lengthy lists of virus hoaxes that have plagued Internet email over the years. These hoaxes have used social engineering techniques to induce users into taking actions that simulate an actual virus or worm, bogging down networks and distracting users. Similar to battling real viruses, one method network administrators have employed to impede the spread of virus hoaxes is educating users about the inability of subject line titles to activate viruses and keeping them informed of hoaxes as they appear.

Internet worms have arguably been more virulent than viruses because they have propagated autonomously without user action. Worms have exploited security vulnerabilities in existing software, often operating systems, and have spread across the global Internet in a matter of minutes. Annually since the millennium, at least one or two major worms have stormed the Internet: the Code Red and Nimbda worms of 2001, the Klez worm of 2002, the Slammer and Blaster worms of 2003, the My Doom, Bagle and Sasser worm variants of 2004, as well as others. When Code Red struck in 2001, it was generally considered the most costly worm or virus to ever hit the Internet, requiring billions of dollars to combat. Often the devastation of worms is not in debilitating or slowing the Internet but in the immense expense of containing attacks. The Code Red worm was characteristic of other worms in several ways. It exploited a vulnerability associated with popular software (Microsoft Windows); it disseminated by scanning networks for other vulnerable systems, and it launched coordinated attacks that overloaded servers. Its primary functional impact was performance degradation on networks, resulting in loss of productivity and requiring administrative and financial resources to fix. Also like other worms, the solution for preventing further attacks was to install a vendor provided software patch.

Technical monocultures seem to have contributed to the destructive breadth and rapid spread of worms are technical monocultures. For example, when a worm exploits a security hole in a Microsoft Windows operating system, the worm has an enormous installed base to attack. The Code Red worm of 2001 affected systems running IIS 4.0 or IIS 5.0 on Windows 2000, Windows NT and beta Windows XP versions [41]. The Nimda worm targeted Microsoft Windows 95, 98, NT and 2000 systems [15]. The Blaster worm similarly targeted Microsoft Windows NT 4.0, Windows 2000, Windows XP, and Windows Server 2003 [16]. The Slammer worm exploited Microsoft SQL Server 2000 and Microsoft Desktop Engine (MSDE) 2000 systems [17]. Microsoft has continuously offered patches to mitigate security vulnerabilities as they arise. It also took an unprecedented step in late

2003 when it publicized a reward of $250,000 for information leading to the identification and arrest of the individual responsible for unleashing either Blaster or SoBig. Worms have often appeared after a vendor issues a patch to fix an identified security hole. Addressing the history of worm appearances, Microsoft's Senior Security Strategist, Scott Culp, testified before the US Congress that "A troubling recent security trend has been the dramatic shortening of the time between the issuance of a patch that fixes a vulnerability and the appearance of a worm exploiting it. In just the past several years, this window has narrowed from hundreds of days in the case of NIMDA, to 26 days for Blaster, to 17 days for the recent Sasser worm" [81].

Some attacks have been "zero-day attacks" in that they have occurred before the software vendor became aware of the software program vulnerability the worm exploited. Only after a worm exploiting a software vulnerability has been released has the software vendor become aware of the problem and developed an associated software patch. The existence of zero-day attacks and the increasingly rapid dissemination of worms raised questions about how information about worms and associated software patches could expeditiously reach Internet users before they become targets. The original objective of CERT in the late 1980s was to provide a single entity that could educate users, issue centralized alerts, and provide security solutions. What has actually materialized is a patchwork of hundreds of distinct CERTs around the world. The U.S. CERT under the U.S. Department of Homeland Security coordinates with the original CERT Coordination Center at Carnegie Mellon University, but is a distinct entity. By 2004, more than 250 CERT organizations blanketed the globe. For example, in January of 2003, India's Ministry of Communications and Information Technology launched the Indian Computer Emergency Response Team (CERT-In) to respond to security incidents in India and help Indian institutions reduce security vulnerabilities. Most countries developed a CERT organization and some formed multiple entities performing similar roles. In effect, hundreds of organizations formed to monitor Internet incident activities which hundreds of vendors issue their own advisories and software patches.

24.10 DDoS WARS

Denial of Service (DoS) attacks have constituted a similar challenge to Internet security. DoS assaults suspend the availability of web sites or other targeted systems by flooding them with enormous volumes of traffic. Distributed Denial of Service (DDoS) attacks bombard a targeted computer simultaneously with traffic from numerous, in some cases, thousands of distributed sites. An analogous scenario would be thousands of individuals simultaneously flooding an emergency dispatcher with calls to such an extent that effectively blocks legitimate calls. These attacks do not necessarily damage systems or steal data but create problems simply by generating outages. DDoS attacks have targeted individual users, commerce web sites, corporate servers, and government sites. Even more damagingly, DDoS assaults have assailed the infrastructures of Internet Service Providers (ISPs), causing performance degradation and service outages affecting all users of the ISP network.

The nature of DDoS attacks starkly confronted the broad Internet community when attacks disrupted service at several high profile web sites between February 7 and 10 of 2000. The first publicly noticeable outage involved Yahoo's web site on February 7. On subsequent days, DDoS attacks targeted such high profile sites as CNN.com, Amazon.com, eBay and ETrade. The attacks also targeted Buy.com on the day it launched its Initial Public Offering (IPO) [29]. The consequences of the February, 2000, coordinated attacks were significant. Targeted companies incurred considerable costs to combat the attacks and lost commerce and advertising revenue during the outages. The targeted companies also suffered negative publicity and the public perception that their systems might not be secure enough for commerce transactions. Additionally, the enormous volumes of traffic flooding the Internet during the series of coordinated attacks resulted in a noticeable degradation of performance among some Internet users. Who was responsible for these attacks on electronic commerce web sites? After almost a year of investigation by the United States Federal Bureau of Investigation (FBI) in conjunction with the Royal Canadian Mounted Police (RCMP), a 15-year old Canadian male known as "Mafiaboy" on the Internet, plead guilty in a Montreal Youth Court to the February, 2000, attacks [75].

How was such a successful DDoS attack able to occur on commerce sites like Amazon and EBay that used encryption services, firewalls, and authentication services? At the time of the incident, DDoS tools, with the sole objective of disabling systems and networks, were freely available on the Internet. available DDoS tools included "Trinoo", "Tribal Flood Network", "mstream", "Trinity" and "Stacheldraht", meaning barbed wire in German. These tools consist of a master program (also called a handler) and agent programs (also called daemons or zombies) installed on compromised systems detected by automated scanning of Internet ports for known vulnerabilities or distributed by worms or Trojan horse viruses. Third party systems not belonging to either the attacker or the target are unwittingly used to launch DDoS attacks and are also victim systems. To launch an attack, the master program issues instructions to the agent programs to repeatedly attempt to make a connection with the targeted computer, often using fake ("spoofed") IP addresses. The high volume of connection requests originating from numerous, coordinated attack systems, flood the target computer with so much traffic it becomes disabled. The owners of the compromised systems used as agents are usually unaware that hackers are using their systems to attack another system. It is difficult to trace the original perpetrator because attacks originate from distributed, third party zombies. The assaults either take advantage of known software vulnerabilities or use existing protocol attributes. Who was responsible for publishing DDoS attack tools like Tribal Flood Network and Stacheldraht? After the 2000 DDoS attacks on ecommerce servers, an anonymous German programmer named "Mixter" reportedly admitted in an interview that he was the author of Tribal Flood Network. Mixter, then a recent high school graduate, claimed to have developed TFN to raise awareness about Internet security vulnerabilities and to enable testing of network security. Mixter also claimed that another German programmer, "Randomizer" wrote the DDoS tool, Stacheldraht [61].

One ongoing area of vulnerability to DDoS attacks has been the Internet itself. The Internet is generally a distributed and redundant system of technologies that would seem inherently difficult to disrupt. However, it includes a centralized system, the Domain Name System (DNS), a hierarchical, massive database management system that translates between the alphanumeric Internet addresses humans understand and the binary IP addresses computers understand. The domain name system is centralized in that there must be a single root for the hierarchical name space. Even though the Domain Name System is distributed across numerous "root" servers, it is a target for attacks because of its essential Internet function. Attacks against the DNS root servers have occurred throughout the years but one particularly coordinated attack occurred on October 21, 2002, when a distributed denial of service attack (DDoS) simultaneously targeted the DNS root servers [86]. Although the attack did not noticeably interrupt Internet functionality for the majority of users, and although it lasted for only an hour, it was disconcerting in that it simultaneously targeted the root servers. Since this attack, the Internet's Domain Name System has become even more distributed, employs greater site mirroring, replication, and load balancing but, because of its importance, is still a vulnerable target for someone seeking to disrupt the Internet's operation.

Has anything stopped DDoS attacks? Preventative measures have helped diminish the risk of becoming a target. At a general level, these measures have included patch management strategies that mitigate software vulnerabilities as they become disclosed, use of monitoring tools to scan for the existence of agent software, distribution of traffic loads across multiple servers, building architectural redundancy, blocking access to certain vulnerable TCP/IP ports, disallowing traffic from broadcast and multicast Internet addresses, and using traffic monitoring tools to identify suspicious flow patterns. But defensive strategies implemented after the major DDoS incidents in 2000 and 2002 failed to prevent further attacks.

High profile incidents have increasingly sought to make an economic or political statement. A well-publicized attack against Al-Jazeera English and

Arabic language web sites occurred shortly after the 2003 inception of the Iraq war. Another episode occurred in 2003, when software vendor SCO announced it was experiencing massive DDoS attacks [58]. During one attack, the company publicized that thousands of servers appeared to be flooding its systems with illegitimate web requests, effectively rendering its corporate Web site unavailable and disrupting internal corporate networks. In its announcement, SCO described the attack as corporate cyberterrorism, "We deplore these activities by those who try to intimidate or harass legitimate businesses through cyber terrorist tactics while hiding their true identity" [90]. Why did SCO assume someone was trying to intimidate them? SCO had launched a lawsuit accusing IBM of intellectual property infringement related to Linux software code, and some in the open source software community believed SCO's tactics were antithetical to the tenets of the open source software philosophy. One attack, according to SCO, occurred two days after IBM's response to the lawsuit. Augmenting the controversy, some in the Linux community accused SCO of fabricating the DDoS attack. On the other hand, some members of the open source community publicly denounced the attacks against SCO. For example, LinuxWorld Magazine issued an urgent appeal calling for an end to the DDoS onslaught against SCO:

> "While many in the Open Source community are not pleased with SCO's lawsuit against IBM, or their proposed legal challenges aimed at Linux users, these DDoS attacks do not promote the Open Source cause, and are not consistent with Open Source values. The Open Source community is based on the notion that principals of free speech should be applied to software development. DDoS attacks clearly deny the victim the ability to communicate freely on the internet. The fact that someone, or some group disagrees with the policies and/or practices of another person or group does not justify this type of attack. We strongly urge whoever is responsible to stop the attack immediately" [6].

After numerous DDoS attacks in 2003, and after implementing protective measures to deter future attacks, SCO confirmed, in 2004, that it was again hit by a crippling assault launched by the MyDoom worm which infected computers and instructed them to bombard SCO's web site with illegitimate requests during a prespecified duration of time [59]. Open source code, which anyone can freely access and modify, has been extolled for a number of reasons including its intrinsic conduciveness to peer review, collaboration, principles of democratization, and freedom to innovate. As Larry Lessig [39] has noted, "From the beginning the trend to enclose code on the Internet has bothered many – some because they believe closed code is less efficient than open code, others because they believe closed code interferes with important values of the Internet". A blurry line sometimes exists between "open source software" and "free software" communities. The Free Software Foundation, founded by Richard Stallman in 1985, stresses a distinction between the two communities based on values: while both are against proprietary software, the open source movement stresses pragmatic values while the free software community stresses ethical issues of freedom [31]. The Free Software Foundation suggests "you should think of "free" as in "free speech" not as in "free beer" [32]. From a different perspective, Sociologist Manuel Castells has suggested that values of the open source movement are the "crucible" of innovation and creativity [11].

Unlike the attacks motivated by political and technical philosophies, the MyDoom worm appeared to target some sites in a manner seemingly motivated by economic issues. For example, on the morning Google announced the details of its initial public offering, Google experienced a slow down in its search service because it was a target of a DDoS attack executed through a variant of the MyDoom worm. Some speculated that the attack was designed to criticize Google's estimated IPO price of $108 to $135 a share. The Google incident was reminiscent of the Buy.com attacks and illustrated that, while perpetrators do not directly profit, some attacks appear to be motivated by economic rationales, whether sullying initial public offerings or allegedly attacking a corporation for unpopular business tactics. Regardless of motivation, Mirkovic et al., in *Internet Denial of Service Attack and Defense Mechanisms*, have stressed that "very few attackers have been caught and prosecuted" [42].

24.11   GROWTH OF WIRELESS INTERNET ACCESS

Securing Internet access via wireless LANs (WLANs) emerged as another specific security challenge. Students began arriving on university campuses with wireless enabled laptops expecting wireless Internet access. Public wireless Internet "hotspots" materialized in coffee shops, hotels and airports. Many businesses augmented traditional local area networks with wireless LANs for cost effectiveness and flexibility. Home Internet users increasingly implemented wireless LANs in homes, accessing the Internet through wireless laptops connected to a network access point in turn connected to the Internet over a high-speed connection. The requirement and expectation for mobility exploded in businesses, homes, universities and public venues. Wireless LAN technologies like the Institute of Electrical and Electronics Engineers (IEEE) 802.11 family of wireless LAN standards, especially 802.11b or "Wi-Fi," have supported this proliferation of wireless Internet access.

Organizations deploying wireless LANs for Internet access and other applications encountered several security challenges. In corporations, employees installed their own inexpensive access points outside the purview of network administrators. These "rogue access points" provided open network entry points for unauthorized access to internal networks. Gaining unauthorized access to data or simply conducting bandwidth piracy was relatively easy to accomplish through "Wi-Fi sniffing", gaining access simply by hunting for unprotected wi-fi hotspots. Another security vulnerability for wireless Internet access, whether via businesses, public venues, or in homes, was the possibility of over-the-air (OTA) interception of information transmitted over WLANs. To protect OTA transmissions, the 1999 version of the 802.11 wireless standard specified the inclusion of an encryption approach known as Wired Equivalent Privacy (WEP) [37]. Vendors built WEP into wireless cards and this standard became the de facto approach for interoperable security in wireless LANs. The main objectives of WEP were to ensure confidentiality of data while transmitted over 802.11

wireless LANs, but it also was intended to provide some access control and data integrity by preventing modification of transmitted messages.

The WEP standard had appropriate objectives, but its inherently weak security became a major impediment to deployments of wireless Internet access. By 2001, a series of independent studies highlighted the various cryptographic vulnerabilities and weaknesses of the WEP specification. The studies determined that WEP had major security flaws and should not be counted on to provide security. One study by UC Berkeley researchers described numerous significant flaws in the protocol that "demonstrate that WEP fails to achieve its security goals" [8]. Another technical paper, developed by Rice University experts in conjunction with AT&T Bell Labs, asserted, "We conclude that 802.11 WEP is totally insecure" [65]. WEP encryption employed a secret key that mobile nodes and an access point would share. The group of researchers was able to intercept and read these keys transmitted over 802.11 networks using only commercially available hardware and software and by implementing an attack method originally described by Fluhrer, Martin and Shamir [28]. The researcher concluded, "We successfully implemented the attack, proving that WEP is in fact completely vulnerable" [65]. The method of attack was passive in that it recovered secret keys as they traversed networks. The validation of WEP's security weaknesses by researchers confirmed concerns that any individual with a Wi-Fi enabled computer within physical proximity to a WEP secured WLAN could gain access as a falsely authenticated user by easily obtaining the LAN's secret key. Additionally, because each network only used one key, the compromised key would pose a security threat to all network nodes.

Widespread concern about the inherent cryptographic weaknesses in WEP prompted businesses to reconsider the speed of planned WLAN deployments. One outgrowth of the cautious adoption of WLAN deployments was a diminishment in WLAN equipment costs, making wireless LAN equipment more accessible to the consumer market. Understanding the immediate market need for greater security than WEP provided, the IEEE's 802.11 working group for wireless local area networks began developing a more robust security approach called 802.11i. The industry expected that the IEEE would finalize this standard by late 2003 or later. In the meantime, Wi-Fi users needed something more powerful than WEP.

The Wi-Fi Alliance, a coalition of wireless vendors formed in 1999 to test and certify the interoperability of products based on the IEEE 802.11 standards, recognized market concerns about encryption weaknesses. As Wi-Fi Alliance chairman Dennis Eaton suggested, "Enterprises, small businesses and home users need a stronger standards-based security solution than WEP and they need it now" [89]. To make data interception much more difficult to accomplish over 802.11 wireless LANs, the Wi-Fi Alliance agreed to incorporate into products a new security standard called Wi-Fi Protected Access, or WPA. The Wi-Fi Alliance agreed upon the interim specification in 2002 and vendors introduced corresponding products by early 2003. The Wi-Fi Alliance considered WPA an interim approach to act as a stopgap measure between the extremely vulnerable WEP and the IEEE's forthcoming security specification, 802.11i, expected to materialize by late 2003. WPA was based on 802.11i draft specifications and the Wi-Fi Alliance worked with the IEEE's 802.11 working group to bring this interim standard to market. The interim WPA encryption upgrade could work on existing wireless hardware rather than requiring new equipment as the forthcoming 802.11i standard was expected to require. WPA provided significantly enhanced encryption over WEP and also added a workable user authentication approach.

In the meantime, the U.S. Department of Defense also acknowledged concerns about unsatisfactory security in wireless LANs and issued specific requirements for wireless LAN security to be included in any commercial wireless products or services deployed in military networks [70]. The DoD mandate, termed Directive 8100.2, specifically prohibited the use of wireless devices for transmitting classified information and required that, for unclassified transmissions, products undergo cryptographic validation under the U.S. National Institute of Standards and Technology's Federal Information Processing Standard (FIPS) 140-2 encryption requirements. The use of the Advanced Encryption Standard (AES) in the forthcoming IEEE wireless security standard would have met the requirements of the DoD mandate.

While the IEEE's 802.11 Working Group continued crafting the much needed 802.11i security standards in 2003, a competing standard emerged. Independent of the international IEEE process, the Standardization Administration of China (SAC) established wireless standards known as GB15629.11 and GB1529.1102, specifying the use of a new WLAN Authentication and Privacy Infrastructure (WAPI) security protocol. It appeared there would be two incompatible security standards, the IEEE's 802.11i security specification and the WAPI standard the People's Republic of China mandated. Outside equipment vendors selling WLAN products in China would require licensing WAPI from one of a number of Chinese companies, effectively creating a barrier to international trade in China's rapidly growing WLAN internet access market. The Chinese standard was based upon the 802.11 standards with the exception of the security protocol, WAPI, which would be essentially a proprietary technology.

As a response to China's new proprietary security protocol mandate, Paul Nikolich, the Chair of the IEEE's 802 LAN/MAN Standards Committee issued a letter to Li Zhonghai, Chairman of the Standardization Administration of China, expressing concern that WAPI would create a balkanization of wireless LAN products [46]. While acknowledging that 802.11 security was problematic and that the IEEE was seeking to alleviate problems through the 802.11i specification, Mr. Nikolich warned that the coexistence of WAPI and 802.11i would prohibit the existence of a unified global security standard for the wireless LAN market. By March of 2004, industry and government groups registered concern about China's proprietary encryption standards and US Secretary of Commerce Donald Evans, US Secretary of State Colin Powell and US Trade Representative Robert Zoellick issued a letter to Chinese Vice Premiers Wu Yi and Zeng Peiyan calling for a resolution of the standards quandary. In June of 2004, the IEEE approved the long anticipated 802.11i security standard solidifying the inclusion of a more powerful cryptographic approach, the Advanced Encryption Standard (AES), into 802.11 technologies. The AES standard provided strong encryption, supporting 128-bit, 192-bit and 256-bit cryptographic keys and requires a hardware upgrade. The incorporation of the Advanced Encryption Standard into WLANs improved wireless Internet access security and helped ameliorate some market concerns about implementing business, home, and public wireless Internet access points. The impasse between China's WAPI and the IEEE's standardization efforts illustrates how security protocols have continued to embody economic and political interests as well as technical requirements.

24.12 SPYWARE INFESTATION

In the opening years of the 21st Century, spyware emerged as an obdurate problem for both individual and institutional Internet users. The general term, spyware, refers to the unauthorized installation of software or unauthorized system configuration changes designed to mine personal information, track online behavior, or deliver pop-up advertisements. Computer users with systems contaminated with spyware experienced degradation in system performance, system crashes, a deluge of pop-up advertisements (usually called adware) generated from monitoring a user's web site visits, tracking search terms, or scanning words written in an email, as well as unwanted modifications such as the reconfiguration of a web browser to direct the user to a different home page. The spyware problem raised many technical, economic and legal questions. Even the definition of spyware itself has been contestable, with many software developers and adware distributors believing their products should not be classified as spyware. Some general questions raised include whether all pop-up ads are necessarily spyware, what specifically should be eliminated by anti-spyware software, what counts as consent to the installation of spyware products, what role should governments play in deterring spyware, and what happens when

a software developer issues an anti-spyware product that classifies some of a competitor's software as spyware. What about "cookies", the files a web server loads on a user's hard drive and accesses the next time the user visits the web server? Are cookies spyware?

In many cases, spyware infects computers with no explicit user consent, but in other cases, Internet users have unknowingly consenting to the installation of spyware on their computers. The rise of peer-to-peer (P2P) file sharing applications for sharing music and movies, as well as other free software for gaming or electronic mail, contributed to the proliferation of spyware. Providers of free software and services bundled spyware programs into free software and ask users to consent to a lengthy licensing agreement that includes broad and often vague language about this bundled software. "Always on" broadband connectivity such as cable access, high-speed wireless, or DSL created greater security challenges for consumers compared to older dial-up connectivity. A 2003 study conducted by the National Cyber Security Alliance (NCSA) found that the vast majority of consumer broadband Internet access lacked even rudimentary security. The findings of the NCSA's broadband security study, conducted by America Online technical staff, apparently demonstrated a gap between the perception of security among broadband consumer users and the reality of their degree of security. Specific findings included the following:

- 67% of Users Do Not Have Properly and Securely Configured Firewalls.
- 62% Do Not Regularly Update Anti-Virus Software.
- 91% of Broadband Users Have Spyware Lurking on Home Computers [45].

The NCSA findings indicated that most broadband users believed they had adequate security but only 11% actually had secure systems. During the 2003 timeframe of the NCSA study, the onus of responsibility for security was on consumers rather than broadband service providers bundling comprehensive security features into service offerings. For example, a consumer purchasing DSL service would have to independently install a firewall and virus and spyware protection software.

Another contributor to the proliferation of spyware was the changing revenue models for Internet content or service providers from subscription to advertising, placing a premium on attracting user web page views containing advertising and implementing highly targeted advertisements.

Who was behind the initial spyware onslaught? A segment of Internet marketing firms and advertising distributors adopted spyware approaches for financial gain, earning commissions when consumers viewed advertisements or for transactions resulting from advertisements. They also benefited by selling so-called spyware removal products for the eradication of the spyware they themselves installed. For example, the United States Federal Trade Commission (FTC) charged Seismic Entertainment Productions, Inc. and Smartbot.net, Inc. with infecting computers with spyware and bombarding computers with pop-up advertisements, beginning in 2003. Both of these companies were small, privately-held Internet marketing companies operated by an individual named Stanford Wallace in the state of New Hampshire [77]. The FTC charged the plaintiffs with downloading and installing software code onto computers without the users' knowledge, resulting in a deluge of pop-up advertisements, creating performance problems, and changing Microsoft Internet Explorer web browser configurations to lead the consumer to a different web page or search engine. The plaintiffs also allegedly sold Spy Wiper and Spy Deleter anti-spyware tools to unwitting consumers who had been impacted by these spyware practices. Similarly, a US District Court, at the request of the Federal Trade Commission, prohibited a Washington Corporation, MaxTheater, Inc., from continuing a marketing scam related to its anti-spyware software [78]. The FTC alleged that this company marketed a free product, "SpywareAssassin", which would claim to detect spyware even if no spyware was present. Then the company would market a software product for $29.95 that it claimed would remove the spyware software, violating the FTC's prohibitions on deceptive claims [27].

Legislative efforts to combat spyware in the United States commenced in 2004. The State of Utah passed anti-spyware legislation, called the *Spyware Control Act*, prohibiting certain types of spyware, including "context based triggering mechanisms" for advertisements, meaning the ability of spyware software to facilitate the serving of an advertisement based on a web site a user is visiting or the content of that web site [84]. California also passed anti-spyware legislation, the *Consumer Protection Against Spyware Act*, prohibiting many spyware tactics such as home page hijacking, system modification, and deceptive software installation practices [64]. Legislative efforts have alternately been criticized for being to broad or too narrow, depending on one's perspective. For example, New York based Internet marketing firm, WhenU.com, Inc., consequently sought and won an injunction preventing Utah from enforcing the Spyware Control Act on the grounds that it violated free speech and the commerce clause of the United States Constitution [83]. On the national front in the United States, the House of Representatives overwhelmingly passed anti-spyware legislation (May, 2005) that would impose up to a two year jail sentence and multimillion dollar fines for such spyware practices as logging user keystrokes or reprogramming the opening web page on a web browser. The House of Representatives had previously passed anti-spyware legislation in 2004 but the bill never reached the Senate for a vote.

The early legislative, law enforcement, and technical efforts to curb spyware have followed similar efforts to battle spam, the unsolicited commercial electronic mail messages that flood both individual and business Internet mail boxes. The preponderance of research reports, surveys, and online statistics by 2003 concurred that spam resulted in billions in lost worker productivity, ranked among individual users as one of the most unwanted Internet intrusions, and resulted in major Internet service providers deleting somewhere between 40–80% of incoming electronic mail [88]. In a matter of seconds at almost no cost, spam marketers could send thousands of electronic mail solicitations. Similar open questions arose in attempts to control spam, such as what counts as spam and what is the efficacy of national and regional laws in an environment that transcends international boundaries.

Prefiguring efforts to prevent spyware, legislative approaches sought to stop spam solicitations. For example, in 2003, the United States Congress enacted the *CAN-SPAM Act of 2003*, short for "Controlling the Assault of Non-Solicited Pornography and Marketing Act of 2003". One outgrowth of anti-spam legislative efforts was the first U.S. felony conviction for spam in April of 2005. Under a Virginia anti-spam statute, Jeremy Jaynes of North Carolina was convicted of inundating America Online (Dulles, Virginia) accounts with tens of thousands of spam e-mail advertisements and initially given a 9-year prison term [2]. Prosecutors had believed Jaynes was actually responsible for sending millions of spam emails a day, grossing $750,000 per month.

24.13 SECURING THE INTERNET: IS IT POSSIBLE?

Some Internet communities have begun a major architectural upgrade to a new network protocol, IPv6, designed to dramatically expand the number of devices able to connect to the Internet. Every device connection to the Internet requires a unique address known as an "IP address". Since the early 1990s [9], the Internet Engineering Task Force (IETF) recognized that a number of circumstances foreshadowed an Internet address shortage and developed a new network protocol known as Internet Protocol Version 6, or IPv6. The anticipated address shortage arose from several circumstances including the original generous allocation of addresses to American institutions when the Internet was primarily a US phenomenon, the subsequent international growth in Internet use and the rapid proliferation of new Internet connected devices like wireless Internet access and Internet telephony (Voice over IP, or VoIP). The IPv6 standard provided an enormous expansion of the number of Internet addresses, and upgrading became a priority especially in Asia and Europe by the opening years of the 21st Century. In the United States, the February, 2003, National Strategy to Secure Cyberspace recommended the US consider the possibility of upgrading to IPv6, noting that Japan, the European Union and China were already upgrading from IPv4 to IPv6 and citing "improved security features" [79], as one of the benefits. Additionally, the United States DoD announced in June of 2003 that it would transition to IPv6 by 2008 as part of its objective to achieve greater "net-centric operations and warfare" [69]. Among the rationales cited in the DoD's IPv6 memorandum was a requirement for end-to-end security. Does IPv6 provide greater security? Many have suggested it does because the standard mandates the use of the IPsec security protocol. Others have noted that IPsec can also be used with IPv4. Still others have recognized that IPv4 and IPv6 will coexist for the foreseeable future and question whether a mixed protocol environment might actually be less secure.

Whatever the future might hold, the statistics maintained by CERT/CC at Carnegie Mellon indicate that Internet security incidents have increased annually. The first significant Internet security incident, the Morris worm of 1988, exploited vulnerabilities in implementations of TCP/IP, the now universal enabler of information exchange between computing devices. From the early days of TCP/IP's rapid and widespread adoption, serious security vulnerabilities inherent to the protocols have been identified and exploited [5]. A quick review of vulnerability and incident reports recorded by various CERTs indicates that protocol vulnerabilities in TCP/IP have presented an ongoing dilemma for Internet security. When each new vulnerability surfaces, vendors issue software patches and recommend additional security approaches. With so many vulnerabilities addressed over the years, has TCP/IP become relatively secure?

Sixteen years after the Morris incident, the British government announced a major security vulnerability in TCP that would allow attacks potentially disabling major segments of the Internet [47]. The morning after this April 20, 2004, announcement, newspapers across the world published headlines such as "Major Internet-Security Hole Found" [87]. This Internet security vulnerability, originally identified by American security researcher, Paul Watson, would impact all systems dependent upon TCP connections, most importantly critical Internet routers using the Border Gateway Protocol (BGP). This particular TCP weakness would allow an attacker the ability to "reset" TCP connections and effectively terminate Internet sessions. Repeated manipulation of this protocol vulnerability would create denial of service attacks against "a large segment of the Internet community" according to the Technical Cyber Security Alert describing this vulnerability [68]. Many network services and applications explicitly rely on the TCP protocol, so any TCP weakness has potentially massive repercussions for the Internet community. Like other discovered TCP vulnerabilities, the solution required downloading of a vendor software patch or implementation of cryptographic solutions. Specific TCP/IP implementations and, in general, TCP/IP as a collection of protocol specifications originally developed under the auspices of the U.S. Department of Defense for relatively trusted users, have required continued vigilance for new security vulnerabilities.

Software specific attacks have similarly continued, with software vendors constantly issuing patches to remedy vulnerabilities and battle virus and worm offensives. Augmenting the traditional security weaknesses of Internet protocols and software are emergent areas of concern, especially the prolific growth of wireless and broadband Internet access. Obviously the shift from a small, relatively trusted network supporting research, academic, and US military applications to a ubiquitous, commercial, and public network has been a major enabler of increased Internet security attacks. There are simply more sites to attack, more is at stake in perpetrating attacks, and more individuals and groups have relatively anonymous and ubiquitous access from which to inexpensively launch attacks. History indicates that the economic, political, and technical tension between securing versus impeding information exchange has endured but yet has not repressed the escalation of the Internet as a global electronic commerce and communications platform.

REFERENCES

[1] J. Abbate, *Inventing the Internet*, MIT Press, Cambridge (1999).

[2] Attorney General of Virginia Press Release, Kilgore announces nation's first felony spam arrest – world's eighth-worst spam kingpin ensnared by tough new Virginia law, December 11 (2003).

[3] A. Bass, *Computer virus halts research*, Boston Globe, November 4 (1988).

[4] A. Bass, *Computer invasion seen as a warning*, Boston Globe, November 5 (1988).

[5] S.M. Bellovin, *Security problems in the TCP/IP protocol suite*, Computer Communication Review 19 (1989), 32–48.

[6] S. Berkowitz et al., *Urgent appeal*, Linux World Magazine, December 10 (2003).

[7] T. Berners-Lee, **Weaving the Web: The Original Design and Ultimate Destiny of the World Wide Web by its Inventor**, HarperCollins, San Francisco (1999), p. 151.

[8] N. Borisov, I. Goldberg, D. Wagner, *Intercepting mobile communications: The insecurity of 802.11*, **Proceedings of the Seventh Annual International Conference on Mobile Computing and Networking**, The ACM Press, New York (2001).

[9] S. Bradner, A. Mankin, *IP: next generation (IPng) white paper solicitation*, RFC 1550, December (1993). Online available at ftp://ftp.rfc-editor.org/in-notes/rfc1550.txt.

[10] D. Cameron, *Student unleashed computer virus*, Sydney Morning Herald, November 7 (1988), 1.

[11] M. Castells, **The Internet Galaxy: Reflections on the Internet, Business, and Society**, Oxford University Press, Oxford (2001), pp. 100–101.

[12] V. Cerf, Letter to the Honorable Timothy Valentine, Committee on Science, Space and Technology, U.S. House of Representatives as a follow-up to his March, 1993, congressional testimony before the Subcommittee on Technology, Environment and Aviation. April 11 (1993). Available at http://www.cpsr-peru.org/cpsr/prevsite/program/clipper/cerf-letter-to-congress.html.

[13] V. Cerf, R. Kahn, *A protocol for packet network intercommunication*, IEEE Transactions on Communications, COM-22 (5) (1974), 637–648.

[14] CERT Advisory CA-90:01, Sun sendmail vulnerability, January 29 (1990).

[15] CERT Advisory CA-2001-26, Nimda worm, September 25, 2001.

[16] CERT Advisory CA-2003-20, W32/Blaster Worm, August 14, 2003.

[17] CERT Advisory CA-2003-04, MS-SQL Server Worm, January 25, 2003.

[18] CERT Advisory CA-1999-04, Melissa macro virus, March 27 (1999).

[19] P. Ceruzzi, **A History of Modern Computing**, 2nd edition, MIT Press, Cambridge (2003), p. 285.

[20] W.J. Clinton, Presidential Executive Order 13026, Washington, DC, November 15 (1996).

[21] DARPA Press Release, DARPA establishes computer emergency response team, December 6 (1988).

[22] D. Denning, *Is cyber terror next?*, Essay for Social Science Research Council, November 1 (2001), available at http://www.ssrc.org/sept11/essays/denning.htm.

[23] W. Diffie, M.E. Hellman, *New directions in cryptography*, IEEE Transactions on Information Theory 22 (1976), 644–654.

[24] P. Edwards, *The Closed World: Computers and the Politics of Discourse in Cold War America*, MIT Press, Cambridge (1996).

[25] M. Eichin, J. Rochlis, With microscope and tweezers: An analysis of the Internet virus of November 1988, Massachusetts Institute of Technology (November, 1988).

[26] T. Eisenberg, *The Cornell commission on Morris and the worm*, Communications of the ACM 32 (6) (1989), 706–709.

[27] Federal Trade Commission Press Release, FTC bars bogus anti-spyware claims. March 11 (2005).

[28] S. Fluhrer, I. Mantin, A. Shamir, Weaknesses in the key scheduling algorithm of RC4, Eighth Annual Workshop on Selected Areas in Cryptography (August, 2001).

[29] L. Garber, *Denial-of-service attacks rip the Internet*, IEEE Computer Magazine (April, 2000), 12–17.

[30] S. Garkinkel, *PGP: Pretty Good Privacy*, O'Reilly and Associates Bejing, Cambridge, Farnham, Koeln, Paris, Sebastopol, Taipei, Tokyo (1995).

[31] GNU Project – Free Software Foundation, Why 'free software' is better than 'open source', available at http://www.gnu.org/philosophy/free-software-for-freedom.html.

[32] GNU Project – Free Software Foundation, The free software definition, available at http://www.gnu.org/philosophy/free-sw.html.

[33] K. Hafner, M. Lyon, *Where Wizards Stay up Late: The Origins of the Internet*, Simon & Schuster, New York (1996), p. 190.

[34] P. Hilts, *Virus hits vast computer network; thousands of terminals shut down to halt malicious program*, Washington Post, November 4 (1988), A1.

[35] T. Hughes, *Rescuing Prometheus: Four Monumental Projects that Changed the Modern World*, Vintage Books, New York (1998), pp. 255–300.

[36] L. Kehoe, A. Cane, *US computer systems attacked by virus*, The Financial Times, November 5 (1988), 1.

[37] LAN MAN Standards Committee of the IEEE Computer Society, Wireless LAN Medium Access Control (MAC) and Physical Layer (PHY) specifications, 1999 Edition, August 20 (1999).

[38] D. Lekkas, S.K. Katsikas, D. Spinellis, P. Gladychev, A. Patel, *User requirements of trusted third parties in Europe*, *User Identification & Privacy Protection: Applications in Public Administration & Electronic Commerce*, S. Fisher-Hübner, G. Quirchmayr and L. Yngström, eds, IFIP WG 8.5 and WS 9.6 (June, 1999), pp. 229–242.

[39] L. Lessig, *Code and Other Laws of CyberSpace*, Basic Books, New York (1999), p. 104.

[40] J. Markoff, *Author of computer 'virus' is son of N.S.A. expert on data security*, The New York Times, November 5 (1988), 1.

[41] Microsoft Security Bulletin MS01-033, Unchecked buffer in index server ISAPI extension could enable web server compromise. June 18, 2001.

[42] J. Mirkovic, S. Dietrich, D. Dietrich, P. Reiher, *Internet Denial of Service Attack and Defense Mechanisms*, Prentice Hall (2005), p. 14.

[43] R.T. Morris, A weakness in the 4.2BSD Unix TCP/IP software, AT&T Bell Laboratories Technical Report #117, Murray Hill, New Jersey (February 25, 1985), available at http://cm.bell-labs.com/cm/cs/cstr.html.

[44] National Computer Security Center, Computer security requirements – guidance for applying the Department of Defense trusted computer system evaluation criteria in specific environments (June 25, 1985), available at http://security.isu.edu/pdf/atcse385.pdf.

[45] National Cyber Security Alliance, Fast and present danger: new study shows majority of broadband users lack basic online protections, Washington, DC, June 4 (2003), available at http://www.staysafeonline.info/press/060403.pdf.L80216-03_19.pdf.

[46] P. Nikolich, Letter to Mr. Li Zhonghi, IEEE 802 LMSC and SAC standards GB15629.11 and GB 1529.1102 (November 23, 2002), available at http://www.ieee802.org/16/liaison/docs/L80216-03_19.pdf.

[47] NISCC Vulnerability Advisory 236929, Vulnerability issues in TCP, April 20 (2004).

[48] Parliament of the Commonwealth of Australia, Parliamentary Joint Committee on the Australian Crime Commission, Cybercrime (March, 2004), available at http://www.aph.gov.au/ senate/committee/acc_ctte/completed_inquiries/200204/cybercrime/report/report.pdf.

[49] J. Postel (ed.), DoD Standard Transmission Protocol, Prepared for Defense Advanced Research Projects Agency, RFC 761 (January, 1980), available at ftp://ftp.rfc-editor.org/in-notes/rfc761.txt.

[50] J. Postel (ed.), Transmission Control Protocol, DARPA Internet Program Protocol Specification, RFC 793 (September, 1981), available at ftp://ftp.rfc-editor.org/in-notes/rfc793.txt.

[51] President's Commission on Critical Infrastructure Protection, Overview Briefing, Washington, DC (June, 1997).

[52] Reuter-AP, *Computer virus unleashed on systems across U.S.*, Toronto Star, November 4 (1988), A1.

[53] J. Reynolds, The Helminthiasis of the Internet, RFC 1135 (December, 1989), available at ftp://ftp.rfc-editor.org/in-notes/rfc1135.txt.

[54] J. Reynolds, The Helminthiasis of the Internet, RFC 1135 (December, 1989). Available at ftp://ftp.rfc-editor.org/in-notes/rfc1135.txt.

[55] R.L. Rivest, A. Shamir, L. Adleman, *A method for obtaining digital signatures and public-key cryptosystems*, Communications of the ACM 21 (2) (1978), 120–126.

[56] R. Saltus, *Computer intruder found doors unlocked*, Boston Globe, November 5 (1988).

[57] K. Schmeh, **Cryptography and Public Key Infrastructure on the Internet**, Wiley, Bochum (2001), p. 365.

[58] SCO Press Release, SCO experiences distributed denial of service attack, PR Newswire, Lindon, Utah, December 10 (2003).

[59] SCO Press Release, SCO experiences massive denial of service attack, Lindon, Utah, February 2 (2004).

[60] D. Seeley, A tour of the worm, Department of Computer Science, University of Utah (1988).

[61] S. Shankland, *German programmer "Mixter" addresses cyberattacks*, C/Net News.com February 14 (2000). Available at http://news.com.com/2100-1023-236876.html.

[62] K. Siyan, C. Hare, **Internet Firewalls and Network Security**, New Riders Publishing, Indianapolis (1995).

[63] E. Spafford, The Internet worm program: an analysis, Purdue Technical Report CSD-TR-823, Department of Computer Sciences, Purdue University, West Lafayette, IN (1988).

[64] State of California, SB 1436, Consumer protection against computer spyware Act, September 28 (2004).

[65] A. Stubblefield, J. Ioannidis, A. Rubin, AT&T Labs Technical Report TD-4ZCPZZ, Using the Fluhrer, Mantin, and Shamir Attack to Break WEP, Florham Park, NJ: AT&T Labs, August 6 (2001).

[66] The United States of America v. Robert Tappan Morris, 928 F. 2d 504, Circuit Judge Jon O. Newman, 2d Circuit Court (1991).

[67] United States Attorney, Northern District of California Press Release, San Jose, CA, January 11, 1996.

[68] U.S. CERT Technical Cyber Security Alert TA04-111A Vulnerabilities in TCP, April 20, 2004.

[69] United States Department of Defense, Memorandum, Internet Protocol version 6 (IPv6), June 9 (2003), available at http://ipv6.disa.mil/docs/stenbit-memo-20030609.pdf.

[70] U.S. Department of Defense, Directive Number 8100.2, Use of commercial wireless devices, services, and technologies in the Department of Defense (DoD), Global Information Grid, April 14 (2004).

[71] U.S. Department of Energy Computer Incident Advisory Capability, CIAC Notes 95-09, April 24 (1995).

[72] United States Department of Homeland Security Press Release, Ridge creates new division to combat cyber threats, June 6 (2003).

[73] U.S. Department of Homeland Security Announcement, Ridge announces the creation of a new computer emergency response center for cyber security, Washington, DC, September (2003).

[74] U.S. Department of Justice Press Release, Creator of "Melissa" computer virus pleads guilty to State and Federal charges, December 9 (1999).

[75] U.S. Department of Justice, Federal Bureau of Investigation Press Release, January 18 (2001).

[76] United States District Court, District of New Jersey, *United States of America v. David Smith*, Criminal No. 99-18 U.S.C. 1030(a)(5)(A) (December 14, 1999).

[77] United States District Court, District of New Hampshire, *Federal Trade Commission, Plaintiff, v. Seismic Entertainment Productions, Inc., Smartbot.net, Inc., and Sanford Wallace*, October 6 (2004).

[78] United States District Court, Eastern District of Washington, *Federal Trade Commission, Plaintiff, v. MaxTheater, Inc., a Washington Corporation, and Thomas L. Delanoy, individually and as an officer of MaxTheater, Inc., Defendants*, FTC File No. 042 3213, December 6 (2005).

[79] U.S. Government, The national strategy to secure cyberspace, Washington, DC, February (2003), p. 30, available at http://www.whitehouse.gov/pcipb/.

[80] U.S. House of Representatives, Subcommittee on Technology, Committee on Science Hearing on Computer Viruses, May 10 (2000).

[81] U.S. House of Representatives Hearing, Cybersecurity and vulnerability management: testimony by Scott Culp, June 2 (2004).

[82] U.S. v. Robert Tappan Morris, Case Number 89-CR-139, U.S. District Judge Howard G. Munson, United States District Court, Northern District of New York, May 16 (1990).

[83] Utah District Court, *WhenU.com, Inc., v. Utah*, Civil Act No. 040907578 (June 22, 2004).

[84] Utah State Legislature, H.B. 323, Spyware regulation (2004), General Session.

[85] M.A. Vatis, Statement before the Senate Judiciary Committee, Criminal Justice Oversight Subcommittee, and House Judiciary Committee, Crime Subcommittee, Washington DC, February 29 (2000), available at http://www.usdoj.gov/criminal/cybercrime/vatis.htm.

[86] P. Vixie, G. Sneeringer, M. Schleifer, Events of 21-Oct-2002, ISC/UMD/Cogent Event Report, November 24 (2002), available at http://d.root-servers.org/october21.txt.

[87] Washington Post, *Major internet-security hole found*, April 21 (2004), E2.

[88] B. Whitworth, E. Whitworth, *Spam and the social-technical gap*, IEEE Computer, October (2004).

[89] Wi-Fi Alliance Press Release, Wi-Fi alliance announces standards-based security solution to replace WEP, October 31 (2002).

[90] P. Yee, Usenet Discussion Posting on comp.protocols. tcp-ip (November 2, 1988).

25

HISTORY OF COMPUTER CRIME

Susan W. Brenner

University of Dayton School of Law
Dayton, Ohio, USA
E-mail: Susan.Brenner@notes.udayton.edu
Website: http://www.cybercrimes.net

Contents

Abstract

This article begins by defining "cybercrime", and then traces the evolution of cybercrime from the criminal misuse of mainframe computers that emerged in the 1960s through the varied and complex types of misuse made possible by the networked personal computer. The article explains how, and why, cybercrime has increased in both incidence and complexity since 1990. It analyzes the more common cybercrimes, e.g., hacking, the dissemination of malware, theft, extortion and child pornography, and notes the continuing possibility of cyberterrorism. The article concludes with an examination and assessment of national and international efforts to deal with cybercrime and a brief speculation on how the phenomenon will evolve over the next few decades.

Keywords: Convention on Cybercrime, cybercrime, cyberterrorism, fraud, hacking, malware, phishing.

25.1 WHAT IS CYBERCRIME?

Any discussion of cybercrime has to begin with the definition of the term, which requires distinguishing 'cybercrime' from 'crime'.

'Crime' consists of engaging in conduct that has been outlawed by a particular society [9]. Crime takes many forms. It encompasses harm to individuals (e.g., murder, rape, assault), harm to property (e.g., arson, theft, vandalism), harm to government (e.g., obstruction of justice, treason, riot) and harm to morality (e.g., obscenity, gambling) [9]. Societies have dealt with crime for millennia and so have well-defined legal principles and procedures that can be used to apprehend and sanction those who commit crimes [9].

"Cybercrime", like crime, consists of engaging in conduct that has been outlawed by a society. Cybercrime differs from crime primarily in the way it is committed: Where real-world criminals use guns to commit crimes, cybercriminals use computer technology to engage in socially outlawed conduct [9]. Most of the cybercrime we see today simply represents the migration of real-world crime into cyberspace [9]. Cybercriminals use computers to commit fraud, theft, stalking, extortion and a variety of other crimes. Their use of computer technology does not fundamentally alter the legal definition of the activity at issue; fraud, after all, is fraud [9]. But cybercriminals' use of computer technology is significant because it can allow them to commit crime on a greater scale and it can make it much more difficult for law enforcement to identify and apprehend them [9].

And not all of cybercrime is simply the computer-facilitated commission of traditional crimes. We have seen the emergence of new types of criminal activity online: A Distributed Denial of Service (DDoS) attack, for example, inflicts harm upon property by shutting down a website and thereby preventing the operators of the website from conducting the commercial or other activity in which they engage [9]. But a DDoS attack does not fit into any of the conceptual categories the law has devised to deal with crime: It is not theft, it is not extortion, it is not vandalism. The law therefore has had to develop legal definitions of this new type of crime [9]. The same is true for another category of cybercrime: the dissemination of viruses, worms and Trojan horses. We have no real-world analogue to this type of activity, and so societies have had to develop specific prohibitions targeting the creation and dissemination of malware [9].

It is probable that the trends we have so far seen in cybercrime will continue to manifest themselves. That is, it is probable (a) that criminals will continue to use computer technology to commit

traditional crimes such as theft and extortion and (b) that they will also devise new ways of inflicting harms upon persons, property, government and morality. Section 25.7 returns to this issue.

25.2 EMERGENCE OF CYBERCRIME (1960s–1990)

This section and the next section trace the evolution of cybercrime from its emergence in the 1960s to the present.

25.2.1 Mainframes and the emergence of hacking

The first published accounts of computers being used illegally appeared in the 1960s, when computers were large mainframe systems [61]. These early crimes tended to involve computer sabotage, computer manipulation or the use of computers for illegal purposes [61]. Since access to mainframe systems was physically limited, and since the systems were not networked with other systems, these crimes tended to be committed by insiders [42]. As is explained below, the legal system did not perceive this as a matter requiring legislative action; the offenders who were apprehended were prosecuted for traditional crimes.

The term 'hacker' emerged in the late 1950s at the Massachusetts Institute of Technology's Artificial Intelligence Laboratory [42]. MIT students had historically used the term 'hack' to refer to creative college pranks, but in the late 1950s it migrated into computer culture and was used to denote a endeavor undertaken not for some instrumental purpose but for pure creativity and intellectual excitement [42]. Those who engaged in such endeavors became known as 'hackers' and their activity became known as 'hacking' [42]. The term 'hacker' originally had only a positive connotation, since it denoted someone who was adept at computer programming and problem-solving [75].

In 1969, the ARPANET, the world's first packet switching network appeared [73]. It linked computers in hundreds of universities, research laboratories and defense contracting companies [57]. In so doing, it linked hackers all over the country and led to the emergence of a distinct hacker culture [57]. This early hacker culture evolved over the next decade; it was limited to a relatively small group until the emergence of the personal networked computer in the 1980s brought it into the mainstream of American society [57]. The 1983 movie "War Games" popularized the image of hackers as creative but benign, but by the end of the 1980s it had become clear that at least some of those who called themselves hackers were up to no good [5]. As a result of activities such as those described in Section 25.2.3, the conception of hacker changed, so that the term now has at least two distinct meanings: hacker as brilliant programmer and hacker as computer criminal [75].

25.2.2 Phone phreaking

Phone phreaking was popular among some in the 1960s [76]. Phone phreaking essentially consisted of manipulating the telephone system to make it do something it normally should not allow, such as providing free long-distance service [76]. Those who engaged in this activity were known as phone phreaks and many phone phreaks moved on to hacking in the 1980s [76]. Kevin Mitnick, for example, who would become one of America's most notorious hackers began as a phone phreak [22]. Like mainframe hacking, phone phreaking was not regarded as a distinct type of criminal activity; it was prosecuted under existing law, usually as either fraud or theft of services [60].

25.2.3 The emergence of 'true' cybercrime

As Section 25.2.1 noted, the appearance of networked personal computers brought hacker culture into the mainstream of American society and, in so doing, produced a distinct change in the types of individuals who could be described as 'hackers'.

When computers were mainframes, 'hackers' were individuals with computer expertise who solved problems and produced innovations in computer programming for sheer intellectual excitement. The ARPANET [73] linked computers around the country, but only those located in defense-funded institutions [50]. In 1978, two computer enthusiasts in Chicago put the first "civilian" bulletin board system online [50]. It created a revolution on online communication, which had been limited to computer scientists and hobbyists; over the next decade thousands of bulletin board systems went online, many of which appealed to a new type of 'hacker' [50].

Bulletin board systems let users interact online with others and share information with them. Much of this activity was innocuous, but some bulletin board users engaged in darker activities, such as sharing phone phreaking tactics and trading pirated software and stolen credit card data [60]. Spurred, perhaps, by the movie "War Games", many of the habitués of the online bulletin boards turned to a new activity: attempting to break into government and other computer systems [60]. In 1981, Ian Murphy, also known as "Captain Zap", became the first person to be prosecuted for hacking in the US; Murphy was charged with theft – not computer crime – for hacking into AT&T's system and changing the clocks that metered billing rates so that subscribers were receiving late-night rates when they called in the middle of the day [17].

In 1983, the year "War Games" came out and the ARPANET became the Internet [50], a group of teen-aged hackersfrom Milwaukee known as the 414s (after the local area code) broke into computer systems at the Los Alamos Laboratory in New Mexico and at the Sloan-Kettering Cancer Center in Manhattan [77]. The 414s claimed to have been acting in the spirit of the original hacker culture, i.e., out of intellectual curiosity; they also claimed they did not realize they were doing anything illegal in breaking into the private computer systems [77]. The 414s were prosecuted federally for computer trespassing; one cooperated and received immunity while the others, all of whom were minors, were given probation [72].

The next year, several soon-to-be infamous hacker groups emerged to exploit the opportunities available in what was now known as 'cyberspace', a term introduced in William Gibson's novel *Neuromancer* [24]. A hacker calling himself Lex Luthor created the Legion of Doom (LOD); when one member, Phiber Optik, was expelled from the LOD for feuding with another member, he and his friends created a rival group – the Masters of Deception (MOD) [63]. A German hacker group, Chaos Computer Club, had already emerged, along with lesser-known hacking gangs [74].

The MOD and the LOD became infamous when the two went to war with each other in the late 1980s [63]. At first, the activities of the two groups were relatively innocuous but as the feud between them heated up the MOD members, in particular, began to cross the line into serious criminal activity [63]. They hacked telephone systems to obtain free telephone services in a fashion reminiscent of the old phone phreakers; but they also hacked information services like TRW to obtain personal information which they sold or used to harass rivals and others [63]. In 1991, five members of the MOD were federally indicted on traditional criminal charges, including conspiracy, wire fraud, computer fraud, computer tampering and wiretapping [63]. Two of the five pled guilty; the others were tried and convicted on all charges [63].

One of the most notable events of this era came late in the 1980s: In 1988, Robert Tappan Morris, a Cornell University graduate student, released the first Internet worm [68]. He released the worm onto the Internet, intending to demonstrate the inadequacy of its security; the worm spread around the country, infecting thousands of computers and causing many thousands of dollars in damage [68]. Morris was prosecuted federally; he was the first person indicted under the recently-revised federal Computer Fraud and Abuse Act [68]. He was convicted and sentenced to three years of probation, four hundred hours of community service and a fine of $10,500 [68]. Morris' exploit resulted in the creation of the CERT computer emergency response center at Carnegie-Mellon University [10].

As Section 25.4 explains, by the end of the 1980s federal legislation had been adopted to deal with hackers who gained unauthorized access to computer systems. States had begun the process of adopting similar legislation, but they were generally far behind the federal effort.

25.3 CYBERCRIME INCREASES IN INCIDENCE AND COMPLEXITY (1990–2004)

As the previous section demonstrates, in its early years cybercrime primarily consisted of hacking – i.e., of gaining unauthorized access to computer systems. By the 1990s personal computers and the network known as the Internet were becoming increasingly sophisticated, and so were cybercriminals. This section notes some of the major developments in cybercrime during this era.

25.3.1 Hacking

Kevin Poulsen – also known as Dark Dante – was a hacker whose career spanned the bridge years between the late 1980s and the early 1990s [45]. Poulsen began as a phone phreak but moved into hacking and was eventually prosecuted for hacking into computers belonging to the Pacific Bell Telephone Company and to several federal government installations [45]. Apprehended in 1991, Poulsen ultimately faced two federal indictments; the charges against him included multiple counts of access device fraud, conspiracy, wiretapping, computer fraud, mail fraud, money laundering and espionage [45]. The espionage charge was based on the allegation that he unlawfully obtained an Air Force order classified as secret [45]. The government ultimately dropped the espionage charge; in 1994, Poulsen pled guilty to seven counts of mail, wire and computer fraud, money laundering and obstruction of justice; he was sentenced to serve 51 months prison and to pay $56,000 in restitution [45]. It was the longest sentence that had been imposed for hacking [45]. Some suggested that the government had made an example of him in an effort to deter others who might be inclined to hack [49].

Similar suggestions were made about the 'other' Kevin – Kevin Mitnick. Mitnick, who became the first hacker to have his face included on an FBI "Most Wanted" poster, was the focus of a multi-state manhunt in the mid-1990s [13]. Like Poulsen, Mitnick began with phone phreaking and then moved into hacking; his first conviction was for hacking a university computer, for which he served six months in jail [13]. In 1987, Mitnick was charged with hacking a software company's computer system; he pled guilty to a misdemeanor and was put on probation [13]. A year later Mitnick faced federal charges of stealing software from the Digital Equipment Corporation; he was arrested by the FBI and eventually convicted [13]. Mitnick served a year in prison on these charges and was then released on probation; in 1992, while he was still on probation, he vanished and became a federal fugitive who was often described as the "most wanted computer criminal in the world" [13]. Federal authorities claimed he had hacked corporate computer systems and had access to software and trade secrets worth millions of dollar [13]. Mitnick was eventually apprehended in 1995 in North Carolina [13]. He was federally indicted on twenty-five counts of access device fraud, computer fraud, wire fraud, computer damage and wiretapping [67]. He eventually pled guilty to five counts and was released from prison in 2000, after having served five years [38].

As the respective careers of the two Kevins demonstrate, hacking at the end of the twentieth century had become primarily an individual activity; the adolescent hacker gangs of the 1980s had disappeared. And this was not only true of American hackers, as two notorious hacks of the 1990s illustrated.

In 1994, the Rome Air Development Center (Rome Labs) at Griffiss Air Force Base in New York was hacked by two hackers who broke into seven computers on the Rome Labs network and installed a password sniffer on each of them [56]. US authorities were concerned because the network contained sensitive Air Force research and development data [56]. They were also concerned because the hackers used the Air Force network to launch attacks on other sensitive systems, including NASA, NATO and the South Korean Atomic Research Institute [56]. Air Force investigators working with Scotland Yard were able to identity the two hackers, both of whom were British adolescents [56]. Both were prosecuted in Britain; one pled guilty, while the other was able to have the charges against him dismissed [56].

In 1996, the U.S. Department of Justice announced that it had charged Julio Ardita, a 21-year-old Argentine citizen, with hacking into Harvard University's computer network and using it to launch attacks on other systems [56]. Like the Rome Labs hackers, Ardita attacked systems belonging to NASA and the U.S. Department of Defense [56]. American and Argentinian authorities cooperated to identify and apprehend Ardita [56]. Ardita ultimately pled guilty the charges brought against him in the United States; hacking was not then a crime in Argentina [56].

These cases exemplify a phenomenon that was established by the 1990s: Attacks tended to be carried out by lone hackers, who were usually adolescent males [56]. The Masters of Deception-style hacker gangs had all but disappeared.

Organized cybercrime, however, did not disappear: As Section 25.7 explains, the beginning of the twenty-first century saw the emergence of a new type of organized cybercrime: adult hacker gangs. These professional hackers are motivated by profit, not by curiosity [54]. They work with organized crime, such as the Russian Mafia, or operate independently as organized cybercrime gangs [54]. Unlike the Masters of Deception, these hacker gangs operate outside the United States in places like Russia, Brazil and China, though they can be almost anywhere [71].

25.3.2 Malware

When Robert Tappan Morris released his worm into the Internet in 1988 (see Section 25.2.3), he 'invented' network malware. The notion of computer viruses was not new: In 1975 John Walker created Pervade, a virus that infected UNIVAC computers through files transferred between systems on magnetic tape [39]. In 1982 a Pittsburgh ninth-grader created the first computer virus to infect personal computers [39]. The Elk Cloner virus, which was spread by infected disks, attacked Apple II computers [39]. In 1984, computer scientist Fred Cohen was the first to use the term "virus" to refer to self-propagating code, and in 1986 two Pakistani brothers released the first virus to infect IBM personal computers [39].

Morris did not invent the concept of a network worm: Two researchers at the Xerox Palo Alto Research Center used the term 'worm' in a 1982 paper to describe a program they used to update an Ethernet application [41]. Morris was the first to release a worm, or a virus, into a networked environment and thereby demonstrate how quickly malware could spread from system to system; his exploit convinced many networks administrators that malware could be a serious threat [41].

And so it has proven to be: About 200 viruses had been identified by 1990, and viruses caused millions of dollars in damage during the remainder of the decade [41]. The Concept virus, which appeared in 1995, was the first to spread in the wild by using security flaws in a macro language but was followed by others, including Melissa; released in 1999, Melissa was the first mass-mailing computer virus [41]. The opening years of the twenty-first century saw the release of the Code Red worm and the Nimda virus, and 2003 saw the first ultrafast, flash worm – Slammer [41]. The Slammer worm infected hundreds of thousands of computers worldwide, shut down cash machines and even impacted on the Internet's nameservers [46]. And in May of 2004 the Sasser worm infected business and government computers, grounded flights and halted rail traffic [36].

Because viruses and worms are not precisely analogous to any type of criminal activity in the real-world, the release of malware tended, at least for a time, not to be addressed by local criminal law. The best example of this is the "Love Bug" virus, which rapidly spread around the world in May of 2000, causing billions of dollars in losses [59]. Investigators quickly traced the virus to the Philippines and quickly identified a student, Onel de Guzman, as the person likely responsible for creating and disseminating the virus [59]. But because the Philippines lacked cybercrime law, de Guzman could not be prosecuted [59].

In the years since the release of the "Love Bug", many jurisdictions, including the Philippines, have adopted laws criminalizing the dissemination of malware [14]. But this has done little to address the underlying problem: Arrests for disseminating viruses and worms are rare, primarily because it is so easy for the architects of malware to remain anonymous [70]. If they release their products from public terminals at universities or cybercafés, it becomes almost impossible to track specific malware to its author [70]. Those who have been caught either left identifying information in their code or called attention to themselves in some other way: David Smith, author of the Melissa virus, was apprehended because the virus contained a unique identifier from Microsoft Word that let authorities locate him [70]. Some have been caught because they included their names or other personal information in their code [43; 44]. Others bragged about their exploits online [44]. Experts agree that the sophisticated authors of malware – like the authors of Code Red, Nimda, Slammer and the original Blaster worm – are not likely to give themselves away and so will remain at large [44].

25.3.3 Theft

The best-known instance of computer theft occurred in 1994: Over a period of two months, Vladimir Levin, a Russian hacker operating from St. Petersburg, siphoned millions of dollars from Citibank accounts hosted on computers located in New York [56]. Working online, Levin transferred the funds to accounts accomplices had set up in various foreign countries [56]. The crime came to light when a South American investment company official noticed suspicious transfers taking place in the middle of the night [56]. The FBI, working with British and Russian authorities, identified Levin, who was arrested in Britain in 1995; he was extradited to the United States, where he faced federal charges of conspiracy to commit wire fraud, bank fraud and computer fraud [26]. Levin eventually pled guilty and was sentenced to serve three years in prison [56]. And Citibank reportedly recovered almost all of the money Levin took from its accounts [56].

The Levin case is a perfect example of "true" online theft, i.e., of the use of computer technology to obtain funds belonging to another. A decade after Levin, a similar case occurred in China: Song Chenglin, a Harbin college student, used a computer at an Internet café to hack into 158 accounts at the Industrial and Commercial Bank of China

(ICBC) [12]. Chenglin transferred over $60,000 into an account he set up for himself; he and his accomplices were apprehended after they went to ICBC branches to withdraw the funds [12].

Prosecuting thefts such as Levin's and Chenglin's generally poses few difficulties for a jurisdiction's criminal law, since they have committed a traditional crime (theft) in a non-traditional manner [14]. Conceptually, their use of computer technology to consummate the theft should be irrelevant; they purposely engaged in conduct which deprived the lawful owner of property of the possession and use of that property [14].

Other types of theft can pose doctrinal problems for the law. In an Oregon case Randall Schwartz was convicted of theft based on his copying a computer password file belonging to Intel, his employer [64]. Schwartz appealed, claiming his action did not constitute theft because he did not "take, appropriate, obtain or withhold" the password file from its rightful owner [64]. Schwartz pointed out that the file and passwords remained on Intel's computers and that the individual users could still use their passwords just as they had before [64]. He contended that since Intel 'had' everything it had before he made the copy, he had not 'taken' anything from Intel [64]. The prosecution argued that by copying the passwords Schwartz stripped them of their value [64]. It claimed that passwords have value only so long as no one else knows what they are and that once Schwartz copied the Intel passwords they were useless for their only purpose, i.e., protecting access to the Intel computers [64]. The state claimed Intel's loss of exclusive possession of the passwords was sufficient to constitute theft [64]. The Oregon Court of Appeals agreed, and affirmed Schwartz's conviction [64].

In the real-world, theft is a zero-sum endeavor: The thief takes physical property from her victim, thereby wholly depriving the victim of its possession and use [14]. Theft statutes have traditionally contemplated this type of all-or-nothing theft [14]. Jurisdictions must therefore ensure that their theft statutes can encompass copying data, as well as the physical extraction of funds and other property.

25.3.4 Fraud

By the beginning of the twenty-first century, online fraud had become the most commonly-encountered variety of cybercrime [51]. Online fraud takes many forms and is very costly. According to the Internet Fraud Complaint Center, the total dollar loss from the cases reported to the IFCC in 2003 was $125.6 million [51].

One of the oldest and still most commonly encountered varieties of online fraud takes the form of advance fee fraud schemes, which have the following characteristics:

- A potential victim receives an email from an 'official' or a relative of an 'official' from a foreign government or agency.
- The official or relative offers to transfer millions of dollars in 'over invoiced contract' funds into the victim's personal bank account.
- The victim is asked to provide blank company letterhead forms, banking account information, telephone/fax numbers.
- The victim receives numerous documents with official looking stamps, seals and logo testifying to the authenticity of the proposal.
- The victim is eventually asked to provide upfront or advance fees for various taxes, attorney fees, transaction fees or bribes, with each fee paid being described as the last that will be needed.
- The victim never receives the promised funds and loses all of the money paid for fees [69].

The most popular version of this scheme is the Nigerian or 419 (after the section of the Nigerian penal code that addresses fraud) scam [69]: The solicitation comes from someone purporting to be a senior civil servant in one of the Nigerian ministries who is seeking assistance in transferring millions of dollars out of the country [69]. Other variants purport to come from Russia or elsewhere, but the dynamic is always the same: The victim parts with advance money for various fees but receives nothing in return [69]. According to the FBI, advance fee fraud schemes inflict some of the highest median dollar losses on those who are victimized by various types of online fraud [51].

Lottery fraud is a variation of the advance fee fraud scams: The victim receives an email announcing the he or she has been chosen as the winner in a European lottery, which pays, say, half a million Euros [62]. The lucky winner, who of course never entered the lottery, which does not exist, is contacted by an 'agent' of the lottery [62]. The 'agent' wants to help the 'winner' receive his or her winnings but the 'winner' will need to pay certain fees to secure the release of the funds, which are allegedly being held by a (non-existent) bank [62]. One British victim of this scam ultimately paid 20,000 Euros in a futile effort to claim his winnings [62].

Internet auction fraud is the most commonly reported variety of online fraud, accounting for 61% of the complaints the Internet Fraud Complaint Center received in 2003 [51]. By 2003, much online auction fraud was originating in Eastern Europe, according to the Internet Fraud Complaint Center [33]. This type of fraud often involves offers to sell what are allegedly factory direct products at low prices [33]. Victims who send purchase money by wire are left without their merchandise and with no recourse, as wire transfer funds are not recoverable [33].

Identity theft emerged as a new type of fraud in the 1990s [20]. Identity thieves misappropriate an individual's personal identifying information – such as name, address, birthdate, Social Security number, bank account and credit card numbers, phone numbers, occupation [20]. Once they have this information they use it to go on spending sprees using the victim's credit cards or using new credit card accounts they open in the victim's name [20]. The identity thieves, of course, do not make payments on the credit card bills they run up on the victim's name [20]. They may even use the victim's credit history to purchase big-ticket items like automobiles or homes, none of which they pay for [20]. All of this destroys the victim's credit history and financial stability [20]. And identity thieves may even give the victim's name to the police if they are arrested; if the thief does not show up for a court date, a warrant can be issued for the victim, who may then be arrested for failing to appear [20]. A Federal Trade Commission survey conduct

in 2003 found that 27.3 million US residents had been the victims of identity theft in the last five years and that identity theft during that period had cost businesses $48 billion and consumers $5 billion in out of pocket expenses [27].

In 2003, a new scheme – phishing – appeared that involves the theft of two different identities: that of a business and that of an individual who will become the victim of the kind of identity theft described in the previous paragraph [21]. In phishing, the perpetrator first creates a website that duplicates the website of a legitimate business or agency; the phisher then sends out mass emails to individuals which tell them there has been some problem with account information they have stored on the duplicated website [21]. The emails send the prospective victims to the duplicated website, where they are encouraged to re-enter their account and other identifying information to 'correct' the problem that has arisen [21]. If they do so, the phisher then harvests their personal information and uses it to commit identity theft as described above [21]. In 2004, phishing scams were dramatically increasing in frequency and in the sophistication of the phishing methodologies being employed [2]. The Anti-Phishing Working Group, which tracks phishing activity, reported that 8,459 new and unique phishing emails were sent out in November of 2004, an increase of 34% since July, 2004 [2]. A 2004 survey found that American consumers had lost $500 million to phishing scams [44].

Legislators in many jurisdictions reacted to the emergence of identity theft by adopting specific statutes criminalizing the fraudulent use of another's identity [14]. Generally, however, specific legislation is not needed to prosecute the types of activity outlined above, as each simply constitutes the commission of traditional fraud in new ways.

25.3.5 Child pornography

Child pornography was not a matter of global concern as long as the material was only available in hard copy. Those interested in such material were generally reluctant to expose themselves to the risk of being apprehended and prosecuted that resulted from inquiring about and purchasing child pornography in a real-world venue [34]. This all changed with the Internet, which lets individuals acquire and distribute child pornography with little risk of being identified and prosecuted [34]. As a result, child pornography has become a highly organized, global subculture, one that takes great pains to avoid being infiltrated by law enforcement [34].

Law enforcement has had some success in apprehending those who traffic in child pornography: In 1996, police in San Jose, California broke up a child pornography ring known as the Orchid Club, which had members in several countries [55]. Three years later, police in Texas raided the home of Thomas and Janice Reedy, who were operating a child pornography website that made them millionaires [37]. The website had 300,000 subscribers in 60 countries and brought in more than a million dollars a month for the Reedys [37]. And in 2001, police from twelve countries cracked the Wonderland Club, arresting 107 individuals and seizing 750,000 digital images of child pornography [55]. These arrests and prosecutions, however, do little to interfere with the global market in child pornography; those who participate in the market are skilled in concealing their identities, and many are in countries where child pornography is not a priority for law enforcement [28].

Child pornography also raises a number of legal issues. One is definitional: What is child pornography? Is it only visual depictions (videos, photographs) of actual children engaging in sexual activity? Or does it also encompass computer-generated images ('virtual' child pornography)? In a controversial decision, the U.S. Supreme Court held that the First Amendment to the US Constitution prevents the criminalization of 'virtual' child pornography because (a) the material qualifies as speech protected by the Amendment and (b) no children are harmed in its creation, so it is only speech [4]. In an earlier decision, the U.S. Supreme Court had held that 'real' child pornography could be criminalized even though it is speech because its creation results in the infliction of harm upon children [4]. The Canadian Supreme Court reached a similar conclusion as to 'textual' child pornography, i.e., text describing sexual activity with children [58]. It held that textual child pornography

could not be criminalized because to do so would violate an individual's right of free expression [58].

Another legal issue that arises as to child pornography is the definition of a 'child' [28]. Countries vary widely in the standards they use to define who is, and who is not, a 'child' [28]. Countries also vary in the extent to which they criminalize child pornography [28].

25.3.6 Extortion

Online extortion takes many forms, but the dynamic is always the same as in the real-world crime: Property is extorted from a victim by threatening him or her with bodily injury or injury to the victim's property [30].

One older but still-common online extortion scam is for hackers to break into the computer systems of online businesses and then threaten to reveal sensitive information unless the business pays the hackers a certain sum of money [54]. Another scam, of more recent vintage, targets online casinos [6]. Hackers use networks of compromised, zombie computers to launch Distributed Denial of Service (DDoS) attacks on casinos; the attacks shut the casinos down, costing them thousands and thousands of dollars in revenue [6]. The attackers usually strike once, let the casino recover, then strike again and demand protection money in exchange for allowing the casino to remain online [6]. These online extortionists attack casinos located in the Carribean, in the United Kingdom, anywhere online casinos are legal [6]. Many attacks come from Russia or Eastern Europe, but they have also been traced to China and to the Middle East [6].

As Section 25.7 notes, many fear that the type of extortion attacks that are now targeting online casinos will eventually be directed government agencies, e-commerce sites, financial institutions and any entity with an online presence [6].

25.3.7 Cyberterrorism

Terrorism is the commission of criminal acts for political motives (Section 2331 of [78]). When Timothy McVeigh blew up the Oklahoma City federal building in 1995, he committed various crimes, including murder and arson, for political reasons [66]. Traditionally, terrorism has involved the use of methods of physical destruction; it is, though, at least theoretically possible that computer technology could be utilized for this purpose [7].

There has and continues to be a great debate as to whether cyberterrorism is a myth or an inevitability [40]. While we have yet to experience a cyberterrorist event, US law has already addressed this issue. In the USA Patriot Act, which was adopted in response to the 9-11 attacks, Congress incorporated the basic federal cybercrime offenses – hacking, cracking, extortion, fraud and malware – into the statutory definition a "federal crime of terrorism", the purpose being to authorize the federal prosecution of cyberterrorists if and when they should strike (Section 2332b of [79]). At least two states have also criminalized cyberterrorism [14].

25.4 CYBERCRIME: THE REACTION

5

As explained above, early varieties of cybercrime involved attacks upon or the misuse of mainframe computers [25]. These cybercrimes were necessarily committed by 'insiders' because only authorized personnel had access to mainframe computers [25]. Cybercrime was therefore a relatively uncommon phenomenon, but it still sparked interest among legislators, as it was clear that traditional criminal law did not address the problem of cybercrime [61].

As a result, early efforts to criminalize cybercrime appeared in the United States in the 1970s, with the introduction of an ultimately unsuccessful federal cybercrime bill in Congress and some similar efforts in a few states [25]. Similar efforts appeared in other countries, as is explained in the next section. The introduction of the personal computer and the rise of the Internet made the adoption of cybercrime legislation far more imperative, since the possibility of misusing computer systems was no longer limited to a few 'insiders' with mainframe access [25]. As our technology evolves, so do our efforts to address the legal issues which result from the misuse of computer and related technologies; the next section reviews the more important efforts that have been taken in this regard.

25.5 DEALING WITH CYBERCRIME: PAST EFFORTS

This section reviews the more significant efforts that have been taken at the national and international level to deal with the challenges cybercrime poses for law enforcement. As the discussion below demonstrates, countries generally agree that an effective strategy against cybercrime combines two features: (a) the criminalization of a core set of offenses, which typically focus on attacks against computer systems and the use of computer systems to commit conventional crimes, such as fraud; and (b) the adoption of legislation which facilitates nation-to-nation cooperation in cybercrime investigations and prosecutions.

25.5.1 National efforts

The first wave of national cybercrime legislation was designed to protect privacy; these laws were a response to the capacity computer technology created for collecting, storing and transmitting data [25]. Legislation was enacted to protect data and citizens' rights to privacy in Sweden (1973); the United States of America (1974); the Federal Republic of German (1977); Austria, Denmark, France and Norway (1978); Luxembourg (1979 and 1982); Iceland and Israel (1981); Australia and Canada (1982); the United Kingdom (1984); Finland (1987); Ireland, Japan and The Netherlands (1988); Portugal (1991); Belgium, Spain and Switzerland (1992); Spain (1995); Italy and Greece (1997) [25]. The concern with privacy also prompted constitutional amendments in Brazil, The Netherlands, Portugal and Spain [25].

The second wave of legislation addressed property crimes. It was precipitated by the inadequacy of existing laws, which were concerned with protecting tangible property [25]. The new laws addressed computer criminals' ability to inflict traditional harms by new means (such as transferring funds from online accounts). The following enacted new laws against computer-related economic crimes: Italy (1978); Australia (state law, 1979); United Kingdom (1981, 1990); United States of America (federal and state legislation in the 1980s); Canada and Denmark (1985); the Federal Republic of Germany and Sweden (1986); Austria, Japan and Norway (1987); France and Greece (1988); Finland (1990, 1995); The Netherlands (1992); Luxembourg (1993); Switzerland (1994); Spain (1995); and Malaysia (1997) [25].

Another wave of legislation came in the 1980s. It was directed toward better protecting intellectual property [25]. The laws adopted in this period imposed criminal penalties for pirating computer software and other types of digital intellectual property [25].

By the 1990s, many countries, especially Western countries, had adopted basic cybercrime legislation that outlawed the more commonly encountered crimes, such as unauthorized intrusions, disseminating malware, computer theft and fraud [25]. A private study published in 2000 found, however, that the laws of many countries did not clearly prohibit cybercrimes [47]. The concern that countries had not acted to prohibit cybercrime prompted the efforts discussed below, particularly the Council of Europe's Convention on Cybercrime [25].

25.5.2 International efforts

Countries soon realized that national legislation, alone, was not adequate to deal with the problem cybercrime because it can so easily transcend national boundaries [25]. As a result, various organizations became involved in an effort to develop transnational solutions to cybercrime [25]. This section outlines the most significant of these efforts in roughly chronological order.

The first effort came from the Organisation for Economic Co-operation and Development (OECD) [25]. Like the other efforts that would follow, this initiative focused on harmonizing computer crime legislation to ensure consistency across countries [25]. The OECD commissioned a study that eventually culminated in a 1986 report; the report recommended that countries criminalize a minimum set of cybercrime offenses [25]. Basically, the OECD recommended that countries criminalize attacks on computer systems, the use of computer systems to commit fraud or forgery, the use of computer systems to infringe software copyrights and

gaining unauthorized access to a computer system [25].

Around the same time, the Council of Europe began a very similar initiative on cybercrime [25]. In 1989, it issued a recommendation which emphasized the need to harmonize cybercrime laws to ensure that countries could respond quickly and adequately to cybercrime [25]. Like the OECD report, this recommendation contained a "minimum list" of crimes to be prohibited; but it also included an "optional list" of offenses on which international consensus would be difficult to reach [25]. The "minimum list" required the criminalization of computer fraud, computer forgery, damage to computer data or programs, computer sabotage, unauthorized access to computer systems, unauthorized interception of data and unauthorized reproduction of a computer program [25]. The "optional list" required criminalizing the alteration of computer data or programs, computer espionage, the unauthorized use of a computer and the unauthorized use of a program [25].

The United Nations has also addressed cybercrime. In 1990, its Congress on the Prevention of Crime and the Treatment of Offenders issued a resolution which called for countries to combat cybercrime by modernizing their law, improving computer security and promoting a comprehensive international framework of standards for preventing, prosecuting and punishing computer-related crime [25]. And in 1995, the United Nations issued its Manual on the Prevention and Control of Computer-Related Crime [25]. It reviewed the need for specific national laws targeting cybercrime and for international cooperation to combat cybercrime [25].

The Group of Eight (G8) has also addressed cybercrime [23]. In 1997, the Justice and Interior Ministers of the G8 adopted ten "Principles to Combat High-Tech Crime"; like the efforts outlined above, these principles focused on the need for cybercrime laws and for international cooperation in cybercrime investigations and for resources [25].

The initiative that has so far yielded the most concrete results in this area began in 1997, when the Council of Europe created a Committee of Experts on Crime in CyberSpace [25]. This Committee was given the task of drafting "a binding legal instrument" dealing with the need to criminalize cybercrime and to ensure international cooperation in cybercrime investigations and prosecutions [25]. The Committee spent the next four years preparing a Convention on Cybercrime [25]. The final version was submitted to the European Committee on Crime Problems in June of 2001, and opened for signature in Budapest on November 23, 2001 [25]. The provisions of the Convention and its status at this writing are discussed below.

In 2002 the Asia Pacific Economic Cooperation (APEC) leaders met and committed themselves to enact "a comprehensive set of laws" that were consistent with the Convention on Cybercrime [3; 35]. In November of 2004, the APEC Ministers issued a statement in which they again agreed to enact "domestic legislation consistent with the provisions of ... the Convention on Cybercrime" [3].

In 2002 the OECD returned to cybercrime, adopting revised Guidelines for the Security of Information Systems and Networks [52]. The new Guidelines were prompted by the increased interconnectivity of networks and an increasing number of threats and vulnerabilities [52]. They articulated a set of principles which, among other things, declared that all participants in the online environment "are responsible for the security of information and networks" [52].

In April of 2004 the Ministers of Justice or Attorneys General of Organization of American States countries issued recommendations on cybercrime [53]. They urged member states to see that "differences in the definition of offenses" did not impede their ability to cooperate in cybercrime investigations and prosecutions [53]. They also encouraged member countries to "evaluate the advisability of implementing the principles" contained in the Convention on Cybercrime [53].

25.5.3 Convention on Cybercrime

The Convention's primary focus is on ensuring that countries adopt substantive cybercrime legislation and procedural laws which will improve international cooperation in cybercrime investigations and prosecutions.

The Convention requires parties to criminalize certain cyber-offenses; like the initiatives outlined in the previous section, it assumes that harmonizing national cybercrime laws will make it difficult for cybercriminals to escape justice and will facilitate the investigation and prosecution of cybercrime offenses. [15]. Specifically, parties must ensure that their domestic law criminalizes: (1) attacks on computer systems and data (illegal access, illegal interception, data interference, system interference and misuse of devices); (2) the use of computer technology to commit fraud or forgery; (3) the use of computer technology to create, possess or distribute child pornography; and (4) the use of computer technology to infringe copyright or related rights [15].

The drafting of the Convention demonstrated how difficult it can be to gain consensus on cybercrime. The drafters considered adding a provision targeting hate-speech but while some countries supported such a measure, others, including the United States, opposed it "on freedom of expression grounds" [19]. The drafters referred this issue to the European Committee on Crime Problems, which drafted a protocol to the Convention that requires parties to criminalize the use of computer technology to disseminate racist and xenophobic material [1]. Racist and xenophobic material is defined as "any representation of ideas or theories, which advocates, promotes or incites hatred, discrimination or violence ... based on race, colour, descent or national or ethnic origin, as well as religion if used as a pretext for any of these factors" [1]. At this writing, the protocol had been signed by twenty-two countries and ratified by one [1].

Article 11 requires that country criminalize attempts to commit these crimes and aiding and abetting their commission [15]. Under Article 12, they must ensure that "corporations, associations and similar legal persons" can be convicted of crimes defined under the Convention [15]. And Article 13 requires countries to ensure that the Convention crimes are "punishable by effective, proportionate and dissuasive sanctions", including incarceration [15].

By 2005, the Convention had been signed by 41 countries but ratified by only 9 [11]: Albania,

Croatia, Cyprus, Estonia, Hungary, Lithuania, Romania, Slovenia and the former Yugoslav Republic of Macedonia. None of the major European countries had ratified it, nor had the United States, Canada or Japan [11]: In November of 2003 President Bush asked the Senate to approve ratification, and a Senate committee held hearings on the Convention in June of 2004, but nothing further had happened by the beginning of 2005 [29]. In September, 2004, the Council of Europe held a conference which was intended to encourage more countries to sign and ratify the Convention; a conference speaker attributed the slow pace of ratification to the complexity of the issues it raises [16].

There is no obvious explanation for the glacial pace at which nations are ratifying the Convention. It is the only affirmative measure that has been taken to enhance law enforcement's ability to deal with cybercrime, both substantively and procedurally, yet it seems not to be a high priority even for the countries that were involved in its creation. The next section considers possible alternatives to the Convention.

25.6 DEALING WITH CYBERCRIME: FUTURE EFFORTS?

The Convention seeks to create an international seamless web of laws that allow for efficient, unproblematic cooperation among nations. In so doing, it treats cybercrime like crime, i.e., as an internal problem that is to be handled unilaterally by an offended nation-state. This continues the localized, decentralized system of law enforcement we have had for centuries, but perhaps we need a new system for cybercrime. Perhaps we cannot adapt our existing approach to deal effectively with cybercrime. Perhaps the otherwise puzzling lack of progress in implementing the Convention attests to the futility of trying to adapt traditional, nationally-based law enforcement to non-territorially based crime. Maybe we should consider an alternative.

There are three ways to structure law enforcement's response to cybercrime: One is the Council of Europe's distributed approach, in which the responsibility for reacting to cybercrime is parsed out among nation-states, each of which defines cybercrime, investigates cybercrime and prosecutes and punishes cybercriminals. The problem with this approach is that it is territorially-based but cybercrime is not.

Another approach is to centralize law enforcement in a single agency that is responsible for controlling cybercrime around the world. In its most extreme form, this option would produce a global agency that is pre-emptively responsible for outlawing cybercrime, for investigating it, for prosecuting cybercriminals and for sanctioning them. Conceptually, this approach would be based on the premise that cyberspace has, in effect, become another jurisdiction, another arena for human activity, and, as such, requires its own law enforcement institutions. Implementing this approach would require us to accept what seems to be inevitable, which is not unknown in the law.

Inevitable or not, counties are unlikely to be willing to surrender complete responsibility for cybercrime in the near future, anyway, which brings us to the third approach, which is a compromise between the first two. Here, the prosecution and sanctioning of cybercriminals would remain the responsibility of discrete nation-states, but the processes of investigating cybercrime and apprehending cybercriminals would be delegated to a central agency, a sort of super-Interpol. Unlike Interpol, this agency would not simply coordinate investigations among law enforcement officers from various countries; it would, instead, be responsible for conducting the investigations and for delivering evidence and offenders to the offended nation-state. In this scenario, the nation-states would presumably bear primary responsibility for defining what is, and is not, a cybercrime; it might be advisable to establish some system for ensuring that their laws were substantially consistent. Aside from expediting the base processes of investigating cybercrime and apprehending cybercriminals, this approach could have the added virtue of improving the overall response to cybercrime. A centralized global agency could track trends in cybercrime, identifying them long before they could become apparent to investigators scattered around the globe, each working within the confines of a separate institutional structure.

Obviously, the last two approaches are mere speculation. The visceral reactions crime (and cybercrime) generate guarantee that our approaches to both will continue to be parochial for the foreseeable future. We are sufficiently invested in territorial reality to find it difficult to surrender control over those who offend our sense of morality to an "outside" entity; we believe crime and punishment are local concerns. The challenge we face is reconciling that belief with the reality that is emerging in cyberspace.

25.7 CYBERCRIME: THE FUTURE

While it can be risky to speculate about the course of future human activity, including criminal activity, trends are emerging that suggest how cybercrime may evolve in the twenty-first century.

One such trend is the professionalization of cybercrime. It is no longer the province of adventurous adolescents or creative computer scientists; cybercrime has become big business. According to one estimate, for example, over $52 billion in goods and services were purchased online in 2004 with fraudulently obtained credit cards [48]. As cybercrime has evolved into big business, it has fallen into the hands of organized crime, of both the traditional and the non-traditional varieties [48]. Large-scale financial cybercrime tends to be controlled by loosely-affiliated 'web mobs' or by more formal gangs often associated with Eastern European organized crime [48]. Web mobs are loosely-affiliated groups of cybercriminals, or hackers, who collaborate in shifting allegiances to carry out cybercrimes for profit; one source reported, for example, that one member of a web mob made $300,000 in 2003 by stealing from businesses in the United States [54]. And American authorities report that the Russian mafia and other Eastern European organized gangs are actively engaged in spamming, phishing, online extortion and the trafficking of stolen goods online [65]. Unlike the web mobs, which are composed of "independent" criminals who commit the cybercrimes themselves, the more traditional organized crime groups often hire individuals to carry out their crimes; they hire those who have computer expertise but reside

in economically depressed countries where gainful employment is rare [65].

A second trend is the increasing automation of cybercrime. Perhaps the best example of this is the use of botnets to commit extortion and other crimes [18]. A botnet is a network of compromised computers, known as zombies; the zombies belong to home users or to businesses, neither of which is aware that their computers have been taken over and are being used to commit crimes [31]. The zombie computers are used to launch distributed denial of service (DDoS) attacks against online businesses; the DDoS attack takes a business offline [18]. The business then receives a message stating that the attack will continue unless the business pays a certain amount ($50,000, for example) to the sender of the message [32]. According to one estimate, in 2004 six to seven thousand businesses were paying online extortionists to prevent their websites from being taken offline [32]. There is concern that this tactic, which is currently used only to attack online businesses, can be used to attack government agencies or parts of a nation's critical infrastructure [32].

These trends indicate that as it matures, cybercrime will conform to our experience with real-world crime in certain aspects but deviate in others. The emphasis on collaboration and organization is nothing new; in this regard cybercriminals are simply replicating a tradition begun by the US Mafia. In the early twentieth-century, Italian-American immigrants and affiliated groups realized they could most efficiently exploit the profit-making opportunities resulting from Prohibition by establishing a hierarchical organization to carry out essential functions, such as producing alcohol, distributing alcohol and preventing encroachments by rivals [8]. This model was eventually applied to later varieties of real-world criminal activity, such as drug-trafficking [8].

Cybercriminals are organizing, but in a different fashion The hierarchical organization utilized by the Mafia and other real-world criminal groups tends to make law enforcement's task that much easier because it provides a focal point; once police identify the organization and its essential personnel, law enforcement officers can concentrate on tracking the activities of that identifiable, localized group [8]. Cybercriminals, on the other hand, are adopting a fluid, more lateral organizational framework; various individuals come together for specific criminal activities and then move on [48]. Membership in the web, for example, tends to be shifting, so there is not a stable core of personnel upon which law enforcement can focus [48]. And the members of these groups are not physically located in the same territory; unlike the members of a Mafia family, the participants in a web mob tend to be from many different countries and tend to remain anonymous [8]. The participants in a web mob or other varieties of online crime may not know the real names of those with whom they collaborate [48]. This makes law enforcement's task that much more difficult. Clever cybercriminals can enjoy the benefits of criminal organization without its disadvantages [8].

Cybercriminals' ability to utilize organization effectively and to combine it with automated crime will create great challenges for law enforcement in the future, challenges that are likely to resist even the improvement in national cooperation sought to be achieved by the Convention on Cybercrime. For real-world crime, enhanced scale tends to be correlated with enhanced vulnerability; that is, the larger a real-world criminal operation is, the more likely it is that the operation, and its participants, will come to the attention of law enforcement [8]. This is not true for online crime. Automated measures like the botnets allow cybercriminals to commit crimes on a heretofore inexperienced scale without, however, running any enhanced risk of apprehension [8]. Indeed, the opposite is true; automated crime tends to be anonymous crime [8]. Botnets come into existence, are used to launch attacks or carry out other crimes, and are then dissolved before law enforcement can track them to an identifiable perpetrator.

The future of cybercrime will almost certainly involve greater reliance on automated methods of commission and on cooperation among groups located in various parts of the world. Indeed, some are already describing web mobs as "multinational enterprises" [48].

REFERENCES

[1] Additional Protocol to the Convention on Cybercrime Concerning the Criminalisation of Acts of a Racist and Xenophobic Nature Committed through Computer Systems (CETS 189), http://conventions.coe.int/Treaty/en/Treaties/Html/189.htm (accessed March 21, 2005).

[2] Anti-Phishing Working Group, Phishing Activity Trends Report – November 2004, http://www.antiphishing.org/APWG%20Phishing%20Activity%20Report%20%20November%202004.pdf (accessed March 21, 2005).

[3] APEC Leaders' Statement on Fighting Terrorism and Promoting Growth (October 26, 2002), http://www.apecsec.org.sg/apec/leaders__declarations/2002/statement_on_fighting.html (accessed March 21, 2005).

[4] Ashcroft v. Free Speech Coalition, 535 US 234 (2002).

[5] C. Avery, Hackers on trial (1999), http://www.english.ucf.edu/publications/enc4932/cassie.htm (accessed March 21, 2005).

[6] S. Baker, B. Grow, Gambling sites, this is a holdup, Business Week, August 9 (2004), http://www.businessweek.com/magazine/content/04_32/b3895106_mz063.htm (accessed March 21, 2005).

[7] S.W. Brenner, M.D. Goodman, In defense of cyberterrorism: an argument for anticipating cyber-attacks, University of Illinois Journal of Law, Technology & Policy 1 (2002), 6–7.

[8] S.W. Brenner, Organized cybercrime: how cyberspace may affect the structure of criminal relationships, North Carolina Journal of Law & Technology 1 (2002), 11–15.

[9] S.W. Brenner, Toward A Criminal Law for Cyberspace: Distributed Security, Boston University Journal of Science & Technology Law 10 (2004), 1–105.

[10] The CERT Coordination Center FAQ, http://www.cert.org/faq/cert_faq.html#A2 (accessed March 21, 2005).

[11] Chart of Signatures and Ratifications, Council of Europe Convention on Cybercrime (CETS No. 185), http://conventions.coe.int/Treaty/Commun/ChercheSig.asp?NT=185&CM=8&DF=12/07/04&CL=ENG (accessed March 21, 2005).

[12] Chinese authorities apprehend online bank robber, ChinaTechNews.com, October 7 (2004), http://www.chinatechnews.com/index.php?action=show&type=news&id=1910 (accessed March 21, 2005).

[13] J. Christensen, The trials of Kevin Mitnick, CNN.com (March 18, 1999).

[14] R.D. Clifford (ed.), The Investigation, Prosecution and Defense of a Computer-Related Crime, Carolina Academic Press, Durham, NC (2005).

[15] Council of Europe Convention on Cybercrime (CETS No. 185), http://conventions.coe.int/Treaty/en/Treaties/Html/185.htm (accessed March 21, 2005).

[16] Council of Europe, *Interview with Henrik Kaspersen*, http://www.coe.int/t/e/com/files/interviews/20040915_interv_kaspersen.asp (accessed March 21, 2005).

[17] M. Delio, *The greatest hacks of all time*, Wired News, February 6 (2001), http://www.wired.com/news/print/0,1294,41630,00.html (accessed March 21, 2005).

[18] M. Desmond, *Attack of the PC zombies!*, PC World, October 19 (2004), http://yahoo.pcworld.com/yahoo/article/0,aid,118208,00.asp (accessed March 21, 2005).

[19] Council of Europe Convention on Cybercrime, Explanatory Report – (CETS No. 185), http://conventions.coe.int/Treaty/en/Reports/Html/185.htm (accessed March 21, 2005).

[20] Federal Trade Commission, *ID theft: what it's all about*, http://www.ftc.gov/bcp/conline/pubs/credit/idtheftmini.htm (accessed March 21, 2005).

[21] Federal Trade Commission, *Identity thief goes "phishing" for consumers' credit information*, http://www.ftc.gov/opa/2003/07/phishing.htm (accessed March 21, 2005).

[22] *Fourth profile: Kevin Mitnick*, ZD Net Australia (2004), http://www.zdnet.com.au/insight/security/0,39023764,39116620-5,00.htm (accessed March 21, 2005).

[23] G8 Justice and Home Affairs Ministers: Communique (May 11, 2004), http://www.usdoj.gov/ag/events/g82004/Communique_2004_G8_JHA_Ministerial_051204.pdf (accessed March 21, 2005).

[24] W. Gibson, *Neuromancer 51*, Ace Books, New York (1995), reissue edition.

[25] M.D. Goodman, S.W. Brenner, *The emerging consensus on criminal conduct in cyberspace*, UCLA Journal of Law & Technology 3 (2002), 12–16.

[26] J. Gould, *Hacker heist*, The Village Voice, December 23 (1997).

[27] G. Gross, *Identity thieves cheat 27 Million in US*, PC World, September 3 (2003), http://www.pcworld.com/news/article/0,aid,112314,00.asp (accessed March 21, 2005).

[28] M.A. Healy, *Child pornography: an international perspective*, Computer Crime Research Center (2004), http://www.crime-research.org/articles/536/ (accessed March 21, 2005).

[29] Hearing Before the US Senate Committee on Foreign Relations: Law Enforcement Treaties (June 17, 2004), http://foreign.senate.gov/hearings/2004/hrg040617a.html (accessed March 21, 2005).

[30] Hobbs Act: 18 US Code §1951.

[31] The Honeynet Project and Research Alliance, *Know your enemy: tracking botnets* (March 13, 2005), http://www.honeynet.org/papers/bots/ (accessed March 21, 2005).

[32] D. Ilett, *Online extortion growing more common*, CNET News (October 8, 2004), http://news.com.com/Expert%3A+Online+extortion+growing+more+common/2100-7349_3-5403162.html (accessed March 21, 2005).

[33] Internet Fraud Complaint Center, Intelligence Note 11/04/2003, http://www1.ifccfbi.gov/strategy/11403RomanianWarning.pdf (accessed March 21, 2005).

[34] P. Jenkins, *Beyond Tolerance: Child Pornography Online*, New York University Press, New York (2001).

[35] Joint Statement: Sixteenth APEC Ministerial Meeting (November 17–18, 2004), http://www.apecsec.org.sg/apec/ministerial_statements/annual_ministerial/2004_16th_apec_ministerial.html (accessed March 21, 2005).

[36] B. Krebs, *"Sasser" worm strikes hundreds of thousands of PCs*, Washington Post (May 3, 2004), http://www.washingtonpost.com/ac2/wp-dyn/A63002-2004May3?language=printer (accessed March 21, 2005).

[37] *Landslide*, The Fifth Estate – CBC News (November 5, 2003), http://www.cbc.ca/fifth/landslide/profile.html (accessed March 21, 2005).

[38] *Legendary computer hacker released from prison*, CNN.com (January 21, 2000).

[39] R. Lemos, *The computer virus – no cures to be found*, ZD Net, November 25 (2003), http://news.zdnet.com/2100-1009_22-5111442.html (accessed March 21, 2005).

[40] R. Lemos, *Cyberterrorism: the real risks*, ZD Net UK, August 27 (2002), http://www.zdnet.co.uk/print/?TYPE=story&AT=2121358-39020369t-10000023c (accessed March 21, 2005).

[41] R. Lemos, *A 20-year plague*, CNET, November 25 (2003), http://ecoustics-cnet.com.com/A+20-year+plague/2009-7349_3-5111410.html (accessed March 21, 2005).

[42] S. Levy, *Hackers: Heroes of the Computer Revolution*, Doubleday, New York (1984).

[43] J. Leyden, *Parson not dumbest virus writer ever, shock!*, The Register, September 1 (2003), http://www.theregister.co.uk/2003/09/01/parson_not_dumbest_virus_writer/ (accessed March 21, 2005).

[44] J. Leyden, *US phishing losses hit $500 Million*, The Register, September 29 (2004).

[45] J. Littman, *The Watchman: The Twisted Life and Crimes of Serial Hacker Kevin Poulsen*, Little Brown, Boston (1997).

[46] M. Loney, *SQL slammer worm wreaks havoc on the Internet*, ZD Net UK (January 26, 2003), http://news.zdnet.co.uk/internet/security/0,39020375,2129330,00.htm (accessed March 21, 2005).

[47] McConnell International, *Cyber crime ... and punishment?* (2000), http://www.mcconnellinternational.com/services/CyberCrime.pdf (accessed March 21, 2005).

[48] J. McCormick, D. Gage, *Shadowcrew: Web Mobs, baseline*, http://www.baselinemag.com/article2/0,1397,1774393,00.asp (accessed March 21, 2005).

[49] C. Morello, *Hacker did time, wants break*, Times – Picayune, September 15 (1996).

[50] C.J.P. Moschovitis et al., *History of the Internet* (1999), http://www.historyoftheinternet.com/chap3.html (accessed March 21, 2005).

[51] National White Collar Crime Center & Federal Bureau of Investigation, IC3 2003 Internet Fraud Report (2004), http://www1.ifccfbi.gov/strategy/2003_IC3Report.pdf (accessed March 21, 2005).

[52] OECD, Guidelines for the Security of Information Systems and Networks (2002), http://www.ftc.gov/bcp/conline/edcams/infosecurity/popups/OECD_guidelines.pdf (accessed March 21, 2005).

[53] Organization of American States, Final Report of the Fifth Meeting of Ministers of Justice or Attorneys General of the Americas (April 28–30, 2004), http://www.oas.org/juridico/english/cybV_CR.pdf (accessed March 21, 2005).

[54] A. Piore, *Hacking for dollars,* Newsweek, December 15 (2003).

[55] *Porn ring "Was real child abuse",* BBC News, January 10 (2001), http://news.bbc.co.uk/1/hi/uk/1109787.stm (accessed March 21, 2005).

[56] R. Power, *Tangled Web*, Que, Indianapolis (2000).

[57] E.S. Raymond, *A brief history of hackerdom* (2000), http://www.hackemate.com.ar/hacking/eng/part_00.htm#toc3 (accessed March 21, 2005).

[58] Regina v. Sharpe, 2001 S.C.C.D.J. 42 (Canada Sup. Ct. 2001).

[59] L. Rohde, *Love bug charges dropped,* PC World, August 21 (2000), http://www.pcworld.com/news/article/0,aid,18146,00.asp (accessed March 21, 2005).

[60] T. Shea, *The FBI goes after hackers,* InfoWorld 38 (March 26, 1984).

[61] U. Sieber, *Legal aspects of computer-related crime in the information society,* Prepared for the European Commission (1998), http://europa.eu.int/ISPO/legal/en/comcrime/sieber.html (accessed March 21, 2005).

[62] J. Scott-Joynt, *How not to win a million*, BBC News, June 22 (2004), http://news.bbc.co.uk/2/hi/business/3808397.stm (accessed March 21, 2005).

[63] M. Slatalla, *Masters of Deception: The Gang that Ruled Cyberspace,* Harper Perennial, New York (1996).

[64] State v. Schwartz, 173 Or. App. 301, 21 P.3d 1128 (Or. App. 2001).

[65] J. Swartz, *Crooks slither into net's shady nooks and crannies,* USA Today, October 21 (2004), http://www.usatoday.com/printedition/money/20041021/cybercrimecover.art.htm (accessed March 21, 2005).

[66] J. Thomas, *Political ideas of mcveigh are subject at bomb trial,* New York Times, June 11 (1997).

[67] United States v. Kevin Mitnick (US District Court – Central District of California 1996), http://www.freekevin.com/indictment.html (accessed March 21, 2005).

[68] United States v. Morris, 928 F. 2d 504 (2d Cir. 1991).

[69] US Secret Service, Public Awareness Advisory Regarding "4-1-9" or "Advance Fee Fraud" Schemes, http://www.secretservice.gov/alert419.shtml (accessed March 21, 2005).

[70] Vamosi, *Why virus writers get away with it,* CNET News, August 23 (2003), http://reviews.cnet.com/4520-3513_7-5068073.html (accessed March 21, 2005).

[71] D. Verton, *Organized crime invades cyberspace,* Computerworld, August 30 (2004), http://www.computerworld.com/securitytopics/security/story/0,10801,95501,00.html (accessed March 21, 2005).

[72] *Who are these crackers?,* ACTLab – University of Texas, http://home.actlab.utexas.edu/~aviva/compsec/cracker/whocrack.html (accessed March 21, 2005).

[73] *ARPANET, **Wikipedia: The Free Encyclopedia**,* http://en.wikipedia.org/wiki/ARPANET (accessed March 21, 2005).

[74] *Chaos Computer Club, **Wikipedia: The Free Encyclopedia**,* http://en.wikipedia.org/wiki/Chaos_Computer_Club (accessed March 21, 2005).

[75] *Hacker, **Wikipedia: The Free Encyclopedia**,* http://en.wikipedia.org/wiki/Hacker (accessed March 21, 2005).

[76] *Phreaking, **Wikipedia: The Free Encyclopedia**,* http://en.wikipedia.org/wiki/Phreaking (accessed March 21, 2005).

[77] *Youth advises house on computer crime,* N.Y. Times, September 27 (1983).

[78] 18 US Code §2331.

[79] 18 US Code Section 2332b(g)(5)(B) and Patriot Act of 2001 §808, P.L. 107-56, 115 Stat. 272.

PART 5

PRIVACY- AND EXPORT REGULATIONS

26

THE EXPORT OF CRYPTOGRAPHY IN THE 20TH AND THE 21ST CENTURIES

Whitfield Diffie and Susan Landau

Sun Microsystems, Inc.
Menlo Park, CA, USA

Contents

26.1 INTRODUCTION

On the 14th of January 2000, the Bureau of Export Administration issued long-awaited revisions to the rules on exporting cryptographic hardware and software. The new regulations, which grew out of a protracted tug of war between the computer industry and the US Government, are seen by industry as a victory. Their appearance, which was attended by both excitement and relief, marked a substantial change in export policy. This paper examines the evolution of export control in the cryptographic area and considers its impact on the deployment of privacy-protecting technologies within the United States.

Before the electronic age, all "real-time" interaction between people had to take place in person. Privacy in such interactions could be taken for granted. No more than reasonable care was required to assure yourself that only the people you were addressing – people who had to be right there with you – could hear you. Telecommunications have changed this. The people with whom you interact no longer have to be in your immediate vicinity; they can be on the other side of the world, making what was once impossible spontaneous and inexpensive. Telecommunication, on the other hand, makes protecting yourself from eavesdropping more difficult. Some other security mechanism is required to replace looking around to see that no one is close enough to overhear: that mechanism is cryptography, the only security mechanism that directly protects information passing out of the physical control of the sender and receiver.

At the turn of the 20th century, cryptography was a labor-intensive, error-prone process incapable of more than transforming a small amount of written material into an encoded *ciphertext* form. At the dawn of the 21st it can be done quickly, reliably, and inexpensively by computers at rates of a billion bits a second. This progress is commensurate with that of communications in general yet the fraction of the world's communications protected by cryptography today is still minuscule. In part this is due to the technical difficulty of integrating cryptography into communication systems so as to achieve security, in part to an associated marketing problem. Proper implementation of cryptosecurity requires substantial up-front expenditure on infrastructure while most of the benefit is lost unless there is nearly ubiquitous coverage, a combination that deters investment. These factors result in a lack of robustness of the market that makes it prey to a third factor, political opposition.

As telecommunication has improved in quality and gained in importance, police and intelligence organizations have made ever more extensive use of the possibilities for electronic eavesdropping. These same agencies now fear that the growth of cryptography in the commercial world will deprive them of sources of information on which they have come to rely. The result has been a struggle between the business community, which needs cryptography to protect electronic commerce and elements of government that fear the loss of their surveillance capabilities. Export control has emerged as an important battleground in this struggle.

26.2 BACKGROUND

In the 1970s, after many years as the virtually exclusive property of the military, cryptography appeared in public with a dual thrust. First came the work of Horst Feistel and others at IBM that produced the U.S. Data Encryption Standard. DES, which was adopted in 1977 as Federal Information Processing Standard 46, was mandated for the protection of all government information legally requiring protection but not covered under the provisions for protecting classified information – a category later called "unclassified sensitive".

The second development was the work of several academics that was to lead to *public-key cryptography*, the technology underlying the security of Internet commerce today. Public-key cryptography makes it possible for two people, without having arranged a secret key in advance, to communicate securely over an insecure channel. Public-key cryptography also provides a digital signature mechanism remarkably similar in func-

tion to a written signature.[1] The effect of new developments in distinct areas of cryptography was to ignite a storm of interest in the field, leading to an explosion of papers, books and conferences.

The government response was to try to acquire the same sort of 'born classified' legal control over cryptography that the Department of Energy claimed[2] in the area of atomic energy. The effort was a dramatic failure. NSA hoped an American Council on Education committee set up to study the problem would recommend legal restraints on cryptographic research and publication. Instead, it proposed only that authors voluntary submit papers to NSA for its opinion on the possible national-security implications of their publication [9: 10].

It did not take the government long to realize that even if control of research and publication were beyond its grasp, control of deployment was not. Although laws directly regulating the use of cryptography in the US appeared out of reach – and no serious effort was ever made to get Congress to adopt any – adroit use of export control proved remarkably effective in diminishing the use of cryptography, not only outside the United States but inside as well.

26.3 EXPORT CONTROL

The export control laws in force today are rooted in the growth of the Cold War that followed World

---

[1]In recognition of the increasing importance of electronic commerce, in June 2000, President Clinton signed into law the Millennium Digital Commerce Act, Public Law 106-229, which establishes the legal validity of 'electronic signatures'. The term is somewhat broader than 'digital signature' but points the way toward a future in which alternatives to the written signature play a central role in commerce.

[2]The courts have never ruled on the constitutionality of this provision of the *Atomic Energy Act of 1946. U.S. Statutes at Large* 60 (1947): 755–775. In 1997, the Progressive magazine challenged it by proposing to publish an article by Robert Morland entitled "The H-bomb Secret, how we got it, why we're telling it". After the appearance of an independent and far less competent article on how h-bombs work, the government succeeded in having the case mooted and leaving the impression in the popular mind that it would have won. In fact, the virtual certainty that it would have lost is undoubtedly why it acted as it did.

War II. In the immediate post-war years the US accounted for a little more than half of the world's economy. Furthermore, the country was just coming off a war footing, with its machinery of production controls, rationing, censorship, and economic warfare. The US thus had not only the economic power to make export control an effective element of foreign policy but the inclination and the regulatory machinery to do so.

The system that grew out of this environment had not one export control regime but two. Primary legal authority for regulating exports was given to the Department of State, with the objective of protecting national security. Although the goods to be regulated are described as *munitions*, the law does not limit itself to the common meaning of that word and includes many things that are neither explosive nor dangerous. The affected items are determined by the Department of State acting, through the *Munition Control Board*, on the advice of other elements of the executive branch, especially, in the case of cryptography, the National Security Agency.[3]

Exports that are deemed to have civilian as well as military uses are regulated by the Department of Commerce. Such items are termed *dual-use* and present a wholly different problem from 'munitions'. A broad range of goods – vehicles, aircraft, clothing, copying machines – are vital to military functioning just as they are to civilian. If the sale of such goods was routinely blocked merely because they might benefit the military of an unfriendly country, there would be little left of international trade. Control of the export of dual-use articles therefore balances considerations of military application with considerations of foreign availability – the existence of sources of supply prepared to fill any vacuum left by US export bans.

The munitions controls are far more severe than the dual-use controls, requiring individually approved export licenses specifying the product and the actual customer as opposed to broad restrictions by product category and national destination. Legal authority to decide which regime is to be applied lies with the Department of State, which can authorize the transfer of jurisdiction to the Department of Commerce, a process called *commodities jurisdiction*.

Assessing whether a product is military or civilian is not always straightforward. Once we leave the domain of the clearly military (such as fighter aircraft), we immediately encounter products that either have both military and civilian uses or products that can be converted from one to the other without difficulty. The Boeing 707, a civilian airplane, was a mainstay of the world's airlines during the 1960s and 1970s. Its military derivatives, the C-135 (cargo, including passengers), the KC-135 (tanker) and RC-135 (intelligence platform) have been mainstays of Western military aviation. Recognizing that civilian aircraft might be put to military use and thus bypass export control, the US government nonetheless permitted their export as a business necessity. The allowability of exports was judged on the basis of how dual-use goods were configured and who was to be the customer. Generally speaking a commercial technology, not explicitly adapted to a uniquely military function, can be sold to a non-military customer without excessive paperwork.

The application of export controls naturally depends heavily on the destination for which goods are bound. Applications for export to US allies, such as the countries of Western Europe, are more likely to be approved than applications for exports to neutral, let alone hostile, nations. Clearly the effectiveness of export controls will be vastly magnified by coordination of the export policies of allied nations. During the Cold War, the major vehicle for such cooperation among the US and its allies was *COCOM*, the *Coordinating Committee on Multilateral Export Controls*, whose membership combined Australia, New Zealand, and Japan with the US and most western European countries.[4] Although COCOM existed primarily to prevent militarily significant exports to non-COCOM countries, that did not mean that the COCOM countries exported freely among themselves. Many products that would not be permitted out of COCOM could be sold to other COCOM countries but still required a burdensome export approval process.

---

[3] US export control regulations are described at length in Root and Liebman's, *United States Export Controls* [10].

[4] Iceland for some reason was not included.

26.4 EXPORT STATUS OF CRYPTOGRAPHY

In the post-WWII period, cryptography was, like nuclear energy, an almost entirely military technology.[5] It is stretching the point only a little to say that insecure analog voice scramblers or hand-authentication techniques that might be found in civilian uses were no more closely related to high-grade military encryption equipment than glowing watch dials or X-ray machines were related to atomic bombs. Not surprisingly, all cryptography, regardless of functioning or intended application was placed in the category of munitions. As the information revolution progressed – particularly as computers began to 'talk' more and more to other computers – the argument for dual-use status slowly improved. Telecommunications between humans can be authenticated by combinations of more or less informal mechanisms: voice recognition, dial-back, request to know the last check written on an account, etc. To achieve high security in communication between computers without human intervention, cryptography is indispensable. Nonetheless, cryptography remained in the 'munition' category long after this seemed reasonable to most observers.

The importance of the munition/dual-use distinction lies in a difference in licensing procedures and a difference in the criteria for export approval. As munitions, cryptographic devices required individually approved export licenses. Two factors combine to make such licenses antagonistic to commercial use of cryptography. One is time: the weeks or months required to get approval often exceed the time commercial organizations allocate to procurement of even major systems. The other is the requirement to identify the end customer. In much of commerce, manufacturers deal with one or more layers of resellers who may either be unaware of the identities of buyers or unwilling to share their information with their suppliers. Munitions are not only more cumbersome to export but more likely to be denied approval outright. The law regulating military exports makes no provision for the probable effectiveness of export policy. If an export is judged militarily imprudent, it is barred regardless of the likelihood that this action will actually prevent the would-be purchaser from obtaining equipment of the type desired.

Even after the business necessity and thus the dual-use character of cryptography had become clear, the problem of distinguishing military from civilian cryptosystems remained elusive. Some cases were straightforward. Systems specially adapted to work with military communication protocols – such as the MK XII IFF[6] devices that identify aircraft to military radars – or those whose implementations were ruggedized for field use or satisfied arcane military specifications against radiation leakage could safely be classified as military. But what about cryptosystems running in ordinary commercial computing equipment in ordinary office environments? Such equipment performs very similarly whether in a general's office or in a banker's.

The challenge of export control is to develop a policy that interferes as little as possible with international trade while limiting the ability of other countries to develop military capabilities that threaten US interests. This requires setting rules to distinguish military uses of technology from civilian ones. In the case of cryptography, the initial attempt was to classify cryptosystems as military or civilian by strength, much as guns might be classified by caliber. Small arms have civilian applications – from hunting and target shooting to personal protection and public safety – whereas artillery is purely military. The distinction, however, proved far harder to make in the case of cryptography than of firearms. A cryptographic system adequate to protect a billion-dollar electronic funds transfer is indistinguishable from one adequate to protect a top-secret message.

---

[5]More precisely, the market for cryptography was almost entirely military. A small number of companies manufactured cryptographic equipment but found their best markets in governmental rather than commercial sales. These were the primary suppliers to nations that did not maintain a domestic capacity to produce cryptographic equipment. Even the United States – where most cryptographic equipment was produced by commercial firms but under exclusive contract to the government – licensed its World War II field cipher system, the M-209, from the Swiss firm Hagelin AG [7: 427].

[6]Identification Friend or Foe.

26.5 THE IMPACT OF EXPORT CONTROL ON CRYPTOGRAPHY

As the US share of the world's economy has declined over the past five decades, export controls have become less effective as a mechanism of US foreign policy. Worldwide growth of manufacturing capacity, particularly in military technology, has made many more products available from non-US sources, while the associated growth of markets outside the US has meant that the cost to US businesses of export controls is far greater. In 1950, it cost US companies little to be prevented from exporting something for which there were few foreign customers. Today, with a majority of potential customers outside the US, a product's exportability can make the difference between success and failure.

This change in impact of export controls has changed their role and export controls on cryptography have come to be used at least as much for their effect on the domestic market as the foreign one. Three factors have made this possible:

- The export market in computer hardware and software is huge. The typical American computer company makes more than half its sales abroad and must manufacture exportable products to be competitive.
- Security is always a *supporting feature*; no system exists for the primary purpose of being secure. To be usable and effective security must be integrated from scratch with the features it supports. Even when it is feasible, adding cryptography to a finished systems is undesirable.
- Making two versions of a product is complicated and expensive, particularly when, as is typically the case, domestic and foreign products must interoperate. Making a more secure product for domestic use, furthermore, points out to foreign customers that you have given them less than your best. These costs would be borne were the domestic demand for security great enough but so far it has not been.

The result of US export controls has thus been to limit the availability of strong cryptography, not merely abroad but at home.

These policies, which put the interests of intelligence and law-enforcement agencies ahead of other national concerns, were made possible by the dominant, though far from invincible, position of US companies in the world market for computer hardware and software. Security, though a small component of most computer systems, is often essential. By forbidding the export of systems with good security, the US risks losing the business of security-conscious customers to foreign competition, thereby accelerating the development of the computer industries outside the US. The fast-growing computer industry in both Europe and Asia have been only too happy to challenge the US position and, as the growth of the world wide web and electronic commerce made the commercial importance of cryptography more obvious, the US government came under more and more pressure to amend its regulations.

26.6 EVENTS AFTER THE COLD WAR

The end of the Cold War at the beginning of the 1990s set the stage for a change in export policy. The first move in industry's direction was a deal struck in 1992 between the National Security Agency, the Department of Commerce, and RSA Data Security, a leading maker of cryptographic software. It provided for streamlined export approval for products using approved algorithms with keys no longer than 40 bits.[7] Two algorithms, both trade secrets of RSA, were approved.

The problem of keylength is not an issue that lends itself well to compromise and the strength represented by 40-bit keys could hardly have pleased either side. In 1992, a message encrypted using a 40-bit key could be cracked by a personal computer using the crudest techniques in a month

---

[7]If the encryption algorithm is properly designed, then the difficulty of unauthorized decryption is determined by the number of bits in the key; an increase of one bit doubles the cost to the intruder. A good encryption algorithm with a 56-bit key is thus 2^{16} or 65,000 times more difficult to crack than one with a 40-bit key. It is often taken for granted that cryptosystems are as strong as their keys suggest and thus it is common to speak of 40-bit cryptography, meaning both that the keys are 40 bits long and that breaking the system takes approximately a trillion encryptions.

or so – hardly sufficient for the lifetime of product plans, let alone personnel records. On the other hand, had such systems been applied to even a few percent of the world's communications they would have created a formidable barrier to signals intelligence. Intercept devices must determine in a fraction of a second whether a message is worth recording. Encryption, broadly applied, seriously interferes with this selection process. If a small enough fraction of messages are encrypted, then being encrypted marks a message as interesting and the message will be recorded. Too many encrypted messages, even weakly encrypted messages, will glut the interceptor's disks and frustrate the collection effort.

At approximately the same time, a case arose that was to demonstrate the difficulty of controlling not only cryptography but open-source software. Philip Zimmerman, a programmer without previous experience in the cryptographic world, wrote an email security program called "Pretty Good Privacy" or PGP that combined several of the most popular cryptographic techniques and employed keys far larger than permitted by export rules. A federal grand jury in San José investigated for over a year before dropping the case. The grand jury gave no explanation for its actions but an event on the other side of the country immediately suggests itself. MIT Press, publishing arm of the Massachusetts Institute of Technology, published the source code of PGP. Even though the type was set in an OCR font for easy conversion to electronic from, the government made no attempt to challenge the implicit claim that a book, no matter how easily read by a computer, was protected by the First Amendment [5].

Government attempts to control cryptography were not limited to its export strategy. In parallel with the keylength-based formula – which it presumably saw as an interim measure – the US government tried to change the rules to give itself a permanent advantage. In early 1993, it moved to replace the fifteen-year-old, 56-bit, Data Encryption Standard with an 80-bit algorithm that provided a special *trap door* for government access.[8]

Although the standard was adopted, it found few takers and was generally counted as a failure.

Looking back over the 1990s, it is hard to judge whether the Clipper program set the stage for the sequence of confrontations and compromises that followed or whether all were merely consequences of the same technological and market forces. The government made several attempts to establish the principle that it had the right to control cryptographic technology in order to guarantee its power to read intercepted messages.[9] Over the same period it restructured the export-control bureaucracy and relaxed the regulations.

While cryptography was classified as a munition, a would-be exporter was required either to seek an export license from "State" or request a transfer of jurisdiction to "Commerce". In 1996 the Department of Commerce Bureau of Export Administration was given direct authority over most cryptographic exports.[10] In the process, however, the personnel to carry out the new role were transferred from the State Department to the Commerce Department, creating a sense that there was likely to be more change of form than substance.

It is during the reorganization of the export-control machinery that Department of Justice personnel were first introduced into the process. In tune with this introduction, though somewhat ahead of it in time, was a move to shape the terms of debate by talking about signals intelligence in terms that were drawn more from law enforcement and less from the military. It was true then and is true now that most US interception of communications is targeted not against criminals (no matter how loosely this term is used) but against other countries – largely countries we recognize and with many of which we are on friendly terms. Spying on your 'friends' is and has always been an uncomfortable activity but much of the discomfort is mitigated by secrecy. A matter never spoken about

---

[8]This was the infamous *Clipper* system, in which the keys were split and escrowed with Federal agencies [12].

[9]In a related move, the government scored a major victory. The Communications Assistance for Law Enforcement Act of 1994 gave it the power to require communications carriers to build wiretapping into their networks.

[10]This was done by adopting Department of State regulations authorizing shippers to go directly to the Department of Commerce for certain categories of goods, rather than submitting their applications first to the Department of State.

creates few awkward pauses in conversation but to engage in a public debate one must have something to say. In the debate about encryption it was necessary for the government to say why it was seeking to expand its powers of interception. The answer was to point to an unholy trinity: terrorists, drug dealers, and pedophiles. Entirely lacking in popular support, these groups were in no position to step forth and speak out against being spied on.[11]

A rationale has its costs. Giving a law enforcement rationale made it hard to maintain the intelligence criteria and as the decade wore on, the government's proposals moved toward the needs of police – individualized court ordered surveillance, perhaps requiring the cooperation of a foreign judicial system – and away from the invisible broad spectrum surveillance that the intelligence community desired. The predictable consequence was that the intelligence agencies, realizing that their needs were not being met, became less vociferous in their support of crypto-control proposals.

In the summer of 1996 the National Research Council released its 18-month study on cryptography policy, *Cryptography's Role in Securing the Information Society* (the *CRISIS* report, conceived at the time of the key-escrow proposal. Acting on a mandate from Congress, the NRC convened a panel of sixteen experts from government, industry, and science, thirteen of whom received security clearances. The panel was heavily weighted towards former members of the government – the chair, Kenneth Dam, for example, had been Under Secretary of State during the Reagan administration – and many opponents of the government's policies anticipated that the NRC report would support the Clinton administration's cryptography policy. It did not.

The report concluded that "on balance, the advantages of more widespread use of cryptography outweigh the disadvantages", and that current US policy was inadequate for the security requirements of an information society. Observing that existing export policy hampered the domestic use of strong cryptosystems, the panel recommended

loosening export controls and said that products containing DES "should be easily exportable".

This was not a message the Clinton administration wanted to hear and no immediate effect on policy was discernible. In the fall of 1996 the government announced that a window of opportunity for export would run for the two years 1997 and 1998. During this window, manufacturers would be allowed to export Data Encryption Standard products quite freely if they had entered into memoranda of understanding with the government promising to develop systems with *key recovery*[12] during the open-window period. This approach did not even survive its own window. In September 1998, the rules were relaxed to permit freer export of products containing DES or other cryptosystems with keys no longer than 56-bits.

It was a classic example of 'too little, too late'. Users around the world had come to feel that cryptographic keys should be 128 bits long. Technical arguments to the effect that there was no point in making the cryptography stronger than the surrounding security system cut little ice with customers. Very strong cryptosystems seem to cost no more to build or run than weaker ones so why not have the strong ones.

The year 1996 also saw the start of Congressional interest in cryptography export. The absurdity of US export controls and the danger that they would have a devastating impact on the growing electronic economy led various members of Congress to introduce bills that would have diminished executive discretion in controlling cryptographic exports. None of the bills – which in their later forms were called SAFE for Security and Freedom through Encryption – was close to having enough votes to override a promised presidential veto. Nonetheless, Congressional support for the liberalization of cryptographic export policy was to grow over the next few years, a policy in keeping with previous Congressional decisions. A decade earlier, the Computer Security Act, contrary to the desires of the Reagan administration, placed civilian computer security research and standards under the control of the National Institute of Standards and Technology, rather than NSA.[13]

---

[11] Whether wiretapping actually plays a significant, let alone indispensable, role in combating any of these phenomena is hard to assess [5: 189–191, 233].

[12]The term "key escrow" had acquired a bad name.

[13]Diffie and Landau [5: supra note 12 at 68–69].

26.7 AMERICA'S INTERNATIONAL STRATEGY

The end of the Cold War, realigned the world and made the "east versus west" structure of COCOM inappropriate. The organization, which had existed since 1949, was replaced by a new coalition, the Wassenaar Arrangement, that included former enemies from the Soviet Union and the Warsaw Pact. The expanded organization, comprising 33 nations, is less unified than its predecessor and its procedures are less formal. Although member nations agree on a common control list, each country performs its own review. In behind-the-scenes negotiations in 1998 the Clinton administration scored a coup: Wassenaar agreed that 'mass-market' cryptography using a key length not exceeding 64 bits would not be controlled.[14] The implication was that anything else would be but the Wassenaar Arrangement is subject to "national discretion", and various nations in the agreement had not previously restricted the export of cryptography. Would they now? The Clinton administration believed so. It looked as if export restrictions would stay. Then evidence surfaced suggesting that the US might be using Cold-War intelligence agreements for commercial spying.

A US signals intelligence network called *ECHELON* that had been in existence for at least twenty years came embarrassingly to light. The Echelon system is a product of the UK–USA agreement, an intelligence association of the English speaking nations dominated by Britain and the United States. According to a report prepared for the European Parliament,[15] Echelon targets major commercial communication channels, particularly satellite systems. Many in Europe drew the inference that the purpose of the system was commercial espionage, and indeed, former CIA Director James Woolsey acknowledged that was at least a partial purpose of the system [16: A18]. Commercial communications play a large and growing role in government communications (both military and nonmilitary) and are thus a "legitimate" target of traditional national intelligence collection. It is the position of the US that it does not provide covert intelligence information to US companies.[16] The potential targets of such spying could hardly be expected to regard US policy as adequate protection under the circumstances. Consternation replaced cooperation in the European community. Nations whose policies had previously ranged from the no-controls stance of Denmark to the relatively strict internal controls of France, were now united on the need to protect their communications from the uninvited ear of US intelligence and cryptography was key to any solution. European policies began to diverge from American ones.

26.8 THE RULES CHANGE

In 1999, a SAFE bill passed the five committees with jurisdiction and was headed to the floor of the House, when it was announced that the regulations would be revised to similar effect. The administration capitulated but avoided the loss of control that a change in the law would have produced.

On 16 September 1999, U.S. Vice President, and Presidential candidate, Albert Gore Jr.[17] announced that the government would capitulate. Beginning with regulations announced for December – and actually promulgated on 14 January 2000 – keylength would no longer be a major factor in determining the exportability of cryptographic products.

In its attempt to make a viable military/civilian distinction, the new regulations take several factors into consideration:

1. They define a concept of *retail* products, similar to the *Mass Market* products defined in the Wassenaar Arrangement.

---

[14]The 64-bit limit was for symmetric, or secret-key, cryptography. This translates to approximately 650 bits for public-key cryptography. (Public key is typically used for key exchange; then the communication is encrypted via a secret-key algorithm using the key just negotiated.)

[15]European Parliament, Directorate General for Research, Directorate A, The STOA Programme, "Development of Surveillance Technology and Risk of Abuse of Economic Information". Prepared by Duncan Campbell, April 1999, PE 168/184/ Part 4/4.

[16]The US government says that it uses intelligence information to assist US business in countering foreign corrupt practices.

[17]The Administration's anti-cryptography policy was inimical to Silicon Valley, whose support was seen as crucial for the Vice President's bid for President.

2. They distinguish sharply between commercial and government customers.
3. They make special provision for software distributed in source code.

In the view of export control, an item is retail if it is:

- sold widely,
- sold in large volume,
- made freely available,
- not customized for each individual user, and not extensively supported after sale,
- not intended explicitly for communications infrastructure protection.

The definition is not entirely in accord with the everyday meaning of 'retail' since many retail items are configured for each customer and some, such as custom tailored clothing, have no wholesale stage.

Retail items are largely free of control. They must be submitted for a "one-time review" that the government is supposed to complete within thirty days. If the would-be exporter has not heard anything within that time, it is free to ship its product. The government can demand additional information or even demand more time because the "review is not proceeding in an appropriate fashion" [4: Section 4g] but the rule is some improvement over the previous versions which required the exporting organization to wait until it received an export license from the government before shipping.

Items that are not retail are regulated primarily on the basis of the customer. For many items, commercial sales are acceptable but government sales are not. The distinction between government and the private sector is far from clear – what is the status of partly-government-owned PTT, for example? – and has been a continuing source of friction.

One especially interesting feature of the regulations is their application to software that may be distributed freely in source form but may not be used for commercial purposes without a license. In this case, the rules for distributing the software (by posting it on the Web, for example) are the same as for open-source software – it is only necessary to inform the Bureau of Industry and Security (formerly the Bureau of Export Administration) when you do it. Export approval is required prior to granting a license to a foreign customer, however.

The new rules go a long way toward achieving the objective enunciated earlier. They are a clever compromise between the needs of business and the needs of the intelligence community. Products employed by individual users, small groups or small companies are fairly freely exportable. Products intended for protecting large communications infrastructures – and it is national communication systems that are the primary target of American communications intelligence – are explicitly exempted from retail status.

26.9 EUROPEAN DECONTROL

In June 2000 the European Council of Ministers announced the end of cryptographic export controls within the European Union and its "close trading and security partners" which include the Czech Republic, Hungary, Japan, Poland, Switzerland and the United States. The liberalized export regulations of January 14 would no longer provide the level playing field the US Administration sought.

On July 17, 2000, in response to the European liberalizations, the Clinton administration adopted similar ones: export licenses would no longer be required for export of cryptographic products to the fifteen EU members and the same additional countries.[18] Furthermore, although companies would have to provide one-time technical reviews to the US Government prior to export, they would be able export products immediately.

26.10 WHY DID IT HAPPEN?

What forces drove the US Government from complete intransigence to virtually complete capitulation in under a decade? Most conspicuous is the Internet, which created a demand for cryptography that could not be ignored and at the same time made it more difficult than ever to control the movement of information but more subtle forces

---

[18]Australia, Canada, Czech Republic, Hungary, Japan, Norway, New Zealand, Poland and Switzerland.

were also at play. One of these was the *open-source* movement.

Ever since software became a big business, most software companies have distributed object code and treated the source code as a trade secret. For many years, the open-source approach to software development – freely sharing the source code with the users – was limited to hobbyists, some researchers, and a small movement of true believers. In the mid-1990s, however, some businesses found that an open-source operating system gave them more confidence and better reliability due to rapid bug fixes and the convenience of customization. Others discovered that they could make good money maintaining open-source software. The fact that sufficiently skillful and dedicated users could get free source code from the Web, compile it, configure it, install it, and maintain it did not mean that there were not other users willing to pay for the same services.

Open-source software has taken its place as a major element in the software marketplace. The consequence is a general decrease in the controllability of software. In particular, a serious threat to effectiveness of the government efforts to stop the export of software containing strong cryptography. A policy predicated on the concept of software as a finished, packaged product, one that was developed and controlled by an identifiably and accountable manufacturer foundered when confronted with programs produced by loose associations of programmers/users scattered around the world.

The problem is not merely one of enforcement. The government has always maintained that it could control the export of information but that view is hard to reconcile with the First Amendment and has never been thoroughly tested. A curious, but widely accepted, convention has grown up under which information of sufficiently limited circulation is not treated as having First Amendment protection. The maintenance manual for an aircraft may be a book but it is treated more like a component of the aircraft than a publication. Proprietary source code was treated in the same way.

By comparison open-source software was widely distributed – arguably published – on web sites.

The Bureau of Export Administration might take the view that publishers of some programs required licenses but the legal basis of their position was doubtful and compliance was low. If a program, such as an operating system, leaves the US without cryptography, foreign programmers can add cryptographic components immeasurably more easily than they could with a proprietary source operating system. US export controls have little influence on this process.

To make that matter more arcane, the government has stopped short of claiming that source code published on paper lacks First Amendment protection, maintaining that only source code in electronic form is subject to export control.

In 1996, Daniel Bernstein, a graduate student at the University of California in Berkeley decided that rather than ignore the law, as most researchers had, he would assert a free-speech right to publish the code of a new cryptographic algorithm electronically. Bernstein did not apply for an export license, maintaining that export control was a constitutionally impermissible infringement of his First Amendment rights. Instead, he sought injunctive relief from the federal courts. Bernstein won in both the district court [2] and the Appeals Court for the Ninth Circuit [3]. Unfortunately for the free-speech viewpoint the opinion of the appeals court was withdrawn in preparation for an *en banc* review – a review by a larger panel of Ninth-Circuit judges – that never took place. The appearance of new regulations provided the government with an opportunity to ask the court to declare the case moot. To the government's delight, the court obliged, indefinitely postponing what the government perceived as the danger that the Supreme Court would strike down export controls on cryptographic source code as an illegal prior restraint of speech.

A final adverse influence on export control came from the government's role as a major software customer and the military desire to stretch its budget by using more *commercial off-the-shelf* software and hardware. If export regulations discouraged the computer industry from producing products that met the government's security needs, the government would have to continue the expensive practice of producing custom products for its own use. This was uneconomical to the point of infeasible; the only way to induce the manufacturers to include sufficiently-strong encryption in domestic products was to loosen export controls.

26.11 THE AFTERMATH OF SEPTEMBER 11TH

On September 11th 2001, the United States was attacked by Al-Qaeda, a terrorist organization. There was no evidence to indicate that encryption played a role in the intelligence lapses that allowed the attack. Indeed, we now know that during the summer of 2001 intelligence garnered from terrorist communications prompted concern amongst the intelligence agencies that a major terrorist attack was imminent 9: 2–3].

Nonetheless, a few weeks after the attacks, New Hampshire Senator Judd Gregg argued for controls on encryption. Neither the Bush administration nor other members of Congress joined Gregg, and, after several weeks, the Senator quietly dropped his efforts. The fact is that in combating terrorism, greater surveillance value appears to come more from traffic analysis than from wiretapping. For example, Osama bin Laden stopped using a cell phone in late 2001 because of tracking by US intelligence in the mountains of Afghanistan. Meanwhile use of "anonymous" cellphones by al Qaeda members led to the arrest of a number of operatives and break-up of some cells [15]. US intelligence agencies were well aware of terrorist threats before the events of September 11th. The government's shift in 2000 on cryptographic export controls occurred in this context and explains the lack of support for Senator Gregg's proposal.

26.12 CONCLUSION

For fifty years the United States used export controls to prevent the widespread deployment of cryptography. This policy succeeded for forty of those years but changes in computing and communications in the last decade of the 20th century increased the private-sector need for security and reduced the policy it to a Cold War relic. Its demise

opens the way for securing the civilian communications infrastructure on which all of society will depend in the 21st century.

26.13 RECOMMENDATIONS

Although the new export regulations in the area of cryptography are a substantial improvement on earlier ones they still leave much to be desired.

- The regulations remain complex. The amendments, exclusive of surrounding procedural and explanatory material, amount to some dozen pages and the material they amend is several hundred.
- Although the burden of timeliness has on its face shifted from the exporter to the government, the conditions that permit the government to require more time are vague and appear to admit of discriminatory application. The use of these extensions should be precisely spelled out.
- The definition of retail is at some variance with the ordinary English use of that term. The regulations should perhaps return to the Wassenaar Arrangement's concept of "mass market".
- The notification requirements for open-source programs, although considerably less onerous than the earlier licensing requirements may still constitute an unconstitutional prior restraint on publication. Considering that they, like all of cryptographic export control, serve the interests of the US signals intelligence organizations, and that those organizations presumably watch the Web anyway, the notification requirements seem to serve little purpose.
- Although understandable from a US intelligence perspective, the restriction on infrastructure protection products may not be compatible with the US desire to protect the critical infrastructure of the industrialized world from terrorist attack.

This issue is fundamentally the same as those faced by the National Research Council CRISIS panel. We remind the readers of their conclusion, "On balance, the advantages of more widespread use of cryptography outweigh the disadvantages". We believe the same holds true for infrastructure protection. On balance, the advantages of more widespread use of cryptography for infrastructure protection products outweigh the disadvantages.

The shortcomings of export law in the cryptographic area are typical of the shortcomings of our export laws in general. Cryptography may therefore point the way toward a fairer export-control regime that balances the broad spectrum of United States interests rather than focusing on military security, which is not currently a major vulnerability. Such a regime, recognizing the importance of international commerce in the post-Cold War world would shift the much of the burden from exporters to the government. Foreign availability tests would be more broadly applied; exporters would be entitled to timely responses; a broader range of export decisions would be appealable to the federal courts; and the effectiveness of export policy would be subject to periodic review.

REFERENCES

[1] AES, available at http://csrc.nist.gov/encryption/aes/

[2] Daniel Bernstein v U.S. Department of State, 922 F. Supp. 1426, 1428–1430 (N.D. Cal. 1996).

[3] Bernstein v U.S. Department of State 176 F. 3d 1132, 1141, rehearing en banc granted, opinion withdrawn, 192 F. 3d 1308 (9th Cir. 1999).

[4] D. Campbell, Interception 2000: Development of surveillance technology and risk of abuse of economic information, Report to the Director General for Research of the European Parliament, Luxembourg (April 1999).

[5] K. Dam, H. Lin, *Cryptography's Role in Securing the Information Society*, National Academy Press (1996).

[6] W. Diffie, S. Landau, *Privacy on the Line: the Politics of Wiretapping and Encryption*, MIT Press (1998).

[7] N. Hager, *Secret Power*, Craig Potton Publishing, New Zealand (1996).

[8] D. Kahn, *The Codebreakers*, Scribners (1996).

[9] S. Landau, *Primes, codes and the National Security Agency,* Notices of the American Mathematical Society (Special Article Series) 30 (1) (1983), 7–10.

[10] National Commission on Terrorist Attacks upon the United States, Threats and Responses in 2001 (Staff Statement No. 10), 13 April (2004).

[11] W.A. Root, J.R. Liebman, *United States Export Controls*, 4th edition, Aspen Law and Business Publications (2002).

[12] United States Department of Commerce, National Bureau of Standards, *Data Encryption Standard*, Federal Information Processing Standard Publication 46 (1977).

[13] United States Department of Commerce, National Institute of Standards and Technology, *Approval of Federal Information Processing Standards Publication 185, Escrowed Encryption Standard*, Federal Register 59 (27), 9 February (1994).

[14] Department of Commerce, Bureau of Export Administration: 15 CFR Parts 734, 740, 742, 770, 772 and 774, Docket No. RIN: 0694-AC11, Revisions to Encryption Items, Effective, 14 January (2000).

[15] U.S. House of Representatives, Select Committee on U.S. National Security, Final Report of the Select Committee on U.S. National Security and Military/Commercial Concerns with the Peoples Republic of China (1999).

[16] Van Natta Jr., D. Butler, *How tiny Swiss cellphone chips helped track global terror web*, New York Times, 4 March (2004), A1.

[17] J.R. Woolsey, *Why we spy on our allies*, The Wall Street Journal, 17 March (2000).

27

HISTORY OF PRIVACY

Jan Holvast

Holvast & Partner, Privacy Consultants
Landsmeer, The Netherlands

Contents

Abstract

Discussion on privacy issues is as old as mankind. Starting with the protection of one's body and home, it soon evolved in the direction of controlling one's personal information. In 1891, the American lawyers Samuel Warren and Louis Brandeis described the right to privacy in a famous article: it is the right to be let alone. In 1967 a new milestone was reached with the publication of Alan Westin's Privacy and Freedom when he defined privacy in terms of self determination: privacy, now, is the claim of individuals, groups, or institutions to determine for themselves when, how, and to what extent information about them is communicated to others.

History of privacy makes clear that there is a strong relationship between privacy and the development of technology. The modern discussion started with the use of cameras and went on to include the development and use of computers in an information society in which personal data on every individual is collected and stored. Not only is it a great concern that privacy is eroding but also that we are entering a surveillance society. This loss of privacy seems to be even more the case since the protection of privacy is strongly dependant upon the political will to protect it. Since 9/11, however, this political will world-wide is oriented more toward the effective and efficient use of technology in the battle against criminality and terrorism than it is toward protecting privacy. Therefore it is time to re-evaluate the use of technology and the protection of privacy. It is not only privacy that is at stake but above all democracy.

Keywords: data protection, information technology, information society, self-regulation, surveillance society, vulnerability.

27.1 INTRODUCTION

"The good news about privacy is that eighty-four percent of us are concerned about privacy. The bad news is that we do not know what we mean". The figures Anne Branscomb [7] mentions are still true for most countries in the Western hemisphere, and the reason for not knowing what we are talking about is primarily because many authors on privacy issues are writing about different aspects of privacy. Some are referring to the need for privacy; whereas, others are referring to the right to privacy, the invasion of privacy, the functions of privacy, or even the (legal) protection of privacy. In this chapter, we start with the need for privacy and attempt to unravel the confusion within that issue. Thereafter, we will give an overview of the concept of privacy, an interpretation of that discussion, and a way of looking at privacy. In addition we will examine the function of privacy in order to clarify the importance of privacy (protection).

The third section is devoted to the attacks on privacy starting with the first publicly discussed cases in 1361 and then focusing on the development during the 20th century until the present day. This section makes clear how strong the relationship is between privacy discussion and technology, in particular information technology as it is called now. It shows the double face of technology, which can help people to master problems and simultaneously can influence people and their conduct in a negative ways. An example of these technologies is the Radio Frequency Identity (RFID). As a pacemaker, the RFID is helpful but as a chip under the skin it can become a tool for tracing all movement of an individual. Another example is ambient technologies which will be present in almost all households in the Western hemisphere.

For some time, there have been ways to protect privacy. In many countries, this protection is included in a country's constitution, and in some cases privacy protection is deliberately translated into privacy and data protection laws. The legal systems are, however, not always the same. In this work we will make a distinction between comprehensive legislation (omnibus laws) and sectoral laws which are intended to protect a particular part of society or areas such as communication technology. In addition to legal measures, self-regulation is used, in particular by industry in the form of codes of conduct or codes of practice. More and more technology itself is used as a means of protection. Security measures are examples but also the often discussed but less implemented Privacy-Enhancing Technologies (PETs) are examples of using technology itself in the protection of privacy. In addition, publicity after privacy has been invaded in an unacceptable way is an important tool of protection, although in an indirect way. We will give some famous examples.

In the fifth section we shall broadly analyse the development of privacy discussions. We start with an overview of discussions on the information society, in particular the increasing vulnerability of individuals and of countries, which is one of the most important characteristics of the present society. Many countermeasures are taken, some of which are aimed at the weakest areas in which humans find themselves. The rise of an information society is not a goal in itself but is seen as an important drive for economic growth. Therefore countries like Japan, the United States and Europe have developed programs to enhance information technologies which also pay attention to social impacts, such as the protection of privacy. This issue raises the question of how information technology can best be regulated. We will present on old regulating model, developed by Dorothy Nelkin [56] which distinguishes among reactive, participatory, and anticipatory control. Looking at the present situation we will see and explain what form of control is most important. The question of regulating will also be discussed in a more general way. Can technology be controlled or is the development deterministic and not able to be influenced by humans? Although this question seems to be the case, in this writer's view modern technology can be controlled.

Returning to the issue of privacy, we will explain how privacy is often invaded. Information has two important characteristics: it is power and it is money. These two reasons drive the collecting, storing, and using of information in the current way that it does. It is also the explanation for the omnipresence of information technology. Everywhere humans walk, sleep, and talk, technology is present. And as humans are increasingly adept at data producing, more and more traces of our daily life will be gathered and known by others, both in government and in industry. Countermeasures will be politically defined, and these measures are, given the power relations, not always oriented towards protecting privacy. Consequently, we must conclude that we are increasingly going to live in a surveillance society in which almost everything about our lives will be known. The consequences of this new society are until now unknown while sociologists seem to have no interest in this new society.

27.2 PRIVACY

27.2.1 The need for privacy

Humans have always had a need for privacy. For Milton Konvitz [49] the feeling of privacy is related to the feeling of shame. Even the bible shows a violation of privacy and its accompanying shame when after Adam and Eve had eaten the fruit of the tree of knowledge, we read the following: "And the eyes of them both were opened, and they knew that they are naked; and they sewed fig-leaves together, and made themselves aprons". In the view

of Richard Hixson [44] Adam and Eve were said to know instinctively both the feeling of privacy and the loss of it. Thus we have been taught that our very knowledge of good and evil is related to a sense of privacy. This privacy-shame dynamic is also demonstrated by another example from Genesis when Noah drunken by the wine lays naked in his tent and was covered by his two sons. In addition to the Bible's passages on privacy, the Qur'an and the sayings of Mohammed at several places show that privacy has always been an important issue.[1] The privacy issue can also be seen in the writings of Socrates and other Greek philosophers [54], when a distinction is made between the 'outer' and the 'inner', between public and private, between society and solitude. Although private life sometimes was seen as an antisocial behaviour, periods of retirement normally were accepted. There always has been a kind of conflict between "the subjective desire for solitude and seclusion and the objective need to depend on others" [44: 5].

An important change took place with the colonisation of America. It appears that issues of privacy were brought along from Europe. The ownership or possession of land in the New World furnished a secure base for the privilege of privacy. Because of the distance between homesteads, in the view of David Flaherty [34], physical privacy became a characteristic of everyday life, and the home itself became the primary place of privacy. The home is still seen in that way since the home is a personal castle, which emphasises the idea that privacy is related to wealth. Historically, poverty and the home meant less privacy, particularly where families share common dwellings with almost no physical separation. Nevertheless the statement of Thomas Pitt shows otherwise in America: "The poorest man may, in his cottage, bid defiance to all the forces of the Crown. It may be frail; its roof may shake; the wind may blow through it; the storm may enter; the rain may enter, but the King of England may not enter; all his forces dare not cross the threshold of the ruined tenement" [1].

---

[1]EPIC [31] mentions from the Qur'an: an-Noor 24:27-28 (Yusufali) and al-Hujraat 49:11-12 (Yusufali) and from Mohammed, Volume I, Book 10, Number 509 (Sahih Bukhari); Book 020, Number 4727 (Sahih Muslim) and Book 31, Number 4003 (Sunan Abu Dawud).

The universal idea of privacy is also stressed by Alan Westin [76] who in his famous book *Privacy and Freedom* shows that the studies of animal behaviour and social organisation suggest that man's need for privacy may well be rooted in his animal origin and that man and animals share several basic mechanisms for claiming privacy among their own kind. Although in the primitive world the feelings of privacy are not comparable with that of the western world, this difference does not prove that there are not universal needs for privacy and that there are not universal processes for adjusting the values of privacy, disclosure and surveillance within each society.

Nowadays it is generally accepted that everybody has a need for privacy, although the way it is appreciated differs from culture to culture and from person to person. At the same time it is clear that a need for privacy can never be absolute and must be balanced against other needs, for example the need for fighting terrorism, criminality and fraud. As we will then see, the discussion on privacy primarily is a political discussion about the way the distinct individual and societal interests can be balanced.

27.2.2 The concept of privacy

In the most fundamental form, privacy is related to the most intimate aspects of being human. Throughout history privacy is related to the house, to family life, and to (personal) correspondence. This relation can be seen as a way of controlling a situation. Since the 14th through the 18th century, people went to court for eavesdropping or for opening and reading personal letters. Since the end of the 19th century, the emphasis shifted more toward personal information with the same intention that is, to control one's own information.

The general discussion on privacy started shortly after the Second World War in the United States. Numerous publications were devoted to the issue of privacy. In these publications attention primarily is paid to a description of the concept of privacy and to the developments of techniques invading privacy, in particular the computer which is seen as primarily responsible for privacy invasion. These publications culminated in the founding in 1962 of the Project *The Impact of Science and Technology*

on Privacy. The project was developed between 1962 and 1966 by the Special Committee on Science and Law of the Association of the Bar of the City of New York. Director of Research was Alan Westin who published extensive details of the results in the Columbia Law Review and in his book *Privacy and Freedom* and laid a profound base for the later discussion [74–76].

In almost all publications from that period, three words are used in relation to privacy: freedom, control and self-determination [2; 9; 42; 47–49; 60; 64]. The concept of privacy is defined in almost the same way as it was in 1891 by Warren and Brandeis. Privacy is described as a right to be left alone and a right of each individual to determine, under ordinary circumstances, what his or her thoughts, sentiments, and emotions shall be when in communication with others. Because of the advancement in technology, privacy will become an ever growing concern. These characteristics of privacy are repeated and elaborated by numerous authors in the beginning of the 1960s.

"Provisionally, the term may be defined as a person's feeling that others should be excluded from something which is of concern to him, and also a recognition that others have the right to do this" [2: 429].

"The essence of privacy is no more, and certainly no less, then the freedom of the individual to pick and choose for himself the time and circumstances under which and most importantly, the extent to which, his attitudes, beliefs, behaviour and opinions are to be shared with or withheld from others" [64: 1189].

"Privacy is an outcome of a person's wish to withhold from others certain knowledge as to his past and present experience and action and his intentions for the future" [47: 307].

In his *Privacy and Freedom*, Alan Westin summarises the discussion and defines privacy based on all of these points. "Privacy is the claim of individuals, groups, or institutions to determine for themselves when, how, and to what extent information about them is communicated to others. Viewed in terms of the relation of the individual to social participation, privacy is the voluntary and temporary withdrawal of a person from the general society through physical or psychological means, either in a state of solitude or small-group intimacy or, when among larger groups, in a condition of anonymity or reserve" [76: 7]. Since 1967, there has almost not been a publication on this subject in which this definition is not presented.

As can be seen from the literature on the subject, two dimensions of privacy can be distinguished: a relational one and an informational one. The first deals with the relation one has to other people, for example controlling who may enter the domestic environment or who is allowed to touch one's body. These aspects sometimes are described as territorial privacy and bodily privacy [31]. The informational dimension is related to the collection, storing and processing of (personal) data.

Common to both dimensions of privacy is the need to maintain control over personal space, the body, and information about oneself; however, it is clear that in certain situations, loss of control is even more important, for example when people lose their consciousness due to an accident. Control can, then, be described in the form of two aspects of freedom: being free to ... and being free from .... The first is the more active part. Within certain borders, humans prefer being free to do what they wish and not be hindered by others or experiences from the past. The second is being free from being watched or eavesdropped on. In both situations the central idea is the concept of self-determination. Although these two freedoms sound rather absolute, it is clear that 'within certain borders' does mean that in all these situations we are depending on others, our neighbours, our co-citizens, and other people. Living in a community means by definition involved with others. But it means at the same time that we must have some free space or sense of freedom since otherwise we would be prisoners of society.

In this writer's view, privacy can be described as the individual's right to self-determination, within certain borders, to his home, body, and information. Although the word 'right' suggests otherwise, the concept of privacy is much more politically determined than legally. This position is more clearly demonstrated by the changing climate of opinions since 9/11. Personal data, such as Passengers Name Records are now made available for governmental use without much debate. A comparable situation shows the discussion on the retention of communication traffic data for at least half a year in order to trace back potential terrorist who have used electronic means of communications. It shows how due to a sudden event the balance between a need for privacy and the need for information can change fundamentally. It is not the right itself that is being discussed but rather the amount of privacy that remains after the government satisfies its need for information. As we will increasingly see, the technical means for collecting and storing information are increasing in an enormous way.

27.2.3 The functions of privacy

It is almost impossible to describe the various ways in which the functions of privacy were seen in the past. Alan Westin has given a comprehensive description of these earlier functions in his study *Privacy and Freedom* [76: 330–338]. He distinguishes among the four functions of privacy which are still important in modern life. The first is a need for personal autonomy, which is vital to the development of individuality and the consciousness of individual choice in anyone's life. Privacy is equally important as it supports normal psychological functioning, stable interpersonal relationship, and personal development. Privacy is the basis for the development of individuality.

In the second place we need privacy as a form of emotional release. Life generates such strong tensions for the individual that both physical and psychological health demand periods of privacy. It supports healthy functioning by providing needed opportunities to relax, to be one's self, to escape from the stresses of daily life, and to express anger, frustration, grief, or other strong emotion without fear of repercussion or ridicule. The consequence of denying opportunities for such privacy can be severe, ranging from increased tension and improvident expression to suicide and mental collapse.

A third function is that of self-evaluation and decision making. Each individual needs to integrate his experiences into a meaningful pattern and to exert his individuality on events. Solitude and the opportunity for reflection are essential for creativity. Individuals need space and time in which to process the information which is coming to them in an enormous amount. Privacy allows the individual the opportunity to consider alternatives and consequences to act as consistently and appropriate as possible.

A fourth function is the need for a limited and protected communication, which is particularly vital in urban life with crowded environments and continuous physical and psychological confrontations. The value of privacy recognises that individuals require opportunities to share confidences with their family, friends and close associates. In short privacy is creating opportunities for humans to be themselves and to stay stable as a person.

Unfortunately since Westin's 1968 study, little attention has been paid to these four functions, and it is still unclear how a significant threat to one's privacy affects psychological growth. Scientists know too little about how people respond under constant surveillance. A concern, however, is that people may become more conformist as they suppress their individuality [28]. On matters related to employees, more information is available. Barbara Garson in *The Electronic Sweatshop* states that there is some empirical prove that for clerical workers whose keystrokes are counted by the minute or airline clerks whose figures are posted daily, electronic monitoring has been linked to pain, stress, and serious disease. Medical reasons then have been some help in limiting the monitoring of employees [39].

27.3 PRIVACY UNDER ATTACK

Literature and court cases show that for a very long time, in one way or another, privacy has always been perceived as attacked. At first the attack on privacy was done by persons with whom individuals have a close contact, such as neighbours and people living in the same village or colony. Later attacks were also accomplished by governmental agencies, industry, or the press. In this chapter we will make a distinction between past situation which lasted until the 1980s and the present situation which covers from the 1980s until 2006 as well as the future situation of which we already now have clear indications of new methods of privacy surveillance.

Although these three periods are distinctive from each other, it is not to say that they can be separated one from the other. An important characteristic of the use of information and information technology is that it is a cumulative process. The beginning of one period does not at all mean that the previous period has concluded. The contrary is the case as, for example, photography and computer uses for privacy invasion show. Many of the earlier techniques are combined with new, even more powerful techniques.

In this overview of (technical) attacks, this contribution will strongly rely on past literature and court cases from the United States since most publications dealing with these discussions and incidents of privacy are published in that country. For the present and future situation, we will use international references, including web-pages. These sources will show that attack on privacy is becoming not only an international but a global problem.

27.3.1 Use of information in the past

Almost all authors on privacy start the discussion with the famous article *The Right to Privacy* of Samuel Warren and Louis Brandeis in the *Harvard Law Review* of December 15, 1890 [71]. Although the effects of this article can not be underestimated this starting point does not mean that there have been no discussions on the invasions of privacy before 1890. As Westin shows in his publications, in the fifteenth century the word 'privacy' was already used in England and historical research shows that colonists in New England were respecting privacy in relation to an individual's home, family, and even written communication. Hixson [44] states that one of the first recorded invasions of privacy by government happened in 1624, when two men protested against interception of their personal letters by Governor William Bradford, who wanted to use them as a prove of conspiracy against him. Hixson shows also that there was opposition against the first US census as early as 1790, although the government required little more than enumeration of persons, both slave and free. This opposition resulted in instructions to census takers in 1840 that individual returns be treated as confidential. It was feared that the citizen was not adequately protected from the danger that private affairs or the secrets of family would be disclosed to the neighbours.

27.3.1.1 *Trespass.* As we have seen, the home and its related physical privacy were, from the beginning, the form of privacy that most vehemently was protected. It is not astonishing that the first cases brought to court had to deal with intrusions of the home, in particular by eavesdropping. EPIC cites James Michael who shows that in 1361 the Justices of Peace Act in England provided for the arrest of peeping toms and eavesdroppers. In 1765, British Lord Camden, striking down a warrant to enter a house and seize papers wrote "We can safely say there is no law in this country to justify the defendants in what they have done; if there was, it would destroy all the comforts of society, for papers are often the dearest property any man can have" [31: 5]. The law of trespass and the constitutional protection of unreasonable search and seizure in the United States as formulated in the Fourth Amendment were interpreted as protections against official and unofficial intrusions.

27.3.1.2 *Correspondence.* In addition to the home, personal mail was seen as a part of private life in need of special protection. Long before this protection was generally accepted, in particular by the use of the telegraph, the first incidents about invasion of reading personal mails are known. One story is from 1624 [44]. Plymouth Plantation was the scene for what Hixson mentions as the first recorded invasion of privacy. Governor William Bradford learned of a plot against the leadership of the small colony. He had intercepted several incriminating letters written by two newcomers and sent to friends in England. When the two men denied any conspiracy the governor produced the letters and asked them to read the content aloud. The men expressed outrage that their private correspondence had been intercepted but did not comply further since they had no legality on which to stand.

27.3.1.3 *The press.* Curiosity has always been an enemy of privacy and is a foible that has stimulated privacy invasion and on which newspapers have exploited individual privacy on a commercial bases. Already in 1873 the first complaints were uttered against the way journalists were using interview techniques. President Cleveland expressed dislike of the way the press treated him on occasion, especially when some journalists followed him and his bride on their honeymoon trip in 1886. Also E.L. Godkin wrote at the end of the 19th century that the chief enemy of privacy in modern life is the curiosity shown by some people about the affairs of other people [44: 29].

Although it is not known in how far Waren and Brandeis were influenced by Godkin, generally the discussion on the attack on privacy starts with the famous article of these two lawyers, published in 1890 in the *Harvard Law Review* under the title *The Right to Privacy* [71]. The reason for publication grew out of a specific situation. Warren had married Miss Mabel Bayard, daughter of Senator Thomas Francis Bayard, Sr. They set up housekeeping in Boston's exclusive Back Bay section and began to entertain elaborately. The *Saturday Evening Gazette,* which specialised in 'blue blood items' reported their activities in lurid details. Warren, together with Louis D. Brandeis, was the first to start a fundamental discussion on his rights not to have his thoughts, statements, or emotions made public without his consent. Since the publication of this famous article, no contribution of the issue of privacy fails to mention it. As later noted, the Warren and Brandeis article added a new chapter to American law [51].

27.3.1.4 *Instantaneous photography.* In their article Warren and Brandeis not only blame the press but also recent inventions and business methods like instantaneous photographs. In combination with the newspaper business, these business methods and new technologies invaded sacred personal and domestic precincts. As predicted in the famous Warren and Brandeis article, these numerous mechanical devices would be the source for "what is whispered in the closet shall be proclaimed from the housetops" [71: 134].

Since 1890, however, the relationship to the use of technical means is new. Already mentioned in the article of Warren and Brandeis, the use of instantaneous photographs makes possible publication for various purposes but without the consent of an individual. A classic type of invasion of privacy is the use without consent of a person's picture to promote a product. The initial test was *Roberson v. Rochester Folding Box Co.,* which startled the New York legal world [52]. A local milling company decided to use a photo of Abigail Rochester, a charming and attractive girl at the time, to promote their product. For that reason the brilliant slogan *The Flour of the Family* was used and, together with the photo, placed in numerous stores, warehouses and saloons. Abigail claimed a 'right of privacy' and brought suit for the sum of $15.000. The New York Court denied the suit, by a 4-3 decision, saying that her claim held no right on grounds that it was yet unknown to common law what had been infringed.

This decision excited much amazement and was strongly influenced later court cases, in particular three years later the leading privacy of *Pavesich v. New England Life Insurance Co.* In that court case, Paolo Pavesich's picture was used, also without his consent, by a life insurance company for an advertisement. The photograph showed a healthy man (Pavesich) who did buy a life insurance policy, in contrast to a sick man who did not and presumably could not make such an 'invaluable' purchase for his future security. In the picture of Pevasich there was a legend underneath: "In my healthy and productive period of life I bought insurance in the New England Life Insurance Co. of Boston Massachusetts, and today my family life is protected". Pavesich had, in fact, never purchased such a life insurance, nor made any such statement as quoted. He found the advertisement distasteful and brought suit for $25,000 damages. In this case the right of privacy was unanimously accepted. The Court found the insurance company subject to damages for invading the privacy of Pavesich [52: 99]. It was a strong precedent for precisely one aspect of personal privacy: the unauthorised use of an individual's picture.

27.3.1.5 *Wiretapping.* An extremely important and much cited case has been *Olmstead v. United States* in 1928 [78]. In this case wiretapping equipment was used by the police as a way of obtaining evidence. However, the complaint was not accepted by five of the nine justices because there had been no actual entry into the houses and nothing tangible had been taken. So the search and seizure amendment did no apply. Even more importantly than the decision, however, had been the dissent of Justice Brandeis, the co-author of the Right to Privacy. In his view, this case indicated that the privacy of the man had been invaded, that is "the right to be let alone – the most comprehensive of rights and the right most valued by civilised men".

Brandeis' reasoning was adopted only forty years later in the *Katz v. United States* case. Federal authorities used electronic listening devices attached to the outside of a telephone booth used by one Charles Katz, whom the authorities suspected of violating gambling laws. Even though the property was not invaded the court found that this method of collecting evidence infringed on the Fourth Amendment rights of Katz. In the view of the court, the constitution protects whatever seeks to be preserved as private. What is most remarkable about this case is the interpretation of what is private within the meaning of the Fourth Amendment. In the view of Justice Harlan private can be defined by the individual's actual, subjective expectation of privacy and the extent to which that expectation was one that society is prepared to recognise as 'reasonable'. This interpretation has since been used in many cases related to homes, business, sealed luggage, and packages. At the same time it is also often criticised and seen to be of limited value since it is restricted to government invasion of privacy and does not apply to objects controlled by third parties such as bank records. Above all, this case is dependent upon what society's expectation of what invasion of privacy is, which is a serious disadvantage since whatever the public views as reasonable tends to evolve more slowly than does information technology [14].

27.3.1.6 *Psychological testing and lie detectors.* Around the 1960s, it was not the single collection of data by means of photography and technical devices that worried people but the mass collection of data with the help of psychological testing, lie detectors, and attitude scales used by social scientists. Not only are these techniques criticised but in particular the philosophy behind the use of them. In his *The Organization Man* William H. Whyte [77] expects that social sciences will become more and more a type of social engineering, the goal of which is to adapt a society to one in which all problems will be solved. In a cynical moment, Whyte promoted a kind of Universal Card with an individuals' fingerprint, IQ and several other personal characteristics attached. To his astonishment the proposal was not criticised but strongly endorsed.

Another criticism came from Vance Packard [60]. In his *The Hidden Persuaders* he shows the strong relationship between techniques that detect hidden personal emotions and feelings and the way this very data are used for advertisement. Packard shows how this technique is used in motivational research and exploited by 'depth boys' who are doing research in subliminal messages and applies the same in ways that a person is not aware.

As a criticism not only of the techniques as discussed but the social sciences in general, Oscar Ruebhausen and Orville Brim [64] are the first to make clear that the development of social research proves that ethical and legal rules are necessary and most especially regulations that allow for the expressed consent of the individual who is willing to cooperate.

27.3.1.7 *Computer as a black box.* At this same point in discussion of privacy rights, a new development was added, that is how the computer should be used as a primary data storage device. Large scale storage of data as well as the processing and exchange of data between organisations are now possible. The computer as a data giant has been seen as frightening by several authors, with as a consequence that during a long time privacy invasion was only related to computers. Numerous publications have appeared with thrilling titles that warn of gigantic invasions of personal privacy, for example *The Assault on Privacy: Computers, Data Banks and Dossiers* [53], *The Data Bank Society* [70] and *The Computer Impact* [68]. The emphasis in this issue is on computers and databases, that

is huge collections of data processed by electronic means.

At the end of the 1970s, a new dimension – telecommunication – was added to the discussion. Telecommunication can be referred to as telematics or as the combination of telecommunication and informatics. It is not only the processing of data which is frightening but above all the distribution of the data to unknown recipients. The combination of computer and telecommunications led, in turn, to a 'tele'-hype of what the future might bring about in society, such as tele-education, tele-work, tele-medication and tele-papers. The future is the human home in which individuals communicate with the outside world exclusively by way of the television. It is a brave new world in which privacy will be strengthened since the home will become even more than ever a castle but at the same time privacy can be attacked by all traces that remain from that type of communication.

27.3.2 Present use of information technology

27.3.2.1 *Video surveillance.* Surveillance video cameras are increasingly being used throughout the public arena [67]. In almost all cities of the western world it is almost impossible to walk around without being recorded nearly every step of the way and it is expected that this surveillance will be expanded in the next years by improved technology, by centralising the surveillance, and by the unexamined assumptions that cameras are providing security.

Cameras in some countries are being integrated into the urban environment in ways similar to the integration of the electricity and water supply at the beginning of the last century [21]. The CCTV market has enjoyed an uninterrupted growth and is in an increasing way integrated into technologies, such as the Internet, face recognition software, and law enforcement databases. CCTV's power is substantially increasing, and it has features that include night vision, computer assisted operations and motion detection facilities.

27.3.2.2 *Biometric identification.* Biometrics[2] is the science and technology of measuring and

---

[2]See http://whatis/techtarget.com/definition/

statistically analysing biological data. In information technology, biometrics refers to technologies for measuring and analysing human body characteristics such as fingerprints, eye retinas and irises, voice patterns, facial patterns and hand measurements, especially for authentication and identification. Biometrics involves comparing a previously captured, unique characteristic of a person to a new sample provided by the person. The information is used to authenticate or verify that persons are who they said they are by comparing the characteristics. New biometric technology attempts to automate the authentication or verification process by converting the provided biometrics into an algorithm, which is then used for matching purposes. This process can mean an attack on one's privacy when the collection takes place without consent or permission and without transparency about the purpose for which these data are used.

27.3.2.3 *Genetic data.*

There is an increase in DNA-analysis for medical testing research and for investigative purposes which are incorporated into routine health [67: 5] testing. Unlike other medical information, genetic data is a unique combination difficult to be kept confidential and extremely revealing about us. Above all it is easy to acquire since people constantly slough off hair, saliva, skin cells, and other samples of our DNA. No matter how hard we strive to keep our genetic codes private, we are always vulnerable to the use of it. The data collected tell about our genetic diseases, risk factors, and other characteristics. For the financial services companies, it would be useful to be able to assess risks on the basis of genes patterns that can indicate an individual's future potential susceptibility to illness and diseases. A specific problem with genetic data is that an individual who discloses his or her genetic information also discloses the genetic data of his or her relatives.

27.3.2.4 *Identity theft.*

Identity theft is one of the fastest growing types of fraud. Identity theft is the assumption of another person's financial identity through the use of the victim's identity information [29]. This information includes a person's name, address, date of birth, social security number, credit card numbers, and checking account information. In literature a distinction is sometimes made between identity theft and identity fraud. Identity theft occurs when someone is using one's personal information to impersonate him or her to apply for new credit accounts in his or her name. Identity fraud involves an unauthorised person using one's credit card number from an existing account to make purchases. Seen from the consequences for the individual, theft normally refers to both.

One of the increasing forms is phishing by which thieves on the Internet pose as legitimate account managers for credit card companies and financial institutions and ask for personal information under the guise of account verification or maintenance. According to survey data released in July 2005 by Chubb Insurance, 20% or one in every five Americans has been a victim of identity theft or fraud [15: 3]. An even more aggressive form is pharming, a word play on farming and phishing. Pharming is a hacker's attack aiming to redirect a website's traffic to another (bogus) website. This website duplicates in every aspect of the look and feel of a bank or other sensitive website. Via this bogus website criminals try to steal, for example, account information.[3]

27.3.2.5 *Data warehousing and data mining.*

Data warehousing is the collation of data into huge, queriable repositories in such a way that they are related to a unique, identifiable natural person. These data are collected in order to make data mining possible, which is a statistical technique enabling analysis of the data in order to find patterns and relations which are nor expected nor predictable. In this way new patterns can be discovered or can confirm already suspected relationships. A famous example is the data mining that marketers show that fathers who buy diapers often pick up beer at the same time. The link prompted some stores to stock the seemingly unrelated items at the same aisle so even more fathers would reach for beer. The underlying expectation is forming profiles of groups of people that make

---

[3] See http://en.wikipedia.org/wiki/pharming

behaviour predictable, for examples potential terrorists or criminals.

That data mining is not only the finding of a relationship but even more an interpretation shows the mistakes a marketer in The Netherlands made. In order to promote the selling of toys the people in an elderly care facility got free diapers, which was not appreciated at all. The individual, naturally, thought the item was to be used against incontinence. The reality, however, was that the data mining had shown, in fact, a strong relationship between that postal code and the buying of toys. The marketers interpreted this as a postal code in which many young families were living rather than a place where grandparents who were buying toys for the grandchildren lived.

27.3.2.6 *Chip or smart cards.* A chip or smart card is a credit card size device with an embedded microprocessor(s), capable of storing, retrieving, and processing a tremendous amount of information relative to one person. This person is obliged to wear and use this card in all contacts he or she has with the distributor or distributors of the cards, since combinations of applications are likely. Examples of these cards are the modern driver license, passport, medical cards, and loyalty cards. The content of the card can be read by making contact with a reader or on a contactless way as is used in public transport.

27.3.2.7 *Global Positioning System (GPS).* With the rapid growth of wireless communications, such as mobile phones, the use of the Global Positioning System and the related Location Based Services (LBS) is increasing. The GPS is a 'constellation' of 24 well-spaced satellites that orbit the Earth and make it possible for people with ground receivers to pinpoint their geographic location. The location accuracy is anywhere from one hundred to ten meters for most equipment. A well-known application is the use of GPS in automobiles to order to pinpoint precisely a driver's location with regards to traffic and weather information. By using mobile telephones it is rather simple to detect the place where the mobile is by using network based technology and/or handset bases technology.

By using the cell of origins method the telephone, once connected, is communicating his position regularly. In this way the user of the telephone always can be traced. Another use also based on tracing is electronic monitoring as an alternative for imprisonment in certain cases.

27.3.2.8 *Internet.* Internet is the most fruitful area for data collection in modern times. It is quite possible to collect tremendous amounts of data on almost all users of the Internet without their knowledge. Using search engines like Google makes clear how elusive the Internet is becoming. Although it is at the same time a mighty instrument in the hands of the consumer or citizen for improving his or her knowledge, it is also an instrument for the service provider for contacting these individuals. The combination of cookies and spam shows in which ways the Internet can be used for advertising purposes. A cookie is a piece of information unique to a user that the user's browser saves and sends back to a web server when the user revisits a website. Cookies form a specific part of the more general area called spyware which extracts information from an individual without the user's knowledge or consent. The purpose of spyware is to gain access to information, to store information, or to trace the activities of the user. Cookies contain information such as log-in or registration information, on line shopping cart information (the user's online buying pattern in a certain retail site), user preferences, and other types of detail gathering. As we have seen phishing (and also pharming) are techniques whereby innocent accountholders are contacted through the Internet and asked for confidential data such as account numbers and passwords.

This detailed information, then, can be used for advertising purposes in the form of spam, which is hundreds of unsolicited junk emails that contain advertising or promotional messages and sent to a large number of people or even to one person at the same time [32]. Spam, therefore, can be described as the electronic distribution of large amounts of unsolicited emails to individuals' email accounts. Spam email is definitely distinctive from the traditional direct mailings in that the costs for such massive mailings fell to the sender. The cost of sending mail through conventional means is very real, including postage costs all paid by the sender. On the other hand, costs of sending bulk emails are very small; furthermore, the receiver, not the sender, pays most of the costs. It is the fact that emails can be sent at low costs and in great quantities that attracts direct marketers and other companies to use spam emails for advertisements.

27.3.2.9 *Key logger.* A key logger system records the key strokes an individual enters on a computer keyboard [31: 39]. Key stroke loggers can be employed to capture every key pressed on a computer keyboard, including information that is typed and deleted. Such devices can be manually placed by law enforcement agents on a suspect's computer or installed remotely by placing a virus on the suspect's computer that will disclose private encryption keys. The question of legitimacy of these methods arose in the case of *United States v. Scarfo* where a key logger was placed in order to capture an individual's PGP encrypted password. The existence was confirmed by the FBI. The key logger did not need physical access to the computer in order to accomplish the desired task of capturing private information.

27.3.3 Technical use in the future

27.3.3.1 *Radio Frequency Identification (RFID).* Use of the RFID is advancing rapidly and, in a sense, is the successor of the chip card. In a similar way, RFID tracks and traces objects and subjects easily. One of the most well known applications is a yellow tag tracing cows in countries of Western Europe. RFIDs are smart tags which make it possible to follow exact movements of the objects wearing it. It is in a type of successor of the barcode with the most important difference being that a barcode is identifying a type of product whereas the RFID is identifying each distinct product. An RFID label consists of two parts: a microchip and an antenna. The chip contains data about the product and a unique code by which the product can be identified. The antenna makes it possible for that data to be sent to a receiver; therefore, one of the most vital differences from past applications is that

the tag can be read from a distance without the wearer of the tag being knowledgeable of the tracing.

This tag can be attached to a product (cow) but can also be implanted in the skin. In the summer of 2004, the RFID application became well known in bars of Barcelona, Spain and Rotterdam, The Netherlands where visitors had the possibility to have an RFID-chip implanted under their skin. This chip recognised people as they entered a bar, knew their preferences for drinks, and knew the bank accounts to be charged for paying the drink bills. This RFID-chip is used during the football World Championship in Germany. On every entrance billet, an RFID-chip was attached thus each visitor could be identified and, in case of incidents, be arrested.

27.3.3.2 *Ambient technology.* In a sense, the RFID-chip is a significant part of a development process called ambient technology or, as it is sometimes referred to, as pervasive or ubiquitous computing. Ambient intelligence is an intelligence system that operates in a surrounding environment, a trend brought about by a convergence of advanced electronic, and particularly wireless, technologies and the Internet [26]. These ambient devices are not personal computers, but very tiny devices, either mobile or embedded, in many objects, including cars, tools appliances, clothing, and consumer goods in such a way that they are become an everyday part of life and reacting to our behaviour as well as participating our human needs.[4] Used in refrigerators they can remind us to use the oldest products and once the item is used automatically adding it to our shopping list. The vacuum cleaner can also be started without human intervention once dust density becomes too high. Utilities are able to monitor the performance of home appliances, sending repairmen or replacements before they break down. Local supermarkets can check the content of customers' refrigerators and make out a shopping list for customers. From desktop computers, office workers can check up on children at home [27].

---

[4]See http://searchnetworking.techtarget.com/sDefinition/

27.3.3.3 *Neurolinguistics.* Neurolinguistics is based on the fact that different people are processing information differently [33]. So, for example, there is a difference between male and female brains with the female brains taking more notice of more cues within a piece of communication and using colours, imagery, and graphics much more to interpret meaning compared with male brains. Neurolinguistics uses knowledge on how information processing styles differ in order to target consumers. It can be used to detect different responses to car designs and to evaluate television commercials. This type of use is called neuromarketing: seeing how different people respond to advertising and other brand-related messages by seeing brain responses. Typically, researchers connect subjects to a functional MRI (Magnetic Resonance Imaging) machine and watch their brain activity throughout an experiment. In one example, neuroscientist Read Montague used a functional MRI to study what he calls the "Pepsi paradox": in a blind taste test, subjects were fairly evenly divided between Pepsi and Coke; however, when the subjects knew what they were drinking, 75% said they preferred Coke. In these tests, Montague saw activity in the prefrontal cortex, indicating higher thought processes. The researcher concluded that it is likely that the subjects were associating the particular drink with images and branding messages from commercials. In another study, at Daimler–Chrysler, researchers found that the 'reward' centres of men's brains were activated by sports cars, similarly to the way those areas respond to alcohol and drugs. Other research, involving United States political campaign messages found that the brains of Democrats and Republicans responded differently to images of the September 11th terrorist attacks on the World Trade Center.

Some anti-marketing activists, such as Gary Ruskin of Commercial Alert, warn that neuromarketing could ultimately be used to manipulate consumers by more effectively playing on their fears or stimulating positive responses. In that sense neurolinguistics will be come part of biophysics.[5]

27.3.3.4 *Memetics.* The science of memetics has recently attracted significant attention [33]. A meme is an idea that is passed from one human generation to another. It is the cultural and sociological equivalent of a gene, the basic element of biological inheritance. In contrast to genetics, a meme acts not vertically through the generations but horizontally. They work as a viral contagion. A good example of the principle is how it is difficult not to start yawning if others are yawning or not applaud when others start to applaud. It is speculated that human beings have an adaptive mechanism that other species don't have. Humans can pass their ideas from one generation to the next, allowing them to surmount challenges more flexibly and more quickly than through the longer process of genetic adaptation and selection. Examples of memes include the idea of God and other forms of belief.[6]

It is believed that changing memes means a change in personality. Therefore it is a concern that others can use memes to influence human behaviour and influence humans both in commercial areas and in political campaigns. The influence might be an unconscious one that might be most enduring if installed at an early stage. In relation to memes it is feared that marketers can use it to infect consumers with a mind virus that is not recognised consciously but which suddenly results in joining a fad or fashion.

27.3.3.5 *Grid technology.* A new way of living will evolve as the Internet morphs into 'the grid'. Wireless tags will be embedded in nearly every object, and even people, linking humans and machines together as 'nodes' on a single global network. By tying all computers together into a single grid, this system will allow any one computer to tap the power of all computers. It is a sort of fourth wave bringing together the power of mainframes, PCs, and the Internet. This grid system will be able to link companies, consumers, and governments together. Biochips might be able to send real-time heart readings to cardiologists by way of the grid. "A smart chip in your convertible could allow the manufacturer to track both the car and your driving

---

[5]See http://searchsmb.techtarget.com/sDefinition/

[6]See http://whatis.techtarget.com/definition/

habit. A digital double of your car might even be parked on the grid, where your mechanic could watch it for engine trouble or the police could monitor your speeding" [38: 67]. The endless reams of data will be too voluminous for human engineers to track. The grid therefore will have to be self-managing, self-diagnosing and self-healing, telling people when things go wrong and instructing us on how to fix them. At the moment there seems to be only one problem: software to make the grid secure does not yet exist. It is said that in a highly networked world the 'castle' model of security with firewalls will not work.

27.3.3.6 *Wireless networking.* Wireless networking has already been in use for several years in the form of Wi-Fi that is, Wireless Fidelity. Wi-Fi was intended to be used for mobile computing devices, such as laptops; however, it is now used increasingly for other applications, including Internet Access, gaming, and basic connectivity of consumer electronics such as television and DVD-players.

A new development in the field of wireless networking is Bluetooth. Bluetooth is an industrial specification for wireless personal area networks (PANs).[7] It provides a way to connect and exchange information between devices like personal digital assistants (PDAs), mobile phones, laptops, PCs, printers, and digital cameras by way of a secure, low cost, globally available short range frequency. The range of Bluetooth depends upon its power class which covers one to one hundred meters; it also includes a low-cost microchip in each device. The difference between Wi-Fi and Bluetooth is that Bluetooth devices can be configured to communicate with any devices even if they are not in the same room.

This flexibility is making Bluetooth vulnerable to interceptions, and the most serious flaws of Bluetooth security may be the disclosure of personal data. Recent research from the University of Tel Aviv in Israel has detected that Bluetooth can be cracked, and these findings have been published in the *New Scientist*. The researchers have shown both active and passive methods for obtaining the

PIN for a Bluetooth Link. The passive attack would allow a suitably equipped attacker to eavesdrop on communication. The active method makes use of a specially constructed message that must be inserted at a specific point in the protocol to repeat the pairing process. After that the first method may be used to crack the PIN.

27.4 THE PROTECTION OF PRIVACY

27.4.1 Introduction

As we have already noted, the first protections of privacy came from the citizen himself (Adam and Eve) or from relatives (the sons of Noah). During the Middle Ages this picture stayed almost the same. However with the rising intrusion of governments into private lives, assistance against privacy intrusion required the help of others, legal legislation and the addition of self-regulation. Later technical instruments like security measures and PET were added.

EPIC distinguishes four models of privacy protection [31: 3]:

- Comprehensive laws: a general law that governs the collection, use, and dissemination of personal information by both the public and the private sector. An oversight body then ensures compliance.
- Sectoral laws: rules in favour of specific laws, governing specific technical applications, or specific regions, such as financial privacy.
- Self-regulation, in which companies and industry establish codes of conduct or practice and engage in self-policing.
- Technologies of privacy: with the development of available technology-bases systems it becomes possible for individuals to protect their privacy and security.

Although there will always be distinctions among countries and cultures in how these four measures will be emphasised, it seems clear that the countries which will protect the data most efficiently will probably use all four of the models simultaneously to ensure data protection.

---

[7] See http://en.wikipedia.org/wiki/Bluetooth

As can be seen from the formulation in this model, emphasis will be on data protection. Nonetheless it is necessary to make the distinction between privacy protection and data protection. The first is a general protection historically oriented towards the home, family life and correspondence while the latter will emphasize the informational dimension.

27.4.2 Comprehensive laws and regulatory agents

The legal interpretation of privacy depends on the way the concept is used. As we have seen, two dimensions can be distinguished: a relational and an informational one. With respect to the relational privacy there has been a long tradition in Europe and in the United States. In terms of protecting data, the regulation is nascent. In 1971 the first privacy act, The Data Protection Act, took effect in the State of Hesse (Germany); shortly thereafter, Sweden and the United States passed privacy legislation. Subsequently, privacy protection has become a part of many constitutions, with the exception of the United States where the protection must be derived from amendments. This contribution will start with a short overview of the situation in the United States followed by a more extensive treatment of the situation in Europe.

27.4.2.1 *Privacy protection.* Although the United States is the country where most of the earliest discussions have taken place, privacy protection has never had a base in the US constitution. An important characteristic of the American constitution, which went into effect in 1789, is that in general it has a negative formulation. The US constitution does not oblige the government to protect but rather to refrain from taking actions. In that sense and in an indirect way, people are protected against government actions. Although more or less all constitutional freedoms are related to privacy, this type of privacy right is not explicitly mentioned in the Constitution. In particular it must be derived from three amendments: the First, Fourth and Fourteenth.

The First Amendment protects the freedom of expression, religion and assembly. The freedom of expression assures the unfettered interchange of ideas for the bringing about of political and social change by the public [78].

The Fourth is centred on the prohibition of unreasonable search and seizure. As can be understood from the history in the United States, it has two deeply rooted concerns: that its citizens' property is protected from seizure by the government and that its citizens' home and person be protected from warrantless and arbitrary searches. This very amendment is the much used, but still unclear, concept of 'reasonable expectation of privacy' which was introduced by Justice Henson in the famous court case *Olmstead v. US*.

The Fourteenth Amendment guarantees due process and nondisclosure of personal information. As this language is more in line with the informational dimension of the computer age, we will treat this amendment in relation to data protection.

Although there has been some form of legal privacy protection for some time now based on case law, international recognition of privacy as a human right can be traced back to the period immediately following the Second World War. Recognition of this particular right emerged as a response to some of the abuses perpetrated by fascist regimes before and during the war. The Universal Declaration of Human Rights was adopted on 10 December 1948 by the United Nations general Assembly. In article 12 the territorial and communications privacy is protected. It states: "No one should be subjected to arbitrary interference with his privacy, family, home or correspondence, nor to attacks on his honour or reputation. Everyone has the right to the protection of the law against such interference or attacks". It was the first international instrument to deal with this right to privacy. As it was in the form of a resolution of the General Assembly it was not legally binding.

In 1950 the European Convention for the Protection of Human Rights and Fundamental Freedoms was drafted. Article 8 of the Convention is still one of the most important international agreements on the protection of privacy: "Everyone has the right to respect for his private and family life, his home and his correspondence". At the same time the second paragraph of the article makes clear that this right to privacy is not absolute. Interference by a public authority is allowed when such is necessary in accordance with the law and is necessary in a democratic society in the interest of national security, public safety and the economic well-being of the country, for the prevention of disorder or crime, for the protection of health and morals or for the protection of the rights and freedoms of others. With this formulation three zones of privacy are defined, that is private and family life, home, and correspondence, although correspondence is very narrowly related to the secrecy of letters.

This Convention has a legal mechanism for its enforcement through the European Commission. It is legally binding on each state that ratifies it and must be put into effect in its domestic laws. This Convention has inspired many countries to create and formulate national laws and constitutions for the protection of privacy that went further than the requirements of this Convention as was already the case in the United States in the First, Fourth and Fourteenth Amendments. In particularly the notion of correspondence has been deepened.

27.4.2.2 *Data protection.*

Although the protection of personal data is dealt with in the Fourteenth Amendment this turned out to be increasingly insufficient in an age in which information became an important force. Therefore in 1974 the Privacy Act was enacted which enforces agencies to process data fairly and limits the disclosure of individual records. The Privacy Act protects in full American tradition primarily against governmental processing of data. In the private sector the emphasis is on self-regulation combined with specific sectoral laws. A few examples out of numerous ones are the Children's Online Privacy Protection Act (COPPA) of 1998, the Fair Health Information Practice Act of 1997, and the Fair Credit Reporting Act of 1997.

As we have seen since the 1960s, the relationship between privacy and the use of data has become closer as has the awareness that this form of privacy should be protected. In addition to the domestic laws on data protection, in 1981 a special Convention was devoted to the use of personal data: the Convention for the Protection of Individuals with Regard to Automatic Processing of Personal Data [20]. In this convention some general guidelines were formulated with regard to data processing and were elaborated in approximately twenty recommendations for specific fields, such as police, medical data and statistical data.

These guidelines in large part are based on the principles of data protection formulated by the Organisation for Economic Co-operation and Development (OECD) which outlines protection is equated to privacy protection [58]. Curiously, the OECD is endorsing the protection of privacy on the one hand, yet they are promoting these principles because there is a danger that disparities in national legislation could hamper the free flow of personal data across the frontiers. Restrictions of these flows could cause serious disruption in important sectors of the economy, such as banking and insurance. For that reason these principles can be seen as guidelines that enhance fair and good practices more than they enhance privacy protection. Nevertheless they have had a big influence on all data protection legislation in Europe and elsewhere.

These principles are:

Collection Limitation Principle: There should be limits to the collection of personal data and any such data should be obtained by lawful and fair means and, where appropriate, with the knowledge or consent of the data subject.

Data Quality Principle: Personal data should be relevant to the purposes for which they are to be used, and, to the extent necessary for those purposes, should be accurate, complete, and kept up-to-date.

Purpose Specification Principle: The purposes for which personal data are collected should be specified not later than the time of data collection and the subsequent use limited to the fulfilment of those purposes or such others as are not incompatible with those purposes and as are specified on each occasion of change of purpose.

Use Limitation Principle: Personal data should not be disclosed, made available or otherwise used for purposes other than those specified in accordance with the specified purpose except: (1) with the consent of the data subject or (2) by the authority of law.

Security Safeguards Principle: Personal data should be protected by reasonable security safeguards against such risks as loss or unauthorised access, destruction, use, modification, or disclosure of data.

Openness Principle: There should be a general policy of openness about developments, practices, and policies with respect to personal data. Means should be readily available of establishing the existence and nature of personal data, and the main purposes of their use, as well as the identity and usual residence of the data controller.

Individual Participation Principle: An individual should have the right: (1) to obtain from a data controller, or otherwise, confirmation of whether or not the data controller has data relating to him; (2) to have communicated to him, data relating to him (i) within a reasonable time; (ii) at a charge, if any, that is not excessive; (iii) in a reasonable manner; and (iv) in a form that is readily intelligible to him; (3) to be given reasons if a request made under subparagraphs (1) and (2) are denied, and to be able to challenge such denial and (4) to challenge data relating to him and, if the challenge is successful, to have the data erased, rectified, completed or amended.

Accountability Principle: A data controller should be accountable for complying with measures which give effect to the principles stated above.

These principles, however, have not been implemented in legislation in all member states of the European Union in the same way. Therefore the fear of hampering the free flow of information remained. In the beginning of 1990 an effect was made to harmonise legislation within the EU. It resulted in a European Directive on Data Protection [23].

This directive enshrines two of the oldest ambitions of the European integration project: the achievement of an Internal Market (the free movement of personal information) and the protection of fundamental rights and freedoms of individuals. It is stated that both objectives are equally important, although this is highly questionable. The status of such a directive is that it binds member states to the objectives to be achieved, while leaving to national authorities the power to choose the form and the means to be used to implement these objectives. EU member states had three years after 1995 to enact or amend their laws so as to comply with the directive.

The directive applies to the public and private sector and covers the processing of personal data by both automated and manual means. Processing includes any operation or set of operations which is performed on personal data, which mean all information relating to an identified or identifiable natural person. This directive elaborates in a way the general OECD principles operationally. These principles are formulated relating to data quality and criteria are given for making data processing legitimate. Special attention is paid to special categories of processing of data of what formerly were called sensitive data: personal data revealing racial or ethnic origin, political opinions, religious or philosophical beliefs, trade-union membership, and the processing of data concerning health or sex life. Processing of data relating to offences, criminal convictions or security means may be carried out only under the control of official authorities, with suitable specific safeguards, or as provided under national law. The controller, that is to say the one who determines the purpose and means of processing is obliged to inform the data subject about the purpose of the processing, except where he already has the information. As already is written into the principles, the data subject has the right of access to his own data as well as the right to rectify, erase or block the processing of data in case the data are not correct or not up to date.

All member states are obliged to comply with the directive and to implement the principles in national laws. With the implementation the member states shall provide that one or more public authorities are responsible for monitoring the application with its territory of the provision. Regarding the transfer of data to third countries (countries outside the European Union) there is the strict rule that this transfer may take place only if the third country in question ensures an adequate level of protection, judged as such by the European Commission. If this level is missing, additional measures have to be taken either in conformity with the directive or in the form of contractual clauses. One of them is the so-called Safe Harbour Principles formulated by the Federal Trade Commission (FTC) in the United Sates.

27.4.2.3 *Regulatory agents.*

An essential aspect of any data or privacy protection is oversight. In most countries with a comprehensive law or an omnibus data protection, there is a data commissioner, sometimes in the person of an ombudsman. Under the Directive 95/46/EC, it is an obligation to have such a data commissioner.

Under article 21 of this directive, all European Union countries, including the new ones, must have an independent enforcement body. These agencies are given considerable power: governments must consult the body when they draw up legislation relating to the processing of personal data; the bodies also have the power to conduct investigations and have a right of access information relevant to these investigations; they may impose remedies such as ordering the destruction of information or ban processing and start legal proceedings, hear complaints, and issue reports. The official is also generally responsible for public education and international liaison in data protection and data transfers. They have to maintain a register of data controllers and databases. They also are represented in an important body at the European Union level through article 29 Working Group which issues reports and comments on technical and political developments.

27.4.3 Sectoral laws

As we have seen, the recommendations based on the Strasbourg Convention form a kind of sectoral legislation in addition to a more comprehensive legislation. In 1997 a special European directive was adopted which is specifically related to the protection of privacy in the telecommunication sector [24]. The development of more advanced digital technologies, such as the Integrated Services Digital Networks (ISDN) gave rise to specific requirements concerning the protection of personal data. Meanwhile this directive is repealed and replaced by a more general directive on privacy and electronic communications [25].

Whereas Directive 95/46/EC formulates general principles of data protection, this directive is oriented towards the use of new advanced digital technologies in public communications networks, in particular the Internet.

The new Directive is a response to two developments that addresses the idea that the private sphere must be protected in a more advanced way. The first is the development of so-called spyware, web bugs, hidden identifiers and other similar devices that can enter the user's terminal unawares. Such devices, for instances cookies, should be allowed only for legitimate purposes and with the knowledge of the user concerned. These cookies may, for example, only be used for analysing the effectiveness of website designs and advertising and in verifying the identity of users engaged in on-line transactions. Users should therefore have the opportunity to refuse to have a cookie or similar device stored on their terminal equipment.

The second development is unsolicited communications. In the EC directive, safeguards are provided against the use of unsolicited communications for direct marketing purposes in particular by means of automated calling machines, telefaxes, e-mail and SMS messages. For these forms of communications the prior explicit consent of the recipient must be obtained before such communications are addressed to them; in short the user must opt in for these communiqués to be legitimate. The only exception is the use of electronic contact details for the offering of similar products or services by the company that has obtained these contact details. The customer should be informed about this use and be given the opportunity to refuse such usage or to opt out.

This directive, then, is meant not only to protect the privacy of the consumer, it allows also for the retention of traffic and location data of all people using mobile telephones, SMS, landline telephones, faxes, e-mails, chatrooms, Internet, and any other electronic communication devices. The traffic data includes all data generated by the conveyance of communications on an electronic communications network, and location data are the data indicating the geographic position of a mobile telephone user, like the GPS. The contents of communications are not covered by these measures.

27.4.4 Protection by self-regulation

Self-regulation appears in different faces and strength. Some are comparable with laws while others are more or less free of obligations. The most comprehensive list of forms is given by Cameron et al. [12: 11,12]. They distinguish:

- Principles which are general statements which may provide international guidance, act as a reference document, or provide a basis for the development of legal instruments in particular jurisdiction.
- Public policies which incorporate aspects of acceptable behaviours, practices, and standards.
- Codes of conduct which incorporate ethical principles but tend to focus on behaviours, outputs, and quality of service which in turn can be used to form the basis for the interpretation of substantive behaviours.
- Guidelines which are legal within a single jurisdiction and used to provide guidance, legal meaning, and relevance even though they are generally not enforceable.
- Legal instruments which generally are the most enforceable provided that they are drafted correctly and the courts are sufficiently qualified to assess the matters brought before them.

In practice the most common forms are the legal instruments, the codes of conduct, or the codes of ethics. These codes are supposed to have five functions [45].

First of all, there is a general feeling that ethical norms for a specific profession make clear that a professional is not only responsible for the product but also for the consequences of introducing it. The idea that a technician has responsibility greater than the mere technical and economical aspects was recognised decades ago. In the 1929 edition of the Encyclopedia Britannica, Alfred Douglas Flinn writes that an engineer is obliged to watch the sociological, economical and cultural consequences of his work and technique in general. He has to

help people in organising their life so that the advantages will be optimal for everyone. This issue of responsibility is especially strong when professionals and laypeople are confronted with negative consequences of technology, such as the risk of nuclear fall-out in the case of a reactor-incident in Chernobyl, Ukraine and Harrisburg, Pennsylvania, USA or the risks of a revolutionary technology, such as DNA-mapping. Looking at some of the principles of ethical norms, this responsibility is broadly seen and moves towards all of society or the common good, employers/clients/system users, as well as colleagues.

A second function of ethical norms originally translated in codes of conduct is seen as supplementary to legal and political measures. As Herbert Maisl [19] states in a report for the Council of Europe, codes of conduct are no substitute for law, but ethics are a legitimate back up to the law. They can be necessary both before and after legislation since legislation cannot cover every detail. Prior to legislation, codes of conduct are a first step towards the introduction of a law; after legislation they can be used as a flexible instrument. Political and legal measures are too sluggish compared to the rapid development which characterises information technology. "The law also has great difficulty keeping up with change", writes Roger Clarke [17]. As an example he points to the Australian Privacy Act. "The Privacy Act, eventually passed in 1988, copes with the Internet technology of the 1970s, but not altogether with the more powerful technologies and applications of the 1990s (...)".

A third function of ethical norms is to make the general public aware in order to ensure that a public debate takes place. Stimulating awareness is the active component of a code in addition to the designation of responsibility for the consequences. As science and technology have collective consequences in an increasing way, we need collective attention for these consequences and, when necessary, collective action. Therefore, codes of conduct are instruments that help clarify what consequences society must be informed about.

A fourth function is the sociological function of 'belonging to a group with the same ideas' or more negatively formulated, a justification for doing things differently from those who are not part of the group [10]. It is an indication to the outer world that the organisation is a mature one and has a type of professional status. "(I)t is a set of guidelines recommended for the benefit of those concerned professionally with computers with the aim of achieving and sustaining a high level of professional behaviour" [57: 22]. In short, it is a part of being defined as professional. This function is the one that is most criticised when too much emphasis is put on the professionalism and less emphasis on its responsibility for society. The fact that norms are formulated is used as an indulgence for bad conduct, as defence of a bad name, or as an excuse for hiding unethical work.

A final function of ethical norms is specific to (international) organisations. A code of conduct can harmonise measures already established within diverse organisations or countries. In many countries political and legal measures are taken to control the impetus of information technology, but the way and extent differ. Not only is there a large difference between developed and developing countries but it is also true that in regional areas such as Western Europe, the differences in a specialised field such as privacy protection are so great that the need for a European directive to harmonise legislation is apparent.

Many organisations have introduced a code of conduct [6]. In nearly all of them the safeguarding of (information) privacy and data integrity is seen as the responsibility of the members. Now, more and more self-regulation is becoming an integral part of legislation.

Regarding this last function, comparing legislation in Europe with the United States shows significant differences. Whereas legislation in Europe is oriented towards both the public and the private sector, in the United States it is primarily oriented toward the public sector. For the US private sector, legislation can be described as a 'patchwork quilt' of various laws since they have chosen to adopt sectoral laws, with a large portion of data protection ceded to a particular industry's self-regulation. Another important difference is that self-regulation in the United States is a substitute for legislation, whereas in Europe self-regulation is seen as an addition to legislation. As formulated in Directive 95/46/EC, the member states and the Commission encourage the drawing up of codes of conduct intended to contribute to the proper implementation of the national provisions. A third difference between Europe and the US is that for the latter in addition to legal codes, there are also special programs are developed to protect on-line privacy, such as the TRUSTe Privacy Program and the Better Business Bureau (BBBOnLine).

27.4.5 Protection by technical means

Related to technological instruments, Charles Raab [61] makes a distinction among the following:

- *Systemic instruments* which are produced by engineers who design the network, the equipment, the computer code, or the technical standards and protocols;
- *Collective instruments,* which result from government policies, such as policy applications in which government and business builds privacy protection into a technical systems for goods and services, such as the development of a public key encryption infrastructure;
- *Instruments of individual empowerment,* required for explicit choices by individuals, such as encryption instruments, devices for anonymity, filtering instruments, and the Platform for Privacy Preferences (P3P).

27.4.5.1 *Systemic instruments.* The systemic approach regulates to the technical rules embedded within the network architecture. The technical standards and protocols as well as the default settings chosen by system developers set threshold information privacy rules. They define the capabilities of networks to invade or protect privacy. As an example anonymous Internet use may be built into the network structure just as surveillance tracking may be built into the network. Cookie management options are developed to allow users greater control over such tracking. Joel Reidenberg [62] calls this kind of regulations the *Lex Informatica.*

27.4.5.2 *Collective instruments.* One example of these instruments is a measure which becomes well known under the name Privacy Enhancing Technologies (PETs). The basic idea behind PET was developed by David Chaum, who published an article in *Communications of the ACM* on security without identification [16]. Although this article and other publications of Chaum got a lot of publicity, the real breakthrough came when the data protection authorities of The Netherlands and Canada published two reports on Privacy-enhancing Technologies [63].

PETs are a coherent system of ICT[8] measures that protect privacy by eliminating or reducing personal data or by preventing the unnecessary or undesirable processing of personal data without losing the functionality of the information system. Eliminating personal data means that adequate measures are taken to prevent identification of a person. Direct and indirect identifiers are removed in such a way that a person can no longer be identified. Reducing personal data means that although identification is possible it is made more difficult and is only allowed in a controlled context, for example by using a Trusted Third Party (TTP). In both situations the starting point is that personal data are not always needed and that the necessity of the personal data must be proved. If that is not possible, either the data are made anonymous or a so-called Identity Protector is used, which converts the actual name to a pseudo-identity.

A very old application of this kind of technology is data security. Data or information security means that a coherent package of measures is taken and maintained for securing the collection and procession of information. These measures are related to availability (information must always be available for the legitimate user), exclusiveness (information may only be used by authorised persons) and integrity (information must be in accordance with reality and be reliable, correct and up to date). To be accurate as to which types of measures are needed, a risk analysis is necessary in which the importance of information is measured as well as the consequences in case the information gets lost. These measures enclose all people and all means that are necessary for data processing. A concrete plan is made including all technical and organisational measures that should be taken, regarding the costs, the state of the technique, and the risks of the processing. Well known security measures are firewalls, which protect against unauthorised access, the use of passwords, and the authorisation for the users of information.

27.4.5.3 *Instruments of individual empowerment.* One peculiar application of PET is offering the individual the means and measures for empowering his own control over the processing of his personal data. It deals with instruments that can be chosen by the individual to enhance his control over the processing and distribution of his data. Sometimes these instruments are called add-ons and well known examples are the cookie killers, the proxy servers, anonymous remailers, and the Platform for Privacy Preferences.[9] A good example of such an instrument in relation to the Internet is the system MUTE, developed by Jason Rohrer [40], in which random access strings are used. Each time a computer (node) is connected with a P2P network that uses software to facilitate the sharing of music, a new IP-address is generated, making it extremely difficult to track the user.

Another example of this type of protection is the research done after PISA: Privacy Incorporated Software Agent. PISA is a specific application of the Intelligent Software Agent Technologies (ISATs). PISA enables the user to control the quality of input of the consumer or the citizen in e-commerce or e-government transactions and communications for protection against loss of control over personal information. A much discussed disadvantage for the protection of privacy is that agents need to maintain a perfect profile of their users in order to know one's likes and dislikes, habits and personal preferences, contact information about friends and colleagues, or the history of websites visited, and many other electronic transactions performed.

---

[8]Information and Communication Technology.

[9]See, e.g., www.anonymizer.com; www.zeroknowledge.com; www.privada.com

27.4.6 Protection through bad publicity

During the last years a strong form of protection has come from the media which publish regularly on all types of misuse of data. In some cases, the public outcry has demonstrated the significant role of the media in informing consumers and facilitating a popular response. The media then highlights not only the effectiveness of protests but also the potential for the technologies such as e-mail and general Internet usage in order for disclosure of information to be used to protect privacy.

In Stockholm a discussion started on February 10, 1986 when the newspaper *Dagens Nyheter* intensified its coverage of the Metropolit study [8]. Metropolit Projects in Copenhagen, Oslo, and Stockholm were initiated based on the same concept. All males born in these three cities were registered from birth certificates by way of regular medical investigation. Age tests, as well as psychological tests, home surveys on military service, and family particulars were carried out. The different files were all identified with a personal identification number which made linkage possible. Discussion of one specific research project rapidly escalated into a general outcry against micro data research methods. The strongest criticism was levelled at the fact that many variables were merged into databases from other sources as well as from paper documents. A subsequent judicial examination proved that no illegal activities had taken place, and that neither data laws nor any other instruction or legal provision had been contravened. Despite the fact that Statistics Sweden had not been in any way involved in the project, the affair had a strong negative influence on the public attitude towards social research in general and Statistics Sweden in particular.

In 1991 Lotus Development Corporation and Equifax abandoned plans to sell Households a CD-ROM database containing names, addresses and marketing information on 120 million consumers, after they received 30,000 calls and letters from individuals asking to be removed from the database. More recently, Lexis-Nexis, has changed plans for P-tracks, a service that provides personal information about virtually every individual in America to "anyone willing to pay a search fee of eighty-five to hundred dollars" [14: 104, 105]. The database includes current and previous addresses, birth dates, home telephone numbers, maiden names and aliases. Lexis was also providing social security numbers but stopped in response to a storm of protest and is honouring the requests of anyone who wishes to be deleted from the database.

So found the Vons chain of supermarkets in Los Angeles [18] itself the recipient of unwelcome front page publicity when it allegedly used data contained in its store-card database to undermine a damages claim from a shopper. The shopper claimed that he slipped on spilt yogurt in the store, shattering his kneecap, but said that when he filed a suit for damages, he was advised by the store's mediator that his club card records showed him to be a frequent buyer of alcohol.

27.5 ANALYSIS

27.5.1 Information society[10]

Looking at the possible attacks on privacy it is interesting to see how information technology is entering our daily life. An information society has replaced former societies and can be seen as a "third wave" as Alvin Toffler [69] has stated. After the agrarian society based on the agriculture and the industrial society based on labour, we now have entered the information society with information as the basic material for workmanship.

Discussion on the information society started roughly with the publication of Daniel Bell's *The Coming of the Post Industrial Society* [3]. Bell's analysis posits that the advanced countries were moving from the industrial stage towards a 'post-industrial stage' of development. He claimed that the majority of economically active people would earn their living from different kinds of post-industrial service sector occupations. In a pre-publication he indicated his notion of a 'knowledge society', characterised by research, development and knowledge. These activities are already responsible for a large proportion of Gross National Product and a large share of employment.

---

[10]This subsection relies heavily on [46].

At the same time a discussion started on the (negative) consequences of this information society, in particular centred on the themes of employment and of privacy. One of the first attempts to bring forward essays on the implications of computer technology together in one volume was *The Computer Impact* [68]. As editor Irene Taviss stated, the essays are intended to present a broad sampling of the major issues raised. They were selected to give the reader a sense of the concrete developments of computer technology and their implications in specific spheres of social activity. Taviss hesitated on the choice of the title although the most appropriate title for a discussion on the social implications of computer technology might appear to be *Computers: Curse or Blessing?* It is clear that the computer generates great uncertainties and great hopes. "It has become a symbol for all that is good and all that is evil in modern society".

In 1976 a milestone was reached with the publication of Joseph Weizenbaum's *Computer Power and Human Reason* [72]. Weizenbaum's critical approach of computer technology, and particularly the way it has been used, was unprecedented. Weizenbaum was shocked by the way people reacted towards the computer program ELIZA, which he had designed to play the role of a psychologist or doctor. This experience led him to attach new importance to the question of the relationship between the individual and the computer, which in turn has led Weizenbaum to the conclusion that the computer, and perhaps technology in general, has been given too much credit. Many problems are seen as technical problems that can be solved by a computer. The computer is seen as more powerful than human beings, and 'common sense' is replaced by science. The consequence is an over-emphasis on rationality and instrumentalism. Those who protest against this development are perceived as anti-technological, anti-scientific and, finally, as anti-intellectual. In reality however the price, which in Weizenbaum's view is actually paid, is servitude and impotence. Therefore human beings, in particular scientists and engineers, have responsibilities that transcend their situation. Every individual must act as if the whole future of the world, of humanity itself, depends on him or her.

Four years later a second milestone was reached with two contributions of by Daniel Bell [4; 5], more or less simultaneously published in two readers about the computer age [22; 36]. In the comprehensive tradition of his earlier publications, Bell gives an overview of the changes on the societal level. What he was calling in 1968 the Knowledge Society and in 1973 the Post Industrial Society becomes the information society, a term that has since been adopted to describe this modern computer-based society. In Bell's view we are living in a society in which information and knowledge are the crucial variables. This information explosion can only be handled through the expansion of computerised and subsequently automated information systems. This utilisation only means that the computer is a tool for managing mass society since it is the mechanism that orders and processes the transactions whose huge numbers have been mounting almost exponentially because of the increase in social interactions. His basic premise is "that knowledge and information" are becoming the strategic resource and transforming agent of the post-industrial society. Inevitably, the onset of far-reaching social changes, especially when those changes proceed as through the medium of specific technologies, confronts a society with "major policy questions". Any technology, like the computer, in his view, is only instrumental, and its impact depends on other social and cultural factors.

In his response Weizenbaum [73] speaks of the 'Computer' Revolution in order to make clear that it is not information that causes the changes but the computer. He agrees that society is transforming into an information society; however, it is not information that is responsible for that but the computer. The central question therefore is not who is responsible for the *information*, but who is responsible for the *actions* based on these computer systems.

27.5.2 Information and economic growth

Eventually, attention to the social consequences of the development towards in an information society is not only coming from social scientists but also from politicians who have become aware of the economic importance of this society. In the 1990s, when information is recognised as an important factor of economic growth, concerned citizens can see a political interest emerging. One of the early actors in this field was the Japanese Ministry of Industry and Trade (MITI), which made Japan the global leader of the development and productions of microelectronics. After the stock market crash in 1987 and the economic recession in the early 1990s many political leaders looked to the digital revolution as a form of salvation. We can cite, as an example, the United States' National Information Infrastructure Program, launched by President Bill Clinton and Vice-President Al Gore.

This concerned position was taken up in the European Commission's report *Europe and the Global Information Society* [43] which was prepared for the European Council meeting in Corfu by the High-Level Group on the Information Society, chaired by Martin Bangemann. The report starts with two key messages. The first is that the advent of the information society is inevitable and will lead to an industrial revolution comparable to that of the 19th century. The second message is that Europe's entry into the information society will be market-driven. As a result of these conclusions, a common regulatory framework must be set up at the level of the European Union in order to maximise the effect of the market while guaranteeing an appropriate level of protection for intellectual property, personal data, and network security.

This last statement is noteworthy insofar as, for the first time at international level, it is accepted and admitted that the development towards an information society is accompanied by risks. "The main risk lies in the creation of a two-tier society of have and have-nots, in which only a part of the population has access to the new technology, is comfortable using it and can fully enjoy its benefits". However this is not the only risk. A regulatory response is also needed in key areas, such as intellectual property, privacy and media ownership. Above all encryption becomes increasingly important, with the proviso that governments need power to override encryption for the purpose of combating crime and protecting national security.

Whether or not this attention to societal risks is purely instrumental in terms of avoiding rejection of the information culture and its instruments by society at large, it is the first time that an international influential body has accepted and confirmed that there are indeed risks. In light of the influence this report has had on various national programs, its relevance cannot be underestimated. All of these programs have since this acknowledgement almost always paid attention to societal and human aspects of the information society.

In a sense the Weizenbaum–Bell dispute is the forerunner of the debate which still dominates discussions today. It is not only a matter of information versus the computer, but it is also the discussion between the optimist and the pessimist, that is between people who see information technology as a societal blessing and those who see the darker side of information technology. The discussion between information and technology was more or less decided in favour of the latter, when Tom Forester presented his next volume on *The Information Technology Revolution* [37]. Since that publication onward, every discussion on information technology includes the new science of collecting, storing, processing, and transmitting information. Although the position seems to be a compromise between Weizenbaum and Bell, in reality the emphasis is on information technology.

After publications by Michael Dertouzos and Joel Moses [22] and by Forester [36; 37], a range of books were published detailing the social implications of the information society; such publications very often are a consequence of a conference devoted to such a theme. This period culminated in The Information Age trilogy of Manuel Castells [13] in which he searches for the social and economic dynamics of the information age. In these three books, Castells sees his main task as analysing informational modes of development of societies, and this analysis revolves around three fundamental axes: the changes that take place in the areas of material production, human experiences and the structures of power. Two main trends are seen as the driving force: globalisation of the economy and the digital revolution. There is, however, a third consequence which is mentioned by Castells in passing even though this phenomenon is as old as any discussion on the information society and that is that society is becoming vulnerable which in turn has consequences for human beings living in that society.

27.5.3 Information society and vulnerability

Since the publication of a report by the Swedish Sarbarhetskommitten (SÅRK) [65] in December 1979, vulnerability within a computerised society has been a significant concern. The main conclusion of the report that "vulnerability is unacceptably high in today's computerised society" has been affirmed by several studies in other countries. Studies on the vulnerability of information technology have resulted in a list of risk factors, and this list clearly shows that vulnerability is strongly related to such matters as complexity, volume of data, centralisation, integration, and above all, interdependency. In this writer's view, vulnerability of information systems can clearly be defined as the state of a system in which the complexity – in the sense of interdependency, volume of data, centralisation, integration and interdependency – is so great that any small incident for a short time can have an enormous effect on the total environment of the system and on the system itself.

The degree of vulnerability depends on several factors, as recent research shows. In official reports published after a disaster, such as at Three-Mile Island, Chernobyl, Ukraine the cause is often ascribed to human failure. This conclusion is in accordance with observations made by many authors. According to Jacques Ellul, the human being will be the greatest source of error in a technological society [30]. A human is unpredictable and since the systems being built are dependent on man, it is logical to conclude that human beings will be the single most frequent source of error. Many other authors arrive at the same conclusion: in a highly technical system, man is becoming the weakest link. He is not compatible to the system and is therefore an important potential danger.

There is, then, a serious need for countermeasures against human vulnerability. Although the already-mentioned studies and reports show that generally there is a low awareness concerning vulnerability, countermeasures can be and are taken as part of security efforts. In almost all systems related to information technology, security measures of external factors are taken against so-called acts of god, which include attending to sprinkler installations, lightening conductors, and in some cases, an emergency plan of some nature. Dependency on other humans is diminished by building and establishing in-house supplies, for example, of electricity (a power station) and computer systems. As a precaution against acts by people outside an organisation, a technical system of advanced access control can be installed, and people, who for instance have to carry out repairs, are screened before they are allowed to perform their tasks. In addition, several countries' governments have accepted and introduced legal measures, in the form of criminal law, against hacking and misuse of information systems.

Internal risk factors are also being combated. The use of automation has, in some cases, been able to remove people from high-risk positions, and the screening of job applicants is now a normal procedure, as is internal and external access control of an organisation. Information is distributed to help make employees more aware of the consequences of the vulnerability of systems that they use. Educational programs are used in an attempt to diminish vulnerability within system-use. In the same manner, attention is being given to deficiencies within an organisation, such as documenting software and hardware, developing emergency plans, and regularly controlling systems. More often now, a total security plan within an organisation is developed.

As this summary shows, most countermeasures are of a technical and organisational nature and are directed against internal and external factors. As experience shows, the complexity itself (and with that, vulnerability) is seldom dealt with. A second observation is that many of the countermeasures are directed against people internal and external to an organisation; consequently, individual freedom can be seen as an enemy with the result that employees are screened, access controlled, and legal measures utilised, usually to the advantage of the owner of the system and to the disadvantage of the employee. This tendency is the reason why some authors are afraid that campaigns against vulnerability sacrifice individual freedoms and, by extension, individual privacy.

27.5.4   Regulating information technology

The reaction of the Bangemann Commission that a regulatory response is necessary is typical for the way control of technological development is seen: from above by government, but at the same time leaving space for a free economy. Departing now from the earlier discussion on control, Dorothy Nelkin [56] has presented a general model for control which can be used for assessing control mechanisms. She uses a well-known method of labelling control at three stages: before, during and after an event; and she calls these control points anticipatory, participatory and reactive, respectively.

Anticipatory control consists of procedures for predicting social, political, and economic consequences of new scientific and technological developments. This method is particularly important when consequences are become obvious, usually once a developed system has matured and changes become almost impossible. The most well-known form of anticipatory control is Technology Assessment that is, identifying the possibilities of applied research and technologies together with the unwanted side-effects. Following the procedures as initiated by Technology Assessment, a Privacy Impact Analysis (PIA) is sometimes suggested. A PIA may be described as a systematic process that evaluates proposed initiatives or strategic options in terms of their impact on privacy [59]. To be effective it is necessary that a PIA be an integral part of a project. The first purpose is to identify the potential effects that a project or a proposal may have upon privacy, for example in the case of introducing RFID and secondly to examine how any detrimental effects upon privacy might be mitigated. In that way decision makers can be cognisant and assured of the privacy dimension within a technical system. In that sense a PIA is comparable to the widely known Environmental Effect Reports (EERs). The Canadian federal government has made PIAs mandatory for new programs and services so that institutions must document their evaluations of privacy risks, the implications of those risks, as well as any discussions of possible remedies, options, and recommendations that avoid mitigating risks. This process can help build more responsible and accountable systems and crucially help foster risk-sensitive information practices.

Participatory control deals with the involvement of citizens in the introduction and regulation of technology. In this capacity the most well-known forms are protest movements and activities aimed at raising awareness. In some sense this method has also been introduced into the labour movement under the heading of participatory design. Another form that is sometimes mentioned is self-regulation, under the condition that (consumer) organisations are involved in the process.

Reactive control is oriented towards the protection of interests of human beings and is a type of control exercised by institutions and governments reacting to a certain development. Well-known forms of reactive control are legal and punitive or disciplinary measures that attempt to prevent consequences that are too negative. Also included in this category are possible claims or complaints.

Looking at the issue of privacy protection, as described earlier, it is clear that most emphasis might be on the reactive element of controlling information systems. Participatory control is practically absent from the information control process with the exemptions of small but influential groups such as Electronic Privacy Information Centre (EPIC) and Privacy International (PI). Whenever a form of participatory control is possible through technical means, either as add-ons or as cryptography, there is strong pressure from government to have final control over the entire process. Technology assessment is still actively studied in some countries, for example the Rathenau Institute in The Netherlands, and within their activities, privacy is a part of a more comprehensive study of the social implications of information technology. Thus as far as can be determined, the Privacy Impact Analysis effort is restricted to Canada.

27.5.5   Can technology be controlled?

Regulation of technology in the sense of reactive control mirrors the way the development is seen by most politicians. The advent of the information society is inevitable as are its consequences. The only thing that can be done is retroactively repairing problems by regulating the effects. It seems a way to promote the idea of technological determinism, and such determinism reduces humankind to powerless individuals who can only accept their fate and wait to see what other people will do to help them. It is our belief that more can be done by human beings themselves than is often admitted.

Discussion on technological determinism is not new. Over the course of time there has always been a vehement discussion between the optimists and pessimists regarding the possibility of controlling technology. It is as Abbe Mowshowitz [55: 6] observed: "The central question is the nature of technology's role in our society. Is it purely instrumental, as most observers believe; or has it become an autonomous, formative element in human affairs?". The pessimists believe that technology is a completely autonomous power in itself that cannot be controlled. In other words, the consequences of technology, both positive as well as negative, have to be accepted as they are. The optimists, like Dorothy Nelkin, believe that in one way or another technology can be influenced and directed.

The discussion on autonomous technology and technological determinism was fuelled, in particular, by the publication in 1954 of a book by the French sociologist and philosopher Jacques Ellul *La Technique ou l'enjeu du siècle*. This book received international attention after its American translation to *The Technological Society* [30]. In Ellul's view technique [technology] as a totality of methods is always striving at absolute efficiency, with the consequence that spontaneous actions disappear and we are left in a completely artificial world. In this world the individual's role will be less and less important in technical evolution. Technique has become a power endowed with its own peculiar force and is for that reason influencing everything: the economy, the state, and the essence of what it is to be a human being. In that sense it even influences human behaviour, which is now oriented to adapting humankind to the technical world. In Ellul's view, in the technological society, there is no place left for a vulnerable human being.

27.5.6 Technology can be controlled

This writer does not, however, believe in technological determinism, or in any other determinism. Discussions on determinism almost always run parallel with the discussion of technological development as a revolution or as an evolution.

As we have seen, Forester [36] speaks first of the microelectronics revolution and later of the information technology revolution [37]. Barry Sherman [66] uses the word 'The New Revolution' and even Castells [13] uses this heavily laden word: "A technological revolution, centered around information technology, is reshaping, at accelerated pace, the material basis of society".

Information is an important factor in an information society but to say that it is a new phenomenon is going too far. Information has always been important. Its importance has increased and it is perhaps more important than ever, but it cannot be seen as the single factor underlying society. This idea means that simultaneously there have always been people working on the collection, processing, and distribution of information. Their numbers have also increased. On the one hand, as a consequence more traditional jobs has been reduced. On the other hand, new information jobs have been created (such as programmers and system analysts). All of these changes are signs that society itself has changed, but it is not the first time that a new technological invention has had societal consequences. The same has been seen with telex, telephone, and television. The most important difference is that we have, for the first time, a convergence of all the technical components with the result that the consequences have been far more rapid and radical. However, the consequences have not been revolutionary in the sense that they are unexpected or unpredictable. One of the proofs of this is that even now, after twenty-five years, no one can tell exactly when the information society made its entrance; and in the same way, it is impossible to say when the industrial revolution came into force.

Therefore the writer prefers to speak of gradual evolution rather than of revolutions. This position means that we are almost never totally surprised by the consequences of the development, but in most cases can more or less predict not only what the consequences are but also to what extent and in what areas they are likely to appear. This change can also be seen in relation to the attacks on privacy. From a small world in which privacy was invaded by neighbours and other villagers to the invasions of (local) government, society's members are now facing attacks not only on an international level but even more on a global level with worldwide Internet and global positioning systems as examples.

The amount of information which is collected, stored, and used is not only startling, but above all the way in which the collection of the information happens is equally startling. From the start of the discussion on the informational-privacy dimension, it is easy to suspect that humans are in the process of losing more and more control over their personal data. From the rather controlled situation from the past in which neighbours and local governments have used our personal data, society now enters a brave, new world where the press-media and the social sciences collects, without our awareness, almost more data than can be imagined. The use of cookies and the methods of phishing, pharming, and bluetoothing are the clear phenomena of this new-order society. When privacy is defined, as this article has done, that is in terms of control and self-determination it is easier to see how individual privacy must be fought for. Looking at future developments, such as the RFID, ambient technology, or grid technology, it becomes increasingly clear that our society is still at the beginning of an on-going process.

27.5.7 Two characteristics of information

The seemingly technological determinism and the feelings of powerlessness are fuelled by the fact that information technology has two characteristics that are so important that it seems that they are following their own course: information is *money* and information is *power*.

27.5.7.1 *Information is money.* Although these two characteristics partly run in parallel with the distinction between the private and public sector, a cross fertilisation appears quite often. The private sector is not only interested in money but very

often influential, as can be seen in the power that insurance companies wield. In addition, the public sector is also interested in influencing people and in money. These two characteristics, then, make it clear that in contrary to general opinion, privacy is not a true juridical issue, but in fact a political one. Making money and having power are not wrong; however, the way such influence is used, and perhaps over-reach, can create problems.

Several tools are used for collecting and analysing personal information: database marketing, relationship marketing, permission marketing, and loyalty programs which all help marketers to find the information they crave. When these collection techniques combine with data warehousing and data mining tools, individual privacy seriously can be at risk. Database marketing is also known as one-to-one marketing, whereas permission marketing acknowledges the need of permission from customers before approaching them with advertising messages, which can stand as one solution to the problem. The philosophy behind this approach is that customers are willing to release personal information if they can profit by doing so, as seen with loyalty cards. The consequence is that direct marketers fill mailboxes; relationship marketers ask for more and more information; telemarketers call at home at dinner time; and spam is a highly used tool for advertisement.

27.5.7.2 *Information is power.* Getting and using power is again a question of balancing several interests and balancing the means and the political choices. Political choices must mean that choices are made in which the privacy is protected as much as is possible. The impetus for information as a means to knowledge and power became visible after the dramatic attacks of terrorists on September 11, 2001. These policy changes were not limited to the United States but also involved most other countries with increasing surveillance powers and minimising oversight and due process requirements. The use of new technologies were incorporated and included which in turn permitted governments to use these powers and formalise its roving powers. In general, the result was a weakening of data protection regimes, an increase in data sharing, and an increase in profiling and identification [31: 25–27] of individuals.

As we have seen, information is power and since the terrorist attacks in New York, Madrid and London this type of power over citizens is becoming more and more a reality. Information is seen as one of the most important weapons in the battle against terrorism and crime. Although measures like the introduction of Passengers Name Records (PNR) and the long retention of traffic communication data are discussed, it is clear that politics in one way or another will win. Data commissioners talk about balancing the interests of privacy protection and the protection of security, but it is clear that one can not speak of a real balance. Laws are used to accept means and measures of data collection which were never accepted without the current political agendas. The introduction of CCTV, the use of Internet data, and the exchange of data among all western countries are clear examples of this untoward development.

Long before the attack of 9/11 intelligence agencies from America, Britain, Canada, Australia and New Zealand jointly monitored all international satellites telecommunications traffic by a system called 'Echelon', which can pick specific words or phrases from hundreds of thousands of messages. In 2000 it was publicly revealed that the America's FBI had developed and was using an Internet monitoring system called 'Carnivore'. The system places a personal computer at an Internet service provider's office and can monitor all traffic data about a user, including e-mail, and browsing. It gives governments, at least theoretically, the ability to eavesdrop on all customers' digital communications.

27.5.8 The omnipresence or technique

Compared with approximately one hundred years ago, the situation has changed dramatically. From an incidental intrusion by humans into each other's lives and, rarely, having the technical means to find out too much, society now has the technical means and capacity of collecting individual data to a serious level. Since technique is omnipresent and, as an old sociological wisdom says, humans are a data producing animal, all tracks and traces left behind by human beings can be and are collected. As we have seen, since information is money and power government and industry are using almost all means at their disposal for this data collection regime.

It is, however, not only the omnipresence of technique which is frightening but also the sheer lack of awareness of its usage. One of the most impressive examples is the way data can be collected from the Internet. Cookies and more general spyware are collecting data without our knowledge. And this lack of transparency increases once data are used. Although in many case we know the purpose of the use, we do not always know for sure whether the actual use is as indicated. Responsible for this lack of transparency in data mining is making clear the distinction between data and information. Data is a collection of details or facts which, as such, are meaningless; however, when those details or facts are placed in a context which makes data usable, serious information can be gleaned. Depending upon the placement-context, the same data can be transformed into different information.

The classical example is the speed of a car. Saying that a car is driving at a speed of forty miles does not mean anything. Depending upon the context, for example in a city or on a highway, the information can be interpreted quite differently: in a city, forty miles per hour can be too fast, especially in a school zone during school in-take hours but on a highway, forty miles per hour may be too slow and seriously jeopardise the flow of traffic.

Another example is the supermarket which introduces loyalty cards and asks patrons to fill in a form in which the sex of the owner of the card and the sex of his of her partner must be filled in. Although it was said that the provided data would only be used for contacting the owner or his or her partner, it is clear that the data can also be used for detecting homosexual relations. Numerous other examples make clear the importance of the distinction between data and information. Knowing for what purpose *data* are used does not mean that for the same purpose the *information* would be used.

27.5.9 Privacy and data protection

In the last century not only have the possibilities of individual intrusion been increased, so have all the forms of protection. From a more general protection based on a country's constitution or its amendments, privacy is said to be protected by a series of legislation by which all data protection acts are shown to be the most visible examples. In Europe in the form of omnibus legislation and in the United States as sectoral protection, data collectors must give the data-subject the idea that the individual's data are used in an ethical way. In addition to this legislation technology can be used as is the case in the stimulation of Privacy Enhancing Technologies.

For showing the correctness of this thinking about individual protection in a data-hungry world, it is necessary to make a critical remark. Privacy rights require making a distinction between privacy protection and data protection. In a classical sense privacy is seen as that part of our intimate life (family life, home, and correspondences) that should be free from intrusion by others (people, industry, government). For that reason privacy protection is part of human rights and these are protected legally. In effect, data protection has become a way of handling data carefully and in accordance with general principles like purpose specification, data quality, and security safeguarding. Once collected the data can be used in a completely different way and can intrude upon individual privacy. The question remains: how far is this intrusion into an individual's privacy acceptable?

All this does not mean that privacy is seen by the consumer as unimportant. On the contrary, all opinion polls show that about eighty to eighty-five percent of the people find privacy and the protection of privacy rather or strongly important, but at the same time see how important it is to use information in the battle against terrorism and crime. They have, however, the naughty feeling that their privacy is not invaded because they are not terrorists or criminals. In relation to some developments like genetic testing a new phenomenon is appearing: the right not to know. While genetic screening has become easier and cheaper, treatment of diseases lags behind. For that reason many people, for example in the case of Huntington's disease, choose not to find out the results of testing

because of the inability to take precautionary measures. A problem in these situations is the dependency from the other family's members who are testing and probably want to rely on a need to know.

27.5.10 Surveillance

The omnipresence of technique and the acceptability of politics and the law to collect, store, and use almost all personal data is making the information society a surveillance society. Simultaneously the number of techniques is increasing so intrusion of privacy is inevitable. Distinct authors [11; 35; 67], agree that surveillance might create conformist actions. People will keep their records clean and will avoid controversial or deviant behaviour, whatever their private views and intentions might really be.

But it is not only surveillance that matters, it is the fact that we are on the way to a *riskless* society in which more and more the policy is oriented toward avoiding risks and errors produced by human beings. Personal data are used for determining the amount of risk a person forms in the eyes of government and industry. In government these figures are used for political reasons, in industry for discriminating between the good and bad consumer. It is this use which makes a consumer into a glass-consumer, in a manner of speaking, for whom there is a deep concern for unfair exclusion, unfair targeting and unfair discrimination [41].

27.6 CONCLUSIONS

The legal measures, and the way political decisions are taken, make clear that this kind of protection is passive. The consumer or citizen plays almost no significant role in the process. It is the government (laws) and industry (laws and self-regulation) who define the way and amount of protection. A more active role can be played when the consumer or citizen is allowed to use technical means, but also in this case it is politics and government who determines when and how these techniques may be used, as we have seen in the discussion around Clipper Chip.

It is the government that wants to control the use of information and refuses to strengthen the position of the individual. For that reason, almost all emphasis is on reactive control of privacy invasions. If some form of participatory control is given, it is always given under the restriction that in the end it is the government who has the ultimate control. Cryptography might be used to secure one's data but the key must stay in the hands of the government. Only in relation to industry does the role of the consumer become legally empowered regarding the use of cookies and spam. At the same time industry is used as a source of information. Traffic and location data must be stored longer then is necessary and must be given to a government in case of suspicion of terrorist actions.

Not only are these techniques empowering governments but they also becoming legal as has been seen in the case of the Patriot's Act in the United States and the Regulation of Investigatory Powers Act in the United Kingdom. The same development can be seen at the Directive on Privacy and Electronic Communications, which opens the possibility to enact from domestic laws the retention of traffic and location data. Most especially, these developments strongly suggest vigilance. Information as a means of power and legislation as legitimising power are dangerous instruments in the hands of unethical politicians who are missing the necessary checks of balances of a democracy. In that case not only is privacy at stake but above all so is democracy. It is time to revisit the use of technology, the law, and the role consumers have in this serious issue. A positive sign comes from the British National Consumer Council in its publication *The Glass Consumer, Life in a Surveillance Society* [50]. Although the title sounds pessimistic, the book ends optimistically with the NCC's agenda and recommendations for the future.

REFERENCES

[1] A. Barth, *The Price of Liberty*, The Viking Press, New York (1961).

[2] A. Bates, *Privacy – a useful concept?*, Social Forces 42 (1964), 429–435.

[3] D. Bell, *The Coming of Post-Industrial Society*, Basic Books, New York (1973).

[4] D. Bell, *The social framework of the information society*, *The Computer Age: A Twenty-Year View*, M.L. Dertouzos and J. Moses, eds, Cambridge, MA, and London, England (1979), pp. 163–212.

[5] D. Bell, *A reply to Weizenbaum*, *The Computer Age: A Twenty-Year View*, M.L. Dertouzos and J. Moses, eds, Cambridge, MA, and London, England (1979), pp. 459–463.

[6] J. Berleur, K. Brunnstein (eds), *Ethics of Computing, Codes, Spaces for Discussion and Law*, Chapman & Hall, London (1996).

[7] A.W. Branscomb, *Who Owns Information? From Privacy to Public Access*, Basic Books, New York (1994).

[8] L. Brantgärde, *Swedish trends in data protection and data access*, *Data Protection and Data Access*, P. de Guchteneire and E. Mochmann, eds, North-Holland, Amsterdam (1990), pp. 105–122.

[9] M. Brenton, *The Privacy Invaders*, Coward-McCann, Inc., New York (1964).

[10] H. Burkert, *The ethics of computing?*, *The Information Society: Evolving Landscapes. Report from Namur*, J. Berleur, A. Clement, T.R.H. Sizer and D. Whitehouse, eds, Springer-Verlag/Captus University Press, New York (1990), pp. 4–19.

[11] D. Burnham, *The Rise of the Computer State. The Threat to Our Freedoms, Our Ethics and Our Democratic Process*, Random House, New York (1983).

[12] J. Cameron, R. Clarke, S. Davies, A. Jackson, M. Prentice, B. Regan, *Ethics, vulnerability and information technology (IT)*, Paper presented at the IFIP's World Conference in Madrid (1992).

[13] M. Castells, *The Information Age, Economy, Society and Culture. Volume I: The Rise of the Network Society*, Blackwell, Oxford, UK (1996).

[14] F.H. Cate, *Privacy in the Information Age*, Brooking Institution Press, Washington, DC (1997).

[15] A. Cavoukian, *Identity Theft Revisited: Security is Not Enough*, Information and Privacy Commissioner, Ontario (2005).

[16] D. Chaum, *Security without identification: Transaction systems to make big brother obsolete*, Communications of the ACM 28 (1985), 1020–1044.

[17] R. Clarke, *Social implications of IT – the professional's role*, The Australian Computer Journal 22 (1990), 27–29.

[18] Computer Business Review, June 2001.

[19] Council of Europe, Committee of experts on data protection (CJ-PD), Legal problems connected with the ethics of data processing, Study by Mr. Herbert Maisl, University of Orléans, Strasbourg (1979).

[20] Council of Europe, Convention for the Protection of Individuals with Regard to Automatic Processing of Personal Data, European Treaty Series No. 108, Strasbourg (1982).

[21] S. Davies, *Big Brother at the box office, electronic visual surveillance and the bog screen*, *Proceedings of*

the 21th International Conference on Privacy and Personal Data Protection, Hong Kong (1999), pp. 151–160.

[22] M.L. Dertouzos, J. Moses (eds), *The Computer Age: A Twenty-Year View*, Cambridge, MA, and London, England (1979).

[23] Directive 95/46/EC of the European Parliament and the Council of 24 October 1995 on the protection of individuals with regard to the processing of personal data and the free movement of such data, Official Journal of the European Communities (1995), 281/31-50.

[24] Directive 97/66/EC of the European Parliament and the Council of 15 December 1997 concerning the processing of personal data and the protection of privacy in the telecommunications sector, Official Journal of the European Communities (1998), 24/1-8.

[25] Directive 2002/58/EC of the European Parliament and the Council of 12 July 2002, concerning the processing of personal data and the protection of privacy in the electronic communication, Official Journal of the European Communities (2002), 201/37-47.

[26] P. Duquenoy, V. Masurkar, *Surrounded by intelligence . . . , Risks and Challenges of the Network Society*, P. Duquennoy, S. Fisher-Hübner, J. Holvast, A. Zuccato, eds, Karlstad University Studies, Karlstad (2004), pp. 121–134.

[27] *The surveillance society*, The Economist, May 1st (1999).

[28] The Economist Technology Quarterly, *Move over, Big Brother*, The Economist, 2 December (2004), 26.

[29] A.J. Elbirt, *Who are you? How to protect against identity theft*, IEEE Technology and Society Magazin (2005), 5–8.

[30] J. Ellul, *The Technological Society*, Vintage Books, New York (1964).

[31] EPIC, Privacy & Human Rights, An international Survey of Privacy Laws and Developments, Electronic Privacy and Information Centre and Privacy International, Washington (2002).

[32] M. Erbschloe, J. Vacca, *Net Privacy, A Guide to Developing and Implementing an Ironclad eBusiness Plan*, McGraw-Hill, New York (2001).

[33] M. Evans, *The data-informed marketing model and its social responsibility*, *The Glass Consumer, Life in a surveillance society*, S. Lace, ed., National Consumer Council (2005), pp. 99–132.

[34] D.H. Flaherty, *Privacy in Colonial New England*, University Press of Virginia, Charlottesville (1972).

[35] D. Flaherty, *The emergence of surveillance societies in the western world: Toward the year 2000*, Government Information Quarterly 4 (1988), 377–387.

[36] T. Forester (ed.), *The Microelectronic Revolution*, Basil Blackwell, Oxford (1980).

[37] T. Forester (ed.), *The Information Technology Revolution*, Basil Blackwell Ltd, Oxford (1985).

[38] R. Foroohar, *Life in the grid*, Newsweek, September 16–23 (2002), 64–67.

[39] B. Garson, *The Electronic Sweatshop, How Computers Are Transforming the Office of the Future into the Factory of the Past*, Penguin Books, New York (1988).

[40] F.S. Grodzinsky, H.T. Tavani, *Verizon vs the RIAA: implications for privacy and democracy*. Paper presented at ISTAS '04, International Symposium on Technology and Society, Globalizing Technological Education (2004).

[41] H. Hall, *Data use in credit and insurance: controlling unfair outcomes*, *The Glass Consumer, Life in a Surveillance Society*, S. Lace, ed., National Consumer Council (2005), pp. 157–186.

[42] A. Harisson, *The Problem of Privacy in the Computer Age, an Annotated Bibliography*, The RAND Corporation, Santa Monica (1967).

[43] High Level Group on the Information Society, *Europe and the Global Information Society*, Cordis Focus, Supplement 2 (1994).

[44] R.F. Hixson, *Privacy in a Public Society. Human Rights in Conflict*, Oxford University Press, New York, Oxford (1987).

[45] J. Holvast, *Codes of ethics: Discussion Paper*, *Ethics of Computing, Codes, Spaces for Discussion and Law*, J. Berleur and K. Brunnstein, eds, Chapman and Hall, London (1996), pp. 42–52.

[46] J. Holvast, P. Duquenoy, D. Whitehouse, *The information society and its consequences: lessons from the past*, *Perspectives and Policies on ICT in Society*, J. Berleur and C. Avgerou, eds, A TC9 Handbook, IFIP, Springer Science & Business Media (2005), pp. 135–152.

[47] S.M. Jourard, *Some Psychological Aspects of privacy*, Law and Contemporary Problems 31 (1966), 307–319.

[48] H. Kalven Jr., *The problem of privacy in the year 2000*, Daedalus 93 (1967), 876–882.

[49] M.R. Konvitz, *Privacy and the law: a philosophical prelude*, Law and Contemporary Problems 31 (1966), 272–281.

[50] S. Lace (ed.), *The Glass Consumer, Life in a Surveillance Society*, National Consumer Council (2005).

[51] A.Th. Mason, *Brandeis, A Free Man's Life*, The Viking Press, New York (1956).

[52] M.F. Mayer, *Right of Privacy*, Law Arts Publishers, Inc., New York (1972).

[53] A. Miller, *The Assault on Privacy: Computers, Dossiers and Data Banks*, The University of Michigan Press, Ann Arbor (1971).

[54] B. Moore Jr., *Studies in Social and Cultural History*, M.E. Sharpe, Inc., Armonk, NY (1984).

[55] A. Mowshowitz, *The Conquest of Will: Information Processing in Human Affairs*, Addison–Wesley Publishing Company, Reading, MA (1976).

[56] D. Nelkin, *Technology and public policy*, *Science, Technology and Society, a Cross Disciplinary Perspective*, I. Spiegel-Rösing and D. de Solla Price, eds, Sage Publications, London and Beverly Hills (1977), pp. 393–443.

[57] G.B.F. Niblett, *Digital Information and the Privacy Problem*, OECD Informatics Studies, no. 2. Organisation for Economic Co-operation and Development, Paris (1971).

[58] OECD, *Guidelines on the Protection of Privacy and Transborder Flows of Personal Data*, Paris (1981).

[59] Office of the Privacy Commissioner for Personal Data, Hong Kong, *E-Privacy: A Policy Approach to Building Trust and Confidence in E-Business, A Management Handbook*, Hong Kong (2001).

[60] V. Packard, *The Hidden Persuaders*, Penguin Books, Midddlesex (1964).

[61] C. Raab, *Regulatory provisions for privacy protection*, *The Glass Consumer, Life in a Surveillance Society*, S. Lace, ed., National Consumer Council (2005), pp. 45–67.

[62] J. Reidenberg, Technologies for Privacy Protection. Paper presented at the 23rd Internation Conference of Data Protection Commissioners, Paris (2001).

[63] H. van Rossem, H. Gardeniers, J. Borking, A. Cavoukian, J. Brans, N. Muttupulle, N. Magistrale, *Privacy-Enhancing Technologies, The Path to Anonymity*. *Volumes I and II*, Registratiekamer, The Netherlands & Information and Privacy Commissioner, Ontario, Canada (1995).

[64] O.M. Ruebhausen, O.G. Brim Jr., *Privacy and Behavioral Research*, Columbia Law Review 65 (1965), 1184–1211.

[65] SÅRK, *The Vulnerability of Computerized Society, Considerations and Proposals*, Liberförlag, Stockholm (1979).

[66] B. Sherman, *The New Revolution, The Impact of Computers on Society*, Wiley, Chicester, New York, Brisbane, Toronto, Singapore (1985).

[67] J. Stanley, B. Steinhardt, *Bigger Monster, Weaker Chains, The Growth of an American Surveillance Society*, American Civil Liberties Union, New York (2003).

[68] I. Taviss (ed.), *The Computer Impact*, Prentice Hall, New Jersy (1970).

[69] A. Toffler, *The Third Wave*, Bantam Books, New York, Toronto, London (1980).

[70] M. Warner, M. Stone, *The Data Bank Society, Organizations, Computers and Social Freedom*, George Allen and Unwin Ltd (1970).

[71] S.D. Warren, L.D. Brandeis, *The Right to Privacy*, A.C. Breckenridge (ed.), *The Right to Privacy*, University of Nebraska Press, Lincoln (1970), pp. 133–153.

[72] J. Weizenbaum, *Computer Power and Human Reason. From Judgment to Calculation*, W.H. Freeman and Co., San Francisco (1976).

[73] J. Weizenbaum, *Once more: The computer revolution*, *The Computer Age: A Twenty-Year View*, M.L. Dertouzos and J. Moses, eds, Cambridge, MA, and London, England (1979), pp. 439–459.

[74] A.F. Westin, *Science, privacy and freedom: Issues and proposals for the 1970's. Part I: The current impact of surveillance on privacy*, Columbia Law Review 66 (1966), 1003–1050.

[75] A.F. Westin, *Science, privacy and freedom: Issues and proposals for the 1970's. Part II: Balancing the conflicting demands of privacy, disclosure, and surveillance*, Columbia Law Review 66 (1966), 1205–1253.

[76] A.F. Westin, *Privacy & Freedom*, The Bodley Head, London, Sydney, Toronto (1967).

[77] W.H. Whyte, *The Organization Man*, Penguin Books, Middlesex (1956).

[78] W. Zelermyer, *Invasion of Privacy*, Syracuse University Press (1959).

28

MUNITIONS, WIRETAPS AND MP3S: THE CHANGING INTERFACE BETWEEN PRIVACY AND ENCRYPTION POLICY IN THE INFORMATION SOCIETY

Andrew Charlesworth

Centre for IT and Law, Law School and Department of Computer Science
University of Bristol
Bristol, UK

Contents

Abstract

The modern history of cryptography, from the early 1970s to the present day, provides a viewpoint upon an often neglected dimension of the modern tensions between personal privacy and public policymaking and, increasingly, between personal privacy and corporate interests. Once national governments, notably in the US, fought to keep strong encryption technologies out of the privacy armoury of the general public, but found themselves thwarted by post-Cold War globalisation and international commercial competition. Now it is the turn of corporations to seek to utilise the power of the State, through the intellectual property regime, both to restrict cryptographic research to legislatively 'approved' ends; and to turn encryption to the service of legally protected privacy invasive technologies.

Keywords: cryptography, encryption, legal regulation, export controls, key escrow, key recovery, key surrender, privacy, surveillance, interception, intellectual property protection, digital rights management.

ACRONYMS

ACLU	American Civil Liberties Union
ACM	Association for Computing Machinery
AEBPR	Advanced eBook Processor

AECA	Arms Export Control Act of 1976 (US)
AES	Advanced Encryption Standard
ATM	Automated Teller Machine
BXA	Bureau of Export Administration (US)
CCL	Commerce Control List (US)
CDPA	Copyrights Designs and Patent Act 1988 (UK)
CDT	Center for Democracy and Technology (US)
CESA	Cyberspace Electronic Security Act (US) (not adopted)
CIPA	Classified Information Procedures Act of 1980 (US)
CoCom	Coordinating Committee for Multilateral Export Controls
CSS	Content Scramble System
DES	Data Encryption Standard
DGSE	Direction Générale de la Sécurité Extérieure / General Directorate for External Security (France)
DMCA	Digital Millennium Copyright Act of 1998 (US)
DRM	Digital Rights Management
DST	Direction de la Surveillance du Territoire / Directorate of Territorial Surveillance (France)
DVD	Digital Versatile Disc
EAR	Export Administration Regulations (US)
ECPA	Electronic Communications Privacy Act of 1986 (US)
EAA	Export Administration Act of 1979 (US)
EES	Escrowed Encryption Standard
EFF	Electronic Frontier Foundation
EU	European Union
IBM	International Business Machines Corporation
IEEE	Institute of Electrical and Electronics Engineers
ITAR	International Traffic in Arms Regulations (US)
KLS	Keystroke Logging Software
LEAF	Law Enforcement Access Field
NBS	National Bureau of Standards (US)
NIST	National Institute of Standards and Technology (US)
NSA	National Security Agency (US)
NSF	National Science Foundation (US)
OECD	Organisation for Economic Co-operation and Development
PCSG	Public Cryptography Study Group (US)
PECSENC	President's Export Council Subcommittee on Encryption (US)
PET	Privacy Enhancing Technology
RG	Direction Centrale des Renseignements Généraux / French Police Intelligence Service (France)
RIAA	Recording Industry Association of America
RIPA	Regulation of Investigatory Powers Act 2000 (UK)
SCSSI	Service Central de la Sécurité des Systèmes Informatiques (France)
SDMI	Secure Digital Music Initiative
SIIA	Software and Information Industry Association (US)
SPA	Software Publishers Association (US)
SSL	Secure Sockets Layer

TLS	Transport Layer Security
TTP	Trusted Third Party
WIPO	World Intellectual Property Organization

28.1 INTRODUCTION

The development of national policies on the distribution and use of cryptographic techniques, and their relationship to, and effect upon, efforts by various groups within society to use encryption to protect individual and group secrets in communications, is a dynamic that has existed for centuries. However, the widespread availability of strong encryption for personal privacy protection for public use, as opposed to its use by the military, spies, diplomats and powerful elites, is a relatively recent phenomenon, driven by the rise of ubiquitous access to personal computing facilities [78; 89; 98; 156].

The initial range of responses by national governments to the post-war developments in strong encryption technologies, made possible by the use of computer ciphers, was entirely predictable. Most research was done under the oversight of national security agencies and the military and the results withheld under the auspices of national security laws. Open research could be deterred in some jurisdictions by withholding of funding, or by the threat of non-disclosure requirements on the fruits of the research, also usually grounded in national security legislation. Only in the 1970s did significant open academic/commercial research begin to be undertaken, and even then both the research and its results were usually subject to some degree of national security restriction. As such, in most, if not all, jurisdictions the impact of strong encryption technologies upon the relationship between the State and its citizens was negligible. While legal protections for privacy in personal information and communications were often available in Western democracies,[1] government bodies and law enforcement agencies were still in a position to legitimately (and illegitimately) access them with little technical difficulty.[2]

As strong encryption technologies began to come widely into the public sphere, governmental legislative and policy responses began to shift from barring disclosure and research, to barring their export to other countries, and restricting their legal use by the public. Until the mid-1990s, such policies remained prevalent. However, the explosion in personal computing, brought about largely by increasing public access to the Internet, was the catalyst for a paradigm shift in the practical uses of cryptographic techniques. The steady technological progress predicted by Moore's Law[3] now placed in the family home a level of processing power that could only be dreamed of by the cryptanalysts of Bletchley Park, and this power could be used to drive viable personal implementations of strong encryption.

The rise of the 'Information Society' did not, of course, just provide the general public with the possibility of greater personal privacy in their personal information and communications – it also exacerbated the need for increased personal privacy/secrecy measures. It multiplied the number of items of information that individuals wished, or

---

[1]The German Land of Hesse enacted what is widely regarded as the first data protection legislation in 1970, Sweden followed with the first national implementation, and by 1980 the majority of the 9 European Union (EU) Member States (excluding the UK, Ireland and Italy) had adopted, or had committed to adopt, national legislation. However, most data protection regimes, both then and now, explicitly exclude data gathered for law enforcement or national security purposes from some or all of the legislative safeguards [23].

[2]From the mid-1980s a series of cases demonstrated that the interception of communications by some European authorities frequently lacked a clear legal basis, and was often subject to minimal oversight to prevent abuses. In the case of *Malone v. United Kingdom* (1984) 7 EHRR 14 the European Court of Human Rights held that this breached Article 8 of the European Convention on Human Rights. As a result, the UK government passed the Interception of Communications Act 1985 to provide the necessary legal framework. See also *Kruslin v. France* (1990) 12 EHRR 547; *Huvig v. France* (1990) 12 EHRR 528; *Valenzuela Contreras v. Spain* (1999) 28 EHRR 483; *Amann v. Switzerland* (2000) 30 EHRR 843.

[3]Moore's Law of Integrated Circuits "The complexity for minimum component costs has increased at a rate of roughly a factor of two per year..." [110].

were obliged, to hold and communicate in conditions of privacy/secrecy; increased the number of social interactions in which private/secret information transfers were necessary or desirable; and continually raised the value for third parties, including government agencies, commercial entities, and criminal elements, in gaining access to such information.

The scene was thus set for an escalating struggle between the individual who sought to retain (and expand) her control over the level of privacy/secrecy available for her information and its communication, and those individuals and entities who sought the means to diminish that capacity. In the Information Society, this struggle has been increasingly expressed not just via traditional socially normative mechanisms such as privacy-enhancing or privacy-decreasing laws, but also via technically normative mechanisms – the architectural design of the information technologies developed to store and communicate information [96; 97].

While the rationales underlying those architectural designs may sometimes fall in the same direction as government policy, or particular legislative goals, this is not a given. For example, the design of information systems for commercial purposes may be driven less by the developers' need to meet particular legal requirements, such as data privacy laws, than by the perceived need to meet certain consumer criteria without the attainment of which preferred e-commerce models cannot be viable. In this circumstance, the self-interest of the designers, government policy and legislation may sometimes coincide, for example, if the level of personal data privacy afforded to consumers within a particular architecture is increased in line with the aims of data protection legislation [24].

However, those interests need not, and often do not, coincide. The encryption systems of varying strengths that are increasingly embedded in consumer software, such as web browsers; and commercial software, such as website payment systems; are not just present because the security of commercial transactions is mandated by law (although such legal requirements are increasingly common, in addition to the legal liabilities owed

by system operators to users under contract and tort law). They are there because browser developers and website owners perceive a need to inculcate a high level of trust in potential users in order to persuade them to move from off-line to on-line commercial activity. However, one result of designing architecture to provide a high level of trust amongst consumers about the privacy of their information and communications is that a former key desire of national governments – to be able to easily surveil their citizens' activities – begins to lose its favoured position to the commercial imperative.

The economic pressures to permit public use of strong encryption in support of widening public take-up of Internet technologies; including e-mail communication, web-browsing and e-commerce; left governments looking for a viable fallback position. It was no longer economically tenable (particularly in *laissez faire* Western economies) to use legal measures to directly ban or inhibit such uses of encryption. Equally, it was politically problematic to be seen to be permitting the widespread use of a technology that could potentially damage national security or hinder law enforcement. How could this circle be squared?

One possible avenue that suggested itself was to mandate a government-approved encryption standard under which national security agencies and law enforcement bodies were provided with a backdoor, or with access via escrowed encryption keys. Attempts to implement this approach dominated government encryption policy and lawmaking in the US and Europe through the late 1990s. Such a policy approach, however, provided the potential for future wholesale public surveillance by government bodies and law enforcement agencies. This possibility, combined in many jurisdictions with the proposed provision of relatively weak protections to deter abuse, was to provoke a furious backlash from privacy organisations and the commercial sector alike.

By the late 1990s, the increasing globalisation of the information technology marketplace, and the ubiquity of the Internet, meant that legal and regulatory measures, such as extensive export bans on strong encryption technology, and the promotion of key escrow systems, were becoming increasingly untenable. The demands of international competition were beginning to force the hand of national governments. Nations that sought to restrict the exports of their national producers, or attempted to mandate national encryption standards, found themselves accused of simply making their industries uncompetitive in the international marketplace. Such arguments were strengthened by the failure of national governments to present convincing evidence that such measures could ever achieve their stated policy goals.

It is obvious that strong encryption has played, and is likely to continue to play, a key role in the contemporary struggle over the appropriate level of privacy protection that should be afforded to individuals in their information and communications. It is equally clear that the influences upon the regulatory framework within which strong encryption is deployed, and within which privacy goals traditionally compete with government national security and law enforcement goals, have radically changed in recent times. There are significant commercial pressures in information technology-rich nations to permit the securing of information and communications via the use of publicly available strong encryption. These include the need to develop public trust in new forms of economic activity, such as e-commerce; and the need to allay the concerns of commercial entities about their exposure to legal risks, to encourage investment and innovation. Such goals have had a powerful effect upon national policies regarding the access of the public to strong encryption technologies. Equally, as the market for such technologies grows, there is a concomitant pressure from producers on their national governments to reduce export restrictions upon strong encryption technologies, to permit them to compete in the globalised information technologies marketplace. The result of this pressure for widening access to strong encryption technologies has been the gradual displacement of the historically dominant concerns of the maintenance of national security and the effectiveness of law enforcement. In theory, at least, this has meant a significant impetus towards strengthening the privacy

rights an individual can exert against the State via a combination of privacy legislation, and modern information and communications architectures.

However, there is a further equally powerful commercial imperative which both threatens development of, and public access to, strong encryption, and is potentially highly invasive of personal privacy. It is grounded in the increasing use of encryption technologies for digital content control. The incorporation of legal protection for digital rights management (DRM) systems into national intellectual property regimes, such as the US Digital Millennium Copyright Act (DMCA), raises two key concerns. First, that research into encryption technologies, and particularly cryptanalysis, might once again be either legally barred, or subject to significant confidentiality and secrecy provisions, not in the interests of national security, but rather because it has the potential to damage the interests of certain types of intellectual property rightsholder. Second, that widespread use of encryption technologies to protect digital information may have the effect, inadvertently or deliberately, of making it difficult, or impossible, for individuals to access such information anonymously. There is thus something of a paradox in the ongoing relationship between encryption and privacy. One set of commercial pressures is helping to increase the ability of the individual to protect their personal privacy via a combination of strong encryption and privacy-enhancing laws, such as data protection legislation. Yet another set is actively seeking to reduce the individual's access to privacy enhancing solutions, via control of the nature, scope and accessibility of encryption research, and the use of potentially privacy-invasive laws, such as laws requiring the protection of technical measures required for DRM.

In short, the rise of ubiquitous computing has resulted in a paradigm shift in the practical uses of cryptographic techniques and has caused marked changes in many national authorities' attitudes and policies towards public use of cryptographic systems. Yet its key impact may yet be that, in key jurisdictions, it has caused the policymaking role of assessing what is deemed to be acceptable public use of cryptography to effectively shift from the public sector to the private sector. This chapter will survey those developments.[4]

28.2 CORKING THE GENIE: RESTRICTING NON-GOVERNMENTAL ENCRYPTION RESEARCH, DEVELOPMENT AND USE

In 1979, David Kahn noted in a famous article in *Foreign Affairs* that, despite the best efforts of national governments to keep cryptology both a secret and a government monopoly, the technology was now increasingly entering the public arena [77: 151]. He identified two key areas of controversy with regard to this development; the first involved whether a nation's ability to decipher foreign codes for intelligence purposes should be regarded as a more important concern than the ability of its citizens to protect their privacy by the use of strong ciphers [77: 151]; the second concerned whether non-governmental research into the area of cryptology should be permitted to proceed unhindered by government regulation designed to protect national security interests [77: 153].

It is clear from the literature of the time that, in the US at least, these issues were highly contentious. The US National Security Agency (NSA)[5] exerted considerable pressure upon other government agencies; academic and commercial institutions; and researchers, in an effort to maintain as broad a control as possible over the strength of encryption products made available to the public. Its concerted and extensive attempts to restrain non-governmental research relied upon a variety of regulatory techniques[6] including:

- straight 'command and control' legal restriction;
- co-regulatory techniques, including voluntary clearance by their authors, via the NSA, of academic papers; and
- regulation by contract approaches, based on the availability, or otherwise, of government research grants [141].

The literature also provides some early hints as to the depths of the mutual distrust between key researchers in the public sphere and the NSA, a distrust that has coloured the debate in the US over the following 30 years.

The initial arena in which control of the strength of encryption product made available to the public came to the fore, was with the development of the cipher system Data Encryption Standard (DES). DES was initially produced by the International Business Machines Corporation (IBM) for the banking industry to reduce the potential vulnerability of communications between automated teller machines (ATMs) and central banking computers, by encrypting traffic between them. In 1973, the US National Bureau of Standards (NBS) sought to adopt a standard cipher which would be used by government agencies to encrypt non-classified computer data, and by private organisations when required to communicate with government agencies securely. It chose DES but, allegedly after pressure from the NSA, adopted a form of DES that was weaker than initially offered, with a 56-bit key rather than the 128-bit key that IBM had envisaged for its original organisational cipher,[7] or the 64-bit key initially proposed by IBM for DES.[8] The NSA,

---

[4]For a similar examination of the early stages of encryption control and the impact on privacy, see [153]; for a similar if shorter, analysis, of the rise in importance of commercial influences on cryptography see [129].

[5]The NSA was formally established in 1952 with the aim of developing US communications intelligence from a military activity divided among the three services to a unified national activity [12]. According to its website "It coordinates, directs, and performs highly specialised activities to protect U.S. government information systems and produce foreign signals intelligence information. ... It is said to be the largest employer of mathematicians in the United States and perhaps the world. Its mathematicians contribute directly to the two missions of the Agency: designing cipher systems that will protect the integrity of US information systems and searching for weaknesses in adversaries' systems and codes". It is a key player in cryptology policy in the US, and has a reputation for aggressively defending its sphere of influence in this area against other government agencies, both military and civilian. A significant portion of civilian cryptology and security research in the US is funded by grants from the NSA. About the NSA, http://www.nsa.gov/about/index.cfm

[6]For further discussion of different approaches to achieving regulatory aims, see [11].

[7]When IBM first described the algorithm that later became the DES, for a system known as Lucifer which was intended for internal use at IBM, a 128-bit key was used [86].

[8]It was suggested by IBM that the reason for reducing the key length from 64 bits to 56 was that the other 8 bits were

which had allegedly been involved in the design of DES, [40; 60] also classified certain aspects of that design (the S-boxes), raising suspicions that the classified elements contained means through which the NSA could break the DES code.[9] These suspicions were partly based on the fact that IBM had been granted a licence by the Office of Munitions Control at the State Department to export computer equipment containing the DES technology, and critics found it hard to believe that the NSA would support the sale of truly unbreakable encryption to other countries [77: 151–153; 86]. Despite, at times, acrimonious debate, and the refusal of some corporations to accept that it was sufficiently secure [86], DES became the official US government civilian cipher in mid-1977 [115].[10]

As the DES controversy was unfolding, it was becoming clear that academic researchers were playing an increasing role in both the development of non-governmental cryptology, and in the criticisms of those systems approved by government organisations for use by the public. Research by Whitfield Diffie and Martin E. Hellman at Stanford University on the theory of public key cryptography [38; 39]; and by Ronald L. Rivest, Adi Shamir and Leonard Adleman in their implementation of public key cryptography in the form of the RSA cryptosystem [142] brought not just exposure of techniques hitherto undiscovered, or undisclosed, by government researchers, but also attracted significant numbers of new researchers into the field.

---

to be used as parity bits for register-to-register transfer of key data [92].

[9]Later disclosures suggest that the NSA-influence on the S-boxes was, in fact, designed to make the cipher more resistant to differential cryptanalysis, a technique not at that time available in the published literature. The NSA did not want to make that technique public knowledge, as it could be used to great effect against other contemporary ciphers [30].

[10]DES was the official federal standard until 2001 when it was replaced by the Advanced Encryption Standard (AES) using 128, 192 or 356-bit key lengths [116]. By this time, serious inroads had been made into the credibility of the DES as a secure standard [44]. Federal Information Processing Standards Publications (FIPS PUBS) are issued by the National Institute of Standards and Technology (NIST) after approval by the Secretary of Commerce under s.5131 of the Information Technology Management Reform Act of 1996 (Public Law 104-106) and the Computer Security Act of 1987 (Public Law 100-235).

This resulted not just in the development of an increasingly highly-skilled network of cryptology researchers based outside traditionally secretive government agencies; but also raised international understandings about the extent to which supposedly secure governmental cipher systems were vulnerable to sophisticated cryptoanalytical attack. As Kahn notes, the explosion in cryptographic research did not just provide new tools for foreign governments to be able protect their information from would-be codebreakers, but it also provided significant impetus for them to adopt those tools. Additionally, unrestricted publication of research materials could potentially provide clues to foreign intelligence agencies as to how to attack US ciphers [39: 154]. Diffie and Hellman were also vocal critics of the DES cipher system, notably with regard to the relatively small key size [70; 86: 438–439]. The dilemma for the US government and its security agencies was how to respond to these issues. In a rapidly growing academic (and commercially sensitive) field, in which researchers like Diffie and Hellman were respected figures, it was increasingly difficult to simply discredit or marginalise academic input into the cryptology debate. Other controls had to be found.

Various methods were posited and some were actually put into practice. One way to undercut academic research was to make certain avenues of research unprofitable or difficult to exploit, for example, by the use of patent secrecy orders. Under the US Secrecy Order Statute:

> Whenever publication or disclosure by [the publication of an application or] by the grant of a patent on an invention in which the Government has a property interest might, in the opinion of the head of the interested Government agency, be detrimental to the national security, the Commissioner [of Patents] upon being so notified shall order that the invention be kept secret and shall withhold [the publication of the application or] the grant of a patent therefor under the conditions set forth hereinafter.[11]

A secrecy order thus causes a patent to be temporarily withheld and further requires that the invention be kept secret. Whilst an order only lasts

---

[11]The Secrecy Order Statute – Chapter 17 of the Patent Act of 1952, 66 Stat. 805 codified at 35 USC §§181–188 (1976). Material in brackets added by amendment in 1999.

for year, it can be renewed indefinitely, subject to appeal to the US Secretary of Commerce. Such orders were used on at least 2 occasions in the late 1970s in an attempt to keep cryptographic inventions out of the public eye. On the first occasion the patent application came about as a result of academic research at the University of Wisconsin, and the secrecy order was rescinded for the reason that (depending on the source one relies upon) it should not have been applied to the outcome of academic research, or because the material had already appeared in the open literature and could not be classified. On the second occasion, a secrecy order was imposed on a patent application for a "phaserphone" voice scrambler created by private inventors for use with citizen's band radios. Here again, the order was quickly lifted. It is unclear from contemporaneous commentary why this occurred, although the inventors' obvious willingness to pursue their rights through the courts, and the perceived shaky constitutionality of the statute, when considered against the freedom of speech requirements of the First Amendment [42; 57; 79; 138; 155], may have played a role [47; 61; 77; 150; 151].

Additional to its efforts to reduce the commercial motivations for cryptology research, it is suggested that the NSA also attempted to cut public funding of research into cryptology. Initially this took the form of attempting to prevent the US National Science Foundation (NSF) from funding any such research at all, on the grounds that the area of research lay solely within the funding scope of the NSA; an approach roundly rejected by the NSF. However, the NSA did come to an arrangement with the NSF, whereby all cryptology research grant applications to the NSF were 'vetted' by the NSA to determine if they posed any threat to national security – which in effect placed the NSA in a prime position to block public cryptology research in areas of the NSA's choice [12: 426–457; 60; 133: 203]. The extent of the effect of that de facto veto power over publicly funded cryptology research remains unclear, although at least one researcher was clearly affected by the vetting, inasmuch as he was refused at least some of his funding by the NSF on the recommendation of the NSA. It

appears that the researcher was then approached by the NSA, which indicated that it would be willing to fund the research, an approach rejected by the researcher because of concerns about the research being classified [87].

It became clear that the direct legal regulation of the outputs of public cryptology research was going to be both fraught with difficulties arising from both the untested constitutional status of the relevant laws, and the poor publicity that swiftly resulted from the attempted use of such measures. As a result, the NSA switched its attention to attempting to co-opt researchers into a less formal regulatory process. It suggested the development of a uniform system of pre-publication review of researchers' work, which could be voluntary, but alternatively, the NSA intimated, could be made mandatory via the application of authorisations in the US International Traffic in Arms Regulations (ITAR).[12] The ITAR were (and are) used to control the import and the export of defence articles and defence services, including "speech scramblers, privacy devices, cryptographic devices", and ancillary equipment. The problem for the NSA of adopting a mandatory pre-publication review process under the ITAR was that the US Department of Justice was not convinced that this approach would be constitutional as it "... establish[ed] a prior restraint on disclosure of cryptographic ideas and information developed by scientists and mathematicians in the private sector".[13]

However, the NSA had not entirely abandoned the path of direct legal regulation of public cryptology research. It sought to address the possible weaknesses in any approach based on the ITAR, by putting forward a legislative proposal to replace the ITAR, as applied to cryptology research, that:

---

[12] Also referred to as the Defense Trade Regulations. The Arms Export Control Act (AECA) 22 U.S.C. 2778 provided statutory authority to control the export of defense articles and services, and gave this authority to the President. The President, via Executive Order 11958, as amended, delegated this statutory authority to the Secretary of State, including the promulgation of implementing regulations, the International Traffic in Arms Regulations (ITAR).

[13] Memorandum from J. Harmon, Department of Justice, to F. Press, Science Advisor to the President (May 11, 1978).

... recommended a system of restrictions based on either pre-publication review, presumably by the NSA, or post-publication criminal sanctions ... both proposals provided for the restraint of publications "likely to have a discernable adverse impact upon the national security". Although the proposals alluded to some form of "judicial review", publication of an article on public cryptography without first obtaining a license from the NSA would be a crime under both proposals [133: 205, 206].

At the request of the NSA, the American Council on Education set up a Public Cryptography Study Group (PCSG) to consider and report on how the development of public cryptology research could be squared with the need to ensure national security [120; 136]. The PCSG were unsupportive of either of the potential mandatory approaches under ITAR and the NSA's legislative proposal [136: 522, 523]. In its final report, the majority of the Group recommended that there should be a two-year experiment in voluntary pre-publication review, by the NSA, of all cryptology research undertaken in the private sector. Should the NSA decide that a paper should not be published, or should be published only after redaction of selected material, it was required to consult with an advisory panel comprised of individuals with top security clearance, but the NSA did not have to follow its decisions [91]. A dissenting opinion published in the PCSG Report argued that even such a voluntary system was an unwarranted interference by government into academic freedom [34].

Landau notes that a report from a subcommittee of the NSF Mathematics and Computer Sciences Advisory Subcommittee, which had been tasked with assessing the NSF's role in supporting cryptology research, published shortly after the PCSG report, emphasised the importance of cryptology to business and private citizens, and disagreed vehemently with the approach taken in the PCSG guidelines. It stated that "the proposed system of prepublication review is unnecessary, unprecedented, and likely to cause damage to the ability and willingness of American research scientists to stay at the forefront of research in public sector uses of cryptology" [91; 167]. In the event, later opinions about the outcome of the PCSG experiment in voluntary

pre-publication review are mixed, with some apparently viewing the guidelines as either a success [33: 267, 268], or at least as not unduly restrictive [93], whilst others saw them as an outright failure [133: 233].

The application of "fear, uncertainty and doubt" tactics, by means of the threatened use of legal measures against public discussion of cryptology, was also a factor at this time, although it is unclear to what extent, if at all, the most widely reported incident was officially authorised by the NSA. Here, a proposed Institute of Electrical and Electronics Engineers (IEEE) conference on cryptology at Cornell University was the subject of a letter to the IEEE, from an NSA employee, which stated that were the conference to go ahead, it might risk breaching the ITAR.[14] This was because encryption technologies required an export license not just for equipment, but also for "technical data" relating that equipment. Both technical data and export were defined very widely by the ITAR and it was just possible that the conference papers could have fallen within their scope, if technical presentations and papers were being made available to foreign researchers. The threat was thus plausible, although the then Deputy Director of the Office for Munitions Control (which is responsible for compliance with the ITAR) at the State Department was quoted as saying that this interpretation sounded unlikely [61: 328; 77: 155–156; 133: 204; 152].

In summary, the NSA, which had dominated cryptological research in the US for decades prior, was, during the late 1970s and early 1980s, clearly fighting a losing battle to keep that area of research out of the public sphere. It was faced with opposition not just from academic researchers, but also, crucially, from commercial interests, such as the banking and telecommunication sectors, and from other branches of government. There seems to have been a growing feeling that the NSA's concerns about the risks to national security were overblown, compared to the risks of not adopting stronger encryption technologies in areas of civilian life, as well as a perception that its policy interventions were sometimes driven primarily by a desire to defend a long-term sphere of influence.

---

[14] 22 CFR §§121–128 (1979).

It is arguable that the NSA might have fought successfully against First Amendment-based legal challenges to its attempts to prevent dissemination of academic cryptology publications, on the grounds of protecting the national security. However, the odds of this approach succeeding appeared to be signally diminished by the increasing insistence, from both inside and outside academia, that strong encryption was essential for the secure development of US communications and computing. By the mid-1980s therefore, the initial regulatory mix of 'command and control' legal restriction; voluntary co-regulation by prepublication control; regulation via government research grants; and influence in the federal standards process, all aimed at restricting or directing the scope of public research, or bringing it back within the NSA fold, was on the wane. Instead, there was a growing acceptance that public cryptological research was not just inevitable, but would also be viewed as broadly socially desirable. The new questions were thus developing into how to ensure that civilian use of cryptology within the US remained under the purview of the NSA and other national security/law enforcement agencies, and how to keep the dissemination of the public cryptographic research outputs within the borders of the US.

28.3 NOW THE GENIE'S OUT – GOVERNMENTS' FIRST WISH: LIMITED CIVILIAN CRYPTOGRAPHY AND EXPORT CONTROLS

Although the debate surrounding cryptology research appears to have quietened through the early to mid-1980s, it appears that the NSA was already planning to regain a key role in the civilian use of encryption. Its new goal was to expand its operations to the supply of new code algorithms to the private sector. This included the replacement for the DES, which the NSA itself was now suggesting was insufficiently secure for its proposed recertification by the NBS, due in 1988. The NSA's stated problems with the DES were that it was too widely used in foreign markets, and by international businesses, to be considered immune from exploitation by foreign government agencies; and

that the increasing power of computer technology was driving down the cost of attacking cryptosystems.

This development was viewed with concern by civil liberties groups and security experts. This was, in part, because the NSA did not propose to disclose how the new codes would operate. Additionally, however, it seemed to presage a continuation of the near-total domination of the civilian encryption market (DES had roughly a 95–98% share of the market in 1985) by a cryptosystem or cryptosystems approved by the NSA [40: 64; 53; 88]. However, the NSA's plans for its Commercial Comsec Endorsement Program (CCEP) to replace the DES appear to have foundered on the unwillingness of industry, notably the banking sector, to give up on the DES. This reluctance was unsurprising, the banks having invested heavily both in equipment, and in obtaining international recognition of the standard, neither of which could cope with the introduction of a new and secretive US code system. As a result, in both 1988 and 1992, and despite NSA qualms, the NBS (by now renamed the National Institute of Standards and Technology (NIST))[15] recertified the DES [10: 13].

In the interim, a further turf war had erupted, in part between the NSA and the NBS, but in large measure between the US Executive branch and Congress. Congress, through legislation known as the Brooks Act,[16] had delegated the power to set computer security standards for unclassified information to the NBS. However, the Brooks Act gave the power to the President to set computer security standards when this was "essential to national defense or national security".[17] The President had largely delegated that power to the NSA. In late 1984 President Reagan issued a National Security Decision Directive 145 (NSDD-145) that allocated control over civilian cryptography designed to protect information that was unclassified, but which had national security implications, to the NSA.

---

[15]The National Bureau of Standards (NBS) became the National Institute of Standards and Technology (NIST) in 1988, to reflect more accurately the involvement of technology in its work.

[16]40 U.S.C. §759(f)(2) (1982).

[17]40 U.S.C. §759(b)(2) (1982).

This would potentially have given the NSA exactly what it had been angling for since the 1970s. It would gain a high degree of control not just over civilian government use of cryptography, but also over elements of its use in the commercial world, not least because the definition of what constituted a 'national security' interest was so vague. But while the NSA was pleased with this new development, Congress and other commentators saw NSDD-145 as an unconstitutional intervention by the Executive:

> It is unconstitutional for the President to issue a directive that supersedes power and authority already delegated by Congress to the National Bureau of Standards. ... If the President believes a change is necessary in the federal computer security structure, action must be initiated in Congress to make the change legally effective [53: 1027].

The net result of NSDD-145 was to stir Congress into action to ensure the re-establishment of the 'proper' constitutional balance between Executive and Congress and to roll back the encroachment of the NSA's involvement into civilian government and the private sector cryptology. In 1987 the passage of the Computer Security Act of 1987 defined the role of NBS in developing civilian computer security standards for the protection of sensitive, but unclassified, information and, in principle, limited NSA to its traditional responsibilities for the protection of classified information [29: 195–198; 40: 68, 69].

This was, however, far from the end of the story, for while the NBS/NIST was given the relevant powers to set standards for civilian computer security, it was required to consult with the NSA on those standards, including those for cryptography. Equally importantly, it was significantly underfunded to actually carry out the task it had now been given. These circumstances meant that the NSA, whose own computer security program was considerably larger and better funded, was in the driving seat when it came to discussions about how the consultation process would work. The resulting Memorandum of Understanding between the two agencies, amongst other things, placed the NSA back in a position to jointly scrutinise with NIST any proposed civilian computer security standards

before they were publicly released, and in excellent position to influence any appeals against approval of proposed standards [10: 14; 40: 71; 59: 410–414].

By this point, it was becoming increasingly obvious that not only was cryptography a useful tool for computer security, but also that it was likely to have practical application in the realm of authentication, in areas such as identification and authentication 'signatures', permission credentials transactions and unforgeable integrity checksums. As such, it was a 'dual use' technology which could bring significant benefits to both military and commercial users [169]. In consequence, the technology would be immensely marketable, and the US appeared to still have a lead in the field. Critics were now arguing that US regulation of cryptographic tools, and particularly the rules on the export of such tools to third countries were likely to harm US commercial interests in a marketplace where international competitors were not as rigorously constrained. Indeed, some went so far as to suggest that export controls, far from aiding national security were, by damaging the nation's economic strength, likely to have the reverse effect [52; 28].[18]

The problem for would-be exporters of cryptography products was the design of the licensing process. US export controls were designed so that dual use licenses for products with both potential military and civilian uses were administered by the Commerce Department, under the Export Administration Act. However, products designated as arms or munitions were licensed by the State Department, under the Arms Export Control Act and the ITAR. The State Department's licensing process tended to demand a licence for every export, and it was able to deny licences without explanation, or vendor recourse to judicial review of the decision. Cryptographic tools had always been viewed as of military application, so fell within the State department remit, which meant, in practice,

---

[18]Although it should be noted that, perhaps contrarily, the authors of the Committee on Science, Engineering, and Public Policy Report appeared to regard the PCSG prepublication review guidelines for cryptology research as an acceptable model for regulation, and one that could be extended to other areas of research.

within the remit of the NSA. The rapid changes in communications technologies and particularly the increasing take up of internet technologies had changed the playing field, such that cryptography technologies could now be legitimately seen as 'dual-use' products, but these changes had failed to make any impact on the allocation of roles within the licensing regime.[19]

Matters finally began to change in 1990 when, following the demise of communism in the former Soviet Union and Eastern Europe, the Western nations began to relax their system of export controls designed to limit the access by the Soviet Union and its allies to Western technology with potential military uses. Prior to 1990, the Coordinating Committee for Multilateral Export Controls (CoCom), an informal non-treaty organisation,[20] which had been set up to co-ordinate these controls, collated three lists of strategic commodities that CoCom members agreed should be the subject of export controls. Two of these dealt with munitions and atomic energy technologies, the third, known as the Industrial List identified products, technologies and technical data with dual military-civilian use potential. Items on the Industrial List were incorporated into the export control program of each of the member countries. While the former two lists proved relatively unproblematic, the Industrial List was always controversial, as the CoCom members tended to have different interpretations of the term 'dual-use', as well as different ideas as to what was appropriate to include on it. Each member nation was responsible for developing a domestic list for enforcement in general compliance with the CoCom lists. By 1991, the CoCom members had rewritten the Industrial List to include only products having "truly significant military applications" [49; 105].

After the revision of the CoCom list, many CoCom members relaxed their controls over cryptographic tools, but under the US export control

regime they were still largely identified as munitions, and were thus remained primarily regulated by the US Department of State rather than the Department of Commerce. In a gesture towards US commercial interests in 1992, a new process for mass market software containing encryption technology was developed, permitting some mass market software to be licensed under the less onerous Department of Commerce regime. An exporter could apply, by means of a commodity jurisdiction request, for a determination as to whether his software should be licensed by the Department of State or Department of Commerce. If software was submitted to this process, it was assigned to several government agencies, including the NSA, for assessment. It was widely believed in the software industry if the NSA could break the encryption relatively quickly, it would approve a license for international sale of the software. In that case the Department of Commerce would be granted jurisdiction, and a licence for export would usually be forthcoming. If the software did not pass the assessment, the exporter would have to apply to the Department of State for a munitions licence, which were rarely granted [49: 478, 479; 68].

The primary difficulty with the US export controls was the fact that, by the early 1990s, as far as cryptographic tools were concerned, they were increasingly being used to attempt to control products that, by their very nature, were simply no longer amenable to traditional export oversight and control means. As cryptographic tools were increasingly software rather than software/hardware based, they were easily transported cross-border by network, or simply on computer systems such as laptops; they were capable of easy and rapid duplication; and identifying the source of infringing software in a commercial market containing grey marketers and software pirates, as well as the primary exporters, was becoming impossible. Additionally, while the US was still imposing a near-absolute bar on the export of products containing the DES and RSA codes, the DES algorithm itself and products based upon it were already in circulation on the Internet, and it was estimated that by this time DES-based encryption products were be-

---

[19]For a detailed overview of US export controls, see [10: Ch. 3].

[20]The CoCom members were Australia, Belgium, Canada, Denmark, France, the Federal Republic of Germany, Greece, Italy, Japan, Luxembourg, The Netherlands, Norway, Portugal, Spain, Turkey, the United Kingdom and the United States.

ing manufactured in a wide range of foreign countries [73: 113].[21]

As the US's CoCom allies did not appear inclined to regulate mass market software with encryption capabilities as munitions; and such controls as there were could clearly be relatively easily bypassed – it was reported that, for example, prior to its invasion of Kuwait, Iraq had purchased cryptographic equipment from the UK [68: 446] – the logic of maintaining a strict export ban on strong cryptography appeared to be being rapidly undermined. It was certainly leading to infuriating inconsistencies for US businesses. Hartzler cites the case of a US company that obtained the cryptographic algorithm for its software program from a book published in the UK, but found it was unable to export the resulting software outside the US [68: 456]. The NSA had already had to give way to the demand for public cryptology research due to lobbying from the commercial sector, and was, in principle, taking a back seat in the creation of civilian computer security standards. Now it was facing pressure to drop its opposition to the export of tools permitting strong encryption, on the grounds that export controls no longer in any meaningful way helped to ensure US national security, and signally damaged the ability of US encryption industry to compete internationally.

However, despite that pressure, and some initial signs that the US might begin to relax its regime, the arrival of the Clinton Administration in 1992 did not bring the expected changes to the export controls relating to cryptographic tools. As will be discussed below, the Administration had its own perceptions about how to most effectively control the risks posed to national security, and to law enforcement activities – through the medium of key escrow. Yet, it appears that this approach was always envisaged as running in tandem with the continuation of export controls, rather than as a replacement for them. While its unilateral efforts to control the use of strong encryption were proving largely ineffective, US policy thinking was now beginning to coalesce on the concept of encouraging development of multilateral controls on cryptography, with the US in the lead. This aim was pursued through two key mechanisms, the revamping of the international export control framework, and the raising of cryptography policy as a key international economic and social policy issue.

The first of those aims was facilitated by the fact that CoCom had been effectively disbanded in March 1994, as the nature of the informal arrangement was no longer seen as a suitable mechanism for achieving an effective multilateral framework for export controls. The interests of the European members in developing economic and trade links with the former Warsaw Pact nations, combined with a decreasing dependence upon the US, rapid developments in technology, and the effects of globalisation, clearly required a different approach. However, it was recognised that some new arrangement would have to replace CoCom as, despite the end of the Cold War, there were new, albeit less polarised, national security threats to be faced by its members. In 1995, negotiations began on what was to become the Wassenaar Arrangement on Export Controls for Conventional Arms and Technologies.

The Wassenaar Arrangement (hereafter Wassenaar)[22] aims to control transfers of conventional

---

[21] According to the Office of Technology Assessment in its report *Information Security and Privacy in Network Environments* (September 1994), the US-based Software Publishers Association (SPA), now known as the Software and Information Industry Association (SIIA), reported having identified 423 US-origin products containing encryption implemented in hardware, software and hardware/software combinations. 245 of these products used the DES and, therefore, were subject to ITAR controls and could not be exported from the US except in very limited circumstances. In total, the SPA identified 763 cryptographic products, developed or distributed by a total of 366 companies (211 foreign, 155 domestic) in at least 33 countries. Evans [49: 489] cites an SPA Report of 1993, which noted that there were 215 foreign hardware, software and combination encryption products produced across 20 countries of which 84 used DES – the statistics may be inconsistent, but the implication is clear. By 1999, Hoffman et al. had identified 805 hardware and/or software products incorporating cryptography manufactured outside the US in 35 countries by 512 companies, including 167 foreign cryptographic products that incorporated strong encryption using algorithms such as Triple DES, IDEA, BLOWFISH, RC5, or CAST-128 [74: 5–7].

[22] The Wassenaar Arrangement, http://www.wassenaar.org/ Parties to the Wassenaar Arrangement included Argentina, Australia, Austria, Belgium, Bulgaria, Canada, the Czech Republic, Denmark, Finland, France, Germany, Greece, Hungary, Ireland, Italy, Japan, Luxembourg, The Netherlands,

armaments and sensitive dual-use goods and technologies that might threaten regional and international security and stability. Like CoCom before it, Wassenaar is not based on any legally binding agreement or treaty and the regime is thus reliant on the voluntary political commitments of its members for its force. The key aim of Wassenaar is to promote transparency in the transfer of arms and sensitive dual-use items, and to act as an information exchange to aid members in identifying undesirable acquisition patterns and clandestine projects, thus achieving "common and consistent export policies while eliminating inadvertent undercuts by other participants" [10: 71–76; 41; 124; 154].

The initial framework of Wassenaar established an export control regime based on two sections or "pillars" covering conventional arms and arms related materials, and dual-use goods. The dual-use pillar is divided into a Tier 1 basic list and a Tier 2 sensitive list that includes a sub-tier of very sensitive items. Cryptography fell within the category of "information security" and was placed within the Tier 2 sensitive list. While the Wassenaar states agreed to be vigilant over exports of items on the list, and to aim to prevent unauthorised transfers or re-transfers, actual policy formation and the scope of authorisation was left up to each state to decide. Wassenaar was also explicit about what fell outside the realm of cryptography subject to its control, including personalised smart cards, digital signature products, banking and financial equipment, and access control equipment. Also excluded was encryption software generally available for retail sale, and software that was in the public domain. Additionally, Wassenaar did not attempt to place a limit on the strength of exported encryption products. From the perspective of US encryption technology developers who had to operate under US export controls that complied with, but went further than required by, the basic Wassenaar requirements; and who faced heavy competition from foreign manufacturers based in states with more liberal approaches to

New Zealand, Norway, Poland, Portugal, the Republic of Korea, Romania, the Russian Federation, Slovakia, Spain, Sweden, Switzerland, Turkey, Ukraine, the United Kingdom and the United States.

encryption, this was not the levelling of the playing field that they might have expected.

Some measures designed to toughen the Arrangement's approach to cryptography were agreed in December 1998, when its members agreed to revise the Dual-Use Control List and implement a maximum bit length of 64-bits on exports of mass-market encryption software. However US-inspired attempts to promote key escrow products were rejected by the other countries, and the new controls did not apply to the "intangible" distribution of cryptography, including downloads from the Internet. Thus, while the US government claimed the changes as a levelling of the competitive barriers between its encryption industry and those of other nations, in practice, the non-binding nature of Wassenaar meant that little real competitive gain accrued to US producers. Indeed, various member governments, such as those of Finland and Denmark, almost immediately began to distance themselves from the controls agreed, due to political pressure at home [107: 439–441].

Any hope of sustaining any meaningful controls via Wassenaar were left in tatters when in January 1999, France, a country which had in theory had stricter cryptographic controls than the United States, declared that it would abandon all controls on cryptography up to 128-bits in strength [46: 27]. Like the US, France had defined encryption hardware and software as munitions. The Law of 18 April 1939 defined eight categories of arms and munitions; Law 73-364 of 12 March 1973 placed cryptography equipment in the second category; Law 86-250 of 18 February 1986 extended the definition of cryptography equipment to include software and specified that all requests be sent to the minister of the PTT with a complete description of the "cryptologic process" and two samples of the equipment; and the Law no. 90-1170 of 29 December 1990 stated that export or use of encryption (defined as "any service aimed at transforming, by means of secret usages, clear information signals that are unintelligible to third parties, or to achieve the opposite, with the aid of devices, materials or software designed for this purpose") had to be approved by the French government. Use of encryption for authentication or integrity purposes had to be declared to the authorities, and its use for private communications had to be authorised. Authorisations were rarely granted, and users were thus effectively forced to use pre-approved products, which the French authorities were in a position to decrypt [112; 147; 153].

While the body responsible for administration of the declarations and authorizations, the Service Central de la Sécurité des Systèmes Informatiques (SCSSI) was said to be keen on strict enforcement of the law, Anderson suggests that few people in France were aware of the laws, and that they were widely ignored [7: 76]. It is possible that the authorities were less concerned about the use of encryption by private individuals than they were by corporations, particularly foreign corporations (see footnote 50). Certainly, the United States Department of Commerce and the National Security Agency in a Report in 1996 were of the opinion that "France has the most comprehensive cryptologic control and use regime in Europe, and possibly worldwide" [37]. Equally, the changes to the Wassenaar Arrangement in December 1998, initially had little impact on French policy because its internal legislation was already much stricter than what had been agreed.

The first shift in French policy had come with the Law of 26 July 1996 which was primarily concerned with telecommunications, but contained a section on encryption. Via a series of implementing decrees,[23] the French government, having recognised that the declaration and authorisation processes for encryption technologies were becoming unworkable, moved to speed up them up. The main changes were:

- No declaration was required for encryption technologies used for authentication/integrity.
- Encryption technologies used to secure privacy/confidentiality did not require authorisation if the user escrowed the keys with a trusted third party (TTP), or the supplier of the technology pre-declared it with the SCSSI.
- Any use, supply, import, and export of encryption using a key of less than 40 bits did not require.

[23]Implementing decrees no. 98-101 and 98-102 of 24 February 1998, the Arrêtés of 13 March 1998 and decrees no 98-206 and 98-207 of 23 March 1998.

- Imports of encryption technologies from EU/EEA countries were not subject to restriction, although use/supply after importation would come under the appropriate scheme.
- Imports of encryption technologies from non-EU/EEA countries required a declaration if for authentication/integrity purposes, or prior authorisation if for privacy/confidentiality purposes.
- Exports of encryption technologies from France required a declaration if for authentication/integrity purposes, or prior authorisation if for privacy/confidentiality purposes [112; 147].

Thus by 1998, France had become the only country to actually begin operating a 'key-escrow' scheme along the lines of the model proposed by the US. While in principle, any person or company could become a TTP, all TTPs had to be approved by the SCSSI. All TTPs were obliged by law to cooperate with public authorities in the hand over of escrowed keys in certain circumstances, including:

- Facilitation of administrative interceptions authorized by the Prime Minister, or judicial interceptions in the scope of a judicial investigation.
- Facilitation of investigations by the judicial police, either for preliminary investigations, or in cases of immediate necessity [112].

In January 1999, however, the French Prime Minister Lionel Jospin abruptly announced that significant changes were to be made to France's encryption policy with the intent of achieving much greater liberalisation. This commenced in March 1999 with Decrees no. 99/199 and 99/200 of 17 March 1999 and an Arrêté dated 17 March 1999. Almost at a stroke, these abolished the complex licensing scheme for cryptographic imports and domestic use, and removed the key escrow requirements for the domestic use of encryption and the need for the system of SCSSI-approved TTPs. The new scheme can be summarised as follows:

- Exempted from regulation – the use and importation of encryption technologies for privacy/confidentiality purposes with an algorithm the key of which is less than or equal to 40 bits, as well as those the key of which is less than or equal to 128 bits, where there has been a prior declaration by their producer, supplier or an importer; or they are for private use.

- Subject to declaration – the supply of encryption technologies for privacy/confidentiality purposes with an algorithm, the key of 40 bits or less; and supply of encryption technologies for privacy/confidentiality purposes the key of which is greater than or equal to 40 bits but less than or equal to 128 bits.
- Subject to authorisation – encryption technologies using an algorithm the key of which is 128 bits, or more [113; 147].

It seems clear that the primary motivation for this move was commercial, with Jospin stating in his initial announcement that the existing law "strongly holds back the usage of cryptography in France, and does not have any impact on allowing the authorities to effectively fight the criminal use of encryption" [107: 442]. There was also significant pressure from the European Commission which saw the French policy as a significant barrier to the promotion of electronic commerce and an integrated European economy.

It is interesting to note suggestions that the development of French encryption policy had been heavily influenced by boundary disputes between the various arms of the French Secret Services, including the DGSE; the Direction de la Surveillance du Territoire (DST); the Secretariat General de Defense National (SGDN); and the Direction Centrale des Renseignements Généraux (RG) echoing the position in the US with the NSA seeking to protect its role. It seems that this was increasingly perceived in French government circles as unhelpful in the new technological environment [65; 147]. In privacy terms, the French public were also unlikely to be sympathetic to arguments based on the need for law enforcement access to communications and other encrypted materials, given the well-documented history of French intelligence agencies illegally intercepting communications.[24]

The approach taken by the French towards access to encrypted material by law enforcement after liberalisation has been geared towards implementing criminal sanctions against suspects who refuse to decrypt data pursuant to a lawful court order.[25]

At roughly the same time as beginning the Wassenaar talks, the US was pursuing the other aspect of its multilateral cryptography control programme through the Organisation for Economic Co-operation and Development (OECD).[26] Here it sought the development of cryptography guidelines by the OECD for recommendation to its member states as considerations as they developed their own encryption policies. Baker and Hurst suggest that this approach was meant to serve two purposes. The first was to raise awareness amongst the OECD members about the possible implications of unregulated encryption, particularly for national security and law enforcement. The second was to send a message to its own encryption industry that once other states started to adopt more restrictive approaches to encryption, the US export controls that they were currently railing against would no longer be such an international competitive disadvantage [10: 41, 42]. If this was the case, then on

---

[24] See *Kruslin v. France*; *Huvig v. France, supra* at n.2.

[25] Law no. 2001-1062 of 15 November 2001 on Daily Security gave the Attorney General the power to require any qualified persons to decrypt or disclose decryption keys if encrypted data are encountered during an investigation. Failure to comply with the request was punishable with a two-year imprisonment sentence and a EUR 30,000 fine. If encryption was used to facilitate the preparation or commission of a crime, the punishment could be increased to three years and EUR 45,000. If decryption could have prevented or mitigated the effects of a crime, the failure to communicate the keys or to decode the data was punished by a 5-year jail sentence and a EUR 75,000 fine. It appears that Law no. 2004-575 of 21 June 2004 for the trust in the digital economy has further increased the maximum punishments for crimes where cryptography was used to prepare or commit the crime or to facilitate the preparation or commission of the crime. The increases do not apply where the perpetrator voluntarily submits the plaintext and private key of encrypted messages. Where it is necessary for the investigation of a crime with a maximum penalty of at least two years' imprisonment, Law no. 2001-1062 permits the police to ask the national security agencies to decrypt relevant encrypted data. To facilitate this process, a Technical Assistance Center was created within the French Ministry of the Interior by Decree 2002-1073 of 7 August 2002.

[26] Organisation for Economic Co-operation and Development (OECD), http://www.oecd.org/home/ At the time of the development of the Guidelines for Cryptography Policy the OECD comprised Austria, Australia, Belgium, Canada, the Czech Republic, Denmark, Finland, France, Germany, Greece, Hungary, Iceland, Ireland, Italy, Japan, Korea, Luxembourg, Mexico, New Zealand, The Netherlands, Norway, Poland, Portugal, Spain, Sweden, Switzerland, Turkey, the United Kingdom and the United States.

some levels, the resulting OECD Guidelines for Cryptography Policy[27] must be seen as a partial success. Certainly, the consciousness raising appears to have inspired the development of cryptographic policies in many of the OECD member states, and resulted in the debate about the use of encryption centring, for a time, primarily upon the kinds of issues with which the US was concerned. However, the resulting guidelines cannot be said to be a resounding endorsement of the US approach to encryption. In summary the principles are:

Principle 1: Cryptographic methods should be trustworthy in order to generate confidence in the use of information and communications systems.

Principle 2: Users should have a right to choose any cryptographic method, subject to applicable law. In order to protect an identified public interest, governments may implement policies requiring cryptographic methods to achieve a sufficient level of protection. Government controls on cryptographic methods should be no more than are essential to the discharge of government responsibilities. Governments should refrain from initiating legislation which limits user choice.

Principle 3: Cryptographic methods should be market driven and be developed in response to the needs, demands and responsibilities of individuals, businesses and governments.

Principle 4: Technical standards, criteria and protocols for cryptographic methods should be developed and promulgated at the national and international level. National standards for cryptographic methods, if any, should be consistent with international standards to facilitate global interoperability, portability and mobility.

Principle 5: The fundamental rights of individuals to privacy, including secrecy of communications and protection of personal data, should be respected in national cryptography policies and in the implementation and use of cryptographic methods.

Principle 6: National cryptography policies may allow lawful access to plaintext, or cryptographic keys, of encrypted data. These policies must respect the other principles contained in the guidelines to the greatest extent possible.

Principle 7: Whether established by contract or legislation, the liability of individuals and entities that offer cryptographic services or hold or access cryptographic keys should be clearly stated. A keyholder should not be held liable for providing cryptographic keys or plaintext of encrypted data in accordance with lawful access. The party that obtains lawful access should be liable for misuse of cryptographic keys or plaintext that it has obtained.

Principle 8: Governments should co-operate to co-ordinate cryptography policies. National key management systems must, where appropriate, allow for international use of cryptography. Governments should avoid developing cryptography policies and practices which create unjustified obstacles to global electronic commerce and creating unjustified obstacles to international availability of cryptographic methods.

It will be immediately apparent from the foregoing that the US and its supporters in the OECD negotiations did not have matters all their own way.[28] While there certainly were concerns amongst the participating states about the possible impact of public access to cryptography upon national security and law enforcement activities, these were necessarily tempered by other considerations. The EU Member States, which also constituted a significant percentage of the OECD member states, had just finished negotiating the EU Data Protection Directive 1995, and were in the processing of implementing its provisions in their national law, so the privacy implications of cryptography could not have been missed [50]. Equally, the EU Commission had also expressed an interest in involving itself in the use of cryptography to facilitate European and global e-commerce [101: 115, 116], an intervention that would not necessarily have received a unanimous welcome from the EU Member

---

[27]OECD Guidelines for Cryptography Policy (1997), http://www.oecd.org/document/11/0,2340,en_2649_201185_1814731_1_1_1_1,00.html

[28]For further discussion of the background to, and the implications of, the Guidelines, see [9; 10: 48–70].

States, some of whom were concerned that such an intervention on the grounds of facilitating commerce would directly, or indirectly, limit the extent to which their national governments could regulate the public use of cryptography for other purposes. While a number of countries were clearly considering restricting cryptography, others, such as Japan, saw any restrictions as an unnecessary fetter on technological development [67].

In the event, as Baker notes [9: 738], the Guidelines are heavily influenced by themes such as public and private sector cooperation, the need for international interoperability and standards, and internationally coordinated and compatible policies – themes which would resonate with those seeking to develop cryptographic products for the commercial environment – as well as public interest themes such as freedom of choice, the need to meet public requirements, and the protection of privacy. The themes that noticeably do not receive heavy emphasis are those of law enforcement and national security. To this point, therefore, the US cryptographic policy initiative appeared to be having little practical effect in changing, as opposed to raising perception of, cryptography policy at the international level [143: 187–193].

To add to the US government's difficulties in maintaining control over the spread of strong encryption tools, its export controls were increasingly under attack from within the US itself. A series of challenges to the regime throughout the 1990s meant that both its rationale, and its legal viability, were thrown into doubt.

The first incident concerned a US cryptographer named Philip Zimmermann, who used the RSA and IDEA algorithms to create a public key encryption software program called Pretty Good Privacy (PGP) to secure the privacy of e-mail and other electronic information [94; 159]. He created the software in 1991, and released the first version (PGP v1.0) as freeware to encourage its distribution, in the US in June of that year. One of the people who received the software made it available on an Internet site, from where it was then downloaded worldwide. It seems clear that making a strong encryption program of that type available,

via the Internet, technically constituted an 'exportation' and that Zimmermann might well be considered to have breached the Arms Export Control Act (AECA) and the International Traffic in Arms Regulations (ITAR). That having been said, the authorities seem to have been fairly slow to react to the matter, as it is reported that the first official contact that Zimmermann received was in February 1993, when he was questioned by United States Customs Service agents about the circumstances of PGP's international release without an export licence. Zimmermann himself claimed that he did not export the software, nor did he intend it to be exported, and that he had clearly included instructions with the code stating that it could only be distributed under circumstances where it was protected from export. He was, however, to remain under investigation by the US Customs Office for the next three years, until the matter was abruptly dropped in January 1996.[29] It is likely that the investigation was terminated in part because of a growing concern in government circles that a high-profile criminal case focusing on the issue would not be a desirable outcome. It would almost certainly have involved examination of the proposition that encryption software was protected speech under the First Amendment, and invited consideration of whether current export restrictions were an unconstitutional prior restraint on free speech. At that time, this was not an argument the government could be at all sure of the courts resolving in its favour.

Shortly afterwards, in February 1996, the State Department tacitly accepted the fact that international civilian use of cryptography was becoming uncontrollable by amending the ITAR rules for personal use of cryptography. Temporary export of products for personal use was exempted from the need of a license, provided the exporter took normal precautions to ensure the security of the product, including locking the product in a hotel room

---

[29]A Justice Department Press Release briefly stated that it "declined prosecution of any individuals involved in connection with the posting to USENET in June 1991 of the encryption program known as Pretty Good Privacy". Department of Justice Press Release from United States Attorney, Northern District of California (Jan. 11, 1996), http://www.eff.org/legal/cases/PGP_Zimmermann/usatty_pgp_011196.announce

or safe; and the product was not intended for copying, demonstration, marketing, sale, re-export, or transfer of ownership or control.

In the event, following the Zimmermann incident, those issues were to be raised anyway, in three key cases brought challenging the constitutionality of the export control regime. In the *Karn* case, Philip Karn had submitted a commodity jurisdiction request in February 1994 to the Department of State with regard to Bruce Schneier's book *Applied Cryptography*. The Office of Defense Trade Controls (ODTC) decided that the book, which contains information on cryptographic protocols, algorithms, techniques, and applications, and includes source code for several cryptographic algorithms, did not fall under the ITAR and was not subject to export control. Having succeeded on that front, Karn then submitted a further commodity jurisdiction request in March 1994, with regard to a companion disk to the book containing a verbatim copy of the source code depicted in the book. This, the ODTC decided, was subject to the jurisdiction of the Department of State pursuant to the ITAR as a 'defense article'. The somewhat unusual outcome of these requests was that Karn required an export licence for the electronic copy of the information, but not for the printed form, even though, given time, a purchaser of the book could type, or scan and OCR, the information into a computer and end up with the identical source code.[30]

Upon exhausting his appeals at the State Department, Karn moved to the courts, claiming that the requirement for a licence for the electronic copy was a restraint on free speech, and thus fell foul of the First Amendment to the US Constitution. In March 1996, however, the District Court disagreed with him. It decided that, under the regime in place at the time, the administrative ruling was a non-reviewable exercise by the State Department of the authority delegated to it by the President under the Arms Export Control Act (AECA). The court did not rule on the issue of whether computer code was protected speech under the First

Amendment, on the basis that the regulation was content neutral, i.e. that it was justified by a significant government interest that was unrelated to the suppression of speech and was tailored towards that end.[31] The significant interest was the Department of State opinion that "encryption source code on machine readable media will make it easier for foreign intelligence sources to encode their communications", and that there was a reasonable fit between what was to be achieved by the regulation and the means used to achieve it. The fact that this led to the absurd situation that a paper copy was exportable, but a disk copy of exactly the same material was not, was not an issue that seems to have unduly troubled the court.[32]

In the meantime, a second case had begun working its way through the legal system. In the *Bernstein* case, Daniel Bernstein, a mathematician, was attempting to publish his work on an encryption algorithm he had designed called 'Snuffle'. He wanted to publish the work in three different forms, an academic paper, the source code of two programs, and instructions on how to program a computer to use his technique. In 1992 he submitted a commodity jurisdiction request to the ODTC at the State Department to determine if his work was subject to the ITAR, and was advised that the source code and instructions were considered controlled defense articles. After a protracted period of negotiation Bernstein brought suit in the federal District Court for the Northern District of California, claiming amongst other things that "the licensing scheme under the ITAR imposes an unconstitutional prior restraint on cryptographic speech", and "that a number of terms make the ITAR vague and overbroad in violation of the First Amendment". The State Department initially moved to dismiss Bernstein's claim, on the ground its determination was not justiciable, whilst informing Bernstein that his academic paper and the other two sets of explanations in English were not subject to export control, but that the two pieces of computer source code were.

In deciding the case, the District Court produced three successive opinions. In the first (*Bernstein I*)

---

[30]As Karn himself put it, "It's old news that the US Government believes only Americans (and maybe a few Canadians) can write C code, but now they have apparently decided that foreigners can't type either!". The Applied Cryptography Case, http://www.ka9q.net/export/

[31]*United States v. O'Brien* 391 U.S. 367 (1968).

[32]*Karn v. U.S. Department of State* 925 F. Supp. 1 (D.Ct. D.C, March 22, 1996). See further [20; 32; 134].

the court held that Bernstein was not challenging the actual decision of the State Department on his commodity jurisdiction request, but rather the constitutionality of the ITAR themselves. In addition it held that computer source code was not just conduct, because while it was functional, it was also a language, and thus expressive and as such, was protected by the First Amendment.[33] In the second (*Bernstein II*) the court considered whether the ITAR licensing scheme constituted unconstitutional prior restraint using the three-prong procedural safeguard analysis previously developed by the U.S. Supreme Court.[34] Under this analysis, the ITAR failed to satisfy all three of the prior restraint safeguards: they did not contain a time limit for a licensing decision, there was no process for prompt judicial review, and they did not require the State Department to go forward and defend the decision in court. The licensing scheme was therefore in violation of the First Amendment. The court also held that certain definitions in the ITAR were unacceptably vague because of "the uncertainty created in scientists about what speech is subject to regulation".[35]

Shortly after the initial decision in *Karn* in March 1996, and just before the ruling in *Bernstein II* in November 1996, President Clinton issued Executive Order 13026 which moved the entire export control scheme for civilian encryption software from the jurisdiction from the State Department under ITAR to the Commerce Department under the Export Administration Regulations (EAR).[36] The move was widely viewed as an attempt by the Clinton Administration to head off yet another potentially damaging loss in the appeal case being brought by Karn before the United States Court of Appeals for the District of Columbia. If this was the government's motivation,

then it was at least partially successful. While both Karn and the Commerce Department requested the court to hear the appeal case as scheduled and to decide the Constitutional issues raised, the court decided to remand to the district court to consider it in the light of the new regulations. At this point Karn made a new commodity jurisdiction request to the Bureau of Export Administration (BXA) at the Commerce Department under the new regulations. The result of Karn's discussions with the BXA was that in December 1997, he was again refused a licence to export the electronic copy of the book, and so he returned to the district court in January 1998.

In the *Bernstein* case, meanwhile, the change of jurisdiction meant that Bernstein's source code was now no longer subject to the ITAR, but rather to the EAR amendments. The court thus permitted him to amend his complaint to include the Commerce Department and the EAR amendments. The resulting decision (*Bernstein III*) on August 25, 1997, on the amended complaint reiterated much of the court's previous analysis, and held that the revised licensing system suffered from much the same procedural failings as its predecessor.[37]

The following year, as the government was pondering its appeal in Bernstein, a third case, *Junger v. Daley* was beginning. Peter Junger taught a course on "Computers and the Law" at Case Western Reserve University Law School in the US and he created a website which he used, amongst other things, to publish class materials and articles for his course. He decided to publish on that web site various encryption programs that he had written. He therefore submitted three applications for thirteen items to the Department of Commerce requesting determinations of commodity classifications for the encryption programs. The Department of Commerce responded that four of the five programs submitted were subject to the Export Regulations and would need a license before he could publish them on his web site.

Junger then filed suit to prevent the Commerce Department from enforcing the EAR against him.

[33]*Bernstein v. U.S. Department of State* 922 F. Supp. 1426 (N.D. Cal. 1996) (*Bernstein I*).

[34]*FW/PBS v. City of Dallas* 493 U.S. 215 (1990).

[35]*Bernstein v. U.S. Department of State* 945 F. Supp. 1279 (N.D. Cal. 1996) (*Bernstein II*).

[36]The Export Administration Act of 1979 (EAA) and the implementing Export Administration Regulations (EAR) establish policies and procedures for the regulation of exports and set out which items need to be licensed for export to which destinations.

[37]*Bernstein v. U.S. Department of Commerce* 974 F. Supp. 1288 (N.D. Cal. 1997) (*Bernstein III*). See further [22; 111; 114; 122; 143].

Like Karn and Bernstein before him, he argued that the regulations violated the First Amendment. The District Court was not, however, impressed with this argument, or with the recent ruling in *Bernstein III*. It held, in July 1998, that the Export Regulations were constitutional because encryption source code could not be considered constitutionally protected writing, as it only carried out the function of encryption and did not contain any expressive content – just because software was written in a 'language' did not mean that it was necessarily expressive and therefore constituted speech. Without expressive content, the export of encryption source code was not constitutionally protected, and thus the EAR licensing system did not constitute unconstitutional prior restraint.[38] As in *Karn* with the ITAR, the court held that the EAR were content neutral, not least because the EAR did not appear to control export of publications on cryptography, only export of source code.[39]

While the *Junger* case appeared to temporarily bolster the government position on export controls,[40] the result of its appeal of the decision in *Bernstein III* did not. A three judge panel of the Court of Appeals (9th Circuit), affirmed by 2 to 1 the District Court's finding that the EAR regulations were facially invalid as a prior restraint on speech, although the dissenting judge took the line that encryption source code was not expression, but simply a functional tool.[41] The government then filed a motion for reconsideration on 21 June 1999 as a result of which the Ninth Circuit granted a rehearing on 30 September 1999, and withdrew its opinion prior to an *en banc* hearing in front of 11 judges.[42]

The unsettling legal position, combined with increasing legislative activity in Congress designed to bolster the position of the US software industry and to safeguard US citizens' use of the Internet[43] [101: 109–114], resulted in the Clinton Administration shifting the ground rules again. In September 1999 the Administration announced a proposal to loosen restrictions on the export of encryption technologies, and in November 1999, the BXA released new draft encryption export regulations. Following consultations, the regulations were adopted in January 2000. Their content meant that, amongst other things:

- encryption products of any key length could be exported to non-government end users after a one time technical review of the product, and to government users under licence;
- 'retail' encryption products could be exported to any end user including foreign governments;
- US firms could export encryption items of any key length to their foreign subsidiaries without a technical review; and
- export controls for source code were relaxed, in that encryption source code freely available to the public could now be exported under a license exemption without a technical review, if the exporter submitted a copy of the source code, or a written notification of its Internet location, to the BXA by the time of export [15].

While not exactly what either civil liberties groups, or US industry, had been looking for – the civil liberties groups felt that the First Amendment issues had still not been fully addressed, as the system could still be seen to require review

---

[38]See further [109] arguing that source code is not speech under the First Amendment.

[39]*Junger v. Daley*, 8 F. Supp. 2d 708, 713 (N.D. Ohio 1998).

[40]See, for example, the discussion in [123].

[41]*Bernstein v. U.S. Department of Justice*, 176 F. 3d 1132 (9th Cir. 1999) (*Bernstein IV*). See [104].

[42]*Bernstein v. U.S. Department of Justice, rehearing en banc granted and opinion withdrawn*, 192 F. 3d 1308 (9th Cir. 1999).

[43]See, for example, Encrypted Communications Privacy Act (S. 1587), Security And Freedom through Encryption (SAFE) Act (H.R. 3011) (both introduced March 5, 1996); Promotion of Commerce Online in the Digital Era (Pro-CODE) Act (S. 1726) (May 2, 1996); Security And Freedom through Encryption (SAFE) Act (H.R. 695) (February 2, 1997); Encrypted Communications Privacy Act (S. 376) Promotion of Commerce Online in the Digital Era (Pro-CODE) Act (S. 377) (both February 27 1997); Secure Public Networks Act (S. 909) (June 16, 1997); Computer Security Enhancement Act (HR 1903) (June 17, 1997); Encryption Protects the Rights of Individuals from Violation and Abuse in CYberspace (E-Privacy) Act (S. 2067) (May 12, 1998); Security And Freedom through Encryption Act (SAFE) (H.R. 850) (February 25, 1999); Promote Reliable On-Line Transactions to Encourage Commerce and Trade (PROTECT) Act (S. 798) (April 14, 1999); Encryption for National Interests Act (H.R. 2616) (July 27, 1999). See for further discussion [10: 18–22; 26; 128].

before researchers could communicate their work to foreign researchers [19; 62]; and US industry had been looking for a considerably simpler set of regulations – the new regulations appear to have addressed many of the issues raised by the *Karn*, *Junger* and *Bernstein* cases. The licensing regime was further relaxed just months later, in July 2000, when it was announced by the Administration that US companies would be able export any encryption products without a license to any end-user in the fifteen nations of the European Union, as well as Australia, Norway, the Czech Republic, Hungary, Poland, Japan, New Zealand and Switzerland, effectively eliminating technical review for such exports.[44]

Karn's case was effectively made moot by the regulations, as both the book and disk at issue could now be exported without licence. Junger's appeal was heard by the Court of Appeals (6th Circuit). In its decision on April 4, 2000, the court reversed the decision of the lower court and held that computer source code should be afforded First Amendment protection, since it was a means of expression among cryptographers: "for individuals fluent in a computer programming language, source code is the most efficient and precise means by which to communicate ideas about cryptography".[45] It then remanded the case back to the District Court for consideration of Junger's constitutional challenge to the amended regulations. Junger settled the case shortly thereafter, apparently to preserve the ruling of the Court of Appeals (6th Circuit) that software could constitute protected speech under the First Amendment.

On April 14, 2000, the Court of Appeals (9th Circuit) also remanded the *Bernstein* case back to the District Court for reconsideration in light of the January 2000 revisions of the export regulations, and in October 2002 the District Court dismissed the case following government assurances that it would not enforce several portions of the regulations. Although the court stated that "If and when there is a concrete threat of enforcement against

Bernstein for a specific activity, Bernstein may return for judicial resolution of that dispute", that dismissal ended Bernstein's long journey through the courts.

The US government's struggle to maintain workable cryptography export controls ebbed and flowed throughout the 1990s, but was fatally undermined by a number of factors. The first, and probably most damaging, was the fact that, with the demise of the Cold War, the rise of an ever-expanding EU as an economic and political bloc, the growth of the Asian economies, and an increasingly globalised marketplace for goods and services, the US no longer had the economic dominance, nor the moral authority, necessary to persuade or to coerce other countries to adopt strict export controls on cryptography. In a dog-eat-dog *laissez faire* global economy, to attempt to restrict research on, and exports of, a commodity that the information marketplace was increasingly viewing as vital to the development of a Global Information Infrastructure, was simply an invitation to other countries nascent cryptographic industries to catch up and cash in. The attempts of the NSA to retain control over the sphere of cryptology thus came to be seen by US corporate developers and implementers, by civilian researchers, and increasingly by legislators, not as a necessary element of maintaining the national security, but rather as a self-interested defence of anachronistic administrative boundaries. The Wassenaar Arrangement and the OECD Guidelines clearly demonstrated both how ineffective non-binding agreements would be for exporting the US policy agenda, and how other nations now expected decisions affecting their national social and economic interests to be a matter of international dialogue and not US diktat.

As civilian cryptography technologies advanced, and the battle to control their form and use within the US was effectively conceded, US export control policy of civilian cryptography became almost impossible to enforce rationally. The ubiquity of international Internet access meant that controls designed to prevent tangible exports were increasingly being targeted at intangible transfers, and the Zimmermann incident demonstrated that this was always going to be a matter of bolting the stable door after the horse had bolted. A successful

---

[44]Press Release, Office of the White House Press Secretary, Administration Updates Encryption Export Policy (July 17, 2000).

[45]*Junger v. Daley* 209 F. 3d 481 (6th Cir. 2000).

prosecution against Zimmermann might have had a temporary deterrent effect on future developers of such software. However, the difficulty of proving that he was in any way responsible for its release on the Internet, and thus internationally, and the impossibility of preventing future releases of other cryptographic software onto the Internet by third parties militated against any lasting effect. Indeed, from early 1996 onwards, the ever increasing number of exemptions from and relaxations of, the export controls showed a government reluctantly, but inevitably, conceding that while export controls could still be used to hinder export of code, this often merely hurt the national interest; and that attempts to place an outright bar on exports of code were a dead letter. The *Karn*, *Bernstein* and *Junger* cases served to point up both the absurdity of the export controls – e.g., that printed source code was exportable, but electronic source code was not – and to open a new front in the cryptography debate – the question of whether code was speech, and therefore protected by the US First Amendment [18; 82; 127; 144].

The issue of civil liberties was present from the start of the civilian cryptography debate. However, during the 1990s, with the development of the Internet and the growth of interest in its political social and economic impacts, an number of existing campaigning groups took an interest in (e.g. the American Civil Liberties Union (ACLU)), and new groups were set up to campaign for (e.g. the Electronic Frontier Foundation (EFF)), what were loosely termed 'digital rights'.

Key amongst those perceived rights were the interlinked rights of freedom of speech and on-line privacy. Privacy rights can broadly be divided into four categories, informational,[46] decisional,[47] physical,[48] and proprietary,[49] of which the first two are of greatest importance in on-line terms. The

early stages of the cryptography debate had not excited wide public debate, with discussion being largely confined to academic circles. However, decisional privacy issues, such as the right of researchers to research on topics of their choice, and to be able to engage in open discussions of those topics with other researchers, without government oversight or prior agreement, were already clearly implicated. The potential for cryptography to secure the communications of the public against intrusion by others, including the State, was also becoming evident. As was seen above, freedom of speech was implicated partly by virtue of whether encryption source code itself was expressive, and partly by virtue of the ability of encryption technologies to facilitate both secret speech and anonymous speech, thereby linking back to privacy.

The gradual decline of the importance of nation-to-nation export controls of cryptography, in terms of checking the spread of strong encryption, as nations developed their own cryptographic capabilities independent of US influence, was probably inevitable in the age of globalisation. The lack of control at that level, did not, however, result in a decline in interest in attempting to regulate encryption's civilian use, particularly in the US and UK.

Firstly, while the military threat of the Cold War had ended, and there was little credible military threat from aggressor nations, economic and commercial tensions remained, with nations seeking to advance the interests of their home industries at the expense of their foreign rivals. Even as international regulatory organisations, such as the World Trade Organisation, were devised to reduce the obvious problems in international trade, such as protectionism, product dumping, and bribery, nations were regearing their intelligence systems to gather economic, rather than military, information. Information which provided an advantage in trade negotiations could be of immense value, economically and strategically, particularly where protecting a civilian industry could also mean boosting potential military innovations and capacity, e.g., the aerospace industry.[50] Ensuring that commercial cryptography remained relatively weak could

---

[46]E.g. access to medical records, employer access to email, on-line anonymity, and executive privilege.

[47]E.g. abortion rights, the right to assisted suicide, the rights of homosexuals and families to make decisions concerning their own lives.

[48]E.g. government search and seizure, peeping toms, and issues concerning bodily integrity.

[49]E.g. publicity rights, identity, and the ownership of the body.

[50]The French policy of banning, or severely restricting the use of encryption technologies, which remained in place until 1999 (see below), is alleged to have been underpinned by just

therefore prove useful, even when dealing with long-term allies.

Secondly, as civilian encryption technologies became more widely available, law enforcement agencies began to become more vocal about concerns that their ability to monitor communications and access stored digital information might be degraded. The ability to draw law enforcement concerns into the debate about cryptography policy, strengthened the hand of those governments keen to maintain controls over civilian use. While it might be difficult to articulate the public policy justifications for maintaining controls on strong encryption for intelligence agency purposes, given a public perception of declining national security threats, and the general undesirability of openly suggesting that it might provide an advantage in spying on one's economic rivals; concerns about law enforcement can be rather more easily and openly communicated to the public. It was this line of thinking which underpinned the other major encryption initiative of the 1990s which, in the US at least, ran in tandem with the policy of export controls – key escrow.

28.4 NOW THE GENIE'S OUT – GOVERNMENTS' SECOND WISH: CRYPTOGRAPHIC KEY ESCROW/KEY RECOVERY[51]

In April 1993, the Clinton Administration announced the other strand of its cryptography policy. In addition to its export controls strategy, it aimed to direct the development of civilian use of encryption by influencing the standards setting agenda. This approach had worked relatively effectively with the DES, despite business and scientific concerns about both the robustness of that standard and the extent to which its adoption might have been influenced by the NSA.

The same standards-based approach was also being used to influence the development of digital signature technologies with the development of the Digital Signature Standard (DSS) during the period 1991–1994. Banisar notes that both the NSA and the FBI had been heavily involved in promoting the proposed standard. Both had put pressure on NIST to adopt a standard that only provided digital signatures, in advance of a move to secure a standard acceptable to the NSA and law enforcement for privacy/confidentiality of communications, thus separating the authentication/confidentiality functions of encryption [13: 269–272]. Despite considerable public concerns, including:

- the DSS standard's lack of privacy protection, its incompatibility with the by then industry-standard RSA algorithm, and the NSA role in designing the algorithm;
- opposition from other parts of the US government, such as the IRS; and
- international concerns about the effect of a US standard that was not compatible with existing international standards, including ISO 9796, the most widely accepted international digital signature standard [116].

---

such considerations. Prior to 1999, foreign companies were required to register their encryption keys with the authorities on 'national security' grounds, a requirement which gave the French Intelligence agencies the opportunity to engage in industrial espionage against them. Shearer and Gutmann note that "The head of the French DGSE (Direction Générale de la Sécurité Extérieure) secret service has publicly stated this organisation helped French companies acquire over a billion dollars worth of business deals from foreign competitors in this way". They go on to suggest that some foreign corporations were aware of these activities and used to deliberately send false information to their French subsidiaries [146; 153: 120, 121].

[51]It is worth noting that the terminology used in the literature, particularly the legal literature, is varied, and the terms key escrow and key recovery are often used interchangeably. Bert-Jaap Koops in his webpage on key recovery

http://rechten.uvt.nl/koops/RECOVERY.HTM makes the following distinctions:

Key escrow: generic term for cryptographic systems that provide government (or user) access to keys by having people deposit their keys with a 'Trusted Third Party'.

Key recovery: originally a term for a cryptographic system which requires people to add a recoverable session key to messages rather than deposit their private keys; but is often used, irrespective of the technology, as a generic term for government access to cryptographic keys, and a replacement for 'key escrow'.

Key encapsulation: a variation on key recovery in its original sense: a cryptographic system which provides data recovery by requiring people to add a recoverable session key to messages rather than deposit their private keys.

NIST was to push ahead and adopt the DSS as a US standard in mid-1994 [118].

In the light of those developments, using the standards process to ensure government access to civilian encrypted communications must have been inviting. However, it was to turn into yet another drawn-out losing battle for the Administration. The announcement in April 2003 was of the Escrowed Encryption Initiative (EEI) (often referred to as Clipper I). The initiative provided for, amongst other things, a hardware implemented solution for encryption, – the Clipper chip – to be incorporated into telecommunications equipment, and the development of a key-escrow scheme, described as "a voluntary program to improve security and privacy of telephone communications while meeting the legitimate needs of law enforcement". The aim was for the Escrowed Encryption Standard (EES) to become a Federal Information Processing Standard (FIPS). The purpose of this was to be able to leverage the US Government's purchasing power by making the Clipper Chip a standard in US Government equipment – a technique known in regulatory theory as 'regulation by contract'. Additionally, of course, should certain key US Government agencies generally adopt Clipper, e.g. the IRS, then anyone wishing to communicate securely with those agencies, e.g., people wishing to file their taxes electronically, would also have to use Clipper, creating a network effect expanding out from the government into wider business and individual use. This was a clever end-run by the Administration around the fact that the Executive actually had no power to control civilian use of encryption, nor to set up a national key escrow system, in the absence of authorisation by Congress: an authorisation which had not been, and was unlikely to be, forthcoming [56: 32].

The EES key escrow system was designed specifically to allow law enforcement access. Any two individuals wishing to exchange information using EES-secured communications would each require a communications device incorporating the Clipper chip. Before transmitting a message the two devices would create an 80-bit session key and pass this to their respective Clipper chips which would then encrypt the session key with a key

unique to that device, and add it to a Law Enforcement Access Field (LEAF), which would also contain an ID for the particular Clipper chip. When the message was sent, the LEAF would be transmitted along with the message. Should a law enforcement agency intercept the encrypted message in the course of a wiretap under warrant, and wish to unencrypt it, the agency would pass the LEAF through a decryption device and obtain the ID for the Clipper chips used. Once the agency had the Clipper chip IDs they could contact the two key escrow agencies provided for in the Initiative in order to request the two halves of the escrowed key for one of the devices used. Using that escrowed key, the agency could then unencrypt the session key held in the LEAF and then decode the encrypted message. Once the law enforcement agency obtained the escrowed key for a device they would be able to unencrypt any other session keys for other communications using that device [35].

The scope of the opposition to the content of the EEI seems to have taken the Administration somewhat by surprise. A large-scale lobbying effort was launched by both industry and public interest groupings which were highly critical of the proposals [66]. Phillips notes that when the EES FIPS was published for comments, of the 320 comments that were received, only 2 were positive [131: 264]. The main concerns focused on the potential for privacy breaches and misuse of escrowed keys, with some pointing to past misuses of interception powers by the FBI, and others noting that if Clipper became widely used in commercial settings, that this would make the key escrow agencies prime targets for criminal attack, particularly by means of coercion or bribery of agency officials. The fact that, in September 1993, the two escrow agencies were named as the NIST and the Treasury Department's Automated Systems Division, both government rather than independent agencies, and in the case of NIST, an agency with rather close links with the NSA, did little to allay concerns about possible governmental abuses. Other concerns included the secretive nature of the escrowed-encryption initiative, the use of a classified algorithm (Skipjack) in the standard, and the requirement that the standard be implemented in hardware only. A longer term concern was that in the event that Clipper became widely used, this would permit the government the latitude to bar other non-escrowed forms of encryption technology.[52]

Supporters of the initiative suggested that Clipper might have positive uses, for example, individuals and businesses that lost their keys would have a way to retrieve data, and employers would be able to crack company-encrypted communications of employees who used company computers and networks in order to commit frauds. However, such suggestions themselves raised further concerns about who would have access to the ability to eavesdrop, and what would constitute "lawful authority" to do so.[53]

Despite the controversy, the Administration formally adopted Clipper as the FIPS for voice communications in February 1994 [117]. This was, however, the zenith of Clipper's success, as from this point onwards, the Administration was in retreat from its original proposal. The extent of the opposition, plus the fact that, in May 1994, the EES was demonstrated to be compromisable [16] forced a rethink, but not an abandonment of the key escrow concept. Instead the Administration moved its support to the development of key escrow systems that "would be implementable in software, firmware or hardware, or any combination thereof, would not rely on a classified algorithm, would be voluntary and would be exportable".[54] Clipper I, the hardware-only solution, absent the Administration's support, died a quiet death in the commercial marketplace.

The next iteration of key escrow, popularly known as Clipper II, combined recognition of the need to overcome some of the key sticking points in Clipper I, with the carrot of export control relaxations. Key escrow was still required, but there was no requirement for specialised hardware or use of a particular algorithm, and keys could be deposited with government-certified private escrow agents rather than with government agencies ("software key escrow" or "commercial key escrow"). The escrow agent would be authorized to provide the keys to either the owner on request, or to law enforcement officials with a court-authorized search warrant. Commercial software makers could export escrowed encryption technologies of up to 64 bits, but also had to ensure that:

- their software code was closed source – which would have made it impossible for third parties to adequately test the implementation of the algorithm;
- their software was designed in such a way that users could not easily alter or adapt it without causing it to fail – a requirement that sat uneasily with the need to provide updates and upgrades for software;
- their software could not be used to provide multiple encryption, such as triple DES;
- their key-escrowed software would not be compatible with, and be unable to communicate with, non-escrowed systems, including products whose key-escrow mechanisms had been altered or disabled.

This proposal met with no warmer enthusiasm from the US software industry than had Clipper I. The 64-bit limit was seen as at odds with the commercial realities of the international marketplace, where both the Swiss 128-bit IDEA algorithm and Triple DES were gaining popularity. Equally, encrypted products whose keys were readily accessible by US government agencies were not especially attractive to international organisations, especially those in competition with US companies in strategic industries and marketplaces.

In May 1996, the Clinton Administration published a further proposal, "Achieving Privacy, Commerce, Security and Public Safety in the Global Information Infrastructure" which was designed to establish a new public key infrastructure for encryption (Clipper III). Under this proposal, US software companies would receive immediate approval to export encryption products

---

[52]See the proposed *Anti-Electronic Racketeering Act* (S. 974, June 1995) designed to considerably restrict encryption, and under which only the use of escrow-like software would have been an affirmative defence for those prosecuted for using cryptography, for an (unsuccessful) example of how this might have played out.

[53]See for further discussion of the Clipper proposals [36; 51; 55; 84; 85; 108; 121; 145].

[54]Vice President's letter to Representative Maria Cantwell, 20 July 1994, reprinted in [72: 236–238].

using 56-bit encryption if they presented a plan to install key recovery in their encryption products for export within two years. Additionally, encryption products of any strength would be eligible for export approval, once the key management infrastructure was in place. The US Government would set policies for handling escrow keys and establish arrangements with other countries to implement the policy. This initiative did have the effect of reducing industry opposition, and garnered some support from commentators [71], and in October 1996 the US software industry grouping Key Recovery Alliance was formed to develop key-recovery techniques, and to lobby for reductions in export restrictions. However, as was outlined in the section above, neither export controls nor key escrow systems were to find much support internationally, as the Wassenaar states, the OECD and the European Commission all failed to provide the type of international co-operation that the US government's plans required in order to have any chance of success.

By early 1998, it was clear that the end was in sight for key escrow, and that the NSA and FBI's attempts to influence the regulation of encryption technologies by ensuring any easy backdoor to the available technologies within the US, and internationally, had largely failed. The attempt to link the issue of key escrow to export controls benefited neither cause, and the US Government's decision in September 1999 to eliminate export controls on encryption after a one-time review, and the further relaxation of export regulations in January 2000, meant that the Administration had largely acceded to industry demands.

It is interesting to speculate just how influential in this long-running policy battle the issue of privacy actually was. Certainly the initial proposal for the Clipper Chip aroused considerable public debate and some high profile opposition, notably from civil liberties groups such as the CPSR and EFF. Yet this opposition, in and of itself, would probably not have been sufficient to derail the proposals in the face of the stated national security and law enforcement concerns. It seems clear, particularly with the later iterations of Clipper, that the most effective opposition came not from the

civil liberties groupings, but from the increasingly powerful software industry lobby, and its allies in the US Congress. Individual privacy was probably less important in influencing the political considerations than commercial concerns about the long term security of financial data and corporate communications, and the development and maintenance of trust in electronic business methods [170]. Individual privacy was certainly a part of that equation, but it was by no means the dominant consideration.

The result of the demise of key escrow as a workable control on the use of encryption technologies, in democratic countries at least, was to throw law enforcement and national security agencies back to more direct routes to obtaining the plaintext of encrypted communications. These might include obtaining the plaintext prior to its encryption, obtaining the plaintext from a recipient who had decrypted it, or simply requiring the person who had encrypted the data to hand over the encryption key. The advantage of these direct methods from a privacy point of view is that they tend to resemble existing/traditional methods of communications interception or data recovery, such as wiretapping. As such they can, in principle, be subjected to similar forms of judicial legal scrutiny, to ensure that the powers granted to law enforcement or government agencies do not go further than are required to meet necessary operational requirements, and that these powers are not then abused.

28.5   NOW THE GENIE'S OUT – GOVERNMENTS' THIRD WISH: MANDATORY KEY SURRENDER AND ENCRYPTION AVOIDANCE

By the early 2000s it was clear that, at the international level, even those states, such as France [107: 441, 442; 112; 113] and the UK [3–6; 8; 45; 81], which had fought a long rearguard action alongside the US to attempt to prevent strong encryption gaining a foothold amongst the general public had, even if they had not totally liberalised their policies on encryption, been forced by market pressures to accept that the use of regulatory tools such as export controls, and the various iterations of key escrow, were no longer effective. However, as was the case in the US, the requirements of national security as a justification for government intervention in the public use of encryption were being rapidly overtaken by the perceived requirements of law enforcement.

The FBI had campaigned ceaselessly during the mid-1990s for new powers to permit them to conduct electronic surveillance more effectively amongst the new communications technologies. While their support for the Clipper Chip went unrewarded, they were successful at mobilising support for their cause in Congress resulting in the passage of the Communications Assistance for Law Enforcement Act of 1994 (CALEA).[55] CALEA required that telecommunications carriers assist law enforcement agencies in executing electronic surveillance pursuant to court order or other lawful authorization, by configuring both new and substantially upgraded telecommunications systems to meet law enforcement interception requirements [173: 138–149].[56] However, CALEA also expressly stated that telecommunications providers were "not responsible for decrypting or ensuring the government's ability to decrypt ... unless the encryption was provided by the carrier and the carrier has the information necessary to decrypt the communication".[57] The FBI and their supporters in Congress and elsewhere [140] have long claimed that encryption was going to make it difficult, if not impossible, to tackle certain types of crime such as racketeering, drug dealing, and money laundering. These claims have, however, been undermined by a paucity of evidence that

encryption has been a significant factor in either the failure of criminal investigations, or the failure to obtain evidence necessary to secure convictions. While the literature suggests that encrypted material is occasionally uncovered during investigations, reports from both the US [14; 99] and UK [135: 107, 108] suggest that it is limited in amount and of minor import. This lack of evidence did not, however, pre-empt suggestions being put forward that in cases where law enforcement agencies were faced with encrypted communications or documents, they should be able to compel production of either plaintext or keys from individuals [139]. In the US, however, adopting this approach was hindered by the Fifth Amendment which provides that "no person shall be compelled in any criminal case to be a witness against himself". The issue of encryption and the Fifth Amendment had been considered at length during the Clipper Chip debate. At that time the preponderance of commentators appeared to conclude that requiring individuals to escrow their encryption keys prior to sending encrypted messages would not violate the Fifth Amendment, following the Supreme Court's judgement in *United States v. Freed*[58] which had laid down the principle that the Fifth Amendment did not protect testimony that was not incriminating at the time it was given, but which might become incriminating at a later date [55; 164; 171]. The question as to whether the government could directly obtain encryption keys, or force individuals to produce non-escrowed keys was a different matter [18; 25; 148].

In September 1999, as the Clinton Administration was significantly reducing export controls on encryption products, and key escrow was seemingly fading as a commercially viable concept, the government was still seeking to signally expand law enforcement powers to access encryption keys directly by proposing legislation known as the Cyberspace Electronic Security Act ("CESA"). Initial drafts of the legislation circulated by the Justice Department in August had suggested that amongst other things, it would permit law enforcement officers to get a sealed warrant from a judge to enter private property, search through computers for

---

[55]Pub. L. No. 103-414, 108 Stat. 4279.

[56]See further, for a history of communications interception in the US [40: Chs 7 and 8]. In August 2005 the FCC, at the request of the US Department of Justice, the FBI and Drug Enforcement Agency (DEA), extended the scope of the CALEA requirements to cover facilities-based broadband Internet access providers and providers of interconnected voice-over-Internet-Protocol (VoIP) services (FCC 05-153, August 5, 2005). At the time of writing this Order was the subject of a legal challenge before the federal appeals court brought by a coalition of public interest, academic and business groups [102].

[57]*Supra* n. 55, at 103 (b3).

[58]*United States v. Freed* 401 US 601 (1971).

encryption keys or passwords, to implant "recovery devices" and to override encryption programs [125]. These highly controversial provisions were dropped before the Act was formally proposed, but the rest of the Act demonstrated that the Administration had not yet given up on key escrow.

CESA was put forward by the Administration as a means of "respond[ing] to both the legitimate and unlawful uses of cryptography, [and] building a legal infrastructure for these emerging issues".[59] It was part of a package of measures including export control relaxations (see above), that aimed to simultaneously pre-empt legislative moves afoot in Congress to liberalise the encryption control regime (notably H.R. 850, the Security and Freedom through Encryption (SAFE) Act), and to increase the abilities of law enforcement and national security agencies to deal with the implications of stronger encryption.

The CESA had three main components:

- it proposed to provide $80 million over four years to fund an FBI Technical Support Center, to develop means of responding to the use of encryption technology by criminals which would then be made available to law enforcement agencies;
- it set out a framework for the compelled disclosure and use of stored recovery information in the possession of recovery agents (in essence, key escrow agencies) which incorporated some privacy protection measures; and
- it made provision for the protection of the secrecy of both:
  - government investigative techniques permitting access to information protected by encryption or other security techniques or devices; and
  - privately held trade secrets (such as encryption algorithms) disclosed to the government to assist it in obtaining access to information protected by encryption.

Of these components, the newly proposed legal framework was the most controversial. It was pro-

moted as providing new federal statutory protections for the privacy of encryption keys, by placing obligations upon recovery agents who provided storage services for keys. These included forbidding recovery agents from disclosing decryption keys or other recovery information, or using those means to decrypt data, except in specific circumstances, e.g., where the person who stored the key had given their consent, or where disclosure was required by a court order. It also prohibited recovery agents from selling or otherwise disclosing their customer lists to other parties. The CESA would also have barred any person from knowingly obtaining stored recovery information from a recovery agent if they knew, or had reason to know, they had no lawful authority to do so – in principle, a means of preventing law enforcement agencies from gaining access to information without going through the proper channels.

However, these new protections were accompanied by new statutory provisions for law enforcement access to decryption information held by third parties. Under these provisions, a governmental agency could compel disclosure of decryption keys via:

- a warrant;
- any process to compel disclosure under federal or state law;
- a court order; or
- an investigative or law enforcement officer's reasonable determination that an emergency situation existed.

Where a court order was required, a court could issue a disclosure order where:

- the use of the stored recovery information was reasonably necessary to allow access to the plaintext of data or communications;
- the access was lawful;
- the governmental agency would seek access within a reasonable time; and
- the plaintext was not subject to a constitutionally protected expectation of privacy, or the privacy interest created by such expectation could be overcome by consent, warrant, order or other authority.

---

[59]The White House, Analysis: The Cyberspace Electronic Security Act of 1999, 13 September 1999, http://www.cdt.org/crypto/CESA/adminstatement.shtml

The CESA would have prohibited recovery agents from disclosing to the person who stored a decryption key the fact that a governmental entity had required them to disclose or use stored recovery information. It would also have permitted the government agency to withhold notice from the person who stored the decryption key that it had receiving stored recovery information, or decrypted data, or communications, from a recovery agent for up to 90 days, or indefinitely if permitted by court order.

By way of a sop to potential recovery agents, and presumably to encourage those organisations still interested in providing key escrow services, CESA provided that where the government compelled disclosure the recovery agent would be both entitled to reasonable compensation for costs incurred in providing stored recovery information or decrypting data or communications. They would also be given immunity from litigation based on actions taken to facilitate compliance with such an order.

There were numerous objections to the CESA framework, both legal and practical. The Justice Department contended that, where encryption keys were stored with a third party there could be no privacy interest in the information encrypted with those keys without a confidentiality agreement. This contention supported its reasoning for the requirement of the specific privacy provisions of the Act.[60] However, it was argued, notably by the non-profit public policy group The Center for Democracy and Technology (CDT), that the CESA framework failed to match the privacy requirements of the Fourth Amendment with regard to the proper issue of warrants. Under the Fourth Amendment, when warrants are issued, law enforcement agencies must show "probable cause" that a crime was committed. The CESA proposed to allow law enforcement agencies to obtain warrants to gain access to decryption keys where "reasonably necessary", a rather less stringent formulation.[61] Additionally, where a federal law enforcement officer

determined that an emergency existed, they could immediately demand a decryption key from a recovery agent thereby, initially at least, avoiding the need for judicial scrutiny of the validity of the demand.

It was also suggested that the withholding of notice by law enforcement agencies from a person whose stored decryption key had been seized potentially breached the Sixth Amendment, inasmuch as they would be unable to exercise their right to establish a defence against accusation because they would be unaware of the "nature and cause of the accusation" [165: 848, 849]. The vagueness of some of the CESA terminology was also criticised – notably the term "constitutionally protected expectation of privacy". The scope of constitutionally protected privacy remains uncertain, to say the least, and the CESA neither defined the scope of the term, nor made clear how and when information became constitutionally protected by an expectation of privacy [165: 854–856].

In practical terms, as the CESA did not require individuals to escrow keys, it was argued that those using encryption to hide criminal activities were, unlike legitimate businesses, unlikely to place their keys where they could be easily accessed by law enforcement, and that this undermined the arguments, for example, for withholding notice of compelled disclosure of keys. Additionally, if US software and encryption services were seen as vulnerable to law enforcement access, it would be a simple matter for users both legitimate and criminal, to simply obtain their software and services from suppliers outside the US. In the event, opposition to the CESA meant that the proposal lacked any influential backers in the House or Senate resulting in it failing to ever be formally introduced as draft legislation. It effectively died when the Clinton Administration left office [15; 126].

However, in the UK, similar legislation was making its way onto the statute books with rather less difficulty. The Regulation of Investigatory

---

[60]See further, *United States v. Miller* 425 U.S. 435, 435 (1976), *Smith v. Maryland* 442 U.S. 735, 737 (1979) and [158].

[61]CDT, Initial CDT Analysis of the Clinton Administration's Proposed Cyberspace Electronic Se-

curity Act (CESA): Standards for Government Access to Decryption Keys, 29 September 1999, http://www.cdt.org/crypto/CESA/cdtcesaanalysis.shtml

Powers Act 2000[62] is legislation that was designed to both update UK surveillance and interception law to take account of new technologies, and to ensure that the UK was in conformity with its obligations under the European Convention of Human Rights following the European Court of Human Rights' decision in the case of *Halford v. United Kingdom.*[63] Amongst other things, the Act makes provision for the investigation of electronic data protected by encryption [137].[64]

The Act applies where any encryption-protected information comes into the possession of any person (or is likely to do so)

- via a statutory power to seize, detain, inspect, search or otherwise to interfere with documents or other property;
- by means of the exercise of any statutory power to intercept communications;
- as a result of having been provided or disclosed in pursuance of any statutory duty; or
- has by any other lawful means not involving the exercise of statutory powers, come into the possession of any of the intelligence services, the police or the customs and excise.[65]

In such circumstances, where permission is granted:

- by court order;
- by warrant;
- by statute;
- or in certain circumstances by the Secretary of State,[66]

that person can require a third party in possession of an encryption key to disclose that key where it is necessary:

- for the exercise or proper performance by any public authority of any statutory power or statutory duty;
- in the interests of national security;

- for the purpose of preventing or detecting crime; or
- in the interests of the economic well-being of the United Kingdom.[67]

The disclosure requirement must be proportionate to achieve the aims to be achieved by its imposition, and the protected information must not be reasonably accessible by other means.[68]

Where a disclosure notice is made, it must:

- be in writing, or in some other permanent record;
- describe the protected information, state the grounds for disclosure, specify the office, rank or position held by the person making it, specify the office, rank or position of the person who gave permission for it; specify a reasonable time period in which it must be complied with; and
- set out the disclosure that is required by the notice and the form and manner in which it is to be made.[69]

Where a disclosure notice is given to an officer or employee of a corporation, it must unless impracticable be given to the most senior employee or officer available, unless this would defeat the purpose of the notice.[70]

When a disclosure notice is issued, if the person receiving it holds both the protected information and a means of obtaining access to the unencrypted information they may use any key in their possession to obtain access to the information, and make a disclosure of the information in an intelligible form; or they may disclose the key to the issuer of the notice. If they hold only the key, they must disclose that. If they no longer hold the key, they must disclose all information that would facilitate the obtaining or discovery of the key or the putting of the protected information into an intelligible form.[71] Failure to make necessary disclosures is a criminal offence carrying a penalty of up to two years imprisonment, and the burden of proof is on the discloser to demonstrate that they no longer had

---

[62]The UK Regulation of Investigatory Powers Act 2000 (RIPA), http://www.opsi.gov.uk/acts/acts2000/20000023.htm

[63]*Halford v. United Kingdom* (1998) 24 EHRR 523.

[64]Regulation of Investigatory Powers Act 2000, Part III: Investigation of Electronic Data Protected by Encryption.

[65]RIPA, s.49 (1)(a)–(e).

[66]RIPA, Schedule 2.

[67]RIPA, s.49 (2)(b), s.49 (3).

[68]RIPA, s.49 (2)(c), (d).

[69]RIPA, s.49 (4).

[70]RIPA, s.49 (5), (6).

[71]RIPA, s.50.

access to the encryption keys when the notice was given.[72]

In certain circumstances, a disclosure notice may require the person to whom the notice is given, and any other person who becomes aware of it or of its contents, and knows nor has reasonable grounds for suspecting that the notice contains a secrecy requirement, to keep secret the giving of the notice, its contents and any actions taken under it. Breaching this requirement ('tipping off') is also a criminal offence carrying a penalty of up to five years imprisonment.[73]

The Act also requires that those who issue disclosure notices, and those who operate on their behalf

- only use disclosed keys for obtaining access to, or putting into an intelligible form, the protected information covered by the notice;
- that the uses to which the keys are put are reasonable and proportionate in the context of the case;
- that the keys are stored in a secure manner and that all records of the key are destroyed when no longer required.[74]

As with the proposed US CESA, persons making disclosures under the RIPA can be recompensed by the government for any costs incurred as result of complying with a disclosure order,[75] although the blanket immunity from liability proposed under the CESA for those disclosing keys is not replicated.

In the UK, the direct constitutional protections afforded to US citizens against the encroachments of government agents are almost wholly lacking, although some limited protection may be obtained via the European Convention of Human Rights, which the UK courts, following the incorporation of the Convention into UK law by the Human Rights Act 1998,[76] are bound to take cognisance of. However, the powers granted to the UK courts

under the Human Rights Act do not extend to an ability to strike down UK legislation that is incompatible with the Convention: rather a court may indicate that the legislation is incompatible, at which point the Government should then act to amend the Act accordingly.[77] The Convention is however designed to allow its Contracting Parties to take actions that would otherwise breach Convention rights. In the case of privacy, Article 8 of the Convention states that:

> There shall be no interference by a public authority with the exercise of this right *except* such as is in accordance with the law and is necessary in a democratic society in the interests of national security, public safety or the economic well-being of the country, for the prevention of disorder or crime, for the protection of health or morals, or for the protection of the rights and freedoms of others.

In this context, it seems clear that the UK Government would have little difficulty in justifying the RIPA's provisions on encryption, except in the case of egregious abuses of the disclosure notice process. Even in such an event, under the Convention processes, a complainant would first have to exhaust their avenues of challenge in the UK courts, and if unsuccessful there, then take an action to the European Court of Human Rights, a process that is both costly and time consuming.

Other states have since adopted a similar approach to obtaining access to encrypted materials. The Council of Europe's Cybercrime Convention requires that all Parties to it "adopt such legislative and other measures as may be necessary to empower its competent authorities to order any person who has knowledge about the functioning of the computer system or measures applied to protect the computer data therein to provide, as is reasonable, the necessary information" to permit search or access to computer systems and computer storage media in their jurisdiction. Citing the Cybercrime Convention as a model, the Australian government introduced a new federal law, the Cybercrimes Act 2001 (Cth), which as one of its measures, amended the federal Crimes Act 1914 (Cth), inserting a new section allowing law enforcement agencies to compel individuals to reveal private encryption keys,

---

[72]RIPA, s.53.

[73]RIPA, s.54.

[74]RIPA, s.55.

[75]RIPA, s.52.

[76]The UK Human Rights Act 1998 (HRA), http://www.opsi.gov.uk/ACTS/acts1998/19980042.htm.

[77]HRA, s.4

ID numbers or passwords for the purpose of prosecuting computer-related offences. Failure to comply is a criminal offence [76].

In the US, the failure of the CESA to gain legislative support, and the limited commercial interest in key escrow schemes, appears to have increased law enforcement agency interest in circumvention of strong encryption by both technological and more traditional means. Research into the efficacy of available commercial encryption software suggests that there are various means by which law enforcement agencies may be able to obtain the plaintext of encrypted documents and e-mails, especially if they have access to suspects' computer systems. In the case *United States v. Scarfo*[78] the FBI obtained the passwords necessary to remove the encryption on certain documents belonging to a member of an organised crime family suspected of running an illegal gambling and loan-sharking operation, by obtaining warrants permitting the installation of keystroke logging software (KLS)[79] on his computer.

The keystroke logger recorded all keyboard input to Scarfo's computer and, when this data was retrieved by the FBI, it included the passphrases which permitted them to access the plaintext of the encrypted files which, the FBI alleged, contained incriminating evidence. After his indictment, Scarfo filed two motions concerning the KLS. The first was a discovery motion which sought information about the KLS including the name and version of the device, its manufacturer, its installation and use instructions, the dates of the retrieval of the captured information in its original form, and a copy of the software and hardware devices used to capture data. The second sought to have the evidence gathered by the KLS suppressed on the grounds that it was "an unlawful general warrant in violation of the Fourth Amendment" and that it "effectively intercepted a wire communication in violation of [ECPA] Title III".[80]

The FBI provided some information regarding the KLS, but was unwilling to divulge details including how it was installed, and how the relevant data was captured, on the grounds that to do so would seriously compromise current and future use of the KLS system. When further pressured by the Court for wider disclosure, the FBI invoked the Classified Information Procedures Act of 1980 ("CIPA"), an unusual step in a case not involving espionage and, after review, the Court permitted the disclosure of only a "Substitute Unclassified Summary Statement" to the defence.

In its assessment of the Fourth Amendment issues raised by the use of the KLS, the Court considered whether the search warrants were too general, in that the KLS recorded considerably more information than simply the passphrases, and that the process of data collection under warrant had continued for a month. In both instances the Court held that the warrants were acceptable in their scope, in that the evidence sought – the passphrases for the encrypted material – was clearly defined, and the nature of the electronic material being searched meant that "law enforcement officers must be afforded the leeway to wade through a potential morass of information".[81]

As regards the ECPA-based challenge, Title III of the Electronic Communications Privacy Act places particularly stringent requirements on law enforcement agencies seeking warrants to intercept electronic communications [2; 21: 202–204]. The defence argued that the fact that the computer on which the KLS was placed contained, or was attached to, a modem, meant that the FBI were potentially intercepting electronic communications, and thus should have obtained a Title III warrant rather than a basic warrant. The FBI offered evidence that the KLS had been configured only to record keystrokes when the modem was not being used, meaning that no electronic communications had been intercepted. This was accepted by the Court, which held that no Title III warrant was required.[82]

---

[78] *United States v. Scarfo* 180 F. Supp. 2d 572 (D.NJ. 2001). The defendant, Nicodemo Scarfo, Jr. had encrypted his gambling files using Pretty Good Privacy (PGP). It transpired that the passphrase was Nds09813-050 – the federal prison identification code for his father.

[79] Often referred to in the literature as a 'Key Logger System', hence 'KLS'.

[80] The Electronic Communications Privacy Act (ECPA) 18 U.S.C. §2510.

[81] Supra n.78 at 578.

[82] Supra n.78 at 582.

The decisions of the Court in the *Scarfo* case with regard to the KLS have been criticised, notably for their potential impact upon individual privacy and their approach to the Fourth Amendment [21; 103]. However, despite indications that Scarfo might have grounds to successfully appeal the decision, he eventually pleaded guilty, and thus removed the possibility of review of these issues by a federal appellate court. To date, there appears to have been no similar case before the US courts, and while it is clear the FBI has continued to advance the KLS technology, with the introduction of its "Magic Lantern" software [172; 69; 166: 348–351], which may potentially be installed without requiring physical access to a suspect's premises or hardware, particularly if a computer is attached to the Internet, the legal framework within which such information capture takes place remains largely untested.

Indeed, there is a growing literature in US legal journals on the relationship between the Fourth Amendment and encryption. This ranges from whether there is a Fourth Amendment right to access to encryption technologies [54; 95]; to whether any form of key escrow is in effect a warrantless search [100] or unduly disturbs the privacy/law enforcement balance [157; 162; 168]; to the degree to which Fourth Amendment requirements should be met by law enforcement agencies seeking access to and decryption of encrypted materials [43; 83; 161]. It is unlikely that greater clarity will emerge in the US situation until such time as the Supreme Court is presented with a case which allows it to re-examine the meaning of a "reasonable right of privacy" in the light of current technologies.

In the meantime, in addition to the possibility of capturing passphrases through the use of KLS, law enforcement agencies, and others, are increasingly looking to exploit other weaknesses in the use of encryption software. While encryption algorithms available to the public may be increasingly hard to defeat via brute force attacks, their effectiveness can often be significantly compromised by other factors. Poor security habits on the part of users, such as short or easily guessable passphrases, poor use of encryption software, lax key security, and

failure to police unencrypted data on media are not uncommon.[83] Also while cryptographic software may often be designed to prevent its processes writing unencrypted data to accessible areas, other programs used to read or print the unencrypted data are unlikely to take the same levels of precaution, and may make numerous backup or temporary copies of the unencrypted data which the user may not then properly remove, or may not even realise are there. It seems likely that, if encryption software becomes more widely used by the general public, law enforcement agency efforts will increasingly focus on techniques for evading the effects of encryption, rather than solely seeking to break it. This, in combination with the suspicions that the discovery of encrypted files often raise ("if they're not doing something wrong, why are these files encrypted") may focus the attention of those seeking to avoid the attention of law enforcement agencies, not on further increasing the strength of encryption, but rather on developing more effective means of camouflaging the results, via the use of techniques such as steganography [130] or dual encryption [63].

28.6 REBOTTLING THE GENIE: COMMERCIAL ATTACKS ON ENCRYPTION RESEARCH, PUBLIC USE AND PERSONAL PRIVACY

Through the early stages of the development of national policies for strong encryption, the interests of national security and law enforcement agencies, and those of the commercial marketplace, appear to have been largely at odds. The former group desired to restrict publicly available research into cryptological techniques and prevent public access to strong encryption. The latter group demanded access to strong encryption to facilitate commercial activities such as banking, e-commerce and Digital Rights Management, and agitated for the removal of governmental interference in the international

---

[83]For example, the Aum Shinri Kyo cult that poisoned the Tokyo subway with nerve gas encrypted their files using RSA, but investigators found the encryption keys on a floppy disk [14: 5].

competition for the supply of such technology [58; 163].

More recently, as some important commercial players have increasingly sought to use encryption techniques as a means to protect, or control access to, their proprietary information and content – and have additionally realised the potential for using encryption-based authentication as a means of precisely monitoring users' information access and use – there has been a shift in their attitudes towards research into cryptology, particularly cryptanalysis. Publicly accessible research that identifies weaknesses in commercial cryptography systems has come, in some circles, to be regarded as a threat to companies' growing ability to identify not just how much of their chargeable 'product' is being consumed, but also who exactly is engaged in the consumption – information that can be mined and combined/mashed with other sources of data to, for example, enhance and precisely target new marketing initiatives. In this environment, the privacy-enhancing potential of encryption is very much secondary to its role in the burgeoning expansion of intellectual property rights protection, and consumer control, to the extent that, ironically, the contemporary commercial use of encryption technology may be yet another nail in the coffin of the individual's right to privacy.

There are several key moments in the changing relationship between cryptanalysis researchers and the content industries. An early pointer was the agreement of the World Intellectual Property Organization (WIPO) Copyright Treaty in 1996. The Treaty, which was designed to take account of the impact of technological developments upon national copyright laws, contained a requirement that the signatories "provide adequate legal protection and effective legal remedies against the circumvention of effective technological measures that are used by authors in connection with the exercise of their rights".[84] This short statement marked a fairly radical change in the nature of copyright law, in that the legal focus expanded from the protection of rights in particular types of works, to the protection of 'fencing' technologies for those works – in

other words, protecting the protection [48]. Even without the benefit of hindsight, it is not difficult to see why rightsholder groups lobbied for the inclusion of such a statement. It must have been clear that planned efforts to protect intellectual property by technical measures were going to have an impact upon activities that the public were long used to engaging in, whether legally or illegally, and that this was likely to result in an escalating technological battle between those seeking to prevent those activities and those seeking to permit the public to continue to engage in them. The WIPO Copyright Treaty effectively provided rightsholders in the member nations with a pre-emptive weapon in that battle [17; 129].

The US implemented the Treaty provisions in the Digital Millennium Copyright Act of 1998 (DMCA).[85] The Act has received considerable criticism on the grounds that it has tipped the copyright regime in the US firmly in favour of rightsholders and against the broader public interest, and has effectively allowed rightsholders to override longstanding public rights in fair use of copyright works. In terms of encryption, it has also provided rightsholders with a significant degree of control over the extent to which encryption research can be effectively carried out and published. The relevant provisions are:

17 U.S.C. §1201(a)(1) No person shall circumvent a technological measure that effectively controls access to a work protected under this title.

1201(a)(2) No person shall manufacture, import, offer to the public, provide, or otherwise traffic in any technology, product, service, device, component, or part thereof that:

(A) is primarily designed or produced for the purpose of circumventing a technological measure that effectively controls access to a work protected under this title;

(B) has only limited commercially significant purpose or use other than to circumvent a technological measure that effectively controls access to a work protected under this title; or

---

[84]WIPO Copyright Treaty (1996), Article 11, http://www.wipo.int/treaties/en/ip/wct/pdf/trtdocs_wo033.pdf

[85]The Digital Millennium Copyright Act of 1998, http://thomas.loc.gov/cgi-bin/query/z?c105:H.R.2281.ENR.

(C) is marketed by that person or another act- ing in concert with that person with that per- son's knowledge for use in circumventing a technological measure that effectively con- trols access to a work.

1201(a)(3) Additional violations – No person shall manufacture, import, offer to the public, pro- vide, or otherwise traffic in any technology, product, service, device, component, or part thereof, that:

(A) is primarily designed or produced for the purpose of circumventing protection af- forded by a technological measure that ef- fectively protects a right of a copyright owner under this title in a work or portion thereof;
(B) has only limited commercially significant purpose or use other than to circumvent pro- tection afforded by a technological measure that effectively protects a right of a copy- right owner under this title in a work or a portion thereof; or
(C) is marketed by that person or another act- ing in concert with that person with that person's knowledge for use in circumvent- ing protection afforded by a technological measure that effectively protects a right of a copyright owner under this title in a work or portion thereof.

Breaches of the provisions can leave individuals liable to civil liabilities or criminal penalties. The DMCA does provide for certain fair use exemp- tions, included amongst which is an exemption for encryption research. It states that

> the term 'encryption research' means activities necessary to identify and analyze flaws and vul- nerabilities of encryption technologies applied to copyrighted works, if these activities are con- ducted to advance the state of knowledge in the field of encryption technology or to assist in the development of encryption products.[86]

It goes on to say that

> it is not a violation of that subsection for a person to circumvent a technological measure as applied

to a copy, phonorecord, performance, or display of a published work in the course of an act of good faith encryption research if:

(A) the person lawfully obtained the encrypted copy, phonorecord, performance, or display of the published work;
(B) such act is necessary to conduct such en- cryption research;
(C) the person made a good faith effort to ob- tain authorization before the circumvention; [...].[87]

To assess whether the exemption criteria are met, the courts must consider:

(A) whether the information derived from the en- cryption research was disseminated, and if so, whether it was disseminated in a manner reasonably calculated to advance the state of knowledge or development of encryption tech- nology, versus whether it was disseminated in a manner that facilitates infringement under this title or a violation of applicable law other than this section, including a violation of pri- vacy or breach of security;
(B) whether the person is engaged in a legitimate course of study, is employed, or is appropri- ately trained or experienced, in the field of en- cryption technology; and
(C) whether the person provides the copyright owner of the work to which the technologi- cal measure is applied with notice of the find- ings and documentation of the research, and the time when such notice is provided.[88]

Further it is not a violation of the DMCA if an individual uses "technological means for research activities" to:

(A) develop and employ technological means to circumvent a technological measure for the sole purpose of that person performing the acts of good faith encryption research [...]; and
(B) provide the technological means to another person with whom he or she is working col- laboratively for the purpose of conducting the acts of good faith encryption research [...] or

---

[86] 17 U.S.C. §1201(g)(1)(A).

[87] 17 U.S.C. §1201(g)(2).
[88] 17 U.S.C. §1201(g)(3).

for the purpose of having that other person verify his or her acts of good faith encryption research [...].[89]

While, in principle, these exemptions may seem quite broad, they are in fact considerably narrower than the fair use rights granted under the Copyright Act of 1976 which provided that an individual could use a copyrighted work without an author's permission if it was used appropriately and legally acquired. The DMCA makes a far larger demand upon the putative researcher, requiring that they make a good faith effort to obtain authorization before circumvention from the author and provide the copyright owner with documentation of their findings, not to mention to prove that they are a *bona fide* researcher, whatever that may be [90].

The effect of these provisions was rapid. In February 2000, the case *Universal City Studios, Inc. v. Reimerdes* (Universal I)[90] was decided. The case revolved around the encryption system, Content Scramble System (CSS), which was used by the movie studios to prevent Digital Versatile Disc (DVD) drives that did not contain appropriate CSS keys from decrypting files on movie DVDs and thus playing the movies. The defendants had obtained, and were making available from their website, a decryption program called DeCSS which decrypted encrypted DVDs, allowing playback on non-compliant DVD players and copying of digital sound and graphics files to hard disk drives. The movie studios filed suit under the DMCA, seeking temporary and permanent injunctions against the defendants, whom they alleged were trafficking in devices that circumvented technological measures that effectively controlled access to copyrighted works, in contravention of the DMCA. The defendants raised a number of defences, including that their actions constituted encryption research and that they were engaged in a fair use under s.107 of the Copyright Act of 1976. The Court, in finding for the plaintiffs, failed to find any evidence that the defendants were engaged in encryption research, and noted that the fair use defence under the Copyright Act of 1976 was for infringement of

copyright, and they were being sued for offering to the public and providing technology primarily designed to circumvent technological measures that control access to copyrighted work.[91]

While it is debatable whether the defendants in *Reimerdes* could reasonably have expected to be considered 'researchers', the defendant in the case *United States v. ElcomSoft Ltd*[92] had a somewhat better claim. Sklyarov was a PhD student researching cryptanalysis and an employee of the Russian software company, ElcomSoft, where he created software called *The Advanced eBook Processor*. The Advanced eBook Processor (AEBPR) was a program that was used to decrypt the proprietary eBook format owned by Adobe Systems, Inc. The eBook format was designed to allow readers to download electronic books from on-line booksellers, but placed restrictions on further editing, copying or printing of the books. AEBPR allowed its users to open an eBook in any Portable Document Format viewer with no restrictions on editing, copying and printing. Sklyarov came to the US to give a presentation on AEBPR at DefCon9 conference in Las Vegas, Nevada whereupon he was arrested, and charged with trafficking in a product designed to circumvent copyright protection measures in violation of Title 17, United States Code, Section 1201(b)(1)(A). While ultimately the personal case against Sklyarov was dropped, and ElcomSoft was found by a jury not to have wilfully violated US law, the case sent a clear message to international cryptology researchers and developers that research into vulnerabilities in proprietary products could lead to threats of, or actual prosecution, should they visit the US [75; 106].

A very clear case of stifled research arose out of a competition sponsored by the Secure Digital Music Initiative (SDMI) Foundation in September 2000 which invited the public to "crack" four watermarked technologies for digital audio files. Professor Edward Felten of Princeton University and some colleagues took up the challenge and duly

---

[89] 17 U.S.C. §1201(g)(4).

[90] *Universal City Studios, Inc. v. Reimerdes*, 82 F. Supp. 2d 211 (S.D.N.Y. 2000).

[91] The judgement was then upheld on appeal in *Universal City Studios, Inc. v. Corley*, 273 F. 3d 429 (2d Cir. 2001).

[92] *United States v. ElcomSoft Ltd* 203 F. Supp. 2d 1111, 1119 (N.D. Cal. 2002). See also *RealNetworks, Inc. v. Streambox, Inc.* 2000 WL 127311 (W.D. Wash. 2000).

decrypted the watermarked files. Foregoing a possible $10 000 reward for assigning SDMI the rights to their research about the watermarked technologies, they decided instead to present their findings at a conference. At this point the Recording Industry Association of America (RIAA), acting on behalf of the SDMI, sent Felten a letter threatening legal action under the DMCA[93] on the grounds that he would be disseminating information on how to circumvent encrypted devices. On receipt of the letter Felten withdrew his paper from the conference. Shortly afterwards, with the assistance of the Electronic Frontier Foundation (EFF) Felton began a legal action against the RIAA requesting a declaratory judgement that the DMCA was unconstitutional in that it breached the First Amendment. The case was dismissed by the Court on the grounds that both the RIAA and government had made (non-binding) declarations that they would not pursue legal action under the DMCA against researchers who published research exposing encryption vulnerabilities, and that there was thus no cause of action [75; 129].

The current situation under the DMCA is that the encryption research rules remain on the statute book, and to date no successful challenge has been brought to the constitutionality of the law either under the First Amendment, or on the basis of the DMCA's diminution of the scope of 'fair use' [75]. It is unclear to what extent the law continues to 'chill' research into encryption vulnerabilities in proprietary products or dissuade 'non-researchers', under the definition in the Act, from exploring the area. What is clear is that attempts to 'step outside' corporately defined copyright licence terms, however restrictively drawn, and even where those attempts would otherwise fall within fair use, or other statutory provisions, can be prevented where the works are protected, be it ever so ineffectively, by an encryption technology. This degree of control exerted by rightsholders has potentially perturbing implications for personal privacy. A key privacy right in the US has been the right to read anonymously, to choose what you read without others, including law enforcement agencies and

other government bodies looking over your shoulder. In a digital environment, without a truly effective method of paying for works anonymously, and in which all works are secured by encryption and no decryption tools are permitted, the rightsholder, in principle, knows exactly who has purchased what and, potentially, where they intend to read or use it. Data aggregated between suppliers of digital works would then permit an increasingly sophisticated picture of a person's preferences in areas such as reading and viewing matter [27; 80]. The fact that this material would be held by the private sector rather than the government is unlikely to be of great comfort in the absence of effective informational privacy laws – private sector companies readily exchange information between themselves and to government agencies.

Since the DMCA was passed, the EU Copyright Directive[94] has been implemented across the EU Member States. Like the DMCA the Directive requires Member States to provide protection for technical measures to protect copyright, but while it states that this should not be permitted to hinder encryption research it provides no provision for how this should occur.[95] In the UK, the relevant legislative provision can be found in s.296ZA(2) of the Copyright Designs and Patent Act (CDPA) 1988 (as amended)[96] which states that the bar against technological circumvention does not apply where "a person, for the purposes of research into cryptography, does anything which circumvents effective technological measures unless in so doing, or in issuing information derived from that research, he affects prejudicially the rights of the copyright owner". While this may be commendably brief, it leaves numerous questions unanswered, not least the types of circumstances in which the interests of the copyright owner might be considered prejudicially affected, and how these might be balanced with public interest questions

---

[93]Letter from Matthew Oppenheim, Legal Counsel, RIAA, to Edward Felten (Apr. 9, 2000), http://www.cs.princeton.edu/sip/sdmi/riaaletter.html

[94]Directive 2001/29/EC of the European Parliament and of the Council of 22nd May 2001 on the harmonisation of certain aspects of copyright and related rights in the information society (O.J. No L167, 22.6.2001, p. 10).

[95]*Ibid.* at Recital 48.

[96]Amended by *The Copyright and Related Rights Regulations 2003* (SI 2003 No 2498), http://www.opsi.gov.uk/si/si2003/20032498.htm

[64; 149]. Here too, therefore, the possibility of a chilling of encryption research, a concomitant reduction in the availability of tools to protect legal rights to access to works, and the possibility of reduced privacy in access to digital works remains, although the Data Protection Act 1998 would, in principle, offer some limited privacy protection.

28.7 CONCLUSIONS

As can be seen here, the laws that relate to encryption in the West have, over time, been heavily influenced by the US. Initially, that influence arose primarily out of the national security agenda, which was driven by the tensions of the Cold War. During that period, which spanned from the end of the Second World War to the mid-1970s, the issue of encryption as a personal privacy tool for the general public was very largely moot. At the end of the Cold War, with the rapid expansion of globalized trade, the US concerns with national security were increasingly overtaken by the speed of technological developments, international commercial competition, and a desire amongst states to set their own national (and, in the case of the EU, supranational) agendas in the new political environment. Even amongst the US's closest allies, there was recognition that the US national security agenda, and in particular the activities of its intelligence agencies, could be as easily turned to commercial as to military ends, and that to accept an encryption regime developed by the NSA might leave the communications of their industries and citizens open to unwelcome scrutiny.[97] The US was not averse to using its political economic and military influence to attempt to sway the encryption debate in the international arena, but those efforts were increasingly undermined by opposition to encryption

---

[97]For example, in 1999, India's Defense Research and Development Organization (DRDO) publicly raised concerns that US encryption software was insufficiently secure, stating that "... no encryption software products can be exported from the U.S. if they are too strong to be broken by the U.S. National Security Agency ..." and that "... when we buy an imported software product that is a 'black box' to us, we cannot be sure that the software package does not contain a time bomb of sorts, to cause havoc to the network when an external command is issued by a hostile nation" [46; 107: 444].

controls at home, and the collapse of international consensus on encryption restrictions abroad.

The 1990s and early 2000s saw increasing concerns about the extent to which encryption would hinder not just national security, but also law enforcement agencies. This resulted in consistent lobbying by those bodies for greater interception and decryption powers. With hindsight, those concerns now appear rather overblown – even today, there is little indication that encryption is unduly hampering either national security or law enforcement activities, or that provision of extra powers in those areas would have prevented any of the serious world-wide national security events of recent years. In many respects, encryption policy appears to have receded from most political and legal agendas, and it is noticeable that there has been a concomitant drop in interest/coverage in practitioner and academic legal journals in the last 2–3 years. Even in privacy law circles, where Privacy Enhancing Technologies (PETs) are still often considered to be an important part of information privacy strategies, encryption receives relatively little attention [132].

It is interesting to speculate why this might be. Undoubtedly, part of the answer is that many of the encryption processes and technologies that members of the public will use are, to all intents and purposes, invisible to them; from the software linking the ATMs that they use on the High Street to their bank, through to the Secure Sockets Layer (SSL) and Transport Layer Security (TLS) protocols that secure their Internet browser communications. In many cases, frankly, that is precisely the way the public wants their encryption services to work, and if encryption services are any more difficult than that to use, then they will probably not be used. Research suggests that most people's concept of the privacy they require is driven not by a strong concern with privacy rights generally, but primarily by the context in which they are operating, and in most contexts encryption of communications or content would be perceived as making normal social interactions unnecessarily complex [1].

The decline in clamour for stricter controls on encryption for national security or law enforcement purposes can perhaps to be seen as a positive development – an acceptance that civilian cryptography can have significant benefits without causing the decline and fall of civilisation as we know it. However, the increasing use of encryption as a tool to lock down, and to permit the hyper-commoditisation of, information goods, when backed by laws which forbid or restrict the rights of citizens to engage in self-help when their legal (and sometimes constitutional) rights are infringed upon by commercial interests, is undoubtedly a negative one. It seems possible therefore that the next round of encryption-focused civil rights battles will not be waged against the ambitions of law enforcement and national security agencies, but rather against those of information society content providers and data aggregators.

REFERENCES

Materials cited

[1] P. 6, *Who wants privacy protection, and what do they want?*, Journal of Consumer Behaviour 2(1) (2002), 80–100.

[2] C.W. Adams, *Legal requirements for the use of keystroke loggers*, **First International Workshop on Systematic Approaches to Digital Forensic Engineering** (SADFE'05) (2005), pp. 142–154.

[3] Y. Akdeniz, *UK Government Policy on Encryption*, 1 Web Journal of Current Legal Issues 1 (1997), Web-only journal, no page numbers, available at http://webjcli.ncl.ac.uk/1997/issue1/akdeniz1.html

[4] Y. Akdeniz et al., *Cryptography and liberty: can the trusted third parties be trusted? A critique of the recent UK proposals*, The Journal of Information, Law and Technology (JILT), 2 (1997), Web-only journal, no page numbers, available at http://www2.warwick.ac.uk/fac/soc/law/elj/jilt/1997_2/akdeniz/

[5] Y. Akdeniz, C. Walker, *UK Government policy on encryption: trust is the key*, Journal of Civil Liberties 3 (1998), 110–116.

[6] Y. Akdeniz, C. Walker, *Whisper who dares: encryption, privacy rights and the new world disorder*, **The Internet, Law and Society**, Y. Akdeniz, C. Walker, D. Wall, eds, Longman, London (2000), pp. 317–348.

[7] R.J. Anderson, *Crypto in Europe – markets, law and policy in cryptography*, **Policy and Algorithms**, E. Dawson, J. Golic, eds, LNCS, Vol. 1029, Springer-Verlag (1996), pp. 75–89.

[8] S. Andrews, *Who holds the key? – A comparative study of US and European encryption policies*, The Journal of Information, Law and Technology (JILT)

2 (2000), Web-only journal, no page numbers, available at http://www2.warwick.ac.uk/fac/soc/law/elj/jilt/2000_2/andrews/

[9] S.A. Baker, *Decoding OECD guidelines for cryptography policy*, International Lawyer 31(3) (1997), 729–756.

[10] S.A. Baker, P.R. Hurst, **The Limits of Trust: Cryptography, Governments and Electronic Commerce**, Kluwer Law International, The Hague (1998).

[11] R. Baldwin, M. Cave, **Understanding Regulation: Theory, Strategy and Practice**, Oxford University Press, Oxford (1999).

[12] J. Bamford, **The Puzzle Palace: A Report on America's Most Secret Agency**, Penguin Books, New York, NY (1983).

[13] D. Banisar, *Stopping science: the case of cryptography*, Health Matrix: Journal of Law–Medicine 9 (1999), 253–287.

[14] W. Baugh, D. Denning, **Encryption and Evolving Technologies: Tools of Organized Crime and Terrorism**, U.S. Working Group on Organized Crime, National Strategy Information Center (1997).

[15] T.E. Black, *Taking account of the World as it will be: the shifting course of U.S. encryption policy*, Federal Communications Law Journal 53 (2001), 289–314.

[16] M. Blaze, *Protocol failure in the escrowed encryption standard*, **Proceedings of the 2nd ACM Conference on Computer and Communications Security**, ACM, ACM Press (1994), pp. 59–67.

[17] B. Bolinger, *Focusing on infringement: why limitations on decryption technology are not the solution to policing copyright*, Case Western Reserve Law Review 52 (2002), 1091–1111.

[18] A.C. Bonin, *Protecting protection: first and fifth amendment challenges to cryptography regulation*, University of Chicago Legal Forum (1996), 495–518.

[19] J.J. Browder, *Encryption source code and the first amendment*, Jurimetrics 40 (2000), 431–444.

[20] J. Camp, K. Lewis, *Code as speech: a discussion of Bernstein v. USDOJ, Karn v. USDOS, and Junger v. Daley in light of the U.S. Supreme Court's recent shift to Federalism*, Ethics and Information Technology 3(1) (2001), 21–33.

[21] N.E. Carrell, *Spying on the Mob:* United States v. Scarfo – *A Constitutional Analysis*, University of Illinois Journal of Law, Technology and Policy (2002), 193–214.

[22] J.J. Carter, *The Devil and Daniel Bernstein: Constitutional flaws and practical fallacies in the encryption export controls*, Oregon Law Review 76 (1997), 981–1025.

[23] A. Charlesworth, *Implementing the European data protection directive 1995 in UK law: the Data Protection Act 1998*, Government Information Quarterly 16 (1999), 203–240.

[24] A. Charlesworth, *The future of data protection law*, Information Security Technical Report 11 (2006), 46–54.

[25] A.M. Clemens, *No computer exception to the constitution: the fifth amendment protects against compelled production of an encrypted document or private key*, UCLA Journal of Law and Technology (2), Web-only journal, no page numbers, available at http://lawtechjournal.com/articles/2004/02_040413_clemens.php

[26] C.A. Cockburn, *Where the United States goes the World will follow – won't it*, Houston Journal of International Law 21 (1999), 491–530.

[27] J.E. Cohen, *A right to read anonymously: a closer look at "copyright management" in cyberspace*, Connecticut Law Review 28 (1996), 981–1039.

[28] Committee on Science, Engineering, and Public Policy, **Scientific Communication and National Security**, National Academy Press, Washington, DC (1982).

[29] Computer Science and Telecommunications Board, **Computers at Risk: Safe Computing In the Information Age**, National Academy Press, Washington, DC (1991).

[30] D. Coppersmith, *The data encryption standard (DES) and its strength against attacks*, IBM Journal of Research and Development 38(3) (1994), 243–250.

[31] C.F. Corr, *The wall still stands – complying with export controls on technology transfers in the post-Cold War, Post-9/11 Era*, Houston Journal of International Law 25 (2002–2003), 441–530.

[32] N.A. Crain, *Bernstein, Karn, and Junger: constitutional challenges to cryptographic regulations*, Alabama Law Review 50 (1999), 869–909.

[33] K.W. Dam, H.S. Lin (eds), **Cryptography's Role in Securing the Information Society**, National Academy Press, Washington, DC (1996).

[34] G.I. Davida, *The case against restraints on nongovernmental research in cryptography* (A minority report of the Public Cryptography Study Group prepared for the American Council on Education), Notices of the American Mathematical Society (October 1981), 524–526; also Communications of the Association for Computing Machinery 24(7), 445–450.

[35] D.E. Denning, *The US key escrow encryption technology*, **Building in Big Brother: The Cryptographic Policy Debate**, L.J. Hoffman, ed., Springer-Verlag, New York, NY (1995), pp. 111–118.

[36] D.E. Denning, W.E.J. Baugh, *Key escrow encryption policies and technologies*, Villanova Law Review 41 (1996), 289–303.

[37] Department of Commerce and the National Security Agency, A Study of The International Market for Computer Software with Encryption, Prepared for the Interagency Working Group on Encryption and Telecommunications Policy (January 11, 1996).

[38] W. Diffie, M.E. Hellman, *New directions in cryptography*, IEEE Transactions on Information Theory 22 (6) (1976), 644–654.

[39] W. Diffie, M.E. Hellman, *Privacy and Authentication: An Introduction to Cryptography*, Proceedings of the IEEE 67(3) (1979), 397–427.

[40] W. Diffie, S. Landau, **Privacy on the Line: The Politics of Wiretapping and Encryption**, MIT Press, Cambridge, MA (1998).

[41] K.A. Dursht, *From containment to cooperation: collective action and the Wassenaar arrangement*, Cardozo Law Review 19 (3) (1997), 1079–1123.

[42] B.S. DuVal Jr., *The occasions of secrecy*, University of Pittsburgh Law Review 47 (1986), 579–674.

[43] S.J. Edgett, *Double-clicking on fourth amendment protection: encryption creates a reasonable expectation of privacy*, Pepperdine Law Review 30(2) (2003), 339–366.

[44] Electronic Frontier Foundation, **Cracking DES**, O'Reilly, Sebastopol, CA (1998).

[45] C. Ellison, *Oppression Net*, Economic Affairs 20(1) (2000), 21–28.

[46] Electronic Privacy Information Center, Cryptography and Liberty 1999: An International Survey of Encryption Policy, EPIC, Washington, DC (1999).

[47] C.B. Escobar, *Nongovernmental cryptology and national security: the government seeking to restrict research*, Computer/Law Journal 4 (1984), 573–603.

[48] B.W. Esler, *Protecting the protection: a trans-Atlantic analysis of the emerging right to technological self-help*, IDEA 43 (2003), 553–606.

[49] C.L. Evans, *U.S. export control of encryption software: efforts to protect national security threaten the U.S. software industry's ability to compete in foreign markets*, North Carolina Journal of International Law and Commercial Regulation 19(3) (1994), 469–490.

[50] A. Fleischmann, *Personal data security: divergent standards in the European Union and the United States*, Fordham International Law Journal 19(1) (1995), 143–180.

[51] S.M. Flynn, *A puzzle even the codebreakers have trouble solving: a clash of interests over the electronic encryption standard*, Law and Policy in International Business 27 (1995), 217–246.

[52] V.M. Fogleman, J.E. Viator, *The critical technologies approach: controlling scientific communication for the national security*, BYU Journal of Public Law 4(2) (1990), 293–394.

[53] R.A. Franks, *The National Security Agency and its interference with private sector computer security*, Iowa Law Review 72(4) (1987), 1015–1039.

[54] J.A. Fraser III, *The use of encrypted, coded and secret communications is an "Ancient Liberty" protected by the United States Constitution*, Virginia Journal of Law and Technology 2(2) (1997),

Web-only journal, no page numbers, available at http://www.vjolt.net/vol2/issue/vol2_art2.pdf

[55] A.M. Froomkin, *The Metaphor is the key: cryptography, the clipper chip, and the constitution*, University of Pennsylvania Law Review 143(3) (1995), 709–897.

[56] A.M. Froomkin, *It came from planet clipper: the battle over cryptographic key escrow*, University of Chicago Legal Forum (1996), 15–75.

[57] R. Funk, *National security controls on the dissemination of privately generated scientific information*, UCLA Law Review 30 (1982), 405–454.

[58] K. Gatien, *How encryption and national security will affect the future of digital film distribution*, Southwestern Journal of Law and Trade in the Americas 8 (2001), 229–249.

[59] K.E. Gegner, S.B. Veeder, *Standards setting and federal information policy: the escrowed encryption standard (EES)*, Government Information Quarterly 11(4) (1994), 403–422.

[60] A. Gersho, *Unclassified summary: involvement of NSA in the development of the data encryption standard*, Communications Magazine, IEEE 16(6) (1978), 53–55.

[61] L.A. Gilbert, *Patent secrecy orders: the unconstitutionality of interference in civilian cryptography under present procedures*, Santa Clara Law Review 22(2) (1982), 325–374.

[62] G. Gordon, *Breaking the code: what encryption means for the first amendment and human rights*, Columbia Human Rights Law Review 32 (2001), 477–516.

[63] D. Grover, *Dual encryption and plausible deniability*, Computer Law & Security Report 20(1) (2004), 37–40.

[64] A. Guadamuz, *Trouble with prime numbers: DeCSS, DVD and the protection of proprietary encryption tools*, The Journal of Information, Law and Technology (JILT) 2002 (3) (2004), Web-only journal, no page numbers, available at http://www2.warwick.ac.uk/fac/soc/law/elj/jilt/2002_3/guadamuz2/

[65] J. Guisnel, *Secret Power*, Craig Potton Publishing, New Zealand (1996).

[66] L.J. Gurak, *Persuasion and Privacy in Cyberspace: The Online Protests over Lotus MarketPlace and the Clipper Chip*, Yale University Press, New Haven, CT (1997).

[67] E.F. Haignere, *An overview of the issues surrounding the encryption exportation debate, their ramifications, and potential resolution*, Maryland Journal of International Law and Trade 22(2) (1998), 319–358.

[68] M.B. Hartzler, *National security export controls on data encryption – how they limit U.S. competitiveness*, Texas International Law Journal 29 (1994), 437–456.

[69] N. Hartzog, *The magic Lantern revealed: a report of the FBI's new "Key Logging" trojan and analysis of its possible treatment in a dynamic legal landscape*, John Marshall Journal of Computer and Information Law 20 (2002), 287–320.

[70] M.E. Hellman, *Letter "Computer Encryption: Key Size*, Science 198(4312) (1977).

[71] C.D. Hoffman, *Encrypted digital cash transfers: why traditional money laundering controls may fail without uniform cryptography regulations*, Fordham International Law Journal 21(3) (1998), 799–860.

[72] L.J. Hoffman (ed.), **Building in Big Brother: The Cryptographic Policy Debate**, Springer-Verlag, New York, NY (1995).

[73] L.J. Hoffman, F.A. Ali et al., *Cryptography policy*, Communications of the Association for Computing Machinery 37(9) (1994), 109–117.

[74] L.J. Hoffman, D.M. Balenson, K.A. Metivier-Carreiro, A. Kim, M.G. Mundy, Growing Development of Foreign Encryption Products in the Face of U.S. Export Regulations, Report No. GWU-CPI-1999-02 Cyberspace Policy Institute, The George Washington University, June 10 (1999).

[75] C. Imfeld, *Playing fair with fair use? The digital millennium copyright act's impact upon encryption researchers and academicians*, Communication Law and Policy 8 (2003), 111–144.

[76] N.J. James, *Handing over the keys: contingency, power and resistance in the context of Section 3LA of the Australian Crimes Act 1914*, University of Queensland Law Journal 23 (2004), 10–21.

[77] D. Kahn, *Cryptology Goes Public*, Foreign Affairs 58 (1979), 141–159.

[78] D. Kahn, **The Codebreakers: The Comprehensive History of Secret Communication from Ancient Times to the Internet**, Scribner, New York, NY (1996).

[79] R.D. Kamenshine, *Embargoes on exports of ideas and information: first amendment issues*, William and Mary Law Review 26 (1985), 863–895.

[80] S.A. Katyal, *Privacy vs. Piracy*, Yale Journal of Law and Technology 7 (2004–2005), 222–345.

[81] G. Kennedy, *Codemakers, codebreakers and rulemakers: dilemmas in current encryption policies*, Computer Law and Security Report 16(4) (2000), 240–247.

[82] J. Kerben, *The dilemma for future communication technologies: how to constitutionally dress the cryptogenie*, CommLaw Conspectus 5 (1997), 125–152.

[83] O.S. Kerr, *The fourth amendment in cyberspace: can encryption create a reasonable expectation of privacy*, Connecticut Law Review 33 (2001), 503–533.

[84] H.R. King, *Big Brother, the holding company: a review of key-escrow encryption technology*, Rutgers Computer and Technology Law Journal 21 (1995), 224–262.

[85] M.I. Koffsky, *Choppy Waters in the surveillance data stream: the clipper scheme and the particularity clause*, High Technology Law Journal 9 (1994), 131–149.

[86] G.B. Kolata, *Computer encryption and the National Security Agency connection*, Science 197(4302) (1977), 438–440.

[87] G.B. Kolata, *Cryptography: a new clash between academic freedom and national security*, Science 209(4460) (1980), 995–996.

[88] G.B. Kolata, *NSA to provide secret codes*, Science 230(4721) (1985), 45–46.

[89] B.-J. Koops, **The Crypto Controversy: A Key Conflict in the Information Society**, Kluwer Law International, The Hague (1999).

[90] V. Ku, *A critique of the Digital Millennium Copyright act's exemption on encryption research: is the exemption too narrow?*, Yale Journal of Law and Technology 7 (2005), 38.

[91] S. Landau, *Primes, codes and the National Security Agency*, Notices of the American Mathematical Society 30(1) (1983), 7–10.

[92] S. Landau, *Standing the test of time: the data encryption standard*, Notices of the American Mathematical Society 47(3) (2000), 341–349.

[93] S. Landau et al., **Codes, Keys and Conflicts: Issues in U.S. Crypto Policy**, Association for Computing Machinery Inc., New York (1994).

[94] E. Lauzon, *The Philip Zimmermann investigation: the start of the fall of export restrictions on encryption software under first amendment free speech issues*, Syracuse Law Review 48(3) (1998), 1307–1364.

[95] T.B. Lennon, *Fourth amendment's prohibitions on encryption limitation: will 1995 be like 1984*, Albany Law Review 58(2) (1994), 467–508.

[96] L. Lessig, *The architecture of privacy*, Vanderbilt Journal of Entertainment Law and Practice 1 (1999), 56–65.

[97] L. Lessig, **Code and Other Laws of Cyberspace**, Basic Books, New York, NY (2000).

[98] S. Levy, **Crypto: How Code Rebels Beat the Government – Saving Privacy in the Digital Age**, Viking, New York, NY (2001).

[99] M. Maher, *International protection of U.S. law enforcement interests in cryptography*, Richmond Journal of Law and Technology 5(3) (1999), available at http://law.richmond.edu/jolt/v5i3/maher.txt

[100] J.C. Mandelman, *Lest we walk into the well: guarding the keys – encrypting the constitution: to speak, search & (and) seize in cyberspace*, Albany Law Journal of Science and Technology 8 (1998), 227–303.

[101] L.H. Marino, *The U.S. export control regime for encryption. Is liberalization on the horizon?*, The Journal of World Intellectual Property 1(1) (1998), 101–119.

[102] C.L. Martin, *Exalted technology: should CALEA be expanded to authorize Internet wiretapping?*, Rutgers Computer and Technology Law Journal 32 (2005), 140–182.

[103] R.S. Martin, *Watch what you type: as the FBI records your keystrokes, the fourth amendment develops carpal tunnel syndrome*, American Criminal Law Review 40 (2003), 1271–1300.

[104] D. McClure, *First amendment freedoms and the encryption export battle: deciphering the importance of*

Bernstein v. United States Department of Justice, 176 F.3d 1132 (9th Cir. 1999), Nebraska Law Review 79(2) (2000), 465–484.

[105] J.F. McKenzie, *Implementation of the core list of export controls: computer and software controls*, Software Law Journal 5 (1992), 1–23.

[106] L. McLaughlin, *After ElcomSoft: DMCA still worries developers, researchers*, IEEE Software 20(2) (2003), 86–91.

[107] F.L. McNulty, *Encryption's importance to economic and infrastructure security*, Duke Journal of Comparative and International Law 9(2) (1999), 427–449.

[108] S.L. Mhlaba, *The efficacy of international regulation of transborder data flows: the case for the clipper chip*, Government Information Quarterly 12(4) (1995), 353–366.

[109] K.A. Moerke, *Free speech to a machine – encryption software source code is not constitutionally protected speech under the first amendment*, Minnesota Law Review 84 (2000), 1007–1049.

[110] G. Moore, *Cramming more components onto integrated circuits*, Electronics Magazine 38 (1965), 114–117, 19 April, 1965.

[111] D.T. Movius, *Bernstein v. United States Department of State: encryption, justiciability, and the first amendment*, Administrative Law Review 49 (1997), 1051–1070.

[112] N. Muenchinger, *The highlights of the new legal regime on encryption in France*, Computer Law and Security Report 15(2) (1999), 86–89.

[113] N. Muenchinger, *The highlights of the new legal regime on encryption in France – Part II*, Computer Law and Security Report 15(4) (1999), 234–237.

[114] R.A. Murr, *Privacy and encryption in cyberspace: first amendment challenges to ITAR, EAR and their successors*, San Diego Law Review 34(3) (1997), 1401–1462.

[115] National Bureau of Standards, **Data Encryption Standard**, Federal Information Processing Standards Publication 46, Department of Commerce: National Technical Information Service, January 15 (1977).

[116] National Institute of Standards and Technology, *The digital signature standard, proposal and discussion*, Communications of the Association for Computing Machinery 35(7) (1992), 36–54.

[117] National Institute of Standards and Technology, **Escrowed Encryption Standard**, Federal Information Processing Standards Publication 185, 9 February (1994).

[118] National Institute of Standards and Technology, **Digital Signature Standard**, Federal Information Processing Standards Publication 186 (FIPS 186), 19 May (1994).

[119] National Institute of Standards and Technology, **Advanced Encryption Standard** (AES), Federal Information Processing Standards Publication 197 (FIPS 197), November 26 (2001).

[120] National Research Council, *Voluntary restraints on research with national security implications: the case of cryptography, 1972–1982*, **Scientific Communication and National Security**, National Academy Press, Washington, DC (1982), Appendix E, pp. 120–125.

[121] K.M. Nelson, *The clipper initiative: fact or fiction in future encryption policy*, Hamline Journal of Public Law & Policy 16 (1994–1995), 291–311.

[122] T. Nguyen, *Cryptography, export controls, and the first amendment in Bernstein v. United States Department of State*, Harvard Journal of Law and Technology 10(3) (1997), 667–682.

[123] Y.C. Ocrant, *A constitutional challenge to encryption export regulations: software is speechless*, DePaul Law Review 48 (1998), 503–557.

[124] P.H. Oettinger, *National discretion: choosing CoCom's successor and the New Export Administration Act*, American University Journal of International Law and Policy 9(2) (1994), 559–595.

[125] R. O'Harrow Jr, *Justice Dept. Pushes for Power to unlock PC security systems. Covert Acts could target homes, offices*, The Washington Post, August 20 (1999), A01.

[126] M.J. O'Neil, J.X. Dempsey, *Critical infrastructure protection: threats to privacy and other civil liberties and concerns with government mandates on industry*, DePaul Business Law Journal 12 (1999–2000), 97–129.

[127] E.J. Park, *Protecting the core values of the first amendment in an age of new technologies: scientific expression vs. national security*, Virginia Journal of Law and Technology 2(3) (1997), Web-only journal, no page numbers, available at http://www.vjolt.net/vol2/issue/vol2_art3.pdf

[128] M.T. Pasko, *Re-defining national security in the technology age: the encryption export debate*, Journal of Legislation 26 (2000), 337–353.

[129] A. Perkins, *Encryption use: law and anarchy on the digital frontier*, Houston Law Review 41 (2005), 1625–1657.

[130] F. Petitcolas et al. (eds), **Information Hiding Techniques for Steganography and Digital Watermarking**, Artech House, Norwood, MA (1999).

[131] D.J. Phillips, *Cryptography, secrets, and the structuring of trust*, **Technology and Privacy: The New Landscape**, P. E. Agre, M. Rotenberg, eds, MIT Press, Cambridge MA (1998), pp. 243–276.

[132] D.J. Phillips, *Private policy and PETs: the influence of policy regimes on the development and social implications of privacy-enhancing technologies*, New Media and Society 6(6) (2004), 691–706.

[133] K.J. Pierce, *Public cryptography, arms export controls, and the first amendment: a need for legislation*, Cornell International Law Journal 17(1) (1984), 197–236.

[134] L.M. Pilkington, *First and fifth amendment challenges to export controls on encryption: Bernstein and Karn*, Santa Clara Law Review 37 (1996), 159–211.

[135] S.A. Price, *Understanding contemporary cryptography and its wider impact upon the general law*, International Review of Law, Computers & Technology 13(2) (1999), 95–126.

[136] Public Cryptography Study Group, Report of the Public Cryptography Study Group, American Council on Education, Washington, DC, February 1981. Reprinted in the Notices of the American Mathematical Society, October (1981), 518–526.

[137] A.S. Reid, N. Ryder, *For whose eyes only? A critique of the United Kingdom's Regulation of Investigatory Powers Act 2000*, Information and Communications Technology Law 10(2) (2001), 179–201.

[138] P.E. Reiman, *Cryptography and the first amendment: the right to be unheard*, John Marshall Journal of Computer and Information Law 14 (1996), 325–345.

[139] P.R. Reitinger, *Compelled production of plaintext and keys*, University of Chicago Legal Forum (1996), 171–206.

[140] P.R. Reitinger, *Encryption, anonymity and markets*, **Cybercrime: Law Enforcement, Security and Surveillance in the Information Age**, D. Thomas, B.D. Loader, eds, Routledge, New York, NY (2000), pp. 132–152.

[141] H.C. Relyea, **Silencing Science: National Security Controls and Scientific Communication**, Ablex Publishing Corporation, Norwood, NJ (1994).

[142] R.L. Rivest, A. Shamir, L. Adleman, *On digital signatures and public key cryptosystems*, Communications of the Association for Computing Machinery 21(2) (1978), 120–126, February 1978.

[143] D.R. Rua, *Cryptobabble: how encryption export disputes are shaping free speech for the New Millennium*, North Carolina Journal of International Law and Commercial Regulation 24(1) (1998), 125–198.

[144] J.M. Ryan, *Freedom to speak unintelligibly: the first amendment implications of government-controlled encryption*, William and Mary Bill of Rights Journal 4 (1996), 1165–1222.

[145] K. Scheurer, *The clipper chip: cryptography technology and the constitution – the government's answer to encryption "Chips" away at constitutional rights*, Rutgers Computer and Technology Law Journal 21 (1995), 263–291.

[146] P. Schweizer, **Friendly Spies: How America's Allies Are Using Economic Espionage to Steal Our Secrets**, Atlantic Monthly, New York (1993).

[147] G.M. Segell, *French cryptography policy: the turnabout of 1999*, International Journal of Intelligence and CounterIntelligence 13(3) (2000), 345–358.

[148] G. Sergienko, *Self-incrimination and cryptographic keys*, Richmond Journal of Law and Technology 2(1) (1996), Web-only journal, no page numbers, available at http://www.richmond.edu/jolt/v2i1/sergienko.html

[149] A. Shah, *The UK's implementation of the anti-circumvention provisions of the EU copyright directive:*

an analysis, Duke Law and Technology Review 0003 (2004), Web-only journal, no page numbers, available at http://www.law.duke.edu/journals/dltr/articles/2004dltr0003.html

[150] D. Shapley, *NSA slaps secrecy order on inventors' communications patent*, Science 201(4359) (1978), 891, 893, 894.

[151] D. Shapley, *Intelligence Agency chief seeks "Dialogue" with academics*, Science 202(4366) (1978), 407–410.

[152] D. Shapley, G.B. Kolata, *Cryptology: scientists puzzle over threat to open research, publication*, Science 197 (4311) (1977), 1345–1349.

[153] J. Shearer, P. Gutmann, *Government, cryptography, and the right to privacy*, Journal of Universal Computer Science 2(3) (1996), 113–146.

[154] K.K. Shehadeh, *The Wassenaar arrangement and encryption exports: an ineffective export control regime that compromises United States economic interests*, American University of International Law Review 15(1) (1999), 271–319.

[155] A.M. Shinn, *The first amendment and the export laws: free speech on scientific and technical matters*, George Washington Law Review 58 (1990), 368–403.

[156] S. Singh, **The Code Book: The Evolution of Secrecy from Mary Queen of Scots to Quantum Cryptography**, Doubleday, New York, NY (1999).

[157] A. Singhal, *The piracy of privacy – a fourth amendment analysis of key escrow cryptography*, Stanford Law & Policy Review 7(2) (1996), 189–210.

[158] D.J. Solove, *Digital dossiers and the dissipation of fourth amendment privacy*, Southern California Law Review 75 (2002), 1083–1167.

[159] R.J. Stay, *Cryptic controversy: U.S. government restrictions on cryptography exports and the plight of Philip Zimmermann*, Georgia State University Law Review 13(2) (1997), 581–604.

[160] R.C. Thomsen II, A.D. Paytas, *US encryption export regulation: US to EU: me too! – the United States amends its export controls on encryption, responding to recent developments in the EU*, Computer Law & Security Report 17(1) (2001), 11–16.

[161] L. Tien, *Doors, envelopes, and encryption: the uncertain role of precautions in fourth amendment law*, DePaul Law Review 54 (2005), 873–908.

[162] C.E. Torkelson, *The clipper chip: how key escrow threatens to undermine the fourth amendment*, Seton Hall Law Review 25(3) (1995), 1142–1175.

[163] A. Torrubia, F.J. Mora et al., *Cryptography regulations for e-commerce and digital rights management*, Computers & Security 20(8) (2001), 724–738.

[164] Z.M. Vedder-Brown, *Government regulation of encryption: the entry of "Big Brother" or the Status Quo?*, American Criminal Law Review 35 (1998), 1387–1414.

[165] H. Victor, *Big Brother is at your back door: an examination of the effect of encryption regulation on privacy and crime*, John Marshall Journal of Computer and Information Law 18 (2000), 825–872.

[166] M.P. Voors, *Encryption regulation in the wake of September 11, 2001: must we protect national security at the expense of the economy*, Federal Communications Law Journal 55 (2002–2003), 331–352.

[167] J. Walsh, *Shunning cryptocensorship*, Science 212 (4500) (1981), 1250.

[168] K.P. Weinberg, *Key recovery shaping cyberspace (pragmatism and theory)*, Journal of Intellectual Property Law 5(2) (1998), 667–700.

[169] C. Weissman, *A national debate on encryption exportability*, Communications of the Association for Computing Machinery 34(10) (1991), 162.

[170] C.R. White, *Decrypting the politics: why the Clinton administration's national cryptography policy will continue to be dictated by the national economic interest*, CommLaw Conspectus 7 (1999), 193–205.

[171] D.F. Wolfe, *The government's right to read: maintaining state access to digital data in the Age of Impenetrable Encryption*, Emory Law Journal 49 (2000), 711–744.

[172] C. Woo, M. So, *The case for magic lantern: September 11 highlights the need for increased surveillance*, Harvard Journal of Law & Technology 15 (2002), 521–538.

[173] J. Yeates, *CALEA and the RIPA: The U.S. and the U.K. responses to wiretapping in an increasingly wireless world*, Albany Law Journal of Science & Technology 12 (2001), 125–166.

Materials reviewed, but not cited

[1] W.A. Ackerman, *Encryption: a 21st century national security dilemma*, International Review of Law, Computers & Technology 12(2) (1998), 371–394.

[2] J.B. Altman, W. McGlone, *Demystifying U.S. encryption export controls*, American University Law Review 46 (1996–1997), 493–510.

[3] B. Barnard, *Leveraging worldwide encryption standards via U.S. export controls: The U.S. government's authority to safeguard the global information infrastructure*, Columbia Business Law Review (1997), 429–484.

[4] G.B. Barrett, *Law of diminishing privacy rights: encryption escrow and the dilution of associational freedoms in cyberspace*, New York Law School Journal of Human Rights 15 (1998), 115–140.

[5] R.C. Barth, C.N. Smith, *International regulation of encryption: technology will drive policy*, **Borders in Cyberspace: Information Policy and the Global Information Infrastructure**, B. Kahin, J.H. Keller, eds, MIT Press, Cambridge, MA (1997), pp. 283–299.

[6] W.P. Berryessa, *Escrowed encryption systems: current public policy may destroy valued constitutional protections*, University of Dayton Law Review 23 (1997–1998), 59–85.

[7] R. Bessette, V. Haufler, *Against all odds: why there is no international information regime*, International Studies Perspectives 2(1) (2001), 69–92.

[8] K. Bharvada, *Electronic signatures, biometrics and PKI in the UK*, International Review of Law, Computers & Technology 16 (2002), 265–275.

[9] C. Crump, *Data retention: privacy, anonymity and accountability online*, Stanford Law Review 56 (2003), 191–230.

[10] M.-T.B. Dinh, *The U.S. encryption export policy: taking the byte out of the debate*, Minnesota Journal of Global Trade 7 (1998), 375–397.

[11] J.C. Erdozain, *Encryption methods and e-commerce. The proposed EU directive*, The Journal of World Intellectual Property 2(4) (1999), 509–516.

[12] A. Etzioni, **How Patriotic is the PATRIOT Act?**, Routledge, New York, NY (2004).

[13] M. Godwin, **Cyber Rights: Defending Free Speech in the Digital Age**, Times Books, New York, NY (1998).

[14] D. Goldstone, *Discussing the constitutionality of regulating the export of encryption products*, New York University Journal of Legislation and Public Policy 3 (1999–2000), 39–49.

[15] P.N. Grabosky, R.G. Smith, *Telecommunications and crime: regulatory dilemmas*, Law & Policy 19(3) (1997), 317–341.

[16] W.A. Hodkowski, *The future of Internet security: how new technologies will shape the Internet and affect the law*, Santa Clara Computer and High Technology Law Journal 13 (1997), 217–275.

[17] O.S. Kerr, *Internet surveillance law after the USA PATRIOT Act: the big brother that isn't*, Northwestern University Law Review 97 (2003), 607–673.

[18] D.R. Klopfenstein, *Deciphering the encryption debate: a constitutional analysis of current regulations and a prediction for the future*, Emory Law Journal 48 (1999), 765–807.

[19] B.-J. Koops, *A survey of cryptography laws and regulations*, Computer Law & Security Report 12(6) (1996), 349–355.

[20] B.-J. Koops, *Crypto regulation in Europe. Some key trends and issues*, Computer Networks and ISDN Systems 29(15) (1997), 1823–1831.

[21] H.S. Lin, *Cryptography and public policy*, Journal of Government Information 25(2) (1998), 135–148.

[22] R.S. Litt, *Crime in the Computer Age: the law enforcement perspective*, Texas Review of Law & Politics 4(1) (1999–2000), 59–68.

[24] W. Madsen, D. Banisar, **Cryptography and Liberty 2000: An International Survey of Cryptography Policy**, EPIC, Washington, DC (2000).

[23] W. Madsen, D.L. Sobel et al., *Cryptography and liberty: an international survey of encryption policy*, John Marshall Journal of Computer and Information Law 16 (1998), 475–527.

[25] D. Masson, *The Genie let loose: ineffectual encryption export restrictions and their deleterious effect on business*, Journal of Technology Law & Policy 3 (1996).

[26] D.L. Morgan, *Digital signatures: will government registration of users mean that anonymity in transactions on the Internet is forever lost*, University of Illinois Law Review (2004), 1005–1031.

[27] J.L. Paik, *The encryption export tax: a proposed solution and remedy to the issues and costs associated with exporting encryption technology*, Cornell Journal of Law and Public Policy 10(1) (2000), 161–193.

[28] V. Pednekar-Magal, P. Shields, *The State and telecom surveillance policy: the clipper chip initiative*, Communication Law and Policy 8 (2003), 429–464.

[29] K.M. Saunders, *The regulation of Internet encryption technologies: separating the wheat from the chaff*, John Marshall Journal of Computer and Information Law 17 (1999), 945–959.

[30] B. Schneier, D. Banisar, **The Electronic Privacy Papers: Documents on the Battle for Privacy in the Age of Surveillance**, Wiley, New York, NY (1997).

[31] C. Sehgal, *The power of the federal government in the Electronic Age*, Texas Review of Law & Politics. 4(1) (1999–2000), 77–84.

[32] S. Singleton, *Encryption policy for the 21st century: a future without government-prescribed key recovery*. Policy Analysis 325, 19 November (1998), available at http://www.cato.org/pubs/pas/pa325.pdf

[33] J.L. Snyder, *U.S. export controls on encryption software*, The Journal of World Intellectual Property 1(1) (1998), 37–54.

[34] J.L. Snyder, J.W. Reed, *Developments in encryption export controls: liberalization at last?*, The Journal of World Intellectual Property 2(6) (1999), 977–990.

[35] J.T. Soma, C.P. Henderson, *Encryption, key recovery, and commercial trade secret assets: a proposed legislative mode*, Rutgers Computer and Technology Law Journal 25 (1999), 97–134.

[36] J.T. Stender, *Too many secrets: challenges to the control of strong crypto and the national security perspective*, Case Western Reserve Journal of International Law 30(1) (1998), 287–337.

[37] M. Wright, *From key escrow to key recovery: variations on a theme*, Computer Fraud & Security 1997(9) (1997), 12–14.

[38] A.W. Yung, *Regulating the Genie: effective wiretaps in the Information Age*, Dickinson Law Review 101 (1996), 95–135.

Cases cited

[1] Amann v. Switzerland (2000) 30 EHRR 843.

[2] Bernstein v. U.S. Department of State 922 F. Supp. 1426 (N.D. Cal. 1996) (Bernstein I).

[3] Bernstein v. U.S. Department of State 945 F. Supp. 1279 (N.D. Cal. 1996) (Bernstein II).

[4] Bernstein v. U.S. Department of Commerce 974 F. Supp. 1288 (N.D. Cal. 1997) (Bernstein III).

[5] Bernstein v. U.S. Department of Justice, 176 F. 3d 1132 (9th Cir. 1999) (Bernstein IV).

[6] Bernstein v. U.S. Department of Justice, 192 F. 3d 1308 (9th Cir. 1999).

[7] FW/PBS v. City of Dallas 493 U.S. 215 (1990).

[8] Halford v. United Kingdom (1998) 24 EHRR 523.

[9] Huvig v. France (1990) 12 EHRR 528.

[10] Junger v. Daley, 8 F. Supp. 2d 708 (N.D. Ohio 1998).

[11] Junger v. Daley 209 F. 3d 481 (6th Cir. 2000).

[12] Karn v. U.S. Department of State 925 F. Supp. 1 (D.Ct. D.C. 1996).

[13] Kruslin v. France (1990) 12 EHRR 547.

[14] Malone v. United Kingdom (1984) 7 EHRR 14.

[15] RealNetworks, Inc. v. Streambox, Inc. 2000 WL 127311 (W.D. Wash. 2000).

[16] Smith v. Maryland 442 U.S. 735, 737 (1979).

[17] United States v. ElcomSoft Ltd 203 F. Supp. 2d 1111, 1119 (N.D. Cal. 2002).

[18] United States v. Freed 401 US 601 (1971).

[19] United States v. Miller 425 U.S. 435, 435 (1976).

[20] United States v. O'Brien 391 U.S. 367 (1968).

[21] United States v. Scarfo 180 F. Supp. 2d 572 (D.NJ. 2001).

[22] Universal City Studios, Inc. v. Reimerdes (Universal I), 82 F. Supp. 2d 211 (S.D.N.Y. 2000).

[23] Universal City Studios, Inc. v. Corley, 273 F. 3d 429 (2d Cir. 2001).

[24] Valenzuela Contreras v. Spain (1999) 28 EHRR 483.

PART 6

INFORMATION WARFARE

The History of Information Security: A Comprehensive Handbook
Karl de Leeuw and Jan Bergstra (Editors)
© Published by Elsevier B.V.

29

THE INFORMATION REVOLUTION AND THE TRANSFORMATION OF WARFARE[*]

Dan Kuehl

Information Resources Management College
National Defense University, USA
E-mail: kuehld@ndu.edu

Contents

29.1 INTRODUCTION

The past two decades have seen an intense debate within defense circles – both in the US and the global military community – over the meaning and impact of the Information Revolution and how it has manifested itself on a range of activities critical to national security, from the evolution of the Revolution in Military Affairs (RMA) to what the U.S. Department of Defense has termed Defense Transformation. Somewhere between the extremes of this debate – "just new wine in old bottles" vs. "the onset of bloodless and non-destructive warfare" – is where the truth will settle. Trying to determine how far on either side of the midpoint this truth will settle, however, is probably much less productive than establishing and exploring some of the major strategic areas that will be affected by this transformation. This essay posits three vital areas in which the Information Revolution is already shaping future national security: protecting a vast and growing number of computer-controlled critical infrastructures from malevolent acts and attacks; the growing importance of Information Operations to US and coalition military affairs; and the struggle for global influence in the "worldwide war of ideas". Several iterations of the US National Security Strategy established this approach to information and national security, and it will serve as the thematic thread around which this paper is organized.[1] First, however, it will be useful to examine this Information Revolution and how it has sparked the emergence of the new operational medium of cyberspace.[2]

29.2 CYBERSPACE

If the Information Revolution is at the heart of an evolving "revolution in military affairs" or RMA, when did this Information Revolution begin? [5].

---

[*]These comments reflect the author's opinions and should not be taken as the official position of the US Government, Department or National Defense University.

[1]See, for example, the last two such documents issued by the Clinton Administration, *A National Security Strategy for a New Century* (1999) and *A National Security Strategy for a Global Age* (2000), both available electronically at http://www.au.af.mil/au/awc/awcgate/nss/nss2000.htm and http://www.au.af.mil/au/awc/awcgate/nss/nss_dec2000_contents.htm

[2]For a particularly insightful perspective see [8].

There are various ways of looking at this revolution. One approach begins with Gutenberg's invention of the printing press in the 1450s, which made possible the widespread dissemination of standardized information. Another, more focused on technology, suggests that this revolution did not begin until the mid-19th century with the invention of the telegraph and then telephone, which made possible real-time information dissemination over long distances, first with symbolic information (dots and dashes), then with actual human voice. Even though this connectivity was extended to intercontinental distances with the advent of undersea telegraph cables, it still depended on a physical link to make the connectivity function. By the mid-20th century the need for a connecting wire had ended, as first the wireless-radio and then television enabled information dissemination via virtual connectivity. The onset of the space age and satellite-enabled connectivity has truly moved this revolution onto a global stage and made possible "omniconnectivity" – the ability for everyone to connect and communicate in every direction simultaneously.

This technological expansion of the means of communication was mirrored by an expansion of the physical environments of warfare. Until the early 20th century warfare was two-dimensional: armies waged war on land, navies waged war at sea, a condition that had lasted since the beginning of war thousands of years ago. The 20th century saw the expansion from two to five physical and operational environments. First, the invention of the internal combustion engine enabled the extension of war at sea to beneath the sea, and it extended warfare into the aerospace above both the land and sea. The last of the three changes, involving spaceflight and the employment of outer space, is closely related to the change with which this chapter is most concerned, the use of cyberspace and the electronic/electromagnetic arena. While the doctrinal aspects of this development, involving Information Operations, will be discussed later in this paper, it is necessary here to define what we mean by cyberspace.

The Department of Defense's "official dictionary", Joint Publication 1-02, defines cyberspace in such a way that there is virtually universal disagreement that it is wrong: "the notional environment in which digitized information is communicated over computer networks". Cyberspace is hardly "notional", and confining it to "digitized and computerized" is far too limiting, failing to reflect the massive technological and social changes with which cyberspace is interwoven. In 2006 the Department of Defense initiated several important efforts that will shape how the US military – and likely many others as well – view and employ cyberspace in military operations. The US Joint Staff in early 2006 began drafting a "National Military Strategy for Cyberspace Operations", and was formally approved by the Joint Chiefs of Staff in December 2006.[3] By summer 2006 another effort was underway, sponsored by the DoD's Undersecretary for Intelligence (USD/I) and focused on a broader and more strategic theory for the role of cyberspace and cyberpower across the entire national security paradigm. Finally, the United States Air Force has rapidly and aggressively stepped forward to lay claim to cyberspace as one of its three operational environments, alongside of the air and outer space. While all three efforts approached the question of defining cyberspace from different perspectives, at the time of the preparation of this chapter the Joint Staff approach closely mirrors that being offered by the USD/I study effort.[4] The USAF approach adds an emphasis on the warfighting aspects of cyberspace, not surprisingly for one of the military services, and they are taking steps to create an operational command which will concentrate on developing capabilities and concepts for military operations in cyberspace.

---

[3] This plan is being developed at the direction of the J-6, who is responsible for communications. A new and temporary office, J6X (headed by a former student of this author) was given the task of writing and coordinating this plan across the DoD.

[4] During the initial meeting of this task force a representative of the Joint Staff effort to develop the National Military Strategy for Cyberspace Operations (J6X) presented a concept for cyberspace that was clearly unacceptable to virtually everyone in attendance. To their credit, the J6X team reworked their approach, perhaps influenced by the approach presented by this author at that initial meeting, to the point where the J6X effort is essentially identical to that presented during the meeting; see footnote 6.

Many authors have offered very useful insights and perspectives on what constitutes cyberspace. Several concepts or issues run through most of them, including the role of telecommunications infrastructures, electronics and information systems.[5] The White House's 2003 "National Strategy to Secure Cyberspace" defined cyberspace as the "nervous system – the control system of the country ... composed of hundreds of thousands of interconnected computers, servers, routers, switches and fiber optic cables that allow our critical infrastructures to work" [28]. These approaches suggest that cyberspace is more than merely computers and digital information. This paper offers a definition that builds upon those threads cited above: "Cyberspace is an operational domain characterized by the use of electronics and the electromagnetic spectrum to create, store, modify and exchange information via networked information systems and infrastructures".[6] These networked and interconnected information systems exist both physically and virtually and within and outside of geographic boundaries. Their users range from nation states and communities to lone individuals who may not profess allegiance to any ideology, state or community. They rely on three distinct yet interrelated dimensions that comprise the information environment as described in American IO doctrine: the global connectivity linking information systems, networks, and human users; the massive amounts of content that can be captured digitally and electronically sent anywhere anytime to virtually anyone; and the human cognition that results from greatly increased access to content and thus impacts human behavior and decision making.[7] While the fundamental aspect of cyberspace that makes it unique from the other environments

occurs naturally in the physical world – electromagnetic energy – it requires man-made technology to use this environment.

Cyberspace has thus become the newest operational environment for warfare and national security affairs. It provides distinctly new ways to conduct military operations and warfare and serves as a vital means of enabling and integrating operations in the other four warfighting domains. It provides crucial new ways to connect people and societies and provides the backbone for how the information exchange and dissemination media – radio, TV, Internet, etc. – have made the world a virtual "fishbowl" in which events anywhere are instantly available everywhere. Finally, it provides the central nervous systems of our societies via the information infrastructures that knit together a wide range of the systems on which our societies and economies depend. The next several segments of this paper will examine these three issues in more detail.

29.3 CRITICAL INFRASTRUCTURE PROTECTION

Infrastructures have been important to national security for a very long time. The Romans, for example, had at least two that were vital to their economic, political and military security and stability: water and transportation, most visible in the form of the networks of aqueducts and roads that encircled the Roman Empire. These systems enabled the growth of Roman cities, the stability of the Roman society, the expansion of commerce across the entire Mediterranean region, and the ability to deploy Roman military power quickly (for that era) and to sustain it in the field. The American concept for strategic airpower in World War II, Billy Mitchell's "industrial web", was a form of infrastructural warfare, and this strategic concept was clearly visible in US air operations throughout the 1990s, against

---

[5] See, in roughly chronological order: [7; 17; 18; 21; 25].

[6] The National Military Strategy for Cyberspace Operations, approved in December 2006, uses a slightly different definition: "Cyberspace is a domain characterized by the use of electronics and the electromagnetic spectrum to store, modify and exchange data via networked systems and associated physical infrastructures". While there are a few differences between this approach and the one offered in the text above they are differences of detail and not of fundamental concept or approach. This author had a major role in the crafting of this definition.

[7] The newly-released (February 2006) Joint Doctrine Pub 3-13 "Information Operations" defines this environment

as "The aggregate of individuals, organizations, and systems that collect, process, or act on information", then goes on to argue that this environment functions via the interrelated effects of three dimensions: the physical (which I have defined as connectivity), the informational (content) and the cognitive.

both Iraq and Serbia.[8] What is critical to any particular society depends on the details of that society and the specifics of its economic, political and military systems.[9] What is dramatically new, however, is the growing use in dozens of countries of internetted and interconnected computer systems to monitor the status and control the operations of these infrastructures, a capability that rests on the widespread and growing reliance on systems employing SCADA – "supervisory control and data acquisition" – technologies which allow us to monitor the status and control the operation of a segment of infrastructure such as a rail network or an electric grid. Nearly any and every capability that supports strategic military, economic and societal strength is linked together in this manner and depends on the smooth and uninterrupted functioning of ICT – "information and communications technologies" – to keep flowing whatever is needed, whether that be electricity, money, a trainload of tanks on their way to a port of embarkation, an air traffic control system, or one of a thousand other critical functions that support and enable all of the different elements and instruments of national power.

This revolution has changed both WHAT an attacker might wish to attack as well as HOW it could be attacked and WHO might need to partner

in its defense. Going back to the WW II model, the means of attacking German industrial infrastructures was massed airpower: hundreds or even thousands of bombers smothering a key target – an oil refinery, or electric generating plant, for example – with hundreds or even thousands of tons of high explosive.[10] With proper planning, adequate force and some luck, sufficient explosive would be delivered to effectively destroy the target so that tomorrow's mission could go on to the next critical target and thus sequentially bring the infrastructure to collapse.[11] The focus on command and control systems and nodes that was highlighted by the American concept of "Command and Control Warfare – C2W" in the early 1990s – also had its origins in previous wars and campaigns. British General Edward Allenby's 1918 Palestine Offensive is just one prominent example, in which he feinted one way to get the Turkish forces out of position (military deception), carefully concealed the evidence of this maneuver (operational security or OPSEC) then used his airpower to strike and destroy the Turkish telegraph nodes along the railroad (C2W) to degrade the Turks' ability to effectively react to

---

[8]This emphasis on attacking (or defending) specific nodes that could degrade entire systems can be seen as early as the first German daylight raid on London in 1917, in which facilities such as banks and railroad centers were intended targets. Air Force Colonel John Warden may have best captured this concept with his 1994 article [26]; available electronically at http://www.airpower.maxwell.af.mil/airchronicles/apj/warden.html

[9]The initial American strategic policy for CIP, Presidential Decision Directive (PDD) 63, issued by President Clinton in 1998, listed six critical infrastructure sectors (telecommunications, energy, banking and finance, transportation, water systems and emergency services); The new National Infrastructure Protection Plan (NIPP) has extended this to between 17 and 20, which is almost certainly too many. See http://www.fas.org/irp/offdocs/pdd/pdd-63.htm for an electronic copy of PDD 63; the Bush Administration expanded this list in Homeland Security Presidential Directive (HSPD) 7, available electronically at http://www.fas.org/irp/offdocs/nspd/hspd-7.html. The NIPP is available at the Department of Homeland Security's website http://www.dhs.gov/dhspublic/display?content=5476

[10]The oft-cited American concept for "precision high-altitude strategic bombardment" is often interpreted as the precision of the bombing, but the reality is that it was tremendously inaccurate in the aggregate: the precision was in the ability to destroy a specific target or facility. In his superb article comparing USAAF and RAF bombing accuracy, W. Hays Parks showed that on a bomber-by-bomber comparison, the RAF was more accurate, although this was more a factor of tactical doctrine and the need to operate in massed formations during daylight for self-protection against German air defenses. What WAS precise was the USAAF's ability – repeatedly demonstrated – to hit and destroy specific industrial installations and thus degrade and eventually fatally weaken Germany's industrial infrastructure. See [9], especially W. Hays Parks 'Precision' and 'Area' bombing: who did which, and when, and Daniel T. Kuehl, Airpower vs. electricity: electric power as a target for strategic air operations.

[11]There is neither time nor space here to explore concepts such as "sequential" or "parallel warfare", which have been at the heart of the airpower debates within American and global defense circles since the Gulf War of 1991, but they are embedded deeply within concepts such as Information Warfare, Decision Superiority, and Command and Control Warfare. See Merrick Krause's short piece Decision dominance: Exploiting transformational asymmetries for a short overview of some of these concepts. Available electronically at http://www.ndu.edu/inss/DefHor/DH23/DH_23.htm

the disaster that was enfolding them.[12] The key addition to this mix of potential targets has been the control mechanism or even software itself, and the growing reliance on SCADA systems has added a new target category to the list of potential targets in strategic warfare.[13] Thus the informationized transformation of warfare has seen us evolve from attacking an infrastructure's physical components to perhaps attacking its informational components instead.

The next step in this transformation focuses on HOW we might attack those components. We understand and have a lengthy history of/experience with ways to physically attack these infrastructures, such as the WW II examples discussed above. But a "computer network attack" on the control systems for an enemy airspace control network would not require the massed forces of WW II, nor even the solitary and precise attacks of Desert Storm. Instead, it might only require a handful of people – who might not even need to be on the same continent, let alone in the same

room – with powerful computer systems and software "weapons" who could cause the simultaneous disruption and even collapse of critical enemy systems, networks and capabilities.[14] In 1943 it took enormous physical and moral courage for American airmen to battle their way for several hours across very hostile skies to strike their targets, but now a similar effect might be achievable in a matter of seconds across intercontinental distances from a setting where the most immediate danger is acute eyestrain. But there are still many unresolved issues surrounding this new and unproven capability. Some of them are legal and ethical, and center on questions such as whether a virtual and non-kinetic cyber "attack" crosses the thresholds of "armed attack and aggression", in the language of the UN Charter. Others involve whether cyberspace has borders that can be crossed and violated, and if so, where are they? Another set concerns the status of those persons conducting the attack: are they criminals, or mercenaries, or uniformed combatants subject to the restrictions and protections of the "Law of Armed Conflict"? Although vigorous and sometimes sensitive debate has taken place within the highest levels of the defense establishment in the US and elsewhere, these issues remain unresolved.[15]

The third focus of this transformation is defensive: WHO is responsible for and capable of protecting and defending our own critical infrastructures against such threats? This is very much an evolving interagency question in which all levels of government – including regional and local – as well as the private sector's business community must be totally involved. Our existing paradigms for national security have been shaped by the battlespaces in which we operate, but the emergence

---

[12]For a short but succinct description see Sir Basil Liddell hart's classic *The Real War, 1914–1918*, esp. pp. 439–448. What General Allenby did was to get well inside of his opponent's "decision cycle" and thus serves as a wonderful example of what Colonel John Boyd called the "observe-orient-decide-act" or OODA loop. For a short description of the OODA Loop concept see http://www.valuebasedmanagement.net/methods_boyd_ooda_loop.html; for a longer examination and analysis of John Boyd's work see either of the two recent biographies of Boyd [6], or [10].

[13]SCADA can be defined as "a *computer system* for gathering and analyzing real time data. SCADA systems are used to *monitor* and control a plant or equipment in industries such as telecommunications, water and waste control, energy, oil and gas refining and transportation. A SCADA system gathers information, such as where a leak on a pipeline has occurred, transfers the information back to a central site, alerting the home station that the leak has occurred, carrying out necessary analysis and control, such as determining if the leak is critical, and displaying the information in a logical and organized fashion. SCADA systems can be relatively simple, such as one that *monitors* environmental conditions of a small office building, or incredibly complex, such as a system that monitors all the activity in a nuclear power plant or the activity of a municipal water system". See http://www.webopedia.com/TERM/S/SCADA.html for this definition and a list of additional links to the term; a Google search of it produced over 700,000 hits!

[14]For a perceptive – and perhaps frightening – set of non-US perspectives on this see several of the chapters in [15], especially Wang Pufeng *The challenge of information warfare*, Wang BaiBaocun and Li Fei *Information warfare* and Wei Jincheng *Information war: a new form of people's war*, available electronically at www.ndu.edu/inss

[15]There is a growing body of literature on these topics, too extensive to cite completely here. The three best books are probably [18; 27] and [19]; in the journal literature [2]; available electronically at http://carlisle-www.army.mil/usawc/Parameters/01spring/bayles.htm

of cyberspace is already complicating this.[16] The critical infrastructures discussed previously are almost exclusively owned and operated by the private sector – they are business enterprises. Who has the responsibility for securing the operation of these infrastructures and protecting their key assets and functions? Although both government and business acknowledge that the owners bear this responsibility, there is one exception: the uniformed military is responsible for the defense of the nation from enemy attack via the air, or land, or sea, or even outer space ... but what of cyberspace? The US Air Force defends the elements of the Northeast Electric Grid against an attack using bombers ... but who defends it against an attack using electrons and malevolent bits and bytes? Segments of the US business and government have been studying this issue for more than a decade, but this "roles and missions" debate is still underway.[17] Critical

national infrastructures cannot be adequately and effectively protected without the integrated and cooperative action of business and government, but these efforts are still underway. We may have to wait until someone decides "the time is right" to make a strategic attack on these infrastructures, probably as an adjunct of a serious geostrategic crisis: finding out then that our interagency efforts have been ineffective might be the cyber equivalent of realizing on Monday, December 8, 1941 that you *can* drop aerial torpedoes in shallow-water harbors.[18]

29.4   INFORMATION OPERATIONS

The Department of Defense has been trying to clearly and effectively define IO for more than a decade, and while this is seen by some as a sign of IO's immaturity, in truth it merely reflects the normal doctrinal process of continuous reflection, evolution and revision. The practitioners and professionals still debate the meaning of "strategic airpower" and "maneuver warfare", even though they were central elements of warfare throughout the 20th century.[19] Most military professionals are quite comfortable with the use of advanced information technology and of information itself, whether as a means of supporting or enabling another activity, such as the need for precision navigational data in the employment of precision weaponry, or the use of information systems such as space-based weather satellites to obtain precise weather data. Most military professionals are equally comfortable with the value of severing or degrading communications and information links

---

[16]I describe cyberspace as a physical environment in which systems that store and exchange information using the electromagnetic spectrum – telegraph, telephone, radio, radar, TV, computer networks, etc. – interact with each other. We have evolved to where military operations are now conducted in five physical media: air, land, water, outer space and cyberspace. See my two 1997 *Strategic Forum* pieces on *Defining Information Power* and *Joint Information Warfare* (www.ndu.edu/inss) for a fuller explanation. I teach my students at the National Defense University that one critical way of looking at "jointness" is not the traditional combination of Services (Army, Navy, etc.) but rather the integration of operational and warfighting environments cited above. See the discussion in the text above and at footnotes 4 and 6.

[17]This is hardly a US-only item of strategic interest: at least a dozen other countries have taken on this issue as vital to their national security. An excellent starting resource is the fine survey and analysis published in Switzerland, *International CIIP [Critical Information Infrastructure Protection] Handbook 2004: An Inventory and Analysis of Protection Policies is Fourteen Countries*, Swiss Federal Institute of Technology, Zurich (2004); available electronically at http://www.isn.ethz.ch/crn/publications/publications_crn.cfm?pubid=224. Also see their newest study, published in 2006, which extends the analysis to 20 countries and several international organizations. An excellent starting point within the US was the publication of the findings of the President's Commission on Critical Infrastructure Protection (PCCIP or Marsh Commission, named after its head, Robert Marsh), *Critical Foundations* (available electronically at http://www.tsa.gov/interweb/assetlibrary/Infrastructure.pdf) in 1997. This was not only the key source document for President Clinton's PDD 63; it served the same role for efforts in several other

countries. For the results of a symposium in which this topic of the shared roles of the partnership see [16], especially Daniel T. Kuehl *The national information infrastructure: the role of the DOD in defending it*, available electronically at http://www.carlisle.army.mil/ssi/pdffiles/PUB224.pdf

[18]One of the earliest – and still best – books on this entire set of issues is [17].

[19]This author still has a vivid memory of the initial class session with the first group of Information Warfare students at NDU, in August 1994, in which they reacted quite strongly to the multiple definitions of IW which we presented them, worrying that "how can you teach us about IW if you can't define it!?" I used the same analogy with them as I have in this article.

and networks as a desirable, perhaps indispensable, enabler for other military operations, such as the attacks on Iraqi strategic C3 systems and nodes in the opening minutes of Operation Desert Storm, 1991.[20] While many are less comfortable or familiar with the criticality of the information itself as a target, electronic warriors have been doing this since the early days of WW II, and many if not most EW tactics and techniques involve distorting or degrading the actual information accessible to the user or defender.[21]

The latest revision of American joint doctrine for IO, Joint Doctrine Publication 3-13, released in February 2006, defines the information environment within which IO is conducted as an operational environment comprised of three distinct but critically-interrelated dimensions: the physical components, the information itself, and the cognitive dimension.[22] The physical dimension is made up of the infrastructures, networks, and systems that store, transfer and use information. These would certainly include any and all systems that operate in cyberspace, as cited previously: radios, radars, TV, the Internet, and more. These can be best understood as the *means* we employ for the use or transmission of information. The information itself – "content" – is the second dimension. Whether that content is a TV broadcast, a radio address, a website, a database, or something else, the information is the raw material that will be used to affect and shape the third dimension, the cognitive.[23] The latest IO doctrine declares this to be the "most important" of the three dimensions because this is where the content delivered by the connectivity impacts human beings: how we think, behave, comprehend and act. Quite obviously, the same information can have different impact on different human audiences, which highlights the indispensable role of cultural awareness and understanding in the analysis of a message and an audience [14]. The image of Saddam Hussein's statue with an American flag draped over the head, or the video of a car bomb attack on an American military vehicle, are two obvious and recent examples of how information content – whether as a TV broadcast, website, blog, or even a cartoon – can have dramatic cognitive impact on a global basis.[24]

The growing military use of the information environment has obvious and critical implications for the warfighter and military strategist. The tremendous expansion of the first dimension, the physical connectivity itself, has had profound impact that is likely to continue to grow. The impact on situational awareness is a case in point. On the blackest day in British military history, the 1st of July 1916 ("first day on the Somme"), British headquarters miles to the rear of the front continued to order attacks forward in the disastrously-erroneous assumption that the initial troops to go "over the top" had achieved their objectives and broken the German defenses. The lack of situational awareness was profound: it took days for British commanders to comprehend the scale of that first day's disaster, which killed nearly 20,000 British troops, left another 40,000 wounded, and scarred the nation for decades. Contrast this with current operations such as occurred during Operation Iraqi Freedom. While ground operations periodically struggled with episodes of uncertainty it is also true that when viewed from a John Boyd – OODA Loop perspective US and UK forces were repeatedly and decisively inside of the Iraqis' ability to assess and react to the pace of our operations. In the air the transformation was even more complete, a prime

---

[20]For a detailed analysis of the planning, execution and impact of these operations see Eliot Cohen and the work of the *Gulf War Air Power Survey* (1993).

[21]See my doctoral dissertation, [11] for perspectives on US use of EW.

[22]The new National Defense Strategy published in spring 2005 describes the information environment as a new "theater of operations". The latest Joint Publication 3-13 "Information Operations" was finally approved and released in February 2006, and the length of time required to gain its approval reflects the intense debate surrounding IO and its place in future military operations.

[23]This is hardly new. Thousands of years ago, for example, Mesopotamian kings had their triumphs immortalized in stone carvings, often depicting the enslavement or execution of captives. Who was the intended audience? It certainly included those rulers' contemporaries, who were thus encouraged to behave and cooperate, lest they suffer the same fate. This was strategic influence, using a stone carving instead of a webpage!

[24]The incident of the violent Islamic protests to the cartoon series that appeared in Danish newspapers in 2005, allegedly defaming the Prophet Mohammed, is a recent example of this.

example being the air strike at the end of the war responding to a report – later proven inaccurate – that Saddam Hussein had been located in a restaurant. If one is willing to set aside that admittedly large gap – he was not there – and focus instead on the rapid response to that report, it took approximately 38 minutes from the initial report ("he's in the restaurant") to the destruction of the target, which was a tremendous improvement over the previous state of operational responsiveness and situational awareness.

This leads to perhaps a better way of looking at how the Department of Defense is defining Information Operations. The current definitional approach to IO as promulgated by the SecDef's October 2003 "IO Roadmap" and the latest Joint Doctrine Pub 3-13 concentrates attention on five "core competencies" and it defines IO in terms of those five activities: Psychological Operations, Operations Security, Military Deception, Electronic Warfare, and Computer Network Operations (CNO).[25] The difficulty with this is that in no other form of warfare – air war, war at sea, etc. – would we define it by means of the things we do there, because those activities change over time and with changes in technology. Instead we define them in terms of the use and control of the operational environment, i.e. the maritime environment, outer space, or the aerospace. So should it be with IO, and a far more effective approach to defining and understanding IO would come from such an approach because we would then characterize IO through the lens of its battlespace, the information environment [12]. Interestingly enough, though probably not coincidentally, this is very similar to the approach that the USAF is taking in its latest doctrine for IO, AF Doctrine Document 2-5, and in other parts of the SecDef's IO Roadmap, which have characterized three distinct yet interrelated aspects of IO: operations that focus on electronics, influence, and networks. The first and third are integral to the information environment's physical domain described earlier in the newest draft joint doctrine for IO, Joint Pub 3-13, and the second or "influence" aspect corresponds almost exactly to the "cognitive" domain.[26]

The Chinese have been writing for more than a decade about the increasingly "informationized" battlefield.[27] What they really mean, however, is more akin to what they have called a "networkization" of the battlespace, because modern warfare is increasingly marked by vastly greater ability than we've ever had to access, store, and transfer information, not just vertically up and down a chain of command but also horizontally across various levels of one or more commands or organizations. This "networkization" has profound implications for doctrine, organizations, operations and even coalitions, and it is at the heart of one of the major lines of US defense transformation, what the US has called "network centric warfare" (NCW, or on a broader scale, net-centric operations). We are fighting adversaries in Iraq and in the larger "war on terror" who are networked in ways that we do not fully understand and are far less technologically based and dependent than we are, so that we

[25] See Chris Lamb's article [13] for a fuller explanation of the Roadmap and how the new definition came about. The intent of the SecDef's Roadmap was to establish IO as a "core military competency" alongside air–land–sea-and-special operations, and I believe this was an admirable and important goal, although I would suggest that the omission of "space" from this list was a serious oversight. This issue of *JFQ* is available electronically at http://www.dtic.mil/doctrine/jel/jfq_pubs/issue36.htm

[26] Air Force Doctrine Document 2-5, *Information Operations*, 11 January (2005); available electronically at http://www.dtic.mil/doctrine/jel/service_pubs/afdd2_5.pdf; General Richard Myers, Chairman, Joint Chiefs of Staff, *Posture Statement* (to the Senate Armed Services Committee), 3 February (2004), available electronically at http://armed-services.senate.gov/statemnt/2004/February/Myers.pdf; see also Chris Lamb's [13]. All four US Services (Army, Navy, Air Force and Marine Corps) understandably view IO from their own unique warfighting perspectives, which naturally lead to different organizational and doctrinal approaches. This is perfectly normal and in fact healthy, but it might also suggest that in order to develop a more comprehensive approach to IO, those organizations and elements that "do" IO should be grouped together and given more organizational and doctrinal autonomy. Even though the Air Force has had (in this author's opinion) the most visionary approach to IO, more organizational autonomy might be someday necessary.

[27] See [15] which contained several excellent articles by Chinese authors and military analysts who offered a range of very interesting and useful insights into evolving Chinese thought in this area. Also see [23], available electronically at http://www.dtic.mil/doctrine/jel/jfq_pubs/issue38.htm, and his [24].

do not fully comprehend both their strengths and vulnerabilities. One of the central tenets of NCW is that it is not solely about the technologies of networking – although that is an indispensable element – but must also include crucial human and organizational dimensions.[28] This increasing "networkization" of warfare is probably the most visible aspect of IO and will have the most significant long-term impact on the transformation of warfare that we will experience in the next decade. It's where IO and other military operations will have the most synergistic impact on each other, and where the warfighter will come into closest contact with it.

29.5 STRUGGLE FOR GLOBAL INFLUENCE

Influence is many things to many people, but while it has attracted a negative connotation for some audiences it is not inherently a bad thing! Influence is, after all, how all peaceful and friendly relationships function, whether on the one-to-one level (a marriage, for example) or at the nation state-to-nation state level. As with the other strategic areas discussed above (infrastructures and IO), the military aspects of influence have been important to warfare since biblical times.[29] There is a lengthy and scholarly literature available on the critical role that influence played in the American Revolution and the Napoleonic Wars [3; 22], and the US government conducted strategic influence operations throughout the decades of the Cold War with the USSR, with the operations of Radio Free Europe being just one visible example.[30] The importance of the struggle for influence in the "global war on terror" (GWOT) or the "worldwide war

of ideas" cannot be overestimated. Although the value of attaining superiority in "influence space" has always been a strategic aspect of warfare and statecraft, there are some new and radically different means available by which to wage what Karen Hughes, Under Secretary of State for Public Diplomacy/Public Affairs, has called a "generational and global struggle of ideas", and our adversaries are availing themselves of these capabilities, too. The last quarter of the 20th century saw a dizzying addition of systems, networks, audiences, and environments. Global satellite TV, the Internet, the "blogosphere", and convergence of information media are all playing a critical role in this struggle.[31] The creation of a common language for virtually any kind of information one can imagine – the bit and byte, one and zero – has dramatically transformed the ways we access, create, store, and transmit information, making it not only possible but routine to create information in quantities that would have been simply unimaginable a few years ago and to make that information instantly available to an audience that has become global. Thus our information reach has become intercontinental, instantaneous, and immense.[32]

But are we using these developments effectively to wage the "war of ideas"? Who are we trying to reach and influence, and do we understand that audience and the cultural contexts through which all information must be filtered? Is there one audience we should focus on, or are there a multiplicity of audiences, all of whom are vital in their own unique way? Conversely, how are our adversaries using the information element of power in

---

[28]The literature on this is large and growing. The fundamental and catalytic contribution was made in 1998 by the late Art Cebrowski, [4]. The DoD's Command-Control Research Project (CCRP) has sponsored several book-length publications on this topic; see, for example, [1]; or any of the several other volumes published at that same source.

[29]Joshua wasn't "blowing down the walls of Jericho"; he was blowing down the morale of those walls' defenders... an early but important example of Psyop, which is the military variant of influence.

[30]See their website at http://www.rferl.org/ for up-to-date information on broadcasts and some of their history.

[31]For one entry point into the blogosphere see http://www.blogospherenews.com/

[32]One survey of Internet use predicts more than 1.2 billion Internet users by 2006, and that only reflects those who are directly connected. If one expands that figure by the number who are touched, affected or influenced by those 1.2 billion, it's obvious that the Internet has become a nearly ubiquitous means of reaching people. Every day, more and more people as individuals, organizations, and countries/societies get connected to this global network, and they do so for one obvious and overriding reason: they have determined that they need to in order to be successful. For current statistics see http://www.clickz.com/stats/web_worldwide/. For graphic depictions of connectivity – an "atlas of cyberspace" – go to http://www.geog.ucl.ac.uk/casa/martin/atlas/isp_maps.html

the global information environment? How do they conduct influence operations, and are there openings we could exploit or dangers against which we must become especially vigilant?

It is not at all certain that we are winning this struggle of ideas and influence. Numerous studies of public diplomacy and strategic communication over the past few years have all concluded that we are *not* doing well and *must* do better.[33] We must recognize that information does have staying power and that its influence persists. Ask anyone anywhere in the world to recall just one photograph from Vietnam and they will instantly cite the famous photo of South Vietnamese Police commander General Loan executing a Viet Cong terrorist with a pistol shot to the head. Regardless of the context or circumstances of that incident, the photo carries an emotional impact. Leap forward 36 years and ask for a photo from Iraq: if the respondent is Arab or Islamic they will almost certainly cite the photo of a hooded Iraqi prisoner standing on a box with his hands apparently connected to electric wires, an image that will create opponents and adversaries in Islamic audiences for decades to come.[34]

---

[33] For the text of her statement at her confirmation hearing before the Senate Foreign Relations Committee see http://www.geog.ucl.ac.uk/casa/martin/atlas/isp_maps.html; the issue of the dismal state of American public diplomacy has been the focus of numerous – several dozen – reports, studies, and surveys over the past few years. One of the most important was the report of the Defense Science Board 2004 summer study, especially the *Report of the Task Force on Strategic Communication*, available electronically at http://www.acq.osd.mil/dsb/reports/2004-09-Strategic_Communication.pdf (truth in advertising requires that I mention that I was a member of this task force).

[34] One of the dangers is the way that technology enables the manipulation and dissemination of imagery. I am not aware that the famous Vietnam photograph cited above received much dissemination at the time, but the Abu Ghraib photos have spawned a series of other such images that have been widely disseminated via the Internet. The issue of faked images provided to major news organizations during the summer 2006 Israeli operation in southern Lebanon provided an excellent case in point. The photojournalist who provided staged or manipulated photos that were published by Reuters was fired, but the images remain on the Internet in a variety of websites and access points, as do the photos of Danish cartoons, hooded prisoners, etc.

Secretary of State Rice recently testified before the Senate Foreign Relations Committee about her planned reorganization within State,[35] but while a realigned department will hopefully smooth its internal coordination process, the need to approach influence from a vastly expanded interagency perspective, indeed from one that includes major parts of the private sector also, grows ever more critical. Under Secretary Hughes spoke to this issue during her confirmation hearings, emphasizing the need to include and integrate the immense capabilities and reach of the private sector in this effort.[36] The "war of ideas" is being waged with all of the modern implements of influence – radio, TV, Internet, and more – and in a million and one critical places, from the TV and computer screens of countless individuals to the souks and madrassas across not only the Islamic world but within Western society as well. USMC General Wallace Gregson recently argued that the most vital audience we must reach is the Moslem who has not yet committed to the side/perspective of the terrorists, a viewed that echoes the conclusions of the 2004 Defense Science Board study on Strategic Communication. This contest for influence – the "war of ideas" – is the most important and decisive aspect of the emerging struggle against violent extremism. It will be as long as the Cold War, it will be as bitter as that contest, and the consequences of losing will be even more profound and disastrous as they would have been had we lost the Cold War.

29.6 CONCLUSION

"Doing strategy" has never been easy, but because of the Information Revolution the difficulties facing the national security strategist have never been more complex or daunting, and the plate of critical strategic issues has never been fuller. While infrastructures have always been important to military and economic power, our growing reliance on

---

[35] See www.state.gov/secretary/rm/2005/50375.htm for the text of her remarks.

[36] See [20]; Secretary Rice's remarks in www.state.gov/secretary/rm/2005/50375.htm; see also Hughes testimony cited earlier.

interconnected computer networks to run our infrastructures presents a vulnerability that a strategic adversary can hardly ignore. Ominously, both Chinese and Al Qa'ida strategists have clearly indicated their awareness of this potential target. The growing networkization of warfare is transforming how we fight, as both conventional and unconventional military capabilities and operations are affected and shaped by the technological, organizational, doctrinal and operational imperatives of the Information Revolution. The struggle for influence may well be the most important and decisive element of our strategic problem for the next several generations. The connectivity between people, organizations and societies continues to grow, creating new avenues and audiences for influence. The national security strategist of the future will have to contend with across-the-board transformations in all three areas – critical infrastructure protection, information operations, and the struggle for global influence – that will strategically affect not only the military but also the economist and diplomat, bringing the partnership of the Public and Private sectors to the fore as a matter of utmost strategic importance. The national security strategist of the future will have to understand these issues and be comfortable in dealing with them. Creating stove piped specialists in each area will not suffice... the creation of information power and its integration with other elements of power require a broad, holistic approach to the synergies and interdependencies among the diplomatic, informational, military and economic elements of power. Exploring these transformations now is the key to effectively exploiting their advantages in the future.

REFERENCES

[1] D.S. Alberts, J.J. Garstka, F.P. Stein, *Network Centric Warfare: Developing and Leveraging Information Superiority*, Command Control Research Project, NDU Press, Washington, DC (1999); http://www.dodccrp.org

[2] W. Bayles, *The ethics of computer network attack*, Parameters, Spring (2001).

[3] C. Berger, *Broadsides and Bayonets: the Propaganda War of the American Revolution*, Presidio Press, San Rafael, CA (1976).

[4] A.K. Cebrowski, J.J. Garstka, *Network-Centric Warfare: Its Origin and Future*, **United States Naval Institute Proceedings**, January (1998), http://www.usni.org/Proceedings/Articles98/PROcebrowski.htm

[5] T.E. Copeland (ed.), *The Information Revolution and National Security*, Army Strategic Studies Institute, Carlisle, PA (2000), http://www.strategicstudiesinstitute.army.mil/pdffiles/PUB225.pdf (accessed October 2006).

[6] R. Corum, *John Boyd; The Fighter Pilot Who Changed the Art of War*, Little–Brown, New York (2005).

[7] D. Denning, *Information Warfare and Security* (1999).

[8] P.F. Drucker, *Beyond the information revolution*, The Atlantic Monthly (October, 1999), http://www.theatlantic.com/doc/prem/199910/information-revolution (accessed October 2006).

[9] J. Gooch (ed.), *Airpower: Theory and Practice*, Cass, London (1995).

[10] G. Hammond, *The Mind of War: John Boyd and American Security*, Smithsonian, Washington, DC (2004).

[11] D. Kuehl, The radar eye blinded: The USAF and EW, PhD thesis, Duke University (1992).

[12] D.T. Kuehl, *Defining information power*, NDU Strategic Forum #115, June (1997), www.ndu.edu/inss

[13] C. Lamb, *IO as a core competency*, Joint Force Quarterly 36 (2004).

[14] M. McFate, *The military utility of understanding adversary culture*, Joint Force Quarterly 38 (2005).

[15] M. Pillsbury (ed.), *Chinese Views of Future Warfare*, NDU Press, Washington, DC (1997).

[16] C. Pumphrey (ed.), *Transnational Threats: Blending Law Enforcement and Military Strategies*, Army War College Strategic Studies Institute, Carlisle, PA (2000).

[17] G. Rattray, *Strategic Warfare in Cyberspace*, MIT Press, Boston (2001).

[18] W.G. Sharp, *Cyberspace and the Use of Force*, Aegis Research, Falls Church, VA (1999).

[19] M.N. Schmitt, B.T. O'Donnell (eds), *Computer Network Attack and International Law*, Naval War College, Newport, RI (2002).

[20] T. Shea, *Transforming military diplomacy*, Joint Force Quarterly 38 (2005).

[21] W. Schwartau, *Information Warfare: Chaos on the Electronic Superhighway*, Thunders Mouth Press, New York (two editions, 1994 and 1996).

[22] P. Taylor, *Munitions of the Mind: a History of Propaganda*, Manchester University Press, Manchester, UK (2003).

[23] T.L. Thomas, *Chinese and American network warfare*, Joint Force Quarterly 38 (2005).

[24] T.L. Thomas, Dragon bytes: Chinese information war theory and practice, Army Foreign Military Studies Office, Fort Leavenworth, KS (2004).

[25] E. Waltz, *Information Warfare: Principles and Operations*, Artech House (1998).

[26] J. Warden, *The enemy as a system*, Air Power Journal, Spring (1995).

[27] T.C. Wingfield, *The Law of Information Conflict: National Security Law in Cyberspace*, Aegis Research, Falls Church, VA (2000).

[28] White House, *National Strategy to Secure Cyberspace*, Washington, DC (2003).

BIOGRAPHIES

A native of New York City, **Matthew M. Aid** has served as a senior executive with a number of large international financial research and investigative companies over the past twenty years. Mr. Aid was the co-editor, along with Dr. Cees Wiebes, of *Secrets of Signals Intelligence During the Cold War and Beyond* (London: Frank Cass, 2001), and is currently completing a multi-volume history of the National Security Agency (NSA) and its predecessor organisations covering the period from 1945 to the present. Mr. Aid is also the author of a number of published articles on intelligence and security issues, focusing primarily on issues relating to Signals Intelligence (SIGINT).

Friedrich L. Bauer is an *emeritus* professor of mathematics and informatics at the Technical University Munich. Professor Bauer has written about numerical analysis, the systematics of program development, and has participated in the committee for ALGOL 60. He received the IEEE computer pioneer award for his contribution to computer science in Germany. He is the author of *Decrypted Secrets*: a handbook about cryptology.

Having begun as a system programmer testing compiler, **Margaret Elinor van Biene-Hershey RE** has moved into developing compilers for PL/1 and Algol as well as other programming tasks that require assembler programming experience. Working with operating systems, data communications, and in particular IBM's IMS, she now works with auditing technical systems, installing large computer centres, and managing these installations for the organisations responsible for them. Publications and presentations on Information Technology (IT) auditing led to setting up a three year post graduate course on IT auditing in the late 1980s. During that same period, she began managing information technology which included the development, installation, and operations for international subsidiaries and daughter companies of a large Dutch bank. Until recently, Professor van Biene has been a partner at Ernst & Young and a lecturer at the VU Amsterdam.

Jeremy Black attended Queens' College, Cambridge; St. John's College, Oxford; and Merton College, Oxford. He lectures in history at the University of Exeter, UK. His research focuses on early modern British and continental European history, with particular interest in international relations, military history, the press, and historical atlases. Professor Black is the single author of forty-six books and the co-author of one. He has edited or co-edited fifteen books. His most recent works include *European Warfare 1494–1660* (2002); *America as a Military Power 1775–1882* (2002); *The World in the Twentieth Century* (2002); and *The Making of Modern Britain: The Age of Edinburgh to the New Millennium* (2001).

Susan W. Brenner is NCR Distinguished Professor of Law and Technology at the University of Dayton School of Law. She is a graduate of the Indiana University School of Law. After obtaining her J.D. degree, she clerked for an Indiana Court of Appeals judge and then for a Federal District Court judge. She then joined the firm of Shellow & Shellow in Milwaukee, Wisconsin where she practised federal criminal law. Professor Brenner left the Shellow firm to join Silets and Martin, Ltd., a Chicago law firm that specialises in defending those charged with federal white–collar crimes. In 1988 she became a professor of law at the University of Dayton School of Law. Her areas of specialisation are cybercrime law and policy, federal grand jury practise and white–collar criminal law. Professor Brenner chaired the International Efforts Working Group for the American Bar Association's Privacy and Computer Crime Committee. She serves on the National District Attorneys

Association's Cybercrimes Committee and chairs the National Institute of Justice – Electronic Crime Partnership Initiative's Working Group on Law & Policy.

Andrew Charlesworth is Senior Research Fellow in IT and Law and Director of the Centre for IT and Law (CITL) at the University of Bristol, a joint venture of the law school and the Department of Computer Science. The CITL is sponsored by Herbert Smith LLP, Vodafone, Barclaycard, Hewlett Packard Laboratories and the Law Society Charitable Trust. He has been teaching and researching in the areas of information technology, internet and e-commerce law for over a decade, and has presented papers on various aspects of IT law at conferences and seminars in Europe, North America, the Middle East and Australia.

After her studies at the University of Amsterdam **Madeleine de Cock Buning** worked as a consultant for the European Commission Directorate General Internal Market in Brussels. At the University of Amsterdam's Institute for Information Law (IViR), she received her PhD on a thesis entitled *Copyright and information technology, on the tenability of technology specific regulation*, published 1998. Presently she works as an attorney-at-law for De Brauw Blackstone Westbroek and is a professor at the University of Utrecht's Molengraaff Institute for Private Law. She is editor of the legal magazine *Intellectual Property and Advertisement Law (IER)*, board member of the Dutch Association for Copyright (VVA), and a panelist on the UDRP for domain names on behalf of the World Intellectual Property Organization (WIPO). She regularly publishes and lectures (inter)nationally in the field of new media, copyright, advertisement and communications law.

B. Jack Copeland received his PhD in Philosophy from the University of Oxford (1979) for research on modal and non-classical logics. He is a professor of philosophy and Director of the Turing Archive for the History of Computing at the University of Canterbury, New Zealand, where he has taught since 1985 and is chair of the School

of Philosophy and Religious Studies. His publications include *The Essential Turing* (Oxford Univ. Press, 2004); *Alan Turing's Automatic Computing Engine* (Oxford Univ. Press, 2005); *Colossus: The Secrets of Bletchley Park's Codebreaking Computers* (Oxford Univ. Press, 2006); *Logic and Reality* (Oxford Univ. Press, 1996); and *Artificial Intelligence* (Blackwell, 1993, 2nd edition forthcoming). In addition, he has published more than one hundred articles on the philosophy and history of computing, the philosophy of mind and philosophical logic. Professor Copeland has held a variety of visiting professorships at international institutions, including the University of Sydney (1997, 2002), the University of Aarhus (1999), the University of Melbourne (2002, 2003) and the University of Portsmouth (1997–2005). During 2000 he was a Senior Fellow in the Dibner Institute for the History of Science and Technology at the Massachusetts Institute of Technology (MIT). Professor Copeland is a member of the Bletchley Park Trust Heritage Advisory Panel and is president of the US based Society for Machines and Mentality. He is the founding editor of *The Rutherford Journal for the History and Philosophy of Science and Technology*, and serves on the editorial boards of various philosophical journals. In 2000 he received the Marsden award from the Royal Society of New Zealand and in 2003 the *Scientific American Sci/Tech Web Award* for his on-line archive www.AlanTuring.net.

Laura DeNardis, a computer networking and security expert residing in Stamford, Connecticut, is currently a visiting fellow (2006–2007) at the Information Society Project at Yale University. Dr. DeNardis studied at Dartmouth College (AB), Cornell University (M Eng), and received a PhD in Science and Technology Studies from Virginia Polytechnic Institute and State University.

Whitfield Diffie began his career in information security as the inventor of the public key cryptography and has made fundamental contributions to many aspects of secure communications. In the 1990s he turned his attention to public policy and played a key role in opposing government key-escrow proposals and restrictive regulations on the export of products incorporating cryptography. He is now the Chief Security Officer at Sun Microsystems and is studying the impact of web services and grid computing on security and intelligence.

Susan Landau is a Distinguished Engineer at Sun Microsystems Laboratories, where she works on security, cryptography and policy, including digital-rights management and surveillance issues. Landau had previously been a faculty member at the University of Massachusetts and Wesleyan University where she worked in algebraic algorithms. She is co-author, along with Whitfield Diffie, of *Privacy on the Line: the Politics of Wiretapping and Encryption* (MIT Press, 1998). She is a member of the National Institute of Standards and Technology's Information Security and Privacy Advisory Board, a member of the editorial board of IEEE Security and Privacy, and she moderates the "researcHers" list, an international mailing list for women computer science researchers. Landau is an AAAS Fellow and received her BA from Princeton, her MS from Cornell and her PhD from MIT.

Joseph Fitsanakis (BA University of Birmingham; MSc, PhD University of Edinburgh) specialises on the politics of government surveillance. He has authored books and articles in Greek on telecommunications policy, as well as numerous articles in English on the conception and implementation of domestic and external intelligence in the US and the UK.

Sylvan Frik was born in Zug, Switzerland in 1976. He studied Political Science, Constitutional, and International Law at the University of Zurich. His PhD thesis is on Switzerland's role in Europe's Security Policy and was submitted in 2002. He has continued as a Senior Research Fellow at the Institute for Political Sciences of the University of Zurich where he lectures on International Organizations. Since 2003, he has been a lecturer on Swiss Foreign Policy at the Swiss Federal Institute of Technology. Since the autumn of 2003 he also worked for Crypto AG as head of the marketing department. He has written various

publications on foreign and security policy topics in military magazines in Switzerland, among them: *Ist die schweizerische Sicherheitspolitik europafähig? Die Neutralität im Lichte der Entwicklungen der ESVP* (Chur, Zürich: Rüegger, 2002); Die Zürcher Kantonsratswahlen 1999: Statistische Analyse des SVP-Wahlsieges mit soziologischer Wahltheorie", in: *Statistische Berichte des Kantons Zürich*, Nr. 1/2000; 79-91 together with Stefan Inderbitzin.

Kees Gispen is Professor of European History and Associate Director of the Croft Institute for International Studies at the University of Mississippi. He received his BA and PhD degrees from the University of California at Berkeley in 1970 and 1981, respectively. He joined the history faculty in Mississippi in 1983 and the Croft Institute in 1998. Gispen served as the Executive Secretary of the Conference Group for Central European History from 1997 to 2005. He is the author of *New Profession, Old Order: Engineers and German Society, 1815–1914* (Cambridge University Press, 1990), and *Poems in Steel: National Socialism and the Politics of Inventing from Weimar to Bonn* (Berghahn Books, 2002) and numerous articles on the social history of technology and the professionalising of the engineering profession. His most recent publication is a chapter on British and American engineers in *Geschichte des Ingenieurs: Ein Beruf in sechs Jahrtausenden*, ed. Walter Kaiser and Wolfgang König (Hanser-Verlag, 2006).

Dieter Gollmann received his degree in engineering mathematics (1979) and his doctorate of technology (1984) from the University of Linz, Austria where he was a research assistant in the Department for System Science. He was a Lecturer in Computer Science at Royal Holloway, University of London and later a scientific assistant at the University of Karlsruhe, Germany where he was awarded the *venia legendi* for Computer Science in 1991. He rejoined Royal Holloway in 1990, where he was the first Course Director of the MSc in Information Security. He was a Visiting Professor at the Technical University of Graz in 1991, an Adjunct Professor at the Information Security

Research Centre, QUT, Brisbane, in 1995 and has acted as a consultant for HP Laboratories Bristol. He joined Microsoft Research in Cambridge in 1998. In 2003, he took the chair for Security in Distributed Applications at the Technical University Hamburg-Harburg, Germany. He is a Visiting Professor with the Information Security Group at Royal Holloway and an Adjunct Professor at the Technical University of Denmark.

Dieter Gollmann is one of the editors-in-chief of the *International Journal of Information Security* link.springer.de/link/service/journals/10207/ and an associate editor of the IEEE *Security & Privacy* magazine http://www.computer.org/security/. His textbook on *Computer Security* recently appeared in its second edition.

Edward Higgs studied history at the University of Oxford, completing his doctoral research in 1978. He was an archivist at the Public Record Office, the national archives in London, from 1978 to 1993 where he was latterly responsible for policy relating to the archiving of electronic records. He was a senior research fellow at the Wellcome Unit for the History of Medicine of the University of Oxford, 1993–1996 and a lecturer at the University of Exeter from 1996 to 2000. He has worked at the University of Essex since 2000 and is now a Reader in History. His early published research was on Victorian domestic service although he has written widely on the history of censuses and surveys, civil registration, women's work, and the impact of the digital revolution on archives. His current research interests include the collection of personal information by the State and the developments of techniques of identification in Britain over the past 500 years.

Jan Holvast (1941) worked at the University of Amsterdam for almost twenty-five years. First in the department of Sociology (Statistics and Methods of Social Research) and then in the Social Science Informatics department where he was responsible for teaching and research of Social Implications of Information Technology. He took his PhD in 1986 based on a study of personal freedom in an information society. From 1991–1994 he was director of the Foundation Privacy Alert and of the

Centre for Privacy Research. In 1994 he started as an independent consultant for Holvast & Partner in the field of privacy protection for ministries, local communities, labour organisations and private companies. He published several books and various articles on the theme of privacy and was chairman of Privacy International and chairman of W.G. 9.2 (Computers and Society) of the International Federation for Information Processing (IFIP). He is editor of a Dutch newsletter on privacy and is editor-in-chief of the Dutch looseleaf *Handbook of Privacy Protection*.

Dan Kuehl is the Director of the Information Operations Concentration, a specialised curriculum on national security in the information age offered to selected senior students at the National Defense University. His courses concentrate on such issues as the information component of national power, information warfare, and public diplomacy. He retired as a Lieutenant Colonel in 1994 after nearly 22 years active duty in the USAF. He holds a PhD in History from Duke University, and his dissertation focused on the Air Force's employment of electronic warfare in the decade after WW II. His publications include a wide range of academic and professional journals, and he has contributed to several books on airpower and information warfare. He is on the editorial boards of *Joint Force Quarterly* and the *Journal of Information Warfare*, is a member of the Public Diplomacy Council and the Cyber Conflict Studies Association, was a member of the Defense Science Board team that wrote the 2004 report on *Strategic Communication*, and is contributing to a major DOD study on cyberspace and cyberpower. He lectures internationally on the subject of information warfare, and his current research focuses on the relationship between the information age and national security.

Karl de Leeuw has a degree in history from the University of Utrecht and in Computer Science from the University of Amsterdam. He wrote his PhD thesis on the history of cryptology in the Netherlands and published extensively about this subject in scholarly journals. He worked as an editor of the correspondence of the Dutch mathematician L.E.J. Brouwer at the Philosophy Department of Utrecht University and as a lecturer in Information Security at the Institute of Informatics of the University of Amsterdam. He is a board member of the Netherlands Intelligence Studies Association. His current research interests are on the crossroad of the history of science and technology, on international relations and on espionage.

Jack Meadows is *emeritus* professor of Information and Library Studies at Loughborough University and was formerly professor of astronomy and of the history of science at Leicester University.

Hans Meijer has worked as a computer programmer and scientist since 1965 until his retirement in 2005, lastly as an associate professor at the Radboud University in Nijmegen. He has specialised in compiler construction and systematic program development and has participated in the development of teaching program for computing science in secondary education.

Jaap-Henk Hoepman is a senior researcher in the Security of Systems (SoS) group at Radboud University. His interests are security and applied cryptography.

Bart Jacobs is professor of software security and correctness in the SoS group and is interested both in technical and societal issues.

Erik Poll is a senior lecturer in the SoS group. His research focuses on software security, specifically for Java.

Robert Plotkin, Esq. is a patent attorney and a lecturer on law at Boston University School of Law. In his law practise, he assists individuals and companies in obtaining, enforcing, and defending against patents on computer technology. At the school of law, he teaches an advanced course entitled "Software and the Law" and has written and spoken on topics including software patents, protection for computer software under the First Amendment to the U.S. Constitution, ethical implications of software source code publication, trademark and domain name disputes, patent licensing, electronic court filing, and electronic privacy in the workplace. He is licensed to practise law in the Commonwealth of Massachusetts and the State

of New York and is registered to practise before the United States Patent and Trademark Office. Attorney Plotkin studied computer science and engineering at Massachusetts Institute of Technology (MIT) and law at the Boston University School of Law.

Bart Preneel received his electrical engineering degree and a doctorate in applied sciences in 1987 and 1993, respectively. He is currently full professor at the Katholieke Universiteit Leuven and has been visiting professor at the Universities of Graz, Bergen-Bochum and Ghent. During the academic year 1993–1994, he was a research fellow at the ECCS Department of the University of California at Berkeley. His main research interests is cryptology and its applications to computer security. He has authored and co-authored more than two hundred scientific publications, has two patents, and has edited ten books. As a consultant, he has been involved in a large number of information security studies for governments and companies and is one of the directors of the research group COSIC. He is vice-president of the International Association for Cryptologic Research and a member of the editorial board of the *Journal of Cryptology* and the *IEEE Transactions on Forensics and Information Security.*

Gerhard F. Strasser was educated at the Universities of Munich (Germany), Grenoble (France), at Rhodes College, Memphis, Tennessee, and at Brown University in Providence, Rhode Island, where he received his PhD in comparative literature. He taught at Northwestern University in Evanston, Illinois and at Pennsylvania State University in University Park until his retirement in 2004. His publications deal primarily with the cultural history of the early modern period and include books and articles on universal languages, cryptology, hieroglyphics, emblematics, mnemonics, and the history of art of the modern era.

Karel Schell (1940) studied physics at the Delft University of Technology. After his graduation he became a lecturer in physics at the Royal Naval College in Den Helder. He then worked for ten years as development manager in various departments for Royal Philips Electronics in Eindhoven before joining the Dutch security printer Royal Enschede & Zn in Haarlem in 1977. In 1988, Mr. Schell became technical director and was responsible for the production of an entire range of security documents, from stock certificates to banknotes as well as postage stamps. In 1993, he left Joh. Enschede and set up Schell Consulting and since that time has more than fifty clients in fifteen countries using his expertise in the area of currency technology.

Robert Verhoogt (1971) is art historian and policy adviser at the department of Cultural Heritage of the Ministry for Education, Culture and Science in the Netherlands. He has written and spoken on topics of nineteenth century (photo-)graphic art reproduction, Dutch romanticism, Lawrence Alma-Tadema, Jozef Israels, Vincent van Gogh, and the history of intellectual property in relation to visual arts. He studied law and history of art at the Amsterdam Free University and wrote his dissertation on *Kunst in reproductie, De reproductie van kunst in de negentiende eeuw en in het bijzonder van Ary Scheffer (1795–1858), Jozef Israëls (1824–1911) en Lawrence Alma Tadema (1836–1912)* at the University of Amsterdam in 2004. A translation of this book in English will be published in 2007. He is a member of the Committee for the History of Intellectual Property.

Chris Schriks (1931) since 1946 is active in the Dutch printing and publishing business. Dr. Schriks started his career at the Boek en Handelsdrukkerij in Helmond, and later was director at the Centrale Drukkerij and Boekbinderij Gebr. Franken in Nijmegen. In 1961 he started the Uitgeverij Walburg Pers in Zutphen which focuses on books about history, architecture, and science. He was also a board member of several publishing houses, a member of the executive committee of directors of the Koninklijke Vereeniging ter bevordering van de belangen des Boekhandels, and chairman of the Dr. P.A. Tiele Stichting. He has written and spoken on topics of the history of publishing, printing, and intellectual property and has published his dissertation, *Het Kopijrecht. 16de tot 19de eeuw. Aanleidingen tot en gevolgen van boekprivileges en boek- handelsusanties, kopijrecht, verordeningen, boekenwetten en rechtspraak in het privaat-, publiek- en staatsdomein in de Nederlanden, met globale analoge ontwikkelingen in Frankrijk, Groot-Brittannie en het Heilig Roomse Rijk.*, which includes an extensive summary in English (Zutphen, 2004). He is now preparing a publication about the history of copyright in The Netherlands and is a member of the Committee for the History of the Intellectual Property.

Jim Wayman is director of the Biometric Identification Research Program of San Jose State University and received his PhD degree in engineering in 1980 from the University of California, Santa Barbara. In the 1980s, under contract to the U.S. Department of Defense, he invented and developed a biometric authentication technology based on the acoustic resonances of the human head. He joined San Jose State University in 1995 to direct the Biometric Identification Research Program, serving as Director of the U.S. National Biometric Test Center at San Jose State from 1997–2000. He has written dozens of book chapters and journal articles on biometrics and is co-editor of *Biometric Systems* (Springer, London, 2005) along with J. Wayman, A. Jain, D. Maltoni and D. Maio. He is a fellow of the British Institution of Engineering and Technology, UK; is a principle expert and head of the delegation for Working Group 1 (Vocabulary Harmonization) on the ISO/IEC JTC1 SC37 standards committee on biometrics; he is a core member of the UK Biometrics Working Group and a member of the Biometrics Experts Group of the UK Home Office; and he is a member of the EC-funded BioSecure Network of Excellence. He is a support contractor for the U.S. Department of Defense Technical Support Working Group, a member of the U.S. Department of Homeland Security Independent Board of External Review and Validation for biometrics, and a committee member of *Whither Biometrics?* for the U.S. National Academies of Science/National Research Council. He previously served on the NAS Authentication Technologies and Its Implications for Privacy committee. He holds four patents in speech processing and has served as a paid biometrics adviser to eight governments.

Pieter E. Wisse is the founder and president of Information Dynamics, an independent company operating from the Netherlands and involved in research and development of complex information systems. He holds an engineering degree (mathematics and information management) from Delft University of Technology and a PhD (information management) from the University of Amsterdam. At the latter university, Pieter is affiliated with the *PrimaVera* research program in information management.

Jeffrey R. Yost is associate director of the Charles Babbage Institute for the History of Information Technology, University of Minnesota. He is the author of a number of books and articles on the business, social, scientific, cultural and intellectual history of computing, software and networking including: *The Computer Industry* (Greenwood, 2005); "Computers and the Internet: Braiding Irony, Paradox and Possibility", in C. Pursell, ed. *Companion to American Technology* (Blackwell, 2005); and "Maximization and Marginalization: A Brief Examination of the History and Historiography of the U.S. Computer Services Industry", in *Enterprises et Histoire* (November 2005). He is currently writing a book on the history of the global computer services industry.

AUTHOR INDEX

Roman numbers refer to pages on which the author (or his/her work) is mentioned. Italic numbers refer to reference pages. Numbers between brackets are the reference numbers. No distinction is made between the first author and co-author(s).

SUBJECT INDEX

Printed and bound by CPI Group (UK) Ltd, Croydon, CR0 4YY

03/10/2024

01040333-0018